PREDATORS WITH POUCHES

THE BIOLOGY OF CARNIVOROUS MARSUPIALS

PREDATORS WITH POUCHES

THE BIOLOGY OF CARNIVOROUS MARSUPIALS

Menna Jones, Chris Dickman and
Mike Archer [Editors]

National Library of Australia Cataloguing-in-Publication entry

Predators with pouches: the biology of carnivorous marsupials.

Bibliography.
ISBN 0 643 06634 9.

1. Marsupials. 2. Carnivora. I. Jones, Menna E.
II. Dickman, Chris R. III. Archer, Michael, 1945– .

599.27

Available from
CSIRO Publishing
150 Oxford Street (PO Box 1139)
Collingwood VIC 3066
Australia

Telephone: +61 3 9662 7666
Freecall: 1800 645 051 (Australia only)
Fax: +61 3 9662 7555
Email: publishing.sales@csiro.au
Web site: www.publish.csiro.au

Cover photographs from top left, clockwise:
Tasmanian Devil, *Sarcophilus laniarius*, ('Eumarrah') photographed at
Trowunna Wildlife Park, Tasmania by Menna Jones
Murine Mouse Opossum, *Marmosa murina*, by Louise Emmons
Brown Four-eyed Opossum, *Metachirus nudicaudatus*, by Louise Emmons
Numbat, *Myrmecobius fasciatus*, by L & O Schick/Nature Focus

Set in Adobe Garamond and Adobe Gill Sans

Cover design by James Kelly
Typeset by Desktop Concepts P/L, Melbourne
Printed in Australia by Ligare

DEDICATION TO PAT WOOLLEY

A hundred years ago the native cat, as it was called, was common around Sydney and became the object of intense study by Professor J.P. Hill, his colleagues and students. For 50 years it was the best-studied species of Australian marsupial, rivalled only by the North American Virginian opossum, studied by Carl Hartman. These two carnivorous marsupials were the pillars on which marsupial reproductive biology and development were raised, and on which all textbooks relied. Neither Hill nor Hartman gave much attention to the ecology of their chosen species, and none to conservation. Knowledge of this sort was only gained after the renaissance in marsupial studies, begun by Adolph Bolliger in the 1940s and 1950s, and greatly fostered by Harry Waring at the University of Western Australia and Francis Ratcliffe in the Wildlife Survey Section of the Commonwealth Scientific and Industrial Research Organisation. Ratcliffe and Waring were responsible for kindling the fire of inquiry in the two people who can justly be recognised as the founders of the modern study of the Dasyuridae: Basil Marlow and Pat Woolley.

In 1961, having decided to study *Antechinus* for her PhD, Pat Woolley collected her first animals in the eucalypt woodlands around Canberra and soon realised that she was dealing with two species: *A. flavipes* and *A. stuartii* (subsequently identified as the southern species, *A. agilis,* by Chris Dickman). She made the astonishing discovery that the males of both species died immediately after the brief mating period and before the females that they had impregnated gave birth. At the first seminar where she presented her conclusions she was met with scepticism or outright disbelief by the senior faculty present. Her discovery triggered a great deal of work on many aspects of dasyurid biology. We now know that the phenomenon of male die off and semelparity is common to all species of this genus and the related genera *Phascogale* and *Dasykaluta* but probably occurs in no other species of marsupial.

As well as this seminal discovery, Pat Woolley was also the first person successfully to breed dasyurids in captivity. Finding and obtaining a nucleus of each species required extensive and arduous fieldwork, followed by meticulous husbandry back in Melbourne. Having studied the reproduction of about half of the 47 species of Australian dasyurids, she turned her attention to the 17 species in New Guinea. During the 1980s and 1990s she made 18 visits to the interior of Papua New Guinea and West Irian, sometimes under extremely arduous conditions. Because of her unrivalled knowledge of the reproduction of dasyurid marsupials she has been a co-author on three major reviews that have attempted to classify the variety of reproductive patterns in the family and to propose evolutionary pathways by which they may have arisen.

Pat Woolley also re-discovered the very rare Julia Creek dunnart, *Sminthopsis douglasi,* which had been described by Mike Archer in 1979 on the basis of three museum specimens, registered between 1911 and 1933, and was presumed to be extinct. With a small nucleus of animals she managed to capture, she established a breeding colony in Melbourne and is now, in association with the Queensland Parks and Wildlife Service and with support from the Australian Academy of Science, studying the movements of animals in the field.

Pat Woolley's signal contributions to Australian mammalogy were recognised by the 1999 Outstanding Achievement Award of the Society of Woman Geographers, USA. Then in 2000 she was elected to Life Membership of the Australian Mammal Society, and in 2001 to Honorary Life Membership of the American Society of Mammalogists, only the second Australian to be so honoured. It is most fitting that this compendium of research on carnivorous marsupials carries a dedication to Pat Woolley, who can justly be recognised as the mother of all dasyurid biology.

Hugh Tyndale-Biscoe

Dedication to Osvaldo Reig

Although he passed away in 1992, Osvaldo Reig is clearly remembered as the leader of South American mammalogists for many years. His research work on evolutionary biology, genetics, mammalogy, and epistemology of the species concept, went far beyond the frontiers of his country and turned him into the most acknowledged Latin American mammalogist. It is impossible to assess his enormous contribution to mammalogy without bearing in mind his strong personality and influence in the formation of young theriologists. Brilliant, outstanding in many fields, but not always an easy man, he was like a huge planet in the Argentinian's scientific solar system. No one studying mammals was able to ignore his monopolic and centripetal gravitational force.

Osvaldo Reig was twice awarded a Guggenheim Fellowship for doing research at Harvard University (1966) and at the British Museum (1971), and in 1973 he received his PhD in Zoology and Paleontology from London University. In 1986 he was appointed as Foreign Member of the Academy of Sciences in the United States of America, of the Academy of Sciences in the former Soviet Union, and member of the Academy of Science of the Third World.

In one of his early works, Reig astonished his colleagues devoted to the study of marsupials with the recognition that *Dromiciops* was a living microbiotheriid. This discovery was a product of his broad knowledge, combining paleontology with the study of living mammals. From the mid-fifties onwards, he worked on Late Cenozoic didelphimorphian marsupials. He described several taxa which are key for the understanding of the Neogene opossum radiations in South America. He set important standards for biostratigraphic considerations on the Pliocene–Early Pleistocene cliffs of Southeastern Pampean region (Didelphoids playing an important role in his conclusions). Some of his outstanding review papers were co-authored with world authorities in evolutionary biology.

During the 1960s he worked on the systematics, ecology, and distribution of living Pampean opossums. By the end of that decade he began working in the genetic and chromosomic evolution of South American marsupials, especially of Southern South American marmosines. Since the fifties and sixties he devoted several papers to theoretical topics in evolutionary biology, and also devoted much of his thoughts to South American mammalian evolution and biogeography. His 1980s contribution, '*Teoría del Origen y desarrollo de la fauna de Mamíferos de América del Sur*', was seminal for our current understanding of mammalian evolution in this and other Southern continents. He was engaged in the initial chromosomic study on New World oppossums, and co-authored the first holistic, integrative, and comprehensive review of didelphid relationships in the first *Carnivorous Marsupials* volume. During the late eighties he contributed, together with J.A.W. Kirsch and L.G. Marshall, to an important review of the phylogenetic affinities of living and fossil South American opossums.

Osvaldo was a passionate, indisputable master of young scientists, and a demolisher of his scientific opponents. The influence exerted by his professorship went beyond the academic realm, becoming a paradigm for the new generations. It was his unbreakable commitment to the democratic ideals that made him return many times to Argentina after every exile. He had over 125 publications including contributions in mammalogy, paleontology, genetics, natural history and evolution.

Adrian Monjeau

Dedication to Rosendo Pascual

Rosendo Pascual could have been a prominent Argentinian geologist. He graduated in geology at the Universidad de La Plata, and obtained his doctoral degree on geological observations at America's highest peak, the Aconcagua, in the Andean Range at Mendoza province. Early in his career, however, he felt intrigued at the numerous fossil mammal remains coming from Patagonia that were housed, mostly unstudied, in the collections of the Museo de La Plata.

In 1963 he was awarded a John Simon Guggenheim Fellowship for the study of Early Tertiary mammals of Patagonia, under the advice of George Gaylord Simpson. Since then, most of his work has been dealing with South American extinct mammals, their evolution, as well as their biochronological, environmental, and biogeographical significance. Chief of the Vertebrate Paleontological Department of the Museo de La Plata since 1959, he has studied most Cenozoic groups of South American mammals: monotremes, sirenians, carnivorans, primates, native 'ungulates' (toxodontids, henricosborniids), xenarthrans, caviomorph rodents (caviids, hydrochoeriids, octodontids, echimyids, neoepiblemids), and marsupials (microbiotherians, sparassodonts, paucituberculatans, and polydolopimorphians). He produced numerous contributions on the probable origins of South American mammals, and on the relation between their evolution and climate, environments, Andean diastrophism, global change, and extinction events. Several of them are still classics on their respective topics. The standard scheme of South American land-mammal ages owes much to his efforts in this field.

As leader of many field expeditions, mostly to Patagonia, Rosendo Pascual discovered several of the most outstanding fossil mammal collections housed at the Museo de La Plata. His interests on South American marsupials, mostly developed in the 1980s, were focused on their taxonomy, adaptations, and environmental significance. Some of the best-preserved specimens of extremely specialised marsupials, such as *Proborhyaena gigantea*, *Groeberia minoprioi* and *Patagonia peregrina*, were collected and/or studied by him and his colleagues at the Museo de La Plata. In recent years, he has been devoted to the study of Mesozoic South American mammals, as well as on the elucidation of the most distinctive episodes in the evolution of our native lineages.

Francisco Goin

Contents

Part 3 Physiology

Part 4 Evolutionary ecology and behaviour

Part 5 Conservation

PREFACE

When *Carnivorous Marsupials* was published in 1982, it provided a major advance in our understanding of these fascinating mammals. It also had a catalytic effect on the research community, stimulating a wealth of studies on the carnivorous marsupials of both the Americas and Australasia. By the late 1990s it was clear that a further synthesis was needed. In 1998, following the initial idea by Menna Jones, the editors approached the Australian Mammal Society to seek support for running a symposium on carnivorous marsupials in concert with the Society's annual scientific meeting. With the generous assistance of Bill Breed and other councillors of the Society at the time, the symposium took place in July 1999 at the Richmond campus of the University of Western Sydney, with over 250 registrants attending. There was high enthusiasm to go to print.

The publication process required several steps. Firstly, while the symposium highlighted recent understanding of many aspects of the biology of carnivorous marsupials, several major players, especially from South America, were unable to attend. We wanted to ensure that a book on carnivorous marsupials would provide up-to-date coverage on the biology of all taxa, so this meant soliciting contributions from additional key people. Fortunately, everybody we contacted agreed to contribute. Secondly, although we wanted a book on carnivorous marsupials to contain overviews on a range of subject areas, we did not want to lose the contributions of researchers who had presented primary data on specific topics at the Richmond symposium. This dilemma was solved with the assistance of David Morton, at CSIRO Publishing, who agreed to publish these contributions in special issues of the *Australian Journal of Zoology* and *Wildlife Research*. After refereeing in the usual way, 12 research papers on carnivorous marsupials were published in issue 48(5) of the *Australian Journal of Zoology* in 2000, and a further nine papers appeared in issue 28(5) of *Wildlife Research* in 2001. Finally, we needed a publisher for the more than 30 overview contributions that we anticipated for the book itself. We are indebted to Kevin Jeans in the early stages of negotiations and then to Nick Alexander of CSIRO Publishing for enthusiastically taking the venture on board. *Predators with Pouches* was now in the pouch, but still with much growth to be achieved.

Authors were encouraged to write synthetic overviews of their topic areas, to be provocative in developing new ideas, and to provide suggestions for future research directions. Much like *Carnivorous Marsupials*, we hope that *Predators with Pouches* will be of use not just to current research students, practitioners and other professionals, but will also stimulate novel directions for the next generation of students and researchers. The three editors each handled about a third of the incoming manuscripts, ensured that each manuscript was reviewed critically by two referees, and liaised with authors about the final typescript. Thirty-three manuscripts survived this process.

Despite great advances in our knowledge of the carnivorous marsupials, especially over the last decade, new species continue to be discovered and the taxonomy of 'known' species often remains contentious. Since 2000 alone, *Antechinus adustus, A. subtropicus* and *Pseudantechinus roryi* have been described as new species from Australia and at least a further two species are known but not yet described, *Gracilinanus ignitus* has been described from Argentina, and Steve Van Dyck has erected the new genera *Micromurexia*, *Paramurexia*, *Phascomurexia* and *Murexechinus* to incorporate known taxa from New Guinea. And these are just the extant taxa; many more extinct forms have been described too. In the face of such change, ongoing debate about whether the kowari is *Dasyuroides byrnei* or *Dasycercus byrnei*, or whether the Tasmanian devil should be *Sarcophilus harrisii* or *S. laniarius* seems almost arcane. Thus, we have not attempted to impose one monotheistic view of nomenclature in this book, but have largely let the opinions of authors stand.

Two other important decisions needed to be made in proceeding with this book. Firstly, how should a carnivorous marsupial be defined? If we took the narrow but popular view that carnivores eat just vertebrate flesh, few of the extant marsupials, at least, would qualify. We have taken the broader view here that carnivory is the consumption of flesh, be it vertebrate or invertebrate. A more difficult question is *how much* flesh needs to be consumed, compared with other foods, before a forager can be considered carnivorous? A pure carnivore clearly will eat entirely animal flesh, but should a species be considered carnivorous if only half the diet comprises flesh, or 70%, or 90%? Most dasyurids and the numbat *Myrmecobius fasciatus* would be classified readily as being toward the 'pure' end of the carnivore spectrum, as would the several now-extinct species of thylacinids, thylacoleonids and borhyaenids. These would be included in any treatment of carnivorous marsupials. However, most South and Central American marsupials have broader diets that include some fruit and other plant materials, as do peramelids, peroryctids, notoryctids and some of the smaller diprotodontids in Australasia. Available evidence suggests that more than half the diet of South and Central American marsupials usually comprises invertebrates or vertebrates, and that the same is probably true for notoryctids. We have used this admittedly rough criterion to include these groups within our coverage, and to exclude peramelids, peroryctids and diprotodontids from consideration.

A second, and easier decision, was to retain our working title *Predators with Pouches* for the finished book. Students of marsupial biology will know, of course, that many marsupials have only a rudimentary pouch; indeed, the tendency toward pouchless-ness is most obvious in some of the carnivorous species. However, we think that 'predators with pouches' neatly encapsulates the subject taxa of the book and should have broad appeal. At the risk of incurring the wrath of purists, we therefore preferred to retain 'predators with pouches' to the end.

OUTLINE OF CONTENTS AND PROGRESS SINCE 1982

The 33 chapters of *Predators with Pouches* are organised under five broad topic areas.

Evolution and systematics

While palaeontologists and morphologists were key contributors to *Carnivorous Marsupials*, many new soft-tissue studies of extant taxa appeared, including those on albumin serology, enzyme electrophoresis and comparative morphology of spermatozoa. Much of this soft-tissue research had been heralded even earlier, in 1968, in the 'Prodromus' of John Kirsch, which had shown that soft-tissue studies could provide a credible test of, or challenge to, phylogenetic conclusions based on considerations of teeth and skulls alone. But the paper in *Carnivorous Marsupials* that caused the most controversy and in the end arguably the most significant transformation in thinking was that given by Fred Szalay based on the comparative morphology of the tarsal bones. As the first thorough, comparative examination of the calcaneum and astragalus, he argued, effectively, that the South American microbiotheriid *Dromiciops australis* had been misinterpreted by soft and hard-tissue studies alike as being part of the American rather than Australian marsupial clade. Several studies, such as that of Vincent Sarich on comparative albumin serology, provided supporting evidence for Szalay's argument. Although there is still disagreement about the precise relationship of *D. australis* to the whole or part of the Australian marsupial radiation, Szalay's 1982 view has been generally accepted.

In the intervening years since 1982 a lot of new discoveries have been made. In Australia, archaic carnivorous marsupials have been found in an early Eocene deposit, at Murgon, south-eastern Queensland, the continent's only known early Tertiary deposit bearing mammalian specimens. One of these, *Djarthia*, is the most plesiomorphic marsupial known from the continent. In this volume, Steve Wroe overviews this Murgon taxon, an enormous number of new thylacinids, dasyurids and other enigmatic carnivorous marsupials that have turned up since 1983 in the very rich late Oligocene to Pliocene deposits of Riversleigh, north-western Queensland, and all other extinct carnivorous marsupials that have been described so far from Australia. In doing so, he has begun to unravel a complex interplay through time between dasyurids, thylacinids and perameloids. Dasyurids, which now fill most of the carnivore niches in Australian ecosystems, have evidently gained ascendancy by outcompeting the formerly more diverse thylacinids and perameloids. Wroe also overviews the history of other carnivorous Australian marsupials, including propleopine kangaroos and thylacoleonids, and concludes that on balance the high diversity and in some cases sizes of carnivores on this continent have not, as others have claimed, been constrained or limited by environmental factors.

In South America the Paleocene deposits of Tiupampa in Bolivia have produced not only the most archaic marsupials of that continent, but many remarkably complete skeletons, such as those of *Pucadelphys*, *Andinodelphys* and *Mayulestes*. These are reported in this volume by Christian de Muizon and Christine Argot. The skeletons reveal that, in contrast to living South American marsupial carnivores, their owners were surprisingly agile and probably arboreal as well as terrestrial, perhaps most resembling Australian dasyurids in their movements. They therefore challenge Szalay's 1982 view that the ancestral marsupial was primarily arboreal. They further argue that *Pucadelphys* and *Andinodelphys* are so generalised that they could present a structural pattern that was ancestral to dasyuroids as well as didelphoids. Francisco Goin reviews in this volume all of the early marsupial carnivore radiations known from South America, and comes up with some very challenging suggestions that go way beyond the phylogenetic overviews presented in *Carnivorous Marsupials*. He challenges presumptions about monophyly for many of the higher taxa of carnivorous marsupials, including Didelphimorphia, Sparassodonta and the diprotodont groups. He also suggests that *Djarthia* and two other Australian late Oligocene taxa, *Ankotarinja* and *Keeuna*, which have been argued about ever since they were described by Archer in 1976, are actually members of Sternbergiidae, an extinct family otherwise known from South America. In a similarly thoughtful and fresh approach, he suggests that Australia's perameloids and the South American derorhynchids may be a clade. This paper in effect suggests that there may have been fewer biogeographical disjunctions across the once-united lands of South America, Antarctica and Australia than we presumed in 1982. Eduardo

Palma in this volume adds further doubt to conclusions reached in 1982, by focusing attention on molecular evidence that appears to relate Australian bandicoots to South American caenolestoids. While these are not strictly speaking carnivorous marsupials, the members of both groups eat meat and have been involved in earlier arguments about relationships on the one hand to dasyuroids (perameloids) and on the other to didelphids (caenolestoids).

Carey Krajewski and Michael Westerman overview the molecular systematics of dasyuromorphian carnivores and conclude that, since 1982, based on studies including albumin micro-complement fixation, DNA hybridisation and DNA sequence data, intrafamilial relationships within Dasyuridae are partly as conceived by Archer in 1982, but in other areas in need of significant revision. On the basis of molecular clock data, they propose that the living descendants of the family are likely the result of an initial radiation that occurred in the middle Miocene, a timing that agrees with the conclusions of Wroe that dasyurids as such are not known prior to the Miocene. Mathew Crowther and Mark Blacket take a more 'why is it so' approach in this volume by trying to understand the reasons for the high species-level diversity of living dasyurids. They make the point that we had only just begun to realise in 1982 that there are sibling species everywhere in Australia, often so many that they significantly confused earlier efforts to relate species to ecological parameters. Why there are so many sibling species and how they form is less clear. These authors explore potential explanations and, in particular, the role of competition. Sergio Solari provides a similar analysis of the diversity of the marmosid species of *Thylamys* in South America. Very little about the diversity of South American didelphoids was overviewed or even considered in *Carnivorous Marsupials*, although these are the most diverse of the South American marsupials. This paper makes a similar point to that of Crowther and Blacket's: that we have significantly underestimated species-level diversity in many groups, and in small carnivorous marsupials in particular, on both continents. The contribution by James Patton and Leonora Costa also focuses on South American didelphoids. These authors take a different approach by using molecular data to analyse phylogeographic patterns across South America. While incidentally *also* noting that species diversity in the groups they examined has been grossly underestimated, they demonstrate the value of this approach for interpreting the history of ecological / geographical barriers that now separate once continuous populations.

Clearly, the new contributions in this volume have moved the frontiers of this field much further forward than where we left them in 1982. The proportionately greater contribution to understanding about relationships coming from molecular data is also clear, although morphological studies are still central to making sense of the relationships of extinct taxa as well as providing the only datasets common to both living and extinct taxa. While in some areas increasing congruence in the interpretation of relationships is evident, such as the now undoubted relationship of microbiotheriids to Australian marsupials, some authors encourage us to be more open-minded than ever about intercontinental relationships. These uncertainties focus in particular on the relationships of perameloids and many of the Oligocene / Miocene taxa from Australia. Many authors are also telling us that we still have a great deal to learn about basic species-level diversity of carnivorous marsupials on both continents. Clearly there will be plenty of fascinating work to do before this research area is revisited in the future. But given the rapid increase in the number of researchers focused on phylogenetics, the increased use of molecular tools and the increasing attention being paid to morphological datasets in addition to teeth and skulls, the need to take stock of progress in the future may come much more quickly than it has up to now.

Reproduction and development

The reproductive and developmental oddities of marsupials have long been a focus of research attention, and this interest is reflected in the five papers in this section. Many carnivorous marsupials breed seasonally, with some showing almost clockwork precision in the inter-annual timing of their reproduction. Bronwyn McAllan emphasises the central role of photoperiod in synchronising ovulation and testicular activity in many of these species, and notes that in some, such as members of the *Antechinus stuartii / A. flavipes* group, the photoperiodic cue is not actual daylength but the daily rate of change in daylength in spring. This unusual response allows animals in eastern Australia to breed later at low rather than at higher latitudes, and thereby to take advantage of the later flush of food resources that occur in more northerly latitudes in spring. McAllan also raises the novel and tantalising possibility that factors such as pheromones may have stimulatory effects. Not all carnivorous marsupials breed seasonally, however, as Pat Woolley demonstrates in her contribution. In this study, the pattern of reproduction is elucidated for the first time for seven species of New Guinea dasyurids. In contrast to their Australian counterparts, all seven species appear capable of breeding year round, and carry relatively small litters of just four or fewer young. The contribution by Bill Breed and co-workers focuses on the structure and development of sperm and eggs. Comparisons of gametes between dasyurids and didelphids reveal large differences in design, as well as in sperm behaviour and storage in the female tract. These findings suggest deep divergences between the two groups, and provide a point of contrast with the phylogenetic interpretations noted above that should be a stimulus for further inquiry. Jose Nogueira and Antonio Castro extend our knowledge of reproductive design by reviewing in detail the structural and functional components and associated glandular elements of the genital system of male didelphids. These authors note that reproduction in female marsupials has historically attracted most research attention, and go on to demonstrate how studies of the male system can increase our understanding of marsupial phylogeny.

Moving beyond reproduction, John Nelson and co-workers provide an overview of perinatal development in marsupials, describing in detail the behaviour, growth and neuro-anatomical stages of the northern quoll, *Dasyurus hallucatus*. At birth, this quoll weighs just 18 mg and is at the earliest recorded stage of development of any marsupial; it therefore provides exceptional insight into the maturation process.

These studies extend the range of carnivorous marsupials for which we now have baseline data on reproduction, and greatly expand our understanding of the proximate and ultimate factors that drive reproductive and developmental patterns. To some extent these advances have been made possible by the use of techniques that were poorly developed in 1982, such as fine-scale radiography or fluorescent labelling. Other advances have been made only after long and arduous travails in the field. As these contributions show, combined field and laboratory studies can provide excellent insight into how carnivorous marsupials reproduce and grow, and yield understanding of the selective regimes that produce the patterns observed.

Physiology

Although this is the shortest of the book's five topic areas, the contributions are diverse in their scope. In his chapter, Ian Hume reviews the range of diets exhibited by carnivorous marsupials and the physiological benefits and costs associated with diets high in vertebrate or invertebrate prey. In general, carnivore diets contain little carbohydrate and variable amounts of fat, but are rich in vitamins, minerals, water and highly digestible protein. However, strict carnivory limits exploitation of non-animal food and consequently confines 'pure' carnivores to a narrow nutritional niche. Diego Astúa de Moraes and colleagues use a different technique to explore the diets of carnivorous marsupials: standardised food preference experiments using captive animals. Using 12 species of didelphids, these authors demonstrate congruence between the species' natural diets and the relative amounts of protein, lipid and carbohydrate present in the foods they selected in the cafeteria trials. Because the foods presented in these experiments were available commercially, the results suggest that the nutritional requirements of a much broader range of carnivorous marsupials may now be studied with relative ease.

In a wide-ranging review, Fritz Geiser outlines the insights that have been gained into the energetics of carnivorous marsupials from studies of (mostly) captive animals. Despite the nutritional benefits derived from a carnivorous diet, the small size (<13 000 g) and consequently high surface-area to volume ratio of carnivorous marsupials means that they often face problems of heat loss or gain. Geiser describes the extraordinary range of behavioural, physiological and morphological mechanisms used by carnivorous marsupials to minimise energy expenditure, and shows that dasyurids and didelphids have often adopted similar solutions to common thermoenergetic problems. Adrian Bradley takes a fresh look at one of the most intriguing problems in carnivorous marsupial biology: the proximate cause of male death after mating in *Antechinus* spp. and some other dasyurids. Using recent research on eutherians, Bradley constructs a new model to explain die-off that incorporates a breakdown in the feedback of corticosteroid hormones that leads to intolerable levels of physiological stress.

While significant advances have been made in the last 20 years, our knowledge of some aspects of the physiology of carnivorous marsupials was already quite advanced in 1982. Much of this knowledge had come from laboratory studies, as can be seen by perusing the six physiology papers presented in *Carnivorous Marsupials*. What has changed is the perception that profound physiological insight can be gained by carrying out comparative studies in the laboratory and on free-ranging animals in the field. This sentiment is conveyed implicitly or explicitly by several of the authors here. Indeed, in making a plea for obtaining more specific nutritional information on carnivorous marsupials, Ian Hume argues that this will allow better understanding of the resource requirements of similar species and thus allow informed management of their habitats; this should maximise our ability to retain species in the wild. Some physiological studies are likely to remain laboratory-based due to the tight environmental control that can be imposed and maintained by the researcher. In other studies, however, significant advances are likely to come from effective integration of laboratory and field observations. This process is under way.

Evolutionary ecology and behaviour

Interest in the ecology and behaviour of carnivorous marsupials has burgeoned since 1982, with exploration of topics such as sperm competition, sex allocation, guild structure and macroecology that were scarcely on the radar when *Carnivorous Marsupials* appeared. More species also have been studied in recent years, allowing more powerful generalisations to be made. To some extent advances have been dependent on developments in technology, such as tracking devices that allow remote eavesdropping of animals in the field, or increased computing power that allows interpretation of complex animal-environment relationships. However, other advances have come from carrying out simple but carefully planned field experiments or from reinterpreting observed behaviours in a more explicitly evolutionary framework. The nine chapters in this section reflect many of these advances and new thinking.

In the first chapter Emerson Vieira and Diego Astúa de Moraes revisit the question of what carnivorous marsupials eat, but their emphasis is on how diet correlates with other aspects of ecology. They show that, among South and Central American taxa, the more strictly carnivorous marsupials tend to be smaller, scarcer and more terrestrial than their more omnivorous counterparts. However, the authors also suggest that the energetic and nutritional benefits of a highly carnivorous diet enable better exploitation of adverse environments. Thus the flesh-eating thick-tailed opossum *Lutreolina crassicaudata* occurs in cooler climates to almost 40° south, while the Patagonian opossum *Lestodelphys halli* occurs in southern Argentina further south than any other living marsupial. Menna Jones also pursues the link between diet and ecology, but her emphasis is on the convergence in trophic morphology and diet among guilds of carnivorous marsupials and placentals. Despite superficial similarities between these two groups, there are differences in skeletal and dental structures – especially the shape of the canine teeth – that appear to reflect different solutions to similar biomechanical problems. Surprisingly, however, both carnivore groups show common patterns of size partitioning to each other within guilds, suggesting that there is convergence in the structuring mechanisms at this level.

In their chapter, Elmer Birney and Adrian Monjeau explore a wide range of patterns in the distribution, diversity, body size and behaviour of all South American marsupials, relating these to latitudinal gradients in temperature and rainfall. Among the many intriguing patterns they reveal are that species diversity correlates most strongly with annual mean minimum temperature, while the more strictly carnivorous, small-bodied and fat-tailed forms occur predominantly in the cooler southern latitudes. These latter results parallel those of Vieira and Astúa de Moraes. Birney and Monjeau conclude their overview by identifying five broad 'life forms', or guilds of South American marsupials that will be of considerable heuristic value to future workers. Chris Dickman's contribution continues the macroecology theme, using field survey data and climatic modelling to describe the distributions and species densities of dasyurids in Australia. In contrast to their South American counterparts, dasyurids show little consistent relationship to broad climatic or latitudinal gradients, but achieve their highest diversity in hummock grass and desert complex habitats of the arid centre. This counter-intuitive result arises because biotic constraints such as competition and predation are less prevalent in arid than in temperate habitats, allowing a higher proportion of species from regional species pools to co-occur in local areas.

The contribution by David Croft switches the emphasis from macroecology to patterns of individual and social behaviour in carnivorous marsupials. Although most species are solitary, they exhibit a broad repertoire of social behaviours and show considerable versatility in the channels used for communication. This chapter makes the important point that, while behaviour needs to be accurately described, it must often be subject to comparative and experimental analyses so that the fitness benefits accruing to individuals are understood. Charlotte Toftegaard and Adrian Bradley take up the analytical challenge in their chapter, which reviews recent studies on chemical communication in dasyurids. Chemical communication among animals can be the most difficult sensory modality for humans to understand because of our limited ability to detect and quantify airborne chemicals, or pheromones. The authors describe both the production and dispersal of pheromones in dasyurids and the poorly known organ of reception, the vomeronasal organ. The increasing availability of techniques that can characterise pheromones (e.g. gas chromatography coupled with mass spectroscopy) and identify their neurological effects (e.g. functional magnetic resonance imaging) suggests that this will be an active and fruitful line for future research.

In their chapter in this volume, Dave Taggart and co-workers assemble a large dataset on reproductive and life history attributes of carnivorous marsupials, and use it to predict species' mating systems and the likelihood that they exhibit sperm competition. In general, sperm competition is most likely to occur in sexually dimorphic species in which behavioural oestrus and copulation are relatively long, testis mass relative to body mass is high, and sperm are stored for several days in the female tract before ovulation and fertilisation occur. Species that share these features also tend to exhibit post-mating male death, thus supporting the intriguing possibility that sperm competition has acted as an evolutionary catalyst for male semelparity. Simon Ward's chapter takes up the story post-mating, and addresses the questions of how and why biased sex ratios occur in the litters of some carnivorous marsupials. Females appear to have the ability to skew the sex ratio of their litters quite strongly toward one sex or the other, but neither the mechanism nor the adaptive significance of this ability are clearly known. If, as suspected, pre-fertilisation mechanisms contribute to sex ratio bias, it is possible that the sexes of sperm respond differently during their period of storage in the female tract.

In the last chapter in this section, Ian Beveridge and Dave Spratt provide an authoritative overview of the endo- and ectoparasites of carnivorous marsupials. The authors demonstrate that the helminth assemblages found in dasyurids are strikingly similar to those in ecologically similar placental mammals, and suggest that this reflects convergence in both the parasite life cycles and feeding behaviour of the hosts. Unfortunately, while dasyurids are shown to harbour a diverse array of parasites, comparisons with other carnivorous marsupials are limited. Neither host-parasite relationships nor impacts of parasites on host health have been well studied, justifying the conclusion of the authors that there will be no shortage of research topics in this field for the foreseeable future.

The contributions presented in this section outline some exciting advances since 1982, especially with respect to interpreting individual behaviour in carnivorous marsupials and understanding patterns in their assemblage structure and the processes that shape them. Because of the breadth of endeavours in this field it is difficult to predict where future advances are most likely to be made. However, there is little doubt that detailed autecological studies will continue to elucidate the behaviour and ecology of individual species, and in turn that these will contribute to better macroecological datasets and meta-analyses. Biotic interactions should also receive more attention, such as those between carnivorous marsupials and their parasites or between carnivorous marsupials and their prey. Several authors made pleas for longer-term (≥3 year) studies so that biological signals can be discerned from background noise; such studies are needed on almost all carnivorous marsupials, but most especially on the more poorly known taxa from South America and New Guinea.

Conservation

The most striking shift in emphasis between *Carnivorous Marsupials* and *Predators with Pouches* concerns conservation issues. In the earlier book just three chapters had any focus on conservation, and two of these dealt with the already-extinct thylacine *Thylacinus cynocephalus*. The seven contributions in the present volume might suggest that the status of carnivorous marsupials has deteriorated rapidly; in reality, however, we now have a much clearer idea of species' status and of the factors threatening carnivorous marsupials than we did in 1982.

Gustavo da Fonseca and co-workers provide a status assessment of Central and South American marsupials. While 23 of these species are listed as threatened on the Red List maintained by the International Union for Conservation of Nature and Natural Resources (IUCN), the authors note that a further 11 species have restricted ranges and require further assessment. Habitat loss is a problem for most or all species, and is acute in 'hotspot' areas that contain up to 80% of the threatened taxa.

Six chapters deal with conservation and management of carnivorous marsupials in the Australasian region. In the first of these, Barbara Wilson and co-workers focus on Australian dasyurids weighing less than 500 g, and show that the larger habitat specialists among them are the most extinction-prone. Habitat loss and modification are key causes of decline, but so too are the impacts of introduced species (principally red foxes, feral cats and rabbits) and changes to fire regimes. The complementary contribution by Menna Jones and colleagues covers the larger Australasian dasyurids and the thylacine, painting a grim picture of their status: two of the eight species are threatened, two are so poorly known that no conservation status can be assigned, and the thylacine, of course, is extinct. Only three species can be considered as being at lower conservation risk. The same gamut of factors that cause losses of smaller dasyurids also threatens the larger species but, in addition, the larger taxa are at risk locally from persecution, poisoning and road mortality.

In the above contributions to the conservation section, most authors provide suggestions for *in situ* management or report on the status of captive breeding programs for carnivorous marsupials, but note that recovery actions taken to date have been scant. In her chapter in this volume, Karen Firestone highlights another important but as yet under-utilised technique for species management: conservation genetics. This has the potential to uncover cryptic species, identify population units that should be priorities for management, and provide solutions to the damaging effects of genetic drift and loss of genetic variability in small populations. Despite the utility of this approach, it has been applied so far to few species of carnivorous marsupials. Keith Morris and colleagues offer perhaps the most optimistic perspective in their chapter by describing the steps taken to successfully recover the chuditch *Dasyurus geoffroii*. Once widespread across much of Australia, this species declined precipitously in the twentieth century and became confined to the far south-west of the continent; it was listed as endangered in 1991. Following detailed autecological research, release of captive-bred animals back into the wild and, most importantly, widespread poison baiting of foxes, the chuditch has been effectively recovered.

The chapters by Tony Friend and Neil Thomas and by Joe Benshemesh and Ken Johnson provide summaries of the conservation status and ecology of some of the most divergent carnivorous marsupials, respectively, the numbat *Myrmecobius fasciatus* and the marsupial moles *Notoryctes* spp. Like the chuditch, the myrmecophilous numbat suffered a dramatic decline in range in the twentieth century, and by the mid-1980s had become confined to two sites in the south-west of Western Australia. A sustained campaign of fox control, coupled with a vigorous reintroduction program, has seen a dramatic increase in numbers, and this species has been recently downgraded from 'endangered' to 'vulnerable'. In contrast, the enigmatic marsupial moles appear to have suffered broadscale declines such that both species are considered endangered. Benshemesh and Johnson review the little that is known of these species and show the extraordinary difficulty of studying them in the field.

Several themes recur in this section. Authors lament declines and losses of their study taxa, but offer detailed prescriptions for identifying species at risk, for demographic or genetic monitoring of small populations, and for diagnosing and managing the factors causing populations to decline. Several authors also lament the lack of resources to implement recovery plans and actions. However, the dramatic recoveries of chuditch and numbat provide hope that recovery should often be achievable, despite the odds, and provide lessons in how to get it right. Such lessons should be heeded by all with an interest in the future of carnivorous marsupials.

Acknowledgements

We are indebted to the Australian Mammal Society for its enthusiastic support of the symposium on carnivorous marsupials in 1999, which provided the impetus for this book. The assistance of then-President Bill Breed and Barry Richardson on the symposium committee is particularly appreciated. We are also indebted to all contributors at the symposium, most particularly to the authors for their papers in the present volume, and to Lisa Akison for assistance in liaising with authors about their contributions. Manuscripts were improved by the generous assistance of a large number of referees, some of whom provided two or even three reviews. They were: Patsy Armati, Steve Austad, Peter Banks, Grant Blackwell, Carolyn Blanchard, Robert Blanchard, Don Bradshaw, Gary Bronner, Andrew Burbidge, Mike Calver, Richard Cifelli, Andrew Cockburn, Steve Cork, Mathew Crowther, Tamar Dayan, Christian de Muizon, Louise Emmons, Judith Field, Hugh Ford, Colin Groves, John Harder, Leon Hughes, Tony Hulbert, Ian Hume, Peter Jarman, Chris Johnson, John Kirsch, David Macdonald, Craig Moritz, Brad Murray, Stuart Nicoll, Meri Oakwood, David Obendorf, Bruce Patterson, Jim Patton, Lisa Pope, Marilyn Renfree, Barry Richardson, John Rodger, Eleanor Russell, Lynne Selwood, Melody Serena, Bill Sherwin, Grant Singleton, Hugh Tyndale-Biscoe, Blaire Van Valkenburgh, Steve Van Dyck, Emerson Vieira, Mike Westerman, Pat Woolley, Steve Wroe, Bruce Wunder and Derek Yalden. Leon Barmuta, Carol McKechnie and Sue Hand provided support throughout the project. Finally, the enterprise would not have been completed without the advice and encouragement of many people at CSIRO Publishing, most notably Kevin Jeans, Briana Elwood and Nick Alexander.

Menna Jones, Chris Dickman and Mike Archer

Part I

Evolution and Systematics

CHAPTER 1

MOLECULAR SYSTEMATICS OF DASYUROMORPHIA

Carey Krajewski[A, B] *and Michael Westerman*[B]

[A] Department of Zoology and Center for Systematic Biology, Southern Illinois University, Carbondale, Illinois, 62901-6501, USA
[B] Department of Genetics, La Trobe University, Bundoora, Victoria 3083, Australia

Molecular phylogenetic studies (albumin microcomplement fixation DNA hybridisation, and DNA sequence analyses) since 1990 have refined the hypothesis of dasyuromorphian relationships inferred from morphological and allozyme data in 1982. DNA sequences weakly support Thylacinidae and Dasyuridae as sister-families apart from Myrmecobiidae, but this problem requires further study. All molecular studies agree on the placement of two endemic New Guinean groups within larger clades of dasyurids. Specifically, *Phascolosorex* and *Neophascogale* ('phascolosoricines') form a monophyletic group with dasyurines (quolls, false antechinuses, and allies), and *Murexia* (including all New Guinean 'antechinuses') forms a monophyletic group with phascogalines (phascogales and antechinuses). Otherwise, the basic phylogenetic structure of Dasyuridae inferred from molecular data is consistent with earlier morphological and allozyme results. DNA sequences place 'phascolosoricines' as sister to a *Dasyurus-Sarcophilus* clade within Dasyurini, and *Murexia* as sister to *Antechinus* within Phascogalini; both results have major implications for morphological evolution and require further evaluation. Within Dasyurini, DNA sequences show several early lineages (including the *Dasyurus-Sarcophilus* group) that may have resulted from an episode of rapid cladogenesis. Within *Antechinus*, DNA sequences recover monophyletic species groups that are identical to those proposed on the basis of allozymes. Within Sminthopsini, *Sminthopsis* and *Ningaui* form a clade apart from *Antechinomys*. Although DNA sequences fail to resolve a monophyletic *Sminthopsis*, they do recover dunnart species groups that are partially congruent with those suggested by morphology. Within Planigalini, *Planigale maculata* and perhaps *Pl. novaeguineae* are sister to other planigale species. Relationships among early dasyurin lineages, *Antechinus* species groups, *Murexia* species, *Sminthopsis* species groups, and most *Planigale* species are currently unresolved. We endorse a revised suprageneric classification of dasyurids that reflects the most robust molecular results. Cladogenic dates estimated from a molecular clock indicate that the four major radiations of modern dasyurids took place in the late mid-Miocene, perhaps in response to climatic drying across Australia. DNA sequence studies concur with recent morphological and allozyme results in suggesting the existence of several (perhaps many) undescribed or cryptic dasyurid species.

INTRODUCTION

The marsupial order Dasyuromorphia consists of three families (Table 1): the extinct Thylacinidae, the monotypic Myrmecobiidae, and the speciose Dasyuridae (Aplin and Archer 1987). Thylacinids are known from the latest Oligocene (Muirhead and Wroe 1998), with some nine described species in six genera, culminating in the recently extinct Tasmanian 'wolf' (*Thylacinus cynocephalus*). Myrmecobiidae includes only a single species, the numbat (*Myrmecobius fasciatus*), with no pre-Pleistocene fossil record and now confined to south-western Australia (Archer 1984). Dasyurid fossils are known since the early to middle Miocene, though no modern genera predate the Pliocene (Wroe 1999). Extant dasyurids currently comprise 66 recognised species, though ongoing revisionary studies promise to increase this number (see below). Dasyurids occur in all major habitat types in Australia and New Guinea, and collectively represent the dominant mammalian carnivores-insectivores in both regions. Their ecological diversity is paralleled by striking variation in life histories (Lee et al. 1982), though in other respects (e.g. karyotypes, general morphology) they have been considered the most primitive Australian marsupials.

Table 1 Traditional classification of modern dasyuromorphian marsupials

Ordinal and familial nomenclature follows Aplin and Archer (1987); dasyurid suprageneric groups are those of Archer (1982a, 1984). Fossil groups are not included.

- Order Dasyuromorphia
  - Family Thylacinidae
    - *Thylacinus*
  - Family Myrmecobiidae
    - *Myrmecobius*
  - Family Dasyuridae
    - Subfamily Muricinae
      - *Murexia*
    - Subfamily Phascolosoricinae
      - *Phascolosorex, Neophascogale*
    - Subfamily Phascogalinae
      - *Phascogale, Antechinus*
    - Subfamily Dasyurinae
      - Tribe Parantechini
        - *Parantechinus, Pseudantechinus, Dasykaluta*
      - Tribe Dasyurini
        - *Myoictis, Dasyuroides, Dasycercus, Dasyurus, Sarcophilus*
    - Subfamily Sminthopsinae
      - Tribe Sminthopsini
        - *Sminthopsis, Antechinomys, Ningaui*
      - Tribe Planigalini
        - *Planigale*

The history of dasyuromorphian taxonomy was ably and thoroughly summarised by Archer (1982a, b) and Aplin and Archer (1987), and we will not recapitulate that information here. Our aim is to describe the contribution that molecular studies since 1982 have made to understanding dasyuromorphian phylogeny and evolution. We begin by describing the systematics of the order as laid out by Archer (1982a, b), and then provide a critical summary of how comparative immunology, DNA hybridisation, and DNA sequencing studies have suggested modifications to this picture. In the course of this summary, we will address key points of conflict between molecular and morphological data, and suggest how those conflicts can be put in perspective by examining their evidential bases. Finally, we describe how integration of the molecular phylogeny with paleontological data has provided a framework for analysing macroevolutionary patterns in Dasyuridae. Readers unfamiliar with the terminology of molecular systematics should consult Hillis et al. (1996).

ARCHER'S 1982 SYNTHESIS

Familial relationships

Aplin and Archer (1987) listed three putative synapomorphies for Dasyuromorphia: incisor number reduced to four uppers and three lowers, enlargement of the epitympanic sinus, and loss of the intestinal cecum. Although Wroe (1997) suggested caution in the interpretation of these characters, there has not been much dispute among recent systematists that at least dasyurids and myrmecobiids ('dasyuroids') form a monophyletic group. The position of thylacinids has been more controversial. Bensley (1903), Sinclair (1906), and Wood (1924) noted the striking dental and basicranial similarity of *Thylacinus cynocephalus* to extinct South American borhyaenids. Could this similarity indicate that thylacinids are part of the American, rather than the Australian, marsupial radiation? Simpson (1941) and Tate (1947) rejected this view and argued that thylacines are part of Dasyuridae – that is, that their closest living relatives are a particular group of modern dasyurids. Archer's (1976b, c) analyses of dental and basicranial data indicated that borhyaenid affinities for thylacines could be supported on phenetic grounds. Marshall (1977) considered the same kinds of evidence in a cladistic framework and concluded that thylacinids are sister to 'dasyuroids'. Lowenstein et al. (1981) applied comparative serology to albumin recovered from a thylacine skin and reached the somewhat surprising conclusions that thylacinids and dasyurids diverged only 7 million years ago (Mya), whereas the numbat lineage was isolated some 24 Mya. Szalay's (1982) landmark study of marsupial tarsal morphology also placed thylacinids with 'dasyuroids'. Archer (1982b) reviewed all these studies, reanalysed his own and Marshall's (1977) morphological data in a cladistic framework, and concluded that how one views thylacine affinities depends on the relative weights assigned to dental, cranial, tarsal, paleontological, and serological evidence. In the end, Archer favoured the hypothesis that thylacinids are sister to 'dasyuroids'.

Dasyurid suprageneric groups

Archer (1982a) considered Dasyuridae to include the Miocene genera *Ankotarinja*, *Dasylurinja*, and *Keeuna*, as well as more recent extinct forms. Wroe (1996, 1997) reviewed the anatomical evidence for the monophyly of the Miocene genera with traditional dasyurids and concluded that all three should be considered Dasyuromorphia *insertae sedis*. Moreover, Wroe (1997) found only three reliable anatomical synapomorphies for Dasyuridae: enlargement of the alisphenoid tympanic wing, development of a distinct periotic tympanic sinus, and presence of a distinct tubal foramen. This list was lengthened somewhat in Wroe's (1999) description of the Miocene *Barinya wangala*, the oldest known fossil dasyurid, which Wroe argued constitutes the sister-group to all modern dasyurids. Thus the monophyly of Dasyuridae seems reasonably well-supported on morphological grounds.

Archer's (1982a, 1984) hypothesis of suprageneric relationships within Dasyuridae is most easily expressed by his suggested classification (Table 1). In erecting this scheme, Archer drew on all previous anatomical studies of dasyurids as well as the allozyme analysis of Australian forms by Baverstock et al. (1982). Excluding the Miocene genera, five putatively monophyletic groups were recognised at the subfamily level, but relationships among them were unresolved. Subfamilies Muricinae and Phascolosoricinae were New Guinean endemics. Archer noted that species of *Murexia* are the most anatomically primitive living dasyurids, whereas phascolosoricines were the most 'phylogenetically enigmatic'. *Phascolosorex* and *Neophascogale* appear to have highly derived dental characteristics, but essentially primitive basicrania (S. Wroe pers. comm.). Archer's Phascogalinae included *Phascogale* and *Antechinus*, the latter genus encompassing both Australian and New Guinean species. It is important to note that Archer's (1982a) placement of New Guinean dasyurids did not have the benefit of allozyme data for comparison with morphology, as New Guinean tissue samples were unavailable to Baverstock et al. (1982). Sminthopsinae was constituted by Archer (1984) to encompass two divergent groups – Planigalini (*Planigale*) and Sminthopsini (*Sminthopsis, Antechinomys, Ningaui*) – identified as monophyletic in Archer (1982a). The union of these in a single subfamily was supported by unpublished immunological comparisons (Archer 1984). Archer's (1982a) Dasyurinae also included two tribes, Dasyurini and Parantechini, the latter comprising three genera of 'false antechinuses' whose affinity with traditional dasyurines was initially demonstrated by penis morphology (Woolley 1982). Monophyly of Parantechini was suggested by allozymes, but Archer (1982a) listed only one anatomical synapomorphy (entoconid reduction on M_4) for the group. The monophyly of Archer's dasyurine tribes was questioned by Kitchener and Caputi (1988).

Genus- and species-level relationships

Antechinus. Archer posited the monophyly of Australian *Antechinus* species apart from New Guinean forms (the monophyly of which was uncertain), and that these two groups formed a clade apart from *Phascogale* (Fig. 1a). Among Australian antechinuses, relationships in Fig. 1a are identical to those estimated from allozymes (Baverstock et al. 1982) and broadly consistent with previous anatomical analyses (e.g. Van Dyck 1980). The phylogeny shows several species groups among which relationships are poorly resolved: (1) *A. bellus-leo-flavipes*, to which (2) *A. stuartii* might be the sister (Van Dyck 1982); (3) *A. swainsonii-minimus*; and (4) *A. godmani*.

Planigalini. Archer (1976a) noted that *P. maculata* is the most anatomically primitive planigale, and that this species closely resembled *P. novaeguineae*. In contrast, *P. gilesi* is distinct from other planigales in lacking a third upper premolar. Allozymes suggested, however, that *P. gilesi* and *P. tenuirostris* form a clade apart from *P. maculata*. Archer (1982a) integrated the morphological and allozyme results by suggesting that *Pl. maculata* and *Pl. novaeguineae* are the sister-clade to *Pl. gilesi-tenurirostris-ingrami*.

Sminthopsini. Archer (1981) maintained that the monotypic *Antechinomys* could not be separated from *Sminthopsis*, though these two groups were monophyletic apart from *Ninguai* (Fig. 1b). Within *Sminthopsis*, Archer (1982a) posited a polytomy relating five species groups whose compositions reflect the anatomical comparisons of Archer (1981) as modified by allozyme results for eight exemplar taxa (Baverstock et al. 1982). It is important to note that seven new *Sminthopsis* species have been described since 1982, and that no morphological analysis of their affinities was available until the work of Van Dyck et al. (1994).

Dasyurini. Archer (1982a) suggested a phylogenetic division between the New Guinean *Myoictis* and all other dasyurins based on the accessory corpora cavernosa and reduced lower third premolars of the former (Fig. 1c). Among the latter, the arid-adapted *Dasycercus* and *Dasyuroides* form the sister-group to an anatomically derived clade of quolls and the devil. In light of interpretations of dental character evolution suggested by the Pliocene *Dasyurus dunmalli*, Archer (1982a) united *D. hallucatus* and *D. albopunctatus* in the genus *Satanellus* apart from other quolls and *Sarcophilus*.

EARLY MOLECULAR STUDIES

Ancient DNA

In 1989–90, three molecular studies were published that impacted our understanding of dasyuromorphian relationships. In the first of these, Thomas et al. (1989) were able to extract 'ancient' DNA from museum specimens of the thylacine. They obtained very short sequences of the mitochondrial cytochrome b (*cytb*) and 12S rRNA (12S) genes (116 bp and 94 bp,

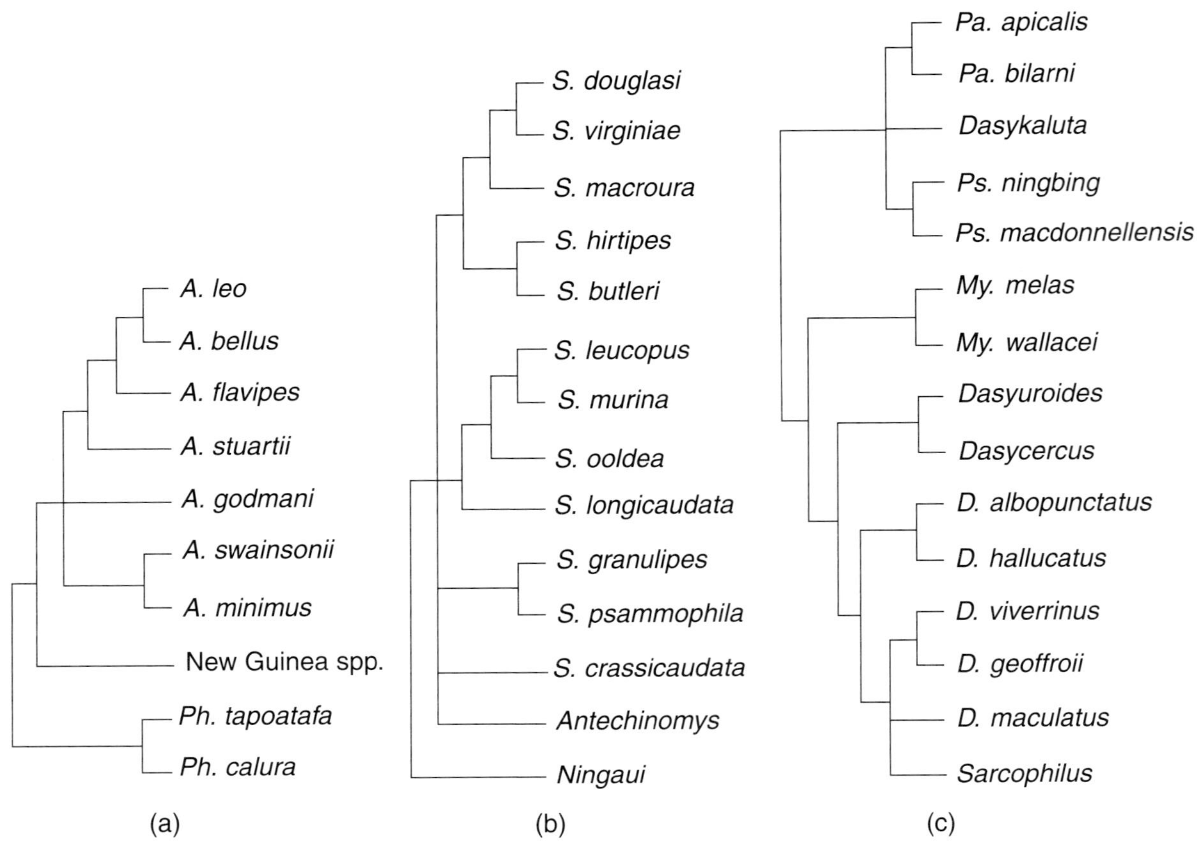

Figure 1 Phylogenetic hypotheses from Archer (1982a) for selected groups of dasyurids. Only extant species are shown. Genus abbreviations: *A.* = *Antechinus*, *D.* = *Dasyurus*, *My* = *Myoictis*, *Pa.* = *Parantechinus*, *Ps.* = *Pseudantechinus*, *S.* = *Sminthopsis*. *(a)* Postulated relationships among phascogalines based on allozyme, cranial, and dental characters. *(b)* Postulated relationships among sminthopsins based on morphology and allozyme characters. Note that Archer (1981) favoured a sister-pairing of *Antechinomys* and *S. crassicaudata*. *(c)* Postulated relationships among dasyurines based on phallic, cranial, dental, and allozyme characters. Note that Archer placed *D. albopunctatus* and *D. hallucatus* in '*Satanellus*'.

respectively), and compared them to homologous sequences from two dasyurids, two diprotodontians, a bandicoot, and a didelphid. Parsimony analysis of 19 informative sites in the 12S gene yielded a tree (rooted with cow) on which *Thylacinus* was sister to *Dasyurus* + *Sarcophilus* with 98% bootstrap resolution. This appeared to constitute strong evidence in favour of dasyuromorphian affinities for the thylacine, but was immediately criticised on the grounds that the small number of informative sites might support the favoured topology by chance (Faith 1990). The ensuing debate fostered development of Faith and Cranston's (1991) topology-dependent permutation tail probability (T-PTP) test, but did not fully resolve questions about the robustness of the thylacine 12S tree.

Microcomplement Fixation (MC′F)

The next major study addressing dasyurid phylogeny was the MC′F survey of Baverstock et al. (1990). MC′F is an immunological technique that measures the cross-reactivity between antisera raised against albumin from one species and antigen (albumin) from another. The index of cross-reactivity (immunological distance) is proportional to the number of amino acid differences between antigens of the species compared (Maxson and Maxson 1986). Several features of the dasyurid phylogeny estimated from MC′F comparisons (Fig. 2a) are significant. First, the single species of *Murexia* included forms a clade with New Guinean antechinuses, while Australian *Antechinus* species are sister to *Phascogale* (contra Archer 1982a). Second, the 'phascolosoricines' (*Phascolosorex* and *Neophascogale*) are associated with dasyurines. Thus MC′F was the first method to clearly suggest relationships between these enigmatic New Guinean groups and larger Australian clades. Third, *Ningaui* and *Sminthopsis* are more closely related to each other than either is to *Antechinomys*. Fourth, the monophyly of *Planigale* with sminthopsin genera (which was initially inferred from immunological data) is not resolved. Fifth, dasyurids are not resolved as monophyletic apart from *Myrmecobius*. The timescale on the

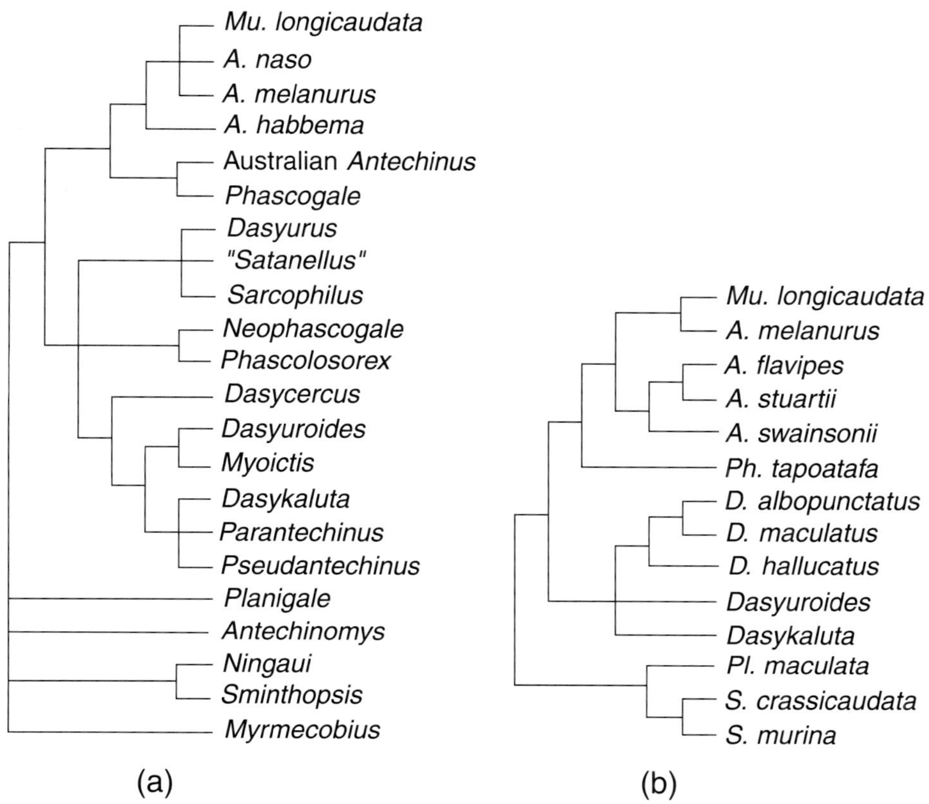

Figure 2 Phylogenetic relationships among dasyurids estimated from early molecular studies. Genus abbreviations as in Fig. 1, as well as *Mu.* = *Murexia, Ph.* = *Phascogale, Pl.* = *Planigale. (a)* Albumin MC′F tree from Baverstock et al. (1990). Note that *A. habbema* was incorrectly labelled '*A. wilhelmina*' in the original paper (P.A. Woolley and P.R. Baverstock pers. comm.). *(b)* DNA hybridisation tree from Kirsch et al. (1990). The bandicoot outgroup has been omitted.

MC′F tree, estimated from an albumin molecular clock, suggested that the diversification of dasyurid and numbat lineages took place some 20 Mya, and that separation of dasyurine (including 'phascolosoricines') and phascogaline (including *Murexia*) clades occurred 10 Mya. Baverstock et al. (1990) recommended caution in the interpretation of these dates, but in light of subsequent discoveries it is worth noting that MC′F suggested a mid-Miocene radiation of modern dasyurids.

DNA hybridisation

The third molecular assessment of dasyurid relationships in this interval was a DNA hybridisation study by Kirsch et al. (1990). DNA hybridisation was pioneered in vertebrate systematics by Sibley and Ahlquist (1990), and the technique owes much of its methodological justification to the correlation between the melting-point depression of artificially hybridised strands of DNA in solution and the average sequence dissimilarity of the species from which the strands were drawn (Springer and Krajewski 1989b). Kirsch et al. (1990) were able to obtain DNA from only 14 dasyurids, but their phylogeny (Fig. 2b) has three important features: (1) it agrees with MC′F in uniting *Murexia* with a New Guinean antechinus; (2) it disagrees with MC′F in that *Murexia* and *Antechinus* form a monophyletic group apart from *Phascogale*; and (3) it resolves *Planigale* and *Sminthopsis* as part of a monophyletic group (vide Archer 1984). Unfortunately, no material was available from numbat or 'phascolosoricines' to address the other points of disagreement between Archer (1982a, 1984) and Baverstock et al. (1990).

DNA SEQUENCE STUDIES: METHODS

The genes

Gene sampling. Molecular analyses of dasyuromorphian relationships published since 1992 have been based on DNA sequences of one or more of four loci. Three of these – *cytb*, 12S and the control region (CR) – are located on the mitochondrial DNA (mtDNA) molecule and are thus completely linked to one another (Fig. 3a). The significance of this linkage cannot be overstated – because mtDNA is essentially nonrecombining in mammals (Moritz et al. 1987, but see Awadalla et al. 1999), the 'gene trees' of *cytb*, 12S, and CR are identical. Any incongruence

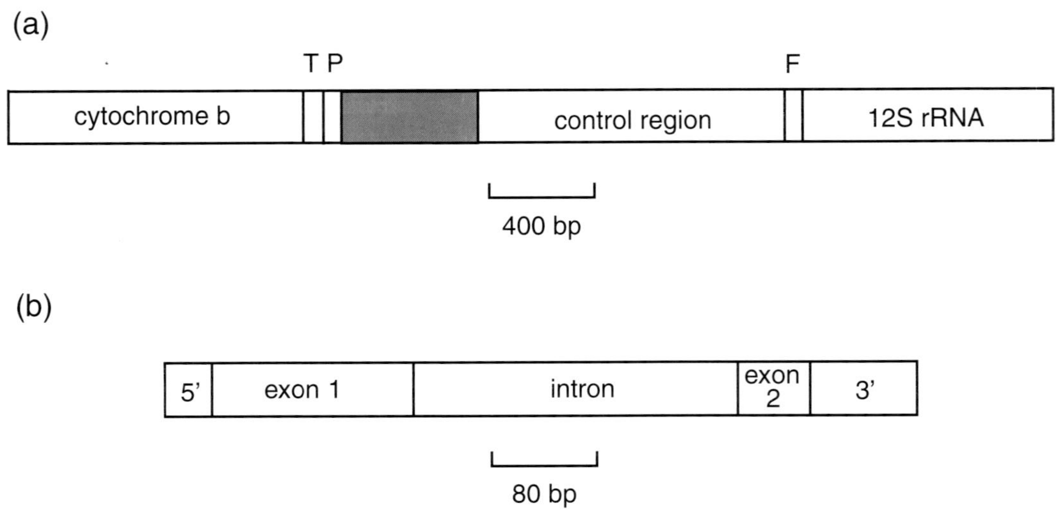

Figure 3 Schematic diagrams of the loci employed to date for DNA sequencing studies of dasyuromorphians. Scale bars are approximate. *(a)* Segment of the mtDNA molecule containing cytochrome b, control region, and 12S rRNA. T, P, and F indicate genes for threonine, proline, and phenylalanine transfer RNAs. The shaded portion of the control region represents Domain I. *(b)* Structure of the protamine P1 locus. 5′ and 3′ indicate upstream and downstream flanking regions.

in phylogenies estimated from these loci for the same set of taxa must be an artifact, though this is not to say that the three loci have the same phylogenetic resolving power (see below). It is also worth remembering, particularly in the interpretation of intraspecific variation, that mammalian mtDNA is inherited primarily through matrilines and that mtDNA trees are maternal genealogies (Avise 1994). To the extent that the structure of genetic variation within species is affected by processes other than female-mediated gene flow, mtDNA results at this level must be interpreted with caution. In contrast, the fourth locus we have studied – protamine P1 (Fig. 3b) – resides on a nuclear chromosome and provides an estimate of species relationships that is independent of mtDNA. Laboratory methods for generating DNA sequences have varied among studies, and original references should be consulted for this information: Krajewski et al. (1992, 1997a), Painter et al. (1995), Retief et al. (1995a), Armstrong et al. (1998), and Blacket et al. (1999). Krajewski et al. (2000) provide a list of GenBank accession numbers for all published sequences.

Cytochrome b. Cytb (1146 bp in dasyurids) is a protein-coding gene whose product is involved in the mitochondrial electron-transport pathway (Howell 1981). The structure and evolutionary properties of the cytb protein have been well-characterised (Esposti et al. 1993), and *cytb* is one of the most frequently employed genes in vertebrate systematics. The evolutionary dynamics of *cytb* are well-known and comparable across vertebrate groups (e.g. Irwin et al. 1991; Krajewski and King 1996). Like most mitochondrial genes, *cytb* evolves with a higher rate of transition substitutions (A↔G and C↔T) than transversions (all other substitution types). The transition bias (ts/tv) is roughly 4:1 in dasyurids. *Cytb* nucleotide frequencies vary more-or-less from uniformity depending on codon position, with third positions showing a low frequency of G (about 3%) and second positions a high frequency of T (about 41%) on the coding strand. Rates of substitution also vary among codon positions, with third positions evolving 20–60 times faster, and first positions 2–6 times faster, than second positions. Transition bias and rate disparity result in a saturation effect at third positions – i.e. these positions begin to accrue multiple transitions at the same site well before such homoplastic events are evident at other positions.

12S rRNA. This locus is one of two mtDNA genes that encode ribosomal RNA subunits. 12S is homologous to the 'small subunit rRNA' of other genomic compartments. Like other rRNA molecules, mature 12S rRNA (965–974 bp in dasyurids) becomes folded into a complex secondary structure of stems (duplex regions formed by intrastrand hydrogen-bonding between runs of complementary nucleotides) and loops (unpaired regions), on which is superimposed a tertiary structure determined by long-distance RNA–RNA and protein–RNA interactions (Springer and Douzery 1996). Secondary structure exerts several evolutionary constraints on 12S gene sequences: (1) values of ts/tv in stems are 9–14 among dasyurids, but only 2–3 in loops; (2) stem base compositions are roughly uniform, whereas loops have low GC content (32-34%); and (3) loops evolve 3–6 times faster than stems. Moreover, constraint on the complementarity of stem sites results in the phenomenon of 'compensatory substitution' in these regions (Dixon and Hillis 1993) – e.g. an A→G transition at one stem site is adaptively compensated by a T→C transition at the complementary site. Such evolutionary correlation among sites violates the assumption of character independence implicit in methods of phylogenetic estimation. Springer et al. (1995) determined that the magnitude of compensation is such that the evidence from a phylogenetically informative stem site is equivalent to only 61% of that from a loop site. However, weighting of stems and loops to adjust for their variable rates has an opposite effect (i.e. loop sites are downweighted) of almost equal magnitude. 12S sequence evolution also includes short insertions and deletions ('indels'), particularly in loop regions. Overall, the rate of 12S evolution is about 25–30% that of *cytb*, making it more useful for resolving deeper nodes of the phylogeny.

Control Region. CR is the only noncoding sequence of substantial length in vertebrate mtDNA (it is about 1.6 kb in marsupials; Janke et al. 1994). Although most sites in this region do not appear to be under much functional constraint, several short stretches of sequence (which probably function in mtDNA replication) are highly conserved. More broadly, the CR can be divided into three domains with different substitution rates (Saccone et al. 1991; Baker and Marshall 1997). The 5′ and 3′ regions (Domains I and III) are typically quite variable and often used as intraspecific markers for population-genetic studies of vertebrates. In contrast, Domain II is more conserved and is often omitted from intraspecific studies. The overall rate of CR evolution is roughly comparable to the silent substitution rate in *cytb*, though this varies among taxonomic groups. CR base compositions show the low GC content typical of mtDNA silent sites. Dasyurid studies (Blacket et al. 1999; Firestone accepted) have so far been confined to Domain I (371–552 bp), but with two surprising observations: (1) rates of evolution are only slightly faster than those observed at *cytb* first positions; and (2) the transition bias is small (ts/tv < 2). The cause of these unusual properties of dasyurid CR sequences is unclear and requires further study. As for 12S, CR evolution involves indels, though in CR these can be quite long (up to 80 bp in dasyurids studied to date). CR sequences also contain repeat motifs of varying lengths that result from block duplication or replication slippage. Alignment gaps caused by indels and repeats create problems for phylogenetic analysis (see below).

Protamine P1. Protamines are small, arginine-rich proteins that replace histones during condensation of the sperm nucleus in the process of spermatogenesis. Most vertebrates have a single protamine gene (*P1*), though some placental mammals have a second (*P2*). The portion of the marsupial *P1* locus (600–650 bp) that has been included in dasyuromorphian studies includes the complete coding sequence (exons 1 and 2), the single intron, and upstream and downstream flanking regions (Retief et al. 1995a). The major features of *P1* sequence evolution are radically different from those of mtDNA sequences. Exon sites, for example, are predominantly constrained by amino acid composition – P1 consists of about 60% arginine residues, and the most frequent type of base changes are silent Arg substitutions (i.e. among CGN and AGR codons). Intron sites evolve much more rapidly than other regions, and are excessively prone to large and small indels, even among closely related species. Introns include short repetitive sequences and block duplications – one such, a 43 bp duplication, is found only in dasyuromorphians and perhaps *Notoryctes* (Krajewski et al. 1997b). In contrast, flanking regions are relatively conserved, though short indels are common. The variable nature of the *P1* intron makes interordinal alignment of sequences difficult. Retief et al. (1995b) employed the alignment algorithm of Lipman and Pearson (1985) to address this problem, and we have used the Retief et al. alignment as a framework in all subsequent dasyuromorphian studies.

Phylogenetic analysis

Overview

There are many methods for estimating phylogenetic relationships from aligned DNA sequences, but recent practitioners emphasise three approaches: parsimony, additive distance, and maximum likelihood (Swofford et al. 1996). Although it is common for individual systematists to champion one or another of these methods on philosophical grounds, we remain unconvinced that any one of them is uniformly superior to the rest. Rather, each method has strengths and weaknesses that will become apparent as we describe its implementation for dasyuromorphian sequences. We will not discuss the statistical justification of these methods, but note that our approach is one of methodological pluralism. This does not mean that we apply generic versions of each technique and emphasise points of consensus among the results. Rather, each technique must be optimised as much as possible for the estimation problem at hand. In some cases it may be altogether inappropriate or infeasible to use one method, but quite reasonable to use another. We think that consensus among methods is much less relevant than finding a good match between an estimation procedure and the properties of the data to which it is applied.

Parsimony strategies

Maximum parsimony is an optimality criterion that identifies the best tree for a set of sequences as that which requires the smallest number of character state changes. If sequence evolution involves only base substitutions, and if all substitution types have the same chance of being homoplastic, it is reasonable to apply uniformly weighted parsimony to estimate the phylogeny. If, on the other hand, some classes of substitution are so prone to homoplasy that they are misleading, it is prudent to exclude them from the analysis. Between these extremes are cases in which some characters are marginally but significantly less reliable than others; these sites are included in the analysis, but given a numerical weight that is lower than the more reliable sites. For dasyurid *cytb* sequences, we have applied several weighting schemes, but the most reliable has been '3TV' weighting (exclusion of third-position transitions); this approach has consistently recovered more 'expected' groupings, and with higher resolution, than other schemes. For 12S, eliminating transitions from intrafamilial comparisons results in a substantial loss of resolution on most nodes. However, such transversion weighting is somewhat more effective at recovering interfamilial relationships. The negible transition bias for CR and *P1* sequences is such that uniform weighting of all substitution types is most appropriate.

Alignment gaps present a problem for phylogenetic analysis. DNA parsimony algorithms assume that each site in an alignment is an independent character. Single-base gaps are not problematic in that they can be assigned a 'fifth state' (i.e. '–' rather than A, G, C, or T), but longer gaps cannot be handled this way. A 2-base gap, for example, presumably arose via a single indel event rather than two one-base events (though the latter is possible). Treating each site in the gap as independent biases the estimation procedure: if taxa X and Y share the gap, this is one shared character, but fifth-state coding treats the gap sites as two characters. On the other hand, both sites may have informative nucleotide variation in other taxa, so eliminating all but one site in the gap discards relevant data. There are several solutions to this problem, none of them entirely satisfactory. For 12S alignments, in which nearly all gaps are short (1–3 bp) and nonoverlapping, we have used fifth-state coding and accepted the small degree of bias that it entails. For CR and *P1*, however, gaps vary greatly in length and often overlap such that treating gap sites as independent would be inappropriate. Instead, we developed 'homologous gap coding' (Retief et al. 1995b). This method assumes a correct alignment and treats colinear gaps as homologous characters. One site (preferably one that also contains nucleotide information) in each homologous gap is coded with a fifth state; all other sites within the gap that contain nucleotide information, or that overlap with such sites in non-homologous gaps, are coded as missing data ('?'). Sites that lack both gap and nucleotide information are eliminated from the analysis. This procedure utilises all phylogenetic information from gaps and nucleotides, but also entails some missing-data codes in the alignment (for the drawbacks of which, see Maddison 1993). We emphasise our solution to the gap problem because phylogenetic analyses of *P1* utterly depend on it – most *P1* trees obtained by other methods are not credible.

Additive distance

Genetic distances estimate numbers of substitutions per site between homologous sequences and require specification of an evolutionary model. It is now clear that trees based on distances are good phylogenetic estimates if the model is accurate (Nei 1991, Yang 1994). The model that we have most often applied to dasyuromorphian mtDNA sequences is Felsenstein's (1993) 'F84' model, the major parameters of which are an overall transition bias, equilibrium base frequencies, and rate categories for groups of sites within the alignment. As noted above, however, groups of structurally or functionally related sites ('partitions') within a gene may differ in parameters other than relative rate. This motivated us to develop an estimator ($\bar{d}$ or d_{MLA}, the weighted-average distance) that allows all model parameters to vary among partitions. Given parameter-values for each partition i, $\bar{d} = \Sigma f_i d_i$, where f_i is the fraction of total sites in i, and d_i is the distance for i. Krajewski et al. (1999) analysed the accuracy and precision of $\bar{d}$, and found that it performs well on both criteria. Intralocus partitions correspond to codon positions in *cytb*, and stems and loops in 12S. Gaps are a major obstacle for distance analysis, in that indel events are not included in standard models of sequence evolution. We have treated gaps in 12S and CR as missing data such that they do not contribute to sequence mismatch, but gaps in *P1* alignments are so extensive that meaningful distance analysis is impossible. Trees are obtained from distance matrices using weighted least-squares (Fitch and Margoliash 1967) and neighbor-joining (Saitou and Nei 1987) methods. Note that these methods are not 'phenetic' – i.e. they are robust to evolutionary rate variation among taxa and do not necessarily produce clusters of maximally similar sequences (Springer and Krajewski 1989a).

Maximum likelihood

Like distance methods, phylogenetic estimation by maximum likelihood (ML) requires a probablistic model of sequence evolution. Once the model and its parameter values are specified, the ML criterion directs us to choose the tree on which branch lengths can be optimised so as to produce the highest probability of observing the actual sequences (Swofford et al. 1996). Rather than taking the intermediate step of estimating pairwise distances, the ML procedure uses character information directly in evaluation of alternative topologies. As a result, the method is more computationally demanding than parsimony or distance analysis, and is not easily amenable to large data sets. On the other hand, ML is known to be more robust than distance analysis to some violations of the assumed model (Yang 1994). We have made use of ML only occasionally in our dasyuromorphian studies and in those cases have used the same F84 model that was employed for distance estimates (see above). Again, however, no generally available model accounts for indel events and thus ML analysis has not been applied to *P1* sequences.

Resolution and the bootstrap

All methods of phylogenetic estimation will return an optimal tree for a set of characters, but this leaves open the question of whether nodes on that tree are significantly better supported than other arrangements. Parsimony analyses may recover several equally short trees, the strict consensus of which collapses unresolved nodes. Similarly, branches lacking resolution on distance and ML trees may have lengths of zero. These indications do not, however, tell us whether bifurcating nodes on an optimal tree are well resolved. Several methods have been developed to address this issue, but the most widely used is Felsenstein's (1985) nonparametric bootstrap. The bootstrap is a means of assessing how frequently a particular parameter-estimate (e.g. an optimal topology) would be obtained if many data sets (e.g. character matrices) were gathered by independent rounds of sampling (Diaconis and Efron 1983). The sampling distribution is approximated by repeated resampling (with replacement) from the observed data set, with the parameter of interest estimated from each pseudoreplicate. The range of parameter-values approximates (or is at least analogous to) a confidence interval on the original estimate. In phylogenetics, bootstrap results are usually portrayed as the frequency with which a particular node on the optimal tree is recovered in a large number of pseudoreplicates. Critical aspects of interpreting phylogenetic bootstraps were discussed by Hillis and Bull (1993), but the most general caution is that bootstrap values are conditional on the data set and method of tree-estimation employed. On the other hand, assessment of confidence is essential if trees from different data sets are to be compared – conflict between two studies on the placement of taxon X is uninteresting if X's positon in one or both studies is poorly resolved.

Multilocus estimates of phylogeny

As noted above, mtDNA genes are linked and share a single mutation history. Trees estimated from *cytb*, 12S, and CR should thus be identical – or, more precisely, they should not conflict over well-resolved nodes. Such conflicts are necessarily due to experimental or analytical error. *P1*, however, is unlinked to mtDNA and phylogenetic conflicts between the two linkage groups could be due to lineage sorting (Avise 1994). Both situations pose a problem if our goal is to combine all DNA sequences in a single phylogenetic analysis. We have used a bootstrap combinable-component criterion to flag instances of such interlocus conflict: relationships that are resolved differently at bootstrap levels ≥90% by each of two genes are considered significantly in conflict. Conflicts between mtDNA and *P1* that could not be resolved by additional lab work or more sophisticated phylogenetic analyses were treated as potential cases of gene-tree incongruence.

DNA SEQUENCE STUDIES: RESULTS

Dasyuromorphian families

Three of our papers have dealt with the question of dasyuromorphian family-level relationships, but these studies addressed distinct questions. Krajewski et al. (1992) reported partial *cytb* sequences from *Thylacinus*, 13 dasyurids, and an outgroup bandicoot. In this paper, we took dasyuromorphian affinities for the thylacine as given, and asked whether the extant sister-group of *Thylacinus* consists of all dasyurids or a specific subgroup of dasyurids (as posited, for example, by Tate 1947); numbats were not included. Our results supported a monophyletic Dasyuridae apart from *Thylacinus* with moderate bootstrap resolution (≤75%). Krajewski et al. (1997b) revisited the issue of interordinal relationships of *Thylacinus* with complete *cytb*, partial 12S, and complete *P1* sequences from 11 dasyurids, numbat, representatives from all other extant marsupial orders, and two outgroup placentals. The motivation for this study was our perception that the limited 12S data supplied by Thomas et al. (1989) were inadequate for testing previous hypotheses. Our results strongly supported dasyuromorphian monophyly (100% bootstrap), but could not resolve a monophyletic Dasyuridae (mtDNA sequences placed *Myrmecobius* within the family). Most recently, we assessed interfamilial relationships based on complete sequences of all three genes for 15 dasyurids, *Thylacinus*, *Myrmecobius*, and outgroups from all other Australian marsupial orders (Krajewski et al. 2000a). This study included a variety of analyses designed to examine the effects of partitioning, and phylogenetic results did vary substantially among analytical treatments. However, the most frequent and best-supported pairing was between *Thylacinus* and a monophyletic Dasyuridae (≤80% bootstrap for all three genes, ≤95% for *cytb* + 12S; Fig. 4). This is inconsistent with Archer's (1984) view that numbats and dasyurids are sisters, but does not conflict with Wroe's (1997) recent evaluation of anatomical data bearing on interfamilial relationships. Thus, although the monophyly of extant dasyuromorphians is resolved, interfamilial relationships require further study.

Dasyurid subfamilies and tribes

Krajewski et al. (1993, 1994) used partial *cytb* sequences to assess the monophyly and relationships of suprageneric taxa recognised by Archer (1982a; Table 1). The most significant results of these studies were: (1) strong support for the affinity of New Guinean 'phascolosoricines' and *Murexia* with dasyurines and phascogalines, respectively, as previously suggested by MC'F and DNA hybridisation; (2) after accounting for the position of New Guinean groups, apparent monophyly of Archer's (1982a) subfamilies Dasyurinae, Phascogalinae, and Sminthopsinae; (3) no resolution of relationships among these three clades; (4) no support for the monophyly of Archer's (1982a, 1984) tribes. The latter point was evidenced by some well-resolved but unexpected pairings, including *Planigale maculata* with *Sminthopsis crassicaudata* and *Pseudantechinus ningbing* with *Myoictis melas*. In addition, some expected groups failed to materialise (e.g. *Dasyurus* species did not form a clade). These early results must be interpreted with care: taxon-sampling was quite limited for sminthopsines and phascogalines; the amount of sequence obtained (657 bp) was relatively small; and phylogenetic analyses were simple. Results (1)–(3) were generally upheld by

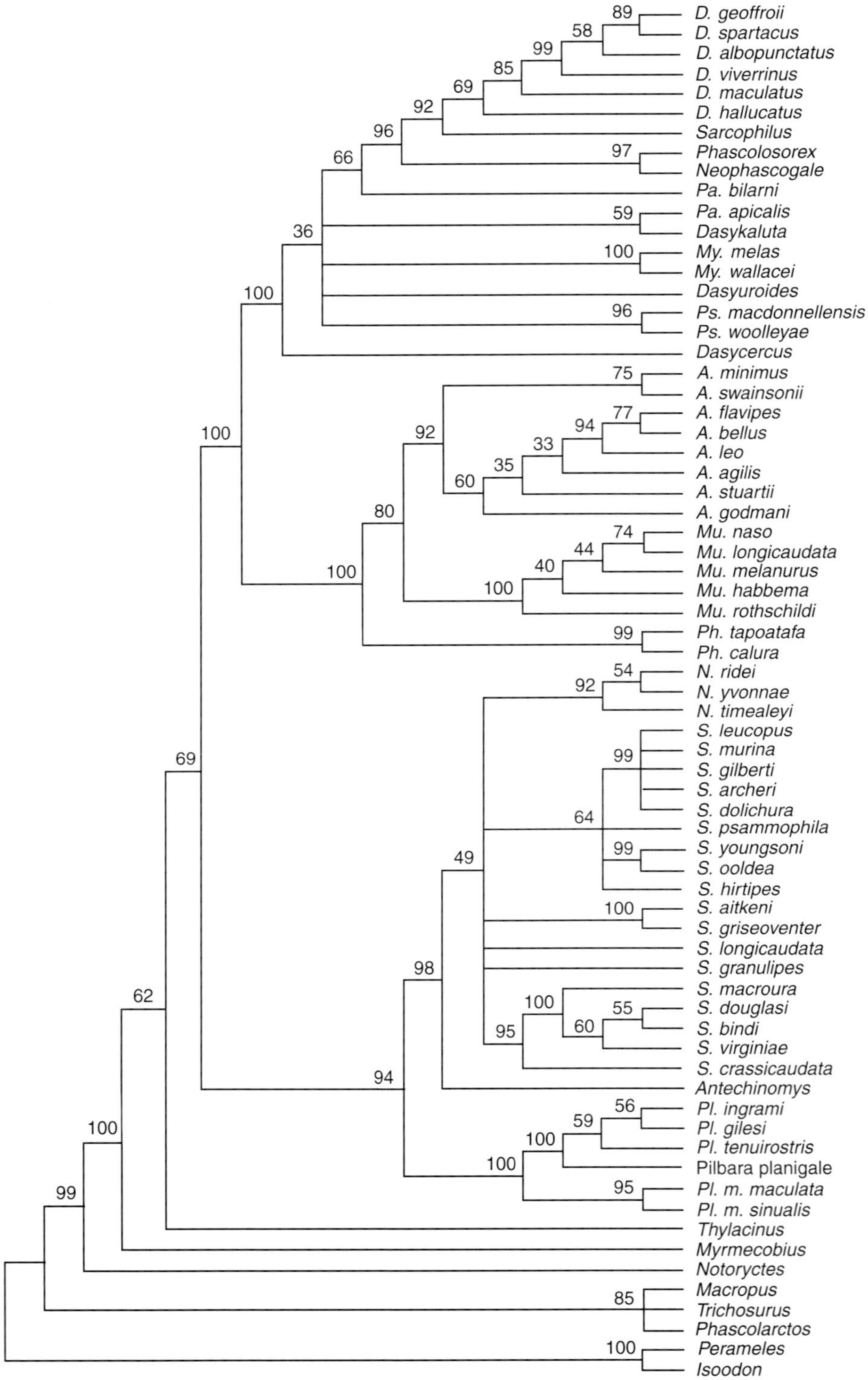

Figure 4 Strict consensus of eight minimum-length trees for combined *cytb*, 12S, and *P1* sequences from dasyuromorphians and selected outgroups (redrawn from Krajewski et al. 2000c). Transitions at third codon positions of *cytb* were given zero weight; indels in *P1* were coded with the homologous gap method of Retief et al. (1995b). Nodal values are bootstrap proportions based on 100 replicates. Genus abbreviations as in Figs 1–2, as well as *N.* = *Ningaui*.

analyses of *P1* sequences from 13 dasyurids (Retief et al. 1995a), but support for tribal monophyly remained equivocal. These issues were not thoroughly revisited until our recent analysis of the complete, three-gene data set from virtually all dasyurid species (Krajewski et al. 2000c). The results of that analysis, however, are robust in showing that Sminthopsini and Planigalini *are* monophyletic (contra Krajewski et al. 1994), but DNA-sequence support for 'Dasyurini' and 'Parantechini' is lacking (Fig. 4). The latter is a negative result – the data neither support nor refute the hypothesis that these groups are monophyletic. More importantly, the three-gene analysis resolves dasyurines and phascogalines as sisters apart from Sminthopsinae, as suggested by DNA hybridisation.

Relationships among dasyurines

Krajewski et al. (1997c) examined the phylogeny of dasyurine species, and their results are consistent with more recent and comprehensive analyses (Krajewski et al. 2000c). Complete sequences of *cytb*, 12S, and *P1* are congruent in resolving three dasyurine nodes: (1) a monophyletic cluster of *Dasyurus* and *Sarcophilus* species; (2) a 'phascolosoricine' clade that is sister to the *Dasyurus-Sarcophilus* group; and (3) a polytomy of seven lineages at the root of Dasyurinae (Fig. 4). Close relationship of *Dasyurus* and *Sarcophilus* is consistent with Archer (1982a), but monophyly of *Dasyurus* apart from *Sarcophilus* is not. Although partial *cytb* analyses (Krajewski et al. 1994) paired *D. maculatus* with *Sarcophilus* (90–98% bootstrap), this association does not persist with more sophisticated analyses of larger data sets (though there is some morphological evidence to support it; S. Wroe pers. comm.). Archer's (1982a) alliance of *D. albopunctatus* and *D. hallucatus* in '*Satanellus*' was recovered by MC'F (Baverstock et al. 1990), but is rejected by DNA sequences which unite *D. albopunctatus* with three other quolls to the exclusion of *D. hallucatus* (Fig. 4). Indeed, *D. hallucatus* appears as sister to all other quolls, as also suggested by the morphocladistic analyses of Van Dyck (1987) and Wroe and Mackness (1998). Domain I CR sequences from all *Dasyurus* and *Sarcophilus* species (Firestone, 2000) gave a tree that is identical to that of Krajewski et al. (1997c) and has comparable levels of resolution. The moderate level of support (79% bootstrap) for *Dasyurus* monophyly from DNA sequences, coupled with anatomical data favouring a *D. maculatus-Sarcophilus* clade, indicate that quoll relationships require further scrutiny.

The dasyurid MC'F tree (Baverstock et al. 1990) clearly associated *Phascolosorex* and *Neophascogale* with dasyurines, but portrayed 'phascolosoricines' as one of three lineages emanating from a trichotomy in the subfamily. This arrangement left open the possibility that 'phascolosoricines' could be sister to a monophyletic group of traditional dasyurines. In contrast, *cytb*, 12S, and *P1* congruently resolve 'phascolosoricines' as sister to the *Dasyurus-Sarcophilus* clade (96% bootstrap; Fig. 4), a phylogeny that requires considerable homoplasy in the evolution of dental and basicranial characters. This unexpected result has prompted us to initiate a more extensive test of 'phascolosoricine' relationships that involves DNA sequences from additional loci as well as anatomical data.

The polytomy among dasyurine lineages may represent an episode of rapid cladogenesis that will be difficult to resolve until much longer DNA sequences are available (Krajewski et al. 1997c). Lineages emanating from the polytomy, however, are interesting in light of groups that do *not* appear (in addition to the 'dasyurin' and 'parantechinin' tribes discussed above). *Dasyuroides* and *Dasycercus*, for example, show no affinity for one another (Fig. 4) despite the fact that some morphologists (e.g. Groves 1993) have considered them congeneric. Similarly, the dibblers *Parantechinus apicalis* and *Pa. bilarni* do not form a clade, indicating that the single putative synapomorphy (three accessory corpora cavernosa) adduced by Archer (1982a) to define the genus may be homoplastic. *Pseudantechinus macdonnellensis* and *Ps. woolleyae* do appear as sisters (this clade also includes *Ps. mimulus*; M. Westerman and J. Young unpubl. data). Unfortunately, we have exhausted our only sample of *Ps. ningbing* and have been unable to obtain complete 12S and *P1* sequences. As in Krajewski et al. (1994), complete *cytb* resolves *Ps. ningbing* as sister to *Myoictis*, whereas partial 12S and *P1* data leave its position unresolved. This suggests that the *cytb* result may be an artifact, though we have tentatively ruled out contamination and pseudogene amplification as possible explanations. Thus, molecular assessment of *Pseudantechinus* monophyly awaits the aquisition of new specimens.

Relationships among phascogalines

Krajewski et al. (1996) reported partial *cytb* sequences from 10 phascogalines (including *Murexia*) and argued that this group consists of three clades: *Phascogale*, Australian *Antechinus*, and New Guinean antechinuses + *Murexia*. Moreover, *cytb* data suggested that antechinuses (Australian and New Guinean) and *Murexia* form a clade apart from *Phascogale* (≤73% bootstrap). Combined *cytb*, 12S, and *P1* analyses (Krajewski et al. 2000c) strongly confirmed the monophyly of the three clades and improved resolution on the *Antechinus-Murexia* pairing (Fig. 4). It is significant that the latter study included outgroups from all other Australian marsupial orders: rooting the phascogaline tree with dasyurid outgroups would be misleading if *Murexia* species were early-branching lineages within the family, as indicated by their primitive morphology (S. Van Dyck pers. comm.). High resolution on the New Guinean clade prompted Armstrong et al. (1998) to reassign all New Guinean antechinuses to *Murexia*, a scheme which we will follow here. Sister-pairing of *Antechinus* and *Murexia* contradicts MC'F results (Baverstock et al. 1990) which favoured *Antechinus* + *Phascogale*, but is consistent with DNA hybridisation (Kirsch et al. 1990) and (excluding the original *Murexia* species) with Archer's (1982a) anatomical results (Fig. 1a). Within *Antechinus*, complete *cytb*, 12S, and *P1* sequences (Armstrong et al. 1998) yielded a tree that is virtually identical to that obtained from allozymes (Baverstock et al. 1982), featuring four species groups whose relationships to one another are unresolved (see above). The recently described *A. agilis* (Dickman et al. 1998) clusters with *A. stuartii*, but with less resolution than expected given that the two had been considered conspecific. Within *Murexia*, not a single node on the DNA phylogeny is highly resolved, and reconstructing the details of this island radiation will require more data and geographic sampling.

Relationships within *Sminthopsini*

Although Archer (1981, 1982a) and Van Dyck et al. (1994) considered *Antechinomys laniger* as a member of *Sminthopsis*, DNA sequence data (Krajewski et al. 1997a, Blacket et al. 1999) resolve *Antechinomys* as sister to a *Sminthopsis-Ningaui* clade (Fig. 4). This resolution is strongly influenced by the inclusion of CR sequences along with *cytb*, 12S, and *P1*, but is congruent with MC'F (Baverstock et al. 1990). Although Krajewski et al. (1997a) recovered a monophyletic *Sminthopsis* (three exemplar species) apart from *Ningaui*, Blacket et al. (1999) were unable to replicate this result with nearly complete species-sampling of dunnarts. Both studies, however, resolved the three *Ningaui* species as a clade (in which *N. ridei* and *N. yvonnae* are sisters). Within *Sminthopsis*, Blacket et al. (1999) recovered four major species groups: (1) a Macroura group (*S. macroura, S. virginiae, S. douglasi, S. bindi*), to which *S. crassicaudata* is the sister; (2) a Murina group (*S. murina, S. leucopus, S. gilberti, S. dolichura, S. archeri, S. butleri*); (3) *S. griseoventer* + *S. aitkeni*; and (4) a weakly resolved clade of *S. psammophila, S. hirtipes, S. youngsoni*, and *S. ooldea*. Inclusion of *S. butleri* in the Murina group is based on unpublished 12S and CR analyses (M. Blacket pers. comm.). The positions of *S. granulipes* and *S. longicaudata* were not resolved, nor were the interrelationships among these and the four species groups. These groups are partially congruent with clades proposed by Archer (1981, 1982a) and Van Dyck et al. (1994), but (as for dasyurines) the polytomy of *Sminthopsis* and *Ningaui* lineages may be the signature of a rapid taxonomic radiation.

Relationships within *Planigalini*

Painter et al. (1995) used partial *cytb* sequences to estimate relationships among *Planigale* species and obtained two major results: (1) *Pl. maculata* is divergent from other planigale species included; and (2) *Pl. gilesi* and *Pl. ingrami* define a clade which, on the basis of a very short (242 bp) sequence, appeared to include *Pl. novaeguineae*. Krajewski et al. (1997a) found strong support for the separation of *Pl. maculata* from a clade containing *Pl. tenuirostris, Pl. gilesi*, and *Pl. ingrami* with complete *cytb*, 12S, and *P1* sequences. Although *Pl. novaeguineae* was not included in the latter study, Blacket et al. (2000) found that this species associated with *Pl. maculata* (62–96% bootstraps) on the basis of complete 12S sequences from numerous planigale populations throughout Australia. Other interspecific nodes on the planigale 12S tree are poorly resolved (as was also true in Krajewski et al. 1997a). Thus, while the branching order depicted in Fig. 4 remains our best estimate of relationships within this genus, the 12S study of Blacket et al. (2000) merits replication with *cytb*, CR, and *P1* sequences.

A PHYLOGENETIC CLASSIFICATION OF DASYURIDS

Based on the results above and consideration of recent fossil discoveries, Krajewski et al. (2000c) proposed the phylogenetic classification of dasyurids given in Table 2. Our scheme follows Wroe (1999) in assigning the Miocene *Barinya* to a distinct subfamily (Barinyainae) that is the primitive sister-group to modern dasyurids. We follow Kirsch et al. (1997) in separating the latter into two subfamilies, Dasyurinae and Sminthopsinae. Dasyurinae is composed of tribes Dasyurini and Phascogalini, the former including *Phascolosorex* and *Neophascogale*, and the latter including *Murexia*. Sminthopsine tribes and their compositions are identical to those given by Archer (1984). We follow Armstrong et al. (1998) in assigning all New Guinean antechinuses to *Murexia*. Should future research identify clades within Dasyurini, the subtribe rank may be used to accomodate these in our classification.

CONFLICTS BETWEEN MOLECULAR AND MORPHOLOGICAL RESULTS

Any group of organisms that has been studied from a variety of different perspectives is bound to have a taxonomic history with incongruent conclusions. Dasyurids are a case in point, as exemplified by the conflicts between molecular and morphological results discussed above. In contrast, DNA sequences, DNA hybridisation, MC'F, and allozyme studies have reached generally similar conclusions, though taxonomic sampling has been limited in all but the first of these. Exceptions include disagreement between MC'F and DNA about the placement of *Murexia*, allozymes versus MC'F and DNA sequences over the positions of *Ningaui* and *Antechinomys* relative to *Sminthopsis*, and allozymes and MC'F versus DNA sequences on the monophyly of 'Parantechini'. It is important to note that neither allozyme (Baverstock et al. 1982) nor MC'F (Baverstock et al. 1990) studies employed an explicit measure of resolution, making it difficult to assess the significance of these discrepancies.

A similar concern about resolution applies to conflicts between DNA sequences and morphology. In general, instances of such conflict can be categorised as those in which: (1) neither DNA nor morphology provide strong evidence for relationships; (2) DNA provides strong, but morphology provides weak, evidence; (3) morphology provides strong, but DNA provides

Table 2 Suprageneric classification of Dasyuridae proposed by Krajewski et al. (2000)

Suprageneric groupings follow Kirsch et al. (1997) and Wroe (1999); species compositions of Antechinus and Murexia are as given by Armstrong et al. (1998). † indicates extinct taxa.

Family Dasyuridae (Goldfuss 1820)
- †Subfamily Barinyainae Wroe 1999
  - *†Barinya*
- Subfamily Dasyurinae (Goldfuss 1820)
  - Tribe Dasyurini Goldfuss 1820
    - *Dasycercus, Dasykaluta, Dasyuroides, Dasyurus, †Glaucodon, Myoictis, Neophascogale, Parantechinus, Phascolosorex, Pseudantechinus, Sarcophilus*
  - Tribe Phascogalini (Gill 1872)
    - *Antechinus, Murexia, Phascogale*
- Subfamily Sminthopsinae Archer 1982
  - Tribe Sminthopsini Archer 1984
    - *Antechinomys, Ningaui, Sminthopsis*
  - Tribe Planigalini Archer 1984
    - *Planigale*

Dasyuridae *incertae sedis*
- *†Ganbulyani*

weak, evidence; and (4) both DNA and morphology provide strong evidence. The first situation is exemplified by our current understanding of relationships among dasyuromorphian families. Examples of the second category are more difficult for non-morphologists to identify, but a potential instance is the monophyly of Sminthopsinae (for which, so far as we are aware, no anatomical evidence has been published). The monophyly of *Sminthopsis* apart from *Ningaui* may be an example of the third category of conflict, though the relevant morphological characters have also been used to include *Antechinomys* within *Sminthopsis*. In any event, these situations are not very interesting, except as examples of nodes that are hard to resolve with DNA, morphology, or both. Cases in the fourth category are much more relevant to our understanding of character evolution and phylogenetic inferences based on that understanding.

There are several examples of strong conflict at the species level, but the most striking cases involve the 'phascolosoricines' and *Murexia*. It is important to be clear on which aspects of the inferred relationships for these taxa are problematic. As noted above, placement of 'phascolosoricines' as sister to all other dasyurins (sensu Table 2) does not entail an excessive amount of homoplasy in cranial characters (because 'phascolosoricine' dentition is autapomorphic and their basicrania are primitive). Nesting them *within* Dasyurini, however, requires that the apparently synapomorphic cranial characters of at least some traditional dasyurins be derived in parallel, or that 'phascolosoricine' basicrania be derived by reversal to primitive states. If the dental, cranial, and external features of *Murexia* (sensu Table 2) are generally plesiomorphic for modern dasyurids, then these species should occupy early branches of the family tree (S. Van Dyck pers. comm.). Nesting *Murexia* within Phascogalini, as indicated by DNA data, implies either that these features are not primitive, that they are primitive states derived by reversal, or that the derived features of *Phascogale* and *Antechinus* (and indeed of dasyurins and sminthopsines) arose in parallel.

What can be done to resolve such conflicts? A traditional approach is to gather more data of both types. Perhaps the evidence of *cytb*, 12S, and *P1* is misleading and other loci will suggest a phylogeny with less radical implications for anatomical evolution. This is conceivable for 'phascolosoricines' in that DNA sequences are the only evidence generated to date for placing them as sister to *Dasyurus-Sarcophilus*. We are currently obtaining sequences from several other mitochondrial and nuclear loci to evaluate and improve resolution on the DNA tree. New sequence data are unlikely to solve the *Murexia* problem, however, because placement of these species within Phascogalini is also robustly supported by MC'F and DNA hybridisation evidence. Anatomical research may result in new characters that give trees more similar to those from DNA sequences, but we think that a more fruitful approach to conflict resolution is to employ Hennig's (1966) strategy of 'reciprocal illumination'. Thus we would like to know exactly what morphological characters or features of molecular data are responsible for conflicting nodes. Have the anatomical traits been correctly homologised, polarised, surveyed for polymorphism and independence, etc.? Do they all come from particular regions (e.g. teeth)? Are relevant molecular characters predominantly from one or a few partitions (e.g. third codon positions)? Do clades united at problematic nodes share unusual molecular features (e.g. base compositions that are similar to one another but different from those of other clades)? Are the models of sequence evolution used to construct molecular trees adequate? These issues have scarcely been mentioned in previous studies, but now that a comprehensive molecular phylogeny for dasyuromorphians is available (Krajewski et al. 2000c; Fig. 4), we are in a position to address them.

MOLECULAR EVIDENCE FOR CRYPTIC SPECIES IN DASYURIDAE

In addition to providing evidence for phylogenetic relationships among recognised dasyurid species, DNA sequence data have also suggested the existence of undescribed or under-ranked taxa. In the former case, sequences derived from specimens assigned to a particular species but collected outside the usual range of that species have sometimes proven quite divergent from specimens within the range. In the latter case, taxa once recognised as full species but later synonymised and/or regarded as subspecies on the basis of morphology have proven to be genetically distinct. Unfortunately, DNA sequences alone rarely provide conclusive evidence for species boundaries. In dasyurids, a few cases of potentially unrecognised species identified by

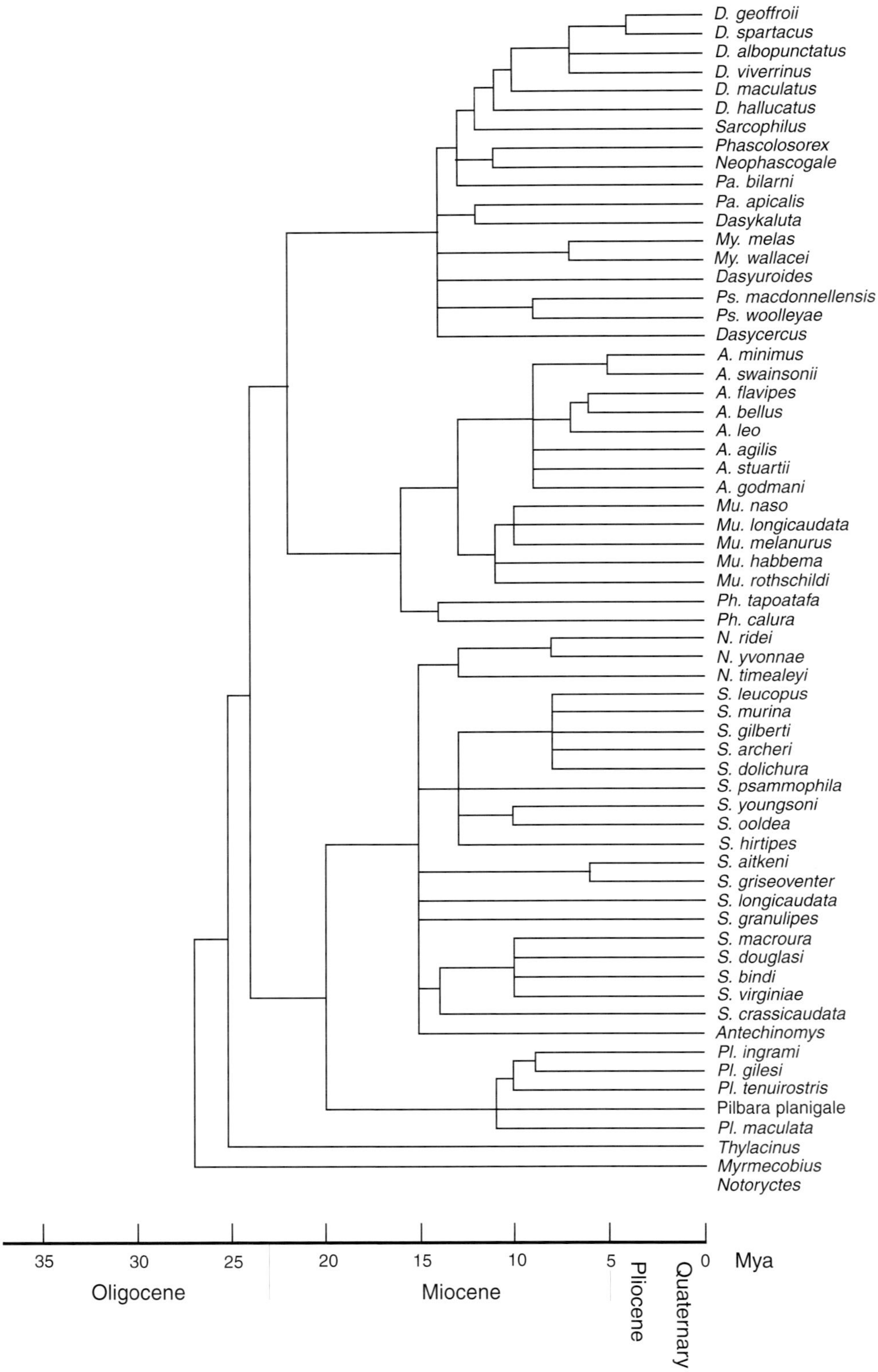

Figure 5 A time-scale depiction of the dasyuromorphian phylogeny in Fig. 4 (redrawn from Krajewski et al. 2000c). Divergence dates were estimated from *cytb* and 12S weighted-average distances and rounded to the nearest million years. Genus abbreviations as in Fig. 4.

sequence data have been followed up with gene-frequency surveys (usually allozymes) and morphological analyses.

Painter et al. (1995) included two planigale specimens from the Pilbara region of Western Australia in their *cytb* survey of the genus. One of these ('Planigale 1') was collected at Millstream Station and identified in the field as *Pl. maculata*. The other ('Planigale 2') was collected near the mouth of the Fortescue River and identified as *Pl. ingrami*. Both localities are outside the putative ranges of *Pl. maculata* and *Pl. ingrami*, and *cytb* sequences of the specimens proved divergent from those of all other *Planigale* species. Krajewski et al. (1997a) discovered that 'Planigale 2' is in fact *Ningaui timealeyi*, but 'Planigale 1' clearly represents a genetically distinct planigale. Blacket et al. (2000) confirmed this result with several additional specimens using both 12S and allozyme data, and showed that an animal collected from the Mt. Tom Price area of the Pilbara is not closely related to any other sampled population. It thus appears that the Pilbara region harbors undescribed planigale species.

Within Sminthopsini, recent mtDNA and allozyme studies (Blacket et al. 2001) have utilised extensive geographic samples from the species groups noted above, particularly the Macroura group. Both mtDNA and allozymes suggest that two taxa (*Sminthopsis froggati* and *S. stalkeri*) that were synonymised with *S. macroura* by Ride (1970) are genetically distinct. Moreover, it appears that *S. macroura* itself comprises distinct lineages from central Western Australia and eastern Australia. Within *S. virginiae*, specimens from the Northern Territory (currently regarded as *S. v. nitella*) show genetic divergences from other recognised *S. virginiae* subspecies that are as large as those between distinct species of dunnarts. The dasyurin genus *Myoictis* currently includes two recognised species (*My. melas* and *My. wallacei*), but DNA sequence and allozyme studies (Westerman et al. in prep.) suggest that members of *My. melas* from widely separated localities in Papua New Guinea may be specifically distinct. A similar situation holds for *Pseudantechinus* in the Pilbara region (Cooper et al. 2000; M. Westerman and J. Young unpubl. data).

THE TIMING OF CLADOGENESIS

In our early papers we used various methods to date cladogenic divergences, but the approach of Krajewski et al. (2000c) is a more rigorous assessment of the timescale of dasyurid evolution (readers should consult the original paper for details and caveats). Here we assumed a sister-group relationship for dasyurids and thylacinids, and used the age of the oldest known thylacinid fossil as a calibration point for interfamilial divergence. This fossil, *Badjcinus turnbulli*, is from the late Oligocene (Muirhead and Wroe 1998) and gives a minimum date of ca. 25 Mya for the separation of thylacinids and dasyurids. Unfortuntately, this is the only calibration point presently available. Although dasyurid *cytb* and 12S do not evolve at uniform rates within or among major clades, we were able to identify rate-uniform subsets of taxa for each gene in each tribe (except sminthopsin *cytb*). We calibrated weighted-average distances from taxa in these subsets to *Thylacinus* using the 25 Mya date. The rates of evolution for *cytb* and 12S obtained (approximately 2%/My and 0.6%/My, respectively) are quite comparable to those estimated for other vertebrate groups, and also to rates assumed in our earlier studies. Calibrated distances through specific nodes on the molecular phylogeny then provided estimates of cladogenic dates (Fig. 5).

The molecular data suggest that subfamilies Dasyurinae and Sminthopsinae arose from a common ancestor in the latest Oligocene or earliest Miocene (ca. 24 Mya), followed by the separation of their constituent tribes in the early Miocene (20–22 Mya). The most striking feature of Fig. 5 is that all four tribes experienced major episodes of cladogenesis in the middle Miocene. Three of them (Dasyurini, Phascogalini, and Sminthopsini) have earliest divergences that date to 14–16 Mya, with Planigalini radiating somewhat later (11 Mya). Krajewski et al. (2000c) speculated on the potential causes of these roughly simultaneous radiations, noting their correlation with a mid-Miocene episode of orogeny that elevated the New Guinean highlands (ca. 15 Mya), increased the rainshadow effect across central and western Australia, and intensified the climatic drying trend of the Miocene. This correlation suggests that modern dasyurids may owe much of their diversity to independent invasions of dry-country habitats that developed during the late Miocene, a hypothesis that remains to be evaluated against more detailed paleoclimatological information. It is also worth noting that, although the molecular clock dates tribal divergences to the middle Miocene, no fossils of modern dasyurids are known from deposits earlier than the Pliocene (Wroe and Muirhead 1999). In contrast, Fig. 5 shows that only the most closely related dasyurid sister-species had common ancestors as recently as the Pliocene, and *no* speciation events date to the Pleistocene. Whether this indicates that the molecular dates are too old, or that there are more closely related dasyurid species than are currently described, remains to be seen.

EPILOGUE

In the 20 years that have passed since publication of the first *Carnivorous Marsupials* anthology, our understanding of dasyuromorphian evolution has expanded considerably. It is worth pausing to appreciate those early volumes, however, because they set the stage for so much of what has been done since. Both of the present authors have spent many hours poring over Archer (1982a) and, as our research leads into new areas, many other chapters have become critically important to us. For example, the compilation of data on dasyurid reproductive strategies by Lee et al. (1982) serves as the basis for a comparative analysis of the reproductive evolution on our molecular phylogeny (Krajewski et al. 2000b). This analysis provides, among other things, further evidence for an association between the ancestors of modern dasyurids and dry-country habitats, but would not have been possible without the detailed information given in the Lee et al. chapter and supplemented with data collected by P.A. Woolley (to whom the current volume is appropriately dedicated).

If the efforts of past workers are acknowledged, so too must those of the present cohort of marsupial biologists. Our phylogeny of dasyuromorphians is based on DNA sequences from virtually all living, and several extinct, species in the order. Many of these animals are rare, and often occur in beautiful but inhospitable or inaccessible localities. Yet field biologists and curators have been diligent in seeking them out and making material available to us. Although the molecular-morphological conflicts described above have occasioned some frustration, a commitment to furthering the common objective of reconstructing marsupial phylogeny has largely prevailed and opened prospects of collaboration between workers from very different research traditions. Indeed, integration of data from many different sources is the most exciting prospect for studying marsupial evolution in the years to come. In the third instalment of *Carnivorous Marsupials*, we can look forward to well-developed hypotheses of phenotypic character evolution, historical biogeography, genome evolution, and causes of speciation, to name just a few of the areas now ripe for comparative study. And, because phylogeny is the framework for comparative biology, systematics will continue to be at the center of this exciting field.

Acknowledgements

We thank Pat Woolley and Steve Wroe for their critical contributions to our research over the years. We also acknowledge the hard work of our students Lori Armstrong, Mark Blacket, Larry Buckley, Brian Cambron, Amy Driskell, Jodie Painter, and Jodie Young. Mark Springer, Steve Cooper, and Jacques Retief initiated us in the acquisition of 12S, CR, and P1 data, respectively. For other valuable input, we thank Ken Aplin, Mike Archer, Peter Baverstock, Mike Braun, Don Colgan, Chris Dickman, Gordon Dixon, Steve Donnellan, Karen Firestone, Max King, John Kirsch, and Steve Van Dyck. We are extremely grateful to all those who so generously provided tissue or DNA samples, without which there would be no molecular phylogeny to discuss. Our research has been funded by the Australian Research Council, the U.S. National Science Foundation, Southern Illinois University, and La Trobe University.

References

Aplin, K.P., & Archer, M. (1987), 'Recent advances in marsupial systematics with a new syncretic classification', in *Possums and Opossums: Studies in Evolution* (ed. M. Archer), pp. xv–lxi, Surrey Beatty, Sydney.

Archer, M. (1976a), 'The basicranial region of marsupicarnivores (Marsupialia), inter-relationships of carnivorous marsupials, and affinities of the insectivorous marsupial peramelids', *Journal of the Linnean Society of London (Zoology)*, **59**:217–322.

Archer, M. (1976b), 'The dasyurid dentition and its relationship to that of didelphids, thylacinids, borhyaenids (Marsupicarnivora) and peramelids (Peramelina: Marsupialia)', *Australian Journal of Zoology Supplemental Series* **39**:1–34.

Archer, M. (1976c), 'Revision of the marsupial genus *Planigale* Troughton (Dasyuridae), *Memoirs of the Queensland Museum'*, **17**:341–65.

Archer, M. (1981), 'Revision of the dasyurid marsupial genus Sminthopsis Thomas', *Bulletin of the American Museum of Natural History* **168**:61–224.

Archer, M. (1982a), 'A review of Miocene thylacinids (Thylacinidae, Marsupialia), thephylogenetic position of the Thylacinidae and the problem of a priorisms in character analysis', in: *Carnivorous Marsupials* (ed. M. Archer), pp. 445–76, Royal Zoological Society of New South Wales, Mosman.

Archer, M. (1982b), 'Review of the dasyurid (Marsupialia) fossil record, integration of data bearing on phylogenetic interpretation and suprageneric classification', in *Carnivorous Marsupials* (ed. M. Archer), pp. 397–443, Royal Zoological Society of New South Wales, Mosman.

Archer, M. (1984), 'The Australian marsupial radiation', in *Vertebrate Zoogeography and Evolution in Australia* (eds. M. Archer & G. Clayton), pp. 633–708, Hesperian Press, Carlisle.

Armstrong, L.A., Krajewski, C., & Westerman, M. (1998), 'Phylogeny of the dasyurid marsupial genus *Antechinus* based on cytochrome-*b*, 12S-rRNA, and protamine-P1 genes', *Journal of Mammalogy*, **79**:1379–89.

Avise, J. (1994), *Molecular markers, natural history, and evolution*, Chapman and Hall, New York, NY.

Awadalla, P., Eyre-Walker, A., & Maynard Smith, J. (1999), Linkage disequilibrium and recombination in hominoid mitochondrial DNA. *Science* 286, 2524–25.

Baker, A.J., & Marshall, H.D. (1997), 'Mitochondrial control region sequences as tools for understanding evolution' in *Avian Molecular Systematics* (ed. D.P. Mindell), pp. 51–82, Academic Press, San Diego, CA.

Baverstock, P.R., Archer, M., Adams, M., & Richardson, B.J. (1982), 'Genetic relationships among 32 species of Australian dasyurid marsupials' in *Carnivorous Marsupials* (ed. M. Archer), pp. 641–50, Royal Zoological Society of New South Wales, Mosman.

Baverstock, P.R., Krieg, M., & Birrell, J. (1990), 'Evolutionary relationships among Australian marsupials as assessed by albumin immunology', *Australian Journal of Zoology* **37**:273–87.

Bensley, B.A. (1903), 'On the evolution of the Australian Marsupialia; with remarks on the relationships of the marsupials in general', trans. *Linn. Soc. London (Zool.)*, **9**:83–217.

Blacket, M.J., Adams, M., Cooper, S.J.B., Krajewski, C., & Westerman, M. (2001), 'Systematics and evolution of the dasyurid marsupial genus *Sminthopsis*: I. The Macroura Species Group', *Journal of Mammalian Evolution*, accepted.

Blacket, M.J., Adams, M., Krajewski, C., & Westerman, M. (2000), 'Genetic variation within the marsupial genus *Planigale*', *Australian Journal of Zoology*, **48**:443–59.

Blacket, M.J., Krajewski, C., Labrinidis, A., Cambron, B., Cooper, S., & Westerman, M. (1999), 'Systematic relationships within the dasyurid marsupial tribe Sminthopsini – a multigene approach', *Molecular Phylogenetics and Evolution*, **12**:140–55.

Cooper, N.K., Aplin, K.P., & Adams, M. (2000), 'A new species of false antechinus (Marsupialia: Dasyuromorphia: Dasyuridae) from the Pilbara region, Western Australia', *Records of the Western Australian Museum*, **20**:115–36.

Diaconis, P., & Efron, B. (1983), 'Computer-intensive methods in statistics', *Scientific American*, **249**:116–130.

Dickman, C.R., Parnaby, H.E., Crowther, M.S., & King, D.H. (1998), '*Antechinus agilis* (Marsupialia: Dasyuridae), a new species from the *A. stuartii* complex in south-eastern Australia', *Australian Journal of Zoology*, **46**:1–26.

Dixon, M.T., & Hillis, D.M. (1993), 'Ribosomal RNA secondary structure: compensatory mutations and implications for phylogenetic analysis', *Molecular Biology and Evolution*, **10**:256–67.

Esposti, M.D., De Vries, S., Crimi, M., Ghelli, A., Patarnello, T., & Meyer, A. (1993), 'Mitochondrial cytochrome *b*: evolution and structure of the protein', *Biochimica et Biophysica Acta*, **1143**:243–71.

Faith, D.P. (1990), 'Chance marsupial relationships', *Nature*, **345**:393.

Felsenstein, J. (1985), 'Confidence limits on phylogenies: an approach using the bootstrap', *Evolution*, **39**:783–91.

Felsenstein, J. (1993), 'PHYLIP (phylogeny inference package)', distributed by the author, Department of Genetics, University of Washington, Seattle.

Firestone, K.B. (2000), 'Phylogenetic relationships among quolls revisited: the mtDNA control region as a useful tool', *Journal of Mammalian Evolution*, **7**:1–22.

Fitch, W.M., & Margoliash, E. (1967), 'Construction of phylogenetic trees', *Science*, **155**:279–84.

Groves, C.P. (1993), 'Order Dasyuromorphia', in *Mammal Species of the World: a Taxonomic and Geographic Reference* (eds. D.E. Wilson & D.A.M. Reeder), pp. 45-62, Smithsonian Institution Press, Washington.

Hennig, W. (1966), 'Phylogenetic systematics', University of Illinois Press, Urbana.

Hillis, D.M., & Bull, J.J. (1993), 'An empirical test of bootstrapping as a method for assessing confidence in phylogenetic analysis', *Systematic Biology*, **42**:182–92.

Hillis, D.M., Moritz, C., & Mable, B.K. (1996), 'Molecular systematics, 2nd edition', Sinauer Associates, Sunderland, Massachusetts.

Howell, N. (1989), 'Evolutionary conservation of protein regions in the protonmotive cytochrome b and their possible roles in redox catalysis', *Journal of Molecular Evolution*, **29**:157–69.

Irwin, D.M., Kocher, T.D., & Wilson, A.C. (1991), 'Evolution of the cytochrome b gene of mammals', *Journal of Molecular Evolution*, **32**:128–44.

Janke, A., Feldmaier-Fuchs, G., Thomas, W.K., von Haesler, A., & Paabo, S. (1994), 'The marsupial mitochondrial genome and the evolution of placental mammals', *Genetics*, **137**:243–56.

Kirsch, J.A.W., Krajewski, C., Springer, M.S., & Archer, M. (1990), 'DNA/DNA hybridization studies of carnivorous marsupials. II: relationships among dasyurids (Marsupialia: Dasyuridae)', *Australian Journal of Zoology*, **38**:673–96.

Kirsch, J.A.W., Lapoint, F.J., & Springer, M.S. (1997), 'DNA-hybridisation studies of marsupials and their implications for metatherian classification', *Australian Journal of Zoology*, **45**:211–80.

Kitchener, D.J., & Caputi, N. (1988), 'A new species of false antechinus (Marsupialia, Dasyuridae) from Western Australia, with remarks on the generic classification within the Parantechini', *Records of the Western Australian Museum*, **14**:35–59.

Krajewski, C., Blacket, M., Buckley, L., & Westerman, M. (1997a), 'A multigene assessment of phylogenetic relationships in the dasyurid marsupial subfamily Sminthopsinae', *Molecular Phylogenetics and Evolution*, **8**:236–48.

Krajewski, C., Blacket, M.J., & Westerman, M. (2000a), 'Effects of partitioning and transversion weighting on phylogenetic precision: familial relationships among dasyuromorphian marsupials', *Journal of Mammalian Evolution*, **7**:95–108.

Krajewski, C., Buckley, L., & Westerman, M. (1997b), 'DNA phylogeny of the marsupial wolf resolved', *Proceedings of the Royal Society of London Series B*, **264**:911–17.

Krajewski, C., Buckley, L., Woolley, P.A., & Westerman, M. (1996), 'Phylogenetic analysis of cytochrome b sequences in the dasyurid marsupial subfamily phascogalinae: systematics and the evolution of reproductive strategies', *Journal of Mammalian Evolution*, **3**:81–91.

Krajewski, C., Driskell, A.C., Baverstock, P.R., & Braun, M.J. (1992), 'Phylogenetic relationships of the thylacine (Marsupialia: Thylacinidae) among dasyuroid marsupials: evidence from cytochrome-b DNA sequences', *Proceedings of the Royal Society of London Series B*, **250**:19–27.

Krajewski, C., Fain, M.G., Buckley, L., & King, D.G. (1999), 'Dynamically heterogenous partitions and phylogenetic inference: an evaluation of analytical strategies with cytochrome *b* and ND6 gene sequences in cranes', *Molecular Phylogenetics and Evolution*, **13**:302–13.

Krajewski, C., & King, D.G. (1996), 'Molecular divergence and phylogeny: Rates and patterns of cytochrome b evolution in cranes', *Molecular Biology and Evolution*, **13**:21–30.

Krajewski, C., Painter, J., Buckley, L., & Westerman, M. (1994), 'Phylogenetic structure of the marsupial family Dasyuridae', *Journal of Mammalian Evolution*, **2**:25–35.

Krajewski, C., Painter, J., Driskell, A.C., Buckley, L., & Westerman, M. (1993), 'Molecular systematics of New Guinean dasyurids (Marsupialia: Dasyuridae)', *Science in New Guinea*, **19**:157–66.

Krajewski, C., Woolley, P.A., & Westerman, M. (2000b), 'The evolution of reproductive strategies in dasyurid marsupials: implications of molecular phylogeny', *Biological Journal of the Linnean Society*, **71**:417–35.

Krajewski, C., Wroe, S., & Westerman, M. (2000c), 'Molecular evidence for the pattern and timing of cladogenesis in dasyurid marsupials', *Zoological Journal of the Linnean Society*, **130**:375–404.

Krajewski, C., Young, J., Buckley, L., Woolley, P.A., & Westerman, M. (1997c), 'Reconstructing the evolutionary radiation of dasyurine marsupials with cytochrome *b*, 12S rRNA, and protamine gene trees', *Journal of Mammalian Evolution*, **4**:217–36.

Lee, A.K., Woolley, P., & Braithewaite, R.W. (1982), 'Life history strategies of dasyurid marsupials', in *Carnivorous Marsupials* (ed. M. Archer), pp. 1–11, Royal Zoological Society of New South Wales, Mosman, Sydney.

Lipman, D.J., & and W.R. Pearson (1985), 'Rapid and sensitive protein similarity searches', *Science*, **227**:1435.

Lowenstein, J.M., Sarich, V.M., & Richardson, B.J. (1981), 'Albumin systematics of the extinct mammoth and Tasmanian wolf', *Nature*, **291**:409–11.

Maddison, W.P. (1993), 'Missing data versus missing characters in phylogenetic analysis', *Systematic Biology*, **42**:576–80.

Marshall, L.G. (1977), 'Cladistic analysis of borhyaenoid, dasyuroid, and thylacinid (Marsupialia: Mammalia) affinity', *Systematic Zoology*, **26**:410–25.

Maxson, R.D., & Maxson, L.R. (1986), 'Micro-complement fixation: a quantitative estimator of protein evolution', *Molecular Biology and Evolution*, **3**:375–88.

Moritz, C., Dowling, T.E., & Brown, W.M. (1987), 'Evolution of animal mitochondrial DNA: relevance for population biology and systematics', *Annual Review of Ecology and Systematics*, **18**:269–92.

Muirhead, J., & Wroe, S. (1998), 'A new genus and species, *Badjcinus turnbulli* (Thylacinidae: Marsupialia), from the Late Oligocene of Riversleigh, Northern Australia, and an investigation of thylacinid phylogeny', *Journal of Vertebrate Paleontology*, **18**:612–26.

Nei, M. (1991), 'Relative efficiencies of different tree-making methods for molecular data' in *Phylogenetic Analysis of DNA Sequences* (eds. M.M. Miyamoto & J. Cracraft), pp. 90–128, Oxford University Press, New York.

Painter, J., Krajewski, C., & Westerman, M. (1995), 'Molecular phylogeny of the marsupial genus *Planigale*', *Journal of Mammalogy*, **76**:406–13.

Retief, J.D., Krajewski, C., Westerman, M., & Dixon, G.H. (1995a), 'The evolution of protamine P1 genes in dasyurid marsupials', *Journal of Molecular Evolution*, **41**:549–55.

Retief, J.D., Krajewski, C., Westerman, M., Winkfein, R.J., & Dixon, G.H. (1995b), 'Molecular phylogeny and evolution of marsupial protamine P1 genes', *Proceedings of the Royal Society of London Series B*, **259**:7–14.

Ride, W.D.L. (1970), 'A guide to the native mammals of Australia', Oxford Univ. Press, Melbourne.

Saccone, C., Pesole, G., & Sbisa, E. (1991), 'The main regulatory region of mammalian mitochondrial DNA: Structure-function model and evolutionary pattern', *Journal of Molecular Evolution*, **33**:83–91.

Saitou, N., & Nei, M. (1987), 'The neighbor-joining method: a new method for reconstructing phylogenetic trees', *Molecular Biology and Evolution*, **4**:406–25.

Sibley, C.G., & Ahlquist, J.E. (1990), *Phylogeny and Classification of Birds*, Yale University Press, New Haven, CT.

Simpson, G.G. (1941), 'The affinities of the Borhyaenidae', *American Museum Novitates*, **578**:1–11.

Sinclair, W.J. (1906), 'Mammalia of the Santa Cruz beds', *Report of the Princeton University Expedition to Patagonia*, **4**:333–60.

Springer, M.S., & Douzery, E. (1996), 'Secondary structure and patterns of evolution among mammalian mitochondrial 12S rRNA molecules', *Journal of Molecular Evolution*, **43**:357–73.

Springer, M.S., Hollar, L.J., & Burke, A. (1995), 'Compensatory substitutions and the evolution of the 12S rRNA gene in mammals', *Molecular Biology and Evolution*, **12**:1138–50.

Springer, M., & Krajewski, C. (1989a), 'Additive distances, rate variation, and the perfect fit theorem', *Systematic Zoology*, **38**:371–75.

Springer, M., & Krajewski, C. (1989b), 'DNA hybridization in animal taxonomy: a critique from first principles', *Quarterly Review of Biology*, **64**:291–318.

Swofford, D.L., Olsen, G.J., Waddell, P.J., & Hillis, D.M. (1996), 'Phylogenetic inference', in *Molecular Systematics, 2nd Edition* (eds. D.M. Hillis, C. Moritz, & B.K. Mable), pp. 407–514, Sinauer Associates, Sunderland, MA.

Szalay, F.S. (1982), 'A new appraisal of marsupial phylogeny and classification', in *Carnivorous Marsupials* (ed. M. Archer), pp. 621–640, Royal Zoological Society of New South Wales, Mosman, Sydney.

Tate, G.H.H. (1947), 'Results of the Archbold Expeditions No. 56. On the anatomy and classification of the Dasyuridae (Marsupialia)', *Bulletin of the American Museum of Natural History*, **88**:97–156.

Thomas, R.H., Schaffer, W., Wilson, A.C., & Paabo, S. (1989), 'DNA phylogeny of the marsupial wolf', *Nature*, **340**:465–67.

Van Dyck, S.M. (1980), 'The cinnamon antechinus, *Antechinus leo* (Marsupialia: Dasyuridae), a new species from the vine-forests of the Cape York Penninsula', *Australian Mammalogy*, **3**:5–17.

Van Dyck, S. (1982), 'The relationships of *Antechinus stuartii* and *A. flavipes* (Dasyuridae, Marsupialia) with special reference to Queensland' in *Carnivorous Marsupials* (ed. M. Archer), pp. 723–766, Royal Zoological Society of New South Wales, Mosman, Sydney.

Van Dyck, S. (1987), 'The bronze quoll, *Dasyurus spartacus* (Marsupialia: Dasyuridae), a new species form the savannahs of Papua New Guinea', *Australian Mammalogy*, **11**:145–56.

Van Dyck, S., Woinarski, J.C.Z., & Press, A.J. (1994), 'The Kakadu dunnart *Sminthopsis bindi* (Marsupialia: Dasyuridae), a new species from the stony woodlands of the Northern Territory', *Memoirs of the Queensland Museum*, **37**:311–23.

Wood, H.E. (1924), 'The position of the "Sparassodonts": with notes on the relationships and history of the Marsupialia', *Bulletin of the American Museum of Natural History*, **51**:77–101.

Woolley, P.A. (1982), 'Phallic morphology of the Australian species of *Antechinus* (Dasyuridae, Marsupialia): a new taxonomic tool?' in *Carnivorous Marsupials* (ed. M. Archer), pp. 767–81, Royal Zoological Society of New South Wales, Mosman, Sydney.

Wroe, S. (1996), '*Murbacinus gadiyuli* (Thylacinidae: Marsupialia), a very plesiomorphic thylacinid from the Miocene of Riversleigh, northwestern Queensland, and the problem of paraphyly for the Dasyuridae (Marsupialia)', *Journal of Paleontology*, **70**:1032–44.

Wroe, S. (1997), 'A reexamination of morphology-based synapomorphies for the families of Dasyuromorphia (Marsupialia). I. Dasyuridae', *Journal of Mammalian Evolution*, **4**:19–52.

Wroe, S. (1999), 'The geologically oldest dasyurid, from the Miocene of Riversleigh, north-west Queensland', *Paleontology*, **42**:501–27.

Wroe, S., & Mackness, B.S. (1998), 'Revision of the Pliocene dasyurid, *Dasyurus dunmalli* (Dasyuridae; Marsupialia)', *Memoirs of the Queensland Museum*, **42**:605–12.

Wroe, S., & Muirhead, J. (1999), 'Evolution of Australia's marsupicarnivores: Dasyuridae, Thylacinidae, Myrmecobiidae, Dasyuromorphia *incertae sedis* and Marsupialia *incertae sedis*', *Australian Mammalogy*, **21**:10–11.

Yang, Z. (1994), 'Statistical properties of the maximum likelihood method of phylogenetic estimation and comparison with distance matrix methods', *Systematic Biology*, **43**:329–42.

CHAPTER 2

EVOLUTION OF AMERICAN MARSUPIALS AND THEIR PHYLOGENETIC RELATIONSHIPS WITH AUSTRALIAN METATHERIANS

R. Eduardo Palma

Centro de Estudios Avanzados en Ecología y Biodiversidad and Departamento de Ecología, Facultad de Ciencias Biológicas, Pontificia Universidad Católica de Chile, Casilla 114-D, Santiago 6513677, Chile

Until now, phylogenetic studies regarding the evolution of marsupials have included morphological, cytogenetic, biochemical and molecular approaches. Within American metatherians, the large-sized didelphids seem to constitute a well supported clade in contrast with mouse opossums, which are not grouped in a natural assemblage. The phylogenetic position of the medium-sized *Metachirus* has long been a point of confusion, but recent molecular studies confirm earlier hypotheses that this genus is a taxon belonging to the large-sized didelphid opossum clade. South American shrew opossums (the caenolestids) appeared as the sister taxon to didelphids, although some molecular reconstructions using different molecular markers show them in close association with the Australian peramelids. Finally, one of the long-standing issues regarding the evolution of Australian and South American marsupials has been the affinities of *Dromiciops gliroides*, a microbiotheriid mouse opossum better known as the 'monito del monte'. According to different data set analyses, this species should be part of a clade that includes the Australasian dasyurids, diprotodontians and notoryctemorph marsupials. Current paleontological calibration, as well as geological and geophysical evidence, would sustain a passage of microbiotheriids from South America to Antarctica with further differentiation in the latter continent. Alternatively, microbiotheriids could constitute the ancestors of Australasian marsupials, having dispersed from Antarctica to Australasia and subsequently differentiating and undergoing further extinction in the latter region.

INTRODUCTION

The major synapomorphic feature that differentiates marsupials from monotremes and eutherian mammals is the mode of reproduction: while monotremes are characterised by being egg-laying mammals, and eutherians have evolved viviparity, marsupials give birth to an embryo-like creature that completes its development attached to the mother's nipple and which may or may not occur in a pouch (Bronson 1989). Another major characteristic of marsupials is their current distribution which is restricted to the American and Australasian regions, having previously been found on all continents including Antarctica (Marshall et al. 1990, Woodburne and Case 1996).

Currently, the major taxonomic diversity of marsupials in the Americas is localised in tropical areas of the Neotropical Region in South America where two orders are found: Didelphimorphia (opossums and mouse opossums), and Paucituberculata (shrew opossums; Gardner 1993, Emmons 1997). A third order, the monotypic Microbiotheria is represented by *Dromiciops gliroides* Thomas, also known as the 'monito del monte' – an endemic taxon inhabitant of temperate forests in southern South America (Mann 1978). The oldest fossil record for marsupials is currently recognised as coming from the late Cretaceous of North America, the Cenomanian of Utah (Cifelli 1990 a, b; 1993 a, b), while the oldest South American record has been ascribed to the early Paleocene Tiupampa fauna of Bolivia (Gayet et al. 1991). Current hypotheses about the appearance of marsupials on the latter continent advocate dispersal events from North America via a proto-Caribbean route that tenuously connected North and South America at the end of the Cretaceous, probably in Maastrichtian times (Eaton 1993). Further records of marsupials have been reported for Europe (middle Paleocene of Belgium; Crochet and Sigé 1983), Africa (Oligocene of Egypt; Bown and Simons 1984), Asia (middle Miocene of Thailand; Ducrocq et al. 1992), and Antarctica (early Eocene; Woodburne and Zinsmeister 1982, 1984; Woodburne and Case 1996). Fossil records for Australasia date back to Oligocene times (Ride 1964).

Although all marsupials became extinct in North America, this region is home to a single species, *Didelphis virginiana*, which represents a southern invader that arrived during the Great American Interchange of fauna during Plio-Pleistocene times (Simpson 1980, Stehli and Webb 1985). Didelphimorphia constitutes the most speciose taxon with 64 currently recognised species, distributed from SE Canada southward to Argentina and Chile (Gardner 1993, Palma and Yates 1998), and grouped into the families Didelphidae and Caluromyidae (Kirsch and Palma 1995). For the Paucituberculata, six species are currently recognised as part of the family Caenolestidae distributed throughout Venezuela, Colombia, Ecuador, Peru, Bolivia and Chile. Finally, the Microbiotheria is composed of the monotypic *Dromiciops gliroides*, restricted to the *Nothofagus* forests of Chile and Argentina (Marshall 1978).

The goal of this chapter is to provide an overview of what is currently known about the relationships between American marsupials, and their linkage (both phylogenetic and biogeographic) to Australian marsupial taxa. The hypotheses about these relationships are the result of different approaches ranging from the morphological to the molecular level. I will provide the results of a review of the major groups recognised in the Americas and their current hypothesised phylogenetic relationships.

MAJOR CHARACTERISTICS OF SOUTH AMERICAN MARSUPIALS

Didelphimorphia: the Didelphidae, opossums and mouse opossums

Three major groups can be distinguished in the Didelphidae based on body size: the large-sized opossums (*Chironectes, Didelphis, Lutreolina*, and *Philander*), the medium-sized opossum *Metachirus* (the 'brown four-eyed opossum'), and the mouse opossums (*Lestodelphys, Monodelphis, Marmosa, Micoureus, Marmosops, Thylamys*, and *Gracilinanus*; Fig. 1). This size-based grouping does not necessarily represent natural groups. The large-sized opossums are distributed from SE Canada, through most of Central America, in the Amazon Basin, and southward to the northern half of Argentina (Gardner 1993). Mouse opossums, on the other hand, range from North America (Mexico) through Central and South America to central Chile and Argentina. Finally, *Metachirus* ranges from Central America to northern Argentina (Gardner 1993).

Figure 1 Didelphimorphia: adult mouse opossum, *Thylamys elegans*, from central Chile (photo by the author).

A monographic revision of *Marmosa* sensu lato, based on morphology, distinguished five species groups that today more or less constitute the genera *Marmosa, Micoureus, Marmosops, Gracilinanus* and *Thylamys*. The generic recognition of these taxa was based on different studies that have included anatomical, chromosomal, biochemical, and molecular analyses (DNA hybridisation and mtDNA sequences; Creighton 1984, Reig et al. 1987, Gardner and Creighton 1989, Kirsch and Palma 1995, Patton et al. 1996, Palma and Yates 1998). Other than body size, didelphids can also be distinguished by their conservative karyotypes composed of three modal numbers: those that have 14 chromosomes (mouse opossums and *Metachirus*), the 2n = 18 (*Monodelphis*), and the 2n = 22 mode (large-sized opossums and *Marmosa canescens*; Engstrom and Gardner 1988, Hayman 1990, Reig et al. 1987, Palma and Yates 1996; Fig. 2).

Didelphimorphia: the Caluromyidae, woolly opossums

Three genera have traditionally been recognised in this family: *Caluromys, Caluromysiops*, and *Glironia* (Kirsch and Palma 1995), most of them being restricted to the Neotropical forests of South America, with the exception *of Caluromys derbianus*, which ranges from Mexico to South America (Emmons 1997). Until now, the only karyotype that has been described is that of

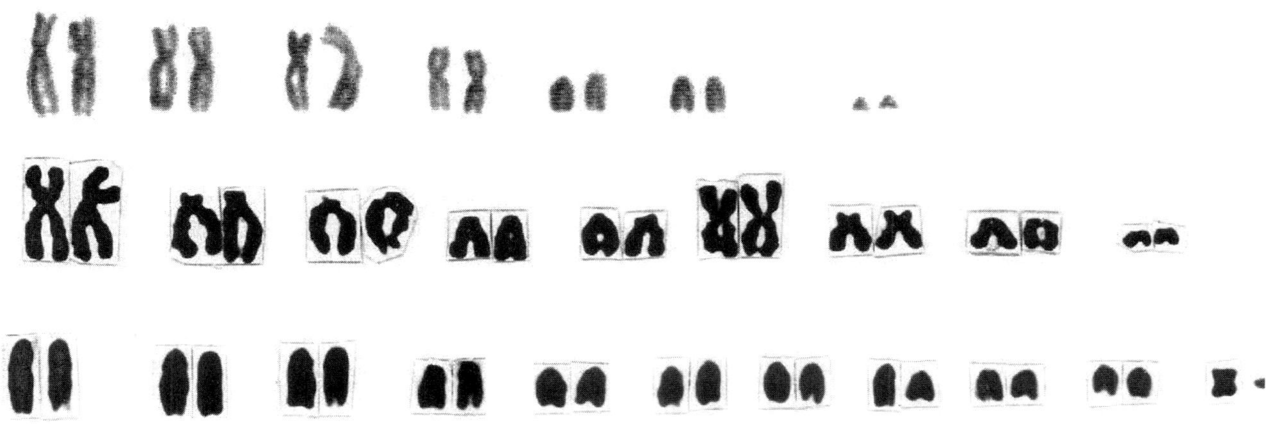

Figure 2 The standard karyotypes of American marsupials: a) 2n = 14 (female *Thylamys elegans*); b) 2n = 18 (female *Monodelphis domestica*); c) 2n = 22 (male *Lutreolina crassicaudata*) (from Palma and Yates 1996).

Caluromys, which also has 2n = 14 (Biggers et al. 1965, Hayman and Martin 1974, Palma and Yates 1996).

Paucituberculata: the Caenolestidae, shrew opossums

Three genera are currently recognised as belonging to this family: *Caenolestes, Lestoros* and *Rhyncholestes,* whose distribution ranges from NW Venezuela (e.g. *Caenolestes*), throughout the Andes of Colombia, Ecuador, Perú, and Bolivia (*Caenolestes, Lestoros*; Gardner 1993, Anderson 1997), and in the temperate forests of southern Chile (e.g. *Rhyncholestes*, Mann 1978; Fig. 3). With regard to karyotypes, shrew opossums are also characterised by having a 2n = 14 complement and a morphology similar to that of mouse opossums.

Some of the major characteristics of the group are that the first lower incisors are enlarged and projected forward (Marshall 1980), a shrew-like external appearance, and a disjunct distribution as seen in the isolated *Rhyncholestes* from the rainforests of southern Chile (Redford and Eisenberg 1992). The disjunct distribution pattern between the southern *Rhyncholestes* and subtropical and tropical caenolestids may be due to the occurrence of a continuous and arid belt that crosses South America from the SE to the NW, separating the southern forests of Chile and Argentina from those of the rest of the continent in what is known as 'diagonal arida' (arid diagonal; e.g. the separation between the Argentinean–Bolivian Yungas and the temperate southern Chilean forests; Villagrán and Hinojosa 1997). Phytogeographic studies have demonstrated the phylogenetic affinities of what was a continuous landscape of forests in southern South America, separated later by the great uplift of the Andes, the Atacama Desert, the Puna and Patagonia among the major open and semi-arid regions. Recent morphological comparisons based on both dental and cranial features between caenolestids and the extinct group of South American marsupials, the Argyrolagidae (this group being convergent in locomotion and diet to bipedal desert rodents e.g. Heteromyidae), confirm both groups as belonging to the order Paucituberculata, which consists in a monophyletic group (Sánchez-Villagra 2001).

Microbiotheria: the Microbiotheriidae, 'monito del monte'

The order Microbiotheria is represented by the single species, *Dromiciops gliroides* (Fig. 4), which is a member of a formerly diverse family, the Microbiotheriidae (Reig 1955, Marshall 1982). Although *Dromiciops* resembles a didelphid mouse opossum (e.g. *Thylamys*), it is differentiated from these by the possession of a pouch, its teeth morphology (e.g. the four lower incisors are evenly spaced), and the entotympanic bone component in the auditory bullae (Hershkovitz 1999). Early serological results showed a relatively early divergence of *Dromiciops* with respect to other American opossums even earlier than the caluromyine and didelphine marsupial (Kirsch 1977; Reig et al.

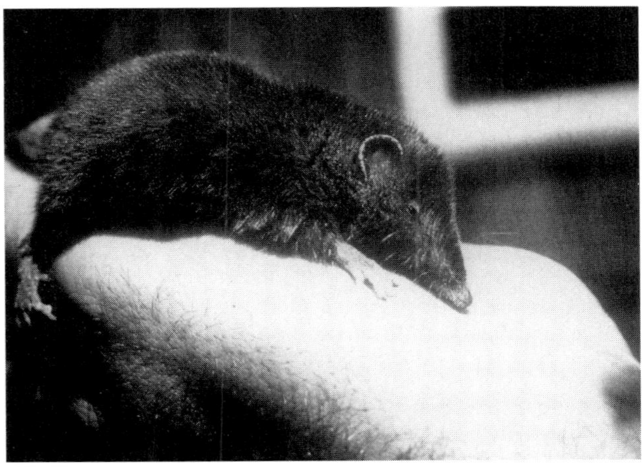

Figure 3 Paucituberculata: adult shrew opossum, *Rhyncholestes raphanurus*, from southern Chile (photo by Bruce D. Patterson).

Figure 4 Microbiotheria: adult 'monito del monte', *Dromiciops gliroides*, from southern Chile (photo by Fabian Jaksic).

1987). The present distribution of *Dromiciops* is restricted to the southern temperate forests of Chile and adjacent Argentina. Chromosomally, *Dromiciops* also exhibits a diploid number of 14 chromosomes, a feature that appears to be found only in the somatic cells of females, since males have 2n = 13 due to the loss of chromosome Y (Gallardo and Patterson 1987; however, see Spotorno et al. 1997).

Phylogenetic relationships among South American marsupial taxa

From a cytogenetic perspective, G-banding patterns have revealed a great homology in autosomal arms for several didelphid taxa. Recent analyses that have considered G and C banding patterns in *Didelphis* and *Philander* have found a high degree of homologies for G bands in both taxa, although C-banding indicated striking differences between both karyotypes (Svartman and Vianna-Morgante 1999). In fact, these authors found that while all autosomes of *Philander opossum* have pericentromeric constitutive heterochromatin, the autosomal C-bands of *D. marsupialis* occur as diffusely stained blocks in the distal portion of the long arms of some chromosomes. Furthermore, variations in the amount of pericentromeric constitutive heterochromatine are the main differences between the karyotypes of *Caluromys philander, C. lanatus, Marmosa murina,* and *Micoureus demerarae* (Svartman and Vianna-Morgante 1999). In general, C and G banding, and genome comparisons through in-situ hybridisation, for several didelphid taxa, revealed a high conservation of the euchromatic portions of the karyotypes, and a high divergence of repetitive sequences as reflected in the heterochromatin of some of the didelphid species.

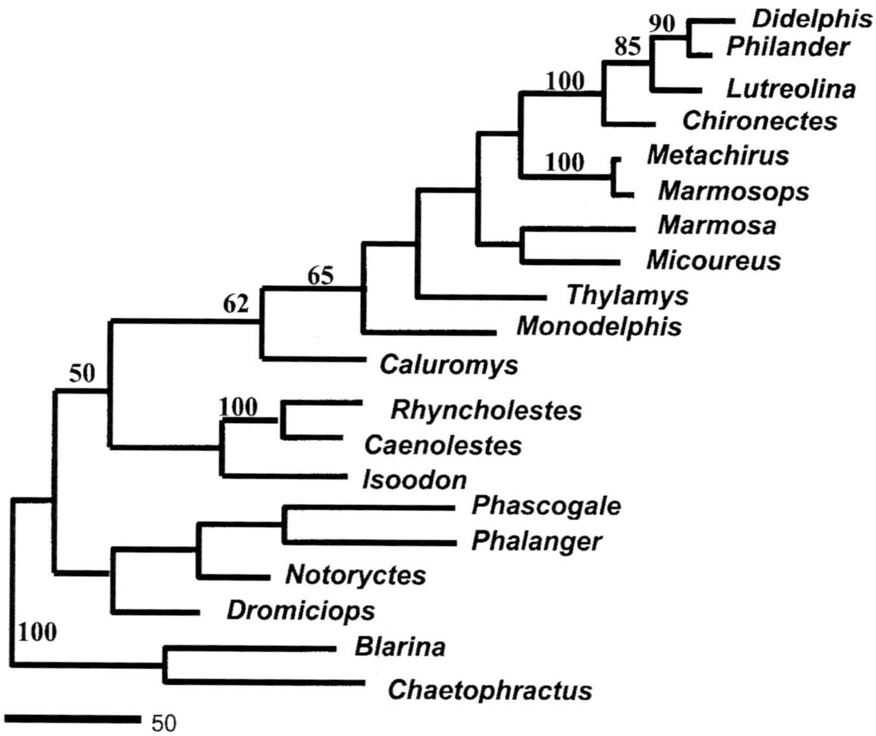

Figure 5 Maximum parsimony analysis (using PAUP 4.0 * implemented by David L. Swoofford) of South American and selected Australian marsupials based on the complete rRNA 12 S gene, CI = 0.4362, 972 steps; numbers on the nodes represent 5000 replicates obtained through bootstrap analysis (values under 50% are not shown) (from Palma and Spotorno 1999).

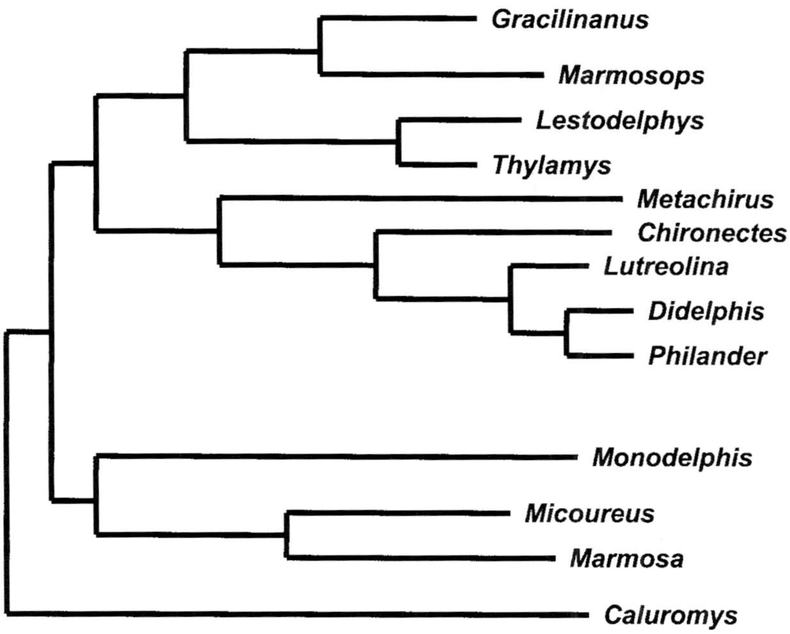

Figure 6 Fitch tree of relationships among didelphid marsupials including mouse and large-sized opossums based on DNA hybridisation studies. (adapted from Kirsch and Palma 1995).

Current phylogenetic relationships within American didelphimorphians seem to be well understood, particularly the phylogeny within large-sized forms (Fig. 5 and Fig. 6). Serological studies, for example, have shown that *Didelphis*, *Philander*, and *Chironectes* form a close union, particularly between these first two genera with respect to the latter (*Lutreolina* was not included in the analysis; Kirsch 1977). Furthermore, molecular studies based on DNA hybridisation, mitochondrial genes such as cytochrome b and 12S, and IRBP nuclear sequences confirm the union between *Didelphis* and *Philander* as sister taxa, with *Lutreolina* as the first outgroup, and *Chironectes* as the most basal taxon within the large-sized didelphids (Kirsch and Palma 1995, Patton et al. 1996, Kirsch et al. 1997, Palma and Spotorno 1999, Jansa and Voss 2000). This reconstruction was earlier proposed by Reig et al. (1987) in a combined phylogenetic analysis that included morphological, anatomical, cytogenetic, and biochemical data. Therefore, based on different data sets, the monophyly of large-sized Neotropical opossums seems to be well supported. However, the relationships of mouse opossums, and the phylogenetic position of the brown four-eyed opossum *Metachirus* has not been consistent. Serological data recovered mouse opossums (e.g. *Marmosa*, *Micoureus*) as closely related to *Monodelphis* with *Metachirus* in an 'intermediate' position between the small and large-sized opossums (Kirsch 1977). Hershkovitz (1992a) based on anatomical traits proposed the family Marmosidae for all mouse opossums including *Metachirus*. However, Kirsch and Palma (1995), based on DNA hybridisation, and Patton et al. (1996) based on mtDNA cytochrome b sequences recovered *Metachirus* as the sister taxon of large-sized American marsupials (Fig. 6). Later, mtDNA 12S sequences recovered *Metachirus* as being closely related to small marsupials (e.g. as the sister taxon to *Marmosops*), although phylogenetically proximate to large-sized forms. However, recent nuclear IRBP sequences confirm previous hypotheses in the sense that *Metachirus* belongs to a clade that includes the large-sized didelphid opossums (Jansa and Voss 2000).

The phylogeny of mouse opossums based on DNA hybridisation, placed the Patagonian *Lestodelphys* as the sister taxon to *Thylamys* as proposed earlier by Creighton (1984) and Reig et al. (1987) using morphological and anatomical characters. DNA hybridisation and allozyme studies also recovered *Marmosops* and *Gracilinanus* as the most related marmosines to the *Thylamys-Lestodelphys* clade (Kirsch and Palma 1995, Palma and Yates 1998; Fig. 6). However, mouse opossums do not constitute a monophyletic group considering that all currently recognised genera were traditionally accepted as species groups of *Marmosa* sensu lato (Tate 1933). Recent didelphid phylogeny has proposed the {[(*Lestodelphys* + *Thylamys*) *Gracilinanus*] *Marmosops*} clade as sister to large-sized opossums, making the [(*Marmosa* + *Micoureus*) *Monodelphis*] node basal to all didelphids (Jansa and Voss 2000).

Further evolutionary studies within didelphids have considered the marmosine genera *Marmosops* (Mustrangi and Patton

1997), and *Thylamys* (Palma and Yates 1998, Palma et al. 2002). The first study, by means of a phylogeographic approach, recognised two instead of a single species of the slender mouse opossum in the Atlantic forest of coastal Brazil. This differentiation was hypothesised to be the result of the intense tectonic activity in the region during the Miocene. The second study proposed the speciation of five species of *Thylamys* from southern South America due to different biogeographic events, mainly including dispersal from subtropical forests to open savanna-like environments, and to Andean and pre-Andean areas during the Plio-Pleistocene period (Palma et al. 2002). In this study, the most plesiomorphic taxon is a form restricted to the subtropical moist forest of eastern Paraguay (*T. macrura*), while the most derived are those distributed in open areas of the Andes and the western slopes of this mountain range (Palma and Yates 1998, Meynard et al. 2002).

Most studies have recovered *Caluromys*, *Caluromysiops* and *Glironia* as either a subfamily within Didelphidae (Caluromyinae; Kirsch 1977, Reig et al. 1987), or as a different family (Caluromyidae; Kirsch and Palma 1995, Kirsch et al. 1997). Serological data analysed through distance methods placed *Caluromys* as a taxon outside other large-and small-sized didelphimorphs (Kirsch 1977), while the combination of different data sets (e.g. cranial, dental, chromosomal) recovered *Caluromys* and *Caluromysiops* as sister taxa, and *Glironia* as the first outgroup to that clade (Reig et al. 1987). Later, molecular data based on DNA hybridisation analysis – which did not include *Caluromysiops* – recovered *Caluromys* and *Glironia* as sister taxa that were different from Didelphidae (Kirsch and Palma 1995, Kirsch et al. 1997). rRNA 12 S sequences also placed the woolly opossums as the sister taxon to all other didelphids (Palma and Spotorno 1999). However, cytochrome b sequences that examined most South American forms in order to hypothesise the didelphimorphian phylogeny failed to recover that caluromyines are a monophyletic assemblage, therefore suggesting *Glironia* and *Caluromys* as belonging to different didelphid clades (Patton et al. 1996).

Relationships between American and Australasian marsupials

The phylogeny of American marsupials and the relationships to their Australasian counterparts has been a topic of strong controversy, particularly since Szalay's (1982) proposition regarding the affinities of *Dromiciops gliroides*. Based on tarsal morphology this author noticed a dichotomy in the astragalar and calcanean articular patterns of marsupials, proposing the existence of characteristically separate ankle joint in American metatherians and a continuous joint in Australasian forms (however, see Hershkovitz 1992b). This allowed Szalay to split marsupials into two cohorts: Australidelphia for all Australasians and *Dromiciops*, and Ameridelphia for all American opossums. Subsequent studies, such as sperm morphology (Temple-Smith 1987) and sexual chromosomes (Gallardo and Patterson 1987, see above), began to support Szalay's hypothesis although these studies – as well as Szalay's itself – did not use strict phylogenetic methodologies. The occurrence of a probable 2n = 13 karyotype in the somatic cells of male *Dromiciops* was earlier proposed as evidence of its relationships to some Australian marsupial taxa, since this condition was also reported for sexual chromosomes in somatic cells of the families Petauridae and Peramelidae (Murray et al. 1979, Hayman 1990). However, similar missing chromosome Y conditions were reported for the Didelphimorphia (e.g. *Thylamys*, *Chironectes*; Palma 1995; Palma and Yates 1996), implying that this missing chromosome might represent a case of parallelism in the evolution of metatherian sexual chromosomes. Interestingly, recent research on *Dromiciops* cytogenetics – in a single young male – has revealed the presence of a very small chromosome in somatic cells, which is, without doubt the Y (Spotorno et al. 1997). At this point, it should be mentioned that other marsupial taxa, particularly didelphimorphs, are characterised by having tiny Y elements (Palma and Yates 1996). Recent morphological phylogenetic analyses between Australasian marsupials and *Dromiciops* confirm the inclusion of this taxon in the Australasian clade. However, these results were not robust enough since different assumptions in the parsimony analysis result in different topologies (Sanchéz-Villagra 2001).

At the beginning of 90s the first molecular, DNA-based, phylogenetic studies on *Dromiciops* relationships started to be published, particularly comparative studies using DNA hybridisation methodology (Kirsch et al. 1991, Westerman and Edwards 1991). These studies showed that marsupial evolution may be represented as a trichotomy of divergence among American, Australian and microbiotheriids (Westerman and Edwards 1991), while Kirsch and his colleagues proposed that the 'monito del monte' constituted the sister taxon of Australian diprotodontian marsupials (Kirsch et al. 1991). This latter hypothesis once again revolutionised the affinities of *Dromiciops* since Kirsch's et al. implied an Australasian origin for the ancestor of diprotodontians and microbiotheriids, with a late dispersal of the latter group to South America via 'island hopping'. In the middle of 90s sequencing studies began to support the affinities of *Dromiciops* to Australian marsupials when comparing partial sequences of the conservative rRNA 12S gene of the mitochondrial genome – studies that included selected American marsupial representatives (Gemmell and Westerman 1994, Springer et al. 1994). The latter authors favoured a relatedness of *Dromiciops* with dasyurids and diprotodontian marsupials, although with a low confidence bootstrap support value (about 50%) in the node that reunited the three lineages (Gemmell and Westerman 1994). Springer et al. hypothesised that the Australasian marsupial mole *Notoryctes* was the sister taxon to *Dromiciops*, however, the authors favoured the relatedness of the microbiotheriid with diprotodontians based on some lines of evidence, such as DNA hybridisation research (Kirsch et al. 1991). Nevertheless, Kirsch et al. (1997) cautioned that the 'monito del monte' lacks many of the synapomorphic features (e.g. morphological) of diprotodontians.

Sequencing studies related to the American-Australasian affinities of marsupials later continued, comparing most of the American taxa (including caenolestids), the microbiotheriid *Dromiciops*, and representatives of the Australian orders Diprotodontia, Dasyuromorphia, Peramelemorphia, and Notoryctemorphia (Colgan 1999, Springer et al. 1998, Palma and Spotorno 1999). Colgan (1999) compared 313 aligned bp of a nuclear coding gene (Pgk), concluding that the monogeneric orders Microbiotheria (*Dromiciops gliroides*) and Notoryctemorphia (*Notoryctes tylops*) were excluded from the Australidelphia as was earlier proposed by Szalay. In the same study, Ameridelphia appeared as monophyletic, and the microbiotheriids and the notoryctemorph marsupials were recovered at the same phylogenetic status as the Ameridelphia. The study by Palma and Spotorno (1999) compared the entire sequence of the rRNA 12S mitochondrial gene in 18 taxa concluding that *Dromiciops* was part of a clade (although with a low bootstrap support; Fig. 5) that included the Australasian forms, with this being the sister taxon of the dasyurid-diprotodontian clade. These data support the earliest studies based on morphology and partial mtDNA sequences. IRBP sequence phylogenetic analyses have failed to recover the Ameridelphia as a monophyletic group since the only paucituberculatan marsupial included in the study – *Caenolestes* – is grouped with Australasian marsupials (Jansa and Voss 2000). Mitochondrial 12S sequences, on the other hand, reunited the peramelid *Isoodon* as the sister taxon to the South American paucituberculatans *Rhyncholestes* and *Caenolestes*, leaving the Australidelphia as paraphyletic (Palma and Spotorno 1999). Interestingly, comparable results were obtained earlier based on partial sequences of the 12S region, in which peramelids appeared as a taxon differentiated from Australasian marsupials (Springer et al. 1994). Further studies based on protamine P1 gene sequences supported the relationships between paucituberculatans and peramelids (Retief et al. 1995), leading Kirsch et al. (1997) to postulate that 'while we regard the evidence as quite strong that microbiotheres are phyletically Australian, peramelids may very well not be.'

Biogeographic considerations for the American and Australasian marsupial affinities

Among the hypothesised biogeographic scenarios proposed to explain the linkage between microbiotheriids and Australasian marsupials, Kirsch and collaborators (1991) proposed a scenario of waif and/or sweepstakes dispersal of microbiotheriids across the Pacific Ocean from Australia to reach Antarctica, with further dispersal to South America. However, this hypothesis is hard to support since no evidence of microbiotheriids has yet been reported for Australasia (but see Kirsch et al. 1997), and the oldest fossil record for the taxon has been ascribed to the early Paleocene Tiupampa fauna of Bolivia in South America, which among other taxa includes caenolestids, didelphids, and a probable microbiotheriid *Khasia cordillerensis* (Gayet et al. 1991, Woodburne and Case 1996). An alternative scenario suggests that some ancient microbiotheriids may have dispersed from South America to Antarctica by corridor dispersal, which may have been facilitated by the occurrence of continuous forests (e.g. *Nothofagus*) while these two continents were still together (Arroyo et al. 1996, Villagrán and Hinojosa 1997). This hypothesis is based on the finding of the fossil microbiotheriids reported for the Antarctic Peninsula, particularly for the Seymour Island from La Meseta Formation, considered to be middle Eocene in age (40 million ybp, Woodburne and Case 1996). Additionally, geophysical and geological evidence have shown that, until the early part of the late Eocene, the Antarctic was united with Australia, and until the late part of the Eocene the Antarctic Peninsula was contiguous with southern South America. The hypothesised microbiotheriid fossils of Antarctica are recognised as highly derived polydolopids, a taxon more derived than *Khasia*. Subsequently, microbiotheriids in Antarctica may have remained isolated after the emergence of the Drake Sea that separated South America from Antarctica about 35 Mya. Consequently, either some ancient forms dispersed to Australasia and differentiated into the ancestors of Australasian marsupials (e.g. of the notoryctemorph [dasyurid-diprotodontian] clade), or the ancestral forms differentiated in Antarctica and dispersed to Australasia giving rise to marsupials of the latter continent (Palma and Spotorno 1999).

Future directions

Most of the studies conducted to understand the evolutionary relationships of South American marsupials, and their relationships to Australasian forms, seem to be well resolved. The impressive development of molecular techniques, using either organellar and/or nuclear markers, has allowed researchers to confirm earlier hypotheses of relationships based on more traditional methods. For example, one of the major issues regarding the phylogenetic relationships of American marsupials is that of the Patagonian 'monito del monte', which has now been fully resolved, leaving no doubt about its Australian affinities. Furthermore, the body of evidence clearly shows that large-sized American opossums constitute a natural group that includes *Metachirus nudicaudatus*. Mouse opossums, however, do not constitute a monophyletic assemblage, and most of them were earlier considered to be part of the single genus *Marmosa*.

Now researchers have a better understanding of the relationships between both major radiations of marsupials from America and Australia, and the phylogeny of major groups in the Neotropical region shows a good resolution. Chromosome evolution is giving the first results using more modern techniques, such as chromosome painting, and these results are confirming the characteristic constancy for major Neotropical marsupial taxa. It would be interesting to include these latter kinds of studies aiming to confirm the link between Australian and American radiation, *Dromiciops gliroides*.

Future studies about marsupial evolution should continue to focus on evaluating the phylogeny at the intrageneric level for some groups, and also make phylogeographic approaches at the intraspecific levels.

Acknowledgements

The author acknowledges Paula E. Neill and Stephen Sanko for reviewing an early version of this chapter. Part of the results presented here were supported by different Chilean grants such as FONDECYT 3950025, FONDECYT 1990156, as well as Fundación Andes 12999/8.

References

Anderson, S. (1997), 'Mammals of Bolivia, taxonomy and distribution', *Bulletin of the American Museum of Natural History*, **231**:1–652.

Arroyo, M.T.K., Riveros, M., Peñaloza, A., Cavieres, L., & Faggi, A.M. (1996), 'Phytogeographic relationships and regional richness patterns of the cool temperate rainforest flora of southern South America', in *High-latitude rainforests and associated ecosystems of the west coast of the Americas: Climate, hydrology, ecology and conservation* (eds. R.G. Lawford, P.B. Alaback, & E. Fuentes), pp. 134–72, Springer-Verlag, Berlin.

Biggers, J.D., Fritz, H.I., Hare, W.C.D., & McFeely, R.A. (1965), 'Chromosomes of American marsupials', *Science*, **148**:1603–03.

Bown, T.M., & Simmons, E.L. (1984), 'First record of marsupials (Metatheria: Polyprotodonta) from the Oligocene in Africa', *Nature*, **308**:447–49.

Bronson, F.H. (1989), *Mammalian reproductive biology*, University of Chicago Press, Chicago.

Cifelli, R.L. (1990a), 'Cretaceous mammals of southern Utah. I. Marsupials from the Kaiparowitz formation (Judithian)', *Journal of Vertebrate Paleontology*, **10**:295–319.

Cifelli, R.L. (1990b), 'Cretaceous mammals of southern Utah. II. Marsupials and marsupials-like mammals from the Wahweap formation (Early Campanian)', *Journal of Vertebrate Paleontology*, **10**:320–31.

Cifelli, R.L. (1993a), 'Early Cretaceous mammals from North America and the evolution of marsupial dental characters', *Proceedings of the National Academy of Sciences USA*, **90**:94l3416.

Cifelli, R.L. (1993b), 'Theria of metatherian-eutherian grade and the origin of marsupials', in *Mammal phylogeny, Vol. 2, Mesozoic differentiation, multituberculates, monotremes, early therians, and marsupials* (eds. F.S. Szalay, M.J. Novacek, & M.C. MacKenna), pp. 205–15, Springer-Verlag, New York.

Colgan, D.J. (1999), 'Phylogenetic studies of marsupials based on phosphoglycerate kinase DNA sequences', *Molecular Phylogenetics and Evolution*, **11**:13–26.

Creighton, G.K. (1984), 'Systematic studies in opossums (Didelphidae) and rodents (Cricetidae)', unpublished PhD thesis, University of Michigan.

Engstrom, M.D., & Gardner, A.L. (1988), 'Karyotype of *Marmosa canescens* (Marsupialia: Didelphidae): a mouse opossums with 22 chromosomes', *The Southwestern Naturalist*, **33**:230–33.

Crochet, J.Y., & Sigé, B. (1983), 'Les mammiferes de Hainin (Paleocen Moyen de Belgique) Part III: Marsupiaux', *Paleovertebrata*, **13**:51–64.

Ducrocq, S.E., Buffetaut, E., Buffetaut-Tong, H., Jaeger, J.J., Jongkanjanasoontorn, Y., & Suteehorn, V. (1992), 'First fossil marsupial from Asia', *Journal of Vertebrate Paleontology*, **12**:395–99.

Eaton, J.G. (1993), 'Marsupial dispersal', *National Geographic Research and Exploration*, **9**:436–43.

Emmons, L. (1997), *Neotropical rainforest mammals: a field guide, 2nd edition*, University of Chicago Press, Chicago.

Engstrom, M.D., & Gardner, A.L. (1988), 'Karyotype of *Marmosa canescens* (Marsupialia, Didelphidae): a mouse opossum with 22 chromosomes', *The Southwestern Naturalist*, **33**:230–33.

Gallardo, M.H., & Patterson, B.D. (1987), 'An additional 14-chromosome karyotype and sex-chromosome mosaicism in South American marsupials', in *Studies in Neotropical mammalogy: essays in honor of Philip Hershkovitz* (ed. B.D. Patterson & R.L. Timm), pp. 111–15, Field Museum of Natural History, Chicago.

Gardner, A.L. (1993), 'Order Didelphimorphia', in *Mammal species of the world: a taxonomic and geographic reference* (eds. D.E. Wilson & D.M. Reeder), pp. 15–23, Smithsonian Institution Press, Washington.

Gardner, A.L., & Creighton, G.K. (1989), 'A new generic name for Tate's (1933) *microtarsus* group of South American mouse opossums (Marsupialia: Didelphidae)', *Proceedings of the Biological Society of Washington*, **102**:3–7.

Gayet, M., Marshall, L.G., & Sempere, T. (1991), 'The Mesozoic and Paleocene vertebrates of Bolivia and their stratigraphic context: a review', *Revista Técnica YPFB, Santa Cruz, Bolivia*, **12**:393–433.

Gemmell N.J., & Westerman, M. (1994), 'Phylogenetic relationships within the Class Mammalia: a study using mitochondrial 12S RNA sequences', *Journal of Mammalian Evolution*, **2**:3–23.

Hayman, D.L. (1990), 'Marsupial cytogenetics', *Australian Journal of Zoology*, **37**:331–49.

Hayman, D.L., & Martin, P.G. (1974), *Mammalia I: Monotremata and Marsupialia, Vol. IV: Chordata 4*, Animal cytogenetics (ed. B. John), Gebruder Borntreger: Berlin.

Hershkovitz, P. (1999), '*Dromiciops gliroides* Thomas, 1894, last of the Microbiotheria (Marsupialia), with a review of the family Microbiotheriidae', *Fieldiana, Zoology*, **93**:1–60.

Hershkovitz, P. (1992a), 'The South American gracile mouse opossums, genus *Gracilinanus* (Gardner and Creighton, 1989): a taxonomic review with notes on general morphology and relationships', *Fieldiana, Zoology*, **70**:1–56.

Hershkovitz, P. (1992b), 'Ankle bones: the Chilean opossum *Dromiciops gliroides* Thomas, and marsupial phylogeny', *Bonner Zoologische Beitruge*, **43**:181–213

Jansa, S.A. and R.S. Voss. (2000), 'Phylogenetic studies on didelphid marsupials I. Introduction and preliminary results from nuclear IRBP gene sequences', *Journal of Mammalian Evolution*, **7**:43–77.

Kirsch, J.A.W. (1977), 'The comparative serology of Marsupialia, and a classification of marsupials', *Australian Journal of Zoology*, **52**:1–152.

Kirsch, J.A.W., Dickerman, A.W., Reig, O.A., & Springer, M.S. (1991), 'DNA hybridization evidence for the Australasian affinity of the American marsupial *Dromiciops australis*', *Proceedings of the National Academy of Sciences, USA*, **88**:10465–69.

Kirsch, J.A.W., & Palma, R.E. (1995), 'DNA/DNA hybridization studies of carnivorous marsupials. V. A further estimate of relationships among opossums (Marsupialia: Didelphidae)', *Mammalia*, **59**:403–25.

Kirsch, J.A.W., Lapointe, F.J., & Springer, M.S. (1997), 'DNA-hybridisation studies of marsupials and their implications for metatherian classification', *Australian Journal of Zoology*, **45**:211–80.

Mann, G. (1978), 'Los pequeños mamíferos de Chile', *Gayana, Zoología.*

Marshall, L.G. (1978), '*Dromiciops australis*', *Mammalian Species*, **99**:1–5.

Marshall, L.G. (1980), 'Systematics of the South American marsupial family Caenolestidae', *Fieldiana, Geology*, **5**:1–145.

Marshall, L.G. (1982), Systematics of the South American marsupial family Microbiotheriidae. *Fieldiana, Geology* **10**:1–75.

Marshall, L.G., Case, J.A., & Woodburne, M.O. (1990), 'Phylogenetic relationships of the families of marsupials', in *Current Mammalogy, Vol. 2* (ed. H.H. Genoways), pp. 433–505, Plenum Press: New York.

Meynard, A.P., Palma, R.E., & Rivera-Milla, E. (2002), 'Filogeografía de las llacas chilenas del género *Thylamys* (Marsupialia, Didelphidae) en base a secuencias del gen mitocondrial citocromo b', *Revista Chilena de Historia Natural*, **75**:299–306.

Murray, J.D., McKay, G.M., & Sharman, G.B. (1979), 'Studies on metatherians sex chromosomes. IX. Sex chromosomes of the greater glider (Marsupialia: Petauridae)', *Australian Journal of Biological Sciences*, **32**:375–86.

Mustrangi, M.A., & Patton, J.L. (1997), 'Phylogeography and systematics of the slender mouse opossum *Marmosops* (Marsupialia, Didelphidae)', *University of California Publications in Zoology*, **130**:1–86.

Palma, R.E. (1995), 'The karyotypes of two South American mouse opossums of the genus *Thylamys* (Marsupialia: Didelphidae), from the Andes, and eastern Paraguay', *Proceedings of the Biological Society of Washington*, **108**:1–5.

Palma, R.E., & Yates, T.L. (1996), 'The chromosomes of Bolivian didelphid marsupials', *Ocassional Papers The Museum of Texas Tech University*, **162**:1–20.

Palma, R.E., & Yates, T.L. (1998), 'Phylogeny of southern South American mouse opossums (*Thylamys*, Didelphidae) based on allozyme and chromosomal data', *Z. Säugetierkunde*, **63**:1–15.

Palma, R.E., & Spotorno, A.E. (1999), 'Molecular systematics of marsupials based on the rRNA 12S mitochondrial gene: the phylogeny of Didelphimorphia and of the living fossil microbiotheriid *Dromiciops gliroides* Thomas', *Molecular Phylogenetics and Evolution*, **13**:525–35.

Palma, R.E., Rivera-Milla, E., Yates, T.L., Marquet, P.A., & Meynard, A.P. (2002), 'Phylogenetic and biogeographic relationships of the mouse opposum *Thylamys* (Didelphimorphia, Didelphidae) in southern South America', *Molecular Phylogenetics and Evolution*, **25**:245–253.

Patton, J.L., Dos Reis, S.F., & Da Silva, M.N.F. (1996), 'Relationships among didelphid marsupials based on sequence variation in the mitochondrial cytochrome *b* gene', *Journal of Mammalian Evolution*, **3**:3–29.

Redford, K.H., & Eisenberg, J.F. (1992), *Mammals of the Neotropics: the southern cone: Chile, Argentina, Uruguay, Paraguay, Vol. 2*, University of Chicago Press, Chicago.

Reig, O.A. (1955), 'Noticia preliminar sobre la presencia de microbiotherinos vivientes en la fauna Sudamericana', *Investigaciones Zoológicas Chilenas*, **2**:121–30.

Reig, O.A., Kirsch, J.A.W., & Marshall, L.G. (1987), 'Systematic relationships of the living and Neocenozoic American "opossum-like" marsupials (Suborder Didelphimorphia), with comments on the classification of these and of the Cretaceous and Paleogene New World and European metatherians', in *Possums and opossums: studies in evolution* (ed. M. Archer), pp. 1–89, Royal Zoological Society of New South Wales, Sydney.

Retief, J.D., Krajewski, C., Westerman, M., Winkfein, R.J., & Dixon, G.H. (1995), 'Molecular phylogeny and evolution of marsupial protamine P1 genes', *Proceedings of the Royal Society of London B*, **259**:7–14.

Ride, W.D.L. (1964), 'A review of Australian fossil marsupials', *Journal and Proceedings of the Royal Society of Western Australia*, **47**:97–131.

Sánchez-Villagra, M.R. (2001), 'The phylogenetic relationships of argyrolagid marsupials', *Zoological Journal of the Linnean Society*, **131**:481–96.

Simpson, G.G. (1980), *Splendid isolation: the curious history of South American mammal*, Yale University Press, New Haven, Connecticut.

Spotorno, A.E., Marín, J.C., Yévenez, M., Walker, L.I., Fernández-Donoso, R., Pincheira, J., Berríos, M.S., & Palma, R.E. (1997), 'Chromosome divergences among American marsupials and the Australian affinities of the American *Dromiciops*', *Journal of Mammalian Evolution*, **4**:259–69.

Springer, M.S., Westerman, M., & Kirsch, J.A.W. (1994), 'Relationships among orders and families of marsupials based on 12S ribosomal DNA sequences and the timing of the marsupial radiation', *Journal of Mammalian Evolution*, **2**:85–115.

Springer, M.S., Westerman, M., Kavanagh, J.R., Burk, A., Woodburne, M.O., Kao, D.J., & Krajewski, C. (1998), 'The origin of the Australasian marsupial fauna and the phylogenetic affinities of the enigmatic monito del monte and marsupial mole', *Proceedings of the Royal Society of London B*, **265**:2381–86.

Stehli, F.G., & Webb, S.D. (eds.) (1985). *The great American biotic interchange*, Plenum Press, New York.

Svartman, M., & Vianna-Morgante, A.M. (1999), 'Comparative genome analysis in American marsupials: chromosome banding and in-situ hybridization', *Chromosome Research*, **7**:267–75.

Szalay, F.S. (1982), 'A new appraisal of marsupial phylogeny and classification', in *Carnivorous marsupials* (ed. M. Archer), pp. 621–640, Royal Zoological Society, New South Wales, Sydney.

Tate, G.G.H. (1933), 'Systematic revision of the marsupial genus *Marmosa*, with a discussion of the adaptive radiation of the murine opossum (*Marmosa*)', *Bulletin of the American Museum of Natural History*, **66**:1–250 + xxvi plates.

Temple-Smith, P. (1987), 'Marsupials and the new biogeography', in *Possums and opossums: studies in evolution* (ed. M. Archer), pp. 171–93, Royal Zoological Society: New South Wales, Sydney.

Villagrán, C., & Hinojosa, L.F. (1997), 'Historia de los bosques de Sudamérica II: fitogeografía', *Revista Chilena de Historia Natural*, **70**:241–67.

Westerman, M., & Edwards, D. (1991), 'The relationships of *Dromiciops australis* to other marsupials: data from DNA–DNA hybridization studies', *Australian Journal of Zoology*, **39**:123–30.

Woodburne, M.O., & Zinsmeister, W.J. (1982), 'Fossil land mammal from Antarctica', *Science*, **218**:284–86.

Woodburne, M.O., & Zinsmeister, W.J. (1984), 'The first land mammal from Antarctica and its biogeographic implications', *Journal of Paleontology*, **58**:913–48.

Woodburne, M.O., & Case, J.A. (1996), 'Dispersal, vicariance, and the Late Cretaceous to Early Tertiary land mammal biogeography from South America to Australia', *Journal of Mammalian Evolution*, **3**:121–61.

CHAPTER 3

EARLY MARSUPIAL RADIATIONS IN SOUTH AMERICA

Francisco J. Goin

Departamento Paleontología Vertebrados, Museo de La Plata, Paseo del Bosque s/n, 1900 La Plata, Argentina. Email: fgoin@museo.fcnym.unlp.edu.ar

The three classic groupings of South American marsupials – sparassodonts, pseudodiprotodonts, and didelphimorphians – may not constitute natural groups. The affinities of borhyaenoids and allies are currently under intense scrutiny and debate. The molar morphology of an early form, *Mayulestes*, suggests peradectian affinities. The monophyly of Paucituberculata + Polydolopimorphia is not supported on the basis of available information. Instead, it is suggested the possible belonging of microbiotherians, glasbiids and polydolopimorphians to a natural group. Didelphimorphia, as currently understood, is probably another waste basket taxon that includes several lineages of still uncertain affinities.

INTRODUCTION

Since the publication of 'Carnivorous Marsupials' two decades ago (Archer 1982a), a series of successive findings of Paleogene South American marsupials, as well as new interpretations on the relationships of extinct and extant taxa, have drastically increased, and notably changed, our perception of their early radiations and affinities. These findings put an end to what can be named the 'George Simpson's era' on the interpretation of metatherian diversity and evolution in South America (see, e.g. Simpson 1980). Among the contributions that helped in the beginning of a new era of the understanding of marsupial relationships, both South American and Australian, fossil or extant, the influential works of Kirsch (1977) and Szalay (e.g. 1982) stand out.

At the beginning of the 80s, Pascual (1980a, 1980b, 1981, 1983) updated knowledge on the early Tertiary marsupials from north-western Argentina. Marshall (1978, 1979, 1980, 1981, 1982a, 1982b) had just finished a series of important contributions on the taxonomy of most extinct South American groups. Marshall et al. (1983), Marshall and Muizon (1988), and Muizon (e.g. 1992) successively described the impressive early Paleocene fauna of Tiupampa (Bolivia). Szalay (1982) introduced a new appraisal of marsupial phylogeny and classification. Several reviews on the relationships of fossil and extant South American marsupials were written by the end of that decade (Aplin and Archer 1987; Reig et al. 1987; Marshall 1987; Marshall et al. 1990). The 90s continued with the production of several dozen papers dealing with punctual aspects of the diversity of Tertiary and Quaternary taxa (not detailed here). More comprehensive contributions, including the relationships of extinct South American groups, were made by Szalay (1993, 1994) and Kirsch et al. (1997). Oliveira (1998) reviewed once again the medial Paleocene fauna of Itaboraí (Brazil). Woodburne and Case (1996) envisioned the land mammal dispersal scenario throughout South America, Australia, and Antarctica (for the latter, see Woodburne and Zinsmeister 1982, 1984; Case et al. 1988; Goin et al. 1999) and supported a chronological framework for trans-Antarctic biotic interchanges. By the turn of the century, several reviews addressed the relationships of most living and some extinct South American marsupial lineages (e.g. Jansa and Voss 2000, Sánchez-Villagra 2001, Szalay and Sargis 2001). A few new, also impressive, marsupial faunas from the medial Paleocene of central Patagonia (Goin et al. 1997, Candela et al. 1998), ?middle–late Eocene of the Peruvian Amazonia (Goin and Candela, in press), and earliest Oligocene Ameghino's 'Astraponoteén plus Supérieur' levels, also at central Patagonia (Goin and Candela 1997, Bond et al. 1997), are still being described.

In this chapter, a new macro systematic arrangement of metatherians is not attempted, but instead focus is given to one aspect of current studies: increasing evidence suggest that the three traditional groupings of South American marsupials – sparassodonts, pseudodiprotodonts, and didelphimorphians – do not connote natural groups.

Abbreviations

AMNH, American Museum of Natural History, New York; DGM, Divisâo de Geologia e Mineralogia do Departamento Nacional de Produçâo Mineral, Rio de Janeiro; MACN, Museo Argentino de Ciencias Naturales 'Bernardino Rivadavia', Buenos Aires; MEF, Museo Paleontológico 'Egidio Feruglio', Trelew; MLP, Departamento Paleontología Vertebrados, Museo de La Plata; MNRJ, Museo Nacional de Historia Natural de Rio de Janeiro; UA, University of Alberta, Edmonton; UCMP, University of California Museum of Paleontology, Berkeley; UPCM, Université Pierre et Marie Curie, Paris; YPFB, Colecciones de Paleontología del Centro de Tecnología Petrolera, Yacimientos Petrolíferos Fiscales de Bolivia, Santa Cruz; ZPAL, Paleozoological Institute of the Polish Academy of Sciences, Warsaw. Dental terminology cited in the text is shown in Fig. 1H. M1, M2, M3, M4, upper molars; m1, m2, m3, m4, lower molars. StA, StB, StC, StD, StE, stylar cusps A, B, C, D, and E, respectively.

DIVERSITY

'Sparassodonts'

Since Marshall et al.'s (1990) review of the phylogenetic relationships of the families of marsupials, the affinities of borhyaenoids and allies have been the subject of intense discussion. Marshall et al. (1990) recognised six families of carnivorous marsupials within the order Sparassodonta: Stagodontidae, Hondadelphidae, Hathliacynidae, Borhyaenidae (Borhyaeninae + Prothylacininae), Proborhyaenidae, and Thylacosmilidae. Later, Marshall and Kielan Jaworowska (1992) suggested close affinities between borhyaenoids and deltatheroidans. Muizon and Lange-Badré (1997) argued against this with an analysis based on the evolution of functional complexes in carnivorous mammals. Muizon (1994, 1999) and Muizon et al. (1997) stated that borhyaenoids are closely related to the early didelphimorphian radiation, and denied special relationships between borhyaenoids and deltatheroidans or stagodontids. Gabbert (1998) discussed several of these previous statements based on basicranial morphology. Rougier et al. (1998) implied in their analysis that Muizon et al.'s (1997) grouping of *Mayulestes* and Borhyaenidae is paraphyletic. Oliveira and Goin (in press) suggested that borhyaenoids evolved from a peradectian stock (see below). It is predictable that the discussion will continue. The lack of cranial materials belonging to several relevant North and South American taxa obscures much of the argumentations. Key to the understanding of the extent and affinities of sparassodonts is the consideration of the basal most taxa.

Most analyses of South American marsupial carnivores (Sparassodonta of Marshall et al. 1990, or Borhyaenoidea of Marshall 1977, Muizon 1994) have assumed their monophyly, also following the traditional view that borhyaenids and allies derived from a didelphid stock, or a didelphid-like ancestor (see, e.g. Simpson 1980). Cranial and dental remains from Tiupampian (early Paleocene) and Itaboraian (medial Paleocene) levels constitute their earliest fossil records in this continent. The molar structure of *Mayulestes* is particularly instructive: protocones and talonids are only slightly reduced, the paracone and metacone twinning is incomplete (i.e. the paracone and the metacone are not coalescent at their bases), stylar cusps are moderately developed, the preparacrista is not reduced in length, the postmetacrista is only moderately enlarged, and, together with the paracristid, is quite transversal to the dental axis; in the lower molars, the paraconid is not reduced or shifted antero-medially, and hypoconulids are high. This combination of features is characteristic of peradectids (i.e. Peradectinae *sensu* Korth 1994), of which *Mayulestes* may constitute a derived lineage of carnivorous or faunivorous feeding habits. Even some minor details, as the crest lingual to StD in the M1-3 of *Mayulestes*, are also present in Paleogene peradectids as *Peradectes* and *Nanodelphys*. It is probable that, had isolated teeth of *Mayulestes* been found in North American late Cretaceous or Paleocene levels, they would have been assigned to the Peradectidae without hesitation. Actually, some isolated dental remains (Fig. 1A, B) belonging to ?Mayulestidae also suggest that representatives of this family were present in North America during Paleocene times (see below). All of Muizon's (1998: 23) diagnostic features of the Borhyaenoidea (including Mayulestidae) exclusively refer to cranial and postcranial features. No peradectids (or peradectians, if stagodontids are excluded from this clade) are yet known from cranial or postcranial materials. The testing of a hypothesis linking *Mayulestes* with peradectids is still lacking on this ground. As far as molar morphology is concerned, the (mostly plesiomorphic) molar structure of *Mayulestes* fits well in the peradectid pattern. Other dental features of *Mayulestes*, as the unreduced incisor formula, agree with this inference.

Allqokirus australis has been regarded as a mayulestid borhyaenoid by Muizon (e.g. 1998). Its upper molar morphology is derived with respect to that of *Mayulestes*: upper molars have much closer paracone and metacone, the postmetacrista is proportionally larger and its distal portion is not labial but postero-labially oriented, stylar cusps C and D are reduced (but not so the stylar shelf), and the lower molars have an enlarged paracristid, a slightly reduced metaconid, and a vestigial entoconid. Most of these features are anticipatory of the pattern seen in hathliacynids and borhyaenids, and are derivable from that of *Mayulestes*. Regardless of the allocation of *Allqokirus* to the Mayulestidae or not, the problems relative to the origin of borhyaenoids still persist.

A second, intimately related problem is whether hathliacynids, borhyaenids (prothylacynines + borhyaenines), proborhyaenids, and thylacosmilids actually belong to a natural group. Even the monophyly of several of these lineages can be subject of debate. For instance, Muizon's (1999) phylogenetic analysis of the Borhyaenoidea imply the paraphyly of the Prothylacyninae as conceived by Marshall (1977, 1978, 1979). Another example comes from the relatively recent discovery of early (middle Miocene) thylacosmilids in northern South America, which also suggests that the origins of this group are far from clear, and may have no direct relation with proborhyaenids or borhyaenids. Goin (1997: 204, Fig. 11.7) reported a thylacosmilid specimen whose mandibular and dental morphology is ancestral to that of *Anachlysictis gracilis*. In turn, *A. gracilis* is slightly generalised regarding the better known late Miocene–Pliocene thylacosmilids. This specimen (a right dentary with the cheek-teeth and an anterior portion of a maxillary) already shows an incipient mandibular flange at the anterior portion of the dentary, the alveolar and ventral borders of this bone are parallel and straight below the cheek-teeth, the masseteric process and the incisor region are notably reduced, the canine is laterally flat, and the molar row is arranged in such a way that the last molar is located not at the middle section of the mandible but quite posterior to it. The upper canine is disproportionally large as compared to the lower one, and is located at the very end of the snout. All these derived features are diagnostic of the Thylacosmilidae. The lower molars show an interesting combination of primitive and derived features: metaconids are reduced but present, talonids are large and widely basined, and the entoconid and pre-entocristid are vestigial, in such a way that the talonid basin is almost lingually open. This molar pattern closely resembles that of another Laventan (middle Miocene) taxon, *Hondadelphys fieldsi*, originally regarded as a specialised didelphid (Marshall 1976) and later as a generalised Sparassodonta (Marshall et al. 1990). The upper molar morphology of *Hondadelphys* is definitely odd for a borhyaenoid: the paracone and metacone are not close to each other, the protocone is wide, and the para-and metaconular inner cristae (post-paraconular and pre-metaconular) are long, almost meeting below the centrocrista. No other borhyaenoid known up to date matches this pattern, which incidentally argues against the concept of 'functional complex' as a highly homoplastic feature in carnivorous marsupials: in *Hondadelphys*, the postvallum-prevallid shearing structures were developed in a quite different molar architecture. Goin (1997) suggested that hondadelphids may represent the plesiomorphic sister-group of thylacosmilids. The molar morphology of *Hondadelphys*, as well of early thylacosmilids, seem to be not derivable from known hathliacynids, borhyaenids or proborhyaenids.

'Pseudodiprotodonts'

The so-called 'pseudodiprotodont' marsupials – e.g. the Pseudiprotodontia of Kirsch et al. (1997), including the Paucituberculata and the Polydolopimorphia – may not constitute a natural group. Among the Paucituberculata are included the Caenolestidae, Palaeothentidae, Abderitidae and, to some authors (but see below), the extremely derived Argyrolagidae, Patagoniidae, and Groeberiidae. To Goin and Candela (1996) the Polydolopimorphia include the Prepidolopidae, Bonapartheriidae, Gashterniidae, Epidolopinae (*Epidolops* spp.) and Polydolopinae (*Polydolops* and allies). Fossil evidence documenting the early phases of their evolutionary histories indicate different origins, homologies, and processes undergone by the Paucituberculata on one side, and the Polydolopimorphia on the other. Oliveira et al. (1996), Oliveira and Goin (in press), Goin et al. (1997, 1998a, 1998b, 1998c) and Goin and Candela (1996, 1997 and in press) report on new Paleogene South American marsupial taxa that actually document these early phases. A summary of their conclusions is as follows:

Previous reviews of 'pseudodiprotodont' taxa had already pointed out several homoplastic resemblances between the Paucituberculata on one side, and the Polydolopimorphia on the other. Most important of them are the different homologies of the enlarged incisor and of the 'plagiaulacoid-like' tooth at the cheek-tooth series (see, e.g. Marshall 1982a). Molar morphology was regarded as essentially identical in both groups. But, paucituberculatans and polydolopimorphians do differ in substantial aspects of their molar structure. Recently discovered Paleogene marsupial faunas, including representatives of both lineages (Goin and Candela 1997), show that the acquisition of a quadrangular molar pattern was differently achieved in each of them. In the upper molars of early paucituberculatans it can be observed that the stylar cusps B and D are already very large, the paracone–metacone difference in size is obvious; the paracone, reduced but distinct, merges at the lingual slope of StB, and the 'hypocone' (enlarged metaconule) has its base much higher than that of the protocone. In the lower molars the entoconid is

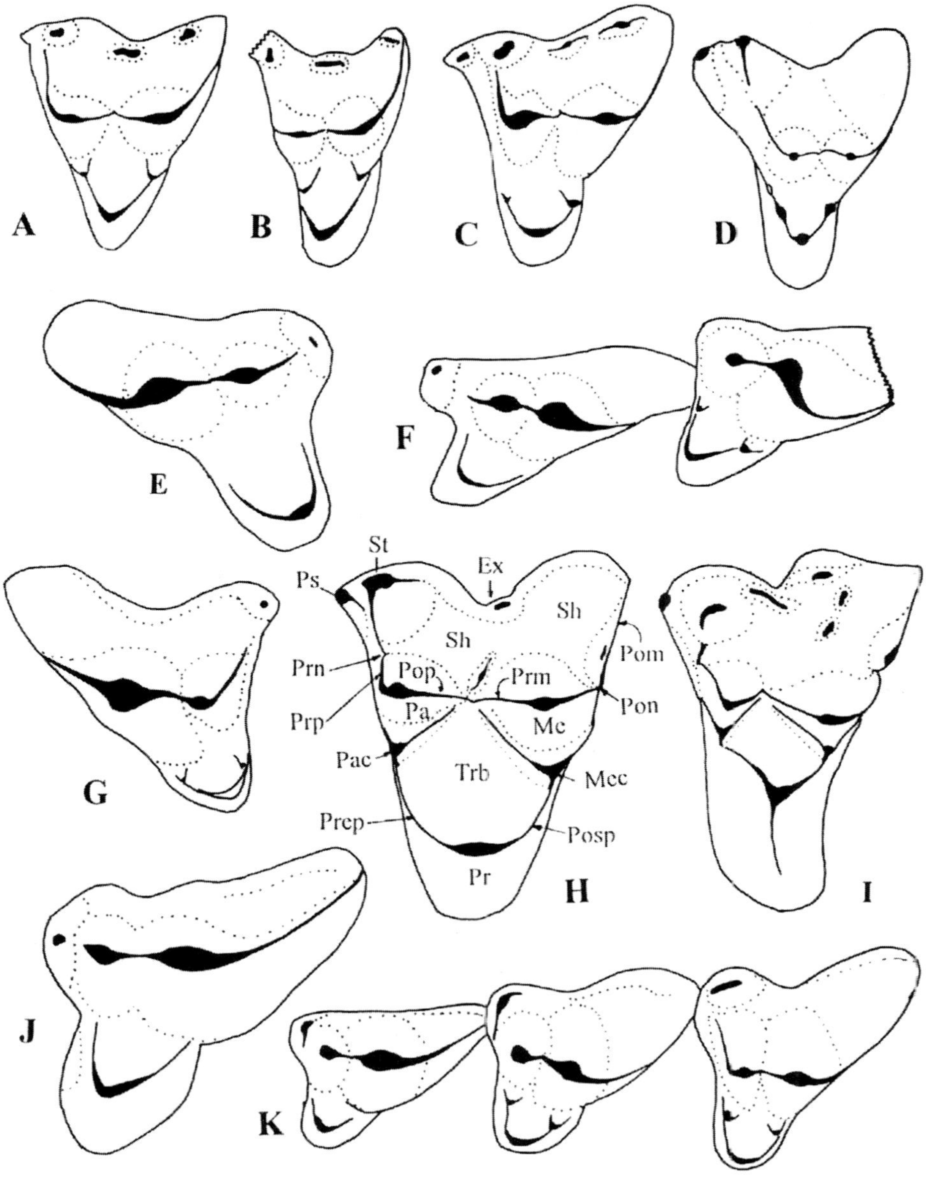

Figure 1 Schematic drawings, in occlusal view, of the upper molars of several carnivorous metatherians. **A**, **B**, uncatalogued UA specimens recovered from Blindman River and Joffre Bridge, Paskapoo Fm., Alberta (late Paleocene, Middle Tiffanian). A, left M2; B, left M3. Specimens were catalogued as *'Peradectes* sp.'. **C**, a reversed image of the holotype of *Allqokirus australis*; YPFB Pal 6104, a right M?3; early Paleocene of Tiupampa (Santa Lucía Fm., Bolivia). Modified from Muizon (1992, Fig. 1b). **D**. *Deltatheridium praetrituberculare*, specimen ZPAL MgM- I/102, late Cretaceous of Asia; modified from Muizon and Lange-Badré (1997, Fig. 1A). *E*, *Notogale mitis*, specimen PU 21875, right M2; middle to late Oligocene (Deseadan age), Patagonia, modified from Marshall (1981, Fig. 55). **F**, *Borhyaenidium musteloides*, part of the type specimen, MLP 57-X-10-153, left M1-2; late Miocene (Huayquerian age) of La Pampa, Argentina; modified from Marshall (1981, Fig. 22b). **G**, *Patene colhuapiensis*, part of the type specimen, AMNH 28448; early Eocene (Casamayoran age), central Patagonia, Argentina; modified from Marshall (1981, Fig. 9b). **H**, *Eodelphis* sp., specimen UA 16104, a left M3; late Cretaceous (Oldman Fm., Alberta). **I**, *Didelphodon vorax*, specimen UCMP 47304, a left M3; late Cretaceous (Lance Fm., Niobrara Co., Wyoming). **J**, *Thylacinus cynocephalus*, specimen UPCM E 558, left M3, Recent; modified from Muizon and Lange-Badré (1997, Fig. 4A). **K**, *Sipalocyon externa*, part of the type specimen, MACN 52-383, left M1-3; early Miocene (Colhuehuapian age), central Patagonia, Argentina; modified from Marshall (1981, Fig. 30b). Drawings are not to scale. Abbreviations for the dental terminology: *Ex*, ectoflexus; *Me*, metacone; *Mec*, metaconule; *Pa*, paracone; *Pac*, paraconule; *Pom*, postmetacrista; *Pon*, posmetacrista notch; *Pop*, postparacrista; *Posp*, postprotocrista; *Pr*, protocone; *Prep*, preprotocrista; *Prm*, premetacrista; *Prn*, preparacrista notch; *Prp*, preparacrista; *Ps*, parastyle (StA); *Sh*, stylar shelf; *St*, Stylocone (StB); *Trb*, trigone basin.

labial–lingually compressed and is not displaced anteriorly on the lingual edge of the talonid; the hypoconid is labially placed, and the twinning of the paraconid and metaconid does not involve the disconnection of the paraconid and the paracristid. Besides differences in size, upper and lower M/m 2–4 are not heterodont, the enamel is thick only in a few taxa, and the crests of the ectoloph complex (preparacrista, centrocrista, postmetacrista) tend to merge on the slopes of the paracone and metacone. Oliveira et al. (1996) argued that the highly modified, 'plagiaulacoid-like' tooth in the cheek-teeth series of the Palaeothentidae and the Abderitidae may not be homologous to the M/m1, but instead represents the heterochronic persistence of the dP/p3 in these lineages.

Oliveira et al. (1996) reported the discovery of medial Paleocene, isolated teeth belonging to a caenolestoid marsupial; they represent the oldest known record of a Paucituberculata. A lower m1 recovered from Itaboraian levels in Brazil shows its metaconid already posteriorly oriented, a wide talonid basin, labially salient hypoconid, and a somewhat labio-lingually compressed entoconid. A second, much smaller specimen of a generalised caenolestid, a fragmentary M?2 coming from the medial Paleocene of Las Flores (Argentina), shows its stylar cusp B already very large, with the paracone almost merging at its base. The molar structure of both specimens seems to be derivable from a generalised didelphimorphian.

A recent phylogenetic review of the Argyrolagidae by Sánchez-Villagra (2001, see also Sánchez-Villagra et al. 2000) supported the monophyly of the Paucituberculata, the latter including caenolestoids and argyrolagids. It should be noted, however, that polydolopimorphians were not included in the analysis.

The polydolopimorphian pattern includes a heterodont molar series with a variable tendency towards unilateral or bilateral hypsodonty, cheek-teeth with a thick enamel layer, upper molars with only minor difference in size between paracone and metacone (which in turn are twinned with the StB and StD, respectively), reduced or vestigial preparacrista that points to StA, open centrocrista (that is, postparacrista not connected to the premetacrista), metaconule large to very large forming, in the more derived forms, a 'hypocone' which is always at the same level as the protocone; lower molars have a very reduced paraconid, a reduced to vestigial hypoconulid, and a high entoconid that is not laterally compressed and is anteriorly shifted, its base being close to that of the metaconid (Goin and Candela 1996; Goin et al. 1998a). One interesting result of this characterisation of the polydolopimorphian cheek-teeth pattern (Goin and Candela 1996) is that the Prepidolopidae and Bonapartheriidae, previously regarded as didelphoids by Pascual (e.g. 1980a, 1980b, 1981), actually represent primitive polydolopimorphians. *Gashternia*, whose metatherian affinities were a matter of discussion (see, e.g. Marshall et al. 1990) also shows this basic pattern (Candela et al. 1998).

The long-thought sister-group relationship between epidolopines and polydolopines – both comprising the Polydolopidae of traditional literature (see, e.g. Pascual and Bond 1981) – was challenged by Goin and Candela (1995). Molars of *Epidolops* share with prepidolopids, bonapartheriids, and *Gashternia* (see Fig. 2G, H), the basic polydolopimorphian pattern described above. Much more derived is that of the Polydolopinae s.s. (*Polydolops* and allies), which includes the alignment of the paraconule, protocone, and metaconule in a lingual row, well-expanded anterior and posterior cingula (which are at the same level as the trigon basin), lingually invasive stylar cusps (especially StC), and a distinct, though variable, tendency towards the development of neomorphic, accessory cuspules at the labial face of the upper molars. Most of these features are already outlined in the early Paleocene polydolopimorphian *Roberthoffstetteria nationalgeographica*.

Although originally described as a member of the Paucituberculata, the recently described *Klohnia charrieri*, from ?late Eocene-early Oligocene levels of central Chile, may constitute instead a very interesting polydolopimorphian; its most conspicuous derived features may also confirm the polydolopimorphian affinities of groeberiids, argyrolagids, and patagoniids. Flynn and Wyss (1999) considered *Klohnia* as the closest known relative of *Patagonia*, and regarded both taxa as the nearest outgroup to the Groeberiidae (see also Koenigswald and Goin 2000). *Klohnia* is derived in many features, such as the possession of an enlarged, gliriform lower incisor, loss of one molar (M/m4?), significant diastema between i?1 and p3, upper and lower cheek-teeth with a thick enamel layer, enlarged and anteriorly placed entoconid, enlarged metaconule, which is at the same level as the protocone, and twinning of the paracone and metacone with the stylar cusps B and D, respectively. The loss of the last molar is infrequent in marsupials; among South American ones, this feature is present only in a few polydolopimorphians, as the polydolopines and the prepidolopid *Punadolops alonsoi* (Goin et al. 1998c). The enlarged, anteriorly shifted entoconid ('isolated lingual talonid cusp' of Flynn and Wyss 1999: 358) characteristic of *Klohnia*, *Groeberia*, and *Argyrolagus*, is already present (though less noticeably) in bonapartheriids and prepidolopids. In the upper molars, the characteristic polydolopimorphian twinning of the paracone and metacone with the stylar cusps B and D, together with the enlargement of the metaconule shaping the posterolingual corner of the teeth, is also present in *Klohnia* (Flynn and Wyss 1999: Fig. 2G). Finally, a thick enamel layer is already present in the cheek-teeth of *Bonapartherium*, *Prepidolops*, *Gashternia*, and *Epidolops*. *Klohnia* is even closer to groeberiids, patagoniids, and argyrolagids, in the great enlargement of the first upper and lower incisors, the latter having an intra-alveolar extension which is lingual to the cheek-teeth series, and in that the last upper and lower premolars are not hypertrophied. In short, (1) the set of derived features in the molar morphology of *Klohnia* suggests its belonging not to paucituberculatans but instead to polydolopimorphians, and (2) it already shares with groeberiids, patagoniids, and argyrolagids, derived features exclusive of this clade (Argyrolagoidea of Marshall 1987).

The question of the origins of the Polydolopimorphia is still under debate. Traditionally regarded as being derived from didelphoids (e.g. Patterson and Pascual 1968, Pascual 1980a, 1980b, 1981), research carried out in the last decade introduced new perspectives and insights. One of them is the probable relationships between microbiotheriids and polydolopimorphians. Microbiotheriids are best known from several Miocene Patagonian taxa (*Microbiotherium* spp.) and from the living 'monito de monte' *Dromiciops gliroides*. Up to 1980, authorities like Simpson regarded microbiotheriids as within the didelphoid radiation (Simpson 1980). Influential studies by Szalay (e.g. 1982) gave a completely different perspective through the study of their tarsal morphology, proposing that microbiotheriids were at the base of the Australian marsupial radiation (his Australidelphia). On the basis of molecular data, Kirsch et al. (1991, see also Kirsch et al. 1997) went further suggesting that, within Australidelphians, microbiotheriids represent the sister-group of the Diprotodontia.

Paleogene microbiotheriids were virtually unknown until quite recently. That implied a strong bias in the recognition of microbiotheriid affinities, as both *Microbiotherium* and *Dromiciops* show a unique combination of generalised and derived features in their dental and cranial morphology. Marshall and Muizon (1988) and Muizon (1992) recognised *Khasia* from the early Paleocene of Tiupampa, in Bolivia. Goin et al. (1998a) mentioned the finding of upper molars referable to microbiotheriids from the middle Paleocene of Las Flores Fm., in central Patagonia, bearing a quite different morphology from those upper molars refered by Marshall (1987) for the same group (see also Oliveira 1998). Upper molars of microbiotheriids from this formation (see Fig. 2A-B) already show enlarged protocones, a reduced stylar shelf with the stylar cusps B and D already close to the paracone and metacone, respectively, enlarged metaconules with respect to the paraconules; in turn, lower molars have wide talonid basins and hypoconulids not twinned with the entoconids. Goin et al. (1998a) described the ?Polydolopimorphia *Palangania brandmayri* (Fig. 2E, F) from the late Paleocene (Riochican Age) of central Patagonia. *Palangania* is peculiar in having a combination of dental features which may be key in the interpretation of the early polydolopimorphian radiation, being microbiotheriids and glasbiids probably basal to this radiation. *Palangania* shares with all polydolopimorphians the incipient twinning of cusps B and D with the para- and metacone, as well as the larger size of the metaconule with respect to the paraconule. In the lower molars the strong reduction of the hypoconulids is also present in early polydolopimorphians as *Prepidolops*, *Epidolops*, and *Bonapartherium*. Additional features support affinities between *Palangania*, early microbiotheriids, and glasbiids: low cusps, strong reduction of M/m4, hypoconulids not twinned to the entoconids but more medially placed on the posterior edge of the talonid, small paraconid, moderate enlargement of cusps B and D, weak centrocrista, preparacrista connecting with cusp A, and a relatively narrow stylar shelf. Taking in account all these features, it was suggested (Goin et al. 1998a, Goin and Candela in press) that microbiotheriids, glasbiids, *Palangania*, and the Polydolopimorphia may constitute a monophyletic group. As stated by Oliveira and Goin (in press), the belonging of microbiotherians and polydolopimorphians to a natural group may give a whole new insight on the evolution of the Australian Diprotodontia.

'Opossum-like' marsupials

A previous consideration on the nature and extent of South American 'didelphimorphs' is that of their alleged source, the Peradectia (or Alphadelphia of Marshall et al. 1990). It is suggested that the whole concept of the Peradectia should be revisited. This basal lineage, presently known from late Cretaceous and/or Tertiary levels in North and South America, Europe, Asia, and Africa, may have undergone an extensive radiation throughout its history, thus reflecting different adaptations, diets, and life histories. It has been argued that the Peradectia may include not only alphadontids, peradectids, and pediomyids, but also caroloameghiniids (Goin et al. 1998b, Oliveira and Goin in press), mayulestids, and, probably, some or all 'borhyaenoid' lineages (see above and Oliveira and Goin in press). The retention of a straight centrocrista may constitute a developmental constraint affecting not only this but a whole character complex in the upper molar morphology. Derived features departing from this pattern may not preclude the recognition of this basic character complex.

The status of several groups that have been regarded as polydolopimorphians since Marshall's (1987) review of the 'opossum-like marsupials' from Itaborai, Brazil (middle Paleocene), has been challenged in recent years. Goin et al. (1998a, 1998b) argued on the peradectian affinities of the Caroloameghiniidae (Fig. 2C, D), and on the didelphimorphian nature of the Protodidelphidae. An upper molar of the caroloameghiniid *Procaroloameghinia*, known from Itaboraian (medial Paleocene) levels at Las Flores, Patagonia (Fig. 2C) shows a straight centrocrista, the para- and metaconule only moderately developed and subequal, StC present, a well-developed postmetacrista, and a protocone that is not eccentric and is notably wide. None of these features are present in the Protodidelphidae, to which caroloameghiniids were previously referred. As was suggested for the carnivorous *Mayulestes*, the molar pattern of caroloameghiniids is

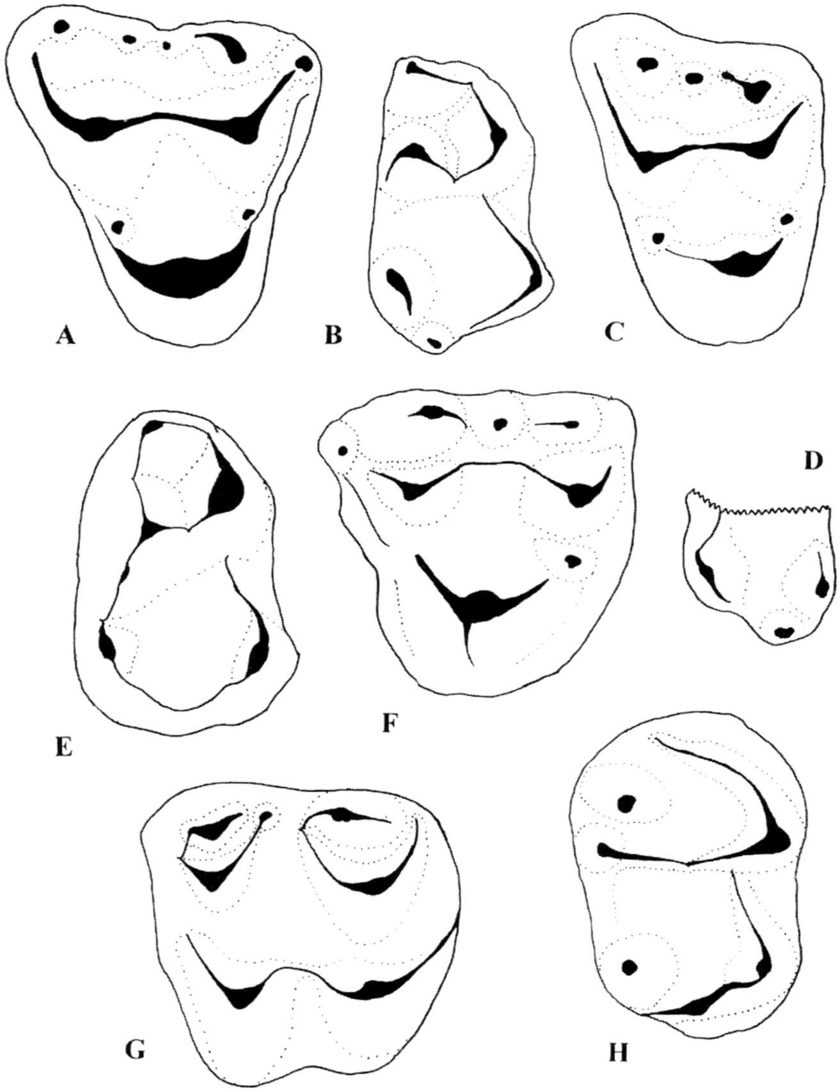

Figure 2 Schematic drawings of upper and lower molars of early South American marsupials, in occlusal view. **A**, **B**, Microbiotheriidae, genus and species indet.; A, MLP 90-II-5-187, right M3; B, MLP 90-II-5-191, right m2. **C**, **D**, *Procaroloameghinia* sp.; C, MLP 90-II-5-138, right M?2; D, MLP 90-II-5-142, left m?4. **E**, **F**, *Palangania brandmayri*; E, MLP 40-VI-20-19, a left m1; F, UNPSJB-PV 114, part of the holotype, a left M3; both specimens come from late Paleocene levels at Pico Salamanca, east-central Patagonia, Argentina. **G**, **H**, *Gashternia* sp.; G, MLP 90-II-5-793, a left M2; H, MLP 90-II-5-895, a right m3. All specimens except those of *Palangania* come from medial Paleocene levels at Yacimiento Las Flores (Las Flores Fm.), in Central Patagonia, Argentina. Drawings are not to scale.

easily derivable from a peradectian stock; in this case, caroloameghiniid ancestors may have developed adaptations towards frugivory while retaining several generalised features (e.g. straight centrocrista). In the second place, protodidelphids lack key synapomorphies of the polydolopimorphia: lower molars have their paraconids normally placed, such that the paracristids are well-developed and straight, hypoconulids are not reduced; upper molars have the paracones and metacones not reduced in size, the para-and metaconules are vestigial or absent, and stylar cusps B and D not twinned with the paracone and metacone, respectively. Goin et al. (1998b) regarded protodidelphids as specialised didelphimorphians.

As happened with the Didelphidae some years ago, the Didelphimorphia as currently understood is probably another waste basket taxon that includes a series of lineages of still uncertain affinities. Comprehensive reviews on the affinities of Paleocene 'opossum-like' marsupials show quite opposing views (cf.

Marshall 1987, Marshall and Muizon 1988, Muizon and Brito 1993, Oliveira 1998). Muizon and Cifelli (2001) stated that resolution for the problem of didelphimorphian extension and affinities '...must await recovery of more complete fossils, particularly crania for taxa from the Cretaceous of North America.' (Muizon and Cifelli 2001: 94).

'Opossum-like' marsupials referred to as the Didelphimorphia, or even the Didelphoidea, include the Pucadelphyidae, Jaskadelphyidae, Sternbergiidae, Derorhynchiidae, Herpetotheriidae, Protodidelphidae, Eobrasiliinae, Didelphidae, Caluromyidae, and Sparassocynidae. Their consideration as a natural group is simply impossible unless plesiomorphic features are involved in the analysis (see also Muizon and Cifelli 2001). Goin (1991, 1993, 1995, 1997) argued against the monophyly of the 'Didelphidae', or even 'Didelphoidea', of the traditional literature. Instead, a unique combination of generalised and derived features support the monophyly of Neogene South American opossums, comprising the Sparassocynidae, Caluromyidae, and Didelphidae – the latter conceived in a very restricted concept, as that of Kirsch et al. (1997). Mandibular and dental diagnostic features of the Didelphoidea (in this restricted concept) include the following combination of generalised and derived features: mandibles relatively short, with the symphysis extending posteriorly to a point below p2/3; alveolar edge of the dentary straight (not sloped downwards at the premolar area); lower incisors subequal in size; p2 larger and higher than p3; posterior crest of the upper and lower premolars not labial but centrally placed on the posterior slope; lower molars with very reduced hypoconulids; upper molars with reduced stylar cusp A, centrocrista moderately invasive in the stylar shelf, well-developed postmetacrista which is postero-labially oriented, and para- and metaconules vestigial or absent. It should be noted that, even if the discussion was restricted to living opossums only, the monophyly of didelphids and caluromyids is sustained by scarce dental and cranial derived features. Kirsch et al. (1997: 212) summarised the conclusions of their DNA-hybridisation studies in living opossums stating that they '...are shown to be as internally divergent as are most members of the order Diprotodontia.'

A number of molecular studies on living 'opossum-like' marsupials have been carried out in the last two decades. Most of them support the monophyly of living didelphoids (Didelphidae + Caluromyidae) (see a review in Kirsch et al. 1997, Springer et al. 1997, Jansa and Voss 2000). The paraphyly of living 'ameridelphians' (Paucituberculata and Didelphimorphia) has been argued by Jansa and Voss (2000).

A review of many early Tertiary, 'opossum-like' marsupials will probably lead to a reconsideration of their family or even ordinal status. One interesting case is that of the North American *Swaindelphys cifellii*, recognised from several upper and lower molars recovered from middle Paleocene levels at Swain Quarry, Wyoming (Johanson 1996). *Swaindelphys* was described as a herpetotheriine differing from all other taxa in the following features: upper molars with greater transverse width to antero-posterior length ratio on M2-3; ectoflexus present on M2, small and symmetrical ectoflexus present on M3; anterior stylar shelf wider on M2-3; posterior stylar shelf more postero-labially directed than posteriorly directed on M2-3; stylar cusps A and B separated by a distinct notch; stylar cusps C and D approximately equal in size, distinct and separate, not doubled or closely associated; cusp B larger than C and D, but no stylar cusp clearly dominant on stylar shelf on M1-3; lower molars have a more posterolabial than directly posterior hypoconulid relative to the entoconid, such that the talonid basin is not opened posteriorly (Johanson 1996). It should be noted that most of this features (with the probable exception of the last one) are plesiomorphic to all known herpetotheriids (see Crochet 1980, Korth 1994), as well as to many other didelphimorphians. In turn, *Swaindelphys* (as well as the early Paleocene, Tiupampian species *Mizquedelphys pilpinenesis* and *Incadelphys antiquus*) is very close to the molar pattern of *Pucadelphys* in its combination of primitive and derived features: retention of well-developed para- and metaconules, complete set of stylar cusps, stylar cusps A and B separated by a notch, unreduced protocone, and long preparacristae which, in the M3, is almost as long as the postmetacrista; among the derived features should be mentioned an incipient 'v'-shaped centrocrista (much better developed in *Swaindelphys*), and the labially (not posterolabially) oriented postmetacrista (especially in M3, where it is almost parallel to the preparacrista). In all this features *Swaindelphys* is basically identical to *Pucadelphys*, thus suggesting its belonging to, or close relation with, the Pucadelphidae. Its fairly deep centrocrista may represent a derived feature regarding other members of this family.

Traditionally refered to the subfamily Didelphinae (Mammalia, Marsupialia, Didelphimorphia, Didelphidae), herpetotheriids were raised to subfamily rank by Reig et al. (1987); later, they were regarded as a full family of didelphimorphians by Kirsch et. al. (1997). Their position and affinities among didelphimorphians are far from clear. In the most comprehensive review of European herpetotheriids, Crochet (1980), while including them in the Didelphinae, suggested the origin of this group (and of all other fossil and recent 'didelphines') from an early offshot of South American late Cretaceous 'opossum-like' marsupials – probably the Laguna Umayo marsupial fauna, as implied in his work (Crochet 1980: Fig. 240). His own concept of Didelphinae was extremely wide, and included taxa presently assigned to different families and even orders (Goin 1991, 1995). Reig el al. (1987) and Marshall (1987) set doubts on the didelphimorphian affinities of herpetotheriids. Even though classifying them among the Didelphidae, they suggested that herpetotheriids could have had an independent origin from a peradectid stock, and that they developed dental adaptations convergent to those of the didelphimorphians. On the contrary Goin (1991) regarded herpetotheriids as true didelphimorphians. In his concept, a clade including the herpetotheriids, as well as the South American Paleocene derorhynchids, protodidelphids, eobrasiliines, and other Paleogene South American taxa, constitute the sister-group of the Didelphoidea.

Herpetotheriids and derorhynchids share a significant combination of derived features: mandibles are anteriorly elongated, the alveolar edge of the dentary slopes downwards in front of the m1, the first two lower incisors are procumbent, and the posterior most two are reduced, p3 is larger than p2, upper molars have a deep centrocrista, the preparacrista is shifted anteriorly at its labial end, and lower molars have reduced paraconids and well-developed, high entoconids (the last two features showing an extreme condition in derorhynchids).

A case can be made that derorhynchids and herpetotheriids actually constitute early Peramelemorphians. Derorhynchids are characterised by the following combination of derived features: short talonids, reduced paraconids, salient hypocones, and large, tall, spire-like entoconids in the lower molars (Fig. 3A, F), and deep centrocristae, metaconules (although small) larger than paraconules, and a variable tendency towards the fusion of stylar cusps C and D (Fig. 3B, E) in the upper ones. A sensible reduction in the size of the posterior most lower incisors has also been noted (Goin et al. 1999 and literature cited). All these derived features fit quite well with the basic peramelemorphian pattern. Moreover, the derorhynchid dental pattern is remarkably similar to that of the generalised peramelemorphian *Yarala burchfieldi* (Muirhead and Filan 1995, Muirhead 2000). Except for the third bilobed lower incisor of *Yarala* (its crown morphology is unknown in derorhynchids), all other derived molar features mentioned by Muirhead (2000), as shared by *Yarala* and all other peramelemorphians, are also present in derorhynchids.

The recent discovery of several new derorhynchid taxa from Antarctic, middle Eocene levels gives empirical evidence on the wide distribution of derorhynchids during Paleogene times. However, one should be cautious in the possible inferences based on this evidence: middle Eocene Antarctic derorhynchids seem to be already too derived for a hypothetical ancestral condition. To Goin et al. (1999: 363) 'none of the presently known Antarctic marsupials accounts for the Australian radiation. Therefore, any dispersals involving taxa precursor to the Australian radiation had to have transpired prior to the medial Paleocene'. Interestingly, Bond et al. (1995) report on the presence of derorhynchid marsupials in early Paleocene levels at Punta Peligro (Peligran age).

Finally, it should be mentioned the probable relationship between several Paleogene South American 'opossum-like' marsupials and the recently described *Djarthia murgonensis*, from the ?early Eocene Tingamarra local fauna of Murgon in southern Queensland, Australia. To Godthelp et al. (1999), *Djarthia* is derived in six features also present in both 'ameridelphian' and 'australidelphian' marsupials: hypertrophy of the M4 preprotocrista relative to that of the M3, marked reduction of the paracones relative to the metacones, dilambdodonty, StC present, slight expansion of the M3 talonid relative to that of the trigonid, termination of the cristid obliqua at a point buccal to the metacristid notch. A quite interesting ?derived feature of *Djarthia* is the presence of a 'central cusp' at the apex of the centrocrista, which led Godthelp et al. (1999) to suggest affinities with *Keeuna woodburnei* and *Ankotarinja tirarensis*. A number of Itaboraian (medial Paleocene) taxa from Itaboraí, Brazil, also show the combination of features diagnostic of *Djarthia*, including the possession of such a cusp. This last feature is observable in specimens DGM 806-Ma (assigned to *Marmosopsis* sp.), MNRJ 2878a and DGM 802-Ma, b, and c, (*Itaboraidelphys* sp.), MNRJ 1429-V, MNRJ 2878-V, and DGM 642-M (*Didelphopsis* sp.), DGM 806-M and DGM 807-M (*Carolopaulacoutoia itaboraiensis*).

The presence of an accessory stylar cusp, together with the remaining diagnostic molar features previously mentioned, suggest the probable belonging of *Didelphopsis*, *Carolopaulacoutoia*, *Itaboraidelphys*, *Marmosopsis*, *Ankotarinja*, *Keeuna*, and *Djarthia* to a natural group (Sternbergiidae Szalay, 1994). If this was the case, then *Djarthia* (as well as the other sternbergiids) may not represent a prototypical Australian marsupial, but instead one more early 'ameridelphian' clade preserving a unique set of generalised as well as derived dental features. To Godthelp et al. (1999, *contra* Woodburne and Case 1996), there is no empirical evidence to dismiss the possibility that both ameridelphians and australidelphians were present in the early Tertiary of Australia.

FUTURE DIRECTIONS

With a few notable exceptions, early marsupial radiations during the late Cretaceous–Paleogene span in South America are still poorly known. Terrestrial faunas corresponding to the Maastrichtian–Danian span are still to be discovered. In equatorial South America, the Paleogene fossil record is mostly a blank. Southern South American fossils, best known from Patagonian localities, still show a strong bias in favour of medium- to large-sized mammals. Most marsupials are, and most probably were, of small size. Since the last decade, the application of screen-washing techniques both in new and in already well-known fossil localities led to the discovery of many small-sized teeth of basal marsupials assignable to most of the currently recognised orders. Several of the new taxa recognised by these methods seem to be key for the interpretation of the evolution of major marsupial lineages.

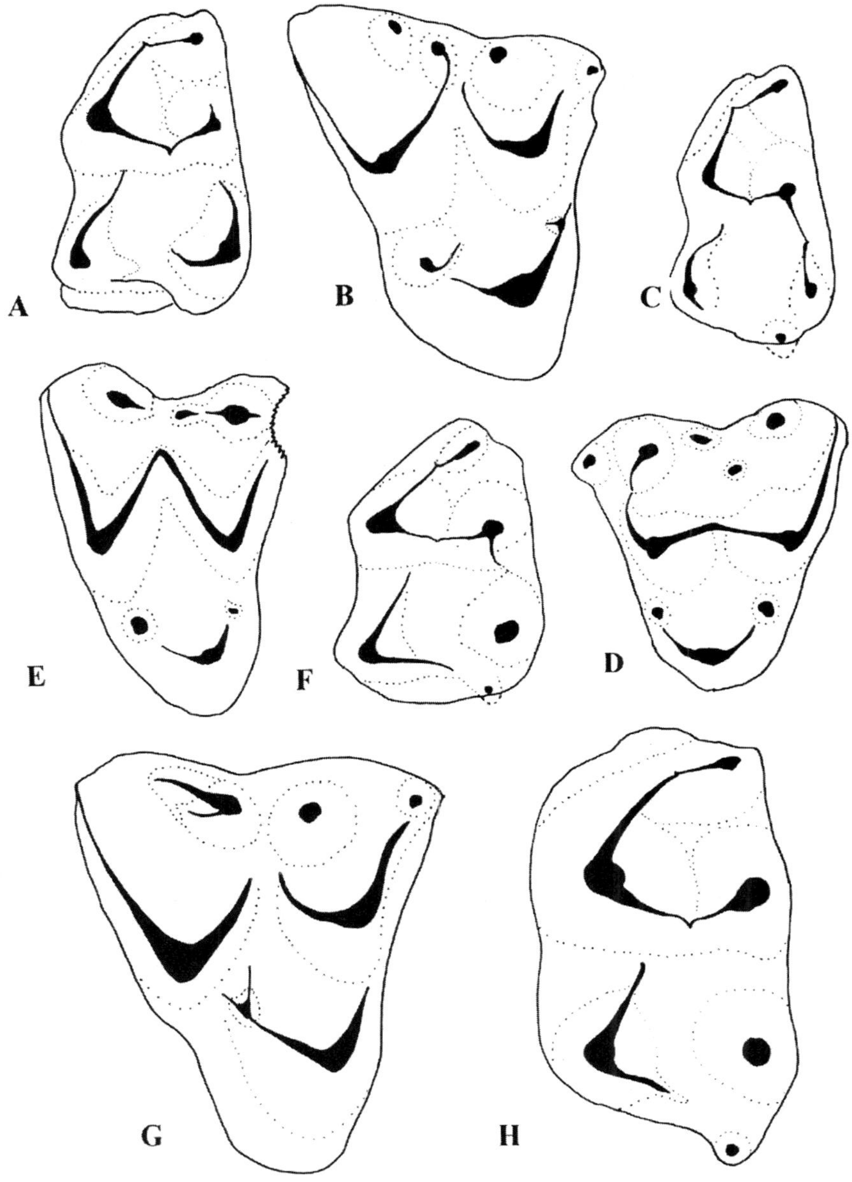

Figure 3 Schematic drawings of upper and lower molars of early South American marsupials, in occlusal view. **A**, **B**, Derorhynchidae, genus and species indet.; A, uncatalogued MEF specimen, a left m2; B, uncatalogued MEF specimen, a right M3. Both specimens come from ?early Eocene levels near Cabeza Blanca, central Patagonia, Argentina. **C**, **D**, Peradectidae, genus and species indet.; C, MLP 90-II-5-166, a left m2 (partially restored); D, MLP 90-II-5-167, a left M2. **E**, **F**, Derorhynchidae, genus and species indet.; E, MLP 90-II-5-342, a right M3; F, MLP 90-II-5-246, a left m2. **G**, **H**, *Guggenheimia* sp.; G, MLP 90-II-5-90, a right M2; H, MLP 90-II-5-72, a left m3. All specimens except the MEF ones come from medial Paleocene levels at Yacimiento Las Flores (Las Flores Fm.), in Central Patagonia, Argentina. Drawings are not to scale.

The recognition of affinities between North and South American marsupials, as well as for South American and Australian, has been probably biased by biogeographic considerations. Land bridges, water barriers, etc., were frequent topics of discussion by the middle of last century in relation to the origin of South American therian mammals from North American ancestors. '…[O]ne or more marsupials, one or more edentates, and one or two or three or more ungulates reached South America about the beginning of the Tertiary (…), and that they came across a water barrier which blocked other mammals. (…) the minimum number of successful landfalls required to account for marsupials, edentates, condylarths and notoungulates is four. (…) The central stock of the order [by then, the Marsupialia], the Didelphidae, could very well (...) have been derived from the North American Cretaceous *Alphadon* (or relatives), with diversification coming after entry into South America' (Patterson and Pascual 1968: 259-260). By the end of that century, a basic dichotomy (excepting microbiotheres) was recognised between South American and Australian marsupial lineages. More recently Kirsch et al. (1997: 257–258) argued against the '...myth of a geographically based dichotomy within the Gondwanan marsupials that we are anxious to deconstruct', a suggestion that was strongly refuted by Szalay and Sargis (2001). Our own perception of marsupial relationships, mostly based on dental evidence, agrees better with Kirsch et al.'s (1997) conclusions on this topic.

As for other mammals, fossil teeth are still the best pieces of evidence at hand for the understanding of early (and not so early) diversity and radiations of marsupial lineages. Exceptionally well-preserved skulls as those of Tiupampa or Riversleigh are, precisely, exceptions. Attempts of reconstructing marsupial phylogeny on the exclusive basis of isolated postcranial remains may be deceptive, as these remains are frequently difficult to assign to taxa diagnosed solely on dentitions. Finally, '...confident resolution of many higher-level systematic problems may be beyond the current capacity of molecular systematic techniques' (Archer et al. 1999: 7). On the contrary, teeth are ubiquous through time and space. They are primary sources for the taxonomy of most extinct (and many extant) taxa. They are complex morphological systems. Character and character complexes analyses have in teeth a wonderful tool for theoretical speculation. Finally, the misleading swamps of homoplasy exist in the analysis of all morphological systems, not only teeth.

Acknowledgements

I thank all colleagues, curators, and chiefs of paleontological departments that facilitated the study of specimens under their care: Marcelo Reguero, Departamento Paleontología Vertebrados, Museo de La Plata; Judd A. Case, Department of Biology, St. Mary's College of California, Moraga; Michael O. Woodburne, University of California, Riverside; Richard Cifelli, Oklahoma Museum of Natural Sciences, Oklahoma; Richard Fox, University of Alberta, Edmonton; Richard Kay, Duke University, Durham; Kenneth Campbell, Los Angeles Museum of Natural Sciences; Christian de Muizon, Muséum National d'Histoire Naturelle, Paris; José Bonaparte, Museo Argentino de Ciencias Naturales 'Bernardino Rivadavia', Buenos Aires; Edison Oliveira, Universidad Federal do Rio Grande do Sul, Porto Alegre; John Flynn, Field Museum of Natural Sciences, Chicago; Wighart von Koenigswald, Institute für Palaeontologie, Universität Bonn. I also thank them, as well as Cornelia Kurz (Hessisches Landesmuseum Darmstadt, Darmstadt, Germany), for their helpful comments while discussing different aspects related to this work. Data dealing with the relationships of herpetotheriids and derorhynchids are presently being studied by C. Kurz and me. Reviewers Richard Cifelli and Stephen Wroe, as well as one of the editors of this volume, Michael Archer, are gratefully acknowledged for their useful critics and comments. The National Sciences Foundation (NSF OPP 9615236) and Dr Michael Woodburne provided financial and logistic support, respectively, for the study of various collections in the United States and Canada in 1997; the Alexander von Humboldt Foundation did so in 1998–1999 for the study of collections at Paris, Bonn, Munich, and Berlin. Finally, many thanks to the editors of this book for their patience in waiting for this manuscript.

References

Aplin, K.O., & Archer, M. (1987), 'Recent advances in marsupial systematics with a new syncretic classification', in *Possums and Opossums: Studies in Evolution* (ed M. Archer.), pp. xv-lixxii, Surrey Beatty & Sons and the Royal Zoological Society of New South Wales, Sydney.

Archer, M. (ed.) (1982a), *Carnivorous Marsupials, Vols. 1 and 2*, Royal Zoological Society of New South Wales, Mosman, New South Wales.

Archer, M., Arena, R., Bassarova, M., Black, K., Brammall, J., Cooke, B., Creaser, P., Crosby, K., Gillespie, A., Godthelp, H., Gott, M., Hand, S.J., Kear, B., Krikmann, A., Mackness, B., Muirhead, J., Musser, A., Myers, T., Pledge, N., Wang Y., & Wroe, S. (1999), 'The evolutionary history and diversity of Australian mammals', *Australian Mammalogy*, **21**:1–45.

Bond, M., Carlini, A.A., Goin, F.J., Legarreta, L., Ortiz Jaureguizar, E., Pascual, R., & Uliana, M.A. (1995), 'Episodes in South American land mammal evolution and sedimentation: testing their apparent concomitance in a Paleocene succession from Central Patagonia', *Actas VI Congreso Arg. Paleont. y Bioestrat., Trelew*, pp. 47–58.

Bond, M., Reguero, M., López, G., Carlini, A.A., Goin, F.J., Madden, R.H., Vucetich, M.G., & Kay, R.F. (1997), 'The "Astraponotéen plus Supérieur" (Paleogene) in Patagonia: report on Project MLP-Duke 1993-1997', *13 Jornadas Argentinas de Paleontología de Vertebrados, La Rioja*, Abstract, p. 12.

Candela, A.M., Goin, F.J., & Pascual, R. (1998), 'Los polydolopimorphia (Mammalia, Marsupialia) de la Formación Las Flores (Paleoceno Medio, Patagonia Central, Argentina)', *Acta Geológica Lilloana* **18**(1):149.

Case, J.A., Woodburne, M.O., & Chaney, D.S. (1988), 'A new genus and species of polydolopid marsupial from the La Meseta Formation, late Eocene, Seymour Island, Antarctic Peninsula', in *Geology and Paleontology of Seymour Island* (eds. R.M. Feldmann & M.O. Woodburne), pp. 505–521, Geol. Soc. Am. Mem, 169, Boulder, CO.

Crochet, J.-Y. (1980): *Les Marsupiaux du Tertiaire d'Europe*. 279 S., 241 fig., 40 tab., 2pl., Singer-Polignac, Paris.

Flynn, J.J., & A.R. Wyss (1999), 'New Marsupials from the Eocene-Oligocene transition of the Andean Main Range, Chile', *Journal of Vertebrate Paleontology* **19**(3):533–49.

Gabbert, S.L. (1998), 'Basicranial anatomy of *Herpetotherium* (Marsupialia: Didelphimorphia) from the Eocene of Wyoming', *American Museum Novitates*, **3235**:1–13.

Godthelp, H., Wroe, S., & Archer, M. (1999), 'A new marsupial from the Early Eocene Tingamarra local fauna of Murgon, Southeastern Queensland: a prototypical Australian marsupial?', *Journal of Mammalian Evolution*, **6**(3):289–313.

Goin, F.J. (1991), 'Los Didelphoidea (Mammalia, Marsupialia, Didelphimorphia) del Cenozoico tardío de la Región Pampeana', unpublished thesis, Facultad de Ciencias Naturales y Museo, Universidad Nacional de La Plata. La Plata, pp. 1–327.

Goin, F.J. (1993), 'Living South American opossums are not living fossils', *6 International Theriological Congress*, Sydney, Abstracts, pp. 112–13.

Goin, F.J. (1995), 'Marsupialia', in *Evolución Biológica y Climática de la Región Pampeana Durante los Últimos Cinco Millones de Años* (eds. M.T. Alberdi, G. Leone, & E. P. Tonni), pp. 165–179, Museo Nacional de Ciencias Naturales, Madrid.

Goin, F.J. (1997), 'New clues for understanding Neogene marsupial radiations', in *A History of the Neotropical Fauna. Vertebrate Paleobiology of the Miocene in Colombia* (eds R.F. Kay, R.H. Madden, R.L. Cifelli, & J. Flynn), pp.185–204, Smithsonian Institution Press, Washington.

Goin, F.J., & Candela, A.M.(1995), 'Una nueva especie de *Epidolops* Paula Couto, 1952 (Marsupialia, Polydolopimorphia, Polydolopidae), Consideraciones sobre el patrón molar inferior de los Epidolopidae', *VI Congreso Arg. Paleont. y Bioestrat.Trelew*, Actas, pp. 143–48.

Goin, F.J., & A.M. Candela, A.M. (1996), 'A new early Eocene Polydolopimorphian (Mammalia, Marsupialia) from Patagonia', *Journal of Vertebrate Paleontology*, **16**(2):292–296.

Goin, F.J., & Candela, A.M.(1997), 'New Patagonian Marsupials from Ameghino's "Astraponoteén Plus Supérieur" (Post-Mustersan–Pre-Deseadan Age)', *56 Annual Meeting of the Society of Vertebrate Paleontology, New York*, Abstracts, Addendum, p. 2.

Goin, F.J.& Candela, A.M.(in press), 'New Paleogene marsupials from the Amazonian basin, Southeastern Perú', *Los Angeles Museum of Natural Sciences, Special volume* (eds. K. Campbell & R. Frailey).

Goin, F.J., Candela, A.M.& Forasiepi, A. (1997), 'New, middle Paleocene marsupials from Central Patagonia', *57 Annual Meeting of the Society of Vertebrate Paleontology, New York*, Abstracts, p. 49A.

Goin, F.J., Candela, A.M., Bond, M., Pascual, R., & Escribano, V. (1998a), 'Una nueva "comadreja" (Mammalia, Marsupialia, ?Polydolopimorphia) del Paleoceno de Patagonia, Argentina, in *Paleógeno de América del Sur y de la Península Antártica*, pp. 79–84, Asociación Paleontológica Argentina, Special Publ. N.5, Buenos Aires.

Goin, F.J., Oliveira, E. V.& Candela, A.M. (1998b), '*Carolocoutoia ferigoloi* n. gen. et sp. (Protodidelphidae), a new Paleocene 'opossum-like' marsupial from Brazil', *Palaeovertebrata*, **27**(3–4):145–54.

Goin, F.J., Candela, A.M.& Lopez, G. (1998c), 'Middle Miocene marsupials from Antofagasta de la Sierra, Northwestern Argentina' *Geobios*, **31**(1):75–85.

Goin, F.J., Woodburne, M.O., Case, J., Vizcaíno, S.F., & Reguero, M. (1999), 'New discoveries of "opossum-like" marsupials from Antarctica (Seymour Island, Middle Eocene)', *Journal of Mammalian Evolution*, **6**(4):335–65.

Jansa, S.A., & Voss, R.S. (2000), 'Phylogenetic studies on Didelphid marsupials I. Introduction and preliminary results from nuclear IRBP gene sequences', *Journal of Mammalian Evolution*, **7**:43–77.

Johanson, Z. (1996), 'New marsupial from the Fort Union Formation, Swain Quarry, Wyoming', *Journal of Paleontology*, **70**:1023–31.

Kirsch, J.A.W. (1977), 'The comparative serology of Marsupialia, and a classification of marsupials', *Australian Journal of Zoology, supplementary series*, **52**, pp. 1–152.

Kirsch, J.A.W., Dickerman, A.W., Reig, O.A., & Springer, M.S. (1991), 'DNA hybridization evidence for the Australasian affinity of the American marsupial *Dromiciops australis*', *Proceedings of the National Academy of Sciences, USA*, **88**:10465–69.

Kirsch, J.A.W., Lapointe, F.-J., & Springer, M. (1997), 'DNA-hybridisation studies of marsupials and their implications for metatherian classification', *Australian Journal of Zoology*, **45**:211–80.

Koenigswald, W.V., & Goin, F.J. (2000), 'Enamel differentiation in South American marsupials and a comparison of placental and marsupial enamel', *Palaeontographica Abt. A.*, **255**:129–68.

Korth, W.W. (1994), 'Middle Tertiary marsupials (Mammalia) from North America', *Journal of Paleontology*, **68**:376–96.

Marshall, L.G. (1976), 'New didelphine marsupials from the La Venta Fauna (Miocene) of Colombia, South America', *Journal of Paleontology*, **50**(3):402–18.

Marshall, L.G. (1977), 'Cladistic analysis of borhyaenoid, dasyuroid, and thylacinid (Marsupialia, Mammalia) affinity', *Systematic Zoology*, **26**:410–25.

Marshall, L.G. (1978), 'Evolution of the Borhyaenidae, extinct South American predaceous marsupials', *University of California Publications in Geological Sciences*, **117**:1–89.

Marshall L.G. (1979), 'Review of the Prothylacyninae, an extinct subfamily of South American "dog-like" marsupials', *Fieldiana n.s.*, **3**:1–50.

Marshall, L.G. (1980), 'Systematics of the South American marsupial family Caenolestidae', *Fieldiana, Geology, n.s.*, **5**:1–145.

Marshall, L.G. (1981), 'Review of the Hathlyacyninae, an extinct subfamily of South American "dog-like" marsupials', *Fieldiana, Geology, n.s.* **7**:1–120.

Marshall, L.G. (1982a), 'Systematics of the extinct South American marsupial family Polydolopidae', *Fieldiana, Geology, n.s.*, **12**:1–109.

Marshall, L.G. (1982b), 'Systematics of the South American Marsupial family Microbiotheriidae', *Fieldiana, Geology, n.s.*, **10**:1–75.

Marshall, L.G. (1987), 'Systematics of Itaboraian (middle Paleocene) age "opossum-like" marsupials from the limestone quarry at Sao José de Itaborai, Brasil', in *Possums and Opossums: Studies in Evolution, Vol. 1* (ed M. Archer), pp. 91–160, Surrey Beatty & Sons Pty Limited and The Royal Zoological Society of New South Wales.

Marshall, L.G., & Kielan-Jaworowska, Z. (1992), 'Relationships of the dog-like marsupials, deltatheroidans and early tribosphenic mammals', *Lethaia*, **25**:361–74.

Marshall, L.G., & Muizon, C. de (1988), 'The dawn of the age of mammals in South America', *National Geographic Research*, **5**(3):268–661.

Marshall, L.G., Muizon, C. de, & Sigé, B. (1983), 'Late Cretaceous mammals (Marsupialia) from Bolivia', *Geobios*, **16**(6):739–45.

Marshall, L.G., Case, J., & Woodburne, M.O. (1990), 'Phylogenetic relationships of the families of marsupials', in *Current Mammalogy, Vol. 2* (ed. H.H. Genoways), pp. 433–505, Plenum Press, New York.

Muirhead, J. (2000), 'Yaraloidea (Marsupialia, Peramelemorphia), a new Superfamily of Marsupial and a description and analysis of the cranium of the Miocene *Yarala burchfieldi*', *Journal of Paleontology*, **74**(3):512–23.

Muirhead, J., & Filan, S. (1995), '*Yarala burchfieldi*, a plesiomorphic bandicoot (Marsupialia, Peramelemorphia) from Oligo-Miocene deposits of Riversleigh, northern Queensland', *Journal of Vertebrate Paleontology*, **18**:612–26.

Muizon, C. de (1992), 'La fauna de mamíferos de Tiupampa (Paleoceno inferior, Formación Santa Lucía), Bolivia', *Revista Técnica de Yacimientos Petrolíferos Fiscales de Bolivia*, **12**(3–4):575–624.

Muizon, C. de (1994), 'A new carnivorous marsupial from the Paleocene of Bolivia and the problem of marsupial monophyly', *Nature*, **370**:208–11.

Muizon, C. de (1998), '*Mayulestes ferox*, a borhyaenid (Metatheria, Mammalia) from the Early Paleocene of Bolivia. Phylogenetic and paleobiologic implications', *Geodiversitas*, **20**(1):19–142.

Muizon, C. de (1999), 'Marsupial skulls from the Deseadan (late Oligocene) of Bolivia and phylogenetic analysis of the Borhyaenoidea (Marsupialia, Mammalia)', *Geobios*, **32**(3):483–509.

Muizon, C. de, & Brito, I.M. (1993), 'Le bassin calcaire de São José de Itaboraí (Rio de Janeiro, Brésil): ses relations fauniques avec le site de Tiupampa (Cochabamba, Bolivie)', *Annales de Paléontologie*, **79**(3):233–69.

Muizon, C. de, & Cifelli, R.L. (2001), 'A new basal "Didelphoid" (Marsupialia, Mammalia) from the early Paleocene of Tiupampa (Bolivia)', *Journal of Vertebrate Paleontology*, **21**(1):87–97.

Muizon, C. de, & Lange-Badré, B. (1997), 'Carnivorous dental adaptations in tribosphenic mammals and phylogenetic reconstruction', *Lethaia*, **30**:353–66.

Muizon, C. de, Cifelli, R.L., & Céspedes Paz, R. (1997), 'The origin of borhyaenoids, South American dog-like marsupials', *Nature*, **389**:486–89.

Oliveira, E.V. (1998), 'Taxonomia, Filogenia e Paleobiogeografia de marsupiais "Poliprotodontes" do Mesopaleoceno da Bacia de Itaboraí, Rio de Janeiro, Brasil', unpublished thesis, Universidade Federal do Rio Grande do Sul, Porto Alegre, pp. 1–327.

Oliveira, E.V., & Goin, F.J. (in press), 'Marsupiais do Início do Terciário da Bacia de Itaboraí: Origem, Irradiação e História Biogeográfica', in *Marsupiais do Brasil* (ed N.C. Caceres), UFRJ, Rio de Janeiro.

Oliveira, E., Goin, F.J., & Candela, A.M. (1996), 'Un nuevo marsupial "Pseudodiprotodonte" del Paleoceno medio de Itaboraí (Brasil), Consideraciones sobre el origen, radiación y heterocronía en los Paucituberculata', *12 Jornadas Argentinas de Paleontología Vertebrados, La Pampa*. Abstr. p. 62.

Pascual, R. (1980a), 'Nuevos y singulares tipos ecológicos de Marsupiales extinguidos de América del Sur (Paleoceno Tardío o Eoceno Temprano) del Noroeste Argentino', *2 Congreso Argentino de Paleontología y Bioestratigrafía, Buenos Aires*, Actas 2, pp. 151–73.

Pascual, R. (1980b), 'Prepidolopidae, Nueva Familia de Marsupialia Didelphoidea del Eoceno Sudamericano', *Ameghiniana* **17**(3):216–42.

Pascual, R. (1981), 'Adiciones al conocimiento de *Bonapartherium hinakusijum* (Marsupialia, Bonapartheriidae) del Eoceno temprano del Noroeste argentino', *Anais II Congreso Latino-americano de Paleontología, Porto Alegre*, pp. 507–20.

Pascual, R. (1983), 'Novedosos marsupiales paleógenos de la Formación Pozuelos (Grupo Pastos Grandes) de la Puna, Salta, Argentina', *Ameghiniana*, **20**(3–4):265–80.

Pascual, R., & Bond, M. (1981), 'Epidolopinae subfam. nov. de los Polydolopidae (Marsupialia, Polydolopoidea)', *Anais II Congreso Latino-americano de Paleontología, Porto Alegre*, pp. 479–88.

Patterson, B., & Pascual, R. (1968), 'The fossil mammal fauna of South America', in *Evolution, Mammals and the Southern Continents* (ed. A. Keast, F.C. Erk, & B. Glass), pp. 247–309, State University of New York Press, Albany, NY.

Reig, O., Kirsch, J.A.W., & Marshall, L.G. (1987), 'Systematic relationships of the living and Neocenozoic American "opossum-like" marsupials (Suborder Didelphimorphia), with comments on the classification of these and of the Cretaceous and Paleogene New World and European metatherians', in *Possums and Opossums: Studies in Evolution* (ed M. Archer), pp. 1–89, Surrey Beatty & Sons and the Royal Zoological Society of New South Wales, Sydney.

Rougier, G.W., Wible, J.R., & Novacek, M.J. (1998), 'Implications of *Deltatheridium* specimens for early marsupial history', *Nature*, **396**:459–63.

Sánchez-Villagra, M.R. (2001), 'The phylogenetic relationships of argyrolagoid marsupials', *Zoological Journal of the Linnean Society*, **131**:481–96.

Sánchez-Villagra, M.R., Kay, R.F., & Anaya-Daza, F. (2000), 'Cranial anatomy and palaeobiology of the Miocene marsupial *Hondalagus altiplanensis* and a phylogeny of Argyrolagids', *Palaeontology*, **43**(2):287–301.

Simpson, G.G. (1980), *Splendid Isolation: The Curious History of South American Mammals*, Yale Univ. Press, New Haven.

Springer, M.S., Kirsch, J.A.W., & Case, J.A. (1997), 'The chronicle of marsupial evolution', in *Molecular Evolution and Adaptive Radiation* (eds. T. Givnish & K. Sytsma), pp. 129–61, Cambridge University Press, New York.

Szalay, F.S. (1982), 'A new appraisal of marsupial phylogeny and classification', in *Carnivorous Marsupials, Vol. 2* (ed M. Archer), pp. 621–40, Royal Zoological Society of New South Wales, Mosman, New South Wales.

Szalay, F.S. (1993), 'Metatherian taxon phylogeny: evidence and interpretation from the cranioskeletal system', in *Mammal Phylogeny, Vol. 1, Mesozoic Differentiation, Multituberculates, Monotremes, Early Therians, and Marsupials* (eds F.S. Szalay, M.J. Novacek and M.C. McKenna), pp. 216–42, Springer-Verlag, New York.

Szalay, F.S. (1994), *Evolutionary History of the Marsupials and an Analysis of Osteological Characters*, Cambridge University Press, New York.

Szalay, F.S., & Sargis, E.J. (2001), 'Model-based analysis of postcranial osteology of marsupials from the Palaeocene of Itaboraí (Brazil) and the phylogenetics and biogeography of Metatheria', *Geodiversitas*, **23**(2):139–202.

Woodburne, M.O., & Case, J.A. (1996), 'Dispersal, vicariance and the Late Cretaceous to early Tertiary land mammal biogeography from South America to Australia', *Journal of Mammalian Evolution*, **3**:121–61.

Woodburne, M.O., & Zinsmeister, W.J. (1982), 'Fossil land mammal from Antarctica', *Science*, **218**:284–86.

Woodburne, M.O., & Zinsmeister, W.J. (1984), 'The first land mammal from Antarctica and its biogeographic implications', *Journal of Paleontology*, **58**:913–48.

CHAPTER 4

COMPARATIVE ANATOMY OF THE TIUPAMPA DIDELPHIMORPHS; AN APPROACH TO LOCOMOTORY HABITS OF EARLY MARSUPIALS

Christian de Muizon and Christine Argot

UMR 8569, Département Histoire de la Terre, 8, rue Buffon, F-75005 Paris, France.
Email: muizon@mnhn.fr; argot@mnhn.fr

The comparison of the postcranial skeletons of the only three subcomplete marsupial skeletons (*Pucadelphys, Andinodelphys*, and *Mayulestes*) from the Palaeocene (Tiupampa, Bolivia) allows an appraisal of their locomotory habits and biology. The anatomy of the cervical, lumbar and caudal vertebrae, shoulder, elbow, hip, knee, and ankle joints is successively analysed. Several features are indicative of arboreal abilities (e.g. prehensile tail, posterodorsal extension of the posterodorsal angle of the scapula, anterior position of the acromion, relatively spherical head of the humerus, prominent medial epicondyle, ulna bent anteriorly, widely opened trochlear incisure, open acetabulum with a concave dorsal edge, large lesser trochanter of the femur, medial orientation of the ectal facet of the calcaneum). However, an increasing gradient of arboreal ability was observed from *Pucadelphys* to *Mayulestes*. Several other features indicate that the three genera were very agile (e.g. anterior position of the anticlinal vertebra, large neural and transverse processes of the lumbar vertebrae, strongly everted ilium, elevated crests of the femoral trochlea). The Tiupampa marsupials were clearly more agile than most living didelphids (except *Metachirus*) and resemble the condition observed in living dasyurids. It is hypothesised that the Tiupampa didelphimorphs could have approached what is regarded as the probable generalised plesiomorphic pattern for marsupial locomotion, i.e. agile terrestrial animals but with good climbing ability. The less specialised taxa (*Pucadelphys* and *Andinodelphys*) could be structurally ancestral to both didelphids and dasyurids.

INTRODUCTION

The anatomy of the postcranial skeleton of mammals is closely related to their locomotion and biology. When dealing with fossil taxa, it generally represents the major (often the only) source of information on their way of life. Because skeletons of fossil mammals are especially rare in early Tertiary (i. e. Palaeocene) and Cretaceous deposits and often poorly preserved (badly crushed, or very incomplete), the biology of early mammals is generally poorly known and the hypotheses on the plesiomorphic locomotory adaptations are highly speculative. Localities that have yielded well-preserved sub-complete mammal skeletons are uncommon. Several specimens of eutherian mammals have been found in several localities of the late Cretaceous of Mongolia (Kielan Jaworowska 1977, 1981; Novacek et al. 1998). Recently, skeletons of non-therian mammal were discovered in the early Cretaceous of China

(2 taxa) (Hu et al. 1997; Ji et al. 1999) and Portugal (1 taxon) (Krebs 1991) and in the Cretaceous of Argentina (1 taxon) (Bonaparte and Rougier 1987). A few eutherian skeletons are known in the Palaeocene (Torrejonian and Tiffanian) of North America (Matthew 1937) and Europe (Russell 1964) but none are recorded in the basal Palaeocene (= Puercan) in North America. Regarding marsupials, a single skeleton has been described from the late Cretaceous of Mongolia (Szalay and Trofimov 1996) and except for the Tiupampa specimens, no marsupial skeletons are known before the Eocene.

The locality of Tiupampa (Santa Lucía Formation, early Palaeocene, Bolivia) has yielded several metatherian partial skeletons that represent the earliest known in the New World. They have been referred to as didelphoids (*Pucadelphys andinus*, *Andinodelphys cochabambensis*) (Marshall and Muizon 1995; Marshall and Sigogneau-Russell 1995; Muizon et al. 1997; Muizon 1998) and borhyaenoids (*Mayulestes ferox*), (Muizon 1998). These taxa are especially well represented in the case of *P. andinus*, for which several well-preserved specimens have been discovered since the 1995 monographs. *A. cochabambensis* is represented by four partial skeletons, while *M. ferox* is still known by one partial skeleton only. The Tiupampa fauna provides unique information on the early evolution of marsupials. The new specimens of *Pucadelphys* and the *Andinodelphys* skeletons (Fig. 1) are not completely prepared yet, but at this stage of the process, several new data are available. The purpose of this paper is to present a preliminary comparative study of the postcranial skeleton of the three taxa in order to approach an appraisal of the primitive condition of the marsupial locomotion and habits.

Figure 1 Block containing several articulated or partly articulated skeletons of *Andinodelphys cochabambensis*.

MATERIAL AND METHODS

Following, we compare several regions of the postcranial skeleton of the three taxa (*Pucadelphys*, *Andinodelphys*, and *Mayulestes*) that are regarded as functionally important in the determination of locomotory habits. Characters of the axial skeleton will be considered first, followed by the forelimb and the hindlimb. Comparisons will be made essentially with the living didelphids since they probably represent the most primitive living marsupials. However, it is noteworthy that the Tiupampa didelphimorphs are not didelphids, and the relation between the bone anatomy and the inferred function does not necessarily resemble the didelphid pattern. This is especially true in the case of *Mayulestes*, which is not even a didelphoid. Furthermore, given the scarcity of marsupial skeletons in the late Cretaceous and early Tertiary and their poor knowledge, the phylogenetic affinities of *Pucadelphys* and *Andinodelphys* are tenuously established and these genera may not have close relationships even with the didelphoids. For this reason, and because *Mayulestes* and *Pucadelphys* have been regarded as partially arboreal (Muizon 1998), the Tiupampa didelphimorphs will also be rodents, in order to have a broader approach of the skeletal features related to arboreality. Dasyurids do not represent a generalised marsupial clade, because they 'show many specialisations of the auditory region that are clearly apomorphic within Dasyuromorphia' (Wroe 1997; 1999: 523). However, dasyurids (and especially Dasyurinae) are in many respects the less derived Australian marsupials. Another important factor to be considered is the incidence of behaviour on the locomotory habits. Small animals are usually eclectic and do not necessarily do what they could do if they do not want to. For instance, *Monodelphis* is usually regarded as poorly adapted to arboreal life and almost exclusively terrestrial (Nowak and Paradiso 1983; Eisenberg 1989; Julien-Laferrière 1991; Grand 1983; Lemelin 1999). However, from captive observations, we know that these didelphids are agile climbers and apparently feel quite comfortable in a tree (R. L. Cifelli, personal observation on *M. domestica*). The same is true for *Lutreolina*, the didelphid species best adapted to life on the pampas (terrestrial), but which can climb well (Nowak and Paradiso 1983). The same is observed in the placental tupaids, in which the limits between arboreality and terrestriality are not clear-cut since one of the most arboreal taxon (*Tupaia minor*) can be found on the ground while the most terrestrial species (*T. tana*) climbs well (Jenkins 1974). Therefore, one should keep in mind that all deductions made from functional anatomy analysis only represent a hypothesis on what an animal could have done, not what it actually did. The myological data referred to in this work are from studies in progress by C. Argot based on dissections of several specimens of seven species of didelphids: *Caluromys philander, Didelphis marsupialis, Marmosa murina, Metachirus nudicaudatus, Micoureus demerara, Monodelphis brevicaudata,* and *Philander opossum.* These dissections helped to understand the relations between osteological features, muscular development and movements performed in the field, and the most relevant results have been reported here. Comparisons were also made with skeletons of the following species: *Monodelphis domestica, Lutreolina crassicaudata, Thylacinus cynocephalus, Dasyurus viverrinus, Antechinomys laniger, Phascogale tapoatafa, Schoinobates volans, Trichosurus vulpecula, Phascolarctos cinereus, Peramales nasuta, Potos flavus, Cryptoprocta ferox, Felis concolor, Nandinia binotata, Bassariscus astutus, Sciurus igniventris, Cyclopes didactylus, Manis tetradactyla, Tupaia gracilis, T. minor, T. tana, T. glis, Nycticebus coucang,* Unfortunately, no postcranial skeletons of *Dromiciops gliroides* (a highly arboreal South American marsupial) were available during this study. In the following descriptions and discussions, reference to a genus actually refers to the species mentioned above. In the case of *Monodelphis*, reference to this genus corresponds to the two species, which we found almost identical in terms of postcranial skeleton. Because the species of *Tupaia* considered in this study have different habits we always refer to species.

The systematic terminology of marsupials follows Muizon et al. (1997) for Didelphimorphia, Marshall et al. (1990) for Didelphoidea and Didelphidae, and Wroe (1999) for Dasyuridae. However, we are conscious of the urgent need for a revision of the three former taxa.

COMPARATIVE ANATOMY

Axial skeleton

Occipital crest and neural process of the axis. On the posterior side of the occipital crest are attached several muscles of the dorsal musculature of the neck (splenius, biventer, complexus, spinalis capitis, and both recti capitis dorsalis), which are involved in vertical and horizontal movements of the head. In the three taxa from Tiupampa the occipital crest is very salient posteriorly. Because the size of the crest increases with ontogenetic age it is difficult to evaluate the relative size of the crest in three genera. However, it is clear that it was more developed than in the living didelphids.

The neural process of the axis is relatively short anteroposteriorly in *Pucadelphys*. It does not project posteriorly and its posterior edge is almost vertical. In contrast, in *Andinodelphys* the neural crest extends posteriorly behind the centrum and postzygapophyses, a condition that is more pronounced in *Mayulestes*. A posteriorly projecting and long neural arch of the axis is present in carnivorous predaceous marsupials (borhyaenoids, thylacinids, some dasyurids). In living didelphids the neural arch is always anteroposteriorly shorter than in *Mayulestes*. In some genera (*Marmosa, Thylamys, Philander, Metachirus*) the posterior process of the neural arch resembles that of *Pucadelphys*, whereas in other genera (*Micoureus, Monodelphis*) it is relatively long approaching (but not reaching) the condition of *Andinodelphys*. The muscle that originates on the sagittal crest of the axis is the obliquus capitis caudalis. It inserts on the transverse processes of the atlas and represents the main rotator of the head. Its rotative action is emphasised by the position of the fibres, oriented anteroventrally rather than longitudinally. Another muscle, more superficial, originates on the posterior process of the neural arch of the axis, the spinalis capitis. The fan shape of this muscle, which inserts on a large part of the occipital crest, would allow it to participate to the rotation of the head. Therefore, the range of these movements is related to the length of the posterior process of the axis which increases the mechanical advantage of these muscles.

The development of the occipital crest and anteroposterior extension of the neural process of the axis indicate a strong musculature of the neck, which is related to predaceous carnivorous habits (Muizon 1998). Rapid lateral shaking of the head is often used by predators to kill small vertebrates. The size of the occipital crest of the Tiupampa genera denotes more predaceous habits than the living didelphids, but the morphology of the neural arch of the axis indicate an increasing gradient of predation from *Pucadelphys* to *Mayulestes*. Because of its smaller size, *Pucadelphys* is likely to have been insectivorous, which may have not required such a strong neck as in *Mayulestes*, and which could explain the lack of posterior extension of the neural arch of the axis. Because of its long posterior process of the axis, *Mayulestes* was the most predaceous of the three genera and may have fed upon other small vertebrates (small marsupials, amphibians, squamates). *Andinodelphys* was probably eating small vertebrates too but could also have been partly insectivorous. The two didelphid genera, which approach the *Andinodelphys* condition of the morphology of the axis are *Micoureus* and *Monodelphis*: although opportunistic feeders, they are also mainly insectivorous (Charles-Dominique et al. 1981). Therefore, the three Tiupampa didelphimorph genera seem to have been more carnivorous than the Recent didelphids, which is corroborated by their dental morphology.

Position of the anticlinal vertebra and neural processes of the lumbar vertebrae. The anticlinal vertebra is the vertebra on which the orientation of the neural process reverses from posterior to anterior. The lumbar vertebrae have a neural process strongly oriented anteriorly in all cursorial and saltatorial mammals and the anticlinal vertebra is located anteriorly (first lumbar or last thoracics). This morphology is related to the

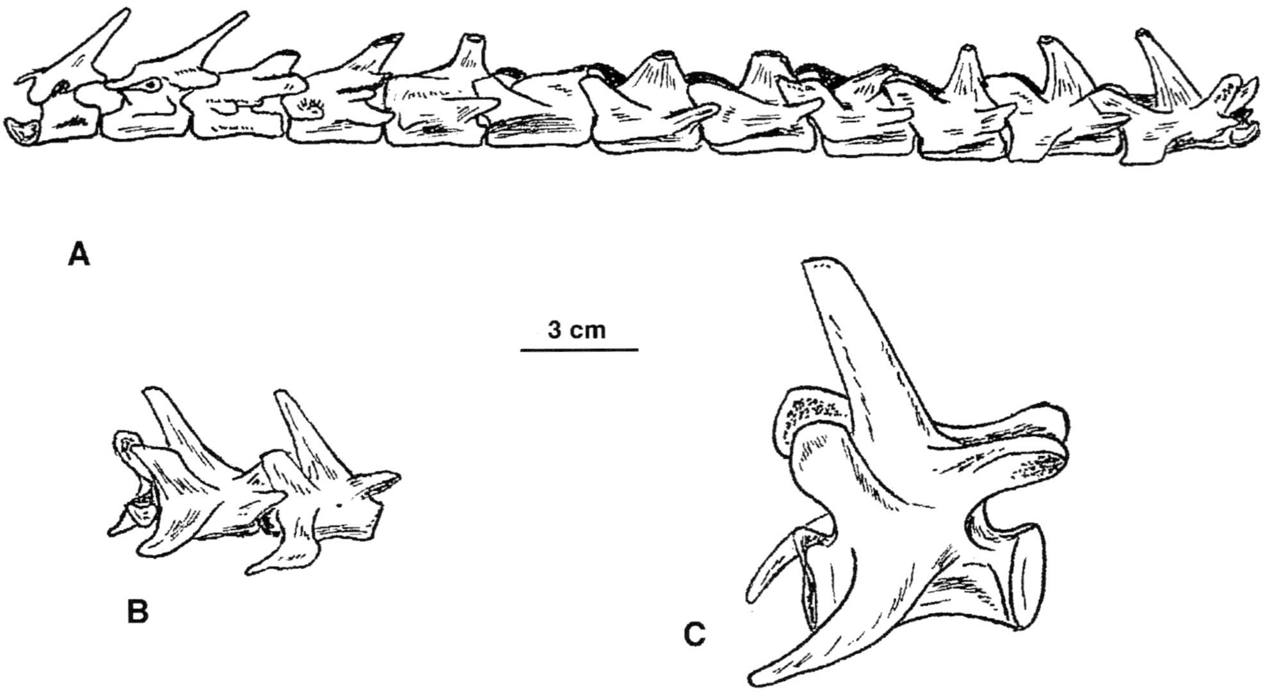

Figure 2 A, Thoracolumbar area of *Pucadelphys andinus* (YPFB Pal 6106), from T?6 to L4; B, Lateral view of L5–L6 of *Pucadelphys andinus* (YPFB Pal 6106); C, Lateral view L5 of *Mayulestes ferox* (MHNC 1249).

strong tractions of the erector spinae, the common epaxial muscular mass which fibres insert on the tip of these processes. The development of the dorsal musculature is related to the powerful extension of the vertebral column during the stroke. It adds propulsive force to that generated by the hindquarter. This common mass does not exist as such in living didelphids, in which the three muscular groups of the back (longissimus dorsi, ilio-costalis and transverso-spinalis) generally remain unfused. In the most agile terrestrial didelphid (*Metachirus*), an aponeurosis runs from the most powerful muscular group of the back (the longissimus dorsi), and inserts on the apex of the last lumbar neural processes, which are anteriorly inclined: in this genus, the anticlinal vertebra is L3. In contrast, in relatively slow climbers (e.g. *Caluromys*, *Phalanger*, *Cyclopes*, *Nycticebus*), the neural processes of the lumbars are vertical or inclined posteriorly, anteroposteriorly long, and low until the last lumbar or first sacral, which generally represents the anticlinal vertebra. In the very agile *Dasyurus*, the anticlinal vertebra is T10, and the neural processes of the lumbar vertebrae are very high (proximodistally) and short (anteroposteriorly). Therefore, an anterior position of the anticlinal vertebra is related to an increased power of the epaxial musculature, due to the migration of the muscular insertions on the apex of the neural processes, their elevation increasing their lever arm.

The last lumbar vertebrae of *Pucadelphys*, *Andinodelphys*, and *Mayulestes* resemble each other in the morphology of their neural processes. They are proximodistally long, anteroposteriorly short, and anteriorly oriented (Fig. 2). They differ from the condition observed in most living didelphids, in which the processes are proximodistally short but anteroposteriorly long. In *Mayulestes*, given the pronounced morphology of the neural process of L5 and L2 (Muizon 1998), it is likely that the anticlinal vertebra was not L4 but more probably L3. This position is similar to that in *Metachirus*, but the L5 process of *Mayulestes* is longer, slenderer, and more inclined anteriorly than in *Metachirus*. In *Pucadelphys* and *Andinodelphys*, the anticlinal vertebra is L2 (probably L2 in the latter). In these genera the neural processes of the postanticlinal lumbars are proximodistally long and even more anteriorly inclined than in *Mayulestes*. The anteroposterior shortness of the neural processes in fossil forms allows an increased flexibility of the lumbar region, an important space being available for interspinous muscular and ligamentous fibres contrary to living didelphids. Moreover, the length of the processes is related to a more powerful back musculature. Therefore, the morphology of the neural processes and the position of the anticlinal vertebra in the Tiupampa didelphimorphs are indicative of a clearly more agile way of locomotion than in the living didelphids (except *Metachirus*). The three of them were probably capable of leaping or bounding and relatively fast

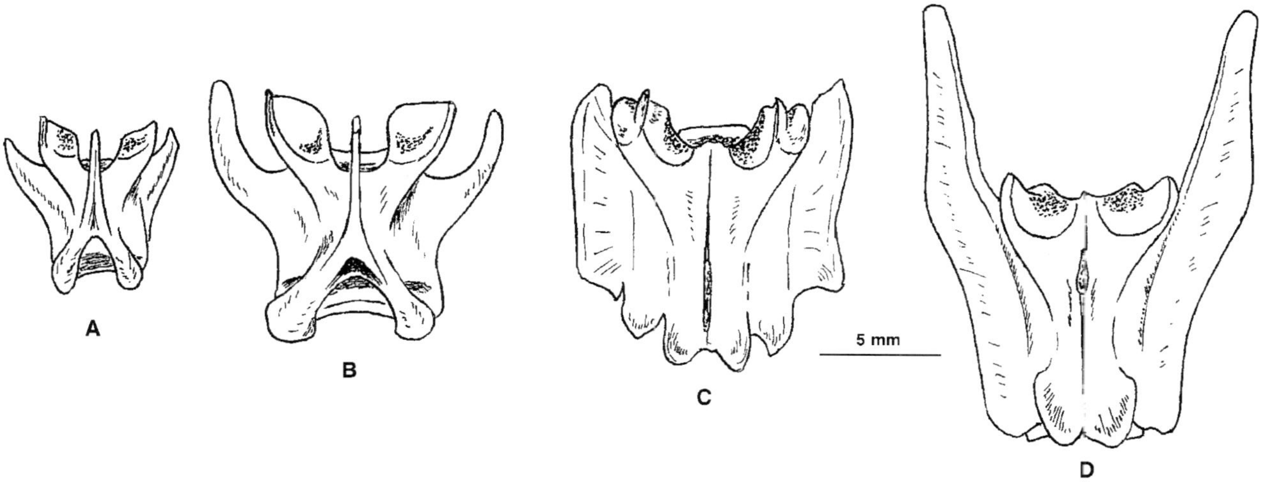

Figure 3 Dorsal view of L5, emphasising the morphological variability of the transverse processes. A, *Pucadelphys andinus* (YPFB Pal 6106); B, *Mayulestes ferox* (MHNC 1249); C, *Caluromys philander*; D, *Metachirus nudicaudatus*.

running: they were probably more similar in this way to living dasyurids than living didelphids.

Transverse processes of the lumbar vertebrae: The transverse processes of both fossils are anteroventrally oriented and slenderer than in living didelphids (Figs 2, 3). This morphology indicates increased lateral flexibility of the vertebral column when the quadrati lumborum act simultaneously, as an increased ventriflexion when the muscular spindles act jointly. The main characteristic of the transverse processes of *Mayulestes* is their strong ventral curvature. The condition of *Andinodelphys* is intermediate between *Pucadelphys* and *Mayulestes* but more resembles the former than the latter. Ventrally and cranially deflected transverse processes provide ample space dorsally for enlargement of epaxial extensor muscles, thus allowing powerful spinal flexions. This condition resembles those of other quadrupedal mammals that use spinal flexion to increase stride length (Shapiro 1995). A condition of the transverse processes approaching that of the Tiupampa marsupials is found in *Dasyurus*, which is quite an agile predator. In contrast, their ventral curvature is not found in didelphids. In *Metachirus*, the transverse processes of the lumbar vertebrae strongly project anteriorly (Fig. 3) and ventrally. This condition indicates powerful sagittal flexion/extension, increasing the mechanical advantage of the quadratus lumborum to flex the vertebral axis. In contrast, horizontal bendings are limited because of the tight imbrication of the anterolaterally oriented transverse processes. Among living didelphids, *Metachirus* is unique because of its locomotory habits. This terrestrial didelphid practises a fast leaping run, which requires more sagittal than horizontal flexions of the column. In agile living taxa (e.g. *Dasyurus* and *Saimiri* (Curtis 1995)) the transverse processes have limited anterior extension but are strongly oriented ventrally. This condition is probably related to an increased lateral flexibility of the column, which could allow easy and fast changes of direction.

In most didelphids and *Pucadelphys*, anapophyses are present on the six lumbar vertebrae although they are usually smaller on L6. In some small and relatively agile didelphids (e.g. *Micoureus*, *Thylamys*), anapophyses are absent on the last lumbars. They are lacking on L5 and L6 of *Andinodelphys* and L5 of *Mayulestes* (L6 of *Mayulestes* is unknown but anapophyses were very probably also lacking on this vertebra). Anapophyses are lacking on L5 and L6 of *Antechinomys* and on L6 of *Dasyurus*. Because a prezygapophysis of a lumbar vertebra imbricates between the corresponding postzygapophysis and anapophysis of the preceding vertebra, the presence of well-developed anapophyses increases the rigidity and reduces the ability of horizontal movements of the lumbar region of the vertebral column. The condition of *Andinodelphys* and *Mayulestes* indicates greater mobility of the posterior lumbar region of the vertebral column than in most living didelphids and *Pucadelphys*. Nevertheless, a potential gradient of vertebral mobility between the three genera is difficult to evaluate. In all cases, the mobility is clearly reduced at the thoraco-lumbar transition, where the anapophyses are well-developed, whereas they are lacking at the level of the last lumbars in *Mayulestes* and *Andinodelphys*. *Pucadelphys* and *Dasyurus* bear anapophyses on the last lumbars; however, their small size probably does not prevent relatively good mobility of this part of the vertebral column.

Morphology and number of the caudal vertebrae. The number of caudal vertebrae and length of the tail are unknown in *Mayulestes*, in which only four caudals are known. In both *Pucadelphys* and *Andinodelphys*, the tail is long and, although incomplete on the best preserved specimens, it is possible to evaluate that it was longer than (at least as long as) the rest of the body (Fig. 4). On

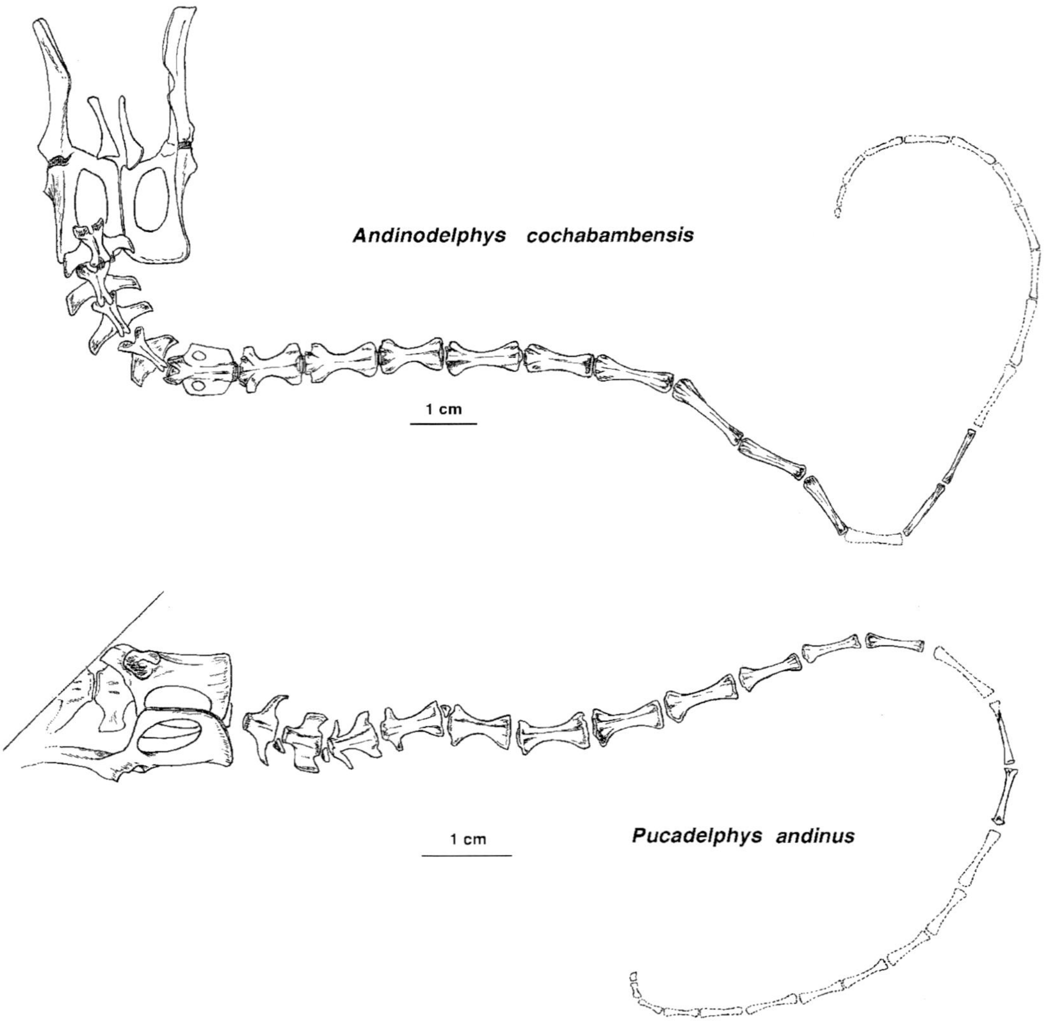

Figure 4 Caudal vertebrae of *Andinodelphys cochabambensis* (MHNC 8308) (top) and *Pucadelphys andinus* (MHNC 8348) (bottom). Dotted lines represent reconstructed vertebrae.

each specimen, 17 vertebrae are preserved and we estimate that 5 to 10 other vertebrae could have been present. The anterior caudal vertebrae of the three genera have a neural process that is reduced on C1–C3 and absent on C4–C5.

The neural processes of the anterior caudals are well-developed in *Caluromys* and *Micoureus*, which are exclusively arboreal didelphids, with a strongly prehensile tail. On these processes inserts the multifidus caudae, an extensor of the tail antagonist to the ventral flexors (the sacro-caudalis and infra-caudalis), very well developed in these forms. The neural processes are small to very small in *Didelphis*, *Philander*, *Metachirus*, and *Thylamys*, which are either generalised or terrestrial forms. In these genera, the tail is long and can have prehensile ability, particularly in *Didelphis* and *Philander*. Therefore, the lack of developed neural processes of the anterior caudals does not preclude the Tiupampa didelphimorphs from having a prehensile tail. The tail of *Metachirus*, although longer than the head and body, is lighter and less muscular than in other genera, especially in its distal part. Its length could be related to the leaping habits of this genus and then to the predominance of the hindquarter (Dor 1937). Nevertheless, the slenderness of the distal caudals indicates that the tail probably cannot bear the weight of the animal as it is able to do in highly arboreal forms. As a matter of fact, such ability is related to the robustness of the distal caudals and to the development of the anterior hemapophyses (on which insert the superficial and deep flexors, essential in prehensile ability). These bony features are much more developed in the arboreal *Caluromys* than in *Metachirus*. Hemapophyses are unknown in the Tiupampa marsupials, except in a new specimen of *Pucadelphys*, not completely prepared yet.

In living didelphids, the anterior part of the tail is rigid and heavy. The anterior caudal vertebrae are probably involved in the lateral movements of the appendix and in the stabilisation of the animal (these movements being in relation to the movements of the pelvis and of the posterior lumbar part of the vertebral axis; Pridmore 1992) rather than in prehensility. The anterior caudal vertebrae are moved by abductors (abductor caudae dorsalis and ischio-caudalis), which insert on their transverse processes, while the tendons of the long flexors/extensors involved in the prehensility insert on the posterior part of the tail. The transverse processes of the anterior caudals are strongly developed laterally and anteroposteriorly in living didelphids, whereas they are slenderer and very posteriorly oriented in mammals which possess a more flexible tail (*Trichosurus*, *Felis*, *Cryptoprocta*, *Bassariscus*). In these latter genera, the posterior orientation is probably due to the tractions of the intertransversarii, deep muscles of the tail involved in the lateral flexions. In the three fossil genera, the transverse processes of C1 protrude laterally. C2–5 are oriented posteriorly in *Pucadelphys* and *Andinodelphys* (Fig. 4), and this suggests a more flexible tail than in living didelphids. The condition of *Mayulestes* is unknown. In the three fossil genera, the strong development of the transverse processes of the posterior caudal vertebrae (i. e. posterior to C5) and the large space between the anterior and posterior transverse processes could be an indication of well-developed intertransversarii muscles.

The prehensile capacities of the tails of the fossils are difficult to evaluate. If the tail of the Tiupampa didelphimorphs was actually prehensile, the slenderness of the very posterior caudal vertebrae (unknown in *Mayulestes*) would indicate a relative weakness as is observed in *Metachirus*. Nevertheless, the tail of the three genera were probably at least partially prehensile as in most living marsupials with climbing capacities (i.e. able to secure a grasp and to grip something, without implying that they were able to support the total body weight of the animal). Although it is difficult to establish a gradient of prehensility, it is likely that they were not as prehensile as in the exclusively arboreal living didelphid, *Caluromys*.

Forelimb

Shoulder joint. On the scapula, the posterodorsal angle morphology, the position of the acromion, the supraglenoid tuberosity, and the orientation and excavation of the glenoid cavity are important elements in understanding the functional anatomy of the shoulder. On the humerus will be considered the morphology and orientation of the head and the tubercles.

The posterodorsal angle of the scapula is extended posterodorsally in arboreal mammals and generally in mammals which present a great mobility of the shoulder and need powerful retractions of the humerus (e.g. fossorial and flying forms). This extension increases the lever arm of the teres major, a powerful adductor and retractor of the limb (Maynard Smith and Savage 1956). Among didelphids, the posterodorsal angle of the scapula of *Caluromys*, the most arboreal didelphid, is more extended than in any other living didelphids. The condition of *Mayulestes* is similar to that of *Caluromys*. In *Andinodelphys* and *Pucadelphys*, this angle is less extended and the infraspinous fossa is smaller than in *Mayulestes* and *Caluromys*, resembling the condition of *Monodelphis* and contrary to fully arboreal forms (Roberts 1974).

The acromion of the scapula of *Caluromys* is large and its ventral edge projects distally, with the result that in medial view, the ventral border of the acromion is widely visible below the glenoid cavity. The well-developed hamatus process extends anteriorly above the supraglenoid process and it is recurved medially, attached to the clavicle by a ligament. In terrestrial didelphids, such as *Metachirus*, *Monodelphis*, and *Lutreolina*, the acromion is smaller and does not project ventrally (with the exception of the processus hamatus, the apex of which is ventral to the glenoid cavity in *Monodelphis*). The hamatus process is slenderer and its apex is located posterior to the supraglenoid process. Its attachment to the clavicle is made through a cartilage, which is still present even in relatively old individuals. In genera with eclectic locomotory habits (*Didelphis*, *Philander*, *Marmosa*) the condition of the acromion is intermediate. A condition similar to that of *Caluromys* is observed in all arboreal mammals. An anteriorly and ventrally projected acromion and a medially recurved apex of the hamatus process (i. e. covering the shoulder joint laterally) increases the stability of the articulation (by the means of a tight muscular wrapping) and, therefore, as stated by Muizon (1998), is related to arboreality. Furthermore, the posterodorsal extension of the posterodorsal angle of the scapula and the anterior position of the acromion could emphasise the action of a scapulothoracic couple of muscles (Fig. 5) which rotate the scapula anticlockwise, increasing the range of extension (and abduction) of the forelimb when extending anteriorly in order to catch any kind of climbing substrate. The condition of the acromion of *Mayulestes* is similar to that observed in *Caluromys* and indicates at least partly arboreal habits. A condition approaching that of *Mayulestes* is also observed in *Andinodelphys* and *Pucadelphys*, although less pronounced. In these genera the strong posterior inclination of the spine does not allow the apex of the acromion to overhang the supraglenoid process. The condition of these genera resembles more those of *Monodelphis* and *Dasyurus*.

The shape of the humeral head plays an important role in the mobility of the shoulder. In *Caluromys*, the head of the humerus is approximately circular in proximal view, indicating good mobility in every direction. In contrast, the humeral head of

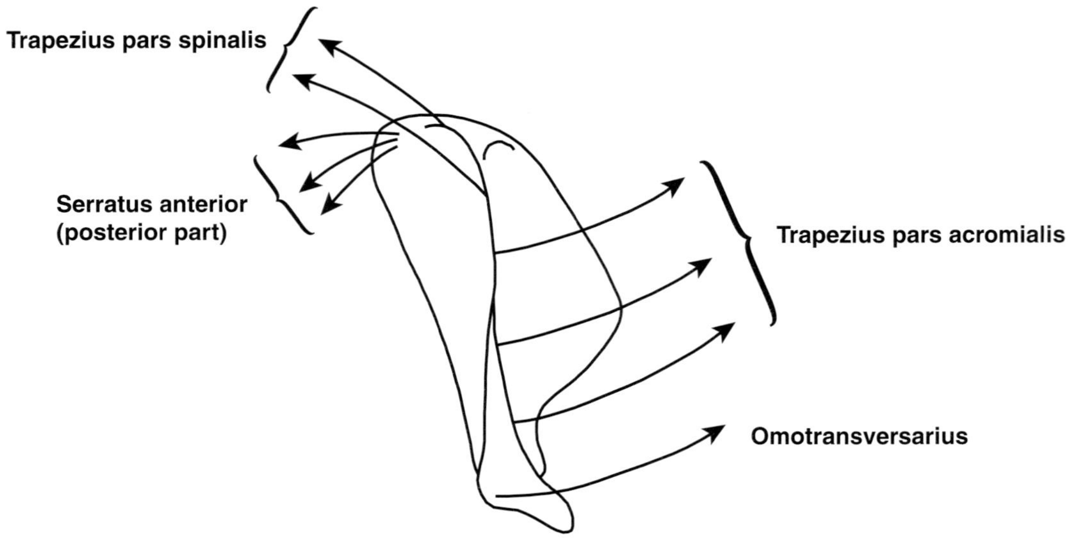

Figure 5 Scapulothoracic muscular couple involved in the rotation of the scapula in *Mayulestes ferox* (see Muizon 1998). The anterior unit is made of the omotransversarius and the pars acromialis of the trapezius; the posterior unit is made of the pars spinalis of the trapezius and the posterior part of the serratus anterior.

Metachirus and *Lutreolina* is distinctly compressed transversely, related to a preference for anteroposterior movements more compatible with their terrestrial habits. In *Monodelphis* (a terrestrial taxon, but which can climb well) the head of the humerus is hemispherical and circular in proximal view. In *Micoureus* (a strictly arboreal taxon) and *Didelphis* (eclectic genus) the head is slightly compressed transversely, whereas in *Marmosa* and *Thylamys* (eclectic genera) the head is almost hemispherical. Among Australian marsupials, a transversely compressed humeral head is present in *Dasyurus* (mainly terrestrial) and *Perameles* (terrestrial and saltatorial), while a circular head is present in *Phascogale* and *Phascolarctos* (arboreal). Therefore, it seems that a hemispherical head of the humerus is related to an increased range of movements in a multidimentional space and a transversely compressed head is related to more sagittal movements and thus to more terrestrial habits. However, this condition is especially difficult to evaluate in the case of eclectic genera. In the fossil genera, the evaluation of this condition is difficult because of the frequent deformation of the specimens. The head of the left (the right is slightly distorted; Muizon 1998) humerus of *Mayulestes* is relatively circular, but more proximally oriented than in living didelphids. Many humerus specimens of *Pucadelphys* are distorted; however, the head of the humerus is also relatively circular in distal view in the two well-preserved specimens (left humeri of YPFB Pal 6106 and MHNC 8348). In *Andinodelphys* one almost undistorted incompletely prepared specimen (MHNC 8308) presents a relatively hemispherical humeral head. Therefore, the shape of the humeral head of *Mayulestes, Andinodelphys*, and *Pucadelphys* is apparently compatible with arboreal habits, although, as noted above, we are conscious that this character must be considered very cautiously.

The elevation of the greater and lesser tubercles is also related to the mobility of the shoulder. A great mobility of the scapulohumeral joint as observed in arboreal taxa (e.g. *Caluromys, Micoureus*) is concomitant with the presence of low tubercles. However, the tubercles of eclectic and terrestrial didelphids are also low and present little difference from those of *Caluromys* and *Micoureus*. The differences observed take place in the lateral protrusion of the tubercles (especially the lower), in relation to the length of the lever arm of the muscles of the 'rotator cuff'. In *Mayulestes*, *Pucadelphys* and *Andinodelphys*, the tubercles are slightly more salient than in the living didelphids, and the greater tubercle is longer anteroposteriorly, approaching the condition observed in arboreal carnivores (e.g. *Cryptoprocta, Nandinia*). This may increase the lever arm of the supraspinatus, a muscle involved in abduction and extension of the arm during the recovery phase of movements; it also stabilises the shoulder joint towards compressional forces, and plays a role in the absorption of kinetic energy generated by the animal in motion (Roberts 1974). Moreover, the lesser tubercle of the fossils also protrudes greatly, increasing the lever arm of the subscapularis, the antagonist muscle. This morphology of both tubercles approaches the condition observed in Australian agile marsupials, like *Dasyurus*, although in this genus the tubercles are clearly more developed than in the Tiupampa didelphimorphs. Therefore, this morphology of the tubercles of *Pucadelphys*,

Andinodelphys, and *Mayulestes*, probably indicates a greater agility than in the living didelphids.

Elbow joint. The elbow joint involves several features of the distal extremity of the humerus (epicondyles, capitulum, trochlea), of the proximal extremity of the ulna (olecranon, flexors fossa, opening of the trochlear notch), and of the morphology of the head of the radius.

The medial epicondyle receives the origin of flexors of the hand and fingers. This epicondyle is very strong and deeply projects medially in arboreal didelphid genera, emphasising the role of these flexors in climbing (for grasping branches). Moreover, the epicondyle of these genera is generally elongated proximodistally (in medial view) and its proximal extremity is tilted posteriorly is such a way that the bridge that overhangs the entepicondylar foramen is distinctly bent posteriorly. This morphology emphasises the torque exerted by the flexors and the power of their tractions occurring during the propulsive phase, when the digits have to encircle the support firmly (Jenkins 1973). It is noteworthy that most (if not all) the epicondyle characters related to arboreal displacements in didelphids are also found in arboreal carnivores (e.g. *Potos, Cryptoprocta, Nandinia*) and in squirrels, especially in the case of the posterior tilting of the epicondyle and bending of the epicondylar bridge. In contrast, the medial epicondyle of the terrestrial *Metachirus* is approximately circular in medial view, does not protrude medially, and the entepicondylar bridge, although thick, is straight. In other terrestrial taxa (but which climb well) such as *Monodelphis* and *Lutreolina*, this epicondyle protrudes medially and the distal extremity of the humerus is much more asymmetrical than in *Metachirus*. In eclectic taxa, the medial epicondyle is not elongated proximodistally and not tilted posteriorly, but the epicondylar bridge is slightly convex. In *Mayulestes*, the medial epicondyle strongly projects medially although to a lesser extent than in *Caluromys*. It is approximately circular in medial view and is not tilted posteriorly. Nevertheless, its medial and distal extension clearly indicates well-developed flexors and thus arboreal abilities. In *Pucadelphys* and *Andinodelphys*, the epicondyle also strongly projects medially; it is tilted posteriorly and the epicondylar ridge is strongly convex posteriorly. In medial view the medial epicondyle of *Andinodelphys* is elongated proximodistally whereas that of *Pucadelphys* is more circular. Like *Mayulestes*, the morphology of the medial epicondyle of both didelphoids indicates clear arboreal affinities.

The major flexor muscle, the flexor digitorum profundus, also originates in a wide fossa on the medial face of the proximal half of the diaphysis of the ulna. This fossa is wide and deep in arboreal forms (*Caluromys, Micoureus*), in relation with the development of this muscle and with its role of grasping, fundamental in their locomotion. This fossa is much shallower in *Metachirus* and *Didelphis*. It is intermediate in *Monodelphis* and *Lutreolina*, which denotes the good potential climbing ability of these taxa. A deep flexor fossa is present in *Mayulestes* and *Andinodelphys*. In *Pucadelphys* it is not so excavated, and the condition approaches that observed in *Didelphis*. Nevertheless, in some specimens of *Pucadelphys* as in *Mayulestes*, the tractions of the flexor digitorum profundus are emphasised by the medial bending of the proximal half of the ulna, a feature not observed in living didelphids.

The terrestrial didelphid *Metachirus* is a very agile cursosaltatorial animal, using more sagittal than transversal movements. The distal articular surface of its humerus presents a relatively deep trochlea, which greatly extends proximally on its anterior and posterior sides. It also extends medially and on the posterior side of the medial epicondyle, it forms a kind of posteromedial capitulum which greatly stabilises the extensions. Anteriorly, the trochlea almost reaches the epicondylar ridge. The posterolateral and anteromedial crests of the trochlea are salient. Anterolaterally the capitulum almost reaches the lateral epicondyle and is bordered by a sharp lateral crest, stabilising the radial head during flexions (Szalay and Dagosto 1980). Posteriorly, a deep and wide olecranon fossa receives the olecranon beak and anteriorly, well-marked radial and coronoid fossae receive respectively the radius and the coronoid process of the ulna. This trochlea articulates with a deep trochlear notch of the ulna, almost semi-circular in lateral view. Therefore, this articulation is well stabilised in all stances. More arboreal mammals also need for a stabilised joint when they are agile and move fast: their distal humerus presents sharp crests, deep fossae and the articular surfaces extend proximally on the diaphysis (*Sciurus, Cryptoprocta, Nandinia*). Nevertheless, the trochlea never extends posteromedially as in *Metachirus*. In this genus, this extension articulates with a robust olecranon beak bearing a concave articular surface. The increased stability provided by this morphology could be related to the need of well-stabilised stances to absorb the excess of energy of impacts at the end of leaps. In contrast, the elbow joint of the arboreal didelphids and especially of *Caluromys* (a relatively slow climber) is very mobile but does not need to be greatly stabilised (especially during extensions, which are mainly used to catch new supports). In such cases, the trochlea is less extended proximally, the crests are low and the olecranon and radial fossae are shallow. Consistently, the trochlear notch of the ulna is shallow, widely open, and the beak of the olecranon is small. Therefore, the morphology of the distal articulation of the humerus is indicative of the stresses exerted on the elbow. In the three Tiupampa didelphimorphs, the trochlear surface extends proximally on the posterior face of the diaphysis of the humerus more than on the anterior; the trochlear crests are relatively elevated and the olecranon and radial fossae are reasonably well developed (although less than in *Metachirus*, the most terrestrial didelphid). Therefore, these genera had a well-stabilised articulation,

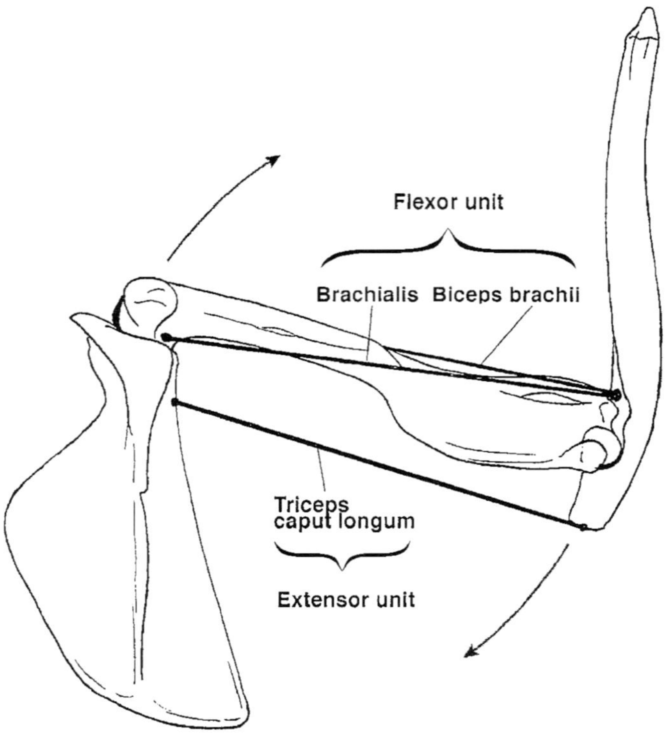

Figure 6 Forelimb muscular couple involved in the lifting of the body weight against gravity in *Mayulestes ferox*, once the manus has secured a new purchase. The flexor unit is made of the Mm. brachialis and biceps; the extensor unit is made of the caput longum of the triceps. The combined action of these muscles exerted upon the proximal part of the ulna explains the posterior convexity of the diaphysis.

which is an indication of agility. In other respects, the widely open and shallow trochlear notch is an arboreal pattern in the fossil taxa with a gradient of increasing arboreality from *Pucadelphys* to *Mayulestes*. The general morphology of the elbow of the fossil genera is consistent with faster movements than practised by the most arboreal didelphids, and also with some degree of arboreality.

The olecranon of the ulna represents the lever arm of the triceps brachii, the major extensor of the elbow. Powerful extensions of the elbow are involved in terrestrial locomotion, especially in cursorial and saltatorial mammals during the propulsive phase. A long olecranon will increase the lever arm of the triceps and provide power to the movement. Conversely, a short olecranon reduces the power of the movement but increases its rapidity. Arboreal didelphids (e.g. *Caluromys*, *Micoureus*) are not extremely agile and their extension of the elbow does not require power but rapidity in order to grasp the nearest support and to secure their position as quickly as possible. Arboreal didelphids generally have a relatively short olecranon. The olecranon of the cursosaltatorial *Metachirus* or of the other terrestrial genus *Lutreolina* is relatively short but is strongly thickened anteroposteriorly and is proportionally much larger than in the other didelphids. Therefore, these didelphids have a fast extension of the elbow and compensate for the lack of power in increasing the volume of the triceps (especially the caput longum). As a matter of fact, the triceps of *Metachirus* is stronger than in any other didelphid. *Monodelphis* and *Didelphis* have a longer olecranon than the strictly arboreal taxa. Therefore, the long olecranon observed in *Mayulestes*, *Pucadelphys*, and *Andinodelphys* would be more related to terrestrial than arboreal habits. However, a long olecranon is not incompatible with arboreal life in the case of a relatively agile animal, which would perform a generalised quadrupedal arborealism as is observed in most tree shrews, squirrels or some viverrids.

The proximal extremity of the olecranon in *Caluromys*, *Micoureus* and *Marmosa*, presents a slight but distinct posterior slope (in lateral view). In contrast, the proximal extremity of the olecranon of the most terrestrial taxa (*Didelphis*, *Metachirus*, *Lutreolina*) is approximately perpendicular to the axis of the bone. A posterior slope of the olecranon is also present in several other arboreal or semi-arboreal eutherians (e.g. *Sciurus*, *Tupaia minor*, *Potos flavus*, *Nasua*). This condition could be related in arboreal forms to the tractions of the triceps acting from semi-flexed stances (Walker 1974), usually performed in these forms in order to increase their stability. The triceps helps to resist surflexions and exerts controlled movements by a braking action.

Therefore, a posterior slope of the apex of the olecranon, which is clearly present in *Mayulestes*, *Andinodelphys*, and *Pucadelphys*, could be related to arboreality. However, this character is perhaps not so constant since such a slope is present in *Monodelphis*, a mostly terrestrial didelphid (but which climbs well), whereas it is absent in *Cryproprocta* and *Nandinia* which are semi-arboreal to fully arboreal carnivores.

More than the absolute length of the olecranon, the proximal convexity of the posterior border of the ulna can help to discriminate arboreal *vs* terrestrial habits. As a matter of fact, the diaphysis of the ulna is distinctly bent anteriorly in the most arboreal didelphids (*Caluromys*, *Micoureus*) but also in *Marmosa*, contrary to what can be observed in *Metachirus*, *Lutreolina*, and *Didelphis*. It is noteworthy that the ulna of *Monodelphis* is more anteriorly bent than that of *Metachirus*, *Lutreolina* or even *Didelphis*, which suggests greater climbing ability in *Monodelphis*. The shape of the proximal ulnar diaphysis could be explained as follows: when climbing, the caput longum of the triceps, which originates on the scapula and inserts on the proximal extremity of the olecranon (therefore, posterior to the humeroulnar articulation) is especially involved in the retraction of the humerus on the scapula. At the same time, the biceps and brachialis are involved in the flexion of the elbow (Fig. 6). The biceps originates on the supraglenoid tuberosity and coracoid process of the scapula. It inserts on the bicipital tuberosity of the radius and on the ulna, distal to the coronoid process. The combined action of both muscles folds the forelimb and pulls the body up. Therefore, the anterior curvature of the ulna can be explained by the fact that both muscles are exerting at the same time a force in a vertical direction respectively posterior and anterior to the humeroulnar articulation, when the hand is firmly grasping the support. The anteriorly bent ulnae of *Mayulestes*, *Andinodelphys* and *Pucadelphys* are then very probably related to arboreal ability. Their long olecranon indicates strong tractions of the triceps, and the massiveness and anteroventral extension of the supraglenoid tuberosity (in *Pucadelphys* and *Andinodelphys*) is related to the tractions of the biceps.

On the anterior face of the ulna (in anterodistal view), the angle between the humeral and radial facets is wide open (approximately 120°), the crest between the two facets is hardly present, and the radial facet is only slightly concave in the arboreal *Caluromys* and *Micoureus*. In *Didelphis* and *Monodelphis*, it approaches 90°, the crest is distinct, and the radial facet is deeper. In *Metachirus* and *Lutreolina*, the angle between the two facets is less than 90°, the crest is distinct, and the radial facet is very concave. The condition of *Caluromys* is also found in most arboreal mammals (when compared to their closest terrestrial relatives) (e.g. *Sciurus*, *Potos*, *Tupaia*) and indicates a relatively unstable articulation. However, given the extreme agility of squirrels, it does not seem to be restricted to relatively slow animals. The arboreal condition is also present in *Mayulestes*. The condition of *Pucadelphys* and *Andinodelphys* is not so specialised and more resembles that of *Didelphis* and *Monodelphis*.

Important movements in arboreal displacements are the pronation and supination of the forearm, which increase the ability to grasp supports in any direction. These movements are induced by the rotation of the radial head against the capitulum of the humerus and the radial notch of the ulna. The regularly concave radial articular facet and almost circular capitulum of *Caluromys* allow the most efficient pronation/supination movements. The radial head of *Metachirus* is relatively circular but the radiohumeral articulation is shallower than in *Caluromys* and is concavoconvex, which greatly reduces pronation/supination movements. In *Didelphis* (an eclectic genus), the radial head is slightly oval-shaped and the pronation/supination movements are probably less efficient than in *Caluromys* (although very effective, as noted by Coues, 1869). However the capitulum, which receives the radial head, is circular and the portion of the humeral facet of the radius, which articulates with the capitulum, is also circular. The oval shape of the radial head is due to a lateral extension of the radiohumeral articular surface, which also involves the tail of the capitulum (the tail of the capitulum is the lateral edge of the articular facet). The result of this condition is a better stabilisation of the flexion/extension of the elbow. The *Didelphis* condition appears more suitable for a terrestrial-scansorial mammal, while that of *Caluromys* is clearly that of a highly arboreal animal. The condition of the Tiupampa didelphimorphs is difficult to evaluate because the radial head, as in the case of the humeral head, is subject to distorsion. For instance, some specimens of *Pucadelphys* and *Andinodelphys* have an oval-shaped radial head, while it is circular in others. Because in both genera the radii with circular heads are apparently less distorted than the others, we believe that the radial head was nearly circular in these genera. This indicates relatively efficient pronation/supination movements, as in arboreal or semi-arboreal didelphid taxa. In the case of *Mayulestes*, the reduced anteroposterior length of the radial head of the single radius known is emphasised by the fact that the anterior edge of the epiphysis was slightly worn prior to fossilisation. Furthermore, it may also have been distorted during fossilisation. Assuming that no distortion occurred, its oval-shaped morphology would be indicative of reduced pronation/supination ability. However, some semi-arboreal mammals have a relatively oval-shaped radial head (e.g. *Cryproprocta ferox*, *Paguma larvata*, *Manis tetradactyla*, *M. tricuspis*, *Lemur fulvus*; MacLeod and Rose 1993), which would indicate that even mammals with limited pronation/supination movements could be agile climbers. In fact, it is noteworthy that not only the global shape of the head, but especially the concavity of the articular facet and the convexity of the radioulnar contact must be taken into account to evaluate the rotative capacities of the forearm.

Hindlimb

Innominate. The ilia of Recent didelphids are usually elongated, narrow and straight, except in the most agile form, *Metachirus*. As mentioned by Muizon (1998: 109-110), this condition is likely to be plesiomorphic since it is present in most primitive non-therian mammals and marsupials. The ilium of the Tiupampa didelphimorphs is shorter and wider than that of the living didelphids, which apparently represents a derived condition. Moreover, one of the most characteristic features of the ilia of *Mayulestes*, *Andinodelphys*, and *Pucadelphys* is their extroversion, which is clearly more pronounced in the latter than in the two former (Fig. 7). This condition is absent in most didelphids but is present in many Recent therians. Among marsupials, this feature is present in the agile *Dasyurus* and *Metachirus*, as also in the highly specialised saltatorial *Perameles* and *Antechinomys*. An extroverted ilium corresponds to a larger insertion area for the longissimus dorsi, the major extensor of the back which originates on the upper part of the sacral face of the ilium and on the iliac crest. Furthermore, on the lateral side of the iliac wing is the origin of the gluteus medius, one of the major extensors of the hip which inserts on the great trochanter of the femur. The short lever arm of the glutei favours fast extensions against powerful ones (Maynard Smith and Savage 1956). In most didelphids (except *Metachirus*) the gluteal fossa, located on the dorsal half of the lateral side of the ilium, is elongated and of similar size as the ventral iliac fossa. In agile forms like *Metachirus*, *Perameles*, but also *Tupaia*, the origin of the gluteus medius is much greater than that of the iliacus, and this development corresponds to a greater trochanter higher than the femoral head. *Mayulestes*, *Pucadelphys* and *Andinodelphys* have a fossa for the gluteus medius smaller than in *Metachirus* and *Perameles* but larger than in the other didelphids. It generally corresponds to a greater trochanter higher than the head, although not so well developed in all specimens. The locomotion of *Metachirus* and *Perameles* is an extremely agile leaping run, especially for *Perameles*. The action of the longissimus dorsi and gluteus medius are essential during the propulsive effort in leaping, jumping, and running. In non-fossorial mammals, the great development of these extensors is, at the least, indicative of agility. Therefore, the three didelphimorphs from Tiupampa, the ilium of which indicates a well-developed and powerful gluteus medius, were certainly more agile than the living didelphids (with the possible exception of *Metachirus*). The morphology of their ilium also indicates that they were capable of leaping or bounding.

Another important element of the innominate in coxofemoral movements is the development of the ischiatic spine because it bears the origin of three powerful extensors of the hip: the biceps femoris, the semimembranosus, and the semitendinosus caput ventrale. These muscles insert on the tibial crest itself, or in an aponeurosis covering both sides of the crus. Like the glutei, their role is critical in any kind of leaping, bounding, and running activity: as a matter of fact, the ischio-pubic complex powerfully acts at the beginning of the stroke to overcome the inertia of the limb, whereas the glutei act after, providing speed to the movements (Maynard Smith and Savage 1956). Therefore, a well-developed and everted ischiatic spine is indicative of agility and rather fast locomotion, and is generally concomitant with an everted iliac wing. This relation is present in the agile *Metachirus*, *Perameles*, and *Antechinomys*, as well as in *Pucadelphys*. However, the ischiatic spine is less everted in *Andinodelphys* and *Dasyurus* (it is more robust in this latter genus). Furthermore, *Mayulestes* has an everted iliac wing but a reduced and non-everted ischiatic spine (Fig. 7). This condition indicates well-developed glutei but a weakly developed ischio-pubic group, possibly emphasising speed and not strength of the movements (i.e., the major part of the work would be done in accelerating the leg). We found no clear explanation for this particular morphology comparing with the locomotor behaviour of living marsupial models.

Hip joint. The mobility of the articulation of the hip is also revealed by the morphology of the acetabulum. In *Caluromys*, the acetabulum is shallower and more open than in the other didelphids, and its dorsal border is markedly concave. This condition allows a greater amplitude of movements of the femur, in flexion/extension but especially in abduction, which is consistent with the arboreal habits of this genus. Concomitantly, this morphology reduces the stability of the joint, which coincides with the rather slow climbing habits of *Caluromys*. Nevertheless, it is noteworthy that a similar condition is found in *Trichosurus*, an arboreal but faster form than *Caluromys*. *Trichosurus* practises walk and half-bounds, this latter gait especially to jump from branch to branch, to move along ground with low cover, and to move up strongly inclined surfaces (Goldfinch and Molnar 1978); such practice is highlighted by the mobility of the lumbar area and the morphology of the ischiatic tuberosities, but not at the level of the hip joint which presents abductive possibility needed for climbing. In contrast, the extremely agile arboreal eutherians *Tupaia gracilis*, and *Sciurus igniventris* have a relatively deep acetabulum which increases the stability of the joint. In *Metachirus* the acetabulum is clearly deeper than in *Caluromys* and its dorsal edge is straight (Fig. 7). The coxofemoral articulation of *Metachirus* is thus well stabilised, which is related to the agile leaping run characterising the locomotor behaviour of this genus. The acetabulum of *Mayulestes* is wide-open and its dorsal edge is very concave, which indicates a great mobility of the hip. This could be a plesiomorphic character, as didelphids and other forms like non-cursorial carnivores are characterised by a relatively shallow acetabulum, and by an abducted femur during the propulsive phase of walking (Jenkins 1971; Jenkins and Camazine 1977). In *Pucadelphys* the acetabulum is relatively open but less than in *Mayulestes*; however, the dorsal edge is strongly concave. No fully prepared innominate of

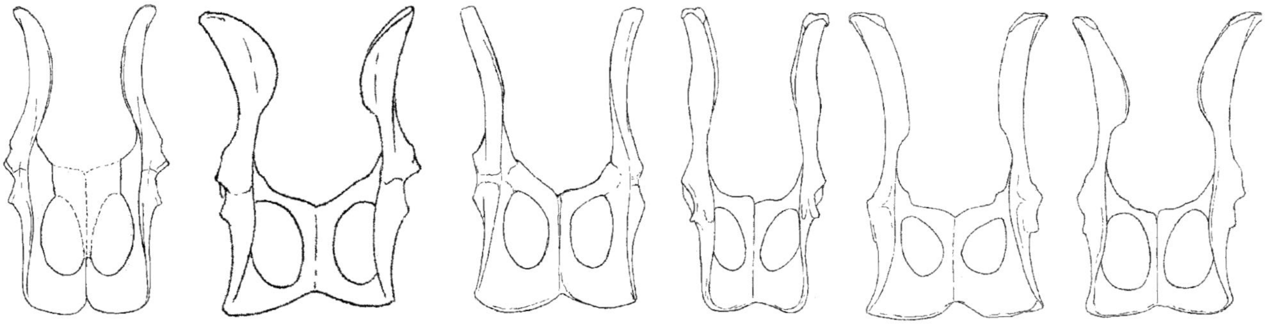

Figure 7 Dorsal view of the innominate in several marsupials: A, *Mayulestes ferox*; B, *Pucadelphys andinus*; C, *Andinodelphys cochabambensis*; D, *Caluromys philander*; E, *Metachirus nudicaudatus*; F, *Dasyurus viverrinus*. Not to scale.

Andinodelphys is available, but one ischiopubis of an immature specimen seems to indicate that the acetabulum was also relatively shallow. Therefore, the coxofemoral articulation of the Tiupampa didelphimorphs appears to have been very mobile, with a gradation from *Pucadelphys* to *Mayulestes*, the former having probably greater leaping ability, as emphasised by the general shape of the innominate. However, the general morphology of the acetabulum indicates a poorly stabilised articulation, when compared with *Metachirus* or agile eutherians. As noted by Jenkins and Camazine (1977), rotation and abduction of the femur, allowed by a relatively open articulation, are not only essential in climbing but also for clambering over uneven, disordered substrates. It is noteworthy that the acetabulum of *Dasyurus viverrinus*, a very agile predator (as are all dasyurids), is very similar to that of *Mayulestes*. If all species of *Dasyurus* are essentially terrestrial (although some are far more so than others) they are agile on a great diversity of substrates and can climb well (Nowak and Paradiso 1983). Moreover, other dasyurids are highly arboreal (*Phascogale*, *Neophascogale*) or semiarboreal (*Antechinus flavipes*). It appears, therefore, that the great mobility of the hip observed in fossil forms does not preclude their agility. Their coxofemoral articulation was probably reinforced by strong ligaments, as suggested in *Mayulestes* by the large fovea capitis on the femoral head (for the insertion of the round ligament), as in *Dasyurus viverrinus*. The round ligament of *Pucadelphys* was probably not as strong as in *Mayulestes*, but the acetabular cavity of this didelphoid is deeper than in the borhyaenoid, although it also has a very concave dorsal border.

The orientation of the acetabulum varies in terrestrial and arboreal taxa. The acetabulum faces laterally in most didelphids (except *Metachirus*), *Sciurus*, *Tupaia gracilis*, *Potos*, *Nandinia*, and generally in highly arboreal forms. It faces ventrolaterally in *Metachirus*, *Dasyurus*, *Antechinomys*, *Perameles* and generally in terrestrial and only occasionally arboreal taxa. This could be related to the reduced capacities of abduction of the femur and to an increased propulsive effort in the latter forms, in which the anterodorsal border of the acetabulum is more robust and prominent laterally. The acetabulum of *Mayulestes*, *Pucadelphys* and apparently *Andinodelphys* (from what can be seen on an incompletely prepared pelvis) faces ventrolaterally but to a lesser extent than in, e.g. *Metachirus*, *Dasyurus viverrinus*, or *Perameles*. Thus the orientation of the acetabulum of the Tiupampa taxa would indicate that they were not strictly arboreal although they could probably have climbed well.

The lesser trochanter of the femur receives the insertion of the iliacus and psoas major. Among didelphids, it is relatively smaller in the terrestrial *Metachirus* and *Lutreolina* than in the other taxa. It is better developed, protrudes more medially, and its distal edge is almost perpendicular to the diaphysis in highly arboreal taxa (*Caluromys* and *Micoureus*); in these forms, the iliacus and psoas major act as flexors but also as external rotators and adductors of the leg. This orientation could also emphasise the rotative movements of the femur during flexion/extension, a component of the movement more important in arboreal than in terrestrial forms (Taylor 1976). A condition similar to that of arboreal didelphids is found in *Sciurus*, *Tupaia gracilis* and *T. minor*. The lesser trochanter of the Tiupampa didelphimorphs is blade-like and is located more posteriorly than medially as observed in some Australian marsupials (*Trichosurus*, *Dasyurus viverrinus*, *D. maculatus*). Its morphology and orientation do not indicate strictly arboreal mammals, but its development could be related to some climbing ability.

Knee joint. The femorotibial articulation is highly relevant to the locomotory habits of mammals. On the femur, the morphology of the distal epiphysis, the femoral trochlea, and the relative proportions of the condyles are especially informative. On the tibia, the morphology of the tibial condyles, the tibial tuberosity, the proximal tibiofibular articulation, and the shape of the tibial diaphysis are of special interest.

The distal epiphysis of the femur is usually anteroposteriorly compressed in slow arboreal mammals, which coincides with a wide and shallow femoral trochlea, with low crests. Among didelphids, the most arboreal taxa (*Caluromys* and *Micoureus*) have the anteroposteriorly shortest distal epiphysis of the femur and the shallowest and widest trochlea. The longest epiphysis and the better developed crests of the trochlea are that of the terrestrial *Metachirus*. *Trichosurus* is similar for this character to *Didelphis* and the agile *Dasyurus* (*D. viverrinus* and *D. maculatus*) resembles *Metachirus*. The morphology of the epiphysis is intermediate in the other didelphid taxa; from the anteroposteriorly shortest to the longest epiphysis: *Marmosa*, *Philander*, *Didelphis*, *Monodelphis*, and *Lutreolina*. This morphological gradient perfectly coincides with the behavioural gradient (arboreal *vs* terrestrial), a relation also observed in other groups, like viverrids (Taylor 1976) and primates (Tardieu 1983). An anteroposteriorly elongated epiphysis, together with posteriorly protruding femoral condyles as can be observed in *Metachirus*, increase the distance between the tendon of the quadriceps femoris and the centre of rotation, which could lengthen the lever arm of this muscle. Moreover, the well-marked crests of the trochlea can better conduct the gliding of the tendon in the groove. And in this terrestrial genus, the role of this extensor of the leg is emphasised by the development of the muscular body (pers. obs. of C. A.). It is noteworthy that extremely agile arboreal mammals (e.g. *Sciurus*, *Tupaia*, lemurs, galagos, indris) more resemble *Metachirus* than the arboreal didelphids in having a narrow, deep trochlea with elevated crests, which conducts the patella (absent in didelphids) in the groove, thus emphasising the role of the quadriceps in the propulsion. Therefore, as expressed by Tardieu (1983), the condition of the distal epiphysis of the femur is more related to the activity of the animal (agile *vs* slow) than to its way of life (arboreal *vs* terrestrial). As a matter of fact, the terrestrial *Metachirus* is also the most agile didelphid and an efficient leaper. The proportions of the distal epiphysis of the femur of *Mayulestes*, *Andinodelphys*, and *Pucadelphys* do not present anteroposterior compression and more resemble that of agile living forms like *Dasyurus* (*D. viverrinus* and *D. maculatus*) and *Metachirus* than that of slower didelphids. The trochlea of *Mayulestes* and *Pucadelphys* is narrower, deeper, and has more elevated crests than in didelphids (although the femoral trochlea of *Andinodelphys* is wider and shallower and has lower crests than those of *Mayulestes* and *Pucadelphys*). Therefore, the general morphology of the distal epiphysis of the femur of the Tiupampa didelphimorphs would be an indication of agility, maybe slightly reduced in *Andinodelphys* relative to the two other genera.

The tibial tuberosity receives the tendon of the quadriceps (two vasti and the rectus femoris), a powerful extensor of the knee. In slow arboreal mammals, the tuberosity is weak and relatively flat, while in terrestrial or in fast arboreal forms (e.g. *Sciurus* or *Tupaia*) it is generally more salient anteriorly, and especially extends distally in a sharp anterior crest. In didelphids there is a clear gradient in the development of the tibial tuberosity from the slow arboreal *Caluromys* and *Micoureus*, to the fast runner-leaper *Metachirus*. In between, the tuberosity varies, from the less salient to the most salient, in *Monodelphis*, *Philander*, *Didelphis* and *Marmosa*. The condition of *Trichosurus* is close to that of *Didelphis*, and that of *Dasyurus* approaches that of *Metachirus* (although it is slightly weaker). Therefore, the size of the tibial tuberosity and especially the sharpening of the anterior crest seems to be related to agility and is probably not a reliable indicator of arboreality *vs* terrestriality. In the Tiupampa didelphimorphs, the tibial tuberosity is usually weak. Therefore, this morphology seems to be indicative of relatively slow movements. This contradicts the observations made on the axial skeleton, on the pelvis and on the distal epiphysis of the femur, which indicated that the Tiupampa didelphimorphs are relatively agile animals. We have no explanation for this contradiction. Nevertheless, it can be noted that the tibial tuberosity and the anterior crest are slightly more developed in *Pucadelphys* than in *Mayulestes*. In the latter, the tibial tuberosity is very wide and the tibial condyles are wider than long, a morphology not observed in any other living marsupial form. The proximal extremity of the tibia bears on its anteromedial side an unnatural ovoid cavity (Muizon 1998), and the anterolateral fossa for the tibialis anterior is reduced to an extent never observed in Recent forms. Because a single tibia of *Mayulestes* is known, a potential distortion during fossilisation can not be excluded.

The femoral condyles of didelphids are strongly asymmetrical, the medial being always distinctly narrower than the lateral. The lateral condyle is the widest in the arboreal taxa (*Caluromys*, *Micoureus*), the narrowest in terrestrial taxa (*Metachirus*, *Monodelphis*, and *Lutreolina*), and intermediate in semiarboreal taxa (*Didelphis*, *Philander*, and *Marmosa*). In *Trichosurus* the lateral condyle is also clearly wider than the medial and approaches the condition of *Didelphis*. This difference is much less pronounced in *Dasyurus*. Moreover, in didelphids, the medial femoral condyle is more convex posteriorly and more extended proximodistally than the lateral, and articulates with a concave medial tibial condyle, whereas the lateral is plane or slightly convex. This general morphology indicates a lateral displacement of the load line, probably in relation to the constant abducted position of the femur during a locomotor cycle as studied by Jenkins (1971) in *Didelphis*. The difference of distal and posterior extension between the femoral condyles, well marked in terrestrial forms, could increase the stabilisation of the joint. In contrast, this difference is very reduced in arboreal taxa like *Caluromys*, which may increase the rotational capacities of the crus with respect to the femur, the articulation being stabilised by the strong collateral ligaments. The difference of the width between femoral condyles may be a plesiomorphic feature (Muizon 1998: 116), related to a non-cursorial pattern and to a very abducted position of the femur. It could also be related in some way to arboreality *vs* terrestriality in marsupials, this difference being emphasised in arboreal taxa. In the Tiupampa didelphimorphs, the medial condyle is narrower than the lateral to an extent intermediate between *Didelphis* and *Dasyurus*, which could indicate a femur slightly less abducted than in didelphids, in relation to a faster locomotion.

Ankle joint. The astragalotibial articulation of didelphids (also observed in phalangerids and petauristids) is very peculiar, permitting in all genera the inversion of the foot, which is a characteristic of most highly arboreal mammals (Jenkins and McClearn 1984). In didelphids, the lateral portion of the distal articular facet of the tibia is helical: the anterior part is more distal than the posterior one, and the whole facet has a crescent form. It coils around the medial portion of the astragalotibial facet, borne by the malleolus, which is located at the anteromedial angle of the epiphysis. Because of the helical nature of the articulation and the obliquity (in distal view) of the great axis of the medial malleolus (relatively to the transverse axis of the femorotibial articulation), a plantar flexion of the foot is accompanied by a combined rotation between tibia and astragalus, in such a way that the plantar surface faces medially to medioventrally (Muizon 1998: 119). In highly arboreal taxa (*Caluromys*, *Trichosurus*), the facet borne by the malleolus is usually reduced, whereas the helical facet is much more developed, increasing the range of rotations. In more terrestrial genera, the helical facet is less developed and the facet of the malleolus protrudes more distally and is more convex: these features stabilise the articulation but reduce the range of rotation. Such a rotation combined with plantar flexion is an adaptation to arboreal grasping, permitted in all positions of the animal (especially during descents head-first). Moreover, the classical peg and socket calcaneocuboid articulation of didelphids also probably has a role in the rotation of the foot, as well as the astragalonavicular articulation: according to Jenkins and McClearn (1984), these movements are more important in the posture of foot reversal than the rotation at the level of the knee, or even the abduction of the hip.

In *Mayulestes*, the lateral part of the distal articular facet of the tibia is not as helical as in didelphids, and is only slightly more distal anteriorly than posteriorly. Moreover, the facet borne by the malleolus is convex and strongly projects distally, as in the most terrestrial didelphids. Therefore, the whole astragalotibial facet does not act as efficiently as the perfectly helical facet of the didelphids. However the medial malleolus is oriented anterolaterally-posteromedially, and the lateral astragalotibial facet is reniform: this morphology indicates that during the extension of the foot, the astragalus can also perform a slight rotation around the malleolus, which tends to orientate the plantar sole medially. Therefore, *Mayulestes* was probably capable of some inversion of the foot but to a lesser extent than in didelphids or phalangerids. The angulation between the two parts of the tibial facet, much more pronounced than in the didelphids, suggests that the two movements (extension and rotation) were probably more independent than in didelphids. Thanks to such movements, *Mayulestes* was probably able to move on uneven, disordered grounds, as suggested by the morphology of its hip. The condition of *Pucadelphys* is similar to that of *Mayulestes*: the lateral astragalotibial facet is incipiently helical and the malleolus is oblique, which indicate some ability of inversion of the foot. Nevertheless, as in *Mayulestes* and terrestrial didelphids, this malleolus strongly projects distally and forms a marked angulation with the lateral part of the facet: this could indicate a better ability to agile terrestrial locomotion than truly arboreal capacity. No distal extremity of the tibia of *Andinodelphys* is available yet. However, the morphology of its astragalus, which presents a well-excavated medial articular surface for the tibial malleolus, approaches the condition observed in *Mayulestes* and *Pucadelphys*.

The calcaneoastragalar articulation provides interesting information on the locomotory habits. As mentioned by Muizon (1998: 120), in the arboreal didelphids, *Caluromys* and *Micoureus*, the ectal facet (in distal view) has a very salient lateral edge and faces more medially than in the terrestrial didelphids. As a consequence, the medial extremity of the posterior ventral articular facet of the astragalus is more salient ventrally in arboreal didelphids than in terrestrial taxa. A medial orientation of the ectal facet produces a medial orientation of the body of the astragalus, which articulates with the tibia. Moreover, the ectal facet is usually larger than the sustentacular one in this 'arboreal pattern', and consequently bears an increased load compared to the other. Therefore, this morphology increases the medial orientation of the plantar face and, then the inversion of the foot, especially during plantar flexion when the astragalus rotates laterally. In *Mayulestes* the orientation of the ectal facet is close to that of *Caluromys* (Fig. 8) and the medial edge of the posterior ventral facet of the astragalus is very salient. In *Pucadelphys*, the orientation of the ectal facet and the ventral extension of the medial edge of the posterior calcaneoastragalar facet is closer to those observed in terrestrial didelphids. The condition of *Andinodelphys* is intermediate between those of *Mayulestes* and *Pucadelphys* (Fig. 8). Summarising, the morphology of these facets indicates an increasing gradient of arboreal ability from *Pucadelphys* to *Mayulestes*.

Foot. The proximal extremity of the fibula is informative on the grasping ability of the foot. It is greatly expanded anteroposteriorly in living didelphids, this character being more pronounced in the arboreal taxa (*Caluromys* and *Micoureus*) than in the terrestrial forms (*Metachirus*, *Monodelphis*, *Lutreolina*). It is also expanded in the highly arboreal Australian marsupials (*Phalanger*, *Phascolarctos*, *Schoinobates*), except *Phascogale* (Fig. 9). On the contrary, it is not expanded in the terrestrial dasyuroids (*Thylacinus*, *Dasyurus*). On the lateral face of the

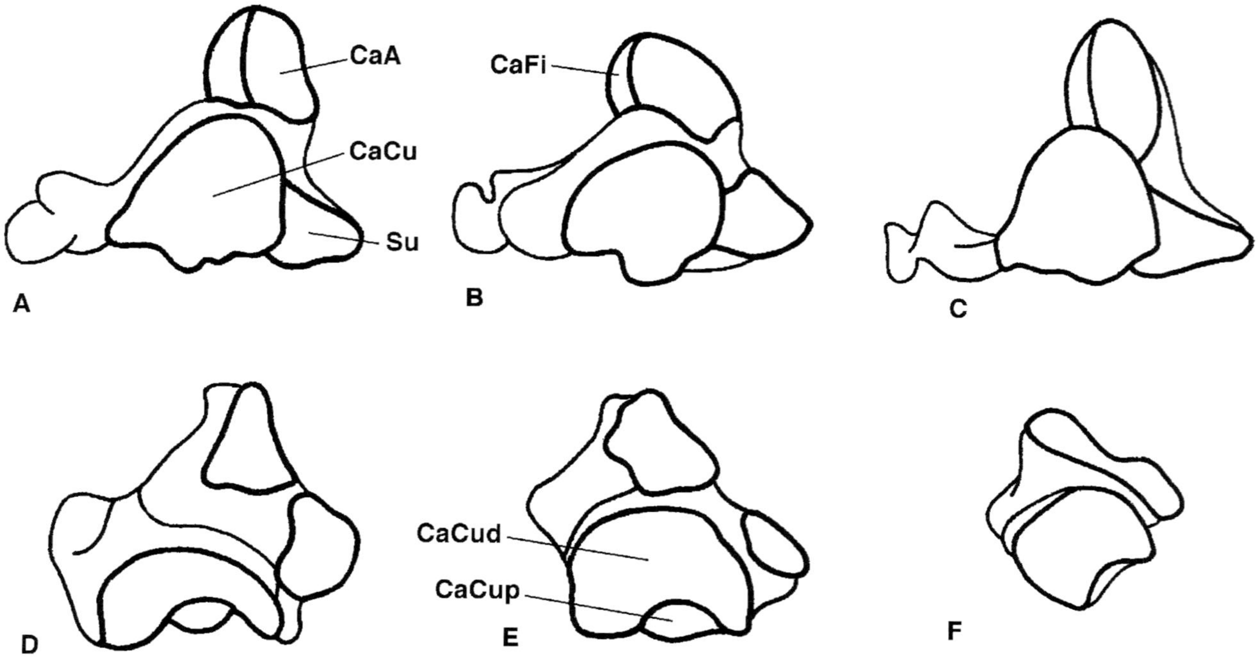

Figure 8 Distal view of the right calcaneum of several marsupials: A, *Mayulestes ferox*; B, *Pucadelphys andinus*; C, *Andinodelphys cochabambensis*; D, *Caluromys philander*; E, *Metachirus nudicaudatus*; F, *Dasyurus viverrinus*. Not to scale. Abbreviations: CaA, calcaneoastragalar facet; CaCu, calcaneocuboid facet; CaCud, dorsal calcaneocuboid facet; CaCup, plantar calcaneocuboid facet; Su, sustentacular facet.

proximal epiphysis of the fibula originates the peroneus longus, the tendon of which inserts on the proximal extremity of the first metatarsal. It is a strong adductor and flexor of MtI, and the development of this muscle is related to the opposability of the hallux. On the medial side of the fibula originates the flexor digitorum fibularis, the major flexor of the digits. Therefore, the development of the area of origin of these muscles on the fibula corresponds to more powerful flexions of the fingers and adduction of the hallux, in relation to efficient grasping as needed in arboreal habits.

In the most terrestrial *Metachirus*, the hallux is still opposable but the head of the fibula is much less expanded and because of the enlargement of the metapodials, the efficiency of the opposability of the hallux is reduced. *Dasyurus viverrinus*, the only species of *Dasyurus* which lacks a first toe (Nowak and Paradiso 1983), also has a narrow proximal extremity of the fibula. In *Phascogale*, an arboreal dasyurid which hallux is reduced comparatively to highly arboreal didelphids, the proximal extremity of the fibula is also reduced, although slightly less than in *Dasyurus*. The proximal extremity of the fibula of *Mayulestes* is unknown but in *Pucadelphys* and *Andinodelphys* it is much less developed than in the arboreal didelphids and resembles that of *Monodelphis* or *Metachirus*. It is also very similar to that of *Phascogale*. In both fossil didelphoids the MtI is proportionally smaller than in the living didelphids (Fig. 10). It is also slender comparing to the other metatarsals, whereas it is very strong and proportionally larger in highly arboreal taxa (*Caluromys, Micoureus*). This indicates clearly less grasping ability in the fossil forms. It is probable that their hallux could have been slightly opposable but obviously much less efficient than in any didelphid, phalangerid or petaurid and it is unlikely that they could have grasped branches as easily and firmly as an opossum or a cuscus do. When climbing, *Andinodelphys* and *Pucadelphys* may have used their hind feet as *Antechinus* or *Phascogale* do. This suggests more terrestrial habits, although not specialised, since their metatarsals are relatively short when compared to those of *Metachirus*, a fully terrestrial genus with saltatorial capacity.

Discussion

Reconstructing the mode of locomotion and displacement of fossil mammals is a difficult exercise, especially when dealing with taxa that are phylogenetically distant from their living representatives. First of all, this requires fairly complete specimens in order to have a global view of the anatomy of the taxa under study. This is especially true for geologically old taxa, which generally have no close relatives in the living faunas. Therefore, the tantalising approach to adapt the pattern observed in living taxa is biased with numerous flaws simply because the fossil generally are very different, genetically remote, and often much more primitive. In the case of the fossil marsupials from

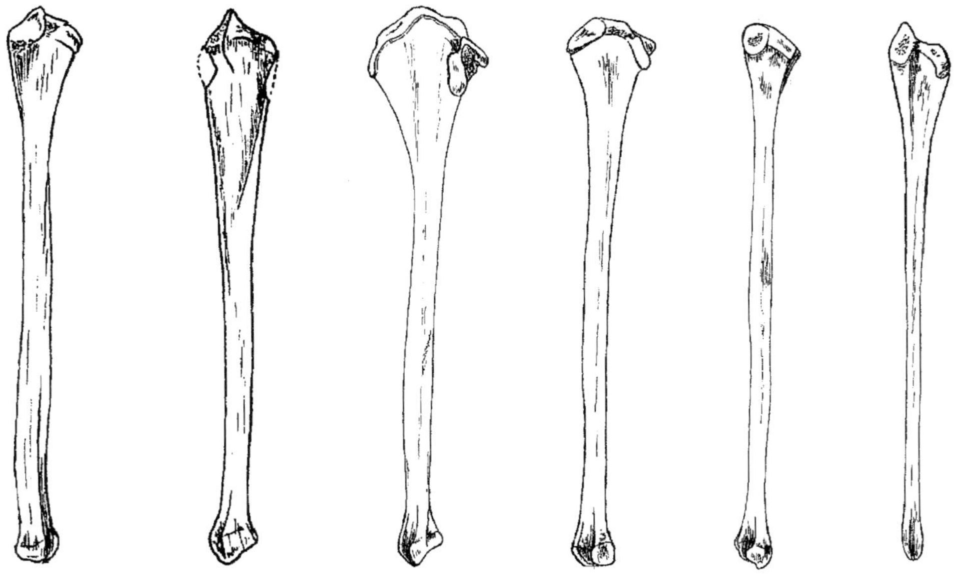

Figure 9 Lateral view of the right fibula of several marsupials: A, *Pucadelphys andinus*; B, *Andinodelphys cochabambensis*; C, *Caluromys philander*; D, *Metachirus nudicaudatus*; E, *Dasyurus viverrinus*; F, *Phascogale tapoatafa*. Not to scale.

Tiupampa, the exceptional completeness of the specimens is a key point for a reconstruction of their locomotory habits, but one must always keep in mind the difficulties mentioned above. Furthermore, as emphasised in the anatomical review above, not all the characters are easy to interpret. Some of them are relatively convincing (at least apparently) to predict the range of locomotor capacities of the animal. Others do not contradict the conclusions drawn from the first characters, but they do not favour them either. Nevertheless, some other characters clearly contradict (at least apparently) the apparent indications provided by the first set. In this case, sometimes a possible explanation can be found, sometimes we simply have no answer to explain the apparent contradiction. The only certitude is that, since an animal lived with this anatomy, its own combination of characters had to be functional, although no equivalent can be found in living models. Consistently, another aspect must be taken into account: the various anatomical elements considered do not provide the same information on the locomotory habits. This state of affairs emphasises the importance of the completeness of the specimens, since the information given by one region of the skeleton can be compared to and tested by that obtained from another. Above all, it is important to recall that any conclusion drawn from the anatomy only represents an interpretation of what an animal was capable of doing, and not necessarily what it was actually doing. Behavioural factors are known to be important in the locomotory habits of Recent mammals and, of course, are not accessible to palaeontologists.

In the anatomical analysis above we have compared the Tiupampa didelphimorphs essentially to two groups of mammals – the American didelphids and the Australian dasyurids – both regarded, in many respects, as the less specialised groups of living marsupials (although, in some respects, dasyurids are clearly more specialised than didelphids). The Tiupampa didelphimorphs are neither didelphids nor dasyurids. However, because their cranial anatomy is clearly more primitive than that of these groups of living marsupials (Marshall and Muizon 1995; Muizon et al. 1997; Muizon 1998), we assume, as a working hypothesis (although we are aware that this is not necessarily true), that their postcranial skeleton is probably essentially primitive relatively to didelphids and dasyurids. In any case, it is certainly closer to the marsupial plesiomorphic condition than dasyurids and didelphids are. The Tiupampa didelphimorphs represent the available fossil marsupials, which permit the best appraisal of the plesiomorphic condition of marsupial postcranial skeleton. The unique skeleton of *Asiatherium* (*A. reshetovi*) from the late Cretaceous of Mongolia is older that the Tiupampa didelphimorphs. However, the preservation of the epiphyses of the limb bones is poor and the specimen is still articulated, which hampers detailed functional studies.

In the anatomical analysis above, three anatomical regions were considered: the axial skeleton, the forelimb and the hindlimb.

The lumbar vertebrae of the Tiupampa didelphimorphs indicate a more important mobility and agility than in didelphids (maybe except *Metachirus*, the most agile didelphid): relative

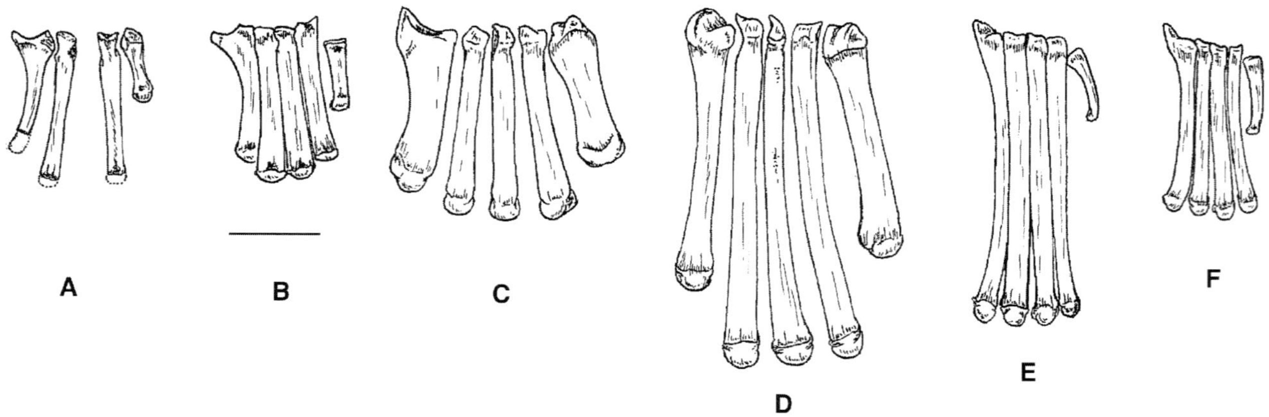

Figure 10 Dorsal view of the metatarsals of several marsupials, showing the relative development of the hallux: A, *Pucadelphys andinus*; B, *Andinodelphys cochabambensis*, C, *Caluromys philander*; D, *Metachirus nudicaudatus*; E, *Dasyurus viverrinus*; F, *Phascogale tapoatafa*. Scale bar: A, 2.5 mm; B–F, 5 mm.

anterior position of the anticlinal vertebra; anteroposterior shortness, proximodistal length, and anterior inclination of the neural processes; anterior and ventral orientation of the transverse processes. Comparisons of these characters with *Dasyurus* indicate that the Tiupampa didelphimorphs approached (probably without reaching) the agility of these living Australian marsupials. The caudal vertebrae indicate some prehensility of the tail (especially the relatively strong development of their transverses processes); however, the relative slenderness of the posterior caudals indicates weakness when compared to highly arboreal models. The Tiupampa didelphimorphs probably had a prehensile tail but it was clearly less efficient than in the highly arboreal extant taxa.

The shoulder of the Tiupampa didelphimorphs was very mobile and indicates a gradient of arboreality from *Pucadelphys* to *Mayulestes*: development and anterior position of the acromion, extended posterodorsal angle of the scapula (in *Mayulestes*), hemispherical head of the humerus. However, the humeral tubercles, slightly higher than in the arboreal didelphids (although not reaching the development observed in *Dasyurus*), are more likely to represent an indication of agility than an argument against mobility of the shoulder. The elbow of the three Tiupampa didelphimorphs reinforces the conclusions drawn from the shoulder. Arboreality is indicated by: the size and orientation of the medial epicondyle, the deep fossa for the flexor digitorum profundus on the lateral side of the olecranon and the medial bending of the ulna, the posterior slope of the olecranon, the proximal convexity of the posterior border of the ulna, the angle between the humeral and radial facet on the anterior face of the ulna, the widely open trochlear notch, and the circular head of the radius (in *Pucadelphys* and *Andinodelphys* only). In *Mayulestes*, the oval-shaped proximal head of the radius, if natural, indicates reduced pronation-supination ability, which does not favour arboreal habits. In other respects, the elevated crest of the humeral trochlea and the reasonably developed olecranon and radial fossae denote a good stability of the elbow and therefore, agility. Summarising, the shoulder and the elbow provide characters relevant to arboreality, although, as expressed above, some (e.g. the shape of the humeral and radial heads) remain difficult to interpret in such a context.

The hindlimb of the Tiupampa didelphimorphs appears to be indicative of both agility and probable arboreality. The short and extroverted ilium indicates agility with some leaping ability, although the relatively weak ischiatic spine of *Mayulestes* seems to preclude this genus from bounding capacities. The open acetabulum with a concave dorsal edge indicates a good mobility of the hip joint and can be related either to arboreality or to locomotion on uneven substrate. The shape of the distal extremity of the femur and the well-marked crests of the femoral trochlea denote agility (although contradicted in *Mayulestes* by the anteroposterior compression of the proximal epiphysis of the tibia). The distal epiphysis of the tibia in *Pucadelphys* and *Mayulestes* and the orientation of the ectal facet on the latter favour some capacity for inversion of the foot, which is related to arboreality. The widened proximal extremity of the fibula, although much less extended than in the arboreal living didelphids and rather slow arboreal Australian forms such as *Schoinobates*, indicates some degree of opposability of the hallux, still less robust in *Andinodelphys* than in highly arboreal didelphid forms. In another way, the relatively short metapodials of the fossil didelphoid are more indicative of arboreality than of a specialised terrestrial locomotor mode.

Conclusion and future directions

The study of the postcranial skeleton of the Tiupampa didelphimorphs indicates that they had arboreal abilities and that they were relatively agile. In this respect, they differ from the living didelphids, in which the arboreal or semi-arboreal taxa are generally relatively slow (when compared with squirrels or tree shrews). A clear gradient of increasing arboreality exists from *Pucadelphys* to *Mayulestes*, although the latter itself was probably not as efficient a climber as the majority of living didelphids (except *Metachirus* and *Chironectes*). In contrast, *Mayulestes* was probably more arboreal than most living dasyurids (except *Phascogale*). In another way, a gradient of decreasing agility is observed from *Pucadelphys* to *Mayulestes*. However, although *Mayulestes* appears to have been probably less agile than *Pucadelphys*, it was certainly more agile than the arboreal didelphids. In terms of agility, the Tiupampa didelphimorphs were probably approaching (without reaching) the living dasyurids (e.g. *Dasyurus, Phascogale, Antechinus, Sminthopsis*). Although most of them are mainly terrestrial, most dasyurids climb well or are semi-arboreal. The association of arboreal abilities to agility suggested by the fossil morphology is, therefore, similar to what is observed in dasyurids. In fact, the locomotion and habits of the Tiupampa didelphimorphs were probably closer to those of the living dasyurids than those of the didelphids, although the latter are regarded as the most primitive living marsupials. Tiupampa didelphimorphs clearly differ from living didelphids or dasyurids, but these two families of Recent marsupials represent the best elements of comparison in order to approach an understanding of their locomotory habits. Given this, there is an urgent need for detailed compared behavioural and anatomical studies of dasyurids, especially focused on osteology, myology and arthrology. Comparative anatomy of the Tiupampa didelphimorphs and didelphids is presently under study by one of us (C. A.).

To conclude, the Tiupampa didelphimorphs were both terrestrial and capable of efficient climbing, the latter adaptation being more emphasised in *Mayulestes* and *Andinodelphys* than in *Pucadelphys*. The Tiupampa didelphimorphs appear to represent a generalised morphotype from which could easily be derived the didelphid and dasyurid morphotypes, both families being more specialised in their locomotory habits than the Tiupampa didelphimorphs. The Tiupampa didelphimorphs could approach a generalised plesiomorphic pattern for marsupial locomotion, which is especially true for *Pucadelphys* and *Andinodelphys*. Therefore, the plesiomorphic condition for marsupials would not be arboreal as hypothesised by Szalay (1994) and Muizon (1998) but terrestrial with good climbing ability, although this ability can vary as is observed in the three Tiupampa didelphimorphs.

ACKNOWLEDGEMENTS

The Tiupampa specimens were collected with funds from the IFEA (Institut Français d'Études Andines), the CNRS (Centre National de la Recherche Scientifique) and the National Geographic Society. Logistical assistance for field vehicle was provided by the IFEA and the IRD (Institut de Recherche pour le Développement). We are greatly indebted to Dr R. Suárez Soruco and Lic. R. Céspedes Paz for their invaluable collaboration during the field expeditions. Specimens from Tiupampa (Bolivia) are property of the 'Museo de Historia Natural de Cochabamba' (Bolivia). We thank R. L. Cifelli and S. Wroe for their careful review which greatly helped to improve the original manuscript. Photograph is by D. Serrette. UMR 8569 of the CNRS (Centre National de la Recherche Scientifique) and line drawings are by C. Argot.

REFERENCES

Bonaparte, J., & Rougier, G.W. (1987), 'Mamíferos del Cretácico Inferior de Patagonia', *IV Congreso Latinoamericano de Paleontología, Santa Cruz, Bolivia*, **1**:343–59.

Dor, M. (1937), *La morphologie de la queue des Mammifères dans ses rapports avec la locomotion*, Impressions Pierre André, Paris, 184 p.

Charles-Dominique, P., Atramentowicz, M., Charles-Dominique, M., Gérard, H., Hladik A., Hladik C. M., & Prévost, M. F. (1981), 'Les Mammifères arboricoles frugivores nocturnes d'une forêt guyanaise: inter-relations plantes-animaux', *Revue d'Ecologie (La Terre et la Vie)*, **35**(3):341–435.

Coues, E. (1869), 'The osteology and myology of *Didelphis virginiana*', *Memoirs of the Boston Society of Natural History*, **2**:41–154.

Curtis, D.J. (1995), 'Functional anatomy of the trunk musculature in the slow loris (*Nycticebus coucang*)', *American Journal of Physical Anthropology*, **97**:367–79.

Eisenberg, J.F. (1989), 'Order Marsupialia', in *Mammals of the Neotropics, Vol. I, The Northern Neotropics*, pp. 20–49, The University Chicago Press, Chicago.

Goldfinch, A.J., & Molnar, R.E. (1978), 'Gait of the brush-tailed possum (*Trichosurus vulpecula*)', *The Australian Zoologist*, **19**(3):277–89.

Grand, T.I. (1983), 'Body weight: its relationship to tissue composition, segmental distribution of mass and motor function. III. The Didelphidae of French Guyana', *Australian Journal of Zoology*, **31**(3):299–312.

Hu, Y., Wang, Y. Luo, Z., & Li, C. (1997), 'A new symmetrodont mammal from China and its implications for mammalian evolution', *Nature*, **390**:137–42.

Jenkins, F.A. Jr. (1971), 'Limb posture and locomotion in the Virginia opossum (*Didelphis marsupialis*) and in other non-cursorial mammals', *Journal of Zoology, London*, **165**:303–15.

Jenkins, F.A. Jr. (1973), 'The funtional anatomy and evolution of the mammalian humero-ulnar articulation', *American Journal of Anatomy*, **137**:281–98.

Jenkins, F.A. Jr., & Camazine, S.M. (1977), 'Hip structure and locomotion in ambulatory and cursorial carnivores', *Journal of Zoology, London*, **181**:351–70.

Jenkins, F.A. Jr., & McClearn, D. (1984), 'Mechanisms of hind foot reversal in climbing mammals', *Journal of Morphology*, **182**:197–219.

Ji, Q., Luo, Z., Ji, S. (1999), 'A Chinese triconodont mammal and mosaic evolution of the mammalian skeleton', *Nature*, **398**:326–30.

Julien-Laferrière, D. (1991), 'Organisation du peuplement de Marsupiaux en Guyane française', *Revue d'Ecologie (La Terre et la Vie)*, **46**(2):125–44.

Krebs, B. (1991), 'Das Skelett von *Henkelotherium guimarotae* gen. et sp. nov. (Eupantotheria, Mammalia) aus dem Oberen Jura von Portugal', *Berliner geowissenschaftliche Abhandlungen* A, **133**:110 p., Berlin.

Lemelin, P. (1999), 'Morphological correlates of substrate use in didelphid marsupials: implications for primate origins', *Journal of Zoology, London* **247**:165–75.

MacLeod, N., & Rose, K.D. (1993), 'Inferring locomotor behavior in Paleogene Mammals via eigenshape analysis', *American Journal of Science*, **293**A:300–55.

Maynard Smith, J., & Savage, R.J.G. (1956), 'Some locomotory adaptations in Mammals', *Journal of the Linnean Society, Zoology*, **42**:603–22.

Marshall, L.G., & Muizon, C. de (1995), 'Part. II: The skull', in *Pucadelphys andinus (Marsupialia, Mammalia) from the early Palaeocene of Bolivia* (ed. Muizon C. de), *Mémoires du Museum national d'Histoire naturelle*, **165**:21–90.

Marshall, L.G., & Sigogneau-Russell, D. (1995), 'Part. III: Postcranial skeleton' in *Pucadelphys andinus (Marsupialia, Mammalia) from the early Palaeocene of Bolivia* (ed. Muizon C. de), *Mémoires du Museum national d'Histoire naturelle*, **165**:91–164.

Muizon C. de (1998), '*Mayulestes ferox*, a borhyaenoid (Metatheria, Mammalia) from the early Palaeocene of Bolivia. Phylogenetic and palaeobiologic implications', *Geodiversitas* **20**(1):19–142.

Muizon C. de, Cifelli R.L., & Céspedes Paz R. (1997), 'The origin of the dog-like borhyaenoid marsupials of South America', *Nature*, **389**:486–89.

Novacek, M.J., Rougier, G.W., Wible, J.R., McKenna, M.C., Dashzeveg, D., & Horovitz, I. (1997), 'Epipubic bone in eutherian mammals from the Creteceous of Mongolia', *Nature*, **389**:483–86.

Nowak, R.M., & Paradiso J.L. (1983), *Mammals of the world, Vols. I & II*, The Johns Hopkins Press, Baltimore, 1362 p.

Pridmore, P.A. (1992), 'Trunk movements during locomotion in the marsupial *Monodelphis domestica* (Didelphidae)', *Journal of Morphology*, **211**:137–46.

Roberts, D. (1974), 'Structure and function of the primate scapula', in *Primate locomotion* (ed. F.A. Jenkins Jr), Academic Press, pp. 171–200.

Shapiro L. (1995), 'Functional morphology of indrid lumbar vertebrae', *American Journal of Physical Anthropology*, **98**:323–42.

Szalay, F.S. (1994), *Evolutionary history of the marsupials and an analysis of osteological characters*, Cambridge, New York, Cambridge University Press, 481 p.

Szalay, F.S., & Dagosto, M. (1980), 'Locomotor adaptations as reflected on the humerus of Paleogene Primates', *Folia Primatologica*, **34**:1–45.

Szalay, F.S., & Trofimov, B.A. (1996), 'The Mongolian Late Cretaceous *Asiatherium* and the early phylogeny and paleobiogeography of Metatheria', *Journal of Vertebrate Paleontology*, **16**(3):474–509.

Tardieu, C. (1983), 'L'articulation du genou. Analyse morphofonctionnelle chez les primates, application aux Hominidés fossiles', *Cahiers de Paléoanthropologie, Editions du CNRS, Paris*, 1–108.

Taylor, M.E. (1976), 'The functional antomy of the hindlimbs of some African Viverridae (Carnivora)', *Journal of Morphology*, **148**(2):227–54.

Walker A. (1974), 'Locomotor adaptations in past and present prosimian primates', in *Primate locomotion* (ed. F.A. Jenkins Jr), Academic Press, pp. 349–81.

Wroe, S. (1997), 'A re-examination of proposed morphology-based synapomorphies for the families of Dasyuromorphia (Marsupialia): Part I Dasyuridae', *Journal of Mammalian Evolution*, **4**:19–52.

Wroe, S. (1999), 'The geologically oldest dasyurid, from the Miocene of Riversleigh, North-West Queensland', *Palaeontology*, **42**:501–27.

CHAPTER 5

MOLECULAR PHYLOGEOGRAPHY AND SPECIES LIMITS IN RAINFOREST DIDELPHID MARSUPIALS OF SOUTH AMERICA

James L. Patton and Leonora Pires Costa

Museum of Vertebrate Zoology, University of California, Berkeley, CA, 94720 USA

We examine the phylogeographic structure of nine genera of rainforest didelphid marsupials (*Didelphis, Philander, Metachirus, Gracilinanus, Marmosa, Marmosops, Micoureus, Monodelphis,* and *Caluromys*) based on sequences from the mitochondrial cytochrome b gene. Multiple geographic representatives of many of the currently recognised species in each genus provide a backdrop to questions concerning the nature of species boundaries and the geographic ranges of these species, as well as the proper application of the available names. We use phylogeographic data to understand the historical connections between the major wet forest biomes of South America, specifically connections between southern Amazonia and the Atlantic Forest of coastal Brazil. Species diversity in several genera is greater than current taxonomy would suggest. While some lineages appear relatively recent (lowland species of *Didelphis*), the majority of extant species are surprisingly divergent, suggesting species formation well before the Pleistocene. Thus, there appears to be little support for the putative role of Pleistocene refuges in generating species diversity in these genera. Geographic samples are limited, but there are areas of strong phylogeographic concordance among lineages within Amazonia and historical connections occur between southern Amazonia and the Atlantic Forest through both the Paraná Basin and around the 'horn' of eastern Brazil.

INTRODUCTION

American opossums are a modestly diverse group of relatively small-bodied marsupials, all but a single species of which are confined to tropical and temperate habitats of the Neotropical Realm. Within this large geographic region, marsupials are common elements of all tropical forest communities and more open habitats (Eisenberg 1989; Eisenberg and Redford 1999; Emmons and Feer 1997; Redford and Eisenberg 1992; Voss and Emmons 1996). Despite this ubiquity, few revisions are available that delimit species boundaries, provide appropriate diagnoses, or even adequately map species distributions. In the past 30 years, only the large-bodied and species-poor genera *Didelphis* and *Philander* have received substantial systematic attention over some, or all, of their respective ranges (Gardner 1973; Hershkovitz 1997). Among the speciose genera, only *Marmosa* (*sensu lato*) has been revised (Tate 1933), but this 70-year-old treatise does not represent current thinking about generic groupings or their included species. Hershkovitz's (1992) review of *Gracilinanus* compiled species in this genus, but did not document their geographic variation or geographic ranges. The most recent taxonomic overview of neotropical marsupials is that of Gardner (1993), who lists 63 species of didelphid marsupials, most of which occur in South America.

Here we review what is known about species boundaries within the polytypic genera. We also emphasise the biogeographic patterns these taxa provide relevant to understanding lowland rainforest phylogeography. We limit our remarks to members of the family that occur in rainforests of South America, both lowland and montane. Finally, we suggest avenues for additional research into the diversity of this lineage of relatively archaic, living mammals.

The generic taxonomy of all didelphids has been reasonably stable for many years, with the exception of the group of murine opossums that Tate (1933) included in the genus *Marmosa.* Gardner and Creighton (1989) suggested that Tate's *Marmosa* was not monophyletic, comprising a series of separate taxa of unknown phylogenetic affinities. They defined the genus *Gracilinanus* to encompass some species, and they suggested that *Marmosops*, *Micoureus*, and *Thylamys* should likewise be elevated to generic status. The lack of monophyly of Tate's *Marmosa* has been substantiated by all recent molecular analyses (Jansa and Voss 2000; Kirsch and Palma 1995; Kirsch et al. 1995; Patton et al. 1996). These molecular data have also clarified some, but not all generic relationships as well as species memberships within these genera.

METHODS AND SAMPLES

We use DNA sequences from the mitochondrial cytochrome b (cyt-b) gene to examine the geographic structure of taxa of rainforest didelphids. The data set for each varies from 450 to 830 base pairs (bp). We supplement data published previously (Mustrangi and Patton 1997; Patton and Costa in press; Patton and da Silva 1997; Patton et al. 1996, 2000) by more recently collected materials. Because of space limitations here, provenance data, museum voucher specimen catalog numbers, and sequences for all specimens examined can be obtained from JLP by request. Our earlier papers provide laboratory methods of DNA extraction, amplification, and sequencing.

We provide synopses of geographic structure based on phylogenetic analyses using maximum parsimony. For each genus, we map the distribution of the species we have examined, plot our sample localities, and provide a generalised cladogram of area relationships. We do not provide detailed phylogenies of individual haplotypes, as we are only interested here in defining gross geographic patterns. We also summarise the strength of particular nodes on these generalised cladograms based on bootstrap analyses, and provide average percentages of Kimura 2-parameter genetic distances for these nodes, as an indication of the depth of lineages. All analyses were performed using PAUP*, version 4.0b2a (Swofford 1998). Tree statistics are given in the legends accompanying each figure.

A word about the nomenclature we use. For simplicity, we consider the family Didelphidae in its traditional context (e.g. Gardner 1993). We make no effort to examine the relationships among the included genera, and thus 'evaluate' competing classifications of didelphids, although various molecular studies have done so (e.g. Jansa and Voss 2000; Kirsch et al. 1995; Patton et al. 1996). In most cases, we examined the voucher of each specimen sequenced and have compared these to other appropriate museum materials. This has not been possible in all cases, however, and we have relied on identifications provided by competent systematists, primarily Alfred L. Gardner, Louise H. Emmons, and Robert S. Voss who have made available some tissues samples. As we emphasise throughout the following, the species-level taxonomy for many genera will certainly change as a consequence of future work, as more molecular data are added and, especially, as appropriate museum revisionary work is undertaken.

SPECIES BOUNDARIES AND MOLECULAR PHYLOGEOGRAPHY

Here we summarise data on phylogeography of the nine widespread genera of rainforest didelphids, defining phylogeography as '...the study of the principles and processes governing the geographic distributions of genealogical lineages, including those at the intraspecific level' (Avise 1994: 233). Extensive sequence data are available for some taxa, from throughout much of their known ranges. Others, however, are poorly represented, and some have not been sampled at all. Of the 16 extant genera of didelphid marsupials, all but three (*Lestodelphys*, *Lutreolina*, and *Thylamys*) have their major distributions and species richness in tropical forests. Two of these (*Caluromysiops* and *Glironia*) are known from only a few scattered records; we provide no data for these. Similarly, we lack data for *Chironectes* and *Hyladelphys*.

Since part of our goal is to hypothesise species units within genera, our species concept needs to be clear. We recognise as species geographically bounded sets of populations that are distinct in morphological traits and that are in some way reproductively isolated from congeners, with or without corresponding molecular divergence. We accept as prima facie evidence for species status sympatry between morphologically diagnosable entities for which there is no evidence of interbreeding. Our concept thus encompasses situations where there are concordant molecular and phenotypic differences, but allows for circumstances where the progenitor of a recently evolved, phenotypically distinct lineage is paraphyletic with respect to the derived species (Patton and Smith 1994). This definition excludes allopatric populations that are divergent for molecular characters but are not clearly distinguishable by morphological traits, populations

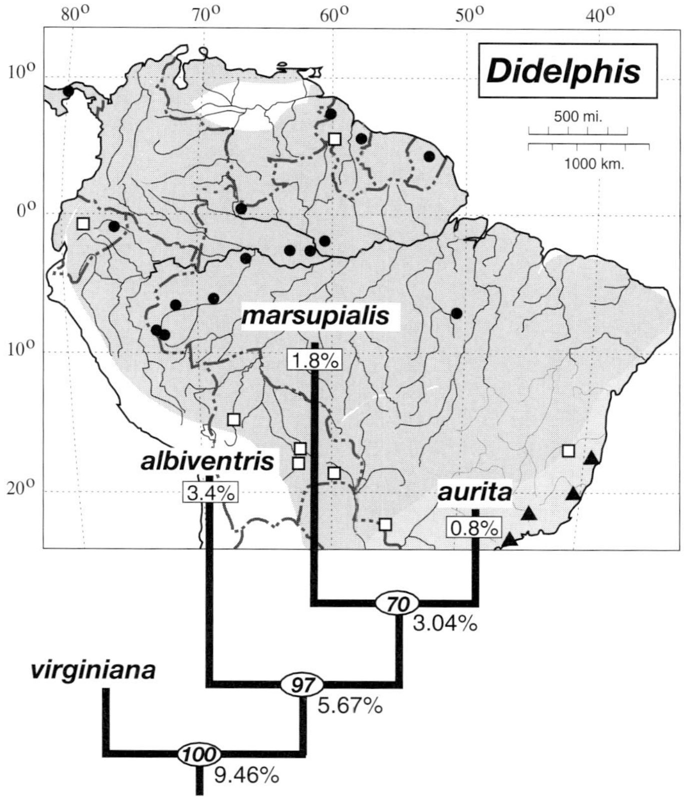

Figure 1 Distribution of the species of common opossum, genus *Didelphis*, in South America. Localities for specimens sampled for 660 bp of the cytochrome b gene are indicated by the symbols (solid circles = *D. marsupialis*; solid triangles = *D. aurita*; open squares = *D. albiventris*). Phylogenetic relationships between all four currently recognised species are indicated by the cladogram, a strict consensus tree of 56 haplotypes based on 120 parsimony informative characters (length = 289 steps; consistency index = 0.633). Bold numbers at nodes are bootstrap values, based on 1000 iterations; percentages at internal and terminal nodes are average Kimura 2-parameter distances.

that would be given species status under a phylogenetic concept (Cracraft 1989).

Subfamily Didelphinae

***Didelphis* Linnaeus, 1758**. Four species are currently recognised in the genus: three in South America (*albiventris*, *aurita*, and *marsupialis*), two in Central America (*marsupialis* and *virginiana*), with only one (*virginiana*) extending into temperate North America. We have data for the first 660 bp for 66 individuals representing each of the four species currently recognised. These include eight individuals of *virginiana* (from localities as far apart as southern Mexico, California, Texas, and eastern Canada); 18 *albiventris* (from southern Brazil to the Andes of Bolivia and Ecuador and the Guianas); 33 *marsupialis* (from Mexico through Costa Rica and Panama in Central America and throughout Amazonia); and seven *aurita* (from eastern Brazil). The phylogenetic relationships among these four are indicated in Fig. 1, as are the respective ranges of the three South American species.

Gardner (1973) hypothesised that *marsupialis* (including *aurita*) and *virginiana* were sister species, with *albiventris* the most distant. However, both DNA/DNA hybridisation data (Kirsch et al. 1993) and mtDNA sequences (Patton et al. 1996) suggest that *virginiana* is basal to a clade that includes *marsupialis* and *albiventris*. The taxa *auritus* and *marsupialis* are reciprocally monophyletic as well as close sister-species (average Kimura 2-parameter distance is 3.04%; Fig. 1). Thus, the molecular data for these taxa support conclusions of species status for *aurita* based on morphological analyses (Cerqueira 1985; Cerqueira and Lemos 2000).

Gardner's (1973) hypothesis that *virginiana* was derived from *marsupialis* suggests that species of common opossums radiated initially in South America and then, following the invasion of Central America after the rise of the Panamanian Isthmus, *virginiana* split from *marsupialis* as it adapted to temperate conditions. The molecular data, however, suggest an initial split between temperate North American and tropical lineages with subsequent differentiation of the latter into the three extant South American species.

Only weak phylogeographic structure is evident in any of the four species of *Didelphis*. This is particularly apparent for *marsupialis*, which has been sampled throughout its range from southern Mexico through Central America, the Guianas, and Amazonia (Fig. 1; Lavergne 1998; Patton et al. 2000). Over this broad distribution, haplotypes from Mexico and Central America generally cluster together; those from the Guianas and eastern Amazonia also form a unit, as do those from western Amazonia (data not shown). Phylogenetic analyses, however, provide little support for each of these haplotype groups. Somewhat greater geographic structure is present in *albiventris*, with samples from Guyana west and south along the Andean arc to northern Bolivia separated from those of southern Bolivia and Brazil by an average of 5.1%. The northern haplotype clade includes both the morphologically marked Guianan subspecies *imperfecta* as well as more typical *albiventris*.

The low levels of sequence divergence within *Didelphis* contrast sharply with that for most other didelphid genera, as presented below. This difference could be due to lower rates of molecular substitution in the common opossum lineage. However, rates of molecular evolution do not differ among the lineages of didelphids examined to date (Lavergne 1998; Patton et al. 1996). Consequently, *Didelphis* as a genus appears to be young, to have speciated relatively recently, and species such as *marsupialis* have attained their widespread distributions throughout the lowland rainforests even more recently. Certainly, *Didelphis* is not the archetypical primitive marsupial, as is often erroneously stated (see Hershkovitz 1969, for example, countered by Clemens 1968, and Gardner 1973).

***Philander* Tiedemann, 1808**. Hershkovitz (1997) reviewed the pouched four-eyed opossums of the genus *Philander* and recognised two species, the widespread grey-coloured *opossum* (from eastern Mexico to coastal Brazil) and the blackish *andersoni* (from the western Amazon). We differ substantially from Hershkovitz in the delineation of species boundaries (Patton and da Silva 1997; Patton et al. 2000). Our samples include 660 bp of sequence for 74 individuals from 29 separate localities from Panama south throughout Amazonia and coastal Brazil.

Patton and da Silva (1997) recognised three species of *Philander* within Amazonia and suggested that populations from coastal Brazil represented a fourth, based on sympatry of both morphologically recognisable and reciprocally monophyletic forms. These authors separated *mcilhennyi*, distributed in western Amazonia south of the Amazon, from *andersoni*, which occurs north of the Amazon in Peru and Ecuador east to southern Venezuela and central Brazil west of the Rio Negro (Fig. 2). These two species differ by colour, colour pattern, and subtle cranial characters (Patton et al. 2000) and are not sisters in the phylogenetic analysis of mitochondrial sequences (Fig. 2; average Kimura 2-parameter distance, 7%). The ranges of *mcilhennyi* and *opossum* overlap over much of eastern Peru and western Brazil (Fig. 2; Gardner and Patton 1972; Hutterer et al. 1995; Patton et al. 2000), and *andersoni* and *opossum* are sympatric in north-eastern Peru (Hice 2001). Patton and da Silva (1997) also argued that the coastal Brazilian taxon *frenatus* should be recognised as a distinct species based on its deeply divergent mitochondrial genome (an average of nearly 14% in comparison to all other taxa; Fig. 2). Finally, the species status of *opossum* itself remains unclear, as the geographic samples of taxa such as the western Amazonian *canus*, Guianan *opossum*, and Middle American *fuscogriseus* form an unresolved polytomy in our analyses with respect to *mcilhennyi*. There is a great deal of morphological diversity within *Philander opossum* throughout its range in Amazonia and on both sides of the Andean slopes, and much remains to be learned of the relationships of these forms.

Hershkovitz (1997) mapped the range of *canus* as continuous from south-eastern Brazil north-west through the Paraná basin to western Amazonia in Brazil, Bolivia, and Peru. There is, however, a major phylogeographic gap between our samples from the coastal Brazil and those from the northern Paraná basin in the Brazilian states of Mato Grosso do Sul and Mato Grosso and adjacent eastern Paraguay. The latter are clearly aligned phylogenetically with those of *canus* from the western Amazonia, differing from them by only 2.5% on average. In turn, all of these samples differ markedly from those of the coastal Brazilian *frenatus* (12.8%; Fig. 2). Analysis of additional samples from the southern Paraná basin, particularly in the Brazilian state of Paraná and Argentina will be especially important to define the geographic extent of both *frenatus* and *canus*, and determine whether the two behave as biological species, as suggested by their very divergent genomes.

***Metachirus* Burmeister, 1854**. A single species of brown four-eyed opossum, *M. nudicaudatus*, is currently recognised (Cabrera 1957; Hall 1981; Gardner 1993). Distributed widely throughout the lowland tropics from southern Mexico to coastal Brazil, this species is restricted to closed canopy forests. We have examined 33 individuals from 21 localities; the data set includes the first 450 bp of the cytochrome b gene. Our samples are geographically extensive in western Amazonia and coastal Brazil. Only single localities in southern, south-eastern, and north-eastern Amazonia are represented (Fig. 3), and we have not examined any specimens from the west coast of South America nor from any Middle American country.

There is surprising sequence diversity across the sampled range of *M. nudicaudatus*, especially so for what has always been considered a single species. Five geographically differentiated and reciprocally monophyletic clades are present, although the resolution at two nodes in a tree of their relationships is limited

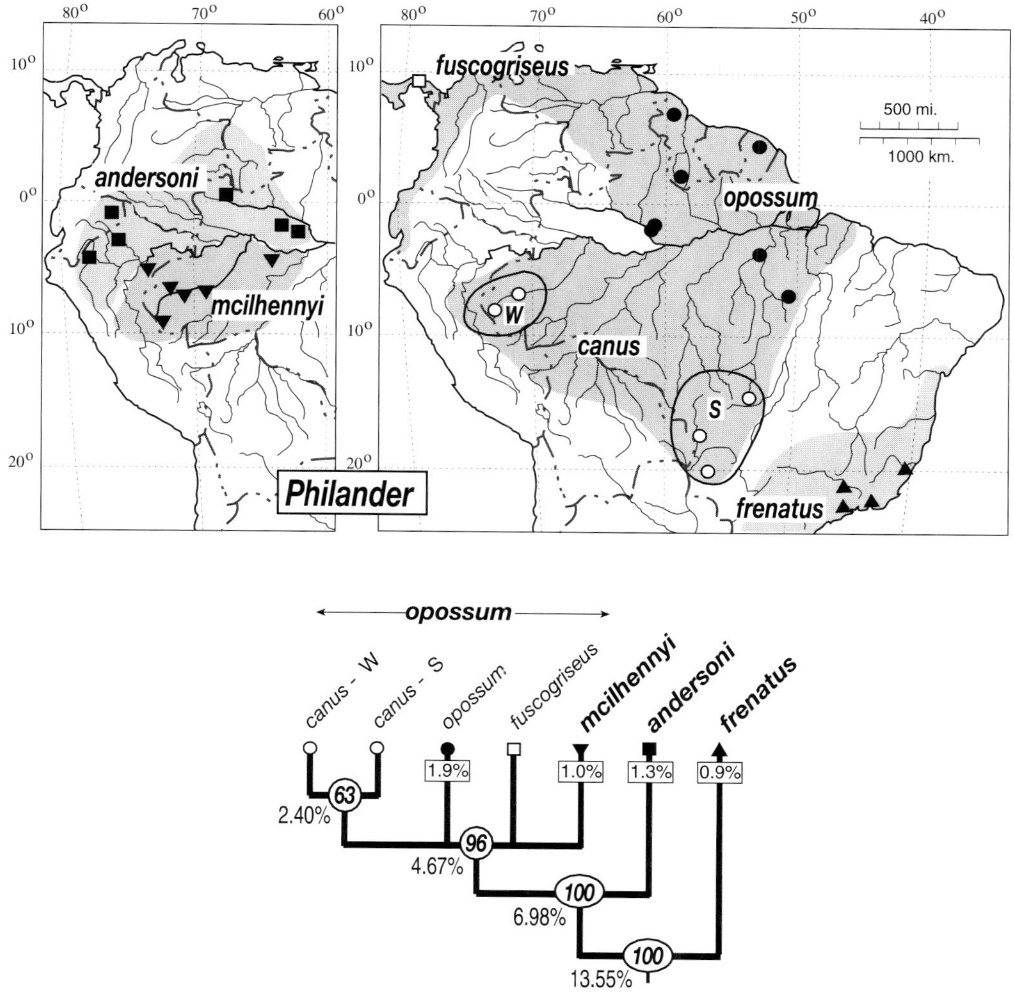

Figure 2 Distribution of the species of pouched four-eyed opossums, genus *Philander*, in South America. Localities for specimens sampled for 660 bp of the cytochrome b gene are indicated by the symbols (solid squares = *P. andersoni*; inverted solid triangles = *P. mcilhennyi*; open squares = *P. opossum fuscogriseus*; solid circles = *P. opossum opossum*; open circles = *P. opossum canus*; and solid triangles = *P. frenatus*). The western and southern clades of *P. o. canus* are enclosed. Phylogenetic relationships between the currently recognised species and geographic units of *P. opossum* are indicated by the cladogram, a strict consensus tree of 48 haplotypes based on 129 parsimony informative characters (length = 369 steps; consistency index = 0.672). Bold numbers at nodes are bootstrap values, based on 1000 iterations; percentages at internal and terminal nodes are average Kimura 2-parameter distances.

(Fig. 3; Patton et al. 2000). Haplotypes from the Atlantic Forest (AF), southern Amazonia (S), and south-western Amazonia (SW) form an unresolved trichotomy of geographic units that are supported by a bootstrap value of 80%. These differ among themselves by an average of 5.1% (Fig. 3). In turn, these three geographic areas link phylogenetically to samples in north-western Amazonia north of the Amazon River. This linkage is somewhat weaker (bootstrap support of 71%), and with substantial sequence divergence of 8.4% between the two geographic groups. Finally, the most divergent geographic clades are those represented by single specimens from French Guiana and the Carajás region of south-eastern Pará state in Brazil. These differ from one another and from all other samples by an average of 13.9%.

The phylogeographic pattern within *M. nudicaudatus* closely links populations from coastal Brazil with those in western Amazonia south of the Amazon. However, large sections of the range of *nudicaudatus* remain to be sampled, particularly across southern Amazonia and the horn of Brazil. Consequently, both the extent of reciprocal phylogeographic structure and area relationships among geographic units must await the analysis of further samples. The high level of sequence divergence, as large or

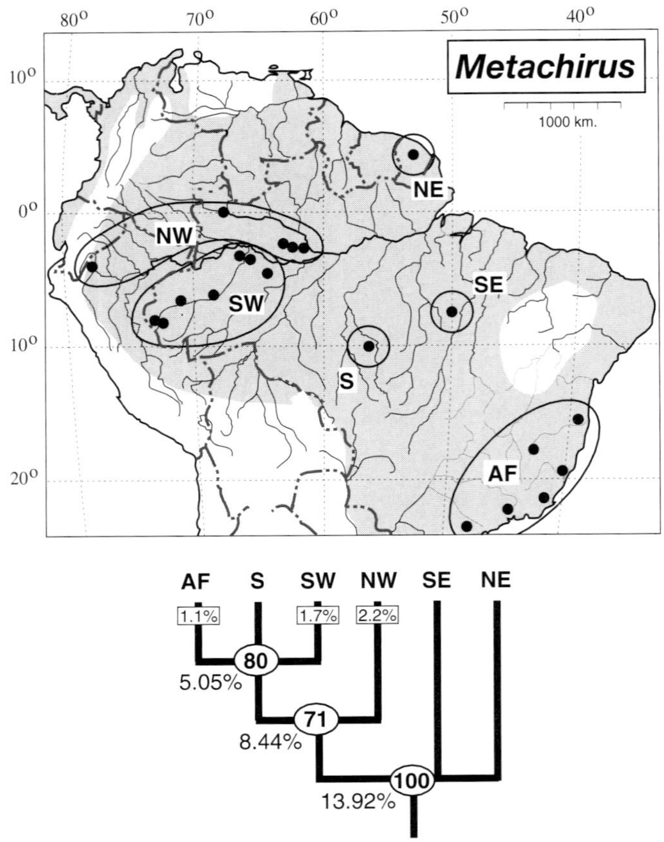

Figure 3 Distribution of the brown four-eyed opossums, *Metachirus nudicaudatus*, in South America. Localities for specimens sampled for 450 bp of the cytochrome b gene are indicated by the symbols; phylogeographic units that are reciprocally monophyletic clusters of similar haplotypes are enclosed. Phylogenetic relationships between these geographic units are indicated by the cladogram, a strict consensus tree of 29 haplotypes based on 141 parsimony informative characters (length = 338 steps; consistency index = 0.740). Bold numbers at nodes are bootstrap values, based on 1000 iterations; percentages at internal and terminal nodes are average Kimura 2-parameter distances.

larger than found between most well-recognised species of all other didelphid genera we have examined, also suggests that more than one species is actually included within the single taxon *nudicaudatus*.

***Marmosa* Gray, 1821**. Tate (1933) included what are now recognised as five different genera in his concept of *Marmosa*, one that is still widely employed by non-systematists in the literature (e.g. Redford and Eisenberg 1992). However, as initially demonstrated by Gardner and Creighton (1989), *Marmosa* should be restricted to a group of species (Gardner 1993) characterised by moderate size, orange-brown dorsal and grey-based ventral colouration, large hind feet, tails with large rhomboid scales arranged in annular rings, and globular auditory bullae without an anterior alisphenoid buttress.

Gardner (1993) lists nine species within the genus, and ranges are mapped in Emmons and Feer (1997) and Hall (1981). Our data for this diverse genus are extremely limited, restricted to 740 bp of sequence for 22 individuals from 16 localities of *murina*, covering much of its known range, as well as three specimens of *lepida* and to one specimen of *rubra*, both from a single (and the same) locality in northern Peru (Fig. 4). *Marmosa lepida*, *M. rubra* and *M. murina* (subspecies *maranii* Thomas) are sympatric in northern Peru. The overall average divergence between three individuals of *lepida* and 16 haplotypes of *murina* is 16.7%; that between the single specimen of *rubra* and *murina* is 20.3%; and that between *lepida* and *rubra* is 20.4% (Fig. 4). Since our data for both *lepida* and *rubra* involve individuals from only a single locality, we cannot comment on the extent of geographic variation in either of these species. However, across Amazonia and along coastal Brazil, haplotypes from individuals allocated to *murina* on morphological grounds comprise four rather distinctive geographic clades that exhibit marked levels of sequence divergence (Fig. 4). Samples from north of the Amazon River in western Amazonia are strongly differentiated from those to the south, at an average level of 9.3%. These two,

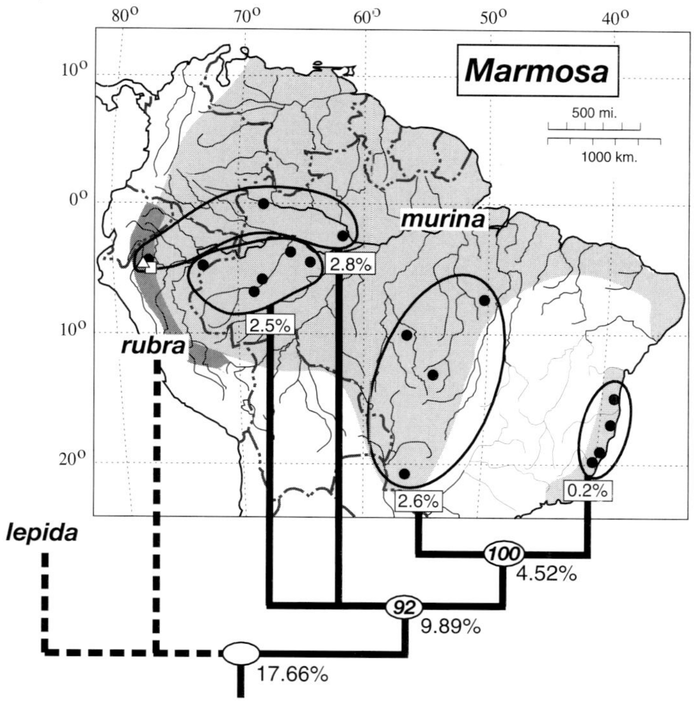

Figure 4 Distribution of the murine opossums, *Marmosa rubra* and *M. murina,* in South America. *Marmosa lepida*, the third species in the genus for which data are available, is known from spotty records in western Amazonia and the Guianan region (mapped by Emmons and Feer 1997). Localities for specimens sampled for 740 bp of the cytochrome b gene are indicated by the symbols (open triangle = *M. lepida*; open square = *M. rubra*; solid circles = *M. murina*). Phylogeographic units of *M. murina* that are reciprocally monophyletic clusters of similar haplotypes are enclosed. Phylogenetic relationships between these species and geographic units are indicated by the cladogram, a strict consensus tree of 20 haplotypes based on 182 parsimony informative characters (length = 469 steps; consistency index = 0.731). Bold numbers at nodes are bootstrap values, based on 1000 iterations; percentages at internal and terminal nodes are average Kimura 2-parameter distances.

in turn, differ from samples from central Brazil, on the one hand, and those from coastal Brazil, on the other, by over 9.9%. Finally, the samples from central Brazil and the Atlantic Forest each comprise separate clades that differ from one another by 4.5%. While all three Amazonian clades individually comprise differentiated haplotypes (2.5% to 2.8%, on average), in each case this degree of divergence is half to one quarter that observed between any two adjacent clades (4.5% to 9.9%; Fig. 4).

Clearly, *M. murina* is a strongly differentiated taxon that is quite likely comprised of more than one biological species. Additional fieldwork will be required to determine if these geographic units (and others remaining to be discovered) contact one another without evidence of intergradiation. Moreover, appropriate museum work will be necessary to associate any of the 18 available names now listed as synonyms (Gardner 1993) to each of these geographic clades.

***Gracilinanus* Gardner and Creighton (1989)**. This group is comprised of delicately built opossums, generally smaller than individuals of *Marmosops* or *Marmosa* with which they are often confused. The tail is slender and of varying length, usually with small caudal scales arranged in rings, and non-petiolate central hairs in each caudal scale triplet. Palates are highly fenestrated and bullae have an anteromedian strut forming a secondary foramen ovale. The dorsal colouration varies considerably, from bright reddish-brown to pale greyish-brown; the ventral pelage is often cream or pale orange with grey-based hairs, but sometimes white or pure cream. The number of species within this genus remains debatable, as do generic boundaries. Gardner (1993) listed six while Hershkovitz (1992), in his review of the genus, recognised the same six species, described three new ones from the Andean slopes of Colombia, Venezuela, and Peru, and recorded what he considered an undescribed additional species from Ecuador. Voss et al. (2001) defined a new genus, *Hyladelphys*, for the new species Hershkovitz (1992) described as *G. kalinowski*.

The genus ranges from the Guiana region, through Venezuela and Colombia, curiously bordering the western limit of the Amazon basin with scattered localities in Peru, Bolivia and

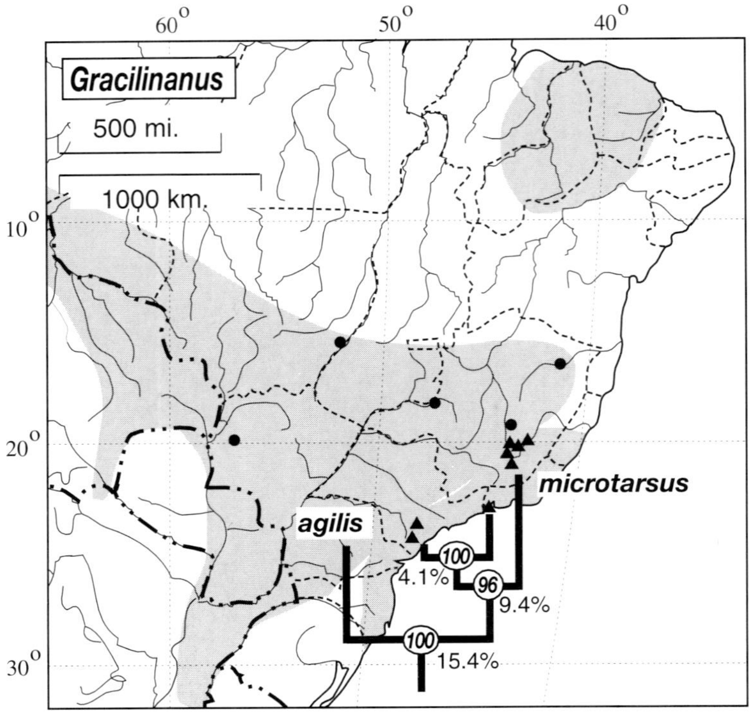

Figure 5 Distribution of the gracile mouse opossum genus *Gracilinanus* in south-eastern South America. Localities for specimens sampled for 750 bp of the cytochrome b gene are indicated by the symbols (solid triangle = *G. microtarsus*; solid circles = *G. agilis*). Phylogenetic relationships between these two species and geographic units within *microtarsus* are indicated by the cladogram, a strict consensus tree of 17 haplotypes based on 159 parsimony informative characters (length = 402 steps; consistency index = 0.741). Bold numbers at nodes are bootstrap values, based on 1000 iterations; percentages at internal and terminal nodes are average Kimura 2-parameter distances.

Paraguay, to the mouth of the Paraná River in Argentina, then north-east along the coast and interior tablelands of Brazil, reaching the south-eastern border of Amazonia. It is apparently absent from all, or at least the majority of the central Amazon Basin in Brazil as specimens recorded from this vast region are either misidentified (see Patton et al. 2000) or of questionable occurrence (Patton and Costa, in press).

Our samples consist only of the two species that occur in coastal Brazil: *microtarsus* from mesic habitats of the Atlantic Forest from Minas Gerais south to Rio Grande do Sul and *agilis*, primarily from dry forests of the interior plateau, an isolated area in the north-east, and through the wet and dry forests of north-eastern Argentina, Paraguay, and Bolivia. We have examined 750 bp of 19 individuals from eight localities of *microtarsus* (including topotypic material) and five of *agilis* (Fig. 5). Samples of *microtarsus* form three phylogeographic groups, one in south-western São Paulo state, a second from Ilha Grande off the Rio de Janeiro coast, and a third from south-central Minas Gerais state. Each clade is highly divergent, with Kimura 2-parameter distances ranging between 4.1 (São Paulo and Rio de Janeiro states) and 9% (these from relative to samples from Minas Gerais). In contrast, samples of *agilis* are quite uniform, averaging only 2.3% divergence among geographic localities sampled over a much larger area (Fig. 5).

One aspect of these data that needs clarification in the future is the correct name to apply to the taxon we, and others, here call *agilis*. Our samples include topotypes of *microtarsus* from the type locality at Ipanema, São Paulo state as well as an individual from Lagoa Santa, Minas Gerais, the type locality of *agilis*. This latter specimen links very closely with other *microtarsus* from nearby localities in southern Minas Gerais (Fig. 5). If our specimen from Lagoa Santa is the same as Burmeister's type of that taxon, then *agilis* would become the name available for the Minas Gerais clade of *microtarsus*, possibly recognised either as a separate species by virtue of the large degree of sequence divergence, or as a synonym of *microtarsus*. In either case, the interior clade which we identify, and to which the name *agilis* is usually applied (Gardner 1993; Emmons and Feer 1997; Eisenberg and Redford 1999), would require another name, presumably one that is currently listed as a synonym of *agilis* (e.g. Gardner 1993). Again, as is true so many other examples, we are in need of a thorough taxonomic revision of this complex of mouse opossums. Hershkovitz's (1992) recent effort made little note of either *microtarsus* or *agilis*.

***Marmosops* (Matschie, 1916)**. These are medium- to small-bodied, pale orange-brown to dark grey-brown murine opossums common in the lower strata of both non-flooded or seasonally flooded lowland forests and mid-elevation Andean forests (Emmons and Feer 1997; Patton et al. 2000). The genus can be distinguished from other murine opossums by the combination of short and straight hair, elongated tail with very small scales arranged in spirals with an enlarged and dark medial hair associated with each scale, relatively small hind feet, and a distinct spine, or buttress, on the alisphenoid portion of the auditory bulla (see Gardner and Creighton 1989).

The systematics of slender mouse opossums is complicated, species limits are poorly understood, and the correct application of available names is equivocal at best for many specimens. The limited molecular data available suggest both greater species diversity than currently appreciated as well as realignments of the names as these are currently allocated to those species that are recognised. Gardner (1993) listed nine species and provided a list of synonyms of each. Our focus has been only on the five species he recognised from Amazonia (*dorothea*, *impavidus*, *noctivagus*, and *parvidens*) and the Atlantic Forest of Brazil (*incanus*). Data for those species we recognise in the Atlantic Forest are extensive, in number of individuals and geographic span of localities from which specimens have been examined (Mustrangi and Patton 1997). However, our effort to date for Amazonian taxa is, like that for other didelphids we summarise here, clearly inadequate in geographic sampling. Our database includes the initial 630 bp of sequence for 38 specimens of Amazonian taxa and 801 bp for 60 specimens of taxa from the Atlantic Forest.

Mustrangi and Patton (1997) revised the Atlantic Forest representatives of *Marmosops* and documented two species, *incanus* and *paulensis*, each diagnosable by a combination of morphological and molecular characters. The level of sequence divergence between them is substantial (16.7%, on average), suggesting a long period of evolutionary independence. *Marmosops incanus* is comprised of seven phylogeographic units that differ by 7 to 11%, which also suggests geographic isolation for a substantial period of time. Each clade is generally limited to separate and parallel mountain chains of the Serra do Mar, coastal Rio de Janeiro, Serra da Mantiqueira, Serra do Espinhaço, plus single localities in Espírito Santo and Bahia states. These values mirror those for clades of *Gracilinanus microtarsus*, distributed over a similar geographic area (Fig. 5). *Marmosops paulensis* occurs in the higher elevations of the coastal ranges from Paraná through São Paulo, Rio de Janeiro, southern Minas Gerais, and Espírito Santo states (Fig. 6). Samples from the forested ridges of coastal São Paulo and Rio de Janeiro states differ from those of the interior ranges of southern Minas Gerais state by an average of 5.9%, again, a substantial level especially considering their close geographic proximity.

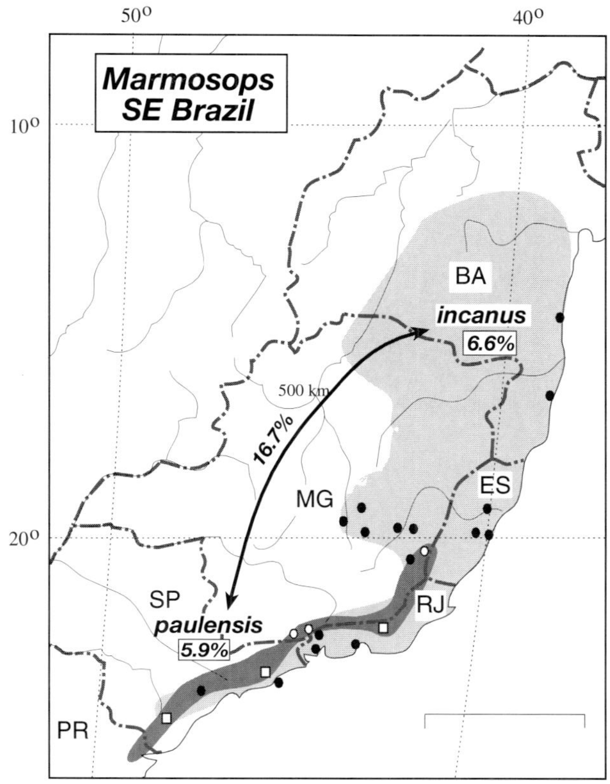

Figure 6 Distribution of the Atlantic Forest species of slender mouse opossums, *Marmosops incanus* (solid circles) and *M. paulensis* (open circles) in south-eastern South America. Open squares represent localities of sympatry of the two species. Data are based on 801 bp of cytochrome b sequence (from Mustrangi and Patton 1997). The two species differ by 16.7% (Kimura 2-parameter distance).

Sequence data for Amazonian and Andean samples also support the existence of several distinct species, each separated by deep levels of divergence (Mustrangi and Patton 1997; Patton et al. 2000). *Marmosops noctivagus*, one of the four Amazonian species recognised by Gardner (1993), is defined by its large size and white venter without lateral grey-based incursions (Patton et al. 2000). It is widely distributed from the mid-elevations of the eastern Peruvian Andes to the central Amazon near Manaus in Brazil, a large range over which *noctivagus* displays little sequence differentiation (Fig. 7A). Our specimens from the upper Andean slopes in southern Peru represent the named forms *albiventris* and *keaysi*, the former which is usually placed as a junior synonym of *M. impavidus* (Cabrera 1957; Gardner 1993). Consequently, some realignment of names allocated to both *noctivagus* and *impavidus* is likely following badly needed museum studies.

The second species we recognise is *impavidus* (Fig. 7B, a taxon with a confused taxonomic history (Tate 1933; Cabrera 1957).

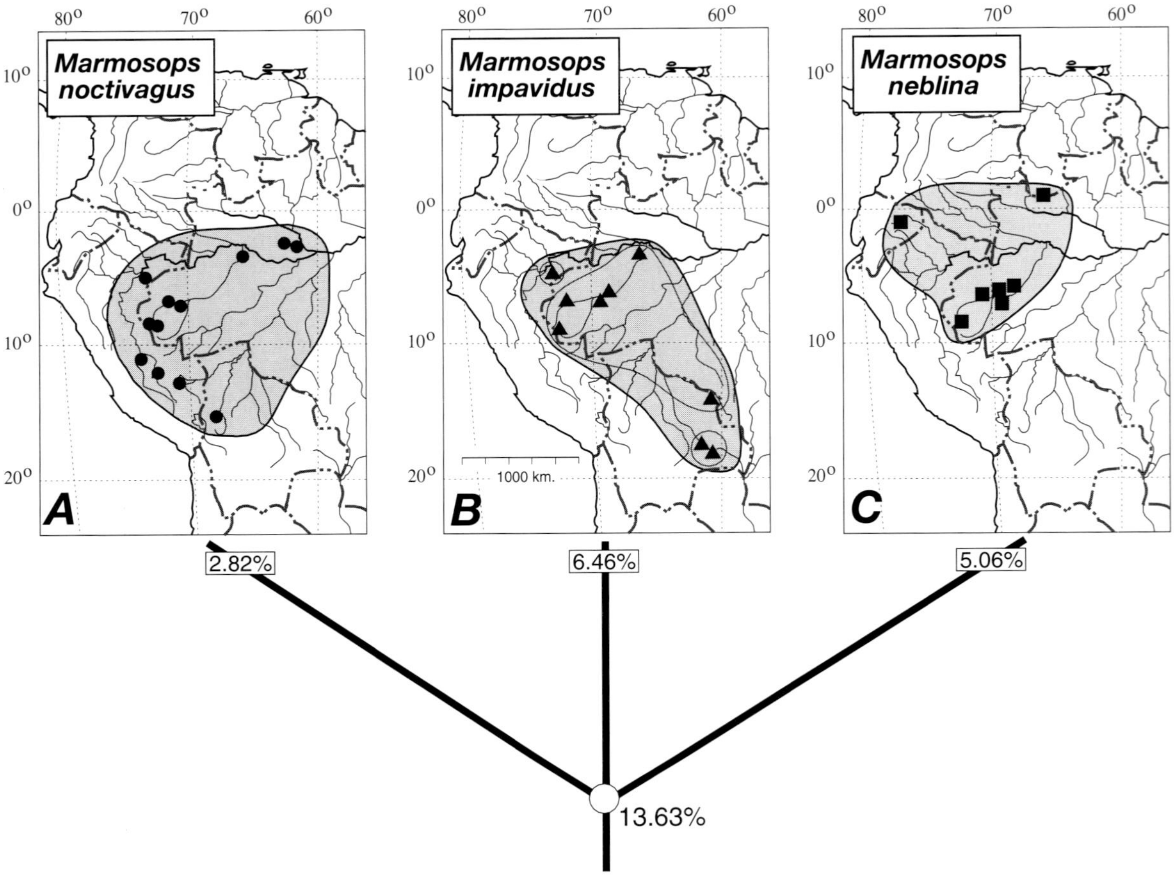

Figure 7 Distribution of three species of slender mouse opossums: *Marmosops noctivagus* (A, solid circles), *M. impavidus* (B, solid triangles), and *M. neblina* (C, solid squares) in the western Amazon for which 630 bp of the cytochrome b gene have been examined. There is no phylogenetic resolution among these three species. The average of 13.6% Kimura 2-parameter distances among them is 13.6%; that within each ranges from 2.8 to 6.5%.

It is generally similar to *noctivagus*, differing in slightly smaller size and with a more narrowed white venter bordered by distinct grey-based lateral incursions (Emmons and Feer 1997; Patton et al. 2000). Our samples from Peru and Brazil match this description, while those from southern Bolivia match the paler colour of the taxon *dorothea*, currently considered a separate species (Gardner 1993). While these samples form a well-supported monophyletic group relative to other species, there is substantial sequence divergence among our samples. The sample from north-eastern Peru differs by nearly 9% from those from Brazil and Bolivia; the Brazilian sample from the Rio Juruá groups closely (2%) with that from eastern Bolivia, but these samples differ from the two from southern Bolivia by more than 5%. It is thus possible that more than one species is represented by what we here call *impavidus*, with *dorothea* being one of them.

The third species of Amazonian slender mouse opossums we recognise is *M. neblina*, a moderate-sized, especially dark taxon with broad grey-based lateral incursions onto the venter that severely constrict the white median stripe. Although originally described (Gardner 1989) as a subspecies of *M. impavidus*, *neblina* clearly deserves species status. The two are sympatric and segregated by habitat over much of the central and upper portion of the Rio Juruá in western Brazil (Fig. 7B, C; Patton et al. 2000). Our samples include the holotype from Cerro Neblina in southern Venezuela, one specimen from eastern Ecuador, and several from the Rio Juruá. Haplotypes from these individuals are intermediate in their level of geographic differentiation between *noctivagus* and *impavidus*, with an average sequence divergence of 5.1%.

Finally, the fourth Amazonian species recognised by Gardner (1993) is *M. parvidens*, a diminutive murine opossum with delicate features, velvety smoky brown dorsal fur, and varyingly grey and/or white venter that ranges widely from the Guianan region and south-eastern Amazonia to the base of the Andes in

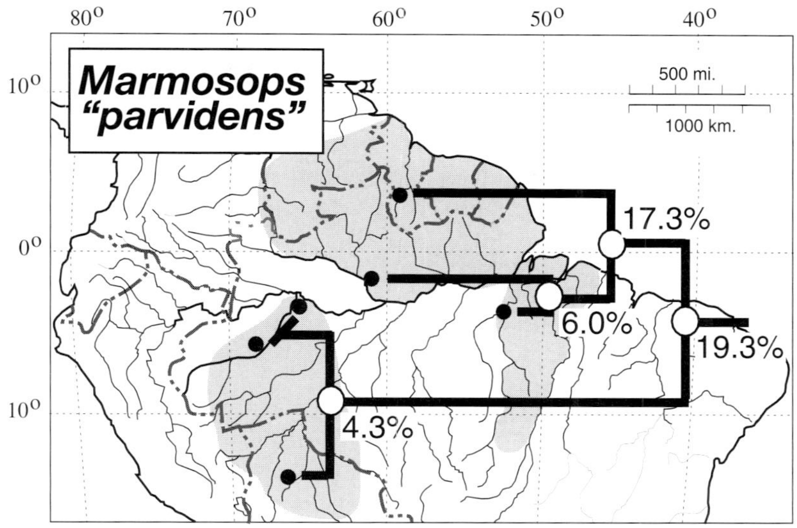

Figure 8 Distribution of *Marmosops 'parvidens'*, as mapped by Emmons and Feer (1997). Our few sampled localities are indicated by solid circles, and the phylogenetic relationships among these by the cladogram, a strict consensus tree based on 630 bp of cytochrome b sequence. This tree, including the three other Amazonian species illustrated in Fig. 7, has a length of 618 steps, a consistency index of 0.494, with 190 parsimony informative characters.

Peru and Bolivia. It was reviewed by Pine (1981), who recognised several subspecies. However, Voss et al. (2001) documented sympatry between two of these races in French Guiana and elevated one of Pine's subspecies (*pinheiroi*) to species status. Our samples are limited to a few individuals from five geographic regions, but the remarkable diversity among them (Fig. 8) fully supports the views of Voss et al. (2001) that multiple species are included under Pine's (1981) concept of '*parvidens*.' The minimal level of divergence between any two areas is that for our samples from western Brazil and northern Bolivia (4.3%). The next most is between samples from the lower Negro and Xingu rivers in Brazil (6.0%). These two pairs of localities, in turn, differ from each other and from the single individual from Guyana by 17 to 19%, levels well above that of most congeneric species of didelphids, even within this highly variable genus. Our specimen from Guyana (Royal Ontario Museum 97938; Patton et al. 2000) is *M. parvidens*, as restricted by Voss et al. (2001), but as yet we are unable to place appropriate names on the other individuals we have examined. Once again, here is another taxon ripe for detailed systematic study.

***Micoureus* Lesson, 1842**. Woolly mouse opossums are arboreal, omnivorous, and common members of the neotropical forest marsupial assemblage. The genus is widely distributed, ranging from Belize in Central America to northern Argentina, and from the lowland Amazonian to mid-elevation elfin forests on both Andean slopes (Emmons and Feer 1997). The number of species recognised is in flux, with Gardner (1993) recognising four species and Emmons and Feer (1997) five. Both sets of authors record three species in lowland South America east of the Andes (*constantiae* from western Brazil and south-eastern Bolivia, *regina* in western Amazonia, and *demerarae* from Amazonia from eastern Peru to the Guianan region and south to the Atlantic Forest of coastal Brazil) and a single species in Central America (*alstoni*). Most maps (e.g. Emmons and Feer 1997) suggest that woolly mouse opossums are absent from the vast cerrado and caatinga of central Brazil; however, recent collections provide new locality records from the gallery forests and brejos of these areas (L. P. Costa and Y. R. Leite unpubl. data).

Our data for *Micoureus* are relatively extensive and include samples of all Amazonian species as well as *M. alstoni* from Panama (Fig. 9). We have examined 63 individuals for at least 630 bp of cytochrome b sequence. Phylogenetic analyses support four distinct, sharply delineated species, although these are not concordant with those listed by Gardner (1993) and Emmons and Feer (1997). The widespread Amazonian *demerarae* is sympatric with *regina* in western Amazonia, so the species status of these two is not in doubt. Patton et al. (2000) provide morphological diagnoses of both and document different habitat preferences where they are sympatric in western Brazil. *Micoureus regina* is relatively uniform over its sampled range, with only 2.3% sequence divergence between localities from western Brazil and northern Peru. However, the usual application of the name *regina* for this western Amazonian clade may be in error, since its type locality ('West Cundinamarca') is probably within the trans-Andean region as defined by Haffer (1975) and not

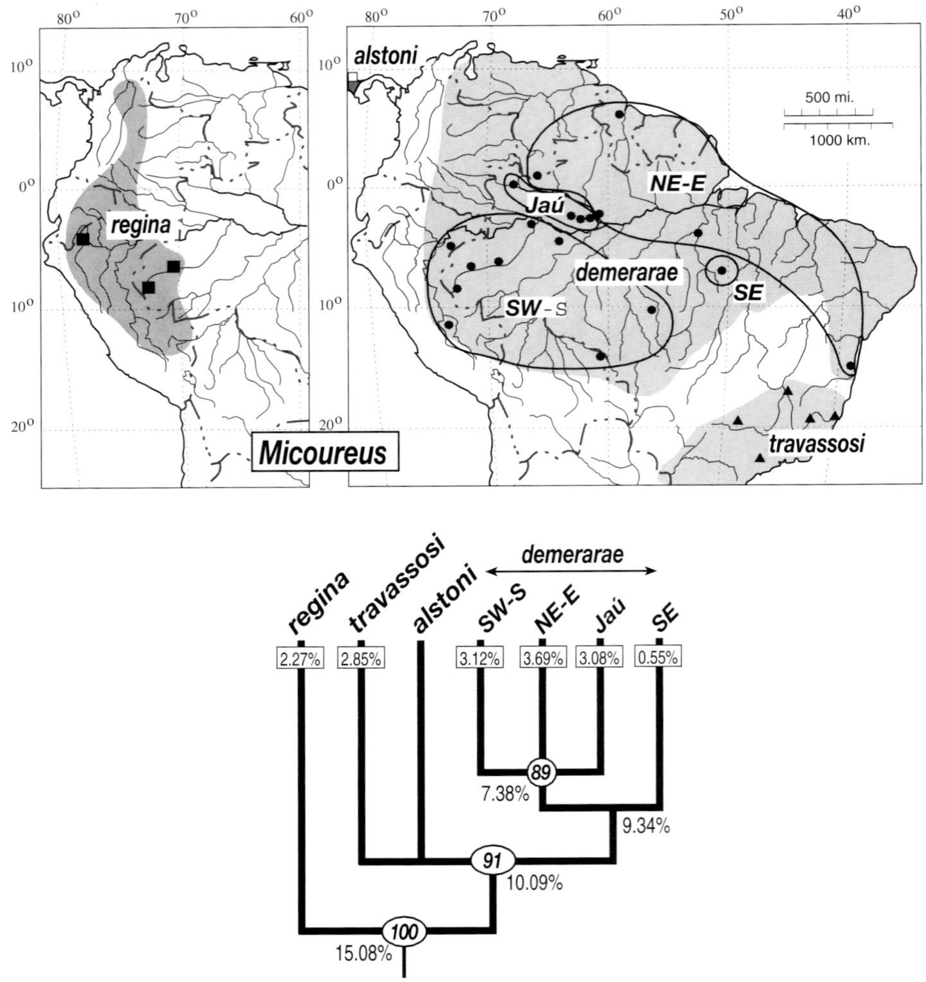

Figure 9 Distribution of the woolly mouse opossum genus *Micoureus* in South America. Localities for specimens sampled for 630 bp of the cytochrome b gene are indicated by the symbols (solid square = *M. regina*; open square = *M. alstoni*; solid circles = *M. demerarae*; and solid triangles = *M. travassosi*). Phylogenetic relationships between these species and geographic units within *demerarae* are indicated by the cladogram, a strict consensus tree of 38 haplotypes based on 163 parsimony informative characters (length = 499 steps; consistency index = 0.544). Bold numbers at nodes are bootstrap values, based on 1000 iterations; percentages at internal and terminal nodes are average Kimura 2-parameter distances.

Amazonia (see Patton et al. 2000). Tate (1933), in his revision of *Marmosa* (*sensu lato*) restricted his concept of *regina* as a species to northern Colombia and used *germana* Thomas as the name for the western Amazonian species we recognise here. Certainly, *regina* (or *germana*) is the most distinctive species of *Micoureus* that we have examined, differing on average by more than 15% from all others, including the Central American *alstoni* and the sympatric *demerarae* (Fig. 9).

Micoureus demerarae is both morphologically polytypic over its broad range and comprised of several highly divergent phylogeographic units. Each clade is relatively uniform, with differentiation among haplotypes from different localities never exceeding 3.7%, even though the geographic distance between them may be more than 1000 kilometers (Fig. 9). The most divergent clade occurs in the Atlantic Forest in the states of Espírito Santo, Rio de Janeiro, Minas Gerais, and São Paulo, and is as distant (10%) from the remainder of the geographic units of *demerarae* as is *M. alstoni* from Panama. Patton et al. (2000) suggested that this clade is a distinct species, applying to it the name *limae* Thomas, which has its type locality in the north-eastern coastal state of Ceará, Brazil. However, more recent collections from the state of Bahia, south of Ceará, link closely to those from the Guianas and eastern Amazonia (Fig. 9). Since the type locality of *demerarae* (in Guyana) is within this region, *limae* is likely a synonym of *demerarae* and another name would apply to the southern Atlantic Forest clade.

For the moment, we use *travasossi* Miranda Ribeiro, with type locality in Rio de Janeiro state, for this taxon, as all of our samples come from central coastal Brazil. However, *paraguayana* Tate, with its type locality in eastern Paraguay, might ultimately prove to be the earliest name available for this taxon.

Micoureus demerarae, as we limit this species here, thus occurs throughout Amazonia, the Guianan region, and around the 'horn' of Brazil to the north coastal state of Bahia (Fig. 9). The four geographic phylogroups include two that are quite widespread and two more restricted, but all are rather divergent from one another. The single sample from the Serra do Carajás in Pará state, south-eastern Amazonia, is the most divergent, averaging more than 9% in sequence from the others, which are equidistant from each other at 7.4%. The question of whether *demerarae* is thus a regionally diverse and polytypic species or whether each geographic unit represents separate species must await further geographic sampling and analyses. Our samples from the SW-S clade (Fig. 9) include localities in eastern Bolivia near the type locality of *constantiae* at Chapada, Mato Grosso, Brazil. Thus, if more than one species is represented, the name *constantiae* would apply to the SW-S clade and *demerarae* to the NE-E clade; the other two (and others to be found?) currently have no available name.

***Monodelphis* Burnett, 1830**. This is the most diverse genus of didelphid in the New World, with at least 15 species currently recognised (Gardner 1993). It is also the most diverse in the variety of habitats occupied, with species found in Andean elfin forests, lowland rainforests, savannas, and grasslands from Panama to Argentina, including the arid Chaco in Bolivia and Paraguay and Caatinga in north-eastern Brazil. Paradoxically, it is also the least known genus, both in the limits of species and their respective geographic ranges. The genus comprises relatively small-bodied terrestrial marsupials with a short tail and small ears. Colour and colour pattern is extremely variable, ranging from the uniform pale grey (*domestica*) to plain brownish (*adusta*) or brownish with distinct black stripes (*americana* or *iheringi*) but also with many combinations of grey, black, and brown with yellow (*dimidiata*) or bright red (*brevicaudata*). In this broad range of colour and colour patterns, *Monodelphis* contrasts sharply with other murine opossums, all species of which tend to be only slight colour variants on the same overall theme.

Some species have received recent revisionary attention (Pine 1976, 1977, 1979; Pine and Abravaya 1978; Pine and Handley 1984; Pine et al. 1985), but only Gomes (1991) has reviewed the entire genus. Gomes also recognised 15 species, but his taxonomic arrangement is quite different from that of Gardner (1993) and others. Clearly, much remains to be done before we achieve a reasonable understanding of the systematics of this genus. In this context, we present our limited molecular data only as a first approximation to untangle the complexities of *Monodelphis*. Our samples are from few localities widely scattered from the Peruvian Andes through Amazonia and the Atlantic Forest, and include materials from the dry forests of interior Brazil and Bolivia. Because Gomes' work is a M.Sc. thesis that has not been published, and thus is not generally available, we will use those names given by Gardner (1993) and Emmons and Feer (1997) as a guide in our discussion. We have 830 bp of sequence from a total of 25 individuals representing eight of the species listed by Gardner. Most of these taxa are represented by only one or two samples, a woefully inadequate coverage of the ranges for the species we examine.

The species units we recognise (Fig. 10) are highly divergent from one another in sequence, averaging 18.4% in all comparisons. As a result, there is no resolution in phylogenetic analyses of these species; rather, all fall together as a basal polytomy. The only exception is our samples of *domestica* and *brevicaudata*. These exhibit somewhat lower levels of differentiation, but still average slightly over 12%, and group together in the parsimony analyses with a bootstrap value of 92. Even the geographic representatives of some species are highly divergent. For example, our *brevicaudata* samples fall into two geographic units, one from central and eastern Amazonia and the other from south-western Amazonia (Fig. 10a). These two differ by 12.9%, a value greater than that in the comparison of either to *domestica* (11.8 and 11.4%, respectively). Similarly, although not to the same degree, our two samples of *americana* (one from São Paulo state and the other from Bahia state, along coastal Brazil; Fig. 10b) differ by 8.8%, and even the two individuals of south-western Amazonian *brevicaudata* differ by 7%. On the other hand, samples of *adusta* from northern to southern Peru (Fig. 10c) show comparably limited divergence, at an average of only 3.4%, as do our samples of *emiliae* from Peru, Brazil, and Bolivia (1.3%; Fig. 10a).

Subfamily Caluromyinae

***Caluromys* J. A. Allen, 1990**. We have sampled only two of the three species of woolly opossums (*lanatus* and *philander*), both occurring in the lowland rainforests of Amazonia and the Atlantic Forest. We lack samples of *derbianus* from Central America and the Pacific coast of north-western South America (Emmons and Feer 1997). While *lanatus* is distributed primarily in western Amazonia and *philander* in eastern regions, the two species are sympatric in the region of Serra da Mesa in northern Goiás in central Brazil, and Voss and Emmons (1996) record sympatry in the central Amazon north of Manaus (Fig. 11). Highly adapted to an arboreal existence, woolly opossums are not frequently trapped or seen in the lower forest strata without the use of canopy traps (Leite et al. 1996; Malcolm 1991; Patton et al. 2000).

We have 607 bp of sequence data from a limited number of individuals and localities of both *C. lanatus* and *C. philander*. Phylogenetic analyses are consistent with the separation of *lanatus* and *philander* as distinct species, which is unsurprising

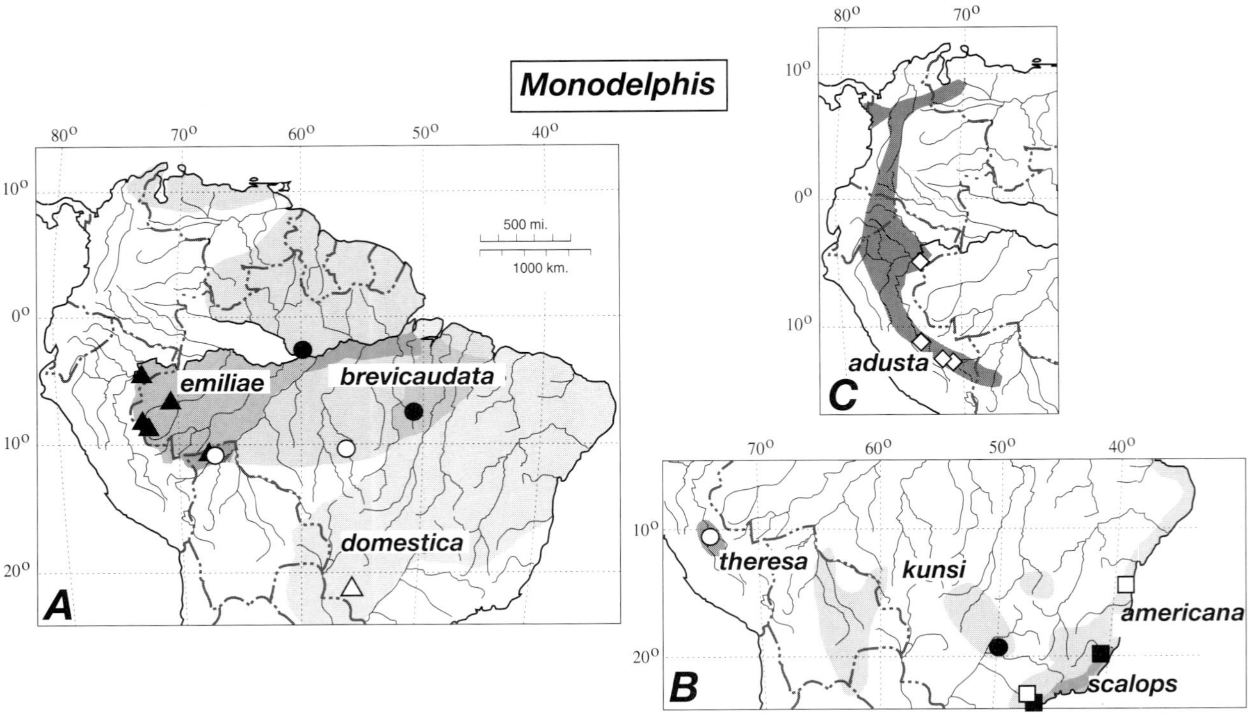

Figure 10 Distribution of some species of short-tailed mouse opossums, genus *Monodelphis*, in South America. Localities for specimens sampled for 830 bp of the cytochrome b gene are indicated by the symbols (A: solid triangles = *M. emiliae*; solid and open circles = *M. brevicaudata*; open triangle = *M. domestica*); (B: open circle = *M. theresa*; solid circle = *M. kunsi*; solid square = *M. scalops*; open squares = *M. americana*); and (C: *M. adusta*).

given their trenchant morphological differences and known sympatry. They differ by an average of 12.5% (Fig. 11). Comparison of these two with *derbianus* will be interesting, especially to determine if the Amazonian taxa are sister-species in comparison to their trans-Andean relative. Sequence differentiation is limited within *lanatus*, averaging only 2.3% among haplotypes distributed from northern Peru to southern Brazil (Fig. 11). Differentiation among the four samples of *philander* averages twice as great, at 4.5% (Fig. 11), with each haplotype nearly equidistant from the others. Given this level of divergence and the limited area examined to date, additional sampling is likely to uncover greater levels of differentiation and phylogeographic structure within *philander*. This contrasts with expectations for *lanatus*, given the high similarity among our samples of this species from localities several thousand kilometers distant.

The temporal age and biogeographic implications of rainforest lineages of didelphid marsupials

We can make several important points regarding the diversification of rainforest didelphid marsupials. First, our data generally support the monophyly of taxa recognised at the species level in current systematic accounts (e.g. Gardner 1993; Emmons and Feer 1997; Voss et al. 2001). However, the extensive phylogeographic structure apparent for many of these taxa suggests that additional species remain hidden within our current nomenclature. Second, a realignment of formal names in some cases will likely result from the combination of additional molecular analyses and revisionary studies that examine appropriate type materials and existing museum collections. Third, sequence divergence is extensive within each genus, suggesting substantial depth to the ages of most taxa. And, finally, shared patterns of phylogeographic structure are beginning to emerge, patterns that can be used to evaluate available biogeographic hypotheses concerning area relationships among the lowland rainforest areas in South and Central America.

Recognising the limitations of our data with respect to taxon and geographic area sampling, we nevertheless provide preliminary observations regarding the age and biogeographic history of these taxa. Even if premature, our observations will be of value if they help direct (and encourage!) future studies.

The age of lineages

We summarise the comparative sequence depth of each of the eight polytypic genera we have examined in Fig. 12, which pro-

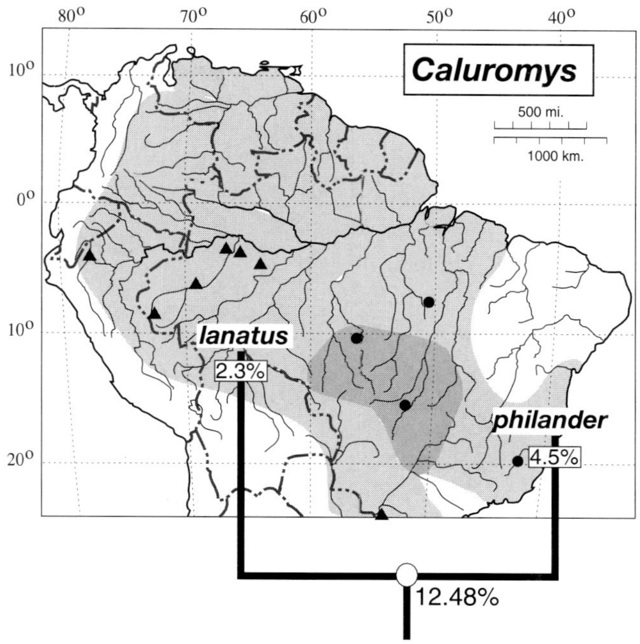

Figure 11 Distribution of the woolly opossum genus *Caluromys* in South America. Localities for specimens sampled for 607 bp of the cytochrome b gene are indicated by the symbols (solid triangles = *C. lanatus*; solid circles = *C. philander*). Phylogenetic relationships between these two species are indicated by the cladogram, a strict consensus tree of 10 haplotypes based on 163 parsimony informative characters (length = 423 steps; consistency index = 0.826). Percentages at internal and terminal nodes are average Kimura 2-parameter distances.

vides histograms of Kimura 2-parameter distances between the most divergent species in each genus relative to all others. We use this approach to avoid non-independence of data points, and to give an overview of maximum divergences within each genus. We do not provide data for *Metachirus*, since by current taxonomy it is monotypic, but we reiterate that the degree of sequence differentiation among the clades of this single species is as large, or larger, than that among many well-recognised species in other genera.

Tempered by differences in the number of species sampled and the short length (630-830bp) of sequence available, a few obvious points are apparent. First, all maximum average distances are large, although intrageneric divergences are significantly different among the sampled genera and range over a factor of two (maximum average distance of 9.5% for *Didelphis* to 19.6% for *Monodelphis*). Second, the maximum average distance generally increases with an decrease in body size. However, since there is no evidence of rate heterogeneity in the cytochrome b gene among these genera (Lavergne 1998; Patton et al. 1996), the differences in divergence can be argued to measure actual differences in the timing of species radiations within each genus. Consequently, we suggest that lineages of the small-bodied murine opossums (especially those of the *Monodelphis* and *Marmosops*) are considerably older than those of *Didelphis*, *Caluromys*, and *Philander*, for example. This hypothesis is supported by the limited fossil record of didelphids, as specimens assigned to *Micoureus* date from the Miocene of Colombia, between 13.5 and 10.1 myr (Goin 1997), while those for *Didelphis* all fall within the past two million years (Gardner 1973; Marshall 1987). The adequacy of the fossil

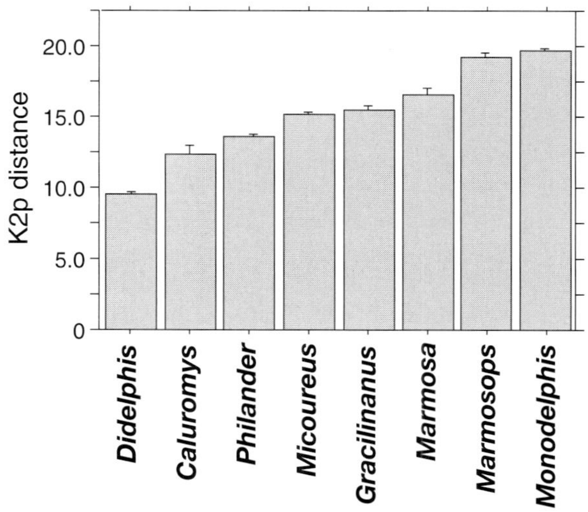

Figure 12 Histograms of average Kimura 2-parameter distances between the most divergent and all other species of eight polytypic genera of rainforest didelphid marsupials. 95% confidence limits are indicated.

record for didelphids is unfortunately poor, and more refined correlations between these molecular distances and that record cannot be made at present. It is important, however, to emphasise the depth of each of these lineages.

If the extant species of didelphids are old, then the timing of cladogenic events within each lineage is old as well. While precise calibration of a molecular clock is not possible with the limited amount of sequence data available for didelphids, a general 'rule of thumb' in molecular evolution of the mitochondrial genome is sequence divergence of about 2% per million years (Brown 1985). Even if we double this estimate, the earliest splitting event for the sequence lineages of extant species in all genera would range from 2.5 myr (*Didelphis*) to 4.9 myr (*Monodelphis*). Moreover, by this estimate, only the species pair *D. marsupialis* (Amazonia) and *D. aurita* (Atlantic Forest) would have had last shared a common ancestor within the last million years. All other species, and even most phylogeographic units within them, appear older. Consequently, the climatic fluctuations of the Pleistocene, commonly viewed to have precipitated species diversification in modern rainforest lineages (e.g. Haffer 1969, 1997), have unlikely played such a role. Moritz et al. (2000) reached this same conclusion in their review of molecular evolution of global rainforest vertebrates on three continents (Africa, Australia, and South America).

Importantly, therefore, the temporal depth of most apparent diversification events for molecular lineages within and between species of these genera also suggests that the lowland neotropical rainforests have been stable for a considerable period of time, if not in actual community composition at least in being a moist, closed-canopy forest. It will be intriguing to learn if a similar temporal depth pertains to temperate lineages (of marsupials or other small mammals) or whether the tropic and temperate zones are fundamentally different in the temporal depth of their component lineages. This comparison awaits sufficient comparative data for a wide range of organisms.

Area relationships among Neotropical rainforests

We direct our final comments to the broad phylogeographic patterns within and among the lowland rainforests of the neotropics uncovered by our data. Since we lack Central American representatives for most of the didelphid genera that occur there, we are unable to evaluate the 'sister' placement of the trans-Andean lowland forests relative to those of Amazonia and the Atlantic coast of Brazil. However, we can address the question: to what degree have lineages within Amazonia and the Atlantic Forest been evolving independently of one another, or to what degree have lineages been shared between these two large forested regions?

To date, the question of area relationships between and within the Atlantic Forest and Amazonia has been addressed primarily by avian taxa (Bates et al. 1998; Cracraft 1988; Cracraft and Prum 1988). In general, each of these two major forest biomes is comprised of independent subsets of lineages, with the Atlantic Forest 'basal' to an area relationship between Amazonia and lowland forests of Middle America. Where area connections are apparent for certain avian taxa between Amazonia and the Atlantic Forest, these occur between south-eastern Amazonia and the north coast of Brazil (Bates et al. 1998). However, for both plants and animals in general, Por (1992) has noted three historical 'pathways' of connection between Amazonia and the Atlantic Forest: a major southern route through the Paraná river basin; a secondary route to the north-east around the horn of Brazil; and a minor route via the gallery forests along rivers of the central Brazilian Cerrado.

In Fig. 13 we illustrate the approximate extent of the lowland rainforests of South America, the four divisions of Amazonia recognised by Alfred Russel Wallace in 1852 (regions bounded by the Amazon, Negro, and Madeira rivers), and the three 'pathways' potentially connecting Amazonia and coastal Brazil. Many modern workers recognise greater subdivision within Amazonia (e.g. Bates et al. 1998; Cracraft and Prum 1988; Cardoso da Silva and Oren 1996), but Wallace's simplified view is adequate for our discussion here. We also provide examples of marsupial taxa that exemplify either a complete reciprocally monophyletic separation of Amazonia from Atlantic Forest clades (Fig. 13, top cladogram) or show connections between those two biomes, either through the northern (Fig. 13, middle cladogram) or southern (Fig. 13, bottom cladogram) routes.

Examples of apparent sister species pairs distributed in Amazonia and the Atlantic Forest include *Marmosops*, *Philander*, *Micoureus*, and *Didelphis*. With the exception of *Didelphis*, divergences between species within the other genera are substantial (10–17%), supporting a period of long separation from a common ancestor. Alternatively, connections between these two global regions are seen in the phylogeography of individual species. For both *Caluromys philander* and *Micoureus demerarae*, populations from coastal Brazil are linked with those in eastern Amazonia, while for *Caluromys lanatus* and *Metachirus nudicaudatus*, the linkage is between coastal Brazil and south-western Amazonia (Fig. 13, middle and bottom cladograms, respectively). For *Marmosa murina* the path of linkage is equivocal as all three routes between the coastal and Amazon forests are supported. The level of sequence divergence in each case is relatively slight (1–5%), suggesting that the connections are more recent, and quite possibly result from Pleistocene vegetation shifts in the region.

Substantial phylogeographic structure also occurs within marsupials of both Amazonian and the Atlantic forests, and typically at divergence levels equivalent to those observed between the two regions. For example, a very deep (19%) eastern versus western division within Amazonia is apparent for *Marmosops 'parvidens'*

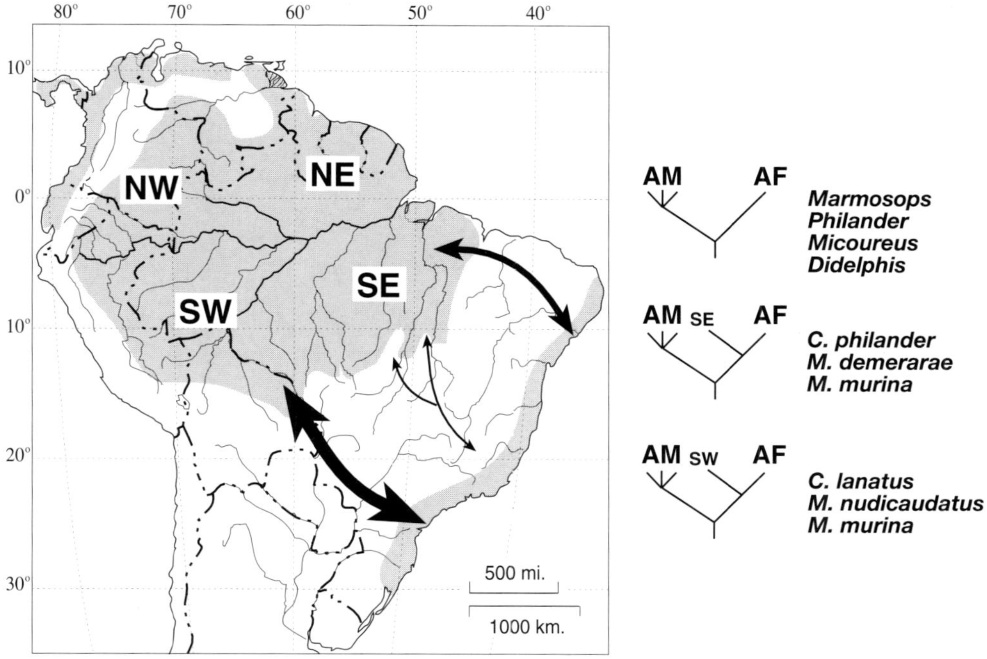

Figure 13 The approximate geographic distribution of lowland rainforest in South America. Amazonia is divided into the four regions recognised initially by Alfred Russel Wallace (1852), and the 'pathways' of connection between Amazonia and the Atlantic Forest of coastal Brazil hypothesised by Por (1992) are indicated. For these, the strength of the arrows indicates the degree of past biogeographic connection. On the right are three hypothetical cladograms depicting area relationships between Amazonia (AM) and the Atlantic Forest (AF): top – reciprocally monophyletic AM relative to AF; middle – south-eastern AM exhibits connection to AF; bottom – south-western AM exhibits connection to AF. Listed beside each cladogram are examples of didelphid genera and/or species for which their respective phylogeography patterns fit one of these three hypotheses.

(Fig. 8). Less deep, but a strongly concordant phylogeographic break is that between north-western and south-western Amazonia, across the upper Amazon. Here, reciprocally monophyletic clades that differ from 7 to 10% are present in *Philander* (*andersoni* versus *mcilhennyi*; Fig. 2), *Metachirus nudicaudatus* (Fig. 3), *Marmosa murina* (Fig. 4), and *Micoureus demerarae* (Fig. 9). This phylogeographic break is also apparent for several murid and echimyid rodent groups (da Silva and Patton 1998). Within the Atlantic Forest, divergences ranging between 15 and 17% occur in pairs of species of *Marmosops* (*incanus – paulensis*; Fig. 6), *Monodelphis* (*americana – scalops*; Fig. 10), and *Gracilinanus* (*agilis – microtarsus*; Fig. 5). The presence of strongly differentiated regional phylogeographic groups, often concordantly placed for multiple independent lineages, only reinforces our conclusions above: rainforest marsupials are old and likely have been stable in geographic position for a substantial period of time.

FUTURE DIRECTIONS

At a time when many organisations, scientific and governmental, need critical information about biodiversity in such threatened areas as the lowland neotropical rainforests, we remain woefully ignorant of even the most basic data on the limits and geographic distributions of individual species. As we have tried to make clear in this review, this is certainly the case for didelphid marsupials. Some might find this surprising, given the more than 200 years of biological investigation of South America, but for scholars with knowledge of this fauna, there is no surprise. If our limited data on phylogeographic structure serve any legitimate purpose, perhaps it will be to spur the kind of critical museum revisionary work that is needed to tie available names to morphological and molecular entities, and the field effort needed to secure additional samples of virtually all taxa to complete the pictures we provide. Our data sets are available to any and all who have the opportunity to add to them. We will send a datafile file of our sequences upon request, or we will generate sequence to add to this data set for anyone who would like their samples examined. In the latter case, we only ask that appropriate voucher material be preserved and maintained in a recognised museum, so that verification of identification can be made now or in the future.

Uncovering the true diversity of rainforest marsupials will require the concentrated effort of multiple individuals, minimally working in parallel but preferably in collaboration. We hope we have made it clear that a molecular phylogeographic approach holds great promise for uncovering the true diversity of rainforest didelphids, identifying the processes by which that diversity was generated, and understanding the history of the neotropical forests that this diversity can tell us.

Acknowledgements

We thank each of the following colleagues for providing specimens used in our research and/or for aiding in our own field collections: Maria Nazareth F. da Silva, Louise H. Emmons, Mark D. Engstrom, Alfred L. Gardner, Lena Geise, Yuri L. R. Leite, Jay R. Malcolm, Rachel T. de Moura, Robert M. Timm, Vera C. S. Vidigal, Robert S. Voss, and Terry A. Yates. Margaret F. Smith and David Pechar provided aid in the laboratory. We also thank Robert S. Voss and two anonymous reviewers for critical comments on the manuscript; any errors in fact or interpretation, however, remain ours. Financial support for either laboratory or field work was provided by the National Geographic Society, Wildlife Conservation Society, World Wildlife Foundation, Museum of Vertebrate Zoology, and National Science Foundation. Permits for our own collection of specimens were kindly provided by agencies of the Peruvian and Brazilian governments. LPC was supported by a fellowship from the Fundação Coordenação de Aperfeiçoamento de Pessoal de Nível Superior (CAPES).

References

Avise, J.C. (1994), *Molecular markers, natural history and evolution*, Chapman & Hall, New York.

Bates, J.M., Hackett, S.J., & Cracraft, J. (1998), 'Area-relationships in the Neotropical lowlands: an hypothesis based on raw distributions of Passerine birds', *Journal of Biogeography*, **25**:783–93.

Brown, W.M. (1985), 'The mitochondrial genome of animals', in *Molecular evolutionary genetics* (ed. R.J. MacIntyre), pp. 95–130, Plenum Press, New York.

Cabrera, A. (1957), 'Catálogo de los mamíferos de América del Sur', *Revista del Museo Argentino de Ciencias Naturales 'Bernadino Rivadavia', Ciencias Zoolólogicas*, **4**:1–308.

Cardoso da Silva, J., & Oren, D.C. (1996), 'Application of parsimony analysis of endemicity in Amazonian biogeography: an example with primates', *Biological Journal of the Linnean Society*, **59**:427–37.

Cerqueira, R. (1985), 'The distribution of *Didelphis* in South America (Polyprotodontia, Didelphidae)', *Journal of Biogeography*, **12**:135–45.

Cerqueira, R., & Lemos, B. (2000), 'Morphometric differentiation between Neotropical black-eared opossums, *Didelphis marsupialis* and *D. aurita* (Didelphimorphia, Didelphidae)', *Mammalia*, **64**:319–27.

Clemens, W.A. (1968), 'Origin and evolution of marsupials', *Evolution*, **22**:1–18.

Cracraft, J. (1988), 'Deep-history biogeography: retrieving the historical pattern of evolving continental biotas', *Systematic Zoology*, **37**:221–36.

Cracraft, J. (1989), 'Speciation and its ontology: The empirical consequences of alternative species concepts for understanding patterns and processes of differentiation', in *Speciation and its consequences* (eds. D. Otte & J.A. Endler), pp. 28–59, Sinauer Associates, Inc., Sunderland, MA.

Cracraft, J., & R.O. Prum. (1988), 'Patterns and processes of diversification: speciation and historical congruence in some neotropical birds', *Evolution*, **42**:603–20.

da Silva, M.N.F., & Patton, J.L. (1998), 'Molecular phylogeography and the evolution and conservation of Amazonian mammals', *Molecular Ecology*, **7**:475–86.

Eisenberg, J.F. (1989), *Mammals of the Neotropics, Vol. 1, The northern Neotropics: Panama, Colombia, Venezuela, Guyana, Suriname, French Guiana*, University of Chicago Press, Chicago.

Eisenberg, J.F., & Redford, K.H. (1999), *Mammals of the Neotropics, Vol. 1: The central Neotropics: Ecuador, Peru, Bolivia, Brazil*, University of Chicago Press, Chicago.

Emmons, L.H., & Feer, F. (1997), *Neotropical rainforest mammals. A field guide. 2nd ed.*, University of Chicago Press, Chicago.

Gardner, A.L. (1973), 'The systematics of the genus *Didelphis* (Marsupialia: Didelphidae) in North and Middle America', *Special Publications, Museum, Texas Tech University*, **4**:1–81.

Gardner, A.L. (1989), 'Two new mammals from southern Venezuela and comments on the affinities of the highland fauna of Cerro de la Neblina', in *Advances in neotropical mammalogy* (eds. K.H. Redford & J.F. Eisenberg), pp. 441–424, Sandhill Crane Press, Gainesville, FL.

Gardner, A.L. (1993), 'Order Didelphimorphia', in *Mammal species of the world: A taxonomic and geographic reference, 2nd ed.* (eds. D.E. Wilson & D.M. Reeder), pp. 15–23, Smithsonian Institution Press, Washington, DC.

Gardner, A.L., & Creighton, G.K. (1989), 'A new generic name for Tate's (1933) *microtarsus* group of South American mouse opossums (Marsupialia: Didelphidae)', *Proceedings of the Biological Society of Washington*, **102**:3–7.

Gardner, A.L., & Patton, J.L. (1972), 'New species of *Philander* (Marsupialia: Didelphidae) and *Mimon* (Chiroptera: Phyllostomidae) from Peru', *Occasional Papers of the Museum of Zoology, Louisiana State University*, **43**:1–12.

Goin, F.J. (1997), 'New clues for understanding Neogene marsupial radiations', in *Vertebrate paleontology in the Neotropics. The Miocene fauna of La Venta, Colombia* (eds. R.F. Kay, R.H. Madden, R.L. Cifelli & J.J. Flynn), pp. 187–206, Smithsonian Institution Press, Washington, DC.

Gomes, N.F. (1991), 'Revisão sistemática do genero *Monodelphis* (Didelphidae: Marsupialia)', Master thesis, Universidade de São Paulo, São Paulo, Brazil.

Haffer, J. (1969), 'Speciation in Amazonian forest birds', *Science*, **165**:131–37.

Haffer, J. (1975), 'Avifauna of northwestern Colombia, South America', *Bonner Zoologische Monographien*, **7**:1–182.

Haffer, J. (1997), 'Alternative models of vertebrate speciation in Amazonia: an overview', *Biodiversity and Conservation*, **6**:451–76.

Hall, E.R. (1981), *The mammals of North America, 2nd ed.*, John Wiley & Sons, New York.

Hershkovitz, P. (1969), 'The evolution of mammals on southern continents. VI. The Recent mammals of the Neotropical Region: A zoogeographical and ecological review', *Quarterly Review of Biology*, **44**:1–70.

Hershkovitz, P. (1992), 'The South American gracile mouse opossums, genus *Gracilinanus* Gardner and Creighton, 1989 (Marmosidae, Marsupialia): a taxonomic review with notes on general morphology and relationships', *Fieldiana Zooogy, n.s.*, **70**, v-56.

Hershkovitz, P. (1997), 'Composition of the family Didelphidae Gray, 1821 (Didelphoidea: Marsupialia), with a review of the morphology and behavior of the included four-eyed pouched opossums of the genus *Philander* Tiedemann, 1808', *Fieldiana Zoology, n.s.*, **86**:1–103.

Hice, C.L. (2001), 'Records of a few rare mammals from northeastern Peru', *Mammalian Biology*, **66**:317–19.

Hutterer, R., Verhaagh, M., Diller, J., & Podooucky, R. (1995), 'An inventory of mammals observed at Panguana Biological Station, Amazonian Peru', *Ecotropica*, **1**:3–20.

Jansa, S.A., & Voss, R.S. (2000), 'Phylogenetic studies on didelphid marsupials. I. Introduction and preliminary results from nuclear IRBP gene sequences', *Journal of Mammalian Evolution*, **7**:43–77.

Kirsch, J.A.W., Bleiweiss, R.E., Dickerman, A.W., & Reig, O.A. (1993), 'DNA–DNA hybridization studies of carnivorous marsupials. III. Relationships among species of *Didelphis* (Didelphidae)', *Journal of Mammalian Evolution*, **1**:75–97.

Kirsch, J.A.W., & Palma, R.E. (1995), 'DNA–DNA hybridization studies of carnivorous marsupials. V. A further estimate of relationships among opossums (Marsupialia: Didelphidae)', *Mammalia*, **58**:402–25.

Kirsch, J.A.W., Dickerman, A.W., & Reig, O.A. (1995), 'DNA/DNA hybridization studies of carnivorous marsupials. IV. Intergeneric relationships of the opossums (Didelphidae)', *Marmosiana*, **1**:57–8.

Lavergne, A. (1998), 'Biologie et structure génétique d'un opossum du genre *Didelphis* en forêt primaire de Guyane', PhD thesis, Universite Monpellier II, Sciences et Techniques du Languedoc, Montpellier, France.

Leite, Y.L.R., Costa, L.P., & Stallings, J.R. (1996), 'Diet and vertical space use of three sympatric opossums in a Brazilian Atlantic forest reserve', *Journal of Tropical Ecology*, **12**:435–40.

Malcolm, J.R. (1991), 'Comparative abundances of Neotropical small mammals by trap height', *Journal of Mammalogy*, **72**:188–92.

Marshall, L.R. (1987), 'Systematics of Itaboraian (Middle Paleocene) age "opossum-like" marsupials from the limestone quarry at São José de Itaborai, Brazil', in *Possums and opossums: Studies in evolution* (ed. M. Archer) pp. 91–160, Surrey Beatty & Sons, Chipping Norton, Australia.

Moritz, C., Patton, J.L., Schneider, C.J., & Smith, T.B. (2000), 'Diversification of rainforest faunas: a molecular view', *Annual Review of Ecology and Systematics*, **31**:533–653.

Mustrangi, M.A., & Patton, J.L. (1997), 'Phylogeography and systematics of the slender mouse opossum *Marmosops* (Marsupialia, Didelphidae)', *University of California Publications in Zoology*, **130**:1–86.

Patton, J.L., & Costa, L.P. (in press), 'Diversidade, limites geográficos e sistemáticos de marsupiais brasileiros', in *Marsupiais brasileiros* (ed. N.C. Cáceres & E.L.A. Monteiro-Filho), Editora da Universidade Federal do Paraná: Curitiba, Brazil.

Patton, J.L., & da Silva, M.N.F. (1997), 'Definition of species of pouched four-eyed opossums (Didelphidae, *Philander*)', *Journal of Mammalogy*, **78**:90–102.

Patton, J.L., da Silva, M.N.F., & Malcolm, J.R. (2000), 'Mammals of the Rio Juruá and the evolutionary and ecological diversification of Amazonia', *Bulletin of the American Museum of Natural History*, **244**:1–306.

Patton, J.L., dos Reis, S.F., & da Silva, M.N.F. (1996), 'Relationships among didelphid marsupials based on sequence variation in the mitochondrial cytochrome b gene', *Journal of Mammalian Evolution*, **3**:3–29.

Patton, J.L., & Smith, M.F. (1994), 'Paraphyly, polyphyly, and the nature of species boundaries in pocket gophers (genus *Thomomys*)', *Systematic Biology*, **43**:11–26.

Pine, R.H. (1976), '*Monodelphis umbristriata* (A. de Miranda-Ribeiro) is a distinct species of opossum', *Journal of Mammalogy*, **57**:785–87.

Pine, R.H. (1977), '*Monodelphis iheringi* (Thomas) is a recognizable species of Brazilian opossum (Mammalia: Marsupialia: Didelphidae)', *Mammalia*, **41**:235–37.

Pine, R.H. (1979), 'Taxonomic notes on *Monodelphis dimidiata itatiaye* (Miranda-Ribeiro), *Monodelphis domestica* (Wagner) and *Monodelphis maraxina* (Thomas) (Mammalia: Marsupialia: Didelphidae)', *Mammalia*, **43**:495–99.

Pine, R.H., & Abravaya, J.P. (1978), 'Notes on the Brazilian opossum *Monodelphis scalops* (Thomas) (Mammalia: Marsupialia: Didelphidae)', *Mammalia*, **42**:379–82.

Pine, R.H. (1981), 'Reviews of the mouse opossums *Marmosa parvidens* Tate and *Marmosa invicta* Goldman (Mammalia: Marsupialia: Didelphidae) with description of a new species', *Mammalia*, **45**:55–70.

Pine, R.H., & Handley Jr., C.O. (1984), 'A review of the Amazonian short-tailed opossum *Monodelphis emiliae*', *Mammalia*, **48**:239–46.

Pine, R.H., Dalby, P.L., & Matson, J.O. (1985), 'Ecology, postnatal development, morphometrics and taxnomic status of the short-tailed opossum, *Monodelphis dimidiata*, an apparently semelparous annual marsupial', *Annals of the Carnegie Museum*, **54**:195–231.

Por, F.D. (1992), *Sooretama, the Atlantic Rain Forest of Brazil*, SPB Academic Publishing, The Hague, The Netherlands.

Redford, K.H., & Eisenberg, J.F. (1992), *Mammals of the Neotropics, Vol. 2: The Southern Cone*, University of Chicago Press, Chicago.

Swofford, D.L. (1998), *PAUP*: phylogenetic analysis using parsimony (*and other methods), version 4.0 Beta*, Sinauer Associates, Inc., Sunderland, MA.

Tate, G.H.H. (1933), 'A systematic revision of the marsupial genus *Marmosa*, with a discussion of the adaptive radiation of the murine opossums (*Marmosa*)', *Bulletin of the American Museum of Natural History*, **66**:1–250.

Voss, R.S., & Emmons, L.H. (1996), 'Mammalian diversity if neotropical lowland rainforests: A preliminary assessment', *Bulletin of the American Museum of Natural History*, **230**:1–115.

Voss, R.S., Lunde, D.P., & Simmons, N.B. (2001), 'The mammals of Paracou, French Guiana: A Neotropical lowland rainforest fauna, Part 2, Nonvolant species'. *Bulletin of the American Museum of Natural History*, **263**:1–236.

Wallace, A.R. (1852), 'On the monkeys of the Amazon', *Proceedings of the Zoological Society of London*, **20**:107–10.

PART I

CHAPTER 6

DIVERSITY AND DISTRIBUTION OF *THYLAMYS* (DIDELPHIDAE) IN SOUTH AMERICA, WITH EMPHASIS ON SPECIES FROM THE WESTERN SIDE OF THE ANDES

Sergio Solari

Departamento de Mastozoología, Museo de Historia Natural, Universidad Nacional Mayor de San Marcos, Aptdo. 14 0434 Lima 14 Peru

Present address: Department of Biological Sciences, Texas Tech University, Lubbock, TX 79409-3131, USA. Email: ssolari@ttacs.ttu.edu

Marsupials of the genus *Thylamys* Gray 1843 (Didelphimorphia: Didelphidae) include small mouse opossums with distinctive morphological traits, mainly distributed on dry open and semi-arid biomes of southern South America. Similar to other Neotropical mammals, its species diversity and distribution are poorly documented. *Thylamys*, with *Didelphis elegans* Waterhouse 1839 as type species, was early on used as a subgenus of *Marmosa*, with a broader definition and including species now in the genera *Marmosops* and *Gracilinanus*. By the 1980s, it became clear that *Marmosa* was not a natural group, and *Thylamys* was raised to full genus and associated to the *elegans* group of Tate (1933). However, the situation for most taxa included in the genus remains obscure because good series are not available.

Peruvian specimens show a wide variation; however, they may represent to two different taxa instead of a single species as it is currently recognised: *Thylamys elegans*. Through comparison and evaluation of discrete morphological traits and morphometric analyses of the variation within and among Peruvian and Chilean populations, I assign Peruvian specimens to *tatei* and *pallidior*. By comparing *tatei* and *pallidior* to *elegans*, I show this taxon to be more restricted than previously thought.

Because *elegans* is the only valid known species from the west of the Andes, it is supposed that *tatei* and *pallidior* should have closer affinities to it. Although *elegans* and *tatei* are alike externally, *pallidior* shares more cranial characters with *tatei*. In addition, the latter two species do not show sexual dimorphism, which is evident in *elegans*. I recommend the use of *tatei* as a valid species, restricted to the western slopes of central Peru. Morphological comparisons included other taxa from the whole geographic range of the genus to update the diversity and distribution of the genus on western South America.

Morphological characters used to distinguish species proved so effective that I have used the same set of characters to group the species. Used in combination with their distributions, I group the seven recognised species in three geographic units, as a first approach to natural groups. I propose a biogeographic scenario to explain the colonisation of the western side of the Andes by the genus, typically found at temperate areas to the east of the Andes. Two different types of dispersal are supposed to have occurred in this migration, favoured by the climatic fluctuations and final uplift of the Andes during the Plio-Pleistocene.

INTRODUCTION

Mouse opossums of the genus *Thylamys* Gray 1843 (Didelphimorphia: Didelphidae) comprise small mouse opossums with nasals of nearly the same width throughout, a distinctive tricolour fur pattern, capacity to store fat in the tail (incrassation), and many other morphological traits in body proportions, skull and dentition (Tate 1933, Creighton 1984, Gardner and Creighton 1989, Hershkovitz 1992b). For a long time it was considered a synonym or subgenus of *Marmosa* (sensu Tate 1933), with a less precise and polyphyletic definition (see Gilmore 1941, Cabrera 1958, Kirsch and Calaby 1977, Marshall 1982, Reig et al. 1985, 1987). Since the monophyly of Tate's *elegans* group was confirmed (Creighton 1984), the name has been used as a full genus (Gardner and Creighton 1989, Gardner 1993).

The oldest fossil record comes from the early Pliocene (Montehermosan) of Buenos Aires, Argentina (Marshall 1982). Other records are from the late Pleistocene of Lagoa Santa, Brazil (Reig et al. 1987), but many of these seem to correspond to *Gracilinanus agilis* (see Hershkovitz 1992b). Only Marshall (1982) and Reig et al. (1985, 1987) have specifically mentioned fossils of *Thylamys*, but their composite definition of the genus does not allow knowing its exact geological range. The report by Palma (1997) based on those references could correspond either to *Gracilinanus* or to *Thylamys*.

Species of *Thylamys* are restricted to South America (Tate 1933, Gardner 1993), with most species limited to the southern portion. They prefer open and temperate areas, such as Pampas, deserts, Andean valleys, Monte desert, Chaco, and even Puna (Creighton 1985). Although there are several names applied to populations along its geographic range (see Cabrera 1958), five (Gardner 1993) to six species (Palma 1994) are now recognised as valid.

HISTORICAL BACKGROUND

The concepts and relationships of *Thylamys* Gray 1843

Formerly, *Thylamys* Gray 1843 was proposed as different from the extant genera *Didelphis*, *Marmosa*, or *Micoureus*, only to include *Didelphis elegans* Waterhouse 1839, which became type species of *Thylamys* by monotypy. Some authors (e.g. Allen and Chapman 1897, Allen 1912, Matschie 1916, Cabrera 1919) applied a broad definition when they considered *Thylamys* should include species currently listed as either *Marmosops* or *Gracilinanus*.

In the first revision of the genus *Marmosa*, Tate (1933) proposed five species groups (*cinerea*, *microtarsus*, *murina*, *noctivaga*, and *elegans*) based on morphological traits, and establishing possible relationships among them. Although he considered *Thylamys* a junior synonym of *Marmosa*, the taxon was not clearly associated to any group. Later, Gilmore (1941) suggested the relevance of some characters to join the *elegans* group and the *microtarsus* section of the *microtarsus* group of Tate (1933) under a subgenus *Thylamys*. These characters were the presence of pectoral mammae, well-developed palatal vacuities, and annular pattern of tail scales. Cabrera (1958) used the name *Thylamys*, following Gilmore (1941), to define a subgenus of *Marmosa*, without regard to other groups of Tate (1933). Despite the uncertainty of Tate's groups, they remained in use as natural taxa equivalent to subgenera. Handley (1956), Petter (1968), and Pine (1981) described and assigned new species of *Marmosa* to those groups.

Marshall (1982) used the names *Thylamys* and *Micoureus* as valid genera in a list of extant and extinct marsupials of South America. Composition of each genus based on the studies by Reig, Kirsch and Marshall, published in 1985 and 1987. Reig et al. (1985, 1987) listed *Marmosa* (with *Marmosops* as subgenus), *Micoure[u]s*, and *Thylamys* (sensu Cabrera 1958) as valid taxa. They used the Tate's groups (1933) to define each genus. *Thylamys* corresponded to an update and amplification of the polyphyletic taxon used by Gilmore (1941) and Kirsch and Calaby (1977).

Creighton (1984) raised *Thylamys* to generic status, but restricted the genus to the *elegans* group of Tate (1933), and included seven species. Gardner and Creighton (1989) rose to full genera to the *noctivaga* group and the *microtarsus* section of the *microtarsus* group (Tate 1933), using the names *Marmosops* and *Gracilinanus*, respectively.

Several hypotheses of phylogenetic relationships have been proposed for the genera of Didelphidae, most of them based on morphological characters and using cladistic methodologies. Creighton (1984) found a closer affinity between *Thylamys* (Tate's *elegans* group) and *Lestodelphys*. From Reig et al. (1985, 1987) it is practically impossible to find true affinities for *Thylamys* because of their composite definition, which includes taxa currently in genera *Thylamys* (*sensu stricto*) and *Gracilinanus*.

Kirsch and Palma (1995) used molecular techniques to group *Thylamys* and *Lestodelphys* in the tribe Thylamyini, which along with *Marmosops* and *Gracilinanus* (tribe Marmosopsini) make up the Thylamyinae. Recently, Jansa and Voss (2000) also recognised this group by comparing gene sequences of all the extant genera of Didelphidae.

Rather than discuss the confidence of proposed phylogenies, it is necessary to establish a framework for the relationships of *Thylamys* (*sensu stricto*) within the Didelphidae. Creighton (1984), Kirsch and Palma (1995), and Jansa and Voss (2000) pointed to *Lestodelphys* as the sister-group of *Thylamys*. For the purposes of this paper, I use *Lestodelphys* to evaluate the character that may help to better define the genus *Thylamys*.

Taxa associated to *Thylamys*

The genus *Thylamys*, like many other small mammals of the Neotropics (e.g. *Cryptotis*, *Monodelphis*), is known from few samples coming from very separate localities. Species descriptions were typically based on incomplete series, with specimens representing only one of the sexes or a single age class (Osgood 1943, Handley 1956, Cabrera 1958).

Although there are morphological characters that appear to define *Thylamys* (Creighton 1984, Gardner and Creighton 1989, Hershkovitz 1992b), there are no taxonomic revisions for many taxa (see Palma 1994, 1995). It is a critical aspect since many of them could belong to species complexes (Creighton 1985, Palma 1997). Often, criteria used to recognise species are not useful for all taxa, or they show high intraspecific variation which prevents valid diagnoses either at the species or subspecies level.

Gray (1843) proposed the generic name *Thylamys* to include *Didelphis elegans* Waterhouse 1839, which becomes the type species by monotypy. However, *elegans* had already been removed from *Didelphis* and assigned to *Micoureus* Lesson 1842 (see Matschie 1916). *Thylamys* was in use either as a subgenus or a junior synonym of *Marmosa* Gray 1821. One of the early names associated with *elegans* was *marmota*, mainly based on the general size and skull shape, in addition to share the 'unique character of not having the nasals expanded posteriorly' (Thomas 1894).

A second species assigned to *Thylamys* was *T. carri*, from Trinidad Island (Allen and Chapman 1897). It expanded the genus to include all the forms 'without postorbital processes, and nasals not expanded posteriorly, but of nearly the same width throughout'. Allen (1900) used this character to describe *T. keaysi*, and to include later (1912) *Marmosa caucae* Thomas 1900. All of these species are currently listed under *Marmosops* (see Gardner 1993).

Thomas's main contribution to *Thylamys* knowledge was the revision of the various forms of *elegans* (1902) where he defined two main 'geographic groups'. In later works he added new species and subspecies (1912, 1921a and b, 1926). The Andean or *elegans* group included *elegans elegans*, *e. venusta*, *e. pallidior*, *e. cinderella*, and *e. sponsoria*. His Paraguayan or *marmota* group included *marmota*, *citella*, *bruchi*, *verax*, and *janetta*.

Like Allen and Chapman (1897), Matschie (1916) regarded *Thylamys* a subgenus of *Marmosa* and included some species now listed in *Marmosops*. Only six out of nine species listed by him are referred to the present concept of *Thylamys*. Cabrera (1919) mentioned the smooth interorbital region and nasals shape as typical of subgenus *Thylamys*. His concept of *elegans* included four current species. Species *carri* and *keaysi* are now under *Marmosops*, and two species of his subgenus *Marmosa* (*pusilla* and *velutina*) actually belong to *Thylamys*.

Table 1 Recognised species of *Thylamys* (*sensu stricto*) listed by Tate (1933), and their correspondence to the current use (Gardner 1993; Palma 1995, 1997).

Use of Tate (1933)	Current use
Genus *Marmosa*	
elegans group	Genus *Thylamys*
a) Section *elegans*	
elegans elegans	*Thylamys elegans elegans*
elegans coquimbensis	*Thylamys elegans coquimbensis*
elegans soricina	*Thylamys elegans soricinus*
marmota marmota (= *grisea*)	*Thylamys macrurus*
marmota verax	*Thylamys pusillus*
janetta	*Thylamys venustus*
pusilla	*Thylamys pusillus*
b) Section *venusta*	
venusta venusta	*Thylamys venustus venustus*
venusta cinderella	*Thylamys venustus cinderella*
venusta sponsoria	*Thylamys venustus cinderella*
pallidior	*Thylamys pallidior*
bruchi	*Thylamys pallidior*
velutina (= *pimelura*)	*Thylamys velutinus*
formosa (= *muscula*)	*Gracilinanus agilis*

Tate (1933) made the first revision of all the species then included in *Marmosa* (s.l.), and grouped them in five natural groups equivalent to subgenera. The *elegans* group included two sections and nine species; all the taxa (except *formosa* and *muscula*) belong to *Thylamys* (Table 1). Although his *elegans* group is the base for present genus *Thylamys*, Tate (1933) did not mention any equivalence between them.

Osgood (1943) treated *Marmosa elegans* with detail. His analysis suggested that differences between taxa, even *pallidior*, are only of subspecific level. The variation within the *elegans* group was explained by geographic gradients, where existing gaps would be artifacts of sampling. He considered *soricina* a valid subspecies of *elegans*. Other subspecies were *coquimbensis*, *soricina*, *venusta* (synonym: *janetta*), *cinderella* (synonym: *sponsoria*), and *pallidior*. He suspected that *pusilla* and *marmota* were variations of the same kind.

Cabrera (1958) followed Gilmore (1941) in recognising *Thylamys* as a valid subgenus of *Marmosa*, and included 'the *elegans* group and the *microtarsus* section of *microtarsus* group of Tate [1933]'. He included 12 species, but just four could be assigned now to *Thylamys* (*sensu stricto*). A main departure from Osgood (1943) is the distinction of *pallidior* from *elegans*. His concept of *pusilla* included subspecies *bruchi*, *pallidior*, and *pusilla* (*marmota* Thomas 1894 as a synonym). For *grisea* he included *marmota* Thomas 1902 as a junior synonym, and for *velutina*, the subspecies *formosa* and *velutina*.

Hershkovitz (1959) proposed the undetermined growth of opossums as the main cause for small and large 'adults'. Therefore, all the taxa within the *elegans* group (Tate 1933) should be conspecifics. The first described name in that group, *Marmosa pusilla* Desmarest 1805, would be the only valid species.

Kirsch and Calaby (1977) followed the arrangement in subgenera proposed by Cabrera (1958), with additions and corrections from later works. Noteworthy changes were the ubication of *Marmosa emiliae* in subgenus *Marmosa*, the recognition of two new species (*karimii* and *tatei*), and the use of *M.* (*Thylamys*) *formosa* as different from *velutina*. The first edition of the World's mammal catalogue (Honacki et al. 1982) recognised three subgenera within *Marmosa* (*Marmosa*, *Thylamys*, and *Stegomarmosa* [Pine 1972]), but did not mention which species would belong to which group. The main reference was Kirsch and Calaby (1977).

Marshall (1982) recognised *Marmosa* as a composite of several genera. Based on unpublished studies of Reig et al. (see below), Marshall differentiated *Thylamys* from *Marmosa*, but did not explain the composition of *Thylamys*, except by the use of common name 'small mouse opossums' (p. 254).

Creighton (1984) presented a detailed phylogenetic evaluation of the intergeneric relationships within subfamily Didelphinae (s.l.), including a formal treatment for the groups of Tate (1933). Creighton proposed to differentiate the genus *Thylamys* from *Marmosa* (*sensu stricto*), and restricted the name to the *elegans* group with some changes based on revision of types and relevant specimens. After adding two recently described species, he listed *elegans*, *grisea*, *karimii*, *pallidior*, *pusilla*, *tatei*, and *velutina*.

Reig et al. (1985) presented the first phylogenetic hypothesis for the genera of Didelphidae, and showed that *Thylamys* (*sensu stricto*) was the closest to *Lestodelphys*. Included species were those proposed by Kirsch and Calaby (1977), with the addition of *lepida*, *emiliae*, and a fossil species. Reig et al. (1987) gave additional details on the employed methodology. This paper included a R.H. Pine's suggestion of recognising Tate's *elegans* group ('i.e. *elegans*, *formosus*?, *griseus*, *karimii*, *pusillus*, *tatei*, *velutinus*') as a natural group.

Gardner and Creighton (1989) recognised the soundness of Tate's (1933) supraspecific groups, as delimited by Creighton (1984). They validated the genera *Micoureus*, *Thylamys*, *Marmosops*, and created a new one for the *microtarsus* section: *Gracilinanus*. The list of species for *Thylamys* was more restricted than that of Creighton (1984), because of synonymy of *karimii* (to *velutinus*) and *tatei* (to *elegans*). Gardner (1993) followed in full this proposal for the second edition of the catalogue of World's mammal species, listing five nominal species in *Thylamys*.

In the first revision of *Thylamys*, Palma (1994) distinguished the eastern (Bolivia and Argentina) populations of *elegans* (sensu lato) as *venustus*. His hypothesis of phylogenetic relationships shows a closer affinity of *elegans* (*sensu stricto*) to the clade *pallidior–pusillus*. Finally, he provided a biogeographic explanation for these patterns. Palma and Yates (1998) used a different set of characters, with allozyme and chromosomal data to re-evaluate the genus. Once again, *T. velutinus* was not included in the analyses. There were no clear conclusions about the affinities of *pallidior*, although *elegans* rather than *venustus* was suggested as its sister-group. However, there was support for a close relationship between *pusillus* and *macrurus*.

The genus *Thylamys* on the western side of the Andes

Here I present the case of Peruvian populations of *Thylamys* as an example of problems involved in the taxonomy and systematics of the genus. *Thylamys elegans* is recognised as the only species of this genus in Peru (Gardner 1993, Pacheco et al. 1995, Palma 1997). Palma (1997) mentioned this species also for Chile, and it could be the only valid species of *Thylamys* on the western side of the Andes.

Authorities disagree about the status of names in the synonymy of *elegans*. Recognition of these taxa relies on geographic range (see Tate 1933, Cabrera 1958) and differences in the colouration of fur (see Osgood 1943). Taxa *coquimbensis* and *soricina* were used as subspecies (Osgood 1943, Palma 1997), but *tatei* and *venusta* were listed as full species (Honacki et al. 1982, Palma 1994). Other authors (Handley 1956, Hershkovitz 1992b) have suggested that more than one species of *Thylamys* would be present on the west side of the Peruvian Andes. None of these authors mentioned any relationship between *elegans* and those species (i.e. *pallidior*, *tatei*).

Didelphis elegans Waterhouse 1839, was originally described from Valparaiso, on the central coast of Chile. Another species, *Didelphys soricina* Philippi 1894, came from Valdivia, southern Chile. The name *soricina* was not in use until Osgood (1943) removed it from the junior synonymy of *elegans* to subspecific status. Other valid names, already in existence but not used, or associated to *elegans*, were *Didelphis pusilla* and *Didelphys macrura*, both from Paraguay, and *Didelphys velutinus*, from Brazil. I will deliberately omit these three species from the following discussion, as they represent taxa geographically isolated from those on the western slope of the Andes (see Palma 1995). Although *pusilla* is sympatric with *venusta* and *pallidior* on northern Argentina and southern Bolivia, it relates only distantly to *venusta* (see Palma and Yates 1998), a lowland inhabitant of the eastern slope of southern Andes.

Thomas (1894) associated *griseus* (a junior synonym for *macrura*) to *elegans*, and eight years later (1902) described several subspecies of *elegans* (s.l.). All these new names had type localities on the Andean region of Bolivia (*pallidior* and *venusta*) and Argentina (*cinderella*). However, Thomas (1902) assigned one specimen from Lima, Peru, to *venusta*, without further details. Tate (1933) recorded the distribution of *venusta*, which considered a true species, with specimens from the highlands of Lima in the subspecies *venusta* (p. 225). Although Tate (1933) was uncertain about their precise identification, he stressed their presence as the northern end of the *elegans* group's distribution, presuming they would occupy a mountain stripe between central Peru and northern Chile (on the western slope of the Andes).

Osgood (1943) updated the distribution of the Chilean populations of *elegans*, including on this the subspecies *coquimbensis* and *soricinus*, along with *venusta*, *cinderella*, and *pallidior* on the eastern slope and the Puna of Bolivia and Argentina. He also mentioned specimens of *Marmosa* (*elegans* group) from southern Peru, which would represent a northern extension of *coquimbensis*. Zuñiga (1942) and Sanborn (1949) reported specimens of *Marmosa* sp. from the 'lomas' of Atocongo, Lima, extending the distribution of the *elegans* group to the central coast of Peru. However, they were never allocated to this group by any posterior work.

Description of *Marmosa tatei*, from Ancash, Peru, represented the northern extension of Tate's *elegans* group (Handley 1956), but there was no mention of the specimens from the Rimac valley. Cabrera (1958) relegated *venusta* as subspecies of *elegans*, including in this taxon the specimens from Lima, and suggesting that these opossums could be a new undescribed subspecies, or part of *M. elegans coquimbensis*. There were no references to *M. tatei*.

Pearson and Pearson (1978) reported *Marmosa elegans* from several habitats (lomas, desert scrub, mountain scrub, and queñual) along an altitudinal gradient (from 60 up to 3900 m) on southern Peru. Mann (1978) recognised *coquimbensis* as a subspecies of *elegans*, but synonymised *soricina* to *e. elegans*. Pine et al. (1979) disagreed with him, recognising specimens from Talca, south Chile, as *elegans soricina*, and others from Tarapacá and Atacama, northern Chile, as an undetermined subspecies of *elegans*. They also mentioned a series from Arequipa, in southern Peru, as very similar to *elegans coquimbensis*, but Hershkovitz (1992b) considered them as *Thylamys pallidior*.

Later, several authors considered *Marmosa tatei* as a valid species (Honacki et al. 1982, Streilein 1982, Creighton 1984) restricting it to the type locality, without any reference to the specimens from Lima (Tate 1933), Arequipa (Pine et al. 1979), or Tacna (Pearson and Pearson 1978). Gardner and Creighton (1989) did not mention *tatei* among recognised species of *Thylamys*, but later *tatei* was regarded as a junior synonym of *elegans* (Gardner 1993). Recently, Palma (1994, 1995) raised *venusta* to the specific level, and reported specimens of *pallidior* from northern Chile (Tarapacá). In addition, he considered *tatei* as a junior synonym of *elegans*, and so all the Peruvian populations (Palma 1997).

Six specific names have been applied to the populations of *Thylamys* on the western slope: *elegans*, *soricina*, *venusta*, *coquimbensis*, *tatei*, and *pallidior*. Most recent authorities (Gardner 1993, Pacheco et al. 1995, Palma 1997) accepted *elegans* (s.l.) as the proper name for these populations. However, the name *pallidior* has been applied to populations from northern Chile (Palma 1995) and southern Peru (Hershkovitz 1992b), suggesting a migration of this species from the Puna of Chile and Bolivia. Previous studies of the genus have obscured its true diversity and, without clear morphological diagnoses, it is not possible to study their biogeographic patterns in South America.

Populations of *Thylamys* display a broad morphological variation along their distribution in Peru. These include samples from arid deserts, desert and mountain scrubs, between the sea level and 3200 m. The habitats are typical of those on the western slope of the Andes (see Cabrera and Willink 1980), and may be related to specific or subspecific variation. However, Pacheco et al. (1995) reported only one species of the genus *Thylamys*. My study aims primarily to assess the variation among the populations from Peru, and compare them with *T. elegans*, suggested by Pacheco et al. (1995) and Palma (1997) as present on western slopes of Peru and Chile. In addition, I will assess the diversity and distribution of the genus *Thylamys*, providing suitable morphological descriptions for species on the western slope of the Andes. I will provide a framework for future research about biogeography and systematics of the species, based on the information obtained from the review of type specimens and representative series.

MATERIALS AND METHODS

Specimens and institutions

Specimens I examined for this research are housed in collections of the following museums (identified throughout the chapter by their respective acronyms):

AMNH	American Museum of Natural History, New York, USA
BM(NH)	The Natural History Museum, London, UK
CBF	Colección Boliviana de Fauna, La Paz, Bolivia
FMNH	The Field Museum, Chicago, USA
MUSM	Museo de Historia Natural, Universidad Nacional Mayor de San Marcos, Lima, Perú
MVZ	Museum of Vertebrate Zoology, University of California, Berkeley, USA

NMNH National Museum of Natural History, Smithsonian Institution, Washington, DC, USA

Methods

Survey of morphological characters

In order to find useful characters to study the relationships within *Thylamys*, I used those considered in previous works on opossums (Gardner 1973, Pine 1981, Pine and Handley 1984, Creighton 1984, Reig et al. 1987, Hershkovitz 1992b), and others mentioned in original descriptions or revisions (Thomas 1902, Tate 1933).

The survey included specimens of both sexes, juveniles and adults, for most taxa within *Thylamys*, and one of *Lestodelphys*. Seventy-four (74) specimens were from Peru, 62 from Chile, and 70 from the Puna or the eastern slope of Bolivia, Argentina, Paraguay, and Brazil, representing twelve *Thylamys* taxa. I will describe each relevant taxon, identify those present on the western side of the Andes, and establish sound morphological characters to define them.

I included type specimens for five taxa previously considered as present on the western side of the Andes, and topotypes for other taxa. I reviewed the holotype and paratype of *tatei*, as well as a series of *soricinus* from Talca, south Chile, at the NMNH, and the type of *coquimbensis*, topotypes of *venusta* (sensu lato), and two specimens of *Lestodelphys* at the FMNH. V. Pacheco took notes of the holotypes of *elegans*, *pallidior*, *venusta*, and *janetta*, and topotypes of *macrurus* at the BM(NH).

Analysis of morphometric variation

I propose the use of 16 measurements for the morphometric study of specimens. They sample a variety of cranial and dental features (Fig. 1), which are informative of size differences between taxa, and allow the comparison with other works (Tate 1933, Pine 1981, Pine et al. 1985). All measurements were taken with digital calipers to the nearest 0.01 mm.

The measurements include: greatest skull length (GSL), condyloincisive length (CIL), palatal length (PL), greatest nasal length (GNL), zygomatic breadth (ZB), postzygomatic width (PZW), postorbital constriction (POC), breadth across bullae (BAB), width of single bulla (WSB), inclusive bulla-petrosal length (BPL), maxillary toothrow (MTR), width at M4-M4 (M4W), M2–M5 length (M2M5), mandibular ramus length (MRL), mandibular toothrow (LTR), m2–m5 length (m2m5).

To properly consider variation usually associated with differences at species level, I assess the variation within (sex and age) and between (geographic) populations of *Thylamys*. Sex and age variation are common sources of heterogeneity for mammals (Pine et al. 1985, Pacheco and Patterson 1992), and should not be considered for defining species.

Age was determined by using a sequence based on tooth replacement (Tribe 1990). The so-called deciduous premolars of Tribe (1990) are true molars (M1) displaced by the larger, late developing third premolar (PM3). However, as Archer (1978) and Hershkovitz (1992a) explained, all the functional adult marsupial teeth are first generation, so there are no replacement teeth. Thus, the fully erupted molar series correspond to the second to fifth molars (M2, M3, M4, and M5). Using this denomination, the sequence of Tribe (1990) is as follows:

Age class 1	M1 functional, M2 erupting.
Age class 2	M1 and M2 functional, M3 erupting
Age class 3	M1–M3 are functional, M4 erupting
Age class 4	M2–M4 functional, and M1 retained, M5 erupting (typical pattern) PM3 and M5 erupting (intermediate pattern)
Age class 5	M2–M5 functional, M1 retained or PM3 erupting
Age class 6	PM3 half to fully erupted, M5 shows little wear
Age class 7	PM3 functional, M5 considerably worn

I consider individuals of age classes 1 to 4 as juveniles, and those of class 6, as adults. Specimens in classes 5 and 7, correspond to young and old adults, respectively. Although there are no similar studies for *Thylamys*, age differences are more conspicuous between males than females of *Monodelphis* (Pine et al. 1985). The same study revealed an 'extreme sexual dimorphism' favouring males of *M. dimidiata*, and probably, by extension, to the Didelphidae (sensu lato). Palma (1997) showed sexual dimorphism in size for *T. elegans*, with males larger than females in 10 out of 12 measurements.

Because sample sizes for most of the localities were too small, analyses required combining them according to geographic proximity, to include specimens from most age classes and both sexes (see Hall 1943, Pacheco and Patterson 1992). Ten taxonomic units were identified from Peru and Chile (Table 2), representing 102 specimens from 36 western localities.

Once the nongeographic variation is evaluated, each sex or age class could be used as comparative units to study geographic variation (see Hall 1943, Pine et al. 1985). Nongeographic variation was assessed by two-way ANOVA tests. I estimated the components of variance (age, sex, interaction, and residual) for each variable. The ANOVA tests were employed only at the larger taxonomic units, before the analysis of geographic variation.

If a significative variation due to age is found, only adults (age class 6) should be included for the analyses of sexual dimorphism. An additional one-way ANOVA was used to assess the variation by sex only. According to the results of this analysis,

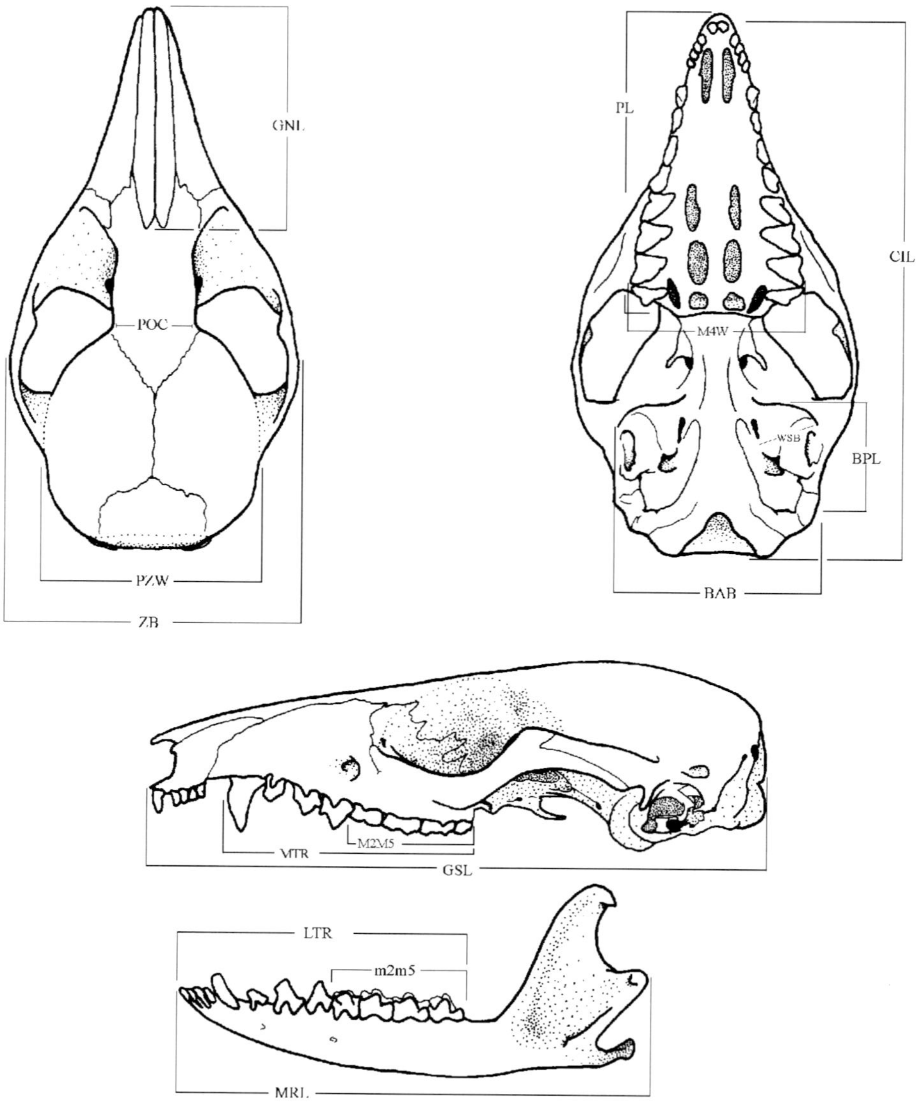

Figure 1 Cranial and dental measurements used for for the morphometric study of specimens

sexes will be treated separately or pooled for geographic analyses. One-way ANOVA's assessed geographic variation within each taxon, and evaluated differences between taxa. All the statistical analyses were performed on SPSS 7.5 (Statistical Package for the Social Sciences) for Windows.

Morphological description of *Thylamys*

The genus *Thylamys* corresponds to the *elegans* group of Tate (1933), and generic definitions by Creighton (1984), Gardner and Creighton (1989), and Hershkovitz (1992b). It is more restricted than the subgenus of Cabrera (1958) or the genus of Reig et al. (1985, 1987). *Thylamys* comprises small to moderate body-sized species, total length less than 310 mm, and ears larger than those in other Marmosinae of similar size.

Dorsal pelage is greyish-brown, darker on the dorsal midline and lighter to the flanks, with ventral pelage white to buffy or grey, making up a tricolour fur pattern (Tate 1933, Creighton 1984). On the back, fur is longer and silky with bases and more than 3/4 of its length dark grey-slate. This pattern also extends over the crown, passes the ears bases, and reaches the area in between the eyes in some species. A patch of the mid-dorsal fur covers the shoulders.

Ventral pelage is highly variable. It may be cream-buffy, whitish, or pure white, extending to all the venter (e.g. *T. macrurus*)

Table 2 Taxonomic units used for study of morphometric variation in *Thylamys* of Peru and Chile. N = number of individuals, f and m are adult females and males (respectively).

Taxonomic unit	N	f	m
Huaraz	10	06	04
Lachay	07	02	03
Central Lima	10	03	01
Central Highlands	05	02	02
Parinacochas	16	01	03
Southern Highlands	18	04	08
Southern Lomas	06	02	02
Valparaiso	20	05	10
Aconcagua	06	01	02
Tierras Blancas	04	02	02

or just to a medial band (*T. elegans*). This colouration might be almost grey for some species (e.g. *T. venustus*), because of the extensive grey-slate bases of the hairs. Hairs may be self coloured, pure white (*T. pallidior*), or grey-based with white-cream tips (*T. elegans*). Delimitation between ventral and dorsal-lateral fur may be sharp, but for some species results in a transitional strip of greyish hairs.

There are no detailed references about presence and number of vibrissae, on head or extremities (see Brown 1971, Brown and Yalden 1973) for *Thylamys* or other Marmosinae. A little variation among specimens was observed. Distribution of vibrissae for *Thylamys* taxa includes (by each body side): 2 superciliary, 5 to 8 genal, 3 submental, 2 interramal, 1 antebrachial, 1 anconeal, and 3 to 4 carpal.

The tail is short, just surpassing 50 to 55% of the total length in most species, except for *velutinus* where it is almost 45% (Petter 1968, Palma 1995, Vieira and Palma 1996). Fat storage (incrassation) is known only for this genus and *Lestodelphys* (Creighton 1984), and it is perhaps related to their survival on highly seasonal habitats (Morton 1980). Tail is covered by small scales, with an annular arrangement (>35/cm). Each scale has three hairs of similar length (2,5-3,5 scales) on its posterior border, giving to the tail a hairy appearance. It may be bicoloured, because of pigmented tips of dorsal hairs, or monocoloured, although it could be affected by incrassation. It is slightly prehensile, using just the tip to grasp thin objects.

A pouch is absent and mammae are arranged in the abdominal region. In addition, there are two pairs of pectoral teats, with a basic formula of 7-1-7 = 15, and occasionally 9-1-9 = 19 (Tate 1933, Creighton 1984, Hershkovitz 1992b) in *Thylamys*. On the contrary, *Lestodelphys* may have up to 15 mammae (Hershkovitz 1992b).

Feet are comparatively small, as well as toes (Creighton 1984), both are densely covered by white or white-cream hairs, increasing the appearance of a small size. Ungual tufts are well developed, reaching the claw tips. However, specimens of *velutinus* in the NMNH lacked these hairs. Thenar and hypothenar pads are not fused with interdigitals (Creighton 1984).

Thylamys is characterised by the shape of the nasals, which are just slightly expanded at the maxilla-frontal suture (Tate 1933, Creighton 1984, Hershkovitz 1992b). Although the nasals may be considered not expanded for some species (i.e. *elegans*), they are variable among species. The nasals may or not narrow after that suture, producing four different patterns by the combination of both characteristics (see below).

Supraorbital processes are not well developed on most species (Tate 1933), although *macrurus* and *janetta*, and old individuals of *pusillus*, *pallidior* and *venustus* (s.l.) might show some degree of beading. Width of the postorbital constriction is variable. Supraorbital processes continue parallel or diverging over braincase, but converge in a sagittal crest in *macrurus*.

The palate is highly fenestrated because of presence of palatine or posteromedial vacuities (Hershkovitz 1992b), enlargement of posterolateral ones, and the occasional development of mesolateral ones in some species (Creighton 1984, Tate 1933). A fenestrated palate is said to characterise *Thylamys* (Gilmore 1941, Cabrera 1958, Reig et al. 1985), however the condition is shared with *Gracilinanus*. Although specimens of *pallidior* presented enlarged vacuities, there are no bases to diagnose any species of *Thylamys* based on the development of some particular fenestra. Premaxillae are rounded (Redford and Eisenberg 1992).

Auditory or tympanic bulla is large and well developed, round shaped, with a slender anteromedial extension of its wall to the alisphenoid floor (Tate 1933, Creighton 1984, Hershkovitz 1992b). Distance between bullae is less than 1.5 times the width of a single bulla, which was considered diagnostic by some authors (Creighton 1984). Presence of a slender anteromedial process bulla-alisphenoid (Tate 1933, Creighton 1984, Hershkovitz 1992b) is another characteristic of *Thylamys*, which is shared with *Lestodelphys*. The slender and narrowed shape of basicranial bridge was suggested by Tate (1933) as typical of his *elegans* group. However, it has a variable shape among *Thylamys* species, therefore lacking taxonomic significance.

Dentition is similar to other Marmosinae, but third upper premolar (PM3) is equal or larger than second one (PM2), in height and length (Tate 1933, Creighton 1984, Hershkovitz 1992b). Molars in general and first upper three in particular, show great compression in length and increase of width. Paracone of second molar (M2) slightly displaced towards the lingual border (Tate 1933). Lower canines are pointed with a well-developed cusp.

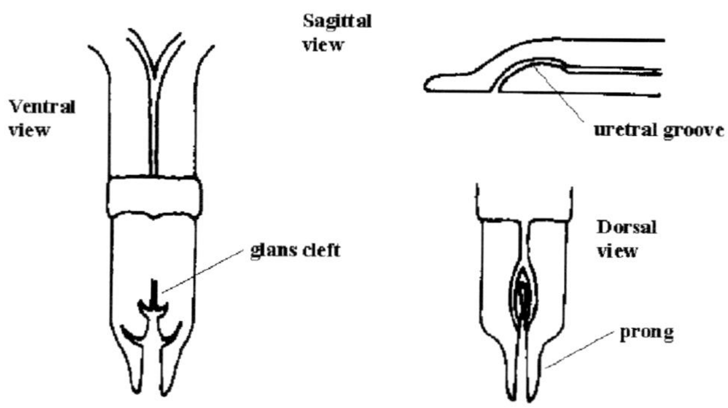

Figure 2 Distinctive characteristics in the penis of *Thylamys*, as shown by *T. pallidior*

For Peruvian and Chilean specimens of *Thylamys* I found a range of 20 to 21 caudal vertebrae. Hershkovitz (1992b) reported 21 vertebrae for a single specimen of *elegans*. In other marmosine genera, number of caudal vertebrae ranged between 22 and 27, apparently related to arboreal habits.

Szalay (1982) and Hershkovitz (1992a) discussed the significance of articular pattern of ankle joint bones for the phylogeny of marsupials. Hershkovitz (1992a) recognised these two patterns and showed both to occur among Didelphimorphia; even they could define natural groups. The separate pattern was reported for astragalus and calcaneus of *Thylamys elegans* and *T. pallidior*. Available specimens confirm this pattern, at least for the astragalus, in *T. pallidior* from Peru, and *T. venustus* from Bolivia.

Although studies of glans penis morphology are just preliminary, they have shown significant variation of taxonomic use to generic level (Reig et al. 1987, Hershkovitz 1992b, Martinelli and Nogueira 1997). In mature individuals, the penis is bifid, with urethral groove extending over internal side of each half allowing its urinary and ejaculatory functions (Hershkovitz 1992b). The most distinctive characteristic in the penis of *Thylamys pallidior* is its short glans cleft, determining shorter halves (Fig. 2). But, short length of glans (6–7 mm), subterminal ending of the urethra, and halves with pointed tips are similar to those of *Gracilinanus* and *Marmosops* (see Martinelli and Nogueira 1997). There is a skin fold on the inner side of each half that could be homologous to that of *Marmosops incanus*, but without more evidence than drawings of Martinelli and Nogueira (1997), I considered this feature particular to *Thylamys*.

The chromosome diploid number for *Thylamys* is 2N=14, which is common with other murine opossums (Palma and Yates 1996, 1998). Sex chromosome variation, as well as mosaicism, has been reported (Palma and Yates 1998).

Variation of morphological characters

Although most of the previously described traits are characteristics for *Thylamys*, some of them showed a variation that could be used with taxonomic purposes. For each character, I describe related anatomical structures, its condition in the studied taxa and *Lestodelphys*, and discuss its use in previous references.

Externals

(01) Colouration of dorsal fur (medial and lateral bands)

The dorsal pattern of colouration on the back, with two dorsal bands, has been considered diagnostic for *Thylamys* (Tate 1933, Hershkovitz 1992b). The mid-dorsal band is very conspicuous, due its darker colouration and width, in *elegans, coquimbensis, soricinus, venustus, sponsoria, cinderella, janetta, pallidior, tatei, macrurus*, and *pusillus*. In *velutinus*, the contrast relative to the lateral bands (of back) is not too clear. The mid-dorsal bands of *sponsoria, soricinus*, and *tatei*, are darker, and with longer hairs. In *Lestodelphys*, the mid-dorsal band is also dark and conspicuous.

(02) Eye-rings

Blackish eye-rings are well developed and projected toward the nose in most species of *Thylamys*, except *velutinus, pusillus*, and *macrurus*. In those species, eye-rings are limited to the area around the eyes. Well-developed and projected eye-rings are present in *Lestodelphys*.

(03) Colour of ventral fur

Ventral fur exhibits a large variation in three of its traits: hair colouration, length and width extension of the mid band. Thomas (1902) used this variation to describe a number of *elegans* subspecies. Ventral hairs are lighter than those on dorsal and lateral bands, ranging from pure white to cream-buffy. Most of these hairs are self-coloured, from the base to the tips. However, in *pallidior* and *coquimbensis*, the hairs are pure snow-white. They are cream-white to whitish in *elegans, tatei, janetta, pusillus, macrurus*, and *velutinus*, but cream-buffy in *soricinus, venustus, cinderella*, and *sponsoria*. Hairs are snow-white in *Lestodelphys*.

(04) Length extension of lighter ventral band

The ventral hairs form a lighter band on the underparts of *Thylamys*, which has a variable extension (Thomas 1902, Osgood 1943). It might extend from the chin to the chest, as occurs in *soricinus, venustus, cinderella*, and *sponsoria*, or to the anus, through the belly, as in *elegans, coquimbensis, janetta, pallidior, tatei, pusillus, macrurus*, and *velutinus*. A band of pure white hairs extends from the chin to the anus in *Lestodelphys*.

(05) Width of the intermediate greyish bands

The presence and width of intermediate greyish bands, determined the width of the lighter ventral band (Tate 1933, Osgood 1943). Intermediate bands include hairs of grey-slate to blackish bases, with grey to cream-buffy tips. In *coquimbensis, pallidior, janetta, pusillus, macrurus*, and *velutinus*, these are very narrow (less than 5 mm each side), so the lighter ventral band extends

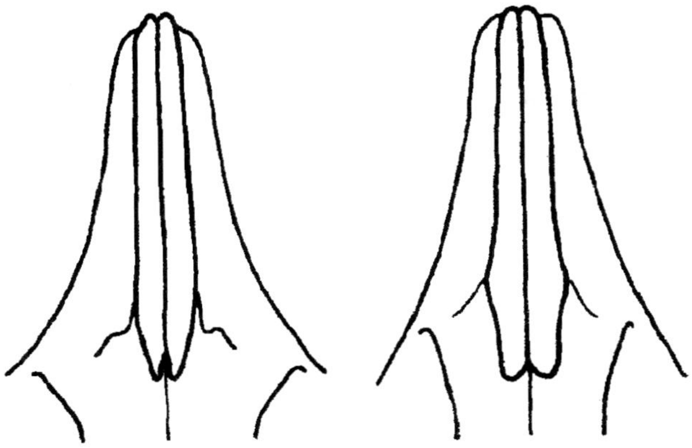

Figure 3 Shape of nasals at the maxilla-frontal suture in *Thylamys*. Left: almost parallel sides, as shown by *elegans*. Right: a little conspicuous expansion, as shown by *tatei*.

to the borders of venter. For other taxa, like *elegans*, *soricinus*, *venustus*, *cinderella*, *sponsoria*, and *tatei*, intermediate bands are wider (7–10 mm), resulting in a narrow ventral band. Ventrally, *Lestodelphys* shows narrow greyish bands.

(06) Relative size of the tail

Among *Thylamys* taxa, there is a little variation in the relative size of the tail, although only *velutinus* has a tail shorter than head and body length. All other taxa have tails longer than the head and body length. A short (and robust) tail is also present in *Lestodelphys*.

(07) Colour of distal end of the tail

Tail colouration in *Thylamys* is typically greyish to brownish on the dorsum, and lighter (white to whitish) on the ventral side, with variations of tone because of colour of tail scale hairs on each side. Nevertheless, *macrurus* and *tatei* present a deviation of this pattern. In *macrurus*, the distal third of the tail is particoloured; for *tatei*, the tip is whitish. Incrassation may affect the typical tail colouration. A whitish tail tip is also present in *Lestodelphys*.

Skull and dentition

(08) Shape of nasals at the maxilla-frontal suture

Nasals shape in *Thylamys* has been used as diagnostic, because of its scarce variation among taxa (Thomas 1894, Allen and Chapman 1897, Hershkovitz 1992b). However, there are two characters related to this shape. One is the nasal width at the maxilla-frontal suture, which may be almost parallel-sided, as occurs in *elegans*, *coquimbensis*, *soricinus*, *venustus*, *cinderella*, and *sponsoria*. However, for *pallidior*, *tatei*, *janetta*, *macrurus*, *pusillus*, and *velutinus*, the expansion is more conspicuous (Fig. 3). Nasals are conspicuously expanded in *Lestodelphys*.

(09) Shape of nasals behind the maxilla-frontal suture

A second character is the narrowing of the nasals just behind this suture, which is more evident by comparing to the nasal width at the suture. A light but conspicuous narrowing of nasals is present in *elegans*, *coquimbensis*, *pallidior*, *tatei*, and *velutinus*. Nasals of almost the same width are presents in *soricinus*, *venustus*, *cinderella*, *sponsoria*, *janetta*, *macrurus*, and *pusillus* (Fig. 4). In *Lestodelphys*, the nasals narrow behind the maxilla-frontal suture.

(10) Frontal-parietal processes

Creighton (1984) suggested a reduced development of the lateral edges of frontals in *Thylamys*, producing rounded superior borders. However, specimens of *janetta*, *macrurus*, and *velutinus* present conspicuous, squared to sharpened edges, which could occur also in other species (see Tate 1933). Old adults (age class VII) presented slightly beaded borders too, as evident in *pusillus*, *venustus*, *cinderella*, *sponsoria*, *pallidior*, and *tatei*. Processes are lacking even among old adults of *elegans*, *coquimbensis*, and *soricinus*. These processes show a light development in *Lestodelphys*.

(11) Development of stylar cusps on the upper molars

Development of a stylar cusp on the second and third upper molars may be used to group some taxa, as they are present only in *macrurus* and *pusillus*. No other *Thylamys* present these cusps. Development of this stylar cusp modifies the shape of the ectoflexus (the labial margin of molars), so it appears serrated in *macrurus* and *pusillus*. The stylar cusp of the anterior upper molars is also developed in *Lestodelphys*.

RECOGNISED SPECIES AND DISTRIBUTION OF *THYLAMYS*

The observed variation in discrete characters allows recognising seven species within *Thylamys*, as well as several of the subspecies mentioned by Tate (1933), Cabrera (1958) and Gardner (1993).

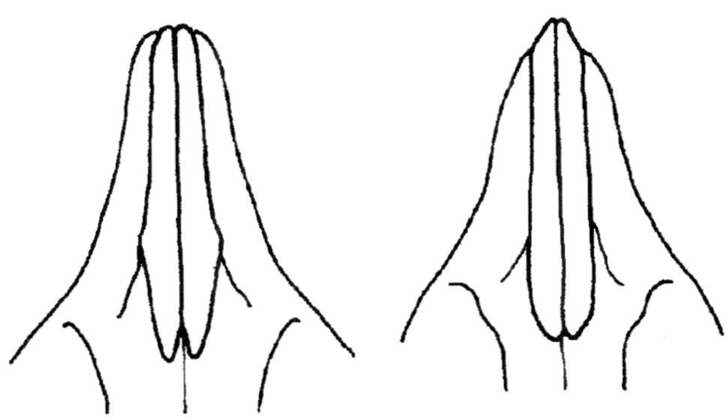

Figure 4 Shape of nasals behind the maxilla-frontal suture in *Thylamys*. Left: nasals just narrowing a little, as shown by *pallidior*. Right: nasals of almost the same width along, as shown by *venustus*.

Table 3 Diagnostic traits of five species of *Thylamys*, from the sampled variation in eight morphological characters (see the text for details)

	pusillus	*venustus*	*elegans*	*pallidior*	*tatei*
Blackish eye-rings	Not well developed	Developed and extended	Developed and extended	Developed and extended	Developed and extended
Ventral colouration	White cream to whitish	Cream-buffy	White cream to whitish	Snow white	White cream to whitish
Ventral lighter band	From the chin to the anus	From the chin to the breast	From the chin to the anus	From the chin to the anus	From the chin to the anus
Width of greyish ventral bands	Less than 4 mm at each side	More than 5 mm at each side	More than 5 mm at each side	Less than 4 mm at each side	More than 5 mm at each side
Shape of nasals at the suture	Slightly expanded	Parallel sided	Parallel sided	Slightly expanded	Slightly expanded
Nasals width behind suture	Almost not narrowed	Almost not narrowed	Conspicuously narrowed	Conspicuously narrowed	Conspicuously narrowed
Supraorbital processes	Slightly beaded to squared	Slightly beaded to squared	Rounded borders	Age related development	Age related development
Stylar cusp C	Developed	Absent	Absent	Absent	Absent

This composition matches closely with the *elegans* group (Tate 1933), with the inclusion of *bruchi*, *verax*, and *marmota* as synonyms of *pallidior*, *pusillus*, and *macrurus*, respectively (see Gardner 1993), and the exclusion of *formosa* (see Gardner and Creighton 1989). The species *venusta* was included as subspecies or synonym of *elegans* by Cabrera (1958) and Gardner (1993). However, it was recognised as a valid species by Palma (1994).

Additional to the six species recognised by Palma (1994), *Marmosa tatei* Handley 1956 shows a particular combination of characters (Table 3) that along its geographical isolation in Peru allow to specifically distinguishing it from *elegans*. Although listed as full species by Honacki et al. (1982), *tatei* was included in *elegans* by Gardner (1993) and Palma (1997). These authors used *elegans* as the only valid name for populations on the western side of the Andes, so including to *coquimbensis*, *soricinus* and *tatei* as subspecies (or synonyms).

According this variation, populations from Peru should be referred to *Thylamys tatei* and *T. pallidior*, and those of Chile to *T. elegans* and *T. pallidior*. As detailed in Table 3, none of the species present in Peru may be confused with *elegans*. Two subspecies of *elegans* are recognised: *coquimbensis* and *soricinus*; both are identifiable based on a particular set of characters that differentiate them from *elegans elegans*. These subspecies are externally similar to *pallidior* and *venustus*, respectively. However, both present the typical parallel-sided nasals of *elegans*, with some degree of variation, and rounded supraorbital borders.

Recognition of *pallidior* in southern and central Peru implies for this species the longest latitudinal distribution among *Thylamys* taxa. It goes from the Patagonia (Birney et al. 1996) of Argentina, to the central western slope of Peru, maybe following the Andean range. Use of *venustus* as a full species by Palma (1994, 1995) did restrict *elegans* to central Chile, on the western slope of the southern Andes. Meanwhile, *tatei* is found at a small area between Lima and Ancash departments on the western slope of north Peru (Fig. 5).

Sympatry between species of *Thylamys* is not extensive or documented. From this revision, it appears that most of them are specialised on certain habitats, but they may extend also to contiguous regions (see Cabrera and Willink 1980). On the western side of the Andes, *pallidior* would be sympatric with *elegans coquimbensis* in northern Chile, and parapatric with marginal populations of *tatei* in northern Peru (Fig. 5). Subspecies of *elegans* occur parapatrically along a latitudinal gradient on central and northern Chile, with *soricinus* on the southern and *coquimbensis* on the northern end (see Mann 1978, Palma 1997).

Other species in the genus are commonly found at the southern part of South America (see below), onto the eastern slope of the Andes. They prefer dry and open habitats, like the Chaco and Pampas (Tate 1933), but also in mountain and desert scrub, or even the yungas. Two species are inhabitants of tropical areas (see Palma 1995, Vieira and Palma 1996): *macrurus* is the only species adapted to the subtropical moist forests of Paraguay, and *velutinus* occurs in the semi-arid habitats of the Cerrado and Caatinga of Brazil (Fig. 5).

Although the status of the name *Thylamys* is stable, as a masculine or neuter noun (Monjeau et al. 1994), the situation is confused among its species. Creighton (1984) listed the names *grisea*, *pusilla*, and *velutina*, which did not agree with the gender of the genus. Gardner (1993) changed the last one to *velutinus*. Following the Code of Zoological Nomenclature (ICZN 1999)

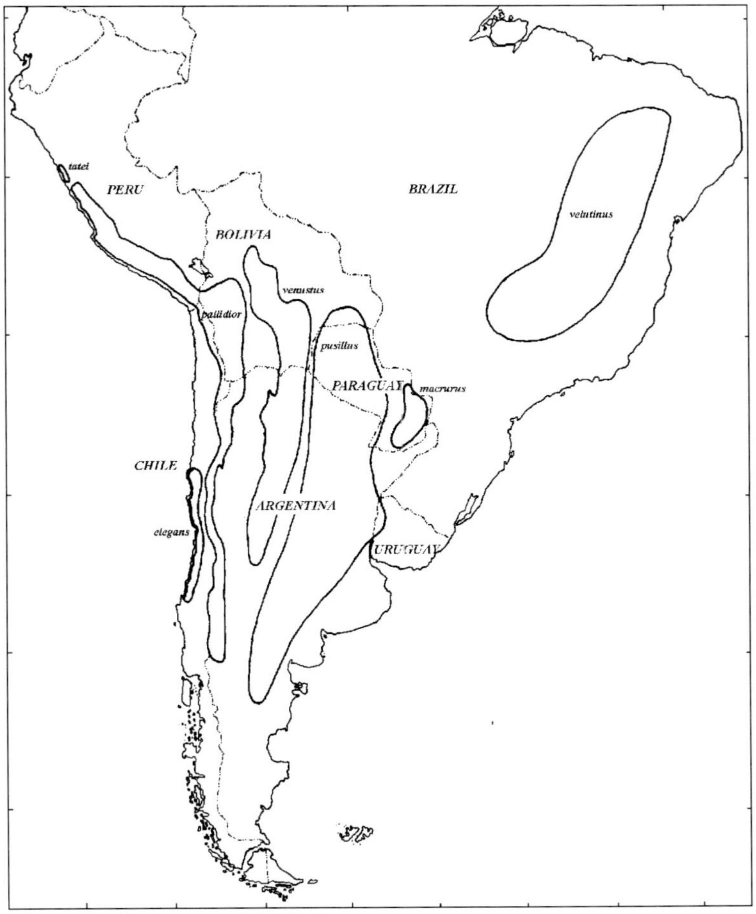

Figure 5 Distribution of *Thylamys* species

in regard to species-group names, if they are adjectives they should agree in gender with the genus (Art. 31.2). This is the case of names (already changed): *velutinus, pusillus, macrurus, venustus, sponsorius*, and *soricinus*. The rule does not apply for *janetta* or *cinderella*, which have not been used as adjectives. It is considered as a name in apposition, and does not need to be changed (Art. 31.2.2). Other specific names, such as *elegans, pallidior, coquimbensis*, and *tatei*, are neuter or formed from personal names (Art. 31.1), and remain unchanged.

Species groups in *Thylamys*

In order to obtain a useful framework to further studies of *Thylamys* species, I group them using their morphological characters (Tables 3 and 4) and geographic distribution (Fig. 5) in three units. These are proposed as equivalent to natural (i.e. monophyletic) groups. Two of them are distributed to the eastern side of the Andes, and the last one (the Andean group) includes all the taxa occurring to the western side of the Andes, although one species (*venustus*) is found only to the east.

Because of my emphasis on species of the Andean region, full details are not given for all the species, but information on names in synonymy is included.

The Brazilian group
Thylamys velutinus
Didelphys velutina Wagner 1842

Archiv für Naturgeschichte, 8: 360

Didelphis pimelura Reinhardt 1849 (Lagoa Santa: Brazil)

Marmosa karimii Petter 1968 (Pernambuco: Brazil)

Type locality: Ipanema, Sao Paulo, Brazil

Description: Medial dorsal band inconspicuous, fur long (>7 mm), greyish, mouse-like; eye-rings not well developed; underparts cream-white, very short hairs; narrow (3 mm) intermediate greyish band. Tail shorter than head and body length (almost 0.75 HB), it is the only species in the genus with this

Table 4 Diagnostic traits of *Lestodelphys* and four species of *Thylamys*, from the sampled variation in nine morphological characters (see the text for details)

	Lestodelphys	*velutinus*	*macrurus*	*pusillus*	*venustus*
Blackish eye-rings	Developed and extended	Not well developed	Not well developed	Not well developed	Developed and extended
Ventral colouration	Snow white	White cream to whitish	White cream to whitish	White cream to whitish	Cream-buffy
Ventral lighter band	From the chin to the anus	From the chin to the anus	From the chin to the anus	From the chin to the anus	From the chin to the breast
Width of greyish ventral bands	Less than 4 mm at each side	Less than 4 mm at each side	Less than 4 mm at each side	Less than 4 mm at each side	More than 5 mm at each side
Relative length of tail	Shorter than head and body	Shorter than head and body	Longer than head and body	Longer than head and body	Longer than head and body
Shape of nasals at the suture	Conspicuously expanded	Slightly expanded	Slightly expanded	Slightly expanded	Parallel sided
Nasals width behind suture	Conspicuously narrowed	Conspicuously narrowed	Almost not narrowed	Almost not narrowed	Almost not narrowed
Supraorbital processes	Slightly beaded to squared	Squared to sharpened	Squared to sharpened	Slightly beaded to squared	Slightly beaded to squared
Stylar cusp C	Developed	Absent	Developed	Developed	Absent

characteristic. Petter (1968) reported incrassation for *M. karimii* in captivity. Tail slightly bicolour. Small feet (12 mm); toes without ungual tuft. Wide nasals little expanded at the maxilla-frontal suture. Then, nasals narrowing to the same width than anterior to the suture. Supraorbital processes well developed in adults, as conspicuous beaded borders, but not projected as lateral edges. Zygomatic arches well expanded. Cusp C not developed, ectoflexus is notch-shaped.

Distribution: Central and Southeast Brazil (Gardner 1993, Palma 1995), including eastern semi-arid habitats of Cerrado and Caatinga (Vieira and Palma 1996).

Specimens reviewed: NMNH 393536-8, from Matto Grosso, Brazil, identified as *Marmosa karimii* by Pine et al. (1970).

Remarks: This is the most distinctive species within *Thylamys*. Preliminary analysis points to an early or basal origin (see below).

The Paraguayan group

The included species are geographically delimited to the west and east of the Paraguay River (Mayr 1982, Creighton 1985, Palma 1995). They occupy the Chaco, and other dry biomes of the western side of the river, as well as the subtropical moist forests to the east (Cabrera and Willink 1980, Palma 1995). Fur colour pattern of underparts is distinctive: short, self-coloured cream-white or whitish hairs, clearly delimited from lateral bands. Other characteristics are: strong and stout skull, presence of stylar cusp C on upper molars, and submetacentric X chromosomes (Palma and Yates 1998). Analyses by Kirsch and Palma (1995) and, Palma and Yates (1998) considered *pusillus* and *macrurus* as closer taxa within *Thylamys*.

Thylamys pusillus
Didelphis pusilla Desmarest 1804

Tabl. Méth. Hist. Nat., in Nouv. Dict. Hist. Nat., 24: 19

Marmosa marmota Thomas 1896

Marmosa citella Thomas 1912 (Corrientes: Argentina)

Marmosa verax Thomas 1921 (Concepción: Paraguay)

Type locality: San Ignacio, Misiones, Paraguay

Description: Dorsal colouration with a well-defined mouse grey medial band, dark bases, short fur; eye-rings poorly defined. Venter cream-white, sharply delimited of dorsal fur. Tail longer than HB length; it is slightly bicolour, fuscous above. Feet covered by short white hairs. Nasals slightly expanded at the maxilla-frontal suture, then narrowing to converge at their posterior end. Well-developed supraorbital processes only in a few adults, probably related to age. Zygomatic arches greatly expanded, giving to the skull a distinctive profile on dorsal view. Medial stylar cusp (C) present on upper molars, very conspicuous on M2 and M3; the ectoflexus on these teeth is serrated.

Distribution: West Paraguay (Chaco), southeast Bolivia, and north to south of Argentina. It would occupy the Chaco of Paraguay, Argentina, and Bolivia (Myers 1982, Anderson 1997), the mountain and desert scrubs, the Patagonia of Argentina

(Birney et al. 1996), and the Pampa of Argentina and Uruguay (Redford and Eisemberg 1992, González and Saralegui 1996).

Specimens reviewed: FMNH 54369,63862, and NMNH 390027-33, from Paraguay, and CBF 012 from Bolivia.

Remarks: Because its small size, some authors (Tate 1933, Cabrera 1958, Petter 1968) suggested a closer affinities of it to *velutinus* and *pallidior*. There is no type specimen for the taxon *pusillus*; its description was based on le micouré nain of Felix D'Azara.

Thylamys macrurus
Didelphys macrura Olfers 1818

in W.L. Eschwege, Journal von Brasilien, Neue Bibliotek Reisen., 15: 205

Didelphis grisea Desmarest 1827

Marmosa marmota Thomas 1912

Type locality: Tapua, Presidente Hayes, Paraguay

Description: Dorsal colouration similar to *pusillus*; eye-rings marked but not extended to the nose; lateral band lighter than medial one, appears grizzled; underparts cream-whitish over all the venter. Tail longer than head and body length. Tail bicoloured and particoloured, white spots on distal third. Large feet with long toes; claws well developed. Slender nasals, just little expanded at the maxilla-frontal suture. After it, nasals narrowing almost to their original width. Supraorbital processes developed as conspicuous sharp borders, projecting to converge as a low sagittal crest. Zygomatic arches well expanded, more than *pusillus*, with stronger bones. Cusp C on first two upper molars, ectoflexus is serrated.

Distribution: Restricted to the east of Paraguay River, but it may be present also in South Brazil (see Gardner 1993). It is the only species adapted to subtropical moist forests (Palma 1995). A single record of the Bolivian Chaco (Anderson 1997) is dubious.

Specimens reviewed: FMNH 26760, and BMNH 3.4.7.21, 99.11.17.1.

Remarks: The largest and stoutest species of the genus. Its taxonomic history is confused because misuse of *Didelphys marmota* a *nomen nudum* made available twice by Thomas (1896, 1912) to refer to actual *pusillus* and *macrurus*. Tate (1933) and Cabrera (1958) used different meanings for this species. Names *macrurus* and *griseus* were both based on le micouré à queue longue of F. D'Azara, so there are no type specimens associated to these species names.

The Andean group

This is the group of greater species richness, and the most uniform in characteristics of its species. Many of them were used as synonym or subspecies of *elegans* in previous works (Thomas 1902, Osgood 1943, Gardner 1993). Studies by Kirsch and Palma (1995), and Palma and Yates (1998) gave support for further relationships between these taxa. Palma (pers. comm.) suggests basal position for *venustus*. Here, diversification is proposed as occurred due to the vegetational fluctuations during the last uplift of the Andes, in the Plio-Pleistocene.

Because of the morphological resemblance between all taxa in this group, I will provide a detailed description just for *elegans* (*sensu stricto*), for each remaining species the descriptions give the most relevant differences with respect to this one. Variation among subspecies is described in the account of each species.

Thylamys elegans
Didelphis elegans Waterhouse 1839

Zool. H.M.S. 'Beagle', Mammalia, p. 95

Didelphis soricina Philippi 1894 (Valdivia: Chile)

Marmosa elegans coquimbensis Tate 1931 (Coquimbo: Chile)

Type locality: Valparaiso, Valparaiso, Chile.

Description: A medium-sized species; wide and conspicuous medial band, greyish or grey-brownish, long (>8 mm) and silky dorsal fur; well-developed eye-rings that extend toward the nose. Lateral bands lighter than dorsal, mouse grey. Underparts from chin to anus, with a narrow band in middle, whitish to cream-whitish, self-coloured hairs. Grey-based hairs with cream-white tips, forming a band wider than 7 mm on each side of belly. Tail only slightly longer than head and body; and seasonally incrassated up to 10 mm of diameter at base. It is bicoloured, grey on dorsum and whitish below. Foot and toes covered by short white hairs. Nasals almost parallel-sided, not expanded at the maxilla-frontal suture. They narrow after they pass the suture, converging on their posterior ends. Nasals look parallel on dorsal view of skull. Supraorbital processes not developed, rounded; without evidence of sagittal crest. Zygomatic arch not very expanded, skull profile elongated. No evidence of cusp C on upper molars, the ectoflexus has a typical notch shape.

Distribution: I consider *elegans* as restricted to the Pacific slope of Chilean Andes, between 32° and 388°S. According to Mann (1978) and Palma (1995), subspecies *soricinus* is found to the south, and *coquimbensis* to the north of this distribution.

Specimens reviewed: BMNH 53.8.29.18 (type of *elegans*), NMNH 269806 (topotype of *elegans*), 541583-6, 541592, FMNH 22330-8, 22666-9, 23302 (type of *coquimbensis*), 23855-6, 23858-60, 23866, 23871-5, 24064, 24395, 119485-7. Also NMNH 541587-91, representing *soricinus*.

Remarks: Specific name *elegans* was suggested to include a complex of Andean taxa, which had been referred as individual species on recent studies. Based on analyses of morphological characters (Table 3) and morphometric variation, I validate *tatei* as different from *elegans*. Taxon *coquimbensis* is only associated with the type specimen, without additional specimens it could be a *nomen dubious*. This specimen is very similar in size and fur colouration to *pallidior*, although the skull is typical of *elegans*, with narrow and little expanded nasals. Subspecies *soricinus* is geographically isolated in southern Chile. Mann (1978) stated the longer fur and darker dorsal colouration were response to the weather of southern Chile. Ventral pelage is more similar to *venustus* (see below). However, the skull is also typical as described for *elegans*, with little expanded but wider nasals and without beaded supraorbital processes.

Thylamys venustus

Marmosa elegans venusta Thomas 1902

Ann. Mag. Nat. Hist., ser. 7, 10: 161

Marmosa elegans cinderella Thomas 1902 (Tucumán: Argentina)

Marmosa elegans sponsoria Thomas 1921 (Jujuy: Argentina)

Marmosa janetta Thomas 1926 (Tarija: Bolivia)

Type locality: Parotani, Cochabamba, Bolivia.

Description: A medium-sized species, similar to *elegans*. Medial band grey-brownish is conspicuously darker than grey-yellowish of lateral bands. Wide blackish eye-rings. Ventral pelage appears tawny-grey, with plumbeous bases and cream tips, except the cream-buffy chest and throat. Tail longer than HB length, incrassated; and distinctly bicoloured, dark fuscous above and whitish below. Foot and digits not densely covered by whitish hairs. Nasals almost parallel throughout, not expanded or narrowed, but they are wider than in *elegans*. The type and topotype specimens have rounded interorbital region, but some old adult specimens present marked edges, even as lateral processes. The cusp C is absent and the ectoflexus looks like a notch.

Distribution: Cochabamba, Bolivia, to Neuquen, Argentina (Gardner 1993, as *elegans*). It was known as the montane representative of *elegans* on eastern Andes, although includes both low and high habitats. In Bolivia, *venustus* is found at lower montane forests and other drier montane habitats, but in Argentina it also occupies moist habitats to the south.

Specimens reviewed: BMNH 2.1.1.120 (type of *venusta*), FMNH 21553 (topotype of *venusta*), 21554;CBF 002-3, and NMNH 290899-900, from Bolivia. FMNH 22352-4 and 41266 (as *sponsoria*), 35014-5 (as *cinderella*), 29168, 30199-203 (as *venustus*), and NMNH 259257-8, from Argentina. In addition, BMNH 26.1.1.167 (type), FMNH 29169, 29170 (topotype), 50972-3, and NMNH 390570, 391293-4, are representing *janetta* from Bolivia.

Remarks: Described as subspecies of *elegans* (Thomas 1902), then raised to species status by Tate (1933). Palma (1994, 1995) supported its status as a different species based in phylogenetic molecular analyses. It would be the basal species within the Andean group (Palma and Yates 1998). Names *cinderella* and *sponsoria* were described as subspecies of *elegans*, but they are a single taxon (Thomas 1921c, Cabrera 1958) and a subspecies of *venustus*. Subspecies *cinderella* would be the lowland and southern representative of *venustus* in Argentina.

The taxon *janetta* is here recognised as a valid subspecies from wet areas of Tarija and Santa Cruz, Bolivia. This is the stoutest subspecies of *venustus*, its ventral pelage is cream-white with plumbeous bases, and general appearance is grey-whitish more than tawny-grey.

Thylamys pallidior

Marmosa elegans pallidior Thomas, 1902

Ann. Mag. Nat. Hist., ser. 7, 10: 161

Type locality: Challapata, Oruro, Bolivia

Description: One of the smallest species in the genus, with long (>11 mm) and silky dorsal pelage, silvery with dark grey bases on medial band; eye-rings blackish. Lateral bands not well defined, greyish with white or cinnamon tips, especially at posterior flank. Face conspicuously paler than dorsal or lateral colouration. Venter pure white throughout, long hairs; intermediate band of grey-based hairs not conspicuous. Tail slightly longer than head and body length, clearly bicolour even when incrassated, dark above. Very small feet (<15 mm) and digits densely covered with short white hairs. Skull elongated, small and delicate. Nasals little expanded at maxilla-frontal suture, then narrow, increasing the contrast. Supraorbital processes not beaded, rounded borders, but variable with age. Molars without cusp C, ectoflexus like a notch.

Distribution: East Argentina, south and east Bolivia (Gardner 1993), and extended to north Chile (Palma 1995) and western slope of Peruvian Andes. The species displays the broadest range of the genus, going from southern Argentina to central Peru, living in deserts, desert and mountain scrubs, and even reaching the Puna. Its wide altitudinal distribution goes from 8000 to 12,000 feet (2400 to 3800 m) on the Andes of Bolivia (Puna) and Argentina, but also reaches the sea level at coastal Peru.

Specimens reviewed: BMNH 2.2.2.116 (type of *pallidior*), CBF 006, 3258-9, FMNH 54255, and NMNH 121157, 271431, from Bolivia. FMNH 41397-8 and NMNH 236331-2 from Argentina. NMNH 391773-7, 54180-2, 541593-600 from Chile. FMNH 24141, 51003-7, 53155, 107398, MUSM 066, 070, 094, 983, 1302, 1749-51, 4754, 5806-14, 5953, 7018, 8374-5, 10722, 10737, 13091-2, 13097, 13104, 16085-

7, and MVZ 116614-5, 119913-5, 137584-5, 136248-50, 137896, 145531, 173937-9, from Peru.

Remarks: Some specimens from central highland of Peru showed variation in the extension of white on venter, with wide grey-based bands bordering a pure white stripe. Tate (1933) also noted this variation when mentioned the specimens from Peru (under *venusta venusta*). Although they were darker dorsally, did not show significant morphometric differences with typical *pallidior* (see below). Other specimens, especially very old males, showed beading of supraorbital processes and incipient sagittal crest; otherwise, they agreed with *pallidior*.

Thylamys tatei
Marmosa tatei Handley, 1956

J. Wash. Acad. Sci., 46: 402

Type locality: Chasquitambo, Bolognesi, Ancash.

Description: A species very similar to *elegans,* but with shorter (8 mm) and darker fur, greyish to grey-slate on medial band; eye-rings blackish. Lateral bands are greyish with cinnamon tones, but not too conspicuous as those of *pallidior*. Venter cream-whitish on medial band, wider at breast and narrow over most of belly, bordered by broad bands of grey-based hairs with cream tips. Tail just slightly longer than HB length, tip whitish for almost 10–15% of tail length. Feet and digits covered by short white hairs. Skull elongated, larger than *pallidior* or *elegans*. Nasals expanded at maxilla–frontal suture and then narrow, some like *pallidior*. Supraorbital processes not beaded in the series of Pariacoto, Ancash, but the type has marked edges, almost convergent on braincase. Molars without stylar cusp C, as typical for the group.

Distribution: This endemic species is just known from Ancash department (Handley 1956), and Lachay, north Lima, in central Peru. Its elevational range goes from 300 m (lomas de Lachay, Lima) to 3000 m (Huaraz, Ancash), including dry habitats such as deserts, lomas, and mountain scrubs of the western slope of the central Andes.

Specimens reviewed: NMNH 302915 (type), 302916 (paratype), FMNH 81443, MUSM 10738, MVZ 135503-12, and others collected by O. Ramírez (University C. Heredia, Lima).

Remarks: Creighton (1984) and Reig et al. (1987) recognised this taxon as a full species, but Gardner (1993) and Palma (1997) listed it as a synonym or subspecies of *elegans*.

ANALYSES OF MORPHOMETRIC VARIATION

Non-geographic variation

The analyses of non-geographic variation were carried out for three units of Peru, one of *tatei* (Huaraz) and two of *pallidior* (Parinacochas and Sierra Sur), and one of Chilean *elegans* (Valparaiso). These units had the largest available series, including juveniles and adults of both sexes. The ANOVA showed the effect of age and sex on the morphometric variation of *pallidior* and *elegans*. Seven and 13 variables, respectively, showed significant variation with age, with average variation over 35% and 59%. Variables most influenced by age were those referred to length of the skull. Sex variation was very low for both species, between 10 and 13%.

Because of the reduced size of these samples, a new ANOVA tested the variation between sexes of adult (age class VI) individuals. The taxonomic units were lumped according to the taxa they represent, to analyse their variation due to age. This variation was significant only for one variable (WSB) in *tatei*, none in *pallidior*, but 11 in *elegans*. Males were larger than females in every case. Three of the non-significant variables for *elegans* were dental measurements (M4W, M2M5, m2m5).

Geographic variation

Due to sexual dimorphism in earlier analyses, adult males and females of *tatei* and *pallidior* were pooled, but sexes were treated separately for *elegans*, in the analysis of variation within taxa. For the between taxa analysis, each sex was compared separately.

Among taxonomic units representing *Thylamys tatei*, only one variable (POC) differed geographically (Table 5). This small general variation may be due to the proximity of these units, less than 120 km apart. For *T. pallidior*, with a larger geographic range (Fig. 5), the units were very homogeneous, with no variation for any variable (Table 5). Finally, males of *T. elegans* showed a larger variation than that found in females, with 13 and five variables (respectively) showing significant differences. This result is worthy because that variation would occur over less than 700 km (Fig. 5).

In order to test the correspondence between taxonomic differentiation and geographic variation, a new ANOVA was carried out. From the results shown in Table 5, it is evident that the variation is equally evident among both males and females of the three species. Only one variable (POC and ZB, respectively) did not show significant variation (Table 5). Post hoc tests (Tukey, Duncan, and Scheffe) grouped to *tatei* and *elegans* as a homogeneous subgroup with larger means than *pallidior*. It is evident that the three species are distinguishable on morphological (Table 3) and morphometric (Table 5) grounds.

BIOGEOGRAPHY

The seven recognised species of *Thylamys* are grouped in three geographic units, proposed as a first approach to natural groups. However, relationships between these units are not conclusive at present. Previous hypotheses (Creighton 1985; Palma 1995) were built on an incomplete knowledge of

Table 5 Results of the analysis of geographic variation within and between taxa, for adults (age class VI) only. Differences are significatives for P < 0.05. See the text for details.

	tatei	*pallidior*	*elegans*	*elegans*	Inter-taxa	
	mm-ff	mm-ff	*mm*	*ff*	mm	ff
BAB	0.889	0.894	**0.007**	0.168	**0.000**	**0.000**
BPL	0.790	0.487	**0.033**	0.226	**0.000**	**0.001**
CIL	0.885	0.440	**0.001**	0.070	**0.000**	**0.004**
GNL	0.424	0.554	**0.021**	0.283	**0.000**	**0.000**
GSL	0.720	0.404	**0.001**	0.136	**0.000**	**0.002**
LTR	0.636	0.272	**0.001**	**0.024**	**0.000**	**0.000**
m2m5	0.484	0.144	0.053	0.063	**0.000**	**0.000**
M2M5	0.671	0.164	0.129	0.059	**0.000**	**0.000**
M4W	0.382	0.275	**0.001**	**0.000**	**0.000**	**0.000**
MRL	0.820	0.414	**0.009**	**0.045**	**0.000**	**0.004**
PL	0.513	0.440	**0.005**	0.258	**0.000**	**0.001**
POC	**0.032**	0.089	**0.037**	0.266	0.111	**0.000**
PZW	0.238	0.747	0.081	**0.010**	**0.000**	**0.009**
WSB	0.827	0.162	**0.020**	0.272	**0.036**	**0.027**
ZB	0.520	0.663	**0.011**	**0.036**	**0.000**	0.084

taxonomy and distributions. In spite of the several taxa proposed to be found on the western slope of the Andes, no one study covered this area.

Most of the biogeographic hypotheses considered *Thylamys* as a genus of subtropical distribution, with the Chaco being the main barrier to taxa ditribution. Thomas (1902) pointed to geographic groups (Andean and Paraguayan) distinguishable by development of supraorbital borders, and distributed at both sides of the Chaco. Creighton (1985) proposed a temperate origin for *pusillus*, with later dispersal to the Andean region and to the subtropical Chaco. In the scenario suggested by Palma (1995), *venustus* would be the most basal taxon, from which all the other taxa originated. New analyses by Palma and Yates (1998) found a closer affinity between *pusillus* and *macrurus*, but did not resolve the relationship between *pallidior*, *elegans*, and *venustus*.

The present separation of *Thylamys* in geographic groups and updated information about distribution of the taxa allow the proposal of a preliminary hypothesis of relationships between species, and a biogeographic scenario for their origin. The following hypothesis is based on morphological similarity as evidence of recent ancestry, but also on previous phylogenies of the genus *Thylamys* (see Palma 1995, Palma and Yates 1998).

Although species of the Paraguayan group share many traits, they differ from each other in geographic ranges. The subtropical Chaco and the temperate regions of southern South America are occupied by *pusillus*, while *macrurus* appears restricted to the subtropical forests to the east of the Paraguay River. Because of its particular characteristics, *Thylamys velutinus* appears as an early divergent taxon, also inhabiting very different habitats (Cerrado and Caatinga forests) on the eastern side of the genus distribution. According to previous hypotheses of tropical lowland ancestry for present taxa of dry, open areas (see Palma 1995), *macrurus* could originate *velutinus* and *pusillus*, by dispersal to the north-east and west, respectively. Both of these events would have occurred during climatic fluctuations of Plio-Pleistocene (5 Mya), but not necessarily at the same time.

From the evaluation of morphological characters, there is a closer similarity between *macrurus* and *pusillus* (Table 4), which would be indicative of a more recent differentiation of these taxa, as compared to the possible isolation of *velutinus*. Palma and Yates (1998) explained this differentiation by changes in habitat, soil, and topography at each side of the river (see Myers 1982), instead of a vicariant effect of the river. Accordingly, *pusillus* should be a successful disperser of this group, but the opportunity to colonise the Andes was reserved for the ancestor of the Andean group, a taxon allied perhaps to *venustus* (see Palma and Yates 1998).

The successful colonisation of highlands could occur when the Andes got their present elevation, during the last two million years (Pleistocene, Simpson 1978). Populations of *venustus* (or an ancestor), residents of open biomes, could be brought passively by the final uplift of the proto-Andes to higher, more temperate habitats. Consequent adaptation and diversification of these populations in highland habitats favoured a later colonisation of lowlands (see Marquet 1994). Ancestors of *elegans* and *pallidior* could be originated by this way. Both habitats, the desert scrub and the Puna, correspond to the semi-arid, temperate conditions of the proposed center of origin for this group (see Potts and Behrensmeyer 1992).

Dispersal of *elegans* and *pallidior* should be more recent (late Pleistocene). Going through the mesic valleys that penetrate the arid coastal desert, *elegans* reached the central part of Chile, being limited to the north by the Atacama desert (see Marquet 1994). On the other hand, *pallidior* succeed in its latitudinal dispersal through the Andean range, actually ranging from 10° to 40° south (Fig. 5). Some cranial characters suggest a closer relationship between *pallidior* and *tatei* (Table 3), and the origin of the latter would be result of a recent isolation of marginal populations of *pallidior*. Altitudinal migration of *pallidior*, from Andean valleys to the desert coast, could be related to the arid conditions of the Holocene. This would be the origin of its intervening populations in the south and central desert of Peru.

SUMMARY AND PROSPECTS

The diversity of small mouse opossums of the genus *Thylamys* has been revised in this chapter. The genus presents a very scattered distribution over its range, with boundaries of species and geographic ranges poorly known. Here I have defined, characterised, and identified two different taxa in Peru, where only one (*T. elegans*) was usually recognised. Based on a set of characters, seven species of *Thylamys* are morphologically and geographically defined.

I group these taxa based on shared morphological characteristics and distribution. Three groups are identified, and a preliminary proposal of biogeographic relationships is given. High variation in most of morphological characters commonly used to identify species prevented evaluation of the whole range of their distributions. Recent molecular analyses by E. Palma (pers. comm.) partially support this hypothesis of relationships, and provide a useful framework for discussion. Inclusion of *velutinus* in these analyses should confirm its relationships to the Paraguayan group, as proposed here. Further molecular analyses, at the population level, should give us a more precise approach to the long history of diversification, colonisation, and extinction of the Andean group, as represented by *elegans* (and its subspecies), *pallidior*, and *tatei*.

ACKNOWLEDGEMENTS

I thank the curators and staff of the several museums that facilitated visits or loans to study their *Thylamys* specimens. I am deeply indebted to Victor Pacheco, Curator of Mammals in the Museo de Historia Natural (MUSM), and professor at University San Marcos, Lima, Peru, for all the time shared with me as advisor. He also checked type specimens at the British Museum, and took measurements of relevant material at other American museums. Special thanks to Bruce Patterson (FMNH), James Patton (MVZ), and Don E. Wilson (NMNH) for their permanent interest in my research. The late Dr Philip Hershkovitz (FMNH) encouraged me to continue beyond my original focus on Peruvian *Thylamys*, so I feel very honored to devote this contribution to his memory.

Some colleagues and friends supported my work in several ways, Alfred Gardner and the late Charles Handley, Jr. (NMNH) shared their expertise on marsupials, and Nuria Bernal (CBF) provided interesting Bolivian specimens for comparison. E. Palma (Pontificia Universidad de Chile) freely shared his first results of molecular analyses, and helped to improve the section on biogeography. D. Silva, J. Santisteban, L. Luna, and two anonymous reviewers made valuable comments and criticisms to previous drafts. N. Salcedo helped my work with her skill in drawings of figures and maps, and in many other ways. A. Alonso and F. Dallmeier (Smithsonian Institution) simplified my research stay at the NMNH. Grants of the Advanced Training Program for the Conservation of the Biological Diversity (MacArthur Foundation) and the Field Museum funded my research work at Chicago.

I want to extend my thankfulness to the editors of this volume, especially to Drs. Menna Jones and Michael Archer for their continue interest. Part of this study was submitted as partial requirement to obtain a Magister in Zoology degree of the University of San Marcos, Lima.

REFERENCES

Allen, J.A. (1900), 'Description of new American marsupials', *Bulletin of the American Museum of Natural History*, **13**:191–99.

Allen, J.A. (1912), 'Mammals from western Colombia', *Bulletin of the American Museum of Natural History*, **31**:71–95.

Allen, J.A., & Chapman, F.M. (1897), 'On a second collection of mammals from the island of Trinidad, with description of new species, and a note on some mammals from the island of Dominica, West Indies', *Bulletin of the American Museum of Natural History*, **9**:13–29.

Anderson, S. (1997), 'Mammals of Bolivia, Taxonomy and distribution', *Bulletin of the American Museum of Natural History*, **231**:1–652.

Archer, M. (1978), 'The nature of the molar-premolar boundary in marsupials and a reinterpretation of the homology of marsupial cheekteeth', *Memoirs of the Queensland Museum*, **18**:157–64.

Birney, E.C, Sikes, R.S, Monjeau, J.A., Guthmann, N., & Phillips, C.J. (1996), 'Comments on Patagonian marsupials from Argentina', in *Contributions in Mammalogy: A memorial volume honoring Dr. J. Knox Jones, Jr.* (eds. H.H. Genoways & R.J. Baker), pp. 149–54, Museum of Texas Tech University, Lubbock.

Brown, J.C. (1971), 'The description of mammals, 1. The external characters of the head', *Mammal Review*, **1**:151–68.

Brown, J.C., & Yalden, D.W. (1973), 'The description of mammals, 2. Limbs and locomotion of terrestrial mammals', *Mammal Review*, **3**:107–34.

Cabrera, A. (1919), *Genera Mammalium. Monotremata, Marsupialia.* Madrid, 232 pp.

Cabrera, A. (1958), 'Catálogo de los mamíferos de América del Sur, I', Revista del Museo Argentino de Ciencias Naturales 'Bernardino Rivadavia', *Ciencias Zoológicas*, **4**(1):1–307.

Cabrera, A.L., & Willink, A. (1980), 'Biogeografía de América Latina', OEA, Serie Biología, Monografía 13: Washington, DC.

Creighton, G.K. (1984), 'Systematic studies on opossums (Didelphidae) and rodents (Cricetidae)', PhD dissertation, University of Michigan, Ann Arbor, Michigan.

Creighton, G.K. (1985), 'Phylogenetic inference, biogeographic interpretations, and the pattern of speciation in *Marmosa* (Marsupialia: Didelphidae)', *Acta Zoologica Fenica*, **170**:121–24.

Gardner, A.L. (1973), 'The systematics of the genus *Didelphis* (Marsupialia: Didelphidae) in North and Middle America', *Special Publications of the Museum, Texas Tech University*, **4**:1–81.

Gardner, A.L. (1993), 'Order Didelphimorphia' in *Mammal species of the World a taxonomic and geographic reference* (eds. D.E. Wilson & D.M. Reeder) pp. 15–23, Smithsonian Institution Press, Washington.

Gardner, A.L., & G.K. Creighton (1989), 'A new generic name for Tate's (1933) *microtarsus* group of South American mouse opossums (Marsupialia: Didelphidae)', *Proceedings of the Biological Society of Washington*, **102**:3–7.

Gilmore, R.M. (1941), 'Zoology', in *The susceptibility to yellow fever of the vertebrates of eastern Colombia. I. Marsupialia* (J.C. Bugher, J. Boshell-Manrique, M. Roca-García, & R.M. Gilmore), pp. 314–319, *American Journal of Tropical Medicine*, **21**:309–33.

Gonzalez, E.M., & A.M. Saralegui (1996), 'La presencia de *Thylamys* Gray 1843 en Uruguay (Mammalia: Didelphimorphia)', *Actas de las IV Jornadas de Zoología del Uruguay*, p. 21.

Gray, J.E. (1843), 'List of the specimens of Mammalia in the Collection of the British Museum', George Woodfall and Son, London.

Hall, E.R. (1943), 'Criteria for vertebrate subspecies, species and genera: the Mammals', *Annals of the New York Academy of Sciences*, **44**:141–44.

Handley, Jr., C.O. (1956), 'A new species of murine opossum (genus *Marmosa*) from Peru', *Journal of the Washington Academy of Sciences*, **46**:402–04.

Hershkovitz, P. (1959), 'Nomenclature and taxonomy of the Neotropical mammals described by Olfers, 1818', *Journal of Mammalogy*, **40**:337–53.

Hershkovitz, P. (1992a), 'Ankle bones: the Chilean opossum *Dromiciops gliroides* Thomas, and marsupial phylogeny', *Bonner Zoologische Beiträge*, **43**:181–213.

Hershkovitz, P. (1992b), 'The South American Gracile mouse opossums, Genus *Gracilinanus* Gardner and Creighton, 1989 (Marmosidae: Marsupialia): A taxonomic review with notes on general morphology and relationships', *Fieldiana, Zoology*, n.s. **70**:1–56.

Honacki, J.H., Kinman, K.E., & Koeppl, J.W. (eds.) (1982), *Mammal species of the world: a taxonomic and geographic reference*, Allen Press, Inc., & The Association of Systematic Collections, Lawrence, Kansas.

International Commission on Zoological Nomenclature (ICZN), (1999), *International Code of Zoological Nomenclature, 4th Ed.*, International Trust for Zoological Nomenclature, London.

Jansa, S. A., & Voss, R. S. (2000), 'Phylogenetic studies on Didelphid marsupials I. Introduction and preliminary results from nuclear IRBP gene sequences', *Journal of Mammalian Evolution*, **7**:43–77.

Kirsch, J.A.W., & Calaby, J.H. (1977), 'The species of living marsupials: an annotated list', in *The Biology of marsupials* (eds. B. Stonehouse & D. Gilmore), pp. 9–26, University Park Press, Baltimore.

Kirsch, J.A.W., & Palma, R.E. (1995), 'DNA/DNA hybridization studies of carnivorous marsupials. V. A further estimate of relationships among opossums (Marsupialia: Didelphidae)', *Mammalia*, **59**:403–25.

Mann, G. (1978), 'Los pequeños mamíferos de Chile (marsupiales, quirópteros, edentados y roedores), Gayana', *Zoología*, **40**:1–342.

Marquet, P.A. (1994), 'Diversity of small mammals in the Pacific coastal desert of Peru and Chile and in the adjacent Andean area: biogeography and community structure'. *Australian Journal of Zoology*, **42**:527–42.

Marshall, L.G. (1982), 'Evolution of South American Marsupialia', *Special Publications Series, Pymatuning Laboratory of Ecology*, **6**:251–72.

Martinelli, P.M., & Nogueira, J.C. (1997), 'Penis morphology as a distinctive character of the murine opossum group (Marsupialia Didelphidae): a preliminary report', *Mammalia*, **61**:161–6.

Matschie, P. (1916), 'Bemer Kungen über die gattung *Didelphis* L.', *Stizungsberichte der Gesellschaft Naturforschender Freunde zu Berlin*, **1916**:259–72.

Monjeau, A.N., Bonino, N., & Saba, S. (1994), 'Annotated checklist of the living land mammals in Patagonia, Argentina', *Mastozoología Neotropical*, **1**:143–56.

Morton, S.R. (1980), 'Ecological correlates of caudal fat storage in small mammals', *Australian Mammalogy*, **3**:81–86.

Myers, P. (1982), 'Origins and affinities of the mammal fauna of Paraguay', *Special Publication Series, Pymatuning Laboratory of Ecology, Pittsburgh*, **6**:85–93.

Osgood, W.H. (1943), 'The mammals of Chile', *Field Museum of Natural History, Zoological series*, **30**:1–268.

Pacheco, V., & Patterson, B.D. (1992), 'Systematics and biogeographic analyses of four species of *Sturnira* (Chiroptera: Phyllostomidae), with emphasis on Peruvian forms', *Memorias del Museo de Historia Natural, UNMSM*, **21**:57–81.

Pacheco, V., de Macedo, H., Vivar, E., Ascorra, C.F. Arana-Cardo, R., & Solari, S. (1995), 'Lista anotada de los Mamíferos peruanos', *Occasional papers in Conservation Biology*, **2**:1–35.

Palma, R.E. (1994), 'Historical relationships of South America mouse opossum (*Thylamys*, Didelphidae): evidence from molecular systematics and historical biogeography', PhD dissertation, University of New Mexico, Albuquerque, New Mexico.

Palma, R.E. (1995), 'Range expansion of two South American mouse opossum (*Thylamys*, Didelphidae) and their biogeographic implications', *Revista Chilena de Historia Natural*, **68**:515–22.

Palma, R.E. (1997), '*Thylamys elegans*', *Mammalian species*, **572**:1–4.

Palma, R.E., & Yates, T.L. (1996), 'The chromosomes of Bolivian Didelphid marsupials', *Occasional papers, The Museum, Texas Tech University*, **162:**1–20.

Palma, R.E., & Yates. T.L. (1998), 'Phylogeny of South American mouse opossums (*Thylamys*, Didelphidae) based on allozyme and chromosomal data', *Zeischrift für Säugetierkunde*, **63**:1–15.

Pearson, O.P., & Peason, C. (1978), 'The diversity and abundance of vertebrates along an altitudinal gradient in Peru', *Memorias del Museo de Historia Natural 'Javier Prado'*, **18**:1–97.

Petter, F. (1968), 'Une sarigue nouvelle du nord-est du Bresil, *Marmosa karimii* sp. nov. (Marsupiaux, Didelphides)', *Mammalia*, **32**:313–16.

Pine, R.H. (1972), 'A new subgenus and species of murine opossum (genus *Marmosa*) from Peru', *Journal of Mammalogy*, **53**:279–82.

Pine, R.H. (1981), 'Review of the mouse opossum *Marmosa parvidens* Tate and *Marmosa invicta* Goldman (Mammalia: Marsupialia: Didelphidae) with description of a new species', *Mammalia*, **45**:55–70.

Pine, R.H., & C.O. Handley, Jr. (1984), 'A review of the Amazonian short-tailed opossum *Monodelphis emiliae* (Thomas)', *Mammalia*, **48**:239–45.

Pine, R.H., Bishop, I.R., & Jackson, R.L. (1970), 'Preliminary list of mammals of the Xavantinha/Cachimbo expedition (Central Brazil)', *Transactions of the Royal Society of Tropical Medicine and Hygiene*, **64**:668–70.

Pine, R.H., Dalby, P.L., & Matson, J.O. (1985), 'Ecology, postnatal development, morphometrics, and taxonomic status of the short-tailed opossum, *Monodelphis dimidiata*, an apparently semelparous annual marsupial', *Annals of the Carnegie Museum*, **54**:195–231.

Pine, R.H., Miller, S.D., & Schamberger, M.L. (1979), 'Contributions to the mammalogy of Chile', *Mammalia*, **43**:339–76.

Potts, R., & Beherensmeyer, A.K. (1992), 'Late Cenozoic terrestrial ecosystems', in *Terrestrial Ecosystems through time: evolutionary paleoecology of terrestrial plants and animals* (eds. A.K. Behrensmeyer, J.D. Damuth, W.A. DiMichelle, R. Potts, H. Sues, & S.L. Wing), pp. 419–451, University of Chicago Press, Chicago.

Redford, K.H., & Eisenberg, J.F. (1992), *Mammals of the Neotropics, vol. 2: The Southern Cone*, University of Chicago Press, Chicago.

Reig, O.A., Kirsch, J.A., & Marshall, L.G. (1985), 'New conclusions on the relationships of the opossum-like marsupials, with an annotated classification of the Didelphimorphia', *Ameghiniana*, **21**:335–43.

Reig, O.A., Kirsch, J.A., & Marshall, L.G. (1987), 'Systematic relationships of the living and Neocenozoic American opossum-like marsupials (suborder Didelphimorphia), with comments on the classification of these and of the Cretaceous and Paleogene New World and European Metatherians', in *Possums and Opossums, Studies in Evolution* (ed. M. Archer) pp. 1–90, Surrey Beatty and Sons Pty. Ltd., & Royal Zoological Society of New South Wales, Sydney.

Sanborn, C.C. (1949), 'A new species of rice rat (*Oryzomys*) from the Coast of Peru', *Publicaciones del Museo de Historia Natural 'Javier Prado', UNMSM*, **3**:1–4.

Simpson, B.B. (1978), 'Quaternary biogeography of the High montane regions of South America', in *The South American herpetofauna: its origin, evolution, and dispersal* (ed. W.E. Duellman), pp. 157–188, University of Kansas, Kansas.

Streilein, K.E. (1982), 'Behavior, ecology, and distribution of South American marsupials', *Special Publications Series, Pymatuning Laboratory of Ecology*, **6**:231–50.

Szalay, F.S. (1982), 'A new appraisal of marsupial phylogeny and classification', in *Carnivorous Marsupials* (ed. M. Archer), pp. 621–640, Royal Zoological Society of New South Wales & Surrey Beatty and Sons Pty. Ltd., Chipping Norton, New South Wales.

Tate, G.H.H. (1933), 'A systematic revision of the marsupial genus *Marmosa* with a discussion of the adaptative radiation of the murine opossums (*Marmosa*)', *Bulletin of the American Museum of Natural History*, **66**:1–250.

Thomas, O. (1894), 'On *Micoureus griseus* Desm., with the description of a new genus and species of Didelph[y]idae', *The Annals and Magazine of Natural History*, **6**(XIV):184–8.

Thomas, O. (1902), 'On *Marmosa marmota* and *elegans*, with descriptions of new subspecies of the latter', *The Annals and Magazine of Natural History*, **7**(X):158–62.

Thomas, O. (1912), 'Three small mammals from South America', *The Annals and Magazine of Natural History*, **8**(IX):408–10.

Thomas, O. (1921a), 'New *Rhipidomys*, *Akodon*, *Ctenomys*, and *Marmosa* from the Sierra Santa Barbara, S.E. Jujuy', *The Annals and Magazine of Natural History*, **9**(VII):183–7.

Thomas, O. (1921b), 'Three new species of *Marmosa*, with a note on *Didelphys waterhousi*, Tomes', *The Annals and Magazine of Natural History*, **9**(VII):519–23.

Thomas, O. (1921c), 'On a further collection of mammals from Jujuy obtained by Sr. E. Budin', *The Annals and Magazine of Natural History*, **9**(VIII):608–17.

Thomas, O. (1926), 'The Spedan Lewis South American Exploration. II. On mammals collected in the Tarija Department, Southern Bolivia', *The Annals and Magazine of Natural History*, **9**(XVII):318–28.

Tribe, C.J. (1990), 'Dental age classes in *Marmosa incana* and other didelphoids', *Journal of Mammalogy*, **71**:566–9.

Vieira, E.M., & Palma, A.R.T. (1996), 'Natural history of *Thylamys velutinus* (Marsupialia: Didelphidae) in Central Brazil', *Mammalia*, **60**:481–4.

Zuñiga, E. (1942), 'Observaciones ecológicas sobre los mamíferos de las lomas', *Boletín del Museo de Historia Natural, UNMSM*, **22–23**:392–9.

PART I

CHAPTER 7

AUSTRALIAN MARSUPIAL CARNIVORES: RECENT ADVANCES IN PALAEONTOLOGY

Stephen Wroe

Institute of Wildlife Research, University of Sydney, School of Biological Sciences, A08, University of Sydney, NSW 2006, Australia
Email: swroe@bio.usyd.edu.au

Over the last two decades, the study of new fossil material has yielded major insights into the evolution of Australia's marsupial carnivores. This material includes the first complete or near-complete crania for Tertiary representatives of Dasyuridae, Thylacinidae and Propleopinae (Hypsiprymnodontidae), as well as the first skulls known for the thylacoleonid genera, *Priscileo* and *Wakaleo*. Also of significance has been the discovery of a new species of marsupial carnivore, *Djarthia murgonensis*, from the Early Eocene Tingmarra Local Fauna. Regarding dasyuromorphian evolution, the study of well-preserved material from Oligocene–Miocene deposits of Riversleigh has been particularly illuminating. Many findings have been unanticipated. For example, it is now clear that during the early and middle Miocene, the now ubiquitous dasyurids were rare, while the recently extinct Thylacinidae were unexpectedly diverse. Furthermore, at present there is no hard evidence for the existence of any extant dasyurid genera or subfamily greater than Pliocene in age. The oldest confirmed dasyurid is from early to middle Miocene deposits and forms a sister clade to the three living subfamilies. Moreover, it is argued that dasyurids are highly specialised among dasyuromorphians, particularly with respect to their basicranial morphology and not 'primitive' Australian marsupials as has often been supposed. On the other hand, from results of analysis of late Oligocene–Miocene cranial material, it is now evident that Thylacinidae constitutes a very conservative lineage with the recently extinct *Thylacinus cynocephalus* little more derived than some late Oligocene–Miocene taxa regarding either cranial or dental features. Propleopinae (giant rat-kangaroos), previously known only from dental remains, are now represented by two skulls. Interpretation of this evidence supports the hypothesis that at least some species included significant amounts of meat in their diets, while phylogenetic analysis hints at the possibility of a special relationship with balbarines. With respect to Thylacoleonidae, the study of cranial material leaves the issue of the family's ordinal level affinities uncertain, while the previously accepted tenet that *Wakaleo* could not be ancestral to *Thylacoleo* is also questioned. On the basis of dental evidence to hand, the Early Eocene *Djarthia murgonensis* can not be placed in any marsupial clade with confidence. Thus, biogeographic scenarios excluding the possibility that ameridelphians ever colonised Australia are considered premature. Growing evidence for a diversity of marsupial carnivores in pre-Pleistocene Australia is considered to diminish, if not contradict the argument that the continent's large terrestrial carnivore niches have long been dominated by reptiles. The flip side of this debate is that arguments for reptilian supremacy are commonly based on assumptions regarding the biology and behaviour of fossil varanid, snake and crocodilian taxa that in many cases are highly speculative. A tendency to consider only estimated maximum dimensions for extinct reptilian species may also have generated false impressions with respect to the their significance in the ecology of their respective communities.

INTRODUCTION

Historically, a polyphyletic group comprising five Australian marsupial families, or members thereof, have been considered as 'carnivorous marsupials'. Three of these comprise the order Dasyuromorphia: Dasyuridae, Thylacinidae and Myrmecobiidae. The remaining two, Thylacoleonidae and Hypsiprymnodontidae, are diprotodontians, members of ?Vombatoidea and Macropodoidea respectively. Since seminal reviews of Australian carnivorous marsupial palaeontology were presented by Archer (1982a, 1982b) and Archer and Dawson (1982), the fossil records for four of the five Australian carnivorous marsupial families have been significantly expanded (Gillespie 1997, Muirhead and Wroe 1998, Wroe 1999a). For taxa of Pliocene age, three new Pliocene species have been added, two dasyurids (Lawson et al. 1999, Wroe and Mackness 2000*a*) and a propleopine (Ride 1993), while discoveries of pre-Pliocene material have been extraordinary. In 1982, the list of carnivorous marsupial taxa exceeding five million years in age then amounted nine species: a single thylacinid (*Thylacinus potens*), three thylacoleonids (*Wakaleo oldfieldi, W. vanderleueri, W. alcootaensis*) and four species then considered to be dasyurids (*Ankotarinja tirarensis, Keeuna woodburnei, Wakamatha tasselli, Dasylurinja kokuminola*). To this list can now be added a further 19 formally described species: 10 thylacinids, two thylacoleonids, two dasyurids, three propleopine kangaroos, one species treated as Dasyuromorphia *incertae sedis* by Wroe (1997a) and *Djarthia murgonensis*, an Early Tertiary marsupicarnivore of uncertain ordinal affinity (Godthelp et al. 1999). Moreover, descriptions of at least seven additional taxa of Oligocene–Miocene age are in preparation, these include: three thylacinids, two thylacoleonids and two dasyurids (Wroe 1999a). Most significantly, whereas no pre-Pleistocene marsupial carnivores were represented by significant cranial material in 1982, complete or near-complete skulls of late Oligocene–Miocene age are now known for at least one species of each of the families listed above, excepting Myrmecobiidae. Investigation of this new material has generated quantum leaps in our understanding of the pre-Pliocene diversity, phylogeny and palaeobiology of Australia's marsupial carnivores. New light has also been shed on the origins and relationships of modern taxa. In some instances, previously accepted dictum has been overturned. For example, Dasyuridae, long held to represent a structurally, even actually ancestral position among Australian marsupials, has been shown to constitute a relatively recent and specialised addition to the continent's fauna (Wroe 1999b). On the other hand, Thylacinidae, monotypic in Recent times, is now thought to have been the more diverse of the three families of Dasyuromorphia during the Miocene (Wroe 1999a).

Dental nomenclature follows Flower (1867) and Luckett (1993) regarding the molar-premolar boundary, where the adult (unreduced) postcanine cheektooth formula of marsupials is P1–3 and M1–4. Dental terminology follows Wroe (1999b). Systematic terminology incorporates amendments to Archer's (1982) classification as suggested by Krajewski et al. (1994), Krajewski et al. (2000a) and Wroe (1996a, 1997b) for Dasyuromorphia, Godthelp et al. (1999) for Marsupialia *incertae sedis* and Aplin and Archer (1987) for other taxa. Institutional abbreviations: QM F = Queensland Museum fossil collection; AM F = Australian Museum fossil collection.

AUSTRALIA'S OLDEST MARSUPIAL CARNIVORE AND OTHER MARSUPIALIA *INCERTAE SEDIS*

On the issue of Australian marsupicarnivore evolution, one of the most significant palaeontological discoveries in recent times is that of the Early Eocene (55 myo) taxon *Djarthia murgonensis* (Godthelp et al. 1999; Fig. 1), from the Tingamarra Local Fauna of south-eastern Queensland. This species is plesiomorphic among marsupials for at least eight of the 18 features considered by Godthelp et al. (1999). Most derived features present in *D. murgonensis* are present in generalised representatives of both ameridelphian and australidelphian clades. Consequently and because the anatomical feature currently given the greatest weight as a potential synapomorphy for Australidelphia, presence of a continuous lower ankle joint, is unknown for this species, Godthelp et al. (1999) considered it prudent to treat *D. murgonensis* as Marsupialia *incertae sedis*. Furthermore, the single derived dental feature common to all extant members of the Australian marsupial radiation, reduction of lower incisor number, is also not preserved. However, *D. murgonensis* does have a potential synapomorphy, presence of a central cusp at the apex of the centrocrista, that could indicate special relationship with two species from the Ngapakaldi Local Fauna of central Australia, *Ankotarinja tirarensis* and *Keeuna woodburnei*. But again, these two taxa, originally described as dasyurids by Archer (1976a), are otherwise plesiomorphic among marsupials for many features. Godthelp et al. (1999) recommend that they too be considered Marsupialia *incertae sedis*. Woodburne and Case (1996) have argued that, excepting microbiotherians, the marsupial faunas of South American and Australia are manifestly distinct. On the basis of the evidence presented by Godthelp et al. (1999) the assertion of Woodburne and Case (1996) is

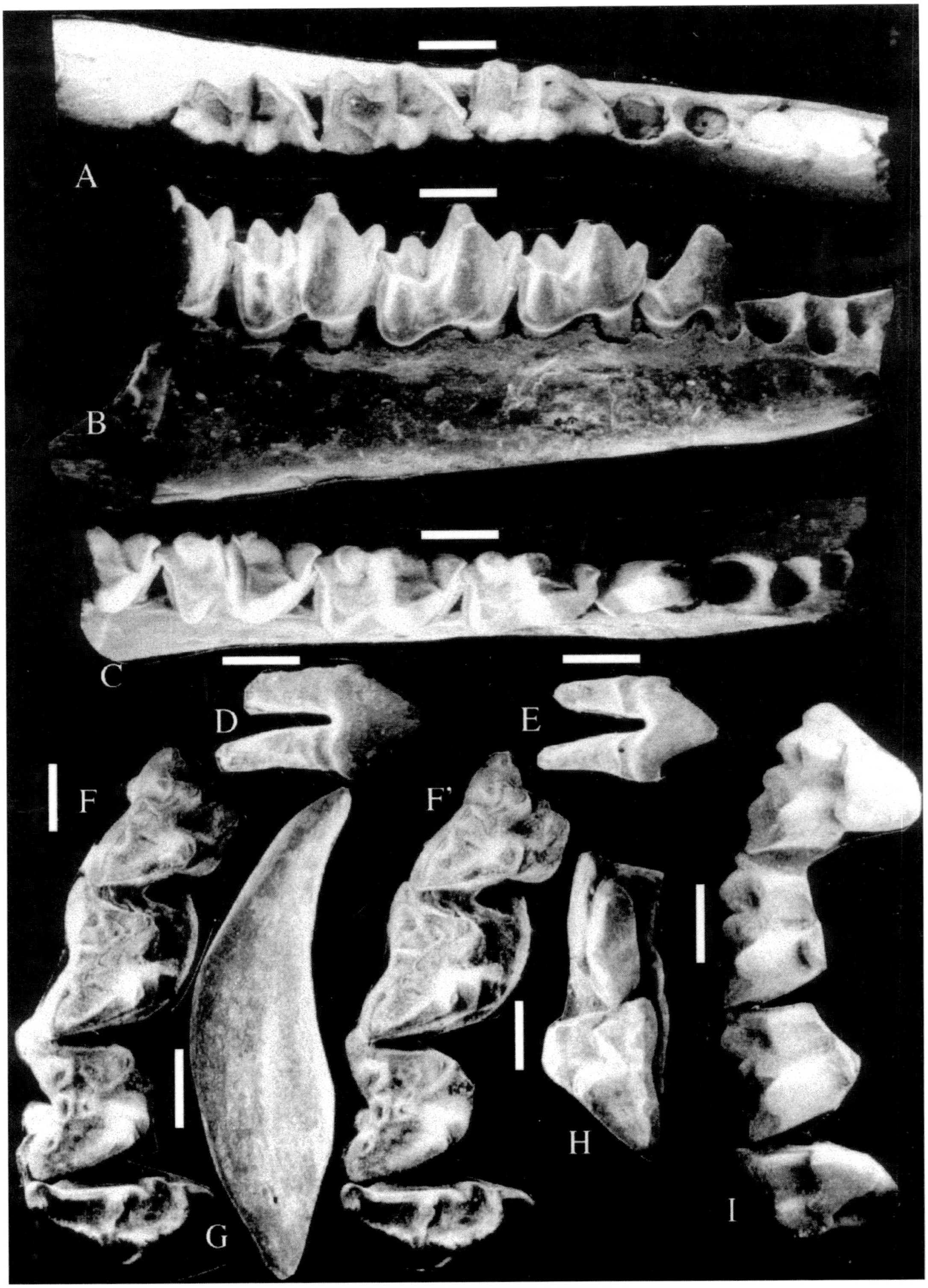

Figure 1 *Djarthia murgonensis*. (A, D–I), QM F31458. (A) M_{1-3}, occlusal view. (B, C), QM F 31460, P_3, M_{1-3} and M_4 trigonid in buccal (B) and occlusal (C) views. (D) right P^3, buccal view. (E) P^1 or P^2, buccal view. (F F') right M^{1-4}, stereo occlusal view. (G) partial right C^1, buccal view. (H) left maxillary fragment with P^3 and M^1, occlusal view. (I)= right M^{1-4}, lingual view. Scale = 1 mm.

regarded as premature. Uncertainty also surrounds the placement of another central Australian fossil species, *Wakamatha tasselli*. This taxon was originally described as a dasyurid by Archer and Rich (1978) on the basis of a partial lower dentary containing M_{3-4}. Addressing the issue of paraphyly in Dasyuridae as then accepted, Wroe (1996a) observed that this taxon, of uncertain age, possessed a number of potential synapomorphies with the most plesiomorphic peramelemorphian known to date, *Yarala burchfieldi* (Muirhead and Filan 1995, Muirhead 2000). These were: the presence of transverse hypocristids, extreme transverse compression of the M_4 talonid, very well-developed anterobuccal cingulids and very low indistinct hypoconulids. *Wakamatha tasselli* shows an intermediate condition for another peramelemorphian synapomorphy, lingual placement of the hypoconulid. If these prove to be actual synapomorphies, then *W. tasselli* represents the least derived bandicoot species known to date. As expanded on below, there are currently no derived dental features defining Dasyuridae and with the possible exception of upper incisor number, none clearly define Dasyuromorphia. In view of these uncertainties, it is posited here that *W. tasselli* too should be treated as Marsupialia *incertae sedis*.

RECENT DEVELOPMENTS IN DASYUROMORPHIAN PALAEONTOLOGY

Interordinal and interfamilial relationships

Despite major new palaeontological discoveries and considerable efforts in the field of molecular biology, both the interordinal and interfamilial relationships of Dasyuromorphia remain the subjects of ongoing debate. At the interordinal level many taxa have been advanced as potential sister taxa to dasyuromorphians. These include Didelphidae (Archer 1976b, Bensley 1903), Peramelemorphia (Kirsch et al. 1977), Notoryctidae (Springer et al. 1997) and Microbiotheriidae plus Diprotodontia plus Notoryctidae (Kirsch et al. 1997). Marshall et al. (1990) and Woodburne and Case (1996) posit that all non-dasyuromorphian Australian marsupials represent the sister clade to Dasyuromorphia and that the Palaeocene Bolivian species *Andinodelphys cochabambensis* is the sister taxon to all Australian marsupials. Various interpretations of higher level phylogeny for these taxa offered in the literature have been reviewed in several recent papers (Aplin and Archer 1987, Godthelp et al. 1999, Kirsch et al. 1997, Luckett 1994, Marshall et al. 1990, Springer et al. 1997). Regarding intrafamilial level relationships, *Myrmecobius fasciatus* has been forwarded as the sister taxon to thylacinids and dasyurids in some previous anatomy based studies (Archer 1984, Aplin and Archer 1987), but not in others (Wroe 1997b). Similarly, among molecular investigations, Lowenstein et al. (1981) placed *M. fasciatus* as the sister taxon to Dasyuridae-Thylacinidae, but at least one (Krajewski et al. 1997) placed the Numbat within Dasyuridae (although these authors treated this poorly supported result as anomalous). Wroe et al. (2000) could not resolve the relative positions of the three dasyuromorphian families, but in the most recent attempts to address this question both Krajewski et al. (2000b) and Wroe and Musser (2001) found dasyurids and thylacinids to be sister taxa. However, support for this relationship in both these latter analyses was weak.

All told, numerical parsimony has been applied on four occasions to anatomical data sets with a view to resolving higher level relationships for dasyuromorphians (Kirsch and Archer 1982, Springer et al. 1997, Wroe et al. 2000, Wroe and Musser 2001). That of Kirsch and Archer (1982), using species as the operational taxonomic unit, produced many anomalous phylogenies. However, their work clearly demonstrated that the inclusion of fossil taxa could significantly impact on results. The morphological investigation presented by Springer et al. (1997), incorporated diprotodontian as well as marsupicarnivore and peramelemorphian taxa, but not Thylacinidae and used the family rather than the species as an operational taxonomic unit. This analysis produced a single most parsimonious tree, with notoryctids the sister to dasyuromorphians and this taxon the sister to a clade inclusive of all other living marsupials, but no interordinal relationships were supported by bootstrap values equal to or greater than 50%. While more recent studies incorporating additional, well represented fossil taxa (Wroe et al. 2000, Wroe and Musser 2001) have far from conclusively solved the major mysteries of dasyuromorphian phylogeny, results constitute a promising basis for further research. The accelerating pace of discovery over the last decade permits the realistic expectation of further significant fossil finds that may well provide deeper insight into questions surrounding the origins, as well as the interfamilial phylogeny of Dasyuromorphia.

Dasyuridae

The most consequential new find in the field of dasyurid palaeontology has been that of near-complete cranial material of early Miocene age from the Neville's Garden Site, Riversleigh, northern Australia. Two skulls, one including both lower jaws, represent the new genus and species, *Barinya wangala* (Figs. 2 and 3). Additional dental material for this taxon has been recovered from the middle Miocene Henk's Hollow Site (AR 6597). Wroe (1999b) identified four dasyurid synapomorphies present in the basicranium of *B. wangala*: development of a tympanic process of the *pars petrosa* to form a distinct periotic hypotympanic sinus, presence of a paroccipital hypotympanic sinus, presence of a deeply invasive sulcus with the posteroventral lip formed by a mesially directed process of the *pars petrosa* for passage of the internal jugular and presence of a distinct tubal foramen for passage of the Eustachian tube. Wroe (1999b) treats *B. wangala* as the sister taxon to modern dasyurids and also forwards an additional four synapomorphies uniting a monophyletic clade comprising all representatives of the three

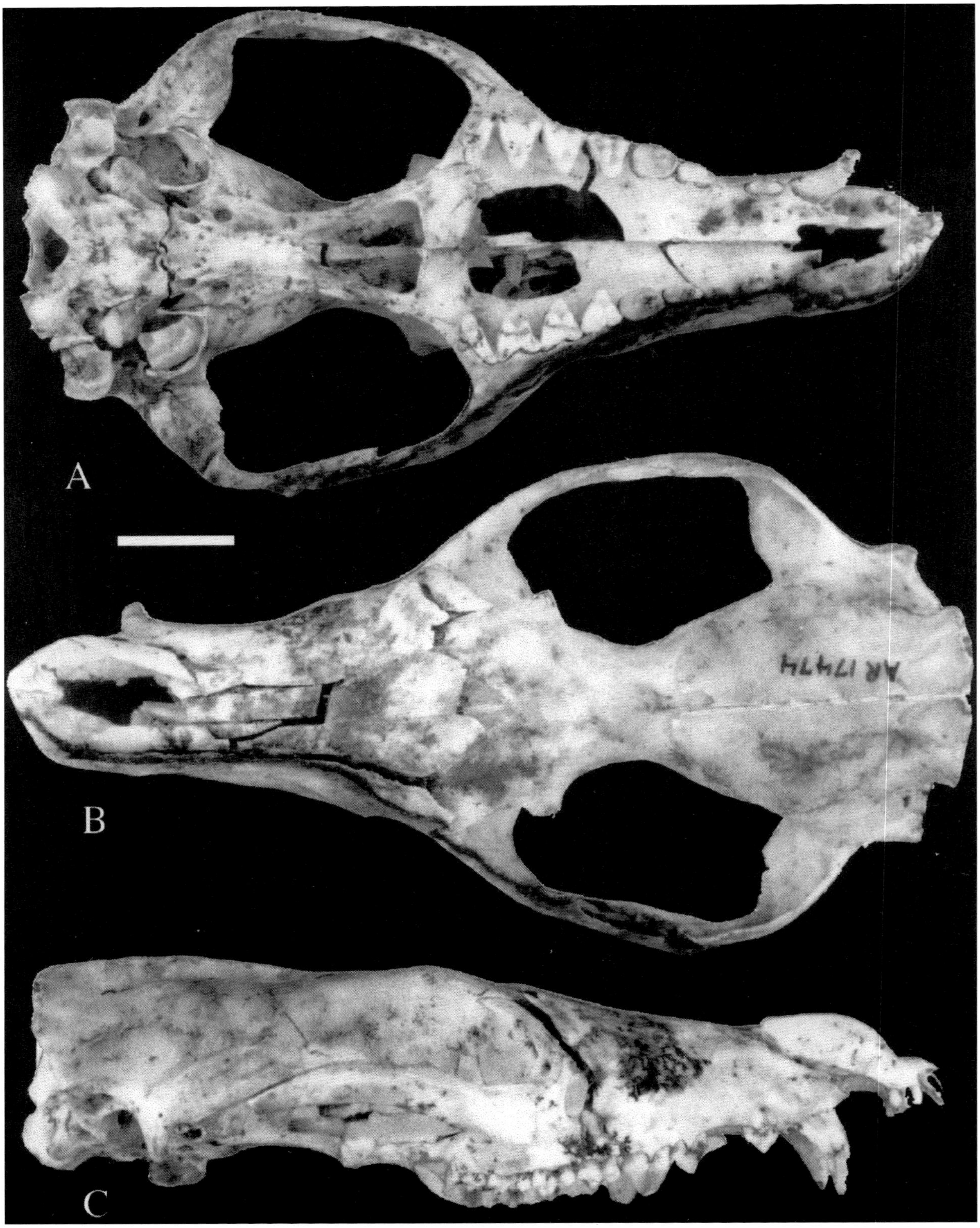

Figure 2 *Barinya wangala*. QM F 31408, skull. (A) ventral view. (B) dorsal view. (C) lateral view. Scale = 2 cm.

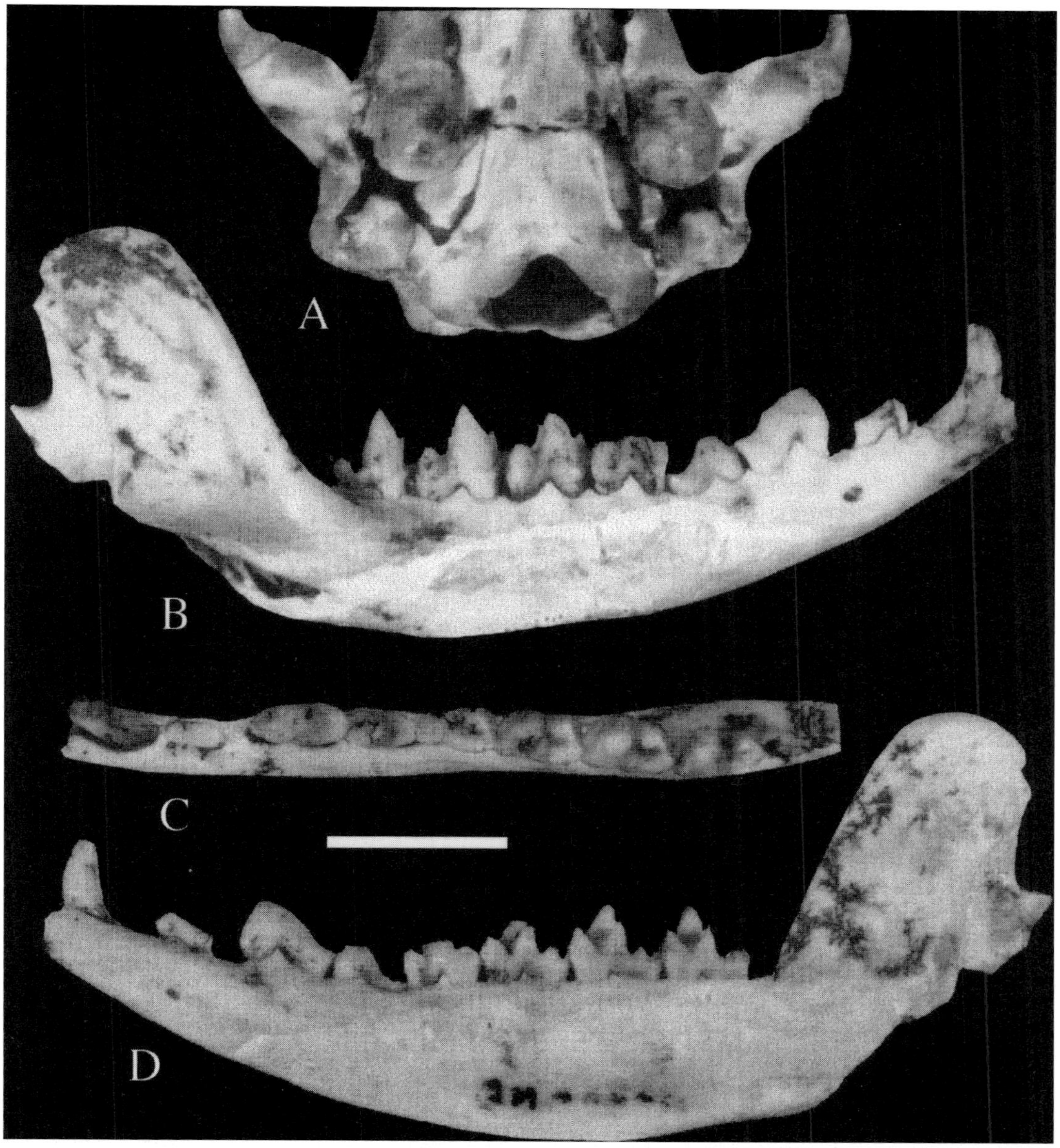

Figure 3 *Barinya wangala*. QM F31409, basicranial region. QM F23889, C_1, P_{1-3}, and M_{1-4} in (B) lateral view, (C) occlusal view and (D) lingual view. Scale = 2 cm.

extant dasyurid subfamilies. These are: presence of a fully enclosed stylomastoid foramen that includes a periotic component; contact between the pars petrosa and a paroccipital tympanic process that that fully encloses the paroccipital hypotympanic sinus ventrally; extensive dorsal enclosure of the internal jugular canal; and contact between the mastoid tympanic process and the pars petrosa. The position of *B. wangala* as sister taxon to all extant subfamilies has been further supported by numerical parsimony analyses (Wroe et al. 2000, Wroe and Musser 2001).

An additional genus and species of Miocene-aged dasyurid, *Ganbulanyi djadjinguli* (see Fig. 4), has been referred to Dasyuridae by Wroe (1998). Dasyurid apomorphies present in this taxon were considered to potentially unite *G. djadjinguli* with either *Sarcophilus* or *Barinya*. Possible synapomorphies with *Sarcophilus* included approximation of the metacone and paracone, anteroposterior orientation of the postmetacrista, and approximation of stylar cusp B and the paracone on M^2, as well as hyper-robusticity of P^2. However, the homology of the then known teeth of *G. djadjinguli* was uncertain and Wroe (1998) also flagged the possibility that the then only upper premolar known for the species might be a P^3, in which case this taxon might be more parsimoniously allied with *Barinya*. The subsequent discovery of a specimen from the Rick's Rusty Rocks Local Fauna supported the latter of these scenarios (Arena et al. 1999).

New dasyurid taxa, as well as additional material of previously known species of Pliocene age, have also been described. '*Dasycercus*' *worboysi* (Lawson et al. 1999; see Fig.4) and *Archerium chinchillaensis* (Wroe and Mackness 2000a; see Fig. 4) from the Big Sink and Chinchilla Local Faunas respectively, represent two new dasyurines of uncertain affinity within the subfamily. Both are known from partial dentitia only and while the presence of derived features shared with most dasyurines permits allocation within Dasyurinae with reasonable confidence, the position of either taxon within the subfamily can not be determined with any precision at present. New material from the Chinchilla Local Fauna has been referred to *Dasyurus dunmalli* (Wroe and Mackness 1998, Wroe and Mackness 2000b). Cladistic analysis incorporating data from these specimens placed *D. dunmalli* within a clade inclusive of *D. albopunctatus*, *D. spartacus*, *D. maculatus* and *Sarcophilus harrisii* (Wroe and Mackness 2000b).

The study of this new material has impacted on preconceptions of dasyurid evolution and phylogeny. While dasyurids have generally been held to represent structural approximations of the archetypal Australian marsupial, it is now clear that the extant subfamilies, in particular, are in fact highly specialised, at least with regard to basicranial features. That said, it is accepted that many dasyurids possess generalised dentitia that do not differ greatly from those of basal ameridelphians. Wroe (1997b) dismisses the only widely accepted dental synapomorphy previously advanced for the family, i.e. a tendency to reduce P_3 (Archer 1982a, Marshall et al. 1990), as an underlying synapomorphy (*sensu* Saether 1986), a practice criticised by a number of authors (Farris 1986, Kitching 1993). Consequently, it is presently impossible to refer specimens of unspecialised fossil marsupials to Dasyuridae on any basis other than overall similarity, unless they are represented by cranial material. A further inference based on palaeontological evidence is that dasyurids represent a relatively recent addition to the Australian fauna. This is particularly so of the monophyletic clade comprising the three extant subfamilies, none of which are represented by diagnosable specimens pre-dating the early Pliocene (Wroe 1999b). Recent molecular evidence indicates an origin for the modern subfamilies of around 24 million years ago (Krajewski et al. 2000a). If this is correct, then the lack of modern dasyurid material in otherwise fossil rich deposits, especially those of Riversleigh in northern Australia, suggests that the modern subfamilies existed in low species abundance during the early Miocene, with the major radiations occurring in the mid Miocene or later. Again this view is not incompatible with recent molecular evidence (Krajewski et al. 2000*a*).

Thylacinidae

The fossil record of Thylacinidae has increased more dramatically than that of any other Australian marsupial carnivore family. Ten new species have been formally described over the last decade: six from late Oligocene–Miocene sites of Riversleigh (*Thylacinus macknessi*, *Wabulacinus ridei*, *Ngamalacinus timmulvaneyi*, *Nimbacinus dicksoni*, *Badjcinus turnbulli*, *Muribacinus gadiyuli*), two from the Bullock Creek Local Fauna of the Northern Territory (*Mutpuracinus archiboldi*, *Nimbacinus richi*), one from the late Miocene Alcoota Local Fauna, *Tjarrpecinus rothi*, and another from the late Miocene Ongeva Local Fauna of central Australia, *Thylacinus megiriani* (Tables 1, 2, 3, Figs 5, 6). Two of these fossil taxa (*Badjcinus turnbulli* and *Nimbacinus dicksoni*) are represented by significant cranial material (Muirhead and Wroe 1998, Wroe and Musser 2001). The rest are known almost entirely from dentitia and jaw fragments. The study of this new material has further corroborated findings based on tarsal morphology (Szalay 1994) and molecular evidence (Lowenstein et al. 1981, Krajewski et al. 1997) that firmly place Thylacinidae within the dasyuromorphian radiation (Muirhead andWroe 1998, Wroe et al. 2000, Wroe and Musser 2001). But, as noted above, where resolution of the interfamilial relationships of thylacinids has been reported in recent studies, it is only weakly supported (Krajewski et al. 2000, Wroe and Musser 2001).

Two analyses using numerical parsimony have addressed the intrafamilial relationships of thylacinids (Muirhead and Wroe 1998, Murray and Megirian 2000). The analysis by Muirhead and Wroe (1998) incorporated five fossil species, as well as *Thylacinus cynocephalus, Thylacinus potens* and five outgroup taxa. In this study six synapomorphies were identified for the family: loss of the mastoid epitympanic sinus; loss of contact between the alisphenoid and petrosal tympanic processes; lateral broadening of the ectotympanic; reduction in size of stylar cusp B; reduction in size of the entoconid; and elongation of P3 in the upper and lower dentitia. However, the validity of some of these synapomorphies seems questionable, the first two of the three middle ear features in particular. This is because, once acquired, it generally accepted that such adaptations are rarely lost (MacPhee 1981). More recently, Murray and Megirian (2000) performed two analyses using Hennig86 (Farris 1988) on data from all 12 described thylacinid taxa. The first of these was performed on a data matrix modified after Muirhead and Wroe (1998) using one cranial and 22 dental characters and the second was based on a set of eight dental features. Murray and Megirian (2000) favoured the results of their second analysis over the first. The basis for this preference appears to be anchored in their assessment of the contentious area of character weighting. Murray and Megirian (2000) dismissed the use of ordered multistate characters by Muirhead and Wroe (1998) opting for another form of character weighting, i.e. character exclusion. The

Table 1 Australian fossil carnivorous marsupial fossil species not represented in modern faunas. Abbreviations for local fauna's are: A = Alcoota; ALL = Allingham; AL = AL 90; B = Bow; BA = Bite's Anntenary; BC = Bullock Creek; C = Chinchilla; CC = Chillagoe Caves; BD = Bluff Downs; BS = Big Sink; CA = Cleft of Ages; CS = Camel Sputum; DT = Dirk's Tower; E = Encore, F = Floraville; FC = Fisherman's Cliff; G = Gag; H = Hiatus; HH = Henk's Hollow; I = Inabeyance; JJ = Jim's Jaw; K = Kutjamarpu; N = Ngapakaldi; O = Ongeva; RRR = Rick's Rusty Rock's; S = Smeaton; T = Tingamarra; TC = Town Cave; US = Upper Site; WC = Wellington Caves; Y = Yanda; [r] = from Riversleigh, northwestern Queensland * = to numerous to list; ? = local fauna unknown; *i. s.* = *incertae sedis*.

Species	Family	Deposit/s	Age	Author/s
Ganbulanyi djadjinguli	Dasyuridae	[r]E, [r]RRR	middle to early-late Miocene	Wroe 1998
Barinya wangala	Dasyuridae	[r]NG, [r]US, [r]JJ, [r]HH, [r]BA	early to middle Miocene	Wroe 1999*b*
Sminthopsis floravillensis	Dasyuridae	F	Pliocene	Archer 1982*a*
Dasyuroides achilpatna	Dasyuridae	FC	Pliocene	Archer 1982*a*
'Dasycercus' worboysi	Dasyuridae	BS	Pliocene	Lawson *et al.* 1999
Archerium chinchillaensis	Dasyuridae	C	Pliocene	Wroe & Mackness 2000*a*
Dasyurus dunmalli	Dasyuridae	B, BD, C,	Pliocene	Bartholomai 1971
Glaucodon ballaratensis	Dasyuridae	S	Pliocene	Stirton 1957
Sarcophilus moornaensis	Dasyuridae	FC	Pliocene	Crabb 1982
Muribacinus gadiyuli	Thylacinidae	[r]G, [r]HH	middle Miocene	Wroe 1996*a*
Badjcinus turnbulli	Thylacinidae	[r]WH	late Oligocene	Muirhead & Wroe 1998
Wabulacinus ridei	Thylacinidae	[r]CS, [r]H	late Oligocene to early Miocene	Muirhead 1997
Thylacinus macknessi	Thylacinidae	[r]NG, [r]G, [r]H, [r]E, [r]WW, [r]DT	late Oligocene to early-late Miocene	Muirhead 1992
Thylacinus megiriani	Thylacinidae	O	late Miocene	Murray 1998
Thylacinus potens	Thylacinidae	A	late Miocene	Woodburne 1967
Ngamalacinus timmulvaneyi	Thylacinidae	[r]I, [r]NG, [r]CS	early Miocene	Muirhead 1997
Nimbacinus dicksoni	Thylacinidae	[r]HH, [r]A, [r]WH, BC, [r]CA	late Oligocene to early-late Miocene	Muirhead & Archer 1990
Nimbacinus richi	Thylacinidae	BC	middle Miocene	Murray & Megirian 2000
Mutpuracinus archiboldi	Thylacinidae	BC	middle Miocene	Murray & Megirian 2000
Tjarrpecinus rothi	Thylacinidae	A	late Miocene	Murray & Megirian 2000
Mayigriphus orbus	Dasyuromorphia *i. s.*	[r]E	early-late Miocene	Wroe 1997*a*
Djarthia murgonensis	Marsupialia *i. s.*	T	early Eocene	Godthelp, Wroe & Archer 1999
Ankotarinja tirarensis	Marsupialia *i. s.*	N	late Oligocene	Archer 1976*a*
Keeuna woodburnei	Marsupialia *i. s.*	N	late Oligocene	Archer 1976*a*
Wakamatha tasselli	Marsupialia *i. s.*	?	?	Archer & Rich 1979
Dasylurinja kokuminola	Marsupialia *i. s.*	Y	late Oligocene	Archer 1982*a*
Priscileo roskellyae	Thylacoleonidae	[r]U	early Miocene	Gillespie 1997
Priscileo pitikantensis	Thylacoleonidae	N	late Oligocene	Rauscher 1987
Wakaleo oldfieldi	Thylacoleonidae	K, [r]CA	early Miocene to early-late Miocene	Clemens & Plane 1974
Wakaleo vanderleueri	Thylacoleonidae	BC,[r]E	middle Miocene to early-late Miocene	Clemens & Plane 1974
Wakaleo alcootaensis	Thylacoleonidae	A	late Miocene	Archer & Rich 1982
Thylacoleo hilli	Thylacoleonidae	B, TC	Pliocene	Pledge 1977
Thylacoleo crassidentatus	Thylacoleonidae	ALL, C, B, BD	Pliocene	Bartholomai 1962
Thylacoleo carnifex	Thylacoleonidae	*	Pleistocene	Owen 1859
Ekaltadeta ima	Hypsiprymnodontidae	r*	late Oligocene to middle Miocene	Archer & Flannery 1985
Ekaltadeta jamiemulvaneyi	Hypsiprymnodontidae	[r]E, CA	early-late Miocene	Wroe 1996*b*
Jackmahoneyi toxoniensis	Hypsiprymnodontidae	B	Pliocene	Ride 1993
Propleopus oscillans	Hypsiprymnodontidae	*	Pleistocene	De Vis 1888
Propleopus wellingtonensis	Hypsiprymnodontidae	WC	Pleistocene	Archer & Flannery 1985
Propleopus chillagoensis	Hypsiprymnodontidae	C	Pleistocene	Archer et al. 1978

Table 2 Character state changes for nodes 1–8 in the first of 24 shortest trees (see Fig. 7).

Node	Character state changes
1	C2 0 ⇒ 1; C3 0 ⇒ 1, C7 0 ⇒ 1; C9 0 ⇒ 1; C11 0 ⇒ 1; C12 0 ⇒ 1; C16 0 ⇒ 1; C19 0 ⇒ 1; C21 0 ⇒ 1
2	C3 1 ⇒ 2; C5 0 ⇒ 1; C6 0 ⇒ 1; C80 ⇒ 1; C10 0 ⇒ 1; C15 0 ⇒ 1; C17 0 ⇒ 1
3	C1 0 ⇒ 1; C12 1 ⇒ 0; C19 1 ⇒ 2; C21 1 ⇒ 2; C22 0 ⇒ 1
4	C10 0 ⇒ 1;C11 1 ⇒ 2; C16 1 ⇒ 2; C18 0 ⇒ 1; C23 0 ⇒ 1
5	C3 2 ⇒ 3; C4 0 ⇒ 1; C6 1 ⇒ 2; C7 1 ⇒ 2; C9 1 ⇒ 2; C11 2 ⇒ 3; C12 2 ⇒ 3; C16 2 ⇒ 3; C17 1 ⇒ 0
6	C3 3 ⇒ 4; C5 1 ⇒ 2; C11 3 ⇒ 4; C16 3 ⇒ 4; C19 2 ⇒ 3; C21 2 ⇒ 3
7	C2 1 ⇒ 2; C16 4 ⇒ 5
8	C2 2 ⇒ 3; C4 1 ⇒ 2; C5 2 ⇒ 3

product was an input data matrix in which four pairs of taxa, including one ingroup and one outgroup taxon, had identical coding for all characters. A deficiency of this procedure is that neither of two resultant shortest trees distinguishes Thylacinidae on the basis of synapomorphy. Moreover, while it is conceded that the use of multistate characters has its problems (Wiley 1981, Lipscombe 1992), results based on the methodology of Muirhead and Wroe (1998) are preferred here primarily because they are far less subject to *apriorism* than those favoured by Murray and Megirian (2000). That said, with most thylacinids currently known primarily from incomplete dental material, it may be imprudent to place too much faith in any hypotheses of intrafamilial relationships at present. Because tables of character state transformations were not provided by either Muirhead and Wroe (1998) or Murray and Megirian (2000) the data presented by these latter authors has been reanalysed and results are given below (Figs. 7–9 and Tables 2–3).

As well as a near-complete skull and skeleton of *Nimbacinus dicksoni* (Wroe and Musser 2001), other thylacinid material discovered from Riversleigh since the description of *B. turnbulli* includes a M_2 (QM F23358, Hiatus Site, late Oligocene) referred here to *Wabulacinus* sp. and a M_1 (QM F 12972, Wayne's Wok Site, late Oligocene) referred here to *Thylacinus* sp. This latter find confirms the prediction by Murray and Megirian (2000) that *Thylacinus* had originated by late Oligocene times. Further extending the temporal range of *Thylacinus* at Riversleigh has been the discovery of a M_2 (QM F 30862) from the early–late Miocene Encore Site. Additional material (AR 10427) representing *Nimbacinus* has also been retrieved from the late Oligocene White Hunter Site. The presence of considerable thylacinid diversity, including specialised *Thylacinus* and *Wabulacinus*, in late Oligocene deposits, indicates that the origin of the thylacinid clade predates estimates forwarded on the basis of molecular data, i.e. 7 to 25 million years (Lowenstein et al. 1981, Sarich et al. 1982).

Dasyuromorphia incertae sedis

One new taxon has been assigned to Dasyuromorphia *incertae sedis*. *Mayigriphus orbus* (Wroe 1997a, see Fig. 4) shares a number of features, derived among dasyuromorphians, that might be construed as indicative of a special relationship with either sminthopsine dasyurids or thylacinids. Specifically, *M. orbus* shares two potential synapomorphies with *Planigale*, a greatly reduced M_1 paraconid concurrent with a moderately reduced M_1 metaconid and reduction of $M_{2\text{-}4}$ metaconids. However, at least three derived features are also shared with thylacinids and/or some specialised dasyurines: lack of a clear differential between reduction of M_1 and $M_{2\text{-}4}$ metaconids; an obtuse angle formed between the paracristid and metacristid; and reduction of the talonid with the cristid obliqua terminating in a lingual position.

Table 3 Character state changes for nodes 1–8 in the first of 2 shortest trees (see Fig. 9).

Node	Character state changes
1	0 length branch
2	C1 0 ⇒ 1
3	C1 1 ⇒ 2
4	C2 0 ⇒ 1
5	C1 2 ⇒ 3; C3 0 ⇒ 1; C8 0 ⇒ 1
6	C1 3 ⇒ 4; C5 0 _ 1
7	C7 0 ⇒ 1
8	C4 0 ⇒ 1; C6 0 ⇒ 1

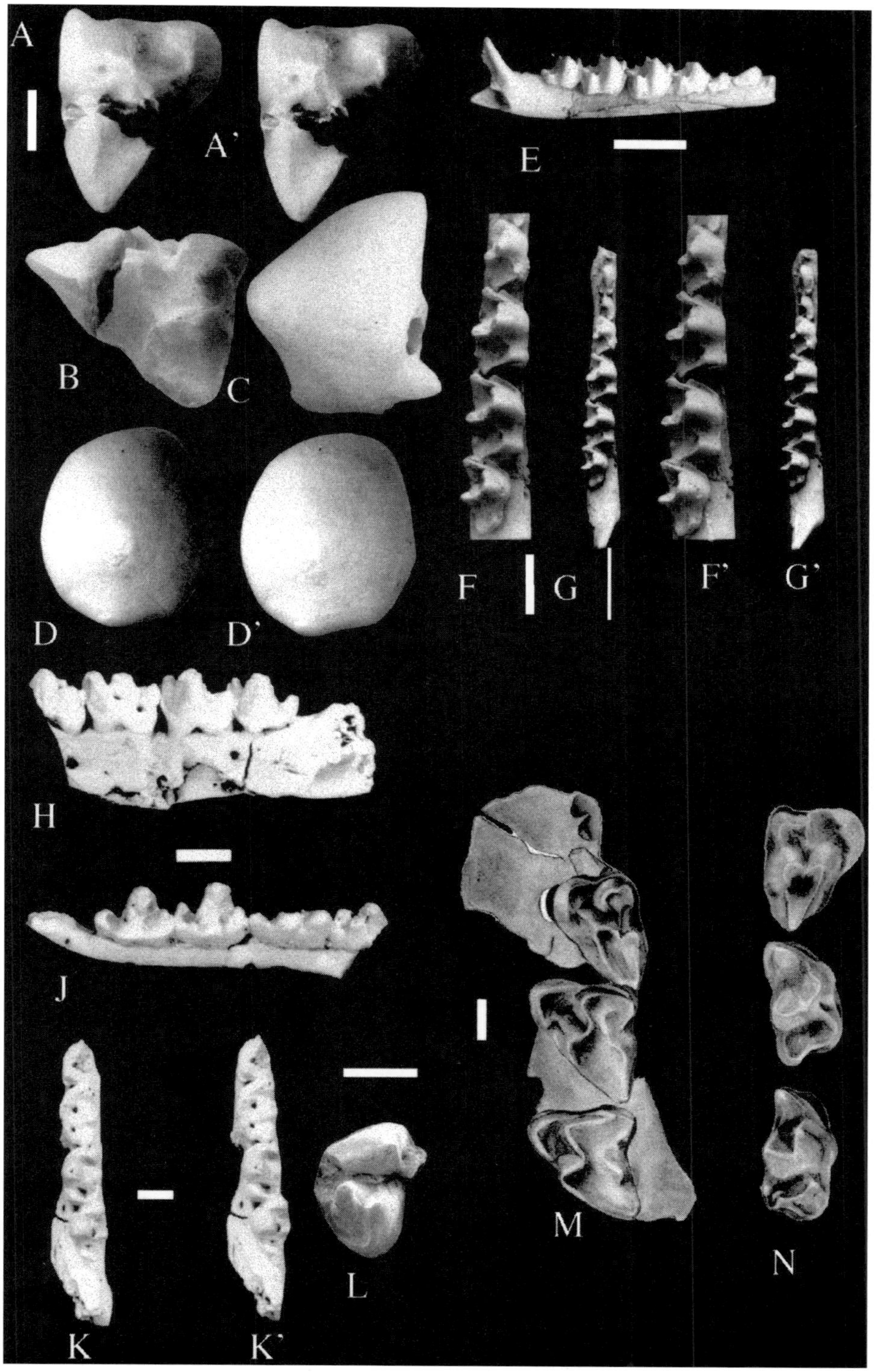

Figure 4 *Ganbulanyi djadjinguli*. QM F24537, right M^1 or M^2. (AA') stereo occlusal and (B) lingual view. QM F24537, P^3. (C) = lateral view, (D, D') = stereo occlusal view. Scale = 2 mm. *Mayigriphus orbus*. QM F23780, P_{2-3}, M_{1-4}. (E) buccal view, (FF' and GG') stereo occlusal views. Scale = 2 mm. '*Dasycerus*' *worboysi*. AM F69805, M_{1-4}. (H) buccal view, (J) lingual view, (KK') stereo occlusal view. AM F69805, right M^1 fragment. (L) = occlusal view. Scale = 2 mm. *Archerium chinchillaensis*. QM F39847, P^3, M^{1-3}. (M) occlusal view. (N) from top to bottom occlusal views of: QM F39848, right M^1; QM F39849, right M_2, QM F39850, right M_2. Scale = 1mm.

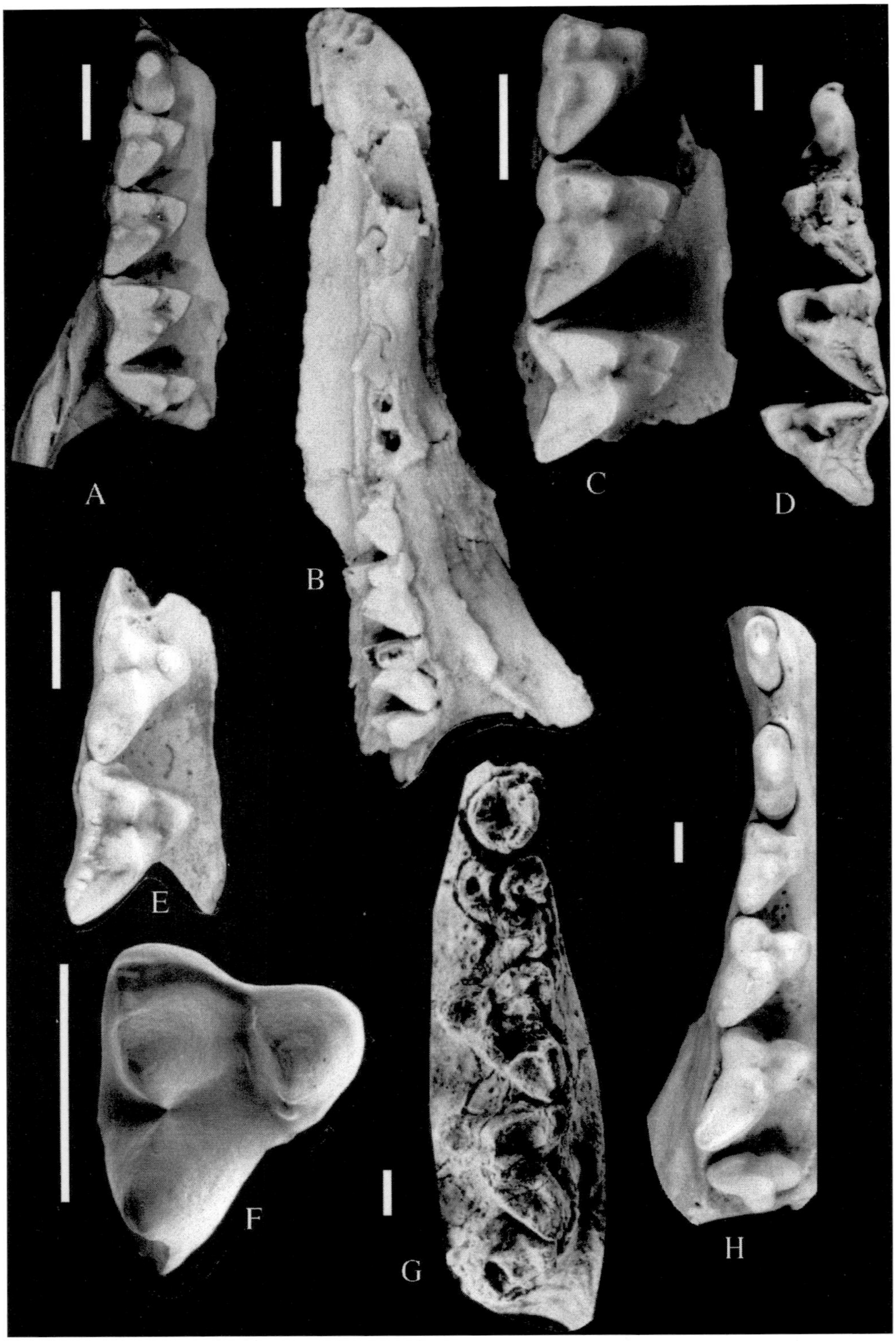

Figure 5 Maxillary fragments. All in occlusal view. (A) *Muribacinus gadiyuli.* QM F 30386, P^3, M^{1-4}. (B) *Badjcinus turnbulli.* QMF 30488, P^{1-2}, M^{1-4}. (C) *Nimbacinus dicksoni.* QMF 16803, M^{1-3}. (D) *Ngamalacinus timmulvaneyi.* QMF 30300, P^3, M^{1-3}. (E) *Wabulacinus ridei.* QMF 16851, M^3. (F) *Thylacinus macknessi.* QMF 16850, M1. (G) *Thylacinus potens.* CPC 6746, P^3, M^{1-4}. (H) *Thylacinus cynocephalus.* AM 217, P^{2-3}, M^{1-4}. Scale = 5 mm.

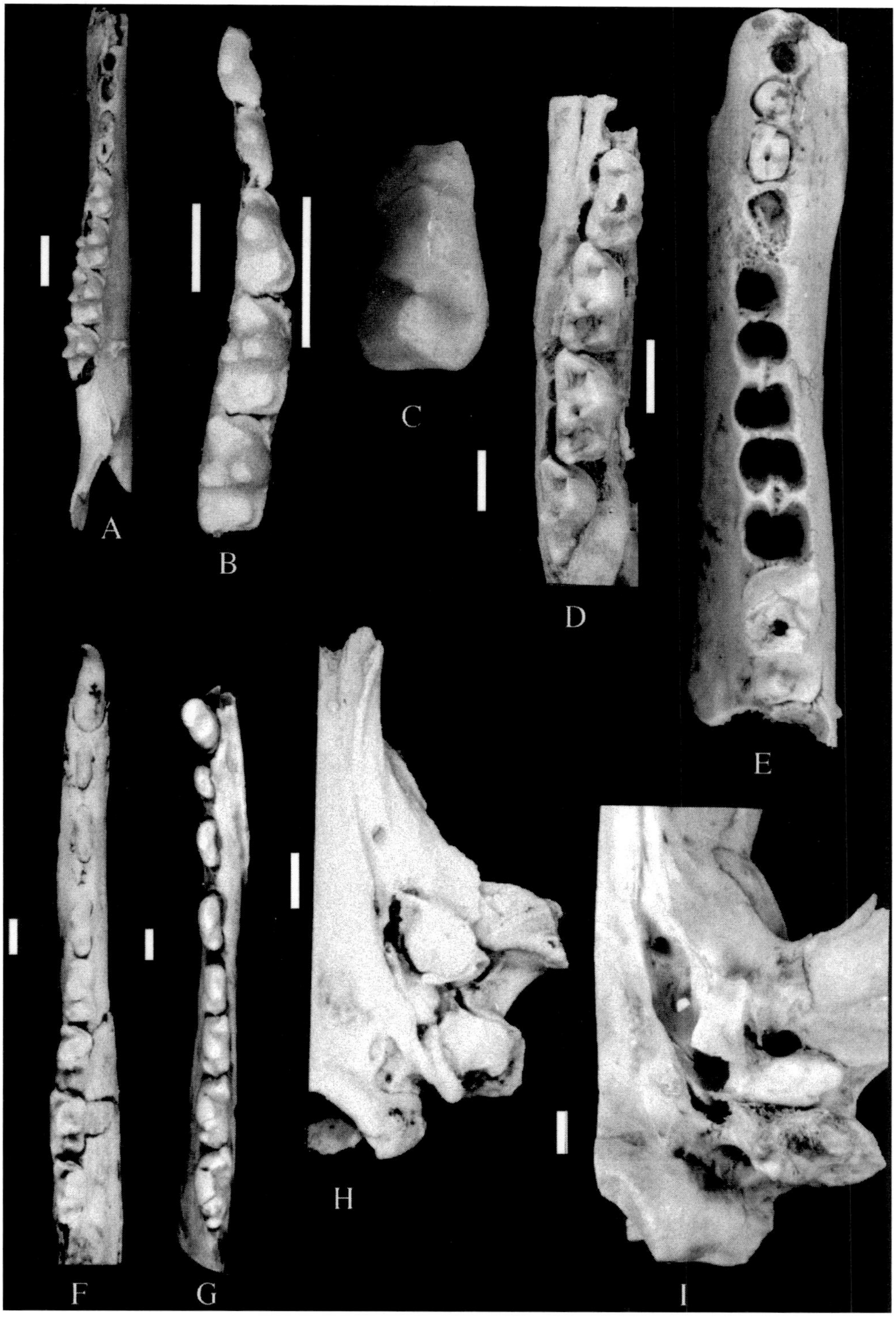

Figure 6 Dentary fragments. All in occlusal view. (A) *Muribacinus gadiyuli*. QMF 30386, P_3, M_{1-4}. (B) *Badjcinus turnbulli*. QMF 30411, P_{2-3}, M_{1-3}. (C) *Nimbacinus dicksoni*. QMF 16802, M_1. (D) *Ngamalacinus timmulvaneyi*. QMF 16853, M_{1-4}. (E) *Wabulacinus ridei* QMF 16852, P_2 (partial), M_3. (F) *Thylacinus macknessi*. QMF 16848, C_1, P_{1-3}, M_{1-4}. (F) *Thylacinus cynocephalus*. AR 8409; C_1, P_{1-3}, M_{2-4}. (G) Left basicranial region of *Badjcinus turnbulli*, QMF 30488. (H) Left basicranial region of *Thylacinus cynocephalus*, QMF 1145. Scales = 5 mm.

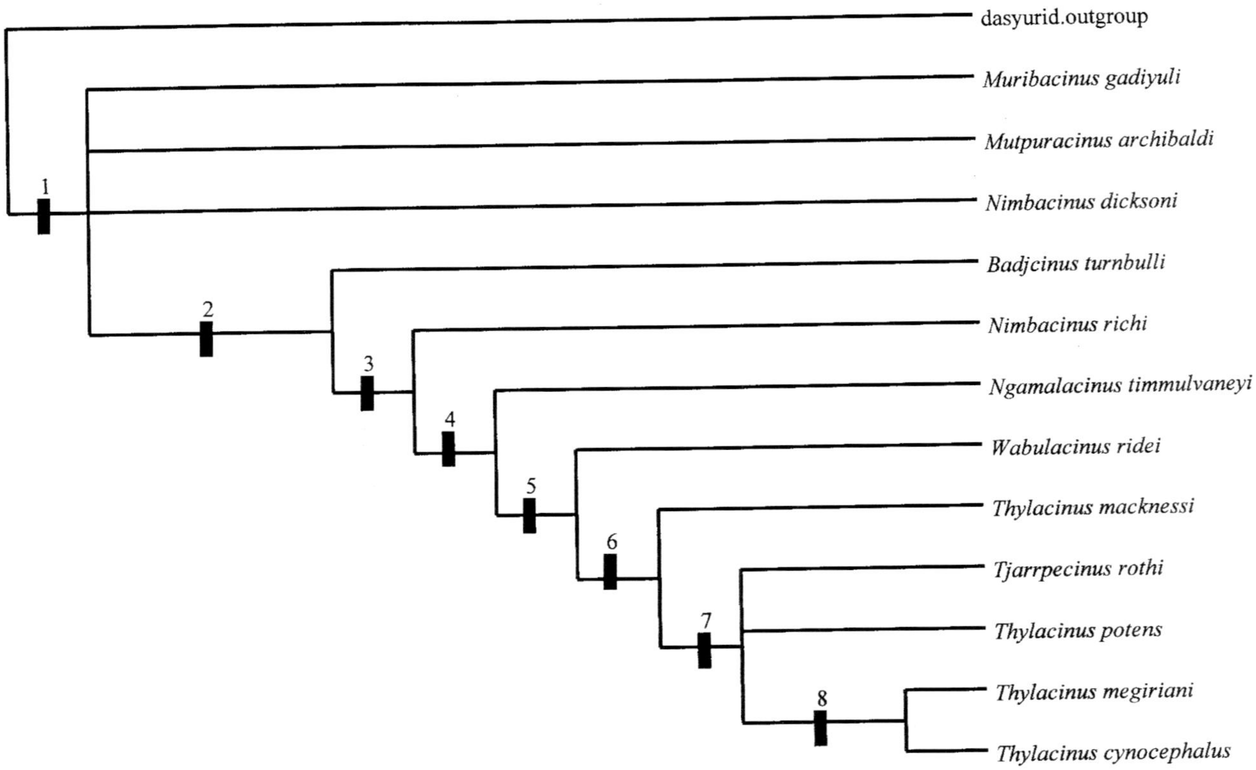

Figure 7 Intrafamilial relationships of Thylacinidae. First of 24 shortest trees of 67 steps based on the treatment of 23 characters modified after Muirhead and Wroe (1998) by Murray and Megirian (2000). See Table 2 for character state changes at nodes.

Recent developments in propleopine palaeontology

Until 1985, Propleopinae was represented by only two species, *Propleopus oscillans* (De Vis 1888) and *P. chillagoensis*, both of Pleistocene age. The pace of discovery began to escalate in 1985 with the description of the first Miocene species, *Ekaltadeta ima* and an additional Pleistocene taxon, *P. wellingtonensis* (Archer and Flannery 1985). Subsequently, a further two species have been described, the early–late Miocene *E. jamiemulvaneyi* (Wroe 1996b) and the Pliocene *Jackmahoneyi toxoniensis* (Ride 1993). Of particular value interest are the first descriptions of cranial material referred to the subfamily: the partial skull of a juvenile *P. oscillans* (Ride et al. 1997) and the near-complete skull of *E. ima* (Wroe et al. 1998, see Fig. 10).

Historically, propleopines have been allied with potoroids in general and the species of *Hypsiprymnodon* in particular (De Vis 1888, Archer and Flannery 1985). Ride (1993) concurred with the conclusion of sister taxon status for Propleopinae and *Hypsiprymnodon*, but considered that both formed a clade of sufficient distinction as to merit higher taxonomic status, hence the erection of a new family, Hypsiprymnodontidae. Monophyly of this clade was based on shared presence of plagiaulacoid upper and lower premolars, specialised trigonids on dP_3 and M_1 with the main crest formed by the parametacristid, retention and withdrawal from the plane of occlusion of P2 (see Wroe and Archer 1995), a deeply invasive masseteric canal and the characteristic hysiprymnodontid mandible. Alternatively, Wroe et al. (1998) regard each of these features as possible macropodoid symplesiomorphies and thus no basis for the postulation of a special relationship. Wroe et al. (1998) further contend that reasonable grounds exist for the consideration of monophyly for propleopines and another extinct macropodoid subfamily, Balbarinae. Features flagged as possible propleopine-balbarine synapomorphies were: very large squamosal epitympanic sinus; ectotympanic roughly linear in lateral view, ectotympanic intimately associated with the glenoid process of the squamosal anteriorly and the squamosal posttympanic and mastoid part of petrosal posteriorly; presence of large ovate lateral carotid fenestra and a greatly expanded frontal sinus. Regarding intrasubfamilial relationships, no studies have included all known taxa and only one has applied numerical parsimony (Wroe 1996b). Balbarines were not included in this latter investigation and it was based on a dental data set only. Work is in progress using additional taxa and a cranial-dental data set (B. Cooke and B. Kear, unpublished data).

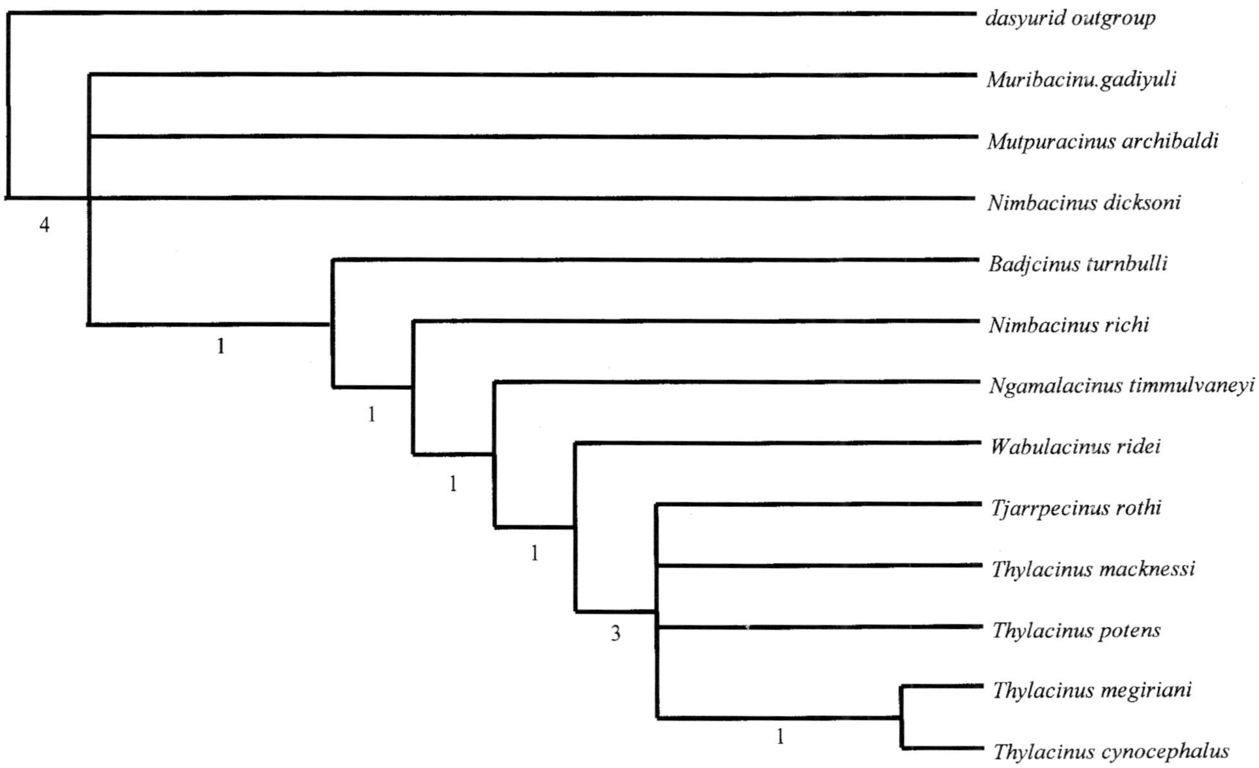

Figure 8 Intrafamilial relationships of Thylacinidae. Strict consensus of 24 shortest trees with Bremer support (Bremer 1994) given below branches. Analysis conducted using PAUP 3.1.1 (Swofford 1993) with multistate characters ordered, ACCTRAN character optimisation and a branch and bound search (CI excluding uninformative characters = 0.73, HI excluding uninformative characters = 0.262, RI = 0.85, RC = 0.64). Polarised character states as abridged by Murray and Megirian (2000) after Muirhead and Wroe (1998): **1**, infraorbital foramen: bound by jugal (0), not bound by jugal (1). **2**, size of paracone: unreduced (0), slight reduction (1), significant reduction (2), extreme reduction (3). **3**, stylar cusp B: well developed (0), slight reduction (1), significant reduction (2), distinct reduction (3), extreme reduction or loss (4). **4**, anterior cingulum: complete (0), incomplete (1). **5**, protocone and conules: well developed (0), slightly reduction (1), significant reduction (2), loss of conules (3). **6**, length of postmetacrista: not elongate (0), significant elongation (1), pronounced elongation (2). **7**, angle of centrocrista: acute (0), obtuse (1), colinear (2). **8**, direction of preparacrista: perpendicular (0), slightly oblique (1), parallel to long axis (2), directly anterior to paracone (3). **9**, angles formed by paracristae and metacristae: narrow (0), significantly wider (1), wide (2). **10**, size of M_1 metaconid relative to those of $M_{2\text{-}4}$: uniform reduction (0), differentially reduced on M_1 relative to $M_{2\text{-}4}$ (1). **11**, metaconid reduction on $M_{2\text{-}4}$: unreduced (0), slight reduction (1), further reduction (2), greatly reduced (3), lost (4). **12**, size of entoconid: unreduced (0), slight reduction (1), further reduction (2), greatly reduced (3), lost (4). **13**, shape of entoconid: uncompressed (0), compressed (1). **14**, diastema between P_1 and P_2: present (0), absent (1). **15**, diastema between P_2 and P_3: present (0), absent (1). **16**, hypoconulid notch: unreduced (0), slight reduction (1), further reduction (2), significantly reduced (3), greatly reduced (4), lost (5). **17**, posterior cingulid and hypocristid: separated (0), joined (1). **18**, carnassial notch of cristid obliqua: absent (0), present (1). **19**, hypocristid: parallel to transverse axis of dentary (0), slight angle (1), moderate angle (2), pronounced angle (3). **20**, carnassial notch in hypocristid: absent (0), present (0). **21**, anterior termination of cristid obliqua: ventral and buccal to carnassial notch of protocristid (0), slightly shifted buccally and dorsally (1), moderately shifted buccally and dorsally (2), markedly shifted shifted buccally and dorsally (3). **22**, height of P_3: higher than P_2 (0), lower than P_2 (0). **23**, length of M_4: shorter than M_2 (0), longer than M_2 (1).

RECENT DEVELOPMENTS IN THYLACOLEONID PALAEONTOLOGY

Since 1982 two new species of thylacoleonid have been placed in a third thylacoleonid genus, *Priscileo*. *Priscileo pitikantensis* (Rauscher 1987), from the late Oligocene Ngapakaldi Local Fauna of central Australia and *P. roskellyae* (Gillespie 1997), from the early Miocene Upper Site Local fauna of Riversleigh. Both have been described on the basis of partial upper dentitia. Near-complete crania have been discovered for both *P. roskellyae* (Gillespie 1999) and *Wakaleo vanderleueri* (Murray et al. 1987). Thus, skulls are now known for representatives of each thylacoleonid genus.

The origin of the thylacoleonid radiation is a contentious issue. Broom (1898) alluded to a special relationship between marsupial 'lions' and burramyids on the basis of P3 morphology. While Bensley (1903) considered the thylacoleonid dentition to

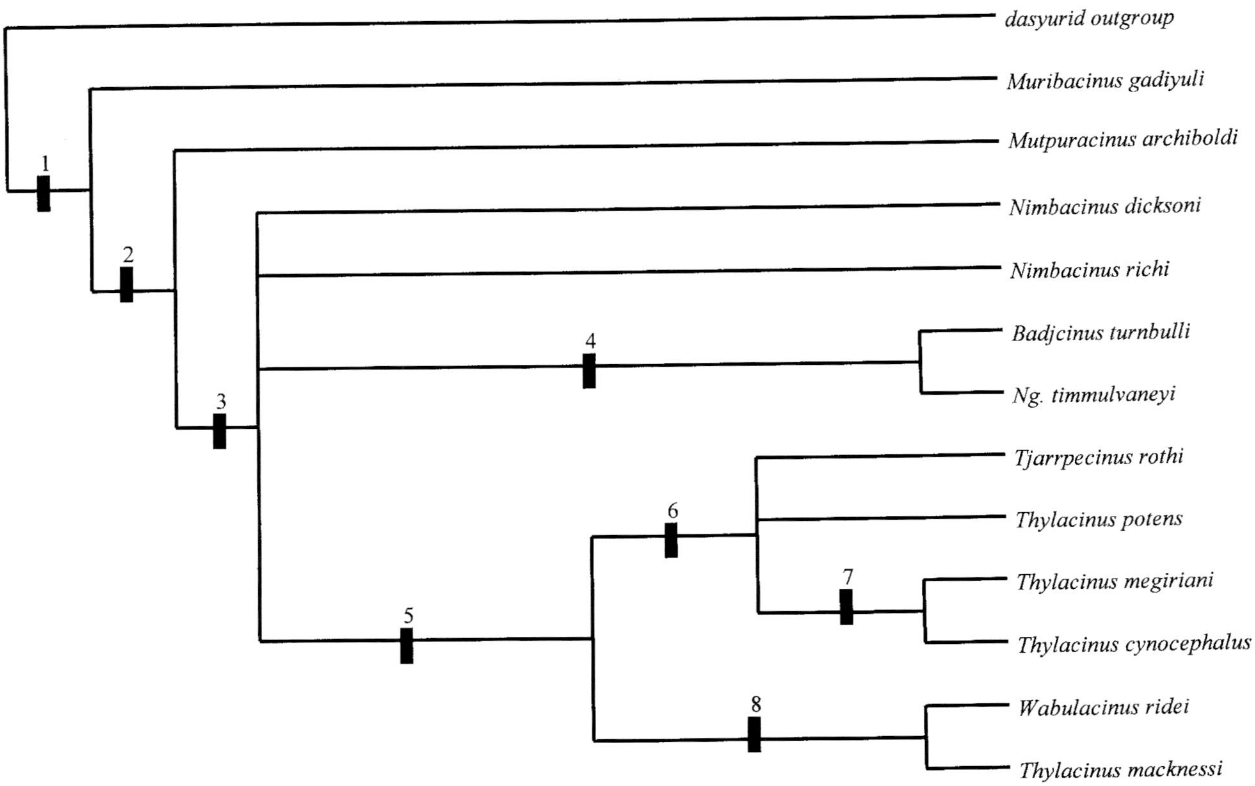

Figure 9 Intrafamilial relationships of Thylacinidae. Least well resolved of two shortest trees of 11 steps from data set provided by Murray and Megirian (2000). Analysis conducted using PAUP 3.1.1 (Swofford 1993) with multistate character (1) ordered, ACCTRAN character optimisation and a branch and bound search (CI excluding uninformative characters = 1.0, HI excluding uninformative characters = 0.0, RI = 1.0, RC = 1.0). Bremer support (Bremer 1994) given below branches. See Table 3 for character state changes at nodes. Character and character states from Murray and Megirian (2000): **1**, pattern of metaconid expression: (0) slight, uniform, (1) slight M_4, (2) differential, marked on M_1, slight M_4, (3) near obsolescence on all molars, (4) entirely lost on all molars. **2**, carnassial notch in hypocristid: (0) absent, (1) present. **3**, elongation of postvallum, shearing surfaces: (0) slight, (1) significant. **4**, stylocone B: (0) present, (1) extremely reduced or lost. **5**, paracone reduction/hypertrophy of metacone: (0) slight–moderate, (1) conspicuous. **6**, ectoflexus: (0) strong expression, (1) weak expression. **7**, precingulum: (0) present, strong (1) reduced–absent. **8**, entoconid: (0) distinct, (1) reduced–absent.

be a modification of the phalangerid type. Phalangeroid affinities for the family seem to have gone unquestioned until 1983, when Aplin and Archer argued for a special relationship with vombatomorphians, a position reiterated by Aplin and Archer (1987) and accepted by at least some others (Marshal et al. 1990, Gillespie 1999, Wroe et al. 1998). The presence of a marsupial apomorphy, i.e. bulla enclosed fully by the squamosal shared by *Thylacoleo carnifex* and other vombatomorphians, is the primary synapomorphy underlying this interpretation. Study of the cranium of *W. vanderleueri* by Murray et al. (1987) led these authors to dispute the phylogeny proposed by Aplin and Archer (1983) and re-advocate phalangeroid affinities with burramyids as their immediate sister taxon. Murray et al.'s (1987) interpretation stands on the acceptance of the bilaminar bullae in *W. vanderleueri*, i.e. with alisphenoid and squamosal contributions, as a diprotodontian plesiomorphy. These authors also treat the absence of a hypocone in thylacoleonids as a retained plesiomorphy. Alternatively, Rauscher (1987) considers the absence of this feature to be an apomorphic reversal to a plesiomorphic state. In Murray et al.'s (1987) scenario, the complete enclosure of the middle ear by the squamosal in *Thylacoleo* and vombatomorphians is treated as homoplasious. The interpretation of thylacoleonid intrafamilial relationships is also in a state of flux. While all authors seem to concur on plesiomorphic sister taxon status for *Prisciléo* within Thylacoleonidae, the relationship of *Wakaleo* to *Thylacoleo* is less certain. The presence of a thylacoleonid apomorphy in then known specimens of *Wakaleo* that was absent in *Thylacoleo*, i.e. loss of P1 or P2, was thought to have precluded the possibility of an ancestor-descendent relationship for the two genera (Clemens and Plane 1974, Rauscher 1987, Archer 1984, Murray et al. 1987). However, because a new species of *Wakaleo* retains a full compliment

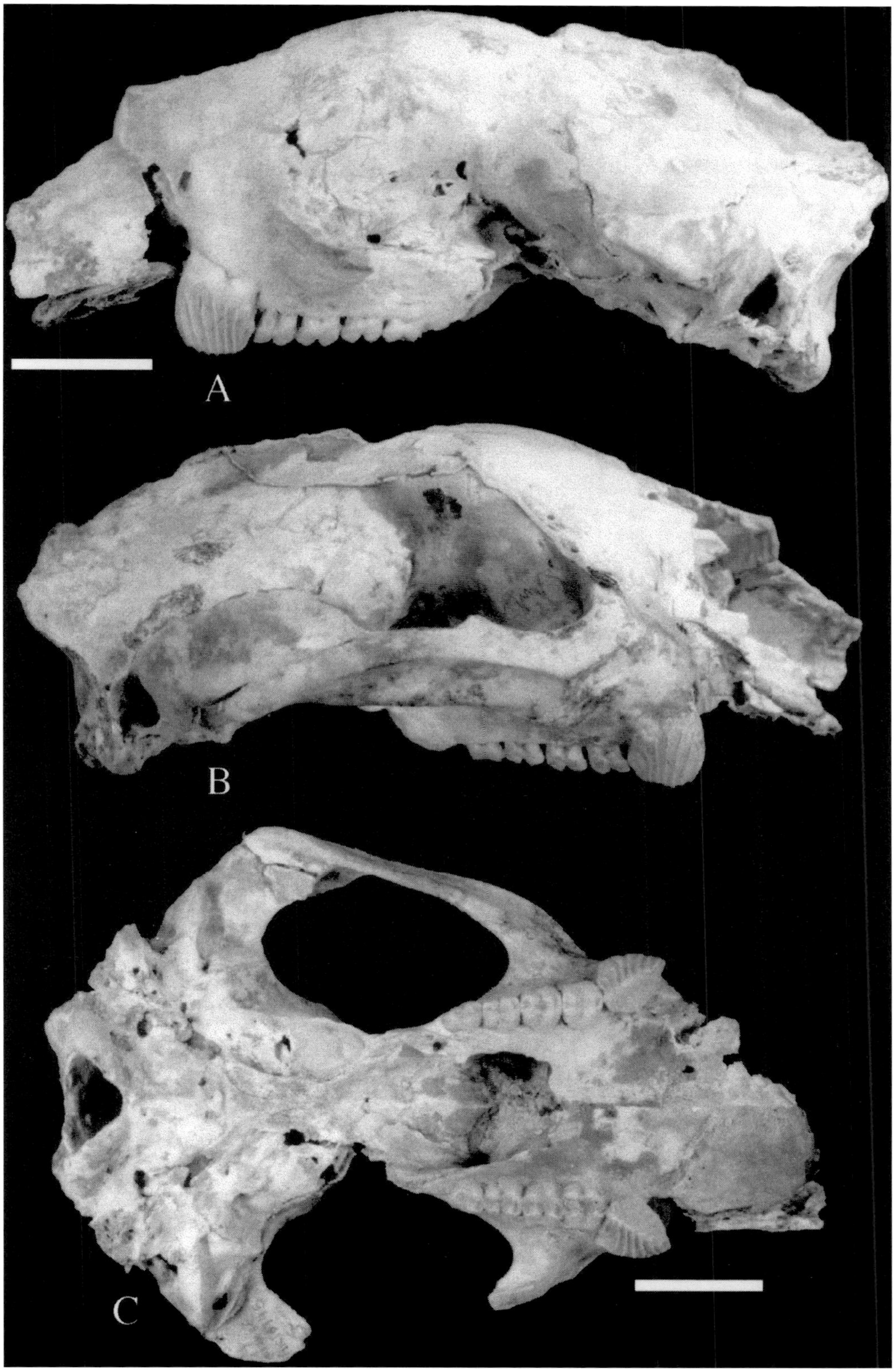

Figure 10 *Ekaltadeta ima*. QM F12426, partial skull. (A) left lateral view. (B) right lateral view. (C) ventral view. Scale = 20 mm.

of three premolars this rationale no longer stands (Gillespie 1999). Cladistic analysis currently under way by A. Gillespie may ultimately resolve these issues.

Palaeobiology

While in recent years progress has been made into the biology and ecology of extant Australian marsupial carnivores (Jones 1996, 1997) and the recently extinct *Thylacinus cynocephalus* (Case 1985, Jones and Stoddart 1998), relatively few detailed analyses have centred on the same for fossil taxa. Exceptions include some species of propleopine (Archer and Flannery 1985, Ride et al. 1997, Sanson 1991, Wroe et al. 1998) and *Thylacoleo carnifex* (Case 1985, Finch and Freedman 1986, Finch and Freedman 1988, Wells et al. 1982, Wroe et al. 1999). The dietary habitus of both giant rat-kangaroos and the Pleistocene marsupial 'lion' have been the subject of much speculation (Wroe et al. 1998, Wroe et al. 1999, Wroe et al. in press). In particular their proclivity for vertebrate flesh has been questioned (Flower 1868, Sanson 1991).

Regarding the question of diet in propleopines, most recent contributors have agreed that vertebrates constituted a significant proportion thereof (Ride et al. 1997, Wroe et al. 1998). However, cranial and dental morphology differs considerably between giant rat-kangaroo species and a polarised debate has emerged over which taxa were more or less 'carnivorous'. Ride et al. (1997) treat *Propleopus oscillans* as primarily predacious, drawing analogy between *P. oscillans* and *Thylacinus*. The same authors consider *Ekaltadeta ima* to be less well adapted to a carnivorous lifestyle, alluding to the fact that in *E. ima* the condyle sits above the molar plane, a feature not typical of carnivorous mammals. Archer and Flannery (1985), Wroe (1997*c*) and Wroe et al. (1998) take a contrary view. Based on the relative proportions of vertical to horizontal shear in their respective dentitions, these authors contend that the teeth of *E. ima* and *Propleopus chillagoensis* were better adapted than *P. oscillans* to slicing through hide, flesh and sinew. Hence they deduce that *P. oscillans* was, in fact, the more omnivorous among propleopines.

With respect to carnivory versus herbivory, controversy over the diet of *T. carnifex* effectively ended following convincing arguments for a meat-eating habit leveled by Wells et al. (1982). However, another contentious issue has arisen, that of body size. As observed by Wroe et al. (1999), weight estimates for *T. carnifex* have been in decline. Owen (1859, 1883) believed that the Pleistocene marsupial 'lion' was comparable in size to *Panthera leo*, but in recent years estimates have dropped from 75–100 kg (Murray 1984) down to 40–60 kg Flannery (1997). More recently still, Webb (1998) gave a weight estimate of 20 kg. Wroe et al. (1999) dispute these figures and in the first quantitative assessment of body weight in *T. carnifex*, posit an average of between 101 and 130 kg for the species. Ramifications of this finding impact directly on the interpretation of lifestyle. Moreover, based on endocranial volume, Wroe et al. (in press) predict an average body mass of 97 kg. Thus, given its robusticity and weight, it appears unlikely that *T. carnifex* was capable of sustaining high speed for any extended period. It also seems improbable that such a large animal was semi-arboreal but increasingly likely that *T. carnifex* regularly preyed on megafauna, as first suggested by Owen (1887).

On reptilian domination of Australia's large terrestrial carnivore niches

The palaeoecological implications of this 'up-sizing' of *T. carnifex* contradict the current consensus view, at least in part and this is considered in detail by Wroe (2002). Over recent decades, many investigators have observed that Australia has been characterised by a paucity of large carnivorous mammal species during Pleistocene and Recent times (Hecht 1975, Archer and Bartholomai 1978, Rich and Hall 1984, Flannery 1991, 1994). Hecht (1975) speculated that the gigantic varanid, *Megalania prisca*, was the dominant, or only predator of Australian megafaunal taxa during the Pleistocene. He suggested (p. 247) that the largest carnivorous marsupial present during this epoch, *Thylacoleo carnifex*, '...could not have filled the big felid niche', although supporting evidence for this hypothesis was not supplied. More recently, Flannery (1991, 1994) expanded on Hecht's argument, positing that a scarcity of large mammalian carnivores, in particular, was symptomatic of long-standing soil nutrient deficiency, further contending that this regime may have extended back until at least early Miocene times. A simple rationale underpinned this hypothesis: poor soils and fluctuating climatic conditions constrained plant biomass; this in turn restricted mammalian herbivore size and species richness; together these limitations culminated in an impoverished large mammalian carnivore fauna in Australia and New Guinea (Meganesia). Flannery (1991, 1994) also suggested that in Meganesia, reptiles, advantaged by lower metabolic requirements, successfully filled large terrestrial carnivore niches taken by mammals elsewhere.

If the estimates forwarded by Wroe et al. (1999) and Wroe et al. (in press) are close approximations of actual body-weight in *T. carnifex*, then the hypothesis of reptilian domination of large carnivore niches is somewhat diminished with respect to Pleistocene Meganesia. Being homeothermic it seems reasonable to conclude that a 100 kg *T. carnifex* would require more food than even the largest *M. prisca* (see below). Wroe et al. (1999) draw attention to additional pertinent facts. For example, *M. prisca* constitutes a relatively rare find in Pleistocene deposits (Rich and Hall 1984) and that, of the two, only *T. carnifex* has been strongly implicated in the butchery of (Horton and Wright 1981, Runnegar 1983) and actual predation on megaherbivores (Scott and Lord 1924).

In addition to the trend towards downsizing *T. carnifex*, another factor that has probably contributed to misconceptions with respect to the roles of reptilian and mammalian carnivores in Pleistocene Australia is a tendency to provide only maximum estimated dimensions for large reptiles. Reliance on maximum dimensions, especially for taxa of indeterminate growth, may lead to heavily skewed interpretations of ecology. Hecht (1975) estimated that *M. prisca* might have reached 7 m and 620 kg, but gave no estimates of minimum, mean or average sizes. He did, however, give the data necessary to determine these values in the form of estimated head-body lengths. Using this data and Hecht's (1975) methodology, the length of the smallest mature specimen can be calculated at less than 2.2 m and the average at around 3.45 m. Ciofi (1999) alludes to a robust male Komodo dragon (*Varanus komodoensis*) of about 2.5 m long weighing 45 kg. With respect to the average weight of *M. prisca*, Auffenburgh (1972) determined that a 3.3 m *V. komodoensis* would weigh around 79 kg. On this basis it is reasonable to postulate that the average Pleistocene *Megalania* would have weighed around 90 kg, assuming no significant difference in shape between 3.3 and 3.45 m long individuals. Using other methods give maximum averages of less than 160 kg (Wroe 2002). Reliance on maximum dimensions also appears to extend to the two remaining large Pleistocene reptiles commonly invoked in support of hypotheses of reptilian domination, the madtsoiid snake *Wonambi naracoortensis* and the mekosuchine crocodile *Quinkana fortirostrum*. Indeed this is commonly the case for many extant reptilian taxa because the data required to determine average body weights is not generally available (Ernst and Zug 1996).

Another factor to be considered in this debate is that doubts persist over fundamental placement of these three Pleistocene reptiles with respect to niche. In fact, it has not yet been clearly established that any were terrestrial. Molnar (1990) raises the possibility that *M. prisca* was semi-aquatic and Molnar (1981) draws attention to the fact that, in the absence of postcranial material, there is no direct evidence to support the interpretation of *Q. fortirostrum* as a terrestrial crocodile. Most fossils of this species are associated with aquatic taxa and the few found in cave or fissure fill deposits are all within 10 miles of major watercourses, an overland distance known to be well within the range of living crocodiles (Cott 1961). These same arguments apply to all Australian crocodilians for which a terrestrial habitus has been suggested, all of which are known from either little or no postcranial material and have been found only in association with aquatic species. Willis (1997a) notes that a high degree of regional endemism among Australian crocodiles of the late Oligocene–Miocene for which a terrestrial lifestyle had been hypothesised may count against the abilities of such species to traverse terrestrial environments. As with both *M. prisca* and *Q. fortirostrum*, the ecology of *W. naracoortensis* also remains enigmatic. Barrie (1990, p. 148) argues that '...large prey capable of struggling vigorously are unlikely to have been taken, since *Wonambi*'s jaws were rather weak' and that '...reduction in lateral flexion would limit its ability to constrict animals, thus implying that it subsisted mainly on small prey'. Indeed, at present there is no firm evidence that *W. naracoortensis* (or, for that matter, any madtsoiid) was in fact a constrictor by habit. The presence of a large number of relatively small weak teeth in this species lead Barrie (1990) to further posit that 'It is possible that *Wonambi* had feeding habits similar to those of *Acrochordus* [aquatic file snakes], fish being available in the lagoons of its habitat.' While *W. naracoortensis* had a wide distribution throughout southern and eastern Australia, it was apparently uncommon throughout this range (Scanlon 1992).

Notwithstanding uncertainty over the size and habits of the largest reptiles, there can be little doubt that there were relatively few large marsupial carnivore species in Pleistocene Australia by way of general comparison with other continents. However, all other continents are much larger and directly contrasting Australian mammalian carnivore diversity with arbitrarily delimited subareas of other continents is inappropriate. Basic principles of island biogeography predict that an isolated landmass will habour fewer species than an area of the same dimensions within a larger continent because of differing balances between rates of immigration and extinction (MacArthur and Wilson 1967, Flessa 1975). Many other factors and conditions unique to the Australian continent may also have contributed to relatively low diversity during the Pleistocene. These include aridity and a lack of geographical barriers such as mountain ranges and large rivers, as well as soil nutrient levels and climatic variables such as temperature. But the relationship between such factors and species richness is unlikely to be simple. For example, the influence of productivity on diversity is not straightforward (Kondoh 2001) and there is strong evidence to suggest that soil nutrient levels do not necessarily effect either productivity or diversity (Jordan and Herrara 1981). In addition to being less diverse, in general, Australia's Pleistocene marsupial carnivores may not have been as large as the largest of presumably analogous taxa elsewhere. This phenomenon may also be explained through the application of biogeographical theory, as mammalian body size clearly correlates with continental area (Marquet and Taper 1998). Also of significance is the possible primacy of phylogeny over vegetation and habitat as a constraint on the distribution of mammalian body size (Seimann and Brown 1999). Finally, and perhaps even more importantly, using differences in species richness as a measure of the relative ecological significance of higher taxa could be inappropriate. Abundances and ranges of individual species may be the real measure of impact. These factors are difficult to quantify with respect to fossil species (Behrensmeyer 1991).

Prima facie evidence for reptilian carnivore supremacy over terrestrial mammalian counterparts during the middle–late Tertiary is perhaps even less convincing than for the Pleistocene. For example, although five crocodile species are known to have existed in the fossil rich Oligocene–Miocene deposits of

Riversleigh (Willis 1993, 1997a, 1997b), whether any were terrestrial is debatable. Four of these five, including *Baru wickeni*, the largest at around 4 m, are known only from late Oligocene deposits. Remaining species are all under 2 m in length and two were less than 1 m long (P.M. Willis pers. comm.). The largest snake, '*Montypythonoides*' *riversleighensis*, was approximately 4 m, large, but not gigantic (J.D. Scanlon pers. comm.). At 2.5 m in length, the madtsoid *Wonambi barriei* was the next largest snake from these deposits (Scanlon and Lee 2000). Two other madtsoiids were less than 1.5 m (Scanlon 1997 and pers. comm.). As with *W. naracoortensis* the habits of these Riversleigh madtsoiids are not well understood. None have been found in cave or fissure fill deposits. With regard to the estimation of size, data on Riversleigh's marsupial carnivores is also lacking. However, most of the 15 marsupial carnivore species known from these Riversleigh local faunas appear to fall within the size range of *Dasyurus maculatus* or larger, the only exceptions being the tiny *Sminthopsis*-sized *Mayigriphus orbus* and *Dasyurus hallucatus*-sized *Barinya wangala*. The biggest are a thylacoleonid at least as large as *Wakaleo vanderleueri* (A. Gillespie pers. comm.), an animal that weighed in at around 44–56 kg (Wroe et al. 1999). At least 10 of these larger species are known from Miocene deposits. On the other hand, in some pre-Pleistocene local faunas marsupial carnivores are rare. Scanlon and Mackness (in press) observe that only two terrestrial marsupial carnivores have been recorded from the early Pliocene Bluff Downs Local Fauna, *Thylacoleo crassidentatus* and *Dasyurus dunmalli*, while three large reptiles are known for which a terrestrial lifestyle has been proposed. An additional, potentially confounding factor effecting the understanding of palaeoecology in Riversleigh and other Tertiary faunas, is the possibility that giant dromornithid birds may have encroached on large terrestrial carnivore-omnivore niches (Wroe 1999*c*). Similarly, being omnivorous, casuariid birds may have competed with generalist marsupial carnivores such as giant rat-kangaroos.

Future directions

While the wealth of new fossil material has already provided considerable insight into the evolution and palaeobiology of Australia's carnivorous marsupial lineages, studies completed to date constitute only a preliminary phase. While numerical parsimony analyses incorporating some of these new specimens have been conducted (Muirhed and Wroe 1998, Wroe et al. 2000), a number of studies are either incomplete, or in early stages (Wroe unpublished data, Cooke and Kear unpublished data, Gillespie unpublished data). Other areas that require more rigorous, quantitative analysis are those of palaeobiology and palaeoecology. In addition to requirements for more data on marsupial carnivore palaeontology is a like need for more data on the ecology of possible competitors among Reptilia and Aves. Ultimately, information gleaned from investigating both the phylogeny and trophic diversity of Australian marsupial carnivore clades and guilds may enable the identification of long-term trends. Correlation with changes in climate and habitat will help to identify factors that have influenced the evolution of communities and their structures.

Acknowledgements

I am indebted to M. Archer, A. Gillespie, H. Godthelp, J.D. Scanlon and P.M.A. Willis, for their constructive criticism and comment on drafts of this manuscript. Funding has been provided to S. Wroe by grants from the following institutions: University of Sydney (U2000 Postdoctoral Research Fellowship), Centre for Research into the Evolution of Australia's Terrestrial Ecosystems (Australian Museum), French Ministry of Foreign Affairs, Linnean Society of New South Wales, Australian Geographic Society, Institute of Wildlife Research, and the University of New South Wales. Support has also been given by the Australian Research Council (to M. Archer); the National Estate Grants Scheme (Queensland) (grants to M. Archer and A. Bartholomai); the Department of Environment, Sports and Territories; the Queensland National Parks and Wildlife Service; the Commonwealth World Heritage Unit (Canberra); the University of New South Wales; ICI Australia Pty Ltd; the Australian Geographic Society; the Queensland Museum; the Australian Museum; Century Zinc Pty Ltd; Mt Isa Mines Pty Ltd; Surrey Beatty and Sons Pty Ltd; the Riversleigh Society Inc.; the Royal Zoological Society of New South Wales; the Linnean Society of New South Wales; and many private supporters.

References

Auffenburgh, W. (1972), 'Komodo dragons', *Natural History*, **81**:52–9.

Aplin, K., & Archer, M. (1983), 'Review of possum and glider phylogeny – the case for a restricted Phalangeroidea', in *Abstracts of the 1983 Possums and gliders symposium of the Australian Mammal Society*, p. 10–11, Armidale.

Aplin, K., & Archer, M. (1987), 'Recent advances in marsupial systematics, with a new, higher level classification of the Marsupialia', in *Possums and Opossums: Studies in Evolution* (ed. M. Archer), pp. xv–lxxii, Surrey Beatty and Sons, Sydney.

Archer, M. (1976a), 'Miocene marsupicarnivores (Marsupialia) from Central South Australia, *Ankotarinja tirarensis* Gen. et. sp. nov., *Keeuna woodburnei* Gen. et. sp. nov., and their significance in terms of early marsupial radiations', *Transactions of the Royal Society of South Australia,* **100**:53–73.

Archer, M. (1976b), 'The dasyurid dentition and its relationship to that of didelphids, thylacinids, borhyaenids (Marsupicarnivora) and peramelids (Peramelina: Marsupialia)', *Australian Journal of Zoology, Supplementary Series,* **39**:1–34.

Archer, M. (1982a), 'Review of the dasyurid (Marsupialia) fossil record, integration of data bearing on phylogenetic interpretation and suprageneric classification', in *Carnivorous Marsupials, Vol. I* (ed. M. Archer), pp. 397–443, Royal Zoological Society of New South Wales, Sydney.

Archer, M. (1982b), 'A review of Miocene thylacinids (Thylacinidae, Marsupialia), the phylogenetic position of the Thylacinidae and the problem of apriorisms in character analysis', in *Carnivorous Marsupials, Vol. 1* (ed. M. Archer), pp. 445–76, Royal Zoological Society of New South Wales, Sydney.

Archer, M. (1984), 'The Australian marsupial radiation', in *Vertebrate Zoogeography and Evolution in Australasia* (eds. M. Archer & G. Clayton), pp. 633–808, Hesperian Press, Sydney.

Archer, M., & Bartholomai, A. (1978), 'Tertiary mammal faunas of Australia: a synoptic review', *Alcheringa,* **2**:1–19.

Archer, M., Bartholomai, A., & Marshall, L.G. (1978), '*Propleopus chillagoensis*, a new north Queensland species of extinct giant rat-kangaroo (Macropodidae: Potoroinae)', *Memoirs of the Natural History Museum of Victoria,* **39**:55–60.

Archer, M., & Dawson, L. (1982), 'Revision of marsupial lions of the genus *Thylacoleo* Gervais (Thylacoleonidae, Marsupialia) and thylacoleonid evolution in the late Cainozoic', in *Carnivorous Marsupials, Vol. 1* (ed. M. Archer), pp. 477–94, Royal Zoological Society of New South Wales, Sydney.

Archer, M., & Flannery, T.F. (1985), 'Revision of the extinct gigantic rat kangaroos (Potoroidae: Marsupialia), with description of a new Miocene genus and species and a new Pleistocene species of *Propleopus*', *Journal of Paleontology*, **59**:1331–49.

Archer, M, & Rich, T. (1979), '*Wakamatha tasselli* Gen. et. sp. nov., a fossil dasyurid (Marsupialia) from South Australia convergent on modern *Sminthopsis*', *Memoirs of the Queensland Museum,* **19**:309–17.

Archer, M., & Rich, T.H. (1982), 'Results of the Ray E. Lemley Expeditions. *Wakaleo alcootaensis* n. sp. (Thylacoleonidae: Marsupialia), a new marsupial lion from the Miocene of the Northern Territory with a consideration of early radiation within the family', in *Vertebrate Zoogeography and Evolution in Australasia* (eds. M. Archer & G. Clayton), pp. 495–502, Hesperian Press, Sydney.

Arena, R., Wroe, S., & Archer, M. (1998), 'Additional material referred to the dasyurid *Ganbulanyi djadjinguli*: phylogenetic and palaeobiological implications', in *Abstracts of the 2nd Riversleigh Symposium*, p. 1, University of New South Wales.

Barrie, J. (1990), 'Skull elements and additional remains of the Pleistocene boid snake *Wonambi naracoortensis*', *Memoirs of the Queensland Museum*, **28**:139–51.

Bartholomai, A. (1962), '*Thylacoleo crassidentatus*',. *Memoirs of the Queensland Museum*, **14**:33–7.

Bartholomai, A. (1971), '*Dasyurus dunmalli*, a new species of fossil marsupial (Dasyuridae) in the upper Cainozoic deposits of Queensland', *Memoirs of the Queensland Museum*, **16**:19–26.

Bensley, B.A. (1903), 'On the evolution of the Australian Marsupialia with remarks on the relationships of marsupials in general', *Transcripts of the Linnean Society of London (Zoology)*, **9**:83–217.

Behrensmeyer, A.K. (1991), 'Terrestrial vertebrate accumulations', in *Taphonomy: releasing the data locked in the fossil record* (eds. P.A. Allison & D.E.G. Briggs), pp. 291–335, Plenum Press, New York.

Bremer, K. (1994), 'Branch support and tree stability', *Cladistics,* **10**:295–304.

Broom, R. (1898), 'On the affinities and habits of *Thylacoleo*', *Proceedings of the Linnean Society of New South Wales,* **23**:57–74.

Case, J.A. (1985), 'Differences in prey utilisation by Pleistocene marsupial carnivores, *Thylacoleo carnifex* (Thylacoleonidae) and *Thylacinus cynocephalus* (Thylacinidae)', *Australian Journal of Mammalogy,* **8**:45–52.

Ciofi, C. (1999), 'The Komodo Dragon', *Scientific American,* **280**:84–91.

Cott, H. B. (1961), 'Scientific results of an inquiry into the ecology and economic status of the Nile Crocodile (*Crocodilus niloticus*) in Uganda and Northern Rhodesia', *Transactions of the Zoological Society of London*, **29**, 1–211.

Clemens, W.A., & Plane, M. (1974), 'Mid-Tertiary Thylacoleonidae (Marsupialia, Mammalia)', *Journal of Paleontology,* **48**:652–60.

Crabb, P.L. (1982), 'Pleistocene dasyurids from southwestern New South Wales', in *Carnivorous Marsupials, Vol. 1* (ed. M. Archer), pp. 511–16, Royal Zoological Society of New South Wales, Sydney.

Ernst, C.H., & Zug, G.C. (1996), *Snakes in question*, CSIRO Publishing, Sydney, 203 p.

Farris, J.S. (1988), Hennig86, Version 1.5, Port Jefferson Station, New York.

Finch, M.E., & Freedman, L. (1986), 'Functional morphology of the vertebral column of *Thylacoleo carnifex* Owen (Thylacoleonidae: Marsupialia)', *Australian Journal of Zoology*, **34**:1–16.

Finch, M.E., & Freedman, L. (1988), 'Functional morphology of the limbs of *Thylacoleo carnifex* Owen (Thylacoleonidae: Marsupialia)', *Australian Journal of Zoology*, **36**:251–72.

Flannery, T.F. (1991), Flannery, T.F. (1991), 'The mystery of the meganesian meat-eaters', *Australian Natural History*, **23**:722–9.

Flannery, T.F. (1994), *The Future Eaters*, New Holland Publishers Pty Ltd: Sydney, 423 p.

Flessa, K.W. (1975), 'Area, continental drift, and mammalian diversity', *Paleobiology*, **1**:189–94.

Flower, W.H. (1868), 'On the affinities and probable habits of the extinct Australian marsupial *Thylacoleo carnifex* Owen', *Quarterly Journal of the Geological Society*, **24**:307–19.

Gillespie, A. (1997), '*Prisciléo roskellyae* sp. nov. (Thylacoleonidae, Marsupialia) from the Oligocene–Miocene of Riversleigh, northwestern Queensland', *Memoirs of the Queensland Museum*, **41**:321–7.

Gillespie, A. (1999), 'Diversity and evolutionary relationships of marsupial lions', *Australian Mammalogy*, **21**:21–2.

Godthelp, H., Wroe, S., & Archer, M. (1999), 'A new marsupial from the early Eocene Tingamarra local fauna of Murgon, Southeastern Queensland: a prototypical Australian marsupial?', *Journal of Mammalian Evolution*, **6**:289–313.

Hecht, M.K. (1975), 'The morphology and relationships of the largest known terrestrial lizard, *Megalania prisca* Owen, from the Pleistocene of Australia', *Proceedings of the Royal Society of Victoria*, **87**:239–50.

Horton, D.R., & Wright, R.V.S. (1981), 'Cuts on Lancefield bones: carnivorous *Thylacoleo*, not humans the cause', *Archaeology and Physical Anthropology in Oceania*, **16**:73–80.

Jones, M. (1997), 'Character displacement in Australian dasyurid carnivores: size relationships and prey size patterns', *Ecology*, **78**:2569–87.

Jones, M.E. (1998), 'The function of vigilance in sympatric marsupial carnivores: the eastern quoll and the Tasmanian devil', *Animal Behaviour*, **56**:1279–84.

Jones, M.E., & Stoddart, D.M. (1998), 'Reconstruction of the predatory behaviour of the extinct marsupial thylacine (*Thylacinus cynocephclus*)', *Journal of Zoology (London)*, **246**:239–46.

Jordan, C.F., & Herrara, R. (1981), 'Tropical rainforests: are nutrients really critical?', *The American Naturalist*, **117**:167–80.

Kirsch, J.A.W. (1977), 'The comparative serology of Marsupialia, and a classification of marsupials', *Australian Journal of Zoology Supplementary Series,* **52**:1–152.

Kirsch, J.A.W., Lapointe, F., & Springer, M.S. (1997), 'DNA–hybridisation studies of marsupials and their implications for metatherian classification', *Australian Journal of Zoology*, **45**:211–80.

Kitching, I.J. (1993), 'The determination of character polarity', in *Cladistics: a Practical Course in Systematics* (eds. P.L. Forey, C.J. Humphries, I.L. Kitching, R.W. Scotland, D.J. Siebert, & D.M. Williams), pp. 22–43, Clarendon Press, Oxford.

Kondoh, M. (2001), 'Unifying the relationships of species richness to productivity and disturbance', *Proceedings of the Royal Society of of London*, **268**:269–71.

Krajewski, C., Buckley, L., & Westerman, M. (1997), 'DNA phylogeny of the marsupial wolf resolved', *Proceedings of the Royal Society of London, Series B*, **264**:911–17.

Krajewski, C., Wroe, S., & Westerman, M. (2000a), 'Molecular evidence for the pattern and timing of cladogenesis in dasyurid marsupials', *Zoological Journal of the Linnean Society*, **130**:375–404.

Krajewski, C., Blackett, M.J., & Westerman, M. (2000b), 'DNA sequence analysis of familial relationships among dasyuromorphian marsupials', *Journal of Mammalian Evolution*, **7**:95–108.

Lawson, L., Muirhead, J., & Wroe, S. (1999), 'The Big Sink Local Fauna: a lower Pliocene mammalian fauna from the Wellington Caves complex, Wellington, New South Wales', *Records of the Western Australian Museum, Supplement,* **57**:265–90.

Lipscombe, D.L. (1992), 'Parsimony, homology, and the analysis of multistate characters', *Cladistics,* **8**:45–65.

Lowenstein, J.M., Sarich, V.M., & Richardson, B.J. (1981), 'Albumin systematics of the extinct mammoth and Tasmanian wolf', *Nature,* **291**, 409–11.

Luckett, P.W. (1994), 'Suprafamilial relationships within Marsupialia: resolution and discordance from multdisciplinary data', *Journal of Mammalian Evolution,* **2**:255–83.

MacArther, R.H., & Wilson, E O. (1967), *The theory of island biogeography*, Princeton University Press, Princeton, 203 p.

MacPhee, R.D.E. (1981), 'Auditory regions of primates and eutherian insectivores: morphology, ontogeny and character analysis', *Contributions to Primatology*, **18**:1–282.

Marquet, P.A., & Taper, M.L. (1998), 'On size and area: Patterns of mammalian body size extremes acrosss landmasses', *Evolutionary Ecology*, **12**:127–39.

Marshall, L.G., Case, J.A., & Woodburne, M.O. (1990), 'Phylogenetic relationships of the families of marsupials', in *Current Mammalogy, Vol. 2* (ed. H.H. Genoways), pp. 433–505, Plenum Press, New York.

Molnar, R.E. (1981), 'Pleistocene ziphodont crocodilians of Queensland', *Records of the Australian Museum,* **33**:803–34.

Molnar, R.E. (1990), 'New cranial elements of a giant varanid from Queensland', *Memoirs of the Queensland Museum,* **29**:437–44.

Muirhead, J. (1992), 'A specialised thylacinid, *Thylacinus macknessi*, (Marsupialia: Thylacinidae) from Miocene deposits of Riversleigh, northwestern Queensland', *Australian Mammalogy,* **15**:67–76.

Muirhead, J. (1997), 'Two new thylacines (Marsupialia: Thylacinidae) from early Miocene sediments of Riversleigh, northwestern Queensland and a revision of the family Thylacinidae', *Memoirs of the Queensland Museum,* **41**:367–77.

Muirhead, J. (2000), 'Yaraloidea (Marsupialia, Peramelemorphia), a new superfamily of marsupial and a description and analysis of the cranium of the Miocene *Yarala burchfieldei*', *Journal of Paleontology,* **74**:512–23.

Muirhead, J., & Archer, M. (1990), '*Nimbacinus dicksoni*, a plesiomorphic thylacine (Marsupialia, Thylacinidae) from Tertiary deposits of Queensland and the Northern Territory', *Memoirs of the Queensland Museum*, **28**:203–21.

Muirhead, J. & Wroe, S. (1998), 'A new genus and species, *Badjcinus turnbulli* gen. et sp. nov. (Thylacinidae: Marsupialia), from the late Oligocene of Riversleigh, northern Australia, and an investigation of thylacinid phylogeny', *Journal of Vertebrate Paleontology*, **18**:612–26.

Murray, P.F. (1998), '*Thylacinus mergiriani*, a new species of thylacine (Marsupialia: Thylacinidae) from the Ongeva Local Fauna of Central Australia', *Records of the South Australian Museum*, **30**:43–61.

Murray, P., Wells, R., & Plane, M. (1987), 'The cranium of the Miocene thylacoleonid, *Wakaleo vanderleuri*: click go the shears – a fresh bite at thylacoleonid systematics', in *Possums and Opossums: Studies in Evolution* (ed. M. Archer), pp. 433–66, Surrey Beatty and Sons, Sydney.

Owen, R. (1859), 'On the fossil mammals of Australia. Part I. Description of a mutilated skull of a large marsupial carnivore (*Thylacoleo carnifex*, Owen), from a calcareous conglomerate stratum, eighty miles SW of Melbourne, Victoria', *Philosophical Transactions of the Royal Society of London,* **149**:309–22.

Owen, R. (1887), 'Additional evidence for the affinities of the extinct marsupial quadrapedal *Thylacoleo carnifex* (Owen)', *Philosophical Transactions of the Royal Society of London*, **178**:1–16.

Pledge, N. S. (1977), 'A new species of *Thylacoleo* (Marsupialia: Thylacoleonidae) with notes on the occurrence and distribution of thylacoleonids in South Australia', *Records of the South Australian Museum* **17**:277–83.

Rauscher, B. (1987), '*Priscileo pitikantensis*, a new genus and species of thylacoleonid marsupial (Marsupialia: Thylacoleonidae) from the Miocene Etadunna formation, South Australia', in *Possums and Opossums: Studies in Evolution* (ed. M. Archer), pp. 423–32, Surrey Beatty and Sons and the Royal Zoological Society of New South Wales, Sydney.

Rich, T.H., & Hall, B. (1984), 'Rebuilding a giant lizard', in *Vertebrate Zoogeography and Evolution in Australasia* (eds. M. Archer & G. Clayton), pp. 291–4, Hesperian Press, Sydney.

Ride, W.D.L. (1993), '*Jackmahoneya* gen. nov. and the genesis of the macropodiform molar', *Memoirs of the Australasian Palaeontological Society*, **15**:441–59.

Ride, W.D.L., Pridmore, P.A., Barwick, R.E., Wells, R.T., & Heady, R.D. (1997), 'Towards a biology of *Propleopus oscillans* (Marsupialia: Propleopinae, Hypsiprymnodontidae)', *Proceedings of the Linnean Society of New South Wales,* **117**:243–328.

Runnegar, B. (1983), 'A *Diprotodon* ulna chewed by the marsupial lion, *Thylacoleo carnifex*', *Alcheringa,* **7**:23–5.

Sanson, G.D. (1991), 'Predicting the diet of fossil mammals', in *Vertebrate Palaeontology in Australasia* (eds. P. Vickers-Rich, J.M. Monaghan, R.F. Baird, & T.H. Rich), pp. 201–28, Pioneer Design Studio Pty Ltd, Lilydale.

Sarich, V., Lowenstein, J.M., & Richardson, B.J. (1982), 'Phylogenetic relationships of *Thylacinus* as reflected in comparative serology', in *Carnivorous Marsupials Vol. I* (ed. M. Archer), pp. 707–9, Royal Zoological Society of New South Wales, Sydney.

Scott, H.H., & Lord, C. (1924), 'Studies in Tasmanian mammals, living and extinct. Notes on a mutilated femur of *Nototherium*', *Papers and Proceedings of the Royal Society of Tasmania for the year 1923*, pp. 56–57.

Springer, M.S., Kirsch, J.A.W., & Case, J. A. (1997), 'The chronicle of marsupial evolution', in *Molecular evolution and adaptive radiation* (eds. T.J. Givnish, & K J. Sytsma), pp. 129–61, Cambridge University Press, Cambridge.

Scanlon, J.D. (1992), 'A new large madtsoiid snake from the Miocene of the Northern Territory', *The Beagle*, **9**:49–60.

Scanlon, J.D. (1997), '*Nanowana* gen. et sp. nov., small madtsoiid snakes from the Miocene of Riversleigh: sympatric species with divergently specialised dentition', *Memoirs of the Queensland Museum*, **41**:393–412.

Scanlon, J.D., & Lee, M.Y. (2000), 'The Pleistocene serpent *Wonambi* and the early evolution of snakes', *Nature*, **403**:416–20.

Siemann, E., & Brown, J.H. (1999), 'Gaps in mammalian body size distributions reexamined', *Ecology*, **80**:2788–96.

Stirton, R.A. (1957), 'Tertiary marsupials from Victoria, Australia', *Memoirs of the National Museum of Victoria*, **21**:129–33.

Swofford, D.L. (1993), PAUP (phylogenetic analysis using parsimony), Version 3.1, Illinois Natural History Survey, Champaign.

Szalay, F.S. (1994), *Evolutionary History of the Marsupialia and an Analysis of Osteological Characters*, Cambridge University Press, New York, pp. 1–455.

Webb, R.E. (1998), 'Marsupial extinction: the carrying capacity argument', *Antiquity*, **72**:46–55.

Wells, R.T., Horton, D.R., & Rogers, P. (1982), '*Thylacoleo carnifex* Owen (Thylacoleonidae): Marsupial carnivore?', in *Carnivorous Marsupials Vol. I* (ed. M. Archer), pp. 573–86, Royal Zoological Society of New South Wales, Sydney.

Willis, P.M.A. (1993), '*Trilophosuchus rackhami* gen. et sp. nov., a new crocodilian from the early Miocene limestones of Riversleigh, northwestern Queensland', *Journal of Vertebrate Paleontology*, **13**:90–8.

Willis, P.M.A. (1997a), 'Review of fossil crocodilians from Australia', *Australian Zoologist*, **30**:287–98.

Willis, P.M.A. (1997b), 'New crocodilians from the late Oligocene White Hunter Site, Riversleigh, northwestern Queensland', *Memoirs of the Queensland Museum*, **41**:423–38.

Wiley, E.O. (1981), *Phylogenetics: The Theory and Practice of Phylogenetic Systematics*, Wiley, New York.

Woodburne, M.O. (1967), 'The Alcoota fauna, Central Australia: an integrated palaeontological and geological study', *Bureau of Mineral Resources Australian Bulletin*, **87**:1–187.

Wroe, S. (1996a), '*Muribacinus gadiyuli* (Thylacinidae, Marsupialia), a very plesiomorphic thylacinid from the Miocene of Riversleigh, Northwestern Queensland, and the problem of paraphyly for the Dasyuridae', *Journal of Paleontology*, **70**:1032–44.

Wroe, S. (1996b), 'An investigation of phylogeny in the giant rat kangaroo *Ekaltadeta* (Propleopinae, Potoroidae, Marsupialia)', *Journal of Paleontology*, **70**:677–86.

Wroe, S. (1997a), '*Mayigriphus orbus*, a new genus and species of dasyuromorphian (Marsupialia) from the Miocene of Riversleigh, northwestern Queensland', *Memoirs of the Queensland Museum* **41**:439–48.

Wroe, S. (1997b), 'A re–examination of proposed morphology–based synapomorphies for the families of Dasyuromorphia (Marsupialia): Part I, Dasyuridae', *Journal of Mammalian Evolution*, **4**:19–52.

Wroe, S. (1997c), 'Stratigraphy, phylogeny and systematics of the giant extinct Rat Kangaroos (Propleopinae, Hypsiprymnodontidae, Marsupialia)', *Memoirs of the Queensland Museum*, **41**:449–56.

Wroe, S. (1998), 'A new genus and species of 'bone–cracking' daysurid (Marsupialia) from the Miocene of Riversleigh, northwestern Queensland', *Alcheringa*, **22**:277–84.

Wroe, S. (1999a), 'Killer kangaroos and other murderous marsupials', *Scientific American*, **280**:68–74.

Wroe, S. (1999b), 'The geologically oldest dasyurid (Marsupialia), from the middle Miocene of Riversleigh, northwestern Queensland', *Palaeontology*, **42**:501–27.

Wroe, S. (1999c), 'The Bird from Hell?', *Nature Australia*, **27**:56–63.

Wroe, S. (2002), 'A review of terrestrial mammalian and reptilian carnivores ecology in Australian fossil faunas and factors influencing their diversity: The myth of reptilian domination and its broader ramifications', *Australian Journal of Zoology*, **50**:1–24.

Wroe, S., & Archer, M. (1995), 'Extraordinary diphyodonty–related change in dental function for a tooth of the extinct marsupial *Ekaltadeta ima* (Propleopinae, Hypsiprymnodontidae)', *Archives of Oral Biology*, **40**:597–603.

Wroe, S., Brammall, J., & Cooke, B.N. (1998), 'The skull of *Ekaltadeta ima* (Marsupialia: Hypsiprymnodontidae?): An analysis of some cranial features among marsupials and a re–investigation of propleopine phylogeny, with notes on the inference of carnivory in mammals', *Journal of Paleontology*, **72**:738–51.

Wroe, S., & Mackness, B.S. (1998), 'Revision of the Pliocene dasyurid *Dasyurus dunmalli* (Dasyuridae, Marsupialia) and a review of the genus *Dasyurus*', *Memoirs of the Queensland Museum*, **42**:605–12.

Wroe, S., & Mackness, B.S. (2000a), 'A new genus and species of dasyurine dasyurid (Marsupialia) from the Pliocene Chinchilla Local Fauna of Southeastern Queensland', *Alcheringa*, **24**:319–25.

Wroe, S., & Mackness, B.S. (2000b), 'Additional material referred to *Dasyurus dunmalli* from the Pliocene Chinchilla Local Fauna of southeastern Queensland: phylogenetic implications', *Memoirs of the Queensland Museum*, **45**:641–45.

Wroe, S., & Musser, A. (2001), 'The skull of *Nimbacinus dicksoni* (Thylacinidae: Marsupialia)', *Australian Journal of Zoology*, **49**:487–514.

Wroe, S., Myers, T.J., Wells, R. T., & Gillespie, A. (1999), 'Estimating the weight of the Pleistocene Marsupial Lion (*Thylacoleo carnifex*: Thylacoleonidae): implications for the ecomorphology of a marsupial super-predator and hypotheses of impoverishment of Australian marsupial carnivore faunas', *Australian Journal of Zoology*, **47**:489–98.

Wroe, S., Ebach, M., Ahyong, S., Muizon, C. de, & Muirhead, J. (2000), 'Cladistic analysis of dasyuromorphian (Marsupialia) phylogeny using cranial and dental features', *Journal of Mammalogy*, **81**, 1008–24.

Wroe, S., Myers, T.J., Seebacher, F., Kear, B., Gillespie, A., Crowther, M., & Salisbury, S. (2003), 'An alternative method for predicting body mass: The case of the Pleistocene marsupial lion', *Paleobiology*, (in press).

CHAPTER 8

BIOGEOGRAPHY AND SPECIATION IN THE DASYURIDAE: WHY ARE THERE SO MANY KINDS OF DASYURIDS?

Mathew S. Crowther[A] *and Mark J. Blacket*[B]

[A]School of Biological Sciences and Institute of Wildlife Research, University of Sydney, New South Wales 2006, Australia
[B]Department of Genetics, La Trobe University, Bundoora, Victoria 3083, Australia

The Dasyuridae of Australia and New Guinea exhibits one of the most spectacular species-level diversity of any marsupial group. This chapter examines the biogeographical areas of Australia and postulates their role in the subspecific variation and speciation of dasyurid taxa. Physical barriers (such as rivers), as well as fluctuations of the arid-zone during the Pleistocene, explain much of the diversity within dasyurid species and genera. However recent molecular work suggests that most speciation within the Dasyuridae was a result of earlier environmental factors during the Miocene and Pliocene. Evidence is also presented that competition between dasyurid taxa lead to variation within species, and in the case of *Antechinus*, sympatric speciation. Finally it is concluded that further studies on dasyurid ecology, systematics and biogeography are needed to quantify their true specific diversity, as well as explain the factors that shaped their evolution.

INTRODUCTION

There are currently at least 64 recognised species within the family Dasyuridae (Flannery 1995, Strahan 1995), occurring in most habitats of Australia and New Guinea. Only one other living group of Australian marsupials – the kangaroos – approximates the spectacular taxonomic diversity of dasyurids. Like most mammals that hunt, dasyurids have a constrained basic body plan and speciation has therefore mostly involved changes in size, physiology and behaviour (Strahan 1995). Dasyurid diversity has consequently sometimes gone unrecognised and it has become obvious that the number of dasyurid forms, be that species, subspecies or races, has often been underestimated in the past.

Many studies of morphological variation within dasyurid species have tried to relate this variation to ecological factors (Archer 1981, Fox 1982, Morton and Alexander 1982, Cockburn et al. 1983). However, much intraspecific variation appears to be due to the presence of cryptic species or subspecies (Baverstock et al. 1984, Kitchener et al. 1984, Dickman et al. 1988, Dickman et al. 1998, Hope et al. 1986, Crowther 2002). The ecological processes thought to produce variation within species may also be involved in the speciation process. In this chapter we examine the literature on dasyurid community ecology and systematics to explain why there are so many forms of dasyurids, and to identify which factors have led to genotypic and phenotypic variability within forms and between closely related taxa. This review concentrates on extant dasyurids; extinct taxa are covered elsewhere (Wroe this volume).

EARLY HISTORY

Despite the superficial similarity in appearance of dasyurids with primitive mammals (see Wroe 1999), dasyurids appear to have had a relatively recent origin. The spectacular modern dasyurid radiation appears to have begun only during the Miocene. Definite dasyurids not been found earlier in the fossil record (Wroe 1999), and their ecological role appears to have largely been occupied by a diverse range of bandicoots and thylacinids (e.g. Wroe 1996, this volume, Wroe and Muirhead 1999). The oldest definite dasyurid is early to middle Miocene in age (Wroe 1999) and the oldest that can be placed in any modern genus is from the early Pliocene (approximately 4.5 million years ago, Archer1982).

The increasing aridity of the late Miocene was probably a major factor that drove the dasyurid radiation and the diversification of modern dasyurid groups (Wroe 1996 1997, Blacket et al. 1999, Krajewski et al. 2000). The early to middle Miocene was a time when rainforest covered much of Australia, but by the late Miocene the northward drift of the continent resulted in New Guinea being uplifted. This led to a decrease in precipitation across Australia and rainforests being replaced by more open forests (White 1994, Archer et al. 1995). There was a brief resurgence of rainforest in the early Pliocene, but this was followed by development of relatively arid environments with loss of forests in the centre and probably west of the continent. Woodland, grasslands, and herbfields had developed by the late Pliocene and early Pleistocene. All of these changes probably aided the diversification of dasyurid genera by creating a mosaic of new habitats and niches.

The Pleistocene saw enormous change in climatic conditions and has been postulated as the time when most modern species and subspecies of Australian mammals (including dasyurids) had their origins (e.g. Archer 1982, Heatwole 1987). Sea levels were raised and lowered, forming and breaking land connections, and drastic changes in climate were accompanied by repeated expansions and contractions of the arid zone with alternating fragmentation and coalescence of peripheral refugia. This was believed to have not only aided the radiation of dasyurid genera with large numbers of arid-dwelling species such as *Sminthopsis*, but also to have caused the separation of more mesic genera in the east and west. However, recent molecular evidence suggests that all modern dasyurid species may have actually originated earlier than the Pleistocene (Krajewski et al. 2000). This also appears to be the case for Australian frogs, whose species origins had similarly been believed to be caused by Pleistocene changes but are now thought to have occurred earlier (e.g. Roberts and Maxson 1985). Indeed, a recent study of relationships between areas of endemism of Australian vertebrates found that many of these areas appear to share a long history and do not simply reflect Pleistocene environmental changes (Cracraft 1991).

RECENT DIVERGENCES

Recent environmental changes often appear to have resulted in divergences within currently recognised dasyurid species. One example of this is Pleistocene changes causing a barrier to reproduction within *Sminthopsis crassicaudata*. The Murray River had a system of large lakes during the Pleistocene. Large-scale divergences between populations of *S. crassicaudata* north and south of the Murray River have been found using allozymes (Hope et al. 1986), mitochondrial DNA control region (d-loop) sequences (Cooper et al. 2000), cytochrome *b* and 12S ribosomal RNA sequences (Blacket et al. 1999, and unpubl. data). Although most of the external variation between previously described subspecies of *S. crassicaudata* has been found to be clinal (Morton and Alexander 1982, Hope and Godfrey 1988), a re-analysis of this variation including cranial morphology may still provide support for this Pleistocene divergence.

Some geographic barriers appear to have formed and disappeared many times in the past. For example, Tasmanian subspecies of *Antechinus swainsonii*, *A. minimus* and *S. leucopus* show small differences in allozymes from their mainland counterparts (Smith 1983) reflecting a relatively recent separation. However, Firestone et al. (1999) found major genetic differences (using the mitochondrial DNA control region and microsatellites) between mainland and Tasmanian *D. maculatus*. Earlier separations of Tasmania from the mainland have been invoked to explain the speciation of *S. leucopus* from *S. murina* and *A. swainsonii* from *A. minimus* (Archer 1982).

Additional examples of water barriers causing relatively recent divergences between island and mainland populations include *Sminthopsis griseoventer boullangerensis* from Boullanger Island, off the coast of Western Australia, which has diverged considerably in morphology and allozymes from the mainland *S. griseoventer griseoventer* (Crowther et al. 1999) but only slightly in mitochondrial DNA control region (Labrinidis et al. 1998) and 12S rRNA (Blacket unpublished). *Sminthopsis aitkeni* from Kangaroo Island has also diverged considerably from mainland *S. griseoventer griseoventer* in morphology (Kitchener et al. 1984) but has the least sequence divergence for the mitochondrial DNA control region (Labrinidis et al. 1998) and 12S ribosomal RNA (Blacket et al. 1999) of any currently recognised dunnart species.

BIOGEOGRAPHY

Inspection of the ranges of Australian terrestrial taxa indicates immediately that Australia can be divided into a number of biogeographical areas. However, researchers working on different taxa have come up with different zones (e.g. Schodde and Calaby 1972, Heatwole 1987), but the following regions are commonly accepted as being generally applicable (Heatwole 1987). These regions are used here as they appear to be important for the distribution of dasyurids: the Timorian region of tropical northern Western Australia and Northern Territory, the Torresian zone of tropical northern Queensland to subtropical northern New South Wales, the Tumbunan zone of the north-eastern Queensland rainforests, the Bassian region of temperate south-eastern Australia, the south-western zone and the Eyrean zone of arid inland Australia. It would, however, be an oversimplification to say that these zones have clear borders and that conditions were homogenous throughout them. Spencer (1896) originally placed the border of the Bassian and Torresian zones at the Clarence River. However, these zones have an extensive area of overlap in northern New South Wales that was coined the 'Macleay-McPherson' overlap zone by Burbidge (1960). Although she defined this area using plants, it appears to apply equally to animal distributions and dasyurids are no exception. It is apparent that the coastal zones have been connected and fragmented throughout geological history and this has led to speciation within the Dasyuridae, and much greater subdivision of biogeographical zones is possible. Indeed, detailed studies of biogeographical relationships of Australian plants (Crisp et al. 1995) and vertebrates (Cracraft 1991) included no less than 14 areas of endemism, and future detailed studies of dasyurid biogeography may also utilise this many, or even more.

The most obvious barrier between the major biogeographical zones, outlined above, is the large gap between the Bassian and the south-western zone. *Sminthopsis murina*, *Antechinus flavipes* and *Phascogale tapoatafa* were long believed to be the same species on both sides of the continent. Recent taxonomic work has shown this to be in error and the amount of time separating these two zones has been sufficient to lead to significant differentiation. Studies of allozymes and morphology have divided Western Australian *Sminthopsis murina* into *S. gilberti*, *S. dolichura* and *S. griseoventer* (Baverstock et al. 1984, Kitchener et al. 1984). *Sminthopsis murina* is genetically similar to both *S. gilberti* and *S. dolichura* (Blacket et al. 1999) and a barrier between these zones could have been responsible for their speciation (*S. griseoventer* appears to represent a different radiation). Although eastern and Western Australian *A. flavipes* are currently recognised as the same species; there are great morphological differences between the subspecies (Crowther et al. 2002) and it has been difficult to crossbreed the two forms in captivity (M. Archer pers. comm.). Additionally, revisionary taxonomic work is currently in progress on *P. tapoatafa* from these two zones (Rhind et al. 2001, P. B. S. Spencer et al. 2001).

Although there is no clearly defined barrier between the Bassian and Torresian regions (as evidenced by the extensive Macleay-McPherson overlap zone), a relatively recent barrier must have existed between these zones. This is the overlap zone of the bats *Rhinolophus megaphyllus ignifer* and *R. m. megaphyllus* (Cooper et al. 1998), subspecies of the grasshopper *Caledia captiva* (Moran and Shaw 1977; Marchant et al. 1988), closely related rock wallaby species (Sharman et al. 1990) and rosella species (Ovenden et al. 1987). The *Antechinus* species that overlap in this region are *A. subtropicus* and *A. stuartii*, which were both originally considered to be *A. stuartii* (Van Dyck and Crowther 2000). The control region mitochondrial DNA differentiation in these latter species is relatively small (Crowther, et al. unpublished) and there are no obvious fixed differences between them (Crowther et al. unpublished), but very striking morphological differences and the presence of two mitochondrial DNA lineages in animals from the same trap line support their status as distinct species. The low levels of divergence shown by these *Antechinus* species as well as other taxa (such as forms of *Sminthopsis leucopus* and *S. murina*; Blacket et al. unpubl. data) provides strong evidence that a recent barrier existed between the Bassian and Torresian zones. Similarly the existence of a recent barrier between the Timorian and Torresian zones is indicated by the presence of subspecies of *S. virginiae* within each. Indeed, these subspecies are highly divergent (based upon mitochondrial DNA sequences, Blacket et al. 1999) and may prove to be different species.

Smaller, distinctive biogeographical areas or refugia may explain much of the species diversity within the Australian fauna and dasyurids in particular. The Pilbara (Western Australia) has a distinctive species of *Ningaui* (*N. timealeyi*), two species of *Planigale* (Painter et al. 1995, Krajewski et al. 1997a, Blacket et al. 2000) and *Dasykaluta rosamondae*, the Kimberley (Western Australia) a distinctive dunnart related to *S. macroura* (Blacket and Crowther unpublished, Blacket et al. 2001), *Pseudantechinus ningbing* and various planigale taxa (such as *P. ingrami subtilissima* Archer 1976, and a number of forms related to *P. maculata* Blacket et al. 2000), the Pilliga (New South Wales) an undescribed dunnart (Maxwell et al. 1996) and unusual forms of *Antechinus* in the Grampians (Victoria) (Crowther 2002).

Opportunities of new habitats

Many species of dasyurids have distributions which are limited to a narrow range of habitat types. For example, *Sminthopsis douglasi* is limited to Mitchell Grass plains (Woolley 1992) and planigales are usually associated with cracking clay soils (Denny 1982). These restrictions can allow speciation to occur if a population expands into a new habitat that other conspecifics do not presently utilise. The importance of new habitats in the speciation process has been alluded to above but is further illustrated with the following example.

The formation of New Guinea appears to have been a major factor in initiating dasyurid diversification not only due to its influence on Australia's post-Miocene climate (see above) but also in the creation of new habitats and extensive geographic isolation. New Guinea was once thought of as a refuge for primitive early forms of dasyurids (e.g. Flannery 1989), but it is now clear that it has a complex history of multiple colonisations from the south and substantial diversification has occurred within New Guinea itself. New Guinea has at least four endemic dasyurid genera (*Murexia, Myoictis, Neophascogale* and *Phascolosorex*) that appear to have originated from various Australian ancestral stocks (e.g. Aplin et al. 1993, Armstrong et al. 1998, Krajewski et al. 1997b). Many new habitats would have been created in the formation of the New Guinea highlands. The appearance and subsequent fragmentation of these habitats associated with the central spine of mountains in New Guinea may have initiated radiations within *Murexia* and *Myoictis*.

Geographic isolation caused by sea-level changes between northern Australia and New Guinea appears to have been a major force in dasyurid speciation both in the distant and recent past. The endemic genera discussed above represent early colonisations of New Guinea, while the two endemic members of the genus *Dasyurus* (*D. spartacus, D. albopunctatus*) appear to have diverged relatively recently from their Australian relatives (Krajewski et al. 1997b). A recent *Planigale* radiation appears to have been initiated by Plio-Pleistocene sea-level changes, with multiple taxa (of uncertain rank) present within what are currently recognised as *P. maculata* (northern Australia) and *P. novaeguineae* (New Guinea) (Blacket et al. 2000). Interestingly, radiation of dasyurids in New Guinea has been mostly limited to larger dasyurids, smaller genera, such as members of the subfamily Sminthopsinae (including *Sminthopsis* and *Planigale*) being present only as recent invaders from the south, the small insectivore niche having already been filled by the microhydromyine rodents (Flannery 1995).

MORPHOLOGICAL VARIATION WITHIN THE DASYURIDAE

Speciation is often accompanied by changes in morphology. Past studies that have tried to examine morphological variation within dasyurid species (e.g. Wakefield and Warneke 1967, Archer 1981, Cockburn et al. 1983) have been largely confounded by the later realisation that least some of this variation was attributable to the presence of cryptic species.

Archer (1981) suggested that the greater size of the alisphenoid tympanic wings (bullae) shown by inland *S. murina* was a response to greater aridity. Although it has subsequently been found that there were actually four species in Archer's (1981) analysis (*S. griseoventer, S. gilberti, S. dolichura* and *S. murina*, Kitchener et al. 1984, Baverstock et al. 1984), the explanation stills holds. Enlarged bullae are also found in inland *A. flavipes* compared to coastal *A. stuartii, A. agilis* and *A. subtropicus*, as well as arid zone *Sminthopsis* and genera such as *Dasycercus*. Larger bullae enhance sensitivity to low frequency sounds produced by owls and snakes (Webster and Webster 1975). This ability is critical for arid zone dasyurids that need to forage further from cover than their forest and heath counterparts. Wroe (1996, 1997) posits that the fully enclosed bullae present even in the oldest known dasyurids may have constituted a successful preadaptation to increasingly arid environments of the late Tertiary.

Coat colour varies greatly in dasyurids and as with many small mammals it appears to correspond highly to substrate colour. Indeed, past taxonomists have often used it to describe subspecies. However, the only species studied in detail in respect to colour is *S. crassicaudata* which shows clinal variation that does not reflect subspecies as currently recognised (Hope and Godfrey 1988). *Sminthopsis macroura macroura* has a dark pelage which matches the dark colour of soils of mid-western New South Wales and Queensland whilst *S. macroura froggatti* of arid Australia is more red in colour. Molecular (Blacket et al. 2001) and morphological (Blacket and Crowther unpublished) data indicate that these forms do not differ substantially and the colour difference is probably a recent response to habitat. The evolution of coat colour to match that of the surrounding substrate can occur relatively quickly, as evidenced by the colours of introduced house mice and rabbits (Crowther unpublished).

A number of papers have detailed variation in nipple number within dasyurid species (Cockburn et al. 1983, Morton 1978, 1982) and concluded that higher nipple numbers are more common in more variable environments. In *Sminthopsis virginiae* nipple number varies between subspecies, apparently according to the environment in which they occur. Distantly related northern Australian subspecies – *S. v. virginiae* and *S. v. nitela* – possess only six nipples, while the New Guinean subspecies – *S. v. rufigenis* (which is most closely related to *S. v. virginiae*, Blacket et al. 1999) – possesses eight. *Antechinus* species appear to show extreme within-species variation, although much of this variation can now be attributed to comparisons among cryptic species (Dickman et al. 1998, Van Dyck and Crowther 2000). Similarly, variation within *Dasycercus* may also be due to the presence of cryptic species (Maxwell et al. 1996, Adams et al. 2000).

Studies of size variation within dasyurid species have been very rare. Wakefield and Warneke suggested that *A. stuartii* decreased in size with latitude (the reverse of Bergmann's Rule). However, it is now known that three species were involved in this cline (*A. subtropicus, A. stuartii* and *A. agilis*). Dickman et al. (1983) found that *A. agilis* and *A. swainsonii* decrease in size with altitude; however, examination of sizes of *A. agilis* and *A. stuartii* also shows an increase of size with latitude (supporting Bergmann's Rule). The only comprehensive study of variation in tail and ear length within a dasyurid was Morton and Alexander's (1982) study of *S. crassicaudata*. Tail length (but not ear length) was strongly correlated with temperature, thus following Allen's rule.

CHARACTER DISPLACEMENT AND SYMPATRIC SPECIATION

Unlike speciation linked with biogeographical breaks (caused by physical, climatic and habitat barriers discussed above), evidence for sympatric speciation is much more limited. Even work supporting character displacement in body size and shape within the Dasyuridae (e.g. Fox 1982) needs to be revisited under taxonomic changes and greater understanding of species ecology. Archer (1981) cited the much larger Victorian *S. leucopus* compared to Victorian *S. murina* as evidence for character displacement between these species. However, Morton et al. (1980) pointed out that these species have different habitats and ranges, and competition cannot be invoked exclusively to explain their size or range differences. Similarly, Fox (1982) used body size differences in *Antechinus, Sminthopsis* and *Planigale* to invoke evidence for interspecific competition. Although these species appeared to show relatively constant ratios in body size, most of the species were not sympatric and there is little evidence to suggest competition between many of these forms. One example that may indicate character displacement is that where *Planigale ingrami* occurs close to *P. tenuirostris*, it shows the flattest skull of anywhere within its range (Archer 1976, Blacket et al. unpublished). Jones (1997), in a more extensive study, has provided evidence for community-wide character displacement for size and carnassial morphology between *Sarcophilus harrisii, Dasyurus maculatus* and *D. viverrinus*.

The most convincing evidence for competition leading to speciation within the Dasyuridae probably comes from the speciation of the recently described *A. agilis* from *A. stuartii* (Dickman et al. 1998). These species are genetically very similar (as indicated by allozyme electrophoresis) and are sympatric only at a small number of areas in southern New South Wales (Dickman et al. 1988, 1998, Crowther 2002). Studies on breeding times have shown that both species ovulate at different rates of change of photoperiod (McAllan and Dickman 1986). This means that in areas where both forms are sympatric (e.g. Kioloa) the two species are reproductively isolated by approximately two weeks (Dickman et al. 1988). Unlike in the speciation of *A. stuartii* and *A. subtropicus*, there is not a clear biogeographical boundary between the two species. Instead the speciation may have been caused by competition with another species, *A. swainsonii*. Interspecific competition between *A. swainsonii* and *A. agilis* has been well demonstrated experimentally (Dickman 1986). *Antechinus swainsonii* is more fossorial whilst *A. agilis* is much more scansorial, and the removal of *A. swainsonii* allows *A. agilis* to spend more time foraging on the ground. Dickman (1986) found that whilst litter invertebrates were abundant throughout the year, invertebrates in the trees were very scarce in winter. Most arboreal invertebrates have their greatest abundance in spring. Dickman (1982) originally attributed the timing of breeding in *A. stuartii* to be delayed when in sympatry with *A. swainsonii*. McAllan and Dickman (1986) have shown that the timing of reproduction in *A. stuartii* and *A. agilis* are regulated by the rate of change in photoperiod, with *A. stuartii* tending to respond to a later change in photoperiod than *A. agilis*. Morphological comparisons of *A. stuartii* and *A. agilis* (Dickman et al. 1998) have shown that *A. stuartii* is a larger species than *A. agilis* with a longer rostrum. *Antechinus swainsonii* is a larger species than either of these species and although it has areas of sympatry with both of these species, the greatest number of sympatric areas and its greatest abundances are within the range of *A. agilis*. The speciation of *A. stuartii* from *A. agilis* could have been caused by competition between *A. stuartii* and *A. swainsonii*. Examination of the effects of rate of change of photoperiod on the breeding time of *A. swainsonii* shows that although it does not respond to a specific rate of change of photoperiod, the time of mating tends to be earlier than in *A. agilis*. This could be taken as evidence that competition over terrestrial resources led to the breeding times of *A. stuartii* within the range of *A. swainsonii* to be delayed. The difference in breeding times between *A. stuartii* and *A. agilis* would be sufficient to stop gene flow between the two forms and cause speciation. It may be no coincidence that *A. subtropicus*, which is a large, long-rostrumed version of *A. stuartii*, occurs in northern New South Wales and south-eastern Queensland where there are few or no *A. swainsonii* (Van Dyck and Crowther 2000).

FUTURE DIRECTIONS

It is apparent that much work is still needed to understand dasyurid species and subspecies diversity in Australia and even more in New Guinea. Determining how many dasyurid species there are might best be accomplished through a combination of field surveys and extensive morphological and molecular analyses. Much of the morphological variation within dasyurid species may actually prove to be due to the presence of cryptic species and previous studies will need to be revisited in order to understand the processes involved in producing this variation. More knowledge is also required on the basic biology of many species (e.g. habitat requirements, distribution and diet) for a more complete picture of their evolution. Phylogeographical studies using mitochondrial DNA genes (e.g. Firestone et al. 1999) appear to be a powerful way to estimate the time and location of barriers. Combined with analyses on other taxa (such as rodents, reptile and invertebrates) with similar dispersal capabilities a more complete biogeographical pattern could be established. Bioclimatic modelling as well as analysis of other variables (soil types etc.), also hold great promise in determining the factors responsible for dasyurid diversity.

ACKNOWLEDGEMENTS

We would like to thanks Chris Dickman and Steve Wroe for their comments and ideas on this chapter. The University of Sydney Postgraduate Award, Australian Museum Collection Fellowship and Postgraduate Award, Ethel Mary Read Fund of the Royal

Zoological Society of New South Wales and Joyce W. Vickery Award of the Linnean Society of New South Wales provided funding for the research that formed the basis for this paper.

REFERENCES

Adams, M., Cooper, N., & Armstrong, J. (2000), *Revision of Dasycercus Systematics*, South Australian Department for Environment and Heritage, Adelaide.

Aplin, K.P., Baverstock P.R., & Donnellan S.C. (1993), 'Albumin immunological evidence for the time and mode of origin of the New Guinean terrestrial mammal fauna', *Science in New Guinea* **19**:131–46.

Archer, M. (1976), 'Revision of the marsupial genus *Planigale* Troughton (Dasyuridae)', *Memoirs of the Queensland Museum*, **17**:341–65.

Archer, M. (1981), 'Results of the Archbold expeditions. No. 104. Systematic revision of the marsupial genus *Sminthopsis* Thomas', *Bulletin of the American Museum of Natural History*, **168**:61–224.

Archer, M. (1982), 'Review of the dasyurid (Marsupialia) fossil record, integration of data bearing on phylogenetic interpretation and suprageneric classification', in *Carnivorous Marsupials* (ed. M. Archer), pp. 397–443, Royal Zoological Society of New South Wales, Sydney.

Archer, M., Hand, S., & Godthelp, H. (1995), 'Tertiary environmental and biotic change in Australia', in *Paleoclimate and Evolution, with Emphasis on Human Origins* (eds. E.S. Vrba, G.H. Denton, T.C. Partridge, & L.H. Burckle), pp. 77–90, Yale University Press, New Haven.

Armstrong L.A., Krajewski C., & Westerman M. (1998), 'Phylogeny of the dasyurid marsupial genus *Antechinus* based on cytochrome *b*, 12S-rRNA, and protamine P1 genes', *Journal of Mammalogy*, **79**:1379–89.

Baverstock, P.R., Adams, M., Archer, M., McKenzie, N., & How, R. (1983), 'An electrophoretic and chromosomal study of the dasyurid marsupial genus *Ningaui* Archer', *Australian Journal of Zoology* **31**:381–92.

Baverstock, P.R., Adams, M., & Archer, M. (1984), 'Electrophoretic resolution of species boundaries in the *Sminthopsis murina* complex (Dasyuridae)', *Australian Journal of Zoology*, **32**:823–32.

Blacket, M.J., Krajewski, C., Labrinidis, A., Cambron, B., Cooper, S., & Westerman, M. (1999), 'Systematic relationships within the dasyurid marsupial tribe Sminthopsini – A multigene approach', *Molecular Phylogenetics and Evolution*, **12**:140–55.

Blacket, M.J., Adams, M., Krajewski, C., & Westerman, M. (2000), 'Genetic variation within the dasyurid marsupial genus *Planigale*', *Australian Journal of Zoology*, **48**:443–59.

Blacket, M.J., Adams, M., Cooper, S.J.B., Krajewski, C., & Westerman, M. (2001), 'Systematics and evolution of the dasyurid marsupial genus *Sminthopsis*: I. The Macroura Species Group', *Journal of Mammalian Evolution*, **8**:149–70.

Burbidge, N.T. (1960), 'The phytogeography of the Australian region', *Australian Journal of Botany*, **8**:75–209.

Cockburn, A., Lee, A.K., & Martin, R.W. (1983), 'Macrogeographic variation in litter size in *Antechinus* (Marsupialia: Dasyuridae)', *Evolution*, **37**:86–95.

Cooper, S.J.B., Reardon, T.B., & Skilins, J. (1998), 'Molecular systematics of Australian rhinolophid bats (Chiroptera : Rhinolophidae)', *Australian Journal of Zoology*, **46**:203–20.

Cooper, S.J.B., Adams, M., & Labrinidis, A. (2000), 'Phylogeography of the Australian dunnart *Sminthopsis crassicaudata* (Marsupialia: Dasyuridae)', *Australian Journal of Zoology*, **48**:461–73.

Cracraft J. (1991), 'Patterns of diversification within continental biotas, hierarchical congruence among the areas of endemism of Australian vertebrates', *Australian Systematic Botany*, **4**:211–27.

Crisp, M.D., Linder, H.P., & Weston, P.H. (1995), 'Cladistic biogeography of plants in Australia and New Guinea: Congruent pattern reveals two endemic tropical tracks', *Systematic Biology*, **44**:457–73.

Crowther, M.S. (2002), 'Morphological variation within *Antechinus agilis* and *A. stuartii* (Marsupialia: Dasyuridae)', *Australian Journal of Zoology*, **50**:339–56.

Crowther, M.S., Dickman, C.R., & Lynam, A.J. (1999), '*Sminthopsis griseoventer boullangerensis* (Marsupialia: Dasyuridae), a new subspecies in the *S. murina* complex from Boullanger Island, Western Australia', *Australian Journal of Zoology*, **47**:215–43.

Crowther, M.S., Spencer, P.B.S., Alpers, D., & Dickman, C.R. (2002), 'Taxonomic status of the Mardo, *Antechinus flavipes leucogaster* (Marsupialia: Dasyuridae): A morphological and molecular approach', *Australian Journal of Zoology*, **50**:627–47.

Denny, M.J.S. (1982), 'Review of Planigale (Dasyuridae, Marsupialia) ecology', in *Carnivorous Marsupials* (ed. M. Archer), pp. 131–8, Royal Zoological Society of New South Wales, Sydney.

Dickman, C.R. (1982), 'Some ecological aspects of seasonal breeding in *Antechinus* (Dasyuridae, Marsupialia)', in *Carnivorous Marsupials* (ed. M. Archer), pp. 139–50, Royal Zoological Society of New South Wales, Sydney.

Dickman, C.R. (1986), 'An experimental study of competition between two species of dasyurid marsupials', *Ecological Monographs*, **56**:221–41.

Dickman, C.R., Green, K., Carron, P.L., Happold, D.C.D., & Osborne, W.S. (1983), 'Coexistence, convergence and competition among *Antechinus* in the Australian high country', *Proceeding of the Ecological Society of Australia'* **12**:79–99.

Dickman, C.R., King, D.H., Adams, M., & Baverstock, P.R. (1988), 'Electrophoretic identification of a new species of *Antechinus* (Marsupialia: Dasyuridae) in south-eastern Australia', *Australian Journal of Zoology*, **36**:455–63.

Dickman, C.R., Parnaby, H.E., Crowther, M.S., & King, D.H. (1998), '*Antechinus agilis* (Marsupialia : Dasyuridae), a new species from the *A. stuartii* complex in south-eastern Australia', *Australian Journal of Zoology*, **46**:1–26.

Firestone, K.B., Elphinstone, M.S., Sherwin, W.B. & Houlden, B. (1999), 'Phylogenetic population structure of tiger quolls *Dasyurus maculatus* (Dasyuridae: Marsupialia), an endangered carnivorous marsupial', *Molecular Ecology*, **8**:1613–25.

Flannery, T.F. (1989), 'Origins of the Australo-Pacific land mammal fauna', *Australian Zoological Reviews*, 1:15–24.

Flannery, T.F. (1995), *Mammals of New Guinea*, Reed, Sydney.

Fox, B.J. (1982), 'A review of dasyurid ecology and speculation on the role of limiting similarity in community organization', in *Carnivorous Marsupials* (ed. M. Archer), pp. 97–116, Royal Zoological Society of New South Wales, Sydney.

Heatwole, H. (1987), 'Major components and distributions of the terrestrial fauna', in *Fauna of Australia. Volume 1A General Articles* (eds. G.R. Dyne & D.W. Walton), pp. 101–35, Australian Government Publishing Service, Canberra.

Hope, R.M., Bennett, J.H., & Chesson, C.M. (1986), 'Genetic variation in natural and laboratory populations of the marsupial *Sminthopsis crassicaudata*', *Biochemical Genetics,* **24**:597–613.

Hope, R.M., & Godfrey, G.K. (1988), 'Genetically determined variation of pelage colour and reflectance in natural and laboratory populations of the marsupial *Sminthopsis crassicaudata* (Gould)', *Australian Journal of Zoology',* **36**:441–54.

Jones, M.E. (1997), 'Character displacement in Australian dasyurid carnivores: size relationships and prey size patterns', *Ecology,* **78**:2569–87.

Kitchener, D.J, Stoddart, J., & Henry, J. (1984), 'A taxonomic revision of the *Sminthopsis murina* complex (Marsupialia, Dasyuridae) in Australia, including descriptions of 4 new species', *Records of the Western Australian Museum,* **11**:201–48.

Krajewski, C., Blacket, M., Buckley, L., & Westerman, M. (1997a), 'A multigene assessment of phylogenetic relationships within the dasyurid marsupial subfamily Sminthopsinae', *Molecular Phylogenetics and Evolution,* **8**:236–48.

Krajewski, C., Young, J., Buckley, L., Woolley, P.A., & Westerman M. (1997b), 'Reconstructing the evolutionary radiation of dasyurine marsupials with cytochrome *b*, 12S, and protamine gene trees', *Journal of Mammalian Evolution,* **4**:217–36.

Krajewski, C., Wroe, S. & Westerman, M. (2000), 'Molecular evidence for the pattern and timing of cladogenesis in dasyurid marsupials', *Zoological Journal of the Linnean Society,* **130**:375–404.

Labrinidis, A., Cooper, S.J.B., Adams, M., & Baczocha, N. (1998), 'Systematic affinities of island and mainland populations of the dunnart *Sminthopsis griseoventer* in Western Australia: data from allozymes and mitochondrial DNA', *Pacific Conservation Biology,* **4**:289–95.

Marchant, A.D., Arnold, M.L., & Wilkinson, P. (1988), 'Gene flow across a chromosomal tension zone. I. Relicts of ancient hybridisation', *Heredity,* **61**:321–8.

Maxwell, S., Burbidge, A.A., & Morris, K. (eds.) (1996), *The 1996 Action Plan for Australian Marsupials and Monotremes*, Wildlife Australia, Canberra.

McAllan, B.M., & Dickman, C.R. (1986), 'The role of photoperiod in the timing of reproduction in the dasyurid marsupial *Antechinus stuartii*', *Oecologia,* **68**:259–64.

Moran, C., & Shaw, D.D. (1977), 'Population cytogenetics of the genus *Caledia* (Orthoptera: Acridinae), III. Chromosomal polymorphism, racial parapatry and introgression', *Chromosoma (Berlin),* **63**:181–204.

Morton, S.R. (1978), 'An ecological study of *Sminthopsis crassicaudata.* III. Reproduction and life history', *Australian Wildlife Research,* **5**:183–211.

Morton, S.R. (1982), 'Dasyurid marsupials of the Australian arid zone: an ecological review', in *Carnivorous Marsupials* (ed. M. Archer), pp. 117–30, Royal Zoological Society of New South Wales, Sydney.

Morton, S.R., & Alexander, F. (1982), 'Geographic variation in the external morphology of *Sminthopsis crassicaudata* (Dasyuridae: Marsupialia)', in *Carnivorous Marsupials* (ed. M. Archer), pp. 695–8, Royal Zoological Society of New South Wales, Sydney.

Morton, S.R., Wainer, J.W., & Thwaites, T.P. (1980), 'Distribution and habitats of *Sminthopsis leucopus* and *S. murina* (Marsupialia: Dasyuridae) in southeastern Australia', *Australian Mammalogy,* **3**:19–30.

Ovenden, J.R., Mackinlay, A.G., & Crozier, R.H. (1987), 'Systematics and mitochondrial genome evolution of Australian rosellas (Aves: Platyceridae)', *Molecular Phylogenetics and Evolution,* **4**:526–43.

Painter, J., Krajewski, C., & Westerman, M. (1995), 'Molecular phylogeny of the marsupial genus *Planigale* (Dasyuridae)', *Journal of Mammalogy,* **76**:406–13.

Rhind, S.G., Bradley, J.S. & Cooper, N.K. (2001), 'Morphometric variation and taxonomic status of brush-tailed phascogales, *Phascogale tapoatafa* (Meyer, 1973) (Marsupialia: Dasyuridae)', *Australian Journal of Zoology,* **49**:345–68.

Roberts, J.D., & Maxson, L.R. (1985), 'Tertiary speciation models in Australian anurans: molecular data challenge Pleistocene scenario', *Evolution,* **39**:325–34.

Schodde, R., & Calaby, J.H. (1972), 'The biogeography of the Australo-Papuan bird and mammal faunas in relation to Torres Strait', in *Bridge and Barrier: The Natural and Cultural History of the Torres Strait* (ed. D. Walker), pp. 257–300, Australian National History Press, Canberra.

Sharman, G.B., Close, R.L., & Maynes, G.M. (1990), 'Chromosome evolution, phylogeny and speciation of the rock wallabies (*Petrogale*: Macropodidae)', *Australian Journal of Zoology,* **37**:351–63.

Smith, A.M.A. (1983), 'The subspecific biochemical taxonomy of *Antechinus minimus*, *A. swainsonii* and *Sminthopsis leucopus* (Marsupialia: Dasyuridae)', *Australian Journal of Zoology,* **32**:753–62.

Spencer, B. (1896), *Report of the work of the Horn Expedition to central Australia. Part 2: Zoology*, Melbourne.

Spencer, P.B.S., Rhind, S.G., & Eldridge, M.D.B. (2001), 'Phylogeographic structure within *Phascogale* (Marsupialia: Dasyuridae) based on partial cytochrome *b* sequence', *Australian Journal of Zoology,* **49**:369–77.

Strahan, R. (1995), *The Mammals of Australia*, Reed, Sydney.

Van Dyck, S., & Crowther, M.S. (2000), 'Reassessment of northern representatives of the *Antechinus stuartii* complex (Marsupialia : Dasyuridae): *A. subtropicus* sp. nov. and *A. adustus* new status', *Memoirs of the Queensland Museum,* 45:611–35.

Webster, D.B., & Webster, M. (1975), 'Auditory systems of the Heteromyidae: Functional morphology and evolution of the middle ear', *Journal of Morphology,* **146**:343–76.

White, M.E. (1994), *After the Greening: The Browning of Australia*, Kangaroo Press, Sydney.

Woolley, P.A. (1992), 'New records of the Julia Creek dunnart, *Sminthopsis douglasi* (Marsupialia: Dasyuridae', *Wildlife Research,* **19**:779–83.

Wroe, S. (1996), '*Muribacinus gadiyuli* (Thylacinidae, Marsupialia), a very plesiomorphic thylacinid from the Miocene of Riversleigh, Northwestern Queensland, and the problem of paraphyly for the Dasyuridae', *Journal of Paleontology,* **70**:1032–44.

Wroe, S. (1997), 'A re-examination of proposed morphology-based synapomorphies for the families of Dasyuromorphia (Marsupialia): Part I. Dasyuridae', *Journal of Mammalian Evolution,* **4**:19–52.

Wroe, S. (1999), 'The geologically oldest dasyurid, from Miocene of Riversleigh, north-west Queensland', *Palaeontology,* **42**:501–27.

Wroe, S. & Muirhead, J. (1999), 'Evolution of Australia's marsupicarnivores: Dasyuridae, Thylacinidae, Myrmecobidae, Dasyuromorphia *incertae sedis* and Marsupialia *incertae sedis*', *Australian Mammalogy,* **21**:10–11.

Part II

Reproduction and Development

CHAPTER 9

SPERM MATURATION AND FERTILISATION IN AUSTRALIAN AND AMERICAN INSECTIVOROUS MARSUPIALS

W.G. Breed[A], *D.A. Taggart*[B], *H.D.M. Moore*[C]

[A]Department of Anatomical Sciences, University of Adelaide SA 5005, Australia
[B]Department of Anatomical Sciences, University of Adelaide SA 5005 and the Royal Zoological Society of South Australia, Frame Road, Adelaide SA 5005, Australia
[C]Department of Molecular Biology and Biotechnology, University of Sheffield, Firth Court, Western Bank, Sheffield SI0 2UH, United Kingdom

The carnivorous and insectivorous marsupials exhibit a number of ancestral morphological features that are not evident in most other Australasian families. In the present review the structural design of the spermatozoon and egg, and their maturation and interaction in the female reproductive tract at the time of fertilisation, are briefly considered for these species.

The evidence suggests that the gametes of the dasyurids and didelphids appear to be highly derived in their structural organisation as well as in being very different from each other. In didelphids sperm pairing occurs, whereas in dasyurids post-testicular sperm maturation is elaborate but no sperm pairing takes place. In the latter group, but not the former, prolonged sperm storage occurs in the higher reaches of the female reproductive tract after mating. Oocytes of dasyurids, but not didelphids, have a large central yolk mass and in the zona pellucida the distribution and abundance of oligosaccharides appear to differ between the species. Thus in both these families of marsupials some unique features of gamete design, organisation and behaviour appear to have evolved, whereas other features are shared by at least a few other marsupial groups.

INTRODUCTION

Over the last 20 years much information has been obtained on the detailed structural organisation of sperm and eggs and the processes of gamete maturation and fertilisation in eutherian mammals. Most data have been accumulated on the gametes of laboratory rodents, domestic livestock, and humans, with many studies facilitated by the development of *in vitro* procedures and the induction of oocyte maturation and fertilisation. Although inter-specific differences clearly occur (e.g. see Yanagimachi 1994), there are common features in the morphology of gametes and their interactions that take place at fertilisation in most, if not all, eutherian mammals that are not evident in other vertebrate groups (see Bedford 1991, 1999). The question thus arises as to whether these processes are specific to eutherians and differ from those in the other large extant group of mammals, the marsupials.

Although the divergence of most eutherian orders occurred several tens of millions of years ago, the most recent common ancestor that eutherians share with marsupials is far more ancient (Janke et al. 1997; Kumar and Hodges 1998; Bromham et al. 1999). Details are debatable, but it is clear that these two major extant mammalian groups diverged well over 100 million years ago. Recent molecular evidence even suggests that marsupials may be more closely related to the monotremes than they are to eutherians (Penny and Hasagawa 1997; Bromham et al. 1999). Did the divergent features of sperm and egg design and the processes of gamete maturation and sperm–egg interaction that are evident in eutherians, but not present in other vertebrates and invertebrates, evolve prior to, or after, the separation of the marsupial lineage? A knowledge of these processes in marsupials should not only shed light on how gamete structure and interaction evolved in this group of mammals, but also, at least indirectly, give some insight into the time of origin of these events in eutherian mammals.

There are a number of families of marsupials living today, most of which occur in Australasia, with fewer being present in South America although the number of species is similar between these two regions (Marshall 1984; Wilson and Reeder 1993). Studies on cranial and dental features suggest that, whereas those of kangaroos and wallabies (macropods), possums and gliders (phalangerids and pseudocheirids), koalas, and wombats are all relatively specialised, the insectivorous and carnivorous marsupials (dasyurids), bandicoots (peramelids), numbats (myrmecobiids) in Australasia, and the opossums (didelphids) in South America are less so (see Marshall 1984; Luckett 1994). Likewise, whereas members of the macropod and phalangerid families ovulate just one egg at any one oestrus, the insectivorous and carnivorous marsupials release many, which is presumably closer to the ancestral condition. Does the morphology of sperm and egg and their interaction at the time of fertilisation show evidence of any ancestral features? With this question in mind, we review sperm maturation and fertilisation in the dasyurids of Australasia and didelphids of the New World, since it is in only these two families of insectivorous marsupials that any significant amount of data are available for the insectivorous and carnivorous marsupials, and even here, relevant information is only available for a very few species.

The first studies on egg maturation were carried out on dasyurids by Hill in 1910 on the native cat, *Dasyurus viverrinus*, and on didelphids by Hartman in 1916 on the American opossum, *Didelphis virginiana*. Subsequently, little was published on this topic until the largely light microscope observations on sperm–egg interactions in *Antechinus* (Selwood 1982) and *Didelphis* (Rodger and Bedford 1982). Since that time, more extensive investigations on various morphological aspects of gametes and their interaction at fertilisation in members of this group have been obtained, particularly on individuals from the laboratory colonies of *Sminthopsis crassicaudata* (Bennett et al. 1990) in Australia and *Monodelphis domestica* in Texas (VandeBerg 1990). This review will summarise relevant data from these species and will highlight the similarities and differences in sperm and eggs and their interactions that occur between these families. Data that pertain to gamete preparation for, and interaction at the time of, fertilisation will be focused on in this present review. For an overview of the detailed structural organisation of the dasyurid and didelphid spermatozoa (see Harding et al. 1982; Temple-Smith 1987, 1994) and for oocyte structure around the time of ovulation (see Selwood 1992; Breed 1996).

Sperm maturation in the male reproductive tract

In eutherian mammals, spermatozoa are neither highly motile nor potentially fertile when they leave the testis, but become so as they travel along the lumen of the epididymis (Moore 1995, 1998; Cooper 1998, Jones 1998). During this migration many of the proteins on, and in, the sperm plasma membrane undergo glycosylation and other alterations such as proteolytic cleavage. Specific proteins are secreted by the principal cells lining the lumen of the epididymis under the control of androgens, and some of these become bound to the surface of the spermatozoa. As a result of interactions with epididymal factors, the spermatozoon retains viability and develops the potential for fertility by the time it reaches the cauda region of the epididymis.

In marsupials, studies on functional changes to spermatozoa as they travel down the epididymis have not yet been carried out, due in part at least to the absence of a repeatable IVF system for most species. However, significant morphological changes occur to the spermatozoon (Cummins 1976; Temple-Smith and Bedford 1976, 1980; Harding et al. 1975; Breed et al. 1994; Lin and Rodger 2000) and these changes appear to be considerably more pronounced in marsupials than in most eutherians. Also, in didelphids, spermatozoa undergo a pairing, or conjugation, during epididymal transit (Biggers and DeLamater 1958; Temple-Smith and Bedford 1980), which involves a rotation of the sperm head in relation to the flagellum followed by a precise alignment and adhesion between two sperm heads (Figs 1a-1c). An electron-dense region develops between the plasma membranes overlying the acrosomes, which resembles that of a tight junction (Taggart et al. 1993b).

No Australian marsupials exhibit sperm pairing, but marked morphological changes take place during epididymal transit, some of which resemble the processes in didelphids. In *Sminthopsis*, when sperm leave the testis, they appear to be in a relatively immature state with not all nuclear chromatin fully condensed. The sperm head lies perpendicular to the tail and a large cytoplasmic droplet occurs around the neck (Figs 2a–c). As sperm travel down the epididymis, the nuclear envelope comes to lie close to the condensed chromatin (Harding et al. 1982; Breed et al. 1994; Soon and Breed 1996) and the large cytoplasmic droplet migrates down the length of the tail and is eventually shed by the time the sperm reaches the cauda region. Perhaps related to this, the sperm head swivels on its axis with the tail so that it comes to lie parallel to it (Temple-Smith and Bedford 1976; Harding et al. 1975; Breed et al. 1994; Soon and Breed 1996) (Figs 2d–h).

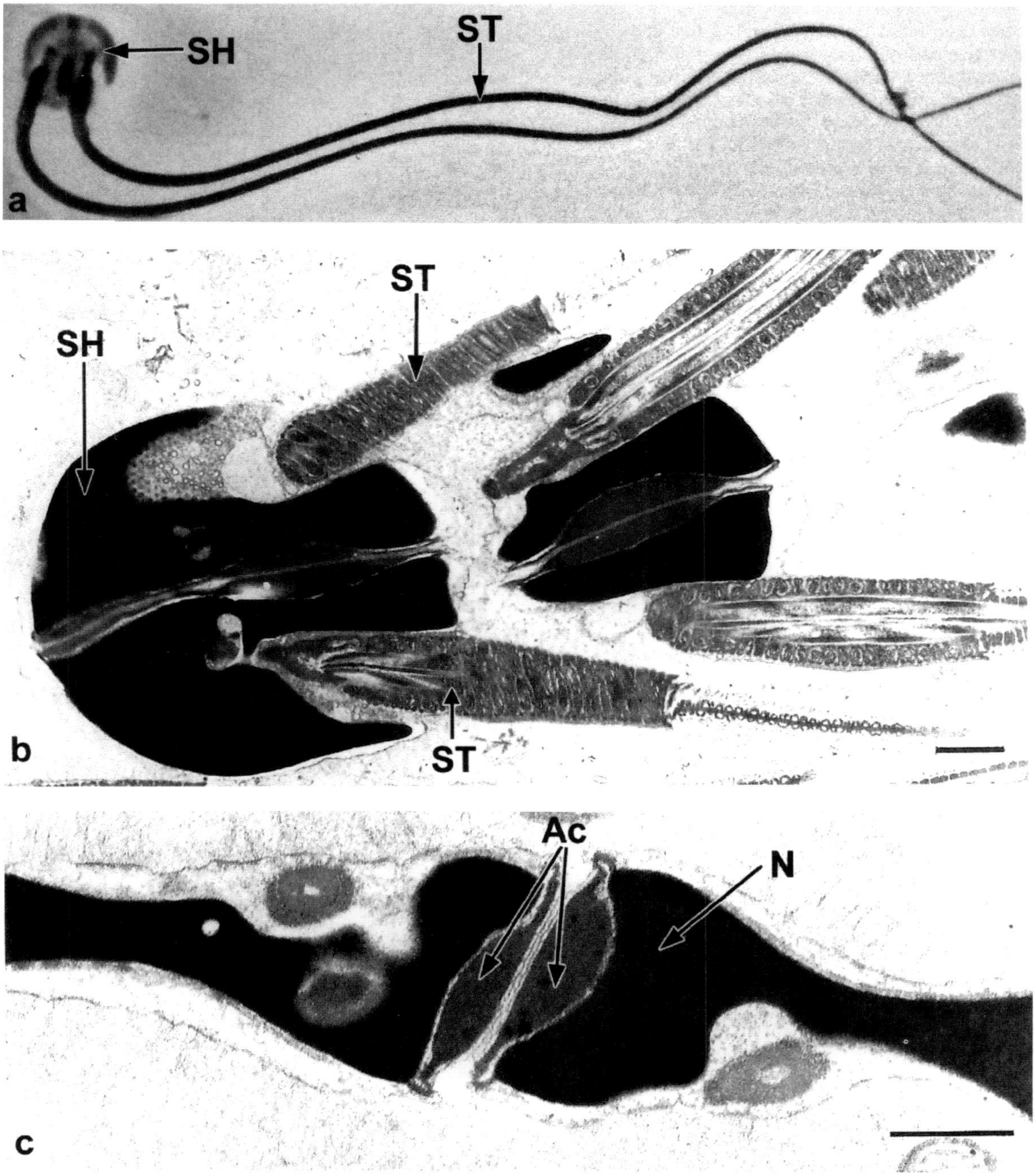

Figure 1 Spermatozoa of *Monodelphis domestica* – a didelphid marsupial. (a) phase contrast, and (b, c) TEM sections. Note attachment of two sperm heads (SH) in region of acrosome (Ac). N = nucleus; ST = sperm tail. From Taggart et al. 1993 a and b. Bar line 0.5 µm.

The molecular changes that occur on the sperm surface are unknown, but preliminary studies with Triton X-100 extracts of spermatozoa have shown that a somewhat different suite of proteins are present on the plasma membrane of cauda sperm compared to those from the caput as analysed by ID gradient SDS PAGE. This suggests that, like tammar (Jones and Clulow 1994) and possum (Lamont et al. 1998; Cooper et al. 1998) spermatozoa, those of *Sminthopsis* undergo considerable changes of the surface molecular organisation as they pass along the epididymis.

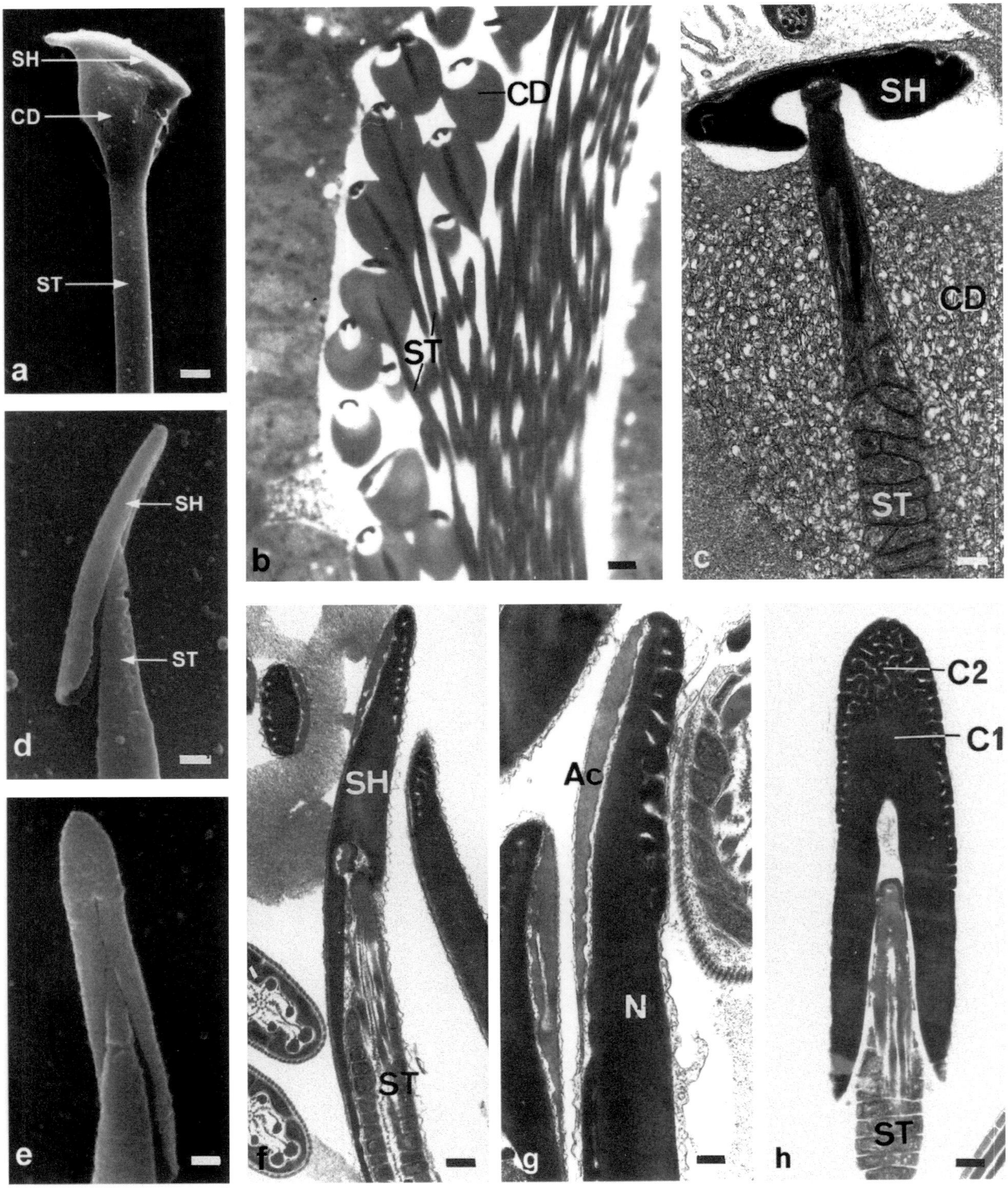

Figure 2 Spermatozoa of *Sminthopsis crassicaudata* – a dasyurid. (a to c) from caput epididymis, and (d to h) from cauda epididymis. Note in caput spermatozoa the sperm head (SH) lies perpendicular to tail (ST) with cytoplasm droplet (CD) in region of attachment of sperm tail to head. In cauda epididymal sperm the sperm head (SH) lies parallel to tail (ST) and chromatin is composed of two regions, C1 and C2. From Breed et al. 1989 and 1994a.

SPERM MATURATION IN THE FEMALE REPRODUCTIVE TRACT

In eutherians, significant functional changes take place to the spermatozoon as it travels along the female reproductive tract. These changes, known as capacitation, eventually bring about potential sperm fertilisability and include removal of inhibitory glycoproteins from the sperm surface, changes in distribution, and glycosylation, of membrane protein and lipid molecules, as well as phosphorylation of a number of intracellular proteins. Capacitation leads to an influx of calcium ions into spermatozoa and a destabilisation of membranes, which permits the acrosome reaction to take place. Probably related to these events is the development of hyperactivated sperm motility (see Yanagimachi 1994), which may facilitate sperm migration from the storage reservoir, that has been shown to exist in the isthmus of the oviduct, to the site of fertilisation (Smith 1998; Suarez 1998).

In didelphid and dasyurid marsupials there appear to be some similarities, and yet some marked differences, in sperm maturation in the female tract to those of eutherians as well as to each other. In didelphids, mating occurs 4 to 24 hours before ovulation and, shortly after mating, sperm accumulate in invaginations of the isthmus epithelium (Rodger and Bedford 1982), where separation of the sperm pairs takes place. The reason why sperm remain in pairs until the time of transport to the site of fertilisation in the ampulla is unclear. It may be a way of protecting the acrosome and/or overlying plasmalemma so that a premature acrosome reaction does not occur (Bedford et al. 1984). Also sperm pairing appears to facilitate progressive motility with the result that there is an increase in forward thrust; this may be important for migration through the viscous environment of the female reproductive tract (Taggart et al. 1993a and b; Moore and Taggart 1995).

In dasyurids, mating takes place up to two weeks before ovulation in *Antechinus* (Selwood and McCallum 1984) and up to three days in *Sminthopsis* (Bennett et al. 1990; Breed and Leigh 1992). Within an hour or so of mating, spermatozoa populate the invaginations of the isthmus crypts (Breed et al. 1989; Taggart and Temple-Smith 1991; Taggart et al. 1997, 1999; Shimmin et al. 1999). Here they become largely immotile and, sometimes at least, become arranged in highly organised sheaths (Breed et al. 1989; Breed 1994a; Bedford and Breed 1994) until around the time of ovulation. Transillumination of the spermatozoa in the isthmus of the oviduct of *Sminthopsis* indicated that, although in close proximity, binding to the plasma membrane of the epithelial cells lining the crypt may not occur. Shortly before ovulation, a reorientation of the sperm head on its axis with the tail takes place so that the spermatozoon migrates forward to the site of fertilisation with the head lying at right angles to the tail. This change may be important to maximise the surface area for contact of the sperm head with that of the outer matrix of the zona pellucida (Bedford and Breed 1994), and thus may be a prerequisite for fertilisation; if this is the case it could be considered to be part of the capacitation process.

EGG COAT COMPOSITION AND SPERM BINDING

All vertebrate eggs when released from the ovary are surrounded by an extracellular coat. In fish and amphibians this is referred to as a vitelline coat, whereas in mammals the homologous structure is the zona pellucida. In eutherians, it is 8–25 µm thick and composed of three, or perhaps four, sulphated glycoproteins, ZPA (=ZP2), ZPB (=ZP1), and ZPC (=ZP3) with, in the laboratory mouse at least, ZPA and ZPC forming heterodimers that apparently polymerise to make up filaments that are cross-linked by ZPB (Wassarman 1988). In eutherians, there is a further vestment, the cumulus oophorus, which is composed of an extracellular matrix that surrounds a mass of ovarian follicle-derived cells (e.g. see Bedford 1991, 1998). After penetrating the cumulus, the capacitated spermatozoon binds to the outer matrix of the zona. In the mouse, the most extensively studied species, primary sperm–zona binding involves receptors on the plasmalemma over the sperm acrosome which bind to the serine/threonine (O-) linked oligosaccharides, perhaps either the α-galactose at the carboxyl terminal end of the ZPC polypeptide (Wassarman 1996), or α-mannose and/or N-acetyl glucosamine residues (see Brewis and Moore 1997; Tulsiani et al. 1997; McLeskey et al. 1998; Brewis and Wong 1999).

That this primary binding may involve several ligands is suggested by other findings that indicate that N-linked carbohydrates may also play a role (Akatsuha et al. 1998; Brewis and Wong 1999). A calcium-dependent exocytosis of the acrosomal contents then takes place as a result of multiple fusions between the outer acrosomal membrane and overlying plasma membrane. The acrosome reaction exposes the inner acrosomal membrane which is likely to be involved in secondary binding of the spermatozoon to the zona matrix. It also promotes fusogenicity of the plasma membrane over the acrosome equatorial segment (Yanagimachi 1994; Brewis and Moore 1997; McLeskey et al. 1998; Brewis and Wong 1999).

The characterisation of egg-binding proteins on the spermatozoon has been controversial with several putative receptor molecules being identified. In the laboratory mouse, two of the likely candidates are β1, 4-galactosyl transferase (GTAse) (Miller et al. 1992) and a 95 kD tyrosine kinase, Zona Receptor Kinase (ZRK) (Leyton and Saling 1989). The GTAse on the sperm surface has been shown to recognise N-acetyl glucosamine residues of the O-linked oligosaccharides of ZPC. Zona pellucida preparations induce phosphorylation of ZRK in mouse and human spermatozoa concomitant with the development of the acrosome reaction, and ZRK is localised to the plasmalemma overlying the acrosomal region of the spermatozoa,

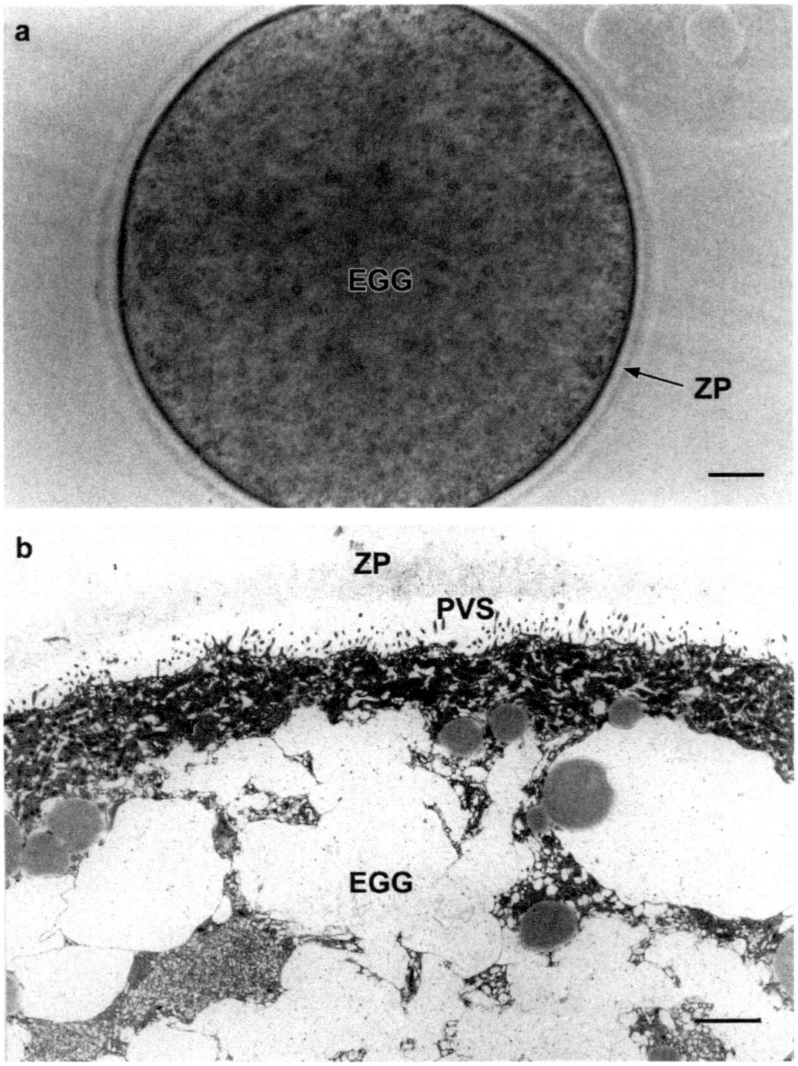

Figure 3 Oocyte from oviduct of *Monodelphis domestica.*(a) Thick plastic section and (b) TEM. Note zona pellucida (ZP) is very thin and egg has very vacuolated cytoplasm. PVS = perivitelline space. From Taggart et al. 1993.

and has been shown to interact with ZPC (Brewis and Moore 1997; Brewis and Wong 1999; McLeskey et al. 1998). After the acrosome reaction, the secondary, stronger, binding of the spermatozoon to the ZPA molecules of the zona pellucida occurs (e.g. see Yanagimachi 1994; McLeskey et al. 1998). This secondary binding may well involve proacrosin/acrosin and/or PH 20 on the sperm surface. Acrosin has a fucose-binding domain and its dispersal may be regulated by a serine protease inhibitor (Moore et al. 1993).

All other eutherians studied to date have been found to have a zona pellucida containing the three glycoproteins ZPA, ZPB and ZPC. The expression of these three genes appears to be ovary-specific. In some species expression is localised to germ cells, whereas in others one or more of the ZP genes appears to be expressed in the surrounding granulosa cells (Dunbar et al. 1991; Grootenhuis et al. 1996; Kolle et al. 1996). Further differences between species are also evident in the location of the primary sperm receptors. Whereas in the mouse these are clearly on the ZPC, results for rabbits and pigs suggest that ZPB, and not ZPC, may be more important (Yurewicz et al. 1993; Yonezawa et al. 1997). Various other putative egg-binding proteins on sperm of other species have also been hypothesised.

In the marsupial, the zona pellucida comes to surround the egg early in follicular development (Frankenberg et al. 1996; Breed 1996; Mate 1998) (Fig. 4). It differs from that of eutherians in being relatively thinner, and more easily digested by proteases

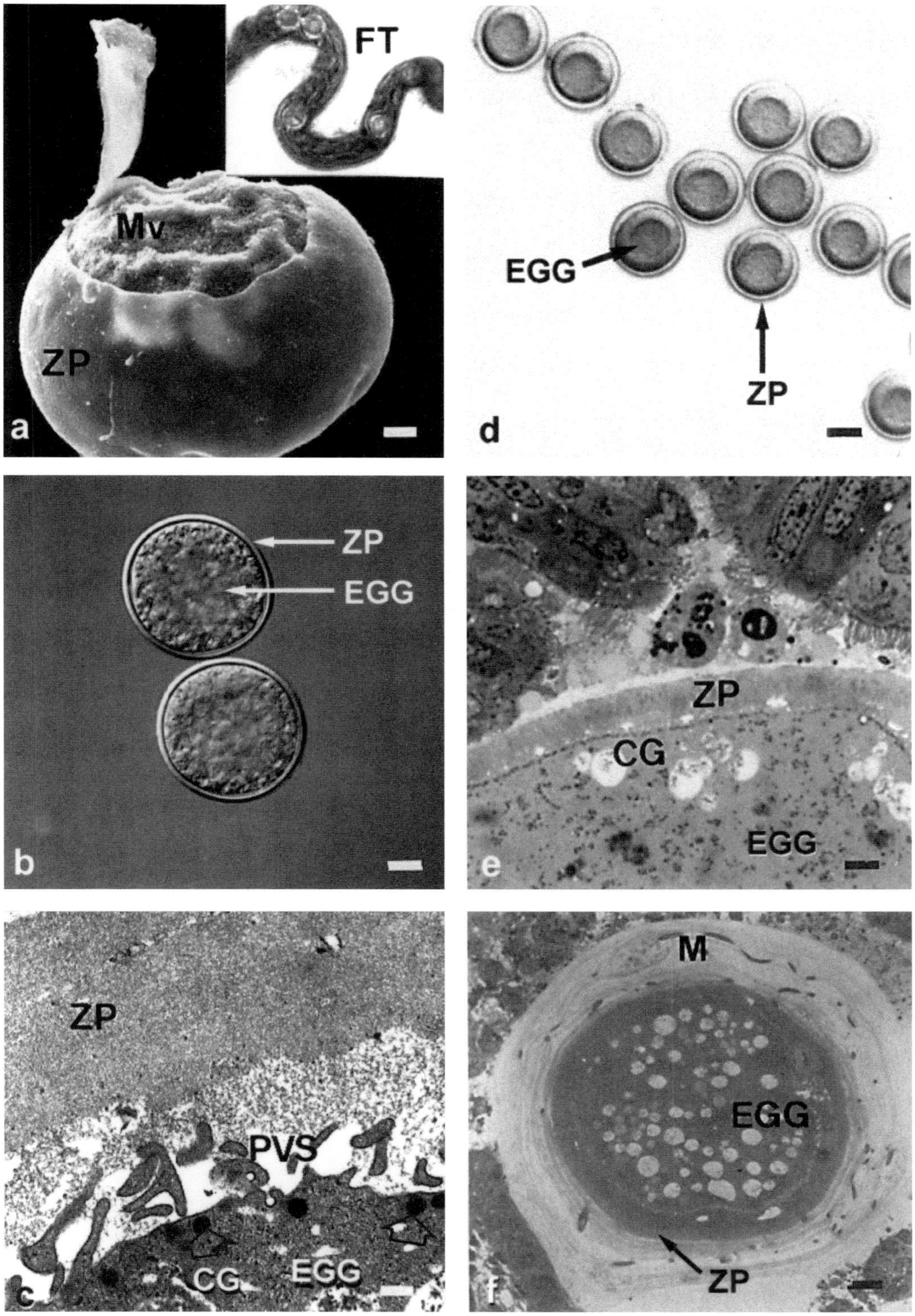

Figure 4 Oocytes from oviduct of *Sminthopsis crassicaudata*. (a) SEM, (b, d) Nomarski, (c, e) TEMs, and (f) a thick plastic section. Note egg is surrounded by the zona pellucida (ZP) and that there are abundant cortical granules (CG) just beneath the oolemma, a large yolk mass in the cytoplasm of the egg. In the oviduct, the egg becomes surrounded by a mucoid (M) coat. Mainly from Breed 1994a, 1996.

(Bedford 1991, 1996) whereas all the cumulus cells are shed prior to ovulation (Selwood 1992; Breed 1996). High resolution scanning and transmission electron microscopy of the zona have, nevertheless, indicated that its composition and fine structural organisation appear to be similar to those of eutherians. Recently, the nucleotide and amino acid sequences of the brush-tail possum zona proteins have been determined and it has been found that, in spite of the solubility differences alluded to above, ZPC (McCartney and Mate 1998), ZPB (Haines et al. 1999), and ZPA (Mate and McCartney 1999; Voyle et al. 1999) are all highly homologous to those of eutherians with the full conservation of the number and location of cysteine residues suggesting a very similar 3-D configuration. Furthermore, *in situ* hybridisation results of ZPA and ZPB have indicated localised expression to the germ cells (Haines et al. 1999; Voyle et al. 1999).

Although full nucleotide and amino acid sequence of the dasyurid and didelphid zona pellucida glycoproteins has yet to be determined, partial sequences of *Sminthopsis* ZPA (Voyle et al. 1999) and ZPB (Voyle, Haines, Rathjen, Hope and Breed unpubl. obs.) show high sequence homology to those of the possum. However, a recent comparative study of oligosaccharides of the zona using fluorescently labelled lectins indicated that the ZP glycoconjugates of all species, except for *Monodelphis,* were masked with sialic acid and that the ZP of *Sminthopsis* was unique in staining intensely with ConA and SBA after prior incubation with potassium hydroxide suggesting an abundance of D-mannose and α-D N-acetyl galactosamine saccharide units (Chapman, Wiebkin and Breed 2000). These preliminary results indicate that, like in eutherians, there are significant differences in sugars of the zona between families of marsupials. Although there is now a little insight into the marsupial zona molecular chemical composition we do not know how the filaments are made up, and nothing is at present known of the complementary molecules on the sperm and how gamete interactions take place at the molecular level.

Sperm penetration of egg coats

Before binding and fusion of the spermatozoon to the oolemma of the egg can occur, the sperm has to penetrate the zona pellucida. In eutherian mammals controversy has existed as to how this takes place, with one view suggesting it occurs largely by enzymatic dissolution, perhaps due to the activity of acrosin or PH 20 on the leading edge of the spermatozoon (Tesarik et al. 1988; Yudin et al. 1999). Another view proposes that zona penetration is largely due to the physical forward thrust of the spermatozoon (Bedford 1991, 1996) and that several features of the eutherian sperm, including its bilateral flattening, sharp apex, rigid disulphide-bonded chromatin, relate to this (see Bedford 1998).

Observations of the zona pellucida around eggs of the didelphids *Didelphis virginiana* (Rodger and Bedford 1982) and *Monodelphis* (Taggart et al. 1993a) indicate that the spermatozoon lies along the outer surface of the zona prior to penetration (Fig. 5), and that, at the time of penetration, at least *in vitro* it makes a large hole in the zona. It has been proposed that this process is markedly different from that of eutherians where only a thin penetration slit in the zona is formed by the spermatozoon (Bedford 1991; Yanagimachi 1994). The larger hole in the marsupial zona is likely to have been brought about by enzymatic dissolution of the zona matrix by the release of acrosomal enzymes (Rodger and Bedford 1982). This contrasts to that of eutherians where physical forward thrust is thought to be equally or more important (Bedford 1991, 1998). However, such clear-cut differences in the processes of zona penetration between eutherians and marsupials are not so apparent from scanning electron micrographs and transmission electron micrographs of *in vivo*-recovered eggs of recently mated animals (see Breed 1996). While some solubilising effect of the acrosomal enzymes on the surrounding zona matrix takes place (Breed and Leigh 1988), the presence of zona matrix around a sperm head in *Monodelphis* (Breed 1996) and packed tightly around a penetrating *Sminthopsis* spermatozoon (Fig. 6) (Breed and Leigh 1988, 1990; Breed 1996) indicates that the forward thrust of the marsupial spermatozoon may be important.

Sperm binding to, and fusion with, the egg plasmalemma

Once the spermatozoon has passed through the zona pellucida it enters the perivitelline space (PVS) where it comes into direct contact with the plasmalemma of the egg (= oolemma). In eutherian mammals the sperm bind to, and fuse with, the microvillous surface of the oolemma by way of receptors on the plasmalemma over the posterior region of the acrosome, the equatorial segment. This differs markedly from that in non-mammalian vertebrates and invertebrates where the inner acrosomal membrane at the apical tip of the sperm binds to, and fuses with, the oolemma (Moore and Bedford 1978; Bedford et al. 1979; see Yanagimachi 1994 for review).

Several fusogenic molecules on the sperm head have been proposed and studies with the monoclonal antibody to the sperm protein PH 30 (= fertilin) have been the most extensive. Fertilin, a heterodimer, is similar to certain viral adhesion molecules, and is synthesised in the testis. It is a member of the ADAM/MDC family of transmembrane proteins and is processed during epididymal migration. It has a disintegrin domain on the β subunit that remains after processing and this is the putative binding domain, whereas a RGD-related peptide on the α subunit possibly acts as a fusogenic peptide (Blobel et al. 1992; Snell and White 1996; Myles and Primakoff 1997). Fertilin was first shown on guinea pig sperm and homologous molecules now having since been found on the sperm head of several other eutherian species. However, recently, when transgenic fertilin 'knockouts' were produced,

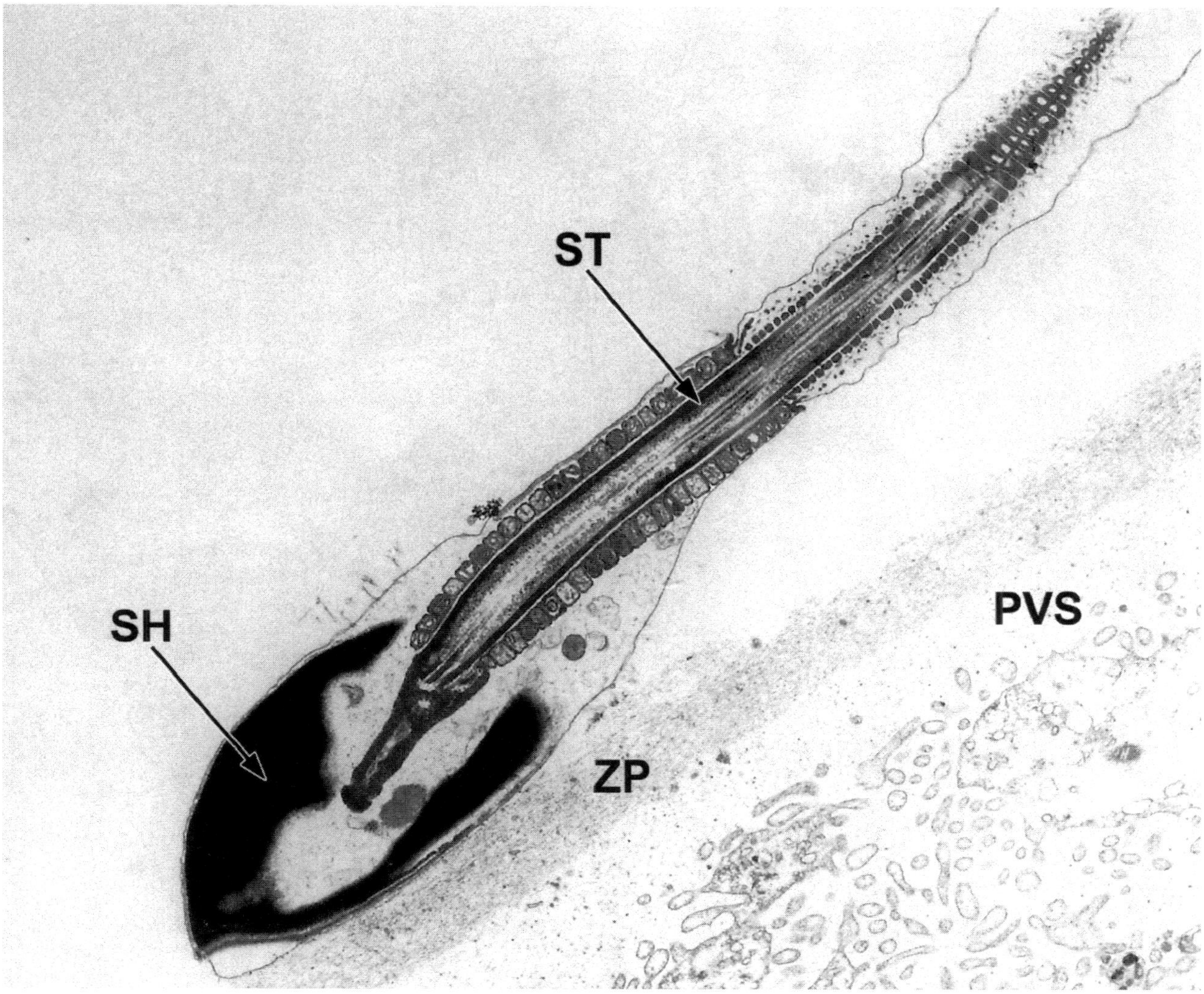

Figure 5 Spermatozoon lying along outer surface of zona matrix of *Monodelphis domestica.* Note acrosomal surface of sperm head (SH) in contract with zona pellucida (ZP). From Taggart et al. 1993a.

sperm transport, as well as sperm–zona binding were reduced. This suggests that fertilin may have other functions apart from facilitating sperm–oolemma binding. Furthermore sperm from a few knockout individuals actually fertilised eggs *in vitro* (Cho et al. 1998) which questions the critical importance of this molecule in this process. At the present time the prevailing view is still however that the disintegrin domain of fertilin β interacts with integrin α6β1 on the oolemma to bring about sperm–oolemma binding although sperm–oolemma fusion is likely to involve other molecules. Other sperm plasmalemma molecules that have been proposed as facilitating sperm–oolemma binding include 2B1 or DE in rats (Rochwerger et al. 1992), 37 kD protein in hamsters which is localised to the apical margin of the spermatozoon (Mat Noor and Moore 1999), and a protein called equatorin which occurs over the equatorial segment of mouse spermatozoa (Toshimori et al. 1998). In marsupials, no studies have yet been published on the molecules involved in sperm–oolemma binding or fusion although preliminary observations suggest that both anti-rabbit PH 30α and anti-rabbit PH 30β will bind to the cell membrane of the possum sperm head (Breed, Neave, Clarke and Holland unpubl. obs.).

In marsupials, significant debate exists in regard to the membrane of the spermatozoon involved in sperm–oolemma binding. From studies carried out on *Didelphis virginiana,* Rodger and Bedford (1982) suggested that sperm binding and fusion took place by way of molecules on the inner acrosomal membrane thus contrasting markedly with the location of these

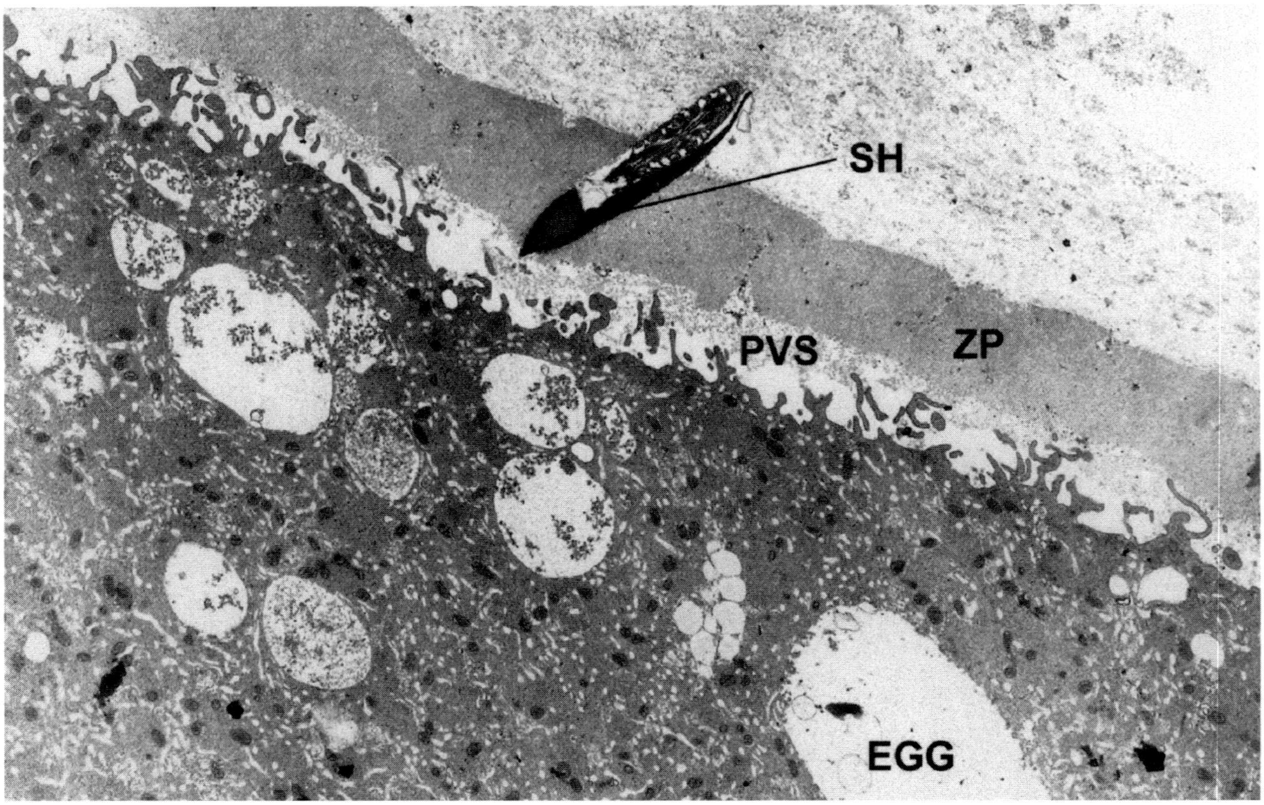

Figure 6 Spermatozoon head (SH) has partly penetrated matrix of zona pellucida (ZP) of *Sminthopsis crassicaudata*. From Breed 1996.

molecules in eutherians. This suggested the retention of an ancestral condition in this group. By contrast, Taggart et al. (1994), working with the didelphid *Monodelphis domestica,* obtained transmission electron microscopic evidence that sperm binding and fusion occurred by way of the plasmalemma over the lateral region of the acrosome in a similar manner to that of eutherians. In order to attempt to throw light on the sperm membrane involved in oolemma binding and fusion of marsupials, a further study was carried out on sperm–egg interactions in *Sminthopsis*. The observations unfortunately did not resolve the question, but they did suggest firstly that initial sperm–oolemma contact occurs by way of the apical tip of the sperm, and that the spermatozoon head comes to lie flat alongside the oolemma in the perivitelline space, before entering the ooplasm. They also suggested that some membrane over the sperm head, perhaps part of the inner acrosomal membrane, is taken into the egg cytoplasm at the time of sperm incorporation (Breed 1994b).

Conclusions

There are two large groups of carnivorous and insectivorous marsupials, the didelphids of America and the dasyurids of Australia. Based on skeletal and dental evidence these species retain many ancestral traits that have been lost by members of most other families of at least Australasian marsupials. They are relatively unspecialised and, although different sets of characters are likely to evolve at different rates, it is possible that the structural organisation of their sperm and eggs as well as the processes of sperm–egg interactions reflect those of the early ancestral marsupials. The relatively high ovulation rate, and hence litter size, and simple design of the pouch in these animals are likely to be closer to the early marsupial condition than the low ovulation rate and elaborate pouch of possums, macropods, koalas, and wombats. The morphology of the gametes, and their maturational changes in preparation for fertilisation, however, clearly show marked differences between the two families of insectivorous/carnivorous marsupials. In didelphids, pairing of spermatozoa occurs, which is presumably a highly derived condition. It is not shown by members of other vertebrate groups and the didelphids share this condition with most other South American, but not Australasian, marsupials. By contrast, in dasyurids a relatively large, and highly morphologically complex, spermatozoon is evident that is quite unlike that of other vertebrates or even most marsupials, except those of bandicoots, honey possums and the numbat (Cummins 1985; Harding et al. 1984; for reviews see Temple-Smith 1987; Harding 1987). Thus, as far as sperm morphology and maturation are concerned, two very different morphotypes have evolved in the didelphids and dasyurids, and in both groups there is little to suggest retention of ancestral features. Rather, in both families, some characteristics are shared by other more advanced marsupials occurring in the two different biogeographic regions, but not by eutherians.

It is clear that in both groups, sperm storage occurs in the isthmus of the oviduct not too unlike that recorded for several eutherian species. The ultrastructure of the cells lining the crypts where this takes place appears to vary between members of the two groups (see Rodger and Bedford 1982; Breed et al. 1989). Such sperm storage has not been found in other marsupial groups (Bedford 1998) but is evident in many extant reptiles and birds as well as some eutherians. This suggests that its existence in the dasyurids and didelphids may be the retention of a primitive condition that has been lost in various other marsupial lineages.

Clearly didelphids and dasyurids ovulate a relatively small oocyte compared to that of non-mammalian species as well as monotremes. However, like that of other marsupials, it is about twice the size of a eutherian egg and has material present within it which is referred to as 'yolk'. The arrangement of this material varies between members of the two groups, with much of that of dasyurids occurring as a central 'yolk mass' (Hill 1910; Selwood 1982; Breed and Leigh 1990). What this material is, and whether it is homologous to the yolk in eggs of other vertebrates, is not clear at the present time. The extracellular matrix, the zona pellucida, that surrounds the oocyte of didelphids and dasyurids is much thinner than that of eutherians and even, it would appear, that of various other marsupials including possums, macropods, koalas and wombats (Tyndale-Biscoe and Renfree 1987; Bedford 1991; Chapman et al. 2000). This thinner zona may reflect a markedly different process of sperm penetration (Bedford 1996, 1998). The recent molecular data, however, suggest similar zona pellucida glycoproteins are present to those of eutherians and that high homology occurs in the nucleotide and amino acid sequences. This is also, to some extent, the case for the egg coat glycoproteins of the amphibian *Xenopus* (Hedrick 1996) and even, to some extent, fish (Lyons et al. 1993; Murata et al. 1995; Chang et al. 1997). Whether the process by which the sperm penetrates the zona in marsupials differs markedly from that in eutherians is not clear but the indirect evidence now suggests that zona penetration may involve both an enzymatic, as well as a physical forward thrust, component of the spermatozoon, like that of eutherians.

Finally for the process of sperm–oolemma binding in marsupials, which is based largely on data from didelphids, different sperm membranes have been proposed for the two species investigated – one for *Didelphis* and the other for *Monodelphis*. Such divergence in the membrane involved in oolemma-binding between two genera within the one family is unlikely to exist. Clearly this process needs to be re-examined to resolve this apparent controversy. Only then will it be possible to determine how similar the processes are in marsupials to those of eutherian mammals.

ACKNOWLEDGEMENTS

We thank Esther Breed for typing the manuscript. Part of the work described in the study was funded by a grant from the ARC to the senior author.

REFERENCES

Akatsuha, K., Yoshida-Komiya, H., Tulsiani, D.R.P., Orgebin-Crist, M.-C., Hiroi, M., & Araki, Y. (1998), 'Rat zona pellucida glycoproteins: molecular cloning and characterisation of the three major components', *Mol Rep Dev*, **51**:454–67.

Allen, C.A., & Green, D.P.L. (1995), 'Monoclonal antibodies which recognise equatorial segment presented *de novo* following the A23187-induced acrosome reaction of guinea pig sperm', *J Cell Sci*, **108**:767–77.

Bedford, J.M. (1991), 'The coevolution of mammalian gametes', in *A comparative overview of mammalian fertilization* (eds. B.S. Dunbar & M.G. O'Rand), pp. 3–39, Plenum Press, New York.

Bedford, J.M. (1996), 'What marsupial gametes disclose about gamete function in eutherian mammals', *Reprod Fertil Dev*, **8**:569–80.

Bedford, J.M., & Breed, W.G. (1994), 'Regulated storage and subsequent transformation of spermatozoa in the fallopian tube of an Australian marsupial, *Sminthopsis crassicaudata*', *Biol Reprod*, **50**:845–54.

Bedford, J.M., Moore, H.D.M., & Franklin, L.E. (1979), 'Significance of the equatorial segment of the acrosome of the spermatozoa in eutherian mammals', *Exp Cell Res*, **119**:1 9–26.

Bedford, J.M., (1998), 'Mammalian fertilization misread? Sperm penetration of the eutherian zona pellucida is unlikely to be a lytic event', *Biol Reprod*, **59**:1275–87.

Bennett, J.H., Breed, W.G., Hayman, D.C., & Hope, R.M. (1990), 'Reproductive and genetic studies with a laboratory colony of the dasyurid marsupial *Sminthopsis crassicaudata*', *Aust J Zool*, **37**:207–22.

Biggers, J.D., & DeLamater, E.D. (1965), 'Marsupial spermatozoa pairing in the epididymis of American forms', *Nature London*, **208**:402–4.

Blobel, C.P., Wolfsberg, T.G., Turich, C.W., Myles, D.G., Primakoff, P., & White, J.M. (1992), 'A potential fusion peptide and an integrin ligand domain in a protein active in sperm-egg fusion', *Nature*, **356**:248–52.

Breed, W.G., & Leigh, C.M. (1988), 'Morphological observations on sperm-egg interactions during in vivo fertilization in the dasyurid marsupial, *Sminthopsis crassicaudata*', *Gam Res*, **19**:131–49.

Breed, W.G., & Leigh, C.M. (1990), 'Morphological changes in the oocyte and its surrounding vestments during in vivo fertilization in the dasyurid marsupial, *Sminthopsis crassicaudata*', *J Morphol*, **204**:177–96.

Breed, W.G. (1994a), 'How does sperm meet egg? – in a marsupial', *Reprod Fert Dev*, **6**:485–06.

Breed, W.G. (1994b), 'Sperm–egg interactions in an Australian marsupial with special reference to changes in acrosomal morphology', *Zygote*, **2**:201–11.

Breed, W.G. (1996), 'Egg maturation and fertilization in marsupials', *Reprod Fert Dev*, **8**:617–43.

Breed, W.G., Leigh, C.M., & Bennett, J.H. (1989), 'Sperm morphology and storage in the female reproductive tract of the fat-tailed dunnart *Sminthopsis crassicaudata* (Marsupialia: Dasyuridae)', *Gam Res*, **23**:61–75.

Breed, W.G., Leigh, C.M., Washington, J.M., & Soon, L.L.L. (1994), 'Unusual nuclear structure of the spermatozoon in a marsupial *Sminthopsis crassicaudata*', *Mol Reprod Dev*, **37**:78–86.

Brewis, I.A., & Moore, H.D.M. (1997), 'Molecular mechanisms of gamete recognition and function at fertilization', *Hum Reprod*, **2**:156–65.

Brewis, I.A., & Wong, C.H. (1999), 'Gamete recognition: sperm proteins that interact with the egg zona pellucida', *Rev Rep*, **4**:135–42.

Bromham, L., Phillips, M.J., & Penny, D. (1999), 'Growing up with dinosaurs: molecular dates and the mammalian radiation', *TREE*, **14**:113–18.

Chang, Y.S., Hsu, C.C., Wang, S.C., Tsao, C.C., & Huang, F.L. (1997), 'Molecular cloning, structural analysis and expression of carp ZP2 gene', *Mol Reprod Dev*, **46**:258–67.

Chapman, J.A., Wiebkin, O.W., & Breed, W.G. (2000), 'Interspecific variation of zona pellucida glycoconjugates in several species of marsupial', *J Reprod Fertil*, **119**:111–20.

Cho, C., Bunch, C. O'D., Faure, J.-E., Goulding, E.H., Eddy, E.M., Primakoff, P., & Myles, D.M. (1998), 'Fertilization defects in sperm from mice lacking fertilin β', *Science*, **281**:1857–9.

Cooper, N.J., Holland, M.K., & Breed, W.G. (1998), 'Extratesticular sperm maturation in the brush-til possum (*Trichosurus vulpecula*)', *J Reprod Fertil Suppl*, **53**:221–6.

Cooper, T.G. (1998), 'Interactions between epididymal secretion and spermatozoa', *J Reprod Fertil Suppl*, **53**:119–36.

Cummins, J.M. (1976), 'Epididymal maturation of spermatozoa in the marsupial *Trichosurus vulpecula*: changes in motility and gross morphology', *Aust J Zool*, **24**:499–512.

Cummins, J.M., Renfree, M.B., & Temple-Smith, P.D. (1985), 'Reproduction in the male honey possum *Tarsipes rostratus*. Marsupialia: the epididymis', *Amer J Anat*, **177**:385–401.

Dunbar, B.S. (1983), 'Morphological, biochemical, and immunochemical characterisation of the mammalian zona pellucida', in *Mechanisms and control of animal fertilization* (ed. J.F. Hartman), pp. 137–75, Academic Press, New York.

Dunbar, B.S., Prasad, S.V., Timmons, T. (1991), 'Comparative structure and function of the mammalian zonae pellucidae', in *A comparative overview of mammalian fertilization* (eds. B.S. Dunbar & M.G. O'Rand), pp. 97–116, Plenum Press, New York.

Evans, J.P., Schultz, R.M., & Kopf, G.S. (1998), 'Roles of disintegrin domains of mouse fertilins α and β on fertilization', *Biol Reprod*, **59**:145–52.

Frankenberg, S., Newell, G., & Selwood, L. (1996), 'A light microscopic study of oogenesis in the brush-tail possum *Trichosurus vulpecula*', *Reprod Fertil Dev*, **8**:541–6.

Grootenhuis, A.J., Philipsen, H.L.A., de Breet-Grijsbach, J.T.M., & van Duin, M. (1996), 'Immunocytochemical localisation of ZP3 in primordial follicles of rabbit, marmoset, rhesus monkey, and human ovaries using antibodies against human ZP3', *J Reprod Fertil Suppl*, **50**: 43–54.

Haines, B.P., Rathjen, P.D., Hope, R.M., Whyatt, L.M., Holland, M.K., & Breed, W.G. (1999), 'Isolation and characterisation of a cDNA encoding a zona pellucida protein (ZPB) from the marsupial *Trichosurus vulpecula* (Brushtail Possum)', *Mol Reprod Dev*, **52**:174–82.

Harding, H.R. (1987), 'Interrelationships of the families of the Diprotodonta – a view based on spermatozoon ultrastructure, in *Possums and opossums: Studies in evolution* (ed M. Archer), pp. 195–216, Surrey Beatty & Sons, NSW.

Harding, H.R, Carrick, F.N., & Shorey, C.D. (1975), 'Ultrastructure changes in spermatozoa of the brushtail possum *Trichosurus vulpecula* (Marsupialia) during epididymal transit. Part II. The Acrosome', *Cell Tiss Res*, **171**:61–70.

Harding, H.R., Woolley, P.A., Shorey, C.D., & Carrick, F.N. (1982), 'Sperm ultrastructure, spermiogenesis, and epididymal sperm maturation in dasyurid marsupials: phylogenetic implications', in *Carnivorous marsupials* (ed. M. Archer), pp. 659–73, Surrey Beatty: Sydney and Royal Zool Soc NSW.

Harding, H.R., Carrick, F.N., & Shorey, C.D. (1984), 'Sperm ultrastructure and development in the honey possum, *Tarsipes rostratus*', in *Possums and gliders* (eds. A. Smith & I. Hume), pp. 451–61, Surrey Beatty & Sons, NSW.

Hartman, C.G. (1916), 'Studies on the development of the opossum *Didelphis virginiana*. I. History of cleavage. II. Formation of the blastocysts', *J Morphol*, **27**:1–83.

Hedrick, J.L. (1996), 'Comparative structural and antigenic properties of zona pellucida glycoproteins', *J Reprod Fertil Suppl*, **50**:9–17.

Hill, J.P. (1910), 'The early development of Marsupialia with special reference to the native cat (*Dasyurus viverrinus*), contributions to the embryology of the Marsupialia', *Q J Microsc Sci*, **56**:1–134.

Janke, A., Gemmell, N.J., Feldmaier-Fuchs, G., von Haeseler, A., & Paabo, S. (1996), 'The complete mitochondrial genome of a monotreme, the platypus, (*Ornithorhynchus anatinus*)', *J Mol Evol*, **42**:153–9.

Jones, R. (1998), 'Plasma membrane structure and remodelling during sperm maturation in the epididymis', *J Reprod Fertil Suppl*, **52**:73–84.

Jones, R.C., & Clulow, J. (1994), 'Interactions of spermatozoa and the reproductive ducts of the male tammar wallaby *Macropus eugenii* (Macropodidae: Marsupialia)', *Rep Fertil Dev*, **6**:437–44.

Kolle, S., Sinowartz, F., Foie, G., Totzauer, I., Amselgruber, W., & Plendl, J. (1996), 'Localisation of the mRNA encoding the zona protein ZP3α in the porcine ovary, oocyte, and embryo by non-radioactive in site hybridisation', *Histochem J*, **28**:441–7.

Kumar, S., & Hodges, B. (1998), 'A molecular timescale for vertebrate evolution', *Nature*, **392**:917–20.

Lamont, A.E., Clarke, H., Cooper, N.J., Holland, M.K., & Breed, W.G. (1998), 'Protein synthesis and secretion by the epididymis of the brush-tail possum, *Trichosurus vulpecula*', *J Reprod Fertil*, **114**:169–77.

Leyton, C., & Saling, P. (1989), '95 kDa sperm proteins bind ZP3 and serve as tyrosine kinase substates in response to zona binding', *Cell*, **57**:1123–30.

Lin, M., & Rodger, J.C. (1999), 'Acrosome formation during sperm transit through the epididymis in two marsupials, the tammar wallaby (*Macropus eugenii*) and the brush-tail possum (*Trichosurus vulpecula*)', *J Anat*, **199**:223–32.

Luckett, W.P. (1994), 'Suprafamily relationships within Marsupialia: resolution and discordance from multidisciplinary data', *J Mam Evol*, **2**:255–83.

Lyons, C.E., Pagette, K.L., Price, J.L., & Huang, R.C.I. (1993), 'Expression and structural analysis of a teleost homology of a mammalian zona pellucida gene', *J Biol Chem*, **268**:21351–8.

Marshall, L.F. (1984), 'Monotremes & marsupials', in *Orders and families of recent mammals of the world* (eds. S. Anderson & J. Knox Jones Jr.), pp. 59–116, John Wiley & Sons, New York.

Mate, K.E. (1996), 'Cytoplasmic maturation of the marsupial oocyte during the periovulatory period', *Reprod Fertil Dev*, **8**:509–19.

Mate, K.E. (1998), 'Timing of zona pellucida formation in the tammar wallaby (*Macropus eugenii*) and brush-tail possum (*Trichosurus vulpecula*)', *Anim Rep Sci*, **53**:237–52.

Mate, K.E., & McCartney, C.A. (1998), 'Sequence and analysis of a zona pellucida 2 cDNA (ZP2) from a marsupial, the brushtail possum (*Trichosurus vulpecula*)', *Mol Reprod Dev*, **51**:322–9.

Mat Noor, M., & Moore, H.D.M. (1999) 'Monoclonal antibodies that recognise epitopes of the sperm equatorial segment region that specifically inhibits sperm-oolemma fusion but not binding', *J Reprod Fertil*, **115**:215–24.

McCartney, C.A., & Mate, K.E. (1999), 'Cloning and characterization of a zona pellucida 3 cDNA from a marsupial, the brushtail possum, *Trichosurus vulpecula*'. *Zygote*, **7**:1–9.

McLeskey, S.B., Dowds, C., Carballada, R., White, R.R., & Saling, P.M. (1998), 'Molecules involved in mammalian sperm egg interaction', *Int Rev Cytol*, **177**:57–113.

Miller, B.J., Macek, M.B., & Shur, B.D. (1992), 'Complementarity between sperm surface β1, 4-galactosyl transferase and egg coat ZP3 mediates sperm-egg binding', *Nature*, **357**:589–93.

Moore, H.D.M., & Bedford, J.M. (1983), 'The interaction of mammalian gametes in the female. Sperm-egg interactions *in vivo*', in *Mechanism and control of animal fertilization* (ed. J.F. Hartman), pp. 453–97, Academic Press, New York.

Moore, H.D.M., & Taggart, D.A. (1993), '*In vitro* fertilization and embryo culture in the grey short-tailed opossum, *Monodelphis domestica*', *J Reprod Fertil*, **98**:267–75.

Moore, H.D.M., & Taggart, D.A. (1995), 'Sperm pairing in the opossum increases the efficiency of sperm movement in a viscous environment', *Biol Reprod*, **52**:947–53.

Murata, K., Sasaki, T., Hasumasu, S., Inchi, I., Enaumi, J., Yasumasu, I., & Yamagaum, K. (1995), 'Cloning of cDNAs for the precursor protein of a low molecular weight subunit of the inner layer of the egg envelope (chorion) of the fish *Oryzias latipes*', *Dev Biol*, **167**:9–17.

Myles, D.G., & Primakoff, P. (1997), 'Why did the sperm cross the cumulus? To get to the oocyte. Functions of the sperm surface pointers PH-20 and fertilin in arriving at, and fusing with, the egg', *Biol Reprod*, **56**:320–7.

Prasad, S.V., Wilkins, B., Skinner, S.M., & Dunbar, B.S. (1996), 'Molecular biology approaches to evaluate species variation in immunogenicity and antigenicity of zona pellucida proteins', *J Reprod Fertil Suppl*, **50**:143–9.

Rodger, J.C., & Bedford, J.M. (1982), 'Separation of sperm pairs and sperm–egg interactions in the opossum, *Didelphis virginiana*', *J Reprod Fertil*, **64**:171–9.

Rochwerger, C., Cohen, D.J., & Cuasnicu, P.S. (1992), 'Mammalian sperm-egg fusion: the rat egg has complementary sites for a sperm protein that mediates gamete fusion', *Dev Biol*, **153**:83–90.

Selwood, L. (1982), 'A review of maturation and fertilization in marsupials with special reference to the dasyurid *Antechinus stuartii*', in *Carnivorous Marsupials* (ed. M. Archer), pp. 65–76, Royal Zoological of NSW, Sydney.

Selwood, L. (1992) 'Mechanisms underlying the development pattern in marsupial embryos. Ch 5', in *Current Topics in Developmental Biology*, **27**:175–233.

Selwood, L., & McCallum, F. (1984), 'Relationship between longevity of spermatozoa after insemination and the percentage of normal embryos in brown marsupial mice (*Antechinus stuartii*)', *J Reprod Fertil*, **79**:495–503.

Shimmin, G.A., Jones, M., Taggart, D.A., & Temple-Smith, P.D. (1999), 'Sperm transport and storage in the agile antechinus (*Antechinus agilis*)', *Biol Reprod*, **60**:1353–9.

Snell, W.J., & White, J.M. (1996), 'The molecules of mammalian fertilization', *Cell*, **85**:629–37.

Soon, L.L.L., & Breed, W.G. (1996), 'Ultrastructure of nuclear condensation and localisation of DNA and proteins in spermatozoa of a dasyurid marsupial *Sminthopsis crassicaudata*', *Mol Reprod Dev*, **43**:217–27.

Smith, T.T. (1998), 'The modulation of sperm function by the oviduct epithelium', *Biol Reprod*, **58**:1102–04.

Suarez, S.S. (1998), 'The oviduct sperm reservoir in mammals: mechanisms of formation', *Biol Reprod*, **58**:1105–07.

Taggart, D.A., & Temple-Smith, P.D. (1991), 'Transport and storage of spermatozoa in the female reproductive tract of the brown marsupial mouse, *Antechinus stuartii* (Dasyuridae)', *J Reprod Fertil*, **93**:97–110.

Taggart, D.A., O'Brien, H.P., & Moore, H.D.M. (1993a), 'Ultrastructural characteristics of *in vivo* and *in vitro* fertilization in the grey short-tailed opossum *Monodelphis domestica*, *Anat Rec*, **237**:21–37.

Taggart, D.A., Johnson, J.L., O'Brien, H.P., & Moore, H.D.M. (1993b), 'Why do spermatozoa of American marsupials form pairs? A clue from the analysis of sperm pairing in the epididymis of the grey short-tailed opossum *Monodelphis domestica*', *Anat Rec*, **236**:465–78.

Taggart, D.A., Selwood, L. & Temple-Smith, F.D. (1997), 'Sperm production, storage, and the synchronisation of male and female reproductive cycles in the iteroparous, stripe-faced dunnart (*Sminthopsis macroura*; Marsupialia): relationship to reproductive strategies within the Dasyuridae', *J Zool Lond*, **243**:725–36.

Taggart, D.A., Shimmin, G.A., McCloud, P., & Temple-Smith, P.D. (1999), 'Timing of mating, sperm dynamics and ovulation in a wild population of agile antechinus (Marsupialia: Dasyuridae)', *Biol Reprod*, **60**:283–9.

Temple-Smith, P.D. (1987), 'Sperm structure and marsupial phylogeny', in *Possums and opossums: studies in evolution* (ed. M. Archer), pp. 171–94, Surrey Beatty & Sons, Sydney.

Temple-Smith, P.D. (1994), 'Comparative structure and function of marsupial spermatozoa', *Reprod Fertil Dev*, **6**:421–35.

Temple-Smith, P.D., & Bedford, J.M. (1976), 'The features of sperm maturation in the epididymis of a marsupial the Brush-tailed possum', *Trichosurus vulpecula*', *Am J Anat*, **147**:471–500.

Temple-Smith, P.D., & Bedford, J.M. (1980), 'Sperm maturation and the formation of sperm pairs in the epididymis of the opossum *Didelphis virginiana*', *J Exp Zool*, **214**:161–71.

Tesarik, J., Drahorad, J., & Peknicora, J. (1988), 'Subcellular immunochemical localisation of acrosin in human spermatozoa during the acrosome reaction and zone pellucida penetration', *Fertil Steril*, **50**:133–41.

Toshimori, K., Saxena, D.K., Tanri, K., & Yoshinaga, K. (1998), 'An MN9 antogenic molecule, equatorin, is required for successful sperm-oocyte fusion on mice', *Biol Reprod*, **59**:22–9.

Tulsiani, D.R.P., Yoshida-Komiya, H., & Araki, Y. (1997), 'Mammalian fertilization: a carbohydrate-mediated event', *Biol Reprod*, **57**:487–94.

Tyndale-Biscoe, C.H., & Renfree, M.B. (1987), *Reproductive physiology of marsupials. Monographs on marsupial biology*, Cambridge University Press, Cambridge.

VandeBerg, J.L. (1990), 'The grey short-tailed opossum *Monodelphis domestica* as a model didelphid species for genetic research', *Aust J Zool*, **37**:235–47.

Voyle, R.B., Haines, B.P., Loffler, K.A., Hope, R.M., Rathjen, P.D., & Breed, W.G. (1999), 'Isolation and characterisation and expression of Zona Pellucida A (ZPA) cDNAs from two species of marsupial', *Zygote*, **7**:239–48.

Wassarman, P.M. (1988), 'Zona pellucida glycoproteins', *Ann Rev Biochem*, **57**:415–42.

Wassarman, P.M. (1990), 'Profile of a mammalian sperm receptor', *Development*, **108**:1–17.

Wassarman, P.M. (1992), 'Mouse gamete adhesion molecules', *Biol Reprod*, **46**:186–91.

Wassarman, P.M. (1999), 'Mammalian fertilization. Molecular aspects of gamete adhesion, exocytosis and fusion', *Cell*, **96**:175–83.

Wilson, D.E., & Reeder, D.M. (1993), *Mammal species of the world. 2nd edn.*, Smithsonian Institute, Washington.

Yanagimachi, R. (1994), 'Mammalian fertilization', in *The physiology of reproduction* (eds. E. Knobil & J.D. Neill), pp. 189–317, Raven Press, New York.

Yonezawa, N., Mitsui, S., Kudo, K., Nakano, M. (1997), 'Identification of an N-glycosylated region of pig zona pellucida glycoproteins ZPB that is involved in sperm binding', *Eur J Biochem*, **248**:86–92.

Yudin, A.I., Vanderort, C.A., Li, M.-W., & Overtheet, J.W. (1999), 'PH 20 but not acrosin is involved in sperm penetration of the macaque zona pellucida', *Mol Reprod Dev*, **53**:350–62.

Yurewicz, E.C., Pack, B.A., Amant, R., & Sacco, A.G. (1993), 'Porcine zona pellucida ZP3α glycoprotein mediates binding of the biotin-labelled M_r 55,000 family (ZP3) to boar sperm membrane vesicles', *Mol Reprod Dev*, **36**:382–9.

CHAPTER 10

TIMING OF REPRODUCTION IN CARNIVOROUS MARSUPIALS

Bronwyn McAllan

Human Biology, School of Biological Sciences, University of New England, Armidale, NSW 2351, Australia

Reproduction in many carnivorous marsupials is seasonal; however the proximal cues for reproduction are known for a minority of species. Species from the genera *Antechinus* and *Sminthopsis* appear to rely on photoperiodic change for timing of reproduction, with other environmental factors such as pheromones (*Antechinus*) and rainfall (*Sminthopsis*) of secondary importance. For most other carnivorous marsupials the proximal cues for reproduction have not been determined, although photoperiod is likely to be the major cue for reproductive timing. Unfortunately, almost nothing is known about reproductive activity for many carnivorous marsupials, especially those from South America, and only further studies can clarify how their reproductive life history is controlled, requiring significant further effort on the part of researchers.

INTRODUCTION

Reproduction in all mammals must be timed to optimise the chance of mating, giving birth, and rearing of young to independence. All mammals have developed physiological and behavioural strategies to ensure the success of reproduction in their environment. Where known, the present summary will focus on the environmental cues that carnivorous marsupials use to ensure successful reproduction. Where little is known about cues for reproduction, a brief summary of published knowledge of reproduction is provided.

Carnivorous marsupials are distributed widely in Papua New Guinea, Australia, and the Americas. The definition of 'carnivory' follows Hume (1999). In Hume's (1999) discussion of the term carnivorous, some deliberation was given to the difficult distinction between carnivorous and omnivorous species. Thus 'carnivorous' refers to those marsupials which, to the best of my knowledge, eat predominantly invertebrate prey, vertebrate prey, or carrion. The brush-tailed possum, *Trichosurus vulpecula*, and the mouse opposum *Marmosa robinsoni*, which are known to devour meat, are not included because of their diverse food preferences (Hume 1999).

The taxonomy of Australian carnivorous marsupials follows that of Strahan (1995) and Krajewski et al. (2000), taxonomy of New Guinean carnivorous marsupials follows that of Flannery (1995) and Krajewski et al. (2000), and taxonomy of South American carnivorous marsupials follows that used by Harder and Fleck (1997). Most of the studies on carnivorous marsupial reproduction concern the dasyurids, and thus much of the following review discusses this group.

Control of reproduction in mammals

Reproduction in many mammals is seasonal. The ability to detect seasonal environmental change, and use of the seasonal cycle to optimise reproduction, is an important survival consideration. It has long been recognised that one proximal cue for reproduction used by many mammals is photoperiod, although rainfall and nutritional status can induce reproduction in some mammals. In response to photic information from the eyes the hormone melatonin is secreted by the pineal gland during the dark phase of the daily photic cycle, and is suppressed during the light phase of the photic cycle (Reiter 1993). Melatonin binds with many parts of the brain, (Williams et al. 1997), resets circadian and circannual biological rhythms, and controls reproductive cycling by suppressing reproductive releasing hormones in the hypothalamus (Reiter 1993).

Most marsupial mammals are seasonal breeders (Tyndale-Biscoe 1980), and the marsupial reproductive strategy is focused more on lactation than pregnancy (Renfree 1983, Hinds and Loudon 1997). There are few detailed studies available on photoperiodism and pineal involvement in seasonal biology in marsupials, with melatonin administration studies the main source of information on pineal activities. Melatonin administration either inhibits reproduction (*Isoodon macrourus*, Gemmell 1987a), brings reproduction forward (*Trichosurus vulpecula*, Gemmell 1987b) or disrupts embryonic diapause depending on time of administration (*Macropus r. rufogriseus*, Loudon and Curlewis 1987). Oral melatonin administration to *Antechinus stuartii* at the winter solstice moves the timing of the breeding period (McAllan et al. 2001). Other studies have looked at the influence of photoperiod on reproductive activity in marsupials. While the role of photoperiod in seasonal cycles has been demonstrated unequivocally in some marsupial species, for others the influence of photoperiod is inferred, or unclear (Gemmell et al. 1993, Hinds and Loudon 1997). Lengthening photoperiods following short-day photoperiods are important for initiating reproduction in *Sminthopsis crassicaudata* and *S. macroura*, (*larapinta*, Godfrey 1969a, 1969b).

Many photoperiodic studies have focused on the tammar wallaby (*Macropus eugenii*), where studies have demonstrated that reproduction is intitiated by the changing photoperiod, which is translated into a reproductive message by the pineal gland and melatonin secretion (McConnell and Tyndale-Biscoe 1985, McConnell et al. 1986). In tammars, the newly formed corpus luteum is inhibited by the frequent suckling stimulus provided by the young (Sharman 1970, Tyndale-Biscoe and Renfree 1987). The quiescent period or embryonic diapause, can also be induced by shortening photoperiod (seasonal diapause, Tyndale-Biscoe 1980, Tyndale-Biscoe 1986). Seasonal embryonic diapause is controlled by the pineal gland, and pineal denervation abolishes seasonal diapause (Renfree et al. 1981). Changes in both amplitude and duration of the pineal hormone, melatonin, have been shown to be the main cue for the pituitary control of seasonal diapause (McConnell 1986, McConnell and Hinds 1985, McConnell and Tyndale-Biscoe 1985, McConnell et al. 1986). The review will now concentrate on what is known about the control of reproduction in the carnivorous marsupials.

Timing of reproduction in carnivorous marsupials

Family Dasyuridae, Subfamily Dasyurinae, tribe Phascogalini

All of the family Dasyuridae are carnivores (Hume 1999, Strahan 1995). Many species within each genus have similar reproductive patterns, and thus reproduction in some genera will be described together.

Genus *Antechinus*

At present there are nine confirmed species of Australian *Antechinus*, (Table 1; Armstrong et al. 1998, Krajewski et al. 2000). Most of the studies on reproduction and the proximal cues that may control reproduction have focused on *A. stuartii* and *A. agilis*. Until recently these two species were thought to be one, *A. stuartii*, but exploration of the role of photoperiod in triggering the mating period in these mammals has prompted further studies into the taxonomic relationship between these species (McAllan and Dickman 1986, Dickman et al. 1988, Dickman et al. 1998). Because many species from the genus *Antechinus* follow similar patterns of reproduction, and many studies have been performed on *A. stuartii*, most of the following discussion refers to this species (Table 1).

Populations of *Antechinus stuartii* have a brief, monoestrus mating period followed by complete male mortality (Lee et al. 1982, Woolley 1966). All males are dead before the females give birth after spontaneous ovulation (Selwood 1980, 1982, Woolley 1966). The male mortality is caused by a failure of the glucocorticoid feedback mechanism (Bradley et al. 1980, this volume, McDonald et al. 1981, McDonald et al. 1986).

All aspects of reproduction are highly synchronised within each local population (Dickman 1982, Lee and Cockburn 1985). Braithwaite (1979) first linked reproductive synchronisation to the photoperiodic cycle. Animals in each population mate within a one- to three-week period, which is at the same time at each geographical location every year. The mating period is shorter at higher latitudes and progressively lengthens the further north the populations are found. The mating period is also later in the north than in the south (Dickman 1982). This is consistent with a rate of change of photoperiod, and not photoperiodic length, being the main promoter of mating activity (McAllan and Dickman 1986).

When the yearly photoperiodic cycle was delayed by two months, beginning in either May or February, it was found that reproductive synchrony was delayed by two months in animals exposed to the different photoperiod in February, but the reproductive cycle of animals exposed in May was delayed, but unsynchronised (McAllan et al. 1991). It appears reproductive synchrony is maintained by the changing photoperiod modulating an underlying endogenous rhythm. Moreover, the oral administration of melatonin from both the autumnal equinox and after the winter solstice, abolishes synchronisation of matings and births, but not reproductive activity in *A. stuartii* (McAllan et al. 2001). The effects of melatonin administration provide further support for the notion that the changing photoperiod is synchronising an underlying endogenous rhythm.

Other factors besides photoperiod are involved in reproductive synchrony. Animals exposed to unchanging photoperiods of different lengths continue to exhibit some of the reproductive changes observed in wild *A. stuartii* (Dickman 1985, Scott 1986). Synchronous mating was not always evident, and exposure to other animals or nesting material increased synchrony, with pheromonal cues important in enhancing synchrony (Dickman 1985, Scott 1986).

In male *A. stuartii*, the synchronous seasonal nature of reproduction is observed in the changing structure of the testis, which includes the cessation of spermatogenesis in mid June (Kerr and Hedger 1983, Taggart et al. 1993, Woolley 1966, this volume). Increasing plasma testosterone levels coincides with spermatogenic failure (Kerr and Hedger 1983), beginning at about the time of the winter solstice, suggesting that there may be hypothalamic interactions with the peripheral endocrine glands by photoperiodic mechanisms. These override local testicular activity causing the subsequent failure of spermatogenesis. In *A. stuartii*, maximum plasma testosterone levels occur two months after spermatogenesis has ceased (Bradley et al. 1980, Kerr and Hedger 1983). The peak plasma testosterone concentrations in *Antechinus* are similar to those found in other marsupials, and some eutherians (Table 2).

Similar plasma testosterone concentrations to those of *A. stuartii* when spermatogenic activity is at its peak are associated with hypogonadism (Crawford et al. 1993, Handelsman 1990), and low testicular size and activity in other mammals (Curlewis 1991, Fuentes et al. 1993, Inns 1982, Kaplan and Mead 1993, Weinbauer and Nieschlag 1990). Maximal spermatogenic activity usually corresponds with the highest plasma testosterone levels in mammals, and the precise mechanism by which spermatogenesis is maintained with low plasma testosterone concentrations, and then ceases before the breeding season when plasma testosterone concentrations are high, has not been determined.

The accessory reproductive glands in other adult marsupials are sensitive to the effects of testosterone (by conversion of testosterone to dihydrotestosterone by 5α-reductase activity), and the administration of testosterone to castrated animals causes an increase in weight and secretory activities of these glands (Cook et al. 1978, Curlewis and Stone 1985, Jones et al. 1988). In pouch young tammar wallabies, a dihydrotestosterone metabolite (5α-androstane-3α, 17β-diol) is important in early differentiation of males (Shaw et al. 2000). In *A. stuartii*, administration of testosterone alone or testosterone and cortisol together to juvenile males increases accessory gland size and secretory activity, and Leydig cell size, but does not affect spermatogenesis, which remains in synchrony with untreated males (McAllan 1998).

Following reproductive activity, the social interactions normally experienced by males are important in intensifying the male 'die-off' (Scott 1987, Wood 1970, Woolley 1966). In captivity, males held alone or with females did not die as rapidly as those held with other males (Wood 1970, Woolley 1966) and the quicker death was the result of increased corticosteroid levels in males that interacted with other males (Scott 1987). Antagonism is common between males (Braithwaite 1974), and social factors including population density may affect the speed at which males in the wild die (Bradley and Monamy 1991).

In females, the spontaneous monoestrous cycle is present in other *Antechinus* species (Calaby and Taylor 1981, Lee et al. 1982, Taylor and Horner 1970, Watt 1997, Wilson 1986, Woolley 1966, Table 1) and some other dasyurids (Calaby and Taylor 1981, Taylor and Horner 1970, Woolley 1988, 1991a, b, c). All members of the genus *Antechinus* experience the synchronised mating period, followed by male mortality, although the timing differs between species (Dickman 1986, Van Dyck 1980, Watt 1997, Wilson and Bourne 1984).

A recent analysis of mating data for the genus *Antechinus* found that the rate of change of photoperiod specific for each species best explained the timing of reproduction in these mammals (McAllan et al. 1999). Absolute photoperiod and ambient temperature could not explain the synchrony or small timespan of reproduction. Of special significance was the separation of the *Antechinus* genus based on rate of change of photoperiod, into two groupings, one of *A. agilis*, *A stuartii* and *A. flavipes*; and the other consisting of *A. swainsonii* and *A. minimus* (McAllan et al. 1999). This is consistent with mitochondrial DNA analysis of the *Antechinus* complex (Armstrong et al. 1998). Interspecific competition may help to maintain the selection of a particular rate of change of photoperiod to ensure staggered demand on food resources and also reproductive and species isolation.

Genera *Murexia*, *Myoictis* and *Neophascogale*

Data for New Guinean *Antechinus* species recently assigned to *Murexia* (Armstrong et al. 1998, Krajewski et al. 2000) suggest that at least some members breed all year around, and that there

Table 1 Family Dasyuridae, subfamily Dasyurinae, tribe Phascogalini. The table below describes the known reproductive biology of the family Dasyuridae, and the source of the information. Under adult body mass the abbreviations are as follows: F = female, M = male. Under seasonal proximal cues, the abbreviations are as follows: P = photoperiod, Ph = pheromones, N = nutrition, and R = rainfall. Under reproductive strategy the abbreviations are as follows: M = monoestrous, P = polyoestrous, +1 = reproduces for more than one year, and F indicates if it is only females which reproduce for more than one year, S = seasonal, NS = non–seasonal, and D = male die–off. The reproductive strategy is also noted in that column, and follows the categorization of Krajewski *et al.* (2000). Spermatorrhoea season follows the description of Bolliger (1942) of sperm discharged into the urine of reproductively mature male marsupials.

Species	Adult body mass (g)	Gestation period (days)	Litter size	Weaning (days)	Sexual maturity (months)	Mating period	Birth season	Sperma-torrhoea season	Seasonal Proximal cues	Life span (years, months)	Repro-ductive strategy	References
Dasyuridae: Subfamily Dasyurini, tribe Phascogalini												
Antechinus adustus	19–38 (F) 23–41 (M)	25–31	6	Approx 120–140	9–10	Early July–early August	Early August–early September			1–2 years (F), 11.5 months (M)	M, +1 (F), S, D Strategy I	Watt 1997
Antechinus agilis	16–25 (F) 25–35 (M)	35	6–8	90–110	9–10	mid July (south)–late August (north)	mid August (south)–mid September (north)	Late June–September	P, Ph	1–2 years (F), 11.5 months (M)	M, +1 (F), S, D Strategy I	Selwood 1980, 1982, Woolley 1966, Scott 1987
Antechinus bellus	Up to 30 (F) Up to 61.5 (M)	28–35	Up to 10	Approx 120	9–10	August	Late September	June–early August		1–2 years (F), 11.5 months (M)	M, +1 (F), S, D Strategy I	Calaby & Taylor 1981
Antechinus flavipes	21–52 (F) 25–75 (M)	Approx. 30	8–12	90–120	9–10	Early July (south)–mid September (north)	Early August (south)–mid October (north)	Late May–September	P? Ph?	1–2 years (F), 11.5 months (M)	M, +1 (F), S, D Strategy I	Smith 1984, Woolley 1973
Antechinus godmani	49.5–62 (F) 70–120 (M)	30?	6	Approx 140	9–10	Late June–early August	Late July–early September			1–2 years (F), 11.5 months (M)	M, +1 (F), S, D Strategy I	Watt 1997
Antechinus leo	30–62 (F) 67–124 (M)		Up to 10	Approx 100–120	9–10	Mid-September–mid October	Mid October–early November	September–October		1–2 years (F), 11.5 months (M)	M, +1 (F), S, D Strategy I	Leung 1999 Van Dyck 1980
Antechinus minimus	35–60 (F) Approx 70 (M)	26–30	6–8	80–90	9–10	3 weeks between mid-June and early August	3 weeks between Mid-July–early September	February–August		1–2 years (F), 11.5 months (M)	M, +1 (F), S, D Strategy I	Wilson 1986, Wilson & Bourne 1984

Antechinus stuartii	18–25 (F) 25–40 (M)	27	6–8	90–95	9–10	Late July (south)–mid September (north)	Late August (south)–mid October (north)	Late June–September	P, Ph	1–2 (F) 11.5 months (M)	M, +1 (F), S, D Strategy I	McAllan et al. 1991, Selwood 1980, 1982, Woolley 1966
Antechinus swainsonii	37–100 (F) 43–175 (M)		8–10	90–95	9–10	May–September	June–October	April–September	P? Ph?	1–2 years (F), 11.5 months (M)	M, +1 (F), S, D Strategy I	McAllan et al. 1999, Woolley 1971c
Phascogale calura	30–48 (F) 39–50 (M)	28–30	6–8		11	3 weeks in July	3 weeks in August		P? Ph?	1–3 years (F), 1 year (M)	M, +1 (F), S, D Strategy I	Bradley 1997
Phascogale tapoatafa	110–190 (F) 175–235 (M)	24–33	4–8	Up to 175	12	May–July	June–August	May–July	P?, R?, N?	1–2 years (F), 1 year (M)	M, +1 (F), S Strategy I	Cuttle 1982, Millis et al. 1999, Soderquist 1993, Soderquist & Ealey 1994

Table 2 Plasma testosterone concentrations in male *A. stuartii* and some other male mammals. Asterisks indicate species that have male post-mating mortality as part of their life history. Some values have been recalculated from ng·mL^{-1}.

Species	Plasma testosterone concentrations	Authors
Antechinus stuartii * (brown Antechinus)	3.5 nmol·L^{-1} (May) to 28.4 nmol·L^{-1} (August)	Bradley et al. 1980
Antechinus swainsonii * (swamp Antechinus)	2-6 nmol·L^{-1} (non-breeding) 17.7 nmol·L^{-1} (August)	McDonald et al. 1981, McDonald et al. 1986
Antechinus flavipes * (yellow-footed Antechinus)	2-6 nmol·L^{-1} (non-breeding) 41.9 nmol·L^{-1} (August)	McDonald et al. 1981
Sminthopsis crassicaudata (fat-tailed dunnart)	16.3 2.4 nmol·L^{-1} (non-breeding), 26.9 nmol·L^{-1} (breeding)	McDonald et al. 1981
Dasyurus viverrinus (eastern quoll)	17.3 nmol·L^{-1} (maximum)	Bryant 1986
Didelphis albiventris (white-belly opossum)	4.2 nmol·L^{-1} (non-breeding), 35.0 nmol·L^{-1} (breeding)	de Queiroz et al. 1995
Isoodon macrourus (bandicoot)	0.4 to 24.3 nmol·L^{-1}	Gemmell et al. 1985
Trichosurus vulpecula (brush-tailed possum)	35-69 nmol·L^{-1} (maximum)	Gemmell et al. 1986
Phascolarctos cinereus (koala)	23.9 nmol·L^{-1} (maximum)	Cleva et al. 1994a
Macropus eugenii (tammar)	3 to 24 nmol·L^{-1}	Inns 1982
Macropus giganteus (grey kangaroo)	7.2 to 29.2 nmol·L^{-1}	Lincoln 1978
Macropus r. rufogriseus (Bennett's wallaby)	16 nmol·L^{-1} (maximum)	Curlewis 1991
Macropus rufus (red kangaroo)	15.4 nmol·L^{-1}	Lincoln 1978
Thylogale thetis (pademelon)	4.0-30.6 nmol·L^{-1}	Lincoln 1978
Aepyceros melampus (impala)	37.8 nmol·L^{-1}	Neaves and Bramley 1972
Lagostomus maximus maximus (vizcacha)	15.6 nmol·L^{-1}	Fuentes et al. 1993
Meriones shawi (Moroccan gerbil)	0.5 to 5.0 nmol·L^{-1}	Zaime et al. 1992
Homo sapiens (human)	23-25 nmol·L^{-1}	Orth et al. 1992
Sus scrofa (pig)	20 nmol·L^{-1}	Schwarzenberger et al. 1993
Spilogale gracilis (spotted skunk)	22.5 nmol·L^{-1}	Kaplan and Mead 1993

are few photopriodic cues similar to those found in southern Australia (Dwyer 1977, Woolley et al. 1991, Woolley this volume). Reproduction in the short-furred dasyure *Murexia longicaudata* and the three-striped dasyure *Myoictis melas* occurs all year around, even when transferred from their equatorial home to a strongly photoperiodic environment in Australia (Woolley this volume). Little is known about the speckled dasyure, *Neophascogale lorentzii* or the broad-striped dasyure *Murexia rothschildi* (Flannery 1995).

Genus *Phascogale*

Phascogales demonstrate the synchronised mating period followed by complete male mortality seen in *Antechinus* and *Dasykaluta* genera (Table 1). The male mortality observed in *P. calura* has been extensively examined by Bradley (1987, 1990, 1997, this volume).

Little is known about the environmental instigators of phascogale reproduction. In the wild and captivity, a mating season of 2–3 weeks duration occurs in May or June (Cuttle 1982, Halley 1992, Bradley 1997, Millis et al. 1999, Soderquist 1993). Phascogales have large home ranges and low population densities (Bradley 1997, Soderquist 1995), which makes determining the proximal cues difficult, although the narrow range of the mating period suggests that photoperiod may be important, as may olfactory cues (see Soderquist and Ealey 1994, Soderquist 1995). The breeding season of *P. tapoatafa* varies little from the Northern Territory to Victoria (Soderquist 1993), but because the mating period occurs when photoperiod is both the shortest and is changing the least in all the sites recorded, photoperiod cannot be discounted as a cue for reproduction. In Victoria other environmental cues are involved, as the date of reproductive activity may be up to two weeks different in adjacent years (Soderquist 1993). The reduced synchrony of oestrous cycles in phascogales (compared to *Antechinus*) has been associated with the lower population densities, which reduce the opportunity for pheromonal cues to intensify synchronised activities (Millis et al. 1999).

Genus *Phascolosorex*

There are two species of *Phascolosorex*, the uncommon *P. doriae*, and the 'moderately common' *P. dorsalis* (Flannery 1995). Small pouch young of *P. dorsalis* have been found in early July, and young adults in October (Flannery 1995). *P. dorsalis* are diurnal, and are believed to be polyoestrous, aseasonal breeders, with no male mortality (Woolley et al. 1991, Krajewski et al. 2000).

Subfamily Dasyurinae, tribe Dasyurini

Genus *Dasycercus*

The single species *Dasycercus cristicaudata* (mulgara) is found in central Australia. In the wild, mulgaras are relatively long-lived, with evidence that some males die after the mating period (Woolley 1971a, 1990a). Much of the information is from captive-bred animals, with mulgaras reproducing for up to six years (Michener 1969, Woolley 1971a; see Table 3). In captivity, two wild-caught females mated on 26 June and 23 June and gave birth approximately seven weeks later (Geiser and Masters 1994). The mulgara appears capable of entering torpor while pregnant, although during lactation no torpor occurred (Geiser and Masters 1994).

Wild females have been found with pouch young between June and September and still lactating in December (Woolley 1990a, Gibson and Cole 1992). The breeding pattern appears to be seasonal, although flexible within the seasonal range (Masters 1998). In a captive population housed in Melbourne the mating period was shorter than that of the wild population, and this may be due to the difference of 15° in latitude. Thus wild and captive studies indicate that photoperiod may be an important cue for reproduction in the mulgara, with both food availability (Masters 1998) and temperature (Geiser and Masters 1994) contributing to the timing of births.

Genus *Dasykaluta*

The genus *Dasykaluta* is also represented by one species, *D. rosamondae*, the little red kaluta. Males die soon after the mating season, although females may breed a second season (Woolley 1991a). Most information concerning reproduction comes from captive studies, although limited evidence from the wild concurs with the information from captive animals (Woolley 1991a; Table 3), where the mating period lasts the first three weeks of September, with young born approximately seven weeks later (Woolley 1991a, Table 3). Isolated, unmated females undergo pseudopregnancy, indicating a spontaneous oestrous cycle (Woolley 1991a). However, in captivity some females bred earlier than September, all but one were second-year females, and the reasons for this are unclear (Woolley 1991a). One can speculate that in the wild the time of mating coincides with the fastest time of photoperiodic change between days, as against the longest photoperiod which occurs in December, although the reproductive behaviour of some of the captive animals confounds this hypothesis.

Genus *Dasyuroides*

The kowari, *Dasyuroides byrnei*, found in central Australia, breeds between May and December, and rears one or two litters in a single season (Woolley 1971a; Table 3). Aslin (1980) concludes that *D. byrnei* are seasonal breeders, and while proximal cues are unknown, they are likely to be photoperiodic. Many breeding laboratory colonies of *D. byrnei* were maintained on the local ambient photoperiod, an important component of maintaining breeding success (Aslin 1980, Fletcher 1989, Woolley 1971a). Pheromonal signaling between males and females may also be important for the success of female reproductive cycles (Aslin 1980), although this has been disputed (Fletcher 1989).

Genus *Dasyurus*

In Australia, the genus *Dasyurus* is represented by four species *D. geoffroii,* the western quoll; *D. hallucatus,* the northern quoll; *D. maculatus,* the spotted-tailed quoll; and *D. viverrinus* the eastern quoll. Two species of *Dasyurus* live in Papua/New Guinea, the New Guinea quoll *D. albopunctatus,* and the bronze quoll *D. spartactus* (see Woolley this volume).

All four Australian species are seasonal breeders, with most breeding from May to June (Fleay 1935a, 1940, Begg 1981a, Bryant 1988, Serena and Soderquist 1988, Dickman and Braithwaite 1992, Braithwaite and Griffiths 1994, Table 3). Most litters are born in June and July, although females may replace a lost litter (Soderquist and Serena 1990, Table 3).

The mating period coincides with significant surges in both testosterone and luteinising hormone secretion in male *D. viverrinus* (Bryant 1992), and peak plasma testosterone values in male *D. hallucatus* (Schmitt et al. 1989). In contrast to the other *Dasyurus* species, in some parts of its range *D. hallucatus* can exhibit post-mating mortality (Dickman and Braithwaite 1992, Braithwaite and Griffiths 1994). Unlike the cycle of complete male mortality found in *Antechinus*, in some localities a few males can survive to a second breeding season (Schmitt et al. 1989, Begg 1981a).

While Fleay (1935a) observed one oestrous cycle per year in *D. viverrinus*, it does appear that *D. viverrinus* is polyoestrous, with females able to replace a lost first litter (Fletcher 1985). Polyoestry has not been recorded for *D. hallucatus*, even though litters may be lost (Begg 1981a). From observations on an isolated female, the oestrous cycle appears to be spontaneous in *D. maculatus* (Fleay 1940). In some localities *D. hallucatus* females have been known to not reproduce more than once, with all not surviving longer than a year (Dickman and Braithwaite 1992, Braithwaite and Griffiths 1994). The proximal cues inducing reproduction in *Dasyurus* are unknown, although for mating coincides with some of the shortest photoperiods of the year, and when photoperiodic change is the least.

Genus *Parantechinus*

The two species of *Parantechinus* – *P. apicalis* or southern dibbler, and *P bilarni*, the northern dibbler – exhibit similar reproductive activity (Table 3). The mating period is somewhat synchronised, and occurs in March in *P. apicalis* (Woolley

Table 3 Family Dasyuridae, subfamily Dasyurinae, tribe Dasyurini. Abbreviations are the same as for Table 1.

Species	Adult body mass (g)	Gestation period (days)	Litter size	Weaning (days)	Sexual maturity (months)	Mating period	Birth season	Spermatorrhoea season	Seasonal Proximal cues	Life span (years, months)	Reproductive strategy	References
Dasyuridae: Subfamily Dasyurinae, tribe Dasyurini												
Dasycercus cristicaudata	60–130 (F), 80–120 (M)	35–44	2–6	120		mid-May to mid-June	June–September	May–August	P?	<6 years	M?, +1, S, D? Strategy II	Masters 1998, Michener 1969; Woolley 1971a 1990a
Dasykaluta rosamondae	20–30 (F), 25–40 (M)	38–62	8	110	10	September	November	July to November		1–2 years (F), 1 year (M)	M?, +1 (F), D?, S Strategy I	Woolley 1991a
Dasyuroides byrnei	70–105 (F), 85–140 (M)	30–35	6	110	9	May–November	June–December	March–December	P?	<5 years	P, +1 Strategy V	Aslin 1980, Woolley 1971a
Dasyurus geoffroii	705–1285 (F) 1175–2075 (M)		4–6	Approx. 150	12	May–June (peak)	May–September			3 years	P, +1, S Strategy V	Serena & Soderquist 1988, Soderquist & Serena 1990
Dasyurus hallucatus	300–850 (F) 400–1000 (M)		6–10	Approx. 125–150	10–11 (F) 12 (M)	May–June	June–July			>2 years (F), <2 years (M)	M, +1, D, S Strategy II	Begg 1981a, Braithwaite & Griffiths 1994, Dickman & Braithwaite 1992, Schmitt et al 1989
Dasyurus maculatus	2000–4000 (F), <7000 (M)	21	4–6	150	12	June–July	July–August			3–4 years	M, +1, S Strategy III	Collins 1973, Conway 1988, Fleay 1940
Dasyurus viverrinus	700–1000 (F), 900–2000 (M)	21	6–8	135–150	12	May–June	May–August	March–August		3–4 years	P, +1, S Strategy III	Bryant 1988, Fleay 1935a, Fletcher 1985
Parantechinus apicalis	40–75 (F) 60–100 (M)	44–53			10–11	March	Late May	April–January		1–4 years (F), 1 year? 2 years? (M)	M, +1, D?, S Strategy II	Dickman & Braithwaite 1992, Woolley 1971b, 1991b
Parantechinus bilarni	12–34 (F) 12–44 (M)	31–47	4–5	Approx. 90	11	Late June–early July	August–September	May–September		1–2 years	M, +1 D?, S Strategy II	Begg 1981b, Woolley 1995a
Pseudoantechinus macdonnellensis	20–40 (F) 25–45 (M)	44–55	5–6	110	11	June–August	Late July–early September	Early May–early September		1–2 years	M, +1 Strategy II	Woolley 1991c, 1995a
Pseudoantechinus ningbing	15–20 (F) 20–25 (M)	Approx. 45–52	4	110–120	11	Late May–early June	Late July–early August	April–July		1–2 years (F), 1 year (M)	M, + 1, S Strategy II	Woolley 1988
Sarcophilus harrisii	>6000 (F) >8000 (M)	31	1–4	Approx. 250	24	March–April	April–May			7–8 years	P?, +1, S Strategy II	Guiler 1970, 1971

1991b, Dickman and Braithwaite 1992) and late June to early July in *P bilarni* (Begg 1981b, Calaby and Taylor 1981; Table 3). Females are monoestrus, and in both captivity and the wild, some males and females survive to breed a second year (Begg 1981b, Woolley 1995a). Unmated female *P. apicalis* undergo a pseudopregnancy (Woolley 1971b).

In contrast to the genus *Antechinus*, spermatogenesis in second year males has been confirmed (Begg 1981b, Woolley 1991b, 1995a). Male mortality is complete in some populations, although facultative die-off has been observed in other populations (Dickman and Braithwaite 1992, Mills and Bencini 2000). The complete and abrupt nature of the post-mating mortality may be dependent on complex interactions between population densities and habitat quality in both species, and when conditions are favourable animals may be able to breed a second year (Dickman and Braithwaite 1992, Mills and Bencini 2000).

Genus *Pseudantechinus*

The genus *Pseudantechinus* is represented by four species, all found in northern or central Australia. Little is known of the biology of *P. mimulus*, and *P woolleyae* is only known from captive observations (Woolley 1995b). *Pseudantechinus macdonnellensis* mates in June in Central Australia, and June to August in Western Australia, with young born between late July and early September (Woolley 1991c; Table 3). *Pseudantechinus ningbing* demonstrates similar breeding seasonality and characteristics as *P. macdonnellensis* (Woolley 1988, Table 3). Some males in both species can breed a second year (Woolley 1988, 1991c). The proximal factors initiating breeding are unknown, although the mating season appears to coincide with short and unchanging photoperiods associated with the year's shortest days.

Genus *Sarcophilus*

The Tasmanian devil, *Sarcophilus harrisii*, although previously found on the Australian mainland, is now restricted to Tasmania (Jones 1995). The Tasmanian devil has a synchronised mating period (Fleay 1935b, Guiler 1970, 1971, Hughes 1982, Table 3). The Tasmanian devil may not be strictly monoestrous, and may be capable of replacing lost litters until the end of September (Guiler 1970). Females are capable of pseudopregnancy (Guiler 1970). Males also appear to undergo cyclic reproductive activity (Guiler 1970). Little is known about the proximal cues that are used to synchronise reproduction in this species.

Subfamily Sminthopsinae, tribe Sminthopsini

Genus *Antechinomys*

The kultarr, *Antechinomys laniger*, is found in central Australia, where adult individuals can be captured all year round, although capture of juveniles is restricted to November and December (Woolley 1984, Table 4). In the wild, breeding occurs from at least August to December. However, in captivity, females first exhibited oestrus in July, and were capable of recurring oestrous cycles until January (Woolley 1984). Captive males followed a seasonal cycle of scrotal enlargement and spermatorrhoea (Woolley 1984, Table 4). Both sexes could reproduce for more than one year and it has been suggested that changing photoperiod is the proximate factor in the timing of reproduction (Woolley 1984).

Genus *Sminthopsis*

At present there are at least 19 described species of the genus *Sminthopsis* (Strahan 1995, Krajewski et al. 2000, Table 4). For most species the breeding season falls between July and December, although the season is shorter for some species (Fox and Whitford 1982, McKenzie and Archer 1982, Aslin 1983, Read et al. 1983, Woolley 1990b, Dickman et al. 1993). Year-round breeding activity has been found for *S. douglasi* and, contentiously, for *S. virginiae* (Taplin 1980, Morton 1987, Woolley 1995c). Nothing is known about the reproductive cycles of almost half of the described species (Krajewski et al. 2000, Table 4).

All species studied have been found to be polyoestrus, and this is believed to allow females to replace a lost litter, or, as in *S. crassicaudata* and *S. macroura*, to produce two litters in a breeding season (Morton 1978, Woolley 1990b). For some species, photoperiod has been proposed as the proximal cue for reproduction (Godfrey 1969a, b, Woolley 1990b, Woolley and Valente 1986). The strict seasonality of the reproductive cycle and modifications to the seasonal cycle length at higher latitudes suggests that photoperiod is also important for reproductive timing for other *Sminthopsis* species. Poor rainfall can curtail the reproductive season; however, severe drought does not eliminate the breeding season altogether, indicating that photoperiod is the overriding cue for reproduction (Morton 1978, Friend et al. 1997).

The influence of photoperiod on reproduction has been comprehensively studied in *S. crassicaudata*. Increased photoperiodic length induces reproductive activity. Refractoriness to 'long day' photoperiod occurs, although reproduction can again be initiated by exposure to 'short day' photoperiods, followed by returning to long photoperiods (Godfrey 1969a, Smith et al. 1978). Exposure of males to 'long day' photoperiods of 16 hours daylight and 8 hours darkness (L:D 16:8) following a regime of L:D 12:12 initiates testicular enlargement, maximising within 3-4 weeks (Holloway and Geiser 1996). Exposure to 'short day' photoperiods (L:D 8:16) following L:D 12:12 initiates testicular regression, again within 4 weeks (Holloway and Geiser 1996). The reproductive stimulation by photoperiod did not extend to use of torpor, indicating that, unlike many torpor-using mammals, seasonal torpor use and seasonal reproduction are uncoupled (Holloway and Geiser 1996).

Table 4 Family Dasyuridae, subfamily Sminthopsinae, tribes Sminthopsini and Planigalini. *Sminthopsis aitkeni, S. archeri, S. bindi, S. butleri, S. gilberti, S. granulipes, S. hirtipes, S. psammophili* and *S. youngsoni* are not included because little information on their reproduction is available. Abbreviations are the same as for Table 1.

Species	Adult body mass (g)	Gestation period (days)	Litter size	Weaning (days)	Sexual maturity (months)	Mating period	Birth season	Spermatorrhoea season	Seasonal Proximal cues	Life span (years, months)	Reproductive strategy	References
Dasyuridae: Subfamily Sminthopsinae, tribe Sminthopsini												
Antechinomys laniger	20 (F) 30 (M)	12	Up to 6	90–100	11.5	July–January	August–December	April–January	P?	1–2 years	P, S Strategy V	Woolley 1984
Sminthopsis crassicaudata	10–20	13–16	7–10	70	4 (F)	July–January	July–February	May–at least November	P	1–2 years	P, S Strategy IV	Godfrey 1969a, Morton 1978, Smith *et al.* 1978, Woolley & Watson 1984
Sminthopsis dolichura	10–17 (F) 11–17 (M)		4–8	90		July–November	August–December			1–3 years	P, S Strategy IV	Friend et al. 1997
Sminthopsis douglasi					6					1–2 years	P, S, Strategy IV	
Sminthopsis grisoventer	9.5–15 (F) 11–17 (M)		Up to 8	70	12	July	August			Up to 2.5 years	P, S Strategy IV	Crowther et al. 1999
Sminthopsis leucopus		Approx. 15	5–8			July–November?	August–December?			1–2 years	P, S Strategy III	Read et al. 1983, Woolley & Gilfillan 1990
Sminthopsis longicaudata	12–24	15–19	1–5			August–December	September–January	From August		1–2 years	P, S Strategy V	Woolley & Valente 1986
Sminthopsis macroura	15–25	12–13	Up to 8	70	3–7 (F), 7–12 (M)	June–January	June–February	Early litters: January–February; late litters: May	P	1–2 years	P, S Strategy IV	Godfrey 1969b, Woolley 1990b,c
Sminthopsis murina					6					1–2 years	P, S, Strategy IV	
Sminthopsis oldea					8–11					1–2 years	P, S Strategy V	Aslin 1983
Sminthopsis virginiae	20	13–20		90	7	June–January	June–February			1–2 years	P, S, Strategy VI	Morton 1987, Taplin 1980

Species	Adult body mass (g)	Gestation period (days)	Litter size	Weaning (days)	Sexual maturity (months)	Mating period	Birth season	Spermatorrhoea season	Seasonal Proximal cues	Life span (years, months)	Reproductive strategy	References
Ningaui ridei	6.5–13	14–21	5–7	70		Early September –January	September –February	August–March	P? R	1–2 ?	P, S Strategy V	Fanning 1982, Kitchener et al. 1983, Kitchener et al. 1986
Ningaui timealyi	2.0–9.4		5–6			"Spring"	September –December	August–March	P? R	1–2 ?	P, S Strategy V	Dunlop & Sawle 1982, Kitchener et al. 1986
Ningaui yvonneae	9.5–11.5		5–7			September	November –December	August–March	P? R	1–2 ?	P, S Strategy V	Carthew & Keynes 2000, Kitchener et al. 1983, Kitchener et al. 1986
Subfamily Dasyurinae, tribe Planigalini												
Planigale gilesi	5–9 (F) 9.5–16 (M)	15.5	3–10		6.5	August–December	Late August–January	July–March	P?	1–4 years	P, S Strategy V	Read 1984
Planigale tenuirostrius	4–7 (F) 4.5–9 (M)	19	4–9		6.5	August–January	September –January	July–March	P?	1–4 years	P, S Strategy V	Read 1984
Planigale maculata	7–15 (F) 6–22 (M)	20	5–11	70	9	All year	All year	All year	P?	2 years	P, NS Strategy V	Aslin 1975 Van Dyck 1979
Planigale ingrami	4.2–4.0 (F) 3.9–4.6 (M)		4–12	90				December–March	P?		NS Strategy V	Taylor et al. 1982

In *S. crassicaudata*, circadian activity patterns are closely associated with the photocycle and are not influenced by feeding times (Crowcroft and Godfrey 1968), similar to *Phascolosorex dorsalis* and *Murexia habbema* (Woolley et al. 1991), but contrasting to *Dasyuroides byrnei* (O'Reilly et al. 1986). Another study observed the photoperiodic entrainment patterns of *S macroura*, and found that, unlike the bandicoots *Isoodon macrourus* and *Permales nasuta*, re-entrainment to shifted photoperiods was swift and synchronous (O'Reilly et al. 1984).

Male cycles follow those of females, with spermatorrhoea observed in *S. macroura* from about May until January (Woolley 1990c, Taggart et al. 1997), although in other species spermatorrhoea could occur for longer (Woolley and Valente 1986, Woolley 1995c). Accessory reproductive (bulbourethral) glands in *S. macroura* males also follow a similar pattern, increasing in size from May, and declining after January (Woolley 1990c, Taggart et al. 1997). Scrotal width follows the seasonal reproductive cycle of females in most species studied, with maximum widths occuring at the beginning of the season, and diminishing at the end of the season (Morton 1978, Woolley and Valente 1986, Woolley 1990c, Friend et al. 1997, Taggart et al. 1997). The extended period of male reproductive maintenance is believed to be associated with the more unpredictable environment experienced by many of the genus (Morton 1978, Woolley and Valente 1986, Friend et al. 1997, Taggart et al. 1997).

In summary, for those members of the genus *Sminthopsis* that have been examined, photoperiod seems to be an important component of timing of reproduction, with proximal cues such as rainfall and temperature playing a less substantial, or even confounding role. It is suspected that photoperiod may be an important proximal cue for all *Sminthopsis*, with increasing photoperiodic length the most likely cue, as most reproductive activity slows in January, after the summer solstice, and when photoperiodic length is declining.

Genus *Ningaui*

Three *Ningaui* species are currently recognised. Most information is available for *N. ridei*, which breeds from October to January. Females are polyoestrus, apparently for replacement of lost litters, rather than for raising two litters per season (Fanning 1982, Table 4). Female *N. timealeyi* can be found with pouch young from September to March, although the breeding season is reduced when rainfall is low (Dunlop and Sawle 1982). Female *N. yvonneae* were found with pouch young in January and March (Kitchener et al. 1983). While rainfall availability affects the length of the breeding season in at least *N. timealeyi,* the shorter breeding season in the more southern species suggests that lengthening photoperiod may also influence reproductive seasonality in ningauis.

In all species male ningauis exhibit similar seasonal spermatogenic cycles (Aslin 1975, Read 1984, Kitchener et al. 1986). Reproductive maturity, indicated by spermiogenesis and increasing prostate weight, occurs in July, with no evidence of later testicular collapse (Kitchener et al. 1986). There survival of ningauis for a second breeding season has been disputed, and while there is some survivorship into the next year, the evidence for a second breeding opportunity for ningauis remains unclear (Kitchener et al. 1986).

Subfamily Sminthopsinae, tribe Planigalini

Genus *Planigale*

Of the five species, only *Planigale maculata* is not uncommon, so data on reproductive activity are patchy (Table 4). In *P. gilesi* and *P. tenuirostris* births are seasonal and both can replace lost litters in the wild (Read 1984). In contrast, *P. maculata* and *P. ingrami* produce litters throughout the year in the Northern Territory (*P. ingrami*, Taylor et al. 1982), Queensland, and in the laboratory (*P. maculata* Aslin 1975, Van Dyck 1979). Museum specimens indicate that *P. maculata* has a more confined breeding period (Van Dyck 1979). Male *P. gilesi* and *P. tenuirostris* demonstrate some seasonality in sperm production, scrotal changes, and mating behaviour in both the wild and captivity (Read 1984). Sparse information for *P. maculata* indicates patterns of male maturity may be similar to those of females (Taylor et al. 1982).

The proximal cues that affect reproduction in planigales are unknown, although photoperiod may be important for initiating spermatogenesis in males, and for the onset of the mating period in female *P. gilesi* and *P. tenuirostris* (Read 1984). Seasonal breeding patterns were observed in captivity in *P. maculata*, contrasting with data from some wild populations from the Northern Territory. The captive population was housed in Adelaide, which has a more distinct yearly photocycle than the Northern Territory (Aslin 1975, Taylor et al. 1982).

Family Thylacinidae

The thylacine or Tasmanian tiger, *Thylacinus cynocephalus,* is now most likely extinct and the only reproductive reports are from government bounty records of 1888–1909 (Guiler 1961). The 148 sub-adults are described either as 'half-grown' or 'pups', and pouch young were not recorded (Guiler 1961). 'Young' numbers peak in May to September, and folklore describes a December breeding season (Guiler 1961, Table 5). Perhaps breeding activity was undertaken during the summer months, but because of the absence of any other information, nothing more can be said of the timing of reproduction in the Tasmanian tiger.

Family Myrmecobiidae

The numbat, *Myrmecobius fasciatus,* breeds from December to April, with young born from January to April/May, and a seasonal cycle of fertility in the males (Calaby 1960, Table 5). Little is known about the proximal cues driving the seasonal cycle, although the breeding season coincides with both the dry season and the longest photoperiod in the year. The tammar wallaby has a similar original distribution to the numbat, and declining photoperiod after the summer solstice reactivates the diapausing blastocyst, with seasonal quiescence initiated before the winter solstice (Hinds and Loudon 1997). *Myrmecobius fasciatus* may be responding to similar proximal cues for reproduction as the tammar wallaby.

Family Notoryctidae

Little is known about the marsupial mole, *Notoryctes typhlops*, with information derived from historical records or scat analyses (Paltridge 1998, Pearson and Turner 2000). A single young has been found in the pouch, which has two teats (Wood-Jones 1923).

Family Didelphidae

Most reproductive studies on the Didelphidae have focused on two omnivores, *Didelphis virginiana* (Virginian opossum) and *Marmosa robinsoni* (murine opossum), and the carnivorous *Monodelphis domestica* (grey short-tailed opossum). Isolated reproductive records are available for other carnivorous species (in Tyndale-Biscoe and Renfree 1987), and sparse information for *Chironectes minimus*, *Lutreolina crassicaudata*, *Marmosa cinerea*, and *Thylamys elegans* can be found in Table 5. *Marmosa cinerea* appears to be a seasonal breeder, with animals in the south of the range breeding during the southern summer (long photoperiod and warmer temperatures). *Thylamys elegans* numbers in the wild increase in response to food availability, and reproductive activity may be seasonal (Meserve et al. 1995, Lima and Jaksic 1999). Nothing is recorded concerning the reproduction of *Lestodelphis halli* (Redford and Eisenberg 1992), except that it has 19 teats (Thomas 1929).

Monodelphis dimidiata is found in central South America. In northern Argentina, young are born between December and January, and females have one large litter (Pine et al. 1985, Table 5). Based on trapping records, adults do not survive until winter, thus this species exhibits complete semelparity (Pine et al. 1985). A similar pattern of reproduction has been deduced from museum records in *Marmosa incana* (*Marmosops incanus*; Lorini et al. 1994).

Some dasyurid marsupials are semelparous, although females may live a second year (Lee and Cockburn 1985). Pine et al. (1985) believe that Australian dasyurids are incompletely semelparous because of female life history, and only *M. dimidiata* truly exhibits semelparity. Adult male *M. dimidiata* disappear from the population in March, two months before the females, and may also exhibit a male mortality syndrome after mating (Pine et al. 1985). It is unclear if male *M. dimidiata* exhibited signs of senescence, and it is tempting to speculate that the same unusual life history pattern as *Antechinus* may occur in *M. dimidiata* (and perhaps *M. incana*), and be under similar environmental control.

The grey short-tailed opossum, *Monodelphis domestica*, has been studied in some detail. Although *M. domestica* is found widely in South America, it is suggested that reproductive timing is completely independent from environmental cues, providing food is available, and females are exposed to male pheromones (Fadem and Rayve 1985, Fadem et al. 1982, Streilein 1982). *Monodelphis domestica* can be found in the arid regions of western Brazil (≅23°–26° 30' S) although many laboratory colonies of *M. domestica* are derived from animals trapped at ≅8°–18° S (Fadem and Rayve 1985, Bergallo and Cerqueira 1994). Photoperiod at higher latitudes could be used as a cue for reproduction, and although laboratory evidence suggests that reproduction in *M. domestica* is uncoupled to environmental cues (Fadem and Rayve 1985), laboratory animals were always kept at L:D 14:10 or a 'summer' photocycle (Fadem and Rayve 1985, Kraus and Fadem 1987).

However, field data contrasts to laboratory studies (Streilein 1982, Bergallo and Cerqueira 1994). Reproduction correlates significantly with the yearly precipitation and photoperiodic change (Bergallo and Cerqueira 1994). Rodents in the same study area were found to respond directly to the irregular rainfall pattern occurring in the region; however, in contrast, *M. domestica* bred in years when there was no rainfall, and were able to breed twice in some years (Streilein 1982, Bergallo and Cerqueira 1994).

Reasonable information is available on reproduction in the grey four-eyed opossums, *Philander opossum* and *P. frenata*. *Philander opossum* has a seasonal, polyestrous reproductive cycle (Fleming 1973, Table 5), and the onset of reproductive activity in Panama (9° N) and Nicaragua (11°–15° N) coincides with the end of the dry season (Biggers 1966, Fleming 1973), which also coincides with the longest photoperiod experienced at these latitudes. In French Guiana, where daylength is only 35 minutes different between the summer and winter solstices, *P. opossum* breeds all year round. Breeding is only reduced by lower litter survival due to food restriction in July and August (Julien-Laferrière and Atramentowicz 1990). Further south (22° 26'–27° 00' S), sketchy information indicates that pouch young can be found from August to January (February in Argentina), and that reproduction is seasonal (Davis 1945, Eisenberg and Redford 1999).

It is difficult to conclude whether the proximal cue for reproduction is photoperiod or rainfall, although photoperiod is believed to be an important cue for a population from south-eastern Brazil (Cerqueira et al. 1993). *Philander opossum* may respond to photoperiod in a similar manner as *D. virginiana* which appears to use photoperiod to cue for reproductive

Table 5 Families Thylacinidae, Myrmecobiidae, Didelphidae, Microbiotheridae, and Caenolestidae. Abbreviations are the same as for Table 1.

Species	Adult body mass (g)	Gestation period (days)	Litter size	Weaning (days)	Sexual maturity (months)	Mating period	Birth season	Spermatorrhoea season	Seasonal Proximal cues	Life span (years, months)	Reproductive strategy	References
Thylacinidae												
Thylacinus cynocephalus			3–4			November –April?	January–October?				S? Strategy ?	Guiler 1961
Myrmecobiidae												
Myrmecobius fasciatus					11	December–April	January–April	October–March	P? R?	>1 year	S Strategy II	Calaby 1960, Krajewski et al. 2000
Didelphidae												
Chironectes minimus	540–790		1–5		10 (F)	November –February	December–January (peak)		P?	Up to 3 years	S	Eisenberg & Redford 1999, Marshall 1978a, Redford & Eisenberg 1992
Lutreolina crassicaudata	200–450	14	7–11								P, S	Eisenberg & Redford 1999, Hunsaker 1977, Redford & Eisenberg 1992
Marmosa cinerea	50–96		Up to 9				September –March		R?		S	Tate 1933; Eisenberg 1989
Monodelphis dimidiata	31–44 (F) 62–102 (M)		Up to 16		9–10		December–February			1 year	S, D, Strategy I	Pine et al. 1985
Monodelphis domestica	80–100 (F) 90–150 (M)	15	Up to 12	42–56	4–5	December–July * all year	December–July * all year		Ph, P, R		P, S Strategy V or Strategy VI?	Bergallo & Cerqueira 1994, Fadem & Rayne 1985, *Streilen 1982
Philander frenata	341–466.5 (F); 340–910 (M)	13–14	Up to 10	70–80	7–11 (F) 9 (M)	July–February	July–March		P	1–3 years	P, S Strategy V	Cerqueira et al. 1993, Hingst et al. 1997

Species	Adult body mass (g)	Gestation period (days)	Litter size	Weaning (days)	Sexual maturity (months)	Mating period	Birth season	Spermatorrhoea season	Seasonal Proximal cues	Life span (years, months)	Reproductive strategy	References
Philander opossum	280–570		2–7	<90	6–7 (F)	Year round	Year round (but seasonal peaks)	Year round?	P, N		P, NS (north) Strategy VI	Davis 1945, Fleming 1973, Julien–Laferrière & Atramentowicz 1990, Cerqueira et al. 1993
Thylamys elegans	18.5–41		Up to 15				September –March				P, S Strategy V?	Eisenberg & Redford 1999
Microbiotheridae												
Dromicops australia	16–42		2–4		Approx. 24	October–December	November –January				S Strategy II (similar to *Sarcophilus*)	Hershkovitz 1999, Marshall 1978
Caenolestidae												
Caenolestes obscurus	14		4				June–July?				S	Kirsch & Waller 1979
Rhynocolestes raphanurus	20.5–32						"Summer"				S	Redford & Eisenberg 1992

timing, with the breeding season becoming shorter as latitude increases. Moreover, the onset of the breeding season is June/ July in Brazil, whereas in Florida (northern hemisphere) it begins in January, indicating a photoperiodic influence (in Tyndale-Biscoe and Renfree 1987).

Captive *Philander frenata* were found to reproduce seasonally when kept under natural changing photoperiod (Hingst et al. 1997), corresponding to field data (Cerqueira et al. 1993, Table 5). Lactation inhibited the oestrous cycle (Hingst et al. 1997). Hingst et al. (1997) concluded that reproduction in *P. frenata* was controlled by changes in the photoperiod.

Family Microbiotheridae

The phyolgenetic affinities of *Dromiciops australis* (Monito del monte, or Colocolo) have been recently explored (Springer et al. 1998), and the little available reproductive information is summarised in Table 5, indicating a seasonal pattern of reproduction (Marshall 1978b, Hershkovitz 1999).

Family Caenolestidae

Six rat-opossums, *Caenolestes caniventer, C. condorensis* (northern shrew opossum), *C. fuliginosus, C. obscurus* (rat opossum), *Lestoros inca* and *Rhynocolestes raphanurus* are carnivorous (Hume 1999). Tyndale-Biscoe and Renfree (1987) reported three papers describing any breeding biology of the Caenolestids. Sadly, although more than a decade has passed since their summary, little more information is available. The paucity of information of the whole family is considered a consequence of limited access to animals, because much their preferred habitats are in remote high elevation and high latitude locations (Albuja and Patterson 1996). Scanty data for *Rhyonocolestes raphanurus* is presented in Table 5.

In one study of *C. obscurus,* enlarged teats, but no young, suggested to the authors that seasonal reproduction preceded the August study period (Kirsch and Waller 1979, Table 5). Osgood (1921) obtained one pregnant and five non-pregnant non-lactating females in February/March. The few data from *C. obscurus* are from equatorial Colombia, (0°–12° N). The seasonality inferred from the few data available is likely induced by poorer weather experienced in the middle of the calendar year in alpine Columbia.

Future directions

Carnivorous marsupials are found throughout the ranges of extant marsupials, and for some the proximal cues timing their reproductive life histories are known. The best studied are the Australian genera, *Antechinus*, and *Sminthopsis*, where photoperiodic cues are important for determining timing of reproduction, with some evidence that other proximal cues can modify the underlying response to photoperiod. While *Sminthopsis* appears to respond to the lengthening photoperiod offered in late winter through to mid-summer, *Antechinus* respond to more precise components of the lengthening photoperiod, the rate of change of photoperiod.

Unfortunately for many of the other carnivorous marsupials, little is known but the most basic reproductive biology. Moreover, for many of the South American and Papua and New Guinean carnivorous marsupials there is little or no information about reproductive biology at all. With such a paucity of information, the first step is to uncover the reproductive biology of many of the species known to be extant. This basic imperative must occur before any experimental evidence of the proximal cues for reproduction can be determined.

Moreover, while for many of the Australian and some of the South American species, basic reproductive biology has been reasonably well described, the inference of the proximal cues needed for reproductive timing has no experimental basis for many species. Clearly for all of the marsupials discussed, significant work needs to be done on determining the reproductive strategies employed by the carnivorous marsupials. Unfortunately, as little is known about some of the species of carnivorous marsupials at the beginning of the new century as was known at the beginning of the last century, and this should be addressed.

References

Albuja, V.L., & Patterson, B.D. (1996), 'A new species of northern shrew–opossum (Paucituberculata: Caenolestidae) from Cordillera del Cóndor, Ecuador', *Journal of Mammalogy,* **77**:41–53.

Armstrong, L.A., Krajewski, C., & Westerman, N. (1998), 'Phylogeny of the dasyurid marsupial genus *Antechinus* based on cytochrome–*b*, 12S–rRNA, and protamine–P1 genes', *Journal of Mammalogy,* **79**:1379–89.

Aslin, H.J. (1975), 'Reproduction in *Antechinus maculatus* Gould (Dasyuridae)', *Australian Wildlife Research,* **2**:77–80.

Aslin, H.A. (1980), 'Biology of a laboratory colony of *Dasyuroides byrnei* (Marsupialia:Dasyuridae),' *Australian Zoologist,* **20**:457–471.

Aslin, H.A. (1983), 'Reproduction in *Sminthopsis ooldea* (Marsupialia:Dasyuridae)', *Australian Mammalogy,* **6**:93–95.

Begg, R.J. (1981a), 'The small mammals of Little Nourlangie Rock, N. T. III. Ecology of *Dasyurus hallucatus*, the Northern quoll (Marsupialia:Dasyuridae)', *Australian Wildlife Research,* **8**:73–85.

Begg, R.J. (1981b), 'The small mammals of Little Nourlangie Rock, N. T. II. Ecology of *Antechinus bilarni*, the sandstone Antehinus (Marsupialia:Dasyuridae)', *Australian Wildlife Research,* **8**:57–72.

Bergallo, H.G., & Cerqueira, R. (1994), 'Reproduction and growth of the opossum *Monodelphis domestica* (Marsupialia:Dasyuridae), in northeastern Brazil', *Journal of Zoology (London),* **232**:551–63.

Biggers, J.D. (1966), 'Reproduction in male marsupials', *Symposia of the Zoological Society of London,* **15**:251–310.

Bolliger, A. (1942), 'Spermatorrhoea in marsupials, with special reference to the action of sex hormones on spermatogenesis of *Tricho-*

surus vulpecula', *Journal of the Royal Society of New South Wales*, **76:**86–92.

Bradley, A.J. (1987), 'Stress and Mortality in the red–tailed phascogale, *Phascogale calura* (Marsupialia: Dasyuridae)', *General and Comparative Endocrinology*, **67**:85–100.

Bradley, A.J. (1990), 'Failure of glucocorticoid feedback during breeding in the male red–tailed phascogale, *Phascogale calura* (Marsupialia:Dasyuridae)', *Journal of Steroid Biochemistry and Molecular Biology*, **37**:155–63.

Bradley, A.J. (1997), 'Reproduction and life history in the red–tailed phascogale, *Phascogale calura* (Marsupialia: Dasyuridae); the adaptive-stress senescence hypothesis', *Journal of Zoology (London)*, **241**:739–55.

Bradley, A.J., & Monamy, V. (1991), 'A physiological profile of male dusky Antechinuses, *Antechinus swainsonii* (Marsupialia:Dasyuridae), surviving the post–mating mortality', *Australian Mammalogy*, **14**:25–7.

Bradley, A.J., McDonald, I.R., & Lee, A.K. (1980), 'Stress and mortality in a small marsupial (*Antechinus stuartii*, Macleay)', *General and Comparative Endocrinology*, **40**:188–200.

Braithwaite, R.W. (1974), 'Behavioural changes associated with the population cycle of *Antechinus stuartii* (Marsupialia)', *Australian Journal of Zoology*, **22**:45–62.

Braithwaite, R.W. (1979), 'Social dominance and habitat utilization in *Antechinus stuartii* (Marsupialia)', *Australian Journal of Zoology*, **27**:517–28.

Braithwaite, R.W., & Griffiths, A.D. (1994), 'Demographic variation and range contraction in the northern quoll, *Dasyurus hallucatus* (Marsupialia: Dasyuridae)', *Wildlife Research*, **21**:203–17.

Bryant, S. (1988), 'Maintenance and captive breeding of the eastern quoll, *Dasyurus viverrinus*', *International Zoological Yearbook*, **27**:119–24.

Bryant, S.L. (1986), 'Seasonal variation of plasma testosterone in a wild population of male eastern quoll, *Dasyurus viverrinus* (Marsupialia: Dasyuridae)', *General and Comparative Endocrinology*, **64:**75–79.

Bryant, S.L. (1992), 'Testosterone–LH response and episodic secretion in the male marsupial *Dasyurus viverrinus*', *General and Comparative Endocrinology*, **87**:410–15.

Calaby, J.H., (1960), 'Observations on the banded Ant–eater *Myrmecobius f. fasciatus* Waterhouse (Marsupialia), with particular reference to its food habits', *Proceedings of the Zoological Society of London*, **135**:183–207.

Calaby, J.H., & Taylor, J.M. (1981), 'Reproduction in two marsupial–mice, *Antechinus bellus* and *A. bilarni* (Dasyuridae), of tropical Australia', *Journal of Mammalogy*, **62**:329–41.

Carthew, S.M, & Keynes, T. (2000), 'Small mammals in a semi–arid community with particular reference to *Ningaui yvonneae*', *Australian Mammalogy*, **22**:103–09.

Cerqueira, R., Gentile, R., Fernandez, F.A.S., & D'Andrea, P.S. (1993), 'A five year study of an assemblage of small mammals in Southeastern Brazil', *Mammalia*, **57**:507–17.

Cleva, G.M., Stone, G.M., & Dickens, R.K. (1994), 'Variation in reproductive parameters in the captive male koala (*Phascolarctos cinereus*)', *Reproduction, Fertility and Development*, **6**:713–19.

Cockburn, A., & Lazenby–Cohen, K.A. (1992), 'Use of nest trees by *Antechinus stuartii*, a semelparous lekking mammal', *Journal of Zoology, London*, **226**:657–80.

Cook, B., McDonald, I.R., & Gibson, W.R. (1978), 'Prostatic function in the brush-tailed possum, *Trichosurus vulpecula*', *Journal of Reproduction and Fertility*, **53**:369–75.

Crawford, B.A., Singh, J., Simpson, J.M., & Handelsman, D.J. (1993), 'Androgen regulation of circulating insulin-like growth factor-I during puberty in male hypogonadal mice', *Journal of Endocrinology*, **139**:57–65.

Crowcroft, P., & Godfrey, G.K. (1968), 'The daily cycle of activity in two species of *Sminthopsis* (Marsupialia: Dasyuridae)', *Journal of Animal Ecology*, **37**:63–73.

Crowther, M.S., Dickman, C.R., & Lynam, A J. (1999), '*Sminthopsis griseoventer boullangerensis* (Marsupialia: Dasyuridae), a new subspecies in the S. murina complex from Boullanger Island, Western Australia', *Australian Journal of Zoology*, **47:**215–43.

Curlewis, J.D. (1991), 'Seasonal changes in the reproductive organs and plasma and pituitary hormone content of the male Bennett's wallaby (*Macropus rufogriseus rufogriseus*)', *Journal of Zoology (London)*, **223**:223–31.

Cuttle, P. (1982), 'Life history strategy of the dasyurid marsupial *Phascogale tapoatafa*', in *Carnivorous Marsupials* (ed. M. Archer), pp. 13–22. Royal Zoological Society of New South Wales, Sydney.

Davis D.E. (1945), 'The annual cycle of plants, mosquitoes, birds, and mammals in two Brazilian rainforests', *Ecological Monographs*, **15**:243–95.

de Queiroz, G.F., Rosa, A.A.M., Ailva, E., & Nogueira, J.C. (1995), 'Testicular sperm reserve and plasma testosterone levels of the South American white–belly opossum (*Didelphis albiventris*), Marsupialia', *Mammalia*, **59**:255–61.

Dickman, C.R. (1982), 'Some ecological aspects of seasonal breeding in *Antechinus* (Dasyuridae, Marsupialia)', in *Carnivorous Marsupials* (ed. M. Archer), pp. 139–50, Royal Zoological Society of New South Wales, Sydney.

Dickman, C.R. (1985), 'Effects of photoperiod and endogenous control on timing of reproduction in the marsupial genus *Antechinus*', *Journal of Zoology (London)*, **206**:509–24

Dickman, C.R. (1986), 'An experimental manipulation of the intensity of interspecific competition: effects on a small marsupial', *Oecologia*, **70**:536–43.

Dickman, C.R., & Braithwaite, R.W. (1992) 'Postmating mortality of males in the dasyurid marsupials, *Dasyurus* and *Parantechinus*', *Journal of Mammalogy*, **73**:143–7.

Dickman, C.R., King, D.H., Adams, M., &. Baverstock, P.R. (1988), 'Electrophoretic identification of a new species of *Antechinus* (Marsupialia:Dasyuridae), in south-eastern Australia', *Australian Journal of Zoology*, **36**:455–63.

Dickman, C.R., Downey, F.J., & Predavec, M. (1993), 'The hairy-footed dunnart *Sminthopsis hirtepes* (Marsupialia Dasyuridae), in Queensland', *Australian Mammalogy*, **16**:69–72.

Dickman, C.R., Parnaby, H.E., Crowther, M.S., & King, D.H. (1998), '*Antechinus agilis* (Marsupialia: Dasyuridae), a new species from the *A. stuartii* complex in south-eastern Australia', *Australian Journal of Zoology*, **46**:1–26.

Dunlop, J.N., & Sawle, M. (1982), 'The habitat and life history of the Pilbara Ningaui, *Ningaui timealyi*', *Records of the West Australian Museum*, **10**:47–52.

Dwyer, P.D. (1977), 'Notes on *Antechinus* and *Cercartetus* (Marsupialia), in the New Guinea Highlands', *Proceedings of the Royal Society of Queensland*, **88**:69–73.

Eisenberg, J.F. (1989), *Mammals of the neotropics, the northern neotropics volume 1*, University of Chicago Press, 449 p.

Eisenberg, J.F., & Redford, K.H. (1999), *Mammals of the neotropics, the central neotropics volume 3*, University of Chicago Press, 609 p.

Fadem B.H., & Rayve, R.S. (1985), 'Characteristics of the oestrous cycle and influence of social factors in grey short-tailed opossums (*Monodelphis domestica*)', *Journal of Reproduction and Fertility*, **73**:337–42.

Fadem, B.H., Trupin, G.L., VandeBerg, J.L., Maliniak, E., & Hayssen, V. (1982), 'Care and breeding of the grey, short-tailed opossum (*Monodelphis domestica*)', *Laboratory Animal Science*, **32**:405–09.

Fanning, F.D. (1982), 'Reproduction, growth and development in *Ningaui sp.* (Dasyuridae: Marsupialia), from the Northern Territory', in *Carnivorous Marsupials*, Volume I (ed, M. Archer), pp. 23–37, Royal Society of New South Wales, Sydney, Australia.

Flannery, T.F. (1995), *Mammals of New Guinea*, Reed Books, Sydney, Australia.

Fleay, D.H. (1935a), 'Breeding of *Dasyurus viverrinus* and general observations on the species', *Journal of Mammalogy*, **16**:10–16.

Fleay, D. (1935b), 'Notes of the breeding of Tasmanian devils', *Victorian Naturalist*, **52**:100–05.

Fleay, D. (1940), 'Breeding of the tiger–cat', *Victorian Naturalist*, **56**:159–63.

Fleming, T.H. (1973), 'The reproductive cycles of three species of opossums and other mammals in the Panama canal zone', *Journal of Mammalogy*, **54**:439–55.

Fletcher, T.P. (1985), 'Aspects of reproduction in the male eastern quoll, *Dasyurus viverrinus* (Shaw), (Marsupialia: Dasyuridae), with notes on polyoestry in the female', *Australian Journal of Zoology*, **33**:101–10.

Fletcher, T.P. (1989), 'Plasma progesterone and body weight in the pregnant and non-pregnant kowari, *Dasyuroides byrnei* (Marsupialia: Dasyuridae)', *Reproduction, Fertility and Development*, **1**:65–74.

Fox, B.J., & Whitford, D. (1982), 'Polyoestry in a predictable coastal environment:reproduction, growth and development in *Sminthopsis murina* (Dasyuridae, Marsupialia)', *Carnivorous Marsupials*, Volume I (ed, M. Archer), pp. 39–48, Royal Society of New South Wales, Sydney, Australia.

Friend, G.R., Johnson, B.W., Mitchell, D.S., & Smith, G.T. (1997), 'Breeding, population dynamics and habitat relationships of *Sminthopsis dolichura* (Marsupialia:Dasyuridae), in semi-arid shrublands of Western Australia', *Wildlife Research*, **24**:245–62.

Fuentes, L.B., Calvo, J.C., Charreau, E.H., & Guzmán, J.A. (1993), 'Seasonal variations in testicular LH, FSH, and PRL receptors; *in Vitro* testosterone production; and serum testosterone concentration in adult male vizcacha (*Lagostomus maximus maximus*)', *General and Comparative Endocrinology*, **90**:133–141.

Geiser, F., & Masters, P. (1994), 'Torpor in relation to reproduction in the mulgara, *Dasycercus cristicaudata* (Dasyuridae:Marsupialia)', *Journal of Thermal Biology*, **19**:33–40.

Gemmell, R.T. (1987a), 'Influence of melatonin on the initiation of the breeding season of the marsupial bandicoot, *Isoodon macrourus*', *Journal of Reproduction and Fertility*, **79**:261–5.

Gemmell, R.T. (1987b), 'The effect of melotonin and the removal of pouch young on the seasonality of births in the marsupial possum, *Trichosurus vulpecula*', *Journal of Reproduction and Fertility*, **80**:301–07.

Gemmell, R.T., Cepon, G., & Barnes, A. (1986), 'Weekly variations in body weight and plasma testosterone concentrations in the captive male possum, *Trichosurus vulpecula*', *General and Comparative Endocrinology*, **62**:1–7.

Gemmell, R.T., Cepon, G., & Sernia, C. (1993), 'Effect of photoperiod on the breeding season of the marsupial *Trichosurus vulpecula*,' *Journal of Reproduction and Fertility*, **98**:515–20.

Gemmell, R.T., Johnston, G., & Barnes, A. (1985), 'Seasonal variations in plasma testosterone concentrations in the male marsupial bandicoot *Isoodon macrourus* in captivity', *General and Comparative Endocrinology*, **59:**184–191.

Gibson, D.F., & Cole, J.R. (1992), 'Aspects of the ecology of the mulgara, *Dasycercus cristicaudata* (Marsupialia: Dasyuridae), in the Northern Territory', *Australian Mammalogy*, **16**:105–12.

Godfrey, G.K. (1969a), 'The influence of increased photoperiod on reproduction in the dasyurid marsupial, *Sminthopsis crassicaudata*', *Journal of Mammalogy*, **50**:132–3.

Godfrey, G.K. (1969b), 'Reproduction in a laboratory colony of *Sminthopsis larapinta* (Marsupialia: Dasyuridae)', *Australian Journal of Zoology*, **17**:637–54.

Guiler, E.R. (1960), 'Breeding season of the Thylacine', *Journal of Mammalogy*, **42**:396–7.

Guiler, E.R. (1970), 'Observations on the Tasmanian devil, *Sarcophilus harrisii* (Marsupialia: Dasyuridae)', *Australian Journal of Zoology*, **18:**63–70.

Guiler, E.R. (1971), 'The Tasmanian devil in captivity', *International Zoo Yearbook*, **11**:32–3.

Halley, M. (1992), 'Maintenance & captive breeding of the brush–tailed phascogale, *Phascogale tapoatafa*', *International Zoo Yearbook*, **31:**71–8.

Handelsman, D.J. (1990), 'Pharmacology of testosterone pellet implants', in *Testosterone: Action, Deficiency, and Substitution* (eds. E. Nieschlag, & H.M. Behre), pp 136–54, Springer-Verlag, Berlin.

Harder, J.D., & Fleck, D.W. (1997), 'Reproductive ecology of new world marsupials', in *Marsupial Biology: Recent Research, New Perspectives* (eds. N. Saunders & L. Hinds), pp. 175–203, UNSW Press.

Hershkovitz, P. (1999), '*Dromicops gliroides* Thomas, 1894, last of the Microbiotheria (Marsupialia), with a review of the family Microbiotheriidae', *Fieldiana, Zoology*, **93**:1–60.

Hinds, L.A., & Loudon, A.S.I. (1997), 'Mechanisms of seasonality in marsupials:a comparative view', in *Marsupial Biology: Recent Research, New Perspectives* (eds. N. Saunders & L. Hinds), pp. 41–70, UNSW Press.

Hingst, E., D'Andrea, P.S., Santori, R., & Cerqueira, R. (1997), 'Breeding of *Philander frenata* (Didelphimorphia, Didelphidae), in captivity', *Laboratory Animals*, **32**:434–8.

Holloway, J.C., & Geiser, F. (1996), 'Reproductive status and torpor of the marsupial *Sminthopsis crassicaudata*:effect of photoperiod', *Journal of Thermal Biology*, **21**:373–80.

Hughes, R.L. (1982), 'Reproduction in the Tasmanian Devil *Sarcophilus harrisii*', in *Carnivorous Marsupials*, Volume I (ed. M. Archer), pp. 49–63, Royal Society of New South Wales, Sydney Australia.

Hume, I.D. (1999), *Marsupial Nutrition*, Cambridge University Press, 434 p.

Hunsaker, II, D. (1977), 'Ecology of New World marsupials', in *The Biology of Marsupials* (ed. D. Hunsaker II), pp. 95–156, Academic Press, New York.

Inns, R.W. (1982), 'Seasonal changes in the accessory reproductive system and plasma testosterone levels of the male tammar wallaby, *Macropus eugenii*, in the wild', *Journal of Reproduction and Fertility*, **66**:675–80.

Jones M. (1995), 'The Tasmanian devil, *Sarcophilus harrisii* (Boitard, 1841)', in *The Mammals of Australia* (ed. R. Strahan), pp. 82–4, Australian Museum/Reed books, Sydney.

Jones, R.C. (1989), 'Reproduction in male Macropodidae', in *Kangaroos, Wallabies and Rat–kangaroos* (eds. G. Grigg, P. Jarman, & I. Hume), pp. 287–305, Surrey Beatty & Sons, New South Wales, Australia.

Jones, R.C., Stone, G.M., Hinds, L.A., & Setchell, B.P. (1988), 'Distribution of 5–reductase in the epididymis of the tammar wallaby (*Macropus eugenii*), and dependence of the epididymis on systematic testosterone and luminal fluids from the testis', *Journal of Reproduction & Fertility*, **83**:779–83.

Julien-Laferrière, D., & Atramentowicz, M. (1990), 'Feeding and reproduction of three didelphid marsupials in two neotropical rainforests (French Guiana)', *Biotropica*, **22**:404–15.

Kaplan, J.B., & Mead, R.A. (1993), 'Influence of season on seminal characteristics, testis size and serum testosterone in the western spotted skunk (*Spilogale gracilis*)', *Journal of Reproduction & Fertility*, **98**:321–6.

Kerr, J.B., & Hedger, M.P. (1983), 'Spontaneous spermatogenic failure in the marsupial mouse *Antechinus stuartii* Macleay (Dasyuridae:Marsupialia)', *Australian Journal of Zoology*, **31**:445–66.

Kirsch, J.A.W. & Waller, P.F. (1979), 'Notes on the trapping and behavior of the Caenolestidae (Marsupialia)', *Journal of Mammalogy*, **60**:390–437.

Kitchener, D.J., Stoddart, J., & Henry, J. (1983), 'A taxanomic apprasial of the genus Ningaui Archer (Marsupialia:Dasyuridae), including description of a new species', *Australian Journal of Zoology*, **31**:361–79.

Kitchener D.J., Cooper, N., & Bradley, A. (1986), 'Reproduction in male *Ningaui* (Marsupialia: Dasyuridae)', *Australian Wildlife Research*, **13**:13–25.

Krajewski, C., Woolley, P.A., & Westerman, M. (2000), 'The evolution of reproductive strategies in dasyurid marsupials: implications of molecular phylogeny', *Biological Journal of the Linnean Society*, **71**:417–35.

Kraus D.B., & Fadem, B.H. (1987), 'Reproduction, development, and physiology of the gray short-tailed opossum (*Monodelphis domestica*)', *Laboratory Animal Science*, **37**:478–82.

Lee, A.K., & Cockburn, A. (1985), 'The evolutionary ecology of marsupials', Cambridge University Press, Cambridge, 274 p.

Lee, A.K., Woolley, P., & Braithwaite, R.W. (1982), 'Life history strategies of dasyurid marsupials', in *Carnivorous Marsupials* (ed. M. Archer), pp. 1–11, Royal Zoological Society of New South Wales, Sydney.

Leung, L.K-P. (1999), 'Ecology of Australian tropical forest rainforest mammals. I. The Cape York antechinus, *Antechinus leo* (Dasyuridae:Marsupialia)', *Wildlife Research*, **26**:287–306.

Lima, M., & Jaksic, F.M. (1999), 'Population dynamics of three neotropical small mammals: Time series models and the role of delayed density-dependence in population irruptions', *Australian Journal of Ecology*, **24**:25–34.

Lincoln, G. (1978), 'Plasma testosterone profiles in male macropodid marsupials', *Journal of Endocrinology*, **77**:347–51.

Lorini, M.L., De Oliveira, J.A., & Persson, V.G (1994), 'Annual age structure and reproductive patterns in *Marmosa incana* (Lund 1841), (Didelphidae, Marsupialia)', *Zeitsschrift für Saugetierkunde/ International Journal of Mammalian Biology*, **59**:65–73.

Loudon, A.S.I., & Curlewis, J.D. (1987), 'Refractoriness to melatonin and short daylengths in early seasonal quiescence in the Bennett's wallaby (*Macropus rufogriseus rufogriseus*)', *Journal of Reproduction & Fertility*, **81**:543–52.

Marshall, L.G. (1978a), '*Chironectes minimus*', *Mammalian species*, **109**:1–6.

Marshall, L.G. (1978b), '*Dromicops australis*', *Mammalian species*, **99**:1–5.

Masters, P. (1998), 'The mulgara *Dasycercus cristicaduata* (Marsupialia: Dasyuridae), at Uluru National Park, Northern Territory', *Australian Mammalogy*, **20**:403–07.

McAllan, B.M. (1998), 'The effect of testosterone and cortisol on the reproductive tract of male *Antechinus stuartii* (Marsupialia)', *Journal of Reproduction & Fertility*, **112**:199–209.

McAllan, B.M., & Dickman, C.R. (1986), 'The role of photoperiod in the timing of reproduction in the dasyurid marsupial *Antechinus stuartii*', *Oecologia*, **68**:259–64.

McAllan, B.M., Joss, J.M.P., & Firth, B.T. (1991), 'Phase delay of the natural photoperiod alters reproductive timing in the marsupial *Antechinus stuartii*', *Journal of Zoology (London)*, **225**:633–46.

McAllan, B.M., Dickman, C.R., & Crowther M.S. (1999), 'Photoperiod as a predictor of the timing of mating in the marsupial genus *Antechinus*', *Proceedings of the 45th Scientific Meeting and Annual General Meeting of the Australian Mammal Society*, Macquarie University, NSW.

McAllan, B.M., Westman, W., & Joss, J.M.P. (2001), 'The effects of oral administration of melatonin on the reproductive cycle of a small marsupial, *Antechinus stuartii*', in *Perspectives in Comparative Endocrinology: Unity and Diversity* (eds. H. Goos, R. Rastogi, H. Vaudry, & R. Pierantoni), pp. 401–07, Monduzzi Editore, Bologna.

Meserve, P.L., Yunger, J.A., Gutiérrez, J.R., Contreras, L.C., Milstead, W.B., Lang, B.K., Cramer, K.L., Herrara, S., Lagos, V.O., Silva, S.I., Tabilo, E.L., Torrealba, M.-A., & Jaksic, F.M. (1995), 'Heterogeneous responses of small mammals to an ElNiño southern oscillation event in the northcentral semiarid Chile and the importance of ecological scale', *Journal of Mammalogy*, **76**:580–95.

McConnell, S.J. (1986), 'Seasonal changes in the circadian plasma melatonin profile of the tammar, *Macropus eugenii*', *Journal of Pineal Research*, **3**:119–25.

McConnell, S.J., & Hinds, L.A. (1985), 'Effect of pinealectomy on plasma melatonin, prolactin and progesterone concentrations during seasonal reproductive quiescence in the tammar, *Macropus eugenii*', *Journal of Reproduction & Fertility*, **75**:433–40.

McConnell, S.J., & Tyndale-Biscoe, C.H. (1985), 'Response in peripheral plasma melatonin to photoperiod change and the effects of exogenous melatonin on seasonal quiescence in the tammar, *Macropus eugenii*', *Journal of Reproduction & Fertility*, **73**:529–38.

McConnell, S.J., Hinds, L.A., & Tyndale-Biscoe, C.H. (1986), 'Change in duration of elevated concentrations of melatonin is the major factor in photoperiod response of the tammar, *Macropus eugenii*', *Journal of Reproduction & Fertility,* **77**:623–32.

McDonald, I.R., Lee, A.K., Bradley, A.J., & Than, K.A. (1981), 'Endocrine changes in dasyurid marsupials with differing mortality patterns', *General & Comparative Endocrinology,* **44**:292–301.

McDonald, I.R., Lee, A.K., Than, K.A., & Martin, R.W. (1986), 'Failure of glucocorticoid feedback in males of a population of small marsupials (*Antechinus swainsonii*), during the period of mating', *Journal of Endocrinology,* **108**:63–8.

McKenzie, N.L., & Archer, M. (1982), '*Sminthopsis youngsoni* (Marsupialia:Dasyuridae), the lesser hairy-footed dunnart, a new species from arid Australia', *Australian Mammalogy,* **5**:267–79.

Michener, G.R. (1969), 'Notes on the breeding and young of the crest-tailed marsupial mouse, *Dasycercus cristicaudata*', *Journal of Mammalogy,* **50**:633–4.

Millis, A.L., Taggart, D.A., Bradley, A.J., Phelan, J., &. Temple-Smith, P.D. (1999), 'Reproductive biology of the brush-tailed phascogale, *Phascogale tapoatafa* (Marsupialia: Dasyuridae)', *Journal of Zoology (London),* **248**:325–35.

Mills, H.R., & Bencini, R. (2000), 'New evidence for facultative male die-off in island populations of dibblers, *Parantechinus apicalis*', *Australian Journal of Zoology,* **48**:501–10.

Morton S.R. (1978), 'An ecological study of *Sminthopsis crassicaudata* (Marsupialia: Dasyuridae), III. Reproduction and life history', *Australian Wildlife Research,* **5**:183–211.

Morton, S.R. (1987), 'The breeding season of *Sminthopsis virginiae* (Marsupialia: Dasyuridae), in the Northern Territory', *Australian Mammology,* **10**:41–2.

Neaves, W.B., & Bramley, P.S. (1972), 'Relationship between blood levels of testosterone and accessory sex gland weights in impala *(Aepyceros melampus)*', *Comparative Biochemistry & Physiology,* **42A**:983–7.

Nowak, R.M., & Paradiso, J.L. (1983), *Walker's Mammals of the World. 4th Ed. Vol. I*, pp. 1–568, *Vol. II*, pp. 569–1362.

O'Reilly, H.M., Armstrong, S.M., & Coleman, C.J. (1984), 'Response to variations in lighting schedules on the circadian activity rhthyms of *Sminthopsis macroura froggatti* (Marsupialia: Dasyuridae)', *Australian Mammalogy,* **7**:89–99.

O'Reilly, H.M., Armstrong, S.M. & Coleman, C.J. (1986), 'Restricted feeding and circadian activity rhthyms of a predatory marsupial, *Dasyuroides byrnei*', *Physiology & Behaviour,* **38**:471–6.

Osgood, W.H. (1921), 'A monographic study of the American marsupial, *Caenolestes*', *Field Museum of Natural History Zoological Series,* **14**:1–162.

Paltridge, R. (1998), 'Occurrence of the marsupial mole (*Notoryctes typhlops*), remains in the faecal pellets of cats, foxes and dingoes in the Tanami desert, N.T.', *Australian Mammology,* **20**:427–9.

Pearson, D.J., & Turner, J. (2000), 'Marsupial moles pop up in the Great Victorian and Gibson Deserts', *Australian Mammalogy,* **22**:115–19.

Pine, R.H., Dalby, P.L., & Matson, J.O. (1985), 'Ecology, postnatal development, morphometrics, and taxanomic status of the short-tailed opossum, *Monodelphis dimidiata*, an apparently semelparous annual marsupial', *Annals of Carnegie Museum,* **54**:195–231.

Read, D.G. (1984), 'Reproduction and breeding season of *Planigale gilesi* and *P. tenuirostris* (Marsupialia: Dasyuridae)', *Australian Mammalogy,* **7**:161–73.

Read, D.G., Fox, B.J., & Whitford, D. (1983), 'Notes on breeding in *Sminthopsis* (Marsupialia: Dasyuridae)', *Australian Mammalogy,* **6**:89–92.

Redford, K.H., & Eisenberg, J.F. (1992), *Mammals of the neotropics, the southern zone volume 2,* University of Chicago Press, 430 p.

Reiter R J. (1993), 'The melatonin rhythm – both a clock and a calendar', *Experientia,* **49**:654–64.

Renfree, M.B. (1983), 'Marsupial reproduction: the choice between placentation and lactation', in *Oxford Reviews of Reproductive Biology 5*, (ed. C.A. Finn), pp. 1–29, Clarendon Press, Oxford.

Renfree, M.B., Lincoln, D.W., Almeida, O.F.X., & Short, R.V. (1981), 'Abolition of seasonal embryonic diapause in a wallaby by pineal denervation', *Nature,* **293**:138–9.

Robinson, A.C. (1995), 'Kangaroo Island Dunnart, *Sminthopsis aitkeni* (Kitchener, Stoddart & Henry, 1984)', in *The Mammals of Australia* (ed. R. Strahan), pp. 123–4, Australian Museum/Reed books, Sydney.

Schmitt, L.H., Bradley, A.J., Kemper, C.M., Kitchener, D.J., Humphreys, W.F., & How, R.A. (1989), 'Ecology and physiology of the northern quoll, *Dasyurus hallucatus* (Marsupialia:Dasyuridae), at Mitchell Plateau, Kimberley, Western Australia', *Journal of Zoology (London),* **217**:539–58.

Schwarzenberger, F., Toole, G.S., Christie, H.L., & Raeside, J.I. (1993), 'Plasma levels of several androgens and estrogens from birth to puberty in male domestic pigs', *Acta Endocrinologica,* **128**:173–7.

Scott, M.P. (1986), 'The timing and synchrony of seasonal breeding in the marsupial, *Antechinus stuartii*: interaction of environmental cues', *Journal of Mammalogy,* **67**:551–60.

Scott, M.P. (1987), 'The effect of mating and agonistic experience on adrenal function and mortality of male *Antechinus stuartii* (Marsupialia)', *Journal of Mammalogy,* **68**:479–86.

Selwood, L. (1980), 'A timetable of embryonic development of the dasyurid marsupial *Antechinus stuartii* (Macleay)', *Australian Journal of Zoology,* **28**:649–68.

Selwood, L. (1982), 'Brown Antechinus *Antechinus stuartii*:management of breeding colonies to obtain embryonic material and pouch young', in *The Management of Australian Mammals in Captivity* (ed. D.D. Evans), pp. 31–7, Zoological Board of Victoria, Melbourne.

Selwood, L. (1985), 'Synchronisation of oestrus, ovulation and birth in female *Antechinus stuartii* (Marsupialia: Dasyuridae)', *Australian Mammalogy,* **8**:91–6.

Serena, M., & Soderquist, T.R. (1988), 'Growth and development of pouch young of wild and captive *Dasyurus geoffroii* (Marsupialia: Dasyuridae)', *Australian Journal of Zoology,* **36**:533–43.

Sharman G.B. (1970), 'Reproductive physiology of marsupials', *Science,* **167**:1221–8.

Shaw, G., Renfree, M.B., Leihy, M.W., Shackleton, C.H.L., Roitman, E., & Wilson, J.D. (2000), 'Prostate formation in a marsupial is mediated by the testicular androgen 5 _–androstane–3 _, 17 _–diol',. *Proceedings of the National Academy of Sciences of the United States of America,* **97**:12256–9.

Smith, G.C. (1984), 'The Biology of the yellow–footed Antechinus, *Antechinus flavipes* (Marsupialia: Dasyuridae), in a swamp forest of Kinaba Island, Cooloola, Queensland', *Australian Wildlife Research,* **11**:465–80.

Smith, M.J., Bennett, J.H., & Cheeson, C.M. (1978), 'Photoperiod and some other factors affecting reproduction in female *Sminthopsis crassicaudata* (Gould), (Marsupialia: Dasyuridae), in captivity', *Australian Journal of Zoology,* **20**:449–63.

Soderquist, T.R. (1993), 'Maternal strategies of *Phascogale tapoatafa* (Marsupialia: Dasyuridae), I. Breeding seasonality and maternal investment', *Australian Journal of Zoology,* **41**:549–66.

Soderquist, T.R. (1995), 'Spatial organization of the arboreal carnivorous marsupial *Phascogale tapoatafa*', *Journal of Zoology (London),* **237**:385–98.

Soderquist, T.R., & Ealey, L. (1994), 'Social Interactions and mating strategies of a solitary carnivorous marsupial, *Phascogale tapoatafa*, in the wild', *Wildlife Research,* **21**:527–42.

Soderquist, T.R., & Serena, M. (1990), 'Occurence and outcome of polyoestry in wild western quolls, *Dasyurus geoffroii* (Marsupialia: Dasyuridae)', *Australian Mammalogy,* **13**:205–8.

Springer, M.S., Westerman, M., Kavanagh, J.R., Burk, A., Woodburne, M.O., Kao, D.J., & Krajewski, C. (1998), 'The origin of the Australasian marsupial fauna and the phylogenetic affinities of the enigmatic minto del monte and marsupial mole', *Proceedings of the Royal Society of London Series B:Biological Sciences,* **265**:2381–6.

Strahan, R. (1995), *The Mammals of Australia,* Australian Museum/ Reed books, Sydney.

Streilein, K.E. (1982), 'The ecology of small mammals in the semiarid Brazilian Caatinga. III. Reproductive biology and population ecology', *Annals of Carnegie Museum,* **51**:251–69.

Taggart, D.A., Johnson, J., & Temple-Smith, P.D. (1993), 'Testicular and epididymal development in the brown marsupial mouse, *Antechinus stuartii* (Dasyuridae, Marsupialia)', *Anatomy & Embryology,* **188**:87–100.

Taggart, D.A., Selwood, L., & Temple-Smith, P.D. (1997), 'Sperm production, storage, and the synchronization of male and female reproductive cycles in the iteroparous, stripe-faced dunnart (*Sminthopsis macroura*; Marsupialia): relationship to reproductive strategies within the Dasyuridae', *Journal of Zoology (London),* **243**:725–36.

Taplin L.E. (1980), 'Some observations on the reproductive biology of *Sminthopsis virginiae* (Tarragon), (Marsupialia:Dasyuridae)', *Australian Zoologist,* **20**:407–18.

Tate G.H.H. (1933), 'A systematic revision of the marsupial genus *Marmosa*, with a discussion of the adaptive radiation of the murine opossums (*Marmosa*)', *Bulletin of the American Museum of Natural History,* **66**:1–248.

Taylor, J.M., & Horner, B.E. (1970), 'Gonadal activity in the marsupial mouse, *Antechinus bellus*, with notes on other species of the genus (Marsupialia: Dasyuridae)', *Journal of Mammalogy,* **51**:659–68.

Taylor, J.M., Calaby, J.H., & Redhead, T.D. (1982), 'Breeding in wild populations of the marsupial-mouse *Planigale maculata sinualis* (Dasyuridae, Marsupialia)', in *Carnivorous Marsupials*, Volume I, (ed. M. Archer), pp. 83–87, Royal Society of New South Wales, Sydney Australia.

Thomas, O. (1929), 'On the Mammals from Patagonia: II The mammals from Señor Budin's Patagonian expedition, 1927–28', *Annals of the Museaum of Natural History 10th Series,* **4**:35–45.

Todhunter, R., & Gemmell, R.T. (1987), 'Seasonal changes in the reproductive tract of the male marsupial bandicoot, *Isoodon macrourus*', *Journal of Anatomy,* **154**:173–86.

Tyndale-Biscoe, C.H. (1980), 'Photoperiod and the control of reproduction in marsupials', in *Endocrinology 1980: Proceedings of the Sixth International Congress of Endocrinology* (eds. I.A. Cumming, J.W. Funder, & F.A.O. Mendelsohn), pp. 277–282, Melbourne, Australia.

Tyndale-Biscoe, C.H. (1986), 'Embryonic diapause in a marsupial:roles of the corpus luteum and pituitary in its control', in *Comparative Endocrinology: Developments and Directions*, pp. 137–155, Alan R. Liss Incorporated.

Tyndale-Biscoe, H., & Renfree, M.B. (1987), *Reproductive Physiology of Marsupials*, Cambridge University Press, Cambridge, 476 p.

Van Dyck, S. (1979), 'Behaviour in captive individuals of the dasyurid marsupial *Planigale maculata* (Gould 1851)', *Memoirs of the Queensland Museum,* **19**:413–29.

Van Dyck, S. (1980), 'The cinnamon Antechinus, *Antechinus leo* (Marsupialia:Dasyuridae), a new species from the vine-forests of Cape York Peninsula', *Australian Mammalogy,* **3**:5–17.

Weinbauer, G.F., & Neischlag, E. (1990), 'The role of testosterone in spermatogenesis', in *Testosterone: Action, Deficiency, Substitution*, (eds. E. Nieschlag, & H.M. Behre), pp. 23–50, Springer-Verlag, Berlin.

Watt, A. (1997), 'Population ecology and reproductive seasonality in three species of *Antechinus* (Marsupialia: Dasyuridae), in the wet tropics of Queensland', *Wildlife Research,* **24**:531–47.

Williams, I.M., Lincoln, G.A., Mercer, J.G., Barrett, P., Morgan, P.J., & Clarke, J.J. (1997), 'Melatonin receptors in the brain and pitutary gland of hypothalamo-pituitary disconnected rams', *Journal of Neuroendocrinology,* **9**:639–43.

Wilson, B.A. (1986), 'Reproduction in the female dasyurid *Antechinus minimus maritimus* (Marsupialia: Dasyuridae)', *Australian Journal of Zoology,* **34**:189–97.

Wilson, B.A., & Bourne, A.R. (1984), 'Reproduction in the male dasyurid *Antechinus minimus maritimus* (Marsupialia: Dasyuridae)', *Australian Journal of Zoology,* **32**:311–18.

Wood, D.H. (1970), 'An ecological study of *Antechinus stuartii* (Marsupialia), in a south-east Queensland rain forest', *Australian Journal of Zoology,* **18**:185–207.

Wood–Jones, F. (1923), *The Mammals of South Australia, Part I Containing the Monotremes and the Carnivorous Marsupials*. 131 p., South Australian Museum.

Woolley, P. (1966), 'Reproduction in *Antechinus* spp. and other dasyurid marsupials', *Symposia of the Zoological Society of London,* **15**:281–94.

Woolley, P.A. (1971a), 'Maintenance of breeding of laboratory colonies:*Dasyuroides byrnei* and *Dasycercus cristicaudata*', *International Zoological Yearbook,* **11**:351–5.

Woolley, P.A. (1971b), 'Observations on the reproductive biology of the dibbler, *Antechinus apicalis* (Marsupialia: Dasyuridae)', *Journal of the Royal Society of Western Australia,* **54**:99–102.

Woolley, P.A. (1984), 'Reproduction in *Antechinomys laniger* ('spenceri' form), (Marsupialia: Dasyuridae): field and laboratory investigations', *Australian Wildlife Research,* **11**:481–9.

Woolley, P.A. (1988), 'Reproduction in the ningbing antechinus (Marsupialia: Dasyuridae): Field and laboratory observations', *Australian Wildlife Research,* **15**:149–56.

Woolley P.A. (1990a), 'Mulgaras, *Dasycercus cristicaudata* (Marsupialia:Dasyuridae: their burrows, and records of attempts to collect live animals between 1966–1979', *Australian Mammalogy,* **13**:61–4.

Woolley P.A. (1990b), 'Reproduction in *Sminthopsis macroura* (Marsupialia: Dasyuridae): I. The female', *Australian Journal of Zoology,* **38**:187–205.

Woolley P.A. (1990c), 'Reproduction in *Sminthopsis macroura* (Marsupialia: Dasyuridae): II. The male', *Australian Journal of Zoology,* **38**:207–17.

Woolley, P.A. (1991a), 'Reproduction in *Dasykaluta rosamondae* (Marsupialia: Dasyuridae): Field and laboratory observations', *Australian Journal of Zoology,* **39**:549–68.

Woolley, P.A. (1991b), 'Reproductive pattern of captive Boullanger Island dibblers, *Parantechinus apicalis* (Marsupialia: Dasyuridae)', *Wildlife Research,* **18**:157–63.

Woolley, P.A. (1991c), 'Reproduction in *Pseudantechinus macdonnellensis* (Marsupialia: Dasyuridae): Field and laboratory observations', *Wildlife Research,* **18**:13–25.

Woolley, P.A. (1995a), 'Observations on reproduction in captive *Parantechinus bilarni* (Marsupialia: Dasyuridae)', *Australian Mammalogy,* **18**:83–5.

Woolley, P.A. (1995b), 'Woolley's Pseudantechinus, *Pseudantechinus woolleyae* (Kitchener & Caputi 1988)', in *The Mammals of Australia* (ed. R. Strahan), pp. 123–124, Australian Museum/Reed books, Sydney.

Woolley, P.A. (1995c), 'Julia Creek dunnart *Sminthopsis douglasi* (Archer 1979)', in *The Mammals of Australia,* (ed. R. Strahan), Australian Museum/Reed books, Sydney, 156 p.

Woolley, P.A. (1995d), 'Butler's dunnart *Sminthopsis butleri* (Arcger 1971)', in *The Mammals of Australia* (ed. R. Strahan), pp. 123–124, Australian Museum/Reed books, Sydney.

Woolley, P.A., & Gilfillan, S.L. (1990), 'Confirmation of polyoestry in captive white-footed dunnarts, *Sminthopsis leucopus*', *Australian Mammalogy,* **14:**137–8.

Woolley P.A., Raftopoulos, S.A., Coleman, G.J., & Armstrong, S.M. (1991), 'A comparative study of circadian activity patterns of two New Guinean dasyurid marsupials, *Phascolosorex dorsalis* and *Antechinus habbema*', *Australian Journal of Zoology,* **39**:661–71.

Woolley P.A., & Valente, A. (1986), 'Reproduction in *Sminthopsis longicaudata* (Marsupialia: Dasyuridae): Laboratory observations', *Australian Wildlife Research,* **13**:7–12.

Woolley, P.A., & Watson, M.R. (1984), 'Observations on a captive outdoor breeding colony of a small dasyurid marsupial, *Sminthopsis crassicaudata*', *Australian Wildlife Research,* **11**:249–54.

Zaime, A., Laraki, M., Gautier, J.-Y., & Garnier, D.H. (1992), 'Seasonal variations of androgens and of several sexual parameters in male *Meriones shawi* in southern Morocco', *General & Comparative Endocrinology,* **86**:289–96.

CHAPTER 11

REPRODUCTIVE BIOLOGY OF SOME DASYURID MARSUPIALS OF NEW GUINEA

P.A. Woolley

Department of Zoology, La Trobe University, Bundoora, Victoria 3086, Australia

Reproductive data has been obtained for seven species of dasyurid marsupials endemic to New Guinea, with the aim of establishing the pattern of reproduction of each. Observations have been made on animals in the field as well as in captivity, and it appears that all seven are capable of year-round breeding. This contrasts with what is known for the Australian species, which, with two exceptions, are strictly seasonal breeders. Aspects of the reproductive anatomy and behaviour of males and females of each species have been documented.

INTRODUCTION

Marsupials of the Family Dasyuridae are found only in the Australasian region – in Australia, New Guinea and other smaller islands lying on the continental shelf. Seventeen of the 70 or so species recognised at the present time occur in the New Guinean region and 15 of the 17 are endemic. While most species of New Guinean dasyurids are well represented in museum collections, it is only in the last 20 years that an attempt has been made to collect animals for study alive in captivity (Woolley 1993). The primary objective was to study their reproductive biology to allow comparison with Australian dasyurids, and the results of observations made on seven endemic species are reported here. Less detailed observations have been made on another four (Woolley 2001). In the course of collecting the animals, information was obtained on breeding in the wild, which supplements data on seasonality of breeding obtained from museum specimens (Woolley 1994).

Six of the seven species studied (*viz* 'Antechinus' habbema, 'Antechinus' melanurus, 'Antechinus' naso, Murexia longicaudata, Myoictis melas and Phascolosorex dorsalis) were collected in the vicinity of Wau (7°22' S, 146°40' E) in Morobe Province, and one (Murexia rothschildi) near Agaun (9°55' S, 149°23' E) in Milne Bay Province, Papua New Guinea. It is now generally recognised that the New Guinean antechinuses are not closely related to members of the genus Antechinus in Australia (see Woolley 1982, 1984 and a discussion of the names applied in Woolley 1989, p 699). Their generic identity has undergone revision (Van Dyck 2002) but for convenience of association of the results presented here with previously published work on the same animals (e.g. Woolley 1984, 1989; Westerman and Woolley 1990; Woolley et al. 1991; Woolley and Valente 1992; Woolley 1993; Westerman and Woolley 1993; Krajewski et al. 1993; Woolley 1994; Krajewski et al. 1996, 1997 and Armstrong et al. 1998) they are referred to as 'Antechinus'. In their paper Armstrong et al. assign the New Guinean antechinuses to the genus Murexia, and this classification has been followed by Krajewski et al. (2000).

MATERIALS AND METHODS

Collection and maintenance of animals

Methods used for the collection of animals, and their maintenance in captivity, are given in detail by Woolley (1993). Trapping in the vicinity of Wau, where six species were obtained, was carried out on 300 days over four periods between September 1981 and March 1983 (Woolley 1993, Table 1). During this time the majority of the animals kept in captivity were collected. Additional specimens were obtained in the course of a study by G. Grossek on the diet of five of the six species (Grossek 1987). He trapped almost continuously between January 1984 and March 1985 and most of the animals captured were marked and released. A small number that were recaptured provided sequential data on reproduction in field animals. The *Murexia rothschildi* were collected near Agaun in February 1986.

Observations

On the day of capture, all individuals were weighed and an attempt made to collect a urine sample to determine if males were showing spermatorrhoea (an indicator of maturity) and if females were in oestrus, as evidenced by the presence of epithelial cells and/or spermatozoa. The width of the scrotum was measured, the appearance of the pouch area (size, hairiness, stained or unstained hair, size of the nipples and mammary tissue, presence of young) and the sternal area (bare glandular area and/or staining of fur) noted. Reproductive tissues were removed from animals that were either found dead in the trap, or died soon after capture. The length and greatest width of the right testis and the prostate (if developed) were measured and the number of pairs and size of the Cowper's (bulbo-urethral) glands noted. Ovaries were examined for the presence of enlarged follicles or corpora lutea, the size of the uteri (greatest width) measured and the appearance of the vaginal complex noted. Uteri that appeared to be enlarged were opened and examined for the presence of eggs or embryos. Tissues were fixed in aqueous Bouin's solution for later histological examination. Physical dimensions (total length, tail length and pes length) were obtained and the carcasses preserved.

The reproductive condition of captive animals was monitored at regular intervals as described for other species of dasyurid marsupials (e.g. *Sminthopsis macroura* Woolley 1990 a, b) to obtain basic information on the pattern of reproduction of each species. Males were examined at weekly intervals. They were weighed, the width of the scrotum measured and a urine sample collected to determine the onset and duration of spermatorrhoea. Females were examined more frequently. They were weighed, the pouch inspected and a urine sample collected to detect the occurrence of epithelial cells indicative of oestrus and establish the duration of pregnancy (or pseudopregnancy). Physical dimensions (see above) were obtained from all animals at the end of the period in captivity, and observations were made on the gross and histological appearance of reproductive organs removed from animals at known stages of the reproductive cycle. Representative sections of the testis, and serial sections of one or both ovaries from each animal were prepared. All sections were cut at 8 μm and stained with haematoxylin and eosin. One diameter of each Graafian follicle, its oocyte and each corpus luteum was determined by counting the number of sections in which it appeared and multiplying this number by eight. The mean diameter of Graafian follicles, etc. in each pair of ovaries was then calculated and all references to size are means for the pair.

Skulls were removed from field specimens that died in the trap or soon after capture, and from some of the animals maintained in captivity to determine the stage of tooth eruption used to assign animals to an age class (see Woolley 1994). Account was also taken of the premolar teeth. All teeth had erupted in adult animals. In juveniles M^4 had not erupted and the deciduous third premolars were in place. None of the animals examined clearly fell into a sub-adult category (M^4 not fully erupted, dp^3_3 lost and p^3_3 unerupted or not fully erupted.

All data collected from every individual at capture, together with terminal data, was collated and the age (juvenile or adult) and reproductive status (sexually immature or mature) of each animal assessed by consideration of all available information. In the case of animals that were released without examination of their teeth or reproductive organs assessment was done by comparison of their body weight, physical dimensions and scrotal size or pouch appearance with those of established age and reproductive status.

RESULTS AND DISCUSSION

Animals examined

Observations were made on 354 trapped animals (296 individuals) and 139 captive animals (Table 1). Some of the wild-caught individuals that were released were recaptured up to a maximum of four times for one male. Of the captive animals, 81 were wild-caught, 40 were laboratory-reared, i.e. either born or conceived in the wild, and 18 were laboratory-bred. Thirty-one of the laboratory-reared and laboratory-bred animals did not survive to weaning, and their sex was not established.

The body weight (Table 2) of sexually mature wild-caught males and females is given as an indication of the size of each species. The large variation in body weight of both males and females suggests that they may continue to grow throughout life. Sexual dimorphism, with males being heavier than females, is apparent in all species except *Murexia rothschildi*, of which only one mature male and female was caught. Captive males of

Table 1 Number of individuals and number of recaptures of each species trapped, and the number of wild-caught, laboratory-reared and laboratory-bred animals maintained in captivity. The laboratory-reared animals were either born or conceived in the wild. M = males, F = females, NW = not weaned

Species	Number trapped		Number maintained in captivity		
	Individuals	Recaptures	Wild-caught	Laboratory-reared	Laboratory-bred
	Total (M/F)	Total (M/F)	Total (M/F)	Total (M/F/NW)	Total (M/F/NW)
'Antechinus' habbema	75 (35/40)	16 (6/4)	18 (7/11)	8 (–/–/8)	–
'Antechinus' melanurus	33 (18/15)	4 (2/1)	16 (6/10)	12 (6/4/2)	–
'Antechinus' naso	36 (24/12)	1 (1/0)	7 (5/2)	4 (–/–/4)	–
Murexia longicaudata	105 (56/49)	35 (17/8)	18 (10/8)	8 (2/3/3)	6 (0/0/6)
Murexia rothschildi	8 (3/5)	–	7 (2/5)	–	4 (3/1/–)
Myoictis melas	7 (2/5)	–	4 (1/3)	–	–
Phascolosorex dorsalis	32 (17/15)	2 (1/1)	11 (5/6)	8 (0/4/4)	8 (2/2/4)

this species, however, reached adult body weights greater than those of females. Such sexual dimorphism is usual among dasyurid marsupials (Lee et al. 1982).

Observations on seasonality of breeding in field animals

Figures 1–7 show the number of males and females in each age class (juvenile or adult), and the number of females known to be in oestrus, pregnant or lactating by month of collection for each of the seven species studied. The crown–rump length of the young of lactating females that were carrying young in the pouch is given. Individuals classed as juvenile were all sexually immature and some of those classed as adult were also sexually immature, i.e. they showed no signs of having come into breeding condition. They are distinguished from the sexually mature adults in the figures, which are presented in the same format as those in Woolley (1994) for ease of comparison of data obtained from field animals and museum specimens. Some individuals of all species except *Murexia rothschildi* and *Myoictis melas* were captured on more than one occasion and data for animals that were recaptured at intervals of four weeks or more are included. In some cases a change in reproductive status occurred between captures.

On the evidence presented it appears that *'Antechinus' habbema*, *'Antechinus' melanurus*, *'Antechinus' naso*, *Murexia longicaudata* and *Phascolosorex dorsalis* may breed at any time of year. The incidence of lactating females, often at different stages of lactation, in most months when adult females were captured, together with the incidence of juveniles in the population supports this conclusion. Too few specimens of *Murexia rothschildi* and *Myoictis melas* were trapped to draw any conclusions about their seasonality (or lack of it). When these results, derived from animals collected in only one location for each species, are combined with data from museum specimens that encompass the entire range of the species (Woolley 1994), and from a field study by Dwyer (1977), it seems likely that all seven species breed throughout the year (Table 3). The incidence by month of pregnant and lactating females is extended from five to ten for *'Antechinus' habbema*; from six to seven for *'Antechinus' melanurus*; from six to eight for *'Antechinus' naso*; from seven to eight for *Murexia longicaudata*; from one to three for *Murexia rothschildi*, and from five to seven for *Phascolosorex dorsalis*. In

Table 2 Body weight of sexually mature wild–caught males and females of each species.

Species	Body weight (g)					
	Males			Females		
	Range	Mean	n	Range	Mean	n
'Antechinus' habbema	30–44	36	23	22–33	27	32
'Antechinus' melanurus	33–61	49	14	32–44	36	12
'Antechinus' naso	42–74	52	8	29–40	37	8
Murexia longicaudata	61–145	102	27	39–65	54	30
Murexia rothschildi	55	–	1	60	–	1
Myoictis melas	110	–	1	61–83	72	3
Phascolosorex dorsalis	43–73	53	14	36–54	42	10

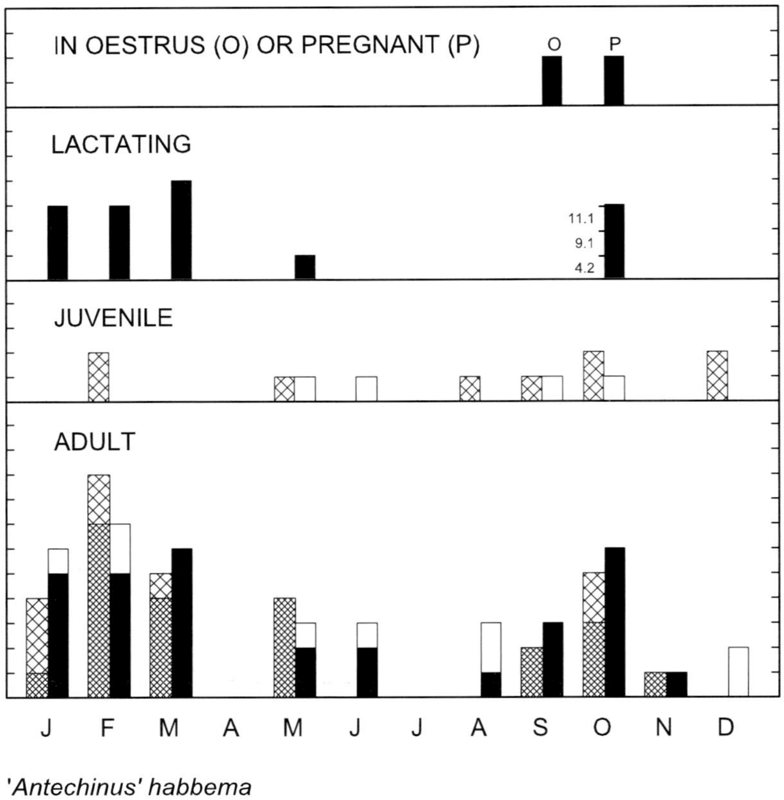

Figures 1–7 Distribution by month of collection of juvenile and adult males and females, and of oestrous, pregnant or lactating females of each species. The crown–rump length of the young is shown beside those lactating females that were carrying young in the pouch. One division on the vertical axis equals one individual. Mature males = fine hatched bars. Immature males = coarse hatched bars. Mature females = black bars. Immature females = white bars

Figure 1 *'Antechinus' habbema* n = 83 (includes single recapture data for each of 4 males and 4 females).

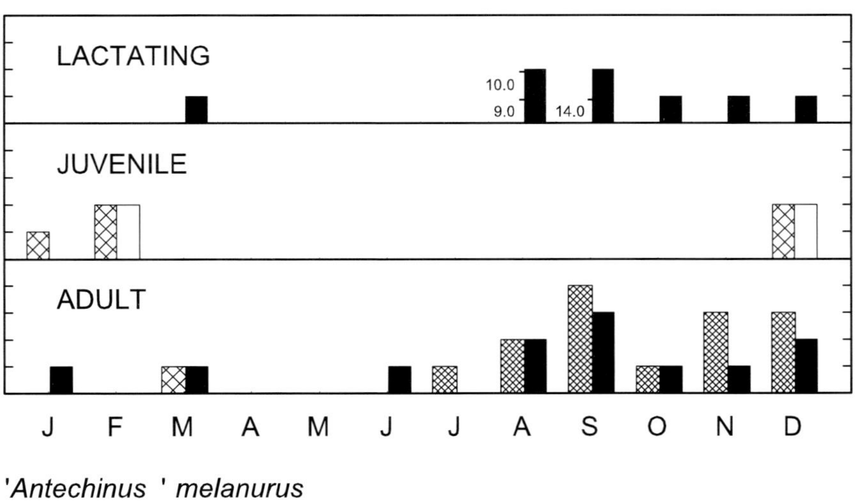

Figure 2 *'Antechinus' melanurus* n = 36 (includes single recapture data for each of 2 males and 1 female).

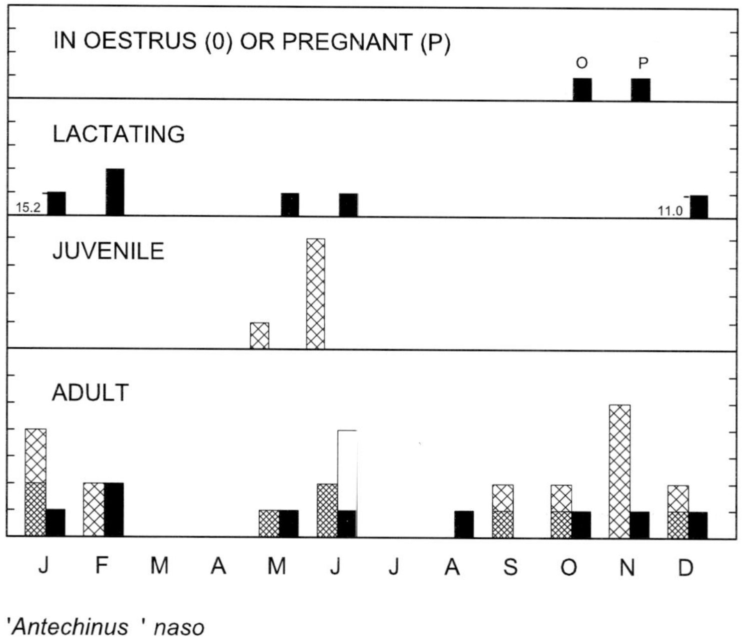

Figure 3 *'Antechinus' naso* n = 37 (includes single recapture data for 1 male).

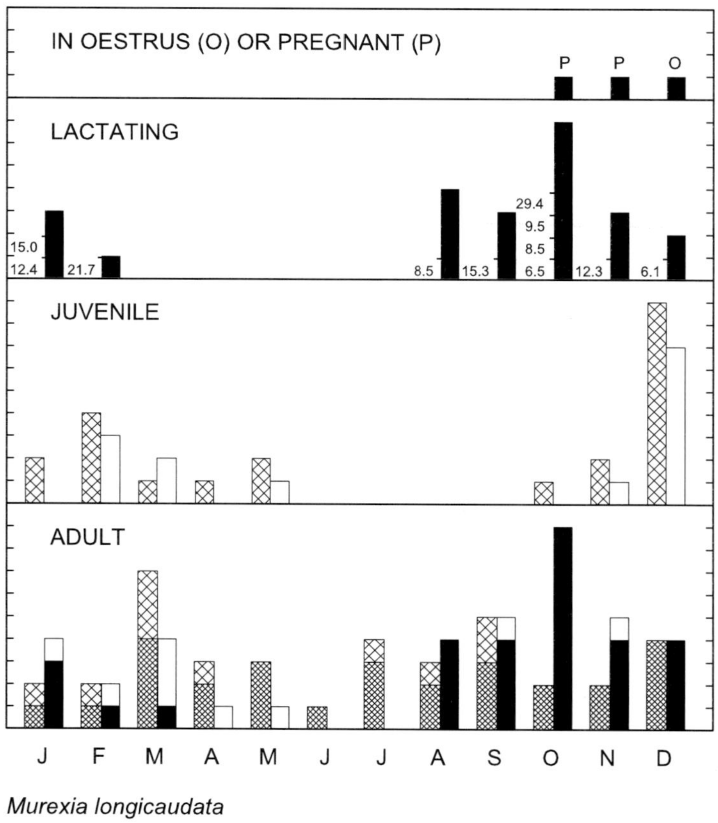

Figure 4 *Murexia longicaudata* n = 113 (includes single recapture data for each of 4 males and 4 females).

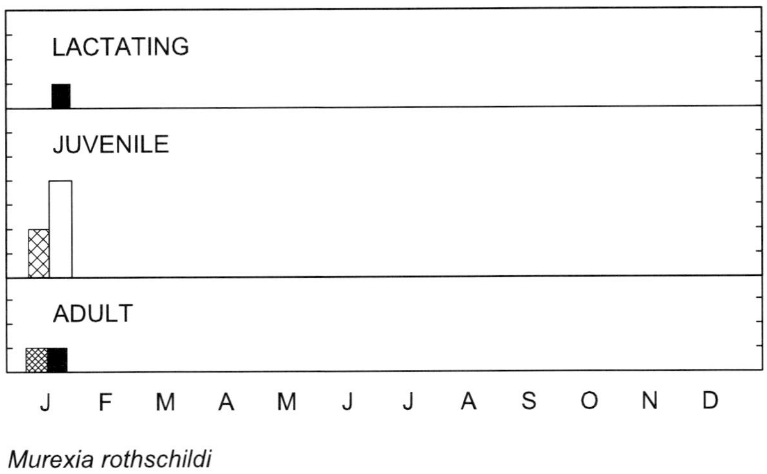

Figure 5 *Murexia rothschildi* n = 8.

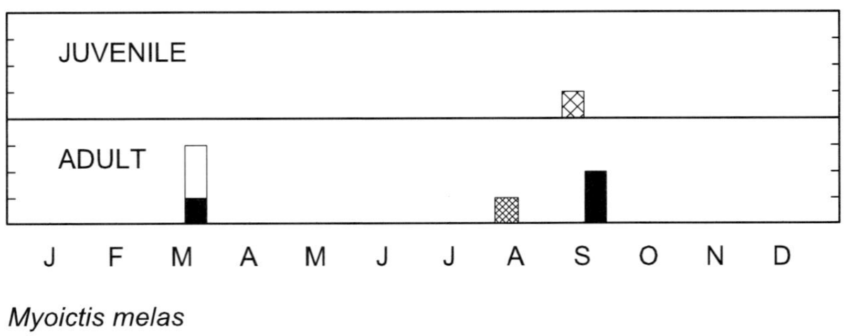

Figure 6 *Myoictis melas* n = 7.

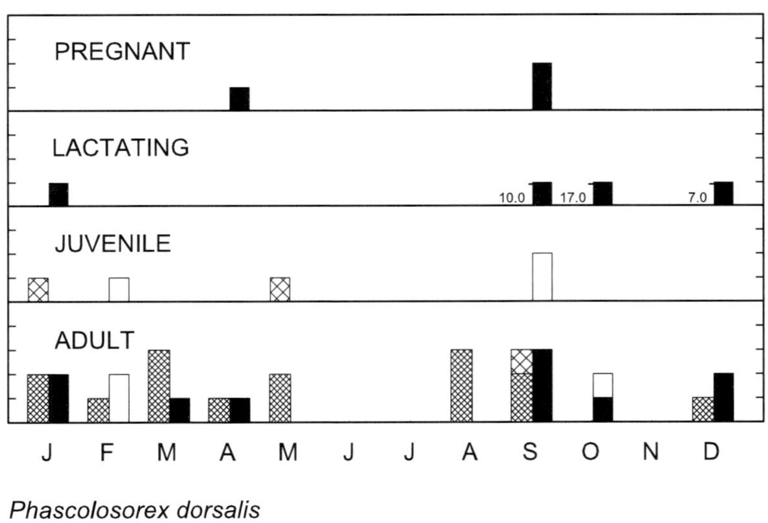

Figure 7 *Phascolosorex dorsalis* n = 34 (includes single recapture data for each of 1 male and 1 female).

Table 3 Summary of the incidence by month of pregnant or lactating females found in this study (■), an earlier study of museum specimens (●) by Woolley (1994) and a field study (◆) by Dwyer (1977).

Species	Month											
	J	F	M	A	M	J	J	A	S	O	N	D
'Antechinus' habbema	■	■	■		■		●	●	●	■●	●	●
'Antechinus' melanurus			■	◆				■●	■◆	■	■	■◆
'Antechinus' naso	■●	■●		●	■●	■●◆			●		■●	■●◆
Murexia longicaudata	■	■●				●		■●	■	■	■	■
Murexia rothschildi	■		●									●
Myoictis melas	●	●	●		●		●		●	●		
Phascolosorex dorsalis	■●			■	●			●	■●	■●		■

addition, the months in which juveniles and sub-adults were found is extended for most species. A reappraisal of the timing of breeding in *Myoictis melas* will be necessary when the revision of the genus, based on morphological and molecular data (Westerman, M., Young, J., Donnellan, S., Woolley, P.A. and Krajewski, C. unpublished) is complete, because the museum specimens upon which the assessment is based are now considered to represent more than one species.

Observations on captive animals

The number of wild-caught animals of each species held in captivity at any one time seldom approached the total number maintained (Table 1) because collection took place, except in the case of *Murexia rothschildi*, over a period of about two years. This sometimes limited the choice of partners for attempts at breeding and only three species *viz Murexia longicaudata, Murexia rothschildi* and *Phascolosorex dorsalis* were bred in captivity. Contrary to the statement in Woolley (1994) *'Antechinus' habbema* was not bred, but young conceived in the wild were born in captivity. They did not survive to weaning. Young of *'Antechinus' melanurus, 'Antechinus' naso, Murexia longicaudata* and *Phascolosorex dorsalis* also born or conceived in the wild were kept in captivity, but not all were weaned (Table 1). Although few animals were bred basic features of reproduction for each species were established by frequent monitoring of captive animals. Some individuals of each species were monitored for periods that were sufficiently long (Table 4) to establish, in conjunction with data obtained from field animals, various life history parameters.

Table 4 Maximum periods for which males and females of each species were monitored in captivity.

Species	Months in captivity	
	Males	Females
'Antechinus' habbema	25.5	12.5
'Antechinus' melanurus	23.5	25.5
'Antechinus' naso	23.0	21.0
Murexia longicaudata	21.0	28.0
Murexia rothschildi	27.0	34.0
Myoictis melas	25.0	31.0
Phascolosorex dorsalis	39.0	36.0

The male

Males of all species have a pendulous scrotum, a 'carrot'-shaped (Rodger and Hughes 1973) disseminate type prostate and either two or three pairs of Cowper's glands. The width of the scrotum and size of the prostate of mature animals (Table 5) could be used as a guide for assessment of the reproductive status of animals captured in the course of other studies. The prostate is macroscopically divisible into two zones, a short proximal semi-translucent zone and a longer distal opaque zone. The prostate of *'Antechinus' habbema* was much shorter than that of other species, and it was unusual in that it appeared short relative to the length of the urethra, but no measurements were made to substantiate this. One of the two or three pairs of Cowper's glands (Table 5) was much larger, usually creamy white in colour, and more opaque than the other pair or pairs, which were usually pinky brown and semi-translucent. The relative sizes of the glands in each species are illustrated diagrammatically in Figure 8. Evidence of a sternal gland, based on external appearance, was seen in only two species (Table 5). Histological sections of skin from the sternal area, which may have revealed the presence of a gland in species in which no staining of the fur occurred, were not prepared.

The males (with the exception of one individual, see below) of all species, once mature, showed spermatorrhoea throughout the period each was maintained in captivity. Also, the scrotum showed no changes in size that would be expected if regression and recrudescence of the testes had occurred. If spermatogenesis was occurring in those maintained for periods greater than one year they were considered to be potentially capable of breeding in more than one year, i.e. to be perennial rather than annual (Lee et al. 1982). One *Phascolosorex dorsalis* male did sire young in two consecutive years. Captive males continued to increase in

Table 5 Attributes of males of each species. Dimensions of the scrotum and prostate are for field animals except where shown in bold type.

Species	Scrotum width (mm)	Prostate length × width (mm)	Cowper's glands No. of pairs	Sternal gland
'Antechinus' habbema	9.0–12.0	6.5 × 3.0–**8.5 × 3.0**	3	no
'Antechinus' melanurus	11.5–12.5	**17.0 × 4.5–23.0 × 5.0**	2	no
'Antechinus' naso	12.0–14.5	12.0 × 3.0–25.0 × 4.5	3	yes
Murexia longicaudata	13.0–15.0	20.0 × 5.5–23.0 × 5.0	2	yes
Murexia rothschildi	15.0	**12.5 × 4.0–15.0 × 5.5**	3	no
Myoictis melas	14.5	**15.5 × 3.5–23.0 × 6.5**	3	no
Phascolosorex dorsalis	12.0–15.0	14.0 × 3.5–17.0 × 5.0	3	no

weight after they matured and the wide range in body weight seen in sexually mature wild-caught males of some species (Table 2) suggests that more than one cohort may be present in the field population, which provides some support for males being perennial. The exceptional *Murexia longicaudata*, one of ten maintained in captivity, may have been physiologically abnormal. An immature adult at capture, after two months it showed spermatorrhoea continuously for seven months and then intermittently for a further 10 months. After a total of 19 months in captivity it was found that the seminiferous tubules contained only Sertoli cells, i.e. it had become reproductively senile.

The female

The females of all species except *Phascolosorex dorsalis* have a Type 1 pouch (Woolley 1974), in which the mammary area has no covering fold of skin and marginal skin folds develop when breeding occurs (Table 6). *Phascolosorex dorsalis* has a Type 3 pouch (Woolley 1974), and the mammary area is covered by a circular skin fold. Females of all species have four nipples and the majority (45 of 56) of wild-caught lactating females appeared to be suckling a full complement of young (Table 6). They either had four young in the pouch or, if the young were not being carried, had four elongated nipples and enlarged mammary glands. Loss of one or two young late in the period of lactation would probably not have been detected, however, because the remaining young would suckle from all nipples. The females of three species *viz 'Antechinus' melanurus, 'Antechinus' naso* and *Murexia longicaudata* showed signs of having a sternal gland (Table 6).

Captive females of all species showed no seasonality or synchronisation in the timing of their oestrous periods. Oestrus was detected by the appearance of cornified epithelial cells in the urine and a transitory increase in body weight (Figure 9) such as occurs in other species of dasyurid marsupials (Woolley 1990a). The number of oestrous periods detected for individual females ranged from one to two (*'Antechinus' habbema*), one to 11 (*'Antechinus' melanurus*), one to six (*'Antechinus' naso*), one to 11 (*Murexia longicaudata*), three to 14 (*Murexia rothschildi*), five to nine (*Myoictis melas*) and two to seven (*Phascolosorex dorsalis*). For most species, some females were in oestrus in most months of the year (Table 7) and thus potentially capable of breeding at any time of the year, i.e. aseasonally rather than seasonally (Lee et al. 1982). Females of all species that were maintained for sufficiently long periods (Table 4) continued to enter oestrus at intervals over consecutive years, suggesting that they may also breed in more than one year.

Some evidence for a lack of seasonality and ability to breed in more than one year was obtained from two *Phascolosorex dorsalis* females that produced young at intervals of 9 months. One gave birth to young in the wild in September, and again in captivity in the following June. The other gave birth to young in December and again the following September. Both litters of this female resulted from matings in the laboratory. One *Murexia*

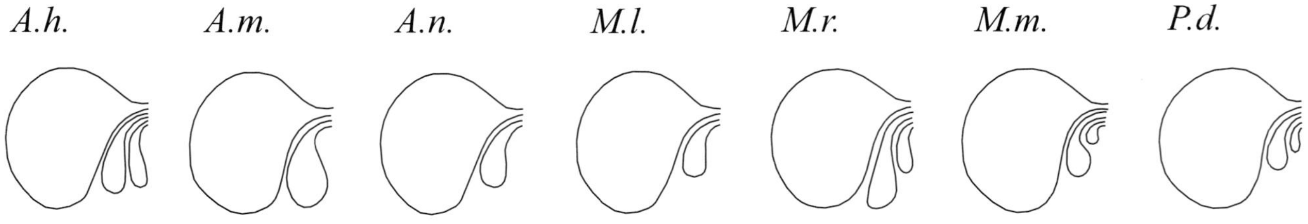

Figure 8 Relative sizes of Cowper's glands (one of each pair) in each species. Diagrammatic representation of the glands arranged in order of decreasing size. *A.h. = 'Antechinus' habbema, A.m. = 'Antechinus' melanurus, A.n. = 'Antechinus' naso, M.l. = Murexia longicaudata, M.r. = Murexia rothschildi, M.m. = Myoictis melas,* and *P.d. = Phascolosorex dorsalis.*

Table 6 Attributes of females of each species. The number of mothers relates to wild-caught lactating females from which the number of young was determined.

Species	Pouch type	No. of nipples	No. of young range (mean)	No. of mothers	No. of corpora lutea	Duration of lactation (months)	Sternal gland
'Antechinus' habbema	1	4	2–4 (3.6)	14	8–11	>3	no
'Antechinus' melanurus	1	4	4 (4)	8	8–13	3.5	yes
'Antechinus' naso	1	4	2–4 (3.7)	6	9–14	–	yes
Murexia longicaudata	1	4	2–4 (3.8)	23	6–16	3.5	yes
Murexia rothschildi	1	4	4 (4)	1	9, 13	3.5	no
Myoictis melas	1	4	–	–	27, 31	–	no
Phascolosorex dorsalis	3	4	3–4 (3.75)	4	8–18	4–4.5	no

longicaudata also had two litters born about 9 months apart, one in the wild in January, and one in captivity in the following September. Two litters were reared over a period of about 18 months by a *Murexia rothschildi* that had been dug from an underground nest with four young which were still suckling. This litter was weaned soon after capture in January 1986, and the second, born in captivity, in March 1987.

The interval between the end of lactation and the return to oestrus was very variable (Table 8). Two species, *Murexia longicaudata* and *Phascolosorex dorsalis*, appeared to return more quickly following premature loss of the young than those that weaned their young. No pattern was seen in other species.

Captive females that failed to mate or suckle young returned to oestrus at intervals (measured by the number of days from peak body weight in one period to peak body weight in the next) that also were very variable (Table 8). Individuals of a species, when first brought into captivity, sometimes showed short intervals, and sometimes long intervals, between oestrous periods. Length of time in captivity had no obvious effect on the interval between successive oestrous periods. In the case of *Phascolosorex dorsalis*, the mean interval between oestrous periods was greater in the six laboratory-reared/-bred females (163 days), than in the six wild-caught females (96.5 days). One laboratory-reared female did not enter oestrus for over one year (427 days) between the third and fourth of five oestrous periods and intervals of 255 days (one) and 200 days (two) were seen in three other laboratory-reared/-bred females. Even if these exceptionally long intervals between oestrous periods are excluded from the calculation, the mean interval is still much longer (140 days) than in wild-caught females. No reason for this difference is apparent. Laboratory-reared and laboratory-bred females

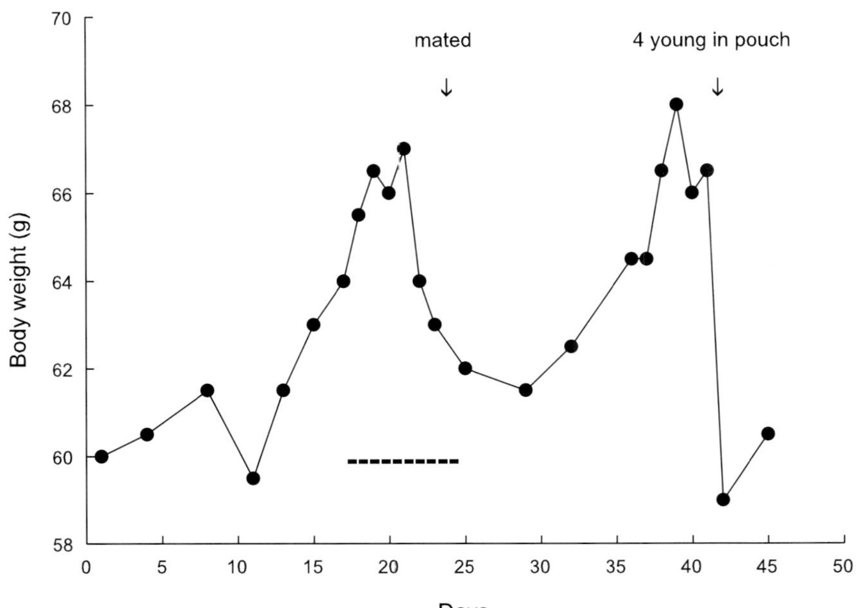

Figure 9 Changes in body weight during oestrus and pregnancy in *Murexia rothschildi*. Interval between mating and parturition is 20 days. Broken line indicates period during which cornified epithelial cells were found in the urine.

Table 7 Incidence by month of oestrus in females of each species. n = number of females in which oestrus was detected

Species (n)	Month											
	J	F	M	A	M	J	J	A	S	O	N	D
'Antechinus' habbema (8)	–	1	–	1	–	–	1	1	5	1	–	–
'Antechinus' melanurus (6)	–	3	2	2	3	2	2	3	4	1	4	1
'Antechinus' naso (1)	1	–	1	–	1	1	1	–	1	–	–	–
Murexia longicaudata (7)	3	5	1	3	4	5	6	6	7	3	6	6
Murexia rothschildi (6)	4	3	5	2	5	2	6	1	5	4	8	3
Myoictis melas (3)	1	2	2	–	3	–	1	2	4	3	2	1
Phascolosorex dorsalis (12)	3	7	4	1	3	5	4	9	3	7	5	5

reached body weights comparable to those of wild-caught animals, and all were maintained under the same conditions.

Attempts were made to mate females when they were in oestrus (Table 9). The male was introduced to the female's cage, often on several consecutive days once the onset of oestrus (cells in urine, increase in body weight) was detected (Woolley 1990a). Pairs were frequently left together overnight if the animals were not showing agonistic behaviour and no mating had occurred during the day, or if mating had started late in the day. Copulation was either observed or its occurrence inferred from the presence of spermatozoa in the urine of the female. Mating was confirmed in only 13 of 96 attempts. It was generally found to have occurred a day or two after the peak in body weight that occurred during oestrus. Not all females that were known to have mated produced pouch young, and in one that did mating was not detected. On the few occasions when copulation was observed it was generally of long duration (Table 9), and within the range known for other species of small dasyurid marsupials in captivity, e.g. from about 30 minutes to 2.5 hours in *Sminthopsis macroura* (Woolley 1990a) and up to 12 hours in *Antechinus stuartii* (Woolley 1966).

The duration of pregnancy, timed from the day of mating to the day of parturition, was established for three species, *viz Murexia longicaudata, Murexia rothschildi* and *Phascolosorex dorsalis*, and an estimate obtained for *'Antechinus' melanurus* from a female that mated but did not produce pouch young (Table 10). During pregnancy body weight increased and birth of the young was accompanied by a sharp fall in body weight (Figure 9). Similar changes in body weight occurred in unmated females and the duration of pseudopregnancy (Woolley 1990a), timed from the day of peak weight at oestrus to the day it dropped after the second increase, was established for females of all species. Given that mating occurred a day or two after peak weight at oestrus was reached, the length of pseudopregnancy as measured provides an estimate of the gestation period for those species in which it was not established. It was of the same order as the length of pregnancy in those species for which this was determined (Table 10).

Ovulation occurs spontaneously in all species. This was confirmed by the presence of corpora lutea in the ovaries and/or eggs in the Fallopian tubes or uteri of females examined after an oestrous period in which they were not paired with a male. The number of eggs shed at ovulation, determined by counting the number of corpora lutea in the ovaries (Table 6), was always in excess of the number of young that could be accommodated in the pouch. The histological appearance of the developing cor-

Table 8 Time to return to oestrus following lactation, and interval between successive oestrous periods in captive females that did not suckle young. Interval between oestrous periods of *Phascolosorex dorsalis* a) = wild-caught females, b) = laboratory-reared and laboratory-bred females

Species	Time to return to oestrus (months)		Interval between oestrous periods (days)		
	Young weaned	Young not weaned	Range	Mean	n
'Antechinus' habbema	–	7, 8	146–166	156	2
'Antechinus' melanurus	1, 3	1, 6	50–99	69.5	18
'Antechinus' naso	–	4	52–68	60	4
Murexia longicaudata	3.5, 5	<1, <1, <1	41–180	68	46
Murexia rothschildi	9	–	43–179	71	41
Myoictis melas	–	–	52–186	100	18
Phascolosorex dorsalis	3, 4, 4.5	<1, <1, <1	a) 58–187	96.5	19
			b) 62–427	163	23

Table 9 Attempts to mate females of each species.

Species	No. of females paired	Total no. of pairings	Copulation		Spermatozoa in urine (no. of pairings)	No. of litters
			No. observed	Duration (hours)		
'Antechinus' habbema	2	2	0	–	0	0
'Antechinus' melanurus	5	9	1	1.5	1	0
'Antechinus' naso	1	3	0	–	0	0
Murexia longicaudata	9	17	5	1.5–4.5	7	2
Murexia rothschildi	7	43	2	0.5, 1.5	1	1
Myoictis melas	2	8	0	–	0	0
Phascolosorex dorsalis	8	14	5	>2	4	4

pora lutea in each species was similar to that described for *Sminthopis macroura* (Selwood and Woolley 1991). Large, fully formed corpora lutea were seen late in the period of pregnancy or pseudopregnancy of *'Antechinus' habbema* (diameter 648 μm), *'Antechinus' naso* (865 μm), *Murexia longicaudata* (841 μm), *Murexia rothschildi* (844 μm) and *Phascolosorex dorsalis* (1207 μm). Fully formed corpora lutea were not seen in any of the *'Antechinus' melanurus* or *Myoictis melas* examined. Based on their histological appearance and size, the corpora lutea of most species regressed quickly, either soon after the end of pseudopregnancy or early in the period of lactation while the young were still being carried in the pouch. Degenerating corpora lutea were found in the ovaries of *'Antechinus' habbema* 5 days after the end of pseudopregnancy (474 μm); in *'Antechinus' naso* with pouch young, crown–rump length 15.2 mm (107 μm); in *Murexia longicaudata* with pouch young, crown–rump length 8.3 mm (355 μm); in *Murexia rothschildi* 42 days after the end of pseudopregnancy (184 μm) and in *Myoictis melas*, in which the corpora lutea were 582 μm and 300 μm at 15 and 30 days respectively after the end of pseudopregnancy. Corpora lutea were no longer recognisable in the ovaries of *'Antechinus' habbema*, *'Antechinus' naso* and *Murexia longicaudata* that were lactating but no longer carrying young in the pouch. In both *'Antechinus' melanurus* and *Phascolosorex dorsalis* the corpora lutea persisted for much longer than in other species. They were large in one *'Antechinus' melanurus* (638 μm) and one *Phascolosorex dorsalis* (946 μm) that were lactating but not carrying pouch young, and still recognisable (215 μm) in another *'Antechinus' melanurus* that had finished lactating. Two generations were recognisable in the ovaries of one *'Antechinus' melanurus*. The most recent generation, formed following oestrus 34 days earlier (and 8 days after the end of pseudopregnancy) were 611 μm in diameter and the older generation formed following oestrus that occurred 110 days earlier, were 196 μm in diameter. A wild-caught *Phascolosorex dorsalis* in an early stage of pregnancy (4–8 cell embryos), had corpora lutea in an early stage of formation, 783 μm in diameter and an older generation 758 μm in diameter. This female had reared a litter previously (based on pouch appearance) but it is not known if the older generation of corpora lutea resulted from the earlier pregnancy, or from a more recent oestrous period that was not followed by pregnancy. Another *Phascolosorex dorsalis* which was in a later stage of pregnancy (blastocysts) had three generations of corpora lutea (Figure 10). The most recent, which were still in a very early stage of formation comparable to those of the previous female, had a diameter of 694 μm. The next most recent generation, based on histological appearance, were 999 μm, and the oldest, 652 μm in diameter. Two Graafian follicles (see below) that

Table 10 The duration of pregnancy (interval between mating and parturition) and pseudopregnancy (interval between peak weight at oestrus and day of body weight drop at the end of pseudopregnancy) in females of each species.

Species	Pregnancy (days)			Pseudopregnancy (days		
	Range	Mean	n	Range	Mean	n
'Antechinus' habbema	–	–	–	31–50	40	6
'Antechinus' melanurus	est. 22	–	1	19–26	23	25
'Antechinus' naso	–	–	–	26–33	28	5
Murexia longicaudata	19, 20	19.5	2	15–29	19.5	51
Murexia rothschildi	20	–	1	11–22	19	45
Myoictis melas	–	–	–	22–28	24.5	21
Phascolosorex dorsalis	18–21	19.75	4	16–26	20.5	47

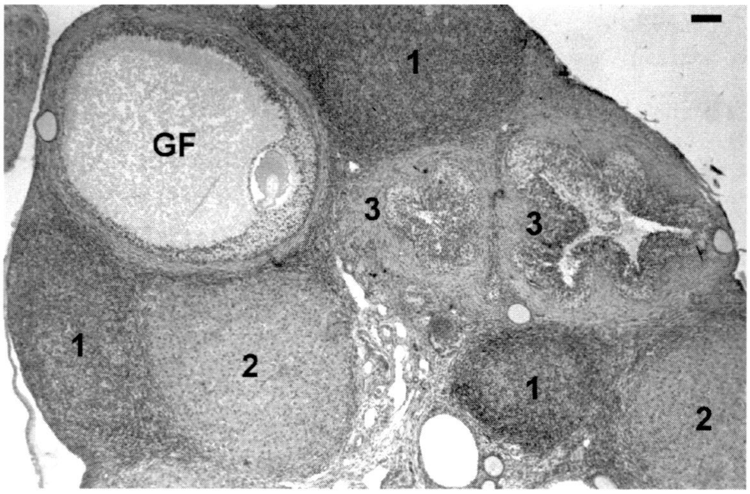

Figure 10 Section through ovary of *Phascolosorex dorsalis* showing three generations of corpora lutea, and a Graafian follicle (GF) that had failed to rupture. 1, first (oldest) generation; 2, second generation; 3, third generation (blastocysts associated with this generation). Scale bar = 100 μm

apparently had failed to rupture were present in one ovary of this female. Three generations of corpora lutea have been found in the ovaries of *Sminthopsis macroura* (Woolley 1990a) but they do not persist for as long as those of *Phascolosorex dorsalis*.

Graafian follicles were seen in only five individuals, one *'Antechinus' habbema*, one *Murexia longicaudata*, two *Murexia rothschildi* and one *Phascolosorex dorsalis*. In *'Antechinus' habbema* the follicles, which may not have been quite fully formed, had a diameter of 358 μm, and the oocytes, 102 μm. In *Murexia longicaudata* they had a diameter of 738 μm, and the oocytes, 193 μm. The oocyte had been shed from one follicle in this individual (Figure 11) and the ruptured follicle had a diameter of 652 μm. The follicles and oocytes of *Murexia rothschildi* were 660 μm and 212 μm respectively in diameter in one individual, and 540 μm and 236 μm in another. In *Phascolosorex dorsalis* the two follicles that had failed to rupture were 1124 μm, and the oocytes 240 μm in diameter.

Development of the young and age at maturity

Two *Murexia longicaudata*, one *Murexia rothschildi* and four *Phascolosorex dorsalis* produced young following mating in captivity. No supernumerary or unattached young were ever found following parturition. One of the *Phascolosorex dorsalis* females was by chance inspected while in the process of giving birth at 1700 hours, when one young was seen moving in a drop of fluid on the fur between the cloaca and pouch opening. The mother

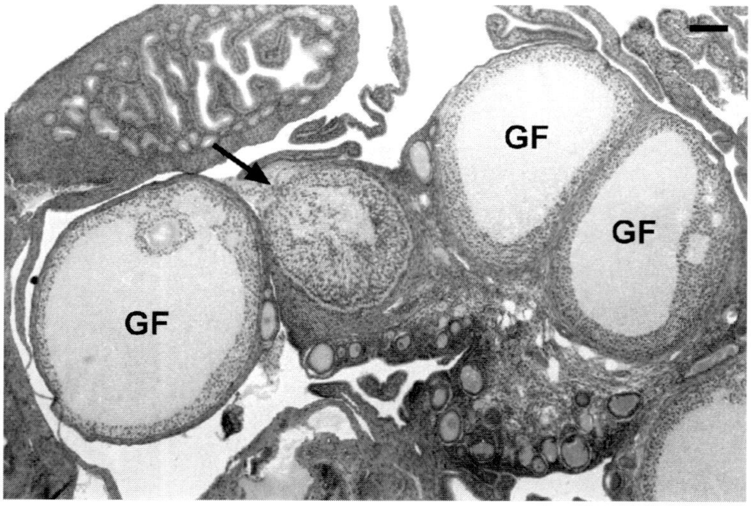

Figure 11 Section through ovary of Murexia longicaudata showing Graafian follicles (GF) and a follicle from which the egg had been shed. The site of rupture (arrow) can be seen. Scale bar = 100 μm

Table 11 Combined body weight of young at weaning, and weight of litter as a percentage of maternal weight for two litters of each species.

Species	Body weight at weaning (g)		Weight of litter as % of maternal weight
	Mother	Young	
'Antechinus' melanurus	51.5	73.5 (4)	143
	41.0	64.0 (4)	156
Murexia rothschildi	60.0	96.0 (4)	160
	72.0	96.0 (4)	133
Phascolosorex dorsalis	53.5	57.0 (3)	106
	42.5	60.0 (4)	141

was immediately returned to her cage, and on the following day this or another young, which was reared for 15 weeks (see below), was found in the pouch. Few observations were made on the young of these females, or of those of wild-caught females in captivity, to avoid disturbance which might have led to loss of the young. Despite this many young did not survive to weaning (Table 1). The *Murexia longicaudata* females mated in captivity each had three young, which were lost within a week of birth. Two of the *Phascolosorex dorsalis* had litters of three, one of which was lost within 30 days and the other weaned. The other two *Phascolosorex dorsalis* each had a single pouch young. One of the singletons was lost at 15 weeks; the other was weaned at 19 weeks. Some measurements of crown–rump length of the young of *Phascolosorex dorsalis* were obtained and used in the estimation of the age of wild-caught pouch young so that age at maturity could be established. Crown–rump length (mm) at 15 days was 8.5–9.0; at 20 days, 9.5; at 30 days, 12.0; at 35 days, 14.0 and at 45 days, 15.0–18.5. The eyes of *Phascolosorex dorsalis* did not open until they were about 95 days old. The duration of lactation in *Phascolosorex dorsalis* (4 to 4.5 months) was longer than in other species (Table 6).

Body weight of the young at weaning was obtained for *'Antechinus' melanurus*, *Murexia rothschildi* and *Phascolosorex dorsalis*. In each species the combined weight of the young in the litter exceeded that of the mother (Table 11) and parental investment (expressed as the weight of the litter as a percentage of weight of the mother) was similar to that given by Russell (1982) for an Australian species of dasyurid marsupial (*Dasycercus cristicauda*) of roughly comparable maternal body weight.

The age at which males and females matured, i.e. when males first showed spermatorrhoea and females first entered oestrus, was established for some species from laboratory-bred and/or laboratory-reared animals (Table 12). For others an estimate was obtained from animals that were juvenile when captured, on the assumption that they were then at least 4 months old (the period of dependence on the mother being of the order of 4 months).

Table 12 Age at maturity (months) for each species.

Species	Males	Females
'Antechinus' habbema	est. 10	–
'Antechinus' melanurus	8	est. 12
'Antechinus' naso	est. 14	–
Murexia longicaudata	11, 12	8, 9
Murexia rothschildi	10	9–10
Myoictis melas	est. 10	–
Phascolosorex dorsalis	11	11–12

PATTERN OF REPRODUCTION

Consideration of all the foregoing information leads to the conclusion that the seven species of endemic New Guinean dasyurids studied are most probably all capable of year-round breeding. The females show aseasonal polyoestry, and births occur throughout the year in the wild. Once mature, the males show spermatorrhoea continuously; the scrotum remains large and they presumably can breed at any time. Both males and females of at least some species breed more than once in a lifetime, and may live for two or more years. Maximum litter size is four, and the young are dependent on the mother for from 3 to 4.5 months. Sexual maturity is reached usually between 8 and 11 months of age.

Further studies, in particular of wild populations, are required to verify various aspects of this simplified overview and to investigate what environmental factors determine when animals breed. Grossek (1987) demonstrated that it is possible to recapture individuals released in the wild, so it should be feasible to monitor reproductive activity in wild populations over an extended period of time, providing a suitable study area free from human interference can be found. The pattern of reproduction of these New Guinean species is unlike that of nearly all the Australian dasyurid marsupials, reproductive data for which has most recently been reviewed by Krajewski et al. (2000). With the exception of *Planigale maculata sinualis* (Taylor et al. 1982) and possibly *Sminthopsis virginiae virginiae* (Taplin 1980, Woolley 1995), both of which occur in the tropical north, the Australian species are strictly seasonal breeders.

ACKNOWLEDGEMENTS

Much of the field work to collect the animals on which this study is based was carried out while the author was on sabbatical leave from La Trobe University. I extend my thanks to all those people, too numerous to mention by name, who assisted me in the field, in the movement of animals from Papua New Guinea to Australia and in the laboratory. My former postgraduate student, George Grossek, kindly allowed me access to his field records. Financial support was largely provided by the Australian Research Grants Scheme (Grant D18315743). Alan Wright assisted with photomicrography and Belinda Lees in the preparation of the manuscript.

References

Armstrong, L.A., Krajewski, C., & Westerman, M. (1998), 'Phylogeny of the dasyurid marsupial genus *Antechinus* based on cytochrome-b, 12S-rRNA, and protamine-P1 genes', *Journal of Mammalogy*, **79**:1379–89.

Dwyer, P.D. (1977), 'Notes on *Antechinus* and *Cercartetus* (Marsupialia) in the New Guinea Highlands', *Proceedings of the Royal Society of Queensland*, **88**:69–73.

Grossek, G.L. (1987), 'The diet of some New Guinean dasyurid marsupials', MSc thesis, La Trobe University.

Krajewski, C., Painter, J., Driskell, A.C., Buckley, L., & Westerman, M. (1993), 'Molecular systematics of New Guinean dasyurids (Marsupialia: Dasyuridae)', *Science in New Guinea*, **19**:157–66.

Krajewski, C., Buckley, L., Woolley, P.A., & Westerman, M. (1996), 'Phylogenetic analysis of cytochrome b sequences in the dasyurid marsupial subfamily Phascogalinae: systematics and the evolution of reproductive strategies', *Journal of Mammalian Evolution*, **3**:81–91.

Krajewski, C., Young, J., Buckley, L., Woolley, P.A., & Westerman, M. (1997), 'Reconstructing the evolutionary radiation of dasyurine marsupials with cytochrome b, 12S rRNA and protamine P1 gene trees', *Journal of Mammalian Evolution*, **4**:217–36.

Krajewski, C., Woolley, P.A., & Westerman, M. (2000), 'The evolution of reproductive strategies in dasyurid marsupials: implications of molecular phylogeny', *Biological Journal of the Linnean Society*, **71**:417–35.

Lee, A.K., Woolley, P.A. & Braithwaite, R.W. (1982), 'Life history strategies of dasyurid marsupials', in *Carnivorous Marsupials* (ed. M. Archer), pp.1–11, Royal Zoological Society of New South Wales, Sydney.

Rodger, J.C., & Hughes, R.L. (1973), 'Studies of the accessory glands of male marsupials', *Australian Journal of Zoology*, **21**:302–20.

Russell, E.M. (1982), 'Patterns of parental care and parental investment in marsupials', *Biological Reviews*, **57**:423–86.

Selwood, L., & Woolley, P.A. (1991), 'A timetable of embryonic development, and ovarian and uterine changes during pregnancy, in the stripe-faced dunnart, *Sminthopsis macroura* (Marsupialia: Dasyuridae)', *Journal of Reproduction & Fertility*, **91**:213–27.

Taplin, L.E. (1980), 'Some observations on the reproductive biology of *Sminthopsis virginiae* (Tarragon) (Marsupialia: Dasyuridae)', *Australian Zoologist*, **20**:407–18.

Taylor, J.M., Calaby, J.H., & Readhead, T.D. (1982), 'Breeding in wild populations of the marsupial-mouse *Planigale maculata sinualis* (Dasyuridae, Marsupialia)', in *Carnivorous Marsupials* (ed. M. Archer), pp. 83–7, Royal Zoological Society of New South Wales, Sydney.

Van Dyck, S. (2002), 'Morphology-based revision of *Murexia* and *Antechinus* (Marsupialia: Dasyuridae)', *Memoirs of the Queensland Museum*, **48**:239–330.

Westerman, M., & Woolley, P.A. (1990), 'Cytogenetics of some New Guinean dasyurids and genome evolution in the Dasyuridae (Marsupialia)', *Australian Journal of Zoology*, **37**:521–31.

Westerman, M., & Woolley, P.A. (1993), 'Chromosomes and the evolution of dasyurid marsupials: an overview', *Science in New Guinea*, **19**:123–30.

Woolley, P. (1966), 'Reproduction in *Antechinus* spp. and other dasyurid marsupials', *Symposia of the Zoological Society of London*, **15**:281–94.

Woolley, P. (1974), 'The pouch of *Planigale subtilissima* and other dasyurid marsupials', *Journal of The Royal Society of Western Australia*, **57**:11–15.

Woolley, P.A. (1982), 'Phallic morphology of the Australian species of *Antechinus* (Dasyuridae, Marsupialia): a new taxonomic tool?', in *Carnivorous Marsupials* (ed. M. Archer) pp. 767–81, Royal Zoological Society of New South Wales: Sydney.

Woolley, P.A. (1984), 'Phallic morphology of the New Guinean species of *Antechinus*', *Bulletin of the Australian Mammal Society*, **8**:182. (Abstract)

Woolley, P.A. (1989), 'Nest location by spool-and-line tracking of dasyurid marsupials in New Guinea', *Journal of Zoology, London*, **218**:689–700.

Woolley, P.A. (1990a), 'Reproduction in *Sminthopsis macroura* (Marsupialia: Dasyuridae) I. The Female', *Australian Journal of Zoology*, **38**:187–205.

Woolley, P.A. (1990b), 'Reproduction in *Sminthopsis macroura* (Marsupialia: Dasyuridae) II. The Male', *Australian Journal of Zoology*, **38**:207–17.

Woolley, P.A., Raftopoulos, S.A., Coleman, G.J., & Armstrong, S.M. (1991), 'A comparative study of the circadian activity patterns of two New Guinean dasyurid marsupials, *Phascolosorex dorsalis* and *Antechinus habbema*', *Australian Journal of Zoology*, **39**:661–71.

Woolley, P.A. & Valente, A. (1992), 'Hair structure of the dasyurid marsupials of New Guinea', *Science in New Guinea*, **18**:29–49.

Woolley, P.A. (1993), 'Collection and laboratory maintenance of New Guinean dasyurid marsupials', in *The Biology and Management of Australasian Carnivorous Marsupials* (eds M. Roberts, J. Carnio, G. Crawshaw, & M. Hutchins), pp. 91–7, American Association of Zoological Parks and Aquariums Monotreme and Marsupial Taxon Advisory Group, Washington, DC.

Woolley, P.A. (1994), 'The dasyurid marsupials of New Guinea: use of museum specimens to assess seasonality of breeding', *Science in New Guinea*, **20**:49–55.

Woolley, P.A. (1995), 'Red-cheeked dunnart', in *The Mammals of Australia* (ed. R. Strahan), pp. 156–7, Reed Books, Chatswood.

Woolley, P.A. (2001), 'Observations on the reproductive biology of *Myoictis wallacei*, *Neophascogale lorentzi*, *Dasyurus albopunctatus* and *Dasyurus spartacus*, dasyurid marsupials endemic to New Guinea', *Australian Mammalogy*, **23**:63–66.

CHAPTER 12

MALE GENITAL SYSTEM OF SOUTH AMERICAN DIDELPHIDS

J.C. Nogueira and A.C.S. Castro

Departamento de Morfologia, Instituto de Ciências Biológicas, Universidade Federal de Minas Gerais, CP 486, 31270-901 Belo Horizonte, Brasil

This chapter presents an overview on the gross anatomy and histology of the male genital system of several South American didelphid species of the genera *Didelphis, Philander, Metachirus, Lutreolina, Caluromys, Monodelphis, Marmosops, Gracilinanus, Marmosa, Micoureus* and *Glironia*, w th emphasis on the white-belly opossum *Didelphis albiventris*. A comparative description of the testes, testicular-epididymal pedicle, epididymides, ductus deferens and spermatic cord, urethra, penis, prostate, two or three pairs of bulbourethral glands, as well as the cloaca and paracloacal glands, are herein presented. Particularly in *Didelphis albiventris*, a few other aspects such as innervation of the male genital system, puberty, biometry and testicular activity during the annual reproductive cycle are also dealt with. Further studies involving other South American marsupials of the genera *Thylamys, Chironectes, Caluromysiops, Caenolestes* and *Dromiciops* are fundamental for a better understanding of the male genital system and phylogeny of American marsupials.

Dedicated to Emeritus Professor Dr Luiz Carlos Uchoa Junqueira, a great enthusiast of comparative morphology, for his 80th birthday.

INTRODUCTION

Marsupials are geographically distributed into two groups that inhabit Australasia and the Americas. The Australasian group consists of about 204 species distributed into 16 families (Groves 1993), some of which have been studied quite extensively. American marsupials comprise about 69 living species (Gardner 1993) grouped in the Families Microbiotheriidae, Caenolestidae and Didelphidae. Brazilian marsupials belong to the Didelphidae which is the most diverse group consisting of 63 living species (Gardner 1993) and including 15 genera (*Didelphis, Philander, Chironectes, Lutreolina, Metachirus, Thylamys, Lestodelphys, Marmosa, Gracilinanus, Micoureus, Marmosops, Caluromys, Caluromysiops, Monodelphis* and *Glironia*). Except for *Lestodelphys*, all the other genera are found in Brazil. Among the 44 Brazilian marsupial species (Fonseca et al. 1996), 24 occur in the State of Minas Gerais of which five are included in the endangered group (Lins et al. 1997). Knowledge of the morphophysiology of the genital systems will provide the fundamental tools for a better understanding of the reproductive biology of marsupials and thus contribute to the establishment of basic criteria and development of improved strategies for the preservation of this peculiar mammalian group of the South American fauna. In addition, it will contribute to the

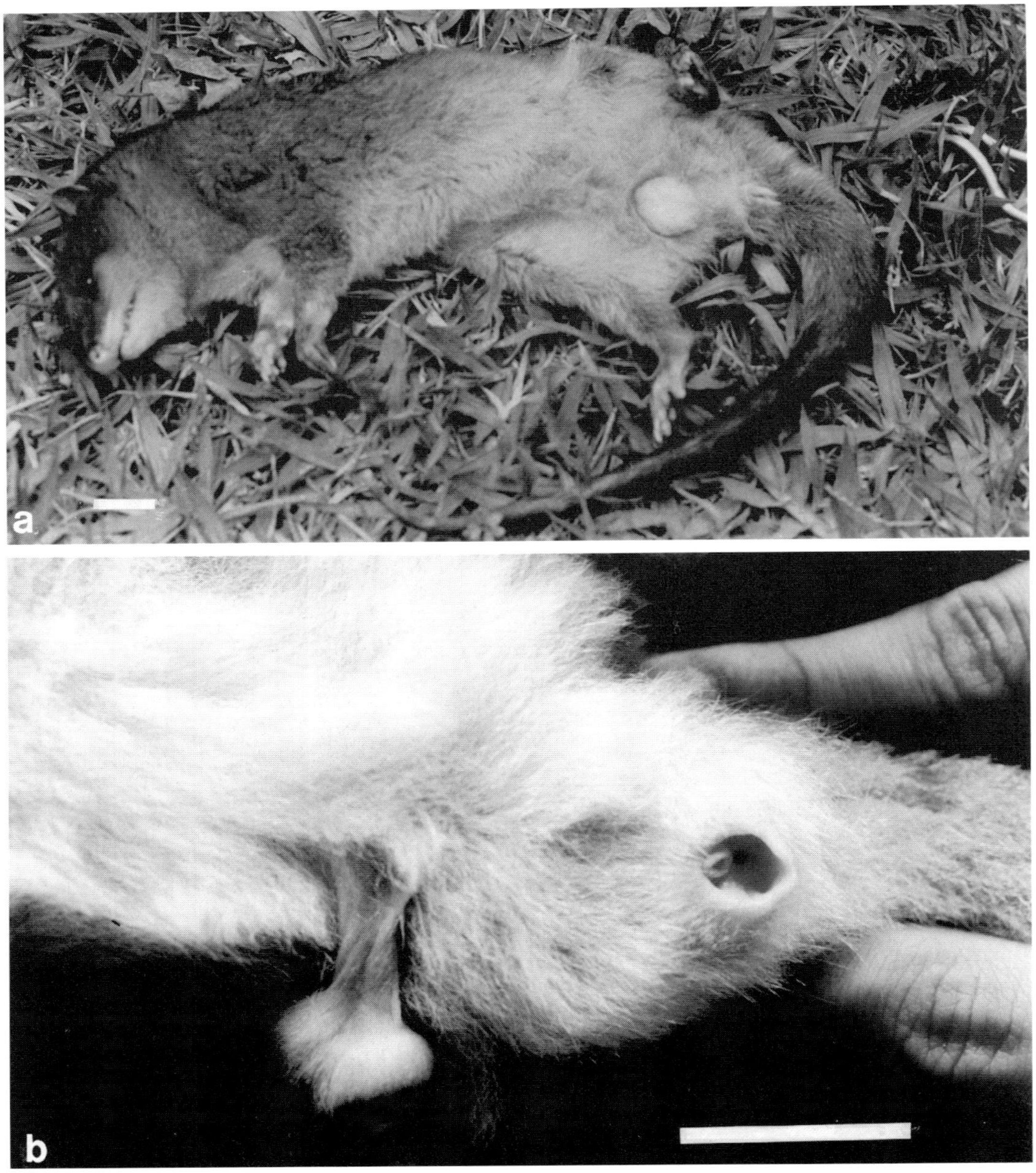

Figure 1 (a) *Lutreolina crassicaudata* under anaesthesia displaying the scrotum and exposed bifid glans penis; (b) Anaesthetised *Caluromys lanatus* showing the cloaca with the genito-urinary and anal openings. The bar corresponds to 2 cm.

comprehension of the phylogenetic and evolutionary aspects of the genital system of the American marsupials.

As the female has always attracted more scientific attention in marsupials than the male, less information on the morphophysiology of the male is available, particularly in didelphids. Therefore, this chapter is concerned with the basic morphology of the male genital system of South American didelphids with emphasis on the reproductive biology of *Didelphis albiventris*.

Didelphidae male genital system

The male genital system (Figs. 1–2) of South American didelphids comprises the testes, epididymides, ductus deferens, urethra, penis and accessory genital glands, the latter consisting of a prostate and of two or three pairs of bulbourethral glands (Cowper's glands). Didelphids lack seminal vesicles, ampullary glands and preputial glands. Biometrical data of some South American didelphids are presented in Table 1.

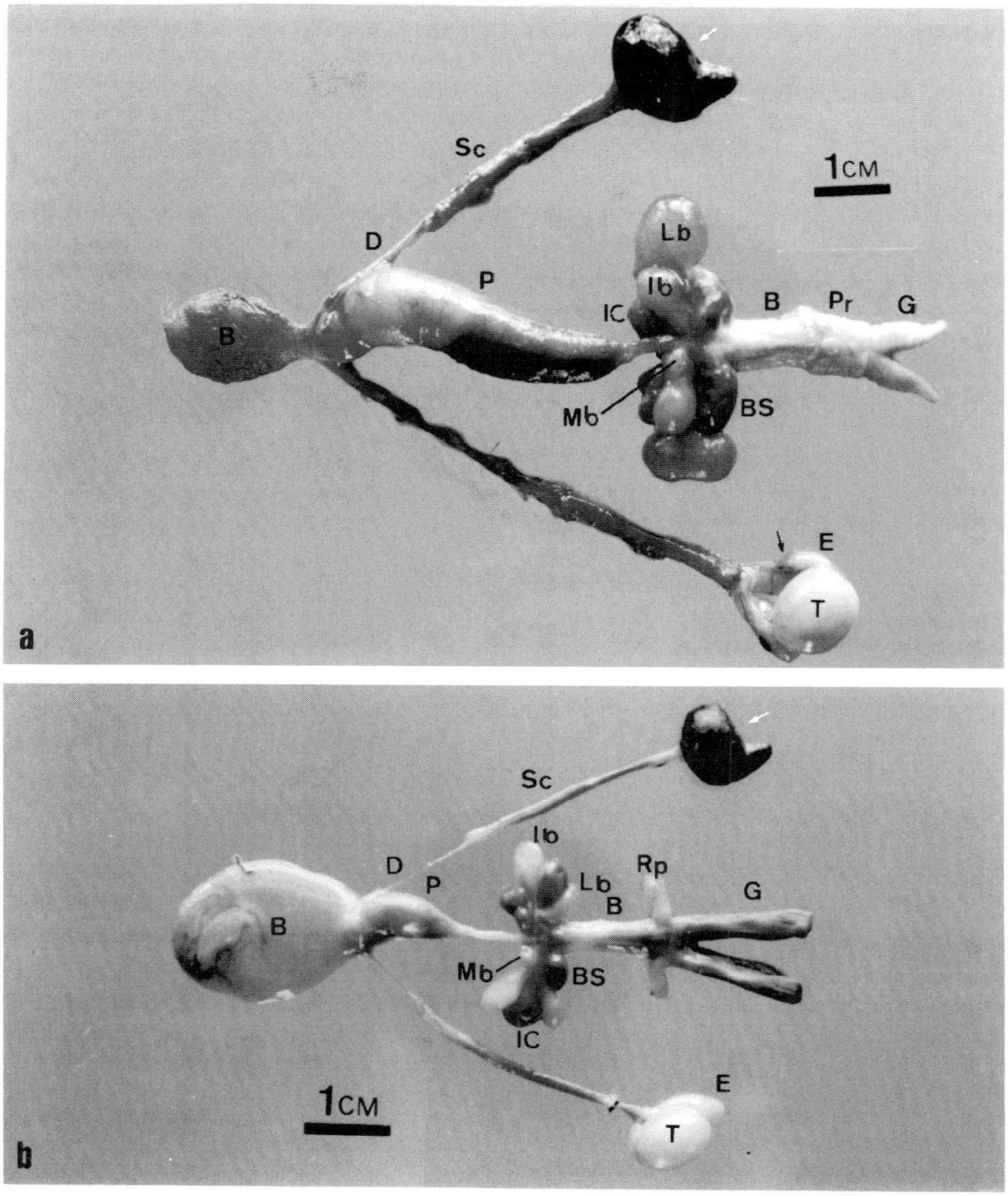

Figure 2 Male genital systems of *Didelphis albiventris* (a) and *Micoureus demerare* (b) collected in pre-mating period: T = testis; E = epididymis; Sc = spermatic cord; D = ductus deferens; P = prostate; Lb = lateral bulbourethral gland; Ib = intermediate bulbourethral gland; Mb = medial bulbourethral gland; Ic = ischiocavernous muscle; Bs = bulbospongiosus muscle; B = body of the penis; Rp = retractor penis muscle; Pr = preputium; G = glans; b = urinary bladder; the black arrow indicates the pigmented (flexuous portion) of the efferent ductules; the white arrow shows testis and epididymis enclosed by the pigmented tunica vaginalis.

Scrotum and scrotal stalk

The pendulous scrotum is located cranial to the penis (Fig. 1a, b) and covers the descended testes and epididymides. The scrotal stalk is long and narrow and contains the spermatic cord and the cremaster muscle. The skin of both the scrotum and stalk has a thin epidermis (3–5 cell layers) containing a fine layer of keratin. The extent of the skin pigmentation is extremely variable among species ranging from virtually none (*Philander opossum, Lutreolina crassicaudata, Metachirus nudicaudatus* and *Glironia venusta*) to small (*Micoureus demerare*) or large (*Didelphis aurita*) patches of pigmentation in other South American marsupials (Table 2). The dense connective tissue of the dermis

Table 1 Biometry (mean ± sd) of the male genital system of South American didelphids (Marsupialia)[1]

Species	N	Age	Length (cm)		Body weight (g)	Weight (mg)					
			Head-body	Tail		Testes	Epididymides	Prostrate	LBUG	IBUG	MBUG
Didelphis albiventris	51	A	36.8±1.2	*	1273.5±161.3	901.5±108.3	340.1±32.4	2167.3±1270.5	349.3±254.4	113.3±53.4	19.2±5.7
Didelphis marsupialis	1	A	41.0	*	1650.0	1485.0±77.8	485.0±7.1	4740.0	555.0±21.2	275.0±21.2	20.0
Philander opossum	4	A	27.5±0.7	27.3±14.4	607.5±83.2	632.5±38.6	217.5±37.7	1510.0±141.4	*	*	*
Metachirus nudicaudatus	5	A	26.5±2.1	32.5±1.5	470.8±116.6	845.0±73.7	287.5±47.7	5622.5±3081.0	*	*	*
Lutreolina crassicaudata	1	A	37.5	27.5	591.3	665.0±21.2	260.0	1760.0	*	*	*
Caluromys lanatus	3	A	29.0±0.0	40.0±0.7	355.7±109.4	441.7±94.7	226.7±52.0	1393.3±554.1	*	**	*
Caluromys philander	2	A	21.7±0.3	24.6±3.7	249.0±76.4	355.0±7.0	170.0±0.0	610.0	*	**	*
Micoureus demerare	8	A	19.8±0.9	*	142.8±24.4	378.6±50.2	100.0±23.8	372.0±161.8	*	*	*
Marmosops incanus	3	A	16.5±0.5	18.8±1.4	103.0±1.48	263.3±10.3	95.0±16.4	260.0±17.3	*	*	*
Gracilinanus agilis	9	A	10.3±0.7	12.9±0.7	38.2±5.45	170.0±27.0	62.2±12.6	610.0±105.0	136.2±55.8	**	23.7±11.5
Marmosa murina	5	A	13.2±1.1	19.2±1.2	53.5±9.0	157.3±20.1	60.4±7.0	97.8±10.4	21.8±3.3	27.8±1.5	2.6±0.5
Monodelphis domestica	2	A	13.7±1.8	*	64.4±19.1	295.0±52.0	130.5±40.0	1210.0±860.0	3.70±0.6	0.01	*
Monodelphis brevicaudata	3	A	15.6±0.7	8.86±0.2	84.3±14.9	178.1±41.3	63.6±11.9	1111.3±571.7	*	*	*
Glironia venusta	1	P	18.0	22.6	104.0	*	*	*	*	**	*

[1] Unpubished data obtained from dissection by Nogueira, JC
N= number of animals; A = adult; P = prepubertal LBUG = lateral bulbourethral glands; IBUG = intermediate bulbourethral glands; MBUG = medial bulbourethral glands; * = structures present, but not measured or weighed; ** = gland absent. The weigth of the testes, epididymides and bulbourethral glands corresponds to the means of the right and left organs.

Table 2 Anatomical characteristics of the male genital system of South American didelphids (Marsupialia)

		Pendulous scrotum and scrotal stalk					
	Age	**Pilous**	**Black pigmented**	**Tunica vaginalis: Black pigmented**	**Prostate: segmented**	**Bulbourethral Glands: pairs**	**References**
Didelphis albiventris	A	xx	Yes	xx	Yes	3	Nogueira 1988
Didelphis marsupialis	A	xx	Yes	xx	Yes	3	van den Broek 1910; Nogueira 1988
Didelphis aurita	A	xx	Yes	xx	Yes	3	Nogueira 2001
Philander opossum	A	x	No	x	Yes	3	Ribeiro 1981
Metachirus nudicaudatus	A	x	No	x	Yes	3	Costa 1995
Lutreolina crassicaudata	A	xx	No	xx	Yes	3	Nogueira 2001
Caluromys lanatus	A	x	No	xx	Yes	2	Carvalho 1996
Caluromys philander	A	xx	No	xx	Yes	2	Carvalho 1996
Micoureus demerare	A	x	Yes	xx	Yes	3	Martinelli 1990
Marmosops incanus	A	xx	Yes	xx	Yes	3	Nogueira 2001
Gracilinanus agilis	A	xx	Yes	xx	Yes	2	Nogueira 2001
Marmosa murina	A	x	Yes	xx	Yes	3	Nogueira 2001
Glironia venusta	P	xx	No	x	No	2	Nogueira et al. 1999a

Legend: A = Adult; P = Prepubertal; x = sparse; xx = dense

contains many sweat glands with prominent mioepithelial cells, hair follicles with small unilobular sebaceous glands and no conspicuous papilar layer. Dense hairs are found predominantly on the ventral surface of the scrotum.

The skin of the scrotal stalk is thin and has sparse and delicate hairs (Ribeiro and Nogueira 1991). It also presents a semilunar fold located at the caudal portion of the stalk which extends into the perineal region. In particular, the hair follicles, sebaceous and sweat glands of the stalk of *Caluromys lanatus* are more abundant than at the scrotum and show structural modifications in the different regions of the cord (Carvalho 1996) which suggests a role for these glands in temperature regulation of the cord, similar to that of some eutherian mammals (Hafez 1995).

Tunica vaginalis

Beneath to the scrotal skin is the tunica vaginalis (an evagination of the peritoneum) which covers each testis, epididymis, testicular-epididymal pedicle (TEP) and the epididymal portion of the ductus deferens, and forms a septum which separates one testis from the other. The unpigmented visceral layer of the tunica vaginalis is intimately adhered to the testis albuginea and to the epididymal capsule. The dense connective tissue of the parietal layer of the tunica vaginalis detaches easily from the overlying scrotal skin and is abundant at the insertion of the cremaster muscle. Occurrence of pigmentation in the tunica vaginalis parietalis has been reported to a variable extent in several species of South American didelphids (Table 1), with the exceptions of the semi-aquatic marsupial *Chironectes minimus* and of the caenolestid, *Caenolestes obscurus* (Rodger 1982). Interestingly, in the microbiotheriid *Dromiciops australis*, the two layers of the tunica vaginalis show black pigmentation (Woolley 1987). The degree of pigmentation intensifies gradually with the advent of sexual maturity, in some cases forming a continuous layer surrounding the testes and epididymides (Figs 2a, b). Biggers (1966) suggests that the pigmentation is involved in testicular temperature regulation by acting as a black body radiator, but its distribution among different species does not seem to correlate with climatic conditions. A more precise role for this pigmentation should be an interesting topic for future investigations.

Testis

Biometry

The testes are generally ovoid in shape with the longer axis positioned horizontally within the scrotum so that the extremitas capitata points cranially and the extremitas caudata, caudally. The average weight of the testes varies within species being about 700 to 900 mg in the larger (*Didelphis albiventris, Metachirus nudicaudatus* and *Lutreolina crassicaudata*) and 150 to 300 mg in the smaller species (*Gracilinamus agilis, Marmosa murina, Marmosops incanus* and *Monodelphis domestica*) (Table 1). Considering testicular weight as a proportion of body weight, it ranged from as low as 0.07% in *Didelhis albiventris* to as much as 0.46% in *Monodelphis domestica*. This is in agreement with the statement that in most marsupial species (R.V. Short cited by Tyndale-Biscoe and Renfree 1987) this relation is less than 0.5%. Interestingly, in adult marsupials (Table 1), there is a negative correlation ($r = 0.58$; $p < 0.01$) between body weight and testicular-body weight ratio, which indicates that smaller species have a higher testicular-body weight ratio. The functional significance of this relation is not fully understood but a similar finding has been reported by Kenagy and Trombulak (1986) for the mammalian testes. These authors state that smaller mammals must allocate a greater proportion of body mass and of energy expenditure to testicular tissue than larger mammals. Also, a functional relationship exists in many mammals between relative size of testes and mating systems. Testes are relatively small in single-male breeding systems (monogamy) and relatively large in multi-male breeding systems (promiscuous). As such, it is likely that the larger testes of *Didelphis albiventris* and *Metachirus nudicaudatus* (Table 1), as compared to the smaller testes of *Monodelphis domestica* and *Gracilinanus agilis*, may be, in part, attributed to the promiscuous way of breeding system of the former species.

Morphology

No intra-testicular septa are seen arising from the tunica albuginea. The delicate interstitial stroma lies between the Leydig cells, the seminiferous tubules (ST) and the intra-testicular excurrent ducts. The stroma represents 4 to 12% of the total testicular volume (Martinelli 1990; Queiroz 1991; Costa 1995; Carvalho 1996). The most conspicuous cells found within the stroma are fibroblasts and macrophages while mast cells are present only in *Caluromys lanatus* (Carvalho 1996). The ST are long, convoluted, well-compacted within the testis and surrounded by a fibroelastic tunica propria. Contrary to most eutherian mammals, the tunica propria of *Didelphis albiventris* (Nogueira and Redins 1987) contains fibroblasts located internally to the myoid cells and to the lamina propria, similar to the situation described in some avian species, which suggests that these cells may participate in the production of intercellular components of the tunica propria.

The ST occupy 65 to 77% of the total testicular volume and their mean diameter ranges from 218 to 250 µm in *Micoureus demerare, Didelphis albiventris, Caluromys lanatus* and *Caluromys philander* (Martinelli 1990; Queiroz and Nogueira 1992; Carvalho 1996). Within the ST are the various germ and somatic (Sertoli) cells (Fig. 3a, b). The ST are directed to the extremitas capitata where an abrupt reduction in diameter occurs (Fig. 3c). At this site, the ST lose germ cells and their wall becomes lined with modified Sertoli cells, whose apices project into the initial portion of the straight tubules (tubuli recti), thus forming a tubular valve-like structure (Fig. 3c). The straight tubules (Fig. 3d) open into a single efferent ductule (Fig. 3e,f) except in *Didelphis albiventris* (Woolley 1987), *Philander opossum* (Ribeiro 1981; Woolley 1987) and *Micoureus demerare* (Martinelli 1990), where they initially open into a branched ductule. The efferent ductule length varies amongst some American marsupials (Woolley 1987; Martinelli 1990; Costa 1995; Carvalho 1996). Leydig cells are the predominant components of the intertubular tissue (Fig. 3 a, b) and occupy 20 and 23% of the total testicular volume in *Micoureus demerare* (Martinelli 1990) and *Caluromys lanatus* (Carvalho 1996), respectively. According to Nogueira and Redins (1988), testicular macrophages of *Didelphis albiventris* are numerous and endocytically active when incorporating exogenous trypan blue. In addition, peculiar junctions with macrophage and Leydig cells are characterised by areas of electron density in the cytoplasm of both cells and electron dense material in the intercellular space. The cytoplasmic dense areas of the macrophages are anchorage points for microfibrils of the interstitial stroma. The association of macrophages with Leydig cells, their endocytic activity and their fixation in the stroma suggest that macrophages may play an important role in testicular function.

The general histological pattern of the testis is similar amongst the South American didelphids. However, the most striking difference resides mainly on the manner of openings of the straight tubules into the intra-testicular efferent ductule and also in the length of the latter ductule. The establishment of morphological patterns and the functional role of the intra-testicular excurrent ducts of American didelphids is essential for comparisons with those South American marsupials of the genera *Dromiciops*, *Caenolestes* and some didephids previously described by Woolley (1987).

Testicular descent

The differentiation of the male gonad in *Didelphis albiventris* takes place early in pouch life when the young are about 9 days old and still small (about 18 mm crown–rump length). As the gonads differentiate, the scrotum begins to delineate externally. Fonseca (1987) and Fonseca and Nogueira (1991) have described the main events related to the descent in *Didelphis albiventris* with emphasis on testicular transabdominal migration, mesonephros involution and pulling of the testis by the gubernaculum. Following gonadal differentiation, migration of the testes towards the inguinal canal takes place and penetration of the scrotum occurs when the young are about 12 cm CR length (around 90 days) and are leaving the pouch in search for the definitive environment.

Vascularisation of the testis

The testis is supplied by the testicular artery. The right and left testicular arteries of *Didelphis albiventris* (Fig. 4) originate from different points of the abdominal aorta (Godinho et al. 1977). After passing through the inguinal canal, the artery gives off numerous thin branches to form the *rete mirabili* of the spermatic cord. The *rete mirabili* is intermingled with veins of

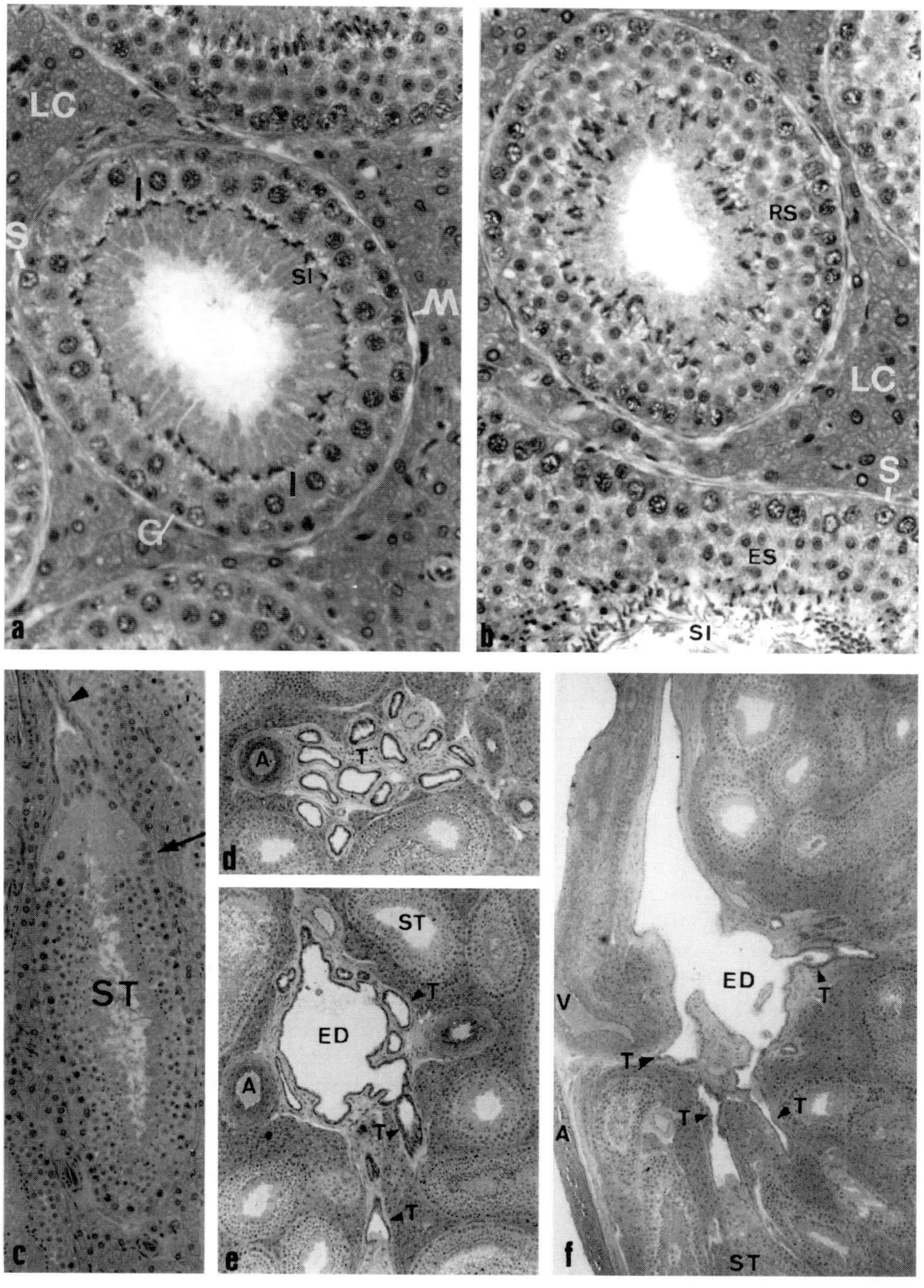

Figure 3 Histological cross-sections (a, b) of the testis of *Metechirus nudicaudatus* at different stages of the cycle of the seminiferous epithelium showing various germ cells, Sertoli cells, myoid cells and intertubular tissue. (a) Sertoli cell (S), spermatogonium (G), primary spermatocytes (I), elongated spermatid (SI) and myoid cell (M). Note the abundance of Leydig cells (LC) in the intertubular tissue. (b) Round spermatid (RS); adjacent to the cross-section, note the presence of spermatid at the initial phase of elongation (ES). HE- 200X; (c–f) intra-testicular excurrent duct system of the *Caluromys lanatus*. (c) A longitudinal section of a seminiferous tubule (ST) depicting the transition to straight tubule (arrow head). The arrow indicates the site of the abrupt disappearance of spermatogenic cells; note the valve-like structure at this transition. HE-120X. (d, e) Testicular cross-sections at the extremitas capitata showing straight tubules (T) enclosed by loose connective tissue; (A) indicates the artery. (e) Straight tubules opening in a single efferent ductule (ED). (f) Longitudinal section of the extremitas capitata depicting the straight tubules opening in the efferent ductule; (A) albuginea; (V) vein. HE-50X.

similar calibre and number. Near the testis the branches reunite into a single vessel that then penetrates the testicular parenchyma. Within the testis, the artery usually divides into two main branches that course towards the extremitas caudata. The *rete mirabili*, before reaching the proximity of the testis sends off a few branches to the epididymis. A similar vascular pattern has been reported in other Brazilian marsupials (Nogueira 2001). Interestingly, the presence of a tunica vasculosa in the testis of *Micoureus demerare* (Martinelli 1990) suggests a different intra-testicular vascularisation pattern for this species.

There is variable abdomino-testicular temperature gradient in conscious and anaesthetised marsupials, the testicular temperature being kept a few degrees below that of the body (Tyndale-Biscoe and Renfree 1987). In anaesthetised *Didelphis albiventris* (Godinho et al. 1977), *Trichosurus vulpecula* (Tyndale-Biscoe and Renfree 1987) and *Macropus eugenii* (Tyndale-Biscoe and Renfree 1987), the mean gradient is 1.6, 2.4 and 5.4°C, respectively. In cryptorchid testis of *Trichosurus vulpecula* and *Macropus eugenii* (Tyndale-Biscoe and Renfree 1987), no such gradient is observed. Since artificial cryptorchidism may cause disruption of spermatogenesis in *Macropus eugenii* (Setchell and Thorburn 1971), as in eutherians (Tyndale-Biscoe and Renfree 1987), it is most likely that a body–testicular temperature gradient be necessary for the normal process of spermatogenesis to occur.

Testicular-epididymal pedicle (TEP)

The testis is attached to the epididymis by the TEP, which is divided into straight and convoluted portions. The TEP emerges at the extremitas capitata of the testis and encloses the straight extra-testicular portion of the efferent ductule as well as blood, lymphatic vessels and nerves (Ladman 1967; Nogueira et al. 1977). The epithelium lining the efferent ductule is composed of secreting cells (neutral mucosubstances) interspersed by ciliated cells and intraepithelial lymphocytes. Ciliated and non-ciliated cells are also present in *Didelphis virginiana* (Ladman 1967; Martan 1983), *Caluromys derbianus* (Martan et al. 1967) and in *Caenolestes obscurus* (Rodger 1982; Woolley 1987). Although neutral mucosubstances are produced in the epithelia of the efferent ductule of American marsupials, variations in the content of acidic or sulphated mucosubstances may occur. In *Didelphis virginiana*, *Caluromys derbianus* and *Marmosa sp* (Martan et al. 1967), phospholipids are also produced. Mast cells are often seen in the TEP connective tissue. The convoluted portion (Fig. 5a) forms an oval-shaped distinct structure enclosed by a dense connective capsule that is continuous with the epididymal capsule at the transition of the head-body of the epididymis. Unlike other genera, in some marsupials such as *Didelphis albiventris* (Fig. 2a), *Didelphis aurita, Didelphis marsupialis, Philander opossum* and *Lutreolina crassicaudata*, the convoluted portion of the efferent ductules shows an intense pigmentation. In *Didelphis virginiana*, serotonin was detected in the epithelium and in mast cells of the efferent ductules (Anderson et al 1979). It also displays marked modifications in epithelial cell height and apex (Fig. 5c) as well as in their secretion. According to Martinelli (1990), the convoluted portion of the efferent ductules enhances the contact of sperm in transit thus suggesting a role in the reabsorption process of testicular fluid or even in the sperm maturation process occurring within the epididymis.

Epididymis

This organ lies dorsally to the testis (Fig. 2a, b), the tail directed ventrally to the extremitas caudata of the testis. It is lobulated (Fig. 5a) and is anatomically divided into head, body and tail. The head is wide, flattened and continuous with the narrow body. The well-developed tail is round (*Monodelphis domestica*) or conical (*Caluromys lanatus, Caluromys philander, Gracilinanus agilis*) in shape, is easily visualised externally and projects beyond the testicular border. The epididymis which is surrounded by a thin capsule is attached to the testis by the mesepididymis. Mean epididymal weight ranges from as low as 60 mg in *Marmosa murina, Gracilinanus agilis*, and *Monodelphis brevicaudata* to as high as 485 mg in *Didelphis marsupialis* (Table 1). The regions of the head, body and tail of the epididymis of marsupials play a role in testicular fluid absorption, sperm maturation and storage, respectively (Setchell 1977; Tyndale-Biscoe and Renfree 1987).

Epididymal parenchyma consists of a coiled, compactly arranged epididymal duct within the lobules. This duct is lined by a columnar epithelium (predominance of principal cells; Fig. 5b) with long apical stereocilia, basal cells, apical cells and intraepithelial lymphocytes (Orsi et al. 1980; Martinelli 1990; Costa 1995). The epididymal duct is divided into various zones based on the morphohistochemical characteristics of its epithelium: six in *Didelphis marsupialis, Didelphis albiventris* and *Philander opossum* (Rodger 1982), seven in *Didelphis albiventris* (Orsi et al. 1981), *Marmosa murina* (Rodger 1982) and *Micoureus demerare* (as *Marmosa cinerea)* (Martinelli and Nogueira 1992), and nine in *Metachirus nudicaudatus* (Costa 1995). Morphofunctional maturation and sperm-pairing takes place during sperm transit in the epididymis (Fig. 5d, e). Taggart et al. (1993) divided the epididymis of *Monodelphis domestica* into 18 segments and described in detail the ultrastructural events related to sperm-pairing in the epididymis. Sperm-pairing also occurs in all South American didelphids studied. The epididymal zone of sperm pairing is characterised by a greater capacity of secretion of neutral mucosubstances, sialomucins, glycogen and/or absorption in the principal cell (Martinelli 1990; Costa 1995). The processes of epithelial secretion, functional maturation and sperm pairing are androgen dependent (Kelce et al. 1987).

An interesting finding in *Didelphis albiventris, Didelphis aurita, Philander opossum, Micoureus demerare, Caluromys lanatus,*

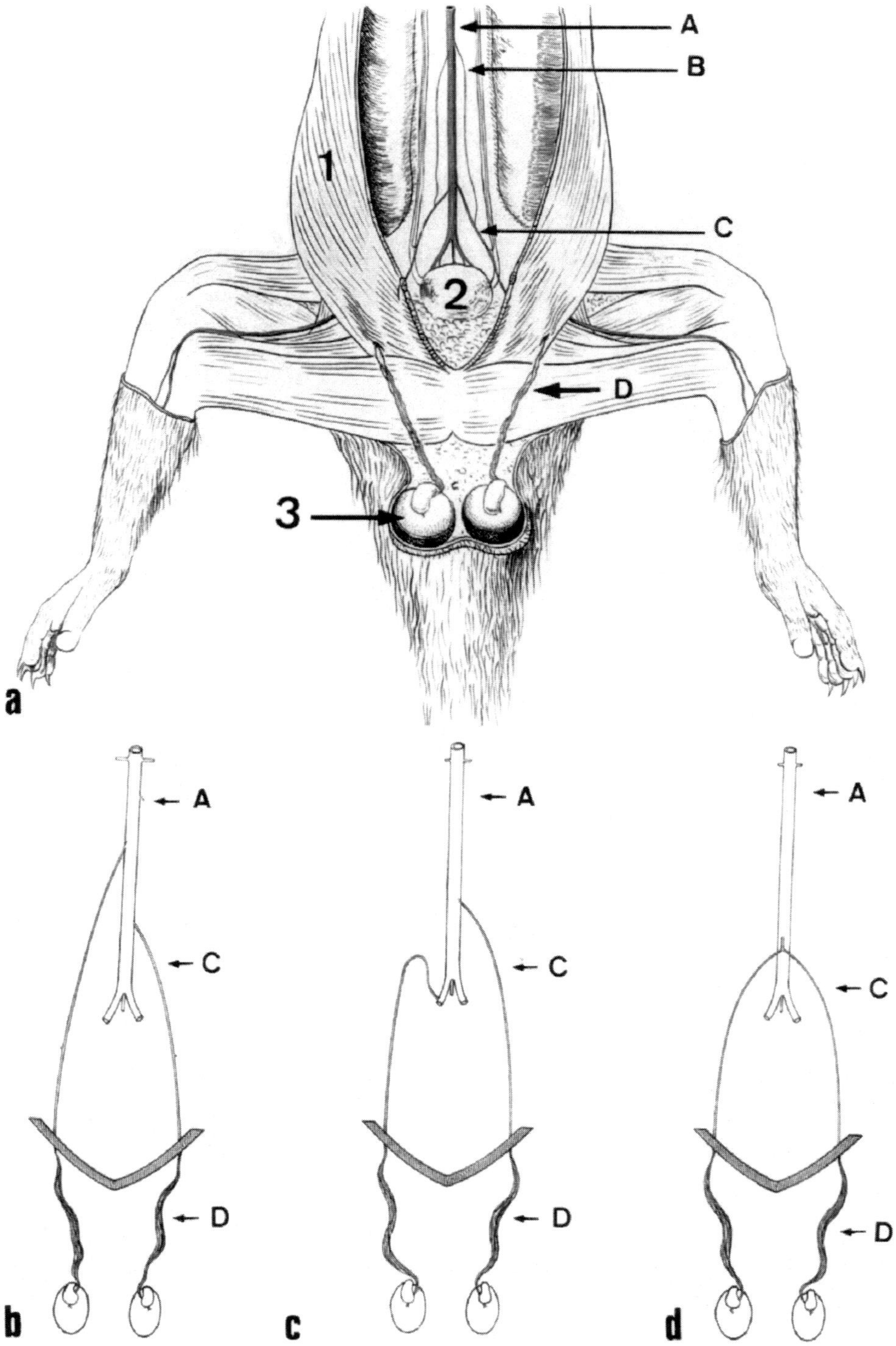

Figure 4 Schematic drawings of the right and left testicular arteries of *Didelphis albiventris* from the origin in the abdominal aorta to the penetration of each testis. 1 = abdominal wall; 2 = urinary bladder; 3 = testis; A = abdominal aorta; B = accessory testicular artery; C = testicular artery; D = *rete mirabile*. (a) The arteries originate independently from the abdominal aorta at about the same level and each of them receives one accessory testicular artery. (b) The origin of the arteries are separated from each other. (c) The right testicular artery originates from the external (common) iliac artery. (d) Both arteries originate by way of a common trunk from the abdominal aorta. With permission of *Acta Anat.*, **99**:204–208 (1977).

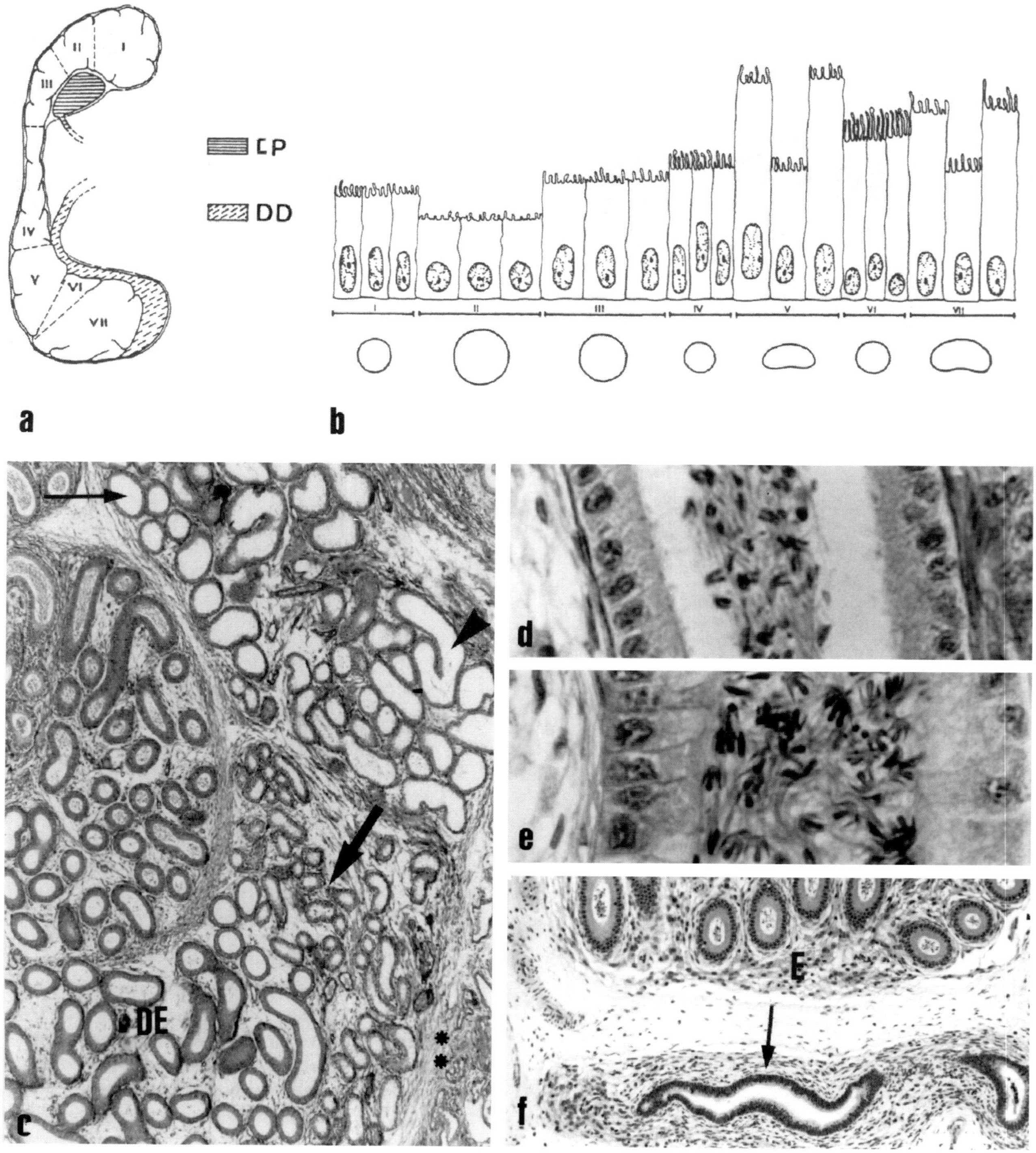

Figure 5 (a) Schematic drawing of the histological zones of the epididymis of *Micoureus demerare* : convoluted portion of the efferent ductules (CP); ductus deferens (DD). (b) Morphological variations in the epithelium, diameter and shape of the lumen of the ductus epididydymis in the different zones of *M. demerare*. With permission of *Rev. Bras. Ciên. Morfol.*, 9(2): 26-31 (1992). (c) Initial (thin arrow), intermediate (arrowhead) and final (thick arrow) regions of the convoluted portion of the efferent ductules of *Caluromys lanatus*. The transition of the efferent ductule to epididymal duct (DE) is indicated by asterix. (d) Unpaired and (e) paired spermatozoa in the epididymal lume of *Metachirus nudicaudatus*. PAS + H – 480X. (f) Duct (arrow) along the lateral border of the epididymis (E).

Caluromys philander and *Metachirus nudicaudatus* is the presence of a duct (Fig. 5f) along the lateral border of the epididymis. Within the columnar epithelium of this duct are slender and compactly arranged cells, which are frequently seen to be eliminating secretory globules at their apical portion. Such a duct in *Metachirus nudicaudatus* branches at the terminal portion of zone VIII giving rise to a tubular gland made up of PAS-positive secretion. According to Martinelli (1990) and Costa (1995), this duct is probably an embryonic remnant of the paramesonephric duct.

Ductus deferens

The ductus deferens is long and divided into four portions: epididymal, funicular, abdominal and intraparietal. The epididymal portion, which lies ventromedial to the body of epididymis, begins at the tail and has a convoluted path. The straight funicular portion runs along the entire length of the spermatic cord and passes through the inguinal ring. The abdominal part extends from the internal inguinal ring to the adventitia of the prostate. The intraparietal portion is the shortest and crosses the cranial segment of the prostate to open into the urethra below to the neck of the bladder. An important difference between eutherians and marsupials lies in the path of the ductus deferens. In eutherian mammals, this duct loops around the ureter before emptying into the urethra (Tyndale-Biscoe and Renfree 1987) while in marsupials it does not cross the ureter. Except for *Caenolestes obscurus* (Rodger 1982) in which an ampulla-like structure has been described, South American didelphids lack an ampulla.

The mucosa of the ductus deferens shows regional variation and has been described in *Philander opossum* (Ribeiro 1981), in *Didelphis albiventris* (Machado et al. 1982) and in other South American marsupial species (Rodger 1982; Martinelli 1990; Costa 1995; Carvalho 1996). It is lined by a slightly acidophilic tall and narrow columnar pseudostratified epithelium with numerous long apical projections. The ductal lumen is irregularly shaped and indented (Fig. 6c) due to the presence of groups of intercalated short columnar cells. Amongst the epithelial cells are small basal and apical cells and lymphocytes. The lamina propria is thin and rich in elastic fibres. The muscular layer is formed mainly by smooth, circularly arranged muscle fibres. Histochemical analysis revealed that both funicular and abdominal portions secrete neutral and acid carboxylated mucosubstances in *Philander opossum, Micoureus demerare, Metachirus nudicaudatus, Caluromys lanatus* and *Caluromys philander* (Ribeiro 1981; Martinelli 1990; Costa 1995; Carvalho 1996). The intraparietal region in *Caluromys lanatus* (Carvalho 1996) secretes neutral mucosubstances. According to Rodger (1982), the ductus deferens of *Caenolestes obscurus*, contrary to didelphids and all other marsupials, is a convoluted ampulla-like structure adjacent to the prostate gland, the epithelial cells of which are PAS and Alcian blue reactive.

Spermatic cord

The morphology of the spermatic cord is similar among South American marsupials (Godinho et al. 1977; Ribeiro 1981; Machado et al. 1982; Martinelli 1990; Costa 1995; Carvalho 1996). It consists of the ductus deferens, testicular artery and vein, lymph vessels, nerves and cremaster muscle (Fig. 6a, b), all of which are enclosed by the funicular stalk. The well-developed cremaster is long and partially surrounds the ductus deferens, vessels and nerves. Its contraction allows the testis to move close to the body wall whenever necessary (under conditions such as stress and defence). In the medial portion of the cord lies a band of dense connective tissue (Fig. 6b), which emerges from the fusion of the parietal and visceral lamina of the tunica vaginalis and obliterates the inguinal canal. This band separates the cremaster from the other funicular components. The canal vaginalis can be partially seen at the proximal and distal regions of the cord. The testicular artery forms along the cord the *rete mirabile* (Fig. 6b, d) which has been implicated in the regulation of testicular temperature leaving both the testicular and scrotal temperature below that of the body (Barnett and Brazenor 1958; Godinho et al. 1977; Setchell 1977; Tyndale-Biscoe and Renfree 1987).

Prostate

This is disseminate and well-developed in marsupials. It is related to the neck of the bladder cranially and with the membranous urethra, caudally (Fig. 2a, b). This carrot-shaped gland is elongated and is more developed at the mating period, at which time its shape changes to being spiral-coiled, and its weight and size increase several times, as seen in *Didelphis albiventris* (Nogueira 1988) and in *Metachirus nudicaudatus* (Costa 1995). The prostate has three anatomical segments, which can be distinguished by colour, length and diameter. The cranial segment is the smallest and pinkish; the medial is the largest and widest and milky in colour; the caudal is the narrowest and dark grey. The parenchyma contains branched glandular tubules (GT) that differ in histology and histochemistry in the the cranial, middle and caudal segments (Fig. 7a). These tubules generally show intense secretory activity during the mating period. The GT produce neutral mucosubstances in all segments; acid carboxylated and sulphated mucosubstances are also found in varying amounts (Nogueira et al. 1985; Martinelli et al. 1991; Costa 1995; Carvalho 1996), proteins and phospholipds (Hruban et al. 1965; Martan and Allen 1965). Interspersed amongst the GT are numerous mast cells and smooth muscle fibres. The latter participate in the contraction of the glandular parenchyma. The prostate also produces glycogen which is the most important energetic component of the semen of American marsupials (Rodger and White 1980). However, this latter statement refers only to the prostate of *Didelphis virginiana* and to date has not been extensively investigated in South American marsupials. The caudal segment is the main producing site of

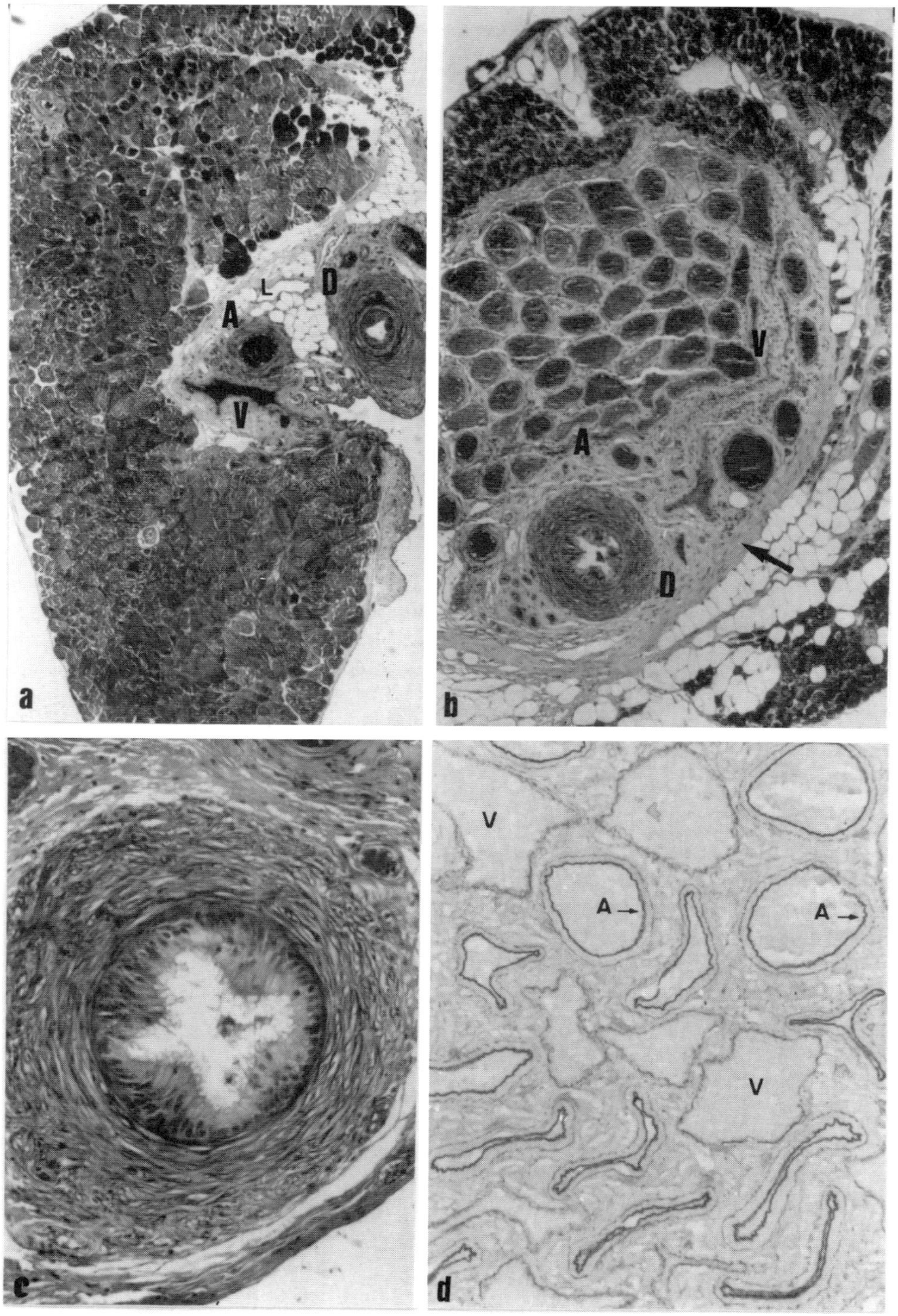

Figure 6 (a) Cross-sections of the proximal (a) and middle (b) portions of the spermatic cord of *Didelphis albiventris*. D = ductus deferens; A = testicular artery; V = testicular vein; L = lymphatic vessel. External to the above mentioned structures is the well-developed cremaster. Gomori's trichrome –30X. (b) Blood vessels and the ductus deferens (D) are separated from the cremaster by dense connective tissue (arrow); numerous branches of the *rete mirabile* of the testicular artery (A) followed by the veins (V). – Gomori's trichrome –52X. The arteries are easily identified in (d) by the presence of their inner elastic membranes – Modified Weigert-Fuchsin- 185X. (c) High magnification of the ductus deferens – Gomori's trichrome. With permission of *Acta Anat.*, **99**:209–219 (1977).

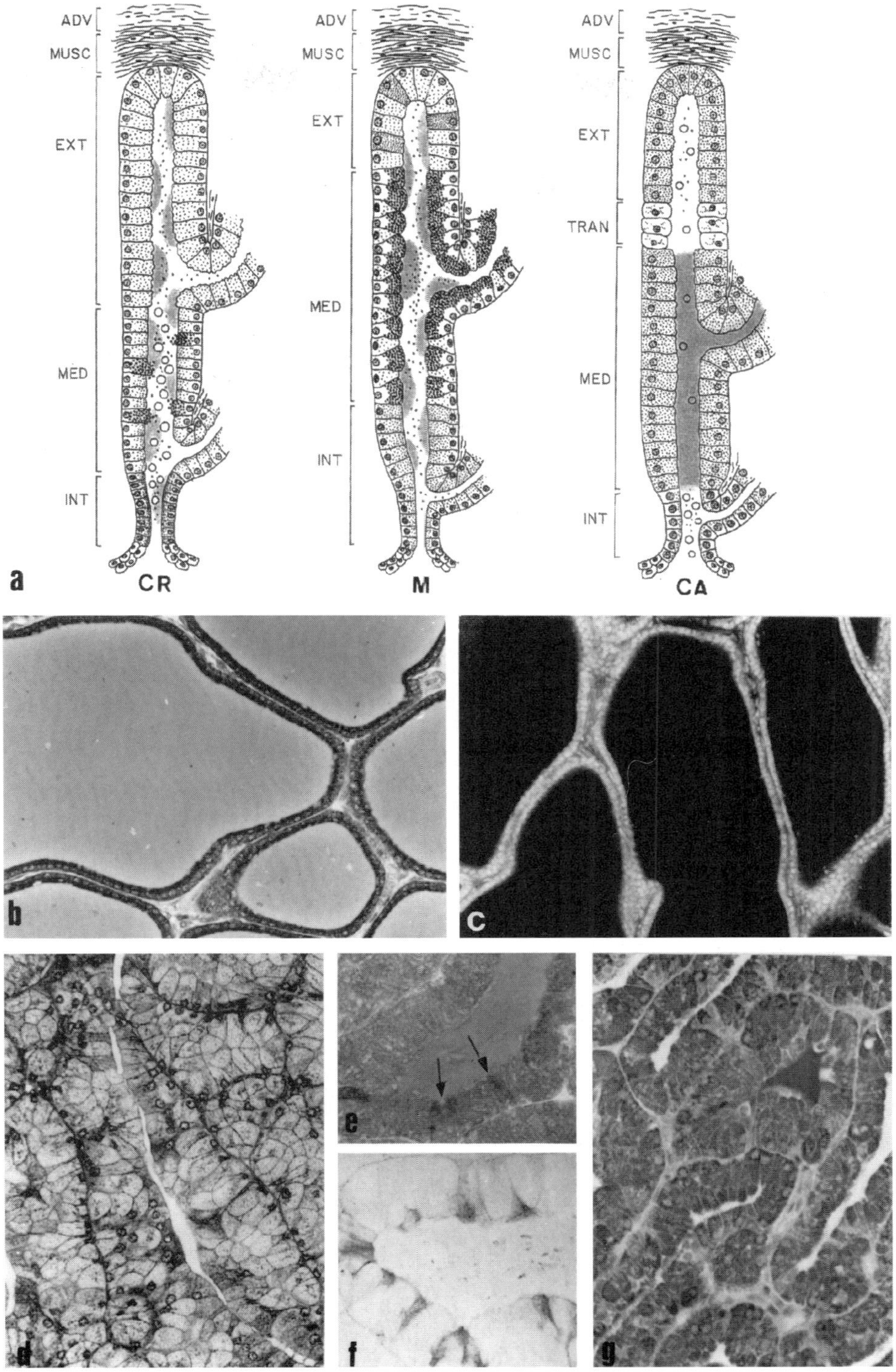

Figure 7 (a) Drawings of the cranial (CR), middle (M) and caudal (CA) segments of the prostate of *M. nudicaudatus* depicting the different cell types of the external (EXT), middle (MED) and internal (INT) zones of tubular glands. ADV= adventitia; MUSC= tunica muscularis. (b–c) Lateral bulbourethral gland with expanded secretory tubules stained with toluidine blue (b) and PAS (c). 190X. (d, e, f) Intermediate bulbourethral gland; stained with toluidine blue (d). The epithelium and its secretion are alcianophilic (e); some cells (arrows) are stained with Alcian blue – pH 2.5) even after acid hydrolisis (f). 190X. (g) PAS positive reaction in the cytoplasm and in the luminal secretion of medial bulbourethral gland. 190X.

glycogen in *Didelphis albiventris* (Garcia and Gonçalves 1984) and in *Philander opossum* (Nogueira et al. 1985). In *Didelphis albiventris*, glycogen concentration in the caudal segment increases markedly during the mating period (Cóser and Nogueira 1988), thus indicating a seasonal activity of the gland.

Bulbourethral glands (Cowper's gland)

Two or three pairs of bulbourethral glands (BG) are present in South American marsupials, which are named lateral, intermediate and medial based on their anatomical relations (Nogueira et al. 1984). Except for *Caluromys lanatus, Caluromys philander, Gracilinanus agilis* and *Glironia venusta* (two pairs), most species (Table 1) of the other genera have three pairs (Table 1). The piriform lateral and the intermediate BGs lie dorsolateral to the urethra and are related dorsally with the ischiocavernous and bulbospongiosus muscles of the radix penis. The lateral BGs are translucent while the intermediate are smaller and opaque except in *Micoureus demerare* (Martinelli 1990). The round medial BGs are the smallest, lie lateral to the urethra and are related caudomedially with the ischiocavernous muscle. Depending on the species, the main ducts of the BGs empty into the initial portion of the penile urethra.

The parenchyma of the bulbourethral gland consists of long branched secretory tubules that greatly expand (Fig. 7b–g) at the mating period (Nogueira 1988). A detailed description of the ultrastructure of this secretory epithelium in *Didelphis albiventris* has been presented by Nogueira and Redins (1989). The main ducts are also secretory in nature and show interspersed endocrine serotonin-producing cells in the epithelium near the urethral opening. Qualitative variations in the content of the BGs secretion, glycogen, neutral mucosubstances and acid carboxylated and sulphated mucosubstances have been detected in several South American didelphids (Nogueira et al. 1984; Ribeiro and Nogueira 1985; Martinelli et al. 1991; Costa 1995; Carvalho 1996). In *Didelphis virginiana*, the three pairs of bulbourethral glands showed no histochemical reaction to glycogen but were reactive to neutral and acid mucosubstances (Martan 1983). The presence of either two or three pairs of BGs and their complex composition in American marsupials strongly suggest a more specific role in their reproductive physiology, other than that of a mere lubricating function of the urethra before ejaculation. Further studies involving biochemical and physiological methods should be conducted in order to elucidate the composition of these secretions as well as to determine the manner in which these are eliminated with respect to the sperm fraction of the semen.

Urethra

The pelvic urethra presents prostatic and membranous segments. The long prostatic segment is related with the prostate tubules, which release their secretion into the urethral lumen. The mucosa is lined by a transitional epithelium and the lamina propria which consists of loose connective tissue and a rich blood and lymph plexus. Sparse endocrine serotonin-producing cells are found along the urethra of *D. albiventris* (Nogueira and Barbosa 2001). The membranous urethra that lies between the caudal segment of the prostate and the radix penis is also lined by a transitional epithelium which lacks glands in its mucosa. After penetrating the penis, the urethra becomes the penile urethra.

Penis

The penis of South American marsupials follows the basic pattern found in other marsupials (van den Broek 1910; Woolley and Webb 1977; Wooley 1982; Ribeiro and Nogueira 1990; Martinelli 1990; Martinelli and Nogueira 1997; Nogueira et al. 1999a, b). This organ occupies a post-scrotal position and attaches to the ischiatic arch by means of the suspensory ligament. In the non-erect state (Fig. 8), the body of the penis displays a pronounced sigmoid flexure (S-shaped curve). When flaccid, it retracts to its normal position, due mostly to the contraction of the retractor muscles. Upon erection, it is projected through the cloaca, thus unfolding the sigmoid flexure.

The major differences in penile morphology between species and genera are associated with the glans and the levator muscles (LM). In all species studied, the voluminous musculature of the radix penis is formed by the ischiocavernous (IC) and bulbospongiosus (BS), which are mostly situated outside of the pelvic cavity. Both muscles converge to join in the radix penis. Encased in the radix penis are the paired levator muscles (LM), which are long, broad or thin, and ventrally situated to the IC and BS muscles and to the main ducts of the bulbourethral glands. The insertion tendon of the LM penetrates the body of the penis, branches at the level of the glans to insert in the albuginea of the cavernous body or in the preputial fornix. In *Metachirus nudicaudatus*, contrary to all other South American marsupials, the LM fuse to each other, thus forming a long muscular band that inserts in the preputial fornix through a small aponeurosis. The paired retractor penis muscles (RP) arise in the sublumbar region and insert on both sides of the dorsal curvature of the sigmoid flexure. Comparative drawings of the major variations of muscles and their tendons are presented in Fig. 8B–M.

The shape and length of the bifid glans penis is extremely variable (Table 3). As the glans bifurcates, the urethra gives rise to two urethral grooves that runs to a variable extent, along the medial surfaces of the hemiglans and terminate at the ventral aspect (*Didelphis albiventris, Didelphis aurita, Didelphis marsupialis, Philander opossum, Lutreolina crassicaudata, Marmosops incanus, Gracilinanus agilis*), at the tip (*Caluromys lanatus, Caluromys philander, Micoureus demerare, Glironia venusta*), very close to the tip (*Marmosa murina*) or at the lateral aspect (*Monodelphis domestica, Monodelphis brevicaudata*) of the glans. In *Metachirus nudicaudatus*, unlike other South American didelphids, the urethral grooves extend a short distance of each glans

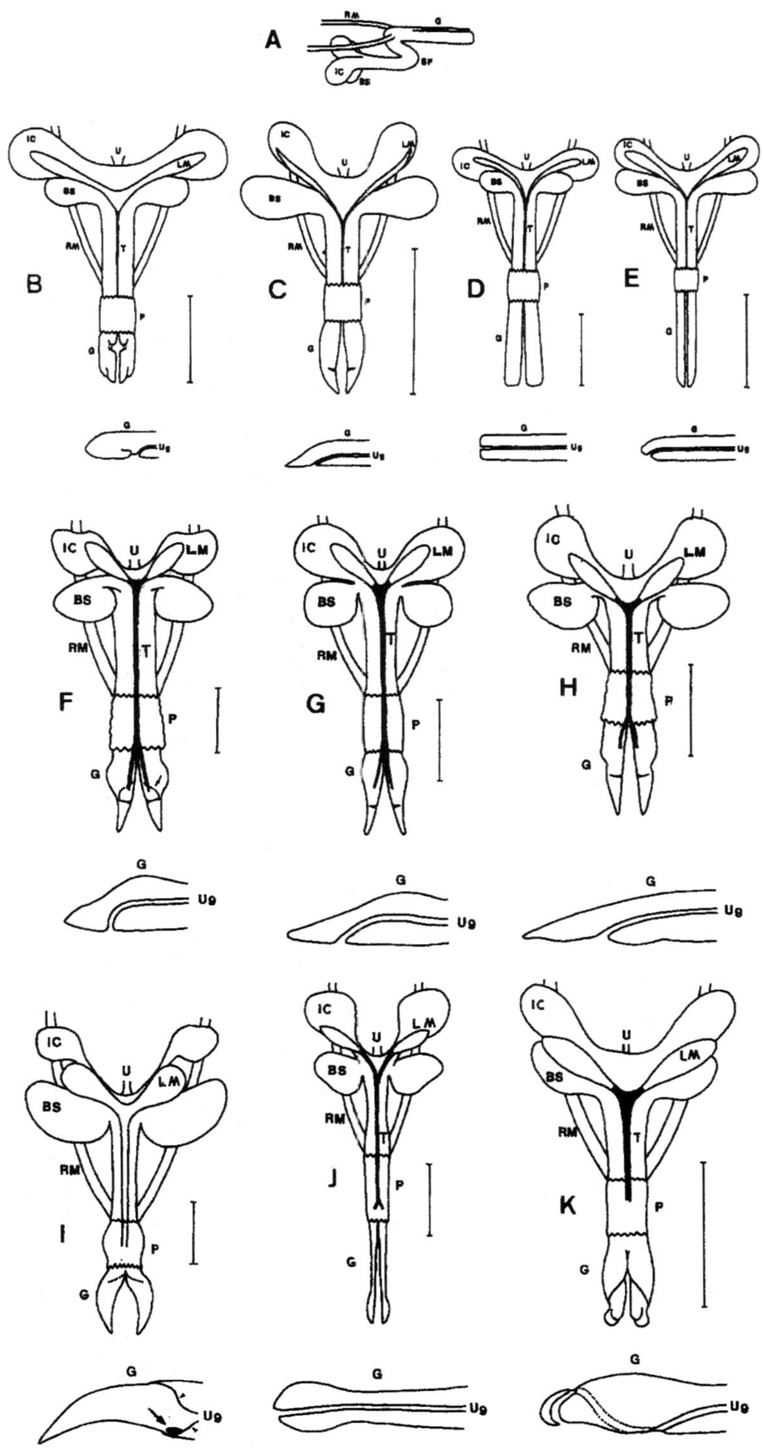

Figure 8 (A) Drawing of the penis (flaccid) of *Micoureus demerare* displaying the prominent sigmoid flexure (SF). (B–K) Drawings of distended penis (ventral view) with the respective hemi-glans (medial view). B = *Marmosops incanus*; C = *Gracilinanus agilis*; D = *Micoureus demerare*; E = *Marmosa murina*; F = *Didelphis*; G = *Philander*; H = *Lutreolina*; I = *Metachirus*; J = *Caluromys lanatus*; K = *Monodelphis*. Bar = 1 cm. Legends: IC= ischiocavernous muscle; BS = bulbospongiosus muscle; LM = levator penis muscle; U = membranous urethra; RM = retractor penis muscle; T = tendon; P = preputium; G = glans; UG = urethral groove. Arrows = opening of the diverticulum; Arrowheads = dorsa and ventral erectile folds.

Table 3 Anatomical aspects of the glans penis of South American didelphids (Marsupialia)

Species	Age	Insertion of LM	Shape	Length	Ending of urethral grooves	Peculiar structures	References
Didelphis albiventris	A	T	Conical	Moderate	Glans halves	Diverticula	Nogueira 1988
Didelphis aurita	A	T	Conical	Moderate	Glans halves	Diverticula	Nogueira et al. 1999b
Didelphis marsupialis	A	T	Conical	Moderate	Glans halves	Diverticula	van den Broek 1910; Nogueira 1988
Philander opossum	A	T	Conical	Moderate	Glans halves	Diverticula	Ribeiro 1981
Lutreolina crassicaudata	A	T	Conical	Moderate	Glans halves	Diverticula	Nogueira et al. 1999b
Metachirus nudicaudatus	A	M	Conical	Moderate	Glans halves	Diverticula; 1	Costa 1995
Caluromys lanatus	A	T	Elliptical	Long	Glans tips	–	Nogueira et al. 1999b
Caluromys philander	A	T	Elliptical	Long	Glans tips	–	Nogueira et al. 1999b
Micoureus demerare	A	T	Not pointed	Long	Glans tips	–	Martinelli and Nogueira 1997
Marmosops incanus	A	T	Tapered	Long	Before tips	2	Martinelli and Nogueira 1997
Gracilinanus agilis	A	T	Pointed	Long	Before tips	–	Martinelli and Nogueira 1997
Marmosa murina	A	T	Rounded	Long	Glans tips	–	Martinelli and Nogueira 1997
Glironia venusta	P	T	Elliptical	Short	Near the tips	–	Nogueira et al. 1999a

Legend: A = Adult; P = Prepubertal; LM = Levator muscles; T = Tendon; M = Muscle; 1 = Dorsal and ventral glans folds; 2 = Ostium-like depressions and lateral appendices half, delimited by dorsal and ventral pronounced folds of erectile tissue. In this species, these structures form two functional channels. Ventral to the urethral grooves are located diverticula of unknown functional significance with different topography and openings in *Didelphis, Philander, Lutreolina* and *Metachirus* (Nogueira et al. 1999 b). All morphological characteristics of the urethral grooves are also summarised in Fig. 8B–M.

Although there are similarities among some species in their phalic morphology, the comparative analysis of this character, allied with other external characters, can be useful in the identification of the species in the field, as well as for the establishment of the phylogenetic relationships among the American marsupials. Further studies of the phallic morphology of other genera and species are needed to determine the consistency of differences among the species of the genera *Chironectes, Thylamys* and *Caluromysiops.* Knowledge of the functional aspects of mating and insemination, as well as morphological studies of the female genital system are also important to provide a more complete understanding of the existing differences in the phallic morphology.

Cloaca and paracloacal glands

This is a cavity in which the digestive and genito-urinary systems open dorsally and ventrally (Fig. 1b), respectively. The prepuce is continuous with the floor of the genito-urinary opening. The cloaca communicates externally through a cloacal cleft. Lateral to this cleft and at the genital opening is a group of cloacal glands (Fig. 9a, b). In *Micoureus demerare*, Martinelli (1990) found piriform glands (Fig. 9a) at each side of the cleft. The largest glands are more ventrally located. The glandular bodies lie among the striated muscle fibres of the cloacal sphincter. This also occurs in *Caluromys lanatus* and *Caluromys philander* (Carvalho 1996). The strategic localisation of these glands is important in the elimination of glandular secretion during contraction of the sphincter thus promoting the lubrication of the cloacal mucosa. Two pairs of paracloacal (anal) glands (Fig. 9a, b) empty in the cloacal mucosa close to the digestive tract opening.

The cloacal mucosa is lined by cornified squamous stratified epithelium (Fig. 9c) which rests in a dense, well vascularised lamina propria containing a few glands. Externally, the cloaca is covered by a folded skin, rich in hair follicles and sebaceous glands (Fig. 9c). The dermis is thin and the dermal–epidermal junction lacks a papillary layer. The hypodermis has small fat lobules intermingled with the coiled parts of sweat glands and hair folicles. The cloacal sphincter separates the deep portion of the hypodermis from the cloacal mucosa. The cloaca of *Caluromys* is the widest amongst South American marsupials studied. Based on morphological and histochemical observations, Martinelli (1990) and Carvalho (1996) state that it is possible that the major and minor cloacal glands of *Micoreus demerare* are modified sebaceous and odoriferous glands, respectively. The presence of a well-developed glandular complex around the cloacal cleft of *Micoreus demerare, Caluromys lanatus* and *Caluromys philander* must be of physiological significance in the processes of penile erection and exposure by lubricating the wall opening to facilitate the sliding of the glans.

Paracloacal glands in marsupial are thought to be a form of scent gland and are often referred to as cloacal, anal, perianal and odoriferous gland. There two pairs of these glands in South American didelphids (*Didelphis marsupialis, Didelphis virginiana, Lutreolina crassicaudata* (Schaffer and Hamperl 1926); *Didelphis aurita* e *Didelphis albiventris* (Munhoz and Merzel 1967); *Philander opossum* (Ferrari et al. 1987); *Micoreus demerare* (Martinelli 1990); *Metachirus nudicaudatus* (Helder-José 1995); *Caluromys lanatus, Marmosops incanus* and *Gracilinanus agilis* (Nogueira, personal information). Two pairs of glands, one on the right and the other on the left of the anal canal, are formed, each consisting of a major and a minor portion. Each gland originates from a single duct which opens in the cloacal mucosa. According to Helder-José (1998) the opening of the glandular ducts in anal mucosa as described in *Didelphis albiventris* (Munhoz and Merzel 1967) and *Philander opossum* (Ferrrari et al. 1987) must be revised. Their wall is made up of three layers: a mucosal (inner), a muscular (intermediate) and a capsule (external). A detailed morphological description of these glands in *Metachirus nudicaudatus* was presented by Helder-José and Freymüller (1995). The inner one is a mucosa the epithelium of which contains holocrine cells, rich in lipid droplets. The surrounding vascular lamina propria contains flattened tubular aprocrine glands whose epithelial cells contain abundant endoplasmic reticulum, prominent Golgi complexes and numerous secretory granules. The middle layer is formed by skeletal striated muscle. The capsule is formed by dense connective tissue. One of these ducts comes from the central cavity, lined by holocrine epithelium, and the other originates from the branched tubular glands of the lamina propria.

In *Metachirus nudicaudatus*, Helder-José (1998) conducted an experiment using hormone replacement (testosterone) to evaluate the possible relationship between testosterone and the paracloacal glands. The author concluded that the development of the major gland but not of the minor appeared to be under the influence of androgen. The size of the major gland was retained throughout the reproductive period but underwent regression during the non-reproductive period. The reduction in the volume of the major glands upon castration during the reproductive period followed by the restoration of the volume to its original size after testosterone treatment suggest that the major glands are androgen dependent.

The role of the paracloacal glands of marsupials is not fully elucidated but they are thought to be involved in defence mechanisms, territorial marking, attraction and sexual dimorphism (Helder-José 1998). Furthermore, the paracloacal glands of

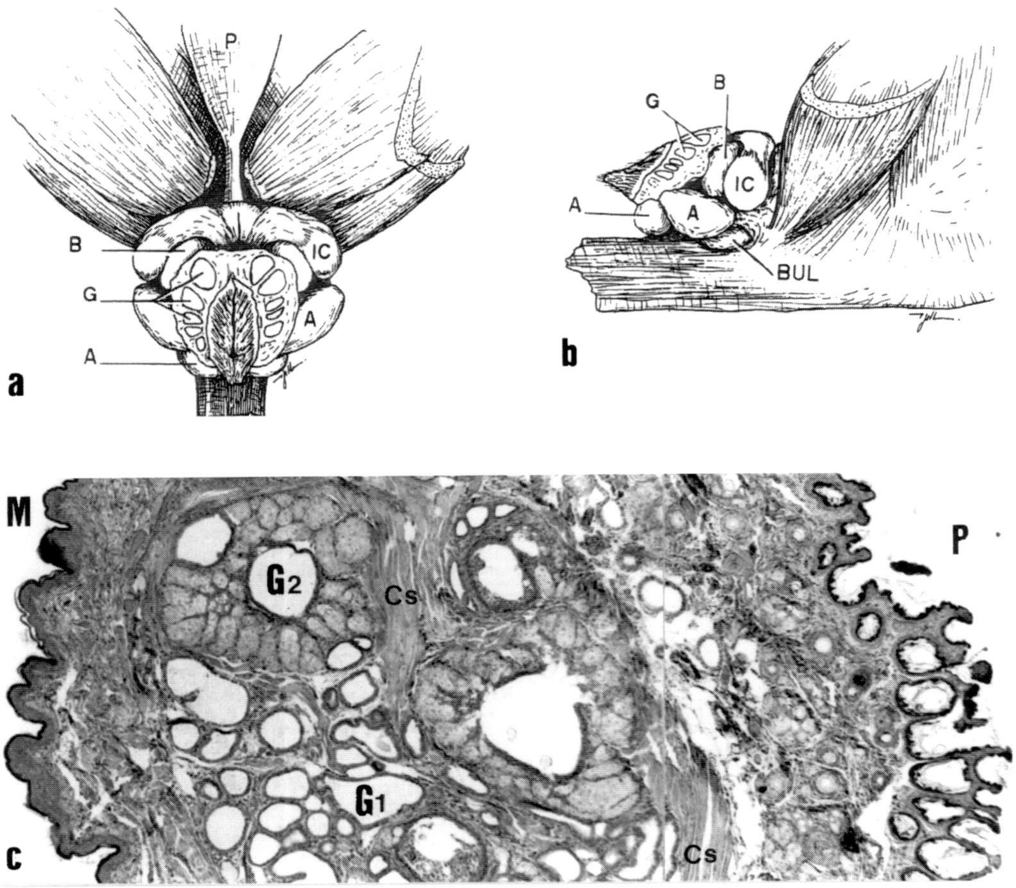

Figure 9 (a, b) Schematic illustrations of the ventral and lateral aspects of the perineal region of *Micoureus demerare*, respectively. A = paracloacal (anal gland); G = cloacal glands; B = bulbospongiosus muscle; IC = ischiocavernous muscle; p = prostate. BUL = lateral bulbourethral gland. (c) View of the cloacal wall of *C. philander*. M= mucosa; minor (G1) and major (G2) cloacal glands; CS = muscular fibres of the cloacal sphincter; P = skin. HE-46X

Didelphis serve as a reservoir of naturally (Naiff et al. 1987) and experimentally (Deane et al. 1984; Lenzi et al. 1984) infected *Trypanosoma cruzi*. Such findings could assume a relevant role in the epidemiology of Chagas' disease being an alternative route of transmission of the disease.

Innervation of the Male Genital System

Histochemical analysis for the demonstration of catecholamines (Tonelli 1982) and histochemical and surgical procedures of denervation (Alves 1986) have been employed in the study of the adrenergic innervation of the male genital system of *Didelphis albiventris*. The organs of the urogenital system are innervated by the sympathetic and parasympathetic autonomic nervous systems. Dissection of nerves and ganglia anatomically related to the pelvic organs revealed the presence of a caudal (inferior) mesenteric ganglion, two testicular ganglia, two hypogastric nerves and two pelvic ganglia and nerves. Histochemical demonstration of catecholamines showed a rich sympathetic innervation in the flexuous portion of the efferent ductuli, in the ductus deferens and in the three segments of the prostate. The testes, epididymidis and the three pairs of bulbourethral glands are supplied exclusively by adrenergic innervation around the vessels. Procedures such as surgical removal of ganglia, neurectomy with or without vessel ligation, and removal of periprostatic tissue accompanied by histochemical detection of catecholamines indicate that short adrenergic neurons are responsible for the adrenergic innervation of the reproductive organ of the male *Didelphis albiventris* (Maruch et al. 1989). Machado et al. (1982) demonstrated that the musculature of the ductus deferens, composed mainly of circular fibres, is richly innervated by adrenergic fibres and by presumed cholinergic fibres. However, under physiological conditions, it did not respond to acetylcholine or catecholamines. The lack of responsiveness to the neurotransmissors in the presence of cholinergic and adrenergic innervation could be accounted for by the absence of longitudinal arranged muscle fibres in the ductus deferens. Based on these morphological and physiological characteristics of the ductus deferens, it is unlikely that this organ is important as a contractile structure, the function of sperm transport being possibly ascribed to the cremaster muscle, which could possibly explain the incidence of spermatorrhoea in marsupials. To our knowledge, no information involving the innervation and physiology of the marsupial penis is available, due possibly to the structural complexity of this organ. Therefore, neurophysiological investigations are needed to fill this gap.

PUBERTY

The literature concerning puberty in South American marsupials is scant. First courtship behaviour in captivity, as indicated by licking and scratching of the scent sternal gland into the corners of a breeding wall and vocalisation when approaching a female, has also been reported at 220 days after birth in *Didelphis aurita* (Motta 1988). In *Didelphis albiventris*, Nogueira (1989) used spermatorrhoea as the basic indicator of puberty followed by testicular histological analysis. Puberty in young raised in captivity begins around 110 to 120 days after leaving the pouch, that is, 200 to 210 days after birth. In this opossum, the appearance of the initial sperm in the urine occurs concomitantly with the detachment of the glans and with the appearance of a yellowish colour of the hairs at the presternal region as consequence of the beginning of secretory activity of the cervical scent gland, also known as supraesternal scent gland. This sebaceous skin gland is also present in *Didelphis aurita* (Motta 1988), *Didelphis virginiana*, *Monodelphis domestica* and *Marmosa robinsoni* (Fadem and Schwartz 1986) and *Marmosops incanus* (Nogueira pers. comm.) and is assumed to be involved in social communication as scent-marking behaviour serves to deposit the secretions of these glands within the environment. In *Marmosa robinsoni*, this gland has an important role in male territorial marking (Boggs 1969) and its secretion is likely to be androgen dependent (Fadem and Schwartz 1986). As such, the gland could serve as a useful biological marker reflecting gonadal hormone activity in a marsupial species.

BIOMETRY OF THE MALE GENITAL SYSTEM THROUGHOUT THE ANNUAL REPRODUCTIVE CYCLE

No significant differences are found in testicular and epididymal weights of *Didelphis albiventris* during along the annual cycle, and both organs contain spermatozoa throughout the cycle (Nogueira 1988). Prostate gland and bulbourethral glands are significantly heavier during the mating period (June to January) and lighter during the resting period (February to May). Increase in the weight of both glands begins in the pre-mating period (June).

TESTICULAR ACTIVITY

Didelphis albiventris is the only Brazilian marsupial in which spermatogenesis has been studied in detail. The seminiferous epithelium is composed of distinct cellular associations (stages) consisting of one or two generations of spermatogonia, spermatocytes and spermatids found at particular phases of development in cross-sectioned tubules. Each cell type of the stage is morphologically integrated with the others in its developmental process. In a cross-sectioned tubule, usually a single cellular association is found at any given time. Sertoli cells are the only somatic cells present within the seminiferous tubule. The cycle of the seminiferous epithelium (CSE) has been studied in *Didelphis albiventris* employing acrosomic (Orsi and Ferreira 1978) and the tubular morphology (Queiroz and Nogueira 1992) systems. Through these systems, ten and eight stages were identified, respectively. The duration of the CSE of *D. albiventris* as obtained by autoradiography after intra-testicular injection of tritiated thymidine is estimated to be 17.3 ± 0.1 days (mean ± s.d). It is noteworthy that this value, although slightly greater than those registered by Setchell and Carrick (1973) for Australian marsupials, is the largest among mammals to date. Quantification of sperm cells based on testicular sperm reserves of *Didelphis albiventris* yielded low values (29.7 ± 1.9 x 10^6 sperm cells/g of testis parenchyma), comparable to that of humans. In the latter, total sperm cell number is considered to be the lowest among mammals. No significant difference was observed in testicular sperm reserves in *Didelphis albiventris* during mating and non-mating periods (Queiroz et al. 1995), thus corroborating the findings of the testicular biometry. Furthermore, among the several testicular parameters analysed through histology in *Didelphis albiventris* during mating and non-mating periods, only the volumetric proportions of Leydig cells differed significantly with an increase of about 40% in the mating period. The circadian variation in plasma testosterone levels indicates that these levels are higher in blood samples collected in the morning during the mating period.

ACKNOWLEDGEMENTS

We are indebted to Dr Gustavo Alberto Bouchardet da Fonseca for his encouragement and incentive. We also thank the anonymous referees for their suggestions and comments. Financial support came from Conselho Nacional de Desenvolvimento Científico e Tecnológico (CNPq), FINEP, Fundação de Amparo à Pesquisa do Estado de Minas Gerais (FAPEMIG), Pró-Reitoria de Pesquisa da UFMG. Special thanks to Célio Murilo Carvalho Valle, Edeltrudes M.V.C. Câmara, Gustavo Alberto Bouchardet da Fonseca, Helder-José, Maria Nazaré Ferreira da Silva and Patrícia Massara Martinelli, for supplying some specimes. The drawings and photomicrographs of *Micoureus demerare, Metachirus nudicaudatus and Caluromys lanatus* were made by Bruno Garzón de Oliveira Câmara, Fernando Val Moro, Humberto

Espírito Santo de Mello, Myrian Morato Duarte and taken from the Master's Thesis of Patricia Massara Martinelli, Suely de Fátima Costa and Godofredo Álvaro de Carvalho under the guidance of Professor Dr José Carlos Nogueira.

REFERENCES

Alves, H.J. (1986), 'Aspectos mesoscópicos e histoquímicos da origem da inervação adrenérgica de órgãos dos sistemas genitais feminino e masculino do gambá *Didelphis albiventris* (Lund, 1841) – Didelphidae-Marsupialia', M.S thesis, Universidade Federal de Minas Gerais, Belo Horizonte, Brasil.

Barnett, C.H., & Brazenor, D.C.W. (1958), 'The testicular *rete mirabile* of marsupials', *Aust J Zool*, **6**:27–32.

Biggers, J.D. (1966), 'Reproduction in male marsupials', in *Comparative Biology of Reproduction in Mammals* (ed. I.W. Rowlands), pp. 251–80, London, Academic Press.

Boggs, J. (1969), 'The general and agonistic behavior of the mouse opossum, *Marmosa robinsoni*', M.S. thesis, San Diego State University, San Diego, California, USA.

Carvalho, G.A. (1996), 'Aspectos morfológicos do sistema genital masculino de *Caluromys lanatus* (Olfers, 1818) e *Caluromys philander* (Linnaeus, 1758) – Marsupialia, Didelphidae', M.S. thesis, Universidade Federeal de Minas Gerais, Belo Horizonte, Brasil.

Côser, A M.L., & Nogueira, J.C. (1998), 'Glycogen, fructose and citric acid levels in the prostate of the white-belly opossum *Didelphis albiventris* (Marsupialia), during mating and non-mating periods', *Braz J Vet & Anim Sci*, **50**:695–8.

Costa, S.F. (1995), 'Morfologia do sistema genital masculino de *Metachirus nudicaudatus (Geoffroy, 1803)* – Didelphidae, Marsupialia', M.S. thesis, Universidade Federal de Minas Gerais, Belo Horizonte, Brasil.

Deane, M.P., Lenzi, H.L., & Jansen, A.M. (1984), '*Trypanosoma cruzi*. Vertebrate and invertebrate cycles in the same mammal host: the opossum *Didelphis marsupialis*', *Mem Inst Oswaldo Cruz*, **79**:513–15.

Fadem, B.H., & Schwartz, R.A. (1986), 'A sexually dimorphic suprasternal scent gland in gray short-tailed opossums (*Monodelphis domestica*)', *J Mammology*, **67**:205-8.

Ferrari, J.A., Helder-José, Musso, F., & Redins, C.A. (1987), 'Aspectos anatômicos e histológicos das glândulas peri-anais da cuíca *Philander opossum* Linnaeus, 1758', *Rev Bras Biol*, **47**:619–23.

Fonseca, C.C. (1987), 'Descenso testicular no gambá *Didelphis albiventris (Lund 1841)*', M.S. thesis, Universidade Federal de Minas Gerais, Belo Horizonte, Brasil.

Fonseca, C.C., & Nogueira, J.C. (1991), 'Morphological aspect of testicular descent in the white-belly opossum *Didelphis albiventris* (Marsupialia)', *Zoolog Jahb/Anatomy*, **121**:115–26.

Fonseca, G.A.B., Herrmann, G., Leite, Y.L.R., Mittermeier, R.A.B., Rylands, A.B., & Patton, J.L. (1996), 'Lista Anotada dos Mamíferos do Brasil', *Occasional Papers in Conservation Biology*, Conservation International & Fundação Biodiversitas, Belo Horizonte, **4**:9–11.

Garcia, P.J, & Gonçalves, R.P. (1984), 'Observações morfológicas da próstata do gambá (*Didelphis azarae*)', *Rev Bras Ciênc Morfol*, **1**:17–23.

Gardner, A L. (1993), 'Order Didelphimorphia, Paucituberculata and Microbiotheria', in *Mammals species of the world 2nd ed.* (eds D.M. Wilson, & D.E. Reeder), pp. 15–23, Washington, Smithsonian Institution Press.

Godinho, H.P., Cardoso, F.M., & Nogueira, J.C. (1977), 'Blood supply to the testis of a Brazilian marsupial *(Didelphis azarae)* and its abdominotesticular temperature gradient', *Acta Anat*, **99**:204–8.

Groves, C.P. (1993), 'Order Dasyuromorphia, Peramelemorphia, Notoryctemorphia, Diprotodondia', in *Mammals species of the world 2nd ed.* (eds D.M. Wilson, & D.E. Reeder), pp. 29–61, Washington, Smithsonian Institution Press.

Hafez, E.S.E. (1995), 'Anatomia da reprodução masculina', in *Reprodução Animal 6th ed.*, Brasileira, pp. 3–20, Editora Manole Ltda, São Paulo.

Helder-José (1998), 'Ação da testosterona sobre as glândulas odoríferas paracloacais do marsupial *Metachirus nudicaudatus* Geoffroy, 1803', PhD thesis, Univesidade de São Paulo, Brasil.

Helder-José, & Freymüller, E. (1995), 'A morphological and ultrastructural study of the paracloacal (scent) glands of the marsupial *Metachirus nudicaudatus* Geoffroy, 1803', *Acta Anat*, **153**:31–8.

Hruban, Z., Martan, J., Slesers, A., Steiner, D.F., Lubran, M., & Reichcigl, M. (1965), 'Fine structure of the prostate epithelium of the opossum (*Didelphis virginiana* Kerr)', *J Exp Zool*, **160**:81–105.

Kelce, W.R., Krause, W.J., & Garjam, V.J. (1987), 'Unique regional distribution of -3 ketosteroid-5-oxireductase, and its relationship to sperm maturation', *Cell & Tissue Res*, **237**:525–35.

Kenagy, G.J., & Trombulak, S.C. (1986), 'Size and function of mammalian testes in relation to body size', *J Mammal*, **67**:1–22.

Krause, W.J. (1991), 'Morphological observations on the paracloacal glands of the North American opossum (*Didelphis virginiana*)', *Zool Anz*, **227**:286–94.

Ladman, A.J. (1967), 'The fine structure of the ductuli efferente of the opossum', *Anat Rec*, **157**:559–75.

Lenzi, H.L., Jansen, A.M., & Deane, M.P. (1984), 'The recent discovery of what might be a primordial scape mecahanism for *Trypanosoma cruzi*', *Mem Inst Oswaldo Cruz*, **79**(Suppl):13–8.

Lins, L.V., Machado, A.B.M., Costa, C.M.R.. & Herrmann, G. (1997), *Roteiro Metodológico para Elaboração de Listas de Espécies Ameaçadas de Extinção. Contendo a Lista Oficial da Fauna Ameaçada de Extinção de Minas Gerais,* Publicações Avulsas Fundação, Biodiversitas, 1:20 e 37, Belo Horizonte, Brasil.

Machado, C.R.S., Calixto, S.L.. & Ladosky, D.W. (1982), 'Morphological and physiological factors involved in the contractility of the spermatic cord and ductus deferens of the opossum (*Didelphis albiventris*)', *J Reprod Fert*, **65**:275–80.

Martan, J. (1983), 'The genital tract of the male opossum *Didelphis marsupialis virginiana* and other marsupials', *Trans Ilini. State Acad Sci*, **76**:3–28.

Martan, J., & Allen, J.M. (1965), 'The cytological and chemical organization of the prostatic epithelium of *Didelphis virginiana* Kerr', *J Exp Zool*, **159**:209–29.

Martan, J., Hruban, Z., & Slesers, A. (1967), 'Cytological studies of the ductuli efferents of the opossum', *J Morphol*, **121**:81–102.

Martinelli, P.M. (1990), 'Morfologia do sistema genital masculino de *Marmosa cinerea* Temminck, 1824 *(Didelphidae, Marsupialia)*', M.S. thesis, Universidade Fedral de Minas Gerais, Belo Horizonte, Brasil.

Martinelli, P.M., & Nogueira, J.C. (1992), 'Epididymal morphology in the South American marsupial *Marmosa cinerea*, Temminck, 1824', *Rev Bras Ciênc Morfol*, **9**:26–31.

Martinelli, P.M., & Nogueira, J.C. (1997), 'Penis morphology as a distinctive character of the murine opossum group (Marsupialia, Didelphidae): a preliminary report', *Mammalia*, **61**:161–6.

Martinelli, P.M., Nogueira, J.C., & Campos, P.A. (1991), 'Morphology, glycogen and mucosubstances histochemistry of the prostate and bulbourethral glands of the *Marmosa cinerea*, Temminck, 1824 (Marsupialia, Didelphidae)', *Rev Bras Ciênc Morfol* **7/8**:3–11.

Maruch, S.G.M., Alves, H.J., & Machado, C.R.S. (1989), 'Sympathetic innervation of the reproductive organs of male opossum, *Didelphis albiventris* (Lund, 1841)', *Acta Anat*, **134**:257–62.

Motta, M.F.D. (1988), 'Estudo do desenvolvimento extra-uterino de *Didelphis aurita* Wied, 1826, em cativeiro – Investigação de critérios para estimativa de idade', M.S. thesis, Universidade Federal do Rio de Janeiro, Rio de Janeiro, Brasil.

Munhoz, C.O.G., & Merzel, J. (1967), 'Morphology and histochemistry study on perianal glands of the opossum', *Acta Anat*, **68**:258–71.

Naiff, D.R., Barret, T.V., & Arias, J.R. (1987), '*Trypanosoma cruzi* nas glândulas anais de *Didelphis marsupialis*: primeiro registro de infecções naturais', *Congr Bras Parasitol*, 10, Salvador, BA. (Abstract)

Nogueira, J.C. (1988), 'Anatomical aspects and biometry of the male genital system of the white-belly opossum *Didelphis albiventris* Lund, 1841 during the annual reproductive cycle', *Mammalia*, **52**:233–42.

Nogueira, J.C. (1989), 'Reprodução do gambá *Didelphis albiventris*', *Ciência Hoje*, **9**:8–9.

Nogueira, J.C. (2001), 'Morfologia do sistema genital masculino de marsupiais brasileiros (Didelphimorphia, Didelphidae)', in *Os Marsupiais do Brasil: Evolução, Biologia e Ecologia* (ed. N.C. Cáceres, & E.L.A. Monteiro-Filho), Imprensa da Universidade Federal do Paraná, Curitiba, Brasil, in press.

Nogueira, J.C., & Barbosa, A.J.A. (2001), 'Distribution of endocrine cells in the male genital system of the *Didelphis albiventris* (Marsupialia)', in preparation.

Nogueira, J.C., & Redins, C.A. (1987), 'Submicroscopic study of the tunica propria of the seminiferous tubules of the Brazilian white-belly opossum (*Didelphis albiventris*)', *Anat Anz*, **163**:349–57.

Nogueira, J.C., & Redins, C.A. (1988), 'Fine structure of macrophages in the testicular interstitial tissue of the white-belly opossum, *Didelphis albiventris*', *Micr Electr y Biol Cel*, **12**:23–34.

Nogueira, J.C., & Redins, C.A. (1989), 'Modificações sazonais na ultra-estrutura das glândulas bulbo-uretrais do gambá *Didelphis albiventris* (Marsupialia)', XII Colóquio Sociedade Brasileira de Microscopia Eletrônica, Caxambu (MG), **1**:163–4.

Nogueira, J.C., Campos, P.A., & Ribeiro, M.G. (1984), 'Histology, glycogen and mucosubstance histochemistry of the bulbo-urethral glands of the *Philander opossum* (Linnaeus, 1758)', *Anat Anz*, **156**:321–8.

Nogueira, J.C., Godinho, H.P., & Cardoso, F.M. (1977), 'Microscopic anatomy of the scrotum, testis with its excurrent ducts system and spermatic cord of *Didelphis azarae*', *Acta Anat*, **99**:209–19.

Nogueira, J.C., Da Silva, M.N.F., & Câmara, B.G.O. (1999a), 'Morphological description of the male genital system of the bushy-tailed opossum *Glironia venusta*. Thomas, 1912 (Didelphimorphia, Didelphidae)', *Mammalia*, **63**:231–6.

Nogueira, J.C., Ribeiro, M.G., & Campos, PA. (1985), 'Histology and carbohydrate histochemistry of the prostate gland of the Brazilian four-eyed opossum *(Philander opossum)*', *Anat Anz*, **159**:241–52.

Nogueira, J.C., Martinelli, P. M., Costa, S.F., Carvalho, G. A & Câmara, B.G. O. (1999b), 'The penis morphology of *Didelphis, Lutreolina, Metachirus* and *Caluromys* (Marsupialia, Didelphidae)', *Mammalia*, **63**:79–92.

Orsi, A M., & Ferreira, A.L. (1978), 'Definition of the stages of the seminiferous epithelium of the opossum (*Didelphis azarae*, Temminck, 1825)', *Acta Anat* **100**:153–60.

Orsi, A M., Mello, R. V. , Ferreira, A. L., & Campos, V.J.M. (1980), 'Morpholgie de céllules epithéliales du canal de l'épididyme de l'opossum (sarigue) sud-américan (*Didelphis azarae*)', *Anat Anz*, **148**:7–13.

Orsi, A M., Ferreira, A.L., Mello, V.R., & Oliveira, M.O. (1981), 'Regional histology of the epididymis in the South American opossum. Light microscope study', *Anat Anz*, **150**:521–8.

Queiroz, G.F. (1991), 'Estudo morfológico e quantitativo da atividade testicular do gambá *Didelphis albiventris* (Marsupialia)', PhD thesis, Universidade Federal de Minas Gerais, Belo Horizonte, Brasil.

Queiroz, G.F., & Nogueira, J.C. (1992), 'Duration of the cycle of the seminiferous epithelium and quantitative histology of the testis of the South American white-belly opossum (*Didelphis albiventris*), Marsupialia', *Reprod Fertil Dev*, **4**:213–22.

Queiroz, G.F, Rosa e Silva, A.M., & Nogueira, J.C. (1995), 'Testicular sperm reserve and plasma testosterone levels of the South American white-belly opossum (*Didelphis albiventris*), Marsupialia', *Mammalia*, **59**:255–61.

Ribeiro, M.G. (1981), 'Aspectos anatômicos e histologia do sistema genital masculino da cuíca *Philander opossum* (Linnaeus, 1758) – Didelphidae, Marsupialia', M.S. thesis, Universidade Federal de Minas Gerais, Belo Horizonte, Brasil.

Ribeiro, M.G., & Nogueira, J.C. (1985), 'Histoquímica de mucossubstâncias nas glândulas bulbo-uretrais (Cowper) de alguns marsupiais brasileiros', pp 664, 37 Reunião S.B.P.C., Belo Horizonte, Brasil.

Ribeiro, M.G., & Nogueira, J.C. (1990), 'The penis morphology of the four-eyed opossum *Philander opossum*', *Anat Anz*, **171**: 65–72.

Ribeiro, M.G., & Nogueira, J.C. (1991), 'Histologia da pele escrotal e túnica vaginal da cuíca *Philander opossum* (Linnaeus, 1758)', *Rev Bras Zool*, **7**:245–50.

Rodger, J.C. (1982), 'The testis and its excurrent ducts in American caenolestid and didelphid marsupials', *Amer J Anat*, **163**:269–82.

Rodger, J.C., & White, T.G. (1980), 'Glycogen not N-acethylglucosamine the prostatic carbohydrate of three Australian and American marsupials, and patterns of these sugars in Marsupialia', *Comp Biochem Physiol*, **67B**:109–13.

Schaffer, J., & Hamperl, H. (1926), 'Über Anal und Zirkumanaldrüsen. 3 Mitt. Marsupialia', *Z Wiss Zool*, **127**:529–69.

Setchell, B.P. (1977), 'Reproduction in male marsupials', in *The biology of marsupials* (eds. B. Stonehouse, & D. Gilmore), pp. 411–57, The Macmillan Press, London.

Setchell, B.P., & Carrick, F.N. (1973), 'Spermatogenesis in some Australian marsupials', *Aust J Zool*, **21**:491–9.

Setchell, B.P., & Thorburn, G.D. (1971), 'The effect of artificial cryptorchidism on the testis and on testicular blood flow in an Australian marsupial *Macropus eugenii*', *Comp Biochem Physiol*, **38A**: 705–8.

Taggart, D.A., Johnson, J.L., O'Brien, H.P., & Moore, H.D.M. (1993), 'Why do spermatozoa of American marsupials form pairs? A clue from analysis of sperm-pairing in the epididymis of the Grey short-tailed opossum, *Monodelphis domestica*', *Anat Rec*, **236**:465–78.

Tonelli, S.M.G.M. (1982), 'Aspectos histológicos e inervação adrenérgica de órgãos do sistema genital masculino do gambá *Didelphis albiven-*

tris (Lund, 1841) – Didelphidae – Marsupialia', M.S. thesis, Universidade Federal de Minas Gerais, Belo Horizonte, Brasil.

Tyndale-Biscoe, C.H., & Renfree, M. (1987), *Reproductive physiology of marsupials*, Cambridge, Cambridge University Press.

van den Broek, A J. P. (1910), 'Untersuchungen über den Bau der männlichen Geschlechtesorgane der Beuteltiere', *Morphol Jahrb*, **41**:347–436.

Woolley, P.A. (1982), 'Phallic morphology of the Australian species of Antechinus (Dasyuridae: Marsupialia): a new taxonomic tool?' in *Carnivorous marsupials* (ed. M. Archer), pp. 767–81, Surrey Beaty & Sons, Sydney.

Woolley, P.A. (1987), 'The seminiferous tubules, rete testis and efferent ducts in Didelphidae, Caenolestidae and Microbiotheriidae', in *Possums and opossums – Studies in evolution* (ed. M. Archer), pp. 217–27, Surrey Beaty & Sons Limited, New South Wales, Australia.

Woolley, P., & Webb, S.J. (1977), 'The penis of dasyurid marsupials', in *The biology of marsupials* (eds. B. Stonehouse, & Gilmore), pp. 307–23, The Macmillan Press, London.

CHAPTER 13

PERINATAL SENSORY AND MOTOR DEVELOPMENT IN MARSUPIALS WITH SPECIAL REFERENCE TO THE NORTHERN QUOLL, *DASYURUS HALLUCATUS*

John Nelson, Richard M. Knight and Craig Kingham

Department of Biological Sciences, Monash University, Clayton 3168, Australia

The sensory and motor development of the early pouch young of the Northern Quoll (*Dasyurus hallucatus*) are described in some detail to give a basis for the relationship between the motor behaviour of the newborn marsupial and its neural and anatomical development. Comparisons are then made with other newborn marsupials and with eutherians at stages of development similar to those of the newborn marsupials. The differences between marsupials and eutherians at these stages are not as great as might be assumed. All have developed parts of the vestibular and somatosensory systems but the trigeminal nerve provides the main sensory input (including a chemical sense) at these stages and it, and the more posterior cranial nerves, are the controllers of the motor behaviour at these stages. Different marsupials are born at different stages and the mothers adopt different birth positions. Thus the journey from the urinogenital sinus to the pouch poses differing challenges for the newborn. Different species have different sensory and motor development and so have different mechanisms to achieve the different challenges involved in reaching the teat.

INTRODUCTION

Marsupials have shorter gestation periods than eutherians and a relatively longer postnatal maturation. Indeed, marsupials are born at stages that are equivalent to eutherian embryonic stages (Nelson 1992) and hence the transitions from embryo through to foetus then to weanling can be observed by the study of young in the pouch. Marsupials and eutherians show similar sequences of development in any one system but there are shifts in the timing of development of some systems relative to others. There is such a shift in the development of the central nervous system relative to the somatic tissues of the craniofacial and other body regions (Nelson 1992; Maier 1992; Smith 1997; Nunn and Smith 1998).

Since pouch young represent embryonic stages, one focus of interest has been on the early development of the various sensory and motor systems (see Mark and Marotte 1992; Saunders 1997) while another has been on the relative development of the sensory and motor systems at birth in marsupials and the senses that are used to find the pouch and then to locate and attach to the teat (e.g. Flynn,1923; Cannon et al. 1976; Renfree et al. 1989).

This chapter will examine the development of the sensory and motor systems at birth in the Northern Quoll, *Dasyurus hallucatus* and compare them with those in other newborn marsupials and some newborn eutherian species. *D. hallucatus* is used because, at birth (see Fig. 1), it is at the earliest recorded stage of development in marsupials according to the Carnegie system

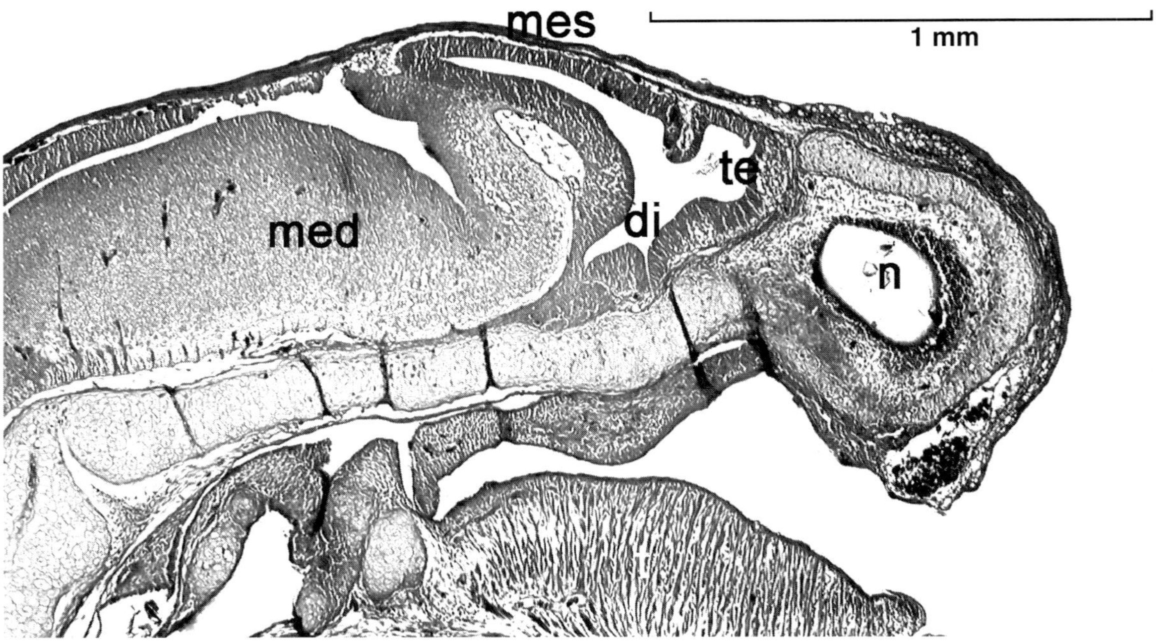

Figure 1 A parasagittal section of the brain of a newborn *Dasyurus hallucatus* (day 0). Abbreviations are: di = diencephalon; med = medulla oblongata; mes = mesencephalon; n = nasal cavity; and te = telencephalon. The bar is 1 mm.

(see Butler 1987; Nelson 1992). We will also correlate sensory and motor development with the reach and grasp motor behaviour in the early pouch young until just before it begins to move from teat to teat in the pouch (about 35–40 days).

Developmental studies of limb use and early motor behaviours have been carried out mainly in eutherians such as mice (Fox 1964 1965), Mongolian gerbils (Cassidy et al. 1992; Cabana et al. 1993), rats (Angulo y Gonzâlez 1932; Narayanan et al. 1971; Altman and Sudarshan 1975), cats (Windle and Griffin 1931), primates (Jeannerod and Decety 1990) and humans (Hooker 1958). Only one marsupial species has been examined, the Brazilian Opossum, *Monodelphis domestica* (Pflieger and Cabana 1996: Pflieger et al. 1996). However, most of these studies have been orientated about the use of the limb in locomotion and there have been few observations of skilled behaviours, such as reaching and grasping (but see Ivanco et al. 1996).

Reaching and grasping is a specialised behaviour that typically involves the extension of a limb towards an object, the orientation and use of the paw to grasp the object, and the subsequent flexion of the limb to bring the object towards the trunk of the animal. In the Quoll, as in many marsupials, reaching and grasping is the only limb movement present at birth and during early postnatal life. It is therefore of interest to examine the change in the form and context of this behaviour and to relate these changes to developments in the sensory and motor systems.

In adult mammals, the movements of the limbs and head can be initiated by tactile stimulation of the surface of the head, limbs or body. Sensory inputs from the peripheral nerves in the forelimb and hindlimb pass first to spinal ganglia and then travel along the dorsal horn of the spinal cord, arriving in the medulla (hindbrain) at the cuneate nucleus (for the forelimb) or gracile nucleus (for the hindlimb). These nuclei have connections to and from the cerebellum and via the medial lemniscus (in the midbrain) and via the ventrobasal nucleus (in the diencephalon) they connect to the somatosensory cortex where the information is processed and passed to the motor cortices for voluntary responses to the stimuli (Crosby et al. 1962; Darian-Smith 1973; Rowe 1982).

Except for a small portion at the angle of the jaw and the back of the head, most of the cutaneous innervation of the head is by the trigeminal nerve. The cell bodies of these sensory nerves are in the trigeminal (Gasserian) ganglion. Their axons pass to three sensory nuclei in the brain: the principal nucleus of the trigeminal in the pons, lateral to the trigeminal motor nucleus; the spinal trigeminal nucleus in the medulla oblongata and upper spinal cord; and the mesencephalic trigeminal nucleus (TMC) in the pons (anterior to the principal) and midbrain. These nuclei have input to the reticular formation of the brainstem and presumably modulate arousal and alerting functions. The TMC also has motor output to the hypoglossal nucleus and to cervical segments C1 and C2. (Crosby et al. 1962; Darian-Smith 1973; Nieuwenhuys et al. 1998).

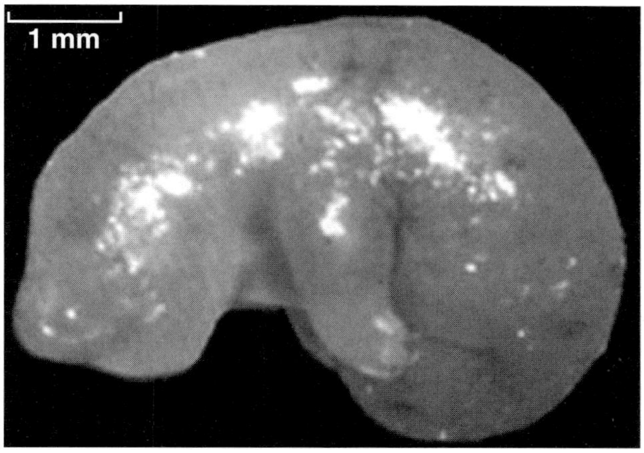

Figure 2 A pouch young on day 0 that has been on the teat for two to three hours. Note the lack of an eye, the well-developed forelimbs and the poorly developed hindlimbs. The bar is 1 mm.

OBSERVATIONS ON THE POUCH DEVELOPMENT OF THE NORTHERN QUOLL

Notes on studies

Dasyurus hallucatus, like other members of the genus, breeds only once a year usually producing about twice the number of young than the available eight teats. It is nocturnal and spends the day in hollows in trees or logs, under rocks, or in the burrows made by other animals. It feeds on insects, spiders, small vertebrates and fruit.

At birth, the young are approximately 18 mg in weight (cf. 100 mg in *Monodelphis*) and organogenesis is not complete until some 30 days after birth (Nelson 1988; 1992). The hindlimbs are little more than undifferentiated buds while the forelimbs are relatively more developed, consisting of the lower forelimb and paw. The lower forelimb is fixed in a position such that the palm points longitudinally along the body axis, which orientates the paw such that the neonate can grasp the hairs within the pouch. The shoulder and upper forelimb are not yet articulated from the body, and the posture is highly ventro-flexed (see Fig. 2).

The relative development of the somatosensory system of the developing Quoll was examined by the injection on successive days of tritiated thymidine, which is incorporated into all cells that are dividing within a few hours of the injection (Altman and Bayer 1964). When the brain is later sectioned and covered with a photographic emulsion, the radioactivity in the labelled cells interacts with the emulsion such that they can be identified by the silver grains in the emulsion above them. The number of silver grains is a measure of the radioactivity. With each division of the labelled cells, the daughter cells receive half of the isotope, so the cells with the densest label are used to determine the birth dates.

Fluorescent labelling of developing nerves was used to visualise the relative development of the sensory and motor innervation of the limbs and of the facial areas at various stages of development. These neural measures were then correlated with the behaviour of the pouch young.

Behavioural observations

At birth (day 0), the Quoll was capable of forelimb-driven movement. The body of the neonate flexed laterally so that the head and neck moved to one side, which allowed the forelimb on the outside of the flexure to advance forwards and grasp a pouch-hair. The body then flexed in the opposite lateral direction turning the head and neck to that side of the body and allowing the other forelimb to move forward and grasp another hair. The swinging of the head from side to side was accompanied by a slight lifting of the head from the body of the mother. Due to the limited movement that is allowed to the elbow joint at this stage, a fully extended reach by the forelimbs is anatomically impossible.

The grasp of the paw was characterised by an initial extension of the digits of the paw and then a flexing of the digits to close the grip. If the grasp was successful, the paw closed around the pouch-hair. If the paw and digits failed to make contact with a hair, the digits still flexed to close the 'grip'. Whether or not the paw contacted a pouch-hair, the digits flexed to close the grasp with the forearm then drawing back to its initial position before beginning another reach,

Even though pouch-hairs were gripped by both forepaws, the young appeared to maintain their position on the teat through the membrane around the lips anchoring them to the teat. Removal of a pouch-hair from the grip did not prompt any immediate reaching or grasping behaviour. Additionally, no spontaneous unclenching and re-grasping of hair was observed at this time. Although these young were never seen re-grasping the hairs during observations, when they were examined the next day they had re-grasped a hair. It is not known if this was stimulated by the mother or by contact with the pouch.

After 12 days, pouch young not only responded to the removal of hairs from the paw but also spontaneously unclenched and re-grasped pouch-hairs. The reaching movements were usually repetitive, with the young spontaneously unclenching and re-grasping a hair several times, even if a grasp was initially successful in securing a pouch-hair. The flexing of the digits coincided with the flexion of the forelimb.

Initial extension of the limb was in the longitudinal plane (i.e. along the body axis). At 12 days, reaching involved only the forelimb and the paw. The forelimb extended anteriorly, and moved through a maximum angle of 45°. There was no rotation of the paw. After 17–20 days, the upper forelimb and shoulder began to add a component to the reach. Reaching was still in the

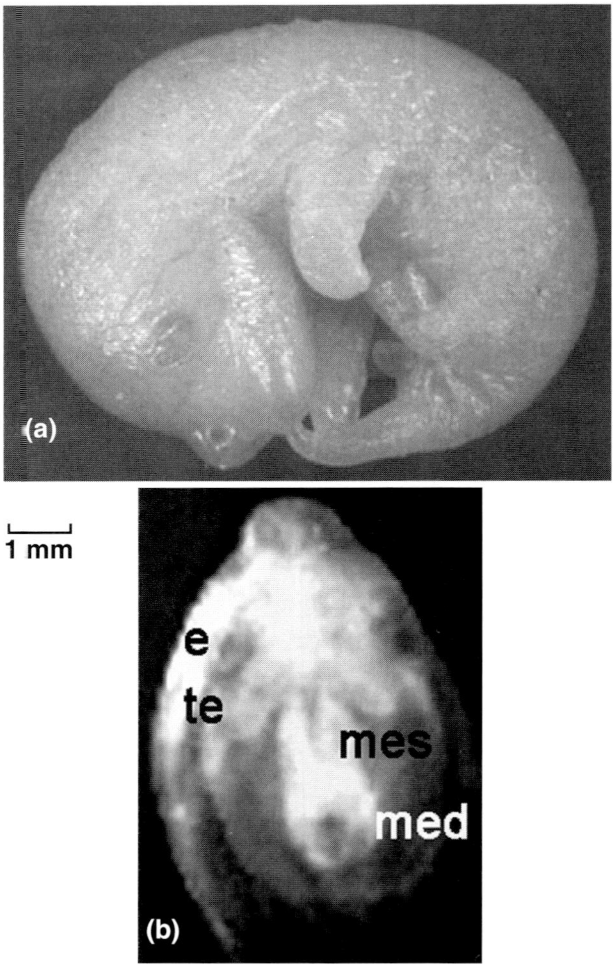

Figure 3 (a) A pouch young on day 13. The eye pigment is present, the forelimb has differentiated into a clearly demarcated upper and lower segment but the hindlimb is still not well developed. The bar is one mm. (b) The same animal that has been partly cleared and photographed in transmitted light to show the state of development of the brain. Abbreviations as in Fig. 1. e = eye. The two cerebral hemispheres of the telencephalon can be seen.

longitudinal plane, but with the ongoing development of the upper forelimb, the arc of movement of the lower forelimb reached 135° relative to the body. The upper forelimb, however, did not become a fully functional component of the reach until around day 30. Prior to this age, most movement remained in the lower limb, and reaching was still in the anterior–posterior plane. At day 30, the paw was capable of rotatory movement around the axis of the limb.

Autoradiographic labelling

Each part of the somatosensory system of the Quoll was examined and scored according to the number of silver grains above each labelled cell and to the number of the variously labelled cells. Densely labelled cells were considered to have ceased mitotic divisions at or close to the day of the dye injection. Lightly labelled cells were considered to be the result of several divisions of the originally labelled cell since the day of the injection. The period of neurogenesis for a nucleus was determined from the day of the first appearance of label to the day when no label was present. However, since it might take several days for these cells to move from the ventricle where they were born to the nucleus, these times should be considered as being the earliest possible time that the nucleus could have been formed and not the absolute time during which the nucleus formed. The nucleus becomes functional at an as yet unknown number of days after the first appearance of label.

The cuneate nucleus can be divided into two parts: lateral (CuL) and the medial (CuM). CuL first showed label on day 5 (see Fig. 4), and no longer showed label after day 15. The densest label (most cells intensely labelled) occurred at day 7. CuM first showed label on day 3, peaked on day 5 and finished around day 12. The gracile nucleus (Gr) showed label from day 3 to day 15 with a peak from day 7 to day 9. Both had completed the majority of their development by day 9, although some new cells were added up to day 15.

The spinal trigeminal nucleus (Tsp) showed label from day 3 to day 19 with a peak around day 5. The paratrigeminal nucleus (Tp) showed label from day 5 to day 12 with a peak around day 7. The principle sensory nucleus of the trigeminal is divided into two parts: dorsal (TsD) and ventral (TsV). TsD showed decreasing amount of label from day 3 to day 7 and occasional subsequent label, while TsV showed label from day 3 to 5, and intermittent label thereafter. TsD, TsV, and especially TM, had very few labelled cells relative to the size of each nucleus when compared with other nuclei. This suggests that the majority of their neurons had been born prior to day 3. The mesencephalic tract of the trigeminal nucleus (TMC) showed label from day 15 to 19 and the medial lemniscus (ML) showed label from day 10 to 19. The motor nucleus of the trigeminal (TM) did not show significant label at any stage.

The ventrobasal nucleus (VB) showed label from day 9 to 26 with a peak around day 12. The somatosensory cortex (SI) showed label from day 10 to day 34, with several peaks. SI is divided into six layers, numbered from the surface of the brain toward the ventricle. Label was present in individual layers in the SI as follows: Layer VI, days 10–12; layer V, days 12–15; layer IV, days 17–22; layer III, days 19–32; layer II, days 22–34. The parts of the somatosensory system in the medulla had birthdates beginning earlier then those in the midbrain, which were in turn earlier than the birthdates in the forebrain. This demonstrated the caudo-rostral development of the somatosensory system that has been seen in other studies (Rowe 1982, 1996).

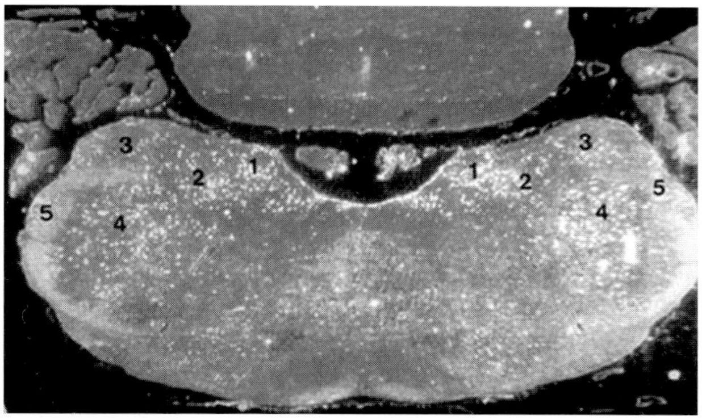

Figure 4 Transverse section of the hindbrain of an adult that had been injected on day 5 with a radioactive tracer. The white spots are over cells that had taken up the label. 1 = gracile nucleus; 2 = medial cuneate; 3 = lateral cuneate; 4 = spinal trigeminal ganglion; and 5 = paraspinal lateral nucleus.

Neuroanatomical observations: the trigeminal

Fluorescent labels from axons running from the maxillary area of the snout to the trigeminal Gasserian ganglion and between this ganglion and the brain were apparent in all specimens examined. The very small size of the brain at birth and at stage 3 makes it difficult to be certain that the sensory nuclei were labelled as the sensory and motor nuclei are then very close to each other. The proportions of the brain with apparent connections to and/or from the trigeminal ganglia (as indicated by the relative amounts of labelled and unlabelled brainstem) varied between the ages examined. The actual size of these regions did not increase substantially between day 3 and 9, although the brain was considerably larger in the day 9 specimens. The size of the area labelled in the day 12 and 13 specimens indicated a sizeable increase in trigeminal innervation when compared with the day 9 specimen. At 23 days the brain resembled the adult form, with the appropriate proportions of label in nuclei associated with the trigeminal nerves. The volume of label within the brainstem showed a greater relative increase within the sensory nuclei compared to the motor nuclei and most of this increase was from day 9 to day 13.

The principal CNS nucleus of the trigeminal receives somatosensory input from the face and also receives gustatory stimuli from the facial (VII), glossopharyngeal (IX) and vagus (X) cranial nerves. The mesencephalic nucleus of the trigeminal projects to the trigeminal motor and hypoglossal motor nuclei. The hypoglossal nerve innervates the intrinsic and extrinsic muscles of tongue and receives afferent proprioception from muscle spindles. The glossopharyngeal nerve has efferents to the stylopharyngeus muscle, and receives tactile, pain and thermal stimuli from mucous membranes of the posterior third of the tongue, tonsils and from the eustachian tube. The vagus nerve is important in swallowing and breathing, and receives taste stimuli from the epiglottis, and visceral sensations from the pharynx, larynx, trachea, and oesophagus as well as thoracic and abdominal viscera. The vagus motor nucleus (nucleus ambiguous) has special visceral efferents to branchiomeric muscles of the pharynx and larynx (Nieuwenhuys et al. 1998).

Somatosensory fibres from the trigeminal also terminate in the (NS). This nucleus also receives gustatory fibres of cranial nerves VII, IX and X, general visceral stimuli of IX and X (from baroreceptors of the carotid sinus, the heart and aorta, the lungs and other viscera) and from the area postrema (which regulates the permeability of the blood-brain barrier). The NS sends efferents to a large number of motor nuclei in the medulla and neck (Nieuwenhuys et al. 1998). The trigeminal thus has input to many of the functional systems in the newborn marsupial.

Neuroanatomical observations: forelimbs and hindlimbs

At day 9 motor transport of carbocyanine dye is seen only from the forelimb. At day 12, a labelled brachial nerve extending from the wrist towards the spinal cord was observed. The nerve was divided into a motor component passing ventrally to the spinal cord and a sensory component passing dorsally (see Fig. 2). It is possible that there may have been a sensory component at day 9 as the one specimen did not show good transport of tracer. The sensory component at day 12 was estimated to be less than half the diameter of the motor component. Label in the forearm increased considerably between days 9 and 23 that indicated an increase in motor input or an increase in sensory output or both. In this period the arm was differentiating into its components so an upper arm was becoming more easily recognised (see Fig. 2). Areas of the hand in the regions where the sensory pads would later develop contained conspicuous label at 23 days.

The ventrobasal nucleus (VB) showed label from day 9 to day 26, peaking around day 12. Both the medial lemniscus (ML) and VB developmentally lag the lateral (CuL) and medial (CuM) parts of the cuneate nucleus and the gracile nucleus (Gr). This demonstrates a pattern of linear development within these related nuclei from hindbrain to forebrain. Extrapolation of this linear sequence indicates that VB should follow ML, but this appears not to be the case. This may be due to the fact that ML contains relatively few cells. Thus ML, with very few neuronal cell bodies, may not have been examined in fine enough detail for the precise determination of nuclei birthdates.

In marsupials, the hindlimbs have a delayed development compared to the forelimbs. The almost simultaneous development of the gracile and cuneate nuclei was thus an unexpected result of our present studies. The gracile nucleus receives input from the hindlimb and also from the lower body (Crosby et al. 1962). Three reasons (not mutually exclusive) could be put forward to explain this phenomenon. Firstly, sensory receptors located on the exterior of the lower body may develop at a similar rate to those on the forearm as a result of the arrival of sensory neurons from the spinal cord. Secondly, the gracile nucleus may develop early and send a limited number of axons to higher centres to establish connections there that might later be pruned as a result of later input of sensory information from the hindlimbs. Thirdly, the nucleus may form early as the distances from the hindlimbs to the hindbrain are much greater than that from the forelimbs and so need a longer time for their growth.

Layer IV of the SI receives afferent fibres from VB. This layer begins to form around 17 days. Output from layer IV goes to layers V and VI, which are mature prior to layer IV. Studies have shown interactions between migrating neurones and established neurones (Shatz 1992). It is theoretically possible, therefore, that migrating neurons destined for layer IV of the SI could be processing information from VB from around 17 days postnatal but certainly not as early as 12 days. Even though it is possible that layers V and VI may be connected to motor areas, the adult pattern of connections, via VB to layer IV then through layers V and VI to the motor centres, cannot be operable before day 17. The effect of this lack of somatosensory input to cortical motor areas before about day 17 is that movements would be restricted to automatic, or reflex, responses unless some sort of circuit exists which bypasses layer IV.

DISCUSSION

Comparison of the quoll to marsupials

Development at birth

At birth, all marsupials use their forearms alternatively to reach and grasp hairs as they move to the pouch and then within the pouch to reach a teat. The head rhythmically moves laterally to one side while the arm is moved forward on the opposite side to reach and grasp a hair (or hairs). This gives an anchor point when the body moves forward on that arm as the head comes to that side and the other arm moves forwards. Our studies show that the Northern Quoll does not have sensory feedback from the hand at this time. This would support the statement of Hughes and Hall (1988) that the movements of marsupials at birth are initiated from a central pattern generator (CPG) within the spinal cord or lower brainstem as did Pflieger and Cabana (1996) in *Monodelphis domestica*.

Knott et al. (1999) state that the development of the pouch young of *Monodelphis* at birth corresponds approximately to a human embryo earlier than 8 weeks of gestation, a rat at E16–17 (embryonic day 16–17) and a mouse at E14–15. This would place *Monodelphis* at about Carnegie stage 21 at birth (Gribnau and Geijberts 1981; Butler 1987.). From published descriptions, Nelson (1992) placed birth in *Didelphis virginiana* at stage 17, *Antechinus stuartii*, *Isoodon macrourus* and *Perameles nasuta* also at stage 17, and *Macropus eugenii* and *M. giganteus* at stage 19. These are all born at a later stage than the Quoll.

Our studies show that at least part of the sensory trigeminal system of the Quoll had formed at birth, as we were able to get fluorescent transport from the maxillary region to the trigeminal ganglion and then to the sensory nuclei of the trigeminal. It is probable that there was some sensory innervation from the maxillary-mandibular region as Merkel cells have been found around the mouth of the newborn Quoll (Gemmell et al. 1988) and on the snout of the newborn *Monodelphis* (Jones and Mungher 1985). Merkel cells are dependent for their survival on innervation by a sensory nerve, as are many other sensory receptors (Purves and Lichtman 1985). Some Merkel cells develop before the arrival of innervation (Mosconi and Rice 1993) and so it is not known if the sensory nerve induces the development of the Merkel cells in the Quoll. Neither hairs nor vibrissae are developed at birth in marsupials but Waite et al. (1998) describe how in the tammar wallaby (*Macropus eugenii*), afferent fibres grow into the brain many weeks before the whisker pattern on the snout develops.

At birth in the Quoll, the olfactory epithelium has formed (Gemmell and Nelson 1988), but our studies on radioactive labelling of the olfactory system indicate that the neurons of the main and the accessory olfactory systems have not formed. The hearing and visual systems were also not developed at birth (Aitkin et al. 1994a, 1994b, 1996; Nelson 1987). The utricles of the vestibular system had formed and were the only parts of the vestibular labyrinth that had a macula (Gemmell and Nelson 1989). This was also found in newborn *Didelphis* (Krause 1991, 1992, 1998), *Monodelphis* (Pflieger and Cabana 1996), brush-tailed possums (*Trichosurus vulpecula*) (Gemmell and Rose 1989), the rat kangaroos *Potorous tridactylus* and *Bettongia gaimardi* (Gemmell and Rose 1989), and the dunnart, *Sminthopsis macroura* (Gemmell and Selwood 1993).

Thus for the Quoll, the main sensory systems controlling movements by the newborn are the trigeminal system and a part of the vestibular system. Most other marsupials are born at later stages of development and may use other systems (for example olfactory) as well.

Development in the pouch

It was not until 12 days in the pouch that the Quoll had sensory feedback from the hand. This was evidenced from our behavioural studies, by the lack of sensory nerves detected in the arm until about 12 days and by the birthdates of many of the neurons in the sensory nuclei of the trigeminal system.

Fentress (1983, 1989) has postulated that the appearance of many motor patterns before they have become associated with sensory input enables the motor pattern to develop without possible perturbations as a result of sensory input. Once the pattern has been perfected, it can then be used to respond to sensory input. His studies showed that many adult motor patterns are composed of units that are at first relatively constant in form that then become variable as sensory input develops. They then become relatively constant in form within larger groupings. The reach and grasp movement of the Quoll did become more variable in form after sensory input, but the forelimb was also differentiating towards the adult form and thus was becoming more able to display variation in movement.

The development of motor control has been described in some detail in the pouch young of *M. domestica* (Wang et al. 1992; Cassidy and Cabana 1993: Qin et al. 1993; Pflieger and Cabana 1996). In *Monodelphis*, the sensorimotor reflex has been established in the cervical region at birth (Lepre et al. 1998) but is not formed in the lumbar region until day 7 (Knott et al. 1999). The authors postulate that sensorimotor reflexes appeared later than the growth of pathways into a motor area because they probably depend on the functional connections within and between motor centres as well as on the formation of the pathway itself. They also found that the expression of motor behaviours was associated with synaptogenesis rather than myelination (Cassidy et al. 1994; Pflieger et al. 1996; Leblond and Cabana 1997).

Our studies found that the ventrobasal nucleus (VB) showed label from day 9 to day 26, peaking around day 12. Afferent fibres from VB are received by layer IV of the SI (this layer begins to form around 17 days). Martin et al. (1988) working on the North American opossum found that thalamic axons (from VB) reach the subplate by 12 days postnatal, but do not innervate an identifiable somatosensory layer IV until estimated postnatal day 46. Molnar et al. (1998) found that there was no such waiting period in *Monodelphis* and that the thalamic fibres had reached the entire cortex by day 9, but those from the cortex had not reached the thalamus until much later. This was also reported in the tammar wallaby (Sheng et al. 1991; Marotte et al. 1997).

Molnar et al. (1995), in an investigation of the development of thalamocortical connections in *Dasyurus hallucatus*, found that a cortical sub-plate was present at around 13 days postnatal and that thalamic fibres did not reach the cortical subplate until around 18 days. Therefore the cortical sub-plate cannot be participating in a 'short circuit' between the thalamus and motor areas in *Dasyurus hallucatus* until at least 18 days. Since there was no waiting period in *Monodelphis*, then it is unlikely that in the Quoll any commands can come down from the cortex until probably day 40 (see also Marotte et al. 1997). There could however be control from the midbrain (red nucleus) or from the diencephalon of the forebrain (Xu and Martin 1989). The red nucleus innervates mainly motor nuclei of V and VII, the nucleus ambiguous and the motor centres of upper spinal cord. It regulates proper muscle tonus, stabilises movements and maintains normal orientation in space.

Comparison of the quoll to eutherians

Early developmental stages

Muller and O'Rahilly in a series of papers (1988, 1989a, 1989b, 1990a, 1990b) have described the Carnegie stages in the human and summarised previous studies. In all cases, motor nuclei were formed before sensory nuclei. Cranial nerves developed during stages 12 and 13 in the order XII, V, VII, VIII, IX/X and XI. They state that in the hindbrain 'an astonishing number of details develop during stage 16' (Muller and O'Rahilly 1989a). By stage 21, the cortical plate has formed and so most of the adult neural structures are present, although many are still in very early stages of differentiation.

Altman and Bayer (1980a, 1980b) found in the rat that cells of the sensory nuclei of the trigeminal were formed from E12 to E16 (stages 13 to 20). We found label of the trigeminal sensory nucleus from day 3 (our earliest animal) to day 7 for the principle component (corresponding to stages 18–19). Label was found from day 19 for the spinal component (corresponding to stages 21–22). Since birth in the Quoll was at stage 15, it is most likely that this nucleus was functional at this time.

In mammals, the sensory trigeminal nucleus receives tactile, pain and temperature information from the face (Crosby et al. 1962). In adult humans, there is also a response to odorous volatiles by fibres in the nasal epithelium (Tucker 1971; Doty et al. 1978). It is possible that this also exits in the newborn although there is no evidence for this. However, Shuleikina-Turpaeva (1986) has show that newborn kittens (which were much further advanced than newborn marsupials) were not able to grasp the teat if they had been deprived of olfactory input directly after birth and before the first feed. It could be expected that some similar chemical sense is also important to newborn marsupials in enabling them to locate and grasp the teat. This cannot be from the main or accessory olfactory systems in the Quoll, as the neurons of the main and the accessory olfactory systems have not formed at birth. Since the hindbrain is essentially the only functional part of the brain at birth, this chemical must be detected by either the trigeminal or the gustatory nerve.

The first reflexes in the human occur at about stage 16 (Hooker 1958). Similar movements and similar sequences of development of these movements have been reported in rats (Angulo y Gonzâlez 1932; Narayanan et al. 1971) and in the cat (Windle and Griffin 1931). These movements are probably general across vertebrates (Bekoff 1985) but there have not been many detailed studies on these first responses to trigeminal stimulation. Recent developments have made it easier to look at these behaviours without disturbing either the mother or the environment of the embryo (Natsuyama 1991).

The general description given below of these reflex movements is mainly from Angulo y Gonzâlez (1932) and Hooker (1958). The first movement, which only occurred in response to a stimulus, involved a slight bending of the neck away from the stimulus and away from the forelimb on that side. The lateral movement then became larger and involved the trunk and the arm that moved forward on the same side as the stimulus. Later movements involved a combination of a retraction of both forelimbs, extension of the head, opening of the mouth and protrusion of the tongue in response to snout stimulation. These later two movements never occurred without head extension at this stage. In other young at the same age, snout stimulation produced a lateral flexion in which the forelimb on the flexed side was bent at the elbow joint and the hand also flexed. The opposite forelimb was extended and the hand also extended. Some animals at the same stage showed a combination of the two patterns in which there was lateral flexion of the trunk with the forelimb on the flexed side slightly retracted and the one on the opposite side raised. The mouth opened and the tongue protruded. This stage was also a time when the embryos began to show a large amount of spontaneous activity.

Thus the pattern of movement seen at birth in marsupials has many features that are seen in eutherians in the uterus. Some trigeminal-based movements are beginning to appear in eutherian embryos at about the same stage of development as similar general patterns are appearing in some newborn marsupials. The movement of the newborn marsupial involves a lateral movement of the head that is slightly raised above the surface (mother's body) and then it is brought down to the surface as it completes its swing. This could then stimulate the trigeminal nerve to produce the reflex seen in humans, kittens and rats in which the head then moves away from the stimulation. It could be that there is still a central pattern generator that produces the rhythmic swinging movement but there could also be some sensory input from the trigeminal. This brief comparison of movements in eutherians and marsupials indicates that there is a need for more detailed studies on the comparative development of these patterns as well as on more detailed observations of the movements of the newborn marsupial. Some obvious questions to be answered are: How do the newborn marsupials respond to trigeminal stimulation? Do they move with mouth opening and tongue protrusion?

What is clear from these eutherian studies is that the first observed reflex involves the trigeminal. The importance of this system has been underestimated because, in adult animals, many of its early functions are under higher inhibition (Teitelbaum et al. 1983) and are obscured by the complexity of the motor functions that are initiated by higher brain centres. The movements, that are seen at birth in marsupials, are not necessarily a newly evolved movement but perhaps just an earlier performance of a pattern that exists in all mammalian embryos.

General considerations

Early development

The maxillomandibular complex of the trigeminal of mammals sends axons to the cervical spinal cord and to the spinal accessory nucleus. The accessory spinal root motor fibres go to the sternocleidomastoid and the upper part of the trapezius muscle. These connections are involved in the first reflexes seen in humans (Crosby et al. 1962; Hooker 1958). The forearm motor system is also at least partially operational at birth in marsupials, since the neonate uses its forearms to crawl toward the nipple and to grasp the pouch-hairs of the mother. Motor commands to the neck are via cervical nerves C1 to C4 and those to the arm are via C5 to T1. In the hindbrain, the main motor nuclei for these structures are in the pons (the lateral vestibular and pontine reticular nuclei) and in the medulla oblongata (medullary reticular nuclei). Extensor motor neurons are facilitated by commands from the nuclei in the pons and inhibited by those in the medulla while it is the reverse for the flexor motor neurons. In the midbrain, the motor centre is the red nucleus which via the rubrospinal tract, facilitates flexor motor neurons and inhibits extensor motor neurons.

The absence of radiographic label in the motor nucleus of the trigeminal nerve in any of the specimens examined suggests that most, if not all, of the development of this structure was completed before birth (we had no animals labelled before day 3). Altman and Bayer (1980b, 1982) found that 80% of the cells of this nucleus in the rat were formed on E12 (stage 13–14). Referring to the cephalic flexure that occurs as the trigeminal is forming in all mammals they state: 'one of the adaptive advantages of this transient morphogenetic process is the spatial approximation of outgrowing trigeminal fibres and their target structures'. The results of their work would lead to the conclusion that all of the cells that would form neurons in this nucleus in the Quoll had been formed before birth.

Apart from the motor nucleus of the trigeminal, which controls mouth opening and sucking, the lateral vestibular nucleus and the pontine nuclei innervate motor neurons in the spinal cord at cervical levels where they supply the muscles that turn the head and move the arm, shoulder and upper extremity. Lesions of this nucleus turn the body to the side opposite to the lesion and the head to the side of the lesion; and section of the spinal cord above L2 produces flexion of the lower legs as a result of loss of vestibulospinal and reticulospinal input (Crosby et al. 1962).

Pflieger and Cabana (1996), in a seminal paper on newborn *Monodelphis*, used DiI to trace the vestibular primary afferents and the vestibular nuclei projections to the cervical spinal cord. They found that some of the fibres of the vestibular nuclei project to the contra-lateral vestibular ganglion and this did not occur in the adult. This raises an important consideration. There are many examples of exuberant growth of axons or collaterals into target areas and then a later pruning of these as a result of the development of function within the system (Purves and Lichtman 1985; O'Leary 1992, O'Leary et al. 1994). Furthermore, during development there is a progressive input from increasingly higher centres so that what was an important neural circuit in an embryo becomes less important as a result of the gigantic increase in the pathways to and from higher centres. Functionally, this system may become 'invisible' as a result of inhibition from higher centres and so only become visible when these higher centres are made non-functional either by chemicals or destruction or by interfering with the feedback from the body senses (Teitlebaum et al. 1983; Bekoff and Kauer 1984).

This emphasises the importance of studies such as Pflieger and Cabana (1996) in which the connections between various neural centres are determined in the embryonic stages and in which it is not assumed that the connections will be as in the adult. It is possible that the embryonic brain has many more connections between various centres in the hindbrain than are visualised in the adult. Pflieger and Cabana (1996) provided evidence for a pathway from the utricle to the lateral vestibular nucleus and then to the cervical cord and suggested that this direct path in this species could provide input via the utricle 'to orient the position of the head and the rhythmic movements of the forelimbs on the mother's belly, against gravity'. Pfaller and Arvidsom (1988) found that a tracer injected into the dorsal root ganglia of C2 had projections in the cuneate nucleus and in all of the trigeminal sensory nuclei and in the spinal, medial and lateral vestibular nuclei.

Later development

In the Quoll, the radioactive label within the hindbrain of the sensory components of the trigeminal nerve decreased in intensity from day 3, suggesting that some parts had been formed before this. (Unfortunately, none of the animals injected with radioactive thymidine before day 3 survived.) There was a progression of development of the somatosensory system from the hindbrain to the forebrain and a similar progression was seen in the motor system. Thus during pouch life, not only was there an increase in the number of neurons in the sensory nuclei, but the sensory information was also being transmitted to higher levels. Higher areas of the brain were also forming at these times and higher motor centres were developing. Thus the behaviours initiated by the trigeminal motor nuclei could progressively be integrated into more complex behaviours that are initiated at higher levels and some of this integration could involve inhibition of the trigeminal motor patterns (Teitelbaum 1967; Davies and Lumsden 1990; Rowe 1996).

It is therefore unlikely that any behaviour prior to 30 days is initiated at the cortical level. Any change in motor patterns must be initiated at lower levels within the brain. However, the reaching behaviour seen in day 12 pouch young, but not previously, indicates some change in circumstance at this time. The fact that the pouch young reach for a hair removed from their grasp indicates 'recognition' of the removal, and 'motivation' to renew the grasp.

At day 23, tracers in forelimb showed that nerves were present in the hand in areas where the sensory pads would later form. In the adult, these pads are ridged and contain large numbers of Meissner corpuscles that are important touch receptors during food manipulation. The ridges on the sensory pads in the hand of the Quoll do not begin to form until about day 40. Vibrissae start to form on the wrist at 35 days. These systems begin to mature around the time of detachment from the nipple at 40 days, suggesting that they have a role in gathering sensory information at this time. These sensory receptors form a large part of the somatosensory spatial map of the adult Northern Quoll, consisting of approximately 36 percent of the somatosensory cortex (Huffmann et al. 1999). The vibrissae are also important in guiding the reach and grasp in the adult Quoll (Nelson, in prep.) and in prey-catching behaviour (Pellis et al. 1992; Pellis and Nelson 1984).

The proportion of the hindbrain that was labelled compared to that which was not labelled following a dye transport from the trigeminal ganglion decreased with age over the period studied. This could be interpreted as indicative of several factors. Firstly, it may indicate that the relative importance of the trigeminal system decreases with age, as more sensory systems develop and neurons born during this period are recruited by these systems. Secondly, this effect may be created by the pruning of collateral projections to other neurons, which may then be incorporated into other systems. Thirdly, neuron death may account for the relative decrease in label. Fourthly, and most probably, this change may be the result of a combination of these factors.

The small amount of development occurring in the midbrain nuclei of the trigeminal was virtually complete by day 7. It would seem, therefore, that most of the increase in size of the trigeminal ganglion that is seen after this time supplies Tsp, Tp, TMC and CuM. These structures are sensory in nature, indicating that the development within the ganglion is sensory, which concurs with the findings of Gasser and Hendrickx (1969). Thus there appears to be a large increase in sensory input to the somatosensory system. The motor development is quite small, suggesting that the motor components of the trigeminal do not increase greatly, but could become responsive to a wider range of stimuli.

Turkewitz and Kenny (1982) propose that the timing and extent of sensory development, regulated by input, is also important for normal behavioural organisation of the righting reflex. 'During the period when only the cutaneous system is functional, embryonic behaviour might more easily be organised around such input than would be the case if the organism had to contend with vestibular input at the same time. Once [righting] behaviour has begun to be organised around cutaneous input, the addition of vestibular input is less likely to disrupt behavioural organisation, but rather would be assimilated into the existing framework' (Turkewitz and Kenny 1982; p 360).

The trigeminal system is involved in the activation of righting behaviour from around day 36 (Pellis et al. 1992). The present study, however, has indicated that the trigeminal system is functional at birth. Similarly, the onset of righting triggered by input from the body commences after day 40 (Pellis et al. 1992) although the present study has suggested that sensory information originating in the forelimb and body could be processed in higher centres of the brain as early as 17 days. The delay in development of righting systems may be interpreted as supportive of the ideas of Turkewitz and Kenny (1982) (but see Pellis 1996). It may be said that the ability to right is superfluous in attached animals, and other systems which are more beneficial to the animal during the pre-detachment period are given a competitive advantage, resulting in a developmental delay for righting systems.

Future directions

This comparison of sensory and motor development, in marsupials and eutherians at developmental stages comparable to those at the time of birth in marsupials, suggests that some aspects of development across marsupials and eutherians are not as different as might first be expected. For example, there are similarities, at these stages, in the relative development of the trigeminal as well as in the form of the early reflexes. There would be great value in comparative studies on such development at these stages, to better understand the evolutionary relationships between the two groups, as important evolutionary changes are brought about by changes in the timing of developmental events (Wolpert et al. 1998).

The detailed study of the sensory and motor system of the Quoll, we hope, indicates that much can be learned from such studies. Marsupials are not born at the same stage of development and hence their sensory systems are not at equivalent stages of development at birth. Nor are their modes of birth the same. Kangaroos and possums crawl upwards to the pouch, bandicoots dangle by the umbilical cord and are swung into the pouch, to some extent, by the rolling from one side to the other by the female, and dasyurids have to move upwards and then horizontally to reach the pouch. Therefore there is not one answer that will explain how the newborn locates the pouch and the teat and more studies are required to determine the functional state of the various sensory systems.

References

Aitkin, L.M., Nelson, J.E., Farrington, M., & Swann, S. (1991), 'Neurogenesis in the brain auditory pathway of a marsupial, the Northern native Cat, *Dasyurus hallucatus*', *Journal of Comparative Neurology,* **309**:250–60.

Aitkin, L.M., Nelson, J.E., & Shepherd, R.K. (1994a), 'Hearing, vocalization and the external ear of a marsupial, the Northern Quoll, *Dasyurus hallucatus*', *Journal of Comparative Neurology,* **349**:377–88.

Aitkin, L., Nelson, J., Farrington, M., & Swann, S. (1994b), 'The morphological development of the inferior colliculus in a marsupial, the Northern Quoll (*Dasyurus hallucatus*)', *Journal of Comparative Neurology,* **343**:532–41.

Aitkin, L., Nelson, J., & Shepherd, R.K. (1996a), 'The development of hearing in a marsupial, the Northern Quoll, *Dasyurus hallucatus*', *Journal of Experimental Zoology,* **276**:394–402.

Aitkin, L., Nelson, J., Martsi–McClintock, A., & Swann, S. (1996b), 'Features of the structural development of the inferior colliculus in relation to the onset of hearing in a marsupial: The Northern Quoll, *Dasyurus hallucatus*', *Journal of Comparative Neurology*, **375**:77–88.

Altman, J., & Bayer, S.A. (1964), 'The use of fine autoradiography in neurological and psychobiological research', in *Response of the Nervous System to Ionising Radiation* (eds. T.J. Haley & R.S. Snide), pp. 336–59, Little Brown, Boston.

Altman, J., & Bayer, S.A. (1980a), 'Development of the brain stem in the rat. I. Thymidine-radiographic study of the time of origin of neurons of the lower medulla', *Journal of Comparative Neurology,* **194**:1–35.

Altman, J., & Bayer, S.A. (1980b), 'Development of the brain stem in the rat. II. Thymidine-radiographic study of the time of origin of neurons of the upper medulla, excluding the vestibular and auditory nuclei', *Journal of Comparative Neurology,* **194**:37–56.

Altman, J., & Bayer, S.A. (1981), 'Development of the brain stem in the rat. V. Thymidine–radiographic study of the time of origin of neurons in the midbrain tegmentum', *Journal of Comparative Neurology*, **196**: 677–716.

Altman, J., & Bayer, S.A. (1982), 'Development of the cranial nerve ganglia and related nuclei in the rat', *Advances in Anatomy, Embryology & Cell Biology,* **74**:1–90.

Altman, J., & Sudarshan, K. (1975), 'Postnatal development and locomotion in the laboratory rat', *Animal Behaviour,* **23**:896–920.

Angulo y Gonzâlez, A.W. (1932), 'The prenatal development of behavior in the albino rat', *Journal of Comparative Neurology,* **55**:395–442.

Bekoff, A., & Kauer, J.A. (1984), 'Neural control of hatching: fate of the pattern generator for the leg movements of hatching in post-hatching chicks', *Journal of Neuroscience,* **4**:2659–66.

Bekoff, A. (1985), 'Development of locomotion in vertebrates: a comparative perspective', in *The Comparative Development of Adaptive Skills: Evolutionary Implications* (ed. E.S. Gollin), pp. 57–94, Lawrence Erlbaum Associates, London.

Butler, H (1987), *An atlas for Staging Mammalian and Chick Embryos*, CRC Press, Boca Raton.

Cabana, T., Cassidy, G., Pflieger, J-P., & Baron, G. (1993), 'The ontogenic development of sensorimotor reflexes and spontaneous locomotion in the Mongolian gerbil (*Meriones unguiculatus*)', in *Proceedings of developmental plasticity and regeneration in the spinal cord: cellular and molecular interactions,* Satellite Symposia of the Third IBRO World Congress of Neuroscience. *Brain Research Bulletin,* **30**:291–301.

Cannon, J.R., Bakker, H.R., Bradshaw, S.D., & McDonald, I.R. (1976), 'Gravity as the sole navigational aid to the newborn quokka', *Nature,* **259**:42.

Cassidy, G., & Cabana, T. (1993), 'The development of the long descending propriospinal projections in the opossum, *Monodelphis domestica*', *Developmental Brain Research,* **72**:291–99

Cassidy, G., Pflieger, J-P., & Cabana, T. (1992), 'The ontogenesis of sensorimotor reflexes in the Mongolian gerbil *Meriones unguiculatus*', *Behavioral Brain Research,* **52**:143–51.

Cassidy. G., Boudrias, D., Pflieger, J.F., & Cabana, T. (1994), 'The development of sensorimotor reflexes in the Brazilian opossum *Monodelphis domestica*', *Brain, Behavior & Evolution,* **43**: 244–53.

Crosby, E.C., Humphrey, T., & Lauer, E.W. (1962), *Correlative Anatomy of the Nervous System*, Macmillan, New York.

Darian–Smith, I. (1973), 'The trigeminal system', in *Handbook of Sensory Physiology* (ed. I. Iggo), pp. 271–314, Springer, Berlin.

Davies, A.M., & Lumsden, A. (1990), 'Ontogemy of the somatosensory system: origins and early development of primary sensory neurons', *Annual Reviews in Neuroscience,* **13**:61–73.

Doty, R.L., Brugger, W.E., Jurs, P.C., Orndorff, M.A., Snyder, P.J., & Lowry, L.D. (1978), 'Intranasal trigeminal stimulation from odorous volatiles: psychometric responses from anosmic and normal humans', *Physiology & Behavior,* **20**:175–85.

Eshkol, N., & Wachmann, A. (1958), *Movement Notation*, Weidenfield & Nicholson, London.

Fentress, J.C. (1983), 'Ethological models of hierarchy and patterning of species-specific behavior', in *Handbook of Behavioral Neurobiology. Volume 6. Motivation* (eds. E. Satinoff & P. Teitelbaum), pp.185–234, Plenum Press, New York.

Fentress, J.C. (1989), 'Comparative coordination (a story of three little p's in behavior)', *Advances in Psychology,* **61**:185–219.

Flynn, T.T. (1928), 'The problem of the birth of the kangaroo', *Workers' Educational Association of Tasmania,* **1**:1–12.

Fox, M.W. (1964), 'A phylogenetic analysis of behavioral neuro-ontogeny in precocial and non-precocial mammals', *Canadian Journal of Comparative Medicine & Veterinary Science,* **28**:197–202.

Fox, M.W. (1965), 'Reflex-ontogeny and behavioural development of the mouse', *Animal Behaviour,* **13**:234–41.

Gasser, R.F., & Hendrickx, A.G. (1969), 'The Development of the trigeminal nerve in baboon embryos (*Papio sp.*)', *Journal of Comparative Neurology,* **136**:159–82.

Gemmell, R.T., & Nelson, J. (1988), 'Ultrastructure of the olfactory system of three newborn marsupial species', *Anatomical Record,* **221**:655–62.

Gemmell, R.T., & Nelson, J. (1989), 'The vestibular system of the newborn marsupial cat, *Dasyurus hallucatus*', *Anatomical Record,* **225**:203–8.

Gemmell, R.T., & Rose, R.W. (1989), 'The senses involved in movement of some newborn Macropodidae and other marsupials from cloaca to pouch', in *Kangaroos, Wallabies and Rat Kangaroos* (eds. G. Grigg, I. Jarman & I. Hume), pp.339–47, Surrey Beatty, Sydney.

Gemmell, R.T., & Selwood, L. (1993), 'Structural development in the newborn marsupial, the stripe-faced dunnart, *Sminthopsis macroura*', *Acta Anatomia,* **149**:1–12.

Gemmell, R.T., Peters, B., & Nelson, J. (1988), 'Ultrastructural identification of Merkel cells around the mouth of the newborn marsupial', *Anatomy & Embryology,* **177**:403–8.

Gemmell, R.T., Veitch, C. & Nelson, J. (1999), 'Birth in the marsupial northern brown bandicoot *Isoodon macrourus*', *Australian Journal of Zoology,* **47**:517–28.

Golani, I. (1976), 'Homeostatic motor processes in mammalian interactions: a chereography of display', in *Perspectives in Ethlogy, Vol. 2* (eds. P.P.G. Bateson, & P.F. Klopfer), pp. 69–134, Plenum Press, New York.

Gottleib, G.E. (1971), 'Ontogenesis of sensory function in birds and mammals', in *The Biopsychology of Development* (eds. E. Tobach, L.R. Aronson, & E.Shaw), pp. 67–128, Academic Press, New York.

Gingras, J., & Cabana, T. (1999), 'Synaptogenesis in the brachial and lumbosacral enlargements of the spinal cord in the postnatal opossum, *Monodelphis domestica*', *Journal of Comparative Neurology,* **414**:551–60.

Gribnau, A.A.M., & Geijsberts, L.G.M. (1981), 'Developmental stages in the rhesus monkey (*Macaca mulatta*)', *Advances in Anatomy Embryology & Cell Biology,* **68**:1–79.

Honig, M.G., & Hume, R.I. (1989), 'DiI & DiO: Versatile fluorescent dyes for neuronal labelling and pathway tracing', *Trends in Neuroscience,* **12**:333–41.

Hooker, D. (1958), *Evidence of Prenatal Function of the Central Nervous System in Man*, James Arthur Lecture on the Evolution of the Human Brain, American Museum of Natural History, New York.

Huffman, K.J., Nelson, J., Clarey, J., & Krubitzer, L. (1999), 'Organization of somatosensory cortex in three species of marsupials, *Dasyurus hallucatus, Dactylopsila trivirgata,* and *Monodelphis domestica*: Neural correlates of morphological specializations', *Journal of Comparative Neurology,* **403**:5–32.

Hughes, R.L., & Hall, L.S. (1984), 'Embryonic development in the common brushtail possum (*Trichosurus vulpecula*)', in: *Possums and Gliders* (eds. A.P. Smith, & I.D. Hume), pp.197–212, Australian Mammal Society, Sydney.

Hughes, R.L., & Hall, L.S. (1988), 'Structural adaptations of the newborn marsupial', in *The Developing Marsupial. Models for Biomedical Research* (eds C.H.Tyndale-Biscoe, & P.A. Janssens), pp. 8–27, Springer, Berlin.

Ivanco, T.L., Pellis, S.M., & Whishaw, I.Q. (1996), 'Skilled forelimb movements in prey catching and in reaching by rats (*Rattus norvegicus*) and Opossums (*Monodelphis domestica*): Relations to anatomical differences in motor systems', *Behavioural Brain Research,* **79**:163–81.

Jeannerod, M., & Decety, J. (1990), 'The accuracy of visuomotor transformations. An investigation into the mechanisms of visual recognition of objects', in *Vision and Action: The Control of Grasping* (ed. M.A. Goodale), pp. 31–48, Ablex Publishing, Norwood, NJ.

Jones, T.E., & Munger, B.L. (1985), 'Early differentiation of the afferent nervous system in glabrous skin of the oppossum (*Monodelphis domesticus*)', *Somatosensory Research,* **3**:169–84.

Knott, G.W., Kitchener, P.D., & Saunders, N.R. (1999), 'Development of motoneurons and primary sensory afferents in the thoracic and lumbar spinal cord of the South American Opossum *Monodelphis domestica*', *Journal of Comparative Neurology,* **414**:423–36.

Krause, W.J. (1991), 'The vestibular apparatus of the opossum (*Didelphis virginiana*) prior and immediately after birth', *Acta Anatomia,* **142**:57–9.

Krause, W.J. (1992), 'A scanning electron microscopic study of the opossum nasal cavity prior to and shortly after birth', *Anatomy & Embryology*, **185**:281–9.

Krause, W.J. (1998), 'A review of histogenesis/organogenesis in the developing North American Opossum (*Didelphis virginiana*)', *Advances in Anatomy Embryology & Cell Biology*, **143**:1–160.

Lepre, M., Fernandez, J., & Nicholls, J.G. (1998), 'Re–establishment of direct synaptic connections between sensory axons and motoneurons after lesions of neonatal opossum CNS (*Monodelphis domestica*) in culture', *European Journal of Neuroscience*, **10**(suppl):2500–10.

Leblond, H., & Cabana, T. (1997), 'Myelination of the ventral and dorsal roots of C8 and L4 segments of the spinal cord at different stages of development in the gray opossum, *Monodelphis domestica*', *Journal of Comparative Neurology*, **386**:203–16.

Maier, W. (1992), 'Cranial morphology of the therian common ancestor, as suggested by the adaptations of neonate marsupials', in *Mammal Phylogeny* (eds. F.S. Szalay, M.J. Novacek, & M.C. McKenna), pp.165–81, Springer, Berlin.

Mark, R.F., & Marotte, L.R. (1992), 'Australian marsupials as models for the developing mammalian visual system', *Trends in Neuroscience*, **15**:51–7.

Marotte, L.R., Leamey, C.A., & Waite, P.M.E. (1997), 'Timecourse of development of the wallaby trigeminal pathway: III. Thalamocortical and corticothalamic projections', *Journal of Comparative Neurology*, **387**:194–214.

Martin, G.F., Cabana, T., & Ho, R.H. (1988), 'The early development of subcortical projections to presumptive somatic sensory-motor areas of neocortex in the North American opossum', *Anatomy & Embryology*, **178**:365–79.

Molnár, Z., Krubitzer, L., Nelson, J.E., & Blakemore, C. (1995), 'Development of thalamocortical innervation in the marsupial northern native cat (*Dasyurus hallucatus*)', *European Journal of Neuroscience*, **8**(suppl):78.

Molnár, Z., Knott, G.W., Blakemore, C., & Saunders, N.R. (1998), 'Development of thalamocortical projections in the South American gray short–tailed opossum (*Monodelphis domestica*)', *Journal of Comparative Neurology*, **398**:491–514.

Mosconi, T.M., & Rice, F.L. (1993), 'Sequential differentiation of sensory innervation in the mystacial pad of the ferret', *Journal of Comparative Neurology*, **333**:309–25.

Müller, F., & O'Rahilly, R. (1988), 'The development of the human brain, including the longitudinal zoning in the diencephalon at stage 15', *Anatomy & Embryology*, **179**:55–71.

Müller, F., & O'Rahilly, R. (1989a), 'The human brain at stage 16, including the initial evagination of the neurohypophysis', *Anatomy & Embryology*, **179**:551–69.

Müller, F., & O'Rahilly, R. (1989b), 'The human brain at stage 17, including the appearance of the future olfactory bulb and the first amygdaloid nuclei', *Anatomy & Embryology*, **180**:353–69.

Müller, F., & O'Rahilly, R. (1990a), 'The human brain at stages 18–20, including the choroid plexuses and the amygdaloid and septal nuclei', *Anatomy & Embryology*, **182**:285–306.

Müller, F., & O'Rahilly, R. (1990b), 'The human brain at stages 21–23, with particular reference to the cerebral cortical plate and to the development of the cerebellum', *Anatomy & Embryology*, **182**:375–400.

Nelson, J. (1987), 'The early development of the eye of the pouch young of the marsupial *Dasyurus hallucatus*', *Anatomy & Embryology*, **175**:387–98.

Narayanan, C.H., Fox, M.W., & Hamburger, V. (1971), 'Prenatal development of spontaneous and evoked activity in the rat (*Rattus norvegicus albinus*)', *Behavior*, **40**:100–34.

Natsuyama, E. (1991), 'In-utero behavior of human embryos at the spinal-cord stage of development', *Biology of the Neonate*, **60**(suppl. 1):11–29.

Nelson, J.E. (1992), 'Developmental staging in a marsupial *Dasyurus halucatus*', *Anatomy & Embryology*, **185**:335–54.

Nieuwenhuys, R., ten Donkelaar, H.J., & Nicholson, C. (1998), *The Central Nervous System of Vertebrates*, Springer, Berlin.

Nunn, C.L., & Smith, K.K. (1998) ,'Statistical analyses of developmental sequences: the craniofacial region in marsupial and placental mammals', *American Naturalist*, **152**:82–101.

O'Leary, D.D.M. (1992), 'Development of connectional diversity and specificity in the mammalian brain by the pruning of collateral projections', *Current Opinion in Neurobiology*, **2**:70–77.

O'Leary, D.D.M., Ruff, N.L., & Dyck, R.H. (1994) 'Development, critical period plasticity, and adult reorganisations of mammalian somatosensory systems', *Current Opinion in Neurobiology*, **4**:535–44.

Pellis, S.M. (1981), 'A description of social play by the Australian magpie *Gymnorhina tibicen* based on Eshkol-Wachmann notation', *Bird Behaviour*, **3**:61–79.

Pellis, S.M. (1996), 'Righting and modular organization of motor programs', in *Measuring Movement and Locomotion: From Invertebrates to Humans* (eds. K-P. Ossenkopp, M. Kavaliers and P.R. Sanberg), pp.115–133, R.G. Landes, Austin.

Pellis, S.M., & Nelson, J.E. (1984) 'Some aspects of predatory behaviour of the Quoll, *Dasyurus viverrinus* (Marsupialia: Dasyuridae)', *Australian Mammalogy*, **7**:5–15.

Pellis, S.M., Pellis, V.C., & Nelson, J.E. (1992), 'The development of righting reflexes in the pouch young of the marsupial *Dasyurus hallucatus*', *Developmental Psychobiology*, **25**:105–25.

Pellis, S.M., Vanderlely, R., & Nelson, J.E. (1992), 'The roles of vision and vibrissae in the predatory behaviour of northern quolls *Dasyurus hallucatus* (Marsupialia: Dasyuridae)', *Australian Mammalogy*, **15**:55–60.

Pfaller, K., & Arvidsson, J. (1988), 'Central distribution of trigeminal and upper cervical primary afferents in the rat studied by anterograde transport of horseradish peroxidase conjugated to wheat germ agglutinin', *Journal of Comparative Neurology*, **268**:91–108.

Pflieger, J.F., & Cabana, T. (1996), 'The vestibular primary afferents and the vestibulospinal projections in the developing and adult opossum, *Monodelphis domestica*', *Anatomy & Embryology*, **194**:75–88.

Pflieger, J.F., Cassidy, G., & Cabana, T. (1996), 'Development of spontaneous locomotor behaviors in the opossum, *Monodelphis domestica*', *Behavioral Brain Research*, **80**:137–43.

Purves, D., & Lichtman, J.W. (1985), *Principles of Neural Development*, Sinauer, Sunderland, Mass.

Qin, Y.Q., Wang, X.M., & Martin, G.F. (1993), 'The early development of major projections from caudal levels of the spinal cord to the brainstem and cerebellum in the gray short–tailed Brazilian opossum, *Monodelphis domestica*', *Developmental Brain Research*, **75**:75–90.

Renfree, M.B., Fletcher, T.P., Blanden, D.R., Lewis, P.R., Shaw, G., Gordon, K., Short, R.V., Parer-Cook, E., & Parer, D. (1989), 'Physiological and behavioural events around the time of birth in macropodid marsupials', in *Kangaroos, Wallabies and Rat-Kangaroos* (eds. G. Grigg, P. Jarman, & I. Hume), pp. 323–7, Surrey Beatty, Sydney.

Rowe, M.J. (1982), 'Development of mammalian somatosensory pathways', *Trends in Neuroscience,* **5**:408–11.

Rowe, M. (1996), 'Sensorimotor cortical organisation: how do marsupials compare with other mammals?', in *Comparison of Marsupial and Placental behaviour* (eds. D.B. Croft, & U. Ganslosser), pp. 3–45, Filander: Fürth.

Saunders, N.R. (1997), 'Marsupials as models for studies of development and regeneration of the central nervous system', in *Marsupial Biology: Recent Research, New Perspectives* (eds. N. Saunders, & L. Hinds), pp. 382–410, University of New South Wales Press, Sydney.

Shuleikina-Turpaeva, K.V. (1986), 'Sensory organization of alimentary behavior in the kitten', *Advances in the Study of Behavior,* **16**:1–37.

Shatz, C.J. (1992), 'How are specific connections formed between thalamus and cortex?', *Current Opinion in Neurobiology,* **2**:78–82.

Sheng, X-M., Marotte, L.R., & Mark, R.F. (1991), 'Development of the laminar distribution of thalamocortical axons and corticothalamic cell bodies in the visual cortex of the wallaby', *Journal of Comparative Neurology,* **307**:17–38.

Shigenaga, Y., Sera, M., Nishimori, T., Suemune, S., Nishimura, M., Yoshida, A., & Tsuru, K. (1988), 'The central projection of masticatory afferent fibres to the trigeminal sensory nuclear complex and upper cervical spinal cord', *Journal of Comparative Neurology,* **268**:489–507.

Smith, K.K. (1997), 'Comparative patterns of craniofacial development in eutherian and metatherian mammals', *Evolution,* **51**:1663–78.

Teitelbaum, P. (1967), *Physiological Psychology: Fundamental Principles*, Prentice-Hall, Englewood Cliffs.

Teitelbaum, P., Schallert, T., & Whishaw, I.Q. (1983), 'Sources of spontaneity in motivated behavior', in *Handbook of Neurobiology, Volume 6 Motivation* (eds. E. Satinoff & P. Teitelbaum), pp. 23–65, Plenum Press, New York.

Troiani, D., Petrosini, L., & Passani, F. (1981) 'Trigeminal contribution to the head righting reflex', *Physiology and Behavior,* **27**:157–60.

Tucker, D (1971), 'Nonolfactory responses from the nasal cavity: Jacobson's organ and the trigeminal system', in *Handbook of Sensory Physiology Volume IV. Chemical Senses. Part I Olfaction* (ed. L.M. Beidler), pp. 151–81, Springer, Berlin.

Turkewitz, G., & Kenny, P.A. (1982), 'Limitations on Input as a Basis for Neural Organisation and Perceptual Development: A Preliminary Theoretical Statement', *Developmental Psychobiology,* **15**:357–68.

Tyndale-Biscoe, H., & Renfree, M. (1987), *Reproductive Physiology of Marsupials: Monographs on Marsupial Biology*, Cambridge University Press, Cambridge.

Waite, P.M.E., Marotte, L.R., Leamey, C.A., & Mark, R.F. (1998), 'Developmental of whisker-related patterns in marsupials: factors controlling timing', *Trends in Neuroscience,* **21**:265–69.

Wang, X.M., Xu, X.M., Qin, Y.Q., & Martin, G.F. (1992), 'The origins of supraspinal projections to the cervical and lumbar spinal cord at different stages of development in the gray short-tailed Brazilian opossum, *Monodelphis domestica*', *Brain Research,* **68**:203–16.

Whishaw, I.Q., Pellis, S.M., Gorny, B.P., & Pellis, V.C. (1991), 'The impairments in reaching and the movements of compensation in rats with motor cortex lesions; an endpoint, videorecording, and movement notation analysis', *Behavioral Brain Research,* **42**:77–91.

Windle, W.F., & Griffin, A.M. (1931), 'Observations on embryonic and fetal movements of the cat', *Journal of Comparative Neurology,* **52**:149–88.

Wolpert, L., Beddington, R., Brockes, J., Jessell, T., Lawrence, P., & Meyerowitz, E. (1998), *Principles of Development*, Oxford University Press, Oxford.

Xu, X.M., & Martin, G.F. (1989), 'Developmental plasticity of the rubrospinal tract in the North American Opossum', *Journal of Comparative Neurology,* **279**:368–81.

PART III

PHYSIOLOGY

CHAPTER 14

NUTRITION OF CARNIVOROUS MARSUPIALS

Ian D. Hume

Institute of Wildlife Research, School of Biological Sciences A08, University of Sydney

Carnivory includes both faunivory (eating of vertebrates) and insectivory (eating of invertebrates). Faunivory is associated with larger body size, insectivory with smaller body size, but not exclusively. When scaled for metabolic body size, the energy requirements of marsupial carnivores are higher than those of most other marsupials but the numbat, an ant eater, is an exception; like eutherian ant eaters it has very low energy requirements. Little is known of the specific nutrient requirements of carnivorous marsupials. Carnivore diets are generally high in water, protein, vitamins and minerals, low in carbohydrate, and highly variable in fat content. Muscle and viscera are generally highly digestible but mammalian hair and teeth, reptilian dermal scales and avian feathers are poorly digested and consequently help in identification of prey remains in carnivore faecal scats. The digestive tract of carnivores is conservative, and the small intestine (the main site of digestion and absorption) dominates. Digesta passage is rapid, but decreases with increasing body size of the carnivore. Strict carnivory is a narrow nutritional niche, with disadvantages as well as advantages. The metabolism of carnivores conforms to the nutrient profile of animal tissues, which eliminates the energetic costs of synthesising redundant enzymes but limits the ability to exploit diets of other than animal material. These general principles of carnivory, derived mainly from eutherians, are applied to marsupials in this chapter. The plea is made for more specific nutritional information on carnivorous marsupials in order to better understand the basis for niche separation among grossly similar species, and thus manage habitats for maximum biodiversity of marsupial carnivores.

INTRODUCTION

Study of the nutrition and metabolism of marsupial carnivores lags behind that of eutherian carnivores, and of marsupial omnivores and herbivores. In many instances it is therefore necessary to review what we know of eutherian carnivores and in some instances to extrapolate across the phylogenetic boundary between eutherian and marsupial mammals.

By definition, a carnivore eats animal material, but rarely do carnivores eat only animal material. Many are seen to eat plant material, either regularly or seasonally when prey species are scarce or unavailable. For instance, although the brown antechinus (*Antechinus stuartii*) is regarded as a carnivore, up to 41% of its diet in a sclerophyll forest in north-eastern New South Wales consisted of flowers of epacrid species during winter when insects, their preferred food, were scarce (Statham 1982). Carnivory includes both faunivory (eating of vertebrates) and insectivory (which is usually taken to mean invertebrates in general, not just insects). There is a general relationship between predator size and prey size among carnivores (Jones 1997). Thus faunivory tends to be the domain of the larger species, and insectivory of the smaller species, but this is only a trend. Often both occur, the relative proportions of vertebrate and invertebrate material in the diet reflecting their respective availabilities. Within the insectivores, larger predators often take larger prey (Fisher and Dickman 1993). Eisenberg (1981) defined the macroniches available to mammals according to their different dietary and habitat specialisations. He reserved the term carnivore for vertebrate eaters, and classified invertebrate eaters as myrmecophages (if they ate mainly or solely ants and/or termites), insectivores or insectivore/omnivores.

Characteristics of carnivore diets

Carnivore diets are characterised by a high content of water, protein, vitamins and minerals, a variable amount of fat and a low level of carbohydrate. The water content of the body less contents of the digestive tract for five domestic mammals and one avian species (domestic chickens) was 54–60%, the protein content was 15–21%, the fat content 17–26% and the mineral (ash) content was 3–5% (Maynard and Loosli 1962). On a dry matter basis, the range of protein contents was 35–49%, of fat 40–60% and of ash 7–12%. Fat contents of wild species are generally lower than these levels. Water, protein, fat and ash contents of invertebrates are more variable. For instance, earthworms (Oligochaeta) contain 82–85% water, but winged termites (Isoptera) only 34% (Redford and Dorea 1984; Hume 1999). On a dry matter basis, the protein content of bee (Hymenoptera) pupae is over 90%, but of waxmoth (Lepidoptera) larvae only 31%. On the same basis fat contents vary from as low as 4% in earthworms to as much as 60% in winged female ants (Hymenoptera), and ash contents from 3% in bee larvae to as much as 61% in termite workers (no doubt because of ingested soil).

The other characteristic of carnivore diets is the high digestibility of muscle and viscera, but low digestibility of the teeth and some bones of vertebrates, the exoskeletons of invertebrates, and reptile dermal scales, bird feathers and mammalian hair. Consequently examination of the scats (faeces) of carnivores for undigested exoskeletons, scales, feathers and hair often is a good way of working out their dietary habits. The scale arrangement and other features of mammalian hair are excellent diagnostic tools, and keys for the identification of mammalian hair are available (e.g. Brunner and Coman 1974).

Problems in determining the diets of carnivores

Although the most common way of estimating the diet of carnivores is to examine the scats, it is much less direct than stomach content analysis because of the substantial modification of ingesta that occurs during its passage through the intestine. However, there are usually good reasons for the indirect approach: it is non-destructive, it allows continual assessment of diet without altering population density, it has no direct influence on other concurrent studies on the same population, and it may allow investigation of ontogenetic changes in the feeding preferences of individuals in a population (Read 1987a). It also has the advantage that invertebrate fragments are more compact in faeces than in the stomach, which reduces the time the investigator spends searching for recogisable parts (Dickman and Huang 1988).

These good reasons for scat rather than stomach content analysis notwithstanding, limitations of scat analysis need to be recognised and acknowledged. For instance, if soft parts are completely digested they will not appear in the faeces. Thus Scott, Hume and Dickman (1999) found no trace in the faeces of long-nosed bandicoots (*Perameles nasuta*) of the skink eggs the animals were observed to consume with relish. Planigales (*Planigale* spp.) use their forepaws to dextrously manipulate the hard exoskeleton of their invertebrate prey so that it can be discarded, leaving virtually no trace in the faeces and a gross underestimation of an important component of the diet. Conversely, any plant structural parts eaten are poorly digested in the carnivore gut and their contribution can be easily overestimated from their proportion in the scats. Dickman and Huang (1988) concluded that faecal analysis can be a relatively reliable method for determining the diet of generalist insectivores that eat hard-bodied prey if results are expressed as the per cent frequency of occurrence over all samples (i.e. the percent of animals in which a prey item is found). Nevertheless, the problem of overlooking soft-bodied prey items remains.

Strict carnivory – a narrow nutritional niche

As carnivore diets are always high in protein, and vitamins are present in their active metabolic form (Morris 1994), it is not surprising that the metabolism of true flesh-eating carnivores, primarily the cats (Felidae), shows some unusual features that are related directly to their specialised diet. For instance, the maintenance protein requirement for most adult mammals is in the range 6–8% of a good quality protein in the diet, but for adult cats it is 13%; for maximal growth of kittens it is 20–30% of the diet. It has been shown that the essential amino acid requirements of growing kittens are largely similar to those of other young mammals, so the higher requirement for dietary protein is to supply nitrogen (Morris 1994). This high requirement for nitrogen is explained by the fact that in cats, as opposed to omnivores, the activities of urea-cycle enzymes and aminotransferases are always high and are not responsive to diets low in protein in order to conserve nitrogen. Rogers, Morris and Freedland (1977) found that even starved cats oxidised amino acids for energy and disposed of the nitrogen at the same rate as fed animals.

The virtual absence of carbohydrate in the diet of strict carnivores means that little hexose is absorbed from the small intestine, and instead the animal's requirements for glucose are met largely from amino acids. Thus, compared with omnivores, hepatic activities of the gluconeogenic enzymes pyruvate kinase and phospho-enol-pyruvate carboxykinase in cats are always high (Rogers et al. 1977). Similar findings were reported by Migliorini et al. (1973) in an avian carnivore, the black vulture (*Coragyps atratus*).

In most mammals the amino acid arginine is not a dietary requirement; citrulline is synthesised in the intestinal mucosa and converted to arginine in the kidney. However, because of the high activity of their urea cycle, cats, ferrets and dogs cannot synthesise enough arginine to supply the urea cycle and thus arginine is an essential amino acid for them. One meal without arginine can lead to hyperammonaemia and death in cats (Morris 1985). On the other hand, cats are less tolerant than omnivores to dietary excesses of some amino acids such as glutamic acid in the diet; plant tissues are higher in glutamic acid than animal tissues. Also, when presented with a choice of purified diets containing either no protein or isolated protein sources, cats do not make appropriate choices to balance their diet in regard to protein, and amino acid imbalanced diets are not avoided (Morris 1994). This is very different from diet selection by omnivores, and reflects the generally constant amino acid composition of the protein consumed by strict carnivores.

Felids (domestic cats, leopard, lion) are also unusual in having a requirement for the β-sulphonic amino acid taurine (Hayes, Carey and Schmidt 1975). Taurine is a metabolite of cysteine oxidation and is present in all animal tissues. Both cats and dogs use taurine exclusively to conjugate bile acids; the taurine is lost in the enterohepatic circulation and degraded by microbes in the gut. This taurine must be replaced, but the rate of synthesis of taurine by cats is limited; hence the need for additional taurine in the diet.

Another adaptation to carnivory seen in cats is in the response of the taste buds of the facial nerve, which is one of four cranial nerves that convey information on taste. Responses are strongest to amino acids, especially those that are described as sweet to humans (proline, cysteine, alanine, lysine, histidine and taurine) (Bradshaw et al. 1996). On the other hand, amino acids with hydrophobic side chains (tryptophan, arginine, isoleucine and phenylalanine), which are bitter to humans, inhibit rates of discharge in the amino acid units of the cat's taste buds. Thus cats prefer solutions of sweet over bitter amino acids. The monophosphate nucleotides, which accumulate in prey tissues after death, also inhibit the amino acid units of the cat's taste buds (Boudreau et al. 1985), which probably explains the preference of cats for fresh meat over carrion. Thus cats are very sensitive to bitter tastes. In contrast, felids are generally insensitive to sugars at any behavioually meaningful concentration (Bartoshuk, Harned and Parks 1971; Beauchamp, Maller and Rogers 1977).

The lack of carbohydrates in the flesh eaters' diet is also reflected in their metabolism. Glucokinase activity in the liver of cats, and in the carnivorous trout, is low or absent, so carnivores are not well adapted to deal with a large glucose load (Cowey et al. 1977). Cats can digest some carbohydrates, especially starch, quite efficiently, but their ability to digest high concentrations of sucrose is limited. Although they can't taste it they show an aversion to sucrose after only six hours of exposure to a concentrated sucrose solution. This is because the limited disaccharidase activity in the cat small intestine is exceeded, allowing undigested sucrose to pass into the large intestine (hindgut) where microbial dysfermentation results in diarrhea and attendant discomfort, triggering an aversion response (Bartoshuk et al. 1971).

There is also evidence that cats require the essential fatty acid arachidonic acid ($C_{20\text{-}4}$) in addition to linoleic acid ($C_{18\text{-}2}$); in most omnivores arachidonic acid can be supplied as linoleic acid, but cats appear to lack the desaturase enzyme involved in conversion of linoleic to arachidonic acid (Hassam, Rivers and Crawford 1977).

As animal tissue contains all the essential vitamins needed by carnivores, their metabolism is modified to the supply of the vitamins in their active form rather than as provitamins, and to removal of excess vitamins. Thus cats, unlike dogs or ferrets, cannot use β-carotene as a provitamin, and require preformed vitamin A (retinol) in the diet. Similarly, cats and dogs have lower rates of synthesis of vitamin D in their skin than do omnivores, reflecting the generally high intake of this vitamin in their meat diet. The high niacin intake of strict carnivores is reflected in cats in the high activity of an enzyme involved in the removal of nicotinamide metabolites that may be undesirable (Morris and Rogers 1989).

Thus the metabolism of eutherian carnivores conforms to the nutrient profile supplied by animal tissues. The adaptations are appropriate as they eliminate the energetic costs of synthesising redundant enzymes and protect against potentially harmful metabolites, but the specialised nature of the adaptations limits the ability of strict carnivores to exploit other diets (Morris 1994), and therefore strict carnivores have a narrow nutritional niche (Hume 1999). There is so little information on the metabolism of marsupial carnivores that the extent to which they mirror their eutherian counterparts metabolically is unknown.

DIGESTION IN CARNIVORES (MARSUPIAL AND EUTHERIAN)

Because of the relatively digestible nature of much of what a carnivore consumes their digestive tract tends to be conservative and uncomplicated. Nevertheless, there are some interesting variations, especially among less strictly carnivorous species. The typical dental formula for the order Carnivora is 3/3, 1/1, 4/4 and 2/3 for a total of 42 permanent teeth (Eisenberg 1981). The canines of the upper jaw are usually enlarged, the premolars tend to be tricuspidate and the molars quadritubercular. The dental formula for dasyurid marsupials is similar, with four pairs of upper and three pairs of lower incisors, well-developed upper and lower canines, two or three pairs of upper and lower blade-like premolars and four pairs of upper and lower molars with sharp, shearing cusps (Strahan 1995).

The upper molars of dasyurids are roughly triangular, resulting in a V-shaped embrasure formed between each upper and lower molar. There are three main cusps on the upper molar (the protocone, metacone and paracone), and two triangular juxtaposed basins on the lower molar (the trigonid and talonid basins) (Archer 1976). Occlusion involves two distinct phases. The first is a puncture-crushing phase in which there is tooth-food-tooth contact as food is caught and pulverised between the protocone of the upper molar and the talonid basin. The second phase is a shearing phase, in which the anterior-most crests of the lower molars shear past the posterior crests of the preceeding upper molars; food is also ground during this phase (Moore and Sanson 1995).

Sanson (1985) suggested that the subtle differences that he observed in dasyurid dentitions may reflect differences in diet selection. This is because of the effort required to breach the barrier imposed by arthropod exoskeletons of different thickness and toughness. Small insectivores tend to chop food finely by many small cutting edges, but such a mechanism may not be appropriate when dealing with larger prey; the cuticle of some larger insects may be too hard for small insectivores to pierce. This is particularly so for some desert beetles. Thus at least some dasyurids cannot afford to be dietary opportunists, but have to be selective in their choice of invertebrate prey (Sanson 1985).

The stomach of marsupial carnivores is usually simple in its morphology (Hume 1999; Hume, Smith and Woolley 2000), although Crisp (1855) noted that the stomach of *Thylacinus* was very muscular and capable of considerable expansion. The only gastric modification among marsupial carnivores is the well-developed compound cardiogastric gland in the South American caenolestid *Caenolestes obscurus* (rat opossum) (Osgood 1921). The gland lies close to the oesophageal opening and consists of thick glandular mucosa with 40-60 openings from unbranched gastric glands. Richardson, Bowden and Myers (1987) thought that the role of the cardiogastric gland in *Caenolestes* may be to supply large quantities of pepsin, hydrochloric acid and mucus in order to process the more-or-less constant intake of high protein food by this small carnivore, but why other small carnivores don't have a similar structure is unexplained.

The small intestine of marsupial carnivores is short, but nevertheless it dominates the digestive tract, being 87% of the total length of the gastrointestinal tract in *Caenolestes* (Richardson et al. 1987). The large intestine is extremely short, only 7% of gastrointestinal tract length. In all but one South American marsupial the small and large intestines are delineated by the presence of a small caecum (Hume 1999). No Australian marsupial carnivore has a caecum, so the gross morphological distinction between small and large intestines is not clear. The one South American species that lacks a caecum is *Dromiciops australis* (Monito del Monte) (Barbour 1977). This is consistent with the inclusion of *Dromiciops* in the cohort Australidelphia rather than the Ameridelphia (Woodburne and Case 1996).

Passage of digesta through the short, uncomplicated carnivore gastrointestinal tract might be expected to be rapid. This is indeed the case. Read (1987b) found that the insect exoskeletons he used as an indigestible marker in two species of planigales (*Planigale* spp.) of 6–9 g body mass maintained on minced meat first appeared in the faeces only 30–60 minutes after a pulse dose, and all were excreted within 3–4 hours. Very similar transit times and 100% excretion times have been reported for shrews, eutherian insectivores of similar small body size (Pernetta 1976). Excretion times generally increase with increasing body size (Table 1). The longest excretion time, in this case mean retention time (the average time an indigestible marker remains in the gastrointestinal tract), recorded for a particle marker in a marsupial carnivore is 13.0 hours in 1 kg eastern quolls (*Dasyurus viverrinus*) on a mealworm diet, and 17.7 hours on a plant diet (D.I. Moyle pers. comm.).

The longer mean retention time (MRT) on the plant diet reflects the increased time that it takes to break down plant material in the digestive system. In the simple carnivore gut this is mainly in the stomach, the antrum of which acts as a grinding pump; food particles are selectively retained, based on their size, then ground by a propulsion/ retropulsion sequence into finer particles (Malagelada and Azpiroz 1989). These fine particles then remain in suspension within the liquid phase by the mixing effect of gastric motility, and pass out through the pylorus, the lumen of which is smaller than the lumen of the antrum, into the duodenum. When a wave of contraction moves over the antrum and approaches the pylorus, the pylorus closes and there is mass contraction of the terminal antrum. The force of this mass contraction against the closed pylorus forces solid particles concentrated by the peristaltic wave through the constricted antral ring (Ehrlein and Akkermans 1984). This ejection of antral contents into the corpus of the stomach produces a shearing effect, with fragmentation of solid particles. Particle breakdown in the stomach is always in a liquid suspension, with the main forces operating being compression and shearing rather than grinding in the way that molar teeth triturate food between hard surfaces. Large indigestible particles do not leave the stom-

Table 1 Digesta passage in six dasyurid marsupials compared

Species	diet	Body mass (g)	Marker	Transit time (hr)	Mean retention time (hr)	Ref.
Planigale gilesi, *P. tenuirostris*	mincemeat	7–12 5–7	Insect exoskeleton	0.5–1.0	3–4 (total excretion time)	1
Sminthopsis crassicaudata	mealworms	18	Stained mealworms	–	0.9	2
S. douglasi	mealworms/ mincemeat	40–70	Co–EDTA Cr–cell walls	1.3 ± 0.4 3.3 ± 0.5	3.3 ± 0.8 3.7 ± 0.9	3
Dasycercus byrnei	mealworms	140	Stained mealworms	–	1.5 (at 21°C) 2.5 (at 32°C)	2
Dasyurus viverrinus	mealworms/small carnivore mix	900–1300	Co–EDTA Cr–cell walls	6.0 (winter) 10.0 (summer)	7.3 (winter) 13.0 (summer)	4

References: 1, Read 1987b; 2, T.J. Dawson and A.C. Paiz (pers. comm. to Hume 1999); 3, Hume, Smith and Woolley (2000); 4, D.I. Moyle (pers. comm. to Hume 1999).

ach until the interdigestive period and the onset of a migrating motility complex over the stomach which clears the organ of residues ready for the next meal.

Because particles are almost invariably retained longer than solutes in the vertebrate stomach (Stevens and Hume 1995), and retention events in the stomach dominate the carnivore digestive system, the whole-tract MRT of particle markers usually exceeds that of solute markers in both eutherian and marsupial carnivores. This is very different from some specialised herbivores, in which there is selective retention of solute rather than particle markers; this is because of a colonic separation mechanism that has evolved in some eutherians and marsupials. In these herbivores particles are still retained longer than liquids and solutes in the stomach, but whole-tract MRTs are dominated by retention events in their complex hindgut (large intestine). Irrespective of which digesta phase is selectively retained, MRTs in the simple carnivore digestive system are usually much shorter than in omnivores and herbivores of similar body size. Thus in omnivorous bandicoots of 1 kg body mass, MRTs of inert digesta markers range from 10 hours for a particle marker on a mealworm diet to 33 hours for a solute marker on a plant diet (Moyle, Hume and Hill 1995; McClelland, Hume and Soran 1999). In the folivorous common ringtail possum (*Pseudocheirus peregrinus*), MRTs ranged from 35 hours for a particle-associated marker to 63 hours for a solute marker (Chilcott and Hume 1985). In both omnivores and herbivores the prolonged retention of digesta aids microbial fermentation of plant cell walls, which is a slow process.

In contrast, even with short digesta retention times, digestion of carnivore diets is high. For instance, the apparent digestibility (apparent assimilation efficiency) of the dry matter of ground house mice by *Antechinus swainsonii* averaged 80%, and of energy 87% (Cowan et al. 1974). Similarly, the apparent digestibility of the dry matter and energy of *Tenebrio* larvae by *A. stuartii* was 84% and 87% respectively (Nagy et al. 1978). Apparent digestibility does not account for faecal material of endogenous or metabolic origin; true digestibility does, and would be even higher than these apparent digestibilities.

The main sites of digestion in carnivores are the stomach and small intestine. There is no information on the gastric secretions of marsupial carnivores, but in eutherian carnivores there is high activity of gastric pepsin, which initiates protein digestion, and also gastric lipase, which initiates digestion of dietary lipids (Stevens and Hume 1995). Gastric chitinase has also been found in some eutherian carnivores such as the fox and dog, but not in the cat or marten (Cornelius, Dandrifosse and Jeuniaux 1975). This difference may be related to the strictly faunivorous habits of felids but more insectivorous nature of canids.

There is some information on small intestinal secretions in marsupial carnivores. The activities of hydrolytic enzymes associated with carbohydrate digestion in the dasyurid small intestine were measured by Kerry (1969). High activites of trehalase were recorded in both *A. stuartii* and *Dasyurus maculatus*, reflecting the reliance of dasyurids on invertebrates as food; trehalose is a storage disaccharide found only in insects. High maltase, isomaltase and sucrase activities suggest that sucrose is also a normal part of their diet. Sucrose could either be ingested directly, for instance in fruit, or indirectly in the contents of the digestive tracts of insect prey that feed on fruit, sap or nectar.

High activities of sucrase, maltase and trehalase were also found in the small intestine of the insectivorous Chilean mouse opossum (*Thylamys elegans*) by Sabat, Bozinovic and Zambrano (1995). That any sucrose consumed by *T. elegans* probably comes mainly from the digestive tracts of insect prey is supported by two findings. First, trehalase activity was constant throughout the year, but sucrase activity was higher in summer (when fruits are more available) than in winter. Second, captive *T. elegans* were not able to meet their maintenance energy requirements on an exclusive diet of fruits, whereas they had no difficulty doing so on a sole diet of insects.

Energy and nutrient requirements of marsupial carnivores

The primary determinant of energy and food requirements of animals is body mass. When scaled for body mass (kg $^{0.75}$), the basal metabolic rates (BMR) of half the marsupials measured to date (see Hume 1999) fall between 65% and 74% of the value expected for a eutherian based on the equation of Kleiber (1961). Most other marsupial BMRs lie below 65% of the expected eutherian value. The highest marsupial BMR recorded is that of the honey possum (*Tarsipes rostratus*), at 158% of the expected eutherian value. This small possum feeds exclusively on pollen and nectar. Other marsupials with relatively high BMRs include three carnivores: *Planigale ingrami* (106%), *Antechinus stuartii* (90%) and *Phascogale tapoatafa* (88%). However, when compared with eutherians of similar dietary habits (the Carnivora), the mean BMR of eight dasyurid marsupials ranging in body mass between 93 and 5050 g (209 ± 12 (SE) kJ. kg$^{-0.75}$. d^{-1}) (Hume 1999) is less than half that of eight species of Carnivora ranging in body mass between 77 and 8750 g (450 ± 49 kJ. kg$^{-0.75}$. d^{-1}) (McNab 1986).

On the other hand, the numbat (*Myrmecobius fasciatus*), an ant-eater, has a BMR only 49% of the expected eutherian value, but in this case it is no lower than those of many eutherian ant- and termite-eaters (McNab 1984). These values for BMR support the contention that there is a basic underlying difference in BMR between eutherians and marsupials, but the influence of other factors is sometimes strong enough to mask any phylogenetic difference (Hume 1999). As well as food habits, factors that may influence BMR include activity level (Lovegrove 2000) and the precision of temperature regulation (McNab 1978, 1986, 1988). McNab concluded that in both marsupials and eutherians, feeding on fruit, tree foliage or invertebrates is associated with low BMRs, especially at large body size. This is because these foods are either seasonally unavailable (fruit, invertebrates), are difficult to digest (tree leaves) or have to be detoxified (tree leaves, some invertebrates). Fossoriality (burrowing), arboreality and nocturnal activity patterns are also associated with low BMR, as are unpredictable environments (Lovegrove 2000).

Field metabolic rate (FMR), or the energy cost of free existence, is more variable than BMR for a species because it includes along with BMR the costs of maintenance, thermoregulation and activity (Nagy 1994), and sometimes other costs associated with tissue growth, fat deposition and reproduction as well. This makes interspecific comparisons difficult (Nagy, Girard and Brown 1999). In addition, comparisons of FMR between marsupial and eutherian carnivores are made complex by the significantly greater slope of the allometric relationship between FMR and body mass in eutherians than in marsupials (Nagy et al. 1999).

FMRs measured for 28 marsupial species are tabulated in Hume (1999). Inspection of these values shows that, after body mass, sex, season and reproductive state are the main sources of variation in FMR. For comparative purposes, a useful measure is the ratio between FMR and BMR (calculated by dividing mass-specific FMR by mass-specific BMR) (Nagy 1987). With one exception the highest ratios are from carnivores, especially very active small insectivores such as *Sminthopsis crassicaudata* (6.6), *Antechinus stuartii* (5.0) and *Marmosa robinsoni* (4.7). The one exception to high ratios in small insectivores is the omnivorous Leadbeater's possum (*Gymnobelideus leadbeateri*) at 6.2 (Smith et al. 1982). Lowest ratios are found in relatively inactive arboreal folivores such as the koala (approximately 2.0), and the greater glider (2.5), and the larger kangaroos (1.8–2.5). Thus although BMRs of marsupial carnivores are generally not higher than the marsupial average, many small species are highly active and as a consequence have higher than average FMRs.

An animal's requirement for water can be determined by measuring its rate of water turnover (WTR). Hume (1999) has tabulated representative values for free-living marsupials. Highest water turnover rates are found in the small carnivorous/insectivorous species, even though some of these have ranges that extend into the arid zone, where water conservation is generally crucial. However, their food of animal tissue contains enough water that special measures for water conservation beyond fossorial and nocturnal habits are not necessary for survival.

In the absence of information to the contrary, it can be assumed that, like their eutherian counterparts, marsupial carnivores have high maintenance requirements for nitrogen. Whether the more strictly faunivorous species have specific requirements for dietary amino acids such as taurine, and require vitamins to be in the active form rather than as pro-vitamins, as do some strict carnivores among the Eutheria, is not known. Similarly, requirements of marsupial carnivores for other specific nutrients among the vitamins, minerals and essential fatty acids are unknown.

Feeding ecology

As suggested by Sanson's (1985) work on small insectivores and Jones and Barmuta's (1998) work on the larger faunivores, diet and niche separation among the marsupial carnivores can be related to often subtle differences in tooth morphology and jaw musculature. This is because there are usually close relationships among diet, feeding structures and digestive tract morphology. However, exceptions to this generalisation are known to exist among extant species, most often because highly selective feeding behaviours can compensate for the lack of a structural specialisation. Also, it is not possible to predict compensations among dentition, gut morphology and feeding behaviours (Sanson 1991). Thus although the dietary habits of extinct forms are often predicted from the dental and masticatory systems of fossils, there is always room for debate. The Pleistocene marsupial lion *Thylacoleo carnifex* is a good example, having been variously described as a carnivore, an omnivore and even a herbivore. The thylacoleonids are complicated because they are of herbivorous phalangeroid stock yet their dentitions have few if any features of extant herbivores. The most prominent feature of the dentition of *Thylacoleo* is the huge size of the sectorial blades formed from the third premolars. The first lower molar contributes to the sectorial blade, but the first upper molar and the second molars are very much reduced, as they are in the Felidae (Sanson 1991). There is no modern equivalent to this dental pattern, hence the wide range of opinions on the diet and feeding niche filled by *Thylacoleo*.

CONCLUSIONS

Carnivore diets are characterised by high contents of water, protein, vitamins and minerals, variable levels of fat, but little carbohydrate. These features are reflected in the metabolism of carnivores, at least in the Eutheria; comparable data from marsupial carnivores are lacking. The muscle and viscera which make up the greatest part of the prey are easily digested in the short and relatively uncomplicated carnivore digestive system, and digesta passage is rapid, especially in small insectivores. Mean retention times of digesta markers increase with body size of the carnivore. In contrast to muscle and viscera, bones, teeth, hair, feathers and scales are poorly digested, and their undigested remains in the faeces provide a means of determining the dietary habits of the carnivore. Much less is known about the nutrition of marsupial carnivores than their eutherian counterparts. This is an area of marsupial biology which deserves greater focus if we are to better understand the basis for niche separation among grossly similar species, and thus manage habitats for the maintenance of maximum biodiversity of marsupial carnivores over the long term.

REFERENCES

Archer, M. (1976), 'The dasyurid dentition and its relationship to that of didelphids, thylacinids, borhyaenids (Marsupicarnivora) and peramelids (Peramelina: Marsupialia)', *Australian Journal of Zoology (suppl.)*, **39**:1–34.

Barbour, R.A. (1977), 'Anatomy of marsupials', in *The Biology of Marsupials* (eds. B. Stonehouse, and D. Gilmore), pp. 237–62, Baltimore University Press, Baltimore.

Bartoshuk, L.M., Harned, M.A., & Parks, L.H. (1971), 'Taste of water in the cat: effects on sucrose preference', *Science*, **171**:699–701.

Beauchamp, G.K., Maller, O., & Rogers, J.G. (1977), 'Flavor preferences in cats (*Felis catus* and *Panthera* sp.)', *Journal of Comparative Physiology and Psychology*, **91**:118–27.

Bourdreau, J.C., Sivakumar, L., Do, L.T., White, T.D., Overek, L.J., & Hoang, N.K. (1985), 'Neurophysiology of the geniculate ganglion (facial nerve) taste systems: species comparisons', *Chemical Senses*, **10**:89–127.

Bradshaw, J.W.S., Goodwin, D., Legrand–Defretin, V., & Nott, H.M.R. (1996), 'Food selection by the domestic cat, an obligate carnivore', *Comparative Biochemistry and Physiology*, **114A**:205–9.

Brunner, H., & Coman. B.J. (1974), *The Identification of Mammalian Hair*, Inkata Press, Melbourne.

Chilcott, M.J., & Hume, I.D. (1985), 'Coprophagy and selective retention of fluid digesta: their role in the nutrition of the common ringtail possum, *Pseudocheirus peregrinus*', *Australian Journal of Zoology*, **33**:1–15.

Cornelius, C., Dandrifosse, G., & Jeuniaux, C. (1975), 'Biosynthesis of chitinases by mammals of the order Carnivora', *Biochemical and Systematic Ecology*, **3**:121–2.

Cowan, I. McT., O'Riordan, A.M., & Cowan, J.S. McT. (1974), 'Energy requirements of the dasyurid marsupial mouse *Antechinus swainsonii* (Waterhouse)', *Canadian Journal of Zoology*, **52**:269–75.

Cowey, C.B., Knox, D., Walton, M.J., & Adron, J.W. (1977), 'The regulation of gluconeogenesis by diet and insulin in rainbow trout (*Salmo gairdneri*)', *British Journal of Nutrition*, **38**:463–70.

Crisp, E. (1855), 'On some points relating to the anatomy of the Tasmanian wolf (*Thylacinus*) and of the Cape hunting dog (*Lycaon pictus*)', *Proceedings of the Zoological Society of London*, **1855**:188–91.

Dickman, C.R., & Huang, C. (1988), 'The reliability of fecal analysis as a method for determining the diet of insectivorous mammals', *Journal of Mammalogy*, **69**:108–13.

Ehrlein, H.J., & Akkermans, L.M.A. (1984), 'Gastric emptying', in *Gastric and Gastroduodenal Motility* (eds. L.M.A. Akkermans, A.G. Johnson, & N.W. Read), pp. 74–84, Praeger, New York.

Eisenberg, J.F. (1981), *The Mammalian Radiations*, University of Chicago Press, Chicago.

Fisher, D.O., & Dickman, C.R. (1993), 'Body size–prey size relationships in insectivorous marsupials: tests of three hypotheses', *Ecology*, **74**:1871–83.

Hassam, A.G., Rivers, J.P.W., & Crawford, M.A. (1977), 'The failure of the cat to desaturate linoleic acid: its nutritional implications', *Nutrition and Metabolism*, **21**:321–8.

Hayes, K.C., Carey, R.E., & Schmidt, S.Y. (1975), 'Retinal degeneration associated with taurine deficiency in the cat', *Science*, **188**:949–51.

Hume, I.D. (1999), *Marsupial Nutrition*, Cambridge University Press, Cambridge.

Hume, I.D., Smith, C., & Woolley, P.A. (2000), 'Anatomy and physiology of the gastrintestinal tract of the Julia Creek dunnart, *Sminthopsis douglasi* (Marsupialia: Dasyuridae)', *Australian Journal of Zoology*, **48**:475–85.

Jones, M.E. & Barmuta, L.A. (1998), 'Diet overlap and relative abundance of sympatric dasyurid carnivores: a hypothesis of competition', *Journal of Animal Ecology*, **67**:410–21

Kerry, K.R. (1969), 'Intestinal disaccharidase activity in a monotreme and eight species of marsupials (with an added note on the disaccharidases of five species of sea birds)', *Comparative Biochemistry and Physiology*, **52A**:235–46.

Kleiber, M. (1961), *The Fire of Life*, Wiley, New York.

Lovegrove, B.G. (2000), 'The zoogeography of mammalian basal metabolic rate', American Naturalist, **156**:201–19.

Malagelada, J.-R., & Azpiroz, F. (1989), 'Determinants of gastric emptying and transit in the small intestine', in *Handbook of Physiology, Section 6: The Gastrointestinal System. Volume 1. Motility and Circulation* (eds. S.G. Schultz, J.D. Wood, & B.B. Rauner), pp. 909–37, American Physiological Society, Bethesda.

Maynard, L.A., & Loosli, J.K. (1962), *Animal Nutrition, 5th ed.*, McGraw–Hill, New York.

McClelland, K.L., Hume, I.D., & Soran, N. (1999), 'Responses of the digestive tract of the omnivorous brown bandicoot, *Isoodon macrourus* (Marsupialia: Peramelidae), to plant- and insect-containing diets', *Journal of Comparative Physiology*, **B169**:411–18.

McNab, B.K. (1978), 'The comparative energetics of neotropical marsupials', *Journal of Comparative Physiology*, **125**:115–28.

McNab, B.K. (1984), 'Physiological convergence amongst ant-eating and termite-eating mammals', *Journal of Zoology, London*, **203**:485–510.

McNab, B.K. (1986), 'Food habits, energetics and the reproduction of marsupials', *Journal of Zoology, London*, **208**:595–614.

McNab, B.K. (1988), 'Complications inherent in scaling the basal rate of metabolism in mammals', *Quarterly Review of Biology*, **63**:25–54.

Migliorini, R.H., Linder, C., Moura, J.L., & Veiga, J.A.S. (1973), 'Gluconeogenesis in a carnivorous bird (black vulture)', *American Journal of Physiology*, **225**:1389–92.

Moore, S.J., & Sanson, G.D. (1995), 'A comparison of the molar efficiency of two insect-eating mammals', *Journal of Zoology, London*, **235**:175–92.

Morris, J.G. (1985), 'Nutritional and metabolic response to arginine deficiency in carnivores', *Journal of Nutrition*, **115**:524–31.

Morris, J.G. (1994), 'Metabolic adaptations of carnivores in relation to diet', in *Nutrition in a Sustainable Environment* (eds. M. Wahlqvist et al.), pp. 502–5, Smith-Gordon, U.K.

Morris, J.G., & Rogers, Q.R. (1989), 'Comparative aspects of nutrition and metabolism of dogs and cats', in *Nutrition of the Dog and Cat* (eds. I.H. Burger, & J.P.W. Rivers), pp.35–66, Cambridge University Press, Cambridge.

Moyle, D.I., Hume, I.D., & Hill, D.M. (1995), 'Digestive performance and selective digesta retention in the long-nosed bandicoot, *Perameles nasuta*, a small omnivorous marsupial', *Journal of Comparative Physiology*, **B164**:552–60.

Nagy, K.A. (1987), 'Field metabolic rate and food requirement scaling in mammals and birds', *Ecological Monographs*, **57**:111–28.

Nagy, K.A. (1994), 'Field bioenergetics of mammals: what determines field metabolic rates', *Australian Journal of Zoology*, **42**:43–53.

Nagy, K.A., Girard, I.A., & Brown, T.K. (1999), 'Energetics of free-ranging mammals, reptiles, and birds', *Annual Review of Nutrition*, **19**:247–77.

Nagy, K.A., Seymour, R.S., Lee, A.K., & Braithwaite, R. (1978), 'Energy and water budgets in free-living *Antechinus stuartii* (Marsupialia: Dasyuridae)', *Journal of Mammalogy*, **59**:60–8.

Osgood, W.H. (1921), 'A monographic study of the American marsupial *Caenolestes*', *Field Museum of Natural History, Zoology Series*, **14**:1–162.

Pernetta, J.C. (1976), 'Diets of the shrews *Sorex araneus* L., & *Sorex minutus* L. in Wytham grassland', *Journal of Animal Ecology*, **45**:899–912.

Read, D.G. (1987a), Diets of sympatric *Planigale gilesi* and *P. tenuirostris* (Marsupialia: Dasyuridae): relationships of season and body size', *Australian Mammalogy*, **10**:11–21.

Read, D.G. (1987b), 'Rate of food passage in *Planigale* spp. (Marsupialia: Dasyuridae)', *Australian Mammalogy*, **10**:27–8.

Redford, K.H., & Dorea, J.G. (1984), 'The nutritional value of invertebrates with emphasis on ants and termites as food for mammals', *Journal of Zoology, London*, **203**:385–95.

Richardson, K.C., Bowden, T.A.J., & Myers, P. (1987), 'The cardiogastric gland and alimentary tract of caenolestid marsupials', *Acta Zoologica (Stockholm)*, **68**:65–70.

Rogers, Q.R., Morris, J.G., & Freedland, R.A. (1977), 'Lack of hepatic enzymatic adaptation of low and high levels of dietary protein in the adult cat', *Enzyme*, **22**:348–56.

Sabat, P., Bozinovic, F., & Zambrano, F. (1995), 'Role of dietary substrates on intestinal disaccharidases, digestibility, and energetics in the insectivorous mouse-opossum (*Thylamys elegans*)', *Journal of Mammalogy*, **76**:603–11.

Sanson, G.D. (1985), 'Functional dental morphology and diet selection in dasyurids (Marsupialia: Dasyuridae)', *Australian Mammalogy*, **8**:239–47.

Sanson, G.D. (1991), 'Predicting the diet of fossil mammals', in *Vertebrate Paleotology of Australasia* (eds. P. Vickers-Rich, J.M. Monaghan, R.F. Baird, & T.H. Rich), pp. 201–28. Pioneer Design Studio, Melbourne.

Smith, A.P., Nagy, K.A., Fleming, M.R., & Green, B. (1982), 'Energy requirements and water turnover in free-living Leadbeater's possums, *Gymnobelideus leadbeateri* (Marsupialia: Petauridae)', *Australian Journal of Zoology*, **30**:737–49.

Statham, H.L. (1982), *Antechinus stuartii* (Dasyuridae, Marsupialia) diet and food availability at Petroi, north-eastern New South Wales', in *Carnivorous Marsupials* (ed. M. Archer), pp. 151–63, Royal Zoological Society of New South Wales, Sydney.

Stevens, C.E., & Hume, I.D. (1995), *Comparative Physiology of the Vertebrate Digestive System, 2nd ed.*, Cambridge University Press, Cambridge.

Strahan, R. (1995), *The Mammals of Australia*, Reed, Sydney.

CHAPTER 15

NUTRITIONAL AND FIBRE CONTENTS OF LABORATORY-ESTABLISHED DIETS OF NEOTROPICAL OPOSSUMS (DIDELPHIDAE)

D. Astúa de Moraes[A, B], *R.T. Santori*[C], *R. Finotti*[A] *and R. Cerqueira*[A]

[A] Laboratório de Vertebrados, Depto. de Ecologia, Universidade Federal do Rio de Janeiro. CP 68020, 21941-590 – Rio de Janeiro, R.J., Brasil

[B] Present Address: Departamento de Zoologia, Instituto de Biociências. USP & Mastozoologia. Museu de Zoologia da Universidade de São Paulo. Av. Nazaré, 481. Ipiranga. 04263-000 São Paulo – SP – Brasil. Email: dmoraes@ib.usp.br

[C] Departamento de Ciências, Faculdade de Formação de Professores, Universidade do Estado do Rio de Janeiro, Rua Francisco Portela, 794. Paraíso. 24435-000 São Gonçalo, RJ. Brasil

Nutritional requirements of 12 species of Neotropical opossums (Didelphidae) were inferred through a laboratory food preference experiment. Nutritional contents of experimental diets follow predictions based on field diet data, with more frugivorous species showing high non-structural carbohydrate proportions and more carnivorous species high protein contents in their selected diets. Mouse opossums selected nutritional and fibre proportions that suggested that they may be more frugivorous and less strictly insectivorous than previously thought. The species form a gradient of differential food specialisation from frugivory to carnivory, and significant differences are found only between extremes of this gradient. Nutritional contents established here are consistent with natural diet patterns and can be used as important complementary data to field diet studies and to prepare diets for captive maintenance of the species.

INTRODUCTION

The diet of a species is an important component of its niche and as such is an important variable in potential competition among species of the same guild (Leite et al. 1994). Furthermore, knowledge on diet choice can play an important role in our understanding of population dynamics (Jensen 1993). Thus, knowing the basic nutritional requirements for a species is one basic and essential step to understand its relationship with the environment.

Knowing the basic nutritional requirements of a species, one can also establish adequate conditions to maintain captive individuals in good health (Périssé et al. 1989). Captive colonies can be established for a variety of purposes, from the establishment of lineages of experimental models for biomedical research to display in public exhibits. Didelphid marsupials have increasingly been used as experimental models, particularly *Didelphis virginiana* and *Monodelphis domestica* (Jurgelsky and Porter 1974; VandeBerg and Robinson 1997). This is especially true in growth and developmental biology studies, as the initial growth stages can be easily obtained and manipulated, without the excessive stress that would be imposed for obtaining early developmental stages in eutherians. Furthermore, the structural simplicity of some morphological and neurological structures, coupled with a morphological similarity, makes didelphids particularly suitable models for varied neurological, physiological (Fournier and Weber 1994), developmental (Smith 1994; Maunz and German 1997; Smith and van Nievelt 1997; Oliveira et al. 1998) and behavioural (Fadem and Corbett 1997;

Kimble 1997) studies. Initially, research was based solely on two model species, the Virginia opossum (*D. virginiana*), as it was the only marsupial naturally present in the US, and the grey short-tailed opossum (*M. domestica*), as many large breeding colonies have been established from individuals collected in Brazil and Bolivia (VandeBerg and Robinson 1997). Nevertheless, other species are now used in biomedical research, such as the southern common opossum, *D. aurita*, or the white-eared opossum, *D. albiventris*. (Perales et al. 1986; Oliveira et al. 1998). Beyond that, there is a need of using species more phylogenetically related for the comparison of morphological patterns.

Breeding species in captivity is not only important for their use in biomedical research, but is also essential for the understanding of the species' natural history. Captive colonies allow for the understanding and establishment of important ecological parameters such as reproductive strategies and behaviour. Exact developmental rates can only be established through captive breeding, and are essential for understanding the evolutionary processes involved in the history of Neotropical mammals. Other bionomic parameters, such as litter size and number, onset of reproduction, among others, are also essential for understanding the ecology of these species.

Thus, preparing a nutritionally balanced diet is a fundamental procedure for setting up a successful breeding colony, and is a primary step towards the use of some of the more than 70 species of Neotropical marsupials as experimental models. Yet, while for small laboratory mammals, most diets are usually known and factory-prepared, knowledge about nutritional needs and diets for wild small mammals in captivity is scarce, especially for Neotropical species. For such species, this knowledge is intrinsically dependent on ecological studies and laboratory food preference experiments (Périssé et al. 1988; Périssé et al. 1989; Fonseca and Cerqueira 1991; Santori et al. 1997; for a review on current knowledge on field data, see Santori and Astúa de Moraes; in press). Studies of diets in the field are important for establishing diet width, resource partitioning among species, and exact trophic relations within a community, but most methods commonly used present some sort of quantitative or qualitative flaw (e.g. Dickman and Huang 1988; Kunz and Whitaker, Jr. 1983; Caron et al. 1985). Furthermore, these methods do not allow for precise nutritional quantification of the diets, as there are no means of establishing precise amounts of nutrients ingested. On the other hand, cafeteria experiments allow us to estimate the relative nutrient content of preferred diets, and can also shed some light on resource use in the field that might go overlooked in field studies, due to the inaccuracy of the methods used.

Definition of dietary categories has always been a controversial issue. With the exception of species with a very narrow diet, there is usually an overlap of categories depending on the species' precise diet composition. For instance, Neotropical marsupials have been classified in categories such as frugivores (Charles-Dominique et al. 1981) or insectivore-omnivores (Leite et al. 1994; 1996), among others. Even though more precise categories have been proposed (Eisenberg 1981; Lee and Cockburn 1985), not all are filled by marsupial species (Hume 1999). Furthermore, many of these are especially useless in classifying opossums, as there are variable amounts of overlap in their diets, resulting in the classification of most species into two or three categories, which is not very helpful for studies such as resource partitioning or community structure.

In this work a food preference experiment for small mammals developed by Périssé et al. (1989) was used to establish the basic nutritional needs of 12 species of Neotropical marsupials. The nutritional contents of the foods chosen were compared among species and discussed in relation to their natural diet as established by traditional means. The relative proportion of the macronutrients used in the diet can give some clues about the food categories that a species consumes in the field, as there is a linear relationship between diet preference and nutritional requirements (Périssé et al. 1988). It can be expected that carnivorous species will select a higher content of protein in their diets, while frugivorous species should select more carbohydrates, related to the higher sugar contents of fruits, and a higher proportion of fibres. The stronger the association between field and experimental diet, the safer the laboratory data will be for diet preparation.

THE FOOD PREFERENCE EXPERIMENT

These tests are done routinely with every animal collected by the Laboratório de Vertebrados, of the Universidade Federal do Rio de Janeiro, always following the same protocol. Tests of food choice were conducted indoors at a 23–29°C mean temperature and 50–80% relative humidity, with animals placed in individual wire-covered polypropylene boxes (45 × 30 × 20 cm).

Experiments consisted of offering 26 different cultivated commercial food items, such as meat, egg, arthropods (shrimps), leaves, roots, seeds and fruits, with previously known and well established nutritional contents (Franco 1987). A complete listing of the items used is presented in Table 1. Although it may seem odd to conduct such an experiment with items unfamiliar to the animals, there are some specific reasons. Such items can be found all year long, which does not necessarily happen with natural food items from the natural habitat of the species. The nutritional contents of commercially established varieties are already known, but would have to be established for natural items. The animals tested come from varied geographical regions, and specific food availability varies geographically. The difficulty of obtaining some of these items would render com-

Table 1 Food items used in the food preference experiments for Neotropical opossums.

Food type	Food items
Fruits	Banana, orange, gumbo, pea, tomato, pimento, fruit of the egg plant, green husk, pumpkin, cucumber, zucchini, chayot
Roots	Carrot, manioc, beet root, turnip, yam
Leaves	Lettuce, cabbage, spinach
Vertebrate protein	Meat, quail egg
Arthropod	Shrimp
Tuber	Sweet potato

parisons impossible. This way, we have the exact same items all year long, for every animal tested, and the results can be compared regardedless of when tests were performed or where animals were collected. Furthermore, as shown below, being presented to apparently unfamiliar food items does not stop the animal from choosing items that reflect their nutritional needs. Even when presented with such unfamiliar items, individuals usually choose a nutritionally balanced diet (Louw 1992).

Water was provided *ad libitum* throughout the experiment. All food items were simultaneously left available to the animal during an 18 h period *ad libitum* (the exact amount of food varied, depending on the body size of the species tested: more food was offered to bigger body-sized species, to be considered *ad libitum*). This way, a hungry animal could satiate itself by taking only the most preferred item without running out of it, therefore not being obliged to use a less preferred one. Those experiments where the animal ran out of a specific item were invalidated and repeated. The experiment is based on the assumption that food consumption by the animals during the test period reflects their own basic nutritional needs at that time. All items were weighed before the experiment, and those tasted by the animals were weighed at the end to estimate consumption. The detailed preparation of the experiment, the calculations of a preference index for each food item and its applications are presented elsewhere (Périssé et al. 1989).

Total quantities and relative proportions of protein, carbohydrate and lipid ingested by each individual, as well as the relative proportion of fibres in the diet (total fibre weight ingested divided by total amount of food ingested), were calculated using the amount of each item consumed and its nutritional composition or fibre contents, previously calculated and published (Franco 1987). Relative proportions of nutrients, as well as relative content of fibre in the diet, were compared between sexes and ages through a Mann-Whitney test, and among species through a Kruskal-Wallis test, followed by a *post-hoc* non-parametric Tukey-like test (Zar 1996). Shannon's diversity index (*H'*) was also calculated for the established diets.

Age classes of tested individuals were established based on the third pre-molar and the number of molars erupted. Individuals were considered young when presenting the third deciduous pre-molar and an incomplete number of molars. Adult animals were those with the permanent third pre-molar and with three to four molars erupted.

SPECIES STUDIED

The species included in this study, with their respective common name and number of individuals analysed are listed in Table 2.

Monodelphis americana, *Chironectes minimus* and *Lutreolina crassicaudata* were not included in statistical comparisons as there was only one individual of each species was available. Their feeding experiments also followed a slightly modified protocol (fish, guts, kidney, coconut, grape, apple, and sunflower were included in the experiment). Nonetheless, the nutritional contents established for these individuals are shown along with the others.

NUTRITIONAL COMPOSITION OF DIETS AND COMPARISONS

The amounts of protein, carbohydrates and lipid ingested and the protein:carbohydrate:lipid (P:C:L) ratios in the diet selected by the adults and young of the 12 species, as well as the fibre contents of the experimental diets, are listed in Tables 3 and 4, respectively. Data are presented separately for young and adults, as it was shown that for some of the species analysed there was a difference in some of the relative proportions of some nutrient ingested, between ages. We present P:C:L ratios as they are in Table 3 and 4 (instead of showing the relative proportions, used in the analysis), because the ratios permit an easy elaboration of any experimental diet, as long as the nutritional contents of the foods used are known. Relative proportions and P:C:L ratios can be interconverted.

Comparisons among sexes

The Mann-Whitney *U* tests showed no significant difference between sexes for the relative proportion of protein (hereafter RPP), carbohydrate (hereafter RPC) or lipid (hereafter RPL), or for the proportion of fibre in the diet, for the nine species analysed. All females used were non-reproductive. Therefore, sexes were pooled for subsequent analysis.

Table 2 Species of opossums included in the study, with their common names and number of individuals examined (y = young; a = adults). Classification follows the proposal of Gardner (1993), modified after Patton & Silva (1997).

Species	Common name	n (y)	n (a)
Family Didelphidae			
Subfamily Caluromyinae			
Caluromys philander (Linnaeus, 1758)	Woolly opossum	4	3
Subfamily Didelphinae			
Chironectes minimus (Zimmermann, 1780)	Water opossum	1	-
Didelphis albiventris Lund, 1840	White-eared opossum	6	7
Didelphis aurita Wied-Neuwied, 1826	Southern common opossum	36	18
Gracilinanus agilis (Burmeister, 1854)	Agile gracile mouse opossum	5	2
Lutreolina crassicaudata (Desmarest, 1804)	Thick-tailed opossum	1	-
Marmosops incanus (Lund, 1840)	Grey slender mouse opossum	3	-
Metachirus nudicaudatus (Desmarest, 1817)	Brown four-eyed opossum	22	4
Micoureus demerarae (Thomas, 1905)	Woolly mouse opossum	7	-
Monodelphis americana (Muller, 1776)	Three-striped short-tailed opossum	1	-
Monodelphis domestica (Wagner, 1842)	Grey short-tailed opossum	18	11
Philander frenata (Olfers, 1818)	Grey four-eyed opossum	51	17

Comparisons among ages

A significant difference was found in nutritional content between the diet selected by young and adult *Didelphis albiventris* (RPP: $U = 5.00$; $p = 0.02$), *Metachirus nudicaudatus* (RPP: $U = 16.00$; $p = 0.05$, and RPC: $U = 16.00$; $p = 0.05$), and *Monodelphis domestica* (RPP: $U = 25.00$; $p = 0.00$, RPC: $U = 21.00$; $p = 0.00$, and RPL: $U = 11.00$; $p = 0.00$). A significant difference was also found in relative fibre content between the diet selected by young and adult *Monodelphis domestica* ($U = 27.00$; $p = 0.00$). Therefore, adults and young were treated separately in all subsequent analyses.

Comparisons among species

The Kruskal-Wallis test indicated significant differences in the relative proportions of nutrients selected by the species studied for both adults (RPP: $\chi^2 = 69.53$; $d.f. = 8$; $p = 0.00$; RPC: $\chi^2 = 63.01$; $d.f. = 8$; $p = 0.00$; RPL: $\chi^2 = 35.07$; $d.f. = 8$; $p = 0.00$) and young (RPP: $\chi^2 = 22.91$; $d.f. = 6$; $p = 0.00$; RPC: $\chi^2 = 22.63$; $d.f. = 6$; $p = 0.00$; RPL: $\chi^2 = 20.80$; $d.f. = 6$; $p = 0.00$). Similarly, a significant difference was found between the relative fibre content in the diets selected by the species studied for both adults ($\chi^2 = 53.04$; $d.f. = 8$; $p = 0.00$) and young ($\chi^2 = 24.49$; $d.f. = 6$;

Table 3 Amount of protein, carbohydrate and lipid ingested, the fibre content (g per 100g of dry matter) and the ratios of protein and carbohydrate to lipid content of diets selected by adult Neotropical opossums. Values represent the mean value for the species (standard deviation in parenthesis).

		Nutritional and fibre content				Ratio		
Species	n	Protein (g)	Carbohydrate (g)	Lipid (g)	Fibre (%)	P	C	L
Caluromys philander	4	0.64 (0.22)	8.19 (3.39)	0.11 (0.04)	2.70 (0.60)	5.87	75.35	1.00
Chironectes minimus	1	7.98 –	7.09 –	0.79 –	2.24 –	10.11	8.96	1.00
Didelphis albiventris	6	20.89 (8.53)	27.90 (7.71)	10.07 (7.74)	1.68 (0.39)	2.07	2.77	1.00
Didelphis aurita	36	16.28 (8.75)	26.70 (19.38)	5.08 (5.40)	1.94 (0.43)	3.20	5.25	1.00
Gracilinanus agilis	5	1.34 (0.97)	3.41 (0.90)	0.32 (0.32)				
Lutreolina crassicaudata	1	27.48 –	16.28 –	7.20 –	1.60 –	3.82	2.26	1.00
Marmosops incanus	3	1.78 (1.45)	4.13 (4.03)	0.45 (0.27)	2.34 (0.64)	4.00	9.28	1.00
Metachirus nudicaudatus	22	8.74 (4.86)	12.88 (10.27)	1.91 (2.30)	1.96 (0.90)	4.57	6.74	1.00
Micoureus demerarae	7	2.30 (1.59)	8.13 (1.53)	0.63 (0.94)	2.81 (0.44)	3.65	12.91	1.00
Monodelphis americana	1	0.62 –	1.70 –	0.01 –	2.31 –	103.13	282.90	1.00
Monodelphis domestica	18	4.37 (1.04)	2.03 (1.31)	0.88 (0.51)	1.03 (0.57)	4.96	2.31	1.00
Philander frenata	51	10.62 (3.75)	11.94 (5.09)	2.31 (1.46)	1.76 (0.45)	4.60	5.17	1.00

Table 4 Amount of protein, carbohydrate and lipid ingested, the fibre content (g per 100 g of dry matter) and the ratios of protein and carbohydrate to lipid content of diets selected by young Neotropical opossums. Values represent the mean value for the species (standard deviation in parentheses).

		Nutritional and fibre content				Ratio		
Species	**n**	**Protein (g)**	**Carbohydrate (g)**	**Lipid (g)**	**Fibre (%)**	**P**	**C**	**L**
Didelphis aurita	18	7.32 (4.63)	10.87 (10.60)	1.53 (1.41)	1.77 (0.41)	4.77	7.09	1.00
Didelphis albiventris	7	5.58 (2.44)	4.69 (4.92)	1.66 (1.49)	1.19 (0.36)	3.36	2.82	1.00
Philander frenata	17	8.82 (3.63)	7.65 (4.44)	1.71 (0.87)	1.54 (0.57)	5.16	4.48	1.00
Metachirus nudicaudatus	4	6.53 (1.42)	3.40 (1.99)	1.04 (0.47)	1.23 (0.44)	6.28	3.27	1.00
Caluromys philander	3	0.56 (0.25)	8.46 (3.20)	0.01 (0.01)	3.09 (0.59)	59.23	890.24	1.00
Gracilinanus agilis	2	1.00 (1.30)	3.09 (2.28)	0.87 (1.21)	3.18 (0.72)	1.16	3.56	1.00
Monodelphis domestica	11	1.24 (0.94)	1.91 (1.62)	0.15 (0.11)	2.13 (0.75)	8.37	12.84	1.00

$p = 0.00$). The results of the Tukey-like test for nutrients and fibres for the adults are summarised in Table 5. For the young, the only differences found were for RPP (*Metachirus nudicaudatus* higher than *Caluromys philander*, $p = 0.05$) and the RPL (*Didelphis albiventris* higher than *Caluromys philander*, $p = 0.05$). The Tukey-like test failed to indicate the location of the difference in the RPC and in the fibre content for young. The diversity indexes of the diets selected by the species are included in Fig. 1.

RELATIONSHIPS BETWEEN RESULTS OF LABORATORY EXPERIMENTS AND FIELD DATA

When ranked according to the nutritional contents of their experimental diets, species were ordered quite similarly to what could be expected in relation to their food habits in nature, at least for those species with some data available (Fig. 1).

Caluromys philander is described as the most frugivorous of the species studied, with its natural diet including 50 to 75% fruit (Julien-Laferrière and Atramentowicz 1990; Leite et al. 1994; 1996). It was also the species with the highest RPC and the lowest RPP and RPL in its diet, for both young and adults. This is a reflection of the consumption of fruits with high carbohydrate contents. The species commonly described as more carnivorous or insectivorous, such as *Philander frenata* and *Monodelphis domestica* (Santori et al. 1995b; Hume 1999), were at the opposite extreme of the distribution of the species, ranked by increasing RPP (Table 3, Fig. 1), with the highest contents of protein, reflecting their more carnivorous habits (these species consumed primarily the meat, eggs and arthropods presented in the experiment).

Although *Lutreolina crassicaudata*, *Chironectes minimus* and *Monodelphis americana* were not included in the statistical comparisons, a simple observation of the relative proportions of nutrients calculated for individuals of these species shows that they are also in agreement with the predictions that more carnivorous species show a higher RPP and a lower RPC in their diets (Fig. 1). *Lutreolina crassicaudata* and *C. minimus are* believed to be the most carnivorous species among these opossums (Marshall 1978a; 1978b), and had a high RPP in their selected diets.

These results indicate that, in broad terms, the choice of food items during the experiments, even unfamiliar items, is consistent with their diet in the field, at least to the extent of our knowledge.

Table 5 Results of the non-parametric Tukey-like test among species (adults). Letters indicate significant difference ($p = 0.05$) among pairs for protein (P), carbohydrate (C), lipid (L) or fibre (F). Standard typeface indicates that the value is higher for species in columns heading; underlined indicates that value is higher for species in row heading.

	C. philander	***M. demerarae***	***G. agilis***	***M. incanus***	***D. aurita***	***D. albiventris***	***M. nudicaudatus***	***P. frenata***	***M. domestica***
C. philander									
M. demerarae									
G. agilis									
M. incanus									
D. aurita	$\underline{L}$								
D. albiventris	$\underline{L}$								
M. nudicaudatus									
P. frenata	$\underline{P}C\underline{L}$	$\underline{P}C$							
M. domestica	$\underline{P}C\underline{L}F$	$\underline{P}C\underline{L}F$	$\underline{P}CF$	$\underline{P}$	$\underline{P}C$	$\underline{P}$	$\underline{P}C$	$\underline{P}C$	

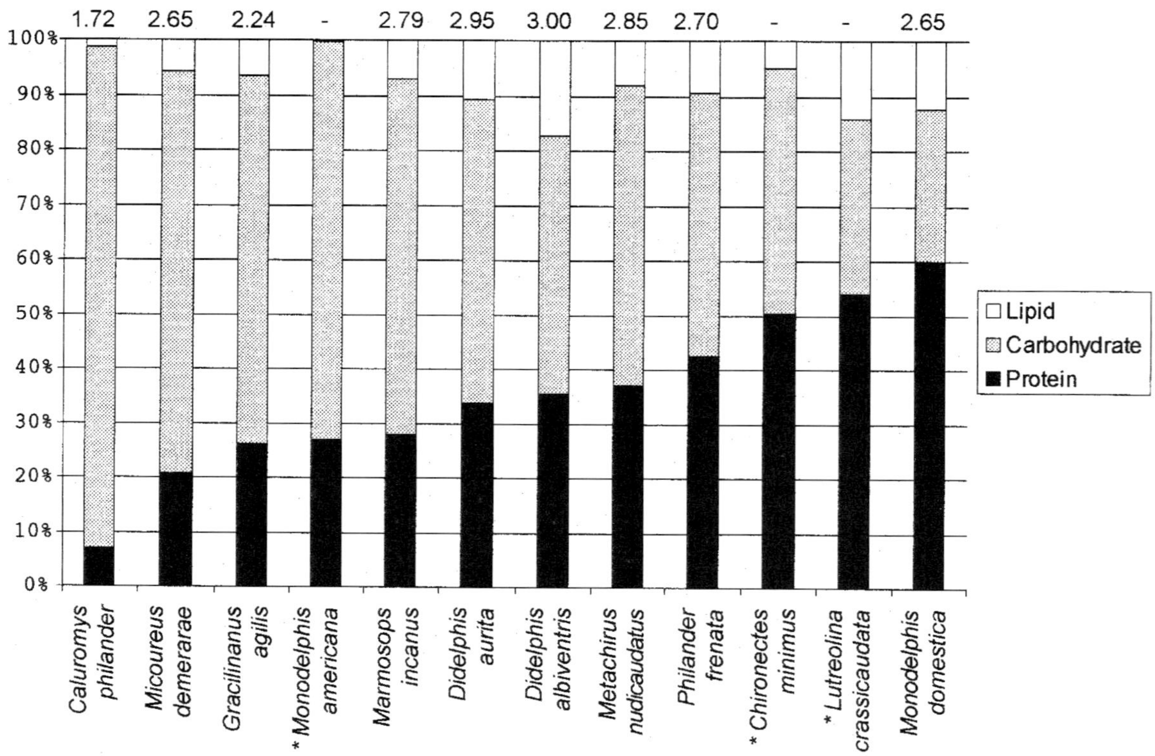

Figure 1 Relative proportions of protein, carbohydrate and lipids in the diets of the 12 species of Neotropical opossums studied. Shannon's diversity index (*H'*) is shown on top of each column. Species were ordered in increasing rank of relative proportionof protein selected (RPP). **Monodelphis americana, Lutreolina crassicaudata* and *Chironectes minimus* are included for visual comparison and had no diversity index calculated.

The more carnivorous the species in the field, the more higher-protein content items were consumed first, and inversely, species with more frugivorous habits essentially consumed fruit items. Regarding the RPP (Table 3), the higher mean ranks for *Metachirus nudicaudatus*, *Philander frenata* and *Monodelphis domestica* are in accordance with field studies indicating relatively high carnivory/insectivory of these species (Santori et al. 1995a; 1995b). At the other extreme, the more frugivorous *Caluromys philander* was followed by the mouse opossums *Micoureus demerarae*, *Gracilinanus agilis*, and *Marmosops incanus*, and in between are the most generalist feeders *Didelphis aurita* and *D. albiventris* (Fig. 1). Conversely, for RPC the higher mean ranks were for the murine opossums and *Caluromys philander*, while in the lower ranks were the more carnivorous species. Finally, with RPL, the lowest mean ranks were again for the mouse opossums and *Caluromys philander*, while the higher were *Didelphis albiventris* and *Monodelphis domestica*. Higher proportions of fibres were found in the diet of *Caluromys philander* and the mouse opossums, and the lower in those of *P. frenata* and especially *M. domestica*. The higher RPC and fibre contents of the diets of the more frugivorous species reflect the high levels of non-structural carbohydrates in the fruits consumed. These species probably meet most of their energetic needs from fruits, while the more carnivorous rely on gluconeogenesis from mainly amino acids in their protein-rich diets (Hume 1999). High levels of fat in the diet are found for species from dry open regions such as *Didelphis albiventris* and *Monodelphis domestica*. These species may be able to satisfy water requirements through fat metabolism, as other carnivorous species do (Edwards et al. 1983; Green et al. 1989).

The diversity index of the experimental diets also reflected data available on the natural diets of these species as it confirmed the position of the *Didelphis* genus as the most generalist, while the remaining ones tended to show less diverse diets, either directed to frugivory (higher RPC) or carnivory (higher RPP) (Fig. 1). The index also confirmed previous observations that *Caluromys philander* seems to be a didelphid with a highly specialised diet.

Successful survivorship and reproduction requires adequate nutrition (Bondrup-Nielsen and Foley 1993). Experimental diets based on estimates of nutritional requirements established with the method described here have proven to be very effective for some species maintained for a long time in our laboratory, such as the didelphid marsupials *Philander frenata* (Périssé et al. 1989; Hingst et al. 1998) and *Didelphis aurita*, as well as the rodents *Akodon cursor, Bolomys lasiurus* and *Calomys* sp. (Cerqueira; unpubl. data). Therefore we are confident that relative proportions of nutrients selected reflect nutritional needs of the animals, and may be useful for ecological comparisons among species, even those based on small sample sizes.

PREDICTIONS ON THE ECOLOGY OF SPECIES

As previously stated, nutritional contents of diets selected in this experiment seem to reflect the natural habits of the species, at least regarding the macronutrient contents of their natural diets. This was established through comparison of nutritional contents of laboratory diets and data available on the natural diet of the most studied species. Thus, based on these relationships, it is possible to suggest the likely nutritional habits of species when there are few or no data on their natural habits available.

The nutritional contents of the mouse opossums' diets reveal that these species may be far more frugivorous than previously thought. These arboreal species are ecologically similar to *Caluromys philander*, as they both use the upper vegetation strata (Hershkovitz 1992; Leite et al. 1996). Thus, they could be expected to share some similarities in feeding habits and nutritional needs. Mouse opossums are usually thought to be mainly insectivorous (Hunsaker II 1977; Leite et al. 1996), but thorough studies have been performed on only a few of these species. An individual *Micoureus demerarae* held in captivity consumed large amounts of fruits daily when offered, and has been maintained in a healthy condition on a diet of small pieces of canine or cat chow, a mixture consisting of oats, cod liver oil, powdered milk, quail egg and banana, and various fruits such as orange, papaya and banana (or fruit baby food), for almost a year (Astúa de Moraes unpubl. data). If fruit was withheld for some days, it was the first item consumed when re-offered, suggesting that a diet with only small amounts of fruit may not be enough to fulfil its needs. Assumptions on the insectivory of these species are usually based on faecal or stomach contents analysis, but we believe that these may be biased towards insects, because not all seeds from fruits consumed are ingested. For instance, Santori (pers. obs.), observed that *P. frenata* rejects big seeds and swallows small ones. Charles-Dominique et al. (1981) had already observed that *Marmosa* fed on the pulp of fruits, and that only very small seeds (less than 0.1 mm^3) were swallowed and found in the digestive tracts. Carvalho et al. (1999) found a great variety of small seeds in scats of *M. demerarae*. This evidence, together with their similar RPC, RPP and a higher relative proportion of fibre in the diet strongly suggests that these species are far more frugivorous than previously thought. Such conclusions would be strengthened with field observations of fruit ingestion or through behavioural studies in the laboratory in order, to determine the extent to which these small species really ingest seeds of consumed fruits.

PROBLEMS OF DIETARY CATEGORIES FOR NEOTROPICAL OPOSSUMS

It has been shown that these species can be ranked in a continuum from the most frugivorous to the most carnivorous species, with omnivorous species in between. However, though statistical differences were found between the extremes of this ranking, there is no clear cut between one species and its neighbours. These species seem to use rather similar food resources, but partition available food items in such a way that there is only a slight overlap between them and the most similar species. This could avoid direct competition when coupled with other ecological variables, such as space use or activity pattern, or with different morphological or physiological capabilities. In fact, patterns in which there is no clear difference between close species, but an overall diversity is present within the whole group, are found repeatedly in other traits of these species, such as gut and cranial morphology.

Although food specialisation is not the rule among the opossums we studied, some characteristics of alimentary tract proportions of these species (Santori et al. unpubl. data) are coupled with food habits. The regions of the digestive tract more closely related to feeding habits of these species are the stomach, caecum and colon relative length (i.e. to total length of digestive tract). In summary, *C. philander* is the species with the best association between digestive tract measurements and food habits. Digestive tract measurements of other species are related, but with a variable degree of morphological differentiation from a generalised form related to omnivory toward a more carnivorous diet. The specimens of *L. crassicaudata* and *M. domestica* examined for gut morphology had measurements which placed them near the *Didelphis* species and murine opossums, respectively. Most striking is the great caecal development in *M. nudicaudatus*. Relative stomach length tends to be greater in the most carnivorous *P. frenata*. This species is more efficient at digesting diets with high protein and fat content (Santori et al. 1995b) and can survive on protein diets (Fonseca and Cerqueira 1991). Carnivores usually have an elastic and voluminous stomach where digestion of proteins begins (Hume 1999). The two omnivorous *Didelphis* species have stomachs of intermediate size. Greater caecum relative length distinguishes *M. nudicaudatus* and *C. philander* from most other opossums. This is in accordance with the feeding habits of *C. philander*, with higher proportions of plant matter in its diet. Microbial fermentation of plant material in the caecum can be inferred from the large size of this region of the hindgut (Hume 1999). However, in *M. nudicaudatus* a large caecum is not clearly related to its apparently more carnivorous food habits (Santori et al. 1995a). In carnivores the caecum as well as the distal colon are thought to be important in the absorption of water and electrolytes (Anderson et al. 1992). Thus the great caecum length of *Metachirus* may be related to water absorption. The relative length of the colon of *C. philander* and *D. albiventris* is remarkable. The highly developed colon in *C. philander* may be related to water absorption from fruits *Didelphis. albiventris* inhabits dry areas of Brazil (Cerqueira 1985), so water absorption from the distal colon may be an important component of its nutritional ecology. Full details on the gut morphology of these species will be published elsewhere.

As for cranial morphology, the Didelphidae are usually considered a morphologically conservative group. Nevertheless, it has been possible to show that, at least among the large body-sized species, morphological variation can sometimes be related to feeding habits, in a proportion that varies directly with the degree of dietary specialisation of the species (Astúa de Moraes 1998). In *L. crassicaudata*, the relative shortening of the rostrum, the high saggital crest and the increased temporal musculature are clearly related to its carnivorous habits (Delupi et al. 1997; Astúa de Moraes et al. 2000), while in *Metachirus nudicaudatus* there is an elongation of the mandible and of the cheek teeth series that can be related to its insectivorous habits (Astúa de Moraes 1998). In contrast, there are less striking differences among genera with different food habits but which are clearly phylogenetically related, such as *Didelphis* and *Philander*, or with similar nutritional requirements but different ecological habits, such as *Philander* and *Chironectes*.

In spite of the patterns described above, all opossums studied here could be considered, to some extent, omnivorous in terms of natural field diet – none of them is a food specialist *sensu stricto* – as they all include fruits and invertebrates in their natural diets, and most big-bodied species also consume small vertebrates (Marshall 1978a, 1978b; Leite et al. 1994, 1996; Santori et al. 1995a, 1997). Differences could only be found among those taxa with the most differentiated diets and consequently, nutritional needs. The remaining species represent transitional states between the extreme taxa, not showing any significant difference. In this sense, the general and usually strict categories widely used for diet classifications (e.g. carnivores, frugivores, insectivores) do not apply well for these species (and probably for many other Neotropical opossum groups), as they may overlook subtle yet important differences in their natural diets. In fact, these species form a continuous gradient of differential food use (see Fig. 1), and boundaries between different diet categories, such as frugivore/omnivore (where *Caluromys* is usually fitted) or insectivore/omnivore (where genera such as *Metachirus* are fitted) are largely arbitrary. The usual diet classification terms should then only be used in a relative meaning in cases like this, stating that, for example, *Caluromys* is one of the most frugivorous of the opossums, or *Lutreolina* and *Chironectes* are more carnivorous genera. Thus altough classification of species into categories as done in reviews (e.g. Streilein 1982) or general species accounts (e.g. Fonseca et al. 1998) is useful, the information contained in these reports should be used in a cautionary way, as it may mask important subtle differences between species.

Acknowledgements

We are thankful to Paula Ceotto for help in performing the experiments, and to all the staff at the Laboratório de Vertebrados – UFRJ, especially Nélio P. Barros, for the maintenance of the animals used, and Angela M. Marcondes for administrative support. Lena Geise, Gabriel Marroig, Marcus V. Vieira, Chris Dickman and an anonymous referee gave helpful advice on the manuscript. This work was supported by Conselho Nacional de Pesquisas (CNPq), Coordenação de Aperfeiçoamento de Pessoal de Nível Superior (CAPES), Fundação de Amparo à Pesquisas do Estado do Rio de Janeiro (FAPERJ), Fundação de Amparo a Pesquisa do Estado de São Paulo (FAPESP), Fundação José Bonifácio (FUJB), Projeto Integrado de Ecologia (PIE) and Projeto de Conservação e Utilização Sustentável da Diversidade Biologica (PROBIO).

References

Anderson, M.D., Richardson, P.R.K., & Woodall, P.F. (1992), 'Functional analysis of the feeding apparatus and digestive tract anatomy of the aardwolf *Proteles cristatus*', *Journal of Zoology*, **228**:423–34.

Astúa de Moraes, D. (1998), *Análise morfométrica do crânio e da mandíbula de marsupiais didelfídeos: implicações ecológicas e funcionais*, unpublished MSc Dissertation, Universidade Federal de Minas Gerais, Belo Horizonte.

Astúa de Moraes, D., Hingst-Zaher, E., Marcus, L.F., & Cerqueira, R. (2000), 'A geometric morphometric analysis of cranial and mandibular shape variation in didelphid marsupials', *Hystrix (Italian Journal of Mammalogy)*, (n.s.), **10**:115–30.

Bondrup-Nielsen, S., & Foley (1993), 'Long-term effects of malnutrition on reproduction: a laboratory study with meadow voles, *Microtus pennsylvanicus*, and red-backed voles', *Clethrionomys gapperi. Canadian Journal of Zoology*, **72**:232–8.

Caron, L., Garant, Y., & Bergeron, J.M. (1985), 'The effect of digestibility on the reliability of food-habit studies from fecal analyses', *Canadian Journal of Zoology*, **63**:2183–86.

Carvalho, F.M.V., Pinheiro, P.S., Fernandez, F.A.S., & Nessimian, J.L. (1999), 'Diet of small mammals in Atlantic Forest fragments in southeastern Brazil', *Revista Brasileira de Zoociências*, **1**:91–101.

Cerqueira, R. (1985), 'The distribution of *Didelphis* in South America (Polyprotodontia, Didelphidae)', *Journal of Biogeography*, **12**:135–45.

Charles-Dominique, P., Atramentowicz, M., Charles-Dominique, M., Gerard, H., Hladik, C.M., & Prevost, M. F. (1981), 'Les mammifères frugivores arboricoles nocturnes d'une forêt guyannaise: inter-relations plantes–animaux', *Revue d'Ecologie (La Terre et La Vie)* **35**:341–435.

Delupi, L.H., Carrera, M.H., & Bianchini, J.J. (1997), 'Morfología comparada de la musculatura craneal en *Lutreolina crassicaudata* (Desmarest, 1804) y *Didelphis albiventris* Lund, 1840 (Marsupialia, Didelphidae)', *PHYSIS (Buenos Aires), Secc. C* **53**:19–28.

Dickman, C.R., & Huang, C. (1988), 'The reliability of fecal analysis as a method for determining the diet of insectivorous mammals', *Journal of Mammalogy*, **69**:108–13.

Edwards, B.A., Donaldson, K., & Simpson, A.P. (1983), 'Water balance and protein intake in the mongolian gerbil (*Meriones unguiculatus*)', *Comparative Biochemistry and Physiology*, **76A**:807–15.

Eisenberg, J.F. (1981), *The mammalian radiations*, University of Chicago Press, Chicago.

Fadem, B.H., & Corbett, A. (1997), 'Sex differences and the development of social behavior in a marsupial, the gray short-tailed opossum (*Monodelphis domestica*)', *Physiology & Behaviour,* **61**:857–61.

Fonseca, C.R.S.D., & Cerqueira, R. (1991), 'Water and salt balance in a South American marsupial, the Gray Four-Eyed Opossum (*Philander opossum*)', *Mammalia,* **55**:422–32.

Fonseca, G.A.B., Herrmann, G., Leite, Y.L.R., Mittermeier, R. A, Rylands, A.B., & Patton, J.L. (1996), 'Lista anotada dos mamíferos do Brasil', *Occasional Papers in Conservation Biology,* **4**:1–38.

Fournier, R.A., & Weber, J.-M. (1994), 'Locomotory energetics and metabolic fuel reserves of the Virginia opossum', *Journal of Experimental Biology,* **197**:1–16.

Franco, G. (1987), *Tabela de composição química dos alimentos, 8th ed.*, Atheneu, Rio de Janeiro.

Gardner, A.L. (1993), 'Order Didelphimorphia', in *Mammal species of the world* (eds. D.E., Wilson & D.M. Reeder), pp. 15–23, Smithsonian Institution Press, Washington.

Green, B., King, D., & Bradley, A. (1989), 'Water and energy metabolism and estimated food consumption rates of free-living Wanbengers, *Phascogale calura* (Marsupialia, Dasyuridae)', *Australian Wildlife Research,* **16**:501–7.

Hershkovitz, P. (1992), 'The South American gracile mouse opossums, genus *Gracilinanus* Gardner & Creighton, 1989 (Marmosidae, Marsupialia): A taxonomic review with notes on general morphology and relationships', *Fieldiana: Zoology,* **70**:1–56.

Hingst, E., D'Andrea, P. S., Santori, R. T., & Cerqueira, R. (1998), 'Breeding of *Philander frenata* (Didelphimorphia, Didelphidae) in captivity', *Laboratory Animals,* **32**:434–8.

Hume, I.D. (1999), *Marsupial nutrition,* Cambridge University Press, Cambridge.

Hunsaker II, D. (1977), 'Ecology of New World marsupials', in *The biology of marsupials* (ed. D. Hunsaker II), pp. 95–156, Academic Press, London.

Jensen, S.P. (1993), 'Temporal changes in food preferences of wood mice (*Apodemus sylvaticus* L.)', *Oecologia,* **94**:76–82.

Julien–Laferrière, D., & Atramentowicz, M. (1990), 'Feeding and reproduction of three didelphid marsupials in two Neotropical forests (French Guiana)', *Biotropica,* **22**: 404–15.

Jurgelski, Jr., W., & Porter, M.E. (1974), 'The opossum (*Didelphis virginiana* Kerr) as a biomedical model. III. Breeding the opossum incaptivity methods', *Laboratory Animal Science,* **24**(2):412–25.

Kimble, D.P. (1997), 'Didelphid behavior', *Neurosciences and Biobehavioral Reviews,* **21**:361–69.

Kunz, T.H., & Whitaker, Jr, J.O. (1983), 'An evaluation of fecal analysis for determining food habits of insectivorous bats', *Canadian Journal of Zoology,* **61**:1317–21.

Leite, Y.L.R.; Costa, L.P., & Stallings, J.R. (1996), 'Diet and vertical space use of three sympatric opossums in a Brazilian Atlantic forest reserve', *Journal of Tropical Ecology,* **12**:435–40.

Leite,Y.L.R., Stallings, J.R., & Costa, L.P. (1994), 'Partição de recursos entre espécies simpátricas de marsupiais na Reserva Biológica de Poço das Antas, Rio de Janeiro', *Revista Brasileira de Biologia,* **54**:525–36.

Lee, A.K., & Cockburn A. (1985), *Evolutionary ecology of marsupials,* Cambridge University Press, Cambridge.

Louw, G. (1992), *Physiological animal ecology*, John Wiley & Sons, New York.

Maunz, M., & German, R.Z. (1997), 'Ontogeny and bone scaling in two New World marsupials, *Monodelphis domestica* and *Didelphis virginiana*', *Journal of Morphology,* **231**:117–30.

Marshall, L.G. (1978a), '*Chironectes minimus*', *Mammalian Species,* **109**:1–6.

Marshall, L.G. (1978b), '*Lutreolina crassicaudata*', *Mammalian Species,* **91**:1–4.

Oliveira, C.A., Nogueira, J.C., & Mahecha, G.A.B. (1998), 'Sequential order of appearance of ossification centers in the opossum *Didelphis albiventris* (Didelphidae) skeleton during develoment in the marsupium', *Annals of Anatomy,* **180**:113–21.

Patton, J.L., & Silva, M.N.F.S. (1997), 'Definition of the species of pouched four-eyed opossums (Didelphidae, *Philander*)', *Journal of Mammalogy,* **78**:90–102.

Perales, J., Muños, R., & Moussatche, H. (1986), 'Isolation and partial characterization of a protein fraction from the opossum (*Didelphis marsupialis*) serum, with protecting property against *Bothrops jararaca* venom', *Anais da Academia Brasileira de Ciências,* **58**:155–62.

Périssé, M., Cerqueira, R., & Sorensen, C.R. (1988), 'A alimentação na separação de nicho entre *Philander opossum* e *Didelphis aurita* (Polyprotodontia, Didelphidae)', *Anais do Seminário Regional de Ecologia, São Carlos*, S.P. **VI**:283–94.

Périssé, M., Fonseca, C.R.S.D., & Cerqueira, R. (1989), 'Diet determination for small laboratory-housed wild mammals', *Canadian Journal of Zoology,* **67**:765–78.

Santori, R.T., Astúa de Moraes, D., & Cerqueira, R. (1995a), 'Diet composition of *Metachirus nudicaudatus* and *Didelphis aurita* (Didelphimorphia, Didelphidae)', *Mammalia,* **59**:511–16.

Santori, R.T., Cerqueira, R., & Kleske, C.C. (1995b), 'Anatomia e eficiência digestiva de *Philander opossum* e *Didelphis aurita* (Didelphimorphia, Didelphidae) em relação ao hábito alimentar', *Revista Brasileira de Biologia,* **55**:323–29.

Santori, R.T., & Astúa de Moraes, D. (in press), 'Alimentação, nutrição e adaptações alimentares de marsupiais brasileiros', in *Marsupiais do Brasil: Avanços em Evolução, Biologia e Ecologia* (eds. E.L.A. Monteiro-Filho, & N.C. Caceres), UFPR, Curitiba.

Santori, R.T., Astúa de Moraes, D., Grelle, C.E.V., & Cerqueira, R. (1997), 'Natural diet at a Restinga forest and laboratory food preferences of the opossum *Philander frenata* in Brazil', *Studies on Neotropical Fauna and Environment,* **32**:12–6.

Smith, K.K. (1994), 'Development of craniofacial musculature in *Monodelphis domestica* (Marsupialia, Didelphidae)', *Journal of Morphology,* **222**:149–73.

Smith, K.K., & van Nievelt, A.F.H. (1997), 'Comparative rates of development in *Monodelphis* and *Didelphis*', *Science,* **275**:683–84.

Streilein, K.E. (1982), 'Behavior, ecology and distribution of South American marsupials', in *Mammalian biology in South America* (eds. M.A. Mares, & H.H. Genoways), Special Publications Series of the Pymatuning Laboratory of Ecology, **6**:251–71.

VandeBerg, J.L., & Robinson, E.S. (1997), 'The laboratory opossum (*Monodelphis domestica*) in laboratory research', *Institute of Laboratory Animal Resources Journal,* **38**:4–12.

Zar, G.H. (1996), *Biostatistical analysis, 3rd ed.,* Prentice-Hall, New York.

PART III

CHAPTER 16

THERMAL BIOLOGY AND ENERGETICS OF CARNIVOROUS MARSUPIALS

Fritz Geiser

Zoology, BBMS, University of New England, Armidale 2351, Australia. Email: fgeiser@metz.une.edu.au

Extant carnivorous marsupials are small (4–13,000 g) and almost all are nocturnal or crepuscular and are active during the coldest part of the day. Heat loss and gain via their relative large surface area is therefore likely to be substantial and to have implications on thermal biology and energy expenditure. In this review, data on thermoenergetics in carnivorous marsupials, with an emphasis on the families Dasyuridae and Didelphidae, are summarised and compared with data on other marsupials and mammals in general.

All carnivorous marsupials have low basal metabolic rates (BMR) when compared to most placental mammals, but similar BMR to those of omnivorous/herbivorous marsupials. Thermal conductances of carnivorous marsupials are similar to those of other similar-sized mammals. Carnivorous marsupials have a high metabolic scope and endogenous heat production is achieved to a large extent by shivering thermogenesis and some poorly understood non-shivering component. During exposure to heat carnivorous marsupials use predominantly panting and licking of fur and appendages for evaporative cooling. Carnivorous marsupials have relatively high field metabolic rates (FMR) and, especially in the small species, high FMR/BMR ratios. To minimise daily energy expenditure many carnivorous marsupials use communal nesting and huddling and torpor extensively and thus can lower energy expenditure substantially. While thermal biology and energetics of carnivorous marsupials generally is well known, most of the information is based on laboratory work. Thus, more fieldwork is needed to put physiological data of carnivorous marsupials into an ecological context.

INTRODUCTION

Most carnivorous marsupials are nocturnal and all are small. Because air temperatures (T_a) are usually coldest during the night when carnivorous marsupials are active, and because convective heat loss presumably is substantial, they face high thermoregulatory costs during foraging. Moreover, even when at rest, carnivorous marsupials have relatively high rates of mass-specific metabolic rate when exposed to low T_a because of their large surface area/volume ratio. However, the high quality protein- and fat-rich diet of insectivorous/carnivorous marsupials contains plenty of energy, but can be difficult to obtain because insect availability fluctuates and vertebrate prey is often hard to catch.

To overcome energy bottlenecks and thermal challenges, carnivorous marsupials have several behavioural and physiological options and some of these have been thoroughly investigated (Martin 1902; MacMillen and Nelson 1969; Dawson and

Table 1 Body mass, basal metabolic rate (BMR), thermal conductance (C) and body temperatures (T_b) of carnivorous marsupials.

FAMILY *Species*	Mass (g)	BMR ($mlO_2/(g\ h)$)	C ($mlO_2/(g\ h°C)$)	T_b (°C)	Source
DIDELPHIDAE					
Marmosa microtarsus	13	1.58	0.26	35	Morrison & McNab 1962
Marmosa robinsoni	122	0.8	0.11	34	McNab 1978
Monodelphis brevicaudata	111	0.68	0.12	33.8	McNab 1978
Monodelphis domestica (WA)	104	0.57	0.14	32.6	Dawson & Olson 1988
Metachirops opossum	751	0.45	0.066	35.8	McNab 1978
Lutreolina crassicaudata	812	0.5	0.053	35.8	McNab 1978
Chironectes minimus	922	0.4	0.031	34.6	Thompson 1988
DASYURIDAE					
Planigale ingrami & tenuirostris	7.1	1.59	0.63	34.5	Dawson & Wolfers 1978
Planigale gilesi	8.3	1.43	0.33	32.6	Geiser & Baudinette 1988
Ningaui yvonneae	11.6	1.35	0.21	34.4	Geiser & Baudinette 1988
Sminthopsis murina	12	1.59		35.6	Geiser et al. 1984, McCarron & Dawson 1984
Planigale maculata	13.1	1.01	0.5	34.2	Morton & Lee 1978
Sminthopsis crassicaudata	14	1.33	0.23	33.8	Dawson & Hulbert 1970, Geiser & Baudinette 1987
Sminthopsis macroura	22	1.07	0.21	33.9	Geiser & Baudinette 1987
Antechinomys laniger	27.4	0.98	0.17	34.8	Geiser 1986b
Antechinus stuartii	36.5	1.0	0.22	34.4	MacMillen & Nelson 1969; Dawson & Hulbert 1970
Pseudantechinus macdonellensis	43.1	0.63	0.15	34.2	MacMillen & Nelson 1969
Antechinus flavipes	61.1	0.89		35.1	Geiser 1985
Dasycercus cristicauda	89	0.52	0.1	37.7	MacMillen & Nelson 1969
Dasyuroides byrnei	118	0.7	0.1	34.3	Geiser & Baudinette 1987
Phascogale tapoatafa	147	0.81	0.08	37.4	MacMillen & Nelson 1969
Dasyurus hallucatus	584	0.51	0.05	38.1	MacMillen & Nelson 1969
Dasyurus viverrinus	910	0.45	0.04	36.7	MacMillen & Nelson 1969
Dasyurus geoffroii	1354	0.42	0.03	35	Arnold & Shield 1970
Dasyurus maculatus	1782	0.3	0.03	36.9	MacMillen & Nelson 1969
Sarcophilus harrisii	6500	0.24	0.014	35	Nicol & Maskrey 1980
MYRMECOBIIDAE					
Myrmecobius fasciatus	400	0.36	0.028	32.5	McNab 1984
NOTORYCTIDAE					
Notoryctes caurinus	34	0.63	0.20	30.8	Withers et al. 2000

WA – warm acclimated; T_b – body temperature of normothermic resting animals; C – minimum thermal conductance below the TNZ

Hulbert 1970; Wallis 1979, 1982; Dawson 1989; Geiser 1994; Green 1997; Hume 1999). Nevertheless, most of the available data on thermal biology and energetics are based on laboratory investigations and much less is known about them in the field.

In this chapter studies of energetics and thermal physiology of insectivorous/carnivorous marsupials are reviewed with emphasis on the Australian Dasyuridae and the South American Didelphidae. To investigate possible energetic consequences of their diet, carnivorous marsupials are compared with similar-sized omnivorous/herbivorous marsupials and mammals in general. The marsupial taxonomy and dietary classification are based on Hume (1999).

BASAL METABOLIC RATE (BMR)

Animals must expend energy to maintain vital functions. In endotherms, maintenance metabolic rate (MR) when measured under standard conditions that exclude thermoregulation and locomotion represents the minimum energy use at high body temperature (T_b) and is termed basal metabolic rate (BMR).

The BMR of carnivorous marsupial are lower than those of most placental mammals (MacMillen and Nelson 1969). BMR of the carnivorous marsupials (28 species, body mass: 7–6500 g; Table 1) appear roughly equivalent to those of similar-sized omnivorous/herbivorous marsupials (Fig. 1) and, as in other organisms,

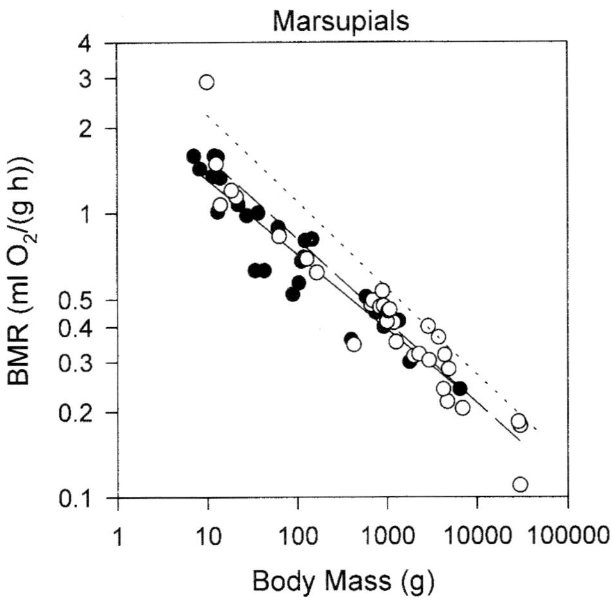

Figure 1 Allometric equations of BMR as a function of body mass (BM) in carnivorous (solid symbols) and omnivorous/herbivorous marsupials (circles) are listed below. The two groups are listed separately although they do not differ significantly (ANCOVA), mainly to provide a separate equation for the carnivorous marsupials; equations for omnivorous/herbivorous marsupials and a combined equation for marsupials are also provided for comparison. The regression line for placental mammals (dotted line) is from Hayssen and Lacy (1985). Regression equations: Carnivorous marsupials (solid line, n = 28): BMR (ml O_2/(g*h)) = 2.39 $BM^{-0.262}$; r^2 = 0.88; Omnivorous/herbivorous marsupials (broken line, n = 32): BMR (ml O_2/(g*h)) = 3.07 $BM^{-0.289}$; r^2 = 0.91; All marsupials (not shown, n = 60): BMR (ml O_2/(g*h)) = 2.58 $BM^{-0.268}$; r^2 = 0.91

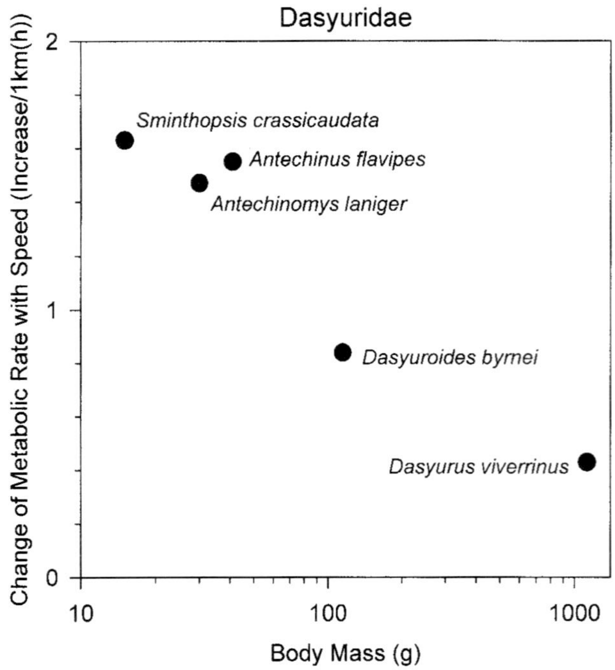

Figure 2 Increase of mass-specific metabolic rate over an increase in speed by 1km/h for several dasyurids. The increase is much more pronounced in the small than in the large species (data from Baudinette et al. 1976).

mass-specific BMR in both groups are negatively correlated with body mass in a logarithmic plot (Fig. 1). Regressions for carnivorous and omnivorous/herbivorous marsupials were indistinguishable (Analysis of Covariance, ANCOVA; $p > 0.1$; data from Table 1 for carnivorous marsupials, and for omnivorous/herbivorous marsupials from Table 1.1 in Hume 1999). This observation and the better than 90% prediction of BMR by the allometric equation for all marsupials (Fig. 1) suggest that maintenance costs of living in thermoneutrality are little or not affected by carnivory.

However, the mass-specific BMR of marsupials differs substantially from the BMR predicted from allometric equations for placentals (Hayssen and Lacy 1985). Especially at low body mass (10 g) the BMR of marsupials is only ~65% of that predicted for placental mammals, whereas the BMR of 10,000 g marsupials is ~80% that predicted for placentals (Hayssen and Lacy 1985). In the past the low BMR in marsupials was often viewed to be a reflection of a primitive thermoregulatory capability (Martin 1902). In recent years it has been suggested that the low BMR of marsupials may be due to ecological rather than, or in addition to, phylogenetic constraints as low BMR in southern hemisphere mammals may reflect the high unpredictability of climate and rainfall (Hulbert and Dawson 1974; Lovegrove 1996).

Thus, in agreement with the original investigations (MacMillen and Nelson 1969; Dawson and Hulbert 1970), maintenance metabolism of carnivorous marsupials is low. However, costs of thermoregulation and locomotion are not considered in this component of energy expenditure and BMR should therefore not be equated with total energy expenditure, as is discussed below.

Locomotion

Dasyurids show similar relationships between body size, running speed, and the associated energetic costs as do other quadrupedal mammals (Baudinette et al. 1976; Baudinette 1982). Moreover, maximum running speed in carnivorous marsupials is similar to that of similar-sized placentals (Garland et al. 1988).

Body size strongly affects cost of locomotion in marsupials (Baudinette et al. 1976). Although the total cost of transport is obviously larger in big than in little animals, the mass-specific energy expenditure in quolls, *Dasyurus viverrinus* (1120 g), when running at a speed of 1 km/h, was only ~1/3 of that in dunnarts, *Sminthopsis crassicaudata* (15 g). In all species, energy expenditure

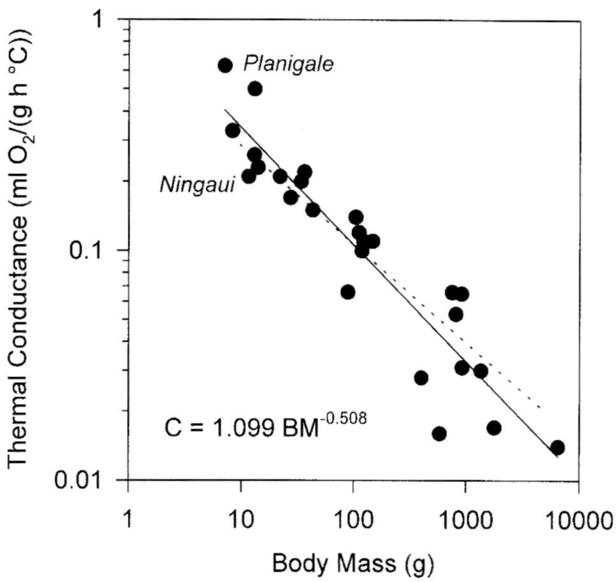

Figure 3 Thermal conductance of carnivorous marsupials (data from Table 1). The broken line is based on the regression equation for all mammals (Bradley and Deavers 1980). *Ningaui* sp. and *Planigale* spp. deviated somewhat from the regression for carnivorous marsupials which is: Carnivorous marsupials (solid line, n = 26):C(ml O_2/(g*h*°C)) = 1.03 BM$^{-0.484}$; r^2 = 0.92

during locomotion increased linearly with speed (Baudinette et al. 1976). Moreover, as in other mammals, the increase of energy expenditure with speed by 1km/h was greater in the small (1.62 times, *Sminthopsis crassicaudata*, 15 g) than in the large species (0.43 times, *Dasyurus viverrinus*, 1120 g)(Fig. 2). Thus, mass-specific costs of locomotion and size in dasyurids are inversely related, which has implications for long distance movements.

Some carnivorous marsupials are arboreal and thus are likely to expend more energy during vertical climbing than horizontal running. To my knowledge the increase in cost of vertical transport in carnivorous marsupials has not been investigated, but it is likely to be similar to that found in squirrels, which, in comparison to horizontal locomotion, increase energy expenditure during vertical climbing by ~2-fold (Wunder and Morrison 1974).

One of the most extreme modes of foraging is the 'swimming' in sand of marsupial moles. In *Notoryctes caurinus*, metabolic rates during sand swimming were ~3.3 times BMR at speeds between 2–18 m/h (Withers et al. 2000), which is lower than those of dasyurid marsupials running at 1 km/h (>5 times BMR; Baudinette et al. 1976).

THERMOREGULATION AND HEAT PRODUCTION

Thermal conductance, insulation and acclimatisation

Since carnivorous marsupials are small, and their relative surface area is large, heat loss during cold exposure should be substantial. However, thermal conductance (C), the inverse of insulation and an indirect measure of heat loss during cold exposure (Bradley and Deavers 1980; Hallam and Dawson 1993), is similar in carnivorous marsupials and mammals in general (Table 1; Fig. 3). Mass-specific thermal conductance in carnivorous marsupials shows a linear decrease with increasing mass on a logarithmic plot (Fig. 3) emphasising the substantial heat loss in the small species when compared to the volume of heat generating tissue. Mass-specific heat loss in a 10 g species is ~10 times greater than in a 1000 g species, which clearly has strong implications on energy expenditure.

Most species appear to be well represented by the general regression equation for conductance of carnivorous marsupials *vs* body mass. Nevertheless, *Ningaui yvonneae* (11 g) and *Planigale* spp. (5–8 g), dasyurids from the Australian arid zone, differ substantially in shape and conductance (Dawson and Wolfers 1978; Geiser and Baudinette 1988) (Fig. 3) and differ from the conductance predicted for carnivorous marsupials. *Ningaui yvonneae*, which has a spherical body and a low thermal conductance (68% of that predicted for carnivorous marsupials), lives on the surface of sandy soils and shelters in spinifex tussocks affording limited thermal buffering. In contrast, *Planigale* spp. are dorsoventrally flattened and have a high thermal conductance (90–158% of that predicted) (Fig. 3), but planigales live in deep soil cracks, which provide both insect prey and a stable thermal environment. Thus, body shape and thermal conductance in these dasyurids are interrelated and appear to be a reflection of where the animals shelter and forage.

Fur is of course also an important component of insulation. *Antechinus stuartii* and *A. swainsonii* show significant increases in fur thickness and insulation in winter in comparison to summer (Wallis 1982). However, some of the change in thermal conductance is likely due to the animals' smaller size in summer, as at that time they are only about half grown. Dunnarts, *Sminthopsis macroura*, and Kowaris, *Dasyuroides byrnei,* also lower their thermal conductance by ~15–20% in winter in comparison to summer (Geiser and Baudinette 1987).

Sminthopsis crassicaudata, show a decreasing reflectance (i.e. animals become darker) with increasing distance from the equator (Hope and Godfrey 1988). As it is known that small dasyurids bask in the sun (see below), the darker fur in the more southerly populations may play an important role in thermoregulation, as it will increase heat absorption.

Acclimation

Thermal acclimation in the laboratory differs from seasonal acclimatisation as usually only one environmental variable is altered and the time period is usually restricted to days or weeks rather than months. Smith and Dawson (1985) reported that cold- and warm-acclimation for 4–6 weeks alters thermoenergetics of

Dasyuroides byrnei substantially. Unlike winter acclimatised *D. byrnei*, which show reduced metabolic rates (Geiser and Baudinette 1987), cold-acclimated individuals showed increased metabolic rates (Smith and Dawson 1985). In contrast, May (1996) reported similar metabolic rates in both cold- and warm-acclimated *D. byrnei*, but observed a lower shivering intensity in cold acclimated individuals. In the short-tailed opossum, *Monodelphis domestica*, cold acclimation increased metabolic rate, but in contrast to placental mammals (Wunder and Gettinger 1996), thermal tolerance was not improved (Dawson and Olsen 1988). Temperature acclimation did not affect metabolism in mouse opossums, *Thylamys (Marmosa) elegans*; however, costs of rewarming from torpor were reduced by cold acclimation (Opazo et al. 1999), perhaps due to a change in insulation. Acclimation to different photoperiods in *Sminthopsis crassicaudata* resulted in change in body mass, but had little effect on thermal biology and energetics (Holloway and Geiser 1996). Thus, acclimation seems to differ from seasonal acclimatisation, and cold acclimation in some carnivorous marsupials appears to differ from placental mammals, which tend to improve non-shivering thermogenesis and thermal tolerance via a proliferation of brown adipose tissue (see below).

Heat production and metabolic scope

Although carnivorous marsupials have a relatively low BMR their ability to increase metabolic rates during cold exposure is pronounced. The metabolic scope (i.e. the ratio of maximum cold-induced metabolism/BMR) in many carnivorous marsupials is greater than in several placental mammals (Dawson 1983; Hinds and MacMillen 1984). The metabolic scope was 9–11 in *Dasyuroides byrnei* and in *Planigale gilesi* and only 3–4 in Australian rodents partially because of the low BMR in the former (Dawson and Dawson 1982; Smith and Dawson 1985). Nevertheless, some northern hemisphere placentals have metabolic scopes similar to those of marsupials (Wunder and Gettinger 1996).

Mechanisms of heat production in carnivorous marsupials, as for marsupials in general, remain controversial. There is general agreement that shivering thermogenesis by rapid movement of antagonistic muscles is a major source of heat in marsupials. In contrast, the mechanism of non-shivering thermogenesis in marsupials, which in placental mammals is to a large extent accomplished in brown adipose tissue (BAT, a specialised thermogenic fat tissue rich in heat-producing mitochondria), has still not been resolved. Despite an intensive search in 38 marsupial species from 18 families no BAT was found using microscopy (Hayward and Lisson 1992). Moreover, noradrenaline, the hormone that activates BAT (Nicol et al. 1997) and elicits a substantial increase of metabolic rate in placental mammals (Wunder and Gettinger 1996), has been reported to show no effect in most carnivorous marsupials (Hulbert and Reynolds 1981; Dawson and Olsen 1988; Opazo et al. 1999). In contrast, Loudon et al. (1985) claimed to have identified BAT in a juvenile wallaby *Macropus rufogriseus* and traces of mitochondrial uncoupling protein (UCP, protein responsible for production of heat in BAT) have been described in *Sminthopsis crassicaudata* (Hope et al. 1997). The increased heat production in *S. crassicaudata* on low doses of noradrenaline (Clements et al. 1998) may however be due to general vasoconstriction and an increase in cardiac output in response to noradrenaline rather than BAT thermogenesis (Nicol et al. 1997). Marsupials are known to use vasoconstrictor-induced non-shivering thermogenesis in skeletal muscle (Eldershaw et al. 1996). Moreover, cold acclimation increases thyroid secretion by 3.1-fold in comparison to warm acclimation in *Antechinus stuartii* (Withers and Hulbert 1988), which also may contribute to a rise in heat production. As all biochemical reactions release heat and marsupials are known to possess UCPs associated with tissues other than BAT (Clements et al. 1998), it is likely that non-shivering thermogenesis plays a role in thermoregulation of marsupials, but further work is required regarding its mechanisms.

Heat exposure

Whereas low temperatures have received much attention with respect to thermoregulation and energetics in carnivorous marsupials, less is known about their responses to heat. Heat exposure is of interest especially in dasyurids, as most members of this family live in arid or semi-arid areas, and despite being nocturnal may occasionally be exposed to physiologically challenging high T_as while resting during the day.

The most extensive study by Robinson and Morrison (1957) showed that exposure to T_a 40°C results in profound hyperthermia (T_b > 40°C) in small (<50 g) dasyurids after ~3 hours, whereas the somewhat larger *Phascogale tapoatafa* (108 g), was able to maintain T_b below 40°C for ~3.5 hours. The arid-zone Mulgara, *Dasycercus cristicauda* (72 g), was able to maintain its T_b at around 38.5°C for more than 5 hours, and similarly quolls, *Dasyurus hallucatus* (650 g), maintained T_b around 39.5°C for 6 hours. In contrast to all the other species, the relatively large Tasmanian devil, *Sarcophilus harrisii* (6700 g), was able to maintain T_b between 37 and 39°C. Robinson and Morrison (1957) concluded that small dasyurids (*Sminthopsis crassicaudata*, *S. macroura* and *Antechinus flavipes*) pant, lick their paws and adopt a 'lizard-like posture' when exposed to heat (they also attempt to increase surface area and maximise contact with the substrate; pers.obs.). The slightly larger species (72–650 g) are able to store heat, show open-mouthed panting and lick paws and/or abdominal fur (Robinson and Morrison 1957). As heat-exposed devils did not change its respiration rate, it was suggested that they sweat (Robinson and Morrison 1957), but sweating could not be confirmed in subsequent studies in resting and exercising devils (Hulbert and Rose 1972; Bell et al. 1983). It thus appears that sweating in marsupials is restricted to

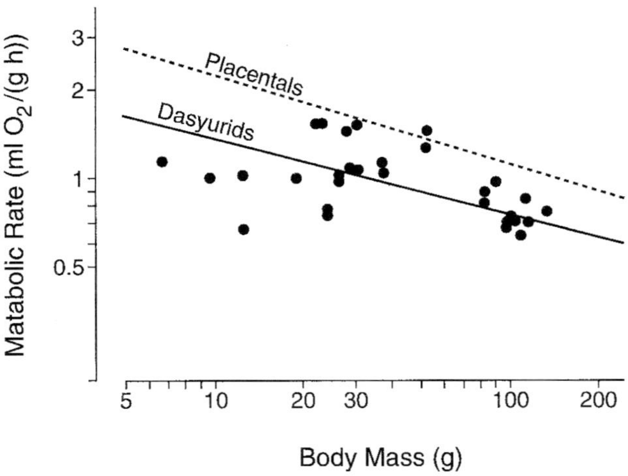

Figure 4 Double-log plot of metabolic rate of the Kowari, *Dasyuroides byrnei*, at T_a 30°C at different ages. The regression lines for BMR of dasyurids is that from MacMillen and Nelson (1969) and that for placentals from Hayssen and Lacy (1985).

diprotodonts (Dawson 1983) and, to cope with heat, dasyurids mainly resort to licking and panting for evaporative cooling.

Development of thermoregulation

Carnivorous marsupials, like other marsupials, are extremely altricial and small at birth (dasyurids ~5–18 mg; Tyndale-Biscoe and Renfree 1987), and develop and grow at a slow rate in their mother's pouch (Morrison and Petajan 1962; Hulbert 1988). Initially, they are permanently attached to a nipple while in later-stage lactation attachment is intermittent (Tyndale-Biscoe 1973). The small size of marsupial neonates, lack of fur, and limited neural capabilities render them incapable of homeothermic thermoregulation. Nevertheless, young macropods and possums are protected within the deep mother's pouch and maternal heat maintains thermal homeostasis of the young (Hulbert 1988, Gemmell and Cepon 1993). In contrast, dasyurid pouches are shallow and open and young are exposed to the outside environment early during development.

Data on development of thermoregulation are available on only two dasyurids. *Dasyuroides byrnei* remain poikilothermic (i.e. unable to maintain a constant T_b) at moderate cold exposure for ~80 days after birth (Geiser et al. 1986). During this time (body mass ~8–25 g), young *D. byrnei* held at T_a 30°C have metabolic rates that are well below those predicted for adult marsupials of the same size (Fig. 4). In neonate Julia Creek Dunnarts, *Sminthopsis douglasi*, metabolic rates are so low that they are able to use their skin for up to 95% of their gas exchange (Mortola et al. 1999). Thus, early in the development of dasyurids, little or no fuel is used for thermoregulation and nutrients can be used almost exclusively for growth. At just over an age of 100 days (body mass ~30–50 g), *D. byrnei* can thermoregulate in the cold, and at T_a 30°C its metabolic rate is higher than predicted for adult marsupials of the same mass reaching values predicted for adult placentals (Fig. 4) because of intensive growth. Only when *D. byrnei* reach adult body mass of ~100 g at ~200 days, is their BMR similar to that predicted for adult dasyurids of their size. Thus, the development of thermoregulation in dasyurids is slow, but the extended developmental period necessitates only small increases in maternal energy requirements, especially in early-stage pouch life (Thompson and Nicoll 1986). This may be important for raising litters especially in an unpredictable desert environment.

NESTS AND HUDDLING

As energy expenditure for thermoregulation in carnivorous marsupials is high, they show behavioural adaptations to minimise energy loss. Many carnivorous marsupials use tree cavities, burrows, soil crevices, caves, tussock grass, or other protected locations as resting places during the day and between foraging bouts. In addition to providing protection from predators, these are important for buffering of thermal extremes (Baudinette 1972). Often the nests of small desert dasyurids, such as those of *Sminthopsis* spp., are not very sophisticated, and they use burrows of bird-eating spiders, scorpions, rodents and lizards, into which they may drag a few leaves (Dickman 1996) or build grass nests under rocks (Morton 1978). Western quolls (*Dasyurus geoffroii*) build extensive nests of eucalypt leaves and some feathers in nursery dens (Serena and Soderquist 1989). Arboreal marsupials, such as *Phascogale tapoatafa*, construct substantial nests in tree hollows consisting to a large extent of stringybark strips, feathers and fur (Soderquist 1993). Similarly, Australian *Antechinus* spp. build substantial nests using eucalypt leaves and grass in artificial nest boxes or tree hollows (Dickman 1991). Small dasyurids from Papua New Guinea build spherical nests of interwoven plant material either in trees or in subterreanean burrows (Woolley 1989). It is well known that nests can lower thermoregulatory costs in small mammals (Vogt and Lynch 1982). In *Sminthopsis crassicaudata* it was estimated that use of a nest reduces daily energy expenditure, which includes costs of foraging, by ~5% (Frey 1991).

Huddling reduces thermal conductance because the surface area/volume ratio of a group is smaller than that of a single animal. Huddling and sharing body heat is common in some but not all dasyurids. Antechinus females and their young huddle extensively and *Sminthopsis crassicaudata* commonly forms groups of two to eight individuals in Autumn/Winter and even share nests with feral mice *Mus musculus* (Morton 1978), which likely results in energetic benefits for both species.

Frey (1991) estimated that huddling in nests reduces the daily energy expenditure of *Sminthopsis crassicaudata* by ~20%. *Antechinus stuartii* nest young at a body mass of ~12 g reduced

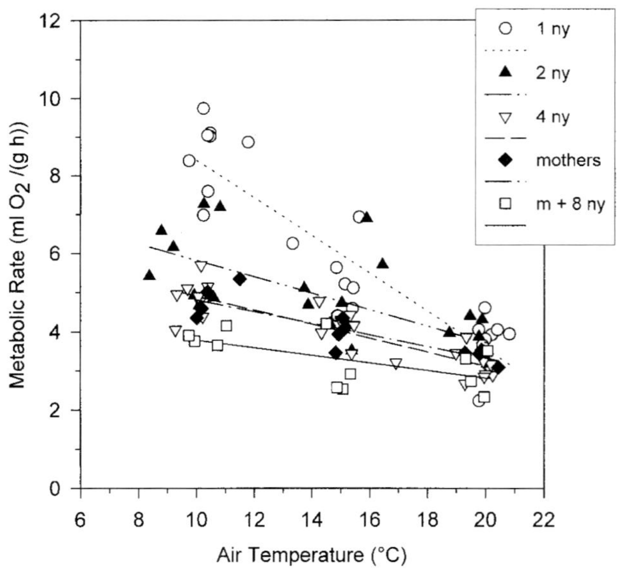

Figure 5 Mass-specific metabolic rates of brown antechinus *Antechinus stuartii* juveniles at a body mass of ~12 g as a function of T_a and group size. Single nest young (1 ny) have significantly higher metabolic rates than groups of nest young (2–4 ny) or nest young with mother (m). (data from Hörner, Westman, Körtner and Geiser unpublished).

resting metabolism at T_a 10°C by >50% when huddling with their mother and eight siblings (Fig. 5). Thus, huddling and nest sharing contributes to limiting energy expenditure in dasyurids. Huddling has the advantage that it can be achieved at normal high T_b of the animal, but it requires the presence of other individuals, and energy expenditure cannot be reduced below maintenance requirements.

Torpor

As most carnivorous marsupials are small, have a high heat loss and have to cope with fluctuating supply of food, many species undergo periods of torpor to lower energy expenditure (Table 2). Torpor is a controlled reduction of T_b and metabolic rate that is widely used by small mammals and birds to reduce energy expenditure during periods of food shortage or adverse environmental conditions. However, torpor in some species is also used to balance energy requirements when food is available or to store fat for future energy bottlenecks (Carpenter and Hixon 1988; Wang 1989; Geiser and Masters 1994; Geiser and Ruf 1995). Torpid animals not only have reduced costs for thermoregulation, but most importantly, as metabolic rate during torpor (TMR) is usually well below BMR (Table 2), their maintenance costs are also substantially lowered (Geiser and Ruf 1995). Unlike in ectothermic organisms, the T_b during torpor is regulated at or above a species-specific threshold by a proportional increase in metabolism (Heller and Hammel 1972) and torpid endotherms can rewarm themselves actively from torpor using endogenous heat production, whereas ectotherms are reliant on uptake of external heat.

Most heterothermic mammals show one of two distinct patterns of torpor. The first, shallow, daily torpor in the 'daily heterotherms' lasts on average for several hours and the mean minimum T_b (determined by the threshold for T_b that is metabolically defended during torpor) is ~17°C. The metabolic rate during daily torpor is reduced to ~30% of the BMR (Geiser and Ruf 1995) and often to less than 10% of the resting metabolic rate during cold exposure (Geiser 1986a). The low metabolic rates in torpid daily heterotherms appear to be largely caused by (i) cessation of heat production for maintenance of high T_b and (ii) temperature effects on metabolic rates caused by the fall of T_b (Heller et al. 1977; Geiser 1988a; Withers 1992; Song et al. 1995; Song et al. 1996; Guppy and Withers 1999).

The second common pattern, prolonged and deep torpor observed in the 'hibernators', is characterised by much longer bouts of torpor that on average last for ~1 to 3 weeks and are interrupted by brief (<24 h) periodic arousals throughout the hibernation season in winter and the mean minimum T_b is ~6°C. The metabolic rate during hibernation is on average reduced to 5% of the BMR (Hock 1951; Geiser and Ruf 1995). However, in small species the metabolic rate may be as low as 1% of BMR, and a fraction of 1% when compared to the resting metabolic rate during cold exposure (Geiser and Ruf 1995, 1996; Song et al. 1997; Guppy and Withers 1999; Geiser and Brigham 2000). The low metabolic rates in torpid hibernators appear to be largely caused by (i) cessation of heat production for maintenance of high T_b, (ii) a physiological inhibition of metabolism, and (iii) temperature effects caused by the fall of T_b (Heller et al. 1977; Geiser 1988a; Song et al. 1997; Malan 1993; Guppy and Withers 1999; Geiser and Brigham 2000).

Species displaying deep and prolonged torpor at low T_a may display short torpor lasting for less than a day at high T_a or at the beginning of the hibernation season. However, these appear to be short bouts of hibernation with metabolic rates well below that during daily torpor even at the same T_b (Geiser 1988a; Song et al. 1997; Geiser and Brigham 2000), and thus differ from torpor in the daily heterotherms. In contrast to the hibernators, the daily heterotherms show daily torpor exclusively both at low and high T_a and often throughout the year (Hudson 1973; Geiser and Baudinette 1987; Coburn and Geiser 1998; Lovegrove 2000).

Daily torpor appears to be most common in carnivorous marsupials, whereas prolonged torpor, common in the omnivorous pygmy-possums (Geiser and Broome 1991; Geiser 1994; Körtner and Geiser 1998), is presently known to occur only in one carnivorous marsupial, *Dromiciops australis*, from South America

Table 2 Summary of body mass and physiological variables in heterothermic carnivorous marsupials

FAMILY Species	Body mass (g)	Minimum T_b (°C)	Torpor duration (h)	TMR (ml O_2/(g h))	TMR/BMR	Source
DIDELPHIDAE						
Marmosa microtarsus	13	16	8	0.25	0.18	Morrison & McNab 1962
Thylamys (Marmosa) elegans	32	14	–	0.4		Opazo et al. 1999
Marmosa robinsoni	122	23	–	–	–	McNab 1978
Monodelphis brevicaudata	40–111	27	–	–	–	McNab 1978
Monodelphis domestica	100	25	–	–	–	Douglas & Nicol 1993, Douglas pers. comm.
MICROBIOTHERIIDAE						
Dromiciops australis	30	7.1 (skin)	120	0.03	–	Rosenmann & Ampuero 1981, Grant & Temple-Smith 1987
DASYURIDAE						
Planigale ingrami	5.8–9.3	–	2–4	–	0.3	Dawson & Wolfers 1978
Planigale tenuirostris	6.6–7.3	–	2–4	–	0.3	Dawson & Wolfers 1978
Planigale gilesi	8.3	14.3	15.25	0.36	0.25	Geiser & Baudinette 1988
Planigale maculata	9.7–16.4	19.6	–	0.4	0.4	Morton & Lee 1978
Ningaui ridei	9	–	–	–	–	Dickman pers. comm. 1999
Ningaui yvonneae	10–13	15.3	12.25	0.3	0.23	Geiser & Baudinette 1988
Sminthopsis youngsoni	10	–	–	–	–	Dickman pers. comm. 1999
Sminthopsis ooldea	11	–	–	–	–	Aslin 1983
Sminthopsis longicaudata	15–20	–	–	–	–	Burbidge et al. 1983
Sminthopsis hirtipes	15	–	–	–	–	Dickman pers. comm. 1999
Sminthopsis crassicaudata	17	13.0	19.5	0.27	0.22	Geiser & Baudinette 1987
Sminthopsis murina	18	15.0	8	0.25	0.22	Geiser et al. 1984
Sminthopsis macroura	20–28	14.0	17.9	0.3	0.29	Geiser & Baudinette 1987
Sminthopsis douglasi	60	16.9	8.8	0.432	0.4	Muller 1996
Antechinomys laniger	27	11.0	16.0	0.14	0.13	Geiser 1986b
Antechinus stuartii	20–60	19.9	9	0.66	0.62	Geiser 1985
Antechinus flavipes	30–70	24.5	5.5	0.48	0.46	Geiser 1985
Antechinus swainsonii	50–100	28.2	–	–	–	Gotts 1976
Dasycercus cristicauda	70–110	14	12	0.12	0.23	Geiser & Masters 1994
Phascogale tapoatafa	110–235	–	–	–	–	Dixon & Huxley 1989
Dasyuroides byrnei	120	20.4	7.5	0.4	0.54	Geiser & Baudinette 1987
Dasyurus geoffroii	1000	23.1	–	–	–	Arnold 1976
Dasyurus viverrinus	1000	25	–	–	–	Moyle unpublished
MYRMECOBIIDAE						
Myrmecobius fasciatus	500	–	–	–	–	Serventy & Raymond 1973
NOTORYCTIDAE						
Notoryctes typhlops	60	–	–	–	–	Wood Jones 1923; Tynda e-Biscoe 1973
Notoryctes caurinus	34	–	–	–	–	Withers et al. 2000

The longest torpor bouts and the lowest metabolic rate during torpor (TMR) are reported.
– no physiological measurements but observations were made.

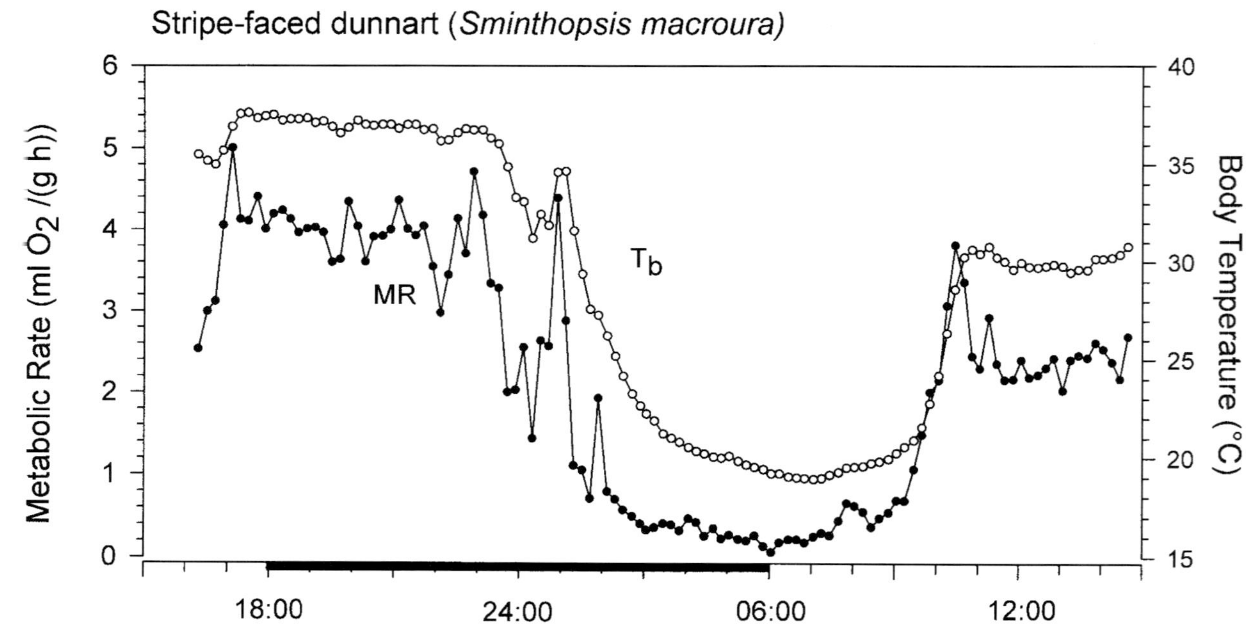

Figure 6 Daily fluctuation of body temperature (T_b) and metabolic rate (MR) of a stripe-faced dunnart *Sminthopsis macroura* at a T_a of 18°C. Torpor entry occurred at ~0100 h and spontaneous arousal at ~0900 h.

(Rosenmann and Ampuero 1981). Details on torpor patterns of families are provided below.

Didelphidae

Carnivorous species of the family Didelphidae tend to be small and it is therefore of no surprise that several display daily torpor (see Table 2). After food deprivation, the murine opossum *Marmosa microtarsus* lowered T_b to ~16°C and torpor bouts lasted up to 8 hours (Morrison and McNab 1962). Daily torpor has also been observed in *Marmosa robinsoni*, *M.* (*Tylamys*) *elegans* and *Monodelphis brevicauda* (McNab 1978; Opazo et al. 1999).

Microbiotheriidae

The only extant species of the Microbiotheriidae, the Monito del Monte or Colocolo, *Dromiciops australis* (*gliroides*, 30 g), which is largely insectivorous (Hume 1999), differs from the other carnivorous marsupials as it is capable of undergoing multiday torpor bouts. Similar to hibernators (Geiser 1988a), Colocolos lowered metabolic rates to ~1% of that in normothermic animals, skin temperatures fell to 7.1°C, and torpor bouts lasted for ~5 days (Rosenmann and Ampuero 1981; Grant and Temple-Smith 1987). This species fattens before winter and apparently disappears during winter (Nowak and Paradiso 1983) suggesting that *Dromiciops australis* hibernates.

Myrmecobiidae

According to an anecdotal report (and supported by unpublished observations on captive individuals; Christine Cooper pers. comm. 2001) the numbat, *Myrmecobius fasciatus* (~500 g) displays torpor in the field. Cold and immobile individuals were found in hollow logs on cold winter mornings and rewarmed after a few hours in the sun (Serventy and Raymond 1973).

Notoryctidae

Tyndale-Biscoe (1973) suggested that observations by Wood Jones (1923) on activity patterns of the marsupial mole, *Notoryctes typhlops* (40–70 g), indicate use of torpor. The second species of the family, *Notoryctes caurinus*, has a low and labile T_b (Withers et al. 2000) suggesting that this species is heterothermic.

Dasyuridae

Torpor has been observed in ~50% of the Australian dasyurid species (Table 2) and it is likely that most members of this family are heterothermic. All dasyurids, which have been investigated with respect to torpor use, displayed daily torpor, usually lasting for several hours from late night or early morning to mid-morning, midday, or afternoon. Torpor has been observed in species ranging in body mass from ~5 g to ~1000 g (Table 2).

Depending on the species, T_b during torpor falls to values between 11°C (Kultarr, *Antechinomys laniger*) and 28°C (*Antechinus swainsonii*; Table 2 – the latter may not represent the minimum T_b of the species). Metabolic rates during torpor of dasyurids fall to ~10–60% of the BMR and torpor bouts last for up to 19.5 hours, but 2–8 hours are more common (Table 2).

The typical daily fluctuation of T_b and metabolism during activity, rest and torpor of a daily heterotherm is illustrated in Fig. 6 for *Sminthopsis macroura*, a commonly studied species in the laboratory. Metabolism fluctuated substantially during the activity phase until entry into torpor, characterised by a reduction of T_b and metabolism, commenced after 0100 h. The animal remained torpid with a T_b around 20°C for ~8 hours. Spontaneous arousal, characterised by a brief metabolic overshoot during the late phase of arousal, commenced at ~0900 h. As can be estimated from Fig. 6, the reduction of metabolic rate during torpor in comparison to activity is substantial and the cost of rewarming incurs only a small energetic penalty.

Torpor in dasyurids can be either spontaneous (food and water *ad libitum*) or may occur after withdrawal or restriction of food (induced torpor). Spontaneous torpor occurs regularly in some species (e.g. *Sminthopsis macroura, Antechinomys laniger, Planigale gilesi*; Godfrey 1968; Geiser 1986b; Geiser and Baudinette 1987, 1988) suggesting, that in these species, torpor is a regular part of their daily cycle to minimise energy expenditure. In other species food and/or water have to be withdrawn or restricted to induce torpor while spontaneous torpor is observed only occasionally (e.g. *Sminthopsis crassicaudata*, and *Antechinus* spp. Godfrey 1968; Wallis 1979, 1982; Geiser and Baudinette 1987; Geiser 1988a) suggesting, that in these species, torpor may mainly be used during acute energetic stress. Torpor in the laboratory reduced average daily metabolic rates of *Sminthopsis crassicaudata* by ~20–40% in comparison to normothermic animals and energy savings were a direct function of torpor bout duration (Holloway and Geiser 1995). Field studies suggest that torpor patterns and energy saving due to torpor in free-ranging *Sminthopsis crassicaudata* are similar to those observed in the laboratory (Frey 1991). However, as daily torpor in free-ranging sugar gliders, *Petaurus breviceps*, is much more frequent, deeper and longer in the field than in the laboratory (Körtner and Geiser 2000), more work on the subject in dasyurids will have to verify this.

Timing of torpor entry in dasyurids appears to be a function of total energy use prior to a torpor episode (Fig. 7; Geiser 1986b). Torpor entry in *Antechinomys laniger* occurred early at low T_a when energy expenditure was high and late when energy expenditure was low. Moreover, induced torpor in dasyurids often begins at night during the second half of the activity phase, whereas spontaneous torpor often begins later in the early morning and is usually restricted to the daily rest phase (Geiser 1986b; Geiser and Baudinette 1987).

Metabolic rates during torpor (TMR) are related to the duration of torpor and mass loss in some dasyurids. *Antechinomys laniger* (Geiser 1986b) and *Planigale gilesi* (Fig. 8) show an increase in torpor bout duration with decreasing TMR in thermoconforming animals (i.e. above the torpor T_b threshold), whereas a decrease in bout duration occurs when thermoregulatory heat production

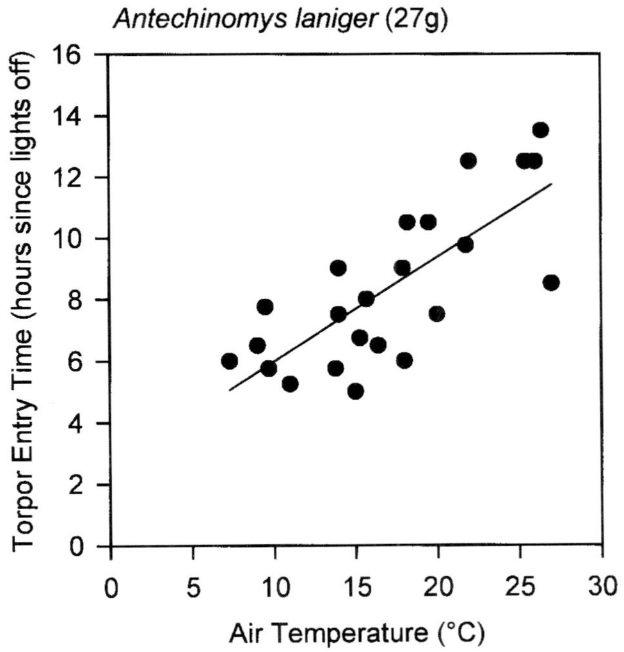

Figure 7 Time of torpor entry in *Antechinomys laniger* as a function of T_a (data from Geiser 1986b). Entries occurred early at low T_a when pre-entry energy expenditure was high and late when T_a was high. Torpor entry time (hours since lights off) = 2.59+0.34 T_a (°C) (r^2 = 0.58)

during torpor commences (Fig. 8a). Thus, the metabolic rate during torpor and duration of torpor appear to be interrelated and energy savings are maximised (and mass loss minimised) at T_a close to or slightly above the T_b threshold during torpor.

Torpor patterns in some dasyurids change with season. Metabolic rates and minimum T_bs during torpor of *S. macroura* (Fig. 9) and *S. crassicaudata* acclimatised to natural outdoors conditions were lower in winter than during summer suggesting a seasonal change of thermal physiology in these species (Geiser and Baudinette 1987). In the field, torpor in *S. crassicaudata* occurred more frequently in late autumn and winter (May–July) when T_a was low, than in early autumn (April) (Frey 1991), and when groups of *S. crassicaudata* displayed social torpor with mice, *Mus musculus* (Morton 1978).

Seasonal changes in the occurrence and depth of torpor in *Antechinus* spp. differ from those in *Sminthopsis* spp. (Geiser 1988b). Antechinus reproduce only once a year in late winter and males die after mating (McAllan and Dickman 1986). Offspring are weaned in summer and grow until the following winter (Woolley 1966; Lee et al. 1982). Torpor in juvenile *Antechinus stuartii* and *A. flavipes* in summer (~50% adult size) was more frequent and deeper than in winter when they had reached adult size (Geiser 1988b). These observations suggest that the seasonal

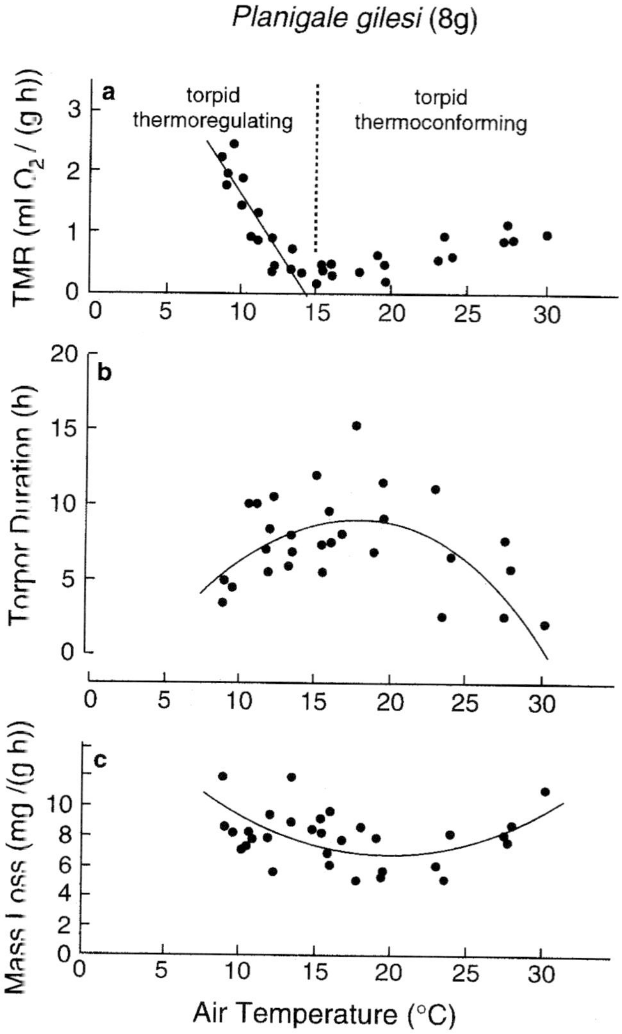

Figure 8 Metabolic rates, torpor duration, and mass loss in torpid planigales, *Planigale gilesi*, as a function of T_a. Duration of torpor increased and mass loss declined over the T_a range in which animals were thermoconforming (i.e. metabolic rate was declining). The opposite was observed in the T_a range in which torpid animals showed a thermoregulatory increase of metabolism (data from Geiser 1985).

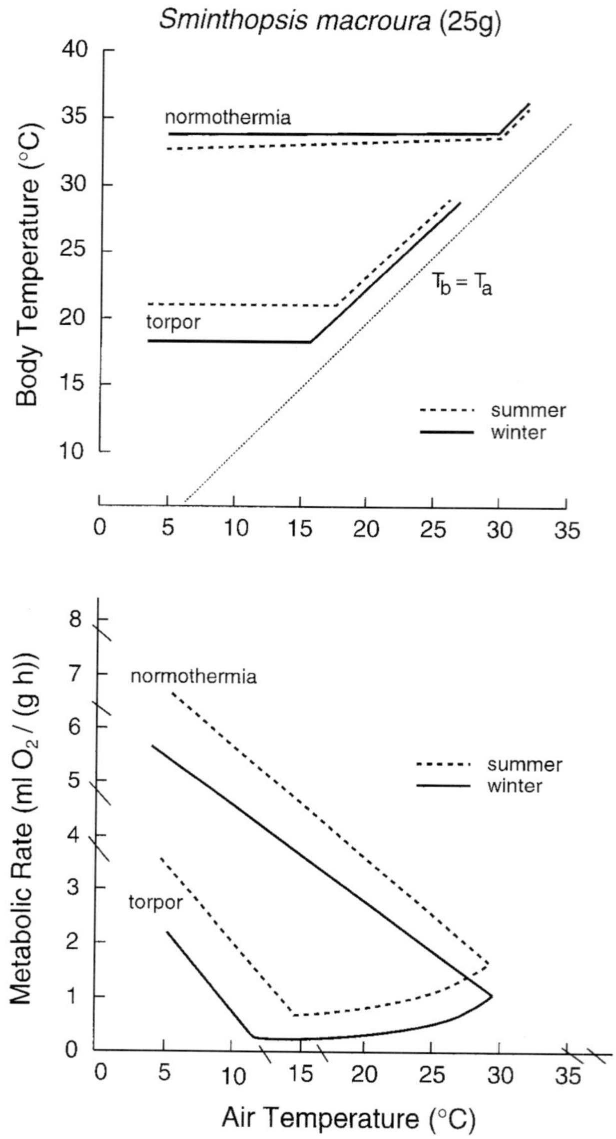

Figure 9 Body temperatures and metabolic rates of stripe-faced dunnarts, *Sminthopsis macroura*, as a function of T_a in summer and winter. In winter body temperatures and metabolic rates in torpid animals were lower than in summer. Energy expenditure of normothermic resting animals in winter was also reduced (from Geiser and Baudinette 1987).

change of torpor patterns in Antechinus is strongly influenced by body size and less by the seasonal change of climate.

Unlike in many rodents, in which reproduction and torpor appear to be mutually exclusive because of hormonal and energetic incompatibilities (Steinlechner et al. 1986; Goldman et al. 1986), several dasyurid marsupials use torpor even when reproductive (Geiser 1996). Torpor has been observed in a wild lactating *Sminthopsis crassicaudata* (Morton 1978). Exposure of *Sminthopsis crassicaudata* to long photoperiod resulted in an increase of testes size, but did not affect torpor patterns (Holloway and Geiser 1996). Female *Dasycercus cristicauda* displayed spontaneous torpor frequently during the period of pregnancy and significantly increased their body mass during this time likely to store fat for the energetically more demanding time of lactation (Geiser and Masters 1994). Thus, it appears that the relatively slow marsupial development allows reproductive females of some species to enter torpor as a small prolongation of gestation may not adversely affect reproductive success, but may improve fitness.

Arousal, often by employing endogenous heat production, ends a torpor episode. The rate of endothermic rewarming from torpor is affected by body mass and maximum rewarming rates are faster in small species (at 10 g ~1°C/min) than in large species (1000 g ~0.25°C/min) (Geiser and Baudinette 1990). Although suggested otherwise, and despite their lack of functional BAT, marsupials do not appear differ in their rate of rewarming from placental mammals (Geiser and Baudinette 1990; Stone and Purvis 1992).

Endothermic arousals can be energetically costly. However, recent evidence suggests that passive rewarming, by the supply of external heat, may be used in endotherms to reduce rewarming costs (Schmid 1996; Geiser et al. 2000; Körtner et al. 2000). This may be especially important in dasyurids that are known to bask (Lovegrove et al. 1999; Geiser et al. 2002). In *Sminthopsis macroura*, passive rewarming reduced costs of rewarming by 36% (Lovegrove et al. 1999) emphasising that the increase in T_a during the time of day dayurids often rewarm, may have an important function in reducing cost of rewarming and thus overall energy expenditure.

Field metabolic rates (FMR)

BMR provides an extensive database for interspecific comparisons of energy use under standard conditions. However, its usefulness for ecological comparisons is somewhat limited, because it is concerned only with one of many aspects of energy expenditure. Moreover, field metabolic rates (FMR), determined by the turnover of doubly-labelled water (DLW), and providing an integrated long-term measure of all metabolic activities of wild animals, often differ considerably from those predicted from BMR measurements (Nagy 1987; Koteja 1991; Degen and Kam 1995; Speakman 1997; Geiser and Coburn 1999).

FMR measurements are available for eight carnivorous marsupials (body mass 13g to 7900 g) and total energy expenditure (kJ/d) increases linearly with body mass on a double-log plot (Green 1997). FMR of carnivorous marsupials, especially in the small species, differ significantly among seasons with the higher values observed in winter (Green 1997). Further, FMR of carnivorous marsupials (*Dasyurus viverrinus*, *Sarcophilus harrisii* and perhaps *Antechinus stuartii*) are significantly raised during late-stage lactation, but not during early lactation (Green 1997). This indicates that increased cost of foraging and milk production during late lactation, when maintenance costs of young become significant, cause an increase in overall energy expenditure in carnivorous marsupials despite the slow developmental rate.

FMR of carnivorous marsupials in comparison to BMR are high (regression equation for marsupials: FMR/BMR = $8.36\ BM^{-0.148}$; $r^2 = 0.66$; data from Green 1997; Hume 1999; compare with placental mammals in Degen and Kam 1995). However, in agreement with the BMR, the FMR/BMR ratio did not differ between carnivorous and omnivorous/herbivorous marsupials (ANCOVA $p > 0.09$). At body masses of ~10 g the FMR is ~6 times BMR whereas at 10,000 g the FMR is only ~2 times BMR. Especially at small body masses, the FMR/BMR ratio of marsupials is well above that of most but not all placental mammals (Degen and Kam 1995; Geiser and Coburn 1999).

Future directions

While laboratory-based knowledge about the thermal biology and energetics of carnivorous marsupials is substantial, most is based on dasyurids and more detailed information on other families of carnivorous marsupials is desirable. Especially data on more South American marsupials (e.g. the Caenolestidae) and the dasyurids from New Guinea are needed. Moreover, the mechanisms of non-shivering thermogenesis in carnivorous marsupials and marsupials in general require further scientific attention. Further, it would be interesting to understand how carnivorous marsupials, in contrast to most other mammals, can overcome apparent hormonal and energetic incompatibilities when entering torpor during reproduction. Finally, more work on thermoenergetics of free-ranging carnivorous marsupials is required to get a better understanding of how their physiological adaptations help them manage energy needs in the wild.

Acknowledgements

I would like to thank Chris Dickman, Tracy Douglas, Gerhard Körtner, Bronwyn McAllan, Louise Streeting and Pat Woolley for help with various aspects of writing this paper, and the Australian Research Council for financial support.

References

Arnold, J.M. (1976), 'Growth and bioenergetics of the Chuditch, *Dasyurus geoffroii*', PhD thesis, University of Western Australia, Perth.

Arnold, J., & Shield, J. (1970), 'Oxygen consumption and body temperature of the Chuditch (*Dasyurus geoffroii*)', *Journal of Zoology, London*, **160**:391–404.

Aslin, H.J. (1983), 'Ooldea dunnart, *Sminthopsis ooldea*', in *Complete Book of Australian Mammals* (ed. R. Strahan), p. 54, Angus and Robertson, Sydney.

Baudinette, R.V. (1972), 'Energy metabolism and evaporative water loss in the California ground squirrel: effects of burrow temperature and water vapor pressure', *Journal of Comparative Physiology*, **81**:57–72.

Baudinette, R.V., Nagle, K.A., & Scott, R.A.D. (1976), 'Locomotory energetics in dasyurid marsupials', *Journal of Comparative Physiology B*, **109**:159–68.

Baudinette, R.V. (1982), 'The energetics of locomotion in dasyurid marsupials', in *Carnivorous Marsupials* (ed. M. Archer), pp. 261–5, Royal Zoological Society NSW, Sydney.

Bell, C.J., Baudinette R.V., & Nicol S.C. (1983), 'Cutaneous water loss at rest and exercise in two species of marsupials', *Australian Journal of Zoology*, **31**:93–9.

Bradley, S.R., & Deavers, D.R. (1980), 'A re-examination of the relationship between thermal conductance and body weight in mammals', *Comparative Biochemistry and Physiology,* **65A**:465–76.

Burbidge, A.A., McKenzie, N.L., & Fuller, P.J. (1983), 'Long-tailed dunnart, *Sminthopsis longicaudata*', in *Complete Book of Australian Mammals* (ed. R. Strahan), pp. 58–9, Angus and Robertson, Sydney.

Carpenter, F.L., & Hixon, M.A. (1988), 'A new function of torpor: fat conservation in a wild migrant hummingbird', *Condor,* **90**:373–8

Clements, F., Hope, P. Daniels, C., Chapman, I., & Wittert, G. (1998), 'Thermogenesis in the marsupial *Sminthopsis crassicaudata*: effect of catecholamines and diet', *Australian Journal of Zoology,* **46**:381–90

Coburn, D.K., & Geiser, F. (1998), 'Seasonal changes in energetics and torpor patterns in the subtropical blossom-bat *Syconycteris australis* (Megachiroptera)', *Oecologia,* **113**:467–73.

Dawson, T.J. (1983), *Monotremes and marsupials: the other mammals*, Edward Arnold, London.

Dawson, T.J. (1989), 'Responses to cold of monotremes and marsupials', in *Animal Adaptation to Cold* (ed. L.C.H. Wang), pp. 255–88, Springer Verlag, Berlin Heidelberg.

Dawson, T. J., & Hulbert, A.J. (1970), 'Standard metabolism, body temperature, and surface areas of Australian marsupials', *American Journal of Physiology,* **218**:1233–8.

Dawson, T.J., & Wolfers, J.M. (1978), 'Metabolism, thermoregulation and torpor in shrew-sized marsupials of the genus *Planigale*', *Comparative Biochemistry and Physiology,* **59A**:305–9.

Dawson, T.J., & Dawson, W.R. (1982), 'Metabolic scope and conductance in response to cold of some dasyurid marsupials and Australian rodents', *Comparative Biochemistry and Physiology,* **71A**:59–64

Dawson, T.J., & Olsen, J.M. (1988), 'Thermogenic capabilities of the opossum *Monodelphis domestica* when warm and cold acclimated: similarities between American and Australian marsupials', *Comparative Biochemistry and Physiology,* **89A**:85–91.

Degen, A.A., & Kam, M. (1995), 'Scaling of field metabolic rate to basal metabolic rate ratio in homeotherms', *Ecoscience,* **2**:48–54.

Dickman, C.R. (1991), 'Use of trees by ground-dwelling mammals: implications for management', in *Conservation of Australia's Forests* (ed. D. Lunney), pp. 125–36, Royal Zoological Society NSW, Sydney.

Dickman, C.R. (1996), 'Vagrants in the desert', *Nature Australia,* **25**:54–62.

Dixon, J.M., & Huxley, L. (1989), 'Mammals of Victoria from the collection and notes of Donald F. Thomson', *Victorian Naturalist,* **106**:4–25.

Douglas, T.A., & Nicol, S. (1993), 'Thermoregulation and the South American grey, short-tailed opossum (*Monodelphis domestica*)', *Australian and New Zealand Society for Comparative Physiology and Biochemistry, Proceedings* (Abstract), **10**:19.

Eldershaw, T.P.D., Ye, J., Clark, M.G., & Colquhoun, E.Q. (1996), 'Vasoconstrictor-induced thermogenic switching in endotherm and ectotherm muscle', in *Adaptations to the Cold: Tenth International Hibernation Symposium* (eds. F. Geiser, A.J. Hulbert, & S.C. Nicol), pp. 311–17, University of New England Press, Armidale, Australia.

Frey, H. (1991), 'Energetic significance of torpor and other energy-conserving mechanisms in free-living *Sminthopsis crassicaudata* (Marsupialia: Dasyuridae)', *Australian Journal of Zoology,* **39**:689–708.

Garland, T. Jr., Geiser, F., & Baudinette, R.V. (1988), 'Comparative locomotor performance of marsupial and placental mammals', *Journal of Zoology, London,* **215**:505–22.

Geiser, F. (1985), 'Physiological and biochemical aspects of torpor in Australian marsupials', PhD thesis, Flinders University, Adelaide.

Geiser, F. (1986a), 'Temperature regulation in heterothermic marsupials', in *Living in the cold* (eds. H.C. Heller, X.J. Musacchia, & L.C.H. Wang), pp. 207–12, Elsevier, New York.

Geiser, F. (1986b), 'Thermoregulation and torpor in the Kultarr, *Antechinomys laniger* (Marsupialia: Dasyuridae)', *Journal of Comparative Physiology B,* **156**:751–7.

Geiser, F. (1988a), 'Reduction of metabolism during hibernation and daily torpor in mammals and birds: temperature effects or physiological inhibition?', *Journal of Comparative Physiology B,* **158**:25–37.

Geiser, F. (1988b), 'Daily torpor and thermoregulation in *Antechinus* (Marsupialia): influence of body mass, season, development, reproduction and sex', *Oecologia,* **77**:395–9.

Geiser, F. (1994), 'Hibernation and daily torpor in marsupials: a review', *Australian Journal of Zoology,* **42**:1–16.

Geiser, F. (1996), 'Torpor in reproductive endotherms', in *Adaptations to the Cold: Tenth International Hibernation Symposium* (eds. F. Geiser, A.J. Hulbert, & S.C. Nicol), pp. 81–6, University of New England Press, Armidale, Australia.

Geiser, F., & Baudinette, R.V. (1987), 'Seasonality of torpor and thermoregulation in three dasyurid marsupials', *Journal of Comparative Physiology B,* **157**:335–44.

Geiser, F., & Baudinette, R.V. (1988), 'Daily torpor and thermoregulation in the small dasyurid marsupials *Planigale gilesi* and *Ningaui yvonneae*', *Australian Journal of Zoology,* **36**:473–81.

Geiser, F., & Baudinette, R.V. (1990), 'The relationship between body mass and rate of rewarming from hibernation and daily torpor in mammals', *Journal of Experimental Biology,* **131**:349–59.

Geiser, F., & Broome, L. (1991), 'Hibernation in the mountain pygmy-possum *Burramys parvus* (Marsupialia)', *Journal of Zoology, London,* **223**:593–602.

Geiser, F., & Masters, P. (1994), 'Torpor in relation to reproduction in the Mulgara, *Dasycercus cristicauda* (Dasyuridae: Marsupialia)', *Journal of Thermal Biology,* **19**:33–40.

Geiser, F., & Ruf, T. (1995), 'Hibernation versus daily torpor in mammals and birds: physiological variables and classification of torpor patterns', *Physiological Zoology,* **68**:935–66.

Geiser, F., & Coburn, D.K. (1999), 'Field metabolic rates and water uptake in the blossom-bat *Syconycteris australis* (Megachiroptera)', *Journal of Comparative Physiology B,* **169**:133–8.

Geiser, F., & Brigham, R.M. (2000), 'Torpor, thermal biology, and energetics in Australian long-eared bats (*Nyctophilus*)', *Journal of Comparative Physiology B,* **170**:153–62.

Geiser, F., Matwiejczyk, L., & Baudinette, R.V. (1996), 'From ectothermy to heterothermy: the energetics of the Kowari, *Dasyuroides byrnei* (Marsupialia: Dasyuridae)', *Physiological Zoology,* **59**:220–9.

Geiser, F., Goodship, N., & Pavey, C.R. (2002), 'Was basking important in the evolution of mammalian endothermy?', *Naturwissenschaften,* **89**:412–414.

Geiser, F., Augee, M.L., McCarron, H.C.K., & Raison, J.K. (1984), 'Correlates of torpor in the insectivorous dasyurid marsupial *Sminthopsis murina*', *Australian Mammalogy,* **7**:185–91.

Geiser, F., Holloway, J.C. Körtner, G., Maddocks, T.A. Turbill, C., & Brigham, R.M. (2000), 'Do patterns of torpor differ between free-ranging and captive mammals and birds?', in *Life in the Cold: Eleventh International Hibernation Symposium* (eds. G. Heldmaier, & M. Klingenspor), pp. 95–102, Springer Verlag, Berlin Heidelberg, Germany.

Gemmell, R.T., & Cepon, G. (1993), 'The development of thermoregulation in the marsupial brushtail possum *Trichosurus vulpecula*', *Comparative Biochemistry and Physiology,* **106A**:167–73.

Godfrey, G.K. (1968), 'Body temperatures and torpor in *Sminthopsis crassicaudata* and *S. larapinta* (Marsupialia: Dasyuridae)', *Journal of Zoology, London,* **156**:499–511.

Goldman, B.D., Darrow, J.M., Duncan, M.J., & Jogev, L. (1986), 'Photoperiod, reproductive hormones, and winter torpor in three hamster species', in *Living in the cold* (eds. H.C. Heller, X.J. Musacchia, & L.C.H. Wang), pp. 331–40, Elsevier, New York.

Gotts, D. (1976), 'Energetics in *Antechinus swainsonii*', MSc thesis, Monash University, Melbourne.

Grant, T.R., & Temple-Smith, P.D. (1987), 'Observations on torpor in the small marsupial *Dromiciops australis* (Marsupilia: Microbiotheriidae)', in *Possums and Opossums* (ed. M. Archer), pp. 273–77, Royal Zoological Society NSW, Sydney.

Green, B. (1997), 'Field energetics and water fluxes in marsupials', in *Marsupial Biology: Recent Research, New Perspectives* (eds. N.R. Saunders & L.A. Hinds), pp. 143–62, University of NSW Press, Sydney.

Guppy, M., & Withers, P. (1999), 'Metabolic depression in animals: physiological perspectives and biochemical generalisations', *Biological Reviews,* **74**:1–40.

Hallam, J.F., & Dawson, T.J. (1993), 'The pattern of respiration with increasing metabolism in a small dasyurid marsupial', *Respiration Physiology,* **93**:305–14.

Hayssen, V.& Lacy, R.C. (1985), 'Basal metabolic rates in mammals: taxonomic differences in the allometry of BMR and body mass', *Comparative Biochemistry and Physiology,* **81A**:741–54.

Hayward, J.S., & Lisson, P.A. (1992), 'Evolution of brown fat: its absence in marsupials and monotremes', *Canadian Journal of Zoology,* **70**:171–79.

Heller, H.C., & Hammel H.T. (1972), 'CNS control of body temperature during hibernation', *Comparative Biochemistry and Physiology,* **41A**:349–59.

Heller, H.C., Colliver, G.W., & Beard, J (1977), 'Thermoregulation during entrance into hibernation', *Pflügers Archiv,* **369**:55–9.

Hinds, D.S., & MacMillen, R.E. (1984), 'Energy scaling in marsupials and eutherians', *Science,* **225**:335–6.

Hock, R.J. (1951), 'The metabolic rates and body temperatures of bats', *Biological Bulletin,* **101**:289–99.

Holloway, J.C., & Geiser, F. (1995), 'Influence of torpor on daily energy expenditure of the dasyurid marsupial *Sminthopsis crassicaudata*', *Comparative Biochemistry and Physiology,* **112A**:59–66.

Holloway, J.C., & Geiser, F. (1996), 'Reproductive status and torpor of the marsupial *Sminthopsis crassicaudata*: effect of photoperiod', *Journal of Thermal Biology,* **21**:373–80.

Hope, R.M., & Godfrey, G.K. (1988), 'Genetically determined variation of pelage colour and reflectance in natural and laboratory populations of the marsupial *Sminthopsis crassicaudata* (Gould)', *Australian Journal of Zoology,* **36**:441–54.

Hope, P., Pyle, D. Daniels, C.B., Chapman I., Horowitz, M., Morley, J.E., & Trayhurn P. (1997), 'Identification of brown fat and mechanisms of energy balance in the marsupial, *Sminthopsis crassicaudata*', *American Journal of Physiology.* **273**:R161–7.

Hudson, J.W. (1973), 'Torpidity in mammals', in *Comparative Physiology of Thermoregulation* (ed. G.C. Whittow), pp. 97–165, Academic Press, New York.

Hulbert, A.J. (1988), 'Metabolism and the development of endothermy', in *The Developing Marsupial* (eds. C.H. Tyndale-Biscoe, & P.A. Janssens), pp. 148–61, Springer-Verlag, Berlin Heidelberg.

Hulbert, A.J., & Rose, R.W. (1972), 'Does the devil sweat?', *Comparative Biochemistry and Physiology,* **43A**:219–22.

Hulbert, A.J., & Dawson, T.J. (1974), 'Standard metabolism and body temperarture of perameloid marsupials from different environments', *Comparative Biochemistry and Physiology,* **47A**:583–90.

Hulbert, A.J., & Reynolds, W. (1981), 'A comparison of strategies of cold acclimation in a marsupial and a eutherian mammal', *Acta Universitatis Carolinae – Biologica,* **1979**:263–5.

Hume, I.D. (1999), *Marsupial Nutrition*, University Press, Cambridge.

Körtner, G., & Geiser, F. (1998), 'Ecology of natural hibernation in the marsupial mountain pygmy-possum (*Burramys parvus*)', *Oecologia,* **113**:170–8.

Körtner, G., & Geiser, F. (2000), 'Torpor and activity patterns in free-ranging sugar gliders *Petaurus breviceps* (Marsupialia)', *Oecologia,* **123**:350–7.

Körtner, G., Brigham, R.M., & Geiser, F. (2000), 'Winter torpor in a large bird', *Nature,* **407**:318.

Koteja, P. (1991), 'On the relation between basal and field metabolic rate in birds and mammals', *Functional Ecology,* **5**:56–64.

Lee, A.K., Woolley, P., & Braithwaite, R.W. (1992), 'Life history strategies of dasyurid marsupials', in *Carnivorous Marsupials* (ed. M. Archer), pp. 1–11, Royal Zoological Society NSW, Sydney.

Loudon, A., Rothwell, N., & Stock, M. (1985), 'Brown fat, thermogenesis and physiological birth in a marsupial, *Comparative Biochemistry and Physiology,* **81A**:815–19.

Lovegove, B.G. (1996), 'The low basal metabolic rates of marsupials: the influence of torpor and zoogeography', in *Adaptations to the Cold: Tenth International Hibernation Symposium* (eds. F. Geiser, A.J. Hulbert, & S.C. Nicol), pp. 141–51, University of New England Press, Armidale Australia.

Lovegrove, B.G. Körtner, G., & Geiser, F. (1999), 'The energetics of arousal from torpor in the marsupial *Sminthopsis macroura*: benefits of summer ambient temperature cycles', *Journal of Comparative Physiology B,* **169**:11–18.

Lovegrove, B.G. (2000), 'Daily heterothermy in mammals: coping with unpredictable environments', in *Life in the Cold: Eleventh International Hibernation Symposium* (eds. G. Heldmaier, & M. Klingenspor), pp. 29–40, Springer Verlag, Berlin Heidelberg, Germany.

Malan, A. (1993), 'Temperature regulation, enzyme kinetics, and metabolic depression in mammalian hibernation', in *Life in the Cold* (eds. C. Carey, G.L. Florant, B.A. Wunder, & B. Horwitz), pp. 241–52, Westview: Boulder.

Martin, C.J. (1902). 'Thermal adjustments and respiratory exchange in monotremes and marsupials. A study in the development of homoeothermism', *Philosophical Transactions Royal Society B,* **195**:1–37.

MacMillen, R.E., & Nelson, J.E. (1969), 'Bioenergetics and body size in dasyurid marsupials', *American Journal of Physiology,* **217**:1246–51.

McAllan, B.M., & Dickman, C.R. (1986), 'The role of photoperiod in the timing of reproduction in the dasyurid marsupial *Antechinus stuartii*', *Oecologia,* **68**:259–64.

McCarron, H.C.K., & Dawson, T.J. (1984), 'Thermoregulatory cost of activity in small dasyurid marsupials', in *Thermal Physiology* (ed. J.R.S.Hales), pp. 327–30, Raven Press, New York.

McNab, B.K. (1978), 'The comparative energetics of neotropical marsupials', *Journal of Comparative Physiology B*, **125**:115–28.

McNab, B.K. (1984), 'Physiological convergence amongst ant-eating and termite-eating mammals', *Journal of Zoology, London* **203**:485–510.

May, E.L. (1996), 'Shivering in cold acclimated kowaris and a role for nonshivering thermogenesis in a marsupial', *Australian and New Zealand Society for Comparative Physiology and Biochemistry, Proceedings* (Abstract), **13**:28.

Morrison, P.R., & McNab, B.K. (1962), 'Daily torpor in a Brazilian murine opossum (*Marmosa*)', *Comparative Biochemistry and Physiology*, **6**:57–68.

Morrison, P.R., & Petajan, J.H. (1962), 'The development of temperature regulation in the opossum, *Didelphis marsupialis virginiana*', *Physiological Zoology*, **35**, 52–65.

Mortola, J.P., Frappell, P.B., & Woolley, P.A. (1999) 'Breathing through skin in a newborn mammal', *Nature*, **397**:660

Morton, S.R. (1978), 'Torpor and nest-sharing in free-living *Sminthopsis crassicaudata* (Marsupialia) and *Mus musculus* (Rodentia)', *Journal of Mammalogy*, **59**:569–75.

Morton, S.R., & Lee, A.K. (1978), 'Thermoregulation and metabolism in *Planigale maculata* (Marsupialia: Dasyuridae)', *Journal of Thermal Biology*, **3**:117–20.

Moyle, D., unpublished, in Reardon, M. (1999), 'Quolls on the run', *Australian Geographic*, **54**:89–105.

Muller, J. (1996), 'Torpor in *Sminthopsis douglasi*', Honours thesis, LaTrobe University, Melbourne.

Nagy, K.A. (1987), 'Field metabolic rates and food requirement scaling in mammals and birds', *Ecological Monographs*, **57**:111–28.

Nicol, S.C., & Maskrey, M. (1980), 'Thermoregulation, respiration and sleep in the Tasmanian devil, *Sarcophilus harrisii* (Marsupialia: Dasyuridae)', *Journal of Comparative Physiology B*, **140**:241–8.

Nicol, S.C., Pavlides, D., & Andersen, N.A. (1997), 'Nonshivering thermogenesis in marsupials: absence of thermogenic response to 3–adrenergic agonists', *Comparative Biochemistry and Physiology*, **117A**:399–405.

Novak, R.M., & Paradiso, J.L. (1983), *Walker's mammals of the world*, Johns Hopkins University Pres, Baltimore.

Opazo, J.C., Nespolo, R.F., & Bozinovic, F (1999), 'Arousal from torpor in the Chilean mouse-opossum (*Thylamys elegans*): does non-shivering thermogenesis play a role?', *Comparative Biochemistry and Physiology*, **123A**:393–7.

Robinson, K.W., & Morrison, P.R. (1957), 'The reaction to hot atmospheres of various species of Australian marsupial and placental animals', *Journal of Cellular and Comparative Physiology*, **49**:455–78.

Rosenmann, M., & Ampuero, R. (1981), 'Hibernación en *Dromiciops australis*', *Archivos de Biologia y Medicina Experimentales (Chile)*, **14**:R294.

Schmid, J. (1996), 'Oxygen consumption and torpor in mouse lemurs (*Microcebus murinus* and *M. myoxinus*): preliminary results of a study in western Madagascar', in *Adaptations to the Cold: Tenth International Hibernation Symposium* (eds. F. Geiser, A.J. Hulbert, & S.C. Nicol), pp. 47–54, University of New England Press, Armidale, Australia.

Serena, M., & Soderquist, T.R. (1989), 'Nursery dens of *Dasyurus geoffroii* (Marsupialia: Dasyuridae), with notes on nest building behaviour', *Australian Mammalogy*, **12**:35–6

Serventy, V., & Raymond, R. (1973), 'Torpidity in desert mammals', *Australia's Wildlife Heritage*, **14**:2233–40.

Smith, B.K., & Dawson, T.J. (1985), 'Use of Helium–Oxygen to examine the effect of cold acclimation on the summit metabolism of a marsupial, *Dasyuroides byrnei*', *Comparative Biochemistry and Physiology*, **81A**:445–9.

Soderquist, T.R. (1993), 'Maternal strategies of *Phascogale tapoatafa* (Marsupialia: Dasyuridae), II juvenile thermoregulation and maternal attendance', *Australian Journal of Zoology*, **41**:567–76.

Song, X., Körtner, G., & Geiser, F. (1995), 'Reduction of metabolic rate and thermoregulation during daily torpor', *Journal of Comparative Physiology B*, **165**:291–7.

Song, X., Körtner, G., & Geiser, F. (1996), 'Interrelations between metabolic rate and body temperature during entry into daily torpor in *Sminthopsis macroura*', in *Adaptations to the Cold: Tenth International Hibernation Symposium* (eds. F. Geiser, A.J. Hulbert, & S.C. Nicol), pp. 63–9, University of New England Press, Armidale, Australia.

Song, X., Körtner, G., & Geiser, F. (1997), 'Thermal relations of metabolic rate reduction in a hibernating marsupial', *American Journal of Physiology*, **272**:R2097–104

Speakman, J.R. 1997. Doubly labelled water: theory and practice. Chapman and Hall, London.

Steinlechner, S., Heldmaier, G., Weber, C., & Ruf, T. (1986), 'Role of photoperiod: pineal gland interaction in torpor control', in *Living in the Cold* (eds. H.C. Heller, X.J. Musacchia, & L.C.H. Wang), pp. 301–7, Elsevier, New York.

Stone, G.N, & Purvis, A. (1992), 'Warm-up rate during arousal from torpor in heterothermic mammals: physiological correlates and a comparison with heterothermic insects', *Journal of Comparative Physiology B*, **162**:284–95.

Thompson, S.D. (1988), 'Thermoregulation in the water opossum (*Chironectes minimus*): an exception that "proves" a rule', *Physiological Zoology*, **61**:450–60.

Thompson, S.D., & Nicoll, M.E. (1986), 'Basal metabolic rate and energetics of reproduction in therian mammals', *Nature*, **321**:690–3.

Tyndale–Biscoe, H. (1973), *Life of Marsupials*, Edward Arnold, London.

Tyndale–Biscoe, H., & Renfree, M.B. (1987), *Reproductive physiology of marsupials*, University Press, Cambridge.

Vogt D.F., Lynch G.R. (1982), 'Influence of ambient temperature, nest availability, hudding and daily torpor on energy expenditire in the white-footed mouse *Peromyscus leucopus*', *Physiological Zoology*, **55**:56–63

Wallis, R.L. (1979), 'Responses to low temperature in small marsupial mammals', *Journal of Thermal Biology*, **4**:105–11.

Wallis, R.L. (1982), 'Adaptation to low environmental temperatures in the carnivorous marsupials', in *Carnivorous marsupials* (ed. M. Archer), pp. 285–90, Royal Zoological Society of New South Wales, Sydney.

Wang, L.C.H. (1989), 'Ecological, physiological, and biochemical aspects of torpor in mammals and birds', in *Advances in Comparative and Environmental Physiology* (ed. L.C.H. Wang), pp 361–401, Springer Verlag, Berlin Heidelberg.

Withers, K.W., & Hulbert, A.J. (1988), 'Cold acclimation in the marsupial *Antechinus stuartii*: Thyroid function and metabolic rate', *Australian Journal of Zoology*, **36**:421–7.

Withers, P.C. (1992), *Comparative animal physiology*, Saunders, Fort Worth.

Withers, P.C., Thompson, G.G., & Seymour, R.S. (2000), 'Metabolic physiology of the north-western marsupial mole, *Notoryctes cauri-*

nus (Marsupialia: Notoryctidae)', *Australian Journal of Zoology*, **48**:241–58.

Wood Jones, F. (1923), *The mammals of South Australia, Parts 1–3*, Government Printer, Adelaide.

Woolley, P.A. (1966), 'Reproduction in *Antechinus* spp., & other dasyurids', *Symposia of the Zoological Society, London,* **15**:281–94.

Woolley, P.A. (1989), 'Nest location by spool-and-line tracking cf dasyurid marsupials in New Guinea', *Journal of Zoology, London,* **218**:689–700.

Wunder, B.A., & Morrison, P.R. (1974), 'Red squirrel metabolism during incline running', *Comparative Biochemistry and Physiology,* **48A**:153–161.

Wunder, B.A., & Gettinger, R.D. (1996), 'Effects of body mass and temperature acclimation on the nonshivering thermogenic response of small mammals', in *Adaptations to the Cold: Tenth International Hibernation Symposium* (eds. F. Geiser, A.J. Hulbert, & S.C. Nicol), pp. 131–9, University of New England Press, Armidale, Australia.

CHAPTER 17

STRESS, HORMONES AND MORTALITY IN SMALL CARNIVOROUS MARSUPIALS

Adrian J. Bradley

School of Biomedical Sciences, Department of Anatomy and Developmental Biology, The University of Queensland, Brisbane, Queensland 4072, Australia

This chapter covers some of the basic ecological, behavioural and physiological changes that have been reported in studies of dasyurid marsupials exhibiting the unusual post-mating male mortality (life-history Strategy I). Rather than attempting to cover all reports, the chapter concentrates upon changes that the author considers to be most relevant in attempting to explain why post-reproductive males in the wild die, yet females survive. Evidence relating to a defect in feedback of corticosteroid hormones in discussed in relation to recent findings for eutherian stress models. The relevance of these studies for explaining the truncated lifespan of Strategy I dasyurid males is discussed and an integrated flow diagram is attempted to synthesise the physiological changes that are likely to occur in the weeks preceding the death of males. Considered together, the adaptive physiological changes that occur during the last few weeks of life of the males are remarkable in enabling them to maximise their reproductive potential prior to a rapid physiological decline that involves stress related dysfunction and pathologies involving renal, gastrointestinal (GI), neuroendocrine and central nervous systems.

INTRODUCTION

While the dramatic post-spawning mortality of Pacific salmon is well known, one of the most remarkable phenomena to occur in the animal kingdom is the annual post-mating death of males that occurs in several species of small dasyurid marsupials in Australia and South America.

Various studies have been carried out during the past 30 years describing the occurrence in several species of small dasyurid marsupials of an annual post-mating mortality of the males (see Woolley 1966; Lee and Cockburn 1985; Lee and McDonald 1985; Cockburn 1989; Tyndale-Biscoe and Renfree 1987). Krajewski, Woolley and Westerman (2000) provide a comprehensive reference list of these small dasyurid species that exhibit the Strategy I life-history pattern (Lee, Woolley and Braithwaite 1982). This is characterised by a monoestrous reproductive pattern, male maturity at 11 months and the disappearance of males from the population within two to three weeks of the commencement of mating. Reproductive strategies in dasyurid marsupials, and the implications of molecular phylogeny, are also treated very comprehensively by Krajewski et al. (2000).

Recent taxonomic reassessments of genus *Antechinus* are relevant to several earlier studies of dasyurid marsupials in Eastern and SE Australia. Studies of *A. stuartii* carried out at Mount Glorious in

SE Queensland would now be regarded as studies of *A. subtropicus* (Van Dyck and Crowther 2000) while studies of *A. stuartii* from SE Australia, such as those ecophysiological studies conducted near Warburton (Bradley et al. 1980), would now be regarded as investigations of *A. agilis* in the light of reassessments by Dickman et al. (1998) and Sumner and Dickman (1998). Behavioural studies carried out on *A. stuartii* in forests near Canberra would be regarded as studies of *A. flavipes* (Sumner and Dickman 1998). To avoid confusion in references to *Antechinus* in this chapter, the new species names will be included in brackets after the species names used in the original citation.

Braithwaite and Lee (1979) observed that the species exhibiting the phenomenon are restricted to coastal regions of Australia where environments are predictable and seasonal. They considered a predictable environment to be important to provide resources for reproduction in species where a prolonged period was required to raise offspring from conception to weaning. Because only one reproductive period was possible under these conditions, there were benefits in investing heavily in their first bout of reproduction. This argument has been extended to provide an explicit physiological model for die-off in *Antechinus* (Lee and Cockburn 1985a, b; Lazenby-Cohen and Cockburn 1988; Cockburn 1992, 1997).

Studies on *Phascogale tapoatafa* (Cuttle 1978, 1982; Soderquist 1993; Soderquist and Ealey 1994) suggest that as with a number of *Antechinus* species, all males die at the conclusion of their first breeding season. Furthermore, Dickman and Braithwaite (1992) have described a post-mating mortality of males in the dasyurid marsupials, *Dasyurus hallucatus* and *Parantechinus apicalis*, the former being the largest dasyurid species to show this unusual phenomenon.

This life-history pattern also occurs in two South American marsupials, *Monodelphis dimidiata* (Pine 1994) and the mouse opossum *Marmosa incana* (Lorini et al. 1994). This life-history pattern has been classified as Strategy 1 (Lee, Woolley and Braithwaite 1982). Examination of West Australian Museum records for *Phascogale calura* in 1980 (Bradley unpublished) indicated that no adult males persisted in the records after the reproductive period, suggesting a post-mating male mortality did occur. This was reported by Kitchener (1981) and subsequently confirmed in field studies (Bradley 1982a; 1990a, b; 1995; 1997). The phenomenon was commented upon by Lee et al. (1982) and Lee and Cockburn (1985).

The Strategy I life-history pattern also occurs in some salmon species in which the death of males closely follows spawning. In these salmon species, an elevation in plasma cortisol concentration has been implicated in the stress related demise of males. In the Strategy I dasyurid males, during the mating period, there also occurs a marked elevation in plasma cortisol concentration however a concurrent rise in plasma testosterone and an androgen induced decrease in the plasma concentration of corticosteroid binding globulin (CBG) results in a marked increase in free (biologically active) cortisol.

This disappearance of males from the population at the end of their first breeding season (Strategy I males) has been shown to be a consequence of stress characterised by an increase in the plasma glucocorticoid (GC) concentration (Barnett 1973; Bradley, McDonald and Lee 1975; Lee, Bradley and Braithwaite 1977; Bradley and Monamy 1991) and exacerbated by an androgen dependent decrease in plasma CBG concentration (Bradley, Mc Donald and Lee 1980; McDonald et al. 1981; Bradley 1987). The phenomenon has been related to negative nitrogen balance (Woollard 1971), anaemia (Cheal et al. 1976; Bradley 1990a), GI haemorrhage and disease (Moore 1974; Bradley 1977, 1985, 1987; Barker et al. 1978; Bradley et al. 1980) and immune suppression and disease (Cheal et al. 1976; Bradley 1977, 1987; Barker et al. 1978, 1991; Bradley et al. 1980; see also review by Lee and McDonald 1985), and to unspecified degeneration of major organs (Williams and Williams 1982). McAllan, Roberts and O'Shea (1996) have also implicated dramatic changes in renal morphology of males in the male mortality.

Wexler and Greenberg (1978) described, in male Sprague-Dawley rats kept in active stud service, the development of abnormal metabolic and pathophysiological changes, many of these involving the cardiovascular system. Furthermore, studies by Von Holst (1972; 1986) have clearly shown the pathophysiological effects of stress in tree shrews *Tupaia belangeri* and in this species renal failure has been reported to be a major cause of death.

QUESTIONS

A. subtropicus males and females captured three months before the natural disappearance of males in the wild and treated to remove ecto-and endo-parasites have survived for up to five years (authors observations), which is in striking contrast with the natural lifespan of approximately 11.5 months for males. How might we explain this difference between realised and potential longevity of males? Why is such a high concentration of free cortisol in the plasma tolerated in males when feedback mechanisms might be expected to terminate the short-term response and allow the plasma free cortisol concentration to return to baseline levels? What is the proximate cause of death and how might this be related to the physiological changes that occur as part of this unusual life-history strategy?

These are but some of the interesting questions that may be raised in respect of the life-history of Strategy I dasyurid marsupial males.

PATHOGENESIS OF GASTRIC ULCERS

A significant advance in our knowledge of the pathogenesis of gastric ulcer in man has been the recognition that the bacteria, *Helicobacter pylori*, was causally related to the development of gastritis and peptic ulceration (see Marshall, McCallum and Guerrant 1991). In pursuit of a bacterial aetiology for GI ulcers in Strategy I dasyurids, numerous samples of gastric tissue from *A. flavipes, A. agilis* and *A. subtropicus* were examined histologically using the Warthin-Starry and Giemsa staining methods. The bacterium was not identified in any of the tissue samples and it seems that a causal association between *Helicobacter* sp. and the pathogenesis of gastric ulcers in these marsupials is unlikely.

Most authors now agree that GU pathogenesis is an example of a multifactorial disease process and many consider the disease to have a 'bio-psychosocial' origin (Leverstein 1998; Overmier and Murison 2000) with a combination of biological and psychological or social factors contributing to GI pathology.

Clinicians have reported that in both human and animal studies, high does of exogenous GCs increase the incidence of gastric ulceration (Bradley et al. 1975; Messner et al. 1983) and based upon this notion it was concluded that the increased release of exogenous GCs observed during the stress response was also ulcerogenic. Indeed, the comparison of experimentally stressed and non-stressed animals will often result in a significant correlation between endogenous steroid production and stress ulceration (Murphy et al. 1979; Filaretova et al. 1999).

While most authors agree that GCs do not have a direct effect upon the gastric mucosa (Kuwayama et al. 1989) there are many views about the ways in which indirect links may operate. GCs may inhibit proliferative repair in *A. flavipes* (Bradley and Richardson unpubl. obs.) and gastric mucosal microcirculation may be compromised (Kuwayama et al. 1989; Caparni de Kaski et al. 1995). The gastric microcirculation is a primary factor in the maintenance of gastric mucosal integrity, and in the protection of the mucosa from *H. pylori*, non-steroidal anti-inflammatory agents, and stress.

Endoscopic studies of GUs carried out on human patients indicate that mucosal blood flow is decreased in nearly all regions of the stomach during the active ulcer stage of the disease (Kawano et al. 1991) compared to normal controls. In contrast, during the ulcer healing stage the blood flow in the area of the ulcer increases markedly. These authors also noted that intractable ulcers, which did not heal within three months of treatment, showed virtually no increase in mucosal blood flow in the ulcerated region. Thus a decrease in the microcirculatory capacity of the stomach would appear to contribute to the development of gastric ulceration, whilst a return to normal levels of blood supply is clearly implicated in ulcer repair.

Decreased mucosal blood flow is a typical aspect of the GI stress response (Svanes et al. 1984). Rats exposed to a water-restraint stress regime are reported to experience a decrease in blood velocity in the submucosal microvessels, accompanied by the appearance of gastric erosions (Filaretova et al. 1999). Curiously it was found in this study that GC deficiency exacerbated the effects of stress on the microcirculation and that GC replacement eliminated the effects of GC deficiency, implying that GC release during stress may have a role in maintaining gastric microcirculation. The effects of stress on the normal structure and function of the gastric microcirculation clearly warrant further investigation.

Because of the ethical issues surrounding the induction of GUs in humans, animal models have greatly aided in our understanding of the pathophysiology of GU disease. A criticism of many earlier studies is that they employed methods to induce ulceration that were unrelated to the natural experience of the animal species under investigation. Some recent studies have recognised this problem and used more naturalistic animal models that allow for experimental analysis of the aetiology and symptomatology of ulcer disease that is more applicable to human GU research (Koolhaas et al. 1997).

During the stress response involving the HPA axis, stimulation of the sympathetic nervous system, a finding that would be expected as part of the stress response, would also contribute to gastric mucosal ischemia. GCs that are clearly elevated during the last weeks of life of Strategy I males are known, in eutherian species, to have a permissive effect that assists catecholamines and other vasoconstrictors to exert their full actions (Bondy 1981; Krakoff 1988).

Haemorrhage from gastric ulcers has been reported in several non-human species. While gastric ulcers do appear in natural populations of wild mammals, death is not the only consequence. Evidence of this was provided by Stemmerman and Hayashi (1969) who described the finding of healing gastric ulcers in individuals in a wild population of Hawaiian Feral Mongoose. Several reports exist for the occurrence of GI ulcers and haemorrhage within the Marsupialia. These include *Sarcophilus harrisii* (Hamerton 1935, 1938) various macropodid marsupials (Hamerton 1934, 1935, 1938, 1939) and *Didelphis virginiana* (Sherwood et al. 1968).

In spite of the current thinking about the absence of a direct link between GCs and gastric ulcer in man, the situation in small marsupials is not so clear. Studies on the pathogenesis, prevention and treatment of gastric ulcer have been examined in a range of mammals that include rodents, lagomorphs, primates and carnivores (Harding and Morris 1977; Kitagawa, Fujiwara and Osumi 1979; Hosoda, Ikedo and Saito 1981; Kivilaakso et al. 1981). These studies, and those carried out in more recent years, indicate that the pathogenesis of stress ulcers is multifactorial and may involve vascular, neurogenic, and central nervous system (CNS) mechanisms (Hernandez 1990).

An obvious question to ask is what evidence exists to support a direct link between glucocorticoid concentration and gastric ulceration in these species. The evidence for a direct link is equivocal. The administration of glucocorticoid for a prolonged period in male *A. stuartii* caused a dose-related mortality (Bradley et al. 1975) with haemorrhage from gastric ulcers being the most common cause of death. However another cause of death was the development of liver abscesses resulting from *Listeria monocytogenes* infection that in turn was most likely aggravated by immunosuppressive effect of the GCs. The administration of GCs to *A. flavipes* (Bradley and Richardson unpublished) resulted in a higher mortality in the glucocorticoid treated groups than in the controls, however the result was again rather equivocal. The GC caused a dose-related decrease in the rate of gastric-cell renewal determined by 3H-thymidine labelling followed by autoradiography (Martin and Menguy 1970). In Strategy I males, starvation plus sympathetic nervous induced gastric mucosal ischemia could effectively combine to threaten both gastric mucosal function and structure, and to compromise repair.

While an increase in parasite/microbiological activity associated with glucocorticoid induced immune suppression in these males has been discounted as the proximate cause of death (Lee et al. 1977), haemorrhagic peptic ulcers seem ubiquitous. In the Strategy I species in which the gastric mucosa has been examined almost all terminal males show some evidence of ulcers, often with evidence of associated acute bleeding. An almost universal finding in post-reproductive males that are found either in a debilitated or hypothermic condition, is the presence of haemorrhagic ulcers (Bradley 1977; Barker et al. 1978; Bradley McDonald and Lee 1980; Bradley 1987; Bradley 1997). While the pathogenesis of these gastric ulcers is unclear, high cortisol concentrations can be shown to reduce the rate at which gastric cell renewal can take place (Bradley and Richardson unpublished) and short periods of hypothermia can predispose males on a restricted food intake to the development of haemorrhagic ulcers (Bradley unpublished).

This observation from studies with rats fits quite well into a model that I envisage to describe the pathogenesis of gastric ulcers in small Strategy I dasyurid marsupials. It is well known that males captured in the wild within about two weeks of the disappearance of males are unlikely to survive, even if housed alone in a quiet environment with adequate shelter, food and water. Females, on the other hand, generally survive with losses being uncommon. Necropsies carried out on males that die invariably reveal extensive haemorrhaging from gastric ulcers. These present as diffuse surface erosions that do not appear to involve larger blood vessels in the adventitia and do not result in perforation of the stomach wall. It is suggested that males, near the time of their disappearance from the natural population, are practising 'physiological brinkmanship' and that the imposition of an additional stressor such as handling, captivity, mild hypothermia or even a threatening olfactory experience (Toftegaard et al. 1999, 2002), may be sufficient to stimulate the stress response and compromise the integrity of an already susceptible gastric mucosa.

Studies of larger dasyurid marsupial species (Dickman and Braithwaite 1992; Oakwood 1999; Oakwood et al. 2001), suggest that life-history variation in dasyurid marsupials is more common than has been previously suspected. My own observations of the GI tract of males of these larger dasyurid marsupials at necropsy is that the incidence of haemorrhage from gastric ulcers is common. Perhaps a larger body mass means that insidious blood loss from a GI lesion is less likely to be lethal and the extra energy reserves available to an animal of larger body mass might become significant for repair and survival. Furthermore, hypothermia would be less likely to develop as body mass increases.

Social stressors have been implicated in combination with hypothermia and restricted diet in the formation of GI lesions (Filaretova et al. 1998). While psychogenic stressors such as restraint are known to promote ulcer formation in other small mammal models (Filaretova et al. 1998) it is suggested that social stressors may be very significant in dasyurid marsupials and, when combined with hypothermia and restricted diet, form a potent ulcerogenic combination. The semiochemical 2,6-dimethylpyrazine, present in the urine of intact *A. subtropicus,* (Toftegaard, Moore and Bradley 1999) is a potent stressor for other males, elevating both plasma cortisol and catecholamines (Toftegaard 1999; Toftegaard and Bradley, in prep; see chapter by Toftegaard and Bradley). The increase in density suggested in Fig. 1 reflects a likely increase in *apparent density* that would be experienced by males during the reproductive period as they move around and deposit urine containing androgen dependent semiochemicals such as 2,6-dimethylpyrazine.

In a study of the pathogenesis of gastric ulceration in *A. subtropicus* (Hanlon 2001) reported that an elevation of free cortisol, combined with elevated testosterone, hypothermia and a psychological stressor (restraint), has a profound effect upon the gastric microvasculature of castrate males. Using intravascular fluorescent-albumin as a marker, ischaemia and leakage of fluid from capillaries into the extracellular compartment occurred, changes that are known to immediately precede the formation of gastric mucosal lesions.

It might therefore be predicted that HPA and sympathetic nervous system activation, following exposure to a range of stressors, would cause ischemia and subsequently promote the formation of gastric lesions (see Fig. 1).

Strategy I dasyurid males, which all appear to exhibit a high incidence of haemorrhage from stomach ulcers, and which are likely to experience psychosocial stress of olfactory origin, may provide very useful new animal models for future studies on the non-microbial pathogenesis of peptic ulcer in mammals.

CONTROL OF CORTICOSTEROID BINDING GLOBULIN CBG

In all Strategy I species that have been studied in some detail, castration causes an elevation in plasma CBG, an effect that is reversed by administration of testosterone (Bradley et al. 1980; McDonald et al. 1981). This inverse relationship between testosterone and CBG does not appear to exist in non-Strategy I species such as *Sminthopsis crassicaudata*. The absence of sex hormone binding globulin (SHBG) from the blood of all dasyurid marsupials (Bradley 1977; Sernia, Bradley and McDonald 1979; Bradley 1982b) would also appear to be important ensuring the exposure of tissues to high concentrations of androgen before and during the reproductive period.

A defect in glucocorticoid regulation

While the pituitary and hypothalamus are major sites involved in the regulation of GC feedback, the hippocampus plays a part by exerting an inhibitory effect on the HPA axis (Wilson 1985). This area of neuroendocrinology has received a great deal of attention, both in man and other mammals, since it was recognised that adrenocortical disruption may also be involved in ageing, depression and neurodegenerative diseases of man such as Alzheimer's disease.

It is now well known that in terminal semelparous dasyurid males *A. swainsonii* and *A. flavipes* (McDonald et al. 1986; Bradley 1990b) and in *A. subtropicus* (Bradley unpublished), a defect develops in glucocorticoid feedback. This defect seems to be an integral part of the physiological breakdown leading to the demise of males however in the absence of further information from studies on marsupials, one must look to studies on other vertebrates to explain why it might occur.

An extensive literature stretching back several decades has shown that prolonged stress or prolonged exposure to GCs, the adrenal steroids secreted during stress, can have adverse effects on the rodent hippocampus. Recent findings suggest a similar phenomenon in the human hippocampus associated with many neuropsychiatric disorders (Sapolsky 2000). GCs typify the double-edged quality of the stress-response. When secreted transiently during the brief physical challenge typical of mammalian stress, GCs aid survival. Glucocorticoids mobilise energy, increase cardiovascular tone, suppress non-essentials such as growth, tissue repair, and reproduction, and potentiate aspects of immunity while preventing activity to the point of autoimmunity (Munck et al. 1984; Sapolsky et al. 2001). However, excessive GCs can have deleterious consequences, including increased risks of hypertension, insulin-resistant diabetes mellitus, amenorrhea, impotency, ulcers, and immune suppression (Munck et al. 1984; Sapolsky et al. 2001). In addition, GCs can also have adverse effects on the nervous system, disrupting learning and memory, and synaptic plasticity (McEwen and Sapolsky 1995). Glucocorticoids can also have adverse morphologic effects, particularly in the hippocampus, a primary neural GC target site, with plentiful GC receptors. These effects include impairing neurogenesis, causing atrophy of dendritic processes and, sometimes, neurotoxic effects. The hippocampus has a variety of functions, but its best-documented function is in the realm of learning and memory. A vast literature shows that the structure plays a critical role in the consolidation of short-term into long-term explicit memory (Squire 1987).

GLUCOCORTICOIDS AND THE HIPPOCAMPUS

There is now considerable evidence that the hippocampus acts as a control site for the HPA axis (Jacobson and Sapolsky 1991; Van Eekelen and De Kloet 1992). Together with the amygdala and septal nuclei, the hippocampus is part of the limbic system that is concerned with olfaction and feeding behaviour (Goya et al. 1995). Furthermore the hippocampus, which is a primary target for GCs, is strongly implicated in the normal functions of mood, memory and learning. Continuous administration of GR antagonist improves cognitive function, while phasic blockade of brain GR function causes a cognitive deficit (Oitzl 1998). Capacity for learning new tasks may effectively be assessed in small mammals using the Morris water maze (Morris 1984) or Y mazes.

While it is accepted that hormones such as CRH or cortisol can be destructive when they are activated for long periods of time, or when the body is unable to terminate their production (Sapolsky 1992; McEwen 1998), another important advance in our knowledge of the stress response has been the introduction of the concept of *allostasis* (McEwen and Stellar 1993). The novelty of this concept of allostasis is that there may be no physiological set-point, with the possibility that any set-point is both fluid and changing (Sterling and Eyer 1981). The maintenance of life depends on the capacity of the organism to sustain its equilibrium via allostasis, in essence the ability to achieve stability through change.

One can envisage, within Strategy I males during the reproductive period, that the 'set-point' for free cortisol is considerably elevated, providing a short-term advantage, but in the longer-term leading to the demise of all males. The alteration of the set point may provide an adaptive advantage but accelerate GC mediated pathologies and senescence.

Sapolsky has significantly advanced our knowledge of the effects of glucocorticoids on the hippocampus (Sapolsky et al. 1984; Sapolsky 1985a, b; 1986a). A more subtle effect of GCs upon the hippocampus may be atrophy of dendritic processes. This may compromise connectivity and function but not result in overt neuronal loss. Complexly arborised dendritic processes are prerequisites for the formation of elaborate neural networks, and GCs can atrophy such processes. Prolonged stress decreases numbers of apical dendritic branch points and the length of apical dendrites in hippocampal CA3 neurons in rodents and nonhuman primates (Sapolsky 1996; Regan and McEwen 1997). This effect is GC-dependent and can emerge after a few weeks of GC overexposure in rodents. Moreover, such atrophy correlates with impaired explicit memory. Thus, transient GC overexposure can alter neuronal morphologic features in a manner deleterious to explicit memory and with the abatement of the stressor or GC exposure, there is re-growth of dendritic processes. Psychosocial stress induced damage to the hippocampal dendritic tree, interneurons, neurogenesis or glia have been suggested (Lucassen et al. 2001) in the three shrew *Tupaia belangeri* where stress differentially affects apoptosis in hippocampal subregions.

Current studies in my laboratory on the brain of *A. subtropicus* employ confocal microscopy and selective staining of dendritic spines to assess the effect of various stressors on dendritic spine density. These changes are related to alterations in cognitive function using open field and radial-arm mazes, and the Morris water maze. Preliminary findings indicate that the short-term memory of *A. subtropicus* males is impaired late in their life in the wild and also with age in captivity. In contrast, the vomeronasal system appears to not undergo any morphological decline (Aland pers. comm.).

AGEING AND NEURODEGENERATIVE DISEASE

The hippocampus is known to be a focus of damage in several neurodegenerative conditions (Scully and Otten 1995). The hippocampus is the only site in the brain where a major loss of corticosteroid receptors has been observed (Angelucci et al. 1980; Sapolsky et al. 1983; Reul et al. 1988; De Kloet 1992). In the progressive age-related depletion, approximately half the corticosteroid receptors are lost. It is also clear that further significant changes occur with ageing while in some animal models, age related changes that are capable of influencing learning and memory, occur in senescent individuals.

The hippocampus is also a major target for damage in Alzheimer's disease in man (Pasquier et al. 1994). Elevated cortisol concentrations, increased urinary free cortisol excretion and defective HPA suppressibility have been reported in many patients with Alzheimer's disease (see review Seckl and Olsson 1995) and correlate with hippocampal damage (de Leon et al. 1988). In Alzheimer's disease, hippocampal GR and MR gene expression are maintained at control levels in surviving neurons, despite a generalised loss of neuronal gene expression and in the face of marked GC hypersecretion (Seckl et al. 1993). Elevated GC concentrations in Alzheimer's disease, perhaps as a consequence of neuronal loss in the hippocampus, might exert an even more potent deleterious effects via the maintained density of GC receptors in the remaining neurons (Goya et al. 1995). The attempt to maintain the hippocampal GR and MR at control levels may reflect a loss of plasticity following disruption of neurotransmitter inputs, since loss of afferent and efferent projections in Alzheimer's disease may isolate the hippocampus neuroanatomically (Hyman et al. 1984).

Preliminary studies (author in prep.) indicate that the hippocampus of *A. subtropicus* does possess saturable Type II glucocorticoid receptors, that the number of these receptors declines during the two weeks preceding the disappearance of males, and that morphological changes consistent with degeneration occur in the dentate gyrus and CA3 region of the hippocampus. It appears that beta-amyloid plaques and neurofibrillary tangles that are associated with neurodegenerative diseases in humans also occur in the brain of some Strategy I dasyurid males late in their life (for *A. stuartii* – McAllan and Norris pers. comm. and for *A. subtropicus* – Bradley unpublished). However, retrospective examination of brains from *A. stuartii* (*A. agilis*) shows that GC administration alone is insufficient to induce the formation of beta-amyloid plaque (Bradley and Masters unpublished).

A MARSUPIAL MODEL FOR STUDIES OF AGING AND SENESCENCE

Koolhaas et al. (1997) point out that our current understanding of the physiological mechanisms underlying stress related disorders is based not only on neuroendocrine and pharmacological studies in human patients but also on experimental studies in a wide variety of animal models. They indicate that while the contribution of social and physical environmental stress to the development of disease is of a general biological phenomenon in animals and man, one may criticise the validity of many of the current stress models. Koolhaas et al. (1997) comment that most of the animal models use stressors that bear little or no relationship to the biology of the species and suggest that more naturalistic models, that allow an experimental analysis of the aetiology and the temporal dynamics of the disease, should be used. Furthermore, in a review of the literature on stress and aging, Stein-Behrens and Sapolsky (1992) considered three model systems to be of particular significance for studies of stress, disease and mortality:

- programmed senescence in marsupial mice and fish as mediated by GC excess
- GC hypersecretion in rats and its role in damaging the aging brain, and
- potential human and primate adrenocortical dysfunction during ageing.

The fact that an adrenal hypersecretion results from ageing of the hippocampus through receptor depletion and/or neuron loss seems to provide an explanation for the inability of aged organisms to cope with stress. Although the HPA axis is manifestly altered with ageing, and the aged hippocampus is apparently involved in this altered response, the question remains: *Are GCs causally related to the process of ageing of the hippocampus?* The observations that chronic GC exposure is associated with neuronal loss in the hippocampus strongly implicate GCs in accelerating some aspects of aging (Goya et al. 1995). Consistent with this concept of hormonal acceleration of the ageing process Bradley (1997) argues that in another small dasyurid marsupial, *P. calura*, which shares the unusual truncated male life-history pattern, the strategy employed by males may be described by an *adaptive-stress senescence model*. This hypothesis draws upon the *senescence hypothesis* (Boonstra 1994) that represents a very significant advance in our understanding of factors controlling small mammal population in the northern hemisphere. In essence the *adaptive-stress senescence hypothesis* advocates an acceptance that for two individuals of the same biological age, one may be senescent as a consequence of experiencing a different hormonal *millieu* during a significant, albeit short part of its life-history. This clearly depends upon how ageing is defined and it is suggested that strict adherence to temporal-age paradigm is too restrictive, particularly when one considers species such as semelparous dasyurid marsupials with the associated hormonally distinct strategies which are employed by females and males.

This hypothesis suggests that stress related changes elevate plasma GCs having short-term beneficial effects upon intermediary metabolism to mobilise energy reserves to enable a heavy investment in a once only reproductive period, but in the long term the elevated GCs have deleterious effects, ultimately promoting pathogenic changes that finally lead to the death of males. While in the short term the physiological stress response is adaptive, in the longer term it leads to adverse changes that are consistent with senescence, in particular those effects that occur in the hippocampus affecting both morphology and function.

Although Barnett (1974) was unable to demonstrate accelerated ageing of male *Antechinus stuartii* relative to females of the same biological age using, as a criterion, the contraction of tail tendon collagen, it is postulated that post-breeding male *P. calura* are effectively senescent relative to females because of damage to critically sensitive brain areas. Males do show an impairment of their glucocorticoid feedback control (Bradley 1990b) and it is suggested that this involves damage to the hippocampal–hypothalamic–pituitary–adrenocortical axis as a consequence of exposure of hippocampal glucocorticoid receptors to high concentrations of free cortisol (Sapolsky 1996) during an adaptive elevation of the free cortisol concentration (Bradley 1987). A relationship between glucocorticoids and hippocampal damage has yet to be thoroughly investigated in marsupials.

Although the weight loss in male *P. calura* is not as profound as that which has been described in other semelparous dasyurid marsupial males, a thinning of the pelage does occur. Because this is accompanied, toward the end of the breeding season, by an increase in the aggressiveness of females toward males, and males which moved greater distance between captures at this time showed a greater incidence of rump and thigh fur loss and evidence of bite marks, it is reasonable to postulate that, at least for *P. calura* and possibly for other species, the female aggression and the male dispersal are causally related. Bearing in mind that these events occur at a time of year when the nights are quite cold, and the males which move long distances are generally trapped at some distance from potential nest trees, it is also reasonable to postulate that these males are exposed to a considerable thermoregulatory challenge.

A consistent observation with *P. calura,* and indeed with all other semelparous dasyurid marsupial males (pers. obs.), is that when individuals are removed from traps in a hypothermic state, they will invariably die within 24 hours in spite of all measures that are taken and the proximate cause of death can usually be identified as acute haemorrhage from gastric ulcers. Similar evidence of bleeding from GI ulcers occuring up to 20 days before the death of male *P. tapoatafa* was reported by Soderquist and Ealey (1994). Furthermore post-mortem investigation of four freshly dead male *P. tapoatafa* also revealed that while ectoparasites were not apparent in the GI tract, lungs or heart, all showed evidence of peptic or intestinal ulceration.

Behavioural changes

In some dasyurid species a change in behaviour between the sexes during the life-history can result in an increase in plasma cortisol in the male. In *Phascogale calura* before and during the mating period the males are clearly dominant in their pursuit of females, however when females are pregnant they become very aggressive toward males (Bradley 1997). In another semelparous species, *Phascogale taqpoatafa*, Soderquist and Ealey (1994) has also reported an increase in female aggression toward males during pregnancy. This is considered to be a potent stimulus for the post-mating terminal dispersal of males (Soderquist and Ealey 1994; Bradley 1995, 1997).

Braithwaite (1974, 1979) reported that behavioural changes occur during the life-history of *A. stuartii* (*A. agilis*). Further behavioural studies on *A. stuartii* (Scott 1987) reported that agonistic encounters between caged males caused in an elevation of plasma cortisol, but that the experience of mating could result in a lowering of plasma cortisol. Lazenby-Cohen and Cockburn (1991) suggested that after a resolution of agonistic interactions early during the life-history, males would cohabit with other males during mating.

While contrived encounters between caged individuals have been criticised and may not give results that can reliably be used to determine the nature of interactions in the field, the difficulty in observing natural interactions explain a general deficit in our knowledge in this area. Further studies that relate social and physiological changes are clearly warranted.

Starvation and negative nitrogen balance

It is curious that at a time when their plasma free cortisol is elevated, *A. stuartii* males do not appear to seek and ingest food during the last days before their disappearance. This may be deduced from a common finding that the stomachs of males are empty. However, stress has been reported to suppress feeding even in food deprived animals (Krahn et al. 1986), an effect that may be mediated by CRH, a potent anorexic agent (Sapolsky et al. 2001). CRH agonists are known to block the anorexic effect of stress (Arase et al. 1988). If hippocampal inhibition of CRH release is diminished in Strategy I dasyurid males CRH release may be sufficient to exert this type of anorexic effect and explain the decreased food intake. During the mating period males undergo a significant reduction in body mass that is consistent with a state of negative nitrogen balance. However the apparent decrease in body mass may be disguised by a glucocorticoid induced increase in total body water TBW (Nagy et al. 1978).

It is reasonable to predict that the survival time for a starving animal should be related to body size and condition at the commencement of fasting. At the commencement of the reproductive period males are in good condition. Because of surface to volume considerations there will be a relationship between body size and development of hypothermia during nitrogen mobilisation/negative nitrogen balance. The threshold would be affected by predictable adverse climatic conditions during the terminal phase and also by altitude and latitude.

It is thought that Strategy I males during the mating period, spend little time searching for food sources, instead devoting their energy to locating females and mating. There is a clear sexual dimorphism in these species, with males being considerably larger than females at the beginning of the mating period. Males therefore have some potential energy reserves but to fully exploit these, males would need to starve themselves. In *A. stuartii* during late winter it is difficult for males to procure food and engage in mating because mating takes place near communal aggregations that are often a considerable distance from male foraging areas (Lazenby-Cohen and Cockburn 1988, 1991; Cockburn and Lazenby-Cohen 1992; Cockburn 1997). There is also evidence that the males enter negative nitrogen balance (Woollard 1971) during which proteins are presumably being utilised for energy sources after carbohydrate and fat reserves have been expended. *A. stuartii* (*agilis*) males exhibit haemoglobinuria, ketonuria, proteinuria, glycosuria, particularly during the terminal two weeks (Bradley 1977). Consistent with a disturbance of intermediary metabolism associated with starvation is the observation that some male *A. stuartii* (*A. agilis*) in this terminal phase show evidence glycosuria, haemoglobinuria , proteinuria, and sometimes, ketonuria (Bradley 1977). The finding of ketonuria is consistent with the metabolic alterations that occur during starvation while the proteinurias and haemoglobinurias are consistent with the renal glomerular damage that has been reported by McAllan et al. (1996) and which may be reproduced in the laboratory by exogenously administering a combination of testosterone and cortisol (McAllan et al. 1997, 1998). A finding of glomerulonephritis and tubular necrosis, affecting in particular the proximal convoluted tubule, is common in male *A. subtropicus* during the last days before their disappearance from the population (author in prep.)

To explain the life-history of semelparous dasyurid species in general, and of *P. calura* in particular, Bradley (1997) has proposed the '*adaptive stress senescence hypothesis*' to explain that males become prematurely senescent as a consequence of exposure to high concentrations of free cortisol for a protracted period prior to their disappearance from the population. The critical factors that seem to predispose males to the development of pathological changes are outlined in Table 1.

Periodic semelparity or plasticity of life-history

Some dasyurid species exhibit a high or even complete male mortality during some years while in other years many males may survive beyond the breeding period and appear to be capable of breeding in the next breeding period. While *Dasyurus hallucatus* may show an incomplete male mortality in some years (Schmitt et al. 1981) they disappear in others (Dickman and Braithwaite 1992) or the phenomenon may occur annually (Oakwood et al. 2001). *D. hallucatus* would appear to be the largest dasyurid species in which a complete male mortality may sometimes occur.

Periodic semelparity may also occur in *Parantechinus apicalis*. While males of this species have been reported to undergo the 'male die-off' (Dickman and Braithwaite 1992), males in the island population may sometimes survive to breed in a second year (Woolley 1991), with survival of males in an island population being recorded in three successive years from 1997 (Mills and Bencini 1999, 2000).

A moribund male *Parantechinus apicalis* found on Boullanger Island toward the end of the breeding period (Bradley and Dickman unpubl. obs.) was found to have acute haemorrhagic ulceration of the gastric mucosa, with a blood sample revealing elevated testosterone (1.49 μg/dl), high cortisol (6.58 μg/dl), low corticosteroid binding globulin (CBG) (0.92 μg/dl) and low albumin.(1.14 g/dl). The free cortisol concentration would be high and this is consistent with the endocrine profile of Strategy I males in their terminal phase.

Curiously the male *D. hallucatus* in the study of Oakwood et al. (2001) did not show cortisol concentrations that were significantly higher than those of the females during the reproductive period. However a significantly higher free cortisol concentration in males could have resulted from decreases in both CBG

Table 1 Critical factors that appear to be significant in allowing Strategy I dasyurid marsupial males to develop their unique adaptive physiological response.

- The absence of sex-hormone–binding globulin in all dasyurid marsupials
- Maintenance of luteinising hormone (LH) production and testosterone release from Leydig cells (→ *consistent with a maintenance of gonadotrophin-releasing hormone release*)
- Failure of spermatogenesis *circa* two months before reproductive period and maintenance of elevated testosterone concentration (→ *consistent with a defect in follicle stimulating hormone (FSH) release and/or target insensitivity*)
- The inverse relationship between plasma testosterone and plasma CBG (→ *rises in plasma testosterone cause decrease in plasma CBG*)
- Androgen-dependent increase in activity of scent glands coinciding with increase in activity of males (→ *stimulates increase in apparent density*)
- A terminal suppression of appetite (→ *reduction in food intake*)
- Major adjustments and sequential change in intermediary metabolism of carbohydrates, fats and proteins in Strategy I males (→ *consistent with mobilisation of energy reserves and starvation*)
- A terminal decrease in plasma albumin in Strategy I males (→ *effect upon steroid partitioning to deliver more free cortisol*)
- Damage to hippocampal neurons (→ *GC receptor decrease and reduction in dendritic connectivity*)
- The development of a defect in glucocorticoid feedback (→ *high concentrations of free cortisol are maintained*)
- Olfactory stimulation of stress pathways also involving hippocampus (→ *activate HPA axis, stimulate increase in free cortisol and cause damage to hippocampal neuron GC receptors and dendritic connectivity*)
- Terminal decline in olfactory function in males with reliance upon vomeronasal system to locate females
- Concurrent terminal deficit in short term-memory and spatial orientation in males.
- Increase in aggressiveness of pregnant females toward males (→ *stimulates terminal dispersal of males*)
- •Vagrant males lacking insulation (both fats and pelage thinning) (→ *more susceptible to hypothermia in cold conditions*)
- Sympathetic stimulation combining with CRH and GCs to cause gastric mucosal ischemia, increased capillary permeability and inhibit repair mechanisms (→ *leads to development of mucosal erosions and acute haemorrhage from ulcers*)
- Androgens and free GC elevated causing renal glomerular damage (→ *proteinuria and haemoglobinuria contributing to anaemia and general debilitated state*)

and albumin concentrations. Such changes, generating an elevated free cortisol concentration and associated immune suppression, would be consistent with the observed weight loss, parasite infections and anaemia. *D. hallucatus*, because it appears to differ in the terminal physiological and pathological changes, clearly warrants further investigation.

Plasticity of life-history strategy in several dasyurid species presents a most interesting prospect. If a failure of glucocorticoid regulation is involved in the male mortality, as current evidence suggests, it might be predicted that in years in which males survive the reproductive period:

- Density would be lower (implications for olfactory communication and consequent stress).
- Plasma testosterone would not be elevated to the same degree as in years when all males disappear.
- Plasma CBG would be higher (CBG depression varies between years in males).
- Plasma free cortisol would be lower.
- Less damage would be caused to hippocampal glucocorticoid receptors and a glucocorticoid feedback defect would not develop.
- Stress-related white blood cell changes would not be found (indicative of general maintenance of immune competence).
- Lesions would not develop in gastric mucosa and renal glomeruli.

At high density male mortality may be more significant than when the population is at low density from which many males may survive the reproductive period. Evidence for these physiological observations would be consistent with the operation of a density dependent population regulation, but it is suggested that density should be regarded in a *behavioural/olfactory* context that would be exacerbated by any reproductive related social aggregation such as occurs during the reproductive period. For example individuals moving more extensively, and using chemical communication, would be expected have a considerable influence on conspecifics (see Fig. 1). It is also important to determine whether males that survive the reproductive period are fertile during the next reproductive-period or whether spermatogenesis ceases during the first year without resumption as in 'obligate' Strategy I dasyurid males.

Single males surviving the reproductive period?

Very occasionally, in populations in which males are regarded as exhibiting a Strategy I life-history, single 'males' may be trapped after the majority of males have disappeared. During extensive trapping for *A. swainsonii* and *A. stuartii* (*A. agilis*), *A. flavipes*, and *A. stuartii* (*A. subtropicus*) the author has captured 2, 1, 2 and 1 respectively of these 'males'. In both cases for the *A. swainsonii* the scrotum either had not formed or had been lost. The absence of a scrotum was also noted in the other cases however one *A. flavipes* had a scrotum but the testes were very small and flaccid. In all cases the plasma testosterone concentration was either very low or undetectable. The absence of a scrotum and testes would explain the low plasma testosterone. For the two *A. swainsonii* the plasma CBG concentration was 9.3 and 10.7 μg/dl while for the two *A. flavipes* the values were 8.2 and 9.9 μg/dl. These blood samples were collected from males approximately two months after the disappearance of all other males and the high CBG values are consistent with the low plasma testosterone concentrations. In magnitude these CBG concentrations resemble those found in pregnant females and these elevated CBG concentrations would provide considerable buffering potential for cortisol, resulting in the exposure of

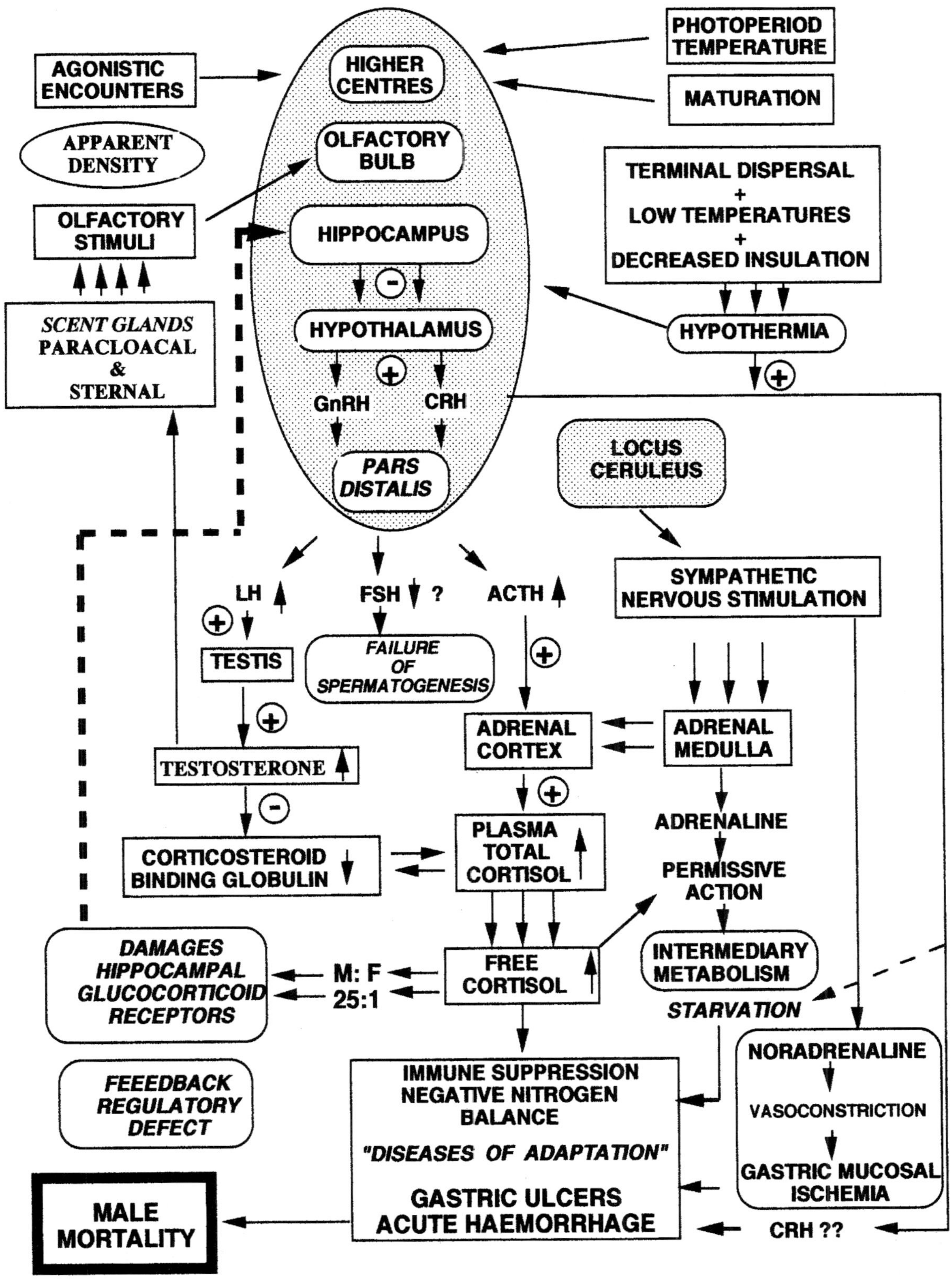

Figure 1 Diagram showing the likely relationships between environmental and physiological factors that may explain why males die at the end of the reproductive period (modified after Bradley 1978; 1985). CNS indicated by stippled areas.

tissues to relatively low free cortisol concentrations even during environmentally stressful periods involving HPA stimulation.

Since cutaneous scent glands in marsupials appear to be controlled by androgens, the absence of testes could indicate that the scent producing glands would be non-functional. This could convey to other males and females the signal that these 'males' were not competitors (Bradley and Monamy 1991). The absence of the characteristic semiochemical from males could explain the persistence of these odd 'males' during and beyond the reproductive period. In *A. flavipes* and *A. stuartii* (*A. subtropicus*) males castrated four months before the breeding season and released may be recaptured beyond the time when all other males have disappeared (Bradley unpubl. obs.). These observations are consistent with an important signalling role for scent glands in males. The survival of these odd 'males' is considered to be independent of the operation of a plasticity of life-history that might explain the persistence of a significant number of adult males in some years.

Why do females survive beyond the breeding period?

The physiological profile of females differs markedly from that of males, particularly during the breeding period and beyond when all females are pregnant. When males undergo an androgen dependent decrease in plasma CBG concentration, the plasma CBG concentration in females gradually increases. This may be attributable to the action of female reproductive hormones, particularly oestradiol. As a consequence of this increase in CBG in females, they are in effect insulated from the effects of an elevation in cortisol – that is they are more *stress resistant* at this time during pituitary-adrenocortical stimulation. This would explain why males in their terminal phase, but not females, exhibit negative nitrogen balance (Woollard 1971; Bradley unpublished). Furthermore, females do appear to consume food when males are reported to reduce their food consumption.

While the abovementioned studies attribute the male mortality to a variety of causes, based upon my experiences working with small dasyurid marsupials for over 25 years I am convinced that the proximate cause of death in the majority of cases is a consequence of haemorrhage from stress induced gastric ulcers. This view is based upon the results of necropsies carried out on many individuals covering several small Strategy I dasyurid marsupial species during this time. Stress induced immune suppression and the development of various pathologies affecting the renal and hepatic systems would also appear to contribute significantly to the generally debilitated physiological state of males leading ultimately to their death.

Future directions

This chapter includes some material that is at present speculative and deserves further investigation. It is included for the specific purpose of providing new directions in which research might proceed based upon exciting studies in other mammals. It is suggested that Strategy I dasyurid marsupials, because of the unique life-history strategy and associated physiological control, present us with a unique opportunity to address fundamental neuroendocrine questions that may be of relevance to all mammals.

Koolhaas et al. (1997) perceptively point out that most animal models use stressors that bear little or no relationship to the biology of the species (to the situations that an animal may meet in its everyday life in a natural habitat). They advocate the use of more naturalistic animal models that allow an experimental analysis. Strategy I dasyurid marsupials, because of their unique biological attributes, present us with exciting opportunities for future research. We would be wise to follow this advice.

References

Angelucci, L., Valeri, P., & Grossi, E. (1980), 'Involvement of hippocampal corticosterone receptors in behavioral phenomena', in *Progress in Psychoneuroendocrinology* (eds. G. Brambilla, G. Rascagni, & D. de Wiede), Elsevier, Amsterdam.

Arase, K., York, D., Shimizu, H., Shargill, N., & Bray, G. (1988), 'Effects of corticotropin-releasing factor on food intake and brown adipose tissue thermogenesis in rats', *Am J Physiol,* **255**:E255–9.

Barker, I.K., Beveridge, I., Bradley, A.J., & Lee, A.K. (1978), 'Observations on spontaneous stress related mortality among males of the dasyurid marsupial *Antechinus stuartii* (Macleay)', *Aust J Zool,* **26**:435–47.

Barker I.K., Carbonell P.L., & Bradley A.J (1981), 'Cytomegalovirus infection of the prostate in the dasyurid marsupials *Phascogale tapoatafa* and *Antechinus stuartii*', *J Wildl Diseases,* **17**:433–41.

Barnett, J.L. (1973), 'A stress response in *Antechinus stuartii* (Macleay)', *Aust J Zool,* **21**:501–13.

Barnett, J.L. (1974), 'Changes in hydroxyproline concentration of the skin of *Antechinus stuartii* with age and hormonal treatment', *Aust J Zool,* **22**:311–18.

Boonstra, R. (1994), 'Population cycles in microtenes: the senescence hypothesis', *Evol Ecol,* **8**:196–219.

Bondy, P. (1981), 'Disorders of the adrenal cortex', in *Williams Textbook of Endocrinology, 7th ed.* (eds. J. Wilson, & D. Foster), Saunders, Philadelphia.

Bradley, A.J. (1977), 'Stress and mortality in *Antechinus stuartii* (Macleay)', PhD thesis, Monash University, Melbourne, Australia.

Bradley, A.J. (1982a), 'The biology of the Red-tailed Phascogale, *Phascogale calura*', *Arid Zone Newsletter:*16–18.

Bradley, A.J. (1982b), 'Steroid binding proteins in the plasma of dasyurid marsupials', in *Carnivorous Marsupial* (ed. M. Archer): 651–7, Surrey Beatty & Sons, Sydney.

Bradley, A.J. (1985), 'Steroid binding proteins and life history in marsupials', in *Current Trends in Comparative Endocrinology Vol. I* (eds. B. Lofts, & W.N. Holmes): 269–70, Hong Kong University Press, Hong Kong.

Bradley, A.J. (1987), 'Stress and mortality in the Red-tailed Phascogale *Phascogale calura* (Marsupialia: Dasyuridae)', *Gen Comp Endocr,* **67**:85–100.

Bradley, A.J. (1990a), 'Seasonal effects on the haematology and blood chemistry in the Red-tailed Phascogale, *Phascogale calura* (Marsupialia: Dasyuridae)', *Aust J Zool,* **37**:533–43.

Bradley, A.J. (1990b), 'Failure of glucocorticoid feedback during breeding in the male Red-tailed Phascogale *Phascogale calura* (Marsupialia: Dasyuridae)', *J Steroid Biochem Molec Biol*, **37**:155–63.

Bradley, A.J. (1995), 'Red-tailed Phascogale *Phascogale calura*', in *Mammals of Australia* (ed. R. Strahan): 102–3, Reed Books, Sydney.

Bradley, A.J. (1997), 'Reproduction and life-history in the red-tailed phascogale, *Phascogale calura* (Marsupialia: Dasyuridae): The adaptive–stress senescence hypothesis', *J Zool Lond*, **241**:739–55.

Bradley, A.J, & Monamy, V.A. (1991), 'A physiological profile of *Antechinus swainsonii* (Marsupialia: Dasyuridae) males surviving the post-mating mortality', *Aust Mammal*, **14**:25–7.

Bradley, A.J., McDonald, I.R., & Lee, A.K. (1975), 'Effect of exogenous cortisol on mortality of a dasyurid marsupial', *J Endocrinol*, **66**:281–2.

Bradley, A.J., McDonald, I.R., & Lee, A.K. (1976), 'Corticosteroid binding globulin and mortality in a dasyurid marsupial', *J Endocrinol*, **70**:323–4.

Bradley, A.J., McDonald, I.R., & Lee, A.K. (1980), 'Stress and mortality in a small marsupial (*Antechinus stuartii* Macleay)', *Gen Comp Endocr*, **40**:188–200.

Braithwaite, R.W. (1974), 'Behavioural changes associated with the population cycle of *Antechinus stuartii* (Marsupialia)', *Aust J Zool*, **22**:45–62.

Braithwaite, R.W. (1979), 'Social dominance and habitat utilization in *Antechinus stuartii* (Marsupialia)', *Aust J Zool*, **27**:517–28.

Braithwaite, R.W., & Lee, A.K. (1979), 'A mammalian example of semelparity', *Am Nat*, **113**:151–5.

Capani de Kaski, M., Rentsch, R., Levi, S., & Hodgson, H.J. (1995), 'Corticosteroids reduce regenerative repair of epithelium in experimental gastric ulcers', *Gut*, **37**:613–16.

Cheal, P.D., Lee, A.K., & Barnett, J.L. (1976), 'Changes in the haematology of *Antechinus stuartii* (Marsupialia), and their association with male mortality', *Aust J Zool*, **24**:299–311.

Cockburn, A, (1989), 'Adaptive patterns in marsupial reproduction', *Trends in Ecology and Environment*, **4**:126–31.

Cockburn, A. (1992), 'Evoltionary ecology of the immune system: Why does the thymus involute?', *Functional Ecology*, **6**:364–70.

Cockburn, A. (1997), 'Living slow and dying young: senescence in marsupials', in *Marsupial Biology: Recent Research, New Perspectives* (ed. N.R. Saunders, & L.A. Hinds), pp. 163–71, Sydney, University of New South Wales Press.

Cockburn, A., & Lazenby-Cohen, K.A. (1992), 'Use of nest trees by *Antechinus stuartii*, a semelparous lekking marsupial', *J Zoology Lond* **226**:657–80.

Cuttle, P. (1978), 'The behaviour in captivity of the dasyurid marsupial *Phascogale tapoatafa* (Meyer)', MSc thesis, Monash University, Melbourne, Australia.

Cuttle, P. (1982), 'Life history strategy of the dasyurid marsupial, *Phascogale tapoatafa*', in *Carnivorous Marsupials* (ed. M. Archer), pp. 13–22, Sydney, Roy. Zool. Soc. NSW.

De Kloet, E.R. (1992), 'Corticosteroids, stress and aging', *Ann NY Acad Sci*, **663**:357–71.

Dickman, C.R., Parnaby, H.E., Crowther, M.S., & King, D.H. (1998), '*Antechinus agilis* (Marsupialia: Dasyuridae), a new species from the *A. stuartii* complex in south-eastern Australia', *Australian Journal of Zoology*, **46**:1–26

Dickman, C.R., & Braithwaite, R.W. (1992), 'Post-mating mortality mortality of males in the dasyurid marsupials, *Dasyurus* & *Parantechinus*', *J Mammal*, **73**:143–7.

Filaretova, L.P., Filaretova, A.A., & Makara, G.B. (1998), 'Corticosterone increase inhibits stress-induced gastric erosions in rats', *Am J Physiol*, **274**:G1024–30.

Filaretova, L.P., Maltcev, N., Bogdanov, A., & Levkovich, Y. (1999), 'Role of gastric microcirculation in the gastroprotection by glucocorticoids released during water-restraint stress in rats', *Chin J Physiol*, **42**:145–52.

Goya, L., Rivero F., & Pascual-Leone, A.M. (1995), 'Glucocorticoids, stress and aging', in *Hormones and Aging* (eds. P.S. Timiras, W.D. Quay, & A. Vernadakis), pp. 249–64, CRC Press, New York.

Hamerton, A.E. (1934), 'Report on deaths occurring in the societys gardens during the year 1933', *Proc Zool Soc Lond*, **104**:389–422.

Hamerton, A.E. (1935), 'Report on deaths occurring in the societys gardens during the year 1934', *Proc Zool Soc Lond*, **105**:433–74.

Hamerton, A.E. (1938), 'Report on deaths occurring in the societys gardens during the year 1937', *Proc Zool Soc Lond*, **108**:489–526.

Hamerton, A.E. (1939). 'Review of mortality rates and report on the deaths occurring in the society's gardens during the year 1938', *Proc Zool Soc Lond*, **108**:281–27.

Hanlon, A.J. (2001), 'The pathogenesis of gastric ulcer in a small marsupial mouse (*Antechinus subrtopicus*)', BSc Hons thesis, University of Queensland.

Harding, R.K., & Morris, G.P. (1977), 'Cell loss from normal and stressed gastric mucosae of the rat', *Gastroenterology*, **72**:859–63.

Hernandez, D.E. (1990), 'The role of brain peptides in the pathogenesis of experimental stress gastric ulcers', *Ann NY Acad Sci*, **597**:28–35.

Hosoda, S., Ikedo, H., & Saito, T. (1981), '*Praemys (Mastomys) natalensis:* Animal model for study of histamine–induced duodenal ulcers', *Gastroenterology*, **80**:16–21.

Hyman, B.T., van Hoesen, G.W., Damasio, A.R., & Barnes, C.L. (1984), 'Alzheimer's disease: cell-specific pathology isolates the hippocampal formation', *Science*, **225**:1168–70.

Jacobson, J.A., & Sapolsky, R.M. (1991), 'The role of the hippocampus in feedback regulation of the hypothalamic–pituitry–adrenocortical axis', *Endocrne Rev*, **12**:118–34.

Kawano, S., Sato, N., Tsuji, S., Hayashi, N., Tsuiji, M., Masuda, E., Takei, Y., Nagano, K., Fusamoto, H., Ogihara, T., Miwa, H., Hamada, T., & Kamada, T. (1991), 'Two dimensional computer color graphics of gastric mucosal blood distribution in normal subjects and ulcer patients', *Endoscopy*, **23**:317–20.

Kitagawa, H., Fujiwara, M., & Osumi, Y. (1979), 'Effects of water immersion stress on gastric secretion and mucosal blood flow in rats', *Gastroenterology*, **77**:298–302.

Kitchener, D.J. (1981), 'Breeding, diet and habitat preference of *Phascogale calura* (Gould, 1844) (Marsupialia: Dasyuridae) in the southern wheat belt, Western Australia', *Rec West Aust Mus*, (Suppl.) **9**:173–86.

Kivilaakso, E., Barzilai, A., Schiessel, R., Fromm, D., & Silen, W. (1981), 'Experimental ulceration of rabbit antral mucosa', *Gastroenterology*, **80**:77–83.

Koolhaas, J.M., Meerlo, P., De Boer, S.F., Strubbe, J.H., & Bohus, B. (1997), 'The temporal dynamics of the stress response', *Neuroscience and Behavioral Reviews*, **21**:775–82.

Krahn, D., Gosnell, B., Grace, M., & Levine, A. (l986), 'CRF antagonist partially reverses CRF- and stress-induced effects on feeding', *Brain Res Bull*, **17**:285–9.

Krajewski, C., Woolley, P.A., & Westerman, M. (2000), 'The evolution of reproductive strategies in dasyurid marsupials: implications of molecular phylogeny', *Biol J Linn Soc*, **71**:417–35.

Krakoff, L. (l988), 'Glucocorticoid excess syndromes causing hypertension', *Cardiology Clinic*, **6**:537–45.

Kuwayama, H., Matsuo, Y., & Eastwood, G.L. (1989), 'Effects of hydrocortisone sodium succinate on the healing of established gastric ulcers in the rat', *Dig Dis Sci*, **34**:A7.

Lazenby-Cohen, K.A., & Cockburn, A. (1988), 'Lek promiscuity in a semelparous mammal, *Antechinus stuartii* (Marsupialia: Dasyuridae)', *Behavioral Ecology and Sociobiology*, **22**:195–202.

Lazenby-Cohen, K.A., & Cockburn, A. (1991), 'Social and foraging components of the home range in *Antechinus stuartii* (Marsupialia: Dasyuridae), *Aust J Ecol*, **16**:301–7.

Lee, A.K., Bradley, A.J., & Braithwaite, F.W. (1977), 'Corticosteroid levels and male mortality in *Antechinus stuartii*', in *Biology and Environment* (eds. B. Stonehouse & D. Gilmore), pp. 209–20, Macmillan, London.

Lee, A.K., & Cockburn, A. (1985a), *The evolutionary ecology of marsupials*, Cambridge University Press, Cambridge.

Lee, A.K., & Cockburn, A. (1985b), 'Spring declines in small mammal populations', *Acta Zool Fenn*, **173**:75–6.

Lee, A.K., & McDonald, I.R. (1985), 'Stress and population regulation in small mammals', in *Oxford reviews of reproductive biology*, pp. 261–304, Oxford University Press, Oxford.

Lee, A.K., Woolley, P., & Braithwaite, R.W. (1982), 'Life history strategies of dasyurid marsupials', in *Carnivorous Marsupials* (ed. M. Archer), pp. 1–11, Royal Society of New South Wales, Sydney, Australia.

de Leon, M.J., McRae, T., Tsai, J.R., George, A.E., Marcus, D.L., Freedman, M., Wolf, A.P., & McEwen, B.S. (1988), 'Abnormal cortisol response in Alzheimer's disease linked to hippocampal atrophy', *Lancet*, **ii**:391–2.

Lorini, M.L., Deoliveira, J.A., & Persson, V.G. (1994), 'Annual age structure and reproductive patterns in marmosa incana (Lund, 1841) (Didelphidae, Marsupialia)', International *J Mammal Biol*, **59**:65–73.

Lucasen, P.J., Vollmann-Honsdorf, G.K., Gleisberg, M., Boldizsar-Czeh, E., De Kloet, R., & Fuchs, E. (2001), 'Chronic psychosocial stress diffrerentially affects apoptosis in hippocampal subregions and cortex of the adult tree shrew', *Eur J Neurosci*, **14**:161–6.

Martin, M., & Menguy, R. (1970), 'Influence of adrenocorticotropin, cortisone, asprin, and phenylbutazone on the rate of exfoliation and the rate of renewal of gastric mucosal cells', *Gastroenterology*, **58**:329–36.

Marshall, B.J., McCallum, R.W., & Guerrant, R.L. (1991), *Helicobacter pylori in peptic ulceration and gastritis*, Boston. Blackwell.

McAllan, B., Roberts, J.B., & O'Shea, T. (1996), 'Seasonal changes in renal morphometry of *Antechinus stuartii* (Marsupialia: Dasyuridae)', *Aust J Zool*, **44**:337–54.

McAllan, B., Roberts, J.B., & O'Shea, T. (1997), 'Effects of testosterone and cortisol on the renal morphology of male *Antechinus stuartii* (Marsupialia)', *Gen Comp* Endocr, **107**:439–49.

McAllan, B., Roberts, J.B., & O'Shea, T. (1998), 'The effects of cortisol and testosterone on renal function in male *Antechinus stuartii* (Marsupialia)', *J Comp Physiol B*, **168**:248–56.

McDonald, I.R., Lee, A.K., Bradley, A.J., & Than, K.A. (1981), 'Endocrine changes in dasyurid marsupials with differing mortality patterns', *Gen Comp Endocrinol*, **44**:292–301.

McDonald I.R., Lee A.K., Than K.A., & Martin R.W.(1986), 'Failure of glucocorticoid feedback in males of a population of small marsupials (*Antechinus swainsonii*) during the period of mating', *J Endocrinol*, **108**:63–8.

McEwen, B.S. (1998), 'Stress, adaptation, and disease. Allostasis and allostatic load', *Ann NY Acad Sci*, **840**:33–44.

McEwen, B. (1999), 'Stress and hippocampal plasticity', *Ann Rev Neurosci*, **22**:105–22.

McEwen, B.S., DeKloet, E.R., & Rostene, W. (1986), 'Adrenal steroid receptors and actions in the nervous system', *Physiol Rev*, **66**:1121–88.

McEwen, B., & Sapolsky, R. (1995), 'Stress and cognitive function', *Curr Opin Neurobiol*, **5**:205–11.

McEwen, B.S., & Stellar, E. (1993), 'Stress and the individual. Mechanisms leading to disease', *Arch Intern* Med, **27**:2093–101.

Mills, H., & Bencini, R. (1999), 'Postmating survival of male dibblers (*Parantechinus apicalis*) in island populations', *Australian Mammal Society Newsletter*, Nov, p. 19.

Mills, H.R., & Bencini, R. (2000), 'New evidence for facultative male die-off in island populations of dibblers, *Parantechinus apicalis*', *Aust J Zool*, **48**:501–10.

Moore, G.H. (1974), 'Aetiology of the die-off of male *Antechinus stuartii*', PhD thesis, Australian National University, Canberra, Australia.

Morris, R.G.M. (1984), 'Developments of a water-maze procedure for studying spatial learning in the rat', *J Neurosci Methods*, **11**:47–60.

Munch, A., Guyre, P., & Holbrook, N. (1984), 'Physiological functions of glucocorticoids in stress and their relation to pharmacological actions', *Endocr Rev*, **5**:25–42.

Murphy, H.M., Wideman, C.H., & Brown, T.S. (1979), 'Plasma corticosterone levels and ulcer formation in rats with hippocampal lesions', *Neuroendocrinol*, **28**:123–30.

Nagy, K.A., Seymour, R.S., Lee, A.K., & Braithwaite, R. (1978), 'Energy use and water budgets in free living *Antechinus stuartii* (Marsupialia: Dasyuridae)', *J Mammalogy*, **59**:60–8.

Oakwood, M. (1999), 'Reproduction and demography of the Northern quoll, *Dasyurus hallucatus*, in the wet–dry tropics of Northern Australia', *Australian Mammal Society Newsletter*, Nov. p. 15.

Oakwood, M., Bradley, A.J., & Cockburn, A. (2001), 'Semelparity in a large marsupial', *Proc R Soc Lond B*, **268**:407–11.

Oitzl, M.S., Fluttert, M., Sutanto, W., & de-Kloet, E.R. (1998), 'Continuous blockade of brain glucocorticoid receptors facilitates spatial learning and memory in rats', *European J Neurosci*, **10**:3759–66.

Pasquier, F., Bail, L., Lebert, F., Pruvo, J.P., & Petit, H. (1994), 'Determination of medial temporal lobe atrophy in early Alzheimer's disease with computed tomography', *Lancet*, **343**:861–2.

Pine, R. (1994), 'Sex and death', *Aust Nat Hist*, **24**:4.

Regan, L.P., & McEwen, B.S. (1997), 'Controversies surrounding glucocorticoid-mediated cell death in the hippocampus', *J Chem Neuroanat*, **13**:149–67.

Reul, J.M., Tonnaer, J.A., & Kloet, E.R. (1988), 'Neurotropic ACTH analogue promotes plasticity of type I corticosteroid receptor in brain of senescent male rats', *Neurobiol Aging*, **9**:253–60.

Sapolsky, R.M. (1985a), 'A mechanism for glucocorticoid toxicity in the hippocampus increased neuronal vulnerability to metabolic insults', *J Neurosci*, **5**:1228–32.

Sapolsky, R.M. (1985b), 'Glucocorticoid toxicity in the hippocampus: temporal aspects of neuronal vulnerability', *Brain Res*, **359**:300–5.

Sapolsky, R.M. (1986), 'Glucocorticoid toxicity in the hippocampus: reversal by supplementation with brain fuels', *J Neurosci*, **2**:2240–7.

Sapolsky, R.M. (1992), 'Neuroendocrinology of the stress response' in *Behavioural Endocrinology* (eds. J.B. Becker, S.M. Breedlove, & D. Grews), Chp 10, MIT Press, Massachusetts, USA, 287 p.

Sapolsky, R.M. (1996), 'Why stress is bad for your brain', *Science*, **273**:749–50.

Sapolsky, R.M. (2000), 'Glucocorticoids and hippocampal atrophy in neuropsychiatric disorders', *Arch Gen Psych*, **57**:925–41.

Sapolsky, R.M., Krey, LC., & McEwen, B.S. (1986), 'The neuroendocrinology of stress and aging; the glucocorticoid cascade hypothesis', *Endocr Rev*, **7**:284–301.

Sapolsky, R.M., Romero, L.M., & Munck, A.U. (2000), 'How do glucocorticoids influence stress-responses? Integrating permissive, suppressive, stimulatory, and preparative actions', *Endocrine Reviews*, **21**:55–73.

Sapolsky, R.M. Krey, L., & McEwen, B.S. (1983), 'Corticosterone receptors decline in a site-specific manner in the aged rat brain', *Brain Res*, **289**:235–40.

Sapolsky, R.M., Krey, L.C., & McEwen, B.S. (1984), 'Stress down-regulates corticosterone receptors in a site-specific manner in the brain', *Endocrinol*, **114**:287–92.

Sapolsky, R.M., Krey, L.C., & McEwen, B.S. (1986), 'The neuroendocrinology of stress and aging: The glucocorticoid cascade hypothesis', *Endocri Rev*, **7**:284–301.

Schmitt, L.H., Bradley, A.J., Kemper, C.M., Kitchener, D.J., Humphreys, W.F., & How, R.A. (1989), 'The demography and ecophysiology of *Dasyurus hallucatus* (Marsupialia: Dasyuridae) in the Mitchell Plateau Area, Kimberley, Western Australia', *J Zool Lond*, **217**:539–58.

Scully, J.L., & Otten, U. (1995), 'Glucocorticoids, neurotrophins and neurodegeneration', *J Steroid Biochem Molec Biol*, **52**:391–401.

Seckl, J.R., French, K.L., O'Donnell, D., Meaney, M.J., Nair, N,.P., Yates, C.M., & Fink, G. (1993), 'Glucocorticoid receptor gene expression is unaltered in hippocampal neurons in Alzheimer's disease', *Brain Res Mol Brain Res*, **18**:239–45.

Seckl, J.R., & Olsson, T. (1995), 'Glucocorticoid hypersecretion and the age impaired hippocampus: cause or effect?', *J Endocrinol*, **145**:201–11.

Scott, M.P. (1987), 'The effect of mating,and agonistic experience on adrenal function and mortality of male *Antechinus stuartii* (Marsupialia)', **68**:479–86.

Semia, C., Bradley, A.J., & McDonald, I.R. (1979), 'High affinity binding of adrenocortical and gonadal steroids by plasma proteins of Australian marsupials', *Gen Comp Endocr*, **38**:496–503.

Sherwood, B.F., Rowlands, D.T., Hackle, D.B., & Le May, J.C. (1968), 'Bacterial endocarditis, glomerulonephritis and amyloidosis in the opossum (*Didelphis virginiana*)', *Am J Path*, **53**:115–26.

Soderquist, T.R. (1993), 'Maternal strategies of *Phascogale tapoatafa* (Marsupialia: Dasyuridae), I. Breeding seasonality and maternal investment', *Aust J Zool*, **41**:549–66.

Soderquist, T.R., & Ealey, L (1994), 'Social interactions and mating strategies of a solitary carnivorous marsupial, *Phascogale tapoatafa*, in the wild', *Wildl Res*, **21**:527–42.

Squire, L. (1987), *Memory & Brain*, New York, Oxford Univ. Press.

Steins-Behrens, B.A., & Sapolsky, R.M. (1992), 'Stress, glucocorticoids and aging', *Aging Clin Exp Res*, **4**:197–210.

Stemmerman, G.N., & Hayashi, T. (1969), 'A survey of the normal and morbid anatomy of the Hawaiian feral mongoose', *Am J Path*, **55**:67a–68a.

Sterling, P., & Eyer, J. (1981), 'Biological basis of stress-related mortality', *Soc Sci Med E*, **15**:3–42.

Sumner, J., & Dickman, C.R. (1998), 'Distribution and identity of specied in the *Antechinus stuartii – A. flavipes* group (Marsupialia: Dasyuridae) in south-eastern Australia', *Aust J Zool*, **46**:27–41.

Svanes, K., Varhaug, J.E., Dzienis, H., & Gronbech, J.E. (1984), 'Gastric mucosal blood flow related to acute mucosal damage', *Scand J Gastroenterol*, **19**:62–6.

Toftegaard, C.L. (1999), 'Morphological and endocrine correlates of chemical communication during the life history of *Antechinus stuartii* (Macleay)', PhD thesis, The University of Queensland, Brisbane, Australia.

Toftegaard, C.L., Moore, C., & Bradley, A.J. (1999), 'Chemical characterisation of urinary pheromones in the brown antechinus, *Antechinus stuartii* (Marsupialia: Dasyuridae)', *J Chemical Ecology*, **25**:527–35.

Toftegaard, C.L., McMahon, K.L.., Galloway, G.J., & Bradley, A.J. (2002). 'Processing of urinary pheromones in *Antechinus stuartii* (Marsupialia: Dasyuridae): Functional magnetic resonance imaging of the brain', *J Mammal*, **83**:71–80.

Tyndale-Biscoe, H., & Renfree, M. (1987), *Reproductive Physiology of Marsupials*, Cambridge University Press, Cambridge.

Van Dyck, S., & Crowther, M.S. (2000), 'Reassessment of Northern representatives of the *Antechinus stuartii* complex (Marsupialia; Dasyuridae): *A. subrtopicus* sp.nov., & *A. adustus* new status', *Mem Qld Museum*, **45**:611–35.

Van Eekelen, J.A., & De Kloet, E.R. (1992), 'Co-localization of brain corticosteroid receptors in the rat hippocampus'. *Prog Histochem Cytochem*, **26**:250–8.

Von Holst (1972), 'Renal failure as a cause of death in *Tupaia belangeri*. exposed to persistent social stress', *J Comp Physiol*, **78**:236–73.

Von Holst (1986), 'Psychosocial stress and its pathophysiological efects in tree shrews (*Tupaia belangeri*)', in *Biological and Psychological Factors in Cardiovascular Disease* (eds. T.H. Schmidt, T.M. Dembroski, & G. Blumchen), Springer-Verlag, Berlin.

Wexler, B.C., & Greenberg, B.P. (1978), 'Pathophysiological differences between paired and communal breeding of male and female Sprague-Dawley rats', *Circulation Research*, **42**:126–35.

Williams, R., & Williams, A. (1982), 'The life cycle of *Antechinus swainsonii* (Dasyuridae: Marsupialia)', in *Carnivorous Marsupials* (ed. M. Archer), pp. 89–95, Royal Zoological Society of New South Wales, Sydney.

Wilson, M. (1985), 'Hippocampal inhibition of the pituitary-adrenocortical response to stress', in *Psychological and Physiological Interactions in Response to Stress* (ed. S. Birchfield), Academic Press, New York.

Woollard, P. (1971), 'Differential mortality of *Antechinus stuartii* (Macleay): nitrogen balance and somatic changes', *Aust J Zool*, **19**:347–53.

Woolley, P.A. (1966), 'Reproduction in *Antechinus spp.*, & other dasyurid marsupials', *Symp Zool Soc Lond*, **15**:281–94.

Woolley, P.A. (1991), 'Reproductive pattern of captive Boullanger Island dibblers, *Parantechinus apicalis* (Marsupialia: Dasyuridae)', *Wildlife Research*, **18**:157–63.

PART IV

EVOLUTIONARY ECOLOGY AND BEHAVIOUR

CHAPTER 18

CARNIVORY AND INSECTIVORY IN NEOTROPICAL MARSUPIALS

Emerson M. Vieira[A] *and Diego Astúa de Moraes*[B]

[A] Laboratório de Ecologia de Mamíferos. Centro de Ciências da Saúde – Centro 2. Universidade do Vale do Rio dos Sinos – UNISINOS. CP 275 Av. Unisinos, 950. São Leopoldo, RS. 93022-000. Brazil. Email: vieira@cirrus.unisinos.br

[B] Laboratório de Vertebrados. Departamento de Ecologia. Universidade Federal do Rio de Janeiro. C.P. 68020. 21941-590. Rio de Janeiro, RJ. Brazil. Present address: Departamento de Zoologia. Instituto de Biociências. USP & Mastozoologia. Museu de Zoologia da Universidade de São Paulo. Av. Nazaré, 481. Ipiranga. São Paulo, SP. 04263-000. Brazil. Email: dmoraes@ib.usp.br

The extent to which the term 'carnivory' can be applied to extant Neotropical marsupials is somewhat difficult to determine clearly. Although no Neotropical marsupial feeds exclusively on plant matter, the relative importance of animals in the diet of each species may vary. At present there is no living counterpart to the extinct carnivorous marsupials once found in many regions of South America, specially members of the Borhyaenidae, the most diverse family. The old explanation for the extinction of all exclusively predatory forms of South American marsupials through competitive exclusion by North American newcoming eutherian carnivores, arrived during the Great American Interchange, is no longer accepted nowadays. In fact, the decline of the marsupial predatory forms began much earlier than the arrival of eutherians, and many were already completely extinct by the time of the rise of the Panama isthmus. The living marsupials possess similar body form and do not show great niche diversity. These animals can be placed in five different classes in relation to their degree of carnivory/insectivory. The most carnivorous genera are *Lestodelphys*, *Lutreolina*, and *Chironectes* that feed mainly on vertebrates and invertebrates. The genera *Philander*, *Metachirus*, *Thylamys*, *Monodelphis*, *Caenolestes*, *Rhyncholestes*, *Lestoros*, and *Dromiciops* feed mainly on invertebrate animals, but also on vertebrates and on plant matter. For Neotropical marsupials there is a negative and significant correlation between arboreal activity and degree of carnivory. Carnivorous/insectivorous marsupials should also have relatively low numerical densities and low biomasses than more omnivorous species. Nevertheless feeding requirements may also provide some advantages for carnivorous/insectivorous marsupials, as high-energy food enables them to survive in less favourable environments at higher latitudes in South America. Basic studies focusing on diet of carnivorous/insectivorous marsupials are still necessary. Moreover, only long-term field studies (i.e. ≥3 years) will be adequately address several relevant ecological aspects of these marsupials.

Introduction

Marsupials have been present in the Neotropical region of South America at least since the Cretacecous–Cenozoic transition, between 70 and 60 Mya (Flynn and Wyss 1998). This group includes the three orders of the magnorder Ameridelphia: Didelphimorphia, Paucituberculata and the extinct order Sparassodonta, following the classification of McKenna and Bell (1997). Members of the order Microbiotheria, from the magnorder Australidelphia, are included as well. The extent to which the term 'carnivory' can be applied to extant Neotropical marsupials is somewhat difficult to outline clearly. These animals have a natural diet based mainly upon small vertebrates, arthropods, fruits and nectar. The relative proportions and importance of each of these items vary from species to species. Furthermore, some species are mainly opportunistic feeders, being able to change their diet according to local food availability. Neotropical marsupials have feeding habits that can be ordered on a continuum from basically frugivorous species to mainly carnivorous ones (Fig. 1). Nonetheless, in the recent Neotropical marsupial fauna there are no truly carnivorous species (i.e. feeding exclusively on vertebrate prey), as the Australasian Thylacine (*Thylacinus cynocephalus*), Tasmanian devil (*Sarcophilus harrisii*), and the spotted-tailed quoll (*Dasyurus maculatus*). On the other hand, contrarily to some of their Australasian counterparts, no Neotropical marsupial feeds exclusively on plant matter, although the relative importance of animals in the diet of each species may vary.

In this chapter we present a general overview of carnivory among Neotropical marsupials. We discuss the morphological, physiological, behavioural and ecological data available for both extinct and recent species. For the latter we consider as 'carnivorous' those species that feed more intensely on animals (invertebrates or vertebrates), that is those belonging to the genera placed in classes IV and V in Fig. 1. By such criteria we include 40 (about 53%) of the currently recognised species. Discussion is mainly focused on those carnivorous/insectivorous marsupials.

The extinct carnivorous marsupials of South America

Ever since the first South American marsupial radiation, it appears that only carnivorous and/or insectivorous forms were present. That is, no strictly herbivorous branches are known, and the ecological role of large terrestrial carnivores was initially filled by marsupials (Hume 1999, Marshall 1978a). The Didelphimorphia, the Paucituberculata and the Microbiotheria include living and fossil species, while the Sparassodonta include only forms now extinct. Evolutionary changes related to carnivory appear in many forms, from morphological to physiological adaptations,

I		II		III		IV		V
Caluromys		*Micoureus*		*Didelphis*		*Philander*		*Lestodelphys*
Caluromysiops	➡	*Marmosa*	➡		➡	*Metachirus*	➡	*Lutreolina*
*Glironia**		*Marmosops*				*Thylamys*		*Chironectes*
		Gracilinanus				*Monodelphis*		
						Caenolestes		
						Rhyncholestes		
						Lestoros		
						Dromiciops		

FRUGIVORY ➡ CARNIVORY

Figure 1 Genera of Neotropical marsupials ranked according to their dietary category. Their feeding habits show a continuum ranging from essentially frugivorous species (I) through more generalistic ones that rely more or less frequently on insects and/or small vertebrates (II–IV) to those that could be considered as genuine carnivores, whose diet is essentially composed of other animals including vertebrates (V). *The monospecific genus *Glironia* was tentatively placed in Class I because of the morphological affinities of the Bushy-tailed opossum *G. venusta* with *Caluromys* and *Caluromysiops*, although there is no available information on the feeding habits of this species.

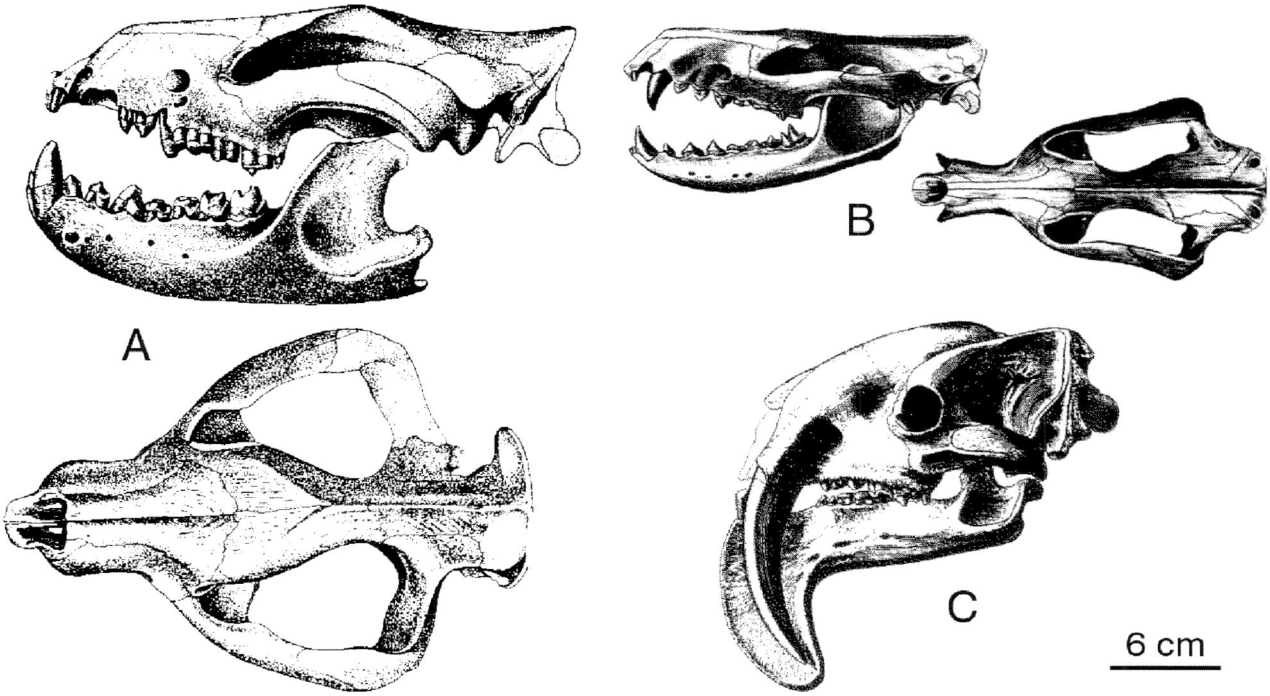

Figure 2 Skulls of the Borhyaenidae *Borhyaena tuberata* (Borhyaeninae) from the lower Miocene of Patagonia, dorsal and lateral view (A), *Cladosictis lustratus* (Hatlyacyninae) from the lower Miocene of Patagonia, dorsal and lateral view (B), and *Thylacosmilus atrox* (Thylacosmilinae) from the lower Pliocene of Catamarca, Argentina, lateral view (C). All three skulls reproduced from Paula Couto (1979), with permission from the Academia Brasileira de Ciências.

and include behavioural changes. Even though living Didelphimorphia and Paucituberculata include species that feed on other animals, it is within the Sparassodonta that the biggest body-sized carnivorous marsupial forms of the Neotropics are found. There is no living counterpart to the extinct carnivorous marsupials once found in many regions of South America, specially members of the Borhyaenidae. On the other hand, several evolutionary changes have occurred even within the living lineages, in the more carnivorous taxa.

The great carnivorous marsupials are found in beds from the Early Paleocene up to the Late Pliocene (Muizon 1999). They originated and went extinct in South America (Simpson 1980), with remains found in beds from Colombia (La Venta) to southern Argentina (Marshall 1978a), and ranging from the size of a weasel (*Mayulestes*) to that of a bear (*Proborhyaena*) (Muizon 1999). It is within the Borhyaenidae (Fig. 2) that the largest bodied extinct carnivorous marsupials, as well as the greatest diversity, are found. Not all Borhyaenidae were exclusively carnivorous, as it included forms thought to be omnivores. However, these shall be reviewed as well, as similar variable degrees of carnivory/insectivory are also found in living forms. As the purpose of this chapter is an analysis of feeding specialisations, it will not deal with major evolutionary relations of South American marsupials that can be found elsewhere (e.g. see Goin, this volume, for a review on diversity and relationships of early South American marsupials).

The fossil forms reviewed here will be focused mainly on the Borhyaenoidea (Borhyaenidae, Hondadelphidae and Mayulestidae) along with some Didelphoidea (the Sparassocynidae). Other more generalist fossil marsupial forms, such as Argyrolagoidea and Polydolopoidea, were left out as they fell outside the scope of this chapter.

Borhyaenidae

The Borhyaenidae was the most diverse family of extinct marsupials, with five subfamilies and 37 genera (McKenna and Bell 1997). According to Marshall's review (1978a), no borhyaenids presented any marked cursorial adaptations, all being quite short legged, but more recent analyses indicate that some species did present morphological modifications towards an increase in terrestrial and cursorial (although limited) locomotion (Muizon 1998). Information on the postcranial skeleton is scarce for many species, but for some genera of the Hathlyacyninae, the smaller body-sized among the Borhyaenidae, such as *Sipalocyon* or *Cladosictis*, the presence of opposable or partially opposable hallux and pollex indicate an arboreal or semi-arboreal mode of life.

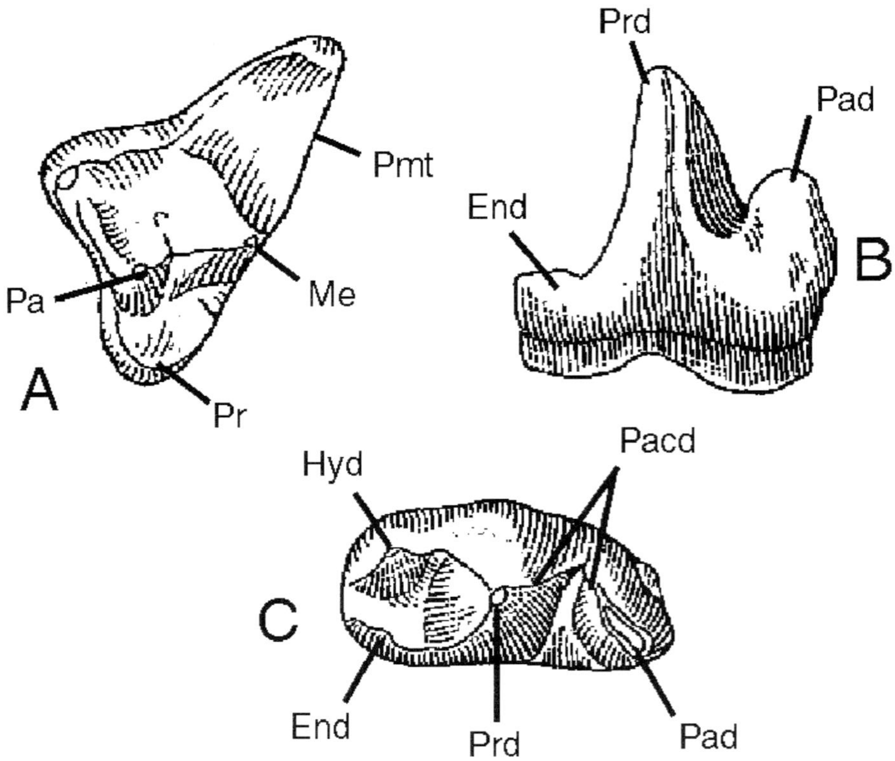

Figure 3 Molars of the boryhaenid *Sypalocyon gracilis* (YPM PU 15373). A. Left M3 in occlusal view. B. Left m3 in lingual and occlusal views. Abbreviations: End, entoconid; Hyd, hypoconid; Me, metacone; Pa, paracone; Pacd, paracristid; Pad, paraconid; Pmt, postmetacrista; Pr, protocone; Prd, protoconid. Reproduced with permission from Muizon and Lange-Bandré (1997).

Other species, such as *Prothylacynus* (Prothylacyninae), with hallux reduction but still enhanced elbow and manus mobility, or *Borhyaena* (Borhyaeninae), with blunt claws and reduced forearm mobility, were probably more terrestrial, comparable to living bears in running and climbing abilities. Most of these carnivorous predators, however, probably did not display chasing strategies for prey capture, but rather acted as ambush predators, stalking their prey (Marshall 1978a, Muizon 1998).

However, knowledge on the feeding habits of fossil species is based essentially on their dental specialisations. Most of the extinct borhyaenoid species present dental features that indicate carnivorous habits. In the Borhyaenidae, these adaptations include clear prevallid-postvallum shear, with the upper blade consisting of developed metacone and postmetacrista, acting against the lower blade composed by the enhanced paraconid and protoconid (with a consequently developed paracristid, with a carnassial notch) (Fig. 3). Along with these, an increased canine size or the presence of robust crushing premolars are also sometimes present (Muizon and Lange-Bandré 1997)

Borhyaeninae. The Borhyaeninae were medium- to large-sized terrestrial carnivores, with strong mandibular rami and dental specialisations indicating a meat diet (Fig. 2). These include the characteristic borhyaenoid dental complex previously mentioned, while some genera such as *Borhyaena* and *Acrocyon* have also developed stout premolars for crushing bone (Marshall 1978a, Muizon and Lange-Bandré 1997).

Proborhyaeninae. The Proborhyaeninae include the largest species among the Borhyaenidae and the biggest marsupial carnivore of South America, *Proborhyaena gigantea*, which attained the size of a large bear, with skulls measuring up to 60 cm (Marshall 1978a). Its canines were highly developed, proportionally more than in any other known mammalian carnivore, with the exception of *Thylacosmilus* and the sabre-toothed feloids (Bond and Pascual 1983, Muizon and Lange-Badré 1997). These teeth apparently were open rooted and long believed to be of continuous growth. However, Bond and Pascual (1983) argued that they were probably of extended but not continuous growth, with their roots closing as the animal aged.

The genus *Arminiheringia* possessed a fused mandibular symphysis, and palatine bones greatly extended posteriorly, interpreted as being a modification in counterpart of greatly hypertrophied pterygoid muscles, in order to allow for breathing during mastication. Other dental specialisations, such as carnassial rotation in order to maintain clear shearing occlusion, are also found in *Arminiheringia* and, in a lesser extent, in some Borhyaeninae (Marshall 1978a).

Thylacosmilinae. The sabre-tooth marsupials of the genus *Thylacosmilus* represent one of the most specialised marsupial carnivores that ever existed. They presented highly developed canines, ever growing and deeply anchored in bony sockets in the maxillary, up to above and behind the eyes (Fig. 2). A well-developed sagittal crest, indicating a strong temporal musculature, and deep flanges on the mandible were present, presumably to protect the massive canines when the mouth was closed, as excessive lateral strain could eventually damage the canines (Marshall 1976, 1982). The canines may have been used for stabbing or slicing the prey, and it is now more widely accepted that Thylacosmilinae were predators and not scavengers as sometimes suggested. The anatomy of the forelimbs also suggest an ambush predation strategy (Marshall 1976).

Prothylacyninae. These medium- to large-sized terrestrial marsupials were probably more omnivorous than the remaining borhyaenids, as judged by dental specialisations, but rodent remains were found in the body cavity of a skeleton of *Lycopsis* (Marshall 1978a).

Hathlyacyninae. These small- to medium-sized borhyaenids (Fig. 2), with sometimes arboreal or semi-arboreal habits, are believed to play an ecological role maybe similar to some actual mustelids or even didelphids, adapted to generalised carnivorous food habits (Marshall 1978a).

Hondaldephidae

Hondadelphys had well-developed shearing modifications, and also presented crushing specialisations. This genus shared some dental specialisations with some Hathlyacyninae (Marshall 1978a, 1982).

Mayulestidae

Based on skeletal morphological features, Muizon (1998) suggests weasel-like predatory habits for *Mayulestes ferox*, with at least partial arboreal habits, but most probably hunting on the ground (Argot 2001). Its dental morphology indicates an increase in carnivory, with some shearing structures adaptations, but not as marked in many borhyaenid species. Being one of the two oldest borhyaenoid known, it probably represents the beginning of the trends toward hypercarnivore adaptation found in later forms (Muizon and Lange-Bandré 1997, Muizon 1998)

Sparassocynidae

Members of the family Sparassocynidae were formerly included within the Didelphidae, but are now placed in a separate family, within the order Didelphimorphia. The only genus, *Sparassocynus*, possessed well-developed carnassial specialisations on the molars, as well as a short, wedge-shaped face, and was probably a small predator (Marshall 1982). With a skull slightly smaller than that of living *Lutreolina*, its snout and mandible were markedly short (Fig. 4). Such reduction of the snout is a recurrent characteristic of carnivores, as it is related to a reduction of torsion forces imposed on it when biting occurs on only one canine, as well as the resistance arm when biting at the anterior teeth (Covey and Greaves 1994). These traits are now found and are most evident in the most carnivorous Didelphidae, *Lutreolina crassicaudata* (Astúa de Moraes et al. 2000 and see Fig. 4). As a whole, most of the cranial features of *Sparassocynus* indicate a powerful closure of the jaws at the molar region (Reig and Simpson 1972).

MARSUPIAL EXTINCTIONS AND THE GREAT AMERICAN INTERCHANGE

Even though there are no extant strict vertebrate-eating marsupials in the Americas, as discussed in the introduction, the marsupial radiation in the Neotropics was once capable of producing predatory forms such as the Borhyaenidae. These forms were confined to South America, where there was a radical and relatively sudden change in faunal composition that occurred in the Pliocene. This event, known as the Great American Interchange, took place mainly after around three Mya (Pascual and Jaureguizar 1990), with the introduction of seventeen families of land mammals into South America and an equivalent invasion of North America by southern immigrants (Webb 1999). This great interchange was preceded by a few procyonids, which arrived in South America 9.0 to 6.0 Mya (Simpson 1980). During the Great American Interchange diverse groups of herbivores and carnivores immigrated from North America across the isthmian land bridge in Panama to South America. The details and consequences of this event have been discussed elsewhere (Marshall and Cifelli 1990; Pascual and Jaureguizar 1990; Webb 1991). Many of these taxa successfully spread and diversified in the new continent, among them most living members of the order Carnivora.

The Great American Interchange certainly resulted in a strong impact on the mammal fauna of South America, including some marsupials. However, the traditional explanation for the decline of the marsupial predatory forms as a consequence of competitive exclusion by 'advanced' eutherian counterparts is no longer accepted (Marshall 1978a). It used to be tempting to relate the striking and complete extinction of marsupial predators to the arrival of placental carnivores. This scenario would include the exclusion of carnivorous marsupials by direct competition or any other indirect interactions, such as the disappearance of potential prey species like extinct edentates, ungulates and rodents that could have been competitively excluded by other newcomers

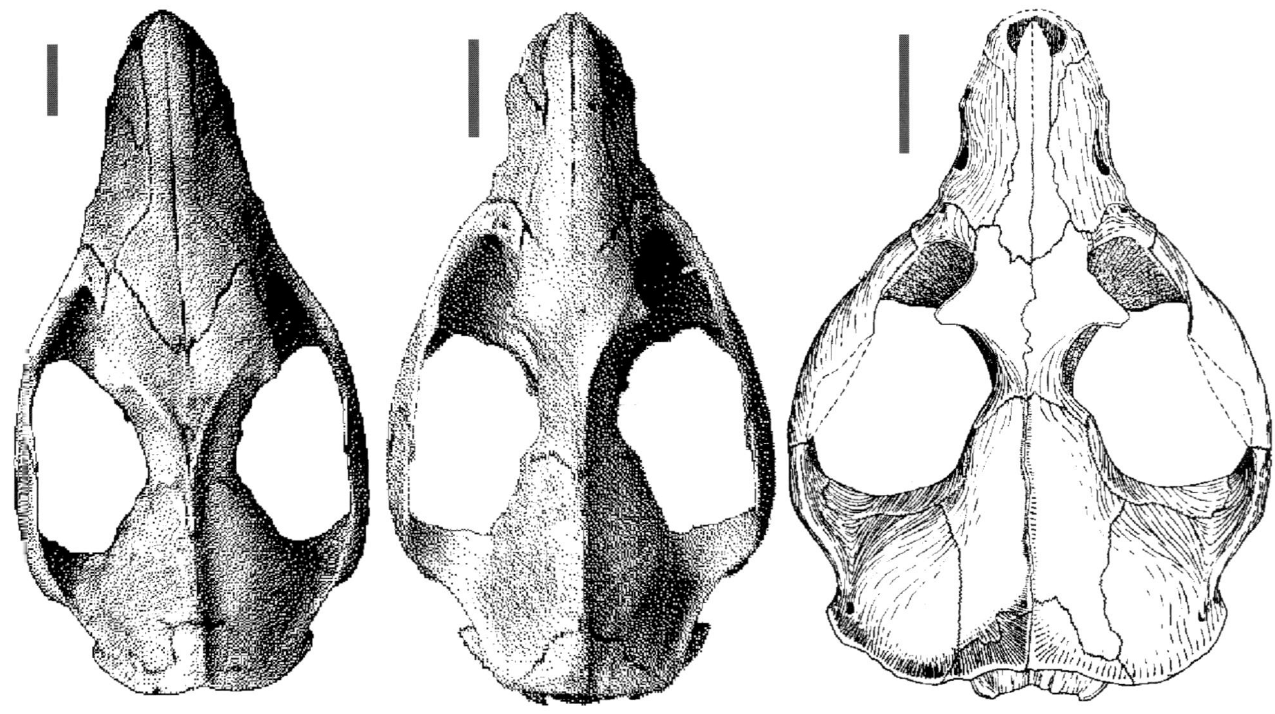

Figure 4 Skulls of the extant marsupials *Didelphis aurita* (left) and *Lutreolina crassicaudata* (center), and also of the extinct *Sparassocynus derivatus* (right). The bar represents 1 cm for each skull. Note the proportional reduction of the snout and enlargement of the temporal fossa in the carnivorous *L. crassicaudata* and *S. derivatus*. Skull of *Sparassocynus* reproduced from Reig and Simpson (1972), with permission from the Zoological Society of London.

(Pascual and Jaureguizar 1990). The marsupials would have been unable to prey successfully on these new herbivores. There is not, however, any strong evidence to support such conjectures. In fact, the decline of the different members of the Borhyaenidae family occurred at different times, and thus, for different reasons. The decline of the Borhyaenidae spans from the Early Oligocene (Deseadan) to Late Pleistocene (Montehermosan) (Marshall 1978a). Indeed there is also a replacement of carnivorous forms from within the Borhyaenidae during this time. One hypothesis proposed for the disappearance of the large borhyaenid forms is competition with large cursorial and carnivorous ground birds, the phororacoids, which would have affected all terrestrial forms (Marshall 1978a).

On the other hand, semi-arboreal forms, as the Hathlyacyninae, present a decline which is simultaneous with the arrival of the first wave of waif-immigrant procyonids, but some may also have been replaced by large didelphids (see Marshall 1978a for more details). Nonetheless, most evidence indicates that carnivorous dog-like marsupials were already extinct prior to the appearance of the first Carnivora in the south American fossil record. The only exception seems to be the sabre-toothed marsupials (*Thylacosmilus*), who were apparently ecologically excluded by the newly arrived machairodonts towards the end of the Pliocene. The latest appearances of marsupial sabre-tooths were immediately followed by the advent of eutherian sabre-tooths in the fossil record.

Marsupial predators are believed to have become extinct by the end of the Pliocene or early Pleistocene (Simpson 1980). The most probable placental competitors (canids, felids, and ursids) did not appear until the Uquian land mammal age, in the Pleistocene, and at that age (or after) there is no fossil record of borhyaenids (Simpson 1980). Nevertheless it should not be forgotten that fossil records generally tell an incomplete history, and that new discoveries can always change the currently proposed scenario. The fact is that extant marsupials show a marked reduction in their kinds of life forms in relation to other epochs (Simpson 1980). The living representatives of the Neotropical marsupial group possess similar body form and do not show great niche diversity, all of them being small to medium-sized omnivorous animals that feed, more or less intensely, on other animals.

Vrba (1992) suggests an alternative theory for extinctions that occurred after the Great American Interchange that she named the Habitat theory. By such theory, these extinctions would have been related mainly to an exceptional cooling of South

America at the time of the appearance of the Panamanian land bridge. This cooling would have caused an increase in open vegetation areas with an equivalent reduction of previously extensive forested areas and consequent disappearance of specialists from such biome. Thus carnivorous marsupials would have been extinguished due to the disappearing of forest or dense woodland resources. Such a theory does not invoke interspecific competition, although both processes are not necessarily mutually exclusive.

The disappearance of most carnivorous marsupial lineages may have also had an influence on the radiation of the actual didelphimorphs. Large-bodied actual Didelphimorphs are recent and all closely related (see Goin, this volume), and their present radiation may be related to the opening of niches previously occupied by smaller extinct boryhaenoid lineages, while early members were all small-bodied, probably due to the use of niches unoccupied by larger borhyaenids.

THE LIVING SPECIES

Living Neotropical marsupials, all of them from the orders Didelphimorphia, Paucituberculata and Microbiotheria, are more homogeneous in general cranial and body shape than the extinct forms, specially the Didelphimorphia. Nevertheless, several species present behavioural, morphological and physiological characteristics adequate for carnivory. As discussed before, the definition of strict trophic categories is imprecise (see Astúa de Moraes et al. in this volume for a more detailed discussion). This sort of delimitation is especially difficult for most South American living marsupial species, for which much still remains to be discovered on the composition and variation in food habits. This classification also may, depending on the author and the concept of carnivory/insectivory adopted, include or exclude some species (Hume 1999). Besides that, the position in the ranking displayed in Fig. 1 might also vary according to local characteristics of the habitat and even due to changes in methods of diet analysis. The position of the mouse opossums (*Gracilinanus*, *Marmosops*, *Micoureus*, and *Marmosa*) in relation to the common opossums (*Didelphis*), for instance, changes if one considers field data from faecal samples or else laboratory data on dietary preferences (see Astúa de Moraes et al. in this volume).

Overall tables indicating the dietary categories of Neotropical marsupials and reviews of the scant information on the feeding ecology of the species have already been published (see Eisenberg 1989; Eisenberg and Redford 1999, Hume 1999, Redford and Eisenberg 1992, Santori and Astúa de Moraes in press). In this section we will focus mainly on those species more commonly accepted as being more carnivorous or insectivorous for which there is morphological, physiological or behavioural data available, while knowing that none of them is a specialist in a strict sense.

Microbiotheriidae

The only living microbiotheriid, the Monito del Monte *Dromiciops gliroides*, does not present any marked morphological feature related to its diet, which can be varied. Analysis of stomach contents indicated that this species show a marked preference for arthropods and insect larvae, which may account for more than 70% of its diet (Meserve et al. 1988) and is able to chase prey on the ground and in trees, as it appears to be a scansorial opossum (Redford and Eisenberg 1992). Although this species feeds mainly on animal matter, plant matter may also be important for it. This species is the exclusive disperser of a lorantaceous mistletoe in Chile (Amico and Aizen 2000), indicating that *D. gliroides* is a constant consumer of this plant.

Caenolestidae

The Caenolestidae, commonly referred to as rat opossums, include three genera and six species. One major striking feature of the family is the excessive development of the first pair of lower incisors, which are frontally projected, while the remaining are reduced. Kirsch and Waller (1979) describe that captive *Caenolestes* and *Lestoros* individuals would readily accept meat in captivity, and when live baby rats were offered, *Caenolestes* would vigorously and repeatedly stab them with the lower incisors while holding them against the ground with its forepaws. This sort of behaviour would indicate (at least potential) predatory habits on small vertebrates, even though invertebrates were more frequently found in stomach remains. Particularly, the stabbing behaviour using the lower developed incisors resembles the proposed attack behaviour of the extinct Australian diprotodont thylacoleonid 'marsupial lions', who would also stab and tear the throat of its prey with their developed lower incisors (Archer 1984). Therefore this behaviour could represent an adaptation toward carnivorous habits. Caenolestids have peculiar lip flaps and Kirsch and Waller (1979) suggest that such structures help to prevent the sensory vibrissae and fur at the side of the mouth from becoming clogged with blood and dirt during feeding.

Barkley and Whitaker, Jr. (1984) report that lepidopteran larvae, centipedes, and arachnids are major foods of *Caenolestes fuliginosus* and that bird flesh is also consumed. The only species of the genus *Rhyncholestes*, *R. raphanurus*, is also primarily a consumer of invertebrates (about 55% of diet overall based on 31 stomach samples) and preys significantly on annelids (Meserve et al. 1988). These food items, along with the soricine-like body form of the latter suggest that *R. raphanurus* search for prey in the soil whereas *Caenolestes* species would chase for prey more often above ground.

Didelphidae

The Didelphidae (*sensu* Gardner 1993, including the Caluromyinae) include at least 68 small- to medium-sized species, among them many with different degrees of specialisations towards carnivory. Although the Didelphidae are usually considered a group with generally conservative morphology, many differences can be found in cranial and mandibular shape, skull musculature, digestive tract proportions, and digestive efficiency, that can be related to carnivory.

Caluromyinae. Although usually considered the most frugivorous among the Didelphidae (Astúa de Moraes et al. this volume, Charles-Dominique et al. 1981, Leite et al. 1996, Carvalho et al. 1999), some authors consider the importance of insects in the diet of species of the genus *Caluromys*. Carvalho et al. (1999) found arthropods in up to 50% of the samples of *C. philander* analysed and Medellín (1991) concluded from the morphological features of *C. derbianus* (developed masseter muscles and reduced teeth) that, while fruit may have greater importance, it includes hard-shelled insects in its diet. It seems that species of the genus *Caluromys* may even prey upon vertebrates. Fleming (1988) observed *C. derbianus* trying to get to netted bats on two distinct occasions and *C. lanatus* was observed preying upon echimyid rodents (D.M. Novaes pers. comm.). Astúa de Moraes (1998) noticed that the canines of adult *C. philander* present almost no wear, specially when compared to the other bigger body-sized species, which may indicate a lesser extent of use, when compared to most carnivorous species.

Feeding habits of the rare bushy-tailed opossum *Glironia venusta* are totally unknown. Nevertheless Emmons and Feer (1997), based on one field observation of what they considered a hunting behaviour, suggest that this species might be an arboreal insectivore. This would make the Bushy-tailed opossum the sole member of such feeding category in the Neotropics. Nevertheless the morphological and cranial similarities with *Caluromys* and *Caluromysiops* (Marshall 1982) suggest similar feeding habits for the entire Caluromyinae group.

Didelphinae. The Didelphinae is the most speciose extant marsupial sub-family, with about 63 species. Most of them present similar body morphology, with a size range of 10 to more than 2000 g. Within the Didelphini tribe, no morphometric differences in the skulls and mandibles of the genera *Didelphis, Philander* and *Chironectes* that could be directly related to the amount of animal prey in their diets have been found (Astúa de Moraes 1998). Though these species are morphometrically clearly distinct, these distinctions are probably more related to their phylogenetic history or to allometric differences. However, Medellín (1991) states that the Water opossum *Chironectes minimus* has bigger teeth (larger molar area and very developed molar cusps) and more voluminous masticatory muscles than the Common grey four-eyed opossum *Philander opossum*, implying a diet including more snails and crustaceans than fish. On the other hand, laboratory experiments have shown that, when compared with the Southern common opossum *Didelphis aurita*, the Southern grey four-eyed opossum *Philander frenata* is more effective in meat digestion, and it also digests meat more effectively than fruit (Santori et al. 1995b). In this sense, it can be said that *Philander* species are more carnivorous than those of the genus *Didelphis*.

The diet of grey four-eyed opossums is quite diverse. According to Santori et al. (1997) *P. frenata* feeds mainly on insects (Formicidae, Coleoptera and Blattariae) and small vertebrates (rodents, reptiles and birds). The animal diet of another congeneric species, *P. opossum*, is composed of big earthworms that emerge from the damp soil after heavy rains, arthropods (mainly Coleoptera, Hymenoptera, Orthoptera), small vertebrates and carrion (Charles-Dominique 1983, Santori and Astúa de Moraes in press). With this species Tuttle et al. (1981) published one of the few studies of didelphid hunting behaviour in field conditions, showing that this opossum locates individual calling frogs acoustically. They also reported a strikingly high capture rate of such leptodactylids (*Physalaemus pustulosus*) with 39 captures during 37 h of observation in only one pond in Panama (Tuttle et al. 1981). These data suggest that, at least under specific conditions (e.g. frog breeding period), vertebrates might be more important in *P. opossum* diet than generally recognised.

Frogs are also part of the diet of the Water opossum *C. minimus* (Fig. 5), which also eats small fish, aquatic insects and other invertebrates such as crabs, although the relative importance of these prey types in the diet is not known (Marshall 1978b and references therein). This is the only didelphid with semi-aquatic habits, being found in rivers and streams with clear water. This species swims by alternate strokes of its webbed hind feet while keeping the forefeet extended in front (Fish 1993). Paddling

Figure 5 The Water opossum *C. minimus* is a Neotropical marsupial with semi-aquatic habits and feeds on small fish, frogs and aquatic invertebrates (e.g. insects and crabs). Photograph by D. Astúa de Moraes

with hind feet enables this marsupial to forage while swimming, feeling for aquatic prey with the forefeet (with its tactile sensitive expanded digital pads) extended anteriorly (Fish 1993).

Another didelphid with essentially carnivorous habits is the Lutrine or Thick-tailed opossum *Lutreolina crassicaudata* (Fig. 6). It is an agile and aggressive hunter whose diet is composed of rodents, marsupials, lagomorphs, birds, snakes, frogs and invertebrates (Marshall 1978c, Monteiro-Filho and Dias 1990, Hunsaker 1977, Hume 1999, Sazima 1992). This marsupial displays several features that can be related to its increased carnivory. A geometric morphometric analysis of skull shape indicated that when compared to the remaining larger body-sized opossums (*Didelphis, Philander, Chironectes, Metachirus* and *Caluromys*), *Lutreolina* has a markedly shorter snout, with a shorter nasal bone (Astúa de Moraes et al. 2000 and Fig. 4). A short rostrum is a feature repeatedly found in carnivores, and it is related to the reduction of torsion forces, when asymmetrical pressures occur on the canines (when only one side occludes on the food and the other stays afloat) (Covey and Greaves 1994).

Figure 6 The Lutrine opossum *Lutreolina crassicaudata* is one of the most carnivorous didelphids. Photograph by D. Astúa de Moraes.

In addition to those characteristics listed above, the Lutrine opossum also presents an increased temporal musculature, when compared to *Didelphis*, as a consequence of the shortening of the rostrum and increase of the area of the temporal fossa, generating a more efficient bite than *Didelphis* (Delupi et al. 1997). Geometric morphometrics also indicated that *L. crassicaudata* has a narrower and more elongated braincase than the other large-sized species, which reflects in a wider temporal fossa, along with a broader and more perpendicular coronoid process, related to the increase in size and volume of the temporal muscle, and its effectiveness (Astúa de Moraes et al. 2000). Broadening of the coronoid process is related to the increased volume of the temporal, as it represents an increased area of insertion of the temporal muscle, while an ascending ramus of the mandible more perpendicular to the horizontal ramus allows a more effective moment arm for the force applied (in this case, by the temporal muscle). All these features undoubtedly characterise *Lutreolina crassicaudata* as the most carnivorous among all Neotropical opossums.

Another relatively large Neotropical opossum, the Brown four-eyed opossum, *Metachirus nudicaudatus*, also presents several morphological features that can be related to its feeding habits. Contrary to the classification of Streilein (1982), field data indicate that this species is one of the most insectivorous of the big opossums (Santori et al. 1995a, Astúa de Moraes et al. this volume) and also preys on small mammals and lizards (Santori et al. 1995a, Medellín et al. 1992). Brown four-eyed opossums possess a broad rostrum, which may reflect an increase in the olfactory bulbs, along with relatively shorter canines and longer molar series resulting in an increase in the crushing and cutting surface (Astúa de Moraes 1998). All of these features are common patterns in insectivores, and may be related to the insectivore habits of *Metachirus*.

The species of the genus *Didelphis* are considered to be the most generalist among the Didelphidae. Diversity of its diet, whether based on field data (Santori et al. 1995a) or laboratory experiments (Astúa de Moraes et al. this volume) is consistently higher than remaining opossums. No clear morphological specialisation could be found in the skull shape and musculature of representatives of the genus (Astúa de Moraes et al. 2000, Medellín 1991). Species of *Didelphis* include in their diets all sorts of items, from vertebrates (marsupials, rodents, lizards, frogs, birds, eggs, fish and carrion) to invertebrates (at least for *D. aurita*, it seems that litter fauna is one major source of resources), nectar, and up to forty species of fruits (Santori et al. 1995a, Monteiro-Filho 1987, Charles-Dominique et al. 1981, see Santori and Astúa de Moraes in press for a major review).

The small murine opossums (genera *Marmosa, Marmosops, Gracilinanus* and *Micoureus*) are also omnivorous species. These marsupials are usually reported as being mainly insectivorous, though seeds have been reported in most field studies, sometimes with an elevated germination success (Leite et al. 1996, Carvalho et al. 1999, Vieira and Izar 1999). These data, coupled with laboratory experiments (Astúa de Moraes et al. this volume), and with the fact that many didelphid have been seen consuming only the pulp of many fruits, rejecting bigger seeds and swallowing only small ones (Charles-Dominique et al. 1981), may indicate that in fact these small species may be including more fruits in their natural diet, and that this data has been overlooked due to methodological limitations.

The other carnivorous/insectivorous opossums from the Neotropics are all of small size, generally with an adult weight of less than 100 g. The Patagonian opossum, *Lestodelphys halli*, is usually considered a carnivorous species that might live like some of

Figure 7 The Fat-tailed mouse opossum *Thylamys velutinus*, a carnivore-insectivore from savannah-like habitats in central Brazil. Note the incrassation at the base of the tail. Photograph by A. Sebben.

Figure 8 The three-striped short-tailed mouse opossum *Monodelphis americana* feeds mainly on invertebrates and inhabits forested habitats in South America. Photograph by D. Astúa de Moraes.

the north temperate weasels, foraging under the snow for small rodents and other animals or, alternatively, enter periods of torpor during winter (Redford and Eisenberg 1992). This animal has exceptionally long canines and strong claws, and a very aggressive and voracious feeding behaviour, being able to eat a vertebrate prey (mouse) half its weight in one night (Hume 1999, Marshall 1977).

A similar ferocious behaviour is shown by the fat-tailed opossum *Thylamys velutinus* (E.M. Vieira pers. obs.). Its common name is derived from a seasonal deposit of fat at the base of the tail (see Fig. 7). Fat-tailed opossums are associated with open areas of South America and feed mainly on invertebrates, but also on small vertebrates and even on small animal carcasses (Simonetti et al. 1984, Vieira and Palma 1996, Palma and Vieira in press). Tail incrassation is also shown by other mouse opossums (e.g. *L. halli* and *D. australis*). This is an adaptation common in small mammals of desert regions (Harris 1987), being associated with life in seasonal or relatively xeric environments (Creighton 1985). Thus this feature helps small carnivorous/insectivorous marsupials to cope with relatively unfavourable environments with seasonal paucity of food, as areas where extended dry seasons (for *T. velutinus*) or else hard winters (for *L. halli* and *D. australis*) occur.

Another group of small carnivorous/insectivorous didelphids is formed by the short-tailed opossums of the genus *Monodelphis*. There are at least 15 species of this genus in the Neotropics (Gardner 1993) generally weighing from 8 to 100 g. These mouse opossums are mainly forest dwellers that search for invertebrates and also small vertebrates or even carrion (Busch and Kravetz 1991, Emmons and Feer 1997) on and under the leaf litter. Some of these animals are brown above with one to three black dorsal stripes (e.g. *Monodelphis americana*, *M. unistriata*, *M. theresa*, see Fig. 8). This colour pattern may provide some concealment for these short tailed opossums that possibly search for prey in daylight.

General patterns related to carnivory/insectivory

The carnivorous/insectivorous diet may impose some ecological constraints to new world opossums. Emmons (1995) states that the pursuit of vertebrate and invertebrate prey may not be compensatory enough for strictly arboreal mammals. Therefore, the more carnivorous Neotropical marsupials generally would not occur mainly in the canopies or at least would be scansorial species foraging frequently on the ground. This pattern was also suggested by Charles-Dominique (1983), who found, based on diet analysis of five species, that arboreal species are more frugivorous than the terrestrial ones. We tested the validity of this assumption for the entire Neotropical marsupial group by comparing the ordination of the genera in diet categories (Fig. 1) with an ordination of genera based on degree of arboreal activity (modified from Vieira in press). The results indicated that there really is a negative and significant correlation between arboreal activity and degree of carnivory (Spearman rank correlation coefficient $r_s = -0.88$, $P < 0.001$; Fig. 9).

Besides being less arboreal than more frugivorous species, carnivorous/insectivorous marsupials also have to cope with other constraints, such as relatively larger areas required to fulfil their nutritional needs. In general, animals that rely on resources at high trophic levels have less energy available in a given area and require a relatively higher range of activity than animals feeding on resources at low trophic levels (Eisenberg 1981). Therefore carnivorous/insectivorous marsupials should have relatively lower numerical densities and lower biomasses than more omnivorous correlates. In order to be able to compare densities of carnivorous/insectivorous marsupials in relation to omnivorous ones it is

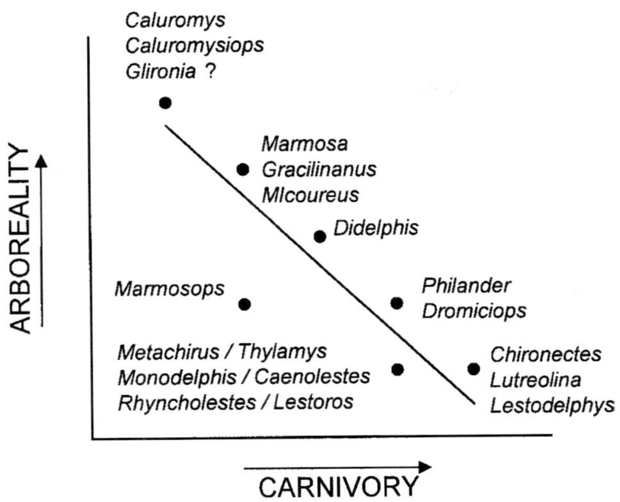

Figure 9 Arboreal activity (from Vieira in press) of Neotropical genera of marsupials as a function of diet category (from Fig. 1).

necessary to analyse similar-sized species studied in the same area with the same field methods, as home range sizes and overall density estimates are influenced by methodological procedures (e.g. trap arrangement). Besides that, patterns of vertical habitat utilisation also must be considered to provide reliable density estimates (Malcolm 1991). Studies that fulfil such requirements in the Neotropics are scarce. One of the few exceptions is the study of Malcolm (1991) in the Brazilian Amazon forest. This author used terrestrial and arboreal trapping and showed that the density of the Red-legged short-tailed opossum *Monodelphis brevicaudata* is lower than the densities of the more omnivorous and similar-sized mouse opossums *Marmosa murina*, *Marmosa cinerea* (= *Micoureus demerarare*), and *Marmosa* (= *Marmosops*) *parvidens.* This pattern appears to hold for the Three-striped short-tailed opossum *Monodelphis americana* when compared to the Brazilian gracile mouse opossum *Gracilinanus microtarsus* in the Atlantic forest in south-eastern Brazil (E.M. Vieira unpublished). Nevertheless it should be considered that reported abundances of *Monodelphis* spp. might be underestimated due to the fact that these animals are not easily captured with standard live traps (Voss and Emmons 1996).

On the other hand, feeding requirements may also provide some advantages for carnivorous/insectivorous marsupials. As these animals are specialised in high-energy food, they are able to survive in less favourable environments like areas at higher latitudes in South America where hard winters occur. In these areas with lower productivity, fruits and other plant food sources also are relatively more scarce. An analysis of the southern limits of geographical distribution of most Neotropical marsupials indicates that there is not a significant difference between distributions of omnivores (classes I, II and III in Fig. 1) and carnivore/insectivores (Kolmogorov-Smirnov test, $P > 0.15$). Both groups show a sharp decrease in number of species after 30° S, probably related to the reduction or disappearance of forested habitats. Nevertheless only the group of carnivorous/insectivorous marsupials has species with southern limits of distribution going beyond 40° S (Fig. 10). The tendency of carnivore/insectivores to reach higher southern latitudes is confirmed by statistical comparison between mean latitudinal values of southern distribution limits of both groups, as carnivore/insectivores show values significantly higher than omnivores ($t = 2.01$, d.f. = 52, $P = 0.05$).

FUTURE DIRECTIONS

About 20–25 years ago the scientific knowledge on Neotropical marsupials was very limited. This situation was recognised by review studies on marsupials published at that time (e.g. Hunsaker 1977, Lee and Cockburn 1985, Streilein 1982). These studies used mainly available data from species of a few genera (mainly *Didelphis* and *Monodelphis*) to discuss the biology and ecology of the group. Fortunately this scenario has changed a lot in the last two decades (see Eisenberg 1989; Eisenberg and Redford 1999, Redford and Eisenberg 1992, Hume 1999 for reviews of the current knowledge on Neotropical marsupials). Nevertheless for several species detailed information on basic life traits is still scarce or non-existent. Consequently population and community processes are also still poorly understood. This scenario holds for the carnivorous/insectivorous species and basic studies focusing on diet of these species are still necessary.

Most available field data on diet of carnivorous/insectivorous marsupials comes from faecal analysis (see reviews in Santori and Astúa de Moraes in press and Hume 1999). Although this method is considered a relatively reliable one (Dickman and Huang 1988) it may be sometimes biased (see Astúa de Moraes

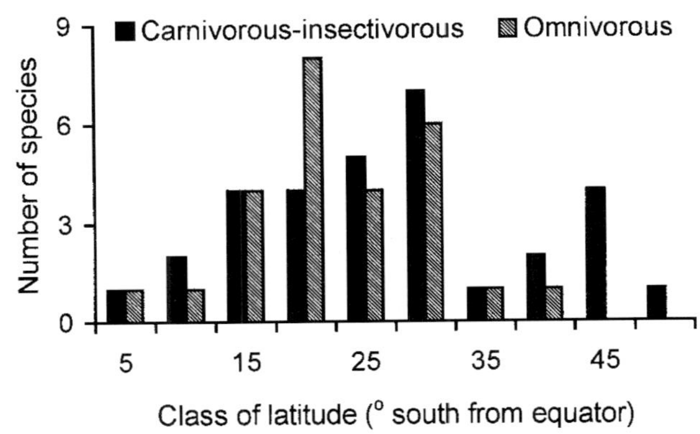

Figure 10 Southern limit of the geographical ranges of Neotropical marsupials (based on distribution maps of Redford and Eisenberg 1992 and Eisenberg and Redford 1999).

et al. this volume). On the other hand, analysis of stomach contents have some disadvantages over scat analysis (see Hume 1999) and is also prone to biases. Nevertheless comparative analysis of both methods for Neotropical marsupials would be very useful. Whenever possible, alternative methods that do not require killing of the animals should be used (such as the method described by Kronfeld and Dayan 1998). Besides that studies conjugating field and laboratory data are very important. These laboratory studies might be conducted not only through cafeteria experiments but also involving observations of preying behaviour on live prey.

Some carnivorous/insectivorous didelphids, such as *Monodelphis* spp. and *Thylamys* spp. are not easily captured with standard live-traps. In order to obtain relevant ecological data on such species other trapping procedures, such as pitfall trapping, should be performed together with standard live-trapping.

Other limitation of our knowledge on carnivorous Neotropical marsupials is the relatively short time length of data collection in the field. Studies of more than two years' duration are very uncommon. Because of that there is a strong paucity of information on medium and long-term population dynamics of predators and prey, types of predator response to seasonal and stochastic food shortages, and temporal variation on potential competitive interactions among predators, among other topics. We believe that only with long term field studies (i.e. ≥3 years) these and many others relevant ecological topics about carnivorous Neotropical marsupials would be adequately addressed.

Acknowledgements

We are grateful to the Academia Brasileira de Ciências, through Dr Diogenes de Almeidas Campos, to the Zoological Society of London, through Dr Linda DaVolls, and to Christian de Muizon, for the authorisation of use of the images of borhyaenids from Paula Couto (1979), of *Sparassocynus* from Reig and Simpson (1972) and of the *Sypalocyon* teeth from Muizon and Lange-Badré (1997), respectively. We are grateful to Jim Patton, Menna Jones, Flávia N. Sá and an anonymous referee, for the many useful suggestions made to improve the text, and to Francisco Goin for providing us with the manuscript of his chapter. DAM is supported by a grant from FAPESP.

References

Amico, G., & Aizen, M.A. (2000), 'Mistletoe seed dispersal by a marsupial', *Nature*, **408**:929–30.

Archer, M. (1984), 'The Australian marsupial radiation', in *Vertebrate zoogeography & evolution in Australasia* (eds. M. Archer & G. Clayton) pp. 633–808, Hesperian Press.

Argot, C. (2001), 'Functional-adaptive anatomy of the forelimb in the Didelphidae, and the paleobiology of the Paleocene marsupials *Mayulestes ferox* and *Pucadelphys andinus*', *Journal of Morphology*, **247**:51–79

Astúa de Moraes, D., Hingst-Zaher, E., Marcus, L.F., & Cerqueira, R. (2000), 'A geometric morphometric analysis of cranial and mandibular shape variation of didelphid marsupials', *Hystrix – Italian Journal of Mammalogy (n.s.)*, **11**:115–30

Astúa de Moraes, D., Santori, R.T., Finotti, R., & Cerqueira, R. 'Nutritional and fibre contents of laboratory-established diets of Neotropical opossums (Didelphidae)', in *Predators with pouches: The biology of carnivorous marsupials* (eds. M. Jones, C. Dickman & M. Archer), pp. 229–37, CSIRO Publishing, Melbourne.

Astúa de Moraes, D. (1998), 'Análise Morfométrica da Mandíbula e do Crânio de Marsupiais Didelfídeos: Implicações Ecológicas e Funcionais', unpublished MSc dissertation, Universidade Federal de Minas Gerais, Belo Horizonte, MG, Brasil.

Barkley, L.J., & Whitaker, Jr., O. (1984), 'Confirmation of *Caenolestes* in Peru with information on diet', *Journal of Mammalogy*, **65**:328–30.

Bond, M., & Pascual, R. (1983), 'Nuevos y eloquentes restos craneanos de *Proborhyaena gigantea* (Marsupialia, Borhyaenidae, Proborhyaeninae) de la Edad Deseadense. Un ejemplo de coevolucion', *Ameghiniana*, **1–2**:47–60.

Busch, M., & Kravetz, F.O. (1991), 'Diet composition of *Monodelphis dimidiata* (Marsupialia, Didelphidae)', *Mammalia* **55**:619–21.

Carvalho, F.M.V., Pinheiro, P.S., Fernandez, F.A.S., & Nessimian, J.L. (1999), 'Diet of small mammals in Atlantic Forest fragments in southeastern Brazil', *Revista Brasileira de Zoociências*, **1**:91–101.

Charles-Dominique, P. (1983), 'Ecology and social adaptation in didelphid marsupials: comparisons with eutherians of similar ecology', in *Advances in the study of mammalian behavior* (eds J. F. Eisenberg & D. G. Kleiman), pp. 395–422, American Society of Mammalogists, Shippensburg, PA.

Charles-Dominique, P., Atramentowicz, M., Charles-Dominique, M., Gérard, H., Hladik, C.M., & Prévost, M.F. (1981), 'Les mammifères frugivores arboricoles nocturnes d'une forêt guyannaise: inter-relations plantes–animaux', *Revue d'Ecologie (La Terre et La Vie)*, **35**:341–435.

Covey, D.S.G., & Greaves, W.S. (1994), 'Jaw dimension and torsion resistance during canine biting in the Carnivora', *Canadian Journal of Zoology*, **72**:1055–60.

Creighton, G.K. (1985), 'Phylogenetic inference, biogeographic interpretations, and patterns of speciation in *Marmosa* (Marsupialia: Didelphidae)', *Acta Zoologica Fennica*, **170**:121–4.

Delupi, L.H., Carrera, M.H., & Bianchini, J.J. (1997), 'Morfología comparada de la musculatura craneal de *Lutreolina crassicaudata* (Desmarest, 1804) y *Didelphis albiventris* Lund, 1840 (Marsupialia, Didelphidae)', *Physis (Buenos Aires) Sección C*, **53**:19–28.

Dickman, C.R., & Huang C. (1988), 'The reliability of faecal analysis as a method for determining the diet of insectivorous mammals', *Journal of Mammalogy*, **69**:108–13.

Eisenberg, J.F. (1981), *The mammalian radiations: An analysis of trends in evolution, adaptation, and behaviour*, University of Chicago Press, Chicago.

Eisenberg, J.F. (1989), *Mammals of the neotropics: The northern neotropics*, University of Chicago Press, Chicago.

Eisenberg, J.F., & Redford, K.H. (1999), *Mammals from the Neotropics: The central Neotropics: Ecuador, Peru, Bolivia, Brazil*, University of Chicago Press, Chicago.

Emmons, L.H. (1995), 'Mammals of rain forest canopies', in *Forest canopies* (eds M.D. Lowman & N.M. Nadkarni), pp. 199–223, Academic Press, London.

Emmons, L.H., & Feer, F. (1997), *Neotropical rainforest mammals: a field guide*, University of Chicago Press, Chicago.

Fish, F.E. (1993), 'Comparison of swimming kinematics between terrestrial and semiaquatic opossums', *Journal of Mammalogy,* **74**:275–84.

Fleming, T.H. (1988), *The Short-Tailed Fruit Bat: A Study in Plant–Animal Interactions*, University of Chicago Press, Chicago.

Flynn, J.J.& Wyss. A.R. (1998), 'Recents advances in South American mammalian paleontology', *Trends in Ecology and Evolution,* **13**:449–54.

Gardner, A.L. (1993), 'Order Didelphimorphia', in *Mammal species of the world* (eds D.E. Wilson & D.M. Reeder), pp. 15–23, Smithsonian Institution Press, Washington.

Goin, F.J. (2003), 'Early marsupial radiations in South America', in *Predators with pouches: The biology of carnivorous marsupials* (eds. M. Jones, C. Dickman & M. Archer), pp. 30–42, CSIRO Publishing, Melbourne.

Harris, J. (1987), 'Variation in the caudal fat deposit of *Microdipodops megacephalus*', *Journal of Mammalogy*, **68**:58–63.

Hume, I.D. (1999), *Marsupial Nutrition,* Cambridge University Press, Cambridge.

Hunsaker, D., II (1977), 'Ecology of new world marsupials', in *The biology of marsupials* (ed D. Hunsaker II), pp. 95–156, Academic Press, New York.

Kirsch, J.A.W., & Waller, P.F. (1979), 'Notes on the trapping and behaviour of the Caenolestidae (Marsupialia)', *Journal of Mammalogy,* **60**:390–437.

Kronfeld, N., & Dayan, T. (1998), 'A new method of determining diets of rodents', *Journal of Mammalogy,* **79**:1198–202

Lee, A.K., & Cockburn, A. (1985), *Evolutionary ecology of marsupials*, Cambridge University Press, Cambridge.

Leite,Y.L.R., Costa, L.P., & Stallings, J.R. (1996), 'Diet and vertical space use of three sympatric opossums in a Brazilian Atlantic forest reserve', *Journal of Tropical Ecology,* **12**:435–40.

Malcolm, J.R. (1991), 'Comparative abundances of neotropical small mammals by trap height', *Journal of Mammalogy,* **72**:188–92.

Marshall, L.G. (1976), 'Evolution of the Thylacosmilidae, extinct sabertooth marsupials of South America', *PaleoBios*, **23**:1–31.

Marshall, L.G. (1977), '*Lestodelphys halli*', *Mammalian species,* **81**:1–3.

Marshall, L.G. (1978a), 'Evolution of the Borhyaenidae, extinct South American predaceous marsupials', *University of California Publications in Geological Sciences,* **117**:1–87

Marshall, L.G. (1978b), '*Chironectes minimus*', *Mammalian species,* **109**:1–6.

Marshall, L.G. (1978c), '*Lutreolina crassicaudata*', *Mammalian species,* **91**:1–4.

Marshall, L.G. (1982), 'Evolution of South American marsupialia', in: *Mammalian biology in South America* (eds M.A. Mares & H.H. Genoways), pp. 251–72, Pymatuning Laboratory of Ecology, Special Publications 6, University of Pittsburgh: Pittsburgh

Marshall, L.G., & Cifelli, R.L. (1990), 'Analysis of changing diversity patterns in Cenozoic land mammal age faunas, South America', *Palaeovertebrata,* **19**:169–210.

McKenna, M.C., & Bell, S.K. (1997), *Classification of mammals above the species level*, Columbia University Press, New York.

Medellín, R.A., Cancino, G., Clemente, A., & Guerrero, R. (1992), 'Noteworthy records of three mammals from Mexico', *The Southwestern Naturalist,* **37**:427–9.

Medellín, R.A. (1991), 'Ecomorfología del cráneo de cinco didélfidos: tendencias, divergencias e implicaciones', *Anales del Instituto de Biología de la Universidad Nacional Autónoma de México, Ser Zool,* **62**:269–86.

Meserve, P.L., Lang, B.K., & Patterson, B.D. (1988), 'Trophic relationships of small mammals in a Chilean temperate rainforest', *Journal of Mammalogy,* **69**:721–30.

Monteiro-Filho, E.L.A. (1987), 'Biologia reprodutiva e espaço domiciliar de *Didelphis albiventris* em uma área perturbada na região de Campinas, Estado de São Paulo. (Mammalia – Marsupialia)', Master thesis, Universidade Estadual de Campinas, São Paulo.

Monteiro-Filho, E.L.A., & Dias, V.S. (1990), 'Observações sobre a biologia de *Lutreolina crassicaudata* (Mammalia: Marsupialia)', *Revista Brasileira de Biologia,* **50**:393–9.

Muizon, C. de (1998), '*Mayulestes ferox*, a borhyaenoid (Metatheria, Mammalia) from the early Palaeocene of Bolivia. Phylogenetic and palaeobiologic implications', *Geodiversitas,* **20**:19–142.

Muizon, C. de (1999), 'Marsupial skulls from the Deseadan (Late Oligocene) of Bolivia and phylogenetic analysis of the Borhyaenoidea (Marsupialia, Mammalia)', *Geobios,* **32**:483–509.

Muizon, C. de & Lange-Badré, B. (1997), 'Carnivorous dental adaptations in tribosphenic mammals and phylogenetic reconstruction', *Lethaia,* **30**:353–66.

Palma, A.R.T., & Vieira, E.M. (in press), 'O gênero *Thylamys* no Brasil: história natural e distribuição geográfica', in *Marsupiais do Brasil: Avanços em Evolução, Biologia e Ecologia* (eds. E.L.A. Monteiro-Filho & N.C. Caceres), IBAMA, Brasília.

Paula Couto, C. de (1979), *Tratado de Paleomastozoologia*, Academia Brasileira de Ciências, Rio de Janeiro.

Pascual, R., & Jaureguizar, E.O. (1990), 'Evolving climates and mammal faunas in Cenozoic South America', *Journal of Human Evolution,* **19**:23–60.

Redford, K.H., & Eisenberg, J.F. (1992), *Mammals of the Neotropics: The southern cone*, Chicago University Press, Chicago.

Reig, O.A., & Simpson, G.G. (1972), '*Sparassocynus* (Marsupialia, Didelphidae), a peculiar mammal from late Cenozoic of Argentina', *Journal of Zoology (London),* **167**:511–39.

Santori, R.T., & Astúa de Moraes, D. (in press), 'Alimentação, nutrição e adaptações alimentares de marsupiais brasileiros', in *Marsupiais do Brasil: Avanços em Evolução, Biologia e Ecologia* (eds. E.L.A. Monteiro-Filho & N.C. Caceres), IBAMA, Brasília.

Santori, R.T., Astúa de Moraes, D., & Cerqueira, R. (1995a), 'Diet composition of *Metachirus nudicaudatus* and *Didelphis aurita* (Didelphimorphia, Didelphidae)', *Mammalia,* **59**:511–16.

Santori, R.T., Cerqueira, R., & Kleske, C.C. (1995b), 'Anatomia e eficiência digestiva de *Philander opossum* e *Didelphis aurita* em relação ao hábito alimentar', *Revista Brasileira de Biologia,* **55**:323–9.

Santori, R.T., Astúa de Moraes, D., Grelle, C.E.V., & Cerqueira, R. (1997), 'Natural diet at a Restinga forest and laboratory food preferences of the opossum *Philander frenata* in Brazil', *Studies on Neotropical Fauna and Environment,* **32**:12–6

Sazima, I. (1992), 'Natural history of the jararaca pitviper, *Bothrops jararaca*, in southeastern Brazil', in *Biology of the pitvipers* (eds J.A.E. Campbell & E.D. Brodie Jr), pp. 199–216, Selva, Tyler.

Simonetti, J.A., Yanez, J.L., & Fuentes, E.R. (1984), 'Efficiency of rodent scavengers in central Chile', *Mammalia,* **48**:608–9.

Simpson, G.G. (1980), *Splendid isolation*, Yale University Press, New Haven.

Streilein, K.E. (1982), 'Behaviour, ecology, and distribution of the South American marsupials', in *Mammalian biology in South America* (eds M.A. Mares & H.H. Genoways), pp. 231–50, Pymatuning Laboratory of Ecology, Special Publications, 6, University of Pittsburgh, Pittsburgh.

Tuttle, M.D., Taft, L.K., & Ryan, M.J. (1981), 'Acoustical location of calling frogs by *Philander opossum*', *Biotropica*, **13**:233–4.

Vieira, E.M. (in press), 'Padrões de uso vertical do hábitat por marsupiais brasileiros', in *Marsupiais do Brasil: Avanços em Evolução, Biologia e Ecologia* (eds. E.L.A. Monteiro-Filho & N.C. Caceres), IBAMA, Brasília.

Vieira, E.M., & Izar, P. (1999), 'Interactions between aroids and arboreal mammals in the Brazilian Atlantic rainforest', *Plant Ecology*, **145**:75–82.

Vieira, E.M., & Palma, A.R.T. (1996), 'Natural history of *Thylamys velutinus* (Marsupialia, Didelphidae) in central Brazil', *Mammalia*, **60**:481–4.

Voss, R.S., & Emmons, L.H. (1996), 'Mammalian diversity in Neotropical lowland rainforests: a preliminary assessment', *Bulletin of the American Museum of Natural History*, **230**:1–115

Vrba, E.S. (1992), 'Mammals as a key to evolutionary theory', *Journal of Mammalogy*, **73**:1–28.

Webb, S. D. (1991), 'Ecogeography of the great American interchange', *Paleobiology*, **17**:226–80.

Webb, S.D. (1999), 'Isolation and interchange a deep history of South American mammals', in *Mammals of the Neotropics* (eds J.F. Eisenberg & K.H. Redford), pp. 13–9, Chicago Press, Chicago.

CHAPTER 19

CONVERGENCE IN ECOMORPHOLOGY AND GUILD STRUCTURE AMONG MARSUPIAL AND PLACENTAL CARNIVORES

Menna E. Jones

School of Zoology, University of Tasmania, Hobart, TAS 7001, Australia. Current address: School of Botany and Zoology, Australian National University, Canberra, ACT 0200, Australia

The extent of ecomorphological convergence in function and form between marsupial and placental carnivores is analysed. Convergence is apparent in trophic functional morphology, with the genera of marsupial carnivores grouping with the families of placental carnivores with which they seem convergent from external appearances. Convergence in trophic morphology or form is superficial, however. A major difference in canine tooth shape relates to different killing behaviour in the marsupial carnivores. Other phylogenetically based differences in skull morphology may represent different solutions to the same biomechanical problems. In terms of morphologies associated with locomotor and hunting behaviour, the marsupial carnivores tend to group together, as does each family of placental carnivore. The skeletal indicators of running speed, activity substrate and prey capture mode appear to be more useful in separating and assigning major groups to their predominant locomotory type than in assigning different species within those groups to their known hunting type and activity substrate classifications. This suggests a strong influence of phylogenetic history on skeletal morphology. Similar morphological size patterns, in the trophic structures that are proximal to prey capture methods, have been found in guilds of both marsupial and placental carnivores. These patterns have an underlying basis in food size partitioning and competitive pressure (evidence from character release). These similar patterns attest to convergence in the structuring mechanisms that influence morphological size relationships within ecological guilds.

INTRODUCTION

The ecology and morphology of species are influenced by ecological forces, both in the abiotic environment and through species interactions such as competition. The end result, however, also can be strongly influenced by phylogenetic history, especially where phylogenetic constraint operates to restrict evolutionary options. Where distantly related species evolve in similar selective environments, some convergence in form and function can be expected. However, important differences in apparently similar environments, differences in environmental history, or phylogenetic constraints mean that the degree of convergence, and similarities and differences, need to be investigated carefully (Wiens 1992). In particular, convergence in function is often not particularly well matched with convergence in form. Behavioural flexibility, in some instances, appears to allow convergence in ecological attributes in species with divergent morphologies (reviewed in Ricklefs and Miles 1994).

One of the classic cases of apparent convergence is that between marsupial and placental mammals, no better exemplified than by the carnivore guilds. The marsupial quolls (*Dasyurus* spp.; Dasyuromorphia: Dasyuridae) from Australia and New Guinea are similar in body size, body form and locomotion to the placental civets and genets (Carnivora: Viverridae) and mongooses (Herpestidae), and to a lesser extent to martens and polecats (Mustelidae), and raccoons (Procyonidae). These are all small to medium-sized carnivores with short legs, somewhat elongate bodies and lithe movements. Tasmanian devils (*Sarcophilus*; Dasyuridae) share the massive, bone-cracking jaws and teeth, and the scavenging habits of the hyaenas (Hyaenidae). Devils resemble small hyaenas even to their sloping hindquarters and cantering gait. And the thylacines (*Thylacinus*; Dasyuromorphia: Thylacinidae) were at least superficially canid-like.

The similarities in general body form between marsupial carnivores and some placental carnivores are quite striking. General similarity in form and function is to be expected as the earliest mammals, predecessors of both marsupial and placental carnivores, were carnivore-insectivores with a body form not too dissimilar from quolls and mustelids-viverrids. That convergence is apparent among several specialised carnivore body forms (mongoose, hyaena and dog-like) is remarkable; however, given the long period of separate evolutionary history of marsupial and placental mammals. This case of apparent convergence can be interpreted as evolutionary adaptation of phenotype to similar selective environments in both faunas, rather than resulting from habitat selection by similar phenotypes (Ricklefs and Miles 1994). The premise that morphology reflects ecology, that is, that morphology and ecology represent the outcome of the ecological and evolutionary relationship of an organism to its environment, is fundamental to the ecomorphological approach (reviewed in Wiens 1989; Ricklefs and Miles 1994). Recent quantitative analyses based on ecomorphology reveal both ecological (convergent) and historical (ancestral) effects on morphology, at the levels of individual species and the biotic assemblage, and imply convergence in community organisation (Schluter and Ricklefs 1993).

Tests to determine whether similar positions in multivariate morphological niche space are occupied by species in different localities with similar environments have produced some support for the idea of community-wide convergence (Ricklefs and Miles 1994). This area of investigation has also led to searches for matched species pairs in communities from different regions, which implies the existence of discrete niches. While there are some compelling examples of closely convergent species pairs (e.g. Cody 1975; Fuentes 1976; Pianka 1986), more rigorous statistical testing has failed to support the notion of species-for-species matching (Ricklefs and Travis 1980; Wiens 1992; Ricklefs and Miles 1994). An extension of this approach to studying community-level convergence combines comparisons of community-wide occupancy of ecological and morphological space with estimates of species packing in the community volume (Ricklefs and Miles 1994). Analyses of community volume and nearest-neighbour distances among similar communities suggest that limiting similarity operates (Ricklefs and Miles 1994), alluding to an inherent structure of communities that is partly a result of interspecific interactions (Ricklefs and Miles 1994). The consideration of species diversity, which is central to this approach, and therefore species packing, is outside the scope of this chapter.

A second type of approach to studying community level convergence is the search for morphological patterns that reveal the ecological forces, specifically competition, that structure communities. The occurrence of convergent ecomorphological structure in communities that have different phylogenetic or historical origins suggests the existence of general structuring rules governing communities (Dayan and Simberloff 1994b). Non-random morphological size relationships among guild members (overdispersion of species means), known as community-wide character displacement (Strong et al. 1979), have generally been interpreted as resulting from interspecific competition (Dayan et al. 1989a; 1989b; 1990; 1992; 1994a). The mechanism may involve ecological resource partitioning by food size, which over evolutionary time results in morphological change or species sorting, and size-structured communities or guilds (Dayan and Simberloff 1998).

Because ecological communities are complex and can include a diversity of life forms, competition is more likely to occur and to be detected at the level of the guild (*sensu* Root 1967; Simberloff and Dayan 1991). As guilds are the functional building blocks of communities, patterns detected at this level are likely to hold also at the community level. Guilds are functional groupings ('a group of species that use the same class of resources in a similar way' Root 1967), that are free of taxonomic association, although they will often, by their nature, reflect taxonomy. The thylacine, devil and quolls can be considered as one guild (all species eat the same class of prey, mammals, using similar locomotory and killing behaviour) (Jones 1997), as can the viverrids and mustelids together, the canids, the felids, the hyaenids and the ursids (e.g. Dayan et al. 1989a; 1990; 1992). However, guilds are far from a precise categorisation, a point that will be demonstrated by the marsupial/placental comparisons below.

In this chapter, I will use ecomorphological analyses of the extant marsupial and placental carnivores to assess the extent of convergence in function and form of the two faunas. I am defining carnivores among the Australasian marsupial fauna, as those species for which vertebrate prey is an important part of the diet. The following questions will be addressed: 1) Do the genera and families, among the marsupial and placental faunas, respectively, that appear convergent, occupy similar positions in multivariate morphological niche space, and do they group into similar functional groups or guilds? Given the broad geographic ranges of most of the marsupial carnivore species, the diversity of habitats across their range in Australia and New Guinea, and the current relative paucity of the Australasian marsupial carnivore fauna, I will compare entire faunas rather than local guilds. 2) Is there convergence in ecomorphological size patterns between marsupial and placental guilds that would indicate common ecological forces, such as competition, structuring the guilds, and general rules that might govern communities?

FUNCTIONAL LINKS BETWEEN MORPHOLOGY AND ECOLOGY

A relationship between the form or morphology of an animal and its function – that is, its ecology, behaviour and physiology – underpins the ecomorphological analytical approach (Wiens 1989; Ricklefs and Miles 1994). For the first set of analyses, that of comparing positions in morphological space, I extended the morphometric database for placental carnivores published by Van Valkenburgh (1985; 1989, unfortunately restricted to carnivores >7 kg in body weight) to incorporate the Australasian marsupial carnivores. In the following subsections, I explain the morphometric ratios, as defined and explored by Van Valkenburgh for placental carnivores (1985; 1989), that were used in these analyses. In some cases, direct equivalence of morphological structures between marsupial and placental carnivores did not exist, so these measurements were adapted for marsupials. These ratios reflect aspects of diet (the relative proportions of meat, invertebrate and bone), prey-killing method, and locomotory and hunting behaviour. In addition, all species of marsupial carnivores were classified after Van Valkenburgh (1985; 1989) according to their dietary category (predominantly meat, meat/bone, meat/nonvertebrate), hunting type (fast pursuit, pounce/pursuit, slow ambush), and activity substrate (arboreal, scansorial, terrestrial, semi-fossorial) (see Table 1). Principal components analysis was used to show which attributes were important in describing convergence or non-convergence between marsupial and placental carnivores in ecomorphological space.

For the analysis of morphological size ratios, I measured five characters that relate directly to the feeding ecology of these predators: condylobasal or skull length (CBL), maximim anterio-posterior diameter of the upper canine tooth (APD), canine strength in bending about the anterio-posterior axis (S_x), an index of the size and strength of the temporalis muscle (SMA), and lower carnassial length (LCL) (Jones 1997). I measured skulls from Tasmania, where until recently four species of marsupial carnivores coexisted, and from three different mainland guilds, including the adjacent mainland where the same species of quolls were until recently extant but where the two larger Tasmanian carnivores had been extinct for a long period of time. In addition, body weight and mean prey mass of each sex and species were determined from a field population.

Skulls and skeletal material from adult animals of equal numbers of both sexes held by the following museums were measured using Vernier calipers (to 0.01 mm accuracy) for use in the analyses: Tasmanian Museum and Art Gallery, Hobart; Queen Victoria Museum, Launceston, Tasmania; Museum of Victoria, Melbourne; the Donald Thomson and the Department of Fisheries and Wildlife Collections housed at the Museum of Victoria; Australian Museum, Sydney; Queensland Museum, Brisbane; South Australian Museum, Adelaide; Western Australian Museum, Perth. Sample sizes and raw data are given in Table 1.

Diet and prey-killing behaviour

Killing behaviour and bone consumption: upper canine shape (CS)

The cross-sectional shape of the upper canine tooth reflects the stresses on the tooth during prey killing or bone consumption and, therefore, the method used to kill or the importance of bone in the diet (Van Valkenburgh 1989). These measurements were taken slightly differently on marsupial carnivore skulls, although not in a way that would affect comparability of the results. Because the canine teeth of marsupial carnivores continue to erupt throughout life (Jones unpubl. data), the medio-lateral (CW) and anterio-posterior (CL) widths of the upper canine tooth, from which canine shape (CW/CL) was calculated, were measured at the level of the gum rather than at the dentine–enamel junction.

Bone and invertebrate consumption: premolar or molar shape and relative size (PMD, RPS, respectively)

Greater width of the largest lower premolar of placental carnivores (PMD) reflects a greater importance of bone or invertebrates in the diet. A larger overall premolar size (RPS) separates the bone-eaters from the invertebrate-eaters (Van Valkenburgh 1989). Because of phylogenetic constraints on tooth eruption patterns in marsupials and the resultant differences in jaw geometry (Archer 1976; Werdelin 1987), the marsupial equivalent of a bone-cracking and invertebrate-crushing tooth is the second molar (M2). The second and first molars are the teeth that become most worn during life in all marsupial carnivores, especially in the bone-eating devil (Pemberton 1990, Jones unpubl. data). The marsupial carnivore second molar also is in an equivalent geometric and biomechanic position in the jaw to the placental largest premolar, being immediately in front of the molar teeth that function as carnassials (see below). In this position in the jaw, large bones can be manipulated and bite strength, while

Table 1 Trophic and foraging mode morphometric values and classifications for marsupial carnivores used in comparative analyses of marsupial and placental carnivores (data for placental carnivores and definitions and ratios from Van Valkenburgh 1985; 1989). Diet category: m = meat, mb = meat/bone, mn = meat/nonvertebrate. Activity substrate: t = terrestrial, s = scansorial. Hunting type: pp = pounce/pursuit, a = ambush. LBW = log body weight. Definitions for CS to ARCH are in the chapter text.

Species	*N* (diet)	*N* (locomotor)	Diet category	Activity substrate	Hunting type	Dental indices						Locomotor indices				
						LBW	CS	RPS	PMD	RBL	RGA	MCP	FMT	OLL	UD	ARCH
thylacine																
Thylacinus cynocephalus	10	2	m	t	pp	1.40	0.72	2.14	0.53	0.81	0.31	2.27	3.31	0.25	2.03	0.15
Tasmanian devil *Sarcophilus* laniarius	10	6	mb	t	pp	0.84	0.98	3.56	0.69	0.89	0.21	2.36	3.42	0.21	1.93	0.16
spotted-tailed quoll *Dasyurus maculatus*	10	8	m	s	a	0.39	0.76	2.84	0.65	0.78	0.34	2.36	3.23	0.22	2.15	0.15
eastern quoll																
D. viverrinus	10	4	mn	t	pp	-0.04	0.67	3.23	0.57	0.73	0.39	2.02	2.41	0.19	2.43	0.14
western quoll																
D. geoffroii	10	2	mn	t	pp	0.04	0.73	2.81	0.58	0.76	0.35	1.7	2.59	0.17	2.17	0.16
northern quoll																
D. hallucatus	10	2	mn	s	pp	-0.26	0.7	2.85	0.58	0.73	0.45	1.71	2.76	0.23	2.02	0.15
New Guinea quoll																
D. albopunctatus	7	2	mn	s?	pp	-0.22	0.73	2.92	0.63	0.76	0.45	1.32	2.75	0.2	2.31	0.15
Bronze quoll																
D. spartacus	2	0	mn	t	pp	-0.07	0.67	3.25	0.56	0.76	0.39					

not as great as more posteriorly, is still strong. PMD (ratio of maximum tooth width to maximum length) and RPS (maximum width divided by the cube root of body weight) were calculated in the same way that Van Valkenburgh (1989) did for placental carnivores, but on M2 instead of the largest premolar.

Vertebrate vs invertebrate diet: relative blade length (RBL) and grinding area (RGA)

Carnivores with a higher proportion of meat in their diet have longer relative cutting blade lengths and smaller relative grinding areas in their molar tooth row than those species for which invertebrate prey or plant matter are important (Van Valkenburgh 1989). Van Valkenburgh's measurements of RBL and RGA were adapted to account for non-equivalence in the molar dentition of marsupial carnivores. Unlike placental carnivore molars, which have a specialised carnassial blade on M1 and specialised grinding surfaces on the posterior molars, the four marsupial molar teeth (see Luckett 1993, for terminology) are similar, each retaining a cutting blade (the paracristid crest connecting paraconid to protoconid) and a grinding platform (the talonid) (Archer 1976). The length of the cutting blade and the size of the grinding area increase and decrease posteriorly, respectively.

Van Valkenburgh's (1989) relative blade length measurement (maximal length of the trigonid, the cutting blade, of the first molar tooth divided by the total length of M1) was adapted for marsupial carnivores. Blade length was taken as the combined length of the cutting blades of both M3 and M4, including the entire paraconid and protoconid, measured on the occlusal surface. Blade length was divided by maximum length of M3 plus M4 (M1 in placentals). Werdelin (1986) viewed M4 as the carnassial equivalent in dasyurids. I consider both M3 and M4 to be important in shearing meat, for two reasons. First, the equivalent position on the jaw relative to the condyle, which affects bite force, of the placental carnassial is halfway between M3 and M4 in the marsupial carnivores (Werdelin 1986). Second, neither M3 nor M4 wear heavily during an animals lifetime and so both retain an effective cutting function. Whether M4 alone or M3 and M4 together are used, the length measurement is not quite equivalent, being slightly less than or more than placental blade lengths, respectively.

Van Valkenburgh's (1989) index of relative grinding area (square root of the total grinding area of the molars divided by the total blade length of the carnassial) was modified for marsupials as follows. The area of the talonid on each molar was measured using a drawing attachment on a Wild dissecting microscope and Mocha Image Analysis Software (to 0.01 mm^2 accuracy). Relative grinding area was then calculated as the square root of the total grinding area of the molars divided by the combined blade length of M3 and M4 (M1 for placentals).

Locomotion and hunting mode

Running speed and activity substrate: hindlimb (FMT) proportions, manus (MCP) proportions and elbow shape (OLL)

Running speed of placental carnivores appears to be loosely correlated with the ratio of femur to metatarsal length (FMT), although this ratio is useful mainly in separating the occasional and semi-fossorial hunters (Van Valkenburgh 1985). Faster runners have longer metatarsals relative to femur length. The ratio of the third metacarpal to the proximal phalanx lengths (MCP) indicates the shape of the manus. Long metacarpals and short digits are characteristic of digitigrade, presumably faster moving animals, while the reverse, short metacarpals and long phalanges, are typical of arboreal animals which grasp branches (Van Valkenburgh 1985). The shape of the elbow, as measured by the ratio of the olecranon process of the ulna divided by the length of the main shaft of the ulna (OLL), indicates activity substrate. The olecranon process is long in semi-fossorial animals which dig, intermediate in terrestrial animals and shortest in arboreal species (Van Valkenburgh 1985). Measurements for these three morphometrics were directly equivalent in marsupial carnivores.

Activity substrate and prey handling behaviour: ungual depth and shape (UD, ARCH, respectively)

Arboreal and scansorial animals which use their claws for climbing tend to have deeper and more curved claws than terrestrial animals. In addition, animals which grasp their prey have more curved claws than those that do not use their forelimbs in prey capture. Claw depth is indicated by the ratio of proximal–distal ungual length measured along the dorsal curve divided by dorso–ventral height at the ungual base (UD) (Van Valkenburgh 1985). Claw curvature is estimated by the ratio of the chord which subtends the dorsal arc and the maximum height of the arc (ARCH) (see Van Valkenburgh 1985). These measurements were directly equivalent in marsupial carnivores.

CONVERGENCE IN TROPHIC FUNCTIONAL GROUPINGS

The genera of marsupial carnivores do indeed group functionally with the families of placental carnivores with which they appear convergent, despite differences in body sizes of the placental and marsupial carnivores compared. Van Valkenburgh (1985; 1987; 1989) confined analyses to animals heavier than 7 kg on the grounds that with fewer predators, the larger carnivores are more likely to experience and demonstrate the effects of competition. Of the Australasian marsupial carnivores (Dasyuridae and Thylacinidae), only the Tasmanian devil and the thylacine are greater than 7 kg in size; all the quolls are less than 7 kg. This size difference could impact on the conclusions, especially in the postcranial analyses, as large body size imposes constraints on locomotor

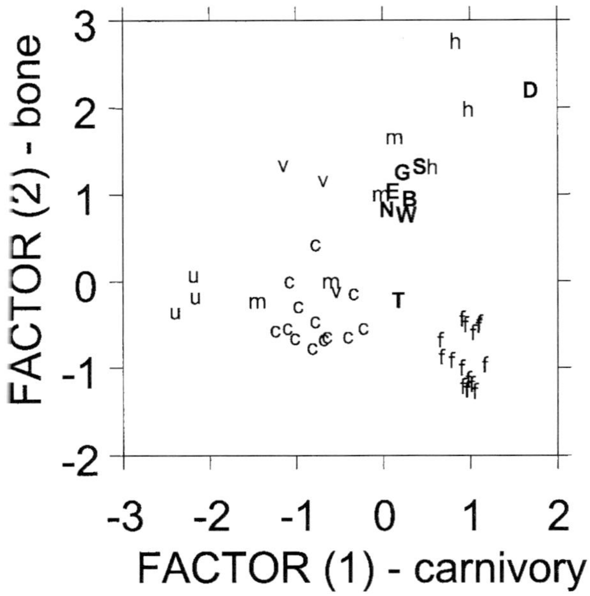

Figure 1 Plot of the first two Principal Components for the analysis of trophic functional groupings among marsupial and placental carnivores. Factor 1 describes the degree of carnivory (relative blade length and grinding area of the molar teeth, positive values indicate more carnivorous) and Factor 2 the importance of bone in the diet (premolar or molar shape and relative size; positive values indicate more bone) (Van Valkenburgh 1989). Capital letters represent species of marsupial carnivores: T = thylacine, D = Tasmanian devil, S = spotted-tailed quoll, W = western quoll, E = eastern quoll, N = northern quoll, G = New Guinea quoll, B = Bronze quoll (see Table 1 for scientific names). Small letters represent families of placental carnivores: f = felids, c = canids, h = hyaenids, m = mustelids, v = viverrids, u = ursids.

adaptations. This mismatch in body sizes could particularly affect comparisons of the quolls, although it must be noted that the devil also is much smaller than the hyaenas with which they appear convergent.

Fig. 1 shows the first two factors of the principal components analysis on the five morphometric ratios related to diet and killing behaviour. The first two factors explained 79% of the total variance. Factor 1 describes the degree of carnivory (proportion of vertebrate prey in the diet) and incorporates relative blade length and relative grinding area (component loadings: 0.91 and –0.95, respectively). Factor 2 includes premolar or molar shape and relative size (component loadings: 0.94 and 0.79, respectively), and indicates the importance of bone in the diet.

The Tasmanian devil, like the hyaenas, has a highly carnivorous dentition and trophic adaptations for bone consumption (D and h, Fig. 1). The quolls group together within the rather looser bounds of the viverrid/mustelid guild. Quolls have molar teeth with intermediate grinding and slicing functions. This attests to their mixed invertebrate/meat diet (Blackhall 1980; Belcher 1995; Oakwood and Eager 1997; Jones and Barmuta 1998). Quolls also have quite a robust second molar tooth. This is the tooth that is used for bone crushing in the devil, but as none of the quolls consume the bones of larger prey, robust molars are probably an adaptation for crushing invertebrates (see Van Valkenburgh 1989).

The thylacine groups with the canids. Their molar teeth are intermediate in grinding and slicing function and are quite slender, with no indications of adaptation for bone consumption. Thylacines have been reported gnawing on bones like a dog, rather than cracking and devouring them whole in the manner of a devil (see Smith 1981). A recent study suggested that thylacines were probably ecologically closer to smaller canids like the coyote, a solitary predator of prey that are small relative to its body size, than to the wolf, a cooperative hunter of very large prey (Jones and Stoddart 1998). Their extremely long rostrum or snout (see Werdelin 1986), very low rates of canine tooth wear and fracture, and limb ratios typical of slow runners, suggest that thylacines hunted prey smaller than their own body size and did not use long, fast pursuits (Jones and Stoddart 1998). Thylacines are associated in sub-fossil cave deposits with small to medium-sized herbivores, weighing from <1 kg to 5 kg (Case 1985). The range of prey reported for the thylacine (wombats, kangaroos, wallabies, pademelons, possums, bandicoots, small mammals and birds, Guiler pers. comm.; Smith 1981; Paddle 2000) suggests that thylacines were generalist predators of prey between <1 kg and 30 kg. It is therefore likely that mean prey size for the thylacine was larger than 5 kg but still small relative to their body size (15–35 kg, Rounsevell and Mooney 1995). Mean prey size of the thylacine probably substantially overlapped mean prey size of the more robustly built but much smaller devil (primarily wombats, wallabies, pademelons and possums, but including species between <1–30 kg Jones and Barmuta 1998), which is large relative to their body size (6–13 kg, Jones, unpub. data). Caution must be heeded, however, as there are no data for either thylacines or devils on the relative proportions of the smaller-bodied juvenile or larger adults of prey species taken. Age of prey individuals will make a substantial difference in comparisons of maximum prey size, particularly for the larger prey species (wombats, kangaroos, wallabies). For example, devils are known to attack adult wombats (30 kg), and to kill juveniles (10 kg), but their success rate is unknown (Jones, unpub. data). Evidence from another indicator of prey size, the strength of the canine teeth and of the muscles that drive them, suggests that devils and thylacines took equivalent-sized prey. Canine tooth strength in male and female devils is equivalent to that of adult male and halfway between adult male and female thylacines, respectively, while temporalis muscle strength in devils is similar to that of female thylacines (Fig. 2). The overall larger body mass of the thylacine may have assisted it in tackling larger prey, though. Thus, the role of top predator in the Tasmanian ecosystem was, at the least, shared equally between thylacines and devils.

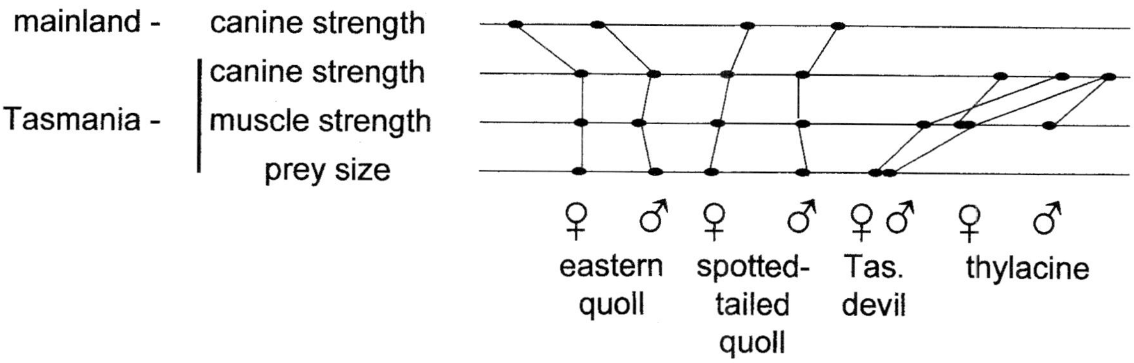

Figure 2 Over-dispersed means in the strength of the canine teeth, an index of strength of the temporalis muscles that drive the canines, and in prey size taken, by sexes and species of the quolls in Tasmania. These patterns are not evident when the devil and thylacine are included. Character release has occurred on the mainland of Australia, where the two larger competitors have been absent for 200+ generations. Figure adapted from (Jones 1997). Scale is logarithmic.

In these trophic analyses (and also in the locomotor analyses later), it is evident that there are no felid- or ursid-like ecomorphs present in the Australasian marsupial fauna. This is certainly true for the Dasyuromorphian radiation. However, the Australasian marsupial radiation has not been without ecological equivalents to the Families Felidae and Ursidae. A number of extinct species and genera of the herbivorous diprotodontian Families Thylacoleonidae (marsupial lions; ?Vombatoidea) and Hypsiprymnodontidae (rat kangaroos; Macropodoidea), are considered to have been carnivorous (reviewed in Wroe, this volume). The thylacoleonids were large-cat-like in form. The hypsiprymnodontids had the body form of a macropod but analyses of their teeth suggest that they ranged from omnivorous to carnivorous (reviewed in Wroe, this volume). The only extant member of this family, the musky rat kangaroo (*Hypsiprymnodon moschatus*) is omnivorous, including fruit, insects and fungi in its diet (Dennis and Johnson 1995).

There are no indicators of close species-for-species matching between marsupial and placental carnivores. Rather, the morphological space occupied by each genus of marsupial carnivore (*Dasyurus*, six species; *Sarcophilus* and *Thylacinus*, one species each) is broadly similar to the space occupied by the placental families with which they appear convergent. This apparent convergence between genera of marsupial carnivores and families of placental carnivores is an interesting case for demonstrating the fuzziness of the guild concept (Simberloff and Dayan 1991). According to Root's definition (Root 1967), the dasyuroids (devil, quolls and thylacine) (Jones 1997), the canids, the felids, and the mustelids, and so on, can be defined as functional guilds (Dayan et al. 1989a; 1990; 1992). Other researchers define guilds more broadly, and include all sympatric carnivores in a multi-species meat eating guild, defining functional groups within the broad guild (e.g. Van Valkenburgh 1985). Convergence between marsupial and placental carnivores in trophic functional groupings seems to sit better with the latter definition.

It is evident from Fig. 1 that the marsupial carnivores have higher scores for carnivory than species from the placental guilds with which they converge. This is probably no more than a consequence of the combined use of M3 and M4 cutting blades in the analysis, however. Had I used only M4, values for carnivory would be less than those for placentals. This point leads us into the next section of the discussion.

SOME NON-CONVERGENCE IN TROPHIC MORPHOLOGY

Although functional convergence in dietary niche is apparent, convergence in trophic morphology is superficial. Phylogenetic history also strongly influences body form. While the third factor in the PCA on trophic morphological attributes, canine shape, explained a much smaller proportion of the total variance (15.2%, eigenvalue 0.76, component loading 0.75), it accounts for an important difference in the feeding ecology of marsupial and placental carnivores.

Canine tooth cross-sectional shape correlates with prey-killing behaviour. The canids and some of the viverrids, including the larger species included by Van Valkenburgh (1987; 1989) in these analyses, have narrow, medio-laterally compressed canines. The larger species of canids kill with shallow slashing bites (Van Valkenburgh and Ruff 1987), and the smaller canid species and the viverrids kill using repeated biting, shaking and tossing of prey (Ewer 1973). These medio-laterally compressed canines are strong in an anterio-posterior direction, the direction where the most force is likely to be placed during prey killing (Van Valkenburgh and Ruff 1987). The canines of the larger felids (>7 kg), which kill using deep, powerful bites that may sever vertebrae or suffocate the prey with a neck, nose or throat hold or crush the skull (Eaton 1970; Schaller 1972; Ewer 1973; Leyhausen 1979), are more rounded (Van Valkenburgh and Ruff 1987). The extra strength provided in a medio-lateral direction probably relates to the potentially tooth-fracturing bending stresses in a medio-

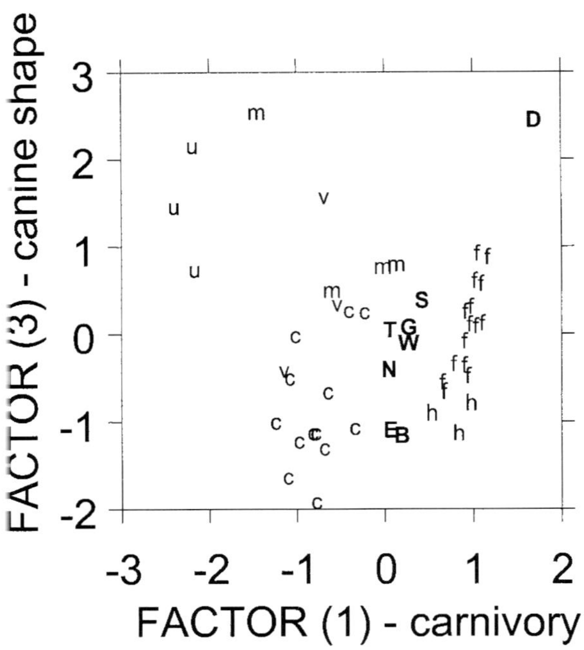

Figure 3 Plot of Factor 3 from the PCA of trophic functional groupings of marsupial and placental carnivores against Factor 1 (degree of carnivory). Factor 3 describes canine shape (positive values have rounder cross-sectional shapes, negative values are more medio-laterally flattened), which is related to killing behaviour (deeper, more penetrating bites *vs* shallow bites, respectively) (Van Valkenburgh 1989). Symbols are as in Fig. 1.

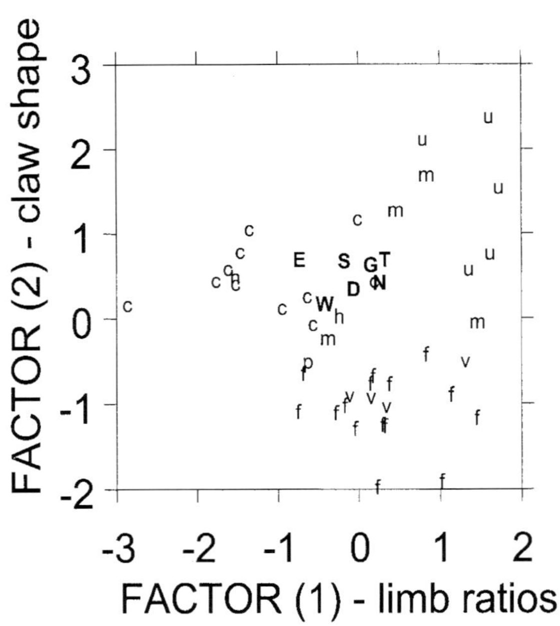

Figure 4 Plot of factors from the PCA of ecomorphological traits associated with locomotor and hunting behaviour of marsupial and placental carnivores. Factor 1 describes limb ratios (femur/metatarsal ratio [FMT], metacarpal/phalanx ratio [MCP], and olecranon or elbow shape [OLL]), which correlate with running speed and activity substrate (positive values are slower and more likely to be arboreal) and claw shape (ungual depth [UD] and shape [ARCH]), which correlates with activity substrate and prey handling behaviour (Van Valkenburgh 1985). Symbols are as in Fig. 1.

lateral direction that may occur if the canine contacts bone during prey killing (Van Valkenburgh and Ruff 1987). Among the smaller placental carnivores, mustelids and herpestids have canine teeth that are more rounded, like felids, and kill their prey using a highly directed nape bite, which drives the long, recurved canines into the back of the skull or between the vertebrae; the diameter of the canines is important in determining what size prey can be killed (see Dayan et al. 1989a; Dayan and Simberloff 1994a; Simberloff et al. 2000).

Marsupial carnivore canines (excluding those of the bone-adapted devil) are oval in cross-section; this is intermediate between the extremes of the dog (narrow canines) and cat (rounded canines) families but broadly overlaps both groups and the civets (Fig. 3). Were the smaller mustelids and herpestids (<7 kg) included in the analyses, it is likely that their canine shape would overlap with that of quolls and the thylacine. Quolls and devils kill prey using a crushing bite to the skull, nape, or chest, depending on the relative sizes of the predator and prey, in which the canines penetrate the skull (smaller prey) or the nape or chest (larger prey, Fleay 1932; Pellis and Nelson 1984; Pellis and Officer 1987; Jones 1995, Jones unpubl. data). This killing method is intermediate between the extremes of the canid and felid species, and perhaps quite similar to that of the smaller mustelids and herpestids (reviewed in Dayan et al. 1989a; Dayan and Simberloff 1994a; Simberloff et al. 2000). The use of crushing, penetrating bites would certainly explain why the canines are more rounded than those of canids.

The small mustelines and the smaller felids, which are more similar in body size to the quolls, are not included in Van Valkenburgh and Ruff's (1987) analysis and data on canine shape were not available at the time of these analyses. Van Valkenburgh and Ruff (1987) covers only three of the larger, semi-fossorial mustelids, whose diets range from being omnivorous with a low proportion of vertebrate prey (*Meles*), to a greater proportion of vertebrate prey (*Mellivora*), to almost exclusively carnivorous species (*Taxidea*) (Ewer 1973). Canine shape of these semi-fossorial species is oval, similar to that of the marsupial carnivores.

That canine shape reflects the chance of potentially tooth-shattering contact with bone, explains why devils have canine teeth that are almost circular in cross-section, even more rounded than any felid. Canine shape in devils is consistent with their strong, crushing, but generally non-penetrating, killing bite to the chest, head or nose of the prey (Jones 1995), similar to that used by some of the larger felids. In addition, canine shape of devils reflects their bone-eating habits. Devils and hyaenas both crack and eat bone and have more rounded canine teeth than expected from their killing method (Fig. 3). Hyaenas kill in a manner similar to canids, yet have canine teeth that are as rounded as felids (Van Valkenburgh and Ruff 1987).

Phylogenetically-based differences in skull morphology also indicate that dasyuromorphs and carnivorans may simply have evolved different solutions to the same biomechanical problems. Dasyuroids have exceptionally long rostrums (most extreme in the thylacine) and narrow, flattened skulls compared with all placental carnivores (Werdelin 1986). Biomechanically, this should mean that their bite strength at the canines is weak. However, the relatively small brain case of marsupial carnivores allows a larger space between the skull and the zygomatic arches for the temporalis muscle. This is exemplified in the devil, which has both a very small brain case and very wide zygomatic arches. Long moment arms for the temporalis and masseter muscles, and large masseter muscle attachment areas on the mandible may also compensate for the long snout. Marsupial carnivores may in fact have bite forces similar to those of equivalent placental carnivores.

ECOMORPHOLOGICAL COMPARISONS OF LOCOMOTOR AND HUNTING MODES

Principal components analysis separated the skeletal ratios into those that reflect running speed and activity substrate (Factor 1; component loadings: FMT = 0.72, MCP = –0.78, OLL = 0.72) and those that reflect activity substrate and prey handling behaviour (Factor 2; component loadings: UD = 0.85, ARCH = –0.79), which together explained 71% of the variance in the data. The dasyuroids clumped together as a reasonably close group with intermediate values on both factors, commensurate with the most common hunting type and activity substrate classifications for the group (Fig. 4, Table 1). Each family of placental carnivore also forms a fairly distinct group, suggesting a strong influence of phylogenetic history on morphology. The skeletal indicators of running speed, activity substrate and prey capture mode appear to be more useful in separating and assigning major groups to their predominant locomotory type than in assigning different species within those groups to their known hunting type and activity substrate classifications.

All species of marsupial carnivores, including the partly arboreal spotted-tailed quoll which also uses its forelimbs to grasp prey and the scansorial northern and New Guinea quolls, have reasonably flattened claws similar to those of terrestrial canids. Van Valkenburgh (1985) notes that while claw-climbing arboreal and scansorial species (e.g. felids) tend to have more curved, deeper claws and those which grasp their prey also have more curved claws, claw shape is not always indicative of climbing behaviour. For example, bears also climb extremely well but

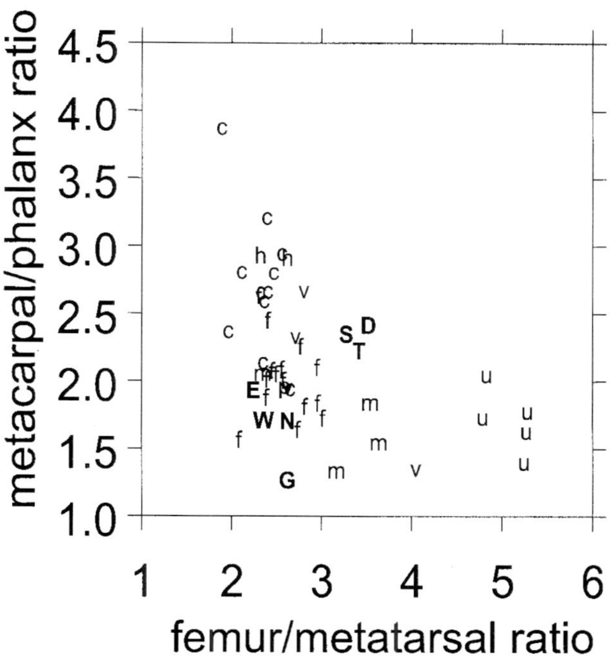

Figure 5 Separate plot of femur/metatarsal (FMT) and metacarpal/phalanx (MCP) ratios for marsupial and placental carnivores. FMT correlates loosely with running speed, MCP with activity substrate (positive values are slower and more likely to be aboreal, Van Valkenburgh 1985). Symbols are as in Fig. 1.

have among the flattest claws of the Carnivora. Claw shape does seem to be strongly influenced by phylogenetic history. Again, there is often more than one evolutionary solution to an ecomorphological problem. Arboreal and scansorial quolls have other morphological traits which may assist with climbing, including striated ridges on the moist foot pads, and a hallux or opposable clawless first digit on the hind foot. An opposable first digit is also well developed in some arboreal herbivorous taxa (e.g. marsupial possums, placental primates). Quolls and devils can also freely rotate the fore and hind limbs inwards, which enhances tree climbing and prey grappling ability.

Marsupial carnivores as a group appear to be intermediate to slow speed runners. High values on Factor 1 indicate slow running species, either arboreal or scansorial, such as bears and some cats, and semi-fossorial mustelids, such as badgers (*Meles, Mellivora, Taxidea*) (Van Valkenburgh 1985). It may be informative to examine two components of Factor 1, the FMT and MCP ratios, separately (Fig. 5).

On the MCP axis, the devil, thylacine and spotted-tailed quoll cluster tightly together with reasonably terrestrial, pounce/pursuit hunting placental species. In addition, values for elbow shape (OLL) are all intermediate (Table 1), indicating terrestrial activity. This is despite the quite arboreal habits and ambushing hunting mode of the spotted-tailed quoll, but see the comments above for other arboreal adaptations. This terrestrial, pounce/pursuit grouping is a fairly accurate description of activity substrate and hunting behaviour for the largely terrestrial devil (that can and does climb, Jones and Barmuta 2000) and is accurate in terms of activity substrate and probably also for hunting type for the thylacine (see Table 1). The smaller quolls all cluster with the arboreal and semi-fossorial placentals. This is curious as these species are either largely terrestrial or scansorial, pounce/pursuit hunters. This apparent mismatch may be an artifact of the lack of smaller carnivores in Van Valkenburgh's (1985) analyses.

On the FMT axis, which is loosely correlated with running speed, all the smaller quolls group with the bulk of the placental pounce/pursuit carnivores. Note that, as Van Valkenburgh (1985) stated, the FMT ratio is not very informative in separating the fast pursuit species from the pounce/pursuit hunters (Fig. 5). However, the spotted-tailed quoll, devil and thylacine are placed with the slow, semi-fossorial carnivores, the badgers. No data on running speed are available to establish whether devils and spotted-tailed quolls are actually slower runners, although I have followed by car devils running along roads at over 30 km/h. It is interesting to note here that the calcaneum or foot pad in dasyuroids extends to the heel and wrist, compared with just the digits in placentals. Dasyuroids do make use of this extended pad, when they are standing quadrupedally or bipedally (this was noted also in the thylacine), although when locomoting, their stance appears to be digitigrade. This does suggest, however, that the functional morphology of the hind limbs of marsupial and placental carnivores may not be strictly equivalent and that locomotory attributes may not be closely convergent; another case of phylogenetic constraint on morphology. How this translates to function and ecological equivalence is not known.

Convergence in morphological size patterns among guilds

Similar morphological size patterns have been found in the trophic structures of the Tasmanian marsupial carnivore guild and in several guilds of placental carnivores. Even size ratios (over-dispersed means) in trophic characters between adjacent size-ranked species and sexes (where species are significantly sexually size dimorphic) have been found in the mustelid/viverrid guild of Israel (Dayan et al. 1989a), mustelid guilds of North America (Dayan et al. 1989a) and Great Britain and Ireland (Dayan and Simberloff 1994a), felid guilds of Israel and Pakistan (Dayan et al. 1990), canid guilds of Israel and North Africa (Dayan et al. 1992), and in the marsupial carnivore guilds of Tasmania and Australia (Fig. 5, Jones 1997), although character displacement was not found in a study of jackals in Africa (van Valkenburgh and Wayne 1994). In addition, similar size patterns have been found in three fossil guilds of canids (Dayan et al. 1992; Dayan et al. 1993) and hyaenids (Werdelin 1996). Such patterns of regularity among guilds, termed community-wide character displacement (*sensu* Strong et al. 1979), have been interpreted as evidence for competition for food by particle (e.g. prey or seed) size effecting either microevolutionary change or species sorting by size (reviewed in Dayan and Simberloff 1998).

An important feature of all of these studies is that the morphological size ratios were sought and found in the trophic structures proximal to the way in which each species kill their prey, which were different for some guilds depending on their killing behaviour (reviewed in Dayan and Simberloff 1998), and which correlates with canine tooth shape (discussed earlier). Over-dispersed means were indeed found in the diameter of the upper canine tooth in mustelids and smaller felids, which use a highly directed bite where the canine is inserted into the space between two cervical vertebrae (Dayan et al. 1989a; Dayan et al. 1990; Dayan and Simberloff 1994a, and references therein). In marsupial carnivores, which use a generalised crushing bite in killing, size patterning was manifest in indices of the strength of the upper canine tooth and in the temporalis muscles that drive them into the prey (Jones 1997). The results for canids, where even spacing was found in carnassial tooth length, were more equivocal (Dayan et al. 1992).

Even spacing in the size of the structures used to capture prey should serve to minimise competition by maximising the differences possible between all species within the range of sizes in the guild (see Dayan and Simberloff 1998). If this holds, and if for example canine tooth strength determines the size of prey that can be caught by a particular sized carnivore, it should be reflected in prey sizes actually taken by the species. This was tested and found to be the case in the Tasmanian marsupial carnivore guild. Over-dispersed means, parallel to the patterns in canine tooth and jaw muscle strength, were found in mean prey size in the diet of the species in the wild (Fig. 2; Jones 1997).

If community-wide character displacement is driven by competition, relaxation of size patterning with an increase in size, a phenomenon termed character release, and an increase in sexual size dimorphism can be expected when one or more competitors are absent and, therefore, presumably resources are relatively more available. Such changes in size patterns have been found in a variety of mammals (reviewed in Dayan and Simberloff 1998), including among carnivores in mustelids (Dayan and Simberloff 1994a), canids (Dayan et al. 1992), and marsupial carnivores (Fig. 2, Jones 1997). Among marsupial carnivores, character release, including an increase in body size and sexual size dimorphism of spotted-tailed quolls, occurred on the adjacent mainland in the absence of the two larger carnivores in Tasmania, the Tasmanian devil and the thylacine. The change in size dimorphism is thought to be in response to the increased importance of intraspecific intersexual competition.

Similarity in the size patterns of trophic structures, with an underlying basis in food size partitioning, and in patterns of character release between marsupial and placental carnivores provides some support for the notion of generality of the ecological forces such as competition that structure communities. They also attest to convergence in the structuring mechanisms that influence morphological size relationships among guilds. Convergence in guild structure of this kind also has been established between Old and New World granivorous rodents. Similar patterns of over-dispersed means of upper incisor width with a basis in food size partitioning have been found among the heteromyid rodents of North America and the gerbilline rodents of Israel (Ben-Moshe et al. 2001).

FUTURE DIRECTIONS

Convergence in trophic functional groupings and in community structure are evident between the genera of marsupial carnivores and some families of placental carnivores despite some differences in form. Canine tooth shape seems to be constrained phylogenetically, and does influence killing behaviour. Other differences in form, such as skull shape, may represent no more than different solutions to the same problem. The implications of differences in locomotory morphology remain enigmatic. They may indicate a strong influence of phylogenetic history on locomotor morphology and that the morphometric ratios are useful only in describing the predominant hunting type and activity substrate within a group and not hunting and locomotor types within ecologically diverse taxonomic groups.

There is much to be learned about the ecomorphology of marsupial carnivores and, by extension, about convergence between marsupial and placental carnivores. This applies equally to the dasyurid marsupial insectivores. Questions of species diversity have barely begun to be addressed. Validation of the functional relationship between morphology and ecology and the fitness consequences of this relationship is needed, including an understanding of how behaviour and physiology compensate for mismatches in convergences in form and function. And, the role of history in phylogenetic constraint is poorly understood.

REFERENCES

Archer, M. (1976), 'The dasyurid dentition and its relationships to that of Didelphids, Thylacinids, Borhyaenids (Marsupicarnivora) and Peramelids (Peramelina: Marsupialia)', *Australian Journal of Zoology* (Suppl), **39**:1–34.

Belcher, C.A. (1995), 'Diet of the Tiger Quoll *Dasyurus maculatus* in East Gippsland, Victoria', *Wildlife Research,* **22**, 341–57.

Ben–Moshe, A., Dayan, T., & Simberloff, D. (2001), 'Convergence in morphological patterns and community organization between Old and New World rodent guilds', *American Naturalist,* **158**:484–95.

Blackhall, S. (1980), 'Diet of the eastern native-cat, *Dasyurus viverrinus* (Shaw), in southern Tasmania', *Australian Wildlife Research,* **7**:191–7.

Case, J.A. (1985), 'Differences in prey utilization by Pleistocene marsupial carnivores, *Thylacoleo carnifex* (Thylacoleonidae) and *Thylacinus cynocephalus* (Thylacinidae)', *Australian Mammalogy,* **8**:45–52.

Cody, M.L. (1975), 'Towards a theory of continental species diversities', in *Ecology and Evolution of Communities* (eds. M.L. Cody, & J.M. Diamond), pp. 214–57, Belknap, Cambridge, Massachusetts.

Dayan, T., & Simberloff, D. (1994a), 'Character displacement, sexual dimorphism, and morphological variation among British and Irish mustelids', *Ecology,* **75**:1063–73.

Dayan, T., & Simberloff, D. (1994b), 'Morphological relationships among coexisting heteromyids: an incisive dental character', *American Naturalist,* **143**:462–77.

Dayan, T., & Simberloff, D. (1998), 'Size patterns among competitors: Ecological character displacement and character release in mammals, with special reference to island populations', *Mammal Review,* **28**:99–124.

Dayan, T., Simberloff, D., & Tchernov, E. (1993), 'Morphological change in Quaternary mammals: A role for species interactions', in *Morphological Change in Quaternary mammals of North America: Integrating Case Studies and Evolutionary Theory* (eds. R.A. Martin, & A.D. Barnosky), pp. 71–83, Cambridge University Press, Cambridge.

Dayan, T., Simberloff, D., Tchernov, E., & Yom-Tov, Y. (1989a), 'Inter- and intraspecific character displacement in mustelids', *Ecology,* **70**:1526–39.

Dayan, T., Simberloff, D., Tchernov, E., & Yom-Tov, Y. (1990), 'Feline canines: community-wide character displacement among the small cats of Israel', *American Naturalist,* **136**:39–60.

Dayan, T., Simberloff, D., Tchernov, E., & Yom-Tov, Y. (1992), 'Canine carnassials: character displacement in the wolves, jackals and foxes of Israel', *Biological Journal of the Linnean Society,* **45**:315–31.

Dayan, T., Tchernov, E., Yom-Tov, Y., & Simberloff, D. (1989b), 'Ecological character displacement in Saharo-Arabian *Vulpes*: Outfoxing Bergmann's rule', *Oikos,* **55**:263–72.

Dennis, A.J., & Johnson, P. M. (1995), 'Musky rat kangaroo *Hypsiprymnodon moschatus* Ramsay, 1876', in *The Mammals of Australia* (ed. R. Strahan), pp. 282–4, Reed Books Australia, Sydney, Australia.

Eaton, R.L. (1970), 'The predatory sequence, with emphasis on killing behaviour and its ontogeny, in the cheetah (*Acinonyx jubatus* Schreber)', *Zeitschrift fur Tierpsychologie,* **27**:492–504.

Ewer, R.F. (1973), *The Carnivores*, Cornell University Press, Ithaca.

Fleay, D. (1932), 'The rare Dasyures (native cats)', *Victorian Naturalist,* **49**:63–9.

Fuentes, E.R. (1976), 'Ecological convergence of lizard communities in Chile and California', *Ecology,* **57**:3–17.

Jones, M.E. (1995), 'Guild structure of the large marsupial carnivores in Tasmania', PhD thesis, University of Tasmania, Hobart.

Jones, M.E. (1997), 'Character displacement in Australian dasyurid carnivores: size relationships and prey size patterns', *Ecology,* **78**:2569–87.

Jones, M.E., & Barmuta, L.A. (1998), 'Diet overlap and abundance of sympatric dasyurid carnivores: a hypothesis of competition?', *Journal of Animal Ecology,* **67**:410–21.

Jones, M.E., & Barmuta, L.A. (2000), 'Niche differentiation among sympatric Australian dasyurid carnivores', *Journal of Mammalogy,* **81**:434–47.

Jones, M.E., & Stoddart, D.M. (1998), 'Reconstruction of the predatory behaviour of the extinct marsupial thylacine', *Journal of Zoology, London,* **246**:239–46.

Leyhausen, P. (1979), *Cat Behaviour: The predatory and social behaviour of wild and domestic cats*, Garland, New York.

Luckett, W.P. (1993), 'An ontogenetic assessment of dental homologies in therian mammals', in *Mammal Phylogeny* (eds. F.S. Szalay, M.J. Novacek, & M.C. McKenna), pp. 182–204, Springer, New York.

Oakwood, M., & Eager, R. (1997), 'Diet of the Northern Quoll, *Dasyurus hallucatus*, in lowland savanna of northern Australia', in *The Ecology of the Northern Quoll,* Dasyurus hallucatus', PhD thesis, Australian National University.

Paddle, R. (2002), *The Last Tasmanian Tiger. The History and Extinction of the Thylacine*, Cambridge University Press, Cambridge, England.

Pellis, S.M., & Nelson, J. E. (1984), 'Some aspects of predatory behaviour of the quoll, *Dasyurus viverrinus* (Marsupialia: Dasyuridae)', *Australian Mammalogy*, **7**:5–15.

Pellis, S.M., & Officer, R.C.E. (1987), 'An analysis of some predatory behaviour patterns in four species of carnivorous marsupials (Dasyuridae), with comparative notes on the eutherian cat *Felis catus*', *Ethology,* **75**:177–96.

Pemberton, D. (1990), 'Social organisation and behaviour of the Tasmanian devil, *Sarcophilus harrisii*', PhD thesis, University of Tasmania.

Pianka, E.R. (1986), *Ecology and Natural History of Desert Lizards: Analysis of the Ecological Niche and Community Structure*, Princeton University Press, Princeton, New Jersey.

Ricklefs, R.E., & Miles, D.B. (1994), 'Ecological and evolutionary inferences from morphology: an ecological perspective', in *Ecological Morphology: Integrative Organismal Biology* (eds. P.C. Wainwright, & S.M. Reilly), pp. 13–41, University of Chicago Press, Chicago and London.

Ricklefs, R.E., & Travis, J. (1980), 'A morphological approach to the study of avian community organization', *Auk,* **97**, 321–38.

Root, R.B. (1967), 'The niche exploitation pattern of the blue–gray gnatcatcher', *Ecological Monographs,* **37**:317–50.

Rounsevell, D.E., & Mooney, N. (1995), 'Thylacine *Thylacinus cynocephalus* (Harris, 1808)', in *The Mammals of Australia* (ed. R. Strahan), pp. 164–5, Australian Museum/Reed Books, Sydney.

Schaller, G.B. (1972), *The Serengeti Lion. A study of predator–prey relations*, University of Chicago Press, Chicago.

Schluter, D., & Ricklefs, R.E. (1993), 'Convergence and the regional component of species diversity', in *Species Diversity in Ecological Communities: Historical and Geographical Perspectives* (eds. R.E. Ricklefs, & D. Schluter), University of Chicago Press, Chicago.

Simberloff, D., & Dayan, T. (1991), 'The guild concept and the structure of ecological communities', *Annual Review of Ecology and Systematics,* **22**:115–43.

Simberloff, D., Dayan, T., Jones, C., & Ogura, G. (2000), 'Character displacement and release in the small Indian mongoose, *Herpestes javanicus*', *Ecology,* **81**:2086–99.

Smith, S. (1981), 'The Tasmanian tiger – 1980', National Parks and Wildlife Service, Tasmania, Technical Report No. 81/1.

Strong, D.R., Szyka, L.A., & Simberloff, D.S. (1979), 'Tests of community-wide character displacement against null hypotheses', *Evolution,* **33**:897–913.

Van Valkenburgh, B. (1985), 'Locomotor diversity within past and present guilds of large predatory mammals', *Paleobiology,* **11**:406–28.

Van Valkenburgh, B. (1989), 'Carnivore dental adaptations and diet: A study of trophic diversity within guilds', in *Carnivore Behaviour, Ecology, and Evolution* (ed. J.L. Gittleman), pp. 410–36, Chapman and Hall, London.

Van Valkenburgh, B., & Ruff, C.B. (1987), 'Canine tooth strength and killing behaviour in large carnivores', *Journal of Zoology, London,* **212**:379–97.

Van Valkenburgh, B., & Wayne, R.K. (1994), 'Shape divergence associated with size convergence in sympatric east African jackals', *Ecology,* **75**:1567–81.

Werdelin, L. (1986), 'Comparison of skull shape in marsupial and placental carnivores', *Australian Journal of Zoology,* **34**:109–17.

Werdelin, L. (1987), 'Jaw geometry and molar morphology in marsupial carnivores: analysis of a constraint and its macroevolutionary consequences', *Paleobiology,* **13**(3):342–50.

Werdelin, L. (1996), 'Community-wide character displacement in Miocene hyaenas', *Lethaia,* **29**:97–106.

Wiens, J.A. (1989), *The Ecology of Bird Communities*, Cambridge University Press, New York.

Wiens, J.A. (1992), *The Ecology of Bird Communities*, Cambridge University Press, Cambridge, UK.

CHAPTER 20

LATITUDINAL VARIATION IN SOUTH AMERICAN MARSUPIAL BIOLOGY

Elmer C. Birney[A] *and J. Adrián Monjeau*[B]

[A] Formerly Professor of Ecology, Department of Ecology, Evolution and Behavior, University of Minnesota, St Paul, MN 55108, USA. Elmer died while we were enjoying writing this paper. I dedicate this chapter to the memory of my friend and colleague.
[B] Instituto de Análisis de Recursos Naturales, Universidad Atlántida Argentina, Mar del Plata, Argentina. Temporary location: Ortega y Gasset 2367 (7600) Mar del Plata, Argentina. Email: amonjeau@copetel.com.ar

We studied latitudinal patterns of species diversity, trophic guild structure, habitat use, body size and use of tail in South American marsupials. We regressed ratios of biological characters on temperature and precipitation parameters as surrogates for latitudinal changes in the physical environment. Our results indicate that mean minimum temperature is positively related with species diversity, the proportion of frugivores, the percentage of marsupials with a long prehensile tail, and the proportion of arboreal and scansorial marsupials forms, and is negatively correlated with the proportion of carnivores, the percentage of marsupials with incrassated tails, the proportion of terrestrial forms, and the proportion of small-sized marsupials. Variability in temperature has higher explanatory power than variability of precipitation. Major life forms in South American marsupials show latitudinal patterns.

INTRODUCTION

The goal of this chapter is to examine a series of macroecological patterns by relating a number of biological features of South America marsupial fauna to latitudinal gradients in climate, but also to the shape of the continent and biome diversity. Our major questions are: What abiotic factors best explain the decrease of species diversity toward the pole? Does the shape of the continent influence latitudinal patterns? Does biome diversity increase the number of marsupial species? Are the marsupial life forms latitudinally patterned? And a fascinating enigma for this book: why do only carnivore species inhabit the southernmost places?

The diversity of Neotropical marsupials, together with the uniqueness of South America as a continuous landmass from the equator to middle southern latitudes, provides a superb opportunity to shed light to these questions.

South American marsupials

New World marsupials are among the oldest living mammals and are an important component in the mammalian communities of South America where they comprise approximately 6.3% of terrestrial species (Patterson 1994). Although living marsupials are more diverse in Australasia, the extant Neotropical marsupials show a great diversity of adaptations. Three orders occur in South America: Didelphimorphia, Paucituberculata and

Microbiotheria (Wilson and Reeder 1993). The didelphimorphs are the most diverse and widely distributed of Neotropical marsupials, with 65 species in three or four families (Galliari et al. 1996, Gardner 1993a, b, c; Hershkovitz 1992). Paucituberculata (6 species) and Microbiotheria (1 species) are restricted to the Andean Cordilleras as remnants of a more diversified and widespread Cenozoic radiation (Albuja and Patterson 1996, Mares and Braun 2000, Reig 1981).

Although Neotropical marsupials have been shown to comprise a significant portion of mammalian biomass in some Neotropical habitats, they are limited to few trophic guilds (e.g. omnivore, frugivore, insectivore/carnivore), rather than the diverse food specialists that characterise Australian marsupial fauna (Mares and Braun 2000). Similarly, the range of body sizes is smaller in Neotropical marsupials compared to Australian forms.

Latitudinal patterns

One of the most striking biogeographic patterns is the gradient of increasing biodiversity from the poles to the equator (Wallace 1878, Darlington 1957). The pattern is a general one, in that it holds true for marine, aquatic and terrestrial organisms as a whole (Rosenzweig 1995), but also for major taxa including marsupials (Brown and Lomolino 1998, Kauffman 1995).

Although the species diversity gradient and other latitudinal patterns have been widely accepted, explanations for these patterns have led to a controversial framework rather than a consensus (Brown and Lomolino 1998; Kauffman 1995; Lyons and Willig 1997, 1999; Pianka 1966, 1978; Rohde 1992; Rosenzweig 1992, 1995). The reason for this lack of consensus is probably because most investigators have focused on a single hypothesis to account for a given latitudinal pattern rather than seeking broader explanations. In order to gain insight about the primary causes of latitudinal patterns, we selected biological features other than biodiversity such as body size, trophic guild structure, use of habitat, and tail structure and function. In order to synthetise the analysis, our selection was based on features in marsupial life forms that seems to be geographically patterned after examining museum specimens and literature.

South American climate

South America lies mainly in the southern hemisphere. The climate is a consequence of its geographical position with respect to latitude, ocean current activity, and major geological features. The presence of the Andes range along the occidental coast modifies the climate of the western regions significantly, establishing an effective meteorological barrier between the west and the east. South America is the only continent that extends as a continuous landmass from the Equator to the middle southern latitudes (Garcia 1994). This combination of geographic position, geological features, and oceanic conditions produced considerable climatic diversity and habitat heterogeneity when marsupial adaptive radiation was occurring.

In terms of ecosystem processes that determine landscape boundaries, most ecologists would agree that climate has the greatest influence (Klijn and Udo de Haes 1994). Monjeau et al. (1997) determined that temperature and precipitation are the best predictors of small mammal community composition in a regional environmental gradient. In this chapter we selected climatic surrogates of environmental variability as suggested by the work of Monjeau et al. (1998) on the responses of small mammal comunities to the hierarchical structure of their environment.

The macroecological approach

After several studies of the ecology and distribution of the southernmost South American marsupials (Birney et al. 1996a, 1996b), and of small mammals in general (Monjeau et al. 1997, 1998), we realised the inherent difficulty in understanding local and regional patterns and processes without a larger perspective. It was this conclusion, together with the influence of Brown's (1995) macroecology book, that prompted us to stand back and take a broader view of latitudinal gradients. The macroecological approach attemps to increase the spatial and temporal scale of ecological inquiry by emphasising analyses of statistical patterns rather than experimental manipulations (Brown 1995). In this chapter we explore relationships between abiotic latitudinal gradients in South America and their associated marsupial life forms.

METHODS AND MATERIALS

Data sources

A list of the species of Marsupialia (Orders Didelphimorphia, Paucituberculata, and Microbiotheria) occurring in South America was compiled from Gardner (1993a, b, c) and modified slightly to account for taxonomic changes published since 1993. The complete list is included in Appendix 1 with a summary of selected aspects of body size, length and tail structure and function, and natural history, including diet, habitat, and place of activity. The distribution of each species was determined from a combination of published distributional maps, as well as distributional records. All of the information presented and later used for our analyses was taken from the literature, primarily from such summary sources such as Anderson (1997), Eisenberg (1989), Emmons and Feer (1997), Mares and Braun (2000), Redford and Eisenberg (1992), and Wilson and Reeder (1993).

The major biome type(s) listed in Appendix 1 for each species and used in our analyses were determined by comparing the known distribution of each species with the map of South American biomes provided by Redford and Eisenberg (1992) (as modified from Udvardy 1978) with the following exceptions: for Bolivia we followed the map provided by Anderson (1997:75, Fig. 470); for Patagonia we followed the landscape classification and nomenclature provided by Monjeau et al. (1998); we considered the Pampas as a single unit instead of recognising a distinction between Argentine and Uruguayan Pampas; we considered the Madeiran to be a part of the more widely known Amazonian Rainforest. Climatological data were taken from the tables provided by Schwerdtfeger (1976).

Selection, treatment, and analysis of data

We initially selected the entire continent of South America as our study area. For both environmental and biological data, we divided the continent into a series of latitudinal bands that extended from the Atlantic or the Caribbean coast to the Panamanian border or the Pacific coast. Each band covers 5° latitude, which is nearly equivalent to 555 km. In order to minimise error in the assignment of marsupial species and biome categories, bands were selected so their north–south boundary corresponded with 0° and latitudes divisible by 5 as determined with reference to a South American map with a Lamberts plane (azimuthal) equal area projection found in a commercial Rand McNally Atlas. The southernmost band included was at 50° S rather than the tip of the continent to avoid inclusion of bands that do not support at least one species of marsupial (the southernmost record of a marsupial in South America is approximately 47° S (Birney et al. 1996a). The northernmost band extended to 10° N rather than the northern margin of the continent, which was necessary to make bands correspond most closely to the latitude designators provided on maps. Thus, the functional study area consisted of the 12 latitudinal bands that lie between 10° N and 50° S. These bands were numbered 1–12 from north to south, and are the units for both data collection and analysis. Two marsupial species (*Marmosa xerophyla* and *Marmosops cracens*) occur north of 10° N but not south of that latitude (Eisenberg 1989). They are included in the data presented in Appendix 1, but were not included in any of our macroecological analysis. For all marsupials, if any part of the known distribution fell within a band, the species was recorded as occurring in that band. In the case of apparently disjunct populations of a single species, the species was recorded as occurring only in those bands of documented occurrence with the exceptions of *Caenolestes convelatus* and *C. fuliginosus*. Localities of record for these species within bands 1 and 3, but not band 2, are provided by Albuja and Patterson (1996). Because these two species are not well studied and because our maps indicate that the specific montane habitats they prefer exist within band 2, we recorded them as present in band 2 throughout our analyses.

Climatic and other environmental data were taken from the chapters by Prohaska (1976) for Argentina, Paraguay, and Uruguay; Miller (1976) for Chile; Johnson (1976) for Perú, Bolivia, and Ecuador; Ratisbona (1976) for Brazil; and Snow (1976) for all of the countries located primarily or entirely north of the Equator in a book on climates of Central and South America (Schwerdtfeger 1976). For each band we selected data for five sites representative of the following areas: A) Pacific coastal; B) the highest elevation site provided in the Andes; C) the westernmost lowland sited east of the Andes; D) the locality nearest the central longitude between the C and E sites; E) Atlantic coastal. No data were available for sites corresponding to C of band 1, D of band 10, C and D of band 11, and B and C of band. For each site selected, we recorded latitude, longitude, elevation, mean annual temperature, maximum annual temperature, minimum annual temperature, mean annual precipitation, mean minimum monthly precipitation, and mean maximum monthly temperature. Latitudinal geographic coordinates were transformed into numerical values by taking the central degree of latitud of each band using hundredth of minutes (e.g. 7.5 rather than 7°30'). Temperature range was calculated by subtracting minimum annual temperature from maximum annual temperature and precipitation range was calculated by subtracting mean minimum monthly precipitation from mean maximum monthly precipitation. We excluded the value from our B (Andean) site for mean band values of all parameters related to temperature. Inclusion of this site greatly increased the variance for the band and greatly influenced the band means owing more to the elevation of each B site than to latitude. The influence of elevation on precipitation is much less profound and predictable than that on temperature, and thus the mean band precipitation values used in our analyses include the data from all selected climatic data sites.

Sixteen biological aspects of individual South American marsupial species and number of species per band were selected for study in our search for macroecological patterns. These were organised into four major natural history super categories as follows: 1) number of marsupial species occurring in each band; 2) diet (carnivorous, frugivorous, and insectivorous); 3) primary place or microhabitat of activity (semiaquatic, terrestrial, scansorial, and arboreal); and 4) characteristics and functions of the tail (long and prehensile, short and prehensile, non-prehensile and non-incrassated, and seasonally incrassated).

Species were classified based on the criteria described below. In the few cases when no information could be found for some super category, less well-known species were assigned to the same categories as better known congeners for which data were available. The interpretation were made very cautiosly.

Diet: Species were categorised based on what we judged to be their primary or preferred diet, despite the fact that most are somewhat omnivorous and do not restrict their choice of food entirely to any one category.

Primary place or microhabitat of activity: Assignment was made to the category that most closely agreed with the literature,

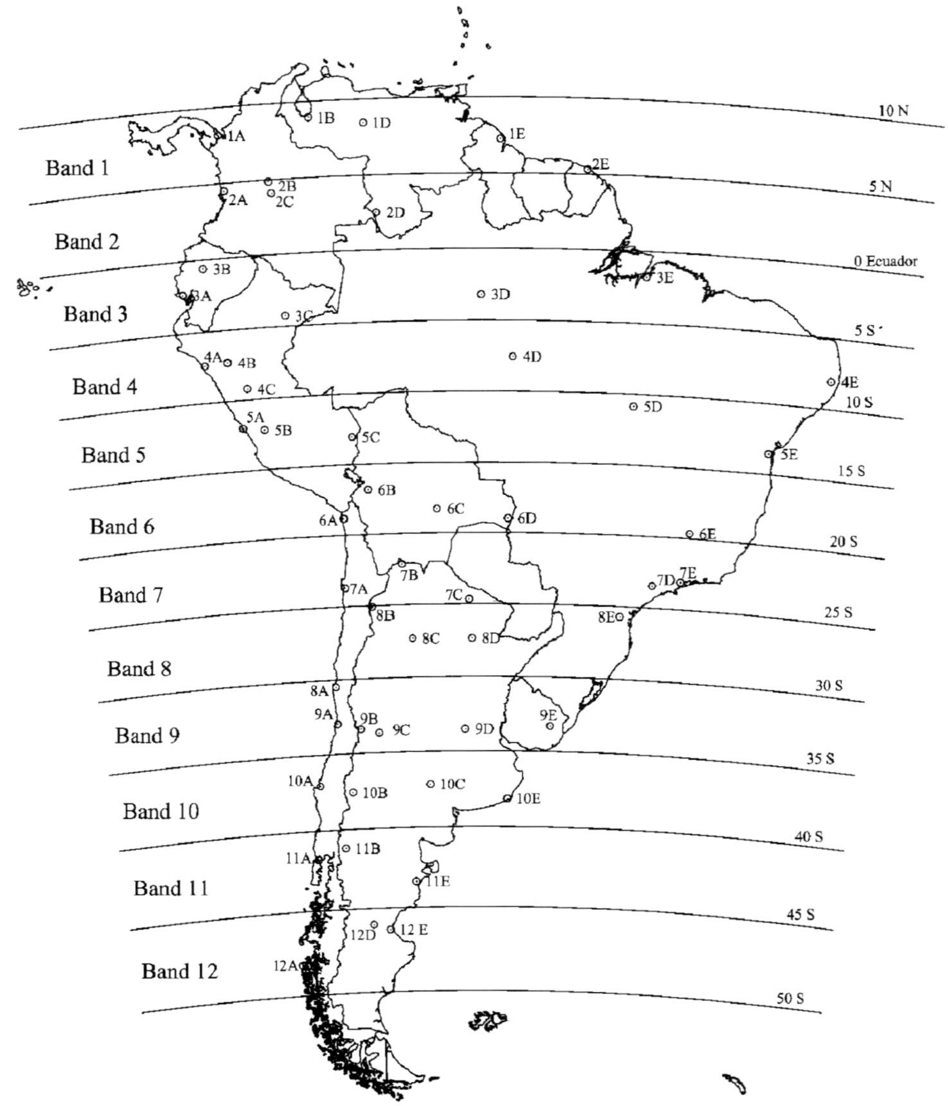

Figure 1 Map of South America showing latitudinal bands and location of the cities used as sources for meteorological data.

which for many species is the place they most often have been observed or trapped.

Characteristics and functions of the tail: Determination of whether or not the tail is prehensile and whether or not it was seasonally incrassated by fat storage was made using a combination of familiarity with the species and museum specimens, comments in the literature, and study of pictures in published works, especially that by Emmons and Feer (1997). Prehensile tails were considered to be long if tail length exceeded length of head and body.

Body size: Data on body mass taken from literature was used to estimate adult body size separately for males and females of each species. Those weighing less than 100 g were categorised as small, those from 100 to 499 g were assigned to a medium category, and all species weighing 500 g or more were considered large.

Correlation analyses were conducted to examine the relationship among variables. Regression analyses, simple and multivariate, were performed for species characteristics versus selected abiotic features of each latitudinal band. Stepwise multiple regression were used to select the best combination of predictors.

RESULTS AND DISCUSSION

As a result of our literature survey, we recorded climatic and other environmental information from selected cities (Fig. 1).

Table 1 Abiotic variables considered in this paper. Band = band number from north to south; Latitude = average latitude per band (in degrees followed by centesims of minutes); MxT = mean annual temperature; MminT = mean minimum extreme temperature; MmaxT = mean maximum extreme temperature; Thrange = average of the thermal range; Pmean = mean annual precipitation; Pdif = average of precipitation differences; Area = area of each band in square kilometers

Band	Latitude	MxT	MminT	MmaxT	ThRange	Pmean	Pdiff	Area (x 10^3)	Biome Div.
1	7.5	24.8	16.6	36.3	19.7	2042	258	1110	6
2	2.5	25.9	16.5	36.5	20	3735	324	1665	8
3	2.5	26	16.8	35.1	18.3	1955	234	2220	10
4	7.5	24.4	11.6	35.9	24.3	1611	204	2819.4	10
5	12.5	23.6	10.3	37	26.7	1277	174	2775	10
6	17.5	22	2.6	37.2	34.6	859	136	1887	7
7	22.5	20.1	1.2	37.3	36.1	726	96	1620.6	8
8	27.5	17.8	-4.2	38.2	42.4	866	110	1176.6	7
9	32.5	16.3	-4.9	39.5	44.4	626	79	999	6
10	37.5	13.8	-8.5	40.2	48.7	726	95	777	6
11	42.5	11	-10.5	34.7	45.2	955	104	466.2	4
12	47.5	10.5	-17.9	34.2	52.1	1618	47	333	3

Appendix 1 summarises biological aspects of individual South American species that were selected for study in our search for macroecological patterns. In Table 1 we calculated the average values of each parameter recorded from cities for each band.

In Table 2 we calculated the species diversity and the proportion of species that belong to a given 'guild' within each super category, per band.

Relationships between latitude and physical variables: searching for surrogates

Climate and latitude

The best surrogate for latitude is temperature (Table 3). Note that within temperature variables, the mean minimum extreme (MminT) correlates with latitude equal than mean annual temperature (MxT), whereas the mean maximum (MmaxT) does not. In addition to this, we found a strong correlation between the MminT and the MxT, but no correlation between MmaxT and the MxT. Therefore, the well known relationship between latitude and mean annual temperature may be driven by the minimum extremes of the thermal range. Given that MminT and MxT are statiscally equivalent, and because we considered this fact to be consistent with observed latitudinal effects in biological patterns, we believe that the MminT, rather than the widely used mean annual temperature, is best surrogate for latitudinal gradients of temperature in this framework. We found that precipitation differences (Pdif) correlates better than mean annual precipitation (Pmean) with latitude. However, the causal mechanisms underlying these correlations is not clear, and this is likely a particular effect related with the combination of climate together with regional geological features.

Band area and latitude

There was a significant correlation, but not causal, between band area and latitude (Table 3). Band area is in turn strongly correlated with biome diversity per band (Table 4).

Species diversity analysis

Mean annual temperature (MxT) was the best single predictor of species diversity (S) in an stepwise multiple regression. Other variables also correlate strongly with S (Table 4), but given the fact that these abiotic variables are highly correlated, there is no combination of variables better than the single MxT. This may result from the strong correlation with mean minimum extreme temperature (MminT, Table 3), which is likely the most relevant limiting factor influencing species richness, as we will discuss in the following section.

What is new about latitudinal patterns of species diversity in South American marsupials?

The pattern of increasing species diversity towards the equator is one of the most strongly supported paradigms of biogeography (Darlington 1957). Ruggiero (1994) and Kauffmann (1995) reported correlations between latitude and species diversity among marsupials in South America. Our results supported the pattern. However, many causal processes have been hypothesised: historical perturbation, productivity, harshness, climatic stability, habitat heterogeneity, competition (Brown and Lomolino 1998). Rohde (1992) suggested that each explanation either contains elements of circularity or is not sufficiently supported. The underlying process explaining the pattern is still obscure and the discussion remains open. We think our findings might shed some new light on the enigma reporting the mean

Table 2 Species diversity and percentage of selected biological features per latitudinal band

Latitudinal band	Species diversity	Trophic guild			Primary place of activity				Characteristics and functions of tail			
		% of carnivore species	% of frugivore species	% of insectivore species	% of semiaquatic species	% of terrestrial species	% of scansorial species	% of arboreal species	% of species with long prehensile tail	% of species with short prehensile tail	% of species with neither prehensile nor fat-filled tail	% of species with fat-filled tail
1 (10 N to 5 N)	28	21.43	25	53.57	3.57	17.86	42.86	35.71	82.14	7.14	10.71	0
2 (5 N to 0)	27	22.22	22.22	55.56	3.70	18.52	48.15	29.63	81.48	7.41	11.11	0
3 (0 to 5 S)	27	20	20	60	2.86	31.43	37.14	28.57	71.43	17.14	11.43	0
4 (5 S to 10 S)	29	20.69	20.69	58.62	3.45	31.03	34.48	31.03	75.86	20.69	3.45	0
5 (10 S to 15 S)	30	20	23.33	56.67	3.33	26.67	33.33	36.67	70	23.33	6.67	0
6 (15 S to 20 S)	30	20	16.67	63.33	3.33	40	23.33	33.33	56.67	20	6.67	16.67
7 (20 S to 25 S)	28	21.43	7.14	71.43	3.57	57.14	17.86	21.43	39.29	35.71	3.57	21.43
8 (25 S to 30 S)	17	29.41	5.88	64.71	5.88	52.94	17.65	23.53	41.18	23.53	5.88	23.53
9 (30 S to 35 S)	8	50	0	50	0	62.50	25	12.50	25	12.50	12.50	50
10 (35 S to 40S)	7	57.14	0	42.86	0	57.14	42.86	0	14.29	14.29	14.29	57.14
11 (40 S to 45 S)	4	25	0	75	0	75	25	0	0	0	0	100
12 (45 S to 50 S)	2	50	0	50	0	100	0	0	0	0	0	100

Table 3 Coeficient of correlation among the abiotic variables considered in this paper. Bold numbers indicate a significant correlation at $p < 0.05$. Latitude variable express the mean latitude of each band; MxT = mean annual temperature; MminT = mean minimum extreme temperature; MmaxT = mean maximum extreme temperature; Thrange = average of the thermal range; Pmean = mean annual precipitation; Pdif = average of precipitation differences; Area = area of each band in square kilometers.

	Latitude	MxT	MminT	MmaxT	ThRange	Pmean	Pdif
Latitude							
MxT	**–0.99**						
MminT	**–0.99**	0.97					
MmaxT	0.11	–0.06	–0.13				
ThRange	0.97	**–0.96**	**–0.99**	0.26			
Pmean	**–0.59**	0.53	0.60	–0.42	**–0.64**		
Pdif	**–0.89**	0.84	0.91	–0.24	**–0.92**	0.84	
Area	**–0.79**	0.80	0.74	–0.05	**–0.73**	0.20	0.51

minimum extreme temperature as a plausible limiting factor explaining why there are fewer marsupial species outside the tropics (*sensu* Blackburn and Gaston 1996), rather than explaining why there are so many species in the tropics.

Given the universality of the latitudinal pattern of species diversity, it is obvious that most of the explanations must rely ultimately on variation in the physical environment towards the poles. Natural forces are hierarchically structured as a series of nested ecosystems in which climate is the least dependent (Klijn and Udo de Haes 1994). Therefore, temperature and related factors seems to be the best explanation for the latitudinal diversity of species richness in terrestrial, marine, and aquatic habitats. Taken together, several hypotheses suggest some common influence on low temperatures limiting the capacity of environments to hold species (Stanley 1979, Brown and Gibson 1983), supporting our proposal of using MminT temperature as a general surrogate for temperature in this context. The correlations obtained between MminT and latitude, mean annual temperature, and species diversity, suggests that the ultimate mechanism causing the pattern is the minimum, rather than the mean temperature.

Although there is little disagreement about the positive correlation between precipitation and species diversity, we found almost no evidence supporting the influence of precipitation in the latitudinal pattern. Whereas mean annual precipitation is not a good surrogate for latitude, precipitation differences are good predictors (Table 3). One plausible explanation is that precipitation differences might be a surrogate for habitat heterogeneity per band, providing physical support to hold species diversity (Brown and Lomolino 1998).

Area effect

Species diversity per band is also correlated to band area (Table 4). This results from the strong correlation between band area and biome diversity mentioned above. Biome diversity is in turn correlated with species diversity ($R^2 = 0.71$; $p < 0.001$), which we consider the biological cause of the area effect in marsupial diversity. Given that the Arrhenius equation (Arrhenius 1921) is the standard form of the species-area curve ($\log S = z\log A + \log C$), we analysed the log-log relationship between area and number of species finding the best correlation between both variables (0.93 in Table 4),

$$\log S = 1.147 \log A - 13.3.$$

The correlation between band area, biome diversity, and species diversity is consistent with the well supported relationship between species diversity and area (Arrhenius 1921, MacArthur and Wilson 1967, Rosenzweig 1995). Band area and latitude also are significantly correlated (–0.79, Table 3), but of course without any possible causality with biological meaning. As Connor and MacCoy (1979) pointed out, area may influence species richness by increasing biome diversity (as supported by our data analysis in Table 4), or by increasing area itself. Because area and latitude are strongly correlated we cannot assess the influence of the area *per se*. We suggest that this particular coincidence, given the shape of the continent, act as a *magnifier* of the worldwide pattern of distribution of species diversity mostly driven by temperature.

Latitudinal patterns in trophic guild structure

Thermal range is the best predictor of the proportion of carnivores and frugivores in a stepwise regression model. There is no additional variable able to add significant explanation to the single thermal range model (Table 4). The biological meaning of this correlation may be the fact that great thermal ranges do not allow year round warranties in food resources for frugivores. As thermal range increases, proportion of frugivores decreases, driving the increase of the proportion of carnivores.

Even masked by the stepwise model, it is interesting to note that the proportion of carnivores was negatively related to either

Table 4 Coeficient of correlation between biotic and abiotic variables per band (n = 12). Bold numbers indicate a significant correlation at p ‘ 0.05. Legends for abiotic variables are the same as Table 3. Biome Div = diversity of biomes; S = species richness; log S = species richness at logaritmic scale; percentages (%) depict the proportion of the total marsupial species per band into a given feature category. %LPT = proportion of species having long prehensile tail; %SPT = proportion of species having short prehensile tail, %NPNI = proportion of species having neither prehensile nor incrassated tail; %IT proportion of species having incrassated tail.

	Latitude	MxT	MminT	MmaxT	ThRange	Pmean	Pdif	Area	Log Area
Biome Div.	**–0.82**	0.83	**0.78**	0.09	**–0.74**	0.22	0.56	0.95	0.96
S	**–0.91**	0.94	0.87	–0.07	**–0.86**	0.35	**0.68**	0.84	0.90
Log S	**–0.90**	0.93	0.87	0.13	**–0.83**	0.27	0.66	0.81	0.93
%carnivores	0.73	**–0.72**	**–0.72**	0.44	0.76	–0.34	**– 0.62**	**–0.65**	**–0.66**
%frugivores	**–0.94**	0.93	0.94	–0.26	**–0.95**	0.60	0.87	0.76	0.77
%insectivores	–0.04	0.03	0.03	–0.39	–0.08	–0.18	– 0.04	0.12	0.13
% semiaquatic	**–0.68**	0.70	**0.61**	–0.01	**–0.60**	0.26	0.50	0.57	0.67
%terrestrial	0.94	**–0.93**	**–0.95**	–0.10	0.91	–0.50	**–0.87**	**–0.72**	**–0.82**
%scansorial	**–0.66**	0.61	0.72	0.24	**–0.66**	0.49	0.78	0.38	0.48
%arboreal	**–0.90**	0.93	0.87	–0.05	**–0.86**	0.36	0.70	0.80	0.86
%LPT	**–0.98**	0.98	**0.97**	–0.08	**–0.95**	0.57	0.87	0.78	0.84
%SPT	–0.35	0.41	0.27	0.40	–0.21	–0.36	– 0.06	0.61	0.69
%NPNI	–0.38	0.39	0.41	0.64	–0.31	0.20	0.39	0.10	0.30
%IT	0.94	**–0.96**	**–0.91**	–0.14	0.87	–0.39	**–0.74**	–0.81	–0.92

mean or minimum temperature whereas the proportion of frugivores was positively related to this environmental parameter (Table 4). As temperature decreases, the proportion of carnivores in the guild increases, whereas the proportion of frugivores decreases. Frugivores dissapear below 30° LS, whereas carnivores remain present to the southernmost band (45–50° LS).

We interpreted this correlation as a result of the decrease in relative proportion of frugivores in the guild, rather than a direct effect of temperature on diversification of carnivores. Rather than a case of replacement, the pattern might be mostly driven by the unavailability of fruits as a reliable resource year round below 30° LS (seasonality is more pronounced), directly affecting the distribution of frugivorous marsupials.

These latitudinal pattern, when precipitation variables were considered as the proportion of carnivores, was negatively related to precipitation variability and proportion of frugivores was positively related (Table 4). As precipitation variability increases, the proportion of carnivores decreases and the proportion of frugivores increases.

Seasonality in rain seems to be an important an direct explanatory variable in the latitudinal changes in frugivore diversity pattern observed. This might result from the sharp differences between the dry and wet season in the tropics which would have allowed more diversification in the trophic dimension of the niceh than more constant precipitation. In fact, some marsupials (e.g. *Caluromiopsis irrupta*, *Caluromys spp.*, *Glironia venusta*) are especialised nectarivores during the dry season (Appendix 1). The percentage of insectivores was related to neither precipitation nor temperature variables.

Latitudinal analysis of vertical use of the habitat

The regression analysis between MMinE and vertical use of the habitat showed a significant correlation ($R^2 = 0.87$; $P < 0.001$). As temperature decreased, terrestrial use of the habitat by the marsupial community increased. Similar results were obtained for scansorial marsupials, but the correlation was weak ($R^2 = 0.47$; $P = 0.03$). The percentage of arboreal marsupials was positively correlated with temperature ($R^2 = 0.87$; $P < 0.001$).

This is because the MminE is strongly related with vegetational types, one of the major factors determining the existance of forests with edible fruits, shrubs, or grasslands. Given this, we can understand why arboreal and terrestrial correlations exhibit similar patterns but in opposing directions.

When precipitation variables were considered, we found significant correlations betwen precipitation variability and the percentage of terrestrial ($R^2 = 0.76$; $P < 0.001$), scansorial ($R^2 = 0.60$; $P < 0.001$), and arboreal marsupials ($R^2 = 0.48$; $P = 0.01$). Precipitation variability is positively correlated with the percentage of arboreal and scansorial marsupials but negatively correlated with the percentage of terrestrial marsupials.

Terrestrial marsupials, such as *Didelphis* and *Lutreolina* seems to prefer places with low precipitation variability, which is typical of savannas, pampas, and open grassland habitats (Appendix 1 and references cited). Precipitation variability seems to favour scansorial marsupials because they can use arboreal resources during the wet season and the ground and underground resources during the dry season.

Latitudinal analysis of body size variation

The regression analysis yielded significant correlations between MMinE and percent of small ($R^2 = 0.45$; $P = 0.02$), and medium marsupials ($R^2 = 0.88$; $P < 0.001$). We found no significant correlation with large marsupials.

The first correlation is weak, but at the lowest extremes of MminE temperatures the proportion of small marsupials reaches 100% of the marsupial guild. All the marsupial species in the bands 11 and 12 are small and have incrassated tails. The trend in body size might be more related with the adaptation of fat storage in the tail rather than to body size itself. The second correlation is strong. There are no medium-sized marsupials south of 30° LS, and the MMinE temperature recorded at that latitude is about 4°C. This temperature seems to be the physiological treshold that limits the distribution of medium-sized marsupial species. We found no correlation between the relative proportion of large marsupials and MminE. However, we think the absence of large marsupials south of 40° LS is of importance. There is a high peak in the relative proportion of large marsupials in the bands 9 and 10, but this does not necessarily indicate a degree of success, but simply results from the absence of medium-sized marsupials in these two bands.

When precipitation variables were considered, we found a significant correlation only between the percentage of medium-sized marsupials and precipitation differences ($R^2 = 0.80$; $P < 0.001$). The greater the differences in the mean precipitational range of the band, the larger the proportion of medium-size marsupials. This seems to be related with the high proportion of scansorials in this size category, and is associated with the ability to use arboreal resources during the wet season, mostly fruits, complementing the diet with resources from the ground during the dry season.

Latitudinal analysis of characteristics and functions of tail

Temperature (as MminE) is significantly correlated with both the percentage of marsupials with long prehensile tails (0.97, Table 4), and the percentage with incrassated tails (0.93, Table 4). However, we found no correlation of temperature with the occurrence of marsupials with either short, prehensile or non prehensile tails.

The southernmost treshold limiting the distribution of long, prehensile-tailed marsupials is surprisingly high, and might be related to the distribution of lianas, a commonly used route for most arboreal tropical mammals (Emmons and Gentry 1983). On the other hand, as temperature decreases, the proportion of marsupials with incrassated tails increases up to 100% in the two southernmost bands (below 40° LS), whereas there are no marsupials with incrassated tails in the warmest portion of the continent. However, the distribution of subtropical *Thylamys* in semideserts (which make up too small a proportion of subtropical marsupials to be noticed as a proportion) shows that incrassated tails are and adaptation to seasonal environments, those with extended periods of resource scarcity. The prehensile tail is an adaptation to arboreality, while a fat storage mechanism is an adaptation to life in either winter cold or winter dry seasons, where increased energy reserves are necessary.

We found precipitation to correlate significantly with both the percentage of marsupials with long, prehensile tails (0.76, Table 4), and the percentage of marsupials with incrassated tails (0.54, Table 4). We found no correlation of precipitation with either short, prehensile or non-prehensile tails. Long, prehensile-tailed marsupials are arboreal, and the distribution of forest is strongly related with precipitation. Marsupials with short, prehensile tails are primarily terrestrial, living in shrubby or open savannas or in the forest. They use the ground as the main foraging space, feeding on mostly insects. The same reason might explain the absence of correlation between the occurrence of marsupials with non-prehensile tails and precipitation differences.

We lack any obvious explanation for the correlation between precipitation differences and the percentage of marsupials with incrassated tails, but it might result from an adaptation to dry winters in deserts (i.e. *Thylamys elegans*).

THE FINAL SYNTHESIS: A TAIL TALE?

Table 2 summarises changes in species diversity and percentage of presence of selected characters in marsupial species considered per latitudinal band and Table 4 shows correlations between the biotic and abiotic variables considered in this analysis.

In order to clarify this complex scenario, we can suggest five major life forms in South American marsupials:

Small, arboreal marsupials with long prehensile tails, mostly frugivores, but some insectivores, inhabiting tropical rainforests: strongly influenced by mean minimum extreme temperatures and to a lesser extent by precipitation variability. The distribution of lianas as the principal way to reach fruit resources might be related strongly to the presence of this life form.

Small, terrestrial marsupials with short prehensile tails, mostly insectivores, less influenced by temperature and precipitation variability because they are adapted to live both in forested and open habitats, using the ground as the primary source of resources.

Medium, marsupials with prehensile tails, generalists and scansorial, less affected by temperature than small marsupials with long prehensile tails, living in a wide variety of habitats but limited by temperature in their southernmost distribution.

Large non-prehensile-tailed marsupials, extremely opportunistic, living in a wide variety of thermal ranges and habitats, mostly terrestrial but also scansorial. Large body size allows them to extend their distribution further south than medium-sized marsupials and to feed on small vertebrates. Note: As B.D. Patterson (reviewer) pointed out, the patterns established in this paper for South American marsupials fail to described North American marsupial variation, where the sole species occupying the northern 30 degrees of latitude within the range of Didelphimorphia is the largest species of all, and lacks an incrassated tail. B.D. Patterson suggested that historical events – specifically the very late re-invasion of North America by marsupials during the Great American Interchange – play a very large role in this case. History then created a very contingent circumstance for Didelphis (virginiana), as he brightly summarised the importance of this phenomenon for mammalian zoogeography in Patterson (1999).

Small marsupials with incrassated tails, insectivorous and/or carnivorous. Perhaps due to the combination of a small body size (cheaper to maintain), and fat storage in the tail, they can reach the southernmost latitudes of the continent. The four southernmost species of marsupials (*Thylamys*, *Lestodelphys*, *Rhyncholestes*, and *Dromiciops*) from three different orders (Didelphimorphia, Paucituberculata, and Microbiotheria, respectively), all have fat-filled tails.

CONCLUDING COMMENTS AND FUTURE DIRECTIONS

Marsupial species diversity change geographically in South America. There exists a latitudinal gradient of replacement of food items in diet, from fruits and nectar in tropical latitudes to insects and meat sources towards the south. Another strong pattern is the replacement of prehensile tails to fat-filled tails towards the south. The latitudinal variation of the minimum extreme temperatures seems to be the primary cause, but also locally influenced by precipitation differences and also magnified by the shape of the continent (band area, not related with latitude). Would the same be true in other parts of the world if one repeated our analyses? The incrassated tail of Australian Burramyidae is noteworthy. In general, we doubt that precipitation differences track latitude as a general, worldwide pattern, but further studies on the influence of minimum extreme temperatures may shed light on the widely known enigma of latitudinal patterns of biological variation. We strongly encourage macroecological research on the relationships between diversity and life forms relative to island shape and latitude in Australasia (e.g. a transect from New Guinea to Tasmania).

ACKNOWLEDGEMENTS

I thank R.S. Sikes for helpful comments on an earlier draft. K. Kramer provided valuable assistance from Minnesota finding and sending to me all sort of files, letters, emails, notes, reprints, copies, diskettes, books, and many other things I needed here in Argentina to put together all the information after Elmer's death. Jorge Marquez helped with the cartography and Fig. 1. B.D. Patterson and another reviewer provided invaluable help with comments, editing, discussion, new references, and good ideas. I am specially grateful to my wife, Ticha Pullol, who provided emotional support, love, and double duty at home, while I was writing this chapter.

REFERENCES

Note: Reference numbers refer to Appendix I.

[1] Aagaard, E.M.J. (1982), 'Ecological distribution of mammals in the cloud forests and paramos of the Andes, Mérida, Venezuela', PhD dissertation, Colorado State University.

[2] Albuja, L., & Patterson, B.D. (1996), 'A new species of Northern shrew-oppossum (Paucituberculata: Caenolestidae) from the Cordillera del Cóndor, Ecuador', *Journal of Mammalogy*, **77**:41–53.

[3] Anderson, S. (1982), 'Monodelphis kunsi', *Mammalian Species*, **190**:1–3.

[4] Anderson, S. (1997), 'Mammals of Bolivia, Taxonomy and Distribution', *Bulletin of the American Museum of Natural History*, **231**:1–652.

[5] Arrhenius, O.(1921), 'Species and area', *Journal of Ecology*, **9**:95–9.

[6] Barkley, L.J., & Whitaker, Jr, J.O. (1984), 'Confirmation of Caenolestes in Peru with information on diet', *Journal of Mammalogy*, **65**:328–30.

[7] Birney, E.C.,Monjeau, J.A., Phillips, C.J., Sikes, R.S., & Kim, I. (1996a), '*Lestodelphys halli*: new information on a poorly known Argentine marsupial', *Mastozoología Neotropical*, **3**:171–81.

[8] Birney, E.C., Sikes, R.S., Monjeau, J.A., Guthmann N., & Phillips, C.J. (1996b), 'Comments on Patagonian marsupials from Argentina', in *Contributions in mammalogy: a memorial volume honoring Dr. J. Knox Jones, Jr.* (eds. H. Genoways, & R.J. Baker), pp. 149–54, Museum of Texas Tech University, Lubbock.

[9] Blackburn, T.M., & Gaston, K.J. (1996), 'A sideways look at patterns in species richness, or why there are so few species outside the tropics', *Biodiversity Letters*, **3**:44–53.

[10] Brown, J.H. (1995), *Macroecology*, University of Chicago Press, Chicago.

[11] Brown, J.H., & Gibson, A.C. (1983), *Biogeography*, Mosby, Toronto.

[12] Brown, J.H., & Lomolino, M.V. (1998), *Biogeography, 2nd ed.*, Sinauer Associates, Sunderland, Massachusetts.

[13] Bucher, J.E., & Hoffmann, R.S. (1980), '*Caluromys derbianus*', *Mammalian Species*, **140**:1–4.

[14] Cabrera A. (1957 [1958] –1961), 'Catálogo de los mamíferos de América del Sur. *Revista del Museo Argentino de Ciencias Naturales Bernardino Rivadavia*', *Ciencias Zoológicas*, **4**(1):i –iv + 1–308 [1958]; **4**(2):v–xxvi + 309–732 [1961].

[15] Cerqueira, R. (1985), The distribution of Didelphys in South America (Polyprotodontia, Didelphidae), *Journal of Biogeography* **12**:135–145.

[16] Charles-Dominique, P. (1983), 'Ecology and social adaptations in didelphid marsupials: comparison with eutherians of similar ecology', in *Advances in the study of mammalian behavior* (eds. J.F. Eisenberg, & D.G. Kleiman), pp. 395–422, American Society of Mammalogists Special Publications No 7.

[17] Charles-Dominique, P., Atramentowicz, M., Charles-Dominique, M., Gerard, H., Hladik, A., Hladic, C.M., & Prevost, M.F. (1981), 'Les mamiféres frugivores arboricoles noctumes d'une forêt guyanaise: Inter-relations plantes–animaux', *Rev Ecol*, **35**:341–435.

[18] Chebez, J.C. (1996), *Fauna misionera. A systematic and zoogeographic catalogue of the vertebrate fauna of the province of Misiones (Argentina)*, L.O.L.A., Buenos Aires, Argentina.

[19] Connor, E.F., & McCoy, E.D. 'The statistics and biology of the species-area relationship', *The American Naturalist*, **113**:791–833.

[20] Contreras, J. (1983), 'La comadreja overa', in *Mamíferos. Fauna Argentina, Vol. I* (eds. G. Montes, & M. A. Palermo), pp. 1–32, Centro Editor de América Latina: Buenos Aires, Argentina.

[21] Creighton, G.K. (1984), 'Systematics and taxonomy of the marsupial family Didelphidae', PhD dissertation, University of Michigan, Ann Arbor.

[22] Creighton, G.K. (1985), 'Phylogenetic inference, biogeographical interpretations, and the patterns of speciation in Marmosa (Marsupialia: Didelphidae)', *Acta Zoologica Fennica*, **170**:121–4.

[23] Crespo, J.A. (1950), 'Nota sobre los mamíferos de Misiones nuevos para la Argentina', *Comunicaciones del Instituto Nacional de Investigación de las Ciencias Naturales anexo al Museo Argentino de Ciencias Naturales Bernardino Rivadavia*, **1**:1–14.

[24] Crespo, J.A. (1964), 'Dos mamíferos nuevos para la provincia de Córdoba', *Neotropica*, **10**:62.

[25] Crespo, J.A. (1974), 'Comentarios sobre nuevas localidades para mamíferos de Argentina y Bolivia', *Revista del Museo Argentino de Ciencias Naturales Bernardino Rivadavia*, **11**:1–31.

[26] da Fonseca, G., Herrmann, G., Leite, Y.L.R., Mittermeier, R.A., Rylands A.B., & Patton, J.L. (1996), 'Lista anotada dos mamíferos do Brasil', *Occasional Papers in Conservation Biology*, **4**:1–38.

[27] Darlington, P.J. Jr. (1957), *Zoogeography: the geographical distribution of animals*, John Wiley & Sons, Inc., New York.

[28] del Valle, A.E., Gader, R., Funes M.C., & Cuerpo de Guardafaunas (1989), *Aves y mamíferos de la Provincia de Neuquén*, Dirección de Ecología de Neuquén: Neuquén, Argentina.

[29] Eisenberg, J.F. (1989), *Mammals of the Neotropics, Vol. 1: The northern tropics: Panamá, Colombia, Venezuela, Guyana, Suriname, French Guiana*, The University of Chicago Press, Chicago.

[30] Emmons, L.H., & Feer, F. (1997), *Neotropical Rainforest Mammals: A Field Guide*, The University of Chicago Press, Chicago.

[31] Emmons, L.H., & Gentry, A.H. (1983), 'Tropical forest structure and the distribution of gliding and prehensile-tailed vertebrates', *The American Naturalist*, **121**:513–24.

[32] Emmons, L.H., & Smith, K.S. (1992), *List of mammals of northeastern Pando and Gazetteer of localities visited in 1992* [parts on an unpublished report 23 Nov 1992, cited in reference 4].

[33] Galliari, C.A.& Pardiñas, U.F.J., & F.J. Goin. (1996), 'Lista comentada de los mamíferos de la Argentina', *Mastozoología Neotropical*, **3**:39–61.

[34] Garcia, N.O. (1994), 'South American Climatology', *Quaternary International*, **21**:7–27.

[35] Gardner, A.L. (1990), 'Two new mammals from southern Venezuela and comments on the affinities of the highland fauna of Cerro de la Neblina' in *Advances in Neotropical Mammalogy* (eds. K.H. Redford, & J.F. Eisenberg), pp. 411–24, The Sandhill Crane Press, Inc., Gainesville, Florida, USA.

[36] Gardner, A.L. (1993a), 'Order Didelphimorphia', in *Mammal Species of the World: A taxonomic and geographic reference, 2nd ed.* (eds. E. Wilson, & D.M. Reeder), pp. 15–23, Smithsonian Institution Press, Washington DC.

[37] Gardner, A.L. (1993b), 'Order Paucituberculata', in *Mammal Species of the World: A taxonomic and geographic reference, 2nd ed.* (eds. E. Wilson, & D.M. Reeder), pp. 25–6, Smithsonian Institution Press, Washington DC.

[38] Gardner, A.L. (1993c), 'Order Microbiotheria', in *Mammal Species of the World: A taxonomic and geographic reference, 2nd ed.* (eds. E. Wilson, & D.M. Reeder), p. 27, Smithsonian Institution Press, Washington DC.

[39] Gardner, A.L., & Creighton, G.K. (1989), 'A new generic name for Tate's (1933) Microtarsus group of South American mouse opossums (Marsupialia, Didelphidae)', *Proceedings of the Biological Society of Washington*, **102**:3–7.

[40] Gardner, A.L., & Patton, J.L.(1972), 'New species of Philander (Marsupialia, Didelphidae) and Mimon (Chiroptera: Phyllostomidae) from Perú', *Occas Pap Mus Zool Louisiana State Univ*, **43**.

[41] Handley, C.O. (1976), 'Mammals of the Smithsonian Venezuelan project', *Brigham Young University Science Bulletin, Biology Series*, **20**(5):1–90.

[42] Handley, C.O. (1996), 'Checklist of the mammals of Panamá', in *Ectoparasites of Panamá R.L.* (eds. R.L.Wenzel & V.J. Tipton), pp. 753–95, Field Museum of Natural History, Chicago.

[43] Handley, C.O., & Gordon, L.K.(1979), 'New species of mammals from northern South America: mouse opossums, the genus Marmosa', in *Vertebrate ecology in the northern Neotropics* (ed. J.F. Eisenberg), pp. 65–71, Smithsonian Institution Press, Washington DC.

[44] Heinonen Fortabat, S.H., & Chebez, J.C. (1997), *Los mamíferos de los parques nacionales de la Argentina*, Monografía Especial, L.O.L.A.

[45] Hershkovitz, P. (1992), 'The South American gracile mouse opossums, genus *Gracilinanus* Gardner and Creighton, 1989 (Marmosidae, Marsupialia): a taxonomic review with notes on general morphology and relationships', *Fieldiana Zoology*, **70**:1–56.

[46] Izor, R.J., &. Pine, R.H. (1987), 'Notes on the black-shouldered opossum, *Caluromiopsis irrupta*', in *Studies in Neotropical Mammalogy: Essays in honor of Philip Hershkovitz* (eds. B.D. Patterson, & R.M. Timm), pp. 117–24, Fieldiana Zoology, Chicago.

[47] Janson, C.H., & Terborgh J., & Emmons, L.H.(1981), 'Non-flying mammals as pollinating agents in the Amazonian forests', *Biotropica, Repro Bot Supp*, 1–6.

[48] Johnson, A.M.(1976), 'The climate of Peru, Bolivia, and Ecuador', in *Climates of Central and Southern America: World Survey of Climatology, Vol 12* (ed. W. Schwerdtfeger), pp. 147–218, Elsevier, Amsterdam.

[49] Kaufman, D.W. (1995), 'Diversity of new world mammals: universality of the latitudinal gradients of species and bauplans', *Journal of Mammalogy*, **76**(2):322–34.

[50] Kelt, D.A., & Martinez, D.R. (1989), 'Notes on distribution and ecology of two marsupials endemic to the Valdivian forests of southern South America', *Journal of Mammalogy,* **70**:220–4.

[51] Kirsch, J.A.W., & Waller P.F. (1979), 'Notes on the trapping and behavior of the Caenolestidae (Marsupialia)', *Journal of Mammalogy,* **60**:390–5.

[52] Klijn, F., & de Haes, U. (1994), 'A hierarchical approach to ecosystems and its implications for ecological land classification', *Landscape Ecology,* **9**:89–104.

[53] Leite, Y.L.R., Costa, L.P.& Stallings, J.R. (1996), 'Diet and vertical space use of the three sympatric oppossums in a Brazilian Atlantic forest reserve', *Journal of Tropical Ecology*, **12**:435–40.

[54] Lorini, M.L., Oliveira, J.A.& Persson, V.G. (1994), 'Annual age structure and reproductive patterns in *Marmosa incana* (Lund 1841) (Didelphidae, Marsupialia)', *Z Saugetierkunde*, **59**:65–73.

[55] Lyons, S.K., & Willig, M.R. (1997), 'Latitudinal patterns of range size: methodological concerns and empirical evaluations for new world bats and marsupials', *Oikos,* **79**:568–80.

[56] Lyons, S.K., & Willig, M.R. (1999), 'A hemispheric assessment of scale dependence in latitudinal gradients of species richness', *Ecology,* **80**:2483–91.

[57] MacArthur, R.H., & Wilson, E.O. (1967), *The theory of island biogeography*, Princeton University Press, Princeton, NJ.

[58] Mares, M.A., & J. K. Braun (2000), 'Systematics and Natural History of Marsupials from Argentina', in *Reflections of a Naturalist: Papers Honoring Professor Eugene D. Fleharty. Fort Hays Studies, Special Issue* (ed. Jerry Choate), pp. 23–45, Stenberg Museum of Natural History, Fort Hays State University, Hays, Kansas, USA.

[59] Mares, M.A. Ojeda, R.A., & Barquez, R.M.(1989), *Guide to the mammals of Salta Province, Argentina (Guia de los mamíferos de la Provincia de Salta, Argentina),* University of Oklahoma Press, Norman, Oklahoma.

[60] Marshall, L.G. (1978), '*Chironectes minimus*', *Mammalian Species,* **109**:1–6.

[61] Marshall, L.G. (1978), '*Dromiciops australis*', *Mammalian Species,* **99**:1–5.

[62] Marshall, L.G. (1978), '*Glironia venusta*', *Mammalian Species,* **107**:1–3.

[63] Marshall, L.G. (1978), '*Lutreolina crassicaudata*', *Mammalian Species,* **91**:1–4.

[64] Massoia, E. (1980), 'Mammalia de Argentina. I. Los mamíferos silvestres de la Provincia de Misiones', *Iguazú,* **1**:15–43.

[65] Massoia, E., & Fornes, A. (1967), 'El estado sistemático, distribución geográfica y datos etoecológicos de algunos mamíferos neotropicales (Marsupialia y Rodentia) con la descripción de *Cabreramys*, género nuevo (Cricetidae)', *Acta Zoologica Lilloana,* **23**:407–30.

[66] Meserve, P.L., & Lang, B.K.& Patterson B.D. (1988), 'Trophic relationships of small mammals in a Chilean temperate rainforest', *Journal of Mammalogy*, **69**:721–30.

[67] Meserve, P.L., Murua, R., Lopetegui, O., & Rau, J.R. (1982), 'Observations on the small mammal fauna of a primary temperate rainforest in southern Chile', *Journal of Mammalogy,* **63**:315–17.

[68] Miles, M.A., De Souza, A.A., Povoa, M.M. (1981), 'Mammal tracking and nest location in Brazilian forest with improved spool-and-line device', *Journal of Zoology (London)*, **195**:331–47.

[69] Miller, A. (1976), 'The climate of Chile', in *Climates of Central and Southern America World Survey of Climatology, Vol 12* (ed. W. Schwerdtfeger), pp. 113–45, Elsevier, Amsterdam.

[70] Monjeau, J.A.(1989), *Ecología y distribución geográfica de los pequeños mamíferos del Parque Nacional Nahuel Huapi y áreas adyacentes*, unpublished PhD dissertation, Universidad Nacional de La Plata, Argentina.

[71] Monjeau, J.A., Birney, E.C., Ghermandi, L., Sikes, R.S., Margutti L., & Phillips, C.J. (1998), 'Plants, small mammals, and the hierarchical landscape classifications of Patagonia', *Landscape Ecology,* **13**:285–306.

[72] Monjeau, J.A., Bonino, N., & Saba, S. (1994), 'Annotated checklist of the living land mammals in Patagonia, Argentina', *Mastozoología Neotropical,* **1**:143–56.

[73]Monjeau, J.A., Sikes, R.S., Birney, E.C., Guthmann, N., & Phillips, C.J. (1997), 'Small mammal community composition within the major landscape divisions of Patagonia, southern Argentina', *Mastozoología Neotropical,* **4**:113–27.

[74] Mustrangi, M.A. (1995), *Phylogeography of Marmosops (Marsupialia, Didelphidae) in the Atlantic forest of Brazil, and the phylogenetic relationships of the Atlantic forest and Amazonian species in the genus,* PhD dissertation, University of California, Berkeley, CA, USA.

[75] Nowak, R.M. (1991), *Walker's mammals of the world, Vol I. 5th ed.*, Johns Hopkins University, Baltimore, USA.

[76] Ojeda, R.A & IUCN's New World Marsupial Specialist Group (2000, in preparation), *The marsupials of Argentina: an annotated checklist of their distribution and conservation*, in http://www.cricyt.edu.ar/INSTITUTOS/iadiza/ojeda/MarsupArg.htm

[77] Olrog, C.C. (1959), 'Notas mastozoológicas. II. Sobre la colección del Instituto Miguel Lillo', *Acta Zoologica Lilloana,* **17**:403–19.

[78] Pacheco, V., De Macedo, H., Vivar, E., Azcorra, C., Arana-Cardo, R., & Solari, S. (1995), 'Lista anotada de los mamíferos peruanos', *Occas Paper Conserv Biol,* **2**:1–35.

[79] Pacheco, V., Patterson, B.D., Patton, J.L., Emmons, L.H., Solari, S., & Ascorra, C.F. (1993), 'List of mammal species known to occur in Manu Biosphere Reserve, Perú', *Publ Mus Hist Nat Univ Nac Mayor de San Marcos, ser A Zoologia,* **44**:1–12.

[80] Palma, R.E., & Yates, T.L. (1998), 'Phylogeny of southern South American mouse opossums (Thylamys, Didelphidae) based on allozyme and chromosomal data', *Zeitschrift fur Saugetierkunde,* **63**:1–15.

[81] Patterson, B.D. (1994), 'Accumulating knowledge on the dimensions of biodiversity: systematics perspectives on Neotropical Mammal', *Biodiversity Letters,* **2**:79–86.

[82] Patterson, B.D., & Gallardo, M.H. (1987), 'Rhyncholestes raphanurus', *Mammalian Species,* **286**:1–5.

[83] Patterson, B.D., Meserve, P.L., & Lang B.K. (1989), 'Distribution and abundance of small mammals along an elevational transect in temperate rainforest of Chile', *J Mammalogy,* **70**:67–78.

[84] Pearson, O.P. (1983), 'Characteristics of a mammalian fauna from forests in Patagonia, southern Argentina', *Journal of Mammalogy,* **64**:476–92.

[85] Pearson, O.P. (1995), 'Annotated keys for identifying small mammals living in or near Nahuel Huapi National Park or Lanin National Park, southern Argentina', *Mastozoología Neotropical,* **2**:99–148.

[86] Pianka, E.R. (1966), 'Latitudinal gradients in species diversity: a review of concepts', *The American Naturalist,* **100**:33–46.

[87] Pine, R.H. (1977), '*Monodelphis iheringi* (Thomas) is a recognizable species of Brasilian opossum (Mammalia: Marsupialia: Didelphidae)', *Mammalia,* **41**:235–7.

[88] Pine, R.H. (1980), 'Taxonomic notes on "*Monodelphis dimidiata itatiayae* (Miranda–Ribeiro)", *Monodelphis domestica* (Wagner) and *Monodelphis maraxina* Thomas (Mammalia: Marsupialia: Didelphidae)', *Mammalia'* **43**:495–9.

[89] Pine, R.H. (1981), 'Review of the mouse opossum Marmosa parvidens and M. invicta with a description of a new species', *Mammalia,* **45**:56–70.

[90] Pine, R.H., & J.P. Abravaya. (1978), 'Notes on the Brazilian opossum *Monodelphis scalops* (Thomas) (Mammalia: Didelphidae)', *Mammalia,* **42**:379–82.

[91] Pine, R.H., & Handley Jr., C.O. (1984), 'A review of the Amazonian short-tailed opossum *Monodelphis emiliae* (Thomas)', *Mammalia,* **48**:239–45.

[92] Pine, R.H., Dalby, P.L., & Matson, J.O. (1985), 'Ecology, postnatal development, morphometrics, and taxonomic status of the short-tailed opossum, *Monodelphis dimidiata*, and apparently semelparous annual marsupial', *Annals of Carnegie Museum,* **54**:195–231.

[93] Pine, R.H., Miller, S.D., & Schamberger, M.L.(1979), 'Contributions to the mammalogy of Chile', *Mammalia,* **43**:339–76.

[94] Prohaska, F. (1976), 'The Climate of Argentina, Paraguay and Uruguay', in *Climates of Central and Southern America. World Survey of Climatology, Vol 12* (ed. W. Schwerdtfeger), pp. 13–122, Elsevier, Amsterdam.

[95] Ratisbona, (1976), 'The Climate of Brazil', in *Climates of Central and Southern America: World Survey of Climatology, Vol 12* (ed. W. Schwerdtfeger), pp. 219–93, Elsevier, Amsterdam.

[96] Redford, K.H., & Eisenberg, J.F. (1992), *Mammals of the neotropics: the southern cone, Vol 2, Chile, Argentina, Uruguay, Paraguay,* University of Chicago Press, Chicago.

[97] Reig, O.A. (1964), 'Roedores y marsupiales del Partido de General Pueyrredón y regiones adyacentes (Provincia de Buenos Aires, Argentina)', *Publicaciones del Museo Municipal de Ciencias Naturales de Mar del Plata,* **1**:203–24.

[98] Reig, O.A.(1981), 'Teoria del origen y desarrollo de la fauna de mamíferos de América del Sur', *Monografie Naturae,* **1**:7–162.

[99] Rohde, K.(1992), 'Latitudinal gradients in species diversity: the search for the primary cause', *Oikos,* **65**:514–27.

[100] Roig, V.G. (1962), 'Aspectos biogeográficos y planteos ecológicos de la fauna de mamíferos de las zonas áridas y semiáridas de Mendoza', *Revista de la Facultad de Ciencias Agrarias,* **9**:58–81.

[101] Rosenzweig, M. (1992), 'Species diversity gradients: we know more and less than we thought', *Journal of Mammalogy,* **73**:715–30.

[102] Rosenzweig, M. (1995), *Species diversity in space and time*, Cambridge University Press, Cambridge.

[103] Ruggiero, A.(1994), 'Latitudinal correlates of the sizes of mammalian geographical ranges in South America', *Journal of Biogeography,* **21**:545–59.

[104] Salazar Bravo, J.A., Campbell, M.L., Anderson, S., Gardner S.L., & Dunnum J.L. (1994), 'New records of Bolivian Mammals', *Mammalia,* **58**:125–30.

[105] Schwerdtfeger, W. (1976), *Climates of Central and Southern America. World Survey of Climatology, Vol. 12*, Elsevier, Amsterdam.

[106] Snow, J.W. (1976), 'The Climate of Central and Northern South America', in *Climates of Central and Southern America World Survey of Climatology, Vol 12* (ed. W. Schwerdtfeger), pp. 295–403, Elsevier, Amsterdam.

[107] Stanley, S.M. (1979), *Macroevolution. Pattern and Process*, W.H. Freeman & Co., San Francisco, CA, USA

[108] Streilein, K.E. (1982), 'Behavior, ecology, and distribution of South American marsupials', in *Mammalian Biology in South America. Pymatuning Symposia in Ecology, Special Publication Series* (eds. M.A. Mares. & H.H. Genoways), pp. 231–50, Pymatuning Laboratory of Ecology, University of Pittsburgh, Pittsburgh, USA.

[109] Tate, G.H.H. (1933), 'A systematic revision of the marsupial genus *Marmosa*', *Bulletin of the American Museum of Natural History,* **66**:1–250.

[110] Udvardy, M.D.F. (1978), 'World biogeographical provinces', *Coevolution Quarterly*, Sausalito California, based on M.D.F. Udvardy, 'International Union for Conservation of Nature and Natural Resources, Occasional Paper', **18**.

[111] Voss, S., & Emmons, L.H. (1996), 'Mammalian diversity in neotropical lowland rainforest: a preliminary assessment', *Bulletin of the American Museum of Natural History,* **230**:1–115.

[112] Voss, R.S., Lunde, D.P., & Simmons, N.B. (in press), 'Mammals of Paracou, french Guiana: A neotropical lowland rainforest fauna, part 2: non-volant species', *Bull Am Mus Nat Hist.*

[113] Wallace, A.R. (1878), *Tropical nature and other essays*, MacMillan, New York.

[114] Wilson, D.E., & Reeder D.M. (eds.), (1993). *Mammal Species of the World: A taxonomic and geographic reference*, 2nd ed., Smithsonian Institution Press, Washington DC.

Appendix 1 Biological aspects of individual South American species selected for study in our search for macroecological patterns. HB = length of head and body, tail = length of tail; Weight = estimated weight in grams and body size category; Distribution = the entire known geographic range of each animal species is described; Band = band number in which the species is present; Ecology = only information from published sources is included on major ecological and habitat features; References = firsthand sources of published information used in the Appendix.

Species account	HB	Tail	Weight (g)	Distribution	Band	Major biomes	Ecology	References
DIDELPHIMORPHIA								
Caluromyopsis irrupta	250–330	310–340	250–300 medium	SE Perú and adjacent Brasil	4, 5	Amazon rainforest, cloud forest	Nocturnal, arboreal, prehensile tail. Habitat: upper levels of the forest.	30, 36, 46, 47
Caluromys derbianus	225–300	384–445	245–370 medium	México, Colombia and W Ecuador	1–3	Central American rainforest region	Nocturnal, arboreal, make nests on tree holes, prehensile tail. Habitat: mature and disturbed evergreen rainforest, dry forest, and plantations.	13, 30, 36
Caluromys lanatus	217–295	330–435	310–410 medium	Colombia, Ecuador, Perú, Bolivia, Brasil, Paraguay, Argentina	1–6	Amazon rainforest	Nocturnal, arboreal, prehensile tail. Habitat: favour dense, vini, midstory and canopy vegetation.	30, 36, 58, 76
Caluromys philander	160–279	250–405	140–390 medium	E Venezuela, Trinidad, Guianas, NE Brazil, Bolivia	1–6	Amazon rainforest	Nocturnal, arboreal. Long prehensile tail. Habitat: it uses middle and upper levels of the forest, dense vini vegetation but also use the high open forest canopy.	16, 17, 30, 36
Glironia venusta	160–205	195–225	150–250 medium	Ecuador, Brazil, Perú, and Bolivia	2–6	Amazon rainforest	Nocturnal, arboreal. Long prehensile tail. It jumps from one branch to another hunting insects. Habitat: upper and middle levels of dense vini vegetation.	30, 36, 62
Chironectes *minimus*	260–298	327–420	590–700 large	Disjunct distribution: a) S. Mexico and Belize to W Venezuela, Colombia, Ecuador, Perú, Bolivia and W. Brasil; b) from mouth of the Amazon to Misiones, Argentina	1–8	Venezuelan coastal rainforest, Central American rainforest, Amazon rainforest, Atlantic Rainforest	Nocturnal, terrestrial and semiaquatic. Long fingers and sandpaper-like palms. Long prehensile tail. Habitat: lives in and near clear rivers and streams in hilly areas.	30, 36, 58, 60, 76
Didelphis albiventris	305–437	290–430	500–2000 large	South America, disjunct distribution: a) a Guyana Highland Isolate in S Venezuela, b) in and east of the Andes around the periphery of rainforest from W Venezuela to Mar del Sur, Argentina.	1–10	Savannas, Cerrado, Caatinga, Chaco, Pantanal, and the Pampas	Nocturnal. Long prehensile tail. Habitat: wide habitat and elevational range in savannas, gallery forest, swamps, cultivated or deforested lands, and in humid forest at high elevations and subtropical latitudes.	20, 28, 30, 58, 76, 97, 108

Didelphis aurita[1]	310–390	310–370	700–1500 large	Coastal Brasil from Bahia to Rio Grande do Sul, west to the lower Rio Paraguay, and Misiones, NE Argentina	5 to 8	Atlantic rainforest and Araucaria forest	Nocturnal, arboreal and terrestrial. Long prehensile tail. Habitat: inhabits humid tropical lowland forest, and *Araucaria* highlands.	15, 20, 30, 36, 44, 53, 58, 76, 96, 111
Didelphis marsupialis	324–425	336–420	565–1610 (males larger than females) large	Trinidad and the Lesser Antillas, México to Bolivia, Paraguay, Brasil, and NE Argentina.	1–8	Amazonian rainforest and northermost of Atlantic rainforest	Nocturnal; arboreal and terrestrial. Long prehensile tail. Habitat: humid forests and outlying gallery forest; they thrive in secondary forest and around dwellings.	15, 30, 36
Gracilinanus aceramarcae			20–30 (inferred) small	Bolivia and Perú. Only two specimens reported	5 6	Probably andean yungas	Nocturnal, arboreal, inferred by similar species. Long prehensile tail. Habitat: No specific or detailed information available.	4, 36, 78
Gracilinanus agilis[2]	90–106	109–150	20–30 small	Perú, Bolivia, E. Brasil, Paraguay, Uruguay, and Argentina	3–9	W Amazonian rainforest, Atlantic forest, N. of Buenos Aires Province	Nocturnal, arboreal. Long prehensile tail. Habitat: frequent the forest understory, where they used slender branches and vines. Found in evergreen gallery forest.	30, 58, 76
Gracilinanus dryas	90–100	130–150	18 small	W Venezuela and E Colombia	1	Colombian montane	Nocturnal, arboreal and terrestrial. Long prehensile tail. Habitat: moist cloud forests, including second growth.	1, 29, 36
Gracilinanus emiliae	72–87	128–143	10 small	Disjunct distribution in Colombia, Surinam, French Guiana, and Brazil only near Belem.	1 and 3	Amazonian rainforest	Nocturnal, inferred by similar species. Extremely long prehensile tail. Arboreal, inferred. Habitat: lowland rainforests.	30, 112
Gracilinanus kalinowskii	89–91	110–117	13–18 small	E. Perú on lower andean slopes, and Guyana and French Guyana.	4	Amazonian rainforest	Nocturnal, probably mostly terrestrial. Relatively much shorter prehensile tail than *G. emiliae*, slightly longer than head and body. Habitat: from lowland and lower montane evergreen forest.	30, 36, 112
Gracilinanus marica	99	134	10 (inferred) small	N coastal Venezuela and Colombia	1	Colombian montane, Venezuelan deciduous forest and Llanos	Nocturnal (inferred), arboreal, prehensile tail. Habitat: found in lowland to montane evergreen and deciduous forest, and savannas.	29, 30, 39, 41
Gracilinanus microtarsus	105–110	145–153	19–29 small	E Brasil and NE Argentina	6–8	Atlantic rainforest	Nocturnal, probably mostly terrestrial. Long prehensile tail. Habitat: Found in wet evergreen coastal forest.	30, 36, 58, 76
Lestodelphys halli	134–144	81–99	50–80 small	Only Argentina: Mendoza, La Pampa, Neuquen, Río Negro, Chubut, and Santa Cruz south to 47° 06 L.S (the southernmost record known for a marsupial species)	9–12	Monte and Patagonian semideserts	Nocturnal, terrestrial, remarkable fat tail. Habitat: mostly in relatively wet patches in a desertic matrix.	7, 8, 36, 58, 71, 73, 76, 91

Appendix 1 *Continued*

Species account	HB	Tail	Weight (g)	Distribution	Band	Major biomes	Ecology	References
Lutreolina crassicaudata	243–400	245–358	350–910 large	Disjunct distribution: a) E. Colombia, Venezuela, Guianas; b) Bolivia, Paraguay, Perú, Brasil, Uruguay and Argentina south to Mar del Sur.	1–2 and 5–10	Savannas	Nocturnal, terrestrial, but a good climber. Tail slighty shorter than head and body. Does not seem to use the tail neither to climb nor to store fat. The female has a very long tail. Habitat: Mainly known from wet savannas near forest in the northern part of the range. Closely associated with aquatic habitats, from flooded grasslands to small streams in the southern part of their geographical range.	30, 36, 58, 63, 76, 97
Marmosa andersoni			10 (inferred) small	Per, Cuzco. Known only from the type locality	5	Andean	Presumably prehensile tail as all the mouse opposums. Presumably nocturnal. Habitat: unknown.	36
Marmosa lepida	97–120	140–150	10 small	Amazon basin of Colombia, Ecuador, Bolivia, Brasil	2–4	Amazon rainforest	Tail very long, prehensile Presumably nocturnal. Habitat: rainforest	30
Marmosa murina	125–150	170–198	43–60 (males larger than females) small	Venezuela, the Guianas, the Amazon basin of Colombia, Ecuador, Perú, Brasil, and Tobago	1–7	Amazon rainforesr and Atlantic coastal forest	Tail long, prehensile Presumably nocturnal. Habitat: Especially common in disturbed roadsides, extremely dense vines, thickets, river edges, and secondary forest.	17, 30
Marmosa robinsoni	128–200	170–239	36–132 (males larger than females) small and medium	Central and South America; Belize, Guatemala, west of the Andes from Colombia to N Perú	2–4	Amazon rainforest	Nocturnal, arboreal and terrestrial. Tail long prehensile. Habitat: found in a wide variety of habitats from evergreen cloud forest to deciduous and thorn forest.	30, 36
Marmosa rubra	121–165	180–220	59–67 small	The Amazon basin of Colombia, Ecuador, and Perú	1–4	Amazon rainforest	Tail long, prehensile Presumably nocturnal. Habitat: found in the ground of lowland rainforest an secondary forest.	30
Marmosa tyleriana	116	166	20–30 (inferred) small	Southern Venezuela	1–2	Guyanan and Madeiran	Nocturnal, arboreal and terrestrial (inferred) prehensile tail. Habitat: inhabits montane regions, especially tepuis.	29, 36
Marmosa xerophila	131	144–181	20–30 (inferred) small	NE Colombia and NW Venezuela	1	Venezuelan dry forest	Nocturnal, arboreal and terrestrial (inferred) prehensile tail. Habitat: found in dry deciduous forest	29, 36, 43
Marmosops cracens	105	132	24–27 small	Known only from the vicinity of the type locality, Falcón, Venezuela	1	Amazonian rainforest	Nocturnal, arboreal, and terrestrial (inferred) Prehensil tail. Habitat: found in moist foothill forests	29, 36, 43

Marmosops dorothea	112–136	144–167	28–48 small	Bolivia and Brazil south to the middle Juruá	5–6	Amazonian rainforest	Nocturnal, terrestrial and arboreal. Uses the ground and understory of the forest in areas with dense vine tangles. Prehensile tail. Habitat: montane forest of la Paz and in the lowlands of southern Amazonian evergreen rainforest and in the adjacent rainforest.	30, 36
Marmosops fuscatus	128–133	152–172	40 small	Northern mountains of Venezuela, Colombia, and Trinidad	1–2	Colombian montane, Venezuelan deciduous forests, and Guyanan	Nocturnal, arboreal and terrestrial. Prehensile tail. Habitat: mainly inhabits montane and cloud forests above 1000 m. Found in evergreen forests and associated clearings.	30
Marmosops handleyi	104–122	129–149	20–30 (inferred) small	Known only from type locality, Antioquia, Colombia	1	Colombian montane	Nocturnal, arboreal, and terrestrial (inferred) Prehensile tail. Habitat: cloud forest.	29, 89
Marmosops impavidus	116–152	146–205	36–51 small	Panamá on high mountains of Darién; Andes from N Colombia to Cochabamba, Bolivia; lowlands near the Rio Amazonas to W Brazil	1–4	Andean and Amazonian rainforest	Nocturnal, arboreal and terrestrial. Prehensile tail. Habitat: montane cloud forest, moist evergreen forest few records in lowland Amazonas.	29, 30, 35, 36, 74
Marmosops incanus	91–194	110–237	13–140 (males much larger than females) small and medium	Brazil, Bahia to San Pablo.	5–7	Atlantic rainforest, Cerrado, Caatinga	Nocturnal, probably uses the ground and low understory. Prehensile tail. Habitat: found in humid lowland coastal forests, montane coastal forest, and semi-deciduous forests of the cerrado and caatinga.	30, 36, 57, 74
Marmosops neblina[3]	112–142	140–178	37–40 (males larger than females) small	Venezuela, on Cerro Neblina, E. Ecuador, Brazil in the lowland along Rio Juruá	2–3	Amazonian rainforest	Nocturnal, mainly terrestrial. Prehensile tail. Habitat: from montane cloud forest on a Venezuelan tepui and lowland evergreen Amazonian forest.	30, 35, 36, 74
Marmosops noctivagus	115–120	164–184	35–37 small	Amazon basin of Ecuador, Peru, W Brazil, S of Amazon River and Bolivia	3–5	Amazonian rainforest	Nocturnal, arboreal and terrestrial. Prehensile tail. Habitat: found in mature, disturbed, and secondaruy forests, usually in fallen trees.	21, 30
Marmosops parvidens	95–105	130–160	15–27 small	East of the Andes in the Amazon Basin of Venezuela, Colombia, the Guianas, Brazil and Perú	1–5	Amazonian rainforest	Nocturnal, arboreal, and terrestrial. Prehensile tail. Habitat: found in mature, closed canopy, evergreen forest, not often in disturbed or secondary forest.	30, 36, 54
Marmosops paulensis	94–153	145–212	16–70 (males much larger than females) small	Brazil, Atlantic montane coastal forests from S Minas Geraes to Sao Paulo	7	Atlantic Coastal Rainforest	Nocturnal, arboreal and terrestrial (inferred) Prehensile tail. Habitat: known only from wet montane and cloud forests.	30, 36

Appendix I *Continued*

Species account	HB	Tail	Weight (g)	Distribution	Band	Major biomes	Ecology	References
Metachirus nudicaudatus	250–280	280–369	300–480 (males slighty larger than females) medium	Costa Rica south to Paraguay and NE Argentina	1–8	Amazonian rainforest, savanna, Central American rainforest region, Atlantic rainforest	Nocturnal, terrestrial. Habitat: seems to favour mature forest with little underground. Also present in dense habitats, secondary forest, and gallery forests.	18, 23, 30, 36, 44, 59, 68, 76
Micoureus alstoni	185–200	230–250	250–300 (inferred) medium	Central and South America: Belize to N Colombia.	1	Colombian montane	Nocturnal, arboreal. Prehensile tail. Habitat: found in lowland and montane evergreen forests	30
Micoureus constantiae	134–150	186–201	68–73 small	Lower Andean foothills in SW Bolivia and NW Argentina (Salta and Jujuy) to Mato Grosso	5–7	Amazonian rainforest and yungas	Nocturnal, arboreal. Prehensile tail. Habitat: inhabits moist forest habitats, including mossy montane forest.	30, 36, 44, 58, 76, 77
Micoureus demerarae	152–210	195–270	62–130 (males larger than females) medium	Colombia to N Argentina, Paraguay, and E. Brazil	1–8	Amazonian rainforest, Central American rainforest region. Atlantic rainforest	Nocturnal, arboreal. Prehensile tail. Habitat: they are usually seen in the middle to upper levels of the forest, but descend to the ground when food is scarce in the dry season.	17, 18, 30, 36, 44, 58, 64, 68, 76
Micoureus phaea[4]	121–179	153–236	63 small	Central and South America: Panama south to W Ecuador	1–3	Central American rainforests	Nocturnal, arboreal. Prehensile tail. Habitat: Known from wet evergreen lowland rainforests and wet montane forests.	30, 36
Micoureus regina	114–151	147–236	35–68 small	The western Amazon Basin and Eastern Andean Slopes from Colombia to S. Perú and Bolivia, Brazil.	1–5	Amazonian rainforest	Nocturnal, arboreal. Prehensile tail. Habitat: Middle to upper levels of the forest, also found in the ground.	30, 36
Monodelphis adusta	84–114	51–67	15–35 small	Central and South America: the northern Andes: Colombia, Ecuador, S Perú, and Bolivia	1–5	Andean and Amazonian rainforest	Terrestrial. Short prehensile tail. Habitat: rainforests, wet grasslands, rocky terrain. A montane species of wet forests at middle elevations.	4, 30, 36, 42, 78, 79, 104
Monodelphis americana	101–105	45–55	23–35 small	Brazil: from Belem to San Pablo	3–7	Atlantic coastal forest	Terrestrial, probably diurnal. Short prehensile tail. Habitat: Usually near water.	30, 36
Monodelphis brevicaudata	134–183	76–105	46–150 small	East of the Andes in Colombia, Venezuela, Guianas, Amazonian Perú, N. Bolivia, Brazil, and probably Argentina	1–6	Amazonian rainforest	Diurnal, terrestrial. Short prehensile tail. Habitat: lives on the ground, near or under fallen brush or ritten logs. Mature, disturbed, and secondary: rainforest.	4, 30, 32, 36, 108

Monodelphis dimidiata	55.0 –151	114 – 231	40 – 84 small	South-eastern Brasil, Uruguay, and Argentina	7–10	Savannas and Argentine pampas	Diurnal and nocturnal. Terrestrial. Short prehensile tail. Habitat: pastures, wetlands, pampa grasslands, and riparian areas next to waterways, grasses that border agricultural areas.	18, 24, 36, 44, 58, 64, 65, 76, 92, 97
Monodelphis domestica	130–191	70–106	36–98 small	Bolivia, Paraguay, Brasil, and northern Argentina	3–7	Chaco forests	Diurnal. Terrestrial. Short prehensile tail. Habitat: Open forest.	36, 44, 58, 76, 108
Monodelphis emiliae	120–158	50–70	52–60 small	Brazil south of the Amazon, and Perú south to the Amazon to Pando, Bolivia	3–5	Amazonian rainforest	Terrestrial, short prehensile tail. Habitat: found around fallen logs in lowland evergreen forest.	4, 30, 36, 91
Monodelphis iheringi	no data	no data	inferred as category 3 small	South-east of Brazil and Argentina	7–8	Atlantic rainforesrs and subtropical forest	Terrestrial, semiaquatic, short prehensile tail. Habitat: found in wet forests, along watercourses.	18, 36, 44, 58, 64, 76, 87
Monodelphis kunsi	71–94	41–42	19 small	Bolivia and Brazil	5–7	Bolivian and Brasilian montane rainforest	Terrestrial, short prehensile tail. Habitat: south-west slope with dense shrubs. Many rocks and fallen logs, leaf mulch and litter.	3, 4, 36, 104
Monodelphis maraxina	130–191	70–106	small	Brazil, Pará, Marajó Island	3	Amazonian (Madeiran)	Terrestrial, short prehensile tail. Habitat: no data.	26, 36, 88
Monodelphis osgoodi	94–100	52–75	14 small	SE Perú and C Bolivia	6	Amazonian rainforest	Terrestrial, short prehensile tail. Habitat: riverine banana field, flat area, forest and shrubs, soil very moist, with organic leaf mulch.	4, 36, 104
Monodelphis rubida	125–137	56–60	45–46 small	E Brazil from Goiáz, Minas Geraes, and San Pablo	5–7	E Amazonian rainforest and Atlantic coastal rainforest	Terrestrial, short prehensile tail. Habitat: lowland forest.	30, 36
Monodelphis scalops	145–146	60–65	70–80 small	SE Brasil in Rio de Janeiro, from Spirito Santo south to Santa Catarina and Misiones, Argentina	7–8	Atlantic coastal rainforest, subtropical rainforest	Terrestrial, short prehensile tail. Habitat: lowland forest.	18, 30, 36, 44, 58, 64, 76, 90
Monodelphis sorex	110–130	65–85	48 small	SE Brasil, S Paraguay, NE Argentina	7–8	Atlantic coastal rainforest, subtropical rainforest	Terrestrial, short prehensile tail. Habitat: lowland forest.	14, 18, 30, 36, 44, 58, 64, 76, 92
Monodelphis theresa	no data	no data	inferred as category 2 small	Brazil: Rio de Janeiro, Minas Geraes, and a disjunct individual from the peruvian Andes	4 and 7	Amazonian and Atlantic rainforest	No data: probably terrestrial. Probably short prehensile tail. Habitat: montane and lowland forest.	30, 36, 78
Monodelphis unistriata	140	60	Inferred as category 3 small	Brazil: San Pablo; Misiones, Argentina	7–8	Atlantic coastal rainforest, subtropical rainforest	Terrestrial, short prehensile tail. Habitat: lowland forest.	18, 30, 36, 58, 64

Appendix 1 *Continued*

Species account	HB	Tail	Weight (g)	Distribution	Band	Major biomes	Ecology	References
Philander andersoni	223–288	267–305	335–600 medium	East of the Andes in the upper Amazon Basin of Venezuela, probably Colombia, and all of Eastern Ecuador	1–3	Amazonian rainforest	Large opossum, long prehensile tail. Nocturnal, arboreal. Habitat: found in mature and disturbed lowland rainforest. Favour dense viny vegetation near water.	30, 36
Philander mcilhennyi [5]	287–307	265–332	396–640 large	Perú: Ucayali Department, Brazil: Acre and Amazonas	2–4	Amazonian rainforest	Large opossum, long prehensile tail. Presumably nocturnal, arboreal and terrestrial. Habitat: lowland evergreen forest.	30, 36, 40
Philander[6] *opossum*	250–302	253–315	200–660 large	Central to South America: S Mexico to Panama, west of the Andes to Paraguay and NE Argentina	1–8	Amazonian rainforest, Atlantic coastal rainforest, subtropical rainforest, and Chaco	Large opossum, long prehensile tail, Nocturnal, arboreal and terrestrial, semiaquatic. Habitat: mature and secondary forests, gardens, plantations, and galery forest. Use the ground to middle vegetation levels: of the forest, common around trefalls and in dense underground near water.	4, 18, 23, 25, 30, 36, 44, 64
Thylamys elegans	approx 115 ?	approx 125 ?	30 small	S of Peru to Central Chile.	6–10	Pacific desert and Chilean sclerophyll	Small opossum. Short tail. Tail seasonally incrassated. Habitat: deserts, semideserts, and matorral scrub in Chile.	36, 85, 96, 114
Thylamys macrurus	126	166	36 small	Paraguay, SE Brasil, and Bolivia	6–7	Amazonian, Chaco, Mato Grosso	Small opossum, short tail. No more data available. Fat tail (inferred). Habitat: no data available.	4, 109
Thylamys pallidior[7]	84–104	93–117	10–17 small	E and S Bolivia, Northern Argentina	6–9	Prepuna and Puna on the Eastern Side of the Andes, semi-arid habitats in northern Argentina south to San Luis	Small opossum, short tail, may become swollen with stored fat. Nocturnal. Habitat: high altitude semideserts, monte shrubby vegetation. High rocky deserts.	4, 8, 36, 44, 58, 59, 109
Thylamys pusillus[8]	85– 106	85–102	40–60 small	South-eastern Bolivia, western Paraguay, southward to the Chaco and Monte desert to Chubut, Argentina	6–12	From Yungas and Chaco in Bolivia to semi-arid deserts in Patagonia	Small opossum, short tail, become swollen with stored fat. Nocturnal. Habitat: seems to inhabit a wide variety of habitats, most semi-arid.	4, 8, 14, 22, 28, 36, 44, 58, 72, 73, 76, 80, 85, 96, 100, 109
Thylamys velutinus	no data	no data	Inferred as category 2 small	SE Brasil	7	Atlantic rainforesr and Serra do Mar	Small opossum. Short tail, seasonally fat. Habitat: No data.	36, 109
Thylamys venustus[9]	81–125	109–140	42 small	SC Bolivia and NC Argentina (Catamarca, Jujuy, Tucuman)	6–8	Yungas (Argentina and Bolivia)	Small opossum, short tail. Fat tail. Habitat: almost restricted to the yungas.	4, 36, 58, 109

PAUCITUBERCULATA								
Rhyncholestes raphanurus	97–128	65–88	20–32 small	SC Chile continental and Chiloe island, and adjacent Argentina	11	Restricted to the southern temperate rainforest	Nocturnal, terrestrial. Store fat in tail during winter. Non prehensile tail. Habitat: in extremely wet *Nothofagus* forest with a dense understory, under or into fallen logs.	8, 37, 44, 50, 58, 66, 67, 71, 72, 73, 76, 82, 83, 85, 93, 96
Lestoros inca	107	128	31 small	S Andean Per and Bolivia	5–6	cloud rainforest in Bolivia and Perú	Nocturnal, terrestrial. Presumably non prehensile tail, as other caenolestids. No data on fat storage, but probably similar to *R.hyncholestes*. Habitat: High altitude Andean mountains.	2, 37
Caenolestes caniventer	90–135	93–139	25–40 small	The Andes of South-western Ecuador and Northern Perú	3–4	Cloud rainforest in Ecuador and Perú	Nocturnal, terrestrial. Presumably non prehensile tail, as other caenolestids. No data on fat storage, but probably similar to *Rhyncholestes*. Habitat: High altitude Andean mountains.	2, 6, 75
Caenolestes condorensis	130	130	48 small	Known only from Cordillera del Condor	3	Cloud rainforest	Nocturnal, terrestrial. Presumably non prehensile tail, as other caenolestids. No data on fat storage, but probably similar to Rhyncholestes. Habitat: ecotone of open and forested habitats, the vegetation strong resembles Venezuelan tepuis. Locality cold and extremely humid.	2
Caenolestes convelatus	no data	no data	inferred as category 2 small	Western Colombia and NW Ecuador	3	Cloud rainforest	Nocturnal, terrestrial. Similar to other species of the genus.	2
Caenolestes fuliginosus	approx 90–135	approx 93–139	25–40 small	The Andes of Northern and Western Colombia, extreme western Venezuela, and Ecuador	2–3	Cloud rainforest	Nocturnal, terrestrial. Presumably non prehensile tail, as other caenolestids. No data on fat storage, but probably similar to Rhyncholestes. Habitat: inhabits cloud forest and páramos.	2, 37, 51, 75
MICROBIOTHERIA								
Dromiciops gliroides	86–122	93–132	16–42 small	SC Chile and adjacent Argentina	10–11	Restricted to the southern temperate rainforest	Small opossum-like species. Short tail, become swollen with stored fat Nocturnal, terrestrial, scansorial. Torpor is reported. Habitat: wet *Nothofagus* forest, usually under or into fallen logs.	8, 38, 44, 50, 58, 61, 66, 70, 71, 73, 76, 83, 84, 85, 96, 108

1. This species was considered as a disjunct population of *Didelphis marsupialis* (see Cerqueira 1985) and is listed as *D. marsupialis* by Redford and Eisenberg (1992), but we follow here the criteria of Gardner (1993), Emmons (1997), and Voss and Emmons (1996).
2. Gardner (1993) considered that the forms *agilis* and *microtarsus* may prove to be conspecific. Emmons (1997) pointed that *microtarsus* and *agilis* are indistinguishable in the field.
3. Included by Emmons and Feer (1997) as a species; Gardner (1993a) as a synonym of impavidus
4. Included by Emmons as a species; W and R as a synonym of *regina*.
5. Included by Emmons and Feer (1997) as a species, Gardner (1993a) as a synonym of *andersoni*.
6. Molecular evidence indicates that the Atlantic forest form is distinct (*P. frenata*). The generic name of this taxon is disputed, given as *Metachirops* (Emmons and Feer 1997).
7. There are uncertainties regarding the distribution and taxonomy of *Thylamys pallidior* and *Thylamys pusillus* (see Mares and Braun 2000:38). We are using here for the southernmost distribution of *pallidior* the localities of the specimens Mares and Braun examined, in San Luis Province near 33 LS, not recognising the re-identification of the species reported in Birney et al. (1996b) as *T. pallidior.*
8. We measured the specimens reported in Birney et al. (1996b) and adults of our sample are larger from the largest individuals of *T. pallidior* reported by Mares and Braun (2000). We will use our adult specimens (all above *largests pallidior*) as the southernmost record of *Thylamys pusillus*: 45 19.78'S 69°50.05' W.
9. Included by Mares and Braun as a species, W and R as a synonym of *elegans*.

PART IV

CHAPTER 21

DISTRIBUTIONAL ECOLOGY OF DASYURID MARSUPIALS

Chris R. Dickman

Institute of Wildlife Research, School of Biological Sciences, University of Sydney, NSW 2006, Australia

Assemblages of small dasyurid marsupials (<500 g) are usually more diverse in arid regions of Australia than in non-arid coastal and sub-coastal areas, whereas the converse is true for larger species. To quantify species overlaps at the regional scale, species density maps were constructed using distributions of all species predicted from bioclimatic modelling. Highest species densities of small dasyurids were predicted to occur in hummock grassland and desert complex habitats (mean species richness 7.0–8.2; maximum 14), and the lowest to occur in forest and heathland (mean species richness 3.6–5.4; maximum 7). Actual field surveys showed that local species richness in most habitats was less than half that predicted at the regional scale, but approached two-thirds the regional level in arid woodland and hummock grassland. Local richness in hummock grassland was the highest of any habitat, averaging 5.3 species with a maximum of eight. Species densities of large (>500 g) dasyurids were greater regionally than locally, but nonetheless ranged from 0–3 at both scales. No large species now occur in arid habitats following the decline of the western quoll, *Dasyurus geoffroii*, and assemblages of three species are restricted to Tasmania.

For small dasyurids, species richness at the local scale depends on the structural complexity of vegetation and other, abiotic components of the environment. Structurally complex habitats allow species to achieve separation of foraging niches and hence reduce dietary overlap. In non-arid habitats interspecific competition appears to maintain foraging niche separation and hence limits the number of species that coexist, but in arid habitats population densities of most species are so low that competition provides little constraint on coexistence. In hummock grassland other potential constraints such as predation appear unimportant; emerging evidence suggests instead that one species, the mulgara *Dasycercus cristicauda*, has a facilitatory effect on the local richness of smaller species. Local richness is further enhanced in arid habitats by climatic events such as fire and rainfall that can stimulate movements of species across the regional landscape. Large dasyurids have been more affected than their smaller relatives by the shocks of European settlement. On mainland Australia there is evidence that remaining species segregate by habitat, while in Tasmania competition appears to drive complementary niche separation along both habitat and prey size axes.

Future work should focus on the enigmatic dasyurids of New Guinea, and attempt to resolve the distributional limits of species with patchy or restricted ranges. The outstandingly rich assemblages of small dasyurids in hummock grasslands also provide a beacon for future research, as do several poorly-known taxa that are restricted to coastal forest and heathland habitats. More long-term and experimental studies are essential to disentangle the factors that influence the distributions and diversity of dasyurids generally.

Introduction

The 69 described species of living dasyurid marsupials (Family Dasyuridae) occur broadly throughout the Australasian region. Fifty-three are restricted to Australia and 14 to New Guinea and surrounding islands; two species are present on both sides of Torres Strait and show some divergence at the subspecific level (Flannery 1995a, b; Strahan 1995). All terrestrial habitats are occupied, with highest population densities of individual species occurring usually in coastal forest and heath. In Australia, for example, exceptional densities of up to 140 individuals ha^{-1} have been recorded for the brown antechinus *Antechinus stuartii* in tall open-forest (Tasker et al. 1999), while the swamp antechinus *Antechinus minimus* reaches a density of 80 individuals ha^{-1} in heath on Great Glennie Island off the coast of Victoria (Wainer 1976). Density estimates have not been made for New Guinea dasyurids, but at least 12 (75%) of the species are restricted largely to forest and could be expected to reach their highest densities in this habitat (Menzies 1991; Flannery 1995a). In contrast, densities of dasyurids in arid environments appear to be uniformly low. Maximum densities of 1–5 individuals ha^{-1} were recorded by Dickman et al. (2001a) for the three most abundant species (of eight) at a site in the Simpson Desert over a period of 10 years. Similarly low densities have been found by other workers elsewhere in the arid zone (Morton 1982; How et al. 1991; Masters 1993, 1998).

While dasyurids achieve highest population densities in forest and heath, fewer species usually co-occur in these habitats than in hummock grassland and other desert environments (Dickman 1989). For example, 2–3 species often co-occur in forest in eastern, northern and western Australia (Braithwaite et al. 1985; Fox 1985), and 3–4 species (possibly as many as six; Gressitt and Nadkarni 1978) sometimes co-occur in New Guinea (Flannery 1995a). Yet, as many as 8–9 species may overlap in parts of arid Australia (Dickman 1989).There is little evidence that dasyurids follow any obvious relationship with productivity (Dickman 1989), as do small mammals in some other world deserts (Abramsky and Rosenzweig 1984). The increased species richness of desert regions thus is counter-intuitive, but could be attributable to several factors. Firstly, opportunities for speciation may have been great in central Australian habitats, as has been the case with scincid lizards (Pianka 1986), allowing many species to co-occur. Possibly too, much dasyurid evolution has taken place in arid environments and taxa are primarily adapted to such conditions. Secondly, biotic processes such as competition and predation may be pervasive in arid but not in forest environments, and could facilitate increased species packing over evolutionary time. Thirdly, biotic and abiotic resources such as food, shelter sites, soil substrates, vegetation and structural microhabitats may be more diverse or abundant in arid environments, thus supporting more species at a local scale. Fourthly, if desert dasyurids are more responsive than their forest counterparts to large scale processes, such as spatial patchiness in climatic conditions (e.g. Dickman et al. 1995), increased interchange with the regional species pool could elevate the richness of local communities more in arid environments than in temperate ones. Finally, it remains possible that the high dasyurid species richness of arid environments is more apparent than real. Relatively few studies have been carried out on desert dasyurids, and the high overlap demonstrated in the distributions of up to 8–9 species (Dickman 1989) may not always translate into high species richness on the ground. Sampling biases and anomalies have been discussed for desert lizards by James and Shine (2000) and more generally for desert vertebrates by Bouskila and Dickman (2003).

In this chapter, I first attempt to describe patterns in the distributions and overlaps of dasyurids, and quantify the numbers of species that co-occur in different arid and non-arid habitats at local and regional scales. These findings are then used to evaluate which of the above factors, or combinations of factors, best account for the distributional patterns discerned. The primary focus is placed on identifying contemporary factors that influence distributions and diversity; radiations and recent phylogeography of dasyurids have been addressed elsewhere (e.g. Krajewski et al. 2000; Crowther and Blacket, this volume). In addition, most attention is focused on the dasyurids of Australia due to the paucity of relevant information on the fauna of New Guinea.

Dasyurid distributions and diversity

In previous analyses the distributions of dasyurids have been taken from generalised maps and superimposed to obtain estimates of regional-level diversity (e.g. Morton 1982; Dickman 1989). While this procedure reveals broad patterns, it is likely to be accurate for mapping occurrences of abundant and well-known species but less so for documenting the distributions of rare, cryptic species or those that occur in poorly-sampled areas. In an attempt to overcome these biases, I collated locality data on all Australian species of dasyurids and used the data to model predicted distributions using the bioclimate prediction system BIOCLIM (Busby 1991). Locality records for all species were obtained from museum records and searches of museum holdings in Australia and overseas, published and unpublished literature, and personal communications from several researchers (see Murray and Dickman 2000). Outlying locality records were checked by examination of specimens, where possible, and discarded from the database if their identity or provenance could not be confirmed. Recent taxonomic changes were incorporated, especially where these involved recognition of cryptic species with separate distributions from their congeners (Dickman et al. 1998; Blacket et al. 2000; Cooper et al. 2000; Van Dyck and Crowther 2000). However, due to a paucity of specimens and uncertainty about likely distributional boundaries, locality records for the mulgara *Dasycercus cristicauda* and ampurta *D. hillieri* were considered together. A single specimen of an apparently new species of *Planigale* from Tom Price, in the Pilbara of Western Australia (Blacket et al. 2000), was also grouped for this analysis with a sympatric but undescribed congener.

The latitudes, longitudes and altitudes of confirmed locality records were obtained from topographic maps. A series of climatic indices was then calculated for each locality record using BIOCLIM, and a profile of the climate prevailing at known localities defined for each species. This bioclimatic profile was then used to predict climatically similar areas outside the known localities where the species could also occur.

In preliminary analyses I compared predicted distributions using all locality records (pre-European; ~1800 to the present) with distributions predicted from records obtained only since 1950. The latter date coincides with a period when several species of dasyurids were last recorded in eastern (Menkhorst 1995), central and western Australia (Burbidge et al. 1988), and was used as a threshold to represent contemporary distributions. With some notable exceptions, however (e.g. the red-tailed phascogale *Phascogale calura* and western quoll *Dasyurus geoffroii*), predicted pre-European and contemporary distributions were largely concordant (Murray and Dickman 2000), justifying retention of all locality records in the analyses. In further explorations I compared predictions based on 'core' distributions, which used climatic values within the range of 10–90% of the minimum and maximum estimates, and predictions based on 'marginal' distributions, which used all climatic values (Lindenmayer et al. 1991). These analyses indicated that up to 55% of known locality records could be omitted if the conservative core distribution approach was used (e.g. Fig. 1a); hence maps used in procedures below are based on core and marginal distributions that use all locality records. Error-checking in all computations used the cumulative frequency method of Lindenmayer et al. (1991). For six species there were insufficient locality records to generate reliable bioclimatic predictions (the Carpentarian pseudantechinus *Pseudantechinus mimulus*, and the dunnarts *Sminthopsis aitkeni, S. archeri, S. bindi, S. butleri* and *S. douglasi*). For these species, distributions were represented instead using simple polygons that enclosed known localities.

Species overlaps are depicted as species density (= richness) maps, and were constructed by direct superimposition of predicted distributions. Dasyurids weighing >500 g (i.e. all species of *Dasyurus* and the Tasmanian devil *Sarcophilus laniarius*) have quite different patterns of distribution from those weighing <500 g; there is also a large gap in body mass between species in both groups. For clarity of presentation, these two groups are therefore considered separately in the analyses below.

To identify the habitats where species co-occur, a simplified map was constructed depicting eight broad structural categories of vegetation, or 'habitat types'. Four habitat types are confined to the arid zone, as defined in Morton (1982) and Dickman (1989), and four to temperate, tropical or alpine/subalpine environments in coastal and sub-coastal areas (the habitat map is depicted in Dickman 1989). Comparison of species density with habitat was then made by taking 25 randomly assigned coordinates within each habitat type and tallying the total number of species predicted to occur there.

Regional pools of species, as predicted from species density maps, are always likely to exceed the numbers of species actually present at any local site (Cornell and Lawton 1992; Holt 1993). Thus, to investigate the relationship between actual species densities and habitat, I reviewed the results of local field surveys carried out in each of the eight defined habitat types and tallied the species present. Data sources included those consulted above for locality records, as well as numerous survey reports listed in Table 1. Surveys by the author in New South Wales, Queensland and Western Australia were also included (C.R. Dickman unpubl. data). Surveys were used only if they had amassed a minimum of 1000 trap-nights over at least two seasons and if they were judged to have used survey techniques appropriate to dasyurids. For example, surveys in areas likely to contain *Ningaui* species were excluded unless pitfall traps had been used.

Distributional patterns

For most species the predicted distribution did not extend more than 200–300 km beyond known localities (Figs. 1a, b), while for others it was contained entirely within the known range (Figs. 1c, d). In addition, for several species (the antechinuses *Antechinus agilis, A. flavipes, A. godmani, A. stuartii, A. subtropicus* and *A. swainsonii*) the results were very similar to those predicted in other studies (Winter 1991; Crowther 2002). These congruences suggest that, at a coarse scale, our understanding of the distributional limits of most dasyurids is quite reliable. However, four species were predicted to occur at distances >500 km from their nearest known localities (Fig. 2).

For two of the exceptional species the predicted range extensions appear reasonable and are accepted here. The Ooldea dunnart *Sminthopsis ooldea* is predicted to extend eastward into western Queensland and north-western New South Wales (Fig. 2a). Although this species has not been recorded east of the

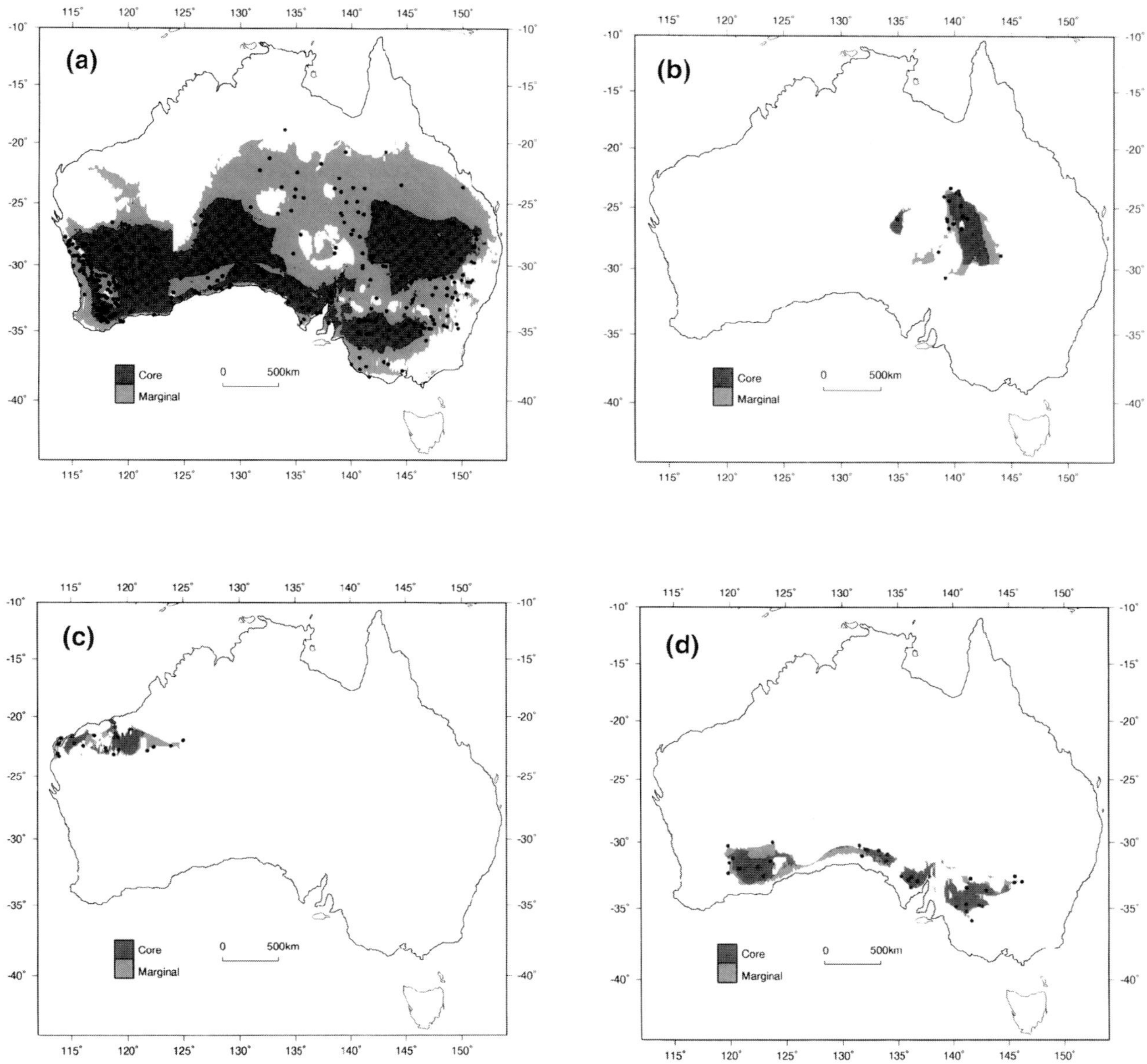

Figure 1 Distributions of dasyurid marsupials predicted by bioclimatic modelling, showing conformity between known and predicted ranges. (a) Fat-tailed dunnart *Sminthopsis crassicaudata*, (b) kowari *Dasycercus byrnei*, (c) little red kaluta *Dasykaluta rosamondae*, (d) southern ningaui *Ningaui yvonneae*. Dark grey areas represent 'core' ranges predicted using climatic values within the range of 10-90% of extreme values; light grey areas represent 'marginal' ranges predicted using all climatic values. Black dots represent known localities from which bioclimatic predictions were made.

South Australian border, surveys in this region have been limited (Ingram and Raven 1991; Dickman et al. 2001b), and a *Sminthopsis* sp. possibly referrable to *S. ooldea* is known from the Pilliga Scrub of north-central New South Wales (Lim 1992). The predicted easterly extension of range of the fat-tailed pseudantechinus *Pseudantechinus macdonnellensis* also encompasses areas that have been little surveyed (Fig. 2b). However, scats attributable to *P. macdonnellensis* have been found at Painted Gorge (23° 15', 138° 06') in western Queensland (C.R. Dickman unpubl. data), and potentially suitable rock outcrops for this species occur further east in the predicted distribution around Mt Isa and as far south as the Grey Range near Quilpie. In contrast, for two species of planigales, the predicted westward extensions of range of >1500 km (Figs. 2c, d) do not appear reasonable. These species usually are associated with deeply cracking soils (Andrew and Settle 1982) that do not occur in the predicted westerly ranges. In addition, extensive surveys have been carried out in potentially suitable habitats in western South Australia and Western Australia, with no planigales recorded (Dell and How 1984, 1985, 1988, 1992; Boscacci et al. 1987;

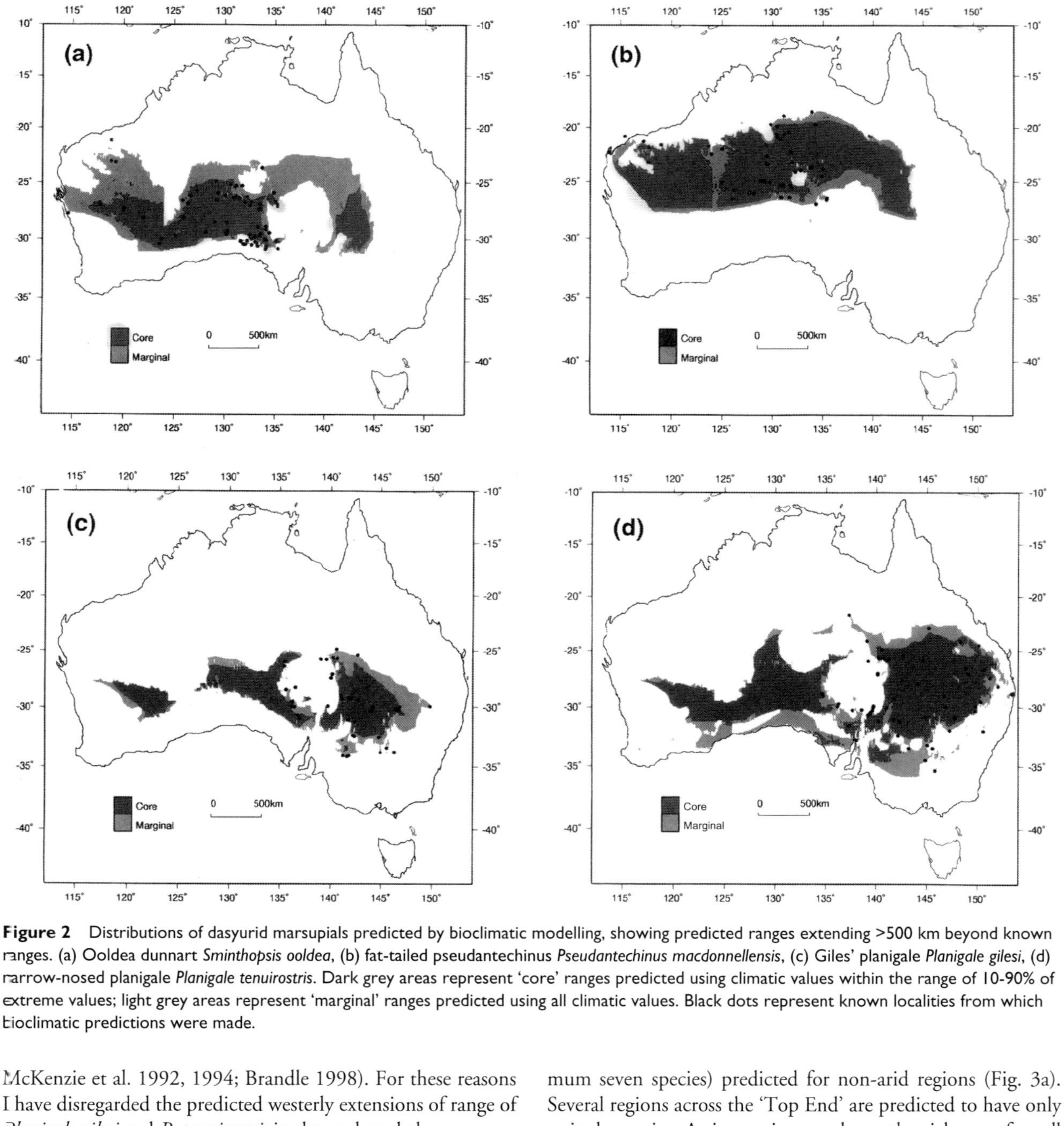

Figure 2 Distributions of dasyurid marsupials predicted by bioclimatic modelling, showing predicted ranges extending >500 km beyond known ranges. (a) Ooldea dunnart *Sminthopsis ooldea*, (b) fat-tailed pseudantechinus *Pseudantechinus macdonnellensis*, (c) Giles' planigale *Planigale gilesi*, (d) narrow-nosed planigale *Planigale tenuirostris*. Dark grey areas represent 'core' ranges predicted using climatic values within the range of 10-90% of extreme values; light grey areas represent 'marginal' ranges predicted using all climatic values. Black dots represent known localities from which bioclimatic predictions were made.

McKenzie et al. 1992, 1994; Brandle 1998). For these reasons I have disregarded the predicted westerly extensions of range of *Planigale gilesi* and *P. tenuirostris* in the analyses below.

OVERLAP PATTERNS

Superimposition of the distribution maps shows that dasyurids can be expected to occur in all parts of Australia, with up to fourteen species <500 g and three species >500 g overlapping. For species <500 g, hotspots with ten or more species are predicted to occur in central Australia, with lower averages (maximum seven species) predicted for non-arid regions (Fig. 3a). Several regions across the 'Top End' are predicted to have only a single species. As in previous analyses, the richness of small dasyurids in arid Australia is particularly outstanding and obvious. For larger dasyurids Tasmania is the only hotspot, where the devil *Sarcophilus laniarius* and two species of quolls *Dasyurus maculatus* and *D. viverrinus* co-occur (Fig. 3b).

Comparison of species densities with major habitat types shows that more small dasyurids are predicted to co-occur regionally in arid compared with temperate or tropical habitats, and that

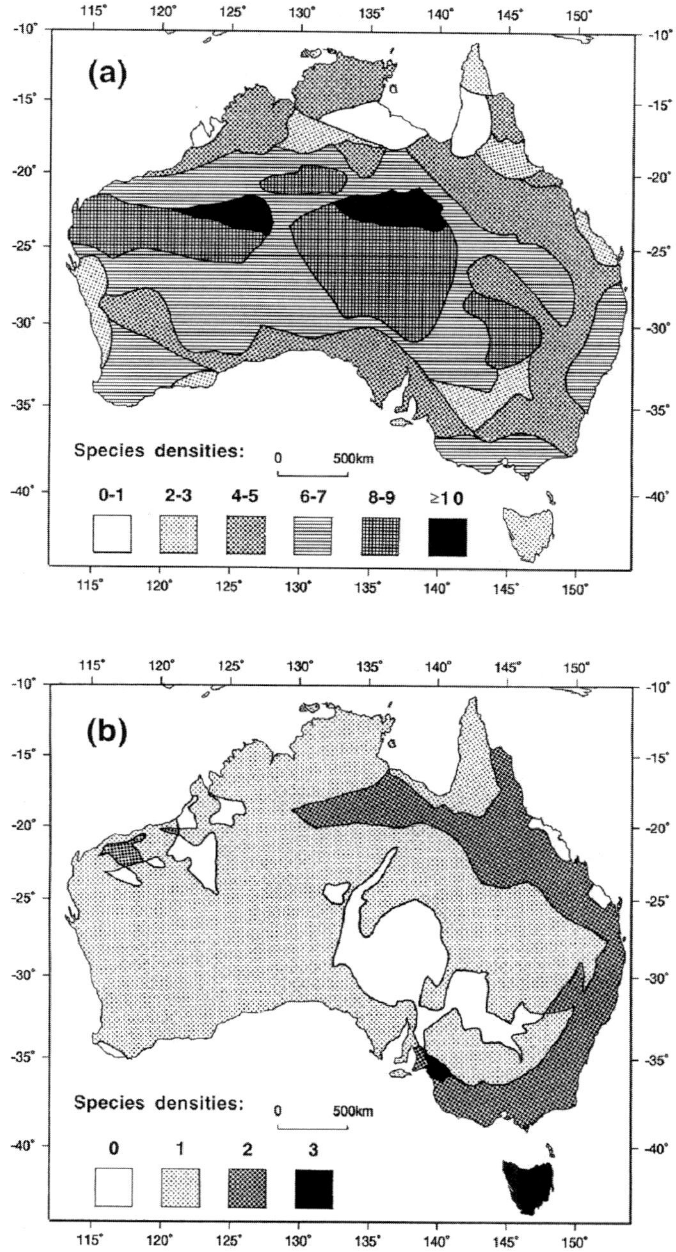

Figure 3 Species density maps of dasyurid marsupials in Australia, constructed by superimposing distributions predicted by bioclimatic modelling. (a) Species <500 g, (b) species >500 g.

hummock grassland and desert complex should be particularly species-rich (Table 1). As expected, local species richness determined from field surveys was always less than that predicted from the regional pool. However, two further results stand out. Firstly, actual species richness in arid habitats is generally greater than that in non-arid ones, with the average number of species in hummock grassland exceeding that in all other habitats. Maximum local richness in hummock grassland was eight, while that in all non-arid habitats was just four. Secondly, the actual species richness found locally in hummock grassland and arid woodland approaches two-thirds of the regional species pools for these habitats, whereas species richness in all other habitats is less than half that of the predicted regional pools (Table 1). For dasyurids >500 g, there were no significant differences in

Table 1 Species richness of small dasyurids (<500 g) in eight major habitat types in Australia. Regional richness values represent the numbers of species predicted to co-occur within habitats from the species density map (Fig. 3a), while local richness values represent the actual numbers of species recorded at different sites within habitats by field surveys. Means are shown ± S.D. with sample sizes in brackets.

Habitat type[1]	Regional richness[2]	Local richness[2,3]	Ratio (%)[4]
Arid			
Hummock grassland	[a]8.22 ± 0.97 (25)	[a]5.31 ± 1.63 (29)	65
Desert complex	[a,b]7.01 ± 0.63 (25)	[a,b]3.40 ± 0.90 (15)	49
Tussock grassland	[b]5.45 ± 1.71 (25)	[b,c]2.30 ± 1.23 (12)	42
Woodland	[b]5.39 ± 1.68 (25)	[a,b]3.28 ± 1.87 (18)	61
Non-arid			
Heath	[b,c]3.57 ± 1.39 (25)	[c]1.74 ± 0.86 (31)	49
Open-forest	[b]5.42 ± 0.40 (25)	[c]1.76 ± 0.85 (93)	32
Rainforest	[b]3.96 ± 1.14 (25)	[c]1.72 ± 0.83 (18)	43
Woodland	[b]4.24 ± 0.83 (25)	[c]1.89 ± 1.00 (35)	45

Notes:

1 Habitat types are described and mapped in detail by Moore (1970), and a simplified map is given in Dickman (1989). Arid woodland is dominated by *Acacia* spp. and includes low-layered, steppe and savanna woodland, as well as some mallee (eucalypt) woodland; non-arid woodland is dominated by *Eucalyptus* spp.

2 Different superscript letters indicate mean richness values that differ between habitat types in *post hoc* tests following significant analyses of variance (regional richness: $F = 4.56$, $P < 0.001$; local richness: $F = 3.12$, $P < 0.01$).

3 Data sources include Calaby (1966); Denny (1975); Posamentier (1976); Braithwaite et al. (1978); Chapman and Kitchener (1978); Norris et al. (1979); Dickman (1980); Gullan and Robinson (1980); Morton (1982); Statham and Harden (1982); Dickman et al. (1983, 1991); Coventry and Dixon (1984); Dell and How (1984, 1985, 1988, 1992); Mollemans et al. (1984); Braithwaite et al. (1985); Dickman and McKechnie (1985); Fox (1985); Friend and Taylor (1985); Morton and Baynes (1985); Lunney and Barker (1986); Wilson et al. (1986); Boscacci et al. (1987); Bradley et al. (1987); Read (1987); Ellis and Henle (1988); Reid and Gillen (1988); Tidemann (1988); How et al. (1988, 1991); Bennett et al. (1989); Braithwaite (1989); Baynes (1990); Bennett (1990); Kemper (1990); Laurance (1990); Friend et al. (1991); Milledge (1991); Moro (1991); Burnett (1992); Dickman and Read (1992); How and Dell (1992); McKenzie et al. (1992, 1993, 1994, 2000); Read (1992); Woolley (1992); Baynes and Jones (1993); Masters (1993); Reid et al. (1993); Barker et al. (1994); Morton et al. (1994); Catling and Burt (1995, 1997); Read (1995); Woinarski and Fisher (1995); Briggs (1996); Woinarski et al. (1996); Catling et al. (1997); Goldingay and Daly (1997); Twyford (1997); Watt (1997); Brandle (1998); Green and Catterall (1998); Murphy (1998); Fisher (1999); Leung (1999); Molsher et al. (1999); Paull and Date (1999); Tasker et al. (1999); Carthew and Keynes (2000); Goosem (2000); Wilson and Wolrige (2000); Moseby and Read (2001); Paltridge and Southgate (2001); Risbey et al. (2001). Unpubl. survey data collected by the author have also been included. Other sources are given in Murray and Dickman (2000).

4 Local richness / regional richness x 100.

local or regional species richness among habitats, but non-arid habitats tended to have the highest species richness (Table 2). Actual numbers of species are again lower than predicted from the regional pool, with little evidence that large dasyurids still persist in any arid habitats.

Determinants of distributions

Dasyurids <500 g

Although the high species richness of small dasyurids in arid Australia has been noted previously (Morton 1982; Dickman 1989), the present analysis quantifies both the surprising paucity of species that co-occur in forest and heathland and the outstanding richness of assemblages in hummock grassland. However, before discussing the factors that produce these differences, some comment on the robustness of the results is warranted.

Firstly, surveys in forest and heathland habitats have generally used Elliott traps to capture dasyurids, whereas those in arid habitats have used pitfall traps. As Elliotts do not efficiently capture very small dasyurids such as *Planigale* spp., could low species richness in forests and heaths be simply a sampling artefact? This is unlikely. Many studies in forest have either used an array of several techniques (e.g. Bennett et al. 1989) or have been sufficiently intense (e.g. Kemper 1990) to have had a high chance of discovering all extant species. If anything, underestimation of species richness may have occurred in arid habitats. For example, Dickman et al. (1993) recorded the hairy-footed dunnart *Sminthopsis hirtipes* in hummock grassland only after 31,000 trap nights of effort, while How et al. (1991) recorded three species of dasyurids for the first time in the Abydos-Woodstock Reserve after >8000 trap nights. Moreover, four of eight species in the latter study were represented by only a single capture, highlighting further the role of chance in identifying this diverse assemblage.

Secondly, bias could arise if dasyurids have declined differently in arid and non-arid regions since European settlement. Although plausible, species declines have been most marked in the inland (Wilson et al. this volume), so that any such bias

Table 2 Species richness of large dasyurids (>500 g) in eight major habitat types in Australia. Regional richness values represent the numbers of species predicted to co-occur within habitats from the species density map (Fig. 3b), while local richness values represent the actual numbers of species recorded at different sites within habitats by field surveys. Means are shown ± S.D. with sample sizes in brackets.

Habitat type[1]	Regional richness[2]	Local richness[2,3]	Ratio (%)[4]
Arid			
Hummock grassland	0.96 ± 0.54 (25)	0.06 ± 0.24 (18)	6.3
Desert complex	0.56 ± 0.51 (25)	0 (9)	–
Tussock grassland	1.20 ± 0.50 (25)	0 (7)	–
Woodland	1.24 ± 0.60 (25)	0 (15)	–
Non-arid			
Heath	1.56 ± 0.87 (25)	0.11 ± 0.32 (19)	7.1
Open-forest	1.92 ± 0.57 (25)	0.59 ± 0.79 (70)	30.7
Rainforest	2.00 ± 0.58 (25)	0.70 ± 0.82 (10)	35.0
Woodland	1.44 ± 0.51 (25)	0.10 ± 0.31 (20)	6.9

Notes:
1 Habitat types are described and mapped in detail by Moore (1970), and a simplified map is given in Dickman (1989). Arid woodland is dominated by *Acacia* spp. and includes low-layered, steppe and savanna woodland, as well as some mallee (eucalypt) woodland; non-arid woodland is dominated by *Eucalyptus* spp.
2 Analyses of variance revealed no differences in species richness between habitat types (regional richness: $F = 0.59$, P ns; local richness: $F = 0.17$, P ns).
3 Data sources are given in Table 1.
4 Local richness / regional richness × 100.

would only have reduced the still-large difference in species richness observed between arid and non-arid habitats. These considerations provide some confidence that the observed results are robust.

In eastern Australian forest and heath, the local richness of small mammal assemblages correlates most strongly with the structural diversity of vegetation (Fox 1985). In tropical woodlands both vegetation structure and floristic complexity are important determinants of local richness, as are indices of primary productivity, such as precipitation and soil fertility (Braithwaite et al. 1985; see also Woinarski et al. 2001). In arid Australia also, emerging evidence suggests that the local richness of small dasyurids depends critically on the complexity of vegetation and other, abiotic components of the environment (Reid et al. 1993; McKenzie et al. 2000). For example, *Ningaui* spp. require spinifex hummocks, while *Planigale* spp. occur primarily in habitats with cracking soils (Read 1987). In an attempt to quantify the importance of habitat complexity, Bouskila and Dickman (2003) removed spinifex hummocks from plots containing six species of dasyurids but left other plots unmanipulated as controls. Species richness remained unchanged on the control plots, but fell by half on the removals; spinifex-dwelling *Ningaui* spp. were affected consistently. Although equivalent experiments have not been carried out in non-arid habitats, it is highly likely that dasyurid species richness is associated generally with structural habitat complexity. Similar species richness-habitat complexity relationships have been described for many taxa (Ricklefs and Schluter 1993).

If there is a positive correlation between local species richness and habitat structure, we might expect that the most complex habitats should occur in arid areas, particularly in hummock grasslands. Although counter-intuitive, there is some evidence that this is so. Thus, counts of microhabitat components such as soil cracks, logs, litter and different vegetation strata were up to 50% greater in 1 ha sites in hummock grassland than in equivalent sites in forest in Western Australia (Bouskila and Dickman 2003). The correlation between species richness and numbers of microhabitat components per site across 39 sites was also very high in this latter study ($r = +0.69$), thus providing further support for the species richness-habitat relationship.

Despite the generality of the association between species richness and habitat complexity, several further factors contribute to patterns of dasyurid richness among habitats. Firstly, because dasyurids are generalist carnivores that can take a broad range of invertebrate and small vertebrate prey, the importance of habitat structure may lie not in complexity *per se*, but rather in the opportunity that it affords for achieving separation of foraging niches and hence reduction in dietary overlap (Dickman 1989). In complex habitat such as undisturbed forest and heath, for example, sympatric dasyurids exploit either a scansorial or soil-fossicking niche; a third species may also occur if the habitat is disturbed (Braithwaite et al. 1978; Fox 1982). In open desert habitats scansorial dasyurids are notably absent (although the red-tailed phascogale would have occupied the arboreal niche in riparian strips and other localised areas), but co-occurring species appear to partition remaining foraging niches quite finely. As noted, soil cracks are used for foraging by *Planigale* spp., as are spinifex hummocks by *Ningaui* spp.; other specialisations include the exploitation of open sand by kultarr *Antechinomys laniger* and several species of dunnarts, dune sides by mulgara and rock outcrops by the fat-tailed pseudantechinus (Fisher and Dickman 1993; Gilfillan 2001). Similar patterns of fine-scale partitioning of foraging niches have been demonstrated among desert lizards (Pianka 1986).

Secondly, while all major habitat types contain relatively large regional pools of species (≥3.6 species), more than half of the pools is present in local assemblages only in hummock grassland and arid woodland (Table 1). This suggests that species in these arid habitats are able to move readily between local areas, or that they stay longer in habitat patches once they have arrived. Two processes have been demonstrated to drive such movements, wildfires and patchy rainfall. Masters (1993) showed that three species of small dasyurids were present primarily in unburnt hummock grassland whereas a fourth species, *Sminthopsis hirtipes,* was numerous only after fire. Dickman et al. (1995) described movements of up to 12 km across drought-stricken hummock grassland by several species of small dasyurids, and noted that the prevailing direction of movement was toward areas that had received local rainfall. In contrast to fire, which likely changes the structure of foraging niches for several years and perhaps favours species that prefer newly-open conditions, rainfall probably just increases food abundance temporarily. Nonetheless, both processes appear to be important regional-level stimuli for movements that increase species richness at local scales. Neither process appears to have large effects in non-arid habitats; *Sminthopsis* spp. often appear locally after fire or other disturbances such as logging, but movements are much more prescribed than in hummock grassland (Lunney and Leary 1989).

Thirdly, there is some evidence that interspecific competition operates in non-arid habitats and contributes importantly to foraging niche separation. Removal and addition experiments have confirmed that competition separates the foraging niches of sympatric *Antechinus* spp. (Dickman 1986a, b), while other studies provide compelling evidence of competition between *Sminthopsis* spp. and other taxa in temperate forest and heath (Fox 1982; Dickman 1988; Righetti et al. 2000). In contrast, while critical experiments have yet to be carried out, competition appears to be less important in arid habitats. Instead, droughts, floods and unpredictable shortfalls in the food supply may often reduce and rarify local populations, hence reducing interspecific encounters so that overlaps in foraging niches are tolerable (Dickman 1989). Intraspecific shifts in both foraging modes and foods taken confirm the flexibility of foraging niches in such situations (Fisher and Dickman 1993; Bos 2001). Risk of predation from owls and larger mammalian carnivores is more likely than competition to constrain foraging niches in open desert habitats such as desert complex or tussock grassland, with larger dasyurids able to tolerate predation risk at the expense of small species (Fisher and Dickman 1993). In more complex arid habitats such as mature hummock grassland, however, predation risk appears minimal due to the abundance and prevalence of shelter (Haythornthwaite and Dickman 2000). Lack of pressure from either competition or predation presumably allows overlap in species' foraging niches in this habitat, and hence contributes to the high richness that characterises it (Table 1).

Finally, one further biotic process may help to account for the extraordinary species richness of small dasyurids in hummock grassland: facilitation by a hummock grass specialist, the mulgara. Local sites with the greatest species richness (7-8 species) almost always contain mulgaras (e.g. How et al. 1991; Reid et al. 1993; Dickman et al. 2001a), whereas less diverse sites often do not. Exclusion of mulgaras from 1.5 ha plots in the Simpson Desert caused the average number of smaller dasyurid species to drop from four to one over the course of a year, whereas no change occurred in open and fence control plots over the same period. The one species that prevailed in the absence of mulgaras, the lesser hairy-footed dunnart *Sminthopsis youngsoni,* was up to 80% more abundant in the exclusion plots than in the controls (C.R. Dickman unpubl. data). The mechanism producing this facilitatory effect is not known. Mulgaras may reduce the population densities of small dasyurids by competition or predation, thus allowing increased overlaps in their foraging niches; they may reduce competition between small dasyurids and rodents by limiting rodent populations; or they may provide sheltering opportunities via provision of burrows. These and other possible explanations await experimental investigation.

In summary, habitat complexity is a primary determinant of the richness of local assemblages of small dasyurids, and accounts in part for the differences in richness between major habitat types in arid and non-arid areas. Complex habitats provide the greatest opportunities for separation of foraging niches, and hence for reducing overlaps in diets. Competition appears to reinforce exclusivity of foraging niches in forest and heathland habitats, hence reducing local richness in these habitats; risk of predation may have similar effects in open desert. In hummock grassland neither competition nor predation appear to be important, but facilitation may occur if mulgaras are present. At the regional scale, changes in the quality of food or habitat resources appear to be driven by climatic events such as fire and rainfall. These changes stimulate movements of dasyurids in arid but not in non-arid environments, and can temporarily increase richness in local areas.

Dasyurids >500 g

The distributions of larger dasyurids have been reduced to a much greater degree by the impacts of European settlement than have the distributions of their smaller counterparts (Wilson et al. this volume). The eastern quoll *Dasyurus viverrinus* disappeared from mainland Australia in the 1960s, and the distributions of both the northern quoll *D. hallucatus* and spotted-tailed quoll *D. maculatus* have become increasingly fragmented (Jones et al. this volume). The western quoll *D. geoffroii* has declined from some 98% of its former range and is now confined to the south west of Western Australia (Morris et al. this volume). These dramatic declines account in large part for the discrepancies in richness between local and regional sites noted earlier (Table 2).

In contrast to the patterns observed for dasyurids <500 g, the larger species occur primarily in non-arid habitats. Indeed, only northern quolls now persist in arid habitats at all (e.g. How et al. 1991). Historical evidence suggests that the once-widespread western quoll occurred at low density in an array of arid habitats and probably occupied a terrestrial, broadly carnivorous niche (Johnson and Roff 1982). This species probably overlapped regionally with the northern quoll in some parts of the northern arid zone (Fig. 3b), but there is little evidence that the two species occurred in close sympatry. For example, Baynes and Jones (1993) recorded both species in cave deposits on Cape Range peninsula in Western Australia, but considered the northern quoll to have been confined to range country and the western quoll to open plains.

It remains unclear, however, why larger dasyurids did not make more extensive use of arid habitats. Dickman (1989) speculated that the large carnivore niche had been pre-empted by organisms such as goannas *Varanus* spp., which are physiologically more tolerant of fluctuating food supplies than similar-sized endotherms. It is also possible that many foraging niches that are partitioned by small dasyurids, such as soil cracks or spinifex hummocks, are not readily available to their larger relatives, hence further reducing opportunities to achieve local coexistence.

Large dasyurids are predicted to have, or have had, highest species richness in open-forest and rainforest (Table 2). Fig. 3b suggests further that two species co-occurred over much of eastern Australia, however syntopy is probably avoided by habitat segregation. Thus the northern-most populations of the spotted-tailed quoll are restricted to rainforest which regionally sympatric northern quolls do not enter, while mainland eastern quolls and southern mainland spotted-tailed quolls apparently preferred dry open-forest and wetter forests, respectively (Caughley 1980; Menkhorst 1995). In the case of the latter two species, some foraging niche separation might also have been expected in any areas of close sympatry. The eastern quoll was probably a terrestrial predator of invertebrates and small vertebrates on the mainland, whereas the spotted-tailed quoll is partly arboreal and includes larger vertebrate prey in its diet (Settle 1978; Caughley 1980).

Eastern quolls, spotted-tailed quolls and Tasmanian devils co-occur in Tasmania, where all three species can occupy the same habitats. As for smaller dasyurids, competition appears to be a major force structuring local communities, with complementary separation occurring between species on prey size and habitat axes (Jones and Barmuta 1998, 2000). Intriguingly, regular displacements in trophic characters occur within and between species, and suggest that competition has influenced size ratios, and presumably diet, over evolutionary time (Jones 1997). An overview and comparison with placental carnivores is provided by Jones (this volume).

FUTURE DIRECTIONS

This review has said very little about the dasyurids of the New Guinea region because almost nothing is known about them. Studies on species boundaries, distributions, habitat and environmental associations are needed before we can speculate on factors that influence local abundance and diversity, and should be a clear priority for future research.

In Australia, a priority for the future is to resolve the distributions of taxa that appear to be highly restricted, and attempt to determine the factors that limit them. Examples include the carpentarian pseudantechinus, the chestnut dunnart *Sminthopsis archeri*, Kakadu dunnart *S. bindi*, Butler's dunnart *S. butleri* and white-footed dunnart *S. leucopus*.

The remarkable regional and local richness of small dasyurids in some arid Australian habitats is not equalled among any other assemblages of insectivorous mammals of which I am aware, although parallels may occur for scincid lizards. Research in arid habitats is also scant, and should be a further priority for future workers. The most fruitful habitats for study are likely to be hummock grasslands. Well-designed experimental studies are needed there to disentangle the importance of biotic factors (i.e. predation, competition, facilitation) in shaping assemblage composition, and detailed autecological research is also needed to delineate foraging niches and how different species use different microhabitat components. Concurrent research on the abundance and availability of invertebrate prey would assist interpretation of foraging niche separation, and also provide much-needed data to evaluate the effects of temporal and spatial fluctuations in the overall food supply. Because of the likely influence of regional events on local assemblages, sampling should be carried out at multiple sites over large (>100 km^2) areas.

In non-arid habitats the factors that drive changes in population size, local distribution and diversity have been studied in some taxa (e.g. some *Antechinus* spp.), but hardly at all in others (e.g. *Parantechinus*, some *Planigale* and *Sminthopsis* spp.). This should be rectified. It is becoming increasingly obvious also that some future research, at least, should be long-term. Different factors may operate at different times on populations; without sampling data that extend over periods of at least 10 years, the suite of factors and their interactions cannot be revealed. An excellent example of the value of long-term research is provided by Woinarski et al. (2001).

Acknowledgements

I thank U. Grott and F.J. Qualls for assistance with running the bioclimatic analyses, M. Ricketts for assistance with preparation of the figures, H.A. Ford and an anonymous referee for helpful comments on the manuscript, and C.A. McKechnie for her support at all times. Funding was provided by the Australian Research Council.

References

Abramsky, Z., & Rosenzweig, M.L. (1984), 'Tilman's predicted productivity-diversity relationship shown by desert rodents', *Nature,* **309**:150–1.

Andrew, D.L., & Settle, G.A. (1982), 'Observations on the behaviour of species of *Planigale* (Dasyuridae, Marsupialia) with particular reference to the narrow-nosed planigale (*Planigale tenuirostris*)', in *Carnivorous marsupials* (ed. M. Archer), pp. 311–24, Royal Zoological Society of New South Wales, Mosman.

Barker, J., Lunney, D., & Bubela, T. (1994), 'Mammal surveys in the forests of the Carrai Plateau and Richmond Range in north-east New South Wales', *Australian Mammalogy,* **17**:19–29.

Baynes, A. (1990), 'The mammals of Shark Bay, Western Australia', in *Research in Shark Bay: Report of the France–Australe Bicentenary Expedition Committee*', (eds. P.F. Berry, S.D. Bradshaw, & B.R. Wilson), pp. 313–25, Western Australian Museum, Perth.

Baynes, A., & Jones, B. (1993), 'The mammals of Cape Range peninsula, north-western Australia', in *The Biogeography of Cape Range, Western Australia* (ed. W.F. Humphreys), pp. 207–25, Western Australian Museum, Perth.

Bennett, A.F. (1990), 'Land use, forest fragmentation and the mammalian fauna at Naringal, south-western Victoria', *Australian Wildlife Research,* **17**:325–47.

Bennett, A.F., Schulz, M., Lumsden, L.F., Robertson, P., & Johnson, P.G. (1989), 'Pitfall trapping of small mammals in temperate forests of southeastern Australia', *Australian Mammalogy,* **12**:37–9.

Blacket, M.J., Adams, M., Krajewski, C., & Westerman, M. (2000), 'Genetic variation within the dasyurid marsupial genus *Planigale*', *Australian Journal of Zoology,* **48**:443–59.

Bos, D.G. (2001), 'Some observations on foraging behaviour in the southern ningaui, *Ningaui yvonneae*', *Australian Mammalogy,* **23**:59–61.

Boscacci, L.J., McKenzie, N.L., & Kemper, C.M. (1987), 'Mammals', in *A biological survey of the Nullarbor region South and Western Australia in 1984* (eds. N.L. McKenzie, & A.C. Robinson), pp. 103–37, South Australian Department of Environment and Planning, Adelaide.

Bouskila, A., & Dickman, C.R. (2003), 'Species diversity of vertebrate communities in arid lands', in *Biodiversity in drylands: toward a united framework for research and management* (eds. M. Shachak, S.T.A. Pickett, J.R. Gosz, & A. Perevolotsky), Oxford University Press, Oxford, (in press).

Bradley, A.J., Kemper, C.M., Kitchener, D.J., Humphreys, W.F., & How, R.A. (1987), 'Small mammals of the Mitchell Plateau region, Kimberley, Western Australia', *Australian Wildlife Research,* **14**:397–413.

Braithwaite, R.W. (1989), 'Shelter selection by a small mammal community in the wet–dry tropics of Australia', *Australian Mammalogy,* **12**:55–9.

Braithwaite, R.W., Cockburn, A., & Lee, A.K. (1978), 'Resource partitioning by small mammals in lowland heath communities of south-eastern Australia', *Australian Journal of Ecology,* **3**:423–45.

Braithwaite, R.W., Winter, J.W., Taylor, J.A., & Parker, B.S. (1985), 'Patterns of diversity and structure of mammalian assemblages in the Australian tropics', *Australian Mammalogy,* **8**:171–86.

Brandle, R. (1998), 'Mammals', in *A biological survey of the stony deserts, South Australia, 1994–1997* (ed. R. Brandle), pp. 147–82, Department for Environment, Heritage and Aboriginal Affairs, Adelaide.

Briggs, S.V. (1996), 'Native small mammals and reptiles in cropped and uncropped parts of lakebeds in semi-arid Australia', *Wildlife Research,* **23**:629–36.

Burbidge, A.A., Johnson, K.A., Fuller, P.J., & Southgate, R.I. (1988), 'Aboriginal knowledge of the mammals of the central deserts of Australia', *Australian Wildlife Research,* **15**:9–39.

Burnett, S.E. (1992), 'Effects of a rainforest road on movements of small mammals: mechanisms and implications', *Wildlife Research,* **19**:95–104.

Busby, J.R. (1991), 'BIOCLIM – a bioclimatic analysis and prediction system', in *Nature conservation: cost effective biological surveys and data analysis* (eds. C.R. Margules, & M.P. Austin), pp.64–8, CSIRO, Melbourne.

Calaby, J.H. (1966), 'Mammals of the upper Richmond and Clarence Rivers, New South Wales', *CSIRO Division of Wildlife Survey, Technical Paper* **10**:1–55.

Carthew, S.M., & Keynes, T. (2000), 'Small mammals in a semi-arid community, with particular reference to *Ningaui yvonneae*', *Australian Mammalogy,* **22**:103–9.

Catling, P.C., & Burt, R.J. (1995), 'Studies of the ground-dwelling mammals of eucalypt forests in south-eastern New South Wales: the effect of environmental variables on distribution and abundance', *Wildlife Research,* **22,** 669–85.

Catling, P.C., & Burt, R.J. (1997), 'Studies of the ground-dwelling mammals of eucalypt forests in north-eastern New South Wales: the species, their abundance and distribution', *Wildlife Research,* **24**:1–19.

Catling, P.C., Burt, R.J., & Kooyman, R. (1997), 'A comparison of techniques used in a survey of the ground-dwelling and arboreal mammals in forests in north-eastern New South Wales', *Wildlife Research,* **24**:417–32.

Caughley, J. (1980), 'Native quolls and tiger quolls', in *Endangered animals of New South Wales* (ed. C. Haigh), pp. 45–8, National Parks and Wildlife Service, Sydney.

Chapman, A., & Kitchener, D.J. (1978), 'Mammals of Durokoppin and Kodj Kodjin Nature Reserves', *Records of the Western Australian Museum* (suppl.), **7**:49–54.

Cooper, N.K., Aplin, K.P., & Adams, M. (2000), 'A new species of false antechinus (Marsupialia: Dasyuromorphia: Dasyuridae) from the Pilbara region, Western Australia', *Records of the Western Australian Museum*, **20**:115–36.

Cornell, H.V., & Lawton, J.H. (1992), 'Species interactions, local and regional processes, and limits to the richness of ecological communities: a theoretical perspective', *Journal of Animal Ecology*, **61**:1–12.

Coventry, A.J., & Dixon, J.M. (1984), 'Small native mammals from the Chinaman Well area, northwestern Victoria', *Australian Mammalogy*, **7**:111–5.

Crowther, M.S. (2002), 'Distributions of species of the *Antechinus stuartii–A. flavipes* complex as predicted by bioclimatic modelling', *Australian Journal of Zoology*, **50**:77–91.

Dell, J., & How, R.A. (1984), 'The biological survey of the eastern Goldfields of Western Australia Part 2: Widgiemooltha–Zanthus study area. Vertebrate fauna', *Records of the Western Australian Museum* (suppl.), **18**:57–81.

Dell, J., & How, R.A. (1985), 'The biological survey of the eastern Goldfields of Western Australia Part 3: Jackson–Kalgoorlie study area. Vertebrate fauna', *Records of the Western Australian Museum* (suppl.), **23**:39–66.

Dell, J., & How, R.A. (1988), 'The biological survey of the eastern Goldfields of Western Australia Part 5: Edjudina–Menzies study area. Vertebrate fauna', *Records of the Western Australian Museum* (suppl.), **31**:38–68.

Dell, J., & How, R.A. (1992), 'The biological survey of the eastern Goldfields of Western Australia Part 6: Youanmi–Leonora study area. Vertebrate fauna', *Records of the Western Australian Museum* (suppl.), **40**:20–32.

Denny, M.J.S. (1975), 'Mammals of Sturt National Park, Tibooburra, New South Wales', *Australian Zoologist*, **18**:179–95.

Dickman, C.R. (1980), 'Ecological studies of *Antechinus stuartii* and *Antechinus flavipes* (Marsupialia: Dasyuridae) in open-forest and woodland habitats', *Australian Zoologist*, **20**:433–46.

Dickman, C.R. (1986a), 'An experimental study of competition between two species of dasyurid marsupials', *Ecological Monographs*, **56**:221–41.

Dickman, C.R. (1986b), 'Niche compression: two tests of an hypothesis using narrowly sympatric predator species', *Australian Journal of Ecology*, **11**:121–34.

Dickman, C.R. (1988), 'Body size, prey size, and community structure in insectivorous mammals', *Ecology*, **69**:569–80.

Dickman, C.R. (1989), 'Patterns in the structure and diversity of marsupial carnivore communities', in *Patterns in the structure of mammalian communities* (eds. D.W. Morris, Z. Abramsky, B.J. Fox, & M.R. Willig), pp. 241–51, Texas Tech University Press, Lubbock.

Dickman, C.R., Downey, F.J., & Predavec, M. (1993), 'The hairy-footed dunnart *Sminthopsis hirtipes* (Marsupialia: Dasyuridae) in Queensland', *Australian Mammalogy*, **16**:69–72.

Dickman, C.R., Green, K., Carron, P.L., Happold, D.C.D., & Osborne, W.S. (1983), 'Coexistence, convergence and competition among *Antechinus* (Marsupialia) in the Australian high country', *Proceedings of the Ecological Society of Australia*, **12**:77–99.

Dickman, C.R., Haythornthwaite, A.S., McNaught, G.H., Mahon, P.S., Tamayo, B., & Letnic, M. (2001a), 'Population dynamics of three species of dasyurid marsupials in arid central Australia: a 10-year study', *Wildlife Research*, **28**:493–506.

Dickman, C.R., Henry-Hall, N.J., Lloyd, H., & Romanow, K.A. (1991), 'A survey of the terrestrial vertebrate fauna of Mount Walton, western Goldfields, Western Australia', *Western Australian Naturalist*, **18**:200–6.

Dickman, C.R., Lunney, D., & Matthews, A. (2001b), 'Ecological attributes and conservation of dasyurid marsupials in New South Wales, Australia', *Pacific Conservation Biology*, **7**:124–33.

Dickman, C.R., & McKechnie, C.A. (1985), 'A survey of the mammals of Mount Royal and Barrington Tops, New South Wales', *Australian Zoologist*, **21**:531–43.

Dickman, C.R., Parnaby, H.E., Crowther, M.S., & King, D.H. (1998), '*Antechinus agilis* (Marsupialia: Dasyuridae), a new species from the *A. stuartii* complex in south-eastern Australia', *Australian Journal of Zoology*, **46**:1–26.

Dickman, C.R., Predavec, M., & Downey, F.J. (1995), 'Long-range movements of small mammals in arid Australia: implications for land management', *Journal of Arid Environments*, **31**:441–52.

Dickman, C.R., & Read, D.G. (1992), *The biology and management of dasyurids of the arid zone in New South Wales*, NSW National Parks and Wildlife Service, Hurstville.

Ellis, M., & Henle, K. (1988), 'The mammals of Kinchega National Park, western New South Wales', *Australian Zoologist*, **25**:1–5.

Fisher, A. (1999), *Wildlife of the mitchell grass downs of northern Australia*, Parks and Wildlife Commission of the Northern Territory, Darwin.

Fisher, D.O, and Dickman, C.R. (1993), 'The body size-prey size relationship in dasyurid marsupials: tests of three hypotheses', *Ecology*, **74**:1871–83.

Flannery, T.F. (1995a), *Mammals of New Guinea*, Reed Books, Chatswood.

Flannery, T.F. (1995b), *Mammals of the south-west Pacific and Moluccan Islands*, Reed Books, Chatswood.

Fox, B.J. (1982), 'A review of dasyurid ecology and speculation on the role of limiting similarity in community organization', in *Carnivorous marsupials* (ed. M. Archer), pp. 97–116, Royal Zoological Society of New South Wales, Mosman.

Fox, B.J. (1985), 'Small mammal communities in Australian temperate heathlands and forests', *Australian Mammalogy*, **8**:153–8.

Friend, G.R., Morris, K.D., & McKenzie, N.L. (1991), 'The mammal fauna of Kimberley rainforests', in *Kimberley rainforests* (eds. N.L. McKenzie, R.B. Johnston, & P.G. Kendrick), pp. 393–412, Surrey Beatty & Sons, Chipping Norton.

Friend, G.R., & Taylor, J.A. (1985), 'Habitat preferences of small mammals in tropical open-forest of the Northern Territory', *Australian Journal of Ecology*, **10**:173–85.

Gilfillan, S.L. (2001), 'An ecological study of a population of *Pseudantechinus macdonnellensis* (Marsupialia: Dasyuridae) in central Australia. I. Invertebrate food supply, diet and reproductive strategy', *Wildlife Research*, **28**:469–80.

Goldingay, R., & Daly, G. (1997), 'Surveys of arboreal and terrestrial mammals in the montane forests of Queanbeyan, New South Wales', *Australian Mammalogy*, **20**:9–19.

Goosem, M. (2000), 'Effects of tropical rainforest roads on small mammals: edge changes in community composition', *Wildlife Research*, **27**:151–63.

Green, R.J., & Catterall, C.P. (1998), 'The effects of forest clearing and regeneration on the fauna of Wivenhoe Park, south-east Queensland', *Wildlife Research,* **25**:677–90.

Gressitt, J.L., & Nadkarni, N. (1978), *Guide to Mt Kaindi: background to montane New Guinea ecology*, Wau Ecology Institute, Wau.

Cullan, P.K., & Robinson, A.C. (1980), 'Vegetation and small mammals of a Victorian forest', *Australian Mammalogy*, **3:**87–95.

Haythornthwaite, A.S., & Dickman, C.R. (2000), 'Foraging strategies of an insectivorous marsupial, *Sminthopsis youngsoni* (Marsupialia: Dasyuridae), in Australian sandridge desert', *Austral Ecology,* **25**:193–8.

Holt, R.D. (1993), 'Ecology at the mesoscale: the influence of regional processes on local communities', in *Species diversity in ecological communities* (eds. R.E. Ricklefs, & D. Schluter), pp. 77–88, University of Chicago Press, Chicago.

How, R.A., & Dell, J. (1992), 'The biological survey of the eastern Goldfields of Western Australia Part 7: Duketon–Sir Samuel study area. Vertebrate fauna', *Records of the Western Australian Museum* (suppl.), **40**:90–103.

How, R.A., Dell, J., & Cooper, N.K. (1991), 'Ecological survey of Abydos-Woodstock Reserve, Western Australia. Vertebrate fauna', *Records of the Western Australian Museum* (suppl.), **37**:78–125.

How, R.A., Dell, J., & Muir, B.G. (1988), 'The biological survey of the eastern Goldfields of Western Australia Part 4: Lake Johnston-Hyden study area. Vertebrate fauna', *Records of the Western Australian Museum* (suppl.), **30**:44–83.

Ingram, G.J., & Raven, R.J. (eds.) (1991), *An atlas of Queensland's frogs, reptiles, birds and mammals,* Board of Trustees of the Queensland Museum, Brisbane.

James, C.D., & Shine, R. (2000), 'Why are there so many coexisting species of lizards in Australian deserts?' *Oecologia*, **125**:127–41.

Johnson, K.A., & Roff, A.D. (1982), 'The western quoll, *Dasyurus geoffroii* (Dasyuridae, Marsupialia) in the Northern Territory: historical records from venerable sources', in *Carnivorous marsupials* (ed. M. Archer), pp. 221–6, Royal Zoological Society of New South Wales, Mosman.

Jones, M.E. (1997), 'Character displacement in Australian dasyurid carnivores: size relationships and prey size patterns', *Ecology,* **78**:2569–87.

Jones, M.E., & Barmuta, L.A. (1998), 'Diet overlap and abundance of sympatric dasyurid carnivores: a hypothesis of competition?', *Journal of Animal Ecology,* **67**:410–21.

Jones, M.E., & Barmuta, L.A. (2000), 'Niche differentiation among sympatric Australian dasyurid carnivores', *Journal of Mammalogy,* **81**:434–47.

Kemper, C.M. (1990), 'Small mammals and habitat disturbance in open forest of coastal New South Wales. I. Population parameters', *Australian Wildlife Research,* **17**:195–206.

Krajewski, C., Woolley, P.A., & Westerman, M. (2000), 'The evolution of reproductive strategies in dasyurid marsupials: implications of molecular phylogeny', *Biological Journal of the Linnean Society,* **71**:417–35.

Laurance, W.F. (1990), 'Distributional records for two 'relict' dasyurid marsupials (Marsupialia: Dasyuridae) in north Queensland rainforest', *Australian Mammalogy,* **13**:215–8.

Leung, L.K.–P. (1999), 'Ecology of Australian tropical rainforest mammals. I. The Cape York antechinus, *Antechinus leo* (Dasyuridae: Marsupialia)', *Wildlife Research,* **26**:287–306.

Lim, L. (1992), *Conservation and biology of the rare Pilliga mouse,* Pseudomys pilligaensis *Fox and Briscoe, 1980 (Muridae: Conilurini),* Zoological Parks Board of New South Wales, Mosman.

Lindenmayer, D.B., Nix, H.A., McMahon, J.P., Hutchinson, M.F., & Tanton, M.T. (1991), 'The conservation of Leadbeater's possum, *Gymnobelideus leadbeateri* (McCoy): a case study of the use of bioclimatic modelling', *Journal of Biogeography,* **18**:371–83.

Lunney, D., & Barker, J. (1986), 'Mammals of the coastal forests near Bega, New South Wales I. Survey', *Australian Zoologist,* **23**:19–28.

Lunney, D., & Leary, T. (1989), 'Movement patterns of the white-footed dunnart, *Sminthopsis leucopus* (Marsupialia: Dasyuridae) in a logged, burnt forest on the south coast of New South Wales', *Australian Wildlife Research,* **16**:207–15.

Masters, P. (1993), 'The effects of fire-driven succession and rainfall on small mammals in spinifex grassland at Uluru National Park, Northern Territory', *Wildlife Research,* **20**:803–13.

Masters, P. (1998), 'The mulgara *Dasycercus cristicauda* (Marsupialia: Dasyuridae) at Uluru National Park, Northern Territory', *Australian Mammalogy,* **20**:403–7.

McKenzie, N.L., Hall, N., & Muir, W.P. (2000), 'Non-volant mammals of the southern Carnarvon Basin, Western Australia', *Records of the Western Australian Museum* (suppl.), **61**:479–510.

McKenzie, N.L., Rolfe, J.K., Hall, N.J., & Youngson, W.K. (1993), 'The biological survey of the eastern Goldfields of Western Australia Part 9: Norseman-Balladonia study area. Vertebrate fauna', *Records of the Western Australian Museum* (suppl.), **42**:33–55.

McKenzie, N.L., Rolfe, J.K., & Youngson, W.K. (1992), 'The biological survey of the eastern Goldfields of Western Australia Part 8: Kurnalpi-Kalgoorlie study area. Vertebrate fauna', *Records of the Western Australian Museum* (suppl.), **41**:37–64.

McKenzie, N.L., Rolfe, J.K., & Youngson, W.K. (1994), 'The biological survey of the eastern Goldfields of Western Australia Part 10: Sandstone-Sir Samuel and Laverton-Leonora study areas. Vertebrate fauna', *Records of the Western Australian Museum* (suppl.), **47**:51–85.

Menkhorst, P.W. (ed.) (1995), *Mammals of Victoria: distribution, ecology and conservation,* Oxford University Press, Melbourne.

Menzies, J.L. (1991), *A handbook of New Guinea marsupials and monotremes,* Kristen Press, Madang.

Milledge, D. (1991), 'A survey of the terrestrial vertebrates of coastal Byron Shire', *Australian Zoologist,* **27**:66–90.

Mollemans, F.H., Reid, J.R.W., Thompson, M.B., Alexander, L., & Pedler, L.P. (1984), 'Biological survey of the Cooper Creek environmental association (8.4.4) north-eastern South Australia', unpublished report to the Department of Environment and Planning, Adelaide.

Molsher, R., Newsome, A.E., & Dickman, C.R. (1999), 'Feeding ecology and population dynamics of the feral cat (*Felis catus*) in relation to the availability of prey in central-eastern New South Wales', *Wildlife Research,* **26**:593–607.

Moore, R.M. (1970), *Australian grasslands,* Australian National University Press, Canberra.

Moro, D. (1991), 'The distribution of small mammal species in relation to heath vegetation near Cape Otway, Victoria', *Wildlife Research,* **18**:605–18.

Morton, S.R. (1982), 'Dasyurid marsupials of the Australian arid zone: an ecological review', in *Carnivorous marsupials* (ed. M. Archer), pp. 117–30, Royal Zoological Society of New South Wales, Mosman.

Morton, S.R., & Baynes, A. (1985), 'Small mammal assemblages in arid Australia: a reappraisal', *Australian Mammalogy,* **8**:159–69.

Morton, S.R., Brown, J.H., Kelt, D.A., & Reid, J.R.W. (1994), 'Comparisons of community structure among small mammals of North American and Australian deserts', *Australian Journal of Zoology,* **42**:501–25.

Moseby, K.E., & Read, J.L. (2001), 'Factors affecting pitfall capture rates of small ground vertebrates in arid South Australia. II. Optimum pitfall trapping effort', *Wildlife Research,* **28**:61–71.

Murphy, M.J. (1998), 'Mammal survey of Seven Mile Beach National Park and Comerong Island Nature Reserve on the south coast of New South Wales', *Australian Zoologist,* **30**:419–25.

Murray, B.R., & Dickman, C.R. (2000), 'Relationships between body size and geographical range size among Australian mammals: has human impact distorted macroecological patterns?', *Ecography,* **23**:92–100.

Norris, K.C., Gilmore, A.M., & Menkhorst, P.W. (1979), 'Vertebrate fauna of south Gippsland',. *Memoirs of the National Museum of Victoria,* **40**:105–200.

Paltridge, R., & Southgate, R. (2001), 'The effect of habitat type and seasonal conditions on fauna in two areas of the Tanami Desert', *Wildlife Research,* **28**:247–60.

Paull, D.C., & Date, E.M. (1999), 'Patterns of decline in the native mammal fauna of the north-west slopes of New South Wales', *Australian Zoologist,* **31**:210–24.

Pianka, E.R. (1986), *Ecology and natural history of desert lizards,* Princeton University Press, New Jersey.

Posamentier, H.G. (1976), 'Habitat requirements of small mammals in coastal heathlands of New South Wales', MSc thesis, University of Sydney, Sydney.

Read, D.G. (1987), 'Habitat use by *Sminthopsis crassicaudata, Planigale gilesi* and *P. tenuirostris* (Marsupialia: Dasyuridae) in semiarid New South Wales', *Australian Wildlife Research,* **14**:385–95.

Read, D.G. (1995), 'Fauna survey in the floodplain of the Great Anabranch of the lower Darling River', *Australian Zoologist,* **30**:57–64.

Read, J.L. (1992), 'Influence of habitats, climate, grazing and mining on terrestrial vertebrates at Olympic Dam, South Australia', *Rangeland Journal,* **14**:143–56.

Reid, J., & Gillen, J. (1988), 'The Coongie Lakes study', unpublished report to the Department of Environment and Planning, Adelaide.

Reid, J.R.W., Kerle, J.A., & Baker, L. (1993), 'Mammals', in *Kowari 4: Uluru fauna. The distribution and abundance of vertebrate fauna of Uluru (Ayers Rock–Mount Olga) National Park, NT* (eds. J.R.W. Reid, J.A. Kerle, & S.R. Morton), pp. 69–78, Australian National Parks and Wildlife Service, Canberra.

Ricklefs, R.E., & Schluter, D. (eds) (1993), *Species diversity in ecological communities,* University of Chicago Press, Chicago.

Righetti, J., Fox, B.J., & Croft, D.B. (2000), 'Behavioural mechanisms of competition in small dasyurid marsupials', *Australian Journal of Zoology,* **48**:561–76.

Risbey, D.A., Calver, M.C., Short, J., Bradley, J.S., & Wright, I.W. (2001), 'The impact of cats and foxes on the small vertebrate fauna of Heirisson Prong, Western Australia. II. A field experiment', *Wildlife Research,* **28**:223–35.

Settle, G.A. (1978), 'The quiddity of tiger quolls', *Australian Natural History,* **19**:164–9.

Statham, H.L., & Harden, R.H. (1982), 'Habitat utilization of *Antechinus stuartii* (Marsupialia) at Petroi, northern New South Wales', in *Carnivorous marsupials* (ed. M. Archer), pp. 165–85, Royal Zoological Society of New South Wales, Mosman.

Strahan, R. (ed.) (1995), *The mammals of Australia,* Reed Books, Chatswood.

Tasker, E., Bradstock, R., & Dickman, C.R. (1999), 'Small mammal diversity and abundance in relation to fire and grazing history in the eucalypt forests of northern New South Wales', in *Bushfire 99: Proceedings of the Australian Bushfire Conference, Albury*, pp. 387–90, Charles Sturt University, Albury.

Tidemann, C.R. (1988), 'A survey of the mammal fauna of the Willandra Lakes World Heritage Region, New South Wales', *Australian Zoologist,* **24**:197–204.

Twyford, K.L. (1997), 'Habitat relationships of small mammals at Port Campbell National Park, Victoria', *Australian Mammalogy,* **20**:89–98.

Van Dyck, S., & Crowther, M.S. (2000), 'Reassessment of northern representatives of the *Antechinus stuartii* complex (Marsupialia: Dasyuridae): *A. subtropicus* sp. nov., and *A. adustus* new status', *Memoirs of the Queensland Museum,* **45**:611–35.

Wainer, J.W. (1976), 'Studies of an island population of *Antechinus minimus* (Marsupialia, Dasyuridae)', *Australian Zoologist,* **19**:1–7.

Watt, A. (1997), 'Population ecology and reproductive seasonality in three species of *Antechinus* (Marsupialia: Dasyuridae) in the wet tropics of Queensland', *Wildlife Research,* **24**:531–47.

Wilson, B.A., Bourne, A.R., & Jessop, R.E. (1986), 'Ecology of small mammals in coastal heathland at Anglesea, Victoria', *Australian Wildlife Research,* **13**:397–406.

Wilson, B.A., & Wolrige, J. (2000), 'Assessment of the diet of the fox, *Vulpes vulpes*, in habitats of the eastern Otway Ranges, Victoria', *Australian Mammalogy,* **21**:201–11.

Winter, J. (1991), 'Mammals', in *Kowari 1: Rainforest animals. Atlas of vertebrates endemic to Australia's wet tropics* (eds.H.A. Nix, & M.A. Switzer), pp. 43–54, Australian National Parks and Wildlife Service, Canberra.

Woinarski, J.C.Z., & Fisher, A. (1995), 'Wildlife of lancewood (*Acacia shirleyi*) thickets and woodlands in northern Australia. I. Variation in vertebrate species composition across the environmental range occupied by lancewood vegetation in the Northern Territory', *Wildlife Research,* **22**:379–411.

Woinarski, J.C.Z., Milne, D.J., & Wanganeen, G. (2001), 'Changes in mammal populations in relatively intact landscapes of Kakadu National Park, Northern Territory, Australia', *Austral Ecology,* **26**:360–70.

Woinarski, J.C.Z., Woolley, P.A., & Van Dyck, S. (1996), 'The distribution of the dunnart *Sminthopsis butleri*', *Australian Mammalogy,* **19**:27–9.

Woolley, P.A. (1992), 'New records of the Julia Creek dunnart, *Sminthopsis douglasi* (Marsupialia: Dasyuridae)', *Wildlife Research,* **19**:779–83.

PART IV

CHAPTER 22

BEHAVIOUR OF CARNIVOROUS MARSUPIALS

David B. Croft

School of Biological Science, University of New South Wales, UNSW, Sydney, NSW 2052, Australia

The behaviour of carnivorous marsupials is reviewed with a particular focus on new knowledge gathered from the 1980s and the interpretation of behaviour in the framework of behavioural ecology. The general life-sustaining behaviour of individuals is first described by examining activity patterns, maintenance activities, nest building and use of shelter, predatory behaviour, predator avoidance and exploration and object play. Carnivorous marsupials are typically small, nocturnal, secretive and exploit a wide range of prey through agile and flexible behaviour. Social behaviour is reviewed with a focus on communication; spacing; agonistic; and sexual behaviour; parental care, and socialisation and play. Although most carnivorous marsupials are solitary, they express a large repertoire of social acts with some unusual behaviour. Knowledge about the behaviour of carnivorous marsupials has progressed well beyond simple description of form and function to evolutionary insights through comparative study and experimentation. There is scope for much more research in this direction, especially with American and New Guinean species.

INTRODUCTION

In the first reviews of the behaviour of carnivorous marsupials (Ewer 1968a; Eisenberg and Leyhausen 1972; Ewer 1973; Eisenberg 1981; Croft 1982), the repertoires of a few well-studied species were assembled and patterns in form, function and evolution were sought. Subsequent reviews have continued this synthesis (Russell 1982; Russell 1984; Russell 1985), or described behaviour in relation to some broader topic (evolutionary ecology – Lee and Cockburn 1985; reproductive physiology – Tyndale-Biscoe and Renfree 1987; nutrition – Hume 1999).

This chapter will continue the synthesis since the behaviour of more species has been studied since the 1980s and some behaviour has been the subject of more sophisticated quantitative analysis with an experimental approach to understanding causation and function. The framework for this approach is behavioural ecology (Krebs and Davies 1997). Behaviour is described by optimality models and evolutionary stable strategies in social contexts where selfish genes and kinship benefits operate. The tools applied are not just inconspicuous observation in natural and contrived situations but the application of technologies breaching our perceptual limitations (e.g. night vision devices), expanding opportunities for data gathering (e.g. light-weight radio transmitters), and illuminating fitness outcomes (e.g. DNA profiling).

In this chapter I will review the individual and social behaviour of carnivorous marsupials within the limits of current knowledge. I will conclude with a brief synopsis of progress in understanding the behaviour of carnivorous marsupials since the 1980s and make recommendations for future research.

INDIVIDUAL BEHAVIOUR

Individual behaviour is a broad category of actions that do not directly involve conspecifics. These actions include activity rhythms, maintenance (grooming/scratching), nest building and the use of shelter, predatory behaviour, predator avoidance, and aspects of behavioural development like exploration and object play. The distinction from social behaviour may be arbitrary since the same actions may be used to pounce on a prey item as on a conspecific rival. Individuals may also gather information about the activities of neighbours in choosing when and where to forage alone.

Activity patterns

The daily activities of an animal predominantly comprise those repetitive actions that maintain its life. Less attention has been paid to constructing a comprehensive time/activity budget of carnivorous marsupials than other aspects of behaviour. The major constraint is that most species are strictly nocturnal and too small and fast-moving to follow their behaviour across several diel cycles in the field. Even so behaviour has been directly observed for the smallest species (e.g. *Ningaui yvonneae* Bos 1999) but more typically behaviour is inferred from indirect methods such as trails of fluorescent pigments, radio tracking (e.g. Fisher and Dickman 1993) or spool-and-line tracking (Carthew 1994). Read (1988) used the daily capture frequency of *Sminthopsis crassicaudata*, *Planigale gilesi* and *P. tenuirostris* to infer that light rain stimulated foraging activity and *Planigale* likewise responded to an increase in air pressure from below the monthly average, but moonlight had no inhibitory effect on the activity of these three arid-zone species.

In captivity, feeding times are quickly learnt and thus may confound the timing and expression of foraging behaviour (Hope et al. 1997). Even so, several studies have shown entrainable circadian activity rhythms by using wheel running (*Dasyurus viverrinus* – Kennedy et al. 1990), activity in plus-shaped mazes (*Sminthopsis macroura* – O'Reilly et al. 1984) or major activities like feeding, drinking and nest box use as well as wheel running (*Phascolosorex dorsalis* and *Antechinus habbema* – Woolley et al. 1991). *Dasyurus* and *Antechinus* were active throughout the night but *Sminthopsis* had a burst of activity at dark onset and thereafter activity was sporadic. The latter has been assumed to be the more typical pattern of the dasyurids. However, *Phascolosorex* was diurnal and so atypical amongst almost exclusively nocturnal dasyrurids.

Activity patterns may be modulated by the presence of potential competitors. Moss and Croft (1988) used direct observation to describe the subterranean and surface activity of *P. gilesi* and *P. tenuirostris* under simulated summer and winter conditions, and the presence or absence of the larger and surface-active *S. crassicaudata*. *Planigale* were active throughout the night with modest variation under all conditions without *Sminthopis*. However, the introduction of *Sminthopis* led to a strong bimodal pattern with the larger mode at dark onset. Righetti et al. (2000) used fluorescent tags and time-lapse colour video recording to examine nocturnal activity of *A. stuartii*, *Antechinus swainsoni* and *Sminthopsis murina*. *Sminthopsis* was markedly less active in short sporadic bouts than *Antechinus* where the larger *A. swainsoni* was significantly more active. *A. swainsoni* decreased activity in the presence of a conspecific but there was no such effect in the other two species. *A. swainsoni* also decreased activity in the presence of *A. stuartii*, and *S. murina* increased activity in the latter's presence but the activity of *A. stuartii* was unaffected in both instances. Dickman (1991) observed *A. stuartii* avoiding *A. swainsoni* in the field, and so the behaviour of sympatric competitors may strongly affect spatio-temporal activity.

There is scant information on the activity of caenolestids, *Dromiciops australis* and those didelphids defined by Hume (1999) as primarily carnivorous. Hunsaker and Shupe (1977) described the activity of a few omnivorous didelphids in similar terms to dasyurids; *viz* nocturnal and crespuscular with either a bimodal pattern with a dark onset peak or unimodal. They noted species like *Didelphis virginiana* did not forage on very cold nights as confirmed by Ryser (1995). Small dasyurids reduce activity under cold temperatures to the extent that they enter daily torpor (reviewed by Geiser 2003).

Maintenance activities

The form and patterning of maintenance activities, like scratching and grooming, have been defined for a few species. Hutson (1972) gives a comprehensive description for a female *Dasyuroides byrnei*. The female washed her face, snout, nape of the neck, throat and chin by wiping with her licked forepaws. Labial glands are known only from *Phascogale tapoatafa* (Russell 1985) and thus face washing does not anoint the head and body with specific social odours. The female licked and nibbled most of her ventrum, progressing in an anterior-posterior direction. The pouch was licked while the female sat but she did not use her forepaws to provide additional access. Pouch cleaning was significantly longer following parturition. Most of the body surface was accessible to scratching by the hind leg. The tail was cleaned either by drawing it over one shoulder towards the mouth or drawing it forwards under the body while sitting. This pattern of grooming and scratching with the hind foot is common to dasyurids, didelphids and morphologically conservative eutherians (Eisenberg 1981). The principal stimulus is cleaning up the muzzle after feeding with additional and specific stimulation from pouch young as part of maternal care.

Nest building and use of shelter

Digging is an uncommon behaviour in American marsupials, except for surface foraging, but several Australian and New Guinean species are accomplished at constructing burrows. *Dasycercus cristicauda* excavates a complex burrow system with one main entrance and one or more 'popholes'. Woolley (1990) describes one example as comprising a large entrance hole leading down 0.5 m to a grass-lined nest. Narrower, near vertical tunnels led up to popholes from the main passageway to the nest. *Dasyurus geoffroii* uses abandonned rabbit warrens or digs its own tunnel leading to a larger chamber (Serena and Soderquist 1989a). The burrows may have one or more entrances. *Antechinus naso* and *A. habbema* nest 65 to 100 cm below ground, gaining access through 4 cm diameter tunnels (Woolley 1989).

Carnivorous marsupials commonly build nests. Dried leaves or grass stems are typically transported in the mouth but *Marmosa, Philander, Caluromys* and *Didelphis* species often carry a bundle in a curl of their prehensile tail. *Dromiciops australis* may modify abandonned bird nests for its own purposes or transport leaves several feet off the ground to construct a nest in clumps of low-growing bamboo (Marshall 1978). The typical nest of smaller dasyurids takes a dome-shaped form secured within a burrow, hollow, rock overhang or some other protective structure. The outer shell is often formed from leaves to provide some rigidity but the core may be softer dried grass. In some species nests are constructed by both sexes and regardless of whether the female is pregnant or has young (e.g. *A. stuartii*, Settle and Croft 1982a). In other species, the female may significantly add to her nest at pregnancy (e.g. *D. cristicauda,* Michener 1969), as young develop during pouch life (e.g. *D. byrnei,* Aslin 1974; *D. geoffroii*, Serena and Soderquist 1989a), or when young leave the pouch (e.g. *P. tenuirostris*, Read 1985). Wardell-Johnson (1986) followed the annual use of nest boxes by *Antechinus flavipes* in regenerating karri forest. Females occupied the nest boxes and constructed large nests of karri leaves during the breeding season from late September through to early January. The nest boxes were again occupied without nest construction during the dispersal of young in February through to April. *Dasyurus geoffroii* of both sexes build more substantial nests in cold weather providing thermal insulation between the occupier and the den entrance whereas in hot weather the nest is simple bedding (Serena and Soderquist 1989a).

Predatory behaviour

Predatory behaviour involves location, capture, killing and ingesting other animals. When the prey size is very small relative to the predator, killing and ingestion become a single act. When the prey is larger, the predator usually has a more specialised killing technique. After an initial approach the predator will bite and/or pin with its forepaws. If the prey shows strong resistance, it may be shaken or tossed and then relocated for a second or third bite. Bites are directed to the anterior of the prey's body and often strike the head or neck (Eisenberg 1985). For example, Kirsch and Waller (1979) described predation on a rat by a captive *Caenolestes obscurus.* The caenolestid seized or pinned down the rat with its forepaws and then killed it with bites from the spear-like incisors.

Sminthopsis crassicaudata typifies the broad range of predatory behaviour of dasyurids. They are adept at catching both terrestrial and arboreal insects and even snatching low-flying moths, beetles and mosquitos out of the air (Ewer 1968b). They stalk and ambush prey, and pursue any moving species of appropriate size but discriminate against noxious species such as adult *Tenebrio.* Even so, Ewer suggested that they were less flexible in their behaviour than eutherian counterparts, a conclusion later disputed with larger species by Pellis and Officer (1987). Predatory efficiency improves significantly with age as shown in comparative observations of juvenile and adult *Planigale maculata* feeding on mouse pups (Van Dyck 1979). Prey-catching sequences vary with size of prey. *Dasyuroides byrnei* take small prey (neonate rats 5–7 g) in a relatively stereotyped manner but predatory sequences for large prey (juvenile rats 44–83 g) are quite variable (Hutson 1975). Experienced *D. byrnei* seize, pin and position fleeing prey and then exact a killing bite, which is a more advanced response than seizing prey in the mouth (Eisenberg and Leyhausen 1972). Pellis and Officer (1987) provide further comparative analysis of the predatory behaviour of *D. byrnei*, two species of quolls (*D. viverrinus* and *D. hallucatus*), *P. tapoatafa* and the domestic cat (*Felis catus*). The *D. byrnei* subjects distinguished themselves by using a frontal attack, which included grasping and pinning the prey before a killing bite to the head. The other species avoided frontal attacks. Two forms of headshake were observed: the snout traversed an arc in space (*F. catus* and *D. hallucatus*) or the sagittal crest traversed an arc in space (remaining three species). Both were equally effective in enhancing the penetration of the canines for a secured prey or disorienting an unsecured one. *Dasyurus hallucatus* use vision to locate and attack mice but blindfolded individuals can compensate with other senses, probably olfaction (Pellis et al. 1992). The vibrissae orient the attack but vision or tactile contact can compensate if the vibrissae are trimmed. Once a mouse is bitten, vision and vibrissae play a secondary role to tactile stimuli for orienting the killing bite and subsequent prey consumption. The texture and lie of the fur may be important cues.

An efficient predator maximises the energy return for the cost of searching for and handling the prey item (Bell 1991). Dasyurids in captivity preferentially consume the most profitable prey and, in the field, forage in the habitat where they are most likely to find them (Fisher and Dickman 1993). Thus their typically generalist diets (although most avoid ants) diverge where larger species preferentially take larger prey. However, there is some debate about the function that best expresses the relationship between prey size

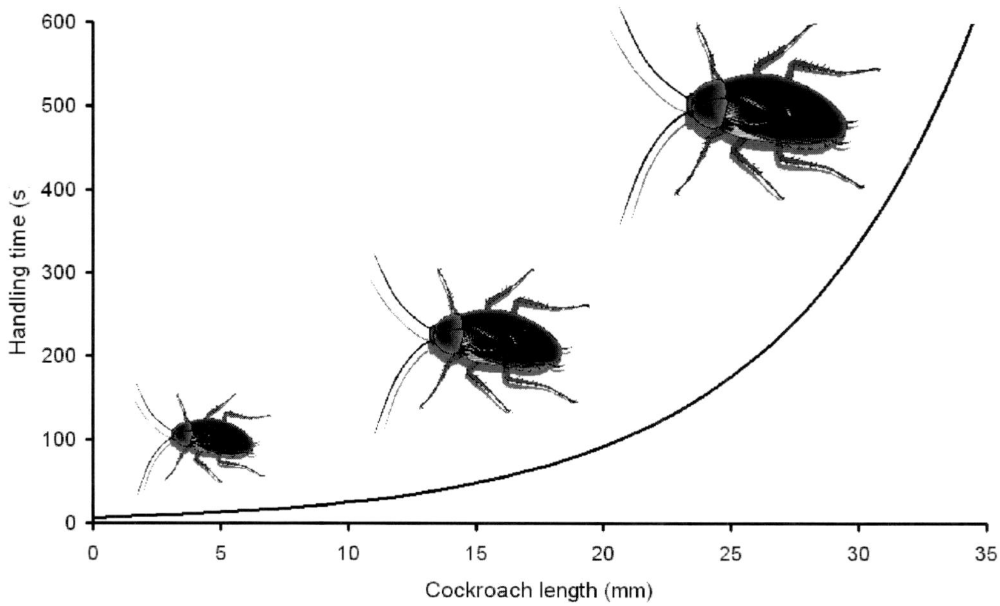

Figure 1 Exponential relationship between handling time (s) and cockroach length (mm) for male *Ningaui yvonnae* (adapted from Woolnough and Carthew 1996).

and handling time. Calver et al. (1989) fed a range of sizes of cockroaches and grasshoppers to *Sminthopsis hirtipes, S. ooldea, Ningaui rideii,* and *N. yvonnae.* Handling time increased linearly with the mass of the prey and at a faster rate with grasshoppers, possibly because they can more effectively defend against predation. Capture success was equivalent across predators, prey types and prey sizes. Woolnough and Carthew (1996) fed various sized cockroaches to *N. yvonnae* and found handling time increased exponentially with prey length (the more typical relationship found with other insectivores) (Fig. 1). The *Ningaui* were optimal foragers selecting those smaller cockroaches that returned the highest energy content for the least handling time. Field tests of observed behaviour against deterministic and stochastic optimality models (derived largely from studies of insectivorous birds) are lacking and are probably precluded by the nocturnality and an opportunistic diet of dasyurids.

The larger carnivorous marsupials, *Sarcophilus laniarius* (= *S. harrisii* Werdelin 1987) and two species of *Dasyurus,* are only sympatric in Tasmania. They are graded in size and differ in arboreal ability. Some scramble competition for resources is a constant possibility, especially among the younger, newly weaned animals. Through such competition they exhibit prey-size preferences, and in terms of canine breaking strength and muscle strength driving the canines – character displacement (Jones 1997). Jones and Barmuta (1998) offer further evidence for diet overlap and the resultant relative abundance of the three extant species. The relative rarity of *Dasyurus maculatus* may derive from competition between young *D. maculatus* and young *Sarcophilus* as well as competition with adult *D. viverrinus.* Killing and hunting behaviour by *Dasyurus, Sarcophilus* and *Thylacinus* is described in Jones and Stoddart (1998) and Jones (2003).

Predator avoidance

The carnivorous marsupials are vulnerable to a range of reptilian, avian and mammalian predators. The largest and most pugnacious species may be at least risk as adults. Individuals presumably minimise predation by foraging at times and under protective cover that lower risk. Oakwood (2000) attributed most predation on *D. hallucatus* to use of open areas with vegetative cover removed by fire. Foraging individuals may detect and avoid areas where non-volant predators are active but may themselves provide scent trails for predators through their own marking behaviour. The nest or den should provide a relatively safe refuge during rest. However, constant use could provide a strong odour cue to goannas, snakes or mammalian carnivores. Many species change nest sites frequently (e.g. *D. hallucatus,* Oakwood and Miles 1998; *S. youngsoni,* Haythornthwaite and Dickman 2000) and so may reduce such risk yet Wardell-Johnson (1986) noted large accumulations of faeces around the entrances of nests used by breeding *A. flavipes* females.

There has understandably been more observation and experimental investigation of the predatory behaviour of the carnivorous marsupials than of their strategies to reduce their risk of falling prey to others. Jones (1998) compared vigilance in *S. laniarius* with no known predators, and the smaller *D. viverrinus* that may be taken by owls and possibly *S. laniarius*. The quoll invested more time in vigilance and sought improved visibility as a function of actual risk and relative risk compared to the devil. Devils reduced vigilance if they were larger, older or in the company of others as expected if the behaviour varied as a function of potential risk. Haythornthwaite and Dickman (2000) studied the response of the much smaller *S. youngsoni* to predation risk in the Simpson Desert. They deployed feeding stations of mealworms in 'open' or 'bush' microhabitat with or without illumination equivalent to moonlight. Risk was assumed to be highest at an illuminated open site and lowest at a non-illuminated bush site. The perception of risk was measured by the giving-up density (number of mealworms left by the forager). No relationship between giving-up density and any of the treatments was found. The null result was ascribed to the broad microhabitat use of *S. youngsoni* and the low predator pressure at the site.

Coulson (1996) reviewed anti-predator behaviour in marsupials and one novel strategy was thanatosis or 'playing dead'. *Didelphis virginiana* will, when seized by the neck, fall into a trance-like state. Although many marsupials become calm if the head is covered with a cloth, the phenomenon exhibited by *D. virginiana* is unusual in that it results in immobility and is displayed most effectively when it is grasped by the loose skin of the neck and shaken slightly. Eisenberg (1981) compared the responses of six didelphid species to grasping by a handler and concluded that the behaviour of *D. virginiana* is unique in the circumstances of its performance and even the congener *D. marsupialis* has a higher threshold for such behaviour. There is no evidence of this behaviour in dasyurids. However, *P. tapoatafa* may deflect attack by piloerection of the tail in response to disturbance (Soderquist 1994). No similar use of the bushy tail of *D. byrnei* has been seen.

Exploration and object play

A characteristic of the dasyurids, especially arid-zone species, is their dietary generalism (Fisher and Dickman 1993), long-range movements (Dickman et al. 1995) and/or drifting home ranges (Read 1984a). Exploration of novel areas is thus likely to be a common behavioural trait. *Antechinus stuartii* (Settle and Croft 1982b) and *D. byrnei* young (Meissner and Ganslosser 1985) explore their mother, siblings and the immediate environment of the nest after eye-opening and venture further afield about 10 days later. However, the safety and limited confines of captivity may induce behavioural artefacts such as a high incidence of mothers carrying dependent young on their backs whilst out 'foraging' (Soderquist and Serena 2000). In the field, *D. geoffroii* young do not explore outside the den until 17 weeks old, several weeks after eye-opening; between 19–21 weeks, bout length and distances travelled outside the den increase rapidly (Soderquist and Serena 2000). At weaning (22–24 weeks), juveniles forage independently up to 500 m from the den following a gradual increase in foraging effort and exploration of the maternal home range.

The next major phase of exploratory behaviour is during dispersal from the natal home range. Sons typically emigrate further than daughters, who may settle in or near their mother's home range (e.g. *A. stuartii*, Cockburn et al. 1985a; *D. geoffroii*, Soderquist and Serena 2000), although male-biased dispersal is not universal (e.g. *D. hallucatus*, Oakwood 2000). Several studies have tried to elucidate those factors leading to post-weaning dispersal, especially proximate causes such as adverse relationships with the mother or siblings. Vestal et al. (1986) staged encounters between mothers and their newly weaned sons or other juveniles in a field population of *A. stuartii*. Mothers were no more aggressive towards sons than daughters but showed significantly more affiliative behaviour towards their own offspring than those of other females. Likewise Soderquist and Lill (1995) observed no maternal aggression towards either sons or daughters in *P. tapoatafa* and hence no support for maternal aggression as a proximate mechanism of male-biased dispersal. The same finding was repeated with *D. geoffroii* (Soderquist and Serena 2000) where mothers may vacate the rearing den in advance of offspring. Soderquist and Lill (1995) also dismissed adverse relationships with siblings, competition for food or nest sites, hyperthermia in a crowded nest, and an ontogenetic switch (e.g. a threshold body weight) as proximate causes of dispersal. Foraging and exploration beyond the natal den took individuals inexorably away but much more so for males suggesting a sex-related genetic determinant for the propensity to explore or move further.

Slow movement with visual and olfactory investigation of the surrounds is one way to explore and learn about a novel area. Young, naïve individuals also frequently manipulate objects and thereby may learn about their characteristics. Repeated interaction with an object may perfect an important skill such as attack or escape behaviour. Such object play takes two forms: (1) the manipulation of objects as part of exploratory behaviour and curiosity as the juvenile familiarises itself with its environment, and (2) functional training in handling objects that are important in adult behaviour. Functional training of predatory skills through play with objects is a relatively poorly documented feature of dasyurid play. *Antechinus stuartii* young engage in tail-pulling and may catch and release, nip and release, pat or toss benign prey like mealworms, cockroaches and moths (Settle and Croft 1982b). *Dasycercus cristicauda* and *Dasyurus maculatus* stalk and make simulated attacks on littermates and other objects, and *S. laniarius* engages in manipulative play simulating predatory behaviour (review in Croft 1982). Smaller dasyurids (e.g. *Sminthopsis* and *Antechinomys* spp.) do not play in this way but also do not stalk prey. Play of any kind is relatively rare in species smaller than 100 g (Byers 1999).

Hunsaker and Shupe (1977) briefly discuss exploratory behaviour of American marsupials and note the absence of reports on play. They emphasise the importance of olfaction as the primary sense used during exploration, with a secondary reliance on sound and movement. Thus most of the information gathered by carnivorous marsupials in the course of their exploration and interaction with their environment is likely to be outside a human observer's perception.

SOCIAL BEHAVIOUR

Social behaviour is simply defined as any action that involves one or more conspecifics. Typically we divide social interactions into three categories: affiliative or aid-giving behaviour, agonistic behaviour and sexual behaviour. A relationship between two individuals may involve a pattern and certain frequency of interactions of all three kinds. One type of interaction may grade into another as when a male entices a female into close proximity by offering a service such as grooming, but then may attempt forced copulation through aggression.

Communication

Communication is the social glue that holds groups of interacting individuals together. Thus the size and complexity of the repertoire of communicatory acts is typically greater in gregarious than solitary species. Most carnivorous marsupials fall into the latter social type (Jarman and Kruuk 1996). However, a solitary individual still needs to learn about the whereabouts of others if it is to stay alone, it needs to find a suitable mate and court with him/her in order to achieve fertilisation, and a mother and offspring will usually interact closely (e.g. suckling of young) until the latter is weaned and independent.

Carnivorous marsupials are typically active only in the dark (nocturnal) or at low light levels (crepuscular). Thus communication is more likely through auditory (hearing) and olfactory channels (scent) than visual (sight) (Croft 1982). However, we should be cautious about making inferences based on our poor night vision for species with a much greater capability than our own. Likewise our colourful environment may be monotone to a nocturnal species. Tactile communication is likely to be confined to parent-offspring and mating behaviour.

Auditory

Mammals produce vocalisations by expelling air from the lungs causing a passive vibration in the glottis lips (Andrew 1963). This produces a sound with a fundamental frequency that may be further modified during passage through the nasopharynx or buccal cavity. The range of sounds produced by marsupials is relatively limited since the glottal structure provides little modulation of air flow (Negus 1962). The calls of very few species have been subjected to sonographic analysis (Eisenberg et al. 1975; Hunsaker and Shupe 1977). Pemberton and Renouf (1993) extended earlier studies on *S. laniarius* by observing behaviour and recording vocalisations of individuals aggregated at carcasses. They described 11 vocalisations mainly used in agonistic encounters and assigned them to one of the four syllable types defined by Eisenberg et al. (1975). The majority were Type II (i.e. mixed tonal and noisy) such as crescendo, growl, whine, snort, bark and yip. A new Type III (i.e. clicks with little harmonic structure), a click train, was identified. The remainder were graded sequences of growls from crescendo to an intense shriek. The latter shriek and humph-growl were Type IV (i.e. noisy syllables with no discrete energy band). Peak energy was at low frequency, 0.1–1.1 kHz, extending to 9–12 kHz, in vocalisations ranging from a 162 ms snort to a 5.5 s growl.

Aitkin et al. (1986) defined 7–12 kHz as the maximal range of sensitivity in the auditory cortex and inferior colliculus of *D. hallucatus* and thus predicted that most vocalisations and behavioural responsiveness to sounds would be in this frequency range. Dempster (1994) identified four vocalisations amongst interacting (hiss, sniff, squawk) and solitary (twitter) quolls. Type II hisses were low frequency sounds in the range 0.3-2.1 kHz and thus well below maximal sensitivity. The less frequent Type IV sniffs (0.6–9.5 kHz) and squawks (1.2–9.3 kHz) better fitted the neurophysiological model. Twitters extended well into the ultrasonic range (1.0–73.5 kHz) but most energy was below 15 kHz. Dempster (1994) concluded that the high neural sensitivity to sound above 10 kHz probably related to prey detection (e.g. rustling sounds) rather than calls given in social contexts where interactants will be in close proximity.

Bishop et al. (1995) have expanded sonographic analysis to include the vocalisations of *S. crassicaudata.* The most common call was the chirp (7–10 kHz), a Type I syllable (i.e. tonal, organised in narrow frequency bands), associated with one individual attending to another in a social context. The remaining calls were described as kiss, fast ch ch ch, pst fz, tchz, a high pitched repetitive sound extending above 12.5 kHz (Type III or IV) and click (Type III). The relatively low frequency 'kiss' vocalisation was associated with exploratory behaviour.

These structural analyses of the vocalisations of carnivorous marsupials have been accompanied by some determination of functional relationships. Since most development is supported in the pouch through lactation in marsupials, there is a long risk of death through losing contact with the mother. A call in response to loss of contact is probably ubiquitous in marsupial young. In dasyurids, the young's call is usually high-pitched such as a squeak, sibilant sisss, tchit-tchit or wheeze, and the mother's call is similar in form at a lower pitch. Of the 22 well-studied dasyurids, mothers call to their displaced young in 23% and young to their mother in 64% (Croft 1982).

In aggressive encounters between individuals, calls by the offensive individual are typical of most species (82%, n = 22). The calls are usually loud and harsh sounding. For example, *Dasyurus hallucatus* hisses and sniffs in association with threat postures, and sniffs while investigating another individual of either sex. Squawks were associated with attack and fighting (Dempster 1995). Didelphids hiss or growl in agonistic contexts. If violent encounters occur, the subordinate may produce a screech-like cry. Defensive calls are rarely reported.

Vocalisations during courtship are given in two contexts: (1) to attract a mate (one or both sexes) from some distance away, and (2) when males and females are enjoined in courtship and, sometimes, mating (usually only male). In didelphids such as *Monodelphis domesticus*, males (and females in some species) may initially produce short click-like calls, which precede further interaction and probably serve to orientate the partners and reduce intersexual aggression (Fadem 1989). Read (1984b) described vocalisations in oestrous female and courting male–female pairs of *P. gilesi* and *P. tenuriostris.* A soft clucking call made by isolated oestrous females was ascribed to 'mate-attracting'. Agonistic interactions between courting males and females were accompanied by 'chits' and affiliative ones by 'tutts'. No 'mate-attracting' call was performed by *Parantechinus apicalis* (Wolfe et al. 2000) but harsh vocalisations by both sexes accompanied courtship and mating. Non-receptive female *P. tapoatafa* deflect advancing males with a sharp, multi-syllabic 'tsk' but ones presumed to be in or near oestrus direct a 'chirp' towards males (Soderquist and Ealey 1994). However, males made no overt response towards 'chirping' females.

An alarm call precedes or coincides with the onset of an attack by a predator. Further vocalisations may be given when an individual is under attack in an attempt to intimidate or retaliate towards a predator. Alarm signals fall into two main categories: (1) 'warning' signals, which serve to communicate the presence, location or behaviour of a predator at some risk to the sender and benefit to conspecific receivers, and (2) 'pursuit deterrent' signals, which are directed to and inform the predator that it has been detected, possibly causing it to give-up an attack (benefits to conspecifics are coincidental). Coulson (1996) found no convincing evidence for 'warning' signals in any marsupial species. Alarm and predator deterrence vocalisations have been recorded from 27% of 22 dasyurid species.

High-frequency vocalisations like a 'twitter' are common to many species (32%, $n = 22$) when exploring a novel environment. There has been no functional explanation of this behaviour performed by solitary individuals but perhaps the calls assist predation (e.g. invoke a detectable response in prey or mask some characteristic of the predator). Non-agonistic interactions with other individuals (novel or familiar) typically contain audible sniffs. The primary sense is probably olfaction and the audible component is a by-product or prelude to a potentially more aggressive vocalisation.

Figure 2 Open-mouthed threat by the South American carnivorous marsupial *Chironectes minimus.*

Chemical

Chemical communication in the carnivorous marsupials is reviewed in a companion chapter by Toftegaard and Bradley (2003), to which the reader is referred.

Visual

Visual signals are best described from macropods, which are large, conspicuous and partially active during the day (e.g. Coulson 1989). This is where we would expect visual signals to be most effective. However, various light-amplifying devices (night-vision devices and very low-light video cameras) have and will reveal hitherto unknown activities of the generally nocturnal carnivorous marsupials. For example, Pemberton and Renouf (1993) described 20 postures in social interactions between feeding *S. laniarius* at dimly illuminated carcasses.

Agonistic interactions are typically accompanied by visual signals that demonstrate body size and weaponry (Fig. 2). Individuals rarely launch straight into a fight without first approaching and testing whether the opponent will hold its ground. The approach may be accompanied by or escalate to a demonstration of strength and aggressive intent, usually referred to as a 'threat display'. Dasyurids display an 'open-mouth threat' that reveals the teeth, especially canines, and is usually accompanied by a harsh vocalisation and raised forepaw (Croft 1982). Similar behaviour is shown in American carnivorous marsupials (Fig. 2). *Sarcophilus laniarius* competing at a carcass stage through quadrapedal (including lying on their bellies with feet extended), tripedal and bipedal postures (Pemberton and Renouf 1993). These are accompanied by a 'gape' (mouth opened for a few seconds and slowly closed). They also 'neck threat', nip in the direction of another's neck, and walk 'stiff-legged'. Thus the pattern of maximal exposure of the weaponry from a posture where an individual may most effectively engage with an opponent in a fight is typical of high intensity threats.

Visual signals are prominent in most species' repertoires, even if they are solely active at night. Most species have a light-coloured ventrum that might enhance the effect of upright postures used in alarm and agonistic interactions. However, few visual signals are not accompanied by a sound or possibly dissemination of an odour. For example, the threat postures of *D. hallucatus* are typically accompanied by a vocalisation, and escalation of one follows the other (Dempster 1995). Likewise some olfactory signals have conspicuous visual components. For example, cloacal-dragging and chest-rubbing behaviour involve specific and obvious motor patterns. Furthermore, many dasyurids defecate on raised objects so that faeces are prominently displayed.

Tactile

Tactile signals must be patterns of touching that are formalised in some way to serve a communicatory function, are relatively constant in form, and are not merely incidental when two individuals touch. This can often be difficult to tease apart. For example, allogrooming is the use of the teeth, lips, tongue and/or forepaws to clean the hair and skin of another individual. It thus provides a service in some relationship but it only becomes interesting in a communicative sense, if it consistently reveals some quality of that relationship; for example, dominance or subordination, a winner or loser of a competitive interaction, a sexually receptive female.

Patterns of touching are an important component of the mother–offspring relationship. Young from birth until weaning transit through the pouch. Mothers keep their pouch and pouch young clean through grooming both. Licking the young's cloaca stimulates it to urinate and defecate and the mother consumes the waste products.

Formalised patterns of touching are common in many carnivorous marsupials during courtship and mating. The males of many species solicit and test a female's receptivity to mounting through tactile signals. 'Paw-on-partner' contact is a typical prelude to a mounting attempt in many dasyurids (Croft 1982). Once mounted, the male of all species grasps the female around the abdomen during copulation. Some palpate the female's abdomen or hindlimbs (e.g. *P. apicalis* Wolfe et al. 2000). A neck-grip is almost universal across the dasyurids. However, it is not typical of American species like *M. domestica* (Trupin and Fadem 1982) or *Marmosa robinsoni* (illustrated in Tyndale-Biscoe and Renfree 1987). These patterns of behaviour serve to restrain the female but probably have no specific signal function. Some males rub their chin on the female's nape but this may be odour deposition from labial or sternal glands.

Spacing behaviour

The dispersion of dasyurids varies from singletons in drifting home ranges (e.g. *S. crassicaudata*, Morton 1978), singletons in defended core ranges (e.g. *D. geoffroii*, Serena and Soderquist 1989) with overlap between sexes, to mixed sex aggregations in undefended ranges (e.g. *Antechinus agilis* and *A. stuartii*, Lee and Cockburn 1985). Some differences in the interpretation of spacing behaviour have arisen from comparison of grid-trapping and radio-tracking (Traill and Coates 1993).

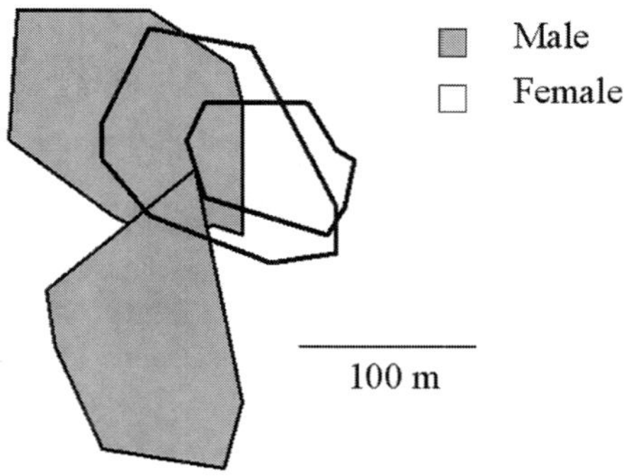

Figure 3 Minimum convex polygons representing the home ranges of two male and two female *Sminthopsis leucopus* in coastal dry heathland at Anglesea, Victoria (adapted from Laidlaw et al. 1996).

Jarman and Kruuk (1996) concluded that the sole pattern of spatial organisation in dasyurids and didelphids was for females to forage alone in an undefended range. For example, *Sminthopsis leucopus* in coastal dry heathland has an average home range of 0.9 ha with extensive overlap within and between the sexes (Fig. 3) (Laidlaw et al. 1996). This contrasts to the Insectivora where only 15% of 345 species show this spatial organisation for adult females and the remainder forage alone but defend the foraging range. The majority of Carnivora (86% of 231 species) show the latter spatial organisation and only 1% the style common to the marsupial carnivores. The marsupial style is conservative and the basal pattern for mammals. Jarman and Kruuk (1996) could find no compelling adaptive reason for female marsupial carnivores to fail to defend their foraging range when so many counterparts in the Insectivora, Carnivora and Macroscelidea do. Perhaps the research effort on the marsupial carnivores has been insufficient to reveal defence. Soderquist (1995) argued that encounters between adult female *P. tapoatafa* are exceedingly rare and they maintain intrasexually exclusive home ranges. Likewise Serena and Soderquist (1989b) drew the same conclusion about female *D. geoffroii* with scent marking used to define and advertise boundaries. These authors have argued strongly for the value of radio tracking coupled with direct observation of compatible species to properly define spatial and social organisation.

At the other end of the spectrum, a number of small dasyurids range over large areas and travel long distances for their diminutive body size. Lunney and Leary (1989) described one 'explorer' male *S. leucopus* travelling 1025 m in 24 h. Some males were sedentary ('resident') with range lengths averaging 105 m while others were more mobile ('explorer') with an average range length of 720 m. High mobility and transience is a characteristic of many small species in drier habitat (e.g. *S. dolichura*, Friend et al. 1997). Morton (1978) coined the term 'drifting home range' to describe the unstable and continually shifting range of *S. crassicaudata* confirmed by Read (1984a) for the same species and noted in *S. hirtipes* and *S. youngsoni* by Dickman et al. (1995). The behaviour is not confined to *Sminthopsis* but also found in *Planigale* (Read 1984). This flexible behavioural pattern is most likely related to the low and unpredictable insect food supply in the arid zone. However, much larger species like *D. cristicauda* are typically sedentary (Masters 1998). This species has a more flexible diet and the opportunity to switch between prey types amongst invertebrates and vertebrates (Chen et al. 1998).

Agonistic behaviour

Agonistic behaviour encompasses elements of both aggression (threat and attack) and retreat (defence and submission) found in two contexts: sexual competition and resource competition (Alcock 1998). Typically in mammals, males compete for copulations and secondarily for resources while females compete for resources to support themselves and the additional costs of production and maintenance of offspring. Agonistic interactions between females to secure copulations are rare.

The typical spatial organisation of most dasyurids is one of males in large home ranges overlapping both sexes and females in smaller, more exclusive areas (see above). In spite of the range overlap, individuals are assumed to meet rarely and thus when encounters are staged in captivity or the field they are socially intolerant. This intolerance extends across both sexes, so that intrasexual and intersexual fights may be equally ferocious (e.g. *D. hallucatus* Dempster 1995). Even so, Dempster (1995) found that male *D. hallucatus* performed more agonistic behaviour than females in same-sex encounters, and females more submissive behaviour in between-sex encounters. Soderquist and Ealey (1994) found no real distinction between intra- and intersexual encounters among *P. tapoatafa* that were observed opportunistically during their foraging or at a supplemental feeding station. Individuals infrequently co-fed, watched each other or approached (if opposite sex) and most often chased each other. Chases rarely resulted in threats and physical contact and few individuals showed any evidence of past wounding. However, these observations were made during the non-breeding season. In other species, such as *D. geoffroii* (Serena and Soderquist 1989b) and *A. stuartii* (Scott 1987) wounding is common between males in the breeding season. Pemberton and Renouf (1993) saw only one wounding in 119 encounters between *S. laniarius* feeding at a carcass but scarring and open wounds were found in 29.5% of a large capture sample (~200).

Dominance relationships, mostly between males, form in captive populations (reviewed in Croft 1982). However, Soderquist and Ealey (1994) found frequent reversals and so little stability between competing male *P. tapotafa* at a food station in the wild. Dominance/subordinance is unlikely to be a useful concept if individuals meet infrequently, especially among males of semelparous species where repeated yielding to another competitor would be maladaptive (Scott 1987). In the longer lived *S. laniarius*, Buchmann and Guiler (1977) showed stable dominance/subordination relationships in captivity but subsequent field observations have failed to support this (Pemberton and Renouf 1993). At carcasses, the intruder most often yielded to the possessor after some interaction at a distance. The first-feeding individual was not necessarily the larger. The possessor only yielded the carcass when apparently satiated.

Sexual behaviour

Wolfe et al. (2000) recently reviewed the behaviour of dasyurids during courtship and copulation in a discussion of the mating behaviour of *P. apicalis*. Some dasyurids (e.g. *Antechinus* and *Phascogale* spp.) are monoestrous but most are polyoestrous like all other marsupials (Tyndale-Biscoe and Renfree 1987). Even so, most species bear only one litter per year and so show a single oestrus. Some species are semelparous and the male has one reproductive season of a few months in which to mate. Sexual behaviour is thus frenetic and a combination of this activity and aggressive encounters with other males leads to a male's early demise from stress, parasitism and poor nutrition (Bradley 2003). Ovulation is spontaneous and oestrus lasts 2–3 days except in *Antechinus* where it may extend over 7–14 days. Since males and females usually live apart, the first phase of courtship involves the male overcoming the female's aggressive response towards him (Croft 1982; Wolfe et al. 2000). Thus interactions include prolonged chasing of the female and some fights. Males frequently vocalise while in pursuit.

When receptive, the female stands and allows the male to approach and investigate her mouth and cloacal region and some allogrooming may occur. In *P. apicalis* cloacal sniffing of the female by the male is significantly more frequent at oestrus than pro-oestrus but other sexual behaviour such as mounting attempts does not significantly differ. Males mount by gripping the scruff of the female's neck in their jaws and clasping the female around the abdomen. Once intromission is achieved the pair may remain in coitus for 1–6 h or longer. While mounted, the males of many species palpate the female with their forelimbs, and some with their hind limbs. *Antechinus* and *Sminthopsis* males rub their chin on the female's nape. This active copulatory phase may be followed by a passive one of several hours where the quietude is broken only by occasional attempts by the female to break free.

Copulation in dasyurids is often characterised by long periods of intromission when compared with other families of mammals (Eisenberg 1981). For example, *Didelphis* may remain in a mounted state for over 60 minutes. *Marmosa* remains in copula for over 30 minutes and may exhibit two long mounts over the course of 6 hours. But *Antechinus* may remain mounted for over 2 hours and *Parantechinus* for up to 6.5 h. Female *A. agilis* frequently have multiple male partners and there may be some siring advantage to a second rather than the first male (Shimmin et al. 2000).

Paternity control, female choice and mating systems

In *D. virginiana* the males range widely during the breeding season (Ryser 1992). Apparently guided by olfactory cues, the male will seek out an oestrous female and attempt to guard her. As many as three males may attend an oestrous female. Mating rights are determined by inter-male dominance. Invariably the largest and most aggressive male will intimidate the others and mating will ensue.

The spatial organisation of dasyurids suggests a similar pattern of mate acquisition. Male ranges typically overlap those of several females and when males compete, as in *Antechinus*, the largest male is typically the winner (Lazenby-Cohen and Cockburn 1988). Thus Lazenby-Cohen and Cockburn (1991) divided the home range of male *A. stuartii* into a small foraging area and a larger social one. These ranges drifted as males were assumed to acquire information about mating opportunities and male competitors. In contrast females showed more fidelity to a presumably well-resourced range. However, inter-male competition at mating attempts is not inevitable as Soderquist and Ealey (1994) observed a few courtship and copulation events in free-ranging *P. tapoatafa* where only one male was present. Male competition may indirectly continue through sperm competition in the female's reproductive tract (Shimmin et al. 2000).

Females are not passive players in the mating game (Walker 1996). Many an ardent male's efforts at fatherhood are thwarted by female tactics before copulation, during copulation, after copulation but before fertilisation, or after fertilisation (Birkhead and Moller 1993). Behaviour plays a primary role in the first two of these periods. While roving males search for receptive females, many of the latter are conspicuously advertising their incipient oestrus. As discussed above, many female dasyurids vocalise and also presumably lay down chemical trails if not follow those of males. A male's ability to gain and retain mating access seems to be the key to its likely mating success. Females often vigorously resist mounting attempts (Wolfe et al. 2000). Thus females seem to exercise choice amongst the potential sires of their offspring by attracting multiple suitors, by inciting intra-male competition, by testing male control of paternity, and, if necessary, by subverting mate guarding.

There is no evidence in carnivorous marsupials that males provide any paternal care. Thus males are likely to be monogamous only if they cannot find additional mates (e.g. females are widely dispersed and show synchronous oestrus – Ostfeld 1990). Females typically forage solitarily in an undefended home range (Jarman and Kruuk 1996). This can lead to three forms of mating system. Firstly, males may defend some area, which habitually attracts roving females (*Resource-defence polygyny*). However, resources are rarely so predictable in most Australian habitats that such a strategy is favoured. Secondly, males may aggregate on notional breeding territories and females move amongst these males, mating preferentially with those in the centre of the aggregation (*Lek polygyny*). Typically females of such species will only mate amongst aggregations of males and will avoid solitary males. The closest strategy is the 'lek promiscuity' of *A. stuartii* discussed in Lazenby-Cohen and Cockburn (1988). Females of many species conspicuously advertise oestrus, which may attract aggregations of males. Males may also aggregate on hotspots where many female ranges overlap. Thus in effect oestrous females incite intra-male competition for mating opportunities. Thirdly, roving males may scramble to mate with roving females (*Scramble competition polygyny*). This would be the likely option for species with 'drifting home ranges' and many transient individuals. This strategy is often modulated by a size-related hierarchy among males and the strength of this may be reflected in the degree of heteromophism in the species (Jarman 1983).

Aid-giving and parental behaviour

There are no reports of altruistic behaviour in carnivorous marsupials and the only service one individual may provide to another is allogrooming. Thus aid-giving behaviour is essentially directed from the parent to the offspring and all of this is maternal. The functions of maternal behaviour are to protect, keep warm and feed the young. The nature and extent of maternal care depends on the pouch anatomy, and shelter habit and type of the species, and the mobility of the permanently emerged young (Russell 1982; Lee and Cockburn 1985).

In some didelphids and most dasyurids, females exhibit incomplete pouches. These vary from simple marginal ridges of skin that develop in the breeding season (e.g. *Antechinus*), partial covering of crescentic antero-lateral folds of skin (e.g. *Sarcophilus*) through to a covering of a circular fold of skin with a central opening (e.g. *Sminthopsis*). Large litters are commonly produced and these are held in the pouch for a relatively short time until the young release the teat. Thereafter, helpless young with little fur, eyes closed and unable to thermoregulate are protected and cared for in a nest. Nests vary in elaborateness from simple depressions in a saucer-shaped structure of dry grass (e.g. *D. byrnei*) through to a superstructure of bark strips with an inner chamber lined with finely teased fibres (e.g. *P. tapoatafa*). The mother regularly returns to the nest to suckle the young and huddle with them to keep them warm. The mother will also actively defend the nest against intruders. If the young are displaced from the pouch or stray from the nest the mother retrieves them in response to their distress calls. Once the young are furred and the eyes are open, they make excursions from the nest with the mother, either following her or riding on her back. Eventually they make independent excursions and return to the mother (in the nest or outside) to suckle until weaned. Russell (1982) argues that young are left in a nest early in development because both the simple pouch offers inadequate protection and the weight of young would hinder efficient foraging and predator escape by the mother. For example, *P. maculata* young are left when the litter is about 70% of maternal body weight (MBW). At weaning the litter is around 300% of MBW.

One of the more interesting questions about parental behaviour is the relative investment in sons and daughters. Adaptive sex ratio variation has been demonstrated in a range of marsupial species (Ashworth 1996). A biased sex ratio at weaning will only arise if one sex is cheaper to rear than the other (Fisher 1930). In most marsupials, higher short-term costs of sons are usually offset by long-term costs of daughters. For example, in the biparous *A. swainsonii* females produce strongly male-biased sex ratios in the first year, but female-biased in the second (Cockburn et al. 1985). This is explained by local resource competition, where sons disperse and do not compete for food and other resources with their mothers whereas philopatric daughters do.

Behaviour is important in the negotiation of supply and demand between the mother and offspring. Trivers and Willard (1973) proposed that in a polygynous mating system large long-lived sons are favoured. Mothers in good condition may preferentially invest in sons if this confers an advantage in the son's reproductive success (i.e. maternal condition is positively correlated with offspring condition at weaning which in turn is correlated with offspring reproductive success). The former is easier to demonstrate than the latter. Ashworth (1996) summarises the evidence for the Trivers-Willard hypothesis in marsupials. Well-fed mothers in good condition produce significantly more sons than daughters in *D. marsupialis, D. virginiana, A. stuartii,* and *P. tapoatafa*. However, there is as yet no evidence that sons are more demanding and that mothers supply this demand in carnivorous marsupials. It is also unclear as to whether mothers directly intervene to bias survival towards one sex. Davison and Ward (1998) suggest not as there is a prenatal bias in the sex ratio of the offspring of *A. agilis.*

Socialisation and play

Since carnivorous marsupials are usually solitary as adults, early socialisation probably helps shape the form of future social interactions but does not lead to enduring social relationships. The principle value of a juvenile interacting with its mother is likely to be the acquisition of essential life skills through social learning (Higginbottom and Croft 1999). The value of interacting with littermates or peers is less clear (Soderquist and Serena 2000). Even so social play is common amongst littermates of the larger bodied carnivorous marsupials (Byers 1999).

Locomotor play is common in many species and serves as physical exercise, which may improve neuro-muscular coordination, cardiovascular function, muscle and bone strength (Fagen 1981). Essential skills such as predator avoidance and, for carnivores and insectivores, pursuit of prey may be exercised by behaviour that consists of sudden running/hopping, leaping, rapid turns and abrupt stops and freezing. In dasyurids, littermates add a social dimension to this behaviour, since much of locomotor play consists of chases, although the roles of pursuer and pursued rarely reverse (Lissowsky 1996).

Social play in carnivorous marsupials includes play-fighting, sex play, play chasing and parallel play. Young *D. byrnei* engage in vigorous social play with their mother or siblings from the same or a different litter (Meissner and Ganslosser 1985). The behaviour typically includes chases, mock attacks and wrestling. Likewise littermates of *D. cristicauda,* all the *Dasyurus* species, and *S. laniarius* engage in wrestling and chasing bouts while young *P. tapoatafa* play 'hide and seek' (Croft 1982). Soderquist and Ealey (1994) describe play of juvenile *P. tapoatafa* as vigorous chasing with role reversal but no social contact (at least outside the nest). Sex play is relatively uncommon (Lissowsky 1996).

Soderquist and Serena (2000) discuss the function of play wrestling in juvenile *D. geoffroii.* They dismiss the hypothesis that it trains skills useful in predatory behaviour since it is a poor mimic of the latter. They support the hypothesis that it provides practice and skill development for fighting since superior fighting skills may be important in a successful outcome in intermale competition for mates. They further suggest that play wrestling may ameliorate any tendency towards siblicide as it fills in hours of close contact between littermates with a vigorous but non-damaging activity.

Conclusion

Our knowledge of the behaviour of carnivorous marsupials has advanced apace since I reviewed the topic in 1982. Researchers have embraced new technologies to better define the structure of behaviour and to reveal its form and function in the field. The advent of reliable, light-weight transmitters has revealed hitherto unknown complexity in the spatial and social organisation of various species. The technique has further provided valuable observations of behaviour in the wild. These often opportunistic samples coupled with those of interacting individuals at point sources of natural or supplemental food sources provide more confidence in understanding natural behaviour and interpreting the results of contrived interactions in captive studies. Functional hypotheses about behaviour have increasingly been tested through comparative or correlative study and experiment. However, many ascribed causes and functions of behaviour remain speculative. Information on the American and New Guinean carnivorous marsupials remains severely limited relative to Australian species. The fascinating behaviour of the well-studied *D. virginiana* and *M. domestica* and the diurnal activity of *P. dorsalis* creates a desire for broader study so that the generality of patterns across these faunas can be explored.

I would like to see further investigation of foraging strategies in the marsupial carnivores and the testing and further development of optimality models. This research could be performed in large arenas with captive populations to investigate foraging efficiency by imposing various handling costs, defining foraging paths by varying prey dispersion, and testing the optimality of patch choice (e.g. tests of the marginal value theorem and ideal free distributions). In the laboratory and field, further study of risk-sensitive foraging would be profitable. Carnivorous marsupials live in a world of odours beyond our perception and we have been quick to speculate on how they use them but slow to determine their chemical structure, to test the ability of individuals to discriminate amongst them and to experimentally investigate functional hypotheses. Likewise, we have characterised some interesting vocalisations, especially the presumed 'mate-attracting' call but have yet to perform playback and other experiments to confirm function. Finally, the marsupial carnivores have some of the most distinctive mating patterns in the marsupials. Molecular genetic techniques will no doubt be applied to unravel the 'truth' about mating systems. However, actions are as equally fascinating as outcomes. Are poorly socialised individuals simply inept and somewhat violent mates or are there novel insights into mate choice to be gained from cleverly contrived encounters and field observations? I hope we will see more imaginative designs in the laboratory than the simple encounter cage and more opportunities taken in the field with low-light video or direct observation at feeding stations to which multiple individuals may be attracted. The latter type of observation will be usefully coupled with an assessment of spatial organisation through other means like radio tracking.

REFERENCES

Aitkin, L.M., Irvine, D.R.F., Nelson, J.E., Merzenich, M.M. & Clarey, J. C. (1986), 'Frequency representation in the auditory midbrain and forebrain of a marsupial, the northern native cat (*Dasyurus hallucatus*)', *Brain Behavior and Evolution,* **29**:17–28.

Alcock, J. (1998), *Animal Behaviour: An Evolutionary Approach. 6th ed.,* Sunderland, Sinauer Associates.

Andrew, R.A. (1963), 'The origin and evolution of calls and facial expressions in primates', *Behaviour,* **20**:1–109.

Ashworth, D.L. (1996), 'Strategies of maternal investment in marsupials: A comparison with eutherian mammals', in *Comparison of Marsupial and Placental Behaviour* (eds. D.B. Croft & U. Ganslosser), pp. 226–51, Fuerth, Filander Press.

Aslin, H. (1974), 'The behaviour of *Dasyuroides byrnei* (Marsupialia) in captivity', *Zietschrift fŸr Tierspychologie,* **35**:187–208.

Bell, W.J. (1991), *Searching behaviour: the behavioural ecology of finding resources,* Cambridge, Chapman and Hall.

Birkhead, T.R., & Moller, A.P. (1993), 'Female control of paternity', *Trends in Ecology & Evolution,* **8**:100–4.

Bishop, N., Bulbert, M., Carr, S., Kroker, S., & Millikan, J. (1995), 'Sonographic analysis of vocalisations in captive dunnarts, *Sminthopsis crassicaudata*', *Australian Mammalogy,* **18**:99–100.

Bos, D.G. (1999), 'Observations of the response of *Ningaui yvonneae* (Marsupialia: Dasyuridae) to pit-fall drift fences', *Australian Mammalogy,* **21**:143–5.

Bradley, A.J. (2003), 'Stress, hormones and mortality in small carnivorous marsupials', in *Predators with pouches: The biology of carnivorous marsupials* (eds. M. Jones, C. Dickman, & M. Archer), pp. 254–67, Melbourne, CSIRO Publishing.

Buchmann, O.L.K., & Guiler, E.R. (1977), 'Behaviour and ecology of the Tasmanian devil, *Sarchophilus harrisiil*', in *The Biology of Marsupials* (eds. B. Stonehouse & D. Gilmore), pp. 155–68, London, Macmillan.

Byers, J.A. (1999), 'The distribution of play behaviour among Australian marsupials', *Journal of Zoology, Lond,* **247**:349–56.

Calver, M.C., Bradley, J.S., & King, D.R. (1989), 'The relationship between prey size and handling time and prey size and capture success in three sympatric species of dasyurid marsupials', *Australian Wildlife Research,* **15**:615–23.

Carthew, S.M. (1994), 'Foraging behaviour of marsupial pollinators in a population of *Banksia spinulosa*', *Oikos,* **69**:133–9.

Chen, X., Dickman, C.R., & Thompson, M.B. (1998), 'Diet of the mulgara, *Dasycercus cristicauda* (Marsupialia: Dasyuridae), in the Simpson Desert, central Australia', *Wildlife Research,* **25**:233–42.

Cockburn, A., Scott, L.K., & Scotts, D.J. (1985a), 'Inbreeding avoidance and male-biased natal dispersal in *Antechninus* spp. (Marsupialia: Dasyuridae)', *Animal Behaviour,* **33**:908–15.

Cockburn, A., Scott, M.P. & Dickman, C.R. (1985b), 'Sex ratio and intrasexual kin competition in mammals', *Oecologia,* **66**:427–9.

Coulson, G. (1989), 'Repertoires of social behaviour in the Macropodoidea', in *Kangaroos, Wallabies and Rat–kangaroos* (eds. G. Grigg, P. Jarman, & I. Hume), pp. 457–73, Sydney, Surry Beatty & Sons.

Coulson, G. (1996), 'Anti-predator behaviour in marsupials', in *Comparison of Marsupial and Placental Behaviour,* (eds. D.B. Croft & U. Ganslosser), pp. 158–86, Fuerth, Filander Press.

Croft, D.B. (1982), 'Communication in the Dasyuridae (Marsupialia): A review', in *Carnivorous Marsupials* (ed. M. Archer), pp. 291–9, Sydney, Royal Zoological Society of New South Wales.

Davison, M.J., & Ward, S. J. (1998), 'Prenatal bias in sex ratios in a marsupial, *Antechinus agilisl*', *Proceedings of The Royal Society Of London Series B Biological Sciences,* **265**:2095–9.

Dempster, E.R. (1994), 'Vocalisations of adult northern quolls, *Dasyurus hallucatus*', *Australian Mammalogy,* **17**:43–9.

Dempster, E.R. (1995), 'The social behaviour of captive northern quolls, *Dasyurus hallucatus*', *Australian Mammalogy,* **18**:27–34.

Dickman, C.R. (1991), 'Mechanisms of competition among insectivorous mammals', *Oecologia,* **85**:464–71.

Dickman, C.R., Predavec, M., & Downey, F.J. (1995), 'Long-range movements of small mammals in arid Australia: Implications for land management', *Journal of Arid Environments,* **31**:441–5.

Eisenberg, J.F. (1981), *The Mammalian Radiations*, Chicago, Chicago University Press.

Eisenberg, J.F. (1985), 'Form and function: The phylogenesis of predatory behaviour', *Australian Mammalogy,* **8**:195–200.

Eisenberg, J. F., Collins, L.R. & Wemmer, C. (1975), 'Communication in the Tasmanian Devil (*Sarcophilus harrisii*) and a survey of auditory communication in the Marsupialia', *Zeitschrift fŸr Tierpsychologie,* **37**:379–99.

Eisenberg, J. F., & Leyhausen, P. (1972), 'The phylogenesis of predatory behaviour in mammals', *Zeitschrift fŸr Tierpsychologie,* **30**:59–93.

Ewer, R.F. (1968a), *Ethology of Mammals*, London, Logos Press.

Ewer, R.F. (1968b), 'A preliminary survey of the behaviour in captivity of the dasyurid marsupial, *Sminthopsis crassicaudata* (Gould)', *Zeitschrift fŸr Tierpsychologie,* **25**:319–65.

Ewer, R.F. (1973), *The Carnivores*, London, Weidenfield and Nicholson.

Fadem, B.H. (1989), 'Sex differences in aggressive behavior in a marsupial, the gray short-tailed opossum (*Monodelphis domestica*)', *Aggressive Behavior,* **15**:435–41.

Fagen, R. (1981), *Animal Play*, Oxford, Oxford University Press.

Fanning, F.D. (1982), 'Reproduction, growth and development in *Ningaui* sp. (Dasyuridae, Marsupialia) from the Northern Territory', in *Carnivorous marsupials, Vol. I* (ed. M. Archer), pp. 23–37, Sydney, Royal Zoological Society of New South Wales.

Fisher, D.O., & Dickman, C.R. (1993), 'Diets of insectivorous marsupials in arid Australia: selection for prey type, size or hardness?', *Journal of Arid Environments,* **25**:397–410.

Fisher, R.A. (1930), *The Genetical Theory of Natural Selection*, Oxford, Clarendon Press.

Friend, G.R., Johnson, B.W., Mitchell, D.S., & Smith, G.T. (1997), 'Breeding, population dynamics and habitat relationships of *Sminthopsis dolichura* (Marsupialia: Dasyuridae) in semi-arid shrublands of Western Australia', *Wildlife Research,* **24**:245–62.

Geiser, F. (2003), 'Thermal biology and energetics of carnivorous marsupials', in *Predators with pouches: The biology of carnivorous marsupials* (eds. M. Jones, C. Dickman & M. Archer), pp. 238–53, Melbourne, CSIRO Publishing.

Hawthornthwaite, A.S., & Dickman, C.R. (2000), ' 'Foraging strategies of an insectivorous marsupial, *Sminthopsis youngsoni* (Marsupialia: Dasyuridae), in Australian sandridge desert', *Austral Ecology,* **25**:193–8.

Higginbottom, K.B., & Croft, D.B. (1999), 'Social learning in marsupials', in *Mammalian Social Learning: Comparative and Ecological Perspectives* (eds. H.O. Box, & K.R. Gibson), pp. 80–101, Cambridge, Cambridge University Press.

Hope, P.J., Wittert, G.A., Horowitz, M., & Morley, J.E. (1997), 'Feeding patterns of *S. crassicaudata* (Marsupialia: Dasyuridae): Role of gender, photoperiod, and fat stores', *American Journal of Physiology,* **272**(1 Part 2):R78–83.

Hume, I.D. (1999), *Marsupial nutrition*, Cambridge, Cambridge University Press.

Hunsaker, D.I., & Shupe, D. (1977), 'Behavior of new world marsupials', in *The Biology of Marsupials* (ed. D.I. Hunsaker), pp. 279–347, New York, Academic Press.

Hutson, G.D. (1972), 'Grooming behaviour and birth in the dasyurid marsupial *Dasyuroides byrnei*', *Australian Journal of Zoology,* **20**:429–34.

Hutson, G.D. (1975), 'Sequences of prey-catching behaviour in the brush-tailed marsupial rat (*Dasyuroides byrnei*)', *Zeitschrift fŸr Tierpsychologie,* **39**:39–60.

Jarman, P.J. (1983), 'Mating system and sexual dimorphism in large, terrestrial, mammalian herbivores', *Biological Reviews of the Cambridge Philosophical Society,* **58**:485–520.

Jarman, P.J., & Kruuk, H. (1996), 'Phylogeny and social organisation in mammals', in *Comparison of Marsupial and Placental Behaviour* (eds. D.B. Croft, &U. Ganslosser), pp. 80–101, Fuerth, Filander.

Jones, M.E. (1997), 'Character displacement in Australian dasyurid carnivores: Size relationships and prey size patterns', *Ecology,* **78**:2569–87.

Jones, M.E. (1998), 'The function of vigilance in sympatric marsupial carnivores: the eastern quoll and the Tasmanian devil', *Animal Behaviour,* **56**:1279–84.

Jones, M.E. (2003), 'Convergence in ecomorphology and guild structure among marsupial and placental carnivores', in *Predators with pouches: The biology of carnivorous marsupials* (eds. M. Jones, C. Dickman & M. Archer), pp. 285–96, Melbourne, CSIRO Publishing.

Jones, M.E., & Barmuta, L.A. (1998), 'Diet overlap and relative abundance of sympatric dasyurid carnivores: a hypothesis of competition', *Journal of Animal Ecology,* **67**:410–21.

Jones, M.E., & Stoddart, D.M. (1998), 'Reconstruction of the predatory behaviour of the extinct marsupial thylacine', *Journal of Zoology, London***, 246**:239–46.

Kennedy, G.A., Coleman, G J., & Armstrong, S.M. (1990), 'Circadian rhythms of wheel-running in the eastern quoll, *Dasyurus viverrinus* (Marsupialia: Dasyuridae)', *Australian Mammalogy,* **13**:11–16.

Kirsch, J.A.W., & Waller, P.F. (1979), 'Notes on the trapping and behavior of the Caenolestidae (Marsupialia)', *Journal of Mammalogy,* **60**:390–5.

Krebs, J.R., & Davies, N. B. (1997), 'The evolution of behavioural ecology', in *Behavioural Ecology: An Evolutionary Approach, 4th ed.* (eds. J.R. Krebs & N.B. Davies), pp. 3–12, Oxford, Blackwell Scientific.

Laidlaw, W.-S., Hutchings, S., & Newell, G.-R. (1996), 'Home range and movement patterns of *Sminthopsis leucopus* (Marsupialia: Dasyuridae) in coastal dry heathland, Anglesea, Victoria', *Australian Mammalogy,* **19**:1–9.

Lazenby-Cohen, C.K.A., & Cockburn, A. (1988), 'Lek promiscuity in a semelparous mammal, *Antechinus stuartii* (Marsupialia: Dasyuridae)', *Behavioral Ecology and Sociobiology,* **22**:195–202.

Lazenby-Cohen, C.K.A., & Cockburn, A. (1991), 'Social and foraging components of the home range in *Antechinus stuartii* (Dasyuridae: Marsupialia)', *Australian Journal of Ecology,* **16**(3):301–7.

Lee, A. K., & Cockburn, A. (1985), *Evolutionary ecology of marsupials*, Cambridge, Cambridge University Press.

Lissowsky, M. (1996), 'The occurrence of play behaviour in marsupials', in *Comparison of Marsupial and Placental Behaviour* (eds. D.B. Croft & U. Ganslosser), pp. 87–207, Fuerth, Filander.

Lunney, D., & Leary, T. (1989), 'Movement patterns of the white-footed dunnart, *Sminthopsis leucopus* (Marsupialia: Dasyuridae), in a logged, burnt forest on the south coast of New South Wales', *Australian Wildlife Research,* **16**:207–15.

Marshall, L.G. (1978), '*Dromiciops australis*', *Mammalian Species,* **99**:1–5.

Masters, P. (1998), 'The mulgara *Dasycercus cristicauda* (Marsupialia: Dasyuridae) at Uluru National Park, Northern Territory', *Australian Mammalogy,* **20**:403–7.

Meissner, K., & Ganslosser, U. (1985), 'Development of young in the kowari *Dasyuroides byrnei* Spencer, 1896', *Zoo Biology,* **4**:351–9.

Michener, G.R. (1969), 'Notes on the breeding and young of the crest-tailed marsupial mouse *Dasycercus cristicauda*', *Journal of Mammalogy,* **50**:633–5.

Morton, S.R. (1978), 'An ecological study of *Sminthopsis crassicaudata* (Marsupialia: Dasyuridae), 2. Behaviour and social organization', *Australian Wildlife Research,* **5**:163–82.

Moss, G.L., & Croft, D.B. (1988), 'Behavioural mechanisms of microhabitat selection and competition among three species of arid zone dasyurid marsupial', *Australian Journal of Ecology,* **13**:485–93.

Negus, V.E. (1962), *Comparative anatomy and physiology of the larynx,* New York, Wagner.

Oakwood, M. (2000), 'Reproduction and demography of the northern quoll, *Dasyurus hallucatus*, in the lowland savanna of northern Australia', *Australian Journal of Zoology,* **48**:519–39.

O'Reilly, H.M., Armstrong, S.M., & Coleman, G.J. (1984), 'Response to variations in lighting schedules on the circadian activity rhythms of *Sminthopsis macroura froggatti* (Marsupialia: Dasyuridae)', *Australian Mammalogy,* **7**:89–100.

Ostfeld, R.S. (1990), 'The ecology of territoriality in small mammals', *Trends in Ecology & Evolution,* **5**:411–15.

Pellis, S.M., & Officer, R.C.E. (1987), 'An analysis of some predatory behaviour patterns in four species of carnivorous marsupials (Dasyuridae), with comparative notes on the eutherian cat *Felis catus*', *Ethology,* **75**:177–96.

Pellis, S.M., Vanderlely, R. & Nelson, J.E. (1992), 'The roles of vision and vibrissae in the predatory behaviour of northern quolls *Dasyurus hallucatus* (Marsupialia: Dasyuridae)', *Australian Mammalogy,* **15**:55–60.

Pemberton, D., & Renouf, D. (1993), 'A field study of communication and social behaviour of the Tasmanian devil at feeding sites', *Australian Journal of Zoology,* **41**:507–26.

Read, D.G. (1984a), 'Movements and home ranges of three sympatric dasyurids, *Sminthopsis crassicaudata, Planigale gilesi* and *P. tenuirostris* (Marsupialia), in semiarid western New South Wales', *Australian Wildlife Research,* **11**:223–34.

Read, D.G. (1984b), 'Reproduction and breeding season of *Planigale gilesi* and *P. tenuirostris* (Marsupialia: Dasyuridae)', *Australian Mammalogy,* **7**:161–73.

Read, D.G. (1985), 'Development and growth of *Planigale tenuirostris* (Marsupialia: Dasyuridae) in the laboratory', *Australian Mammalogy,* **8**:69–78.

Read, D.G. (1988), 'Weather and trap response of the dasyurid marsupials *Sminthopsis crassicaudata, Planigale gilesi* and *P. tenuirostris*', *Australian Wildlife Research,* **15**:139–48.

Righetti, J., B.J. Fox and D.B. Croft (2000), 'Behavioural mechanisms of competition in small dasyurid marsupials', *Australian Journal of Zoology,* **48**:561–76.

Russell, E.M. (1982), 'Patterns of parental care and parental investment in marsupials', *Biological Reviews,* **57**:423–86.

Russell, E.M. (1984), 'Social behaviour and social organization of marsupials', *Mammalian Review,* **14**:101–54.

Russell, E.M. (1985), 'The metatherians: Order Marsupialia', in *Social Odours in Mammals* (eds. R.E. Brown, & D.W. Macdonald), pp. 45–104, Oxford, Clarendon Press.

Ryser, J. (1992), 'The mating system and male mating success of the Virginia opossum (*Didelphis virginiana*) in Florida', *Journal of Zoology, London,* **228**:127–39.

Ryser, J. (1995), 'Activity, movement and home range of Virginia opossums', *Bulletin of the Florida Museum of Natural History,* **38**:177–94.

Scott, M.P. (1986), 'The timing and synchrony of seasonal breeding in the marsupial, *Antechinus stuartii*: interaction of environmental and social cues', *Journal of Mammalogy,* **67**:551–60.

Scott, M.P. (1987), 'The effect of mating and agonistic experience on adrenal function and mortality of male *Antechinus stuartii* (Marsupialia)', *Journal of Mammalogy,* **68**:479–86.

Serena, M., & Soderquist, T.R. (1989a), 'Nursery dens of *Dasyurus geoffroii* (Marsupialia: Dasyuridae), with notes on nest building behaviour', *Australian Mammalogy,* **12**:35–6.

Serena, M., & Soderquist, T.R. (1989b), 'Spatial organization of a riparian population of the carnivorous marsupial *Dasyurus geoffroii*', *Journal of Zoology, London,* **219**:373–83.

Settle, G.A., & Croft, D.B. (1982a), 'Maternal behaviour of *Antechinus stuartii* (Dasyuridae, Marsupialia) in captivity', in *Carnivorous marsupials, Vol. 2* (ed. M. Archer), pp 365–81, Sydney, Royal Zoological Society of New South Wales.

Settle, G.A., & Croft, D.B. (1982b), 'The development of exploratory behaviour in *Antechinus stuartii* (Dasyuridae, Marsupialia) young in captivity', in *Carnivorous marsupials, Vol. 2* (ed. M. Archer), pp. 383–96, Sydney, Royal Zoological Society of New South Wales.

Shimmin, G.A., Taggart, D.A., & Temple-Smith, P.D. (2000), 'Sperm competition and genetic diversity in the agile antechinus (Dasyuridae: *Antechinus agilis*)', *Journal of Zoology, London,* **252**:343–50.

Soderquist, T. (1994), 'Anti-predator behaviour of the brush-tailed phascogale (*Phascogale tapoatafa*)', *Victorian Naturalist,* **111**:22–4.

Soderquist, T.R. (1995), 'Spatial organization of the arboreal carnivorous marsupial *Phascogale tapoatafa*', *Journal of Zoology, London,* **237**:385–98.

Soderquist, T.R., & Ealey, L. (1994), 'Social interactions and mating strategies of a solitary carnivorous marsupial, *Phascogale tapoatafa*, in the wild', *Wildlife Research,* **21**:527–42.

Soderquist, T., & Lill, A. (1995), 'Natal dispersal and philopatry in the carnivorous marsupial *Phascogale tapoatafa* (Dasyuridae)', *Ethology,* **99**:297–312.

Soderquist, T., & Serena, M. (2000), 'Juvenile behaviour and dispersal of chuditch (*Dasyurus geoffroi*) (Marsupialia: Dasyuridae)', *Australian Journal of Zoology,* **48**:551–60.

Stoddart, D.M. (1976), *Mammalian Odours and Pheromones*, London, Edward Arnold.

Toftegaard, C.L., & Bradley, A.J. (2003), 'Chemical communication in dasyurid marsupials', in *Predators with pouches: The biology of carnivorous marsupials* (eds. M. Jones, C. Dickman & M. Archer), pp. 347–57, Melbourne, CSIRO Publishing.

Traill, B.J., & Coates, T.D. (1993), 'Field observations on the brush-tailed phascogale *Phascogale tapoatafa* (Marsupialia: Dasyuridae)', *Australian Mammalogy,* **16**:61–5.

Trivers, R.L., & Willard, D.E. (1973), 'Natural selection of parental ability to vary the sex ratio of offspring', *Science,* **179**:90–2.

Trupin, G.L., & Fadem, B.H. (1982), 'Sexual behavior of the gray short-tailed opossum (*Monodelphis domestica*)', *Journal of Mammalogy,* **63**:409–14.

Tyndale-Biscoe, C.H., & Renfree, M.B. (1987), *Reproductive physiology of marsupials*, Cambridge, Cambridge University Press.

Van Dyck, S. (1979), 'Behaviour in captive individuals of the dasyurid marsupial *Planigale maculata* (Gould 1851)', *Memoirs of The Queensland Museum,* **19**:413–29.

Vestal, B.M., Lee A.K., & Saxon, M.J. (1986), 'Interactions between adult female and juvenile *Antechinus stuartii* (Marsupialia: Dasyuridae) at the time of juvenile dispersal', *Australian Mammalogy,* **9**:27–33.

Walker, L. (1996), 'Female mate-choice', in *Comparison of Marsupial and Placental Behaviour* (eds. D.B. Croft, & U. Ganslosser), pp. 208–25, Fuerth, Filander.

Wardell-Johnson, G. (1986), 'Use of nest boxes by mardos, *Antechinus flavipes leucogaster*, in regenerating karri forest in south Western Australia', *Australian Wildlife Research,* **13**:407–17.

Werdelin, L. (1987), 'Some observations on *Sarcophilus laniarius* and the evolution of *Sarcophilus*', *Records of the Queen Victoria Museum,* **90**:1–27.

Wolfe, K.M., Robertson, H. & Bencini, R. (2000), 'The mating behaviour of the dibbler, *Parantechinus apicalis*, in captivity', *Australian Journal of Zoology,* **48**:541–50.

Woolley, P.A. (1988), 'Nest location by spool-and-line tracking of dasyurid marsupials in New Guinea', *Journal of Zoology, London,* **218**:689–700.

Woolley, P.A. (1990), 'Mulgaras, *Dasycercus cristicauda* (Marsupialia: Dasyuridae); their burrows, and records of attempts to collect live animals between 1966 and 1979', *Australian Mammalogy,* **13**:61–4.

Woolley, P.A., Raftopoulos, S.A., Coleman, G.J., & Armstrong, S.M. (1991), 'A comparative study of circadian activity patterns of two New Guinean dasyurid marsupials, *Phascolosorex dorsalis* and *Antechninus habbema*', *Australian Journal of Zoology,* **39**:661–71.

Woolnough, A.P., & Carthew, S.M. (1996), 'Selection of prey by size in *Ningaui yvonneae*', *Australian Journal of Zoology,* **44**:319–26.

CHAPTER 23

CHEMICAL COMMUNICATION IN DASYURID MARSUPIALS

C.L. Toftegaard[A,B] *and A.J. Bradley*[B]

[A] Department of Medical Physiology, The Panum Institute, University of Copenhagen, Blegdamsvej 3, 2200 Copenhagen K, Denmark.
[B] School of Biomedical Sciences, Department of Anatomy and Developmental Biology, The University of Queensland, Brisbane, Queensland 4072, Australia. Email: A.Bradley@mailbox.uq.edu.au

This chapter gives a brief overview of studies that have been carried out to describe the way in which dasyurid marsupials communicate by chemical means. This involves the production of chemical substances from both cutaneous scent glands and glands associated with the reproductive tract, the dispersal of these substances by various morphological adaptations, and the detection of these airborne chemicals (pheromones) by the special sensory system, the vomeronasal organ (VNO). Gas chromatography coupled with mass spectroscopy GC-MS is increasingly being applied to identify substances used by animals in chemical communication. While many of the structures used by animals to detect pheromones have been known for many years, it is only now with the availability of sensitive techniques such as functional magnetic resonance imaging (fMRI) that we are able to visualise regional changes in brain activity in temporal sequence in response to these pheromones.

INTRODUCTION

Within the last two decades, the role of olfaction in the mediation and control of reproductive behaviour and in modulation of reproductive physiology in many eutherian species has been extensively studied. In many species, chemosignals or pheromones of male origin may alter the timing of adult oestrus cycles and accelerate the onset of puberty in females (Vandenbergh 1969, Carter et al. 1980, Drickamer 1983), and numerous studies have described the effect of female urinary pheromones on testosterone, LH and gonadotropin levels in males of many species (Clancy et al. 1988, Waring et al. 1996). Although the complexity of social communication in marsupials appears fully equivalent to that observed in ecologically similar eutherian taxa (Gansglosser 1982, Fadem 1986), it is not well understood and remains relatively unexplored. Pheromonal effects on ovulation and oestrus have been recorded in only four marsupial species: the woolly opossum *Caluromys philander* (Perret and M'Barek 1991), the grey short-tailed opossum *Monodelphis domestica* (Fadem 1985, 1987), and the dasyurids *S. crassicaudata* (Smith et al. 1978), and *A. stuartii* (Scott 1986). Fadem (1987) reports that adult female grey short-tailed opossums normally remain in anoestrus when housed in single-sex groups, and only enter oestrus when exposed to male pheromones (Fadem 1989b, Stonerook and Harder 1992). In contrast, Scott (1986) found that isolated female *A. stuartii* ovulate synchronously when placed with grouped females suggestive of an oestrus-stimulated female chemosignal. Based on the findings of Scott (1986) and the observations of Cockburn and Lazenby-Cohen (1992) that male *A. stuartii* may use olfactory cues to locate females during the mating period, further studies are warranted to explore the importance of olfactory cues in reproduction and social communication in this species.

The brains of eutherian and metatherian mammals differ very little in both their structure and function (Johnston 1977). As is the case in eutherian mammals, marsupial sensory systems are well-developed, particularly the olfactory bulb which is prominent in all marsupials (Croft 1982). Furthermore the nasal cavity of marsupials is covered with extensive areas of olfactory epithelium, and may contain a functionally intact vomeronasal organ (Croft 1982) that is believed to be specialised for reception of chemosignals (Jacobson 1811, Meredith and O'Connell 1979, Taniguchi et al. 1992a, 1992b).

Because of several problems associated with the use of the term *pheromone*, the term *semiochemical* is now more commonly used in studies of chemical communication. A pheromone is generally regarded as a substance that elicits a stereotypic response whatever the circumstances associated with its release and reception. Such a substance may have dual effects, acting as an attractant to members of one sex and at the same time eliciting an aggressive response from members of the other sex. A *semiochemical* encompasses any form of chemical communication, from a single compound to a complex mixture. While the chemical may be an odour, it does not have to be. The response that an individual makes to the semiochemical will depend upon the context, physiological state and previous experience of the recipient (Mykytowycz 1972). Semiochemistry is thus much broader than the concept of a pheromone with the former broad term including the latter (Albone 1984).

Taxonomic reassessments

Taxonomic reassessments of genus *Antechinus* are relevant to several earlier studies of dasyurid marsupials in eastern and SE Australia. Studies of *A. stuartii* carried out at Mount Glorious in SE Queensland would now be regarded as studies of *A. subtropicus* because of a recent reassessment (Van Dyck and Crowther 2000) while studies of *Antechinus stuartii* from SE Australia, such as those ecophysiological studies conducted near Warburton (Bradley et al. 1980), would now be regarded as investigations of *A. agilis* in the light of a reassessment by Sumner and Dickman (1998). Behavioural studies carried out on *A. stuartii* in forests near Canberra would be regarded as studies of *A. flavipes.*

In spite of the apparent morphological and genetic differences, most researchers would agree that apart from slight differences in the timing of reproduction all these species share a common life history strategy. To avoid confusion in references to *Antechinus* in this chapter, the new species names will be included in brackets after the species names used in the original citation.

Scent-marking behaviour

Scent marking and subsequent olfactory investigation of chemosignals are fundamental components of social interactions within the majority of mammalian species. In mammals, a variety of scent marking mechanisms are found ranging from urine, faeces, vaginal secretions and saliva to specialised glands such as the temporal gland of elephants, supraorbital and inter-digital glands of many ungulates, anal glands of carnivores, and the sternal gland of the marsupial sugar glider (Strahlendorff 1987, Stoddart et al. 1994). Research into the chemistry of mammalian pheromones, which builds on the success of chemical investigations of insect pheromones, has mainly focused on social conditions which influence sexual maturation and reproduction in rodents (Drickamer 1984, Jemiolo et al. 1994), but is now increasingly employed in a wide range of other species (e.g tigers, beavers, moles, badgers and primates).

Within the Metatheria, little is known about the chemistry of chemosignals although the involvement in reproduction and social organisation is well documented (Fadem 1989a, Fadem et al. 1989). Biggins (1979, 1984) reported on the importance of chemical communication to the brushtail possum *Trichosorus vulpecula.* Woolhouse and co-workers (1994) have described a possible paracloacal and sternal gland pheromone in the *Trichosorus vulpecula* which may act as an intraspecific attractant. Furthermore, partially racemic compounds have been identified as possum urinary metabolites (Carman and Klika 1992), the function of which, however, is unknown. To our knowledge, only one study has identified chemical compounds with possible pheromonal effects in a dasyurid marsupial. In *Antechinus subtropicus* from Mount Glorious in SE Queensland males possess both a sternal gland and a cluster of paracloacal glands which are believed to produce chemical signals for intraspecific communication (Toftegaard 1999). Females also possess paracloacal glands, the function of which is unknown, but they may also serve in chemical communication.

Using a solid phase microextraction technique, urinary volatiles from male, female and castrated *A. subtropicus* were extracted and analysed using GC-MS. Fourteen volatile compounds were identified, some of which were gender-specific (Table 1).

The GC-MS profile of the males were distinguished by two pyrazine compounds (2,6-dimethylpyrazine and 2-ethenyl-6-methylpyrazine) and a series of methyl ketones which were not detected in the profiles of females, nor in the castrate (Toftegaard, Moore and Bradley 1999). Urine from females, however, contains several aldehydes not present in the profile of males. The apparent sexual dimorphism in urinary constituents such as pyrazine derivatives in *A. subtropicus* is of particular interest because such compounds

Table 1 Structure of volatile compounds identified in urine of *Antechinus stuartii* using GC-MS.

Urinary volatile	Mol. Wt. (Da)	Elemental formula	Present in males	Present in females	Present in castrates
2,6-dimethylpyrazine	108.14	$C_6H_8N_2$	X		
2-ethenyl-6-methylpyrazine	120.15	$C_7H_8N_2$	X		
2-heptanone	114.19	$C_7H_{14}O$	X	X	
2-octanone	128.22	$C_8H_{16}O$	X		
2-nonanone	142.24	$C_9H_{18}O$	X		
2-decanone	156.27	$C_{10}H_{20}O$	X		
2-hexanone	100.16	$C_6H_{12}O$	X		
2-undecanone	170.29	$C_{11}H_{22}O$	X		
Nonanal	142.24	$C_9H_{18}O$		X	
Decanal	156.27	$C_{10}H_{20}O$		X	
Undecanal	170.30	$C_{11}H_{22}O$		X	
Benzaldehyde	106.13	C_7H_6O			
Decanol	158.29	$C_{10}H_{22}O$			
2,4-dithiapentane	108.23	$C_3H_8S_2$	X		X
N-butyl benzene sulfonamide	213.30	$C_{10}H_{15}NO_2S$	X	X	X
limonene	136.24	$C_{10}H_{16}$	X	X	

have also been reported in tree shrews *Tupaia belangeri* (Strahlendorff 1987) where male-specific 2,5-dimethylpyrazine was identified. The presence of this compound was interpreted to be directly related to the activation of male chinning behaviour. The identification of urinary pyrazines as key regulatory components in reproduction has shown that long-term exposure of female mice to male-specific 2,5-dimethylpyrazine may inhibit their overall reproductive fitness (Jemiolo and Novotny 1993). Pyrazines have been found in the adrenal gland of mice (Novotny et al. 1986), and adrenalectomy, but not ovariectomy, eliminates the biological activity of excreted urine (Drickamer and McIntosh 1980). Finally, it has been proposed that tetraethylpyrazine and 2,5-dimethylpyrazine may modulate the gonadotropin releasing hormone and/or gonadotropic hormone release from the hypothalamic-pituitary axis in juvenile rats (Yamada et al. 1989). The role of these sexually dimorphic urinary constituents in social interactions and during the life history of *A. subtropicus* is not yet fully understood. It may be that male-specific pyrazines serve as female attractants during the breeding period or conspecific male aggression signals. Further studies are needed to address the involvement of urinary pheromones during social interactions in *A. subtropicus.*

Olfactory stimuli in the form of pheromones, semiochemicals or 'social odours', may represent specific metabolites in the animals' waste products such as urine or faeces, or may be released from specialised glandular regions on the body. Chemosignals may be actively deposited by rubbing the glandular area against the substrate known as scent marking, or passively through indiscriminate liberation. Scent glands have been described in 15 of the 19 orders of mammals (Thiessen and Rice 1976). Within the Metatheria, scent glands have been identified in the sugar glider *Petaurus breviceps* (Stoddart and Bradley 1991), *Phascogale calura* (Bradley 1997), and grey short-tailed opossum *Monodelphis domestica* (Fadem and Schwartz 1986, Fadem et al. 1989, Fadem 1990). Several different forms of scent marking behaviour have been reported in dasyurids (Table 2).

Most species studied are known to use cloacal dragging to deposit chemosignals. Sternal and chin rubbing are also frequently used marking mechanisms.

Sexual dimorphism has been reported in marking behaviour and physiology. During the mating period, for example, male *A. subtropicus* secrete copiously from their cutaneous sternal glands (Bradley pers. obs.), and have been found to frequently scent mark their nesting area using cloacal rubbing (Braithwaite 1974). Although females of this species also possess cloacal glands, they do not appear to engage in marking behaviour. In *Planigale maculata*, only males use sternal marking (Van Dyck 1979), and in *Sminthopsis crassicaudata*, only males use cloacal marking in female–male encounters (Ewer 1968a).

Morphological adaptations to pheromone dispersal

The hair covering the surface of specialised integumentary glands located around the body of various mammals are known to serve special functions. They may occur in the form of well-developed tufts as in the case of the flank gland of the shrew (Balakrishnan 1987), and the tarsal and metatarsal glands of

Table 2 Modes of chemical deposition used by dasyurids in olfactory communication. (+) indicates observed; (-) shown not to occur.

Species	Urine dribble	Cloacal drag	Chin rub	Cheek rub	Sternal rub	Reference
Planigale maculata		+	+			Van Dyck 1979
Planigale tenuirostris		+			+	Andrew and Settle 1982
Ningaui spp.		+	+			Fanning 1982
Antechinus flavipes	+	+			+	Ewer 1968
Antechinus stuartii		+	+		+	Rigby 1972, Braithwaite 1974
Sminthopsis crassicaudata	-	+			+	Ewer 1968a+b
Sminthopsis macroura	+	+				Van Dyck 1979
Antechinomys laniger		+	+			Happold 1972, Eisenberg and Golani 1977
Dasycercus cristicauda		+				Ewer 1968b, Sorenson 1970
Dasyuroides byrnei	+	+	+			Sorenson 1970, Aslin 1974
Dasyurus viverrinus	+				+	Eisenberg and Golani 1977
Sarcophilus harrisii	+	+			+	Ewer 1968b, Eisenberg and Golani 1977

deer (Müller-Schwarze et al. 1977) functioning as visual stimuli during intraspecific interactions, or they may be structurally different from hairs of other body regions and modified for holding and releasing materials of olfactory relevance. Such structural modifications have been reported in the Crested rat, *Lophiomys imhausi* (Stoddart 1979) and some species of pteropodid and molossid bats (Hickey and Fenton 1987). The term 'osmetrichia' has been proposed for structurally specialised mammalian scent hair (Müller-Schwarze et al. 1977) that is capable of retaining material containing olfactory information.

As far as we are aware, the presence on the hair surface of specialisations for holding material that may provide chemical signals in social communication has not been described in dasyurid marsupials. Bradley (1997) has discussed the apparent importance of the sternal gland and its secretions in *Phascogale calura* in which the sternal gland of males becomes very active when the plasma testosterone concentration rises during the breeding period. In ovariectomised females of another marsupial species the grey short-tailed opossum *Monodelphis domestica* the development of suprasternal glands and chest marking may be stimulated by testosterone administration but not by oestradiol (Fadem 1990). Similarly in castrated males testosterone, but not oestradiol, stimulated chest marking (Fadem et al. 1989a, b). This androgen dependence of cutaneous scent gland function has also been described (Stoddart and Bradley 1991; Stoddart, Bradley and Mallick 1994) in the sugar glider *Petaurus breviceps*.

The production of sternal gland secretion in *A. stuartii* (*A. agilis*), but also in *A. subtropicus* in SE Queensland, is at a maximum at the time of breeding coinciding with male peak androgen concentration (Bradley et al. 1980) however individual differences are apparent. In light of this relationship between onset of breeding and activity of the gland and the apparent morphological difference between hair covering the sternal gland and hair from other body regions, the hair of the sternal gland was studied using the scanning electron microscope (SEM) (Toftegaard and Bradley 1999). In this study comparisons were made between the surface ultrastructure of sternal gland hair and hair from the dorsum of sexually intact males (n = 3), castrated males (n = 2) and females (n = 3) to determine whether hairs that overlie the scent producing gland would show surface specialisation to promote scent retention. Morphological differences were evident between the sternal gland hairs and back hairs of the intact males that had active sternal glands. The specialised sternal gland hairs were only about one-third of the length of the control hairs from the dorsal body region in contrast with the relatively flat surface of the long dorsal hair created by the smooth, even cuticular scales. The scales of the sternal gland hairs project outward from the cortex, creating angled chambers between the scales.

At magnifications of x7000 numerous ridges and grooves were found lining the distal edges of the scales (Fig. 1); however, only hairs from the sternal gland of the two actively secreting males had large numbers of these grooves. Hairs taken from both the sternal and dorsal areas of females lacked visible grooves and no comparable grooves were identified on the sternal hairs of the castrate males (hairs examined 6 months after castration) and males which were not actively secreting. The scales from hair of actively secreting males appeared very efficient in retaining lipid material. Some sternal hairs were still covered with sebum after a period of 72 hours in solvent, thus making examinations of grooves difficult. The general appearance of the hair shaft changed from root to tip, the scales becoming elongate and narrowly conical towards the tip. This pattern was identical for all hair samples examined.

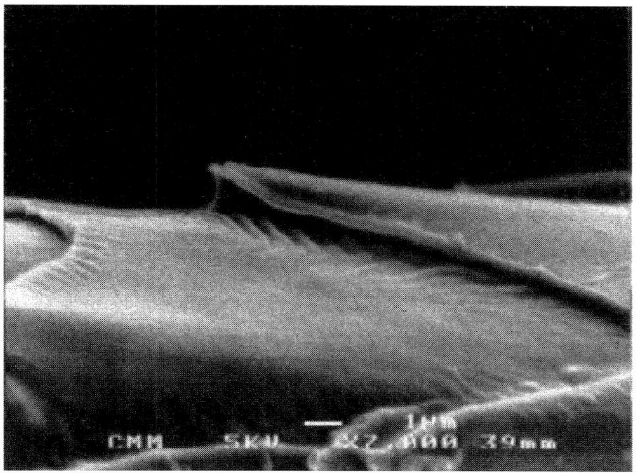

Figure 1 Scanning electron micrograph of a sternal gland hair from an actively secreting male showing the pattern of arrangement of cuticular angular scale 'chambers' displaying distal ridge-like grooves. x7000 (white scale bar = 1 μm).

The present study revealed that the specialised fur overlying the male sternal gland surface exhibits structural modifications in the form of rough keratinised chambers which act as reservoirs for the glandular secretions. Structurally these modifications resemble the sebum-storing flank gland hairs of the musk shrew, *Suncus murinus viridescens* (Balakrishnan 1987) and common shrew, *Sorex araneus L.* (Kapischke and Mühle 1988), which have been recognised as important instruments in scent-marking in these species. However, this specialisation is found only in the sternal gland area of the male *A. stuartii*, in contrast to the musk shrew in which both sexes possess the modification.

Pheromones have been chemically identified in the ventral gland sebum of the jird *Meriones tristrami* (Kagen et al. 1983) and *Meriones unguiculatus* (Thiessen et al. 1974). From ongoing chemical and behavioural studies in our lab we suggest that male *A. stuartii* may discharge a continuous, but highly volatile scent from the sternal gland during the breeding season for which the glandular sebum acts as a carrier. The presence of these modified scent gland hairs as special sebum storing chambers may facilitate the continuous availability of the volatile pheromone as well as delay its oxidation and degradation upon exposure to air. Since the chemical substances of olfactory relevance are of low molecular weight (C. Moore pers. comm.), it is proposed that the lipid secretions produced by the sternal glands and retained in the modified hair chambers act as a medium for olfactory information required for intraspecific communication during the breeding season. While cloacal marking is common during behavioural encounters (Braithwaite 1974) male *A. stuartii* do not appear to use their sternal gland for scent-marking. It is proposed that the glandular sebum contains a pheromone that may act as a status signal during the breeding season, which occurs in many mammal species (Boero 1995). We propose that the functional significance of the sternal hair modifications is to stabilise and delay oxidation of this pheromone, an effect that has been suggested by Allen (1975; 1982) to occur with paracloacal secretions in *Trichosurus vulpecula*.

Quite clearly the hair overlying the sternal gland in *Antechinus stuartii* possesses surface features that would appear to assist both in the storage and propagation of odour.

STERNAL GLAND MORPHOLOGY AND CONTROL

All adult *A. stuartii* males possess a sparsely haired patch of skin approximately 3 mm in diameter in the sternal region which is covered by a thin oily film. This is particularly prominent during the breeding period in September. Histological examination of biopsies (Bradley and Stoddart 1991) taken from the sternal gland (Fig. 2a) revealed that this organ consists of two distinct layers of glandular tissue (Fig. 2a), a deep layer of apocrine tissue (Fig. 2b) overlain by a layer of holocrine sebaceous tissue (Fig. 2c). The sebaceous secretion appears colourless when produced by the gland; however, with prolonged oxidation, an orange staining of the fur surrounding the glandular tissue is observed.

The sternal gland of adult male *A. stuartii* contains both apocrine and sebaceous glandular tissues, the latter of which may be under androgen control (Toftegaard and Bradley 1999). This is in agreement with a study by Dixon (1976) on the Greater Galago *Galago crassicaudatus crassicaudatus*, in which it was found that low levels of testosterone are insufficient to maintain sebaceous glands and result in atrophy of the entire sternal gland. The low plasma testosterone levels observed in the castrate male *A. stuartii* are presumably caused by the continuous secretion of androgens from the zona reticularis of the adrenal cortex.

Sebaceous gland secretions may act as a vehicle for the volatile substances produced by the apocrine glands (Mykytowycz 1972). This type of cooperative action has been suggested for civetone, the crude secretion from the glands of the civet cat. This substance is weel known for its use as a fixative by perfumers to extend the valuable components of the perfume. It acts as a *fixative*, binding to the molecules with high volatility and, in so doing, slowing their release rate. This ensures that the more ephemeral odours produced by the animal persist in the scent for much longer than would otherwise be expected (Whitten 1969).

Experiments on different mammalian species using castration, and castration followed by subsequent hormone replacement, have shown that androgens greatly increase sebaceous gland size, whereas oestrogen may reduce them (Dryden and Conway 1967, Jannett 1978, Bradley and Stoddart 1993, Helder and Freymuller 1995, Romo et al. 1996). The same has been demonstrated in paracloacal glands (Bradley and Stoddart 1993). In the marsupial sugar glider *Petaurus breviceps*, plasma androgens

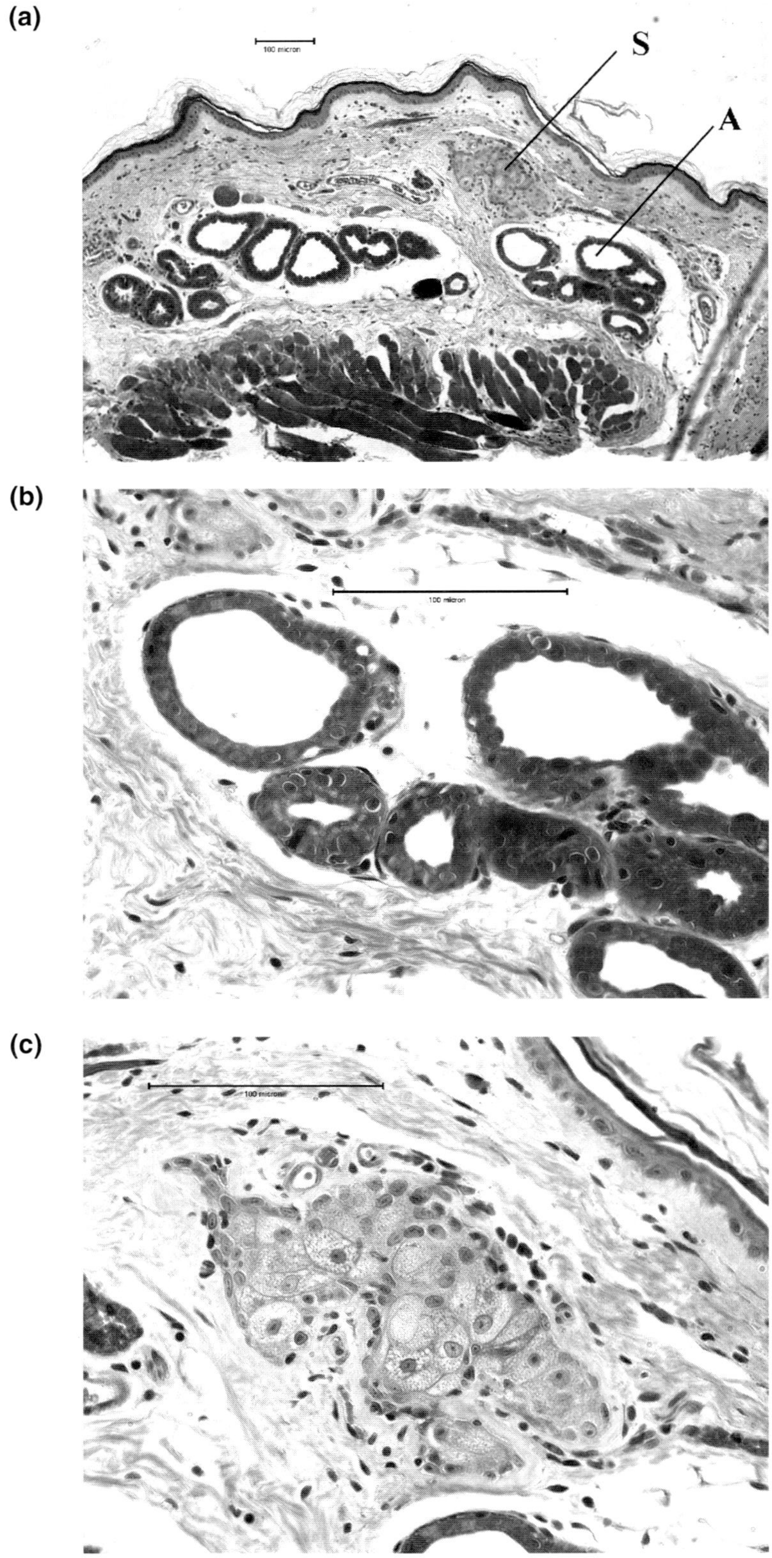

Figure 2 Longitudinal section through a gland biopsy core sample taken from a male *A. substropious* (a) showing the two distinct glandular zones (A = apocrine tissue zone, S = sebaceous tissue zone). (b) apocrine tissue. (c) sebaceous tissue. [Scale bars = 100 μm].

have been found to play an essential role in the development and activity of cutaneous scent-producing glands (Stoddart and Bradley 1991, Mallick et al. 1994, Stoddart et al. 1994). Although such correlations between androgens and gland activity have yet to be described in dasyurids, the production of sternal gland secretions in male *A. stuartii* (*A. agilis*) is at a maximum with the onset of the mating period, which coincides with a peak androgen concentration in the blood of males (Bradley et al. 1980).

Androgen replacement treatment increases sebaceous gland size, mitotic activity and lipogenesis (Ebling et al. 1971, Stoddart and Bradley 1991) in many mammalian species. Although the apocrine tissue within the side glands of the shrew *Suncus murinus* (Dryden and Conway 1967) and the skin glands of the hare and rabbit have been found to respond in the same way to the withdrawal and replacement of testosterone, this was not observed in male *A. subtropicus*. In this species only apocrine cell height was observed to increase following testosterone implantation, whereas apocrine cell nuclear diameter was found to decrease as the plasma testosterone concentration increased.

Knowledge of the specific behavioural contexts in which the sternal gland of *A. subtropicus* is used is at present limited. It is known, however, that specialised osmetrichia are found (Toftegaard and Bradley 1999) overlying the sternal gland that may prevent rapid oxidation of the gland exudate thus prolonging the release of a chemical signal. From studies on the sugar glider, *Petaurus breviceps* (Stoddart and Bradley 1991), it is known that the frontal gland secretions may contribute to the whole-body odour which, when released by dominant males, can effect the elevation of plasma cortisol and catecholamine levels as well as heart rate in subordinates. Unlike sugar gliders, *A. stuartii*, have no distinct hierarchical system; however, social communication during lekking displays at the time of breeding may be facilitated by whole-body odour in which sternal gland secretions may play a vital part. The fact that testosterone levels increase prior to the onset of the breeding season (Cockburn and Lazenby-Cohen 1992) supports the theory that sternal gland secretions may be involved in the triggering of sexual behaviour.

PERCEPTION OF CHEMOSIGNALS – OLFACTORY SYSTEMS

Two systems are currently recognised as perceiving chemosensory input: the main olfactory system and the accessory olfactory system. The main olfactory system is suggested to serve a general function as a 'molecular' analyser for environmental chemicals without predetermined meaning. In this capacity, it would serve a role in the association between odours and contexts. The accessory olfactory system, which consists of chemoreceptor neurons in the vomeronasal organ, and their central pathway through the accessory olfactory bulb, amygdala and basal forebrain, has been implicated in pheromone detection and chemical communication in several species (Pfeiffer and Johnston 1993, Jackson and Harder 1996). Although used extensively in the literature, the proposed segregation of function between the main and accessory olfactory system is not absolute (Meredith 1991).

First described by Jacobson (1811), the vomeronasal organ (VNO) has been shown to be of particular importance in the mediation of chemical signals such as pheromones, and play a crucial role in biologically significant functions related to feeding, and more specifically breeding (Johnston and Rasmussen 1984, Johnston 1985, Pfeiffer and Johnston 1993). VNOs are two paired crescent-shaped tubular chemosensory organs that run caudally in the upper palate at the base of the nasal septum. Depending on the particular species, they open either to the nasal cavity, the mouth or both (Schilling et al. 1990). The VNO is encased by either a bony or cartilaginous capsule, and contains a specialised neuroepithelium that differs from the main olfactory epithelium in that the apical portions of the bipolar receptor cells are lined with microvilli instead of cilia. The VNO is derived embryologically from the olfactory placode, and vomeronasal sensory neurons, like olfactory neurons, regenerate continuously throughout life (Liman 1996). The olfactory nerves of the main olfactory system amalgamate in the main olfactory bulb, from which projections are sent to the primary olfactory cortex, the nucleus of the lateral olfactory tract, the olfactory tubercle, and the periamygdaloid region. In contrast, the vomeronasal nerves containing axons of bipolar neurons project to the accessory olfactory bulb, from which secondary neurons extend into the bed nucleus of the stria terminalis, the medial amygdala, and the hypothalamus, thereby facilitating direct pheromonal influences on reproductive physiology and behaviour (Wysocki and Meredith 1991, Kennedy and Anholt 1997). Relatively few studies have described the anatomy and function of the accessory olfactory system in metatherian species (Kratzing 1978, Shammah-Lagnado and Negrao 1981, Kratzing 1982a, b, 1984; Jackson and Harder 1996), and information regarding VNO anatomy and functional significance in chemical communication in dasyurids is at present lacking.

NEW DIRECTIONS IN OLFACTORY RESEARCH

Our knowledge of the central neural mechanisms of marsupial olfaction is limited because of the lack of objective methods to evaluate it. Positron emission tomography (PET) is one useful method for investigating brain structure-function relationships. However, since PET uses radioactive substances, subjects may suffer from radiation overexposure. The advent of magnetic resonance imaging (MRI) has enabled researchers to use a non-radioactive high-resolution *in vivo* method to study the auditory, visual, motor and sensory regions of the mammalian brain. Recent studies of human olfactory function (Koizuka et al. 1994, Yousem et al. 1997, Sobel et al. 1998) have demonstrated that it is possible to use MRI to detect alterations of brain activation. This technique of functional MRI (fMRI) is particularly

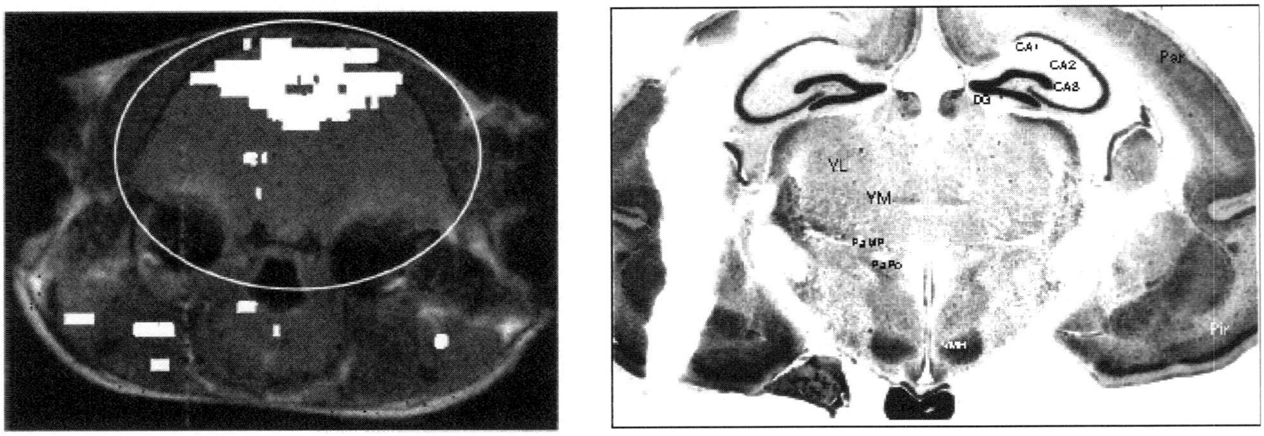

Figure 3 White pixels indicate signal activation in coronal section through the brain of a female *A. subtropicus* during exposure to male urine. The brain tissue is circled in white. Tissue lying outside this circle is bone and soft tissue. Activation is evident in the frontal (area 1–3) cortex, hippocampal fissures (CA1–3), dentate gyrus (DG), anterior paraventricular, ventrolateral (VL) and ventromedial thalamic (VM) nuclei, and paraventricular (PaMP) nucleus of the hypothalamus. A corresponding coronal histological section through the brain of *Antechinus subtropicus* is indicated on the right-hand side. Note that section on the right-hand side contains only brain tissue and is equivalent to the tissue outlined on the left-hand side.

useful because it is possible to tune the instrument to detect different target molecules. General brain activity can be inferred by looking at differences in the number of oxygen molecules with a defined tissue before and after a particular experimental treatment. Regional changes in vascular perfusion may be indicated within the hippocampus and hypothalamus (Fig. 3) of an isofurane anaesthetised marsupial mouse *Antechinus subtropicus* (Toftegaard et al. 2002).

This picture of altered brain activity was obtained by computing the difference between images collected approximately 70 seconds apart before and after the presentation of a pheromonal stimulus (urinary odour) to the experimental subject. Concern that global vascular change may occur in the brain of an anaesthetised animal and obscure the effect ascribed to the pheromone may be answered by pointing out that pixels indicated in red highlight differences between two otherwise identical brain images.

This technique shows particular promise when it is realised that fMRI can be arranged to target specific and discrete neurotransmitter molecules such as glutamate. This makes it possible to monitor animals during longitudinal experimental studies in which subjects may be investigated using a non-invasive, non-destructive technique. In addition to the intrinsic scientific merit of monitoring progressive change in individuals, the use of fMRI also enables experimental designs using fewer animals. The benefit for animal ethics in this conservative approach is obvious.

Future directions

The use of functional MRI promises to revolutionise the study of chemical communication in mammals. In combination with standard methods for describing cytoarchitecture, such as tract tracing, fMRI allows researchers to describe the temporal sequence of changes that occur in response to the presentation of an odour of biological significance. In *A. subtropicus* 2,4 dimethylpyrazine, that appears in the urine of intact males, has been shown to cause a significant physiological response and it is likely that other compounds may also cause physiological change. Undoubtedly different species will be shown to produce their own unique chemical signatures that can be detected using GC-MS. In many cases these semiochemicals are readily available from commercial sources. When compounds are not commercially available, research arms of some large companies can often supply small quantities for testing (e.g. pyrazines sourced from the tobacco industry). The future seems very promising for studies of the role of chemical communication in behaviour in marsupials. Studies carried out using marsupial models in the US, New Zealand and Australia promise to significantly advance our knowledge in this area of biology during the next few years and the intrinsic and practical benefits will certainly become more apparent both for metatheria and eutheria. While the dasyurid marsupials provide interesting models for the role of pheromones in the pituitary-adrenocortical axis and pathological change, studies of Petauridae and Phalangeridae, in which clear social structures exist, promise to provide important new information in the future.

Acknowledgements

We thank the Zoological Society of London for permission to use the scanning electron micrograph of osmetrichia.

REFERENCES

Albone, E.S., Eglinton, G., Walker, J.M., & Ware, G.C. (1974), 'The anal sac secretion of the red fox (*Vulpes vulpes*); its chemistry and microbiology. A comparison with the anal sac secretion of the lion (*Panthera leo*)', *Life Sci*, **14**:387–400.

Allen, N.T. (1975), 'A new mechanism for scent dispersal by the anal glands of the Possum, *Trichosurus vulpecula*', unpublished Honours thesis, Zoology Department, University of Western Australia, Perth.

Allen, N.T. (1982), 'A study of the hormonal control, chemical constituents, and functional significance of the paracloacal glands in *Trichosurus vulpecula*', unpublished MSc thesis, Zoology Department, University of Western Australia.

Balakrishnan, M. (1987), 'Sebum-storing flank gland hairs of the musk shrew, *Suncus murinus viridescens*', *J Zool Lond*, **213**:213–20.

Biggins, J.G. (1979), 'Olfactory communication in the brush-tailed possum *Trichosurus vulpecula* Kerr, 1792 (Marsupialia: Phalangeridae)', unpublished PhD thesis, Department of Zoology, Monash University, Melbourne.

Biggins, J.G. (1984), 'Communications in possums: a review', in *Possums and Gliders* (eds. A.P. Smith, & I.D. Hume), pp. 35–57, Australian Mammal Society & Surrey Beatty & Sons, Sydney.

Boero, D.L. (1995), 'Scent-deposition behaviour in the alpine marmot (*Marmota marmota* L): Its role in territorial defence and social communication', *Ethology*, **100**:26–38.

Bolliger, A., & Whitten, W.K. (1948), 'The paracloacal (anal) glands of *Trichosurus vulpecula*', *Proceedings of the Royal Society of New South Wales*, **82**:36–43.

Bradley, A.J. (1997), 'Reproduction and life-history in the Red-tailed phascogale, *Phascogale calura* (Marsupialia: Dasyuridae): The adaptive-stress senescence hypothesis', *J Zool Lond*, **241**:739–55.

Bradley, A.J., McDonald, I.R., & Lee, A.K. (1980), 'Stress and mortality in a small marsupial (*Antechinus stuartii*, Macleay)', *Gen & Comp Endocrinol*, **40**:188–200.

Bradley, A.J & Stoddart, D.M. (1991), 'A non-destructive small scale biopsy method for obtaining repetitive tissue samples from skin glands in a marsupial', *Aust Mammal*, **14**:151–3.

Bradley, A.J., & Stoddart, D.M. (1993), 'The dorsal paracloacal gland and its relationship with seasonal changes in cutaneous scent gland morphology and plasma androgen in the marsupial sugar glider (*Petaurus breviceps*; Marsupialia: Petauridae)', *J Zool Lond*, **229**:331–46.

Braithwaite, R.W. (1974), 'Behavioural changes associated with the population cycle of *Antechinus stuartii* (Marsupialia)', *Aust J Zool*, **22**:45–62.

Carman, R.M., & Klika, K.D. (1992), 'Partially racemic compounds as Brushtail Possum urinary metabolites', *Aust J Chem*, **45**:651–7.

Carter, C.S., Getz, L.L., Gavish, L., McDermott, J.L., & Arnold, P. (1980) 'Male-related pheromones and the activation of female reproduction in the prairie vole *Microtus ochrogaster*', *Biol Reprod*, **23**:1038–45.

Clancy, A.N., Singer, A.G., Macrides, F., Bronson, F.H., & Agosta, W.C. (1988), 'Experiential and endocrine dependence of gonadotropin responses in male mice to conspecific urine', *Biol Reprod*, **38**:183–91.

Cockburn, A., & Lazenby-Cohen, K.A. (1992), 'Use of nest trees by *Antechinus stuartii*, a semelparous lekking marsupial', *J Zool Lond*, **226**:657–680.

Dixon, A.F. (1976), 'Effects of testosterone on the sternal cutaneous glands and genitalia of the male Greater Galago (*Galago crassicaudatus crassicaudatus*)', *Folia Primatol*, **26**:207–13.

Drickamer, L.C. (1983), 'Male acceleration of puberty in female mice (*Mus musculus*)', *J Comp Psychol*, **97**:191–200.

Drickamer, L.C. (1984), 'Urinary chemosignals from mice (*Mus musculus*): Acceleration and delay of puberty in related and unrelated young females', *J Comp Psychol*, **89**:414–20.

Drickamer, L.C., & McIntosh, T.K. (1980), 'Effects of adrenalectomy on the presence of a maturation–delaying pheromone in the urine of female mice', *Horm Behav*, **14**:146–52.

Dryden, G.L., & Conway, C.H. (1967), 'The origin and hormonal control of scent production in *Suncus murinus*', *J Mammal*, **18**:420–8.

Ebling, F.J., Ebling, E., McCafferty, M., & Skinner, J. (1971), 'The response of the sebaceous glands of the hypophysectomized castrated male rat to 5 alpha-dihydrotestosterone, androstenedione, dehydroepiaandrosterone and androsterone', *J Endocrinol*, **51**:181–90.

Ewer, R.F. (1968a), 'A preliminary survey of the behaviour in captivity of the dasyurid marsupial, *Sminthopsis crassicaudata* (Gould'), *Z.f. Tier-psychol*, **25**:319–65.

Fadem, B.H. (1985), 'Evidence for the activation of female reproduction by males in a marsupial, the gray short–tailed opossum (*Monodelphis domestica*)', *Biol Reprod*, **33**:112–16.

Fadem, B.H. (1986), 'Chemical communication in gray short-tailed opossums (*Monodelphis domestica*) with comparisons to other marsupials and with reference to monotremes', in *Chemical Signals in Vertebrates IV. Ecology, Evolution and Comparative Biology* (eds. D. Duvall, D. Müller-Schwarze, & R.M. Silverstein), pp. 587–607, Plenum Press, New York.

Fadem, B.H. (1987), 'Activation of estrus by pheromones in a marsupial: Stimulus control and endocrine factors', *Biol Reprod*, **36**:328–32.

Fadem, B.H. (1989a), 'The effects of pheromonal stimuli on estrus and peripheral plasma estradiol in female gray short-tailed opossums (*Monodelphis domestica*)', *Biol Reprod*, **41**:213–17.

Fadem, B.H. (1989b), 'Sex differences in aggressive behavior in a marsupial, the Gray Short–Tailed opossum (*Monodelphis domestica*)', *Aggres Behav*, **15**:435–41.

Fadem, B. (1990), 'The effects of gonadal hormones on scent marking and related behaviour and morphology in female gray short-tailed opossums (*Monodelphis domestica*)', *Horm Behav*, **24**:459–69.

Fadem, B.H., & Schwartz, R.A. (1986), 'A sexually dimorphic suprest-ernal scent gland in gray short-tailed opossums (*Monodelphis domestica*)', *J Mammal*, **67**:205–8.

Fadem, B.H., Erianne, G.S., & Karen, L.M. (1989), 'The hormonal control of scent marking and precopulatory behavior in male gray short-tailed opossums (*Monodelphis domestica*)', *Hormones & Behavior*, **23**:381–92.

Gansglosser, V.U. (1982), 'Social structure and communication in marsupials', *Zool Anz*, **209**:294.

Helder, J., & Freymuller, E. (1995), 'A morphological and ultrastructural study of the paracloacal (scent) glands of the marsupial *Metachirus nudicaudatus* Geoffroy, 1803', *Acta Anat*, **153**:31–8.

Hickey, M.B., & Fenton, M.B. (1987), 'Scent dispersing hairs (Osmetrichia) in some pteropodidae and molossidae (Chiroptera)', *J Mammal*, **68**:381–4.

Jackson, L.M., & Harder, J.D. (1996), 'Vomeronasal organ removal blocks pheromonal induction of estrus in Gray Short-Tailed opossums (*Monodelphis domestica*)', *Biol Reprod*, **54**:506–12.

Jacobson, L. (1811), 'Description anatomique d'un organe observè dans les mallifieres', *Ann Mus Hist Nat (Paris)*, **18**:412–24.

Jannett, F.J. (1978), 'Dosage response of the vesicular, preputial, anal, and hip glands of the male vole, *Microtus montanus*, to testosterone propionate', *J Mammal*, **59**:772–79.

Jemiolo, B., & Novotny, M. (1993), 'The long-term effect of a urinary chemosignal on reproductive fitness in female mice', *Biol Reprod*, **48**:926–9.

Jemiolo, B., Gubernick, D.J., Yoder, M.C., & Novotny, M. (1994), 'Chemical characterization of urinary volatile compounds of *Peromyscus californicus*, a monogamous biparental rodent', *J Chem Ecol*, **20**:2489–500.

Johnston, R.E. (1977), 'The causation of two scent marking behaviors in female golden hamsters (*Mesocricetus auratus*)', *Anim Behav*, **25**:317–27.

Johnston, R.E. (1985), 'Olfactory and vomeronasal mechanisms of communication', in *Taste, Olfaction and the Central Nervous System* (ed. D.W. Pfaff), New York, Rockefeller University Press, 346 p.

Johnston, R.E., & Rasmussen, K. (1984), 'Individual recognition of female hamsters by males: Role of chemical cues and of the olfactory and vomeronasal systems', *Physiol Behav*, **33**:95–104.

Kagan, R., Ikan, R., & Haber, O. (1983), 'Characterization of a sex pheromone in the jird (*Meriones tristrami*)', *J Chem Ecol*, **9**:775–83.

Kapischke, H, & Mühle, H. (1988), 'Zur morphologie der seitendrüsenhaare bei der waldspitzmaus (*Sorex araneus*) (Mammalia, Insectivora, Soricidae)', *Zool Abhandl*, **44**:71–4.

Kennedy, S.W., & Anholt, R.R.H. (1997), 'Pheromone regulated production of inositol-(1, 4, 5)-triphosphate in the mammalian vomeronasal organ', *Endocrinol*, **138**:3497–504.

Koizuka, I., Yano, H., Nagahara, M., Mochizuki, R., Seo, R., Shimada, K., Kubo, T., & Nogawa, T. (1994), 'Functional imaging of the human olfactory cortex by magnetic resonance imaging', *Otalaryngol*, **56**:273–75.

Kratzing, J.E. (1978), 'The olfactory apparatus of the bandicoot (*Isoodon macrourus*): Fine structure and presence of a septal olfactory organ', *J Anat*, **125**:601–13.

Kratzing, J.E. (1982), 'The anatomy of the rostral nasal cavity and vomeronasal organ in *Tarsipes rostratus* (Marsupialia: Tarsipedidae)', *Aust Mammal*, **5**:211–19.

Kratzing, J.E. (1984), 'The anatomy and histology of the nasal cavity of the koala (*Phascolarartos cinereus*)', J Anat, **138**:55–65.

Liman, E.R. (1996), 'Pheromone transduction in the vomeronasal organ', *Curr Opin Neurobiol*, **6**:487–93.

Mallick, J., Stoddart, D.M., Jones, I., & Bradley, A.J. (1994), 'Behavioral and endocrinological correlates of social status in the male sugar glider (*Petaurus breviceps* Marsupialia: Petauridae)', *Physiol Behav*, **55**:1131–4.

Meredith, M. (1991), 'Sensory processing in the main and accessory olfactory systems: Comparisons and contrasts', *J Steroid Biochem Molec Biol*, **39**:601–14.

Meredith, M., & O'Connell, R.J. (1979), 'Efferent control of stimulus access to the hamster vomeronasal organ', *J Physiol*, **286**:301–16.

Müller-Schwarze, D., Volkman, N.J., & Zemanek, K.F. (1977), 'Osmetrichia: Specialized scent hair in black–tailed deer', *J Ultrastruct Res*, **59**:223–30.

Mykytowycz, R. (1972), 'The behavioural role of the mammalian skin glands', *Naturwissenschaften*, **59**:133–9.

Novotny, M., Jemiolo, B., Harvey, S., Wiesler, D., & Marchlewska-Koi, M. (1986), 'Adrenal-mediated endogenous metabolites inhibit puberty in female mice', *Science*, **231**:722–5

Perret, M., & M'Barek, S.B. (1991), 'Male influence on oestrous cycles in female woolly opossums (*Caluromys philander*)', *J Reprod Fert*, **91**:557–66.

Pfeiffer, C.A., & Johnston, R.E. (1993), 'Hormonal and behavioral responses of male hamsters to females and female odors: Roles of olfaction, the vomeronasal system, and sexual experience', *Physiol Behav*, **55**:129–38.

Romo, E., de Miguel, M.P., Arenas, M.I., Frago, L., Fraile, B., & Paniagua, R. (1996), 'Histochemical and quantitative study of the cloacal glands of *Triturus marmoratus marmoratus* (Amphibia: Salamandridae)', *J Zool Lond*, **239**:177–86.

Schilling, A., Serviere, J., Gendrot, G., Perret, M. (1990), 'Vomeronasal activation by urine in the primate *Microcebus murinus*: A 2 DG study', *Exp Brain Res*, **81**:619–25.

Scott, M.P. (1986), 'The timing and synchrony of seasonal breeding in the marsupial, *Antechinus stuartii*: Interactions of environmental and social cues'. *J Mammal*, **67**:551–60.

Shammah–Lagnado, S.J., Negrao N. (1981), 'Efferent connections of the olfactory bulb in the opossum (*Didelphis marsupialis aurita*): A Fink-Heimer study', *J Comp Neurol*, **201**:51–63.

Smith, M.J., Bennett, J.M., & Chesson, C.M. (1978), 'Photoperiod and some other factors affecting reproduction in female *Sminthopsis crassicaudata* (Marsupialia: Dasyuridae) in captivity', *Aust J Zool*, **26**:449–63.

Sobel, N., Prabhakaran, V., Desmond, J.E., Glover, G.H., Goode, R.L., Sullivan, E.V., & Gabrieli, J.D.E. (1998), 'Sniffing and smelling: Separate subsystems in the human olfactory cortex', *Nature*, **392**:282–6.

Stoddart, D.M. (1979), 'A specialised scent-releasing hair in the Crested rat *Lophiomys imhausi*', *J Zool Lond*, **189**:553.

Stoddart, D.M., & Bradley, A.J. (1991), 'The frontal and gular dermal scent organs of the marsupial sugar glider (*Petaurus breviceps* Waterhouse)', *J Zool Lond*, **225**:1–12.

Stoddart, D.M., Bradley, A.J., & Mallick, J. (1994), 'Plasma testosterone concentration, body weight, social dominance and scent-marking in male marsupial sugar gliders (*Petaurus breviceps;* Marsupialia: Petauridae)', *J Zool Lond*, **232**:595–601.

Stonerook, M.J., & Harder, J.D. (1992), 'Sexual maturation in female gray short-tailed opossums, *Monodelphis domestica*, is dependent upon male stimuli', *Biol Reprod*, **46**:290–4.

Stralendorff, F.V. (1987), 'Partial chemical characterization of urinary signalling pheromone in tree shrews (*Tupaia belangeri*)', *J Chem Ecol*, **13**:655–79.

Sumner, J., & Dickman, C.R. (1998) 'Distribution and identity of specied in the *Antechinus stuartii – A. flavipes* group (Marsupialia: Dasuyuridae) in south-eastern Australia', *Aust J Zool*, **46**:27–41.

Taniguchi, K., Matsusaki, Y., Ogawa, K., & Saito, T.R. (1992a), 'Fine structure of the vomeronasal organ in the common marmoset (*Calilithrix jacchus*)', *Folia Primatol*, **59**:169–76.

Taniguchi, K., Taniguchi, K., Arai, T., & Ogawa, K. (1992b), 'Enzyme histochemistry of the olfactory and vomeronasal sensory epithelia in the golden hamster', *J Vet Med Sci*, **54**:1007–16.

Thiessen, D.D., Regnier F.E., Rice, M., Goodwin, M., Isaacks, N., & Lawson N. (1974), 'Identification of the ventral scent marking pheromone in the male Mongolian gerbil (*Meriones unguiculatus*)', *Science,* **184**:83–4.

Thiessen, D., & Rice, M. (1976), 'Mammalian scent gland marking and social behavior', *Psychol Bull,* **83**:505–39.

Toftegaard, C.L. (1999) 'Morphological and endocrine correlates of chemical communication during the life history of *Antechinus stuartii* (Macleay)', PhD thesis, University of Queensland, Brisbane, Australia

Toftegaard, C.L., & Bradley, A.J. (1999), 'The structure of specialised osmetrichia in the brown antechinus, *Antechinus stuartii* (Marsupialia: Dasyuridae)', *J Zool Lond,* **248**:27–33.

Toftegaard, C.L., Moore, C., & Bradley, A.J. (1999) 'Chemical characterisation of urinary pheromones in the brown antechinus, *Antechinus stuartii* (Marsupialia: Dasyuridae)', *J Chemical Ecology,* **25**:527–35.

Toftegaard, C.L., McMahon, K.L., Galloway, G.J., & Bradley, A.J. (2002), 'Processing of urinary pheromones in *Antechinus stuartii* (Marsupialia: Dasyuridae): Functional magnetic resonance imaging of the brain', *J Mammal,* **83**:71–80.

Vandenbergh, J.G. (1969), 'Male odor accelerates female sexual maturation in mice', *Endocrinol,* **84**:658–60.

Van Dyck, S. (1979), 'Behaviour in captive individuals of the dasyurid marsupial *Planigale maculata* (Gould 1851)', *Mem Qld Museum,* **19**:413–31.

Van Dyck, S., & Crowther, M.S. (2000) 'Reassessment of Northern representatives of the *Antechinus stuartii* complex (Marsupialia; Dasyuridae): *A. subrtopicus* sp.nov., & *A. adustus* new status', *Mem Qld Museum,* **45**:611–35.

Waring, C.P., Moore, A., & Scott, A.P. (1996), 'Milt and endocrine responses of mature male Atlantic salmon (*Salmon salar* L.) parr to water-borne testosterone, 17, 20 beta-dihydroxy-4-pregnen-3-one 20 sulfate, and the urines from adult female and male salmon', *Gen Comp Endocrinol,* **103**:142–9.

Whitten, W.K. (1969), 'Mammalian pheromones', in *Olfaction and Taste III* (ed. C. Phaffmann), pp. 252–7 Rockefeller University Press, New York.

Woolhouse, A.D., Weston, R.J., & Hamilton, 3.H. (1994), 'Analysis of secretions from scent-producing glands of Brushtail Possum (*Trichosorus vulpecula* Kerr)', *J Chem Ecol,* **20**:239–53.

Wysocki, C.J., & Meredith, M. (1991), 'The vomeronasal system', in *Neurobiology of Taste and Smell* (eds.T.E. Finger, & W.L. Silver), pp 125–50, John Wiley & Sons, New York.

Yamada, K., Itoh, R., & Otha, A. (1989), 'Influence of pyrazine derivatives on the day of vaginal opening in juvenile female rats', *Japan J Pharmacol,* **49**:529–30.

Yousem, D.M., Williams, S.C.R., Howard, R.O , Andrew, C., Simmons, A., Allin, M., Geckle, R.J., Suskind, D., Bullmore, E.T., Brammer, M.J., & Doty, R.L. (1997), 'Functional MR imaging during odor stimulation: Preliminary data', *Radiology,* **204**:833–8.

PART IV

CHAPTER 24

REPRODUCTIVE BIOLOGY OF CARNIVOROUS MARSUPIALS: CLUES TO THE LIKELIHOOD OF SPERM COMPETITION

D.A. Taggart[A, B, C], *G.A. Shimmin*[B], *C.R. Dickman*[D] *and W.G. Breed*[C]

[A]Department of Zoology, University of Melbourne, Parkville, Vic 3011 and Conservation and Research Unit, Royal Zoological Society of South Australia, Frome Rd, Adelaide, SA 5000, Australia.
Email: david.taggart@adelaide.edu.au
[B]Department of Environmental Biology, Adelaide University, SA 5005, Australia
[C]Department of Anatomical Sciences, Adelaide University, SA 5005, Australia
[D]School of Biological Sciences, University of Sydney, Sydney, NSW 2006, Australia

The limited relevant data that are available which may pertain to the occurrence of sperm competition and mating systems in carnivorous marsupials are presented. A positive allometric relationship was found between body mass and testis mass (n = 46 species) with some variation around the mean. There was also a positive allometric relationship between body mass and numbers of stored sperm in the male reproductive tract (n = 8). The limited data available suggest that length of copulation varies markedly between species and the sperm, once deposited in the female tract, undergo efficient transport to the higher reaches of the female tract, with storage taking place in the isthmus of the oviduct until ovulation. Females ovulate many oocytes and, where investigated in *Antechinus*, multiple paternity was found to occur in the animals in both laboratory experiments and in the wild. These data are discussed in relation to the apparent breeding system, and the possibility of inter-male sperm competition.

INTRODUCTION

In eutherian mammals there is now a considerable body of data that suggests that males, living in multi-male breeding groups where a female is likely to mate with several males during the one oestrous period, have relatively larger testes than those of species that occur as monogamous pairs, or as single male breeding units. This difference in relative testis size is thought to be due to differences in intensity of sperm competition (Short 1979; Harcourt et al. 1981; Kenagy and Trombulak 1986).

The absolute and relative testis size, and the potential for sperm competition in various marsupial species, have recently been considered and found to vary markedly between species (Dickman 1993; Rose et al. 1997; Taggart et al. 1998). In these studies it was established that relatively large testes occurred in some marsupial species and correlated with the production of abundant sperm. This was not the case in the few dasyurids investigated where relatively small testes and few sperm were found to occur. In these studies, like those in eutherians, there was some evidence which suggested that relative testis size may be associated with a specific breeding system and hence potential for inter-male sperm competition.

The carnivorous marsupials comprise around 112 extant species that occur in at least 29 genera in five families. These include the Dasyuridae with 69 species in eighteen genera, the New World Didelphidae with around four species in three genera; the Caenolestidae with about seven species in two genera, the Marmosidae with up to 31 species in at least five genera, and the Microbiotheriidae which consists of one species only. Previous studies on selected species of dasyurid and didelphid marsupial have indicated some unusual morphological and behavioural, reproductive characteristics (such as low sperm numbers, relatively long sperm, occurrence of sperm pairing, lengthy copulation and sperm storage in the female reproductive tract). In studies on a variety of other vertebrate and invertebrate species some of these characteristics have been correlated with an increased likelihood of inter-male sperm competition (Short 1979). Further investigation of the reproductive biology of the carnivorous marsupials may thus provide some clues as to the likelihood of sperm competition occurring in species within this group. In this chapter we present data on relative testis mass and various other reproductive biological characteristics of species of carnivorous marsupials, and then consider these data in relation to the possible breeding system that may occur and hence the potential for inter-male sperm competition.

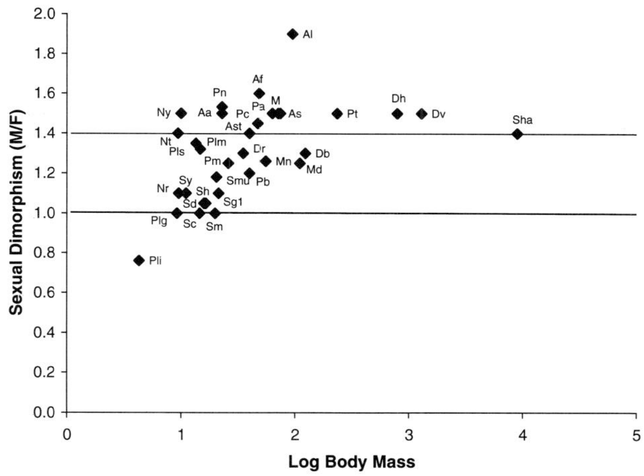

Figure 1 Sexual dimorphism (male/female) against Log body mass for carnivorous marsupials. For explanation of species codes refer to Table 1. Note: A sexual dimorphism of 1 denotes that males and females are of similar body size for that species.

SEXUAL DIMORPHISM IN BODY WEIGHT

Sexual dimorphism in favour of males is likely to indicate competition between males for mating access to females. In almost all of the carnivorous marsupials examined, male body mass exceeded that of females (Fig. 1, Table 1). Among the small species examined (<40 g male body mass) the *Ningaui* species, *Antechinus agilis* and *Pseudantechinus ningbing* showed the largest sexual dimorphism of body mass. In *Planigale gilesi*, *Sminthopsis crassicaudata*, *Sminthopsis griseoventer* and *Sminthopsis macroura* a significant difference in body mass between the sexes does not appear to occur, whereas in *Planigale ingrami* moderate female-biased dimorphism is evident (Fig. 1). Of the medium-sized carnivorous marsupials (40 to 250 g male body mass), the species in the genus *Antechinus* and *Phascogale*, and also *Parantechinus apicalis* and *Dasycercus cristicauda*, showed a strong male dimorphism (Fig. 1). Of the carnivorous marsupials with the largest body mass (>250 g male body mass) males from species in the genus *Dasyurus* and *Sarcophilus harrisii* are all considerably larger than females (Fig. 1).

REPRODUCTIVE DATA

Published, and unpubl., morphological and behavioural reproductive data were compared for all available carnivorous marsupials (~47 species) (Table 1). Where appropriate, morphological and behavioural, reproductive data were collected from wild-caught animals in order to eliminate potential bias associated with the housing and management of animals in captivity (e.g. weight gain due to captivity; abnormal social structure; unusual timing of reproductive activity due to unnatural environmental conditions). However, where data on particular topics from field animals were not available (for example; the maximum length of copulation, and sperm storage in the female reproductive tract) those from captive colonies were used.

To the best of our knowledge, data on testis mass and epididymal sperm number were collected during the breeding season, except for the semelparous dasyurids in which spermatogenic failure occurs prior to the onset of mating (Lee et al. 1982; Kerr and Hedger 1983). In these species (*Antechinus agilis, A. swainsonii, A. stuartii, Phascogale tapoatafa, P. calura*) body and testis mass were recorded around the time of maximal testicular activity, which is approximately 1–2 months prior to the commencement of the mating season (Woolley 1966; Taggart and Temple-Smith 1990a; Millis et al. 1999). Data presented on testis and body mass for some other *Antechinus* species (e.g. *A. flavipes, A. godmani* and *A. minimus,* and also *Dasyurus hallucatus*) must be viewed with caution as these were most likely recorded during the breeding season, well after testicular activity has peaked (Woolley 1966; Kerr and Hedger 1983; Taggart and Temple-Smith 1994; Oakwood 2000).

TESTES MASS

A positive allometric relationship between body mass and testis mass was found in the species studied (Table 1; Fig. 2; $y = 0.7612x - 1.7812$; $r^2 = 0.86$, n = 46). Examination of individual data (Fig. 2; Table 1) show that among species of small body mass (<40 g) relatively large testis mass occurred in *Sminthopsis murina* (residual +0.21), *Sminthopsis macroura* (residual +0.17), *Sminthopsis crassicaudata* (residual +0.21), *Sminthopsis granulipes* (residual +0.26), *Sminthopsis leucopus* (residual +0.12), *Antechinus agilis* (residual +0.13) and *Pseudantechinus macdonnellensis* (residual +0.16). By contrast, relatively small testis mass

Table 1 Summary of available reproductive data as it relates to sperm competition for carnivorous marsupials. Data were derived from a variety of published material (listed in the bibliography), unpublished material belonging to the authors (refer to text) and where indicated unpublished data of Dr Menna Jones (*) and Dr Pat Woolley (+). Species symbol (Sym); Male body weight (M-Wt g); Combined testes mass in grams (TM); Relative testes mass (RTM); Residual Log Testes mass*Log Body Mass (RES); Combined total epididymal sperm number $\times 10^6$ (ESN); Maximum length of copulation (Cop max); Maximum length of behavioural oestrus in days (MLO); Maximum length of sperm storage, in days, in the female (Sp. St); Sexual dimorphism M:F (M:F); Maximum number of teats (Teats); Maximum number of oocytes (Oocytes); Monoestrus, Facultative monoestry or Polyoestrus (Oest). Predicted likelihood of inter-male sperm competition occurring based on available data (Sp Comp): Ratings – Unclear (?); Unlikely (-); Low probability (+); Low-moderate probability (++); Moderate / moderate-high probability (+++); Highly likely (++++).

Species	Common Name	Sym.	M-Wt g	TM (g)	RTM	RES	ESN ($\times 10^6$)	Cop max (hrs)	MLO (d)	Sp. St (d)	M:F	Teats	Oocytes	Oest	Sp Comp
A.agilis	Agile antechinus	Aa	23.0	0.240	1.044	0.125	7	12	11	15	1.50	10	19	M	++++
A.flavipes	Yellow-footed Antechinus	Af	48.8	0.340	0.697	0.027	8.22	11	7	13	1.60	12		M	++++
A.stuartii	Brown antechinus	Ast	40.0	0.640	1.600	0.368		12		14	1.40	8		M	++++
A.swainsonii	Dusky antechinus	As	71.4	0.630	0.882	0.169	7.2	9.5	7	8	1.50	8		M	++++
A.leo	Cinnamon Antechinus	Al	95.0							14	1.90	10		M	++++
A.godmani	Atherton antechinus	Ag	105.0	0.196	0.187	-0.465									++++
A.minimus	Swamp antechinus	Am	51.0	0.460	0.902	0.144									++++
P.macdonnellensis	Fat-tailed pseudantechinus	Pm	26.0	0.288	1.108	0.164		2		~15	1.25	6	12	M	++?
P.ningbing	Ningbing pseudantechinus	Pn	23.0	0.094	0.409	-0.282		6		12	1.53	4	20	FM	++?
S.crassicaudata	Fat-tailed dunnart	Sc	14.6	0.166	1.139	0.116	1.22	2.5-11	3	4	1.00	8	14	P	++-+++
S.macroura	Stripe-faced dunnart	Sm	20.0	0.240	1.200	0.171	1.06	2.5	2 or 3	5	1.00	6	39	P	++-+++
S.dolichura	Little long-tailed dunnart	Sd	16.0	0.142	0.888	0.017					1.05	8			+?
S.virginiae	Red-cheeked dunnart (QLD)	Sv	31.0	0.251	0.810	0.046								P	+?
S.virginiae nitela	Red-cheeked dunnart NT	Sv1	33.5	0.257	0.767	0.030								P	+?
S.virginiae (PNG)	Red-cheeked dunnart (PNG)	Sv2	48.0	0.298	0.621	-0.024									+?
S.griseoventer	Grey-bellied dunnart	Sg	20.0	0.151	0.755	-0.030									+?
S.griseoventer boull'	Grey-bellied dunnart	Sg1	21.4	0.221	1.033	0.113		3			1.10	8		M	++?
S.granulipes	White -tailed dunnart	Sgr	18.5	0.280	1.514	0.264									++++?
S.hirtipes	Hairy-footed dunnart	Sh	16.5	0.155	0.939	0.045		2.5			1.05	6		P	++
S.murina	Common dunnart	Smu	20.5	0.266	1.298	0.208		2		4	1.18	10		P	++
S.aitkeni	Kangaroo Is Dunnart	Sa	18.0	0.110	0.611	-0.133									-
S.bindi	Kakadu dunnart	Sb	15.3	0.106	0.690	-0.097									-
S.leucopus	White -footed dunnart	Sl	22.5	0.232	1.031	0.117									++
S.ooldea	Ooldea dunnart	So	12.0	0.106	0.883	-0.015									+?
S.youngsoni	Lesser hairy-footed dunnart	Sy	11.0	0.115	1.045	0.049		3.2			1.10	6		P	++
D.rosamondae	Little-red kaluta	Dr	35.3	0.126	0.357	-0.297					1.30	8			++?
D.byrnei	Kowari	Db	123.6	1.381	1.117	0.329	1.7	2.0-3.0	1 to 3		1.30	6	11	P	+++?
D.hallucatus	Northern quoll	Dh	796.0	2.486	0.312	-0.032					1.50	8			++?

D.viverrinus	Eastern quoll	Dv	1300.0	3.000	0.231	-0.112	4		3	14	1.50	6	35	P	++
*D.maculatus**	Spotted-tailed quoll	Dm						8 to 12	3		1.75	6		P	+++
*S.harrisii**	Tasmanian devil	Sha	9000.0	24.120	0.268	0.152		1.5			1.40	4	56	M	+++
P.tapoatafa	Brush-tailed phascogale	Pt	235.0					8	5	14	1.50	8		M	+++
P.calura	Red-tailed phascogale	Pc	47.2								1.45	8	>13	M	+++?
P.bilarni	Northern dibbler	Pb	40.0							2	1.20	6	10	M	-
P.bilarni	Northern dibbler	Pb1	22.5	0.061	0.271	-0.463								M	-
P.apicalis	Southern dibbler	Pa	63.6	0.480	0.755	0.090		7.7		10	1.50	8		M	+++
D.cristicauda	Mulgara	M	74.5	0.770	1.034	0.243		3			1.50	6		M	++++?
N.ridei	Wongai ningaui	Nr	9.5	0.087	0.916	-0.024		4.5	1 to 3	3	1.10	7		P	-
N.yvonneae	Southern ningaui	Ny	10.0	0.074	0.740	-0.111					1.50			P	-
N.timealeyi	Pilbara ningaui	Nt	9.4	0.047	0.500	-0.287					1.40			P	-
P.tenuirostris	Narrow-nosed planigale	Plt	9.3						1					P	-
P.ingrami	Long-tailed planigale	Pli	4.3								0.76	10		P	-
P.gilesi	Giles planigale	Plg	9.2						3		1.00			P	-
P.maculata	Common planigale	Plm	14.8					2.5	2		1.32	12		P	-
P.m.sinualis	Common planigale	Pls	13.5								1.35	12	17	P	-
T. cynocephalus	Tasmanian tiger	Tc	35000												?
M. melas+	Three stripe dasyure	Mm	117.5	0.733	0.624	0.071									++ -+++?
M. wallacei+		Mw	240.8	0.866	0.360	-0.094									?
M. longicaudata+	Short-furred dasyure	Ml	78.2	0.370	0.473	-0.092					3.1				?
M. aspersa+		Ma	222.0	0.560	0.252	-0.257									?
P. dorsalis+	Narrow-striped dasyure	Pd	52.6	0.346	0.658	0.010									?
N. lorentzii+	Speckled dasyure	Nl	237.0	0.714	0.301	-0.173									?
M. habbema+	Habbema dasyure	Mh	35.0	0.231	0.660	-0.031								M?	?
M. naso+	Long-nosed dasyure	Mn	55.7	0.393	0.706	0.047					1.26			M?	?
M. melanurus+	Black-tailed dasyure	Mme	45.2	0.250	0.553	-0.081								M?	?
M. domestica	Grey short-tailed opossum	Md	110.0	0.570	0.518	-0.017	4.2	0.1	1.5	1	1.25	14	16	P	-
M. robinsoni	Robinson's mouse oposuum	Mr						3	3			15		P	+++?
C. obscurus	Common shrew opossum	Co	40				1.8								?

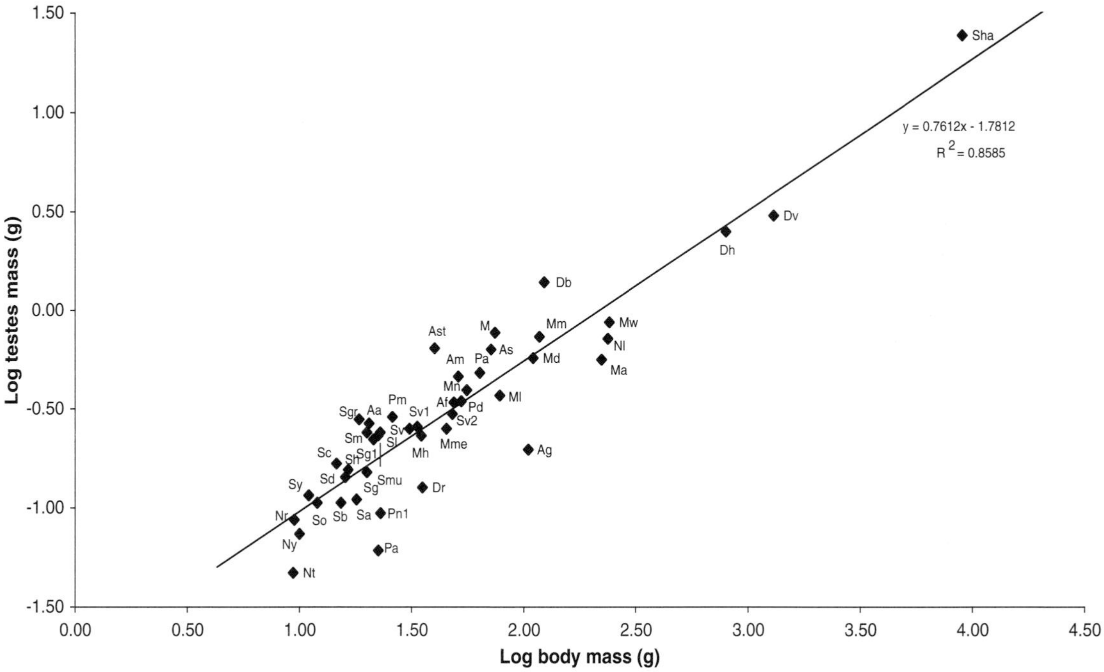

Figure 2(a) Regression of testis mass against body mass for all available data on carnivorous marsupial species. For explanation of species codes refer to Table 1. $R^2 = 0.86$. Note: The higher a data point sits above the regression line (the average), the larger the testes mass was relative to body mass and the greater the likelihood that inter-male sperm competition may occur. The lower a data point sits below the regression line, the smaller the testes mass was relative to body mass and the less likely inter-male sperm competition may occur.

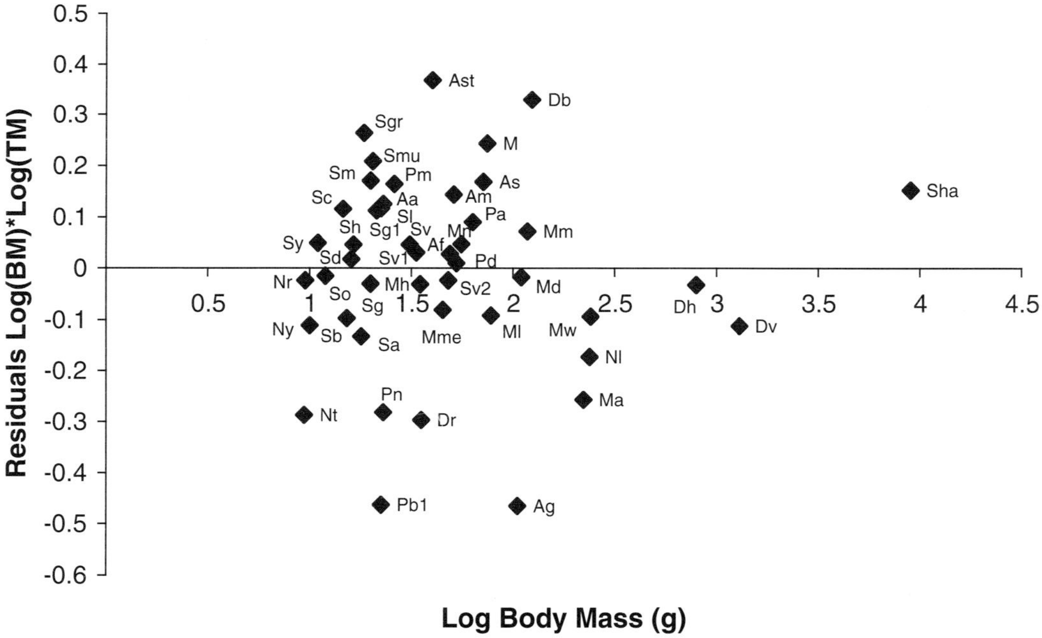

Figure 2(b) Residuals of Log Body Mass*Log Testes Mass plotted against Log Body Mass for all available data on carnivorous marsupial species. For explanation of species codes refer to Table 1. The higher the residual is above zero, the greater the likelihood that inter-male sperm competition may occur. The lower the residual is below zero, the less likely inter-male sperm competition may occur.

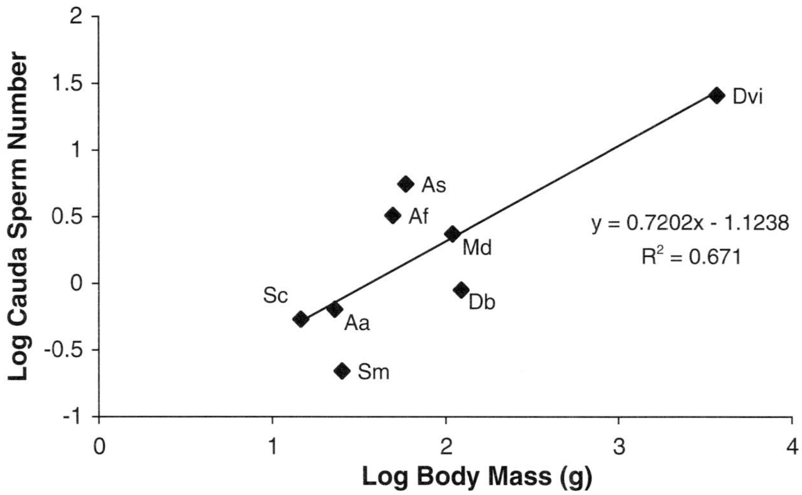

Figure 3 Regression of sperm number in cauda epididymides against body mass for all available data on carnivorous marsupial species. For explanation of species codes refer to Table 1.

was evident in *Sminthopsis bindi* (residual -0.10), *Sminthopsis aitkeni* (residual -0.13), *Pseudantechinus ningbing* (residual -0.28), *Parantechinus bilarni* (residual -0.46), *Dasykaluta rosamondae* (residual -0.30), *Ningaui timealeyi* (residual –0.29) and *Ningaui yvonneae* (residual -0.11) (Fig. 2; Table 1).

Of the carnivorous marsupials with a body mass of between 40 g–250 g, a relatively large testes mass occurred in *Antechinus stuartii* (residual +0.37), *A. minimus* (residual +0.14) and *A. swainsonii* (residual +0.17), *Dasycercus cristicauda* (residual +0.24), and *Dasycercus byrnei* (residual +0.33), whereas, a relatively small testes mass was evident in *Sminthopsis virginiae* (residual -0.024), *Antechinus godmani* (residual –0.46) and most of the New Guinean species (Fig. 2; Table 1). As indicated above, where relatively small testis mass is evident and little information is available on the timing of data collection relative to the breeding season (e.g. *Antechinus godmani and A. flavipes;* and the New Guinean Spp), the information may not represent the maximum relative testis size for the species.

Among the larger carnivorous marsupials (>250 g) only *Sarcophilis harrisii,* the Tasmanian devil, has clearly large testes for body mass (residual +0.15) whereas those of *Dasyurus viverrinus* (residual –0.11) are small for body mass. The testes of *Dasyurus hallucatus* (residual –0.03) also appear small to average for body mass, although this may relate to the timing of data collection relative to peak testicular activity (refer above; Oakwood 2000).

SPERMATOGENIC FAILURE PRIOR TO BREEDING SEASON AND MALE-DIE OFF

Spermatogenic failure prior to the breeding season together with male die-off following a 'rut', has been reported in all species of *Antechinus* and *Phascogale* examined (Woolley 1966; Taylor and Horner 1970; Inns 1976; Kerr and Hedger 1983; Taggart and Temple-Smith 1990a; 1994; Taggart et al. 1997; Millis et al. 1999). The northern quoll, *Dasyurus hallucatus,* also exhibits a male die-off following the rut, however total spermatogenic failure has not been recorded, although a reduction in scrotal diameter between pre breeding and breeding periods has (Oakwood 2000; Oakwood et al. 2001). Studies on testis size in eutherians suggest that the energetic costs associated with sperm production may be significant, particularly in small mammals (Kenagy and Trombulak 1986). Recent studies by Taggart et al. (1997) have extended this hypothesis in an attempt to explain testicular failure in semelparous dasyurids prior to mating. These authors suggested that it might be advantageous for semelparous dasyurids living in an environment of intense competition, to store sperm in the epididymis and then to cease production of sperm in the testes with a consequence that testis size becomes reduced. The energy saved by such a strategy could then be directed into securing mates which, in semelparous dasyurids, places huge stress on males and often results in their death before females have even ovulated (Woolley 1966; Lee et al. 1977; Williams and Williams 1982).

A potential additional benefit associated with restricting sperm production to the period prior to mating is that, following production of spermatozoa, androgen levels can be elevated dramatically to enhance an individual's competitive ability (increased weight, aggression, and/or energy) without suppressing spermatogenesis (Sun et al. 1989; McLachlan et al. 1994). Evidence that this occurs within the semelparous, but not iteroparous, dasyurids can be found by comparing monthly changes of both plasma androgen levels and prostatic weight (Woolley 1966; Wilson and Bourne 1984; Kerr and Hedger 1983; Taggart and Temple-Smith 1989; 1990a; Taggart et al. 1997).

The intense reproductive effort of the semelparous males, and their associated death, may thus have evolved as a direct consequence of intense competition between males for mates (Dickman 1993). Support for this hypothesis comes from studies of *Parantechinus* and *Dasyurus* (Woolley 1991; Dickman and Braithwaite 1992) where male die-off occurs only in some years, or in some populations. In *Parantechinus apicalis* total male die-off has been reported in high density island populations but not in low density mainland populations (Woolley 1991; Dickman and Braithwaite 1992; Mills and Bencini 2000). In iteroparous species, like *S. macroura*, where individual densities are usually lower and sperm storage is of relatively short duration, the opportunities for multiple matings are more limited and intense inter-male competition therefore less likely. This, combined with the possibility of some males breeding for several years, suggests that the reproductive effort of males for iteroparous species is likely to be less in each breeding season than that in the semelparous species (Morton 1982; Dickman 1993).

Number of sperm in the epididymidis

It is clear that compared to eutherian mammals of similar size relatively few sperm are present within the epididymis of the carnivorous marsupials for which data are available (Table 1). In the Australasian species at least, there appears to be limited capacity for sperm storage in the cauda epididymidis (Taggart and Temple-Smith 1990a; 1994). Data on caudal epididymal sperm number relative to body mass for the eight species for which there are data show an allometric relationship (Fig. 3; $y = 0.7202x - 1.1238$; $r^2 = 0.67$). Within this group *Antechinus agilis* and *Antechinus flavipes* have relatively large numbers of sperm in the cauda epididymides, whereas *Dasycercus byrnei* and *Sminthopsis macroura* have relatively few epididymal sperm.

In semelparous dasyurids the sperm available to inseminate females may be limiting since spermatogenesis ceases prior to the mating season (Kerr and Hedger 1983). Consequently a finite, and non-replenishable, sperm reserve exists from the time of the onset of breeding activity. In addition to a finite number of sperm being available for insemination, sperm are also continually lost in the urine (spermatorrhoea) at a rate of 2.9×10^5 sperm per day (Taggart and Temple-Smith 1990a) leaving even unmated males with as few as 4.0×10^5 sperm by the end of the mating season. This constitutes only about 4% of the sperm available at the commencement of the mating period (Taggart and Temple-Smith 1990a).

Length of copulation and copulatory behaviour

Information on mating behaviour in carnivorous marsupials is restricted to data obtained from individuals housed in captivity. Data from the *Antechinus* species, *Phascogale tapoatafa* (Dickman 1993; Millis et al. 1999), *Parantechinus apicalis* (Dickman 1993; Wolfe et al. 2000), *Dasyurus maculatus* (Dickman 1993) and didelphids (Barnes and Barthold 1969) indicate that intromission can last for as long as 7 to 12 hours (Table 1). In mating trials on *Antechinus agilis,* where multiple males had access to a female, it was found that the presence of a male competitor did not appear to influence the mating duration (Shimmin et al. 2002). By contrast, duration of copulation in *Sminthopsis* species, *Dasycercus byrnei*, *Pseudantechinus macdonnellensis*, *Planigale maculata*, *Dasycercus cristicauda* and *Sarcophilis harrisii* appears to only last up to 3.2 hours (Table 1), although an early study by Ewer (1968) on *S. crassicaudata* suggested prolonged oestrus from dusk to dawn. Preliminary data however, have found that, in this species ejaculation normally occurs within the first 30 minutes of the onset of mating (Lampard, Sarma and Breed unpubl. obs.). A loose exponential relationship was observed between the maximum length of copulation and the maximum length of oestrus ($y = 1.9599e^{0.1064x}$, $R^2 = 0.6544$) in the species examined. Interestingly, all high values are for semelparous species and all low values are for iteroparous species (Fig. 4).

In *Antechinus agilis* mating by a male given first access to a female at oestrus typically lasted around 7 hours, but males given second mating access to females generally mated for a shorter period of time than did the first male (Shimmin 1999; Shimmin et al. 2002). In addition, males mating closer to the time of ovulation mated for a shorter duration irrespective of whether they were, or were not, the first male to mate. Behaviour during the mating time was variable with periods of thrusting, pelvic side to side movements, and kneading of the female's flanks, interspersed with periods of rest or apparent inactivity. Thrusting and pelvic side to side movements were most prevalent during the first half of the mount period with female repelling behaviour becoming more frequent towards the end with the eventual forced dismount of the male (Shimmin 1999; Shimmin et al. 2002).

As with *S. crassicaudata,* data on presence of sperm in the female reproductive tract of *A. agilis* indicate that ejaculation occurs early in the mating sequence and that thrusting during the latter half of mating does not result in ejaculation and the deposit of additional spermatozoa (Shimmin et al. 1999). In *S. crassicaudata* too we have found that, although ejaculation occurs soon after initial mounting, bouts of multiple intromissions subsequently occur (Lampard, Sarma and Breed unpubl. obs.).

This behavioural repertoire of copulatory behaviour appears similar for many of the carnivorous marsupials (e.g. other species of *Antechinus*, *Sminthopsis, Phascogale tapoatafa* and *Monodelphis domestica;* Taggart, Breed Shimmin and Moore unpubl. data).

Mate guarding

Little information is available on mate guarding in carnivorous marsupials due in part to the cryptic nature of mating behaviour

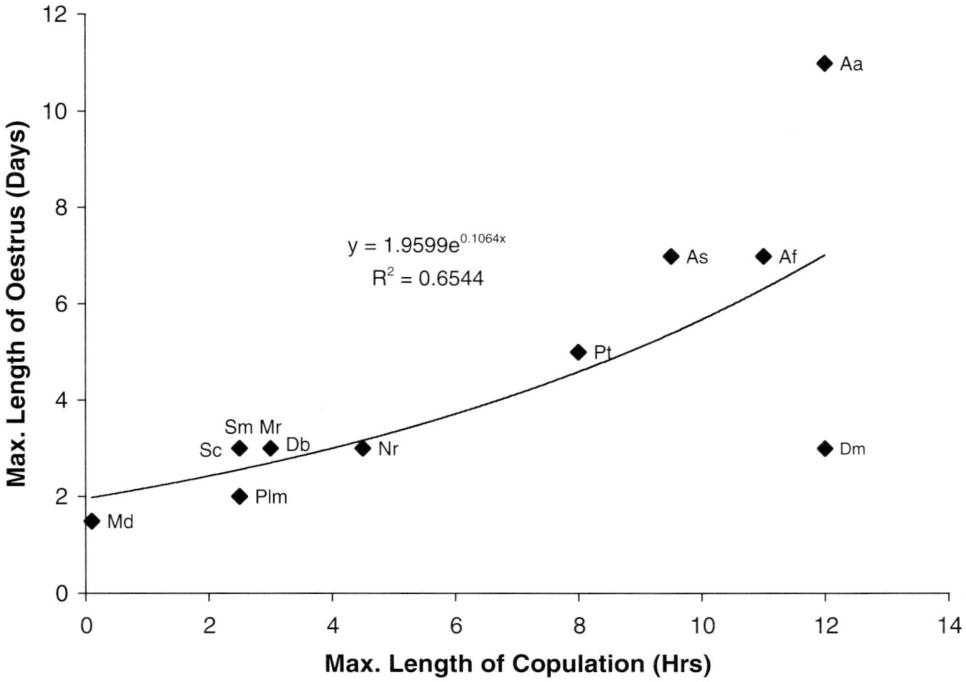

Figure 4 Regression of the maximum length of copulation against the maximum length of oestrus for all available data (N = 9) from the carnivorous marsupials. For explanation of species codes refer to Table 1. Note separation of semelparous (high values) and iteroparous species (low values).

and the lack of knowledge of the social organisation of most species. In *A. agilis* (Dickman 1993; Shimmin et al. 1999; 2002) and *S. crassicaudata* (Lampard, Sarma and Breed unpubl. obs.) however, as sperm are delivered very early in the prolonged mating sequence (Table 1) it has been concluded that a large proportion of each copulation probably serves as a form of contact mate guarding, preventing access to the female by other males during this period and helping to ensure the successful transport and colonisation of spermatozoa in the female reproductive tract.

A number of species in the genera *Antechinus, Sminthopsis, Marmosa* and *Didelphis* have also been observed to lock during copulation (Barnes and Barthold 1969; Shimmin et al. 2002; Taggart, Breed and Moore unpubl. obs.). In *Antechinus,* whilst locked in copulation the dominant male is able to turn through 180° from his normal mounted position and fight off the approaches of other subdominant males whilst maintaining intromission (Shimmin et al. 2002). Other forms of mate guarding, however, have not been reported.

SPERM NUMBERS EJACULATED

Data on sperm number in the ejaculate are available for three species of carnivorous marsupial (*Antechinus agilis,* 80-480x10^3, Taggart and Temple-Smith 1991, Taggart et al. 1999, Shimmin 1999; *Didelphis virginiana,* 2.98x10^6, Bedford et al. 1984; *Sminthopsis crassicaudata,* about 2x10^5, Breed et al. 1989). The measured values for these three species are in the order of 10^2 to 10^3 fewer sperm than those typical for eutherian mammals and most other marsupials (Taggart 1994; Taggart et al. 1998). Since few sperm are also found in the cauda epididymides of the other carnivorous marsupials, it is likely that these species too would deliver relatively few sperm at ejaculation (Table 1).

SPERM TRANSPORT IN THE FEMALE REPRODUCTIVE TRACT

Upon ejaculation, semen is deposited in the upper part of the urogenital sinus and sperm travel rapidly through the cervix (Hughes and Roger 1971; Tyndale-Biscoe and Rodger 1978; Taggart 1994; Breed 1994). Copulatory plugs similar to those occurring in various macropods (Tyndale-Biscoe and Rodger 1978), vombatids (Taggart et al. 1998) and phalangerids (Hughes and Rodger 1971), have been found in some opossums in the families Didelphidae and Marmosidae (Hartman 1924; McCrady 1938; Barnes and Barthold 1969), whereas histological sections of recently mated *Sminthopsis crassicaudata* have shown that an eosinophilic intravaginal plug is present (Breed 1994). These copulatory plugs may act as a temporary physical barrier to subsequent mating by other males and/or help to retain the ejaculate in the vagina in close proximity to the cervix

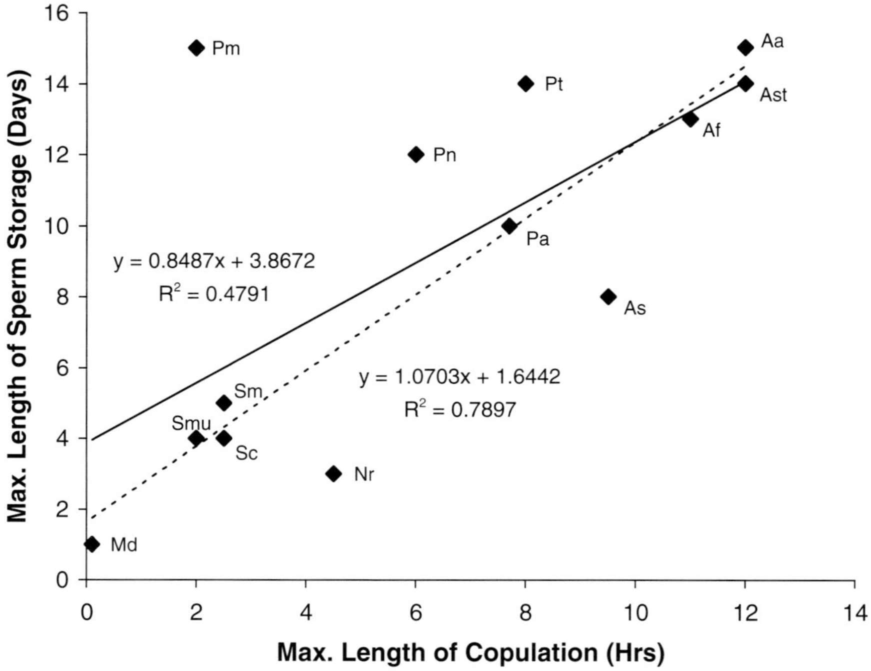

Figure 5 Regression of the maximum length of copulation against the maximum length of sperm storage in the female reproductive tract for available data (N = 13) from the carnivorous marsupials. For explanation of species codes refer to Table 1. Note: There appears to be no relationship between these parameters (solid regression line); however, if plotted without the outlying point for *Planigale maculata* (Pm) a strong, positive allometric relationship is observed (dashed regression line).

to facilitate sperm transport through the lower regions of the female reproductive tract.

In most mammals, barriers to sperm transport occur at the cervix and utero-tubal junction with the result that as few as 0.01% of ejaculated sperm reach the isthmus of the oviduct (e.g. in rabbit – Overstreet and Cooper 1978; 1979). By contrast, in the carnivorous marsupials studied 5–100% of ejaculated sperm reach the lower isthmus of the oviduct (*Antechinus agilis*, 15–100%, Taggart and Temple-Smith 1991; *Sminthopsis crassicaudata*, about 10%, Breed et al. 1989; *Didelphis virginiana*, about 5%, Bedford et al. 1984). In three *S. crassicaudata* euthanased about 1 hour after mating several thousand spermatozoa had already reached the isthmus of the oviduct with similar numbers of sperm being present in the lateral vagina and isthmus 3 hours after mating (Breed unpubl. obs.). This suggests that, in these species at least, there are few, if any, barriers to sperm transport and that there is rapid migration of the sperm population ejaculated to the isthmus of the oviduct. The extremely high rates of sperm transport success may be due, in part, to the unusual sinusoidal mode of sperm movement and the morphology of the female tract (Taggart and Temple-Smith 1990b; 1991). In the American opossums, the occurrence of sperm pairing during epididymal transit may also enhance the efficiency of sperm transport (Moore and Taggart 1995).

Sperm storage and release in the female reproductive tract

In dasyurids and didelphids, extended periods of sperm storage in the lower isthmus of the female reproductive tract occur with the duration of storage in the 13 species studied varying from 2 to 15 days (Table 1). Sperm storage in the isthmus also occurs for at least 24 hours in *Dasycercus byrnei* (Payne and Breed unpubl. obs.). At first glance there does not appear to be a relationship between length of copulation and maximum duration of sperm storage in the female reproductive tract in the carnivorous marsupials studied ($y = 0.8487x + 3.8672$, $r^2 = 0.48$; Fig. 5); however, a strong, positive allometric relationship ($y = 1.0703x + 1.6442$, $r^2 = 0.79$; Fig. 5) is observed if the outlying point for *Planigale maculata* is omitted. In general it appears that in species where copulation length is most prolonged the duration of sperm storage is relatively longer.

The storage capacity of the isthmic crypts in *Antechinus agilis* may become maximal before ovulation (Taggart et al. 1999, Shimmin et al. 1999). Males, therefore, may not only compete for mating access to females, but their sperm may also compete for storage space in the crypts (Dickman 1993; Taggart and Temple-Smith 1991; Taggart et al. 1998; Shimmin et al. 2000a). These crypts, which appear to be lined by an epithelium of a distinctive cellular structure (Breed et al. 1989), facilitate long-term sperm storage as sperm are continually shed from the lumen of the isthmus and are voided in the urine between the time of insemination and ovulation (Selwood and McCallum 1987). However, since sperm are continuously lost from the isthmus up until the time of ovulation, repopulation of crypts by sperm of males that mate subsequently may continue up until the time of ovulation.

Limited data are available on the release of sperm from the isthmic crypts. However it has been suggested that, in *Sminthopsis crassicaudata* at least, a vanguard population of sperm located in the crypts closest to the ovary are the first to be released following ovulation and thus are the most likely ones to fertilise the oocytes (Bedford and Breed 1994).

NUMBER OF OOCYTES OVULATED AND SEX BIAS IN LITTERS

Females of most dasyurids appear to ovulate more oocytes (eggs) than there are teats available for the attachment of pouch young (Hughes 1982; Breed and Leigh 1992; Shimmin et al. 2000b). This greater number of ovulated oocytes compared to teat number probably results in an excessive number of embryos and foetuses. The large numbers of eggs ovulated, and consequently of young born, provide the possibility of competition for teat access between young of different paternity. The chances of success could relate, at least in part, to the birth sequence and /or the fitness of the individual males involved. The maximum number of oocytes ovulated and the teats available within the pouch of several species are indicated in Table 1.

Studies of pouch young within populations of three species of *Antechinus* and several *Didelphids* (Cockburn et al. 1985; Austad and Sunquist 1986; Sunquist and Eisenberg 1993; Davison and Ward 1998) have reported biased sex ratios within litters. Within semelparous *Antechinus* species, first-year animals have a female-biased sex ratio (e.g. *A. agilis*), whereas male-biased litters tended to occur in the first year of breeding in the more iteroparous species. Pre-fertilisation mechanisms have been implicated in the generation of sex-biased litters in *A. agilis* (Cockburn 1990; Davison and Ward 1998), with the most likely mechanism relating to the proportion, assortment, and/or viability of X- and Y-bearing sperm during the period of sperm storage in the female reproductive tract between copulation and fertilisation (Cockburn 1990; Davison and Ward 1998).

MULTIPLE PATERNITY

Multiple paternity has been demonstrated in the agile antechinus (*A. agilis*) (Shimmin et al. 2000a; Kraaijeveld-Smit et al. 2002a; 2002b) and the brush-tailed phascogale (*Phascogale tapoatafa*; Millis 1995). In both species, studies have examined paternity within litters associated with competitive mating trials between two males (Shimmin et al. 2000a; Millis 1995). In addition, in *A. agilis*, data have also recently been collected from wild populations where mixed paternity was also found to occur (Kraaijeveld-Smit et al. 2002a; 2002b). Studies in *A. agilis* indicate that spermatozoa from more than one male can occupy the isthmic sperm storage crypts concurrently prior to ovulation, and that many males (~5) can sire young within the one litter (Selwood and McCallum 1987; Shimmin et al. 2000a; Kraaijeveld-Smit et al. 2002a; 2002b).

In studies on *A. agilis* (Shimmin et al. 2000a), of the 61 young to which DNA paternity was assigned, 72% were sired by the second male that mated when both matings occurred early in oestrus, 62% were sired by the second mating male when one mating occurred early and one in mid-oestrus, and 58% were sired by the second mating male when both matings occurred in mid-oestrus. Overall the second mating male sired 64% of young. In a field study on the same species in which radionuclide labels were used to identify individual mating males, a promiscuous mating strategy was suggested. This has subsequently been verified and the extent of promiscuity detailed using molecular techniques (Kraaijeveld-Smit et al. 2002a; 2002b).

The high incidence of litters sired by multiple males (Shimmin et al. 2000a) may have evolved so that the female could increase her reproductive fitness by effectively outbreeding her own litter (Arnqvist 1989; Loman et al. 1988).

LIKELIHOOD OF SPERM COMPETITION BASED UPON AVAILABLE REPRODUCTIVE DATA

The following data relate to the possible mating system and likelihood of inter-male sperm competition in the various genera of carnivorous marsupials based upon, morphological and behavioural, reproductive data:

Genera *Antechinus* and *Phascogale*

In *Antechinus* (e.g. *Antechinus agilis*), sexual maturity generally occurs at around 11 months of age and mating takes place within a short, highly synchronised period each year (Woolley 1966; Lee and Cockburn 1985). As a consequence of spermatogenic failure prior to the mating season (Woolley 1966, Kerr and Hedger 1983), males must rely on stored epididymal spermatozoa for fertilising females (Taggart and Temple-Smith 1989; 1990a). By contrast, in *P. tapoatafa* (Millis et al. 1999) and *P.calura* (Bradley 1997) spermatogenic activity, although substantially reduced, continues throughout the mating season. In all species examined in both genera, a strong male-biased sexual dimorphism is apparent and all males die within a few weeks of mating (Lee et. al. 1977; Bradley et al. 1980; Cuttle 1982; Lee and Cockburn 1985; Bradley 1997). Although territoriality is weak, it is likely that the heaviest males occupy the most resource-rich habitat, mate first, and subsequently are the first to die (Braithwaite 1979; Williams and Williams 1982; Lee and Cockburn 1985; Bradley and

Monamy 1990). Monoestry and female receptivity lasting 5 to 11 days have been reported for species in both of these genera (Marlow 1961; Woolley 1966; Bradley 1997; Millis et al. 1999). In *A. agilis* a lek mating system has been proposed with males aggregating in a few communal nests, which females visit for mating (Lazenby-Cohen and Cockburn 1988; Cockburn and Lazenby-Cohen 1992). Male and female *A. agilis* and *P. tapoatafa* are known to be promiscuous (Wittenberger 1979; Scott and Tan 1985; Lee and Cockburn 1985; Shimmin et al. 2000a; Millis 1995; Millis et al. 1999; Kraaijeveld pers. comm.). Information presented on the extended periods of copulation, efficient sperm transport in the female, lengthy isthmic sperm storage (Selwood 1980; Selwood and McCallum 1987; Taggart and Temple-Smith 1991; Millis et al. 1999) and paternity studies (Millis 1995; Shimmin et al. 2000a; Kraaijeveld pers. comm.) indirectly support a mating system that is promiscuous. These data, together with the values presented here for relative testis mass and sexual dimorphism, suggest that sperm competition between males is highly likely to occur. Although the data are incomplete, similar reproductive and mating patterns are suggested for other species in these two genera (Inns 1976; Leung 1995; 1999; Bradley 1987; 1995; 1997; Van Dyck 1995). The low relative testis mass values for *A. godmini* and *A. flavipes* may be due to obtaining material during the breeding season after testicular regression has been initiated (Woolley 1966; Inns 1976; Kerr and Hedger 1983; Taggart and Temple-Smith 1990a).

Genus *Sminthopsis*

Social organisation in the iteroparous species of *Sminthopsis*, like that of *S. crassicaudata*, contrasts with that of *Antechinus* in that males generally do not die after mating, so that in the wild a proportion of males, as well as females, survives to breed in a second year (Morton 1978a; 1978b). In all *Sminthopsis* species examined, sperm production in the testes occurs throughout the breeding season (Woolley 1990; Friend et al. 1997; Taggart et al. 1997). Populations are generally less dense than those of *Antechinus* species and the social organisation is 'loose' by comparison (Morton 1978a). Male *S. crassicaudata* (Morton 1978a) and probably *S. youngsoni* and *S. hirtipes* (Dickman et al. 2001) do not appear to defend a territory or home range.

Seasonal, or extended seasonal, breeding patterns have been observed in *S. crassicaudata, S. murina, S. leucopus, S. ooldea, S. griseoventer, S. hirtipes and S. dolichura* that inhabit mesic and semi-arid regions. Others, like *S. virginiae,* that occur in the wet–dry tropics, exhibit a more flexible breeding strategy, including seasonal breeding (Morton et al. 1987; see Strahan 1995) as well as year-round breeding (Taplin 1980).

Females are polyoestrus, receptivity is relatively short (1 to 3 days), and sperm storage within the female tract occurs for up to 3 days (Morton 1978c; Breed et al. 1989). As only a single adult male has been found with pro-oestrous or oestrous females, litters in most species are probably sired by a single male (Morton 1978a; Lee and Cockburn 1985), although when two captive males were tested with a single female at oestrus mating by both males occurred in one out of four tests carried out (Lampard, Sarma and Breed unpubl. obs.). Values for relative testis mass vary from being large in *S. crassicaudata, S. murina, S. granulipes, S. leucopus, S. griseoventer* (Boullanger Island) and *S. macroura* to small in *S. bindi* and *S. aitkeni*. Interestingly the species that fall above the regression line inhabit arid or semi-arid regions, whereas those that fall below the regression line are found in more temperate or tropical regions.

Sperm competition in the female reproductive tract probably sometimes occur in *S. crassicaudata, S. murina, S. granulipes, S. leucopus, S. griseoventer* (Boullanger Is) and *S. macroura,* but perhaps not in *S. virginiae, S. dolichura, S. hirtipes, S. virginiae nitela, S. virginiae* (PNG), *S. ooldea, S. bindi* and *S. aitkeni*.

Genus *Ningaui*

Radio tracking suggests that individuals in this genus do not have fixed home ranges and they occur at low density. Extended seasonal breeding would appear to be the norm (Coventry and Dixon 1984; Kitchener et al. 1986), although long-term field studies of *N. ridei* indicate that oestrus and parturition only occur in the spring (Dickman et al. 2001). Males and females do not often appear to survive into a second year (Dunlop and Sawle 1983). There is no evidence of spermatogenic failure before, or during, the mating season, or of male die-off following mating (Kitchener et al. 1986).

Both males and females produce mate-seeking calls before and during the breeding season (Fanning 1982) and captive adults are intolerant of one another outside the breeding season. Males have a sternal gland and actively mark bark and other surfaces as they pass along them (Fanning 1982). In male *N. ridei,* sternal gland activity appears to be closely associated with the reproductive season (Dickman et al. 2001).

Behavioural oestrus may last for only one to three days (Fanning 1982), the maximum length of copulation is 4.5 to 5 hours (Fanning 1982; Dickman unpubl. obs.). Sperm are stored for 2 to 3 days. Relative testis mass in all species examined was near the regression line or below it. Thus the data on testis size, reproductive behaviour and social structure suggest that the likelihood of sperm competition occurring is low and that the mating system may involve a mate defence polygyny, where a male defends an oestrous female during her short period of receptivity.

Genus *Parantechinus*

Both *P. apicalis,* which occurs in the south-west of Western Australia, and *P. bilarni* in tropical regions of the Northern Territory, appear to have restricted breeding seasons (Woolley 1995; Woolley and Begg 1995). Total male die-off has been reported in the island, but not mainland, population of *P. apicalis* (Dickman and Braithwaite 1992; Woolley 1991; Mills and Bencini 2000), and it does not occur in *P. bilarni* (Begg 1981). Like species in the genera *Antechinus* and *Phascogale,* both *P. apicalis* and *P. bilarni* are monoestrus (Woolley 1991; 1994). Spermatogenesis occurs throughout the mating season in both species (Begg 1981; Woolley 1991; Dickman and Braithwaite 1992), with relative testis mass being considerably greater in *P. apicalis.* Maximum length of copulation, and sperm storage in the female reproductive tract, of *P. apicalis* are relatively long, but in *P. bilarni* they are short (Woolley 1995; Wolfe et al. 2000). A strong male-biased sexual dimorphism in body mass has been reported in *P. apicalis,* whereas only a small male bias body mass has been found to occur in *P. bilarni* (Woolley 1995; Woolley and Begg 1995). Based on the available evidence, it would appear that sperm competition is likely in *P. apicalis* and perhaps less likely in *P. bilarni.*

Genus *Pseudantechinus*

The field data available suggest that both *P. ningbing* and *P. macdonnellensis* are monoestrous, and that in *P. macdonnellensis* at least, there is a degree of site fidelity displayed by females (Woolley 1995). Laboratory and field data suggest that the breeding season in *P. ningbing* and *P. macdonnellensis* is short and highly seasonal and that spermatogenic failure does not occur prior to the breeding season (Woolley 1988; 1991; Gilfillian 2001). Males in this genus are significantly larger than females and in *P. macdonnellensis* at least, may survive to reproduce in 2 years and females up to 3 (Gilfillian 2001).

Relative testis mass is high for *P. macdonnellensis,* and low for *P. ningbing.* Data from captive breeding studies suggest that length of copulation is short in *P. macdonnellensis* and moderate in *P. ningbing.* Length of sperm storage in both species is up to about 12 days. A high relative testes mass and long periods of sperm storage in the female reproductive tract point toward the occurrence of sperm competition in *P. macdonnellensis,* at least.

Genus *Planigale*

Planigales occur in relatively low densities, often in areas close to permanent water, or in areas that are periodically flooded on cracking soils (Denny 1982; Denny et al. 1979). Data available suggest that animals have shifting home ranges, and sometimes occupy communal nests, although olfactory marking has been observed (Morrison 1975; Van Dyck 1979; Andrews and Settle 1982; Read 1984a; 1984b; 1987). The presence of a dominance hierarchy has been reported in captive colonies of several species (Morrison 1975; Van Dyck; 1979; Andrews and Settle 1982). In two of these studies females were found to be dominant to males (Morrison 1975; Van Dyck 1979).

Extended seasonal breeding appears to occur (Aslin 1975; Read 1984b; 1995; Van Dyck 1995). All species are polyoestrous and can potentially rear several litters in a year (Aslin 1975; Taylor et al. 1982). Testicular regression prior to breeding does not appear to occur (Aslin 1975). Behavioural oestrus lasts for one day in *P. tenuirostris* (Read 1995), 2 days in *P. maculata,* and 3 days in *P. gilesi,* with the maximum length of copulation observed being about 2.5 hours in *P. maculata* (Aslin 1975; Whitford et al. 1982). No sexual dimorphism is apparent in *P. gilesi,* but in the other species a small to moderate male-biased sexual dimorphism has been found (Denny 1982; Taylor et al. 1982; Dickman unpubl. obs.).

The available information, including low animal densities, short copulation length, short length of behavioural oestrus, and little sexual dimorphism suggests that sperm competition may not occur.

Genus *Dasycercus*

In the mulgara, *Dasycercus cristicauda* little is known of the social organisation and reproductive biology (Woolley 1971; 1995; Masters 1998). Females are seasonal breeders, producing one litter per year (Masters 1998), and are probably monoestrous (Dickman unpubl. obs.). The maximum length of copulation reported is 3 hours (Dickman et al. 2001). Males and females may survive for several years (Masters 1998). The relative testis mass is high and a strong male-biased sexual dimorphism is apparent (Dickman unpubl. obs.).

The kowari, *Dasycercus byrnei,* has overlapping home ranges of several kilometres. Animals appear to be solitary only coming together briefly to mate (Aslin and Lim 1995). Extended seasonal breeding has been reported (Woolley 1971; Aslin and Lim 1995). Females are polyoestrous, mating is relatively short (<3 hours), and behavioural oestrus lasts for 1 and 3 days (Woolley 1971). A moderate sexual dimorphism in favour of males has been observed (Table 1) and testis mass is large relative to body mass. Total male die-off does not occur in species within this genus, nor is there spermatogenic cessation before the mating season (Woolley 1971; Masters 1998; Dickman et al. 2001).

The available data on the mulgara including a strong male-biased sexual dimorphism, large relative testes mass and monoestry suggest sperm competition may occur in this species. In the kowari, however, the data appears to conflict as information on relative testis mass and male sexual dimorphism suggest that sperm competition is likely but their solitary existence, low densities, and relatively short length of copulation and behavioural oestrus perhaps suggest otherwise.

Genus *Dasykaluta*

The breeding season in *Dasykaluta rosamondae* is highly seasonal, with one litter being produced per season (Ride 1964; Woolley 1991). Males of this species die shortly after the mating season (Ride 1964; Woolley 1991). Although males can breed in only one season, females may survive to breed over 2 or more years (Woolley 1991). Relative testis mass may be low in this species (see Tyndale-Biscoe and Renfree 1987) but, based on similarities in life history to those of *Antechinus* and *Phascogale,* testis and body weight data may have been collected during the breeding season when testes mass may be somewhat reduced. A moderate male-biased sexual dimorphism is apparent.

Based on the available data and the similarities in life history between this species and those of *Antechinus* and *Phascogale,* sperm competition is considered likely.

Genus *Dasyurus*

All four Australian species are polyoestrous and display a seasonal breeding pattern (Fletcher 1985; Soderquist and Serena 1990; Edgar and Belcher 1995; Oakwood 2000; Oakwood et al. 2001). Although home ranges of individuals overlap considerably, both males and females tend to be solitary and actively defend core home range areas (Godsell 1982; 1995; Dempster 1995). Olfactory marking is common in all species (Dempster 1995). No complete post-mating mortality of males has been observed in *D. viverrinus* or *D.maculatus* (Fletcher 1985; Schmitt et al. 1989; Edgar and Belcher 1995), although it may sometimes occur in *D. hallucatus* (Dickman and Braithwaite 1992; Oakwood 2000). Sperm production continues throughout the breeding season in *D. viverrinus* (Fletcher 1985; Bryant 1986) and *D. hallucatus* (Oakwood 2000; Oakwood et al. 2001), although in *D. hallucatus* scrotal width peaks 2 months prior to the mating season and then declines significantly (Oakwood 2000; Oakwood et al. 2001).

The only information available on length of copulation is that for *D. maculatus* in which males mate for 8 to 12 hours (Edgar and Belcher 1995; Jones pers. comm.). Behavioural oestrus in this species, and in *D. viverrinus,* can last for up to 3 days (Hill and O'Donoghue 1913; Edgar and Belcher 1995) with sperm storage in the female tract of *D. viverrinus* occurring for up to 7 to 14 days prior to ovulation (Hill and O'Donoghue 1913). Relative testis mass is low–average in *D. viverrinus,* but possibly considerably higher in *D. hallucatus.* A strong male-biased sexual dimorphism is evident in all species examined (Begg 1981; Fletcher 1985; Dempster 1995). The available data suggests that the likelihood of inter-male sperm competition is low -moderate in *D. viverrinus,* and *D.maculatus,* but possibly greater in *D. hallucatus.*

Genus *Sarcophilus*

The Tasmanian devil, *Sarcophilus harrisii,* has individual, extensively overlapping home ranges (Pemberton and Renouf 1993; Jones 1995). In this monoestrous species, breeding is highly synchronised with mating occurring in March each year (Hughes 1982; Jones 1995). Males do not die after mating and the testes are active throughout the breeding season (Hughes 1982). Relative testis mass is large, and a moderate male-biased sexual dimorphism is apparent (Hughes 1982). Mating is thought to be promiscuous and lasts for about 1.5 hours (Jones pers. comm.). The available information suggests that sperm competition is likely in this species.

Genera *Myoicitis, Murexia, Phascolosorex* and *Neophascogale*

With the exception of the three-striped dasyure, *Myoicitis melas,* which has larger than average testes for body mass, relative testes mass of the other five species examined was average or below average for body mass (Woolley pers. comm.) (Table 1). Data on male and female body weight in the short-furred dasyure, *Murexia longicaudata* (Flannery 1995), suggest a very strong male-biased sexual dimorphism. Due to the lack of information on reproduction and life history in these species (Woolley 1994), no predictions are made about the likelihood of inter-male sperm competition.

Genera *Monodelphis* and *Marmosa*

Of the 74 marsupial species that inhabit South America, detailed reproductive and life history data are available for only three species. The genus *Monodelphis* comprises approximately 15 species, all of which appear to be carnivorous (Moore 1992; Bergallo and Cerqueira 1994; Nowak 1999). Captive studies on *M. domestica* have demonstrated that females will mate with multiple males within the one oestrus (Moore and Taggart unpubl. obs.). Based on captive studies, behavioural oestrus lasts for 1 to 1.5 days and concludes with ovulation (Moore 1992). Male and females lock during mating, which lasts for approximately 5 minutes. Females ovulate 18 to 24 hours after their first mating (Moore 1992). No mating occurs post-ovulation. Sperm storage also occurs in isthmic crypts and lasts for up to one day (Moore 1992).

M. domestica is thought to have a solitary, semi-arboreal, lifestyle in the wild (Collins 1973; Emmons 1999), supported by the observations in captivity that both sexes are very aggressive and will fight to the death if a male is introduced to a female when she is not in behavioural oestrus (Moore and Taggart unpubl. obs.). It has been demonstrated that oestrus is pheromonally induced by the presence of an unfamiliar male (Fadem and Rayve 1985; Baggott et al. 1987; Moore 1992). A moderate male-biased sexual dimorphism occurs as it does in other species within the genus *Monodelphis* (Pine et al. 1985; Bergallo and Cerqueira 1994). Sperm production takes place throughout the year in captivity and relative testis mass is about average for body weight. Females are polyoestrous. In the wild *M. domestica* has an extended seasonal breeding pattern (Bergallo and Cerqueira 1994), with both sexes surviving to reproduce over several seasons (Bergallo and Cerqueira 1994). This, however, may not be the case for all *Monodelphis* species, as both *M. dimidiata* and *M. henseli* are reported to exhibit semelparity (Pine 1994; Pine et al.

1985). A post-mating mortality of males has also been described in *M. dimidiata* (Pine 1994).

Together these data suggest that sperm competition is unlikely in *M. domestica,* due to their relatively short oestrus, short periods of sperm storage in the female tract prior to ovulation and solitary existence. It is possible that this is the case for at least some of the other species within the genus; however, as semelparity has been described in two species (Pine et al. 1985), post-mating mortality of males in one and a large interspecific variation in extent of sexual dimorphism in others (Ventura et al. 1998), it is likely that interspecies differences in mating system and the likelihood of sperm competition will also occur.

The genus *Marmosa* contains nine species (Nowak 1999). Robinson's mouse opossum, *M. robinsoni,* is the only species that has been studied in any detail and is reputedly solitary in nature, nomadic and occurs at low densities (Fleming 1972; 1973; Nowak 1999). Life history studies have suggested that while there is overlap between home ranges of males and between opposite sexes, females do not have overlapping home ranges (Fleming 1972; 1973). Copulation may last for up to 3 hours in the wild, and up to 7 hours in captivity (Barnes and Barthold 1969; Hunsaker and Shupe 1977). The length of behavioural oestrus appears to be about 3 days (Barns and Barthold 1969). A strong male-biased sexual dimorphism has been reported (Harder 1992). Females are polyoestrous and can produce up to two litters annually, although the average lifespan is probably less than one year (Hunsaker 1977). Vaginal plugs have been reported.

Although data are incomplete for *M. robinsoni,* the longer periods of copulation and behavioural oestrus, together with the presence of copulatory plugs, suggest that sperm competition may occur.

Of the other South American species, the only relevant data we could find include that for the *Marmosops* genus (Family Marmosidae), in which a post-mating mortality of all males has been reported (Lorini, de Oliveira and Persson 1994; Emmons 1999), and for *Dromisiops australis* (Family Microbiotheriidae), which has been reported as living in pairs, at least during the mating season.

FUTURE DIRECTIONS

This review has shown that there is a correlation between relative testis mass and body mass within the carnivorous marsupials, as has been found for eutherian mammals (Harcourt et al. 1981; Kenagy and Trombulak 1986) and the Marsupialia (Rose et al. 1997; Taggart et al. 1998). When data for species within a genus are examined, similar trends are apparent, and for some groups, there is a marked deviation from the mean for body mass.

Data on breeding systems, copulatory behaviour, paternity guarding and social organisation are relatively sparse but, in some cases, the occurrence of a multi-male breeding system is evident with a high probability for inter-male sperm competition. In general, these species tend to have a large relative testis mass, long periods of behavioural oestrus, lengthy copulations, prolonged periods of sperm storage within the female reproductive tract prior to ovulation, strong male-biased size dimorphism, and higher population densities. These features, along with others like testes that produce few, but very large sperm, total spermatogenic failure prior to the mating season and highly efficient sperm transport within the female reproductive tract, suggest that particular species of marsupial carnivore would serve as excellent model species for studies of sperm competition in mammals.

The study of this group of marsupials therefore offers a unique opportunity to test many of the predictions associated with sperm competition theory that remain unresolved – for instance, issues such as multiple paternity within litters, differential investment by males in association with sperm competition, the control and effect of mating order on paternity success and the concept that females might exert some physiological control over the paternity of their offspring.

We present these data as a starting point for discussions on sperm competition and the mating systems of the carnivorous marsupials. The strength of the predictions made are only as good as the data from which they are drawn and hence, as more information becomes available, the accuracy of these assessments on the likelihood of inter-male sperm competition will improve.

ACKNOWLEDGEMENTS

We thank Dr Pat Woolley for access to dasyurid material from New Guinea and Ms Rachel Norris for assistance in calculating residuals for relative testes mass.

REFERENCES

Andrew, D.E., & Settle, G.A. (1982), 'Observations on the behaviour of the planigale with particular reference to the narrow-nosed planigale (*Planigale tenuirostris*)', in *Carnivorous Marsupials* (ed. M. Archer), pp. 311–24, Roy. Zool. Soc. NSW Sydney, Australia.

Arnqvist, G. (1989), 'On multiple mating and female fitness: comments on Loman et al. (1988)', *Oikos,* **54**:248–50.

Aslin, H. (1975), 'Reproduction in *Antechinus maculatus* Gould (Dasyuridae)', *Aust Wildl Res,* **2**:77–80.

Aslin, H.J., & Lim, L. (1995). 'Kowari', in *The Mammals of Australia* (ed. R. Strahan), pp. 59–61, Reed Books, Chatswood, NSW, Australia.

Austad, S.N., & Sunquist, M.E. (1986), 'Sex ratio manipulation in the common opossum', *Nature,* **324**:58–60.

Baggott, L.M., Davis-Butler, S., & Moore, H.D.M. (1987), 'Characterisation of oestrus and timed collection of oocytes in the grey short-tailed opossum, *Monodelphis domestica*', *J Reprod Fert,* **79**:105–14.

Barnes, R.D., & Barthold, S.W. (1969), 'Reproduction and breeding behaviour in an experimental colony of *Marmosa mitis* Bangs (Didelphidae)', *J Reprod Fert, (Suppl),* **6**:477–82.

Bedford, J.M., & Breed, W.G. (1994), 'Regulated storage and subsequent transformation of spermatozoa in the Fallopian tubes of an Australian marsupial, *Sminthopsis crassicaudata*', *Biol Reprod,* **50**:845–54.

Bedford, J.M., Rodger, J.C., & Breed, W.G. (1984), 'Why so many mammalian spermatozoa – a clue from marsupials?', *Proc R Soc Lond,* **221**:221–33.

Begg, R.J. (1981), 'The small mammals of Little Nourlangie Rock, N.T. III. Ecology of *Dasyurus hallucatus,* the northern quoll (Marsupialia: Dasyuridae)', *Aust Wildl Res,* **8**:73–85.

Bergallo, H.G., & Cerqueira, R. (1994), 'Reproduction and growth of the opossum, *Monodelphis domestica* (Mammalia: Didelphidae) in north-eastern Brazil', *J Zool,* **232**:551–63.

Bradley, A.J. (1987), 'Stress and mortality in the red–tailed phascogale, *Phascogale calura* (Marsupialia: Dasyuridae)', *Gen Comp Endocrin,* **67**:85–100.

Bradley, A.J. (1997), 'Reproduction and life history in the red–tailed phascogale, *Phascogale calura* (Marsupialia: Dasyuridae): the adaptive-stress senescence hypothesis', *J Zool (Lond),* **241**:739–55.

Bradley A.J. (1995), 'Red-tailed phascogale', in *The Mammals of Australia* (ed. R. Strahan), pp. 102–3, Reed Books, Chatswood, NSW, Australia.

Bradley, A.J., & Monamy, V. (1990), 'A physiological profile of male dusky antechinuses, *Antechinus swainsonii* (Marsupialia: Dasyuridae) surviving post-mating mortality', *Aust Mammal,* **14**:25–7.

Bradley, A.J., McDonald, I.R., & Lee, A.K. (1980), 'Stress and mortality in a small marsupial (*Antechinus stuartii*, Macleay), *Gen Comp Endocrin,* **40**:188–200.

Braithwaite, R.W. (1979), 'Social dominance and habitat utilisation in *Antechinus stuartii* (Marsupialia)', *Aust J Zool,* **27**:517–28.

Breed, W.G. (1994), 'How does sperm meet egg? – in a marsupial', *Rep Fertil Devel,* **6**:485–506.

Breed, W.G., Leigh, C.M., & Bennett, J.H. (1989), 'Sperm morphology and storage in the female reproductive tract of the fat-tailed dunnart, *Sminthopsis crassicaudata* (Marsupialia: Dasyuridae)', *Gamete Res,* **23**:61–75.

Breed, W.G., & Leigh, C.M. (1992), 'Marsupial fertilization: some further ultrastructural observations on the dasyurid *Sminthopsis crassicaudata*', *Mol Rep Dev,* **32**:277–92.

Bryant S.L. (1986), 'Seasonal variation of plasma testosterone in a wild population of male eastern quoll, *Dasyurus viverrinus* (Marsupialia: Dasyuridae) from Tasmania', *Gen Comp Endocrin,* **64**:75–9.

Cockburn, A. (1990), 'Sex ratio variation in marsupials', *Aust J Zool,* **37**:467–79.

Cockburn, A., & Lazenby-Cohen, K.A. (1992), 'Use of nest trees by *Antechinus stuartii,* a semelparous lekking marsupial', *J Zool (Lond),* **226**:657–80.

Cockburn, A., Scott, M.P., & Dickman, C.R. (1985), 'Sex ratio and intrasexual kin competition in mammals', *Oecologia,* **66**:427–29.

Collins, L.R. (1973), *Monotremes and marsupials. A reference for zoological institutions*, Smithsonian Institution Press, Washington.

Coventry, A.J., & Dixon, J.M. (1984), 'Small native mammals from the Chinaman Well area of North-western Victoria', *Aust Mammal,* **7**:111–15.

Cuttle P. (1982), 'A preliminary report on aspects of the behaviour of the dasyurid marsupial *Phascogale tapoatafa*', in *Carnivorous marsupials* (ed. M. Archer), pp. 325–32, Royal Zoological Society of New South Wales, Sydney.

Davison, M.J., & Ward, S.J. (1998), 'Prenatal bias in sex ratios in a marsupial, *Antechinus agilis*', *Proc R Soc Lond,* **265**:2095–9.

Dempster, E.R. (1995), 'The social behaviour of captive northern quolls, *Dasyurus hallucatus*', *Aust Mamm,* **18**:27–34.

Denny, M.J.S. (1982), 'Review of planigale (Dasyuridae, Marsupialia) ecology' in *Carnivorous marsupials* (ed. M. Archer), pp. 131–8, Royal Zoological Society of New South Wales, Sydney.

Denny, M.J.S., Gibson, D., & Read, D. (1979), 'Some results of field observations on planigales', *Aust Mamm Soc Bull,* **5**:27.

Dickman, C.R. (1993), 'Evolution of semelparity in male dasyurid marsupials: A critique and an hypothesis of sperm competition', in *The biology and management of Australasian carnivorous marsupials* (eds. M. Roberts, J. Carnio, G. Crawshaw, & M. Hutchins), pp 25–38, American Association of Zoological Parks and Aquariums Publishers, Toronto.

Dickman, C.R., Haythornthwaite, A.S., McNaught, G.H., Mahon, P.S., Tamayo, B., & Letnic, M. (2001), 'Population dynamics of three species of dasyurid marsupials in arid central Australia: a 10-year study', *Wildlf Res,* **28**:493–506.

Dickman, C.R., & Braithwaite, R.W. (1992), 'Postmating mortality of males in the dasyurid marsupials, *Dasyurus* and *Parantechinus*', *J Mamm* **73**:143–7.

Dunlop, J.N., & Sawle, M. (1983), 'The habitat and life history of the Pilbara Ningaui, *Ningaui timealeyi*', *Rec West Aust Mus,* **10**:47–52.

Edgar, R., & Belcher, C. (1995), 'Spotted-tailed quoll', in *The Mammals of Australia* (ed R. Strahan), pp. 67–9, Reed Books, Chatswood, NSW, Australia.

Emmons, L.H. (1999), *Neotropical rainforest mammals* (ed, L.H. Emmons), University of Chicago Press, London, UK.

Ewer, R.F. (1968), 'A preliminary survey of the behaviour in captivity of the dasyurid marsupial *Sminthopsis crassicaudata*', *Z Tierpshchol,* **25**:319–65.

Fadem, B.H., & Rayve, R. (1985), 'Characteristics of the oestrous cycle and influence of social factors in grey short-tailed opossums (*Monodelphis domestica*)', *J Reprod Fert,* **73**:337–42.

Fanning, F.D. (1982), 'Reproduction, growth and development in *Ningaui* sp (Dasyuridae, Marsupialia) from the Northern Territory', in *Carnivorous Marsupials* (ed. M. Archer), pp. 23–37, Royal Zoological Society of NSW, Sydney.

Flannery, T. (1995), 'Short-furred dasyure', in *Mammals of New Guinea* (ed. T. Flannery), pp. 87–9, Reed Books, Chatswood, NSW, Sydney, Australia.

Fleming, T.H. (1972), 'Aspects of the population dynamics of three species of opossums in the Panama Canal Zone', *J Mammal,* **53**:619–23.

Fleming, T.H. (1973), 'The reproductive cycles of three species of opossums and other mammals in the Panama Canal Zone', *J Mammal,* **54**:439–55.

Fletcher T.P. (1985), 'Aspects of reproduction in the male eastern quoll, *Dasyurus viverrinus* (Shaw) (Marsupialia: Dasyuridae), with notes on polyoestry in the female', *Aust J Zool,* **33**:101–10.

Friend, G.R., Johnson, B.W., Mitchell, D.S., & Smith, G.T. (1997), 'Breeding, population dynamics and habitat relationships of *Sminthopsis dolichura* (Marsupialia: Dasyuridae) in semi arid shrublands of Western Australia', *Wildlf Res,* **24**:245–62.

Gilfillian, S.L. (2001), 'An ecological study of a population of *Pseudantechinus macdonnellensis* (Marsupialia: Dasyuridae) in central Australia. I. Invertebrate food supply, diet and reproduction', *Wildlf Res*, **28**:469–80.

Godsell, J. (1982), 'The population ecology of the eastern quoll, *Dasyurus viverrinus*, in southern Tasmania', in *Carnivorous Marsupials* (ed. M. Archer), pp. 199–207, Roy. Zool. Soc. NSW Sydney, Australia.

Godsell, J. (1995), 'Eastern quoll', in *The Mammals of Australia* (ed. R. Strahan), pp. 70–1, Reed Books, Chatswood, NSW, Australia.

Harcourt, A.H. (1991), 'Sperm competition and the evolution of non-fertilising sperm in mammals', *Evolution*, **45**:314–28.

Harcourt, A.H., Harvey, P.H., Larson, S.G., & Short, R.V. (1981), 'Testis weight, body weight and breeding system in primates', *Nature, London*, **293**:55–7.

Harder, J.D. (1992), 'Reproductive biology of South American marsupials', in *Reproductive Biology of South American Vertebrates* (ed W.C. Hamlett), pp. 211–28, Springer-Verlag, New York.

Hartman, C.G. (1924), 'Observations on the motility of the opossum genital tract and the vaginal plug', *Anat Rec*, **27**:293–303.

Hill, J.P., & O'Donoghue, C.H. (1913), 'The reproductive cycle in the marsupial *Dasyurus viverrinus*', *Q J Microscopical Sci*, **59**:133–74.

Hughes, R.L. (1982), 'Reproduction in the Tasmanian devil', in *Carnivorous Marsupials* (ed. M. Archer), pp. 49–63, Roy. Zool. Soc. NSW, Sydney, Australia.

Hughes, R.L., & Rodger, J.C. (1971), 'Studies on the vaginal mucus of the marsupial, *Trichosurus vulpecula*', *Aust J Zool*, **19**:19–33.

Hunsaker, D. (1977), 'Ecology of New World marsupials', in *The Biology of Marsupials*, (ed D. Hunsaker), pp. 95–158, Academic Press, New York.

Hunsaker, D., & Shupe, D. (1977), 'Behaviour of New World marsupials', in *The Biology of Marsupials*, (ed D. Hunsaker), pp. 279–347, Academic Press, New York.

Inns, R.W. (1976), 'Some seasonal changes in *Antechinus flavipes* (Marsupialia: Dasyuridae)', *Aust J Zool*, **24**:523–31.

Jones, M. (1995), 'Tasmanian devil', in *The Mammals of Australia* (ed. R. Strahan), pp. 82–4, Reed Books, Chatswood, NSW, Australia.

Kenagy, G.J., & Trombulak, S.C. (1986), 'Size and function of mammalian testes in relation to body size', *J Mamm*, **67**:1–22.

Kerr, J.B., & Hedger, M.P. (1983), 'Spontaneous spermatogenic failure in the marsupial mouse *Antechinus stuartii* MacLeay (Dasyuridae: Marsupialia)', *Aust J Zool*, **31**:445–66.

Kitchener D.J., Cooper, N., & Bradley, A.J. (1986), 'Reproduction in male *Ningaui* (Marsupialia: Dasyuridae)', *Aust Wildlf Res*, **13**:13–25.

Kraaijeveld-Smit, F.J., Ward, S.J., & Temple-Smith, P.D. (2002a), 'Multiple paternity in a field population of a small carnivorous marsupial, the agile antechinus, *Antechinus agilis*', *Behav Ecol Sociobiol*, **52**:84–91.

Kraaijeveld-Smit, F.J., Ward, S.J., Temple-Smith, P.D., & Paetkau, D. (2002b), 'Factors influencing paternity success in *Antechinus agilis*: last-male sperm precedence, timing of mating and genetic compatibility', *J Evol Biol*, **15**:100–107.

Lazenby-Cohen, K.A., & Cockburn, A. (1988), 'Lek promiscuity in a semelparous mammal, *Antechinus stuartii* (Marsupialia: Dasyuridae)', *Behav Ecol Sociobiol*, **22**:195–202.

Lee, A.K., Bradley, A.J., & Braithwaite, R.W. (1977), 'Corticosteroid levels and male mortality in *Antechinus stuartii*', in *The Biology of Marsupials* (eds. B. Stonehouse, & D. Gilmore), pp. 209–20, MacMillan, London.

Lee, A.K., & Cockburn, A. (1985), *Evolutionary Ecology of Marsupials*, Cambridge University Press, Cambridge.

Lee, A.K., Woolley, P., & Braithwaite, R.W. (1982), 'Life history strategies of dasyurid marsupials', in *Carnivorous Marsupials* (ed. M. Archer), pp. 1–11, Roy. Zool. Soc. NSW Sydney, Australia.

Leung, L.K.P. (1995), 'Cinnamon antechinus', in *The Mammals of Australia* (ed R. Strahan), pp. 91–2, Reed Books, Chatswood, NSW, Australia.

Leung, , L.K.P. (1999), 'Ecology of Australian tropical rainforest mammals. I. The Cape York antechinus, *Antechinus leo* (Dasyuridae: Marsupialia)', *Wildlf Res*, **26**:287–306.

Loman, J., Madsen, T., & Hakansson, T. (1988), 'Increased fitness from multiple matings and genetic heterogeneity: a model of a possible mechanism', *Oikos*, **52**:69–72.

Lorini, M.L., De Oliveira, J.A., & Persson, V.G. (1994), 'Annual age structure and reproductive patterns in *Marmosa incana* (Lund, 1841) (Dideliphidae, Marsupialia)', *Z Saugetierkd*, **59**:65–73.

McCrady, E. (1938), 'The embryology of the opossum', *Am Anat Memoirs*, **16**:1–233.

McLachlan, R.I., Wreford, N.G., de Kretser, D.M., & Robertson, D.M. (1994), 'The effects of testosterone on spermatogenic cell populations in the adult rat', *Biol Reprod*, **51**:945–55.

Marlow, B.J. (1961), 'Reproductive behaviour of the marsupial mouse, *Antechinus flavipes* (Waterhouse) (Marsupialia) and the development of pouch young', *Aust J Zool*, **9**:203–18.

Masters, P. (1998), 'The mulgara, *Dasycercus cristicauda* (Marsupialia: Dasyuridae) at Uluru National Park, Northern Territory', *Aust Mamm*, **20**:403–4.

Millis A.L. (1995), 'Reproductive and genetic studies of the brush-tailed phascogale, *Phascogale tapoatafa* (Marsupialia: Dasyuridae)', BSc (Honours) thesis, Department of Anatomy, Monash University, Clayton Victoria.

Millis, A., Taggart, D.A., Bradley, A.J; Phelan, J., & Temple-Smith, P.D. (1999), 'Reproductive biology of the brush-tailed phascogale, *Phascogale tapoatafa* (Marsupialia: Dasyuridae)', *J Zool (Lond)*, **248**:325–35.

Mills, H.R., & Bencini, R. (2000), 'New evidence of facultative male die-off in island populations of dibblers, *Parantechinus apicalis*', *Aust J Zool*, **48**:501–10.

Moore, H.D.M. (1992), 'Reproduction in the grey short-tailed opossum, *Monodelphis domestica*', in *Reproductive Biology of South American Vertebrates* (ed. W.C. Hamlett), pp. 229–41, Springer-Verlag, New York.

Moore, H.D.M., & Taggart, D.A. (1995), 'Sperm pairing in the opossum increases the efficiency of sperm movement in a viscous environment', *Biol Reprod*, **52**:947–53.

Morrison, R.G.B. (1975), 'Emergence of the pygmy antechinus', *Aust Nat Hist*, **18**:164–7.

Morton, S.R. (1978a), 'An ecological study of *Sminthopsis crassicaudata* (Marsupialia: Dasyuridae), I. Distribution, study areas and methods', *Aust Wildlf Res*, **5**:151–62.

Morton, S.R. (1978b), 'An ecological study of *Sminthopsis crassicaudata* (Marsupialia: Dasyuridae), II. Behaviour and social organization', *Aust Wildlf Res*, **5**:163–82.

Morton, S.R. (1978c), 'An ecological study of *Sminthopsis crassicaudata* (Marsupialia: Dasyuridae), III. Reproduction and life-history', *Aust Wildlf Res*, **5**:183–211.

Morton, S.R. (1982), 'Dasyurid marsupials of the Australian arid zone: An ecological review', in *Carnivorous Marsupials* (ed. M. Archer), pp. 117–30, Roy. Zool. Soc. NSW Sydney, Australia.

Morton, S.R., Armstrong, M.D., & Braithwaite, R.W. (1987), 'The breeding season of *Sminthopsis virginiae* (Marsupialia: Dasyuridae) in the Northern Territory', *Aust Mammal*, **10**:41–2.

Nowak, R.M. (1999), *Walker's Mammals of the World* (ed. R.M. Nowak), pp. 17–68, John Hopkins University Press, London, UK.

Oakwood. M. (2000), 'Reproduction and demography of the northern quoll, *Dasyurus hallucatus*, in the lowland savanna of Northern Australia', *Aust J Zool*, **48**:519–39.

Oakwood, M., Bradley, A.J., & Cockburn, A. (2001), 'Semelparity in a large marsupial', *Proc R Soc Lond B*, **268**:407–411.

Overstreet, J.W., & Cooper, G.W. (1978), 'Sperm transport in the reproductive tract of the female rabbit I. The rapid transit phase of transport', *Biol Reprod*, **19**:101–14.

Overstreet, J.W., & Cooper, G.W. (1979), 'Effect of ovulation and sperm motility on the migration of rabbit spermatozoa to the site of fertilization', *J Reprod Fert*, **55**:53–9.

Pemberton, D., & Renouf, D. (1993), 'A field study of communication and social behaviour of the Tasmanian devil at feeding sites', *Aust J Zool*, **41**:507–26.

Pine, R.H. (1994), 'Sex and death', *Aust Nat Hist*, **24**:4.

Pine, R.H., Dalby, P.L., & Matson, J.O. (1985), 'Ecology, postnatal development, morphometrics and taxonomic status of the short-tailed opossum, *Monodelphis dimidiata*, an apparently semelparous annual marsupial', *Ann Carneg Mus*, **54**:195–231.

Read, D.G. (1984a), 'Reproduction and breeding season of *Planigale gilesi* and *P. tenuirostris* (Marsupialia: Dasyuridae)', *Aust Mamm*, **7**:161–73.

Read, D.G. (1984b), 'Movements and home ranges of three sympatric dasyurids, *Sminthopsis crassicaudata*, *Planigale gilesi* and *P. tenuirostris* (Marsupialia: Dasyuridae), in semi arid western New South Wales', *Aust Wildlf Res*, **11**:223–34.

Read, D.G. (1987), 'Habitat use by *Sminthopsis crassicaudata*, *Planigale gilesi* and *P.tenuirostris* (Marsupialia), in semi arid New South Wales, Australia', *Aust Wildlf Res*, **14**:385–96.

Read, D.G. (1995), 'Narrow-nosed Planigale', in *The Mammals of Australia* (ed. R. Strahan), pp. 113–15, Reed Books, Chatswood, NSW, Australia.

Ride, W.D.L. (1964), '*Antechinus rosamondae*, a new species of dasyurid marsupial from the Pilbara district of Western Australia; with remarks on the classification of *Antechinus*', *West Aust Nat*, **9**:58–65.

Rose, R.W., Nevison, C.M., & Dixon, A.F. (1997), 'Testes weight, body weight and mating systems in marsupials and monotremes', *J Zool (Lond)*, **243**:523–32.

Schmitt, L.H., Bradley. A.J., Kemper, C.M., Kitchener, D.J., Humphreys, W.F., & How, R.A. (1989), 'Ecology and physiology of the northern quoll, *Dasyurus hallucatus* (Marsupialia, Dasyuridae) at Mitchell Plateau, Kimberly, Western Australia', *J Zool*, **217**:539–58.

Scott, M.P., & Tan, T.N. (1985), 'A radiotracer technique for the determination of male mating success in natural populations', *Behav Ecol Sociobiol*, **17**:29–33.

Selwood, L. (1980), 'A timetable for embryonic development of the dasyurid marsupial, *Antechinus stuartii* (Macleay)', *Aust J Zool*, **28**:649–68.

Selwood, L., & McCallum, F. (1987), 'Relationship between longevity of spermatozoa after insemination and the percentage of normal embryos in brown marsupial mice (*Antechinus stuartii*)', *J Reprod Fert*, **79**:495–503.

Shimmin, G.A. (1999), 'Sperm competition in brown marsupial mice (*Antechinus stuartii;* Southern form); an investigation into reproductive fitness, mating behavior sperm transport and siring success', PhD thesis, Dept of Anatomy, Monash University, Clayton, Victoria, Australia.

Shimmin, G.A., Taggart, D.A., & Temple-Smith, P.D. (2000a), 'Sperm competition and genetic diversity in the agile antechinus (Dasyuridae: *Antechinus agilis*)', *J Zool (Lond)*, **252**:343–50.

Shimmin, G.A., Taggart, D.A., & Temple-Smith, P.D. (2000b), 'Variation in agile antechinus (*Antechinus agilis*) reproductive surpluses at different teat number locations', *Aust J Zool*, **48**:511–17.

Shimmin, G.A., Taggart, D.A., & Temple-Smith, P.D. (2002), 'Mating behaviour in the agile antechinus (*Antechinus agilis*, Marsupialia: Dasyuridae)', *J Zool (Lond)*, **258**:39–48.

Shimmin, G.A., Jones, M., Taggart, D.A., & Temple-Smith, P.D. (1999), 'Sperm transport and storage in the agile antechinus (*Antechinus agilis*)', *Biol Reprod*, **60**:1353–9.

Short, R.V. (1979), 'Sexual selection and its component parts, somatic and genital selection as illustrated by man and the great apes', *Adv Study Behav*, **9**:131–58.

Soderquist, T.R., & Serena, M. (1990), 'Occurrence and outcome of polyoestry in wild western quolls, *Dasyurus geoffroii* (Marsupialia: Dasyuridae)', *Aust Mammalogy*, **13**:205–8.

Strahan, R. (1995), *The Mammals of Australia, 2nd ed*, Reed Books, Chatswood, NSW, Australia.

Sun, Y.T., Irby, D.C., Robertson, D.M., & de Kretser, D.M. (1989), 'The effects of exogenously administered testosterone on spermatogenesis in intact and hypophysectomized rats', *Endocrinology*, **125**:1000–10.

Sunquist, M.E., & Eisenberg, J.F. (1993), 'Reproductive strategies in female *Didelphis*', *Bull Florida Mus Nat Hist*, **36**:109–40.

Taggart, D.A. (1994), 'A comparison of sperm and embryo transport in the female reproductive tract of marsupial and eutherian mammals', *J Reprod Fertil Dev*, **6**:1–22.

Taggart, D.A., & Temple-Smith, P.D. (1989), 'Structural features of the epididymis in a dasyurid marsupial (*Antechinus stuartii*)', *Cell Tiss Res*, **258**:203–10.

Taggart, D.A., & Temple-Smith, P.D. (1990a), 'The effects of breeding season and mating on total number and relative distribution of spermatozoa in the epididymis of the brown marsupial mouse, *Antechinus stuartii*', *J Reprod Fert*, **88**:81–91.

Taggart, D.A., & Temple-Smith, P.D. (1990b), 'An unusual mode of progressive motility in spermatozoa from the dasyurid marsupial, *Antechinus stuartii*', *Reprod Fert Dev*, **2**:107–14.

Taggart, D.A., & Temple-Smith, P.D. (1991), 'Transport and storage of spermatozoa in the female reproductive tract of the brown marsupial mouse, *Antechinus stuartii* (Dasyuridae)', *J Reprod Fert*, **93**:97–110.

Taggart, D.A., & Temple-Smith, P.D. (1994), 'Comparative epididymal morphology and sperm distribution studies in dasyurid marsupials', *J Zool (Lond)*, **232**:365–81.

Taggart, D.A, Selwood, L., & Temple-Smith, P.D. (1997), 'Sperm production, storage and the synchronization of male and female reproductive cycles in the iteroparous stripe faced dunnart, *Sminthopsis macroura* (Marsupialia): Relationship to reproductive strategies within the Dasyuridae', *J Zoology (Lond)*, **243**:725–36.

Taggart, D.A., Breed, W.G., Temple-Smith, P.D, Purvis, A, & Shimmin, G. (1998), 'Sperm competition and mating strategies in marsupials and monotremes', in *Sperm Competition and Sexual Selection* (eds. T.R. Birkhead & A.P. Moller), pp. 667–752, Academic Press, London..

Taggart, D.A., Shimmin, G.A., McCloud, P., & Temple-Smith, P.D. (1999), 'Timing of mating, sperm dynamics and ovulation in a wild population of agile antechinus (Marsupialia: Dasyuridae)', *Biol Reprod*, **60**:283–9.

Taplin, L.E. (1980), 'Some observations on the reproductive biology of *Sminthopsis virginiae* (Tarragon), (Marsupialia: Dasyuridae)', *Aust Zool*, **20**:407–18.

Taylor J.M., & Horner B.E. (1970), 'Gonadal activity in the marsupial mouse, *Antechinus bellus*, with notes on other species of the genus', *J Mammal*, **51**:659–68.

Taylor, J.M., Calaby, J.H., & Redhead, T.D. (1982), 'Breeding in wild populations of the marsupial-mouse *Planigale maculata sinualis* (Dasyuridae, Marsupialia)', in *Carnivorous Marsupials* (ed. M. Archer), pp. 83–7, Royal Zoological Society of New South Wales, Sydney.

Tyndale-Biscoe, C.H., & Rodger, J.C. (1978), 'Differential transport of spermatozoa into the two sides of the genital tract of a monovular marsupial, the tammar wallaby (*Macropus eugenii*)', *J Reprod Fert*, **52**:37–43.

Tyndale-Biscoe, C.H., & Renfree, M.B. (1987), *Reproductive Physiology of Marsupials*, Cambridge University Press, Cambridge.

Van Dyck S. (1979), 'Behaviour in captive individuals of the dasyurid marsupial, *Planigale maculata* (Gould 1851)', *Memoirs of the Queensland Museum*, **19**:413–29.

Van Dyck S. (1995), 'Atherton antechinus', in *The Mammals of Australia* (ed. R. Strahan), pp. 89–90, Reed Books, Chatswood, NSW, Australia.

Ventura, J., Pérez–Hernánzez, R., & López–Fuster, M.J. (1998), Morphometric assessment of the *Monodelphis brevicaudata* group (Didelphimorpha: Didelphidae) in Venezuela. *J. Mammal.* **79**:104–117.

Whitford, D., Fanning, D., & White, A.W. (1982), 'Some information on reproduction, growth and development in *Planigale gilesi* (Dasyuridae, Marsupialia)', in *Carnivorous Marsupials* (ed. M. Archer), pp. 77–81, in Royal Zoological Society of New South Wales, Sydney.

Williams, R., & Williams, A. (1982), 'The life cycle of *Antechinus swainsonii* (Dasyuridae, Marsupialia)', in *Carnivorous Marsupials* (ed. M. Archer), pp. 89–95, Roy. Zool. Soc. NSW, Sydney.

Wilson, B.A., & Bourne, A.R. (1984), 'Reproduction in the male dasyurid *Antechinus minimus maritimus* (Marsupialia: Dasyuridae)', *Aust J Zool*, **32**:311–18.

Wittenberger, J.F. (1979), 'The evolution of mating systems in birds and mammals', in *Handbook of Behavioral Neurobiology, Vol. 3* (eds. P. Marler, J.G. Vandenbergh), pp. 271–349, Plenum Press, New York.

Wolfe, K.M., Robertson, H., & Bencini, R. (2000), 'The mating behaviour of the dibbler, *Paraantechinus apicalis*', *Aust J Zool*, **48**:541–50.

Woolley, P. (1966), 'Reproduction in *Antechinus spp*., and other dasyurid marsupials', *Symp Zool Soc Lond*, **15**:281–94.

Woolley, P. (1971), 'Observations on the reproductive biology of the dibbler, *Antechinus apicalis* (Marsupialia: Dasyuridae)', *Journal of the Roy Soc W Aust*, **54**:99–102.

Woolley, P. (1988), 'Reproduction in the *Ningbing Antechinus* (Marsupialia: Dasyuridae), Field and laboratory observations', *Aust Wildlf Res* **15**:149–56.

Woolley, P. (1990), 'Reproduction in *Sminthopsis macroura* (Marsupialia: Dasyuridae), II. The male', *Aust J Zool*, **38**:207–17.

Woolley, P. (1991), 'Reproductive pattern of captive Boullanger Island dibblers, *Parantechinus apicalis* (Marsupialia: Dasyuridae)', *Wildlf Res*, **18**:157–63.

Woolley, P. (1994), 'The dasyurid marsupials of New Guinea: Use of museum specimens to assess seasonality of breeding', *Sci New Guinea*, **20**:49–55.

Woolley, P. (1995), 'Southern dibbler', in *The Mammals of Australia* (ed. R. Strahan), pp. 72–3, Reed Books, Chatswood, NSW, Australia.

Woolley, P., & Begg, R.J. (1995), 'Northern dibbler', in *The Mammals of Australia* (ed R. Strahan), pp. 74–5, Reed Books, Chatswood, NSW, Australia.

PART IV

CHAPTER 25

BIASED SEX RATIOS IN LITTERS OF CARNIVOROUS MARSUPIALS: WHY, WHEN AND HOW?

Simon J. Ward

Department of Zoology, University of Melbourne, Parkville, VIC 3052, Australia

Several studies have now established that biased sex ratios occur in litters of some species of dasyurid and didelphid marsupials. It is presumed that these biased ratios have an adaptive significance and they have been interpreted within various theoretical frameworks, predominantly the Local Resource Competition Hypothesis (LRC), the Trivers-Willard Hypothesis (TWH), and the First Cohort Advantage Hypothesis (FCAH). No single framework has been successful in explaining all cases observed, even within a single species, and none is likely to. One barrier to providing adaptive explanations for sex-biased litters is that we know of no low-cost mechanism of sex ratio manipulation. Recent studies with *Antechinus agilis* have shown a bias of 2F:1M that is generated prior to birth, and the full level of bias cannot be explained by sex-selective embryo loss alone. Pre-fertilisation mechanisms must contribute to the generation of these sex-biased litters. As we continue to narrow down the stage(s) at which sex bias is generated, we come closer to determining the mechanism(s) that generate sex bias in litters, but at present we can only speculate. One possibility, in *A. agilis* at least, is that sexes of sperm respond differently to the period of sperm storage between copulation and fertilisation.

INTRODUCTION

There is now no doubt that sex ratios in litters or clutches of mammals and birds can be biased towards one sex or the other. A clear example comes from a recent study of *Antechinus agilis*, which showed that of 215 neonates born to 19 mothers from a single population, 146 were females (68%) and only 69 were males (32%), a ratio that differs significantly from parity (x^2 = 24.2, $p < 0.001$) (Davison and Ward 1998). Much energy has been put into developing theories to explain such biased sex ratios within an adaptive framework, but only recently has there been investigation of the developmental stages at which such biases are generated. Identification of the stage is only the first step in understanding the mechanisms that generate them. The absence of a known low-cost mechanism of sex ratio manipulation is a major barrier to providing adaptive explanations for sex-biased litters (Clutton-Brock and Iason 1986; Krackow 1995).

This review concerns itself mostly with biases in sex ratio of young in litters, but this is only one part of sex allocation. A mother also has biased sex allocation if she produces an equal number of sons and daughters but invests differently in them according to sex.

Table 1 Examples of biased sex ratios in carnivorous marsupials (all cases here are statistically significant).

Species	Sex ratio	Conditions	Reference
Didelphis marsupialis	0.57	supplementally fed mothers	Austad and Sunquist 1986
D. virginiana	0.57	first cohort	Wright et al. 1995
Sarcophilus laniarius	0.43	all mothers	Guiler 1970, Hughes 1982
Dasyurus hallucatus	0.65	first year mothers	Oakwood 2000
Antechinus agilis	0.32	first year mothers	Davison and Ward 1999
A. swainsonii	0.57	first year mothers	Cockburn et al. 1985
	0.61	first year mothers	Hansen and Ward in prep.
A. flavipes	0.43	first year mothers	Coates 1995
	0.61	second year mothers	

Much of the theory on sex allocation has concerned itself with this kind of post-natal investment (e.g. Cockburn 1994) because it is much easier to monitor, but also there was an overwhelming view that the primary sex ratio is set by simple Mendelian separation of sex chromosomes. However, there is now considerable evidence (such as that above for antechinus, and other studies) that pre-natal factors can bias sex ratios, including factors operating pre-fertilisation (Krackow 1995).

Most theories on biased sex allocation in mammals are based on studies of long-lived monotocous eutherian species such as primates or deer, but such species do not lend themselves to experimental manipulation because of the long period or very large sample sizes needed to show a significant effect. Small mammals with multiple young per litter are much better suited to such studies, and mechanisms of sex ratio bias are more likely to be discovered in such species. Carnivorous marsupials have been the focus of important studies of sex ratio bias, including rare experimental manipulations of wild populations (Austad and Sunquist 1986, Dickman 1988, Sunquist and Eisenberg 1993), and are also being used in studies of the mechanisms.

Biased sex ratios have been recorded in litters of several species of carnivorous marsupials, including both American didelphids and Australian dasyurids (Table 1), and also in other marsupials (e.g. brushtail possum, Hope 1972; various macropods, Johnson and Jarman 1983, Poole et al. 1985, Johnson 1989, Sunnucks and Taylor 1998, Fisher 1999). In some cases the bias favours males, in others females, and different cohorts of the same population may show different, even opposite biases. This list led Cockburn (1990) to suggest that sex ratio bias may be more prevalent in marsupials than in eutherian mammals or in birds. In this paper I will review the studies of sex ratio bias in litters of carnivorous marsupials and discuss the 'why', 'when' and 'how'. The section on 'why' will discuss the theories that have been used to explain biased sex allocation in marsupials, 'when' will review the evidence on the developmental stage at which bias is generated, and 'how' will discuss possible mechanisms, though is largely speculative.

WHY: THEORIES OF ADAPTIVE BIASED SEX ALLOCATION IN CARNIVOROUS MARSUPIALS

Fisher (1930) was one of the first to model patterns of sex allocation and his conclusions underpin much of sex allocation theory today. He argued that in populations of sexually-reproducing species the total parental investment in males must be the same as that in females, so natural selection will favour the primary sex ratio that results in equal parental investment in the sexes. This does not, however, mean that a 1:1 sex ratio is always the optimum sex ratio for a parent. Adaptive biased sex ratios are now considered to occur under a number of conditions, particularly where the costs of producing sons and daughter are not equivalent and where mothers in a population differ in their ability to invest in reproduction. Two theoretical frameworks have received most attention in mammalian studies, though they deal with sex ratios at two different levels. These are the Local Resource Competition (LRC) Hypothesis, which addresses population sex ratios, and the Trivers-Willard Hypothesis (TWH), which considers the sex ratio of litters produced by specific individuals in a population. A third framework, the First-Cohort Advantage Hypothesis (FCAH), has more recently been proposed to account for sex ratios of specific litters within a multi-litter breeding season, and is based on studies of American opossums.

The Local Resource Competition Hypothesis was first invoked by Clark (1978) to explain male-biased sex ratios in prosimian primates. She argued that where sons and daughters have different patterns of natal dispersal, mothers will face different levels of competition with the two sexes of offspring in the future. In the sense of Fisher's (1930) model, competition with the sex that does not disperse (or between litter mates of that sex) increases the costs of investment in that sex. Under these conditions a mother may bias the sex ratio of her litter towards the dispersing sex (so reducing future competition and costs). This is likely to be more pronounced early in a female's breeding life, and may be relaxed or reversed later. In most mammal systems it is males that disperse, so a bias towards males is expected in earlier litters.

Cockburn et al. (1985) reported a positive relationship between the sex ratio of litters of first year female *Antechinus* (across three species) and the degree of iteroparity in populations, and interpreted these results in the light of LRC. They argued that in more iteroparous populations (typified by *Antechinus swainsonii*), females were likely to survive to breed again and so would encounter competition with their philopatric daughters, and were likely to bias their first litter towards males. Females in more semelparous populations (typically *A. agilis* – previously referred to as *A. stuartii*) were unlikely to breed again, and would produce female-biased litters. One corollary of this interpretation is that females in iteroparous populations should show a shift or reversal in sex ratio between their first and second litter. Oakwood (2000) found this to be the case in *Dasyurus hallucatus*, with ratios of 0.65 and 0.3 for first and second year females, respectively. Cockburn et al. (1985) report evidence for such a shift, but Hansen and Ward (in prep.) found no shift in *A. swainsonii*, though there was a significant increase in the variance of the sex ratio in second litters (mean sex ratio ± variance in first litters, 0.65 ± 0.05, n = 4; second litters, 0.61 ± 0.18, n = 10). The reasons for such an increase are unclear.

Antechinus flavipes was represented in Cockburn et al.'s (1985) relationship by a single population that was intermediate between *A. swainsonii* and *A. agilis* in both sex ratio (~0.5) and degree of iteroparity. More recently, Coates (1995) reported patterns of sex ratio bias in this species that run directly counter to the predictions of Cockburn et al. (1985). Coates (1995) found that mothers breeding for their first time produced significantly female-biased litters (0.43), and second-year mothers produced significantly male-biased litters (0.61) (Table 1). Overall in the population there was no significant sex ratio bias in litters.

The Trivers Willard Hypothesis (also referred to as the Maternal Condition Hypothesis) proposes that the sex allocation may be biased in polygynous species, depending on the condition of the mother (Trivers and Willard 1973). It argues that if the potential increase in fitness gained from extra maternal investment is greater in one sex than in the other, then a mother with greater ability to invest in offspring (i.e. in better condition) should bias her investment towards the sex that will gain more. In most mammalian polygynous systems it is considered that larger males have greater competitive ability, and therefore fitness, so sons are likely to gain more fitness benefits from greater maternal investment than daughters, so mothers in good condition should invest more in sons. In some cases this will be expressed in male-biased sex ratios, in others it will be in greater provisioning or care of sons compared to daughters, perhaps resulting in differential mortality.

Four experimental field studies of this theory have been carried out on carnivorous marsupials. In all four studies one group of females was supplementally fed (provisioned) in an attempt to increase maternal body condition, and sex ratios in their pouch litters were compared to those of a group of control (non-supplemented) females. Austad and Sunquist (1986) performed this experiment on *D. marsupialis* in Venezuela. After supplementary feeding, 13 of 17 (76%) mothers produced male-biased litters, producing an overall sex ratio of 0.57 (n = 270), in comparison to 55% of control mothers with male-biased litters and a 0.50 (n = 256) overall sex ratio. (This difference in overall sex ratio is statistically significant, but see below.) Supplemented mothers also weaned larger young with better post-weaning survival, but did not produce larger litters and did not skew somatic investment in male and female offspring. The results from a parallel experiment with *D. virginiana* in Florida are less clear-cut (Sunquist and Eisenberg 1993). Here bias was apparent in one year, but the following year was a drought year and there was no bias.

Austad and Sunquist's (1986) study of sex ratio in provisioned *D. marsupialis* was the first experimental field test of the TWH, and some consider it to provide the best empirical support for the hypothesis. However, there has been some debate on the interpretation of the results. Austad and Sunquist (1986) and Sunquist and Eisenberg (1993) found significant differences in the pooled numbers of male and female offspring in litters of provisioned and control mothers, consistent with the TWH. However, Wright et al. (1995) argue that since the TWH predicts selection at the individual level, not at the population level, analyses should focus on the numbers of biased litters, not on pooled data. They found that male-biased litters were not significantly more numerous in provisioned mothers (using Sunquist and Eisenberg's (1993) data) than in control ones. Following up this conflict in interpretation, Hardy (1997) acknowledged the 'level of analysis' problem highlighted by Wright et al. (1995), but was wary of their new analysis that used only categorical classification of litters (as male-biased or not). He used generalised linear statistical models in a reanalysis of the same data using sex ratios of individual litters or of young produced by individual mothers, and found that litters of provisioned mothers were more male-biased than were those of control mothers, again supporting TWH.

In a broader study of competition between *A. agilis* and *A. swainsonii*, Dickman (1988) investigated the sex ratios in litters of *A. agilis* when provisioned and when competition with *A. swainsonii* was reduced or eliminated. These species are sympatric in many wet forest areas in south-eastern Australia, with *A. agilis* the smaller more scansorial species. Sex ratios were female-biased in control areas (0.34–0.36), but were male-biased in provisioned areas (0.59) and where interspecific competition was reduced (0.53), and maternal body mass was significantly greater in these areas. These results conform with the predictions of TWH. However, there was no significant change in maternal body mass or sex bias (0.44), in areas where *A. swainsonii* was removed completely. This complete removal may have led to increased intraspecific competition, so negating the benefits to maternal condition of reduced interspecific competition (Dickman 1988).

Coates (1995) studied sex ratios in *A. flavipes*, using both population-wide comparisons and experimentation. In this species, females typically produce two litters in a lifetime (one each year) and he found that mothers breeding for their first time produced significantly female-biased litters, and second-year mothers produced significantly male-biased litters (see above). Population-wide there was a relationship between maternal mass and sex ratio in litters, but this was confounded by second-year mothers being heavier than first-years, and there was no such relationship within either cohort. When he experimentally provisioned part of the population there was a shift, but not significant, towards producing more males by first-year mothers, and no change at all in the sex ratio in litters of second-year mothers. Coates (1995) concluded that his results provide little support for the TWH in this species.

Whilst three of these four food provisioning experiments provide evidence of biased sex ratios consistent with the TWH, other data are equivocal. Sunquist and Eisenberg (1993) reported that female *D. marsupialis* that gained weight during lactation (therefore considered to be in good condition), whether provisioned or control, tended to produce male-biased litters. Dickman (1988) also reported that the sex ratio within litters of *A. agilis* was greater in heavier mothers. However, Davison and Ward (1998), also working on *A. agilis*, and Coates (1995), with *A. flavipes* (see above), found no relationship between maternal body mass and sex ratio in unmanipulated litters. Perhaps the level of provisioning in the first three experimental studies was sufficient to elevate maternal condition to a level at which such effects can be detected statistically.

The First-Cohort Advantage Hypothesis, proposed by Wright et al. (1995), applies to populations in which two or more litters are produced within an annual breeding season. In such populations, offspring produced in the first litter of a season (first cohort) will be older at the start of the next breeding season, and probably bigger. In polygynous species, bigger males are more likely to be successful reproducers, so this hypothesis predicts that first litters will be male-biased. In a two-year study of *D. virginiana* in Florida, Wright et al. (1995) followed the breeding of 105 females and their offspring. More first litters were male-biased than were female-biased. Female-biased litters were more numerous in the second cohort, but not significantly so, and overall there was no bias in sex ratio across the population. Following the male offspring into the next breeding season showed that those from first litters were significantly heavier at the start of that season than were males from second litters. Wright et al. (1995) used these results to support their FCAH, and reported other data from the study counter to the predictions of both LRC and TWH.

Wright et al.'s (1995) FCAH is similar to theories used to explain changes in sex ratio within clutches as breeding seasons progress in some bird of prey species (Daan et al. 1996). However, the FCAH cannot provide explanation for all cases of sex bias in carnivorous marsupials since it can only be applied to species with multiple litters per year, so is inappropriate for *Antechinus* spp. Furthermore, it doesn't apply to all cases of species with multiple litters per year: Hardy's (1997) reassessment of Sunquist and Eisenberg's (1993) data on sex ratios of *D. marsupialis* found no cohort effect. Presumably the biologies and selective pressures on the two *Didelphis* species are different (Hardy 1997).

My discussion of the major theories of biased sex allocation above has treated each separately, yet it is clear that they are not mutually exclusive and within a single species evidence may support more than one theory. For example, biased sex ratios in litters of *A. agilis* have been interpreted in the light of LRC (Cockburn et al. 1985) and TWH (Dickman 1988). This is because these theories were not formulated to address the same issues – one is at the population level and the other at the individual level. Research by Cockburn (1994) on biased sex allocation by brood reduction during late lactation in *A. agilis* (after obligate attachment to teats and before weaning) has shown that LRC and TWH may influence sex allocation conjointly. It is also likely that different factors become important under different social and environmental conditions (Dittus 1998). Certainly, the factors influencing sex ratio and sex allocation are many and varied, and we should not expect any single theoretical framework to explain all cases of biased sex allocation.

WHEN: AT WHAT DEVELOPMENTAL STAGE IS BIAS GENERATED?

In most cases, reports of biased sex ratios in carnivorous marsupials are based on counts of male and female pouch young in the field, so after the stage when the presence of a scrotum or pouch/teats is apparent on inspection (Cockburn et al. 1985, Austad and Sunquist 1986, Dickman 1988). In these species, few pouch young (<5%) are lost between when they attach to teats and when they reach this sexable stage (Cockburn et al. 1985, Austad and Sunquist 1986, Dickman 1988), so these reports indicate the sex ratio soon after birth. Many carnivorous marsupials produce supernumerary young, which provides the potential for sex ratios to change between birth and attachment through differential success of neonates in reaching and attaching to teats; in *D. virginiana*, up to 25 young are born, but there is a maximum of 13 viable teats (Hartman 1952); 17 and 10 in *A. agilis* (Davison and Ward 1998). However, Davison and Ward (1998) found that in *A. agilis*, the sex ratio of all young born (0.32) was the same as that of pouch young that successfully attached to teats (0.32) or of supernumerary neonates that would have died (0.34). These data indicate that there is no change in sex ratio at this stage, and the bias in sex ratio must be generated prior to birth.

There are two broad stages prior to birth in which these biases could be generated: pre- and post-fertilisation, though they are not mutually exclusive. Selwood (1983) showed considerable embryo loss during development of captive-mated *A. agilis*. She used the number of corpora lutea in the ovaries to indicate the number of eggs ovulated, and therefore the maximum number of embryos fertilised, and found that 77% of embryos survived to the bilaminar blastocyst stage and only 60% reached implantation. Davison and Ward (1998) found slightly higher survival in a sample of 10 wild-mated *A. agilis*, with 75% of embryos reaching birth. If this loss of embryos was sex-specific it could contribute to the sex-bias observed in litters at birth. However, even if all the embryos that were lost in the litters reported by Davison and Ward (1998) had been male, the minimum sex ratio this would have produced was 0.46, and the actual sex ratio at birth in these 10 litters was 0.26. So, at most, only two-thirds of the observed bias could have been achieved through extreme bias against male embryos in this case. Some of the bias in sex ratio must have been generated pre-fertilisation.

An investigation of the sex ratio in embryonic litters at various stages through gestation in *A. agilis* is currently underway. Preliminary data suggest that there is no change in sex ratio (Ward, unpublished data), indicating that pre-fertilisation mechanisms are likely to be of major importance in this species. The potential for maternal manipulation of sex ratio during pregnancy is perhaps greater in other carnivorous marsupials as rates of embryonic loss can be much higher. As few as 32% of embryos survive to the blastocyst stage in *Sarcophilus laniarius* (Hughes 1982), and in *D. viginiana,* Hartman (1919) concluded that approximately one-third of eggs were unfertilised or abnormal in the early stages, and at blastocyst and later stages, 16% of embryos were abnormal.

How: What mechanisms might generate sex bias?

Mechanisms for producing sex bias in litters of marsupials, or eutherians, are still largely speculative, and little can be added to the treatments given the topic by Cockburn (1990) and Krackow (1995), other than to narrow the focus. We now know that, at least in *Antechinus* species and probably in other carnivorous marsupials, the biases in sex ratios are generated prior to birth. As a result, mechanisms must operate to bias the ratio of X and Y sperm reaching (or penetrating) the ova (i.e. pre-fertilisation), and or to produce sex-selective loss of embryos during gestation. Krackow (1995) believes that gonadotrophin and or steroid hormonal mechanisms are most likely to operate in mammals and birds prior to birth, and provides considerable discussion on the topic. However, he notes that all are speculative and there is no empirical evidence for any mechanism operating in a system of adaptive biased sex allocation in mammals or birds.

Pre-fertilisation

A mechanism operating at this stage must either vary the ratio of X and Y sperm reaching ova, or vary the relative ability of the two sexes of sperm to penetrate and fertilise the ova. Vaginal pH may influence mucus viscosity and sperm motility, and is influenced by gonadotrophin or steroid hormones (Krackow 1995). If the sexes of sperm differ in motility, hormone-induced changes in pH could exaggerate or minimise differences between the sexes. The Y chromosome is typically smaller than the X, and this lead to the thought that X sperm are heavier and may be slower swimmers, however, attempts to show such effects have generally failed (but see Roberts 1972).

The relative times of mating and ovulation have frequently been considered as potential mechanisms influencing sex ratio. In a number of eutherians, including humans, the relative time of insemination within the oestrous cycle has been found to affect sex ratio at birth, but it is not clear why or how (Krackow 1995). Periods of sperm storage between copulation and ovulation/fertilisation have been reported in multiple species (across multiple genera) of dasyurids and didelphids (and at least one bandicoot) (Selwood 1982, Rodger and Bedford 1982). In these carnivorous marsupials, storage occurs in crypts in the isthmus of the oviduct, and may be for periods up to two weeks (Taggart and Temple-Smith 1991). If the two sexes of sperm differ in their patterns of senescence, the length of the period of sperm storage could influence the ratio of competent X and Y sperm reaching the ova. Biased sex ratios could then occur through behavioural control of the timing of mating within the period of oestrus, or physiological mechanisms controlling the timing of ovulation.

It is possible that males produce biased ratios of X and Y sperm, or their seminal fluids may have pH or viscosity affects similar to those outlined above, but these are unlikely to be widespread mechanisms, for a couple of reasons. A female marsupial invests much more effort in her offspring than does the father, so she is expected to have more control over sex allocation. A result of this is that all of the theories applied to biased sex allocation relate to maternal factors rather than paternal ones. This is also a function of the observations of patterns of biased sex allocation, whereby sex ratios change with cohorts (of litters or mothers) within a population, but mating patterns are considered to be constant. This, however, has never been tested empirically. We also now know that multiple paternity often occurs within single litters of *A. agilis* (Kraaijeveld-Smit et al. 2002), which must decrease the impact of any one male on the sex ratio within a litter.

Embryonic stages

Eutherian studies have confirmed that sex-biased loss of embryos can be a mechanism for producing adaptive biased sex ratios (e.g. in coypus, Gosling 1986), but generally the examples show it occurring at a developmental stage after that at which birth occurs in marsupials. Specifically, it occurs after the gonads have differentiated and have started producing sex-specific hormones, and such primary sex hormones provide a likely key to gender recognition. Didelphid and dasyurid neonates are born after very short gestations at very altricial stages (Hughes and Hall 1988). In the absence of primary sex hormones during marsupial embryonic development, more subtle influences must operate. In mice and cattle, male blastocysts develop more rapidly than female ones, such that they reach developmental stages, such as implantation, at different times (Krackow 1995). Since implantation requires complex interactions between the embryo and the uterus, a degree of synchrony is required, which is amenable to manipulation by variation in steroid hormone levels (Krackow 1995). However, this would result in bias being generated at or after implantation, yet in marsupials, the post-implantation period is very short and is not necessarily the period during which most embryo loss occurs (Selwood 1983, pers. obs.).

FUTURE DIRECTIONS

A great deal is still to be learnt on nearly all aspects of biased sex ratios, and biased sex allocation, in mammals generally. The frequency of reports of sex bias in carnivorous marsupial litters, and the large litter sizes of these animals, make them very useful target species for future research of the topic. More careful definition of the stage(s) at which biases are generated is needed and is an important precursor to understanding the mechanism(s). Once the mechanism(s) is known, we will have a much better concept of how much variation is possible, and a more solid background for building theories of adaptive biased sex allocation. Carefully-planned experimental studies to test theories of biased sex allocation and to identify those members of a population most likely to produce sex-biased litters are also required, as are longer-term studies to follow the reproductive success of offspring to determine the fitness gains to mothers that bias sex ratio in their litters.

ACKNOWLEDGEMENTS

My thanks to Hugh Tyndale-Biscoe for discussions on the topic and, in particular, his reluctance to accept adaptive explanations of sex ratio bias without consideration of the mechanisms that may produce uneven sex ratios. My thanks also to Chris Johnson and an anonymous reviewer for their comments on the original manuscript.

REFERENCES

Austad, S.N., & Sunquist, M.E. (1986), 'Sex-ratio manipulation in the common opossum', *Nature,* **324**:8–60

Clark, A.B. (1978), 'Sex ratio and local resource competition in a prosimian primate', *Science,* **201**:163–5.

Clutton-Brock, T.H., & lason, G.R. (1986), 'Sex ratio variation in mammals', *Quarterly Review of Biology,* **61**:339–73.

Coates, T. (1995), 'Reproductive ecology of the yellow–footed antechinus, *Antechinus flavipes* (Waterhouse), in north east Victoria', PhD thesis, Monash University, Clayton.

Cockburn, A. (1990), 'Sex ratio variation in marsupials', *Australian Journal of Zoology,* **37**:467–79.

Cockburn, A. (1994), 'Adaptive sex allocation by brood reduction in antechinuses', *Behavioral Ecology and Sociobiology,* **35**:53–62.

Cockburn, A., Scott, M.P., & Dickman, C.R. (1985), 'Sex ratio and intrasexual kin competition in mammals', *Oecologia,* **66**:427–9.

Davison, M.J., & Ward, S.J. (1998), 'Prenatal bias in sex ratios in a marsupial, *Antechinus agilis*', *Proceedings of the Royal Society of London* B, **265**:2095–9.

Dickman, C.R. (1988), 'Sex ratio variation in response to interspecific competition', *American Naturalist,* **132**:289–97.

Daan, S., Dijkstra, C., & Weissing, F.J. (1996), 'An evolutionary explanation for seasonal trends in avian sex ratios', *Behavioral Ecology,* **7**:426–30.

Dittus, W.P.J. (1998), 'Birth sex ratios in toque macaques and other mammals: integrating the effects of maternal condition and competition', *Behavioral Ecology and Sociobiology,* **44**:149–60.

Fisher, R.A. (1930), *The genetical theory of natural selection*, Clarendon Press, Oxford.

Fisher, D.O. (1999), 'Offspring sex ratio variation in the bridled nailtail wallaby, *Onychogalea fraenata*', *Behavioral Ecology and Sociobiology,* **45**:411–19.

Gosling, L.M. (1986), 'Selective abortion of entire litters in the coypu: adaptive control of offspring production in relation to quality and sex', *American Naturalist,* **127**:772–95.

Guiler, E.R. (1970), 'Observations on the Tasmanian devil, *Sarcophilus harrisii* (Marsupialia: Dasyuridae), II. Reproduction, breeding and growth of pouch young', *Australian Journal of Zoology,* **18**:63–70.

Hardy, I.C.W. (1997), 'Opossum sex ratios revisited: significant or nonsignificant?', *American Naturalist,* **150**:420–4.

Hartman, C.G. (1919), 'Studies in the development of the opossum (*Didelphis virginiana* L.), Parts III & IV', *Journal of Morphology,* **32**:1–140.

Hartman, C.G. (1952), *Possums*, University of Texas Press, Austin.

Hope, R.M. (1972), 'Observations on the sex ratio and the position of the lactating mammary gland in the brush-tailed possum *Trichosurus vulpecula* (Ker) (Marsupialia)', *Australian Journal of Zoology,* **20**:131–7.

Hughes, R.L. (1982), 'Reproduction in the Tasmanian devil, *Sarcophilus harrisii* (Dasyuridae, Marsupialia)', in *Carnivorous Marsupials, Vol. 1* (ed. M. Archer), pp. 49–63, Royal Zoological Society NSW, Sydney.

Hughes, R.L., & Hall, L.S. (1988), 'Structural adaptations of the newborn marsupial', in *The Developing Marsupial: Models for Biomedical Research* (eds. C.H. Tyndale-Biscoe, & P.A. Janssens), pp. 8–27, Springer-Verlag, Heidelberg.

Johnson, C.N. (1989), 'Dispersal and philopatry in the Macropodoids', in *Kangaroos, Wallabies and Rat Kangaroos, Vol. 2* (eds. G. Grigg, P. Jarman, & I. Hume), pp. 593–601, Surrey Beatty & Sons, Sydney.

Johnson, C.N., & Jarman, P.J. (1983), 'Geographic variation in offspring sex ratios in kangaroos', *Search,* **14**:152–4.

Kraaijeveld-Smit, F.J., Ward, S.J., & Temple-Smith, P.D. (2002), 'Multiple paternity in a field population of a small carnivorous marsupial, the agile antechinus, *Antechinus agilis*', *Behavioral Ecology and Sociobiology,* **52**:84–91.

Krackow, S. (1995), 'Potential mechanisms for sex-ratio adjustment in mammals and birds', *Biological Reviews,* **70**:225–41.

Oakwood, M. (2000), 'Reproduction and demography of the northern quoll, *Dasyurus hallucatus*, in the lowland savanna of northern Australia', *Australian Journal of Zoology*, **48**:519–539.

Poole, W.E., Merchant, J.C., Carpenter, S.M., & Calaby J.H. (1985), 'Reproduction, growth and age determination in the yellow-footed rock-wallaby, *Petrogale xanthopus* Gray, in captivity', *Australian Wildlife Research,* **12**:127–36.

Roberts, A.M. (1972), 'Gravitational separation of X and Y spermatozoa', *Nature,* **238**:23–5.

Rodger, J.C., & Bedford, J.M. (1982), 'Separation of sperm pairs and sperm–egg interactions in the opossum, *Didelphis virginiana*', *Journal of Reproduction and Fertility,* **64**:171–9.

Selwood, L. (1982), 'A review of maturation and fertilization in marsupials with special reference to the dasyurid: *Antechinus stuartii*', in *Carnivorous Marsupials, Vol. 1* (ed. M. Archer), pp. 65–76, Royal Zoological Society NSW, Sydney.

Selwood, L. (1983), 'Factors influencing pre-natal fertility in the brown marsupial mouse *Antechinus stuartii*', *Journal of Reproduction and Fertility,* **68**:317–24.

Sunnucks, P., & Taylor, A.C. (1998), 'Sex of pouch young related to maternal weight in *Macropus eugenii* and *M. parma* (Marsupialia: Macropodidae)', *Australian Journal of Zoology,* **45**:573–8.

Sunquist, M.E., & Eisenberg, J.F. (1993), 'Reproductive strategies in female *Didelphis*', *Bulletin of the Florida Museum of Natural History, Biological Sciences,* **36**:109–40.

Taggart, D.A., & Temple-Smith, P.D. (1991), 'Transport and storage of spermatozoa in the female reproductive tract of the brown marsupial mouse, *Antechinus stuartii*', *Journal of Reproduction and Fertility,* **93**:97–110.

Trivers, R.L., & Willard, D.E. (1973), 'Natural selection of parental ability to vary the sex ratio of offspring', *Science,* **179**:90–2.

Wright, D.D., Ryser, J.T., & Kiltie, R.A. (1995), 'First-cohort advantage hypothesis: a new twist on facultative sex ratio adjustment', *American Naturalist,* **145**:133–45.

CHAPTER 26

PARASITES OF CARNIVOROUS MARSUPIALS

I. Beveridge[A] *and D.M. Spratt*[B]

[A] Department of Veterinary Science, University of Melbourne, Parkville, Victoria 3052, Australia
[B] CSIRO Sustainable Ecosystems, GPO Box 284, Canberra, ACT 2601, Australia

The parasites of dasyurid marsupials are far better known than those of other carnivorous Australian or of South American marsupials. The known parasite fauna of dasyurids is diverse and includes representatives of eight genera of trematodes, nine genera of cestodes, 33 genera of nematodes, a single species of acanthocephalan, 17 genera of fleas, one genus of lice, two genera of ticks, 27 genera of mites and five species of protozoans. The families of parasites represented are often cosmopolitan but there is a high level of endemicity at the genus and particularly at the species level. Several genera of helminths and arthropods provide links between the South American and Australian parasite fauna, but there are many instances of acquisitions of parasites from other host groups. There is a remarkable similarity between types of parasites present in dasyurid marsupials and those found in eutherian carnivores or insectivores, illustrating the importance of diet and parasite life cycles as well as host phylogeny in determining the structure of parasite assemblages. Many species of parasites cause observable lesions in their hosts, but their overall effect on health and population parameters is poorly understood. The study of parasites of carnivorous marsupials is embryonic, but offers considerable scope for future investigations.

INTRODUCTION

As is the case with many other groups of mammals, carnivorous marsupials harbour a diverse fauna of both internal and external parasites. Fleas, ticks, lice and mites are commonly encountered while handling small dasyurids, particularly in high-rainfall areas of eastern Australia, while careful dissection and microscopic examination of internal organs often reveals a multiplicity of parasitic nematodes, cestodes, digeneans, acanthocephalans, mites and protozoans, not simply restricted to the gastrointestinal tract, but occurring in virtually all organs of the body as exemplified by *Antechinus agilis* (Fig. 1).

The array of parasites encountered poses a number of biological questions. From the point of view of the parasites, is the parasite fauna endemic, is it host-specific, that is restricted to carnivorous marsupials, and given the relatively basal position of the dasyurids within the Australian marsupial radiation, do the parasites provide any insights into their relationships with South American marsupials? From the point of view of the hosts, the obvious questions which arise are how hosts become infected, what are the ecological factors which influence parasite burdens and do the parasites adversely affect the health of the hosts? While knowledge of the parasites of Australian marsupials is still

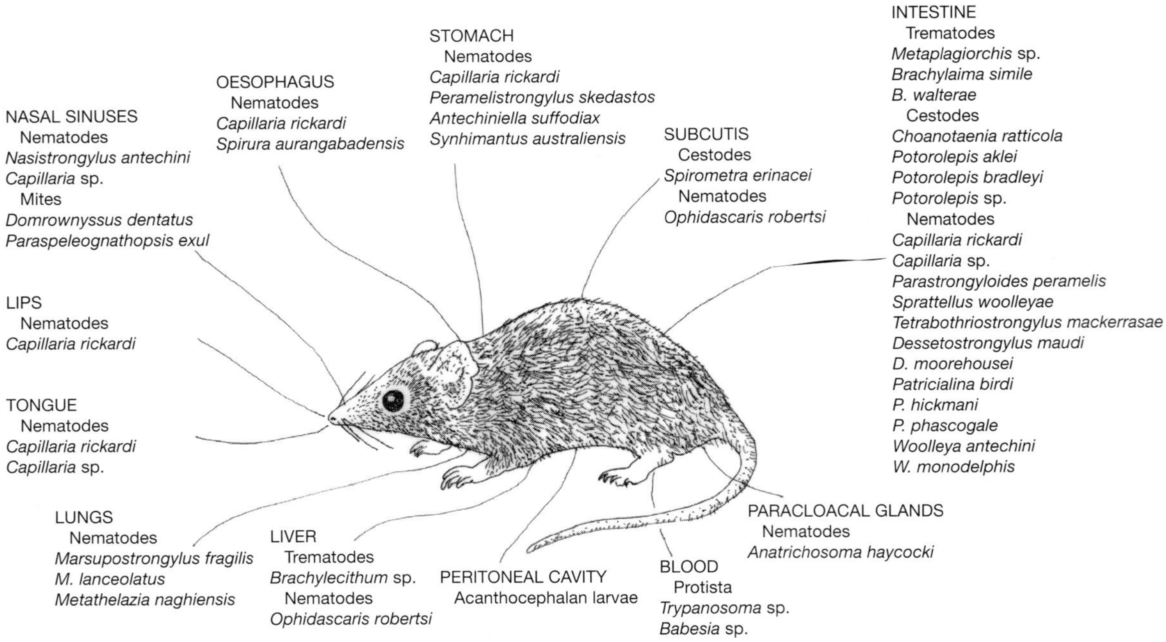

Figure 1 Diagram of *Antechinus agilis* showing the principal internal parasites found and the sites in which they occur.

relatively fragmentary, an attempt is made here to address some of these more general questions. Most attention is devoted to the dasyurids since the parasite fauna of representative species of this group is better known. Of the 65 species of dasyurids currently named, there are some parasite records for 32 species (see for example Spratt et al. 1991) , while only a few species, *Antechinus agilis, A. stuartii, A. swainsonii, Dasyurus hallucatus, D. maculatus, D. viverrinus* and *Sarcophilus harrisii* have been studied in detail. The parasite fauna of thylacinids and myrmecobiids is extremely poorly known (Spratt et al. 1991). Perhaps suprisingly, the helminth fauna of most South American marsupials is also very poorly documented, limiting the comparisons which can be drawn between the two groups of mammals.

Detailed lists of helminth parasites known from particular Australian carnivorous marsupial species are available (Spratt et al. 1991) as are lists for ectoparasites (ticks: Roberts 1970; fleas: Dunnet and Mardon 1974, Mardon 1976 1978, Calder 1996; lice: von Keler ,1971, Palma and Barker 1996; mites: Domrow 1987, 1992, Domrow and Lester 1985) and parasitic protozoa (O'Donoghue and Adlard 2000). A review of the diseases of dasyurids (Obendorf 1993) includes those caused by parasites.

Types of parasites found in or on dasyurid marsupials

Trematodes or flukes

The trematodes are represented by only a small number of species in dasyurids. The eight genera known (*Brachylaime, Coelomotrema, Dasyurotrema, Dicrocoelium, Mehlisia, Pharyngostomoides, Neodiplostomum, Strzeleckia*) (15 species) belong to cosmopolitan families (Cribb 1998), with three of the genera (*Coelomotrema, Dasyurotrema, Strzeleckia*) endemic and restricted to dasyurids (Cribb 1992; Cribb and Spratt 1991,1992; Cribb and Pearson 1993). Most are inhabitants of the intestine, while *Coelomotrema*, a large, oblong species, inhabits the abdominal cavity of its host (Angel 1970) (Fig. 2). All trematodes require molluscs as the first intermediate host. To date, only one of the life cycles of a trematode found in carnivorous marsupials, a member of the genus *Brachylaime*, has been partially elucidated (Peisley and Howell 1975).

Cestodes or tapeworms

Cestodes likewise are represented by relatively few taxa in carnivorous marsupials, with nine genera and 14 species reported from dasyurids. In the case of three species, *Choanotaenia ratticola* (Dilepididae), *Mirandula parva* (Dilepididae) and *Paralinstowia semoni* (Anoplocephalidae), the primary hosts are rodents or bandicoots, and infections of dasyurids are probably fortuitous (Beveridge and Spratt 1996). Species of *Paralinstowia* also occur as parasites of didelphid marsupials in South America (Gardner and Campbell 1992a). Members of the linstowiine genus *Oochoristica* (Anoplocephalidae) are primarily parasites of reptiles and the single species found in dasyurids, *O. eremophila* in *Dasykaluta rosamondae*, may therefore also represent a host-switch from reptiles (Beveridge 1977; Beveridge and Spratt 1996). Only two species of the linstowiine genus *Mathevotaenia*

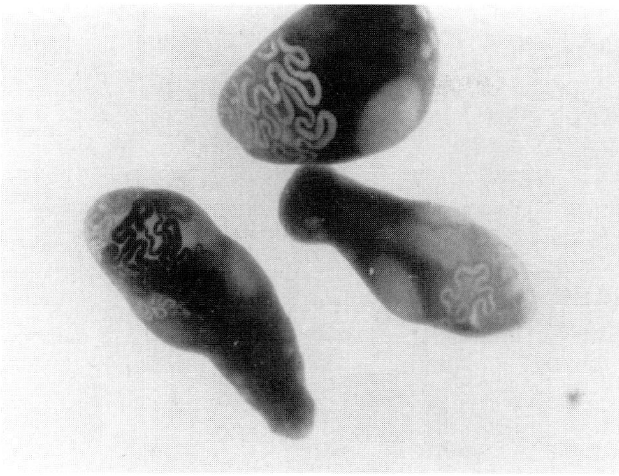

Figure 2 The trematode *Coelomotrema antechinomes* found in the abdominal cavity of species of *Antechinus*, *Antechinomys* and *Sminthopsis*. (Photograph courtesy of Dr P.A. Woolley.)

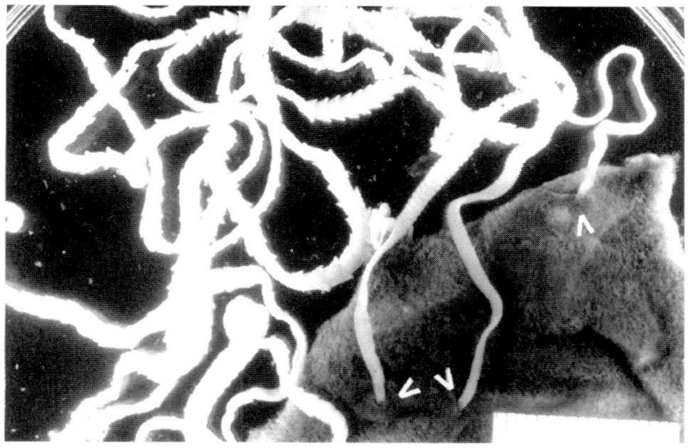

Figure 3 The cestode *Dasyurotaenia dasyuri* in the intestine of *Dasyurus maculatus* from Tasmania. Note that the anterior ends of the cestodes disappear into the wall of the small intestine. (Photograph courtesy of Dr D.L. Obendorf.)

(Anoplocephalidae) are known from Australia – one, *M. antechini*, in a dasyurid (*Pseudantechinus macdonnellensis*) and the second in a bat – while overseas the genus occurs in a wide array of eutherians (Beveridge 1994). *Mathevotaenia* may be rare in Australia, or, equally likely, its apparent rarity may reflect the lack of intensive collecting of suitable hosts.

By contrast, the three genera *Anoplotaenia*, *Dasyurotaenia* and *Potorolepis* are endemic and are restricted to marsupials. *Anoplotaenia dasyuri* is a common parasite of the Tasmanian devil, *Sarcophilus harrisii*, occurring in numbers of up to 15,600 in the small intestine (Gregory et al. 1975). Larval stages of the parasite are found in the skeletal and cardiac muscle of potoroos, pademelons and wallabies in Tasmania (Gregory et al. 1975) and the life cycle is completed when the larval stage is ingested by scavenging devils. The larval stage, curiously for a cestode, has also been found in the musculature of potential definitive hosts, *Dasyurus maculatus* and *D. viverrinus* (see records in Spratt et al. 1991), and probably also in the thylacine, *Thylacinus cyanocephalus*, in which it was reported from an animal that died in an American zoo under the name *Dithyridium cynocephali* (see Obendorf and Smith 1989).

Dasyurotaenia is represented by two named species, one in the Tasmanian devil and one in *Dasyurus* spp. (Beveridge 1984) (Fig. 3). Additional species from *Dasyurus* spp. remain undescribed (Spratt et al. 1991). The life cycle of members of this genus has not been elucidated but the larva of one species has been found in a potoroo (Beveridge 1984). This cestode genus is unusual in that the scolex is deeply embedded in the muscular outer layers of the small intestine, its position of attachment being marked on the outer surface by a prominent elevation of the serosa (Beveridge 1984) (Fig. 4). Again, the taxonomic affinities of the genus are unknown. *D. robusta*, from the Tasmanian devil, holds the dubious honour of being the only tapeworm currently listed as an 'endangered' species by a state government agency.

Potorolepis is an hymenolepidid genus currently found exclusively in dasyurids, peramelids and potoroids in Australia and New Guinea (Vaucher et al. 1984; Vaucher and Beveridge 1997). Hymenolepidid cestodes are common in insectivorous eutherians throughout the world and utilise arthropods as intermediate hosts. However, no life cycles of Australian species have been elucidated. While the taxonomic independence of the Australian hymenolepidids has recently been confirmed (Vaucher and Beveridge 1997), their phylogentic associations are uncertain, with none described from South American marsupials (Schmidt 1986). However, material apparently belonging to

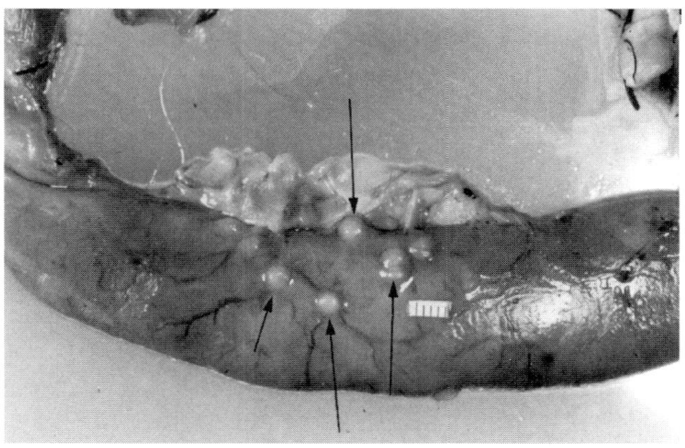

Figure 4 Small intestine of *Dasyurus maculatus* from Tasmania showing nodular projections on the external surface of the intestine induced by the scoleces of the cestode *Dasyurotaenia dasyuri* embedded in the mucosa (Photograph courtesy of Dr D.L. Obendorf.)

this genus has recently been recovered from preserved specimens of the South American marsupial, *Dromiciops,* held in the Australian National Wildlife Collection, Canberra (unpubl. obs.), opening up the possibility of a connection with South America.

Spirometra erinacei (Diphyllobothriidae), in contrast to all the other cestodes discussed thus far, utilises dasyurids as well as other mammals, reptiles and amphibians as intermediate or paratenic hosts. The second intermediate or plerocercoid stage of this cosmopolitan species, commonly known as a 'sparganum', is found either subcutaneously or in body cavities and develops into the adult cestode when ingested by a cat, dog or fox. The first intermediate stage or procercoid develops when the egg, deposited in faeces, hatches in water and the released coracidium is ingested by a copepod (Sandars 1953). It is assumed that the cestode was introduced to Australia with eutherian carnivores. Although spargana are usually regarded as non-pathogenic, heavy infestations have been observed in the inter-muscular tissues and body cavities of *Dasyurus hallucatus* at Cannon Hill, NT (D.M. Spratt, pers. obs.) and in six of 19 *Parantechinus bilarni* captured at Mt Brockman, NT, some of which died after capture with clinical signs suggesting the involvement of the spargana in their demise (P.Woolley, pers. comm.).

Nematodes or roundworms

Nematodes are by far the most diverse group of helminth parasites found in dasyurid marsupials. Because of the diversity not only of nematode families but also of life cycles, they are dealt with here on the basis of major taxonomic groupings.

Trichostrongyles

The trichostrongyles represent the most abundant internal parasites of dasyurids comprising nine endemic and largely host-specific genera with 23 species. Two of the genera (*Peramelistrongylus, Mackerrastrongylus*) are also common in peramelids. Most are parasitic in the small intestine, where they coil around villi, but one species, *P. skedastos,* occurs in the stomach, and another remarkable species, *Nasistrongylus antechini,* clearly related to species inhabiting the intestines (Durette-Desset and Beveridge 1981a) is found in the nasal cavities of *Antechinus agilis* and *A. swainsonii.* This large nematode causes remarkably little pathological change within the nasal cavity (Beveridge and Barker 1975a), a site it shares with the mites, *Domrownyssus dentatus* and *Ascoschoengastia rattus.* Very few strongylid nematodes are known to inhabit nasal cavities, and in its occupation of this site, *Nasistrongylus* resembles the metastrongyloid genus *Skrjabingylus,* which is a common parasite of the nasal cavities of mustelids in the Holarctic region (Anderson 1992).

Taxonomic diversity within the trichostrongyloid parasites is high, with three families represented. *Mackerrastrongylus, Sprattellus* and *Tetrabothriostrongylus* belong to the subfamily Mackerrastrongylinae (family Mackerrastrongylidae), a group thought to have evolved from nematode parasites of echidnas (Durette-Desset 1985). Similarly, *Copemania* (Nicollinidae), represented by a single species *C. obendorfi* in *Dasyurus maculatus,* is also thought to have evolved by host-switching from an ancestral genus *Nicollina* parasitic in echidnas (Durette-Desset and Beveridge 1981b). *Dessetostrongylus, Nasistrongylus, Patricialina* and *Woolleya* by contrast belong to the family Herpetostrongylidae, which shares a common ancestry with the Viannaiidae, a family of trichostrongyloid nematodes which has diversified in South American marsupials and rodents (Durette-Desset 1985). In this case, the Herpetostrongylidae are probably derived from ancestors in South American marsupials and may have reached Australia with the original marsupial immigrants (Humphery-Smith 1983). Of the few parasites known from the thylacine, *Thylacinus cynocephalus,* one is a trichostrongyloid, *Woolleya sprenti,* which also occurs in *Dasyurus* spp. (Mawson 1973).

The only known nematode parasites of the numbat, *Myrmecobius fasciatus,* are two species of trichostrongyle nematode, *Beveridgiella calabyi* and *B. inglisi,* found in the small intestine (Spratt et al. 1991).

No life cycles of the trichostrongyle nematodes of marsupials are known (Anderson 1992), but based on studies of related groups, life cycles are expected to be direct, with either oral infection or skin penetration being the principal modes of infection.

Metastrongyles or lung-worms

Metastrongyles are prominent parasites of the smaller dasyurid marsupials, with heavy accumulations of nematodes and their eggs or larvae leading to gross discolouration of the margins of the lung lobes of their hosts. The most striking example recorded to date is of infection with *Filaroides pilbarensis* (Filaroididae) in *Dasykaluta rosamondae* in which nematodes may occupy a large proportion of the lung mass (Fig. 5). Two species

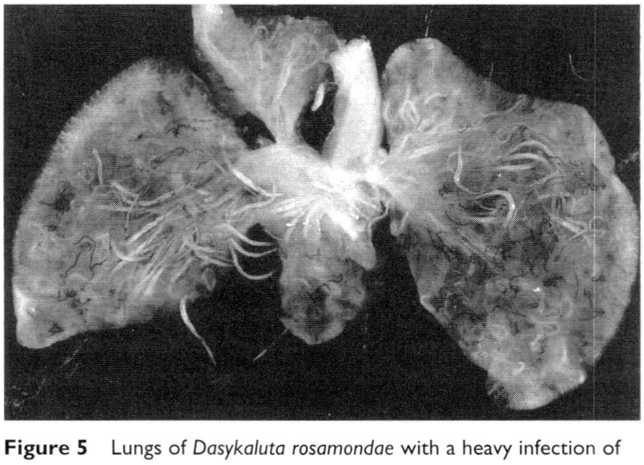

Figure 5 Lungs of *Dasykaluta rosamondae* with a heavy infection of the nematode *Filaroides pilbarensis.*

of metastrongyles frequently co-occur in the lungs of *Antechinus agilis* in south-eastern New South Wales – *Marsupostrongylus fragilis* (Angiostrongylidae), occupying the major bronchi and bronchioles, and *M. lanceolatus*, occurring in 'nests' in the terminal bronchioles, alveoli and lung parenchyma (Spratt 1979).

Many metastrongyles utilise molluscs as intermediate hosts while others have direct life cycles (Anderson 1992). The only known life cycle of a metastrongyloid occurring in Australian marsupials is that of *M. fragilis*. Infections were established experimentally in laboratory-reared *Antechinus agilis* infected with third-stage larvae from wild gastropods *Nitor pudibundus* and *Helicarion virens* (Spratt 1979). The prepatent period was less than 20 days. In addition, third-stage larvae developed in laboratory-reared *H. virens* within 12 days of infection.

The monotypic genus *Antechinostrongylus* (Angiostrongylidae) occurs, as the name implies, exclusively in *Antechinus* spp. while *Marsupostrongylus* occurs in a wide range of both carnivorous and herbivorous groups of marsupials (Spratt 1979). By contrast with these endemic genera, *Filaroides* is a large, cosmopolitan genus with representatives in pinnipeds, canids, mustelids and primates (Anderson 1978) and a single species in dasyurids. The phylogenetic associations of Australian lungworms with other metastrongyloids are unclear.

Spiruroid nematodes

Spiruroids are generally large, robust nematodes that occur in the stomach or oesophagus of their hosts and utilise arthropods, often coleopterans or orthopterans, as intermediate hosts. Three species, belonging to the families Spiruridae and Spirocercidae, are known from Australia.

The sole member of the Spiruridae, *Spirura aurangabadensis* is found in the oesophagus of *Antechinus*, *Planigale* and *Sminthopsis* spp. in south-eastern Australia, and also occurs in chiropterans and primates in south-east Asia. The hypothesis which most simply explains this range of hosts is that the parasite occurs primarily in mammalian hosts in Asia and that it has been introduced to Australia by bats migrating from south-east Asia since the spirurid nematodes are able to utilise a variety of arthropod as well as mammalian hosts (Spratt 1985).

The family Spirocercidae is represented by *Cyathospirura seurati* and *Cylicospirura heydoni*, which are common parasites of the stomachs of larger dasyurids, occurring in tumour-like nodules (Mawson 1968). *Cyathospirura seurati* also occurs in feral eutherians such as the cat, fox and dingo as well as in eutherians in Asia (Hasegawa et al. 1993). *C. heydoni* is restricted to marsupials, occurring in *Dasyurus albopunctatus, D. hallucatus* and *D. maculatus* (Mawson 1968; Oakwood and Spratt 2000) and, surprisingly, in nodules in the viscera of wallabies such as *Macropus rufogriseus* (see Spratt et al. 1991). Phylogenetic relationships within this family are not well established. However, Chabaud and Bain (1981a) suggested that the origin of the family was neotropical, including the genus *Didelphonema*, found in *Didelphis* in the southern United States. They did not consider the specific relationships of the Australian species.

Acuarioid nematodes

Acuarioids (family Acuariidae) are represented by two species in small dasyurids (*Antechinus* spp.), *Antechiniella suffodiax* and *Synhimantus australiensis*, both occurring as parasites of the stomach of these hosts. Acuarioids have arthropods as intermediate hosts and all related taxa are parasitic in birds leading to the suggestion that these parasites of dasyurids have originated by host-switching from aquatic birds (Beveridge and Spratt 1996).

Seuratoid nematodes

The diversity of seuratoid nematodes within dasyurids has only been demonstrated relatively recently (Smales 1998, 1999a, b, c, e; Smales and Rossi 1999) with four genera and 10 species recognised. Several genera occur exclusively in dasyurids, while the genus *Linstowinema* is shared with peramelids (Smales 1999a). The subfamily Echinonematinae, to which all of these these genera belong, is endemic and its relationships are obscure. Chabaud et al. (1980) suggested on the basis of larval morphogenesis that the echinonematids were related to *Seuratum* in bats. The phylogenetic relationships of the genus therefore currently remain uncertain, but recent publications suggest a diverse fauna exists in dasyurids, with many taxa yet to be discovered.

Gnathostomatoid nematodes

Gnathostoma spp. (Gnathostomatidae) are primarily parasitic in eutherian mammals. The life cycle includes a free-swimming larva which is ingested by a copepod. In the copepod, the larva develops to the third stage. A wide range of vertebrates such as small amphibians and reptiles ingest infected copepods and larvae develop to an advanced third stage. The life cycle is completed when the second intermediate host is ingested by the appropriate mammalian definitive host. If ingested by an unsuitable mammalian host, the nematode encysts in the stomach, oesophagus or kidney (Daengsgvang 1982).

Two species of *Gnathostoma* are known from Australia – *G. spinigerum*, which is parasitic in cats and dogs, and *G. doloresi*, which is found in pigs (Smales 1999d). Third-stage larvae of *G. spinigerum* have been found encysted in *Dasyurus hallucatus* and *Phascogale tapoatafa*, while a single specimen of *G. doloresi* has been found in *Isoodon macrourus* (Smales 1999d). These findings are not surprising given the low level of host specificity exhibited by the larval stages of *Gnathostoma* spp.

Physalopteroid nematodes

The genus *Physaloptera* (family Physalopteridae) is cosmopolitan and highly speciose, occurring in the stomachs of carnivores and birds of prey. Intermediate hosts are usually coleopterans. The genus is represented in Australia by species in bandicoots (Norman and Beveridge 1999), rodents (Johnston and Mawson 1941), birds (Mawson et al. 1986), dogs and feral cats (Barton and McEwan 1993; O'Callaghan and Beveridge 1996) and a single species, *P. sarcophili*, in large dasyurid marsupials (*Sarcophilus, Dasyurus*) (Johnston and Mawson 1940; Spratt et al. 1991; Oakwood and Spratt 2000). Associations within this genus and phylogenetic origins remain obscure. A related genus, *Turgida*, occurs in neotropical marsupials (Potkay 1970; Alden 1995), but the phylogenetic relationships between the genera have not been investigated.

Thelazioid nematodes

The Thelazioidea consists of three families – the Thelaziidae, containing the eyeworms of birds and mammals, the Rhabdochonidae of fish and the Pneumospiruridae of carnivorous mammals. The superfamily is represented by a single species within the Australian marsupial helminth fauna and is the only pneumospirurid nematode known from marsupials. *Metathelazia naghiensis* occurs in the lungs of *Antechinus stuartii, A. flavipes* and *A. bellus*, and the southern brown and long-nosed bandicoots (Spratt 1980). Fewer than 10 species of *Metathelazia* are known from carnivorous mammals in America, Europe, the Middle East and from the primate, *Tupaia glis*, in Malaysia. No life cycles are known and phylogenetic relationships are unresolved.

Filarioid nematodes

The Filarioidea occur in the tissues and tissue spaces of all classes of vertebrates except fishes (Anderson and Bain 1976). Members of the Filariidae occur in subcutaneous sites in mammals and create skin lesions which attract dipteran intermediate hosts. Members of the Onchocercidae occur in all of the organ systems of the body and in most tissues; all are transmitted by blood- or lymph-feeding arthropods. Filariidae are not known from Australian marsupials but four genera of Onchocercidae are present, three of which occur in dasyurids (Spratt and Varughese 1975). *Cercopithifilaria johnstoni* occurs in the sub-cutis of Tasmanian devils and a spectrum of other marsupial, monotreme and murid hosts and is transmitted by the tick *Ixodes trichosuri* (Spratt and Haycock 1988). The broad host range of this species and the 'zoologically incoherent' host range of the genus around the world is attributable to the role of ixodid ticks as intermediate hosts and vectors (Spratt and Haycock 1988; Chabaud and Bain 1994). *Sprattia capilliforme* occurs in the superior mesenteric and portal veins and venous spaces of the liver of *Dasyurus hallucatus*; other species occur in *Trichosurus caninus, T. vulpecula* and *Isoodon macrourus* (Spratt et al. 1991).

The genus *Breinlia* has radiated widely in Australasian marsupials, particularly the Macropodoidea, and is known also from phalangerids, petaurids, peramelids and the echidna (Spratt et al. 1991). *Breinlia dasyuri* is the only species known from dasyurids and has been recorded only in *Dasyurus maculatus*. Species in which life cycles have been studied are transmitted by mosquitoes (see Bain et al. 1981). The genus *Breinlia* occurs in lemurs and in rodents in the Indo-Malaysian region and was interpreted as being of recent origin there, having invaded from Australia (Chabaud and Bain 1994). However, species of *Breinlia* are now known to occur in indigenous rodents in northern Australia, suggesting that the genus may be Indo-Malaysian in origin (Beveridge and Spratt 1996).

Ascaridoid nematodes

The Ascaridoidea (family Ascarididae) is represented by only two species within the Australian marsupial helminth fauna with both occurring in dasyurids. One occurs as the adult nematode, while the other occurs exclusively as the larval stage.

Baylisascaris tasmaniensis is an ascaridoid nematode found commonly in the intestine of the Tasmanian devil, *Sarcophilus harrisii*, as well as in *Dasyurus* spp. in Tasmania (Sprent 1970; Sprent et al. 1973). Unidentified specimens of *Baylisacaris* have also been found in the New Guinea dasyurid, *Dasyurus albopunctatus* (see Spratt et al. 1991). Larval stages are common in the tissues of common wombats, *Vombatus ursinus* (Munday and Gregory 1974), as well as in the Tasmanian pademelon, *Thylogale billardierii* (Spratt et al. 1991). The relationships of this remarkable ascaridoid are unclear since congeners occur in ursids, mustelids and procyonids primarily in the nearctic region (Sprent 1968). Sprent (1970) suggested that if the unlikely possibility that *Bayslisascaris* had been introduced to Australia by bears in circuses or zoos could be eliminated, then it may indicate an 'heirloom', a link with an ancient group of carnivores no longer present on the Australian continent.

Ophidascaris robertsi is an ascaridoid which occurs in the adult stage in the stomachs of pythons, primarily *Morelia amethistina* and *M. spilota* (Pichelin et al. 1999). The larval stages are exteremely large (up to 15 cm) and occur in the viscera and subcutaneous tissues of a wide variety of eutherians and marsupials (Sprent 1963; Spratt et al. 1991). They are common in small dasyurid marsupials (Spratt et al. 1991), sometimes occurring in large numbers (Fig. 6). Completion of the life cycle depends upon the mammalian intermediate host being preyed upon by a python.

A curious, temporary addition to the ascaridoid fauna of marsupials was the description by Sprent (1971) of a new genus, *Cotylascaris*, from the thylacine, *Thylacinus cyanocephalus*. The nematodes had been passed in the faeces of a thylacine in captivity in the Regent's Park zoo in London. Subsequently (Sprent 1972), it was discovered that the nematodes described were in fact *Ascaridia columbae*, a cosmopolitan parasite of pigeons. The suggestion was made that the thylacine had caught and eaten a

Figure 6 Subcutaneous tissues of *Antechinus agilis* with a heavy infection of the larval nematode *Ophidascaris robertsi*. The adult nematodes occur in the stomachs and intestines of pythons.

pigeon and that the parasites of the pigeon had been passed undigested in the faeces of the thylacine. The incident is worth noting as it indicates the potential hazards associated not only with the collection of parasites from captive animals, but also the collection of parasites passed in the faeces and not obtained from their normal habitat in the gastro-intestinal tract.

Rhabdiasoid nematodes

Rhabdiasoids are microscopic nematodes inhabiting the small intestines of their hosts and having direct life cycles. A single species of *Parastrongyloides* is found in the intestine of small dasyurids, with related species in bandicoots and possums (Spratt et al. 1991). Additional species are known from insectivores in Europe (Anderson and Bain 1982).

Trichinelloid nematodes

Trichinelloid nematodes differ from all of the nematode parasites considered thus far in lacking posterior sensory organs termed phasmids and in possessing a distinctive oesophageal structure in which the central oesophageal tube is surrounded by specialised cells termed stichocytes. Three genera are known to occur in dasyurids, *Anatrichosoma* and *Capillaria* from the family Trichuridae and *Trichinella* from the family Trichinellidae.

Anatrichosoma is a genus of small nematodes which occur in the gastric, nasal and buccal mucosa of tupaids, primates and murids as well as in the paracloacal glands of *Antechinus* spp. (Spratt 1982; Spratt et al. 1991). Relationships within the genus are uncertain.

Capillaria (*sensu lato*) is a cosmopolitan, highly speciose genus of nematodes which occurs in almost all parts of the bodies of mammals, birds, fish and reptiles. In Australia, only one endemic species from marsupials has been named, *C. rickardi*, from the gastrointestinal tract of *Antechinus* spp. and *Sminthopsis leucopus* (Beveridge and Barker 1975b). However, numerous undescribed species have been collected from dasyurid marsupials (Spratt et al. 1991, Zhu et al. 2000).

The nematode genus *Trichinella* is of considerable economic and human health significance because of a complex of species, formerly aggregated under the name *T. spiralis*, which are common in wildlife but also infect rats, pigs and occasionally horses. Humans eating undercooked meat from these hosts may also become infected and infections may be fatal (Clausen et al. 1996). *T. spiralis* has never been found in Australia.

By contrast, the related species *T. pseudospiralis* was reported from *Dasyurus maculatus* and *Sarcophilus harrisii* from Tasmania by Obendorf et al. (1990). First described from Russia by Garkavi (1972), *T. pseudospiralis* is unusual in the genus in that it utilises birds of prey as well as mammals as definitive hosts (Obendorf and Clarke 1992). In Tasmania, Obendorf et al. (1990) postulated that the principal means of transmission was by cannibalism as they found both the larval stages in muscle and adult nematodes in the intestines of the same host species. No larval stages have been found in the muscles of herbivorous marsupials or domestic herbivores (Obendorf et al. 1990). The origins of the population of *T. pseudospiralis* in Tasmania remain undetermined.

Acanthocephalans or 'thorny-headed worms'

Acanthocephalans are uncommon in marsupials and the phylum is represented by a single endemic genus, *Australiformis*, occurring primarily in bandicoots (Spratt et al. 1991). Its larval stage (cystacanth) is found occasionally in the abdominal cavity of *Antechinus* and *Phascogale* (Spratt et al. 1991).

Fleas (Siphonaptera)

Sixteen genera of fleas with 32 species have been reported from dasyurid marsupials (Mardon 1976 1978; Calder 1996). Many fleas exhibit a low degree of host specificity with their larval and pupal stages occurring off the host. Thus a variety of cosmopolitan species and genera such as *Ctenocephalides*, *Echidnophaga*, *Pulex* and *Xenopsylla*, occurring most frequently on dogs, cats, chickens and rodents, are also found on dasyurids. A number of endemic flea genera, such as *Macropsylla*, *Stephanopsylla* and *Stephanocircus*, are common parasites of Australian rodents and/or bandicoots and are also parasites of dasyurids. Of the remaining genera, *Acanthopsylla* is parasitic primarily on dasyurids, while *Coronapsylla* is endemic and restricted to *Antechinus* spp. The most unusual genus of fleas is *Uropsylla* which is restricted to larger dasyurids in Tasmania. In this species the larval and pupal stages occur in the subcutaneous tissues of the host, one of the few known fleas to have all life cycle stages on the host (Dunnet and Mardon 1991). Many of the flea genera and families present are cosmopolitan, providing few clues as to the phylogenetic origins of the Australian representatives. However, the family Stephanocircidae or 'helmeted fleas' is one group that occurs primarily on South American and Australian marsupials (Traub and Dunnet 1973), suggesting a common origin. The Pygiopsyllidae by contrast have diversified greatly in the Australasian region, occurrences in the African and Asian regions being interpreted as emigrants from Papua New Guinea by Mardon (1978). The latter example indicates that the Australasian region may be the origin of some groups of parasitic arthropods. Larvae of pygiopsyllid fleas were found commonly in the nests of the dasyurids *Murexia longicaudata* and *Antechinus habbema* in Papua New Guinea by Woolley (1989).

Lice (Phthiraptera)

All lice known from Australian marsupials belong to the suborder Amblycera, one of the groups of so-called 'biting lice'. This group attains its greatest diversity in birds but has one family, the Boopiidae, which, with a single striking exception, is restricted to Australasian marsupials. The exception is *Heterodoxus spiniger* which is found on dingoes and on canids worldwide (von Kéler 1971). Within this family, the type genus *Boopia* has seven species restricted to dasyurid marsupials. Lice are highly host-specific and complete their entire life cycles on the host. The boopid lice are restricted to Australasian marsupials (von Kéler 1971). They probably originated from lice parasitic on birds and are phylogenetically distinct from the Trimenoponidae, which are found on South American marsupials (von Kéler 1971).

Ticks (Ixodoidea)

Six species of ticks have been reported from dasyurid marsupials, belonging to the genera *Ixodes* and *Haemaphysalis*, both of which are members of the family Ixodidae (Roberts 1970). While the genera are cosmopolitan, the species found on dasyurids are endemic. Species of *Haemaphysalis* found on dasyurids are also common on rodents while three of the species of *Ixodes* (*tasmani*, *fecialis*, *holocyclus*) have broad host ranges, a common feature of ticks. One species, *I. antechini*, is a prominent and distinctive parasite of *Antechinus* spp. While the latter is suspected to be the vector of protozoan parasites of *Antechinus* spp., its role has not been proven (Barker et al. 1978). *I. holocyclus*, the so called 'paralysis tick' of eastern Australia, is of particular significance. This tick induces a fatal paralysis in previously unexposed hosts (Stone et al. 1989), is a vector of the filarioid nematode *Cercopithifilaria johnstoni*, found in various marsupials including dasyurids (Spratt and Haycock 1988), and is also a vector of Queensland tick typhus, *Rickettsia australis*, causing clinical disease in humans along the east coast of Australia (Campbell and Domrow 1974).

Mites

Dasyurid marsupials are host to a wide variety of parasitic mites including 32 species of trombiculids, in which only the larval stage is parasitic (Domrow and Lester 1985) and 15 species of mesostigmatid mites, in which all stages of the parasite are found on the host (Domrow 1987). The species of trombiculds usually exhibit a low level of host specificity and are found on a wide range of rodents, bandicoots and other marsupials. They are of considerable medical importance as they also act as vectors of scrub typhus, *Orientia tsutsugamushi*.

The largest and therefore the most readily collected mites are members of the mesostigmatid mites of the families Laelapidae and Rhinonyssidae. Woolley (1989) found that species of *Mesolaelaps* and *Ornithonyssus* were common inhabitants of the nests of *Murexia* and *Antechinus* in Papua New Guinea.

The astigmatid mites are represented by 10 genera which exhibit a higher degree of host specificity than many other mites. Several genera of 'fur mites', *Campylochirus* and *Dasyurochirus* (12 species), are restricted to dasyurids, while the family Sarcoptidae, which includes the causative agent of human scabies, *Sarcoptes scabiei*, has three endemic species in the genera *Diabolicoptes* and *Satanicoptes* restricted to large dasyurids. The marsupial-inhabiting genera were considered as basal genera in a phylogenetic study of the Sarcoptidae by Klompen (1992). No comparable genera occur on South American marsupials. *Caenolestocoptes*, parasitic on caenolestid marsupials in South America, was considered to be among the most primitive members of the family Sarcoptidae by Fain (1982b); however, the genus was transferred to the family Rhyncoptidae by Klompen (1992). It was used by him as an outgroup to the Sarcoptidae and the possibility still remains of a connection between the rhyncoptid mites of South American marsupials and the basal sarcoptids present in dasyurids.

Among the remaining mites parasitic on or in marsupials, however, several groups do have affinities with mites parasitic on South American marsupials. The myobiid subfamily Archemyobiinae, parasitic on didelphids and microbiotherids in South America has affinities with the tribe Australomyobiini, with three species of *Australomyobia* restricted to dasyurids (Fain 1982a, p. 79, 1982b, p. 112; Fain 1978; Fain and Lukoschus 1979; Domrow 1992). The atopomelid genus *Didelphoecius* on didelphids in South America has affinities with *Dasyurochirus* on dasyurid marsupials in Australia (Fain 1982b, p. 114). Among Psoroptids, *Petauralges*, parsasitic on *Phascogale* is related to the genus *Listropsoralges* on didelphid marsupials (Domrow 1988). Additional genera such as *Cytostethum* on macropodoids and *Austrochirus* on peramelids have diversified within the Australasian region but precise phylogentic relationships remain to be determined (see Domrow 1992).

Protozoan parasites

The protozoan parasites of dasyurids have been little studied. Haemoparasites reported to date include the haemogregarine *Haemogregarina dasyuri* from *Dasyurus* spp., which exhibits no obvious pathogenic effects, and an undescribed species of *Babesia* found in *Antechinus agilis*, which causes severe, fatal anaemia and haemoglobinuria in male antechinuses during the mating season (Barker et al. 1978). An unidentified species of *Cryptosporidium* has been identified on the surface of enterocytes in the small intestine of *A. agilis* during the period of male mortality (Barker et al. 1978). Unidentified species of *Sarcocystis* have been found in the musculature of *Antechinus stuartii, A. swainsonii, Dasykaluta rosamondae, Pseudantechinus bilarni, P. macdonnellensis, P. ningbing, Planigale maculata, Sarcophilus harrisii* and *Dasyurus hallucatus* (Mackerras 1958; Munday et al. 1978; Attwood and Woolley 1982; Oakwood and Spratt 2000). The stage reported is an intermediate stage in the life cycle of the parasite and completion of the life cycle requires ingestion by as yet undetermined predators.

Possibly the most significant protozoan parasite reported to date from dasyurids is *Toxoplasma gondii* which is a common intestinal parasite of felids with intermediate stages occurring in the tissues of many mammals and birds (Obendorf 1993). It is particularly common in the muscles of sheep in Australia (Speare 1985) and when fed to small dasyurids causes a fulminating encephalitis and death with neurological signs as the most common clinical manifestations (Attwood and Woolley 1972, 1982; Attwood et al. 1975).

DIVERSITY AND RELATIONSHIPS OF PARASITES

The above brief survey of the parasites of dasyurid marsupials provides adequate evidence in support of the hypothesis that they harbour a diverse fauna of helminth, arthropod and protozoan parasites. Most of the parasite taxa encountered are endemic to Australia and while many are restricted to dasyurid marsupials, a number are also shared by rodents and peramelids (Beveridge and Spratt 1996), hosts whose habitats overlap to some degree with those of dasyurids. This overlap provides circumstances in which parasites can switch from one host group to another either because hosts utilise similar pathways through vegetation in which haematophagous arthropods occur, or they utilise similar 'nesting sites' which are inhabited by a range of non-specific blood-feeding arthropods, or they feed on the same arthropods which harbour the intermediate stages of heteroxenous helminth parasites.

Of those parasites which are restricted to dasyurid marsupials, some such as the herpetostrongylid nematodes (*Woolleya, Beveridgiella, Dessetostrongylus, Patricialina*) and the cestode genus *Paralinstowia* have clear associations with the helminth groups occurring in South American marsupials. Unfortunately, the parasites of microbiotheriid marsupials, the South American family most likely to indicate associations with the Australian helminth fauna, remain largely uninvestigated. The nematode genera *Sprattellus* and *Tetrabothriostrongylus* appear to have been derived from the subfamily Tachynematinae, parasitic in echidnas, as does the genus *Copemania*, a genus derived from nicollinids which are also parasitic in echidnas. This diversity of phylogenetic origins in the trichostrongyloids is mirrored in the spirurid nematodes with taxa apparently acquired from bats and waterfowl. The helminth fauna present in dasyurids therefore appears to have complex phylogenetic origins including elements shared with South American marsupials as well as other taxa acquired by host-switching from monotremes, bats and birds.

Similar conclusions can be drawn from the arthropod parasite fauna found on or in dasyurid marsupials. The fauna is diverse and there is a high degree of endemicity at least at the species level. There are many fleas, mites and ticks which exhibit a relatively low level of host specificity and therefore many of these arthropod parasites of dasyurids are shared with rodents and bandicoots, a pattern which was also evident in the case of the helminths. The lice present on dasyurids belong to a group restricted to marsupials and probably derived from ancestors on birds (von Kéler 1971). Many of the flea genera and families present are cosmopolitan, providing few clues to the phylogenetic origins of the Australian representatives. However, some families apparently have a Gondwanan origin while others have diversified in the Australasian region, and extended their range to the African and Asian regions. Likewise, there are several groups of mites with apparent connections to related taxa occurring on South American marsupials. However, as with the helminths, the precise phylogenetic affiliations of many groups of arthropods parasitic on dasyurid marsupials remain to be investigated.

The best studied family of South American marsupials, from a parasitological point of view, is the family Didelphidae. Potkay (1970) and Alden (1995) reviewed studies of the helminths of *Didelphis virginiana and D. marsupialis* and listed a range of nematodes, trematodes, cestodes and acanthocephalans. Genera which are also present in Australian dasyurids included *Gnathostoma* and *Trichinella*, while the genera *Gongylonema, Trichuris* and *Spirocerca*, though present in Australia, occur primarily in domestic animals and have not been found in dasyurids. *Aspidodera, Cruzia*, and *Viannaia*, which are common in American marsupials, have not been found in Australia. The wide range of trematodes found in *Didelphis* includes a number of species which are endemic to North America as well as a few cosmopolitan species. The only completely identified cestodes known from *Didelphis* are species of *Mesocestoides* occurring in the small intestine, a genus not found in Australia, and plerocercoids of *Spirometra mansonoides* in body cavities. Likewise the acanthocephalan fauna consists of North American endemics. The helminth fauna of *Didelphis* therefore includes many of the genera found in omnivorous, carnivorous or insectivorous mammals globally with only one distinctive genus, the trichostrongyloid *Viannaia*. Recently, Gardner and Hugot (1995) described a new genus of oxyurid nematode, *Didelphoxyuris thylamysis* from the didelphid *Thylamys elegans* from Bolivia. No oxyurid nematodes are known from Australian carnivorous marsupials (Spratt et al. 1991).

The helminth fauna of microbiotheriids would be of particular interest, but is poorly known. *Dromiciops australis* harbours the rictulariid nematode genus *Pterigodermatites*, but the genus is also common in edentates and rodents in South America (Navone and Suriano 1992). The related genus *Quentius*, described by Chabaud and Bain (1981b) from *Marmosa* sp. in Colombia, was considered to be the most primitive rictulariid known. Chabaud and Bain (1981b) suggested that the rictulariids arose in South America but did not diversify until South and North America were joined and rictulariids were able to colonise hosts in the northern hemisphere. These hosts included muroid rodents. In Australia, the related genus *Rictularia* is found only in rodents (Mawson 1971a, b), but the hypothesis of Chabaud and Bain (1981b) suggests that this is an invasion from Asian rodents rather than a connection with the parasites of South American marsupials. Their hypothesis indicates the potential complexities involved in establishing the biogeographical relationships between parasites and their mammalian hosts. The current state of ignorance and of complexity has been exemplified in a recent study of the cestodes of marsupials from Bolivia by Gardner and Campbell (1992b). They suitably crystallised the state of knowledge of this topic when they commented in a concluding statement that 'the recognition of *Linstowia schmidti* n.sp. from Bolivian marsupials, nearly a century after the original description of (a related species in marsupials) *Linstowia iheringi*, is further indication of how little is known of the helminth fauna of mammals in the neotropics'.

The Effects of Parasites on Their Hosts

To date, the potential effects of parasites on dasyurid hosts are limited to observations of pathological changes visible grossly or at the level of light microscopy. No experimental studies have been undertaken, so the possibility that parasites which exhibit few visible pathological changes in natural infections might be involved in the regulation of populations cannot currently be eliminated.

Several species of parasites have been observed to exhibit deleterious effects on their dasyurid hosts. An undescribed species of *Babesia* was considered to be responsible for mortality in some males of *Antechinus agilis* during the breeding season (Barker et al. 1978). The parasite is probably present throughout the year (Cheal et al. 1976; Barker et al. 1978) but significant parasitaemias are associated with reduced haematocrits and haemoglobinuria only in males during the breeding season (Barker et al. 1978).

Gastric lesions are associated with two species of parasitic nematodes, *Antechiniella suffodiax* and *Capillaria rickardi*. *A. suffodiax* burrows deeply into the stomach wall producing digitiform projections on the external surface of the stomach, each of which contains the enlarged anterior extremity of a nematode. The tails of the nematodes lie free in the lumen of the stomach. The lesions induced are of a chronic nature and are composed of mature fibrous tissue with infiltrates of mononuclear cells (Beveridge and Barker 1975b). *C. rickardi* occurs in tunnels in the epithelium of the gastric mucosa or duodenum (Beveridge and Barker 1975b). Occasionally it is associated with ulceration of the mucosa and haemorrhage (Beveridge and Barker 1975b; Barker et al. 1978). Gastric and duodenal haemorrhage is one of the major immediate causes of mortality in males of *A. agilis* during the breeding season and it is possible that *C. rickardi* is involved in the initiation of ulcers (Barker et al. 1978). The spirurid nematodes *Cyathospirura seurati* and *Cylicospirura heydoni* are found in nodular lesions in the stomachs of *Dasyurus* spp. The lesions are chronic in nature and the effects on the host are not known.

Metastrongyloid nematodes induce a significant interstitial pneumonia in dasyurids (Barker et al. 1978; McColl and Spratt 1982). The most spectacular indication of the potential pathogenicity of metastrongyloids is provided by *Filaroides pilbarensis* in *Dasykaluta rosamondae*. In this instance, the pathogenic effects are due primarily to the elimination of most functional pulmonary tissue by large numbers of worms and their eggs.

Studies of the seasonal dynamics of helminths of *Antechinus agilis* have indicated an increase in the numbers of helminths in male hosts during the breeding season (Beveridge and Barker 1976). The differences were interpreted as being due to increased susceptibility to infection resulting from suppression of the lympho-reticular system under the influence of high concentrations of circulating corticosteroids (Barker et al. 1978).

Toxoplasma gondii is a protozoan parasite which utilises felids as definitive hosts and a wide range of mammals including humans as intermediate hosts. Attwood and Woolley (1972) and Attwood et al. (1975) have reported *T. gondii* as a significant cause of mortality in small captive dasyurids maintained on a diet including raw sheep meat, a common source of bradyzoites. The significance of toxoplasmosis in wild dasyurids has not been investigated. The disease has been incriminated as a significant cause of mortality in wild bandicoots, *Perameles gunnii,* in Tasmania (Obendorf and Munday 1990) and in a critically small population of *P. gunnii* at Hamilton, Victoria (Lenghaus et al. 1990). Toxoplasmosis does not appear to be a contributing factor to mortality in *Dasyurus hallucatus* in the Kakadu region of the Northern Territory (Oakwood and Pritchard 1999).

Heavy infestations with mites are common causes of dermatitis in domestic animals and the numerous species found in or on dasyurids also have the potential to cause disease. Currently, only the hair follicle and sebaceous gland inhabitant *Demodex antechini* has been reported as a cause of disease in dasyurids (Nutting and Woolley 1965).

FUTURE DIRECTIONS

The study of the parasites of carnivorous marsupials is very much in its infancy. Although there are clear indications that these marsupials harbour an extremely diverse parasite fauna, much more collecting and basic descriptive work remains to be done. This is not only true of the Australian mammals but also of South American representatives, particularly groups such as the microbiotheriids. While there are some obvious similarities beween helminths from Australian and South American marsupials, these relationships need to be tested primarily by establishing more robust phylogenies for the parasites involved since most parasites are simple organisms and provide relatively few morphological characters for phylogenetic analyses. The phylogenetic relationships of many of the helminth and arthropod groups discussed here are unknown, but the fragmentary data currently available suggest tantalising possibilities for future research, particularly in the case of the parasites of larger dasyurids. Life cycles of parasites are extremely poorly understood and the ecological associations between host and parasite have scarcely been studied. Similary, the impact of parasites on host populations is currently largely a matter of speculation. Numerous species of parasites cause observable lesions, but the effects of parasitism are often more subtle than the production of visible lesions, influencing the metabolism and ultimately the reproductive success of hosts (Scott 1987). Experimental studies on the effects of parasites on carnivorous marsupials remain a totally unexplored area of investigation.

Thus, there are many reasons which one can advance in favour of studying the parasites of carnivorous marsupials. The one feature of any such enterprise appears to be that there will be no shortage of research topics in this field in the forseeable future.

ACKNOWLEDGEMENTS

We wish to thank Dr P.A. Woolley for the provision, over the years, of meticulously preserved parasites from dasyurids, for the contribution of Fig. 2, for critical comments on a draft of the manuscript and for the provision of the unpublished data included in it. Dr D. Obendorf is also thanked for providing material for study and for the provision of photographs used in this paper.

REFERENCES

Alden, K.J. (1995), 'Helminths of the opossum, *Didelphis virginiana*, in Southern Illinois, with a compilation of all helminths reported from this host in North America', *Journal of the Helminthological Society of Washington,* **62**:197–208

Anderson, R.C. (1978), 'Keys to genera of the superfamily Metastrongyloidea', in *Commonwealth Institute of Helminthology Keys to the Nematodes Parasites of Vertebrates, No. 5* (eds. R.C. Anderson, A.G. Chabaud, & S. Willmott), Commonwealth Agricultural Bureaux, Farnham Royal, England.

Anderson, R.C. (1992), *Nematode Parasites of Vertebrates, their Development and Transmission,* Commonwealth Agricultural Bureaux, Wallingford.

Anderson, R.C., & Bain, O. (1976), 'Keys to genera of the order Spirurida. Part 3. Diplotriaenoidea, Aproctoidea and Filarioidea', in *Commonwealth Institute of Helminthology Keys to the Nematodes Parasites of Vertebrates, No. 3* (eds. R.C. Anderson, A.G. Chabaud, & S. Willmott), Commonwealth Agricultural Bureaux, Farnham Royal, England.

Anderson, R.C., & Bain, O. (1982), 'Keys to the superfamilies Rhabditoidea, Dioctophymatoidea, Trichinelloidea and Muspiceoidea', in *Commonwealth Institute of Helminthology Keys to the Nematodes Parasites of Vertebrates, No. 9* (eds. R.C. Anderson, A.G. Chabaud, & S. Willmott), Commonwealth Agricultural Bureaux, Farnham Royal, England.'

Angel, L.M. (1970), '*Coelomotrema antechinomes*, gen. et sp. nov., a prosthogonomid trematode from *Antechinomys* and *Antechinus* (Marsupialia: Dasyuridae) from Australia', *Australian Journal of Zoology,* **18**:19–124.

Attwood, H.D., & Woolley, P.A. (1972), 'Toxoplasmosis in dasyurid marsupials', *Pathology,* **7**:50.

Attwood, H.D., & Woolley, P.A. (1982), 'Histopathology of captive dasyurid marsupials', in *The Management of Australian Mammals in Captivity* (ed. D.D. Evans), pp. 27–30, Zoological Board of Victoria, Melbourne.

Attwood, H.D., Woolley, P.A., & Rickard, M.D. (1975) 'Toxoplasmosis in dasyurid marsupials', *Journal of Wildlife Diseases,* **11**:543–51.

Bain, O., Petit, G., Ratanaworabhan, N., Yenbutra, S., & Chabaud A.G. (1981), 'Une nouvelle filaire d'Ecureuil en Thailande, *Breinlia (B.) manningi* n. sp. et son developpement chez *Aedes*', *Annales de Parasitologie Humaine et Comparée,* **56**:193–201.

Barker, I.K., Beveridge, I., Bradley, A.J., & Lee, A.K. (1978), 'Observations on spontaneous stress-related mortality among males of the dasysurid marsupial *Antechinus stuartii* Macleay', *Australian Journal of Zoology,* **26**:435–47.

Barton, M.A., & McEwin, D.R. (1993), 'Spirurid nematodes in dogs and cats from central Australia', *Australian Veterinary Journal,* **70**:270.

Beveridge, I. (1977), 'On two new linstowiid cestodes from Australian dasyurid marsupials', *Journal of Helminthology,* **51**:31–40.

Beveridge, I. (1984), '*Dasyurotaenia robusta* Beddard, 1912 and *D. dasyuri* sp. nov., from carnivorous marsupials', *Transactions of the Royal Society of South Australia,* **108**:185–95.

Beveridge, I. (1994), 'Family Anoplocephalidae Cholodkowsky, 1902', in *Keys to the Cestode Parasites of Vertebrates* (eds. L.F. Khalil, A. Jones, & R.A.Bray), pp. 315–74, Commonwealth Agricultural Bureaux, Wallingford.

Beveridge, I., & Barker I.K. (1975a), '*Nicollina antechini* sp. n. (Nematoda: Amidostomatidae) from the nasal cavity of the dasyurid marsupial *Antechinus stuartii* Macleay, 1841, and associated pathology', *Journal of Parasitology,* **61**:489–93.

Beveridge, I., & Barker I.K. (1975b), 'Acuariid, capillariid and hymenolepid parasites of the dasyurid marsuypial *Antechinus stuartii*', *Journal of Helminthology,* **49**:211–27.

Beveridge, I., & Barker, I.K. (1976), 'The parasites of *Antechinus stuartii* Macleay from Powelltown, Victoria with observations on seasonal and sex-related variations in numbers of helminths', *Australian Journal of Zoology,* **24**:265–72.

Beveridge, I., Rickard, M.D., Gregory, G.G., & Munday, B.L. (1975), 'Studies on *Anoplotaenia dasyuri* Beddard, 1911 (Cestoda: Taeniidae), a parasite of the Tasmanian devil: observations on the egg and metacestode', *International Journal for Parasitology,* **5**:257–67.

Beveridge, I., & Spratt, D.M. (1996), 'The helminth fauna of Australasian marsupials: origins and evolutionary biology', *Advances in Parasitology,* **37**:135–254.

Calder, A.A. (1996), 'Siphonaptera. *Zoological Catalogue of Australia vol. 28. Neuroptera, Strepsiptera, Mecoptera, Siphonaptera* (ed. A. Wells), pp. 137–81,185–97, 222–26, CSIRO Publishing, Melbourne.

Campbell, R.W., & Domrow, R. (1974), 'Rickettsiosis in Australia: infection of *Rickettsia tsutsugamushi* and *R. australis* from naturally infected arthropods', *Transactions of the Royal Society of Tropical Medicine and Hygiene,* **68**:397–402.

Chabaud, A.G., & Bain, O. (1981a), 'Description de *Spirobakerus weitzeli* n.g., n. sp., et remarques sur les nematodes Spirocercidae', *Annales de Parasitologie Humaine et Comparée,* **56**:173–8.

Chabaud, A.G., & Bain, O. (1981b), 'Quentius kozeki n.g., n. sp., nematode rictulaire parasite d'un marsupial americain', *Annales de Parasitologie Humaine et Comparée,* **56**:73–80.

Chabaud, A.G., & Bain, O. (1994), 'The evolutionary expansion of the Spirurida', *International Journal for Parasitology,* **24**:1179–201.

Chabaud, A.G., Seurau, C., Beveridge, I., Bain, O., & Durette-Desset, M.-C (1980), 'Sur les nématodes Echinonematinae', *Annales de Parasitologie Humaine et Comparée,* **55**, 427–33.

Cheal, P.O., Lee, A.K., & Barnett, J.L. (1976), 'Changes in the haematology of *Antechinus stuartii* (Marsupialia) and their association with male mortality', *Australian Journal of Zoology,* **24**:299–311.

Clausen, M.R., Meyer, C.N., Krantz, T., Moser, C., Gomme, G., Kayser. L., Albrectsen, J, Kapel, K.M.O., & Bygbjerg, I.C. (1996), 'Trichinella infection and clinical disease', *Quarterly Journal of Medicine,* **89**:631–36.

Cribb, T.H. (1992), 'The Brachylaimidae (Trematoda: Digenea) of Australian native mammals and birds, including descriptions of *Dasyurotrema* n.g., and four new species of *Brachylaima*', *Systematic Parasitology,* **22**:45–72.

Cribb, T.H. (1998), 'The diversity of the Digenea of Australian mammals', *International Journal for Parasitology,* **28**:899–911

Cribb, T.H., & Spratt, D.M. (1991), '*Strzeleckia major* n.g., n.sp., & *S. minor* n.sp. (Digenea: Hasstilesiidae) from *Antechinus swainsonii* (Marsupialia: Dasyuridae) in Australia', *Systematic Parasitology,* **19**:73–9.

Cribb, T.H., & Spratt, D.M. (1992), 'Dicrocoeliidae (Trematoda: Digenea) of Australian mammals with description of *Dicrocoelium antechini* n.sp., *Athesmioides aiolos* n.g., n.sp., & *Platynosomum burrman* n. sp.', *Systematic Parasitology,* **21**:211–22.

Cribb, T. H., & Pearson, J. C. (1993), '*Neodiplostomum spratti* n. sp. (Digenea: Diplostomidae) from *Antechinus* spp. (Marsupialia: Dasyuridae) in Australia, with notes on other diplostomids from Australian mammals', *Systematic Parasitology,* **25**:25–35.

Daengsgvang, S. (1982), 'Gnathostomiasis', in *Handbook Series in Zoonoses. Section C: Parasitic Zoonoses, Vol. 2* (ed J.H. Steele), pp.147–80, CRC Press, Boca Raton.

Domrow, R. (1987), 'Acari Mesostigmata parasitic on Australian vertebrates: an annotated checklist, keys and bibliography', *Invertebrate Taxonomy,* **1**:817–948.

Domrow, R. (1988), '*Petauralges mordax* sp. n. (Acari: Psoroptidae) from an Australian dasyurid marsupial', *Journal of the Australian Entomological Society,* **27**:87–90.

Domrow, R. (1992), 'Acari Astigmata (excluding feather mites) parasitic on Australian vertebrates: an annotated checklist, keys and bibliography', *Invertebrate Taxonomy,* **6**:1459–606.

Domrow, R., & Lester, L.N. (1985), 'Chiggers of Australia (Acari: Trombiculidae): an annotated checklist, keys and bibliography', *Australian Journal of Zoology Supplementary Series,* **114**:1–111.

Dunnet, G.M., & Mardon, D.K. (1974), 'A monograph of Australian fleas (Siphonaptera)', *Australian Journal of Zoology (Suppl.),* **30**:1–273.

Dunnet, G.M., & Mardon, L.N. (1991), 'Siphonaptera (Fleas)' in *The Insects of Australia, 2nd edn*, pp. 705–16, Melbourne University Press, Melbourne.

Durette-Desset, M.-C. (1985), 'Trichostrongyloid nematodes and their vertebrate hosts: reconstruction of the phylogeny of a parasitic group', *Advances in Parasitology,* **24**:239–306.

Durette-Desset, M.-C., & Beveridge, I. (1981a), 'Deux genres aberrants de nématodes trichostrongyloïdes parasites de marsupiaux australiens: *Asymmetracantha* Mawson, 1960, et *Nasistrongylus* n. gen.', *Bulletin du Muséum National d'Histoire Naturelle, Paris, 4ème série,* **3**:1053–9

Durette-Desset, M.-C., & Beveridge, I. (1981b), '*Copemania obendorfi* n. gen., n. sp. nématode trichostrongyloïde parasite d'un marsupial australien', *Annales de Parasitologie Humaine et Comparée,* **56**:63–6.

Fain, A. (1982a), 'Spécificité et évolution parallèle hôtes-parasites chez les Myobiidae (Acari), *Mémoires du Muséum National d'Histoire Naturelle, Paris, n.s. A, Zoologie,* **123**: 77–85.

Fain, A. (1982b) 'Notes et discussions sur les marsupiaux et leurs parasites', *Mémoires du Muséum National d'Histoire Naturelle, Paris, n.s. A, Zoologie,* **123**:111–14.

Fain, A., & Lukoschus, F.S. (1979), 'Two new parasitic mites (Acari: Astigmata) from the skin of Australian vertebrates', *Mitteilungen aus dem Hamburgischen Zoologischen Museum und Institut,* **76**:387–93.

Gardner, S.L., & Campbell, M.L. (1992a), 'Parasites as probes for biodiversity', *Journal of Parasitology,* **78**:596–600.

Gardner, S.L., & Campbell, M.L. (1992b), 'A new species of *Linstowia* Zschokke, 1899 (Cestoda: Anoplocephalidae) from marsupials in Bolivia', *Journal of Parasitology,* **78**:795–99.

Gardner, S.L., & Hugot, J.P. (1995), 'A new pinworm, *Didelphoxyuris thylamisis* n. gen., n. sp. (Nematoda: Oxyurida) from *Thylamys elegans* (Waterhouse, 1939) (Marsupialia: Didelphidae) in Bolivia', *Research and Reviews in Parasitology,* **55**:139–47.

Garkavi, B.L. (1972), 'Species of *Trichinella* isolates from wild animals', *Veterinariya,* **10**:90–1.

Gregory, G.G., Munday, B.L., Beveridge, I., & Rickard, M.D. (1975), 'Studies on *Anoplotaenia dasyuri* Beddard, 1911 (Cestoda: Taeniidae), a parasite of the Tasmanian devil: life cycle and epidemiology', *International Journal for Parasitology,* **5**:187–91.

Hasegawa, H., Arai, S., & Shiraishi, S. (1993), 'Nematodes collected from rodents on Uotsini Island, Okinawa, Japan', *Journal of the Helminthological Society of Washington,* **60**:39–47.

Humphery-Smith, I. (1983), 'An hypothesis on the evolution of Herpetostrongylinae (Trichostrongyloidea: Nematoda) in Australian marsupials, and their relationships with Viannaiidae, parasites of South American marsupials', *Australian Journal of Zoology,* **31**:931–42.

Johnston, T.H., & Mawson, P.M. (1940), 'New and known nematodes from Australian marsupials', *Proceedings of the Linnean Society of New South Wales,* **65**:468–76.

Johnston, T.H., & Mawson, P.M. (1941), 'Some parasitic nematodes in the collection of the Australian Museum', *Records of the Australian Museum,* **21**:9–16.

Kéler, S. von (1971) 'A revision of the Australasian Boopiidae (Insecta: Phthiraptera) with notes on the Trimenoponidae', *Australian Journal of Zoology (suppl.),* **6**:1–126.

Klompen, J.S.H. (1992), 'Phylogenetic relationships in the mite family Sarcoptidae (Acari: Astigmata)', *Miscellaneous Publications of the Museum of Zoology, University of Michigan,* **180**:1–154.

Lenghaus, C., Obendorf, D.L., & Wright, F.H. (1990), 'Veterinary aspects of *Perameles gunnii* biology with special reference to species conservation', in *Management and Conservation of Small Populations* (eds. T.W. Clark, & J.H. Seebeck), pp. 89–108, Chicago Zoological Society, Brookfield Illinois.

Mackerras, M.J. (1958), 'Catalogue of Australian mammals and their recorded internal parasites. Part I. Monotremes and marsupials', *Proceedings of the Linnean Society of New South Wales,* **83**:101–25.

Mardon, D.K. (1976) 'A revision of the genus *Papuapsylla* Holland (Siphonaptera: Pygiopsyllidae)', *Australian Journal of Zoology (suppl.),* **45**:1–69

Mardon, D.K. (1978) 'On the relationships, classification, aedeagal morphology and zoogeography of the genera of Pygiopsyllidae (Insecta: Siphonaptera)', *Australian Journal of Zoology, (suppl.),* **64**:1–69.

Mawson, P.M. (1968), 'Two species of Nematoda (Spirurida: Spiruridae) from Australian dasyurids', *Parasitology,* **58**:75–8.

Mawson, P.M. (1971a), 'Two new species of *Rictularia* (Nematoda) from Australian rodents', *Transactions of the Royal Society of South Australia,* **95**:61–4.

Mawson, P.M. (1971b), 'Pearson Island Expedition 1969, 8. Helminths', *Transactions of the Royal Society of South Australia,* **95**:169–83.

Mawson, P.M. (1973), 'Amidostomatinae (Nematoda: Trichostrongyloidea) from Australian marsupials and monotremes', *Transactions of the Royal Society of South Australia,* **97**:257–79.

Mawson, P.M., Angel, L.M., & Edmonds, S.J. (1986), 'A checklist of helminths from Australian birds', *Records of the South Australian Museum,* **19**:219–325.

McColl, K.A., & Spratt, D.M. (1982), 'Parasitic pneumonia in a koala (*Phascolarctos cinereus*) from Victoria, Australia', *Journal of Wildlife Diseases,* **18**:511–12.

Munday, B.L., & Gregory, G.G. (1974), 'Demonstration of larval forms *Baylisascaris tasmaniensis* in the wombat (*Vombatus ursinus*)', *Journal of Wildlife Diseases,* **10**:241–2.

Munday, B.L., Mason, R.W., Hartley, W.J., Presidente, P.J.A., & Obendorf, D.L. (1978), '*Sarcocystis* and related organisms in Australian wildlife: I. Survey findings in mammals', *Journal of Wildlife Diseases,* **14**:417–33.

Navone, G.T., & Suriano, D.M. (1992), '*Pterogodermatites* (*Paucipectines*) *spinicaudatis* n. sp. (Nematoda: Rictulariidae) from *Dromiciops australis* (Marsupialia: Microbiotheriidae) in Bariloche, Rio Negro, Argentina. Biogeographical distribution and host parasite relationships', *Memorias do Instituto Oswaldo Cruz, Rio de Janeiro,* **87**:533–8.

Norman, R.J. deB., & Beveridge, I. (1999), 'Redescriptions of the species of *Physaloptera* Rudolphi, 1819 (Nematoda: Spirurida) parasitic in bandicoots (Marsupialia: Perameloidea) in Australia', *Systematic Parasitology,* **43**:103–21.

Nutting, W., & Woolley, P. (1965), 'Pathology in *Antechinus stuartii* due to *Demodex* sp.', *Parasitology,* **55**:383–9.

Oakwood, M., & Pritchard, D. (1999), 'Little evidence of toxoplasmosis in a declining species, the northern quoll (*Dasyurus hallucatus*)', *Wildlife Research,* **26**:329–33.

Oakwood, M., & Spratt, D.M. (2000), 'Parasites of the northern quoll, *Dasyurus hallucatus* (Marsupialia: Dasyuridae) in tropical savanna, Northern Territory', *Australian Journal of Zoology,* **48**:79–90.

Obendorf, D.L. (1993), 'Diseases of dasyurid marsupials', in *The Biology and Management of Australasian Dasyurid Marsupials* (eds. M. Roberts, J. Carnio, G. Crawshaw, & M. Hutchins), pp. 29–46, Monotreme and Marsupial Advisory Group of the American Association of Zoological Parks and Aquariums, Metropolitan Toronto Zoo.

Obendorf, D.L., & Clarke, K.P. (1992), '*Trichinella pseudospiralis* infection in free-living Tasmanian birds', *Journal of the Helminthological Society of Washington,* **59**:144–7.

Obendorf, D.L., & Smith, S.J. (1989), 'A re-examination of *Dithyridium cynocephali* Ranson, 1905, a metacestode parasite from the thylacine *Thylacinus cynocephalus*', *Papers and Proceedings of the Royal Society of Tasmania,* **123**:133–6.

Obendorf, D.L., & Munday, B.L. (1990), 'Toxoplasmosis in wild eastern barred bandicoots, *Perameles gunnii*', in *Bandicoots and Bilbies* (eds. J.H.Seebeck, P.R.Brown, R.L.Wallis, & C.M. Kemper), pp. 193–7, Surrey Beatty & Sons, Sydney.

Obendorf, D.L., Handlinger, J.H., Mason, R.W., Clarke , K.P., Forman, A.J., Hoooper, P.T., Smith, S.J., & Holdsworth, M. (1990), '*Trichinella pseudospiralis* infection in Tasmanian wildlife', *Australian Veterinary Journal,* **67**:108–10.

O'Callaghan, M.G., & Beveridge, I. (1996), 'Gastro-intestinal parasites of feral cats in the Northern Territory', *Transactions of the Royal Society of South Australia,* **120**:175–6.

O'Donoghue, P.J., & Adlard, R.D. (2000), 'Catalogue of protozoan parasites recorded in Australia', *Memoirs of the Queensland Museum* (in press).

Palma, R.L., & Barker, S.C. (1996), 'Phthiraptera', in *Zoological Catalogue of Australia, vol. 26: Psocoptera, Phthiraptera Thysanoptera* (ed. A. Wells), pp. 81–247, 333–61, 373–96, CSIRO Publishing, Melbourne.

Peisley, E., & Howell, M.J. (1975), 'A brachylaimid trematode from marsupial mice with observation on its life cycle', *International Journal for Parasitology*, **5**:441–7.

Pichelin, S., Thomas, P.M., & Hutchinson, M.N. (1999), 'A checklist of helminth parasites of Australian reptiles', *South Australian Museum Monograph Series,* **5**:1–61.

Potkay, S. (1970), 'Diseases of the opossum (*Didelphis marsupialis*): a review', *Laboratory Animal Care*, **20**:502–11.

Roberts, F.H.S. (1970), *Australian Ticks*, CSIRO, Melbourne.

Sandars, D.F. (1953), 'A study of Diphyllobothriidae (Cestoda) from Australian hosts', *Proceedings of the Royal Society of Queensland*, **63**:65–70.

Schmidt, G.D. (1986), *Handbook of Tapeworm Identification*, CRC Press, Inc. Boca Raton.

Smales, L.R. 1998. 'New species of *Seurechina* (Nematoda: Seuratidae) parasitic in dasyurid marsupials from Australia', *Transactions of the Royal Society of South Australia,* **122**:179–84.

Smales, L.R. 1999a. '*Linstowinema* Smales, 1997 (Nematoda: Seuratidae) from dasyurids (Marsupialia: Dasyuridae) from Australia', *Systematic Parasitology,* **43**:29–39.

Smales, L.R. 1999b. '*Bainechina rossiae* gen. et sp. nov. (Nematoda: Seuratidae) from Australian dasyurid marsupials', *Transactions of the Royal Society of South Australia,* **123**:37–41.

Smales. L.R. 1999c. '*Chabaudechina* n.g. (Nematoda: Seuratidae), with the description of two new species from dasyurid marsupials from Australia', *Systematic Parasitology,* **44**:145–51.

Smales, L.R. (1999d), 'Species of *Gnathostoma* (Nematoda: Spiruroidea) from bandicoots and dasyurids (Marsupialia) from Australia', *Transactions of the Royal Society of South Australia,* **123**:83–4.

Smales, L.R. (1999e), '*Linstowinema breve* n.sp. (Nematoda: Seuratidae), a parasite of *Antechinus agilis* (Marsupialia: Dasyuridae) in Australia', *Parasite,* **6**:329–32.

Smales, L.R., & Rossi, P.R. (1999), '*Inglechina virginiae* sp. n. (Nematoda: Seuratidae) from *Sminthopsis virginiae* (Marsupialia: Dasyuridae) from Northern Australia', *Journal of the Helminthological Society of Washington,* **66**:33–6.

Speare, R. (1985), 'The veterinary aspects of man as an intermediate host for *Toxoplasma gondii*', *Australian Veterinary Practitioner,* **15**:11–18.

Spratt, D.M. (1979), 'A taxonomic revision of the lungworms (Nematoda: Metastrongyloidea) from Australian marsupials', *Australian Journal of Zoology (suppl.),* **67**:1–45.

Spratt, D.M. (1980), '*Metathelazia naghiensis* sp. n. (Nematoda : Pneumospiruridae) from the long-nosed bandicoot, *Perameles nasuta* (Marsupialia)', *Journal of Parasitology,* **66**:1032–5.

Spratt, D.M. (1982), '*Anatrichosoma haycocki* sp. n. (Nematoda: Trichuridae) from the paracloacal glands of *Antechinus* spp., with notes on *Skrjabinocapillaria* Skarbilovitsch', *Annales de Parasitologie Humaine et Comparée,* **57**:63–71.

Spratt, D.M. (1985), '*Spirura aurangabadensis* (Ali & Lovekar) (Nematoda: Spiruridae) from small Dasyuridae (Marsupialia)', *Transactions of the Royal Society of South Australia,* **109**:25–9.

Spratt, D.M., Beveridge, I., & Walter E.L. (1991), 'A catalogue of Australasian marsupials and their recorded helminth parasites', *Records of the South Australian Museum, Monograph Series,* **1**:1–105.

Spratt, D.M., & Haycock, P. (1988), 'Aspects of the life history of *Cercopithifilaria johnstoni* (Nematoda: Filarioidea)', *International Journal for Parasitology,* **18**:1087–92.

Spratt, D.M., & Varughese, G. (1975), 'A taxonomic revision of filarioid nematodes from Australasian marsupials', *Australian Journal of Zoology (suppl.),* **35**:1–99.

Sprent, J.F.A. (1963), 'The life history and development of *Amplicaecum robertsi*, an ascaridoid nematode of the carpet python (*Morelia spilotes variegatus*), II. Growth and host specificity of larval stages in relation to the food chain', *Parasitology,* **53**:321–37.

Sprent, J.F.A. (1968), 'Notes on *Ascaris* and *Toxascaris*, with a definition of *Baylisascaris* gen. nov.', *Parasitology,* **58**:185–98.

Sprent, J.F.A. (1970), '*Baylisascaris tasmaniensis* sp. nov. in marsupial carnivores: heirloom or souvenir?', *Parasitology,* **61**:75–86.

Sprent, J.F.A. (1971), 'A new species and genus of ascaridoid nematode from the marsupial wolf (*Thylacinus cynocephalus*)', *Parasitology,* **63**:37–43.

Sprent, J.F.A. (1972), '*Cotylascaris thylacini:* a synonym of *Ascaridia columbae*', *Parasitology,* **64**:331–2.

Sprent, J.F.A., Lamina, J., & McKeown, A. (1973), 'Observations on migratory behaviour and development of *Baylisascaris tasmaniensis*', *Parasitology,* **67**:67–83.

Stone, B.F., Binnington, K.C., Gauci, M., & Aylward, J.H. (1989), 'Tick-host interactions for *Ixodes holocyclus*: the role, effects, biosynthesis, and nature of its toxic and allergenic secretions', *Experimental and Applied Acarology,* **7**:59–69.

Traub, R., & Dunnet, G.M. (1973) 'Revision of the siphonapteran genus *Stephanocircus* Skuse, 1893 (Stephanocircidae)', *Australian Journal of Zoolog (suppl.),* **20**:1–128.

Vaucher, C., & Beveridge, I. (1997), 'New species of *Potorolepis* Spasskii (Cestoda: Hymenolepididae) parasitic in dasyurid marsupials from New Guinea',. *Transactions of the Royal Society of South Australia.* **121**:95–102.

Vaucher, C., Beveridge, I., & Spratt, D.M. (1984), 'Cestodes du genre *Hymenolepis* Weinland, 1858 (sensu lato) parasites de marsupiaux australiens et descriptions de cinq espèces nouvelles', *Revue Suisse de Zoologie,* **91**:443–58.

Woolley, P.A. (1989), 'Nest location by spool-and-line tracking of dasyurid marsupials in New Guinea', *Journal of Zoology (London),* **218**:689–700.

Zhu, X., Spratt, D.M., Beveridge, I., Haycock, P., & Gasser, R.B. (2000), 'Mitochondrial DNA polymorphism within and among species of *Capillaria sensu lato* from Australian marsupials and rodents', *International Journal for Parasitology,* **30**:933–8.

Part V

Conservation

CHAPTER 27

MARSUPIALS OF THE NEW WORLD: STATUS AND CONSERVATION

Gustavo A.B. da Fonseca[A, B], *Adriano Pereira Paglia*[B, C], *James Sanderson*[A] *and Russell A. Mittermeier*[A]

[A]Center for Applied Biodiversity Science, Conservation International, 1919 M Street NW suite 600, Washington, DC 20036
[B]Departamento de Zoologia, Universidade Federal de Minas Gerais, Av. Antonio Carlos 6627, Belo Horizonte, MG Brazil 31275-000
[C]Conservation International do Brasil, Av. Getúlio Vargas, no. 1300, 7.o Andar – Savassi – 30112-021 Belo Horizonte – Minas Gerais, Brazil

Marsupials are well represented in the New World with three orders. Here we discuss the conservation status of New World marsupials by considering their geographic ranges, association with major biomes and vegetation types, and the latest IUCN's Species Survival Commission Red List assessment. We also discuss marsupial geographical distributions with respect to the global conservation strategy that focuses on global biodiversity hotspots. The IUCN Red List considers 23 New World marsupials, or about a third of all extant species, as threatened with extinction. We highlight that an additional 11 species with restricted ranges should also require timely attention. New World biodiversity hotspots contain 78% of threatened species recognised by IUCN's Red List, as well as 80% of the remaining species whose conservation status may be in need of further revision. These hotspots comprise ecosystems that have already lost 70% or more of their native habitat, and are in most need of conservation attention.

INTRODUCTION

Marsupials are known from South America since the upper Cretaceous, with records of approximately 100 million years of age being found in Peru and Bolivia (Eisenberg and Redford 1999). Fossils that can be traced to the South American autochtonous fauna are also known from the Eocene of Antarctica. Marsupials in the Western Hemisphere now occur from southern Canada to southern Mexico, and south throughout Central and South America, with maximum richness occurring between the equator and 20° south latitude. Diversity declines toward the Southern Cone.

Patton and Costa (in press) reviewed the distribution, geographical limits and systematics of Brazilian marsupials. Several authors such as Gardner (1973, 1993) have presented additional taxonomic information, and Patton and his co-authors (Patton et al. 1996, Patton and Costa, in press) have discussed phylogenetic relationships based on genetic analysis. In this chapter, we have followed the arrangement proposed by Gardner (1993), which groups New World marsupials in three orders: Didelphimorphia, Paucituberculata and Microbiotheria. Despite not being a strictly valid phylogenetic arrangement, we use the term 'marsupials' to refer collectively to members of these three orders.

Species of the order Didelphimorphia occur widely over the Neotropics and occupy nearly all habitat types, except high elevations and deserts (Redford and Eisenberg 1992). The species of the two other orders occur only in western South America. The contemporary New World marsupial fauna is composed basically of species of small body size, the largest of them in the genus *Didelphis*. Judging from the current fauna, the origins of which can be traced to the original South American stock, marsupials stood best the invasion of North American mammals via the Panamanian land bridge, and also coped well in the face of other rapid environmental changes. Notwithstanding the disappearance of many forms from the New World at the Pliocene/Pleistocene transition, including all representatives of five families (Eisenberg and Redford 1999), marsupial standing diversity can still be considered relatively high in relation to other mammalian taxa, particularly considering its ancient origin. Amongst New World mammals, only rodents, bats and primates are more speciose (Fonseca et al. 1999). Together with rodents and bats, marsupials are a dominant feature of small mammal communities of South America, particularly those of forest biomes (Fonseca et al. 1989; Stallings 1989; Fonseca and Robinson 1990).

Here we discuss the conservation status of New World marsupials, especially with respect to the current IUCN Red List of Threatened Species (Hilton-Taylor 2000), compiled by the Species Survival Commission (SSC). The patterns of species distribution and diversity are also interpreted in light of their conservation ramifications. We also traced species distribution to major ecosystem, biome type, and global biodiversity hotspot (Myers et al. 2000), the latter representing severely threatened regions that have exceptional concentrations of endemic plant and vertebrate species. It is worth noticing, however, that much remains to be done in clarifying alpha taxonomy of New World marsupials, that distribution ranges are constantly being revised, and new forms are still being routinely described (see Patton et al. 1996; Patton et al. 2000; Patton and Costa, in press).

New World marsupial diversity and distribution

We are still far from determining the true range of marsupial diversity in the New World. Current described forms vary from 70 to over 80 species, depending on the taxonomic scheme one chooses to follow. Our analysis derives from Wilson and Reeder (1993), Eisenberg (1989), Redford and Eisenberg (1992) and Eisenberg and Redford (1999), complemented by Hershkovitz (1992), Patton et al. (1996), Perez-Hernandez et al. (1994), Rodrigues-Mahecha et al. (1995), Pacheco et al. (1995), and Fonseca et al. (1996; 1999). We concluded, based on this body of literature, that 76 species can at this point be considered valid, distributed in 19 genera and three families (Table 1).

Monodelphis, Marmosops, and *Marmosa* are the most diversified genera, comprising some 55% of all marsupial species. Nine out

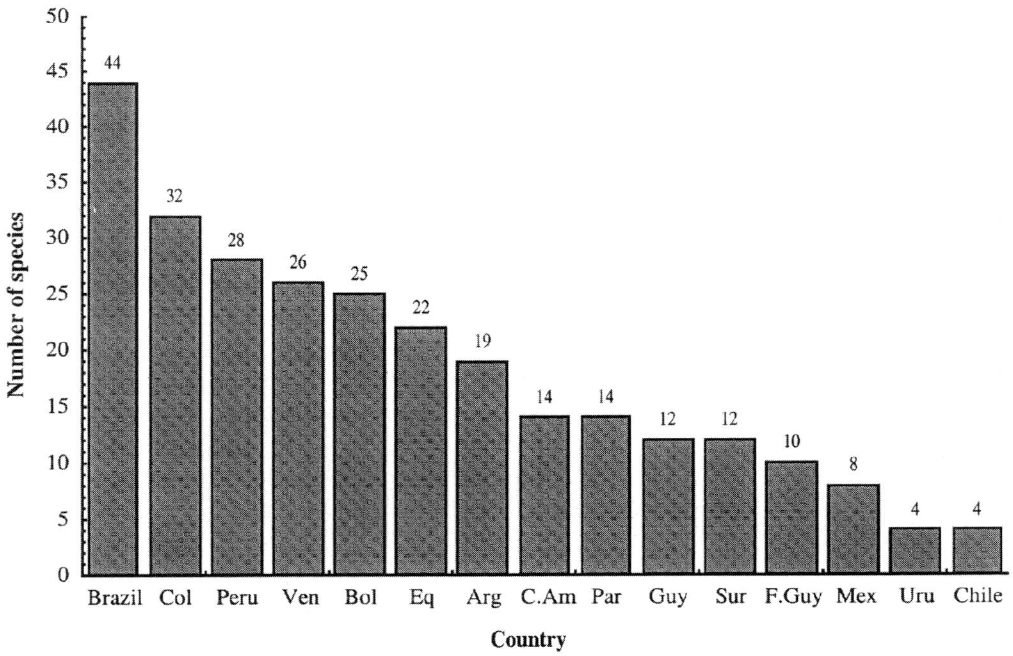

Figure 1 Marsupial species richness by country. For the purpose of this analysis, Central American countries have been lumped in a single category (C.Am).

Table 1 Marsupials of the New World.

ORDER DIDELPHIOMORPHIA
Genus *Caluromys*
C. derbianus; C. lanatus; C. philander
Genus *Caluromysiops*
C. irrupta
Genus *Chironectes*
C. minimus
Genus *Didelphis*
D. albiventris; D. aurita; D. marsupialis; D. virginiana
Genus *Glironia*
G. venusta
Genus *Gracilinanus*
G. aceramarcae; G. agilis; G. dryas; G. emiliae; G. marica; G. microtarsus
Genus *Lestodelphys*
L. halli
Genus *Lutreolina*
L. crassicaudata
Genus *Marmosa*
M. andersoni; M. canescens; M. lepida; M. mexicana; M. murina; M. robinsoni; M. rubra; M. tyleriana; M. xerophila
Genus *Marmosops*
M. cracens; M. dorothea; M. fuscatus; M. handleyi; M. impavidus; M. incanus; M. invictus; M. neblina; M. noctivagus; M. parvidens; M. paulensis
Genus *Metachirus*
M. nudicaudatus
Genus *Micoureus*
M. alstoni; M. constantiae; M. demerarae; M. regina
Genus *Monodelphis*
M. adusta; M. americana; M. brevicaudata; M. dimidiata; M. domestica; M. emiliae; M. iheringi M. kunsi; M. maraxina; M. orinoci; M. osgoodi; M. rubida; M. scalops; M. sorex; M. theresa; M. unistriata
Genus *Philander*
P. andersoni; P. frenata; P. mcilhennyi; P. opossum
Genus *Thylamys*
T. elegans; T. macrura; T. pallidior; T. pusilla; T. velutinus
ORDER PAUCITUBERCULATA
Genus *Caenolestes*
C. caniventer; C. condorensis; C. convelatus; C. fuliginosus; C. tatei
Genus *Lestoros*
L. inca
Genus *Rhyncolestes*
R. raphanurus
ORDER MICROBIOTHERIA
Genus *Dromiciops*
D. gliroides

of 19 genera are monotypic, and Microbiotheridae is represented by a single species, *Dromiciops gliroides*, found in Chile and Argentina, and phylogenetically quite distant from all other New World marsupials. *Lestodelphis halli*, a Patagonian endemic, has the southernmost range of any taxa. *Didelphis virginiana* reaches the furthest north, into southern Canada.

We analysed the patterns of species richness by country (Fig. 1). Brazil has the largest number of extant species (44). Colombia has the second highest richness, followed closely by Peru, Venezuela, Bolivia and Ecuador. As expected, species richness is closely related to the size of the country, a trend that is somewhat confounded by a noticeable latitudinal gradient (Fig. 2), with northern and southern countries having less diversity than expected for their size. On the other hand, Amazon basin countries hold exceptional diversity.

Species distribution was also discriminated at the level of particular biomes in search of patterns of species richness (Table 2). The three tropical forest ecosystems hold the largest number of species, with 32 in Amazonia, 24 in the Atlantic Forest, and 16 in the moist forests of Central America and the Caribbean. The Chaco and the Pantanal harbor the lowest number of species (Table 2). After tropical forests, open habitats follow second in richness, such as the savannas of Central Brazil, better known as the Cerrado.

Amazonia shows the highest degree of marsupial endemism, followed by the much smaller Atlantic Forest (Fig. 3). In the former, 17 endemics (59%) belong to only three genera: *Gracilinanus, Marmosa* and *Marmosops*. In the latter, most of the species-level endemism is due to the genus *Monodelphis*, with six species (46%) restricted to this region. The Andean region also harbors several endemic species, and the entire genus *Caenolestes* is confined to the northern portion of Cordillera.

Species were further classified into four main typologies: forest, open area or xeric biomes, as well as Andean ecosystems (Table 3). As with other Neotropical mammals (see Fonseca et al. 1999), open habitat biomes have very few (Cerrado and Lhanos of Northern Venezuela) or no endemics (Pampas, Chaco and Pantanal – Table 3).

CONSERVATION STATUS

The 2000 IUCN Red List of Threatened Species (Hilton-Taylor 2000) recognises 17 New World marsupials as Vulnerable, three as Endangered, and another three as Critically Endangered, collectively representing nearly one-third of all extant New World species (Fig. 4, Table 6). There has not been any change in status for any species listed as threatened in the 1996 Red List (IUCN 1996; Hilton-Taylor 2000). Only two out the 23 threatened marsupials are not endemic to the biome where they occur: *Caluromys derbianus*, which is distributed in humid

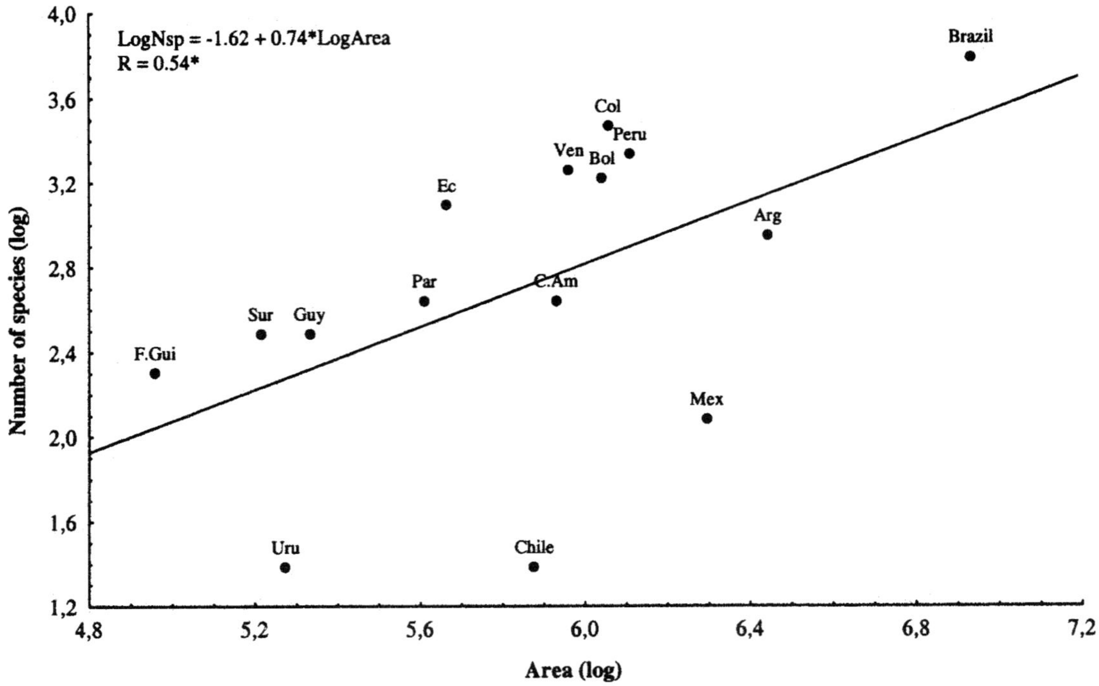

Figure 2 Relationship between species richness and country size (* significative to α = 5%)

Table 2 Marsupial species richness by biome type.

Biome type	Number of species
Forest biomes	
Amazon Forest	32
Atlantic Forest	24
Humid Forests of Mexico, Central America and Caribbean	16
Temperate Forests of Chile	2
Open habitat formations	
Savannas of Central South America (Cerrado)	16
Chaco of Bolivia and Paraguay	13
Pantanal	13
Lhanos of Venezuela and Colombia	7
Pampas of Argentina and Uruguay	4
Andean ecosystems	
Northern Andes (Peru, Ecuador, Colombia and Venezuela)	8
Central Andes (Peru, Bolivia, Chile and Argentina)	5
Xeric biomes	
Caatinga	4
Patagonia	2
Xeric region of northern Venezuela	2
Coastal Chile	1

Table 3 Number of endemic and threatened species by biome type.

Habitat	Number of species	Endemic species	Threatened species
Forest biomes	56	37 (66.1%)	14 (25.0%)
Open area biomes	21	4 (19.0%)	2 (9.5%)
Andean ecosystems	13	11 (84,6%)	5 (38.5%)
Xeric biomes	9	3 (33.3%)	4 (44.4%)

forests of Mexico and Central America, expanding into northwestern South America west of the Andes, and *Monodelphis rubida*, reported to occur in the Atlantic Forest and also in the Cerrado savannas of Central Brazil. Given their high levels of endemism, as well as the highest totals for threatened species, Amazonia, the Atlantic Forest, and the Central Andes region should be considered 'marsupial hotspots', and given the highest priority for conservation of this ancient mammalian order.

There are also 18 species listed by IUCN as Lower Risk: Near Threatened, a category reserved for taxa which do not qualify for threatened status, but which are close to qualifying or likely to qualify in the near future. It is difficult to assess, however, whether such classification is indeed an indication of impending threat to particular taxa. For instance, marsupials now ranked as Near Threatened – such as *Caluromys lanatus* and *C. philander* – have large geographic ranges, and based on extensive field sur-

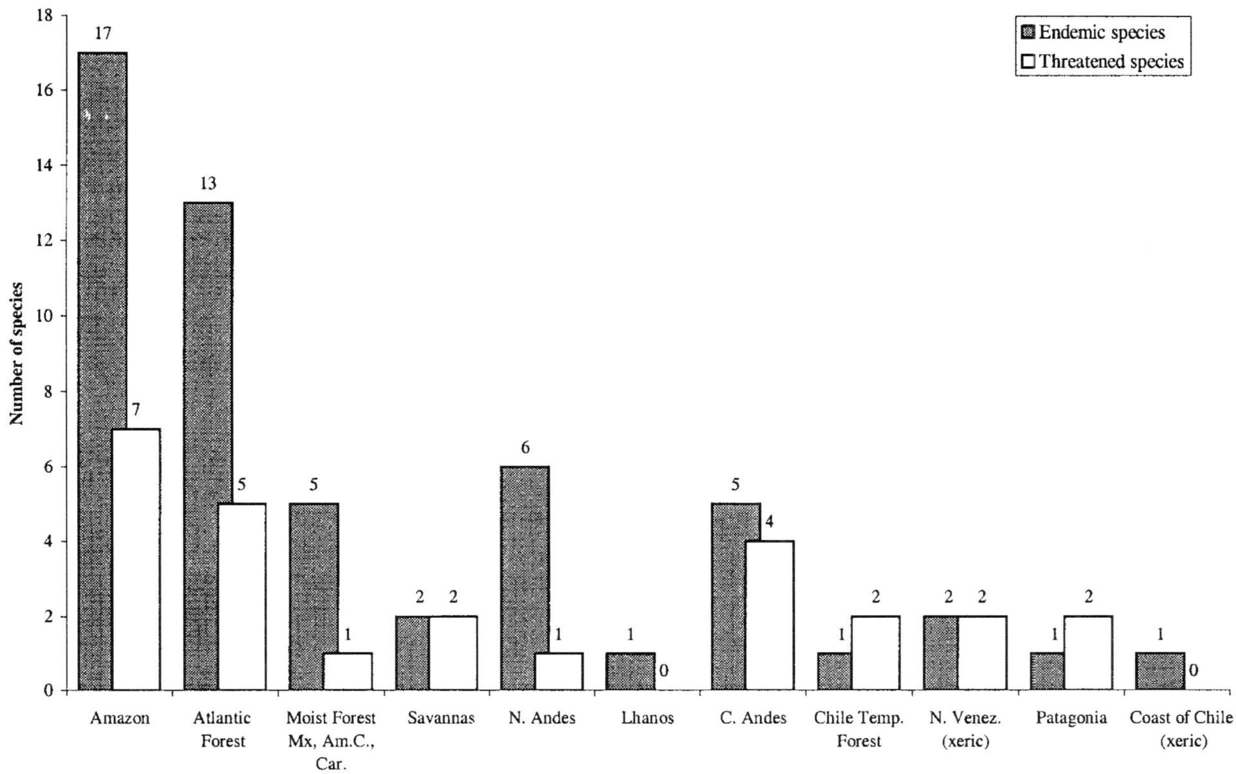

Figure 3 Number of endemic and threatened species by major biome.

veys (G. Fonseca and A. Paglia, unpublished data), these two species are frequent and rather abundant throughout most of their area of distribution. Another five are classified as Data Deficient (of which only two have been considered here as valid species). Further analysis, based on geographic range sizes and distribution, suggests that an additional nine species, currently not considered in any of the risk categories of the IUCN Red List, might need to be closely monitored to assess their status in the wild (Table 5).

Threatened species

Myers et al. (2000) have identified 25 global biodiversity hotspots, which are severely threatened ecosystems harboring very high levels of species endemism. Their analyses also highlighted the fact that 57% and 82% of mammals and birds, respectively, which are considered by IUCN as threatened, are found in these very same areas (Mittermeier et al. 1999). Not unexpectedly, six New World ecosystems identified as global biodiversity hotspots were found to hold 17 (73.9%) of the threatened marsupials identified in the IUCN Red List. These hotspots are the Tropical Andes, the Atlantic Forest, Central Chile, the Brazilian Cerrado, Mesoamerica, and the Chocó-Darién-Western Ecuador area. Table 4 summarises these results for all species considered as threatened by the IUCN. With little remaining habitat and increasing pressure, the hotspots – and the species found in them – are of the highest conservation concern.

Regional endemism is often associated with threatened status. Of the three marsupial species restricted to Chile and a small area in adjacent Argentina, two are threatened and mostly confined to the Central Chile hotspot. The known geographic range of *Rhyncholestes raphanurus* is restricted to Chiloé Island and the Valdivian rainforest around Puerto Montt, in addition to one record in southern Argentina (Ojeda and Giannoni 2000), the species showing preference for moist forests with dense understory. *Dromiciops gliroides* is also confined to the Notophagus forest but is recorded slightly further north than *R. raphanurus*, also in southern Argentina. With such a small geographic range, ongoing habitat destruction associated with logging constitutes the main threat to the species' continued existence. Still in the southern cone, the distribution of the threatened Patagonian opossum, *Lestodelphys halli*, is poorly known, since it is mostly derived from a few specimens from Patagonia and Central Argentina. This range of *L. halli* extends further south than that of any other marsupial.

Five of the six most threatened New World marsupials (*Marmosops handleyi*, *Marmosa andersoni*, *Gracilinanus aceramarcae*, *Marmosops cracens*, *Marmosa xerophila*, *Monodelphis kunsi*) are all

Table 4 New World marsupials considered as threatened by IUCN's Red List, and their occurrence in biodiversity hotspots. Of the 23 threatened species, 17 (73.9%) occur within hotspots.

Species	IUCN Category	Hotspot
Marmosops handleyi	Critically Endangered	Tropical Andes
Marmosa andersoni	Critically Endangered	Tropical Andes
Gracilinanus aceramarcae	Critically Endangered	Tropical Andes
Marmosops cracens	Endangered	Tropical Andes
Marmosa xerophila	Endangered	-
Monodelphis kunsi	Endangered	Tropical Andes
Rhyncholestes raphanurus	Vulnerable	Central Chile
Dromiciops gliroides	Vulnerable	Central Chile
Lestodelphys halli	Vulnerable	-
Monodelphis emiliae	Vulnerable	-
M. maraxina	Vulnerable	-
M. osgoodi	Vulnerable	Tropical Andes
M. rubida	Vulnerable	Cerrado, Atlantic Forest
M. scalops	Vulnerable	Atlantic Forest
M. sorex	Vulnerable	Atlantic Forest
M. theresa	Vulnerable	Atlantic Forest
M. unistriata	Vulnerable	Brazilian Cerrado
Caluromys derbianus	Vulnerable	Chocó-Darién, Mesoamerica
Caluromysiops irrupta	Vulnerable	-
Gracilinanus dryas	Vulnerable	Tropical Andes
G. emiliae	Vulnerable	-
Marmosops dorothea	Vulnerable	Tropical Andes
Glironia venusta	Vulnerable	Tropical Andes

Table 5 New World marsupial species with ranges included in New World biodiversity hotspots, and not listed by IUCN's 2000 assessment in any threatened category.

Species	Hotspot
Micoureus regina	Tropical Andes
Marmosa rubra	Tropical Andes
Marmosops paulensis	Atlantic Forest
Caenolestes caniventer	Tropical Andes
Caenolestes condorensis	Tropical Andes
Caenolestes convelatus	Tropical Andes
Caenolestes fuliginosus	Tropical Andes
Caenolestes tatei	Tropical Andes
Lestoros inca	Tropical Andes

known from the richest hotspot on Earth – the Tropical Andes. Of the three critically endangered New World marsupials, two are only known from the Tropical Andes: *Marmosa andersoni* has only been recorded for its type locality near Cuzco, Peru, at an altitude of 600m (Eisenberg and Redford 1999; Nowak 1999),

Table 6 New World marsupials threatened with extinction, according to the IUCN 2000 Red List (Hilton-Taylor 2000), including those classified as Lower Risk: Near Threatened or Data Deficient.

Critically endangered

Gracilinanus aceramarcae (Tate, 1931)

Marmosa andersoni (Pine, 1972)

Marmosops handleyi (Pine, 1981)

Endangered

Monodelphis kunsi (Pine, 1975)

Marmosa xerophila (Handley and Gordon, 1979)

Marmosops cracens (Handley and Gordon, 1979)

Vulnerable

Caluromys derbianus; Caluromysiops irrupta; Dromiciops gliroides; Glironia venusta; Gracilinanus dryas; Gracilinanus emiliae; Lestodelphys halli; Marmosops dorothea; Monodelphis emiliae; Monodelphis maraxina; Monodelphis osgoodi; Monodelphis rubida; Monodelphis scalops; Monodelphis sorex; Monodelphis theresa; Monodelphis unistriata; Rhyncolestes raphanurus

Near Threatened

Caluromys lanatus; C. philander; Chironectes minimus; Gracilinanus agilis; G. marica; G. microtarsus; Marmosa lepida; Marmosops fuscatus; Marmosops impavidus; Marmosops incanus; Marmosops invictus; Marmosops parvidens; Micoureus alstoni; Micoureus constantiae; Monodelphis americana; Monodelphis dimidiata; Monodelphis iheringi; Thylamys macrura

Data Deficient

*Gracilinanus kalinowskii**; *Gracilinanus longicaudis**; *Gracilinanus perijae**; *Marmosa canescens; Marmosa tyleriana*

* In the present paper, these were not considered as valid species.

and *Marmosops handleyi* was described by two specimens collected in Antioqua, Colombia (Pine 1981). *Gracilinanus aceramarcae*, also listed as critically endangered, is a little known species, having been collected at Rio Aceramarca, Bolivia, and in the Andes of southern Peru at an altitude of 2700 m.

The ranges of the two endangered *Marmosops* adjoin the northern-most extent of the Tropical Andes hotspot. *Marmosops cracens* has a highly restricted range in the state of Falcón, Venezuela. All specimens have been taken below 170 m in moist foothills (Eisenberg 1989). *Marmosa xerophila* is highly specialised for living in xeric environments and is only known to occur north of the Andes at elevations below 90 m at the mouth of Lake Maracaibo in Colombia and Venezuela. *Monodelphis kunsi* is known only from specimens collected in savannas at two localities in Bolivia and another two in Brazil. *Monodelphis osgoodi, Gracilinanus dryas* and *Marmosops dorothea* are all confined to the Tropical Andes, and are ranked by IUCN as vulnerable, partly on the basis of their limited distributions.

The type locality of *Gracilinanus emiliae* is located in the state of Pará, Brazil, but its range can be extended to the three Guianas,

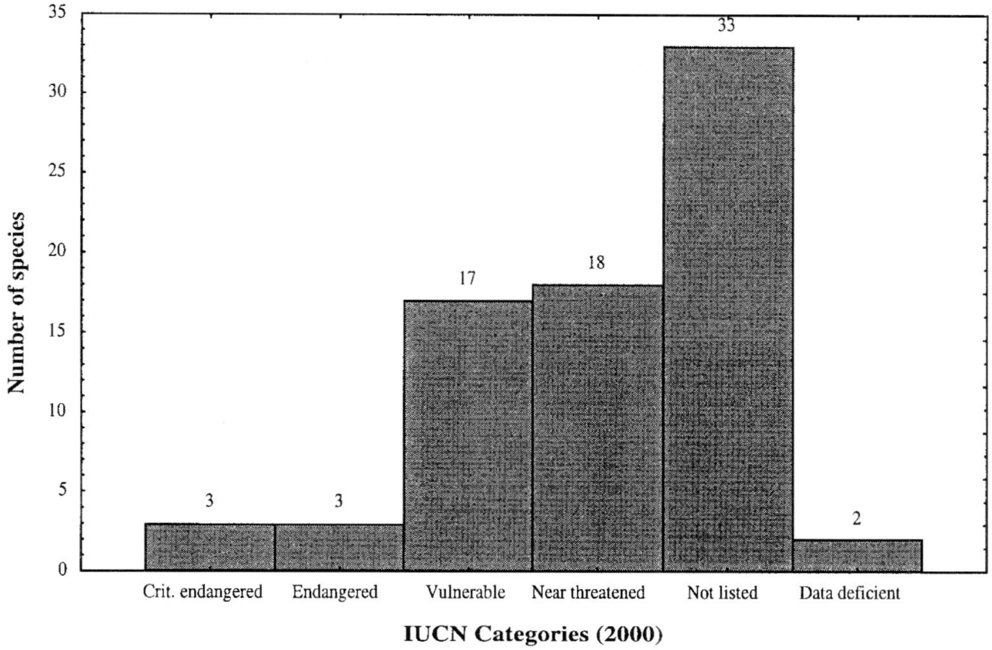

Figure 4 Number of New World marsupial species into the IUCN threatened categories.

Venezuela and Colombia (Voss et al. 2001). Few specimens of *Monodelphis maraxina* are known, all collected in the Marajó Island in Pará, Brazil (Wilson and Reeder 1993). *Monodelphis rubida* and *M. unistriata* occur in the Brazilian Cerrado, and *M. scalops* in south-eastern Brazil and adjacent Argentina (Eisenberg and Redford 1999). *Monodelphis emiliae* extends from north-eastern Amazonia in Brazil to north-eastern Peru in Iquitos, with some records in central Brazil (Patton et al. 2000). *Monodelphis sorex* is known to occur in south-eastern Brazil and Misiones Province, Argentina (Redford and Eisenberg 1992). It is listed as vulnerable by the IUCN and its range extends into the Atlantic Forest hotspot. *Monodelphis theresa* is confined to the Atlantic Forest of Rio de Janeiro, Brazil. *Glironia venusta* occurs in the lowland forests of Amazonia, extending into northern Bolivia.

Caluromys derbianus occurs in western Ecuador and north-eastward to Veracruz, Mexico. The range of *C. derbianus* is restricted to the lower elevations of moist evergreen tropical rainforests. Ranked as vulnerable by IUCN, its geographic distribution transverses two biodiversity hotspots: Mesoamerica and Chacó-Darién Western Ecuador. *Caluromysiops irrupta*, the black-shouldered opossum, is known to occur in the extreme south-eastern part of Colombia and in the Peruvian Amazon west to Acre, Brazil, and is listed as vulnerable

OTHER SPECIES OF CONSERVATION CONCERN

Several species with highly restricted ranges occurring in biodiversity hotspots, but which are currently not listed as threatened by the IUCN, may need further consideration as of their conservation status. Eight of these species are in the Tropical Andes hotspot (Table 5). Information on many of these forms is typically quite limited. From the great wealth of information assembled by Eisenberg (1989) on the mammalian fauna of the Neotropics, all that he can tell us of *Micoureus regina* is that this species 'has been described from the region of Bogotá, Colombia.' *Lestoros inca,* the Peruvian shrew opossum, has arguably the smallest geographic range of any living marsupial, confined to habitats between 2100 and 3600 m in the southern Peruvian Andes (Eisenberg and Redford 1999). Most other species of the order Paucituberculata are also restricted to the Tropical Andes, yet not listed as threatened by the IUCN: *Caenolestes fuliginosus* occurs in the Andes of western Venezuela and northern Colombia above 2300 m, a continuous but highly restricted mountain range. *Caenolestes caniventer* is restricted to the Andes of Ecuador and has a comparatively small range. *C. condorenis, C. convelatus, C. fuliginosus* and *C. tatei* are also restricted to the Tropical Andes hotspot. *Marmosa rubra* also has a restricted geographic range within the Tropical Andes hotspot of eastern Ecuador, extending down into the Peruvian Amazon. *Marmosops paulensis* surely has one of the smallest geographic ranges of any marsupial, confined to the southern coastal Atlantic Forest.

CONCLUSIONS

Marsupials are still well diversified in the New World, usually being well represented in small mammal communities of quite a diverse range of habitats (Fonseca and Kierulff 1989; Fonseca et al. 1999). Given a number of factors, particularly the loss of native habitat, approximately one-third of all species are now recognised as having poor conservation status in the wild. However, given the paucity of data, we suspect many more taxa may be facing demographic hardships, especially those inhabiting hotspots where over 70% of all native vegetation has already been degraded. Lack of comprehensive field studies currently prevent a more clear picture of the conservation status of the full extent of marsupial diversity in the New World, particularly those with more restricted ranges. Moreover, many genera, particularly *Marmosa, Marmosops* and *Micoureus,* are undergoing major taxonomic revisions that will have direct consequences for the reassessment of the conservation status of a large fraction of the New World marsupial fauna. Likewise, the quite speciose genus *Monodelphis* would greatly benefit from further and comprehensive systematic work.

Acknowledgements

We thank Menna Jones for the invitation to contribute to this volume, and two anonymous reviewers who made important corrections in the earlier draft.

References

Eisenberg, J.F., & Redford, K. (1999), *Mammals of the Neotropics, Vol. 3, The Central Neotropics: Ecuador, Peru, Bolivia, Brasil,* Chicago, University of Chicago Press.

Eisenberg, J.F. (1989), *Mammals of the Neotropics, Vol. 1, Mammals of Northern Neotropics: Panama, Colombia, Venezuela, Guyana, Suriname, French Guiana,* Chicago, University of Chicago Press.

Fonseca, G.A.B., & Kierulff, M.C. (1989), 'Biology and natural history of Brazilian Atlantic Forest small mammals', *Bulletin of the Florida State Museum, Biological Sciences,* **34**(3):99–152.

Fonseca, G.A.B., & Robinson, J.G. (1990), 'Forest size and structure: Competitive and predatory effects on small mammal community structure', *Biological Conservation*, **53**(4):265–94.

Fonseca, G.A.B., Hermann, G., & Leite, Y.L.R. (1999), 'Macrogeography of Brazilian mammals', in *Mammals of the Neotropics, Vol. 3, The Central Neotropics: Ecuador, Peru, Bolivia, Brasil* (eds. J.F. Eisenberg, & K. Redford.), pp. 549–63, Chicago, University of Chicago Press.

Fonseca, G.A.B., Herrmann, G., Leite, Y.L.R., Mittermeier, R.A., Rylands, A.B., & Patton, J. (1996), 'Lista anotada dos mamíferos do Brasil', *Occasional Papers in Conservation Biology, Vol 4,* Conservation International & Fundação Biodiversitas.

Gardner, A.L. (1973), 'The systematics of the genus *Didelphis* (Marsupialis: Didelphidae) in North and Middle America', *Special Publications, The Museum, Texas Tech University,* **4**:1–81.

Gardner, A.L. (1993), 'Order Didelphimorphia', in *Mammal Species of the World* (eds. D.E. Wilson, & D.M. Reeder), pp. 15–23, Washington, DC, Smithsonian Institution Press.

Hershkovitz, P. (1992), 'The South American gracile mouse opossums, genus *Gracilinanus* Gardner & Creighton, 1989 (Marmosidae, Marsupialia): A taxonomic review with notes on general morphology and relationships', *Fieldiana: Zoology,* **70**:1–56.

Hilton-Taylor, C. (2000), *2000 IUCN Red List of Threatened Species,* Gland, Switzerland, The World Conservation Union, IUCN.

IUCN The World Conservation Union (1996), *IUCN Red List of Threatened Animals,* Gland, Switzerland, The World Conservation Union, IUCN.

Mittermeier, R.A., Myers, N., Gil, P.R., & Mittermeier, C. (1999), *Hotspots: Earth's Biologically Richest and Most Endangered Terrestrial Ecosystems,* Agrupación Sierra Madre, México, CEMEX and Conservation International.

Myers, N., Mittermeier, R.A., Mittermeier, C.G., Fonseca, G.A.B., & Kent, J. (2000), 'Biodiversity hotspots for conservation priorities', *Nature,* **403**:853–8.

Nowak, R.M. (ed.), (1999). *Walker's Mammals of the World, 6th ed.,* Baltimore, The Johns Hopkins University Press.

Ojeda, R.A., & Giannoni, S.M. (2000), *Lista de marsupials de Argentina/ The marsupials of Argentina: An annotated checklist of their distribution and conservation*, Grupo de Especialistas de Marsupiales del Nuevo Mundo (UICN)/IUCN's New World Specialist Group.

Pacheco V., Macedo, H., Vivar, E., Ascorra, C., Arana-Cardó, R., & Solari, S. (1995), 'Lista anotada de los mamiferos Peruanos', *Occasional Papers in Conservation Biology, Vol. 2*, Washington, DC, Conservation International.

Patton, J.L., & Costa, L.P. (In press), Diversidade, limites geográficos e sistemáticos de marsupiais brasileiros. In *Marsupiais Brasileiros* (eds. N.C. Cáceres & E.L.A. Monteiro-Filho), Curitiba, Brasil: Editora da Universidade Federal do Paraná.

Patton, J.L., Da Silva, M.N.F., & Malcolm, J.R. (2000), 'Mammals of the Rio Jurua and the evolutionary and ecological diversification of Amazonia', *Bulletin of the American Museum of Natural History,* **224**:1–306.

Patton, J.L., Reis, S.F., & da Silva, M.N.F. (1996), 'Relationships among didelphid marsupials based on sequence variation in the mitochondrial cytochrome B gene', *Journal of Mammalian Evolution,* **3**:3–29.

Pérez-Hernández, R.P., Soriano, P., & Lew, D. (1994), *Marsupiales de Venezuela,* Caracas, Venezuela, Editora Arte, S.A.

Pine, R. (1981), 'Review of the mouse opossum *Marmosa parvidens* and *M. invicta* with a description of a new species', *Mammalia,* **45**:56–70.

Redford, K.H., & Eisenberg, J.F. (1992), *Mammals of the Neotropics, Vol. 2, The Southern Cone,* Chicago, University of Chicago Press.

Rodríguez-Mahecha, J., Hernandez-Camacho, J.I., Defler, T.R., Alberico, M., Mast, R.B., Mittermeier, R. A., & Cadena, A. (1995), 'Mamíferos Colombianos: Sus nomes comunes e indigenas', *Occasional Papers in Conservation Biology, Vol. 3*, Washington, DC, Conservation International.

Stallings, J.R. (1989), 'Small mammal inventories in an eastern Brazilian park', *Bulletin of the Florida State Museum, Biological Sciences,* **34**(3):153–200.

Voss, R.S., Lunde, D.P., & Simmons, N.B. (2001), 'The mammals of Paracou, French Guiana: A Neotropical lowland rainforest fauna – part 2. Nonvolant species', *Bulletin of the American Museum of Natural History,* **263**:3–236.

Wilson, D.E., & Reeder, D.M. (1993), *Mammals Species of the World, 2nd ed,* Washington, DC, Smithsonian Institution Press.

CHAPTER 28

DASYURID DILEMMAS: PROBLEMS AND SOLUTIONS FOR CONSERVING AUSTRALIA'S SMALL CARNIVOROUS MARSUPIALS

B.A. Wilson[A], *C.R. Dickman*[B] *and T.P. Fletcher*[C]

[A]School of Ecology and Environment, Deakin University, Geelong, Victoria 3217, Australia
[B]Institute of Wildlife Research, School of Biological Sciences, University of Sydney, New South Wales 2006, Australia
[C]Co-operative Research Centre for Conservation and Management of Marsupials, Perth Zoo, South Perth, Western Australia 6151, Australia

Dasyurid marsupials are distributed throughout the major terrestrial environments of Australia, but since European settlement have suffered local and regional extinctions, range reductions and population declines. In this paper we examine the conservation status of small dasyurids (<500 g) and the threats they face. We also evaluate recovery procedures for threatened taxa and assess their success. Twenty-four percent of smaller dasyurids are classified as vulnerable, endangered or data deficient. Large body size and occupancy of one or two habitat types are correlated strongly with endangerment; species currently considered as 'low risk, near threatened' group closely with vulnerable and endangered species, indicating a risk of further declines. The processes contributing most to declines include habitat loss and fragmentation, altered fire regimes and predation. As of April 2001, no Recovery Plans had been adopted by the Commonwealth Government for any small dasyurid species. There is much information on the reproduction and development of smaller dasyurids, making them suitable for captive breeding. However, captive breeding programs have been limited, the dibbler *Parantechinus apicalis* being the only species bred systematically for reintroductions. There is a need for integration between captive breeding programs and recovery planning, as well as for more information on the population viability and metapopulation structures of small dasyurids, genetic diversity of populations and inbreeding depression. We suggest a program of survey, research, management and education to improve conservation outcomes for all small dasyurids.

INTRODUCTION

Although the biology, evolution and ecology of dasyurid marsupials (Family Dasyuridae) was reviewed extensively in the two volumes arising from the Carnivorous Marsupial symposium held in 1980 (Archer 1982), there was little reference then to the status and conservation of any taxa. Since that time, the decline of many species and threats affecting them have been documented and assessed. The conservation of large dasyurid species is addressed in other chapters in this volume. Here, we assess extinctions and declines of the smaller dasyurids, those weighing less than 500 g, together with the problems of conserving and recovering threatened species.

Presently, 10 genera of small dasyurids are recognised in Australia (Table 1), ranging in size from the very small (<10 g) species of *Ningaui* and *Planigale*, to medium-sized, robust animals belonging to *Dasycercus*. The different species are distributed throughout

Table 1 Ecological attributes and conservation status of smaller dasyurid marsupials. *Maxwell et al. (1996)

Taxa Family Dasyuridae	*Status	Weight class	Diet	Habit	Habitat	Region
Antechinomys						
Antechinomys laniger	DD	2	1	G	cd	E
Phascogale						
P. calura	En	2	1	TG	f	E S
P. tapoatafa pirata	LR(nt)	3	1 2	TG	efg	Ti To Tu I
P. t. tapoatafa	LR(nt)	3	1 2 4	TG	ef	K S
Antechinus						
A. adustus	LR(lc)	2	1	TG	g	Tu
A. agilis	LR(lc)	2	1	TG	ef	K
A. bellus	LR(lc)	2	1	TG	ef	Ti
A. flavipes flavipes	LR(lc)	2	1 2	TG	bdefg	E K I
A. flavipes leucogaster	LR(lc)	2	1	TG	bef	S
A. f. rubeculus	LR(lc)	2	1 3 4	TG	efg	To Tu
A. godmani	LR(nt)	2	1 4	TG	g	Tu
A. leo	LR(nt)	2	1	TG	g	Tu
A. minimus maritimus	LR(nt)	2	1	G	cfk	K I
A. minimus minimus	LR(lc)	2	1	G	cefgk	T I
A. stuartii stuartii	LR(lc)	2	1 2	TG	def	K
A. swainsonii insulanus	LR(nt)	2	1 3	G	def	K
A. s. mimetes	LR(lc)	2	1 3	G	def	K
A. s. swainsonii	LR(lc)	2	1 3	G	efg	T I
A. subtropicus	Vu	2	1	TG	g	Tu
Dasycercus						
D. byrnei	Vu	3	1 4	G S	cd	E
D. cristicauda	Vu	2	1 4	G S	cd	E
D. hillieri	En	3	1 4	G S	cd	E
Dasykaluta						
D. rosamondae	LR(lc)	2	1 4	G S	cd	E
Ningaui						
N. ridei	LR(lc)	1	1	G S	cde	E
N. timealeyi	LR(lc)	1	1	G S	bcde	E
N. yvonneae	LR(lc)	1	1	G S	cde	E
Parantechinus						
P. apicalis	En	2	1 2 4	G	d	S I
Pseudantechinus						
Ps. bilarni	LR(lc)	2	1	T G	befg	Ti
Ps. macdonnellensis	LR(lc)	2	1	T G	be	E
Ps. mimulus	Vu	2	1	T G	be	TiI
Ps. ningbing	LR(lc)	2	1	T G	e	Ti
Ps. woolleyae	LR(lc)	2	1	T G	bcef	E
Planigale						
P. gilesi	LR(lc)	2	1	G S	cd	E
P. ingrami	LR(lc)	1	1	G S	cde	ETiTo
P. maculata	LR(lc)	2	1	G	bcdefgk	ETiToTuKI
P. tenuirostris	LR(lc)	1	1	G S	cd	EK
Sminthopsis						
S. aitkeni	En	2	1	G	de	I
S. archeri	DD	2	1	T G	e	To
S. bindi	LR(lc)	2	1	G S	e	Ti
S. butleri	Vu	2	1	G	e	Ti
S. crassicaudata	LR(lc)	2	1	G S	cde	EK
S. dolichura	LR(lc)	2	1 4	G S	cdef	ES
S. douglasi	En	2	1 4	G S	c	E

Table 1 Ecological attributes and conservation status of smaller dasyurid marsupials. *Maxwell et al. (1996) *(Continued)*

Taxa Family Dasyuridae	*Status	Weight class	Diet	Habit	Habitat	Region
***Sminthopsis* (cont.)**						
S. gilberti	LR(lc)	2	1	G	def	ES
S. granulipes	LR(lc)	2	1	G	de	S
S. griseoventer boullangerensis	CR	2	1 4	G	d	I
S. griseoventer griseoventer	LR(lc)	2	1	G	dek	S
S. hirtipes	LR(lc)	2	1	G S	cde	E
S. leucopus	DD	2	1	G	cdefg	TuKTI
S. longicaudata	LR(lc)	2	1	T G S	bcde	E
S. macroura	LR(lc)	2	1	G S	bcd	ETi
S. murina murina	LR(lc)	2	1	G	defg	EK
S. murina tatei	LR(nt)	2	1	G	g	Tu
S. ooldea	LR(lc)	2	1	G S	cde	E
S. psammophila	En	2	1	G S	cd	E
S. virginiae nitela	LR(lc)	2	1	G	e	TiI
S. virginiae virginiae	LR(lc)	2	1	G	e	To
S. youngsoni	LR(lc)	1	1	G S	cde	E

Status categories: DD – data deficient; Vu – vulnerable, En – endangered, CR – critically endangered, LR – low risk, lc – least concern, nt – near threatened.
Antechinus subtropicus is considered vulnerable here; it was not identified in Maxwell et al. (1996).
Weight class: 1 = 1–10 g, 2 = 11–100 g, 3 = 101–1000 g, 4 = 1001–10,000 g
Diet: 1 = terrestrial invertebrates, 2 = nectar/pollen/plant exudates, 3 = fruit, 4 = terrestrial vertebrates
Habit: T = tree, G = ground, S = subterranean
Habitat: b = rock outcrop, c = grassland, d = shrubland, e = woodland, f = open forest, g = closed forest, k = wetland
Region: E = Eyrean, Ti = Timorian, To = Torresian, Tu = Tumbunan, K = Kosciuszkan, S = South Western, T = Tasmanian, I = Islands

Australia and occur in a wide variety of environments, from arid deserts to shrublands, heathlands, woodlands, forests and grasslands (Dickman, this volume). A further eight genera containing at least 16 species occur in New Guinea, with most taxa being restricted to forest (Flannery 1995).

In this chapter we review current knowledge of the conservation status of the smaller dasyurid taxa and assess extinctions, range reductions and population declines. We examine the effects on populations of major disturbance factors such as land clearance, habitat fragmentation, disease, changed fire regimes and introduced predators, and assess population responses to determine if there are any ecological correlates to endangerment. We then review management and recovery processes that have been implemented in recent times, assess their success and associated problems, and finally identify gaps in knowledge and future areas of research. The review is focused primarily on Australian dasyurids due to the paucity of relevant information on any New Guinean species.

CONSERVATION STATUS AND ECOLOGICAL ATTRIBUTES OF THREATENED AND NON-THREATENED DASYURIDS

Conservation status

The conservation status of Australian dasyurids was reviewed recently by Maxwell et al. (1996). We have used the evaluations of status of these authors in the analyses below, modifying them with respect only to the addition of *Antechinus subtropicus* and the inclusion of *A. adustus* as a full species rather than as a sub species of *A. stuartii* (Van Dyck and Crowther 2000) (Table 1). The status of several taxa has been subject to ongoing review since 1996 (e.g. *Sminthopsis douglasi*, *S. leucopus*, *Planigale* spp.). We have preferred to retain the evaluations of Maxwell et al. (1996) for consistency rather than introduce ad hoc changes here. Maxwell et al. (1996) also identified the conservation status of species and subspecies, and we have followed this distinction. Of the 58 taxa weighing <500 g, 14 have been classified as vulnerable, endangered or data deficient, and one as critically endangered. Three of the six larger taxa of dasyurids are classified as threatened; the Eastern Quoll *Dasyurus viverrinus*, Northern Quoll *D. hallucatus* and Tasmanian Devil *Sarcophilus laniarius* are low risk.

Ecological attributes

Recent studies have suggested that a wide range of ecological attributes can influence the likelihood that any given species will become threatened. Such attributes include life history traits, distribution and abundance, degree of ecological specialisation, and body size (McKinney 1997). Analyses of the extinction-proneness of native mammals have evaluated the importance of some of these attributes, and indicate that non-flying taxa above 35 g are at particular risk (Burbidge and McKenzie 1989; Cardillo and Bromham 2001). There is some evidence also that ground-dwelling mammals in arid and semi-arid environments have suffered declines (Burbidge and McKenzie 1989; Copley et al. 1989; Dickman et al. 1993), especially in the southern parts of the Australian mainland. Little attention has focused specifically on regional dasyurid faunas, although Dickman et al. (2001) have shown recently that large body size and occupancy of a single region are risk factors for the dasyurids of New South Wales.

In this section, we adopt the five categories of ecological attributes used by Dickman et al. (2001), and compare the attributes of threatened versus non-threatened species using the status categories of Maxwell et al. (1996). The attribute categories are body weight, region, habit, habitat and diet (Table 1). For ease of analysis, categories were subdivided. The *body weight* category was subdivided into four levels, 1–10 g, 11–100 g, 101–1000 g, and 1001–10 000 g, using mean body weight data taken from Burbidge and McKenzie (1989), Strahan (1995) and Van Dyck and Crowther (2000). *Region* was subdivided into Eyrean, Timorian, Torresian, Tumbunan, Kosciuszkan, South Western, Tasmanian, and islands. These regions were defined by Heatwole (1987), except for the inclusion of the Tumbunan element (north-eastern closed forests) and islands off the coast of the mainland, exclusive of Tasmania. Species were assigned to a region if the region contained ≥10% of the species' entire geographical range, or if the species occurred, or had occurred, in ≥10% of the region. Assignations to region were based on detailed distribution maps for all species (Murray and Dickman 2000). *Habit* was split into arboreal, ground-dwelling and subterranean, and *habitat* into rock outcrop, grassland, shrubland, woodland, open forest, closed forest and wetland. These subdivisions follow Lunney et al. (1997), while assignations of species to the different levels were based on Strahan (1995) and primary references. *Diet*, finally, was subdivided into the food categories, terrestrial invertebrates, terrestrial vertebrates, fruit, and nectar, pollen and other plant exudates. Data for placement of species within dietary subdivisions were again taken from Strahan (1995) and primary literature. The species-by-attribute matrix is shown in Table 1.

To compare the attributes of species in the different status categories, we used chi-squared contingency analysis and non-metric multidimensional scaling (nMDS). For contingency analysis, frequencies of species in each status group were tallied separately for each category of ecological attribute, and tested for association using chi-squared. For nMDS, species were scored for presence/absence in the different subdivisions of all attribute categories, and ordinated to produce two-dimensional plots. Significance testing used one-way analysis of similarities (ANOSIM) on 5000 random permutations of the data, and the contribution of individual species to overall differences between status groups evaluated using similarity-percentage (SIMPER) analysis (Clarke 1993; Legendre and Legendre 1998). A more detailed discussion of these procedures is given by Dickman et al. (2001).

In all analyses, the three species listed as 'data deficient' by Maxwell et al. (1996), *Antechinomys laniger*, *Sminthopsis archeri* and *S. leucopus*, were grouped with species listed as vulnerable. We considered this reasonable as few records of these species may mean that they are represented by small or diffuse populations that are likely to be of some conservation concern. The critically endangered dunnart, *Sminthopsis griseoventer boullangerensis*, was grouped with 'endangered' species. All analyses were run with larger species of dasyurids (≥500 g) included and omitted, to determine the influence of the larger species on overall patterns of vulnerability. Because initial nMDS ordinations using all status groups were associated with high statistical stress values and were visually difficult to interpret, we re-ran analyses by collapsing the status groups 'low risk, least concern' and 'low risk, near threatened' into a single group, and then collapsing the groups 'vulnerable' and 'endangered' into another single group. However, these procedures failed to clarify ordinations, and they are presented showing the four original status groups. Joint use of contingency analysis and nMDS was expected to identify individual ecological attributes, and combinations of attributes, that contribute most to explaining differences in species status.

In contingency analyses, body weight and status were associated when all dasyurids were included ($\chi^2_{(9)} = 21.58$, $P < 0.05$) and when species ≥500 g (*Dasyurus* spp., *S. laniarius*) were omitted ($\chi^2_{(6)} = 12.95$, $P < 0.05$). This was due largely to the over-representation of larger species (≥101 g) in the categories 'vulnerable' and 'endangered'.

Comparisons of status by diet, habit, habitat and region were not significant whether larger dasyurids were included or not. There was also no association between status and the numbers of different categories of food in the diet or between status and numbers of regions occupied. However, status and numbers of habitats occupied were associated both when *Dasyurus* spp. and *S. laniarius* were included in the comparison ($\chi^2_{(9)} = 20.29$, $P < 0.05$) and when they were not ($\chi^2_{(9)} = 20.22$, $P < 0.05$). Further analysis showed that low risk, least concern species contributed heavily to this result. Including dasyurids ≥500 g, 25 of 37 species (67.6%) in the low risk, least concern category occupy three or more habitats compared with only 9 of 28 species (32.1%) in all the remaining status categories ($\chi^2_{(3)} = 9.20$, $P < 0.05$). Similarly, with larger dasyurids excluded, 24 of 36 low risk, least concern species (66.7%) occupy three or more habitats whereas only 4 of the remaining 22 species (18.2%) do so ($\chi^2_{(3)} = 14.43$, $P < 0.01$). All endangered species occupy only one or two habitat types.

Two-dimensional ordinations produced by nMDS are shown in Figs. 1a and 1b. With all dasyurids included (Fig.1a) clear overall differences were detected between the four status groups ($R = 0.24$, $P < 0.001$), with the strongest pairwise differences evident between 'low risk, least concern' and 'low risk, near threat-a

b

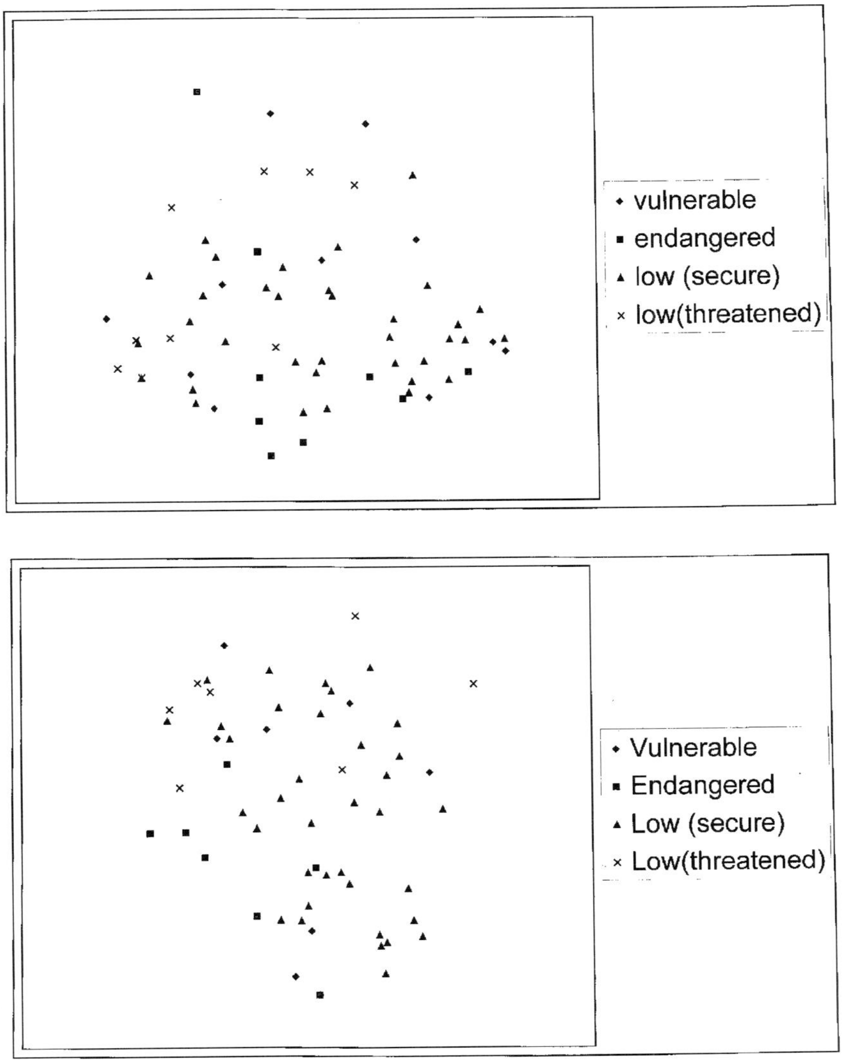

Figure 1 Ordinations produced by nMDS of the ecological attributes of vulnerable, endangered, low-risk least concern and low-risk near threatened dasyurids: (a) all species included; (b) only species <500 g included.

ened' (R = 0.37, P < 0.001), 'low risk, least concern' and 'vulnerable' (R = 0.20, P < 0.01) and 'low risk, least concern' and 'endangered' (R = 0.19, P < 0.05). No individual species contributed strongly (>6%) to discrimination between the groups. With only smaller (<500 g) dasyurids included (Fig. 1b), there were again strong differences overall between status groups (R = 0.21, P < 0.001). The group 'low risk, least concern' again differed markedly from species in the 'low risk, near threatened' (R = 0.38, P < 0.001) and vulnerable (R = 0.14, P < 0.05) status groups, but in this comparison 'low risk, near threatened' and endangered species differed also (R = 0.32, P < 0.01). No individual species contributed more than 8% to differences between status groups. Some separation of the status groups can be seen in Figs. 1a and b, especially the 'low risk, least concern' group, but both ordinations were associated with high stress values of 0.22–0.23 which reduce the reliability of visual interpretation alone.

In summarising this section, large body size and occupancy of one or two habitat types stand out as the strongest correlates of endangerment in dasyurids. These associations remain whether all species or only those <500 g are considered. It is clear also that species considered currently as 'low risk, near threatened' group more closely with vulnerable and endangered species than they do with 'low risk, least concern' species, especially when all attributes are considered jointly. It would be simplistic to conclude that small habitat generalists are always at less risk than larger specialists, because there are many exceptions. However, it is reasonable to suggest that attention be focused on larger species, especially those weighing ≥101 g, and that there be no complacency with respect to species listed currently 'low risk, near threatened' due to their ecological similarity with their vulnerable and endangered counterparts.

MAJOR THREATS AND IMPACTS

Habitat loss, fragmentation and decline

The loss of natural vegetation in Australia has been most severe in areas developed for agriculture such as the plains of western Victoria, central New South Wales, South Australia and the wheatbelt of Western Australia. Indeed, historical disruption of these regional landscapes for settlement and farming has caused dramatic changes in the species richness and composition of fauna in remaining remnants (Kitchener et al. 1980; Bennett 1987, 1990a, b; Lunney and Leary 1988; Laurance 1997). Some species are more vulnerable to fragmentation and can decline or disappear very rapidly (Bennett 1987), particularly habitat specialists such as the rainforest-dwelling *Antechinus godmani*, *A. adustus* and *A. leo* (Laurance 1991, 1993, 1994; Leung 1999). Conversely, Bennett (1990a, b) found that the generalist *A. agilis* was one of the most abundant and widespread mammals in remnant forest patches at Naringal, western Victoria occurring in 26 of the 39 patches assessed. Lunney and Leary (1988) found that, although larger dasyurids such as *D. maculatus* and *P. tapoatafa* had become extinct in the Bega region of south-eastern New South Wales, species such as *A. agilis*, *A. swainsonii* and *S. leucopus* remained in remnant forests.

Habitat fragmentation due to extensive land clearance in the wheatbelt of Western Australia has been a major factor in the decline of *Phascogale calura* in that region, while *P. tapoatafa* has also suffered declines in south-eastern Australia from land clearance and habitat alteration by timber harvesting and mining activities (Kitchener et al. 1980; Lunney and Leary 1988; Humphries and Seebeck 1996). Habitat clearing has been implicated in the decline of *Parantechinus apicalis*, while *Antechinus minimus* is prone to habitat fragmentation due to roading and housing developments in coastal south-western Victoria (Wilson et al. 2001).

Several traits and factors that underlie the variable responses of species to fragmentation have been identified (Laurance 1991, 1997). These include species' natural rarity, their tolerance of the habitat matrix and edge effects, and their dispersal capacity. Species that are most susceptible to habitat fragmentation and are first to disappear are usually those that occur naturally at low densities (Diamond 1984). Small populations are more vulnerable to stochastic processes such as random variation in demographic parameters, loss of genetic diversity and catastrophic events such as fire or storms. Large animals, species high in the food chain (owls, carnivorous mammals) and species with specialised food and habitat requirements also contribute to a species' susceptibility to fragmentation, as suggested by the analyses above.

Species richness and composition of marsupial communities generally are influenced strongly by habitat patch size, but habitat diversity, the level of disturbance, the shape of patches and time since isolation are also important (Bennett 1987; Williams 1997). In rainforest, *A. godmani* and *A. adustus* have disappeared or declined sharply in disturbed fragments under 100 ha in area (Laurance 1997), while in urban bushland *A. stuartii* occurs only in patches of 400–2000 ha or larger (Matthews et al. 1999). However, in forest fragments in southern Victoria *A. agilis* occurs in fragments of less than 100 ha (Bennett 1987), perhaps because this species makes some use of the surrounding habitat matrix.

The destruction of vegetation communities via 'dieback', caused by the plant pathogen *Phytophthora cinnamomi*, is also strongly associated with reductions in small mammal abundance and species diversity (Wilson et al. 1990; Wilson et al. 1994; Newell and Wilson 1993; Newell 1994). Species such as *A. agilis* and *S. leucopus* are less abundant in diseased heathland than in healthy areas, or utilise them less frequently (Laidlaw and Wilson 1989; Laidlaw 1997), but reasons for this remain to be found.

Introduced predators and herbivores

The role of the cat, *Felis catus*, and the red fox, *Vulpes vulpes*, in the extinction and decline of many Australian mammal species has been recognised relatively recently. Cats rely on small to medium-sized prey (Catling 1988; Paton 1993; Paltridge et al. 1997), with small mammals <220 g often preferred (Dickman 1996; Mifsud 2000; Risbey et al. 1999, 2000). Foxes prey upon a wide variety of animals including small to medium-sized mammals (Seebeck 1978; Wallis and Brunner 1986; Risbey et al. 1999, 2000).

Interestingly, although foxes have been shown to deplete populations of medium-sized native mammals (e.g. Kinnear et al. 1988, 1998), there is a lack of data on the impact of fox or cat removal on small mammals (<200 g) (Dickman 1996). Removal of foxes and cats at two sites in the Simpson Desert failed to produce any increase in dasyurid populations, although native rodents did respond (P. Mahon pers. comm.). However, in semi-arid habitat at Shark Bay, Western Australia, captures of native small mammals, particularly small rodents, increased significantly when cats were controlled (Risbey et al. 2000). Cats have recently been shown to prey heavily on the Julia Creek Dunnart, *S. douglasi*, and there is evidence that they have extirpated the species from one area of its limited range in recent times (Mifsud 2000). Introduced predators have also been implicated in the declines of *D. cristicauda, D. hillieri, S. butleri, S. psammophila,* and *P. calura* (Maxwell et al. 1996).

Introduced herbivores have likely contributed to extinctions and declines of dasyurids through the effects of overgrazing, including removing shelter, competition for food and soil compaction (Recher and Lim 1990; Morton 1990). For smaller dasyurids the major threats are removal of vegetation shelter and the destruction of burrows (Dickman and Read 1992). Declines of *D. cristicauda* , *D. hillieri, D. byrnei, S. douglasi* and *A. laniger* have been associated with effects of introduced herbivores (Maxwell et al. 1996).

Fire

The effect of fire on Australian small mammals has been a focus of many studies over the past two decades (Catling and Newsome 1981; Suckling and McFarlane 1984; Friend 1993; Wilson 1996), with evidence mounting that declines of several species are due to changes in fire regimes since European settlement (Burbidge and McKenzie 1989; Kitchener et al. 1980; Wilson and Friend 1999). Fire-effects vary depending on the fire regime, fire intensity, frequency and season of occurrence. Fires of high intensity generally reduce small mammal populations substantially, with some animals surviving in small, unburnt patches; conversely where fires are of low intensity and large areas remain unburnt, many animals survive (Christensen and Kimber 1975; Recher et al. 1974; Newsome et al. 1975; Catling and Newsome 1981; Fox 1982).

Small dasyurid species exhibit varied recolonisation responses to fire, entering seral stages of vegetation as their specific requirements are met (the 'habitat accommodation' model, Fox 1982). Species such as *S. murina* and *S. leucopus* achieve maximal abundance in mid-successional stages (3–5 years), *A. stuartii* prefers mid- to late succession (5–6 years) (Fox 1982, 1990; Wilson et al. 1990), while species such as *A. swainsonii* and *A. minimus* that require dense ground cover, exhibit low population numbers up to 6 years after fire and depend on late successional habitat (6 to >10 years) (Newsome et al. 1975; Lunney et al. 1987; Wilson et al. 1990; Aberton 1996; Wilson et al. 2001). In central Australia *Sminthopsis youngsoni* is more abundant on older areas (11–15 years), while *S. hirtipes* is usually more abundant on areas 1–4 years post-fire (Masters 1993).

Some smaller dasyurid species have probably suffered declines from inappropriate fire regimes and are currently threatened by fires that are too frequent or infrequent. *Dasycercus cristicauda* is considered to have been affected by changed fire regimes in hummock grasslands of the central deserts (Gibson and Cole 1992), as is the Sandhill Dunnart *S. psammophila* in more southerly parts of the arid zone (Pearson and Robinson 1990). Changes in burning regimes likely contributed to the decline of *Parantechinus apicalis*, particularly on the mainland. All modern records of this species are from long unburnt vegetation, and management regimes to prevent burning of habitat have been recommended (Muir 1985). Frequent burning of woodland remnants in the Western Australian wheatbelt is probably a major factor in the decline of *Phascogale calura* (Kitchener et al. 1980).

Timber harvesting

The impacts of logging on small dasyurids have been assessed to a limited degree. Lunney et al. (1987) found that numbers of *A. agilis* were unaffected for 15 years after logging but those of *A. swainsonii* increased in the regenerating forest. There was evidence that *S. leucopus* increased after logging, but declined in regenerating forest when regrowth was dense (Lunney and Ashby 1987). Dasyurids such as *A. agilis* and *A. swainsonii* may survive and inhabit linear areas of habitat within logged areas (Friend 1982; Bennett 1987, 1990a, b; Lindenmayer et al. 1994). Such survival may provide a source of animals for recolonisation of logged, regenerating areas (Friend 1979; Lindenmayer et al. 1994). A number of variables was found to be important for the occurrence of *A. agilis* in linear habitat strips including the density, cover and structural characteristics of the vegetation at ground level, the presence of *Acacia* trees, the continuity of habitat within the linear strips, and the topography of the strips, for example whether the midslope joined to ridges or gullies (Lindenmayer et al. 1994).

Ground-dwelling and scansorial dasyurids also have requirements for large mature trees in forests for foraging and nesting (Dickman 1991). Species such as *A. agilis, A. stuartii, A. flavipes* and *P. calura* principally use tree surfaces and hollows for foraging, particularly under loose bark. They also use hollows in trees, and in logs and leaf litter, for shelter sites, as do species such as *S. griseoventer* and *S. dolichura* (Dickman 1991). The presence of *A. agilis* in montane Ash forest is associated with proximity to hollow-bearing trees in linear habitat strips (Lindenmayer et al. 1994). This species nests in tree cavities and its social behaviour includes movements between nests in hollow-bearing trees (Cockburn and Lazenby-Cohen 1992). Declines of phascogales have been correlated both with the decline in quality of habitat in production forests, and with the decrease in availability of nest sites and foraging habitat.

Population viability and metapopulations

While the causes of declines of wildlife populations due to habitat clearance and fragmentation are deterministic, remnant populations are vulnerable to processes intrinsic to small populations such as demographic changes (e.g. birth rate), loss of genetic variability and environmental and catastrophic events (Soulé 1987; Burgman et al. 1988). Small, isolated populations face high risks of extinction, but may become more viable if some individuals can move between patches; this is the concept of the metapopulation (Gilpin and Hanski 1991). Although models of metapopulation structure and dynamics have been developed (Gilpin and Hanski 1991), few have been applied to Australian mammal species (e.g. Lindenmayer and Lacy 1995). We have little understanding of the viability, metapopulation structures and dynamics of any small dasyurids.

RECOVERY AND MANAGEMENT

Action Plans, Recovery Plans and Action Statements

Action Plans are documents which review the conservation status of major taxonomic groups against IUCN categories, identify threats and recommend actions to minimise those threats. The 1996 Action Plan for Australian Marsupials and Monotremes (Maxwell et al. 1996) proposed Recovery Outlines for 21 of the 58 small dasyurid species including both threatened and non-threatened taxa (Table 2). Recovery Plans are more detailed documents based on the Recovery Outline and give steps to be taken to ensure survival of the nominate species. Similarly, Action Statements outline the actions to be taken to ensure long-term survival of species under the Flora and Fauna Guarantee Act 1988 of Victoria.

By April 2001, Final/Interim Recovery Plans had been developed by State authorities for only two of the small dasyurids (Kowari *Dasycercus byrnei*, Lim 1992 and the dibbler *Parantechinus apicalis*, Start 1998) listed in Maxwell et al. (1996), and that for one species, *D. byrnei*, is out of date (Table 2). Draft Recovery Plans are in preparation for the Mulgara *D. cristicauda* and Ampurta *D. hillieri*, and the Kowari Recovery Plan is being revised by the South Australian Department of Environment and Heritage (P. Copley pers. comm.). The Victorian Dept of Natural Resources and Environment has produced Action Statements for *Planigale gilesi* and *Phascogale tapoatafa* which are not among the 15 species identified in Maxwell et al. (1996) as data deficient, critically endangered, endangered or vulnerable.

The actual status or existence of Recovery Plans was difficult to assess until a database documenting all Recovery Plans approved under the Commonwealth Government Environment Protection and Biodiversity Conservation (EPBC) Act 1999 was developed by Environment Australia (www.ea.gov.au/biodiversity/threatened/recovery/list.html). This database lists Recovery Plans adopted by the Commonwealth Government for threatened species and ecological communities and incorporates plans adopted under the Endangered Species Protection Act 1992 which the EPBC Act replaced. State Government websites which are linked to the Environment Australia website are variably informative regarding the threatened species recovery process and status of progress of species within their jurisdiction. Only the Victorian Government website details the full Action Statement for any of the small dasyurid species whose status is summarised in Table 2.

The Action Plan for Monotremes and Marsupials (Maxwell et al. 1996) can be found at the same Environment Australia website. Most actions within the plans involve management of taxa *in situ* including undertaking surveys to confirm or define distributions, and identify threatening processes. Management actions in the Action Plan identify captive breeding for only three species (*P. apicalis, S. douglasi and S. griseoventer boullangerensis*), and reintroduction or translocation for three species with a fourth to be assessed (Table 2).

Captive breeding

Many dasyurid species have been bred in captivity, predominantly for research on reproduction and development (physiology, fertilisation, mating behaviour, etc). The basic reproductive biology of many of the small dasyurids is summarised in Tyndale-Biscoe and Renfree (1987) and Taggart et al. (this volume), but data on breeding season, gestation, litter size, development of young is incomplete for many species. Some species information is limited to field observations of females with pouch young (Strahan 1995).

There is much additional information available on mating strategies, mating times, gestation lengths, oestrus, endocrinology, development stages, fertilisation reactions (e.g. Woolley 1966, 1971a, b, 1984, 1991, 1995; Fletcher 1985, 1989a, b; Wilson 1986; Wilson and Bourne 1984; Selwood 1980, 1981, 1982a, b, 1983; Taggart et al. 1997, 1998). Based on this extensive research the application of assisted reproductive technologies to dasyurids, i.e. induced ovulation, IVF, may be possibilities for the future. At the present, though, it is hard to identify which species could need this level of intervention in the near future.

The Australasian Regional Association of Zoological Parks and Aquaria (ARAZPA) is the zoo industry body which manages the Australasian Species Management Program (ASMP). The ASMP organises species management and planning at a regional level for the industry. ASMP is organised into categories which indicate the status of a taxon in the regional collection, whether management is occurring and at what level (Johnson et al. 2001). A Conservation Program includes specific conservation objectives that would initiate or assist efforts to support *in situ* populations of the target species. A Population Management Program is established for species which would benefit from some regional population management and should be held in more than one institution with a population greater than 20. The 2001 Regional Census and Plan (Johnson et al. 2001) shows a Conservation Program for the dibbler, *P. apicalis* and a Population Management program for brush-tailed phascogale, *P. tapoatafa*. Clearly this is an area where strategic research could be undertaken to fill gaps in knowledge which is aligned with the conservation mission of these institutions and the conservation needs of threatened fauna.

Table 2 Conservation status, Recovery Plans and captive husbandry of small dasyurids

Species	Conservation status	Recovery outline	EPBC listed	Recovery Plan/ Action Statement	Management action	Held in captivity*
Antechinomys						
A. laniger	DD	+				
Antechinus						
A. adustus	LR(lc)					
A. agilis	LR(lc)					6
A. bellus	LR(lc)					11
A. flavipes flavipes	LR(lc)					4
A. flavipes leucogaster	LR(lc)					
A. flavipes rubeculus	LR(lc)					
A. godmani	LR(nt)	+				
A. leo	LR(nt)	+				
A. minimus maritimus	LR(nt)					
A. minimus minimus	LR(lc)	+				
A. stuartii stuartii	LR(lc)					6
A. swainsonii insulanus	LR(nt)	+				
A. swainsonii mimetes	LR(lc)					
A. swainsonii swainsonii	LR(lc)					
A. subtropicus	Vu					
Dasycercus						
D. byrnei	Vu	+	+	Expired RP		2,7
D. cristicauda	Vu	+	+			2,11
D. hillieri	En	+	+		?	
Dasykaluta						
D. rosamondae	LR(lc)					
Ningaui						
N. ridei	LR(lc)					
N. timealeyi	LR(lc)					
N. yvonneae	LR(lc)					
Parantechinus						
P. apicalis	En	+	+	Interim RP	C, R	9
Phascogale						
P. calura	En	+	+		R	2
P. tapoatafa pirata	LR(nt)	+				
P. tapoatafa tapoatafa	LR(nt)	+		AS		1,3,4,5,6,9,10
Planigale						
P. gilesi	LR(lc)			AS		
P. ingrami	LR(lc)					
P. maculata	LR(lc)					4,6,11
P. tenuirostris	LR(lc)					
Pseudantechinus						
Ps. bilarni	LR(lc)					
Ps. macdonnellensis	LR(lc)					2
Ps. mimulus	Vu	+	+		R	
Ps. ningbing	LR(lc)					
Ps. woolleyae	LR(lc)					

Table 2 Conservation status, Recovery Plans and captive husbandry of small dasyurids *(Continued)*

Species	Conservation status	Recovery outline	EPBC listed	Recovery Plan/ Action Statement	Management action	Held in captivity*
Sminthopsis						
S. aitkeni	En	+	+			
S. archeri	DD	+				
S. bindi	LR(lc)					
S. butleri	Vu	+	+			
S. crassicaudata	LR(lc)					1,3,6,8,11
S. dolichura	LR(lc)					
S. douglasi	En	+	+		C	
S. gilberti	LR(lc)					
S. granulipes	LR(lc)					
S. griseoventer boullangerensis	CR	+	+		C	9
S. griseoventer griseoventer	LR(lc)					
S. hirtipes	LR(lc)					
S. leucopus	DD	+				
S. longicaudata	LR(lc)					2
S. macroura	LR(lc)					
S. murina murina	LR(lc)					
S. murina tatei	LR(nt)	+				
S. ooldea	LR(lc)					2
S. psammophila	En	+	+			
S. virginiae nitela	LR(lc)					
S. virginiae virginiae	LR(lc)					
S. youngsoni	LR(lc)					2

Conservation status: see caption Table 1.
Recovery outline: Recovery outline in Maxwell et al. (1996).
EPBC listed: Listed under Environment Protection and Biodiversity Conservation Act 1999 (Commonwealth of Australia) as Critically Endangered, Endangered or Vulnerable
Management action: C = captive breeding, R = translocation/reintroduction
Held in captivity: On display in an institution affiliated with the Australasian Regional Association of Zoological Parks and Aquaria: 1, Adelaide Zoo (SA); 2, Alice Springs Desert Park (NT); 3, Australia Zoo (QLD); 4, Currumbin Sanctuary (QLD); 5, Dreamworld (QLD); 6, Healesville Sanctuary (VIC); 7, Melbourne Zoo (VIC); 7, Monarto Zoological Park (SA); 8, Pearcedale Conservation Park (VIC); 9, Perth Zoo (WA); 10, Taronga Zoo (NSW); 11, Territory Wildlife Park (NT). Data from the Australasian Species Management Program Census and Plan (Johnson et al. 2001).

The role of zoos and wildlife parks in captive breeding of smaller dasyurids has been somewhat limited. Given the extensive amount of research on dasyurid reproduction and the conservation mission of zoos, this is puzzling. Small nocturnal species are difficult and expensive to display given this almost always means nocturnal house-type buildings with reverse lighting. Nevertheless species such as the long-tailed dunnart *Sminthopsis longicaudata* are spectacular and virtually unknown to the general Australian public, and education is one of the major purposes of a modern zoo. The educative value of maintaining displays where the public can see some of the smaller, more cryptic, endangered species cannot be underestimated. Clearly there needs to be more integration of zoo and wildlife park captive management programs with State agencies conservation planning where captive breeding is deemed necessary in a Recovery Plan. Zoos after all do have significant expertise in management of small populations on sound genetic principles. Zoos in Australia use studbooks to manage populations of captive animals. Studbooks contain a comprehensive record on species including breeding stocks, their locations and genetic information to assist in the regional management of species, and are maintained by experienced staff who liaise with participating organisations and make recommendations for the appropriate pairings of breeding animals.

Despite the availability of the breeding data outlined above, there is little information on the longevity of captive colonies of dasyurids, and whether there may be problems with inbreeding depression. A colony of *Sminthopsis crassicaudata* has been maintained at the University of Adelaide for many generations, and a colony of kowaris was maintained at the Institute for Medical and Veterinary Science (Adelaide, South Australia) for many years. Data from these colonies might give some indication whether long-term captive colonies do express genetic defects. Zoos and wildlife parks are ideally placed to address some of these issues.

The dibbler *P. apicalis*, is the only small dasyurid species that has been systematically bred for reintroductions/translocation. Research into the breeding biology of the species was undertaken by Dr P. Woolley at Latrobe University (Woolley 1971b, 1991, 1995). In 1985 dibblers were found on Boullanger and Whitlock Islands in Jurien Bay, and at Fitzgerald River National Park, Western Australia. In 1997 four pairs of dibblers were collected from the two islands to establish a captive breeding colony at Perth Zoo as part of the dibbler Interim Recovery Plan (Start 1998) and the dibbler was bred in captivity for the first time at Perth Zoo in 1996. Between 1997 and 1999, three more animals were brought in from the wild, a total of approximately 131 individuals were born in captivity and 98 were released on Escape Island in Jurien Bay (1998). In 2000, four dibblers were brought in from Fitzgerald River National Park to commence breeding of animals from the southernmost part of their range.

Translocations

Translocations have been defined by the IUCN (1998) as the movement of living organisms from one area for free release to another and include introductions (outside historical range), reintroductions (into previous native range where organisms have been extirpated in historic times) and re-stocking (building up numbers in original habitat).

There have been relatively few reintroductions of smaller dasyurids. In Western Australia there have been 82 translocation events since 1971, most have been since 1990 and include 59 (72%) reintroductions, 13 (16%) re-stockings and 10 (12%) introductions (Morris 2000). Most translocations have been mammals (21 of the 24 taxa), but only one has involved a small dasyurid, *Parantechinus apicalis* (Morris 2000). All other species have been larger taxa such as the numbat, woylie, banded hair wallaby, euro (Morris 2000; Fletcher and Morris 2003). Captive-bred dibblers (n = 27) were released onto Escape Island in 1998 and further introductions (n = 50) occurred from 1998–2001. Animals released have been recaptured in 1999 and 2000 and unmarked animals have been captured in 2000, confirming that the species has bred (J.A. Friend pers. comm.). Reintroductions to mainland sites were proposed for 2001, as part of the Interim Recovery Plan (Start 1998).

In Victoria a translocation/reintroduction of *Antechinus minimus*, was trialled in the Eastern Otways in southern Victoria where populations of the species were small and spatially fragmented across the landscape (Wilson et al. 1986; Wilson et al. 1990; Aberton 1996). There was evidence that the species was no longer extant in the area following the 1983 'Ash Wednesday' fire which burnt some 40,000 hectares (Wilson et al. 1990; Aberton 1996). A re-introduction of the species was thus trialled in 1992-94, but was judged not to have been successful (Aberton 1996).

As noted above, metapopulation structures and habitat distributions for smaller dasyurid species are poorly known, yet understanding of these features provides a solid basis for planning and implementation of translocations. The distribution of high quality habitat for many species is patchy, with poorer habitat intervening. An important aspect of the dynamics of the metapopulation is thus to assess the size, quality and spatial position of suitable patches (Gilpin and Hanski 1991) before re-introductions of species proceed. Identification of suitable habitat for one dasyurid, the swamp antechinus, has provided the basis for developing a spatially-based habitat model to determine the distribution of high quality habitat for protection, and the possibility of translocations or re-introductions (Wilson et al. 2001).

Conservation genetics

Assessment of genetic variability in captive colonies, source and re-introduced populations is important for future recovery of several species of dasyurids. The relatively limited research that has been carried out is summarised by Firestone (this volume).

FUTURE DIRECTIONS

Based on the findings of our review we suggest an extensive program of research, management and education that should assist broadly in the conservation of the remaining species of small dasyurids. These suggestions are adapted from detailed recommendations advanced for management of dasyurids in western New South Wales (Dickman and Read 1992):

1 Surveys should be carried out to identify the distributions of all threatened, data deficient and 'low risk, near threatened' species. Surveys should also target poorly surveyed areas, including western Cape York, the Gulf of Carpentaria, Arnhem Land, the Kimberley, Pilbara, and sandridge and upland deserts of the continental interior.

2 Because dasyurids are morphologically conservative, taxonomic research should be carried out to identify cryptic species and resolve species boundaries. Recognition and description of four new species since 1998 emphasises the importance of this recommendation (Dickman et al. 1998; Cooper et al. 2000; Van Dyck and Crowther 2000). Small amounts of tissue collected during surveys, together with re-examination of museum material, would provide the necessary raw material for taxonomic appraisals. *Planigale* spp., the *Sminthopsis murina* complex, *S. macroura* complex and *Phascogale tapoatafa* are candidate groups for evaluation.

3 Detailed studies should be initiated on the biology, demography and resource requirements of selected species to help reveal factors that limit their distribution and abundance. Such studies may also form the basis for mensurative or manipulative experiments that critically reveal the intensity of, and interactions among, threatening processes. Candidate species include *D. hillieri*, *A. laniger*, *S. psammophila*, *S. archeri*, *S. butleri* and *S. douglasi*.

4 Threatening processes should be managed as they become known for each species. Management may require reservation or some other form of land protection, manipulation of fire regimes, control of introduced species, especially predators, and captive breeding and translocation.

5 Long-term monitoring should be established to confirm the effectiveness of management regimes, and to identify and guide any changes that would improve conservation outcomes. Given the limited information on population viability, metapopulation structure and genetic diversity of most small dasyurids, monitoring should be designed to obtain spatial replication as well as robust estimates of demographic parameters and genetic sampling.

6 Programs of education should be developed to improve the low public profile of small dasyurids, and ensure continued support for conservation and management initiatives. Campaigns to increase public awareness of species, such as *Sminthopsis douglasi* and *Dasycercus byrnei*, can produce very beneficial results. Longer-term programs based on schools and using print and electronic media, could be expected to have longer-lasting, broadscale value.

7 There is extensive information on the reproduction and development of smaller dasyurids, making them suitable for *ex situ* breeding. However, captive breeding and reintroductions have been limited; reintroductions have been either experimental (*Antechinus minimus maritimus*), or onto offshore islands (*Parantechinus apicalis*). Zoos and wildlife parks could offer extra resources should their capabilities be integrated into the conservation and recovery process.

8 All suggestions from 1–7 apply to the entire dasyurid fauna of New Guinea. Dasyurids here remain the most poorly known of the Australasian region.

REFERENCES

Aberton, J.G. (1996), 'Succession of small mammal communities after fire and reintroduction of Swamp Antechinus, *Antechinus minimus*', PhD thesis, Deakin University, Geelong.

Archer, M. (ed.) (1982), *Carnivorous Marsupials, Vol. 1, 2*, Royal Zoological Society of New South Wales, Mosman, New South Wales.

Bennett, A.F. (1987), 'Conservation of mammals within a fragmented forest environment: The contributions of insular biogeography and autecology', in *Nature Conservation: The Role of Remnant Native Vegetation* (eds. D.A. Saunders, G.W. Arnold, A.A. Burbidge, & A.J.M. Hopkins), pp. 41–52, Surrey Beatty & Sons Pty. Ltd, Sydney, New South Wales.

Bennett, A.F. (1990a), 'Habitat corridors and the conservation of small mammals in a fragmented forest environment', *Landscape Ecology*, **4**:109–22.

Bennett, A.F. (1990b), 'Land use, forest fragmentation and the mammalian fauna at Naringal, south-western Victoria', *Australian Wildlife Research*, **17**:325–47.

Burbidge, A.A., & McKenzie, N.L. (1989), 'Patterns in the modern decline of Western Australia's vertebrate fauna: causes and implications', *Biological Conservation*, **50**:143–98.

Burgman, M.A., Akcakaya, H.R., & Loew, S.S. (1988), 'The use of extinction models for species conservation', *Biological Conservation*, **43**:9–25.

Cardillo, M., & Bromham, L. (2001), 'Body size and risk of extinction in Australian mammals', *Conservation Biology*, **15**:1435–40.

Catling, P.C. (1988), 'Similarities and contrasts in the diets of foxes *Vulpes vulpes* and cats *Felis catus* relative to fluctuations in prey populations and droughts', *Australian Wildlife Research*, **15**:307–17.

Catling, P.C., & Newsome, A.E. (1981), 'Responses of the Australian vertebrate fauna to fire: an evolutionary approach', in *Fire and the Australian Biota* (eds. A.M. Gill, R.H. Groves, & I.R. Noble), pp. 273–310, Australian Academy Science, Canberra.

Christensen, P., & Kimber, P. (1975), 'Effect of prescribed burning on the flora and fauna of south-west Australian forests', *Proceedings of the Ecological Society of Australia*, **9**:85–107.

Clarke, K.R. (1993), 'Non-parametric multivariate analyses of changes in community structure', *Australian Journal of Ecology*, **18**:117–43.

Cockburn, A., & Lazenby-Cohen, K.A. (1992), 'Use of nest trees by *Antechinus stuartii*, a semelparous lekking marsupial', *Journal of Zoology London*, **226**:657–80.

Cooper, N.K., Aplin, K.P., & Adams, M. (2000), 'A new species of false antechinus (Marsupialia: Dasyuromorphia: Dasyuridae) from the Pilbara region, Western Australia', *Records of the Western Australian Museum*, **20**:115–36.

Copley, P.B., Kemper, C.M., & Medlin, C. (1989), 'The mammals of north-western South Australia', *Records of the South Australian Museum*, **23**:75–88.

Diamond, J.M. (1984), '"Normal" extinctions of isolated populations', in *Extinctions* (ed. M.H. Nitecki), pp. 191–246, University of Chicago Press, Chicago, Illinois.

Dickman, C.R (1991), 'Use of trees by ground-dwelling mammals: implications for management', in *Conservation of Australia's Forest Fauna* (ed. D. Lunney), pp. 125–136, Royal Zoological Society of New South Wales, Mosman, New South Wales.

Dickman, C.R. (1996), *Overview of the impacts of feral cats on Australian native fauna*, Australian Nature Conservation Agency, Canberra.

Dickman, C.R., & Read, D.G. (1992), 'The biology and management of the dasyurids of the arid zone in New South Wales', *Species Management Report No. 11*, New South Wales National Parks and Wildlife, Hurstville. 112 p.

Dickman, C.R., Pressey, R.L., Lim, L., & Parnaby, H.E. (1993), 'Mammals of particular conservation concern in the Western Division of New South Wales', *Biological Conservation*, **65**:219–48.

Dickman, C.R., Parnaby, H.E., Crowther M.S., & King, D.H. (1998), '*Antechinus agilis* (Marsupialia: Dasyuridae), a new species from the

A. stuartii complex in south eastern Australia', *Australian Journal of Zoology*, **46**:1–26.

Dickman, C.R., Lunney, D., & Matthews, A. (2001), 'Ecological attributes and conservation of dasyurid marsupials in New South Wales', *Pacific Conservation Biology*, **7**:124–33.

Flannery, T. (1995), *Mammals of New Guinea*. Reed Books, Sydney.

Fletcher, T.P. (1985), 'Aspects of reproduction in the male eastern quoll *Dasyurus viverrinus* (Shaw) (Marsupialia: Dasyuridae) with notes on polyoestry in the female', *Australian Journal of Zoology*, **33**:101–10.

Fletcher, T.P. (1989a), 'Luteinising hormone in the Kowari *Dasyuroides byrnei* (Marsupialia: Dasyuridae) during oestrous cycles and pregnancy, and effects of gonadectomy in males and females', *Reproduction Fertility and Development*, **1**:55–64.

Fletcher, T.P. (1989b), 'Plasma progesterone and body weight in the pregnant and non-pregnant Kowari *Dasyuroides byrnei* (Marsupialia: Dasyuridae)', *Reproduction Fertility and Development*, **1**:65–74.

Fletcher, T.P and Morris, K. D. (2003), 'Captive breeding and predator control: a successful strategy for conservation in Western Australia', in *Reproductive Sciences and Integrated Conservation* (eds. W.V. Holt, A.R. Pickard, J.C. Rodger, & D.E. Wildt), Cambridge University Press, Cambridge.

Fox, B.J. (1982), 'Fire and mammalian secondary succession in an Australian coastal heath', *Ecology*, **63**:1332 – 41.

Fox, B.J. (1990), 'Changes in the structure of mammal communities over successional time scales', *Oikos*, **59**:321–29.

Friend, G.R. (1979), 'The response of small mammals to clearing and burning of eucalypt forest in south-eastern Australia', *Australian Wildlife Research*, **6**:151–64.

Friend, G.R. (1982), 'Mammal populations in exotic pine plantations and eucalypt forests in Gippsland, Victoria', *Australian Forestry*, **45**:3–18.

Friend, G.R. (1993), 'Impact of fire on small vertebrates in mallee woodlands and heathlands of temperate Australia – a review', *Biological Conservation*, **65**:99–114.

Gibson, D.F., & Cole, J.R. (1992), 'Aspects of the ecology of the Mulgara', *Australian Mammalogy*, **15**:105–12.

Gilpin, M., & Hanski, I. (1991), *Metapopulation dynamics: empirical and theoretical investigations*, Academic Press, London.

Heatwole, H. (1987), 'Major components and distributions of the terrestrial fauna', in *Fauna of Australia, Vol. 1A, General Articles* (eds. G.R. Dyne, & D.W. Walton), pp.101–135, Australian Government Publishing Service, Canberra.

Humphries, R.K., & Seebeck, J.H. (1996), 'Brush-tailed Phascogale, Action Statement', Department of Natural Resources and Environment, Victoria.

Johnson, K., Wilcken, J., & Barker, J. (eds.) (2001), *Australasian Species Management Program Regional Census Plan, 11th ed.* Australasian Regional Asssociation of Zoological Parks and Aquaria, Sydney.

Kinnear, J.E., Onus, M.L., & Bromilow, R.N., (1988), 'Fox control and rock-wallaby population dynamics', *Australian Wildlife Research*, **15**:435–50.

Kinnear, J.E., Onus, M.L., & Sumner, N.R. (1998), 'Fox control and rock-wallaby population dynamics – II. An update', *Wildlife Research*, **25**:81–8.

Kitchener, D.J., Chapman, A.J., Muir, B.H., & Palmer, M. (1980), 'The conservation value for mammals of reserves in the Western Australian wheatbelt', *Biological Conservation*, **18**:179–207.

Laidlaw, W.S. (1997), 'The effects of *Phytophthora cinnamomi* on the flora and fauna of the Eastern Otways', PhD thesis, Deakin University, Geelong.

Laidlaw, W.S., & Wilson, B.A. (1989), 'Distribution and habitat preferences of small mammals in the eastern section of the Angahook-Lorne State Park', *Victorian Naturalist*, **106**:224–36.

Laurance, W.F. (1991), 'Ecological correlates of extinction proneness in Australian tropical rainforest mammals', *Conservation Biology*, **5**:79–89.

Laurance, W.F. (1993), 'The pre-European and present distributions of *Antechinus godmani* (Marsupialia: Dasyuridae), a restricted rainforest endemic', *Australian Mammalogy*, **16**:23–7.

Laurance, W.F. (1994), 'Rainforest fragmentation and the structure of small mammal communities in tropical Queensland', *Biological Conservation*, **69**:23–32.

Laurance, W.F. (1997), 'Responses of mammals to rainforest fragmentation in tropical Queensland: a review and synthesis', *Wildlife Research*, **24**:603–12.

Legendre, P., & Legendre, L. (1998), *Numerical Ecology*, Elsevier, Amsterdam.

Leung, L. (1999), 'Ecology of Australian tropical rainforest mammals. I. The Cape York Antechinus, *Antechinus leo* (Dasyuridae: Marsupialia)', *Wildlife Research*, **26**:287–306.

Lim, L. (1992), 'Recovery plan for the kowari *Dasyuroides byrnei* Spencer, 1896 (Marsupialia: Dasyuridae)', Australian National Parks and Wildlife Service, Canberra.

Lindenmayer, D.B., Cunningham, R.B., Donnelly, C.F, Triggs, B.J., & Belvedere, M. (1994), 'Factors influencing the occurrence of mammals in retained linear strips (wildlife corridors) and contiguous stands in montane ash forests in the Central Highlands of Victoria, south-east Australia', *Forest Ecology and Management*, **67**:113–33.

Lindenmayer, D.B., & Lacy, R.C. (1995), 'Metapopulation viablility of Leadbeater's Possum, *Gymnobelideus leadbeateri* in fragmented old-growth forests', *Ecological Applications*, **5**: 164 – 82.

Lunney, D., & Ashby, E. (1987), 'Population changes in *Sminthopsis leucopus* and other small mammal species in forest regenerating from logging and fire near Bega, New South Wales', *Australian Wildlife Research*, **14**:275–84.

Lunney, D., Cullis, B., & Eby, P. (1987), 'Effects of logging and fire on small mammals in Mumbulla State Forest near Bega, New South Wales', *Australian Wildlife Research*, **14**:163–81.

Lunney, D., & Leary, T. (1988), 'The impact on native mammals of land-use changes and exotic species in the Bega district, New South Wales, since settlement', *Australian Journal of Ecology*, **13**:67–92.

Lunney, D., Curtin, A.L., Fisher, D., Ayers, D., & Dickman, C.R. (1997), 'Ecological attributes of the threatened fauna of New South Wales', *Pacific Conservation Biology*, **3**:13–26.

Masters, P. (1993), 'The effects of fire-driven succession and rainfall on small mammals in spinifex grassland at Uluru National Park, Northern Territory', *Wildlife Research*, **20**:803–13.

Matthews, A., Dickman C.R., & Major R.E. (1999), 'The influence of fragment size and edge on nest predation in urban bushland', *Ecography*, **22**:349–56.

Maxwell, S., Burbidge, A.A., & Morris, K. (eds) (1996), 'The 1996 Action Plan for Australian Marsupials and Monotremes', Wildlife Australia, Canberra.

McKinney, M.L. (1997), 'Extinction vulnerability and selectivity: combining ecological and paleontological views', *Annual Review of Ecology and Systematics,* **28**:495–516.

Mifsud, G. (2000), 'The ecology of the Julia Creek Dunnart (*Sminthopsis douglasi*)', MSc thesis, LaTrobe University, Melbourne.

Morris, K.D., (2000), 'Fauna translocations in Western Australia 1971–1999: a review', in *Biodiversity and the Re–introduction of Native Fauna at Uluru–KataTjuta National Park* (eds. J.S. Gillen, R. Hamilton, W.A., Low & C. Creagh), Bureau of Rural Sciences, Canberra.

Morton, S.R. (1990), 'The impact of European settlement on the vertebrate animals of arid Australia: a conceptual model', *Proceedings of the Ecological Society of Australia,* **16**:201–13.

Muir, B.G. (1995), 'The Dibbler *Parantechinus apicalis*: Dasyuridae found in the Fitzgerald River National Park, Western Australia', *The Western Australian Naturalist,* **16**:48–51.

Murray, B.R., & Dickman, C.R. (2000), 'Relationships between body size and geographical range size among Australian mammals: has human impact distorted macroecological patterns?', *Ecography,* **23**:92–100.

Newell, G.R. (1994), 'The effect of *Phytophthora cinnamomi* on the habitat utilization of *Antechinus stuartii* in a Victorian woodland', PhD thesis, Deakin University, Geelong.

Newell, G.R., & Wilson, B.A. (1993), 'Small mammals and cinnamon fungus (*Phytophthora cinnamomi*) in the Brisbane Ranges, Victoria', *Wildlife Research,* **20**:251–60.

Newsome, A., McIlroy, J., & Catling, P. (1975), 'The effects of an extensive wildfire on populations of twenty ground vertebrates in south-east Australia', *Proceedings of the Ecological Society of Australia,* **9**:107–23.

Paltridge, R., Gibson, D., & Edwards, G. (1997), 'Diet of the feral cat *Felis catus* in central Australia', *Wildlife Research,* **24**:67–76.

Paton, D.C., (1993), 'Impacts of domestic and feral cats on wildlife', in *Proceedings of Cat Management Workshop* (ed. G. Siepen & C. Owens), pp. 9–15, Queensland Department of Environment and Heritage, Brisbane.

Pearson, D.J., & Robinson, A.C. (1990), 'New records of the Sandhill Dunnart *Sminthopsis psammophila* (Marsupialia: Dasyuridae) in South and Western Australia', *Australian Mammalogy,* **13**:57–9.

Recher, H.F., & Lim, L. (1990), 'A review of current ideas of the extinction, conservation and management of Australia's terrestrial vertebrate fauna', *Proceedings of the Ecological Society of Australia,* **16**:379–94.

Recher, H.F., Lunney, D., & Posamentier, H. (1974), 'Effects of wildfire on small mammals at Nadgee Nature Reserve, New South Wales', in *Proceedings of the Third Fire Ecology Symposium*, pp. 30–6, Monash University, Melbourne.

Risbey, D., Calver M.C., & Short, J.C. (1999), 'The impact of cats and foxes on small vertebrate fauna of Heirisson Prong, Western Australia. I Exploring the potential impact using diet analysis', *Wildlife Research,* **26**:621–30.

Risbey, D., Calver, M.C., Short, J.C., Bradley J.S., & Wright I. W. (2000), 'The impact of cats and foxes on small vertebrate fauna of Heirisson Prong, Western Australia. II Field experiments', *Wildlife Research,* **27**:223–36.

Seebeck, J.H. (1978), 'Diet of the fox *Vulpes vulpes* in a western Victorian forest', *Australian Journal of Ecology,* **3**:105–8.

Selwood, L. (1980), 'A timetable of development of the dasyurid marsupial *Antechinus stuartii* (Macleay)', *Australian Journal of Zoology,* **28**:649–68.

Selwood, L. (1981), 'Delayed embryonic development in the dasyurid marsupial, *Antechinus stuartii*', *Journal of Reproduction and Fertility (suppl.),* **29**:79–82.

Selwood, L. (1982a), 'Brown antechinus *Antechinus stuartii*: management of breeding colonies to obtain embryonic material and pouch young', in *The Management of Australian Mammals in Captivity*, (ed. D.D. Evans), pp. 31–7, Zoological Board of Victoria, Melbourne.

Selwood, L. (1982b), 'A review of maturation and fertilization in marsupials with special reference to the dasyurid: *Antechinus stuartii*', in *Carnivorous Marsupials*, (ed. M. Archer), pp.65–76, Royal Zoological Society of New South Wales, Mosman, New South Wales.

Selwood, L. (1983), 'Factors influencing pre-natal fertility in the brown marsupial mouse *Antechinus stuartii*', *Journal of Reproduction and Fertility, Supplement* **68**:317–24.

Soulé, M.E. (ed.) (1987), *Viable populations for conservation,* Cambridge University Press, Cambridge.

Start, A. (1998), 'Interim Recovery Plan No. 18. The Dibbler *Parantechinus apicalis* 1998–2000', Department of Conservation and Land Management, Western Australia.

Strahan, R. (ed.) (1995), *The Mammals of Australia*, Reed Books, Sydney.

Suckling, G., & McFarlane, M. (1984), 'The effects of fire on forest fauna – A review', in *Fighting Fire with Fire* (ed. E.H.M. Ealey), pp. 107–28, Graduate School Environmental Science, Monash University, Melbourne.

Taggart, D.A., Selwood, L., & Temple-Smith, P.D. (1997), 'Sperm production, storage and the synchronization of male and female reproductive cycles in the iteroparous, Stripe–faced Dunnart (*Sminthopsis macrourao*: Marsupialia) – relationship to reproductive strategies within the Dasyuridae', *Journal of Zoology,* **243**:725–36.

Taggart, D.A., Breed, W.G., Temple-Smith, P.D., Purvis, A., & Shimmin, G. (1998), 'Sperm competition and mating strategies in marsupials and monotremes', in *Sperm Competition and Sexual Selection* (eds. T.R. Birkhead, & A.P. Moller), pp. 623–56, Academic Press, London.

Tyndale-Biscoe, H., & Renfree, M. (1987), *Reproductive physiology of marsupials,* Cambridge University Press, Cambridge.

Van Dyck, S., & Crowther, M.S. (2000), 'Reassessment of northern representatives of the *Antechinus stuartii* complex (Marsupialia: Dasyuridae): *A. subtropicus* sp. nov. and *A. adustus* new status', *Memoirs of the Queensland Museum,* **45**:611–35.

Wallis, R.L., & Brunner, H., (1986), 'Changes in mammalian prey of foxes, *Vulpes vulpes* (Carnivora: Canidae) over 12 years in a forest park near Melbourne, Victoria', *Australian Mammalogy,* **10**:43–4.

Williams, S.E. (1997), 'Patterns of mammalian species richness in the Australian tropical rainforests: Are extinctions during historical contractions of the rainforest the primary determinants of current regional patterns in biodiversity?', *Wildlife Research,* **24**:513–30.

Wilson, B.A. (1986), 'Reproduction in female *Antechinus minimus maritimus*', *Australian Journal of Zoology,* **34**:189–97.

Wilson, B.A. (1996), 'Fire effects on vertebrate fauna and implications for fuel reduction burning and management', in *Fire and Biodiversity: the effects and effectiveness of fire management,* pp. 131–47, Biodiversity Series, paper No. 8, Biodiversity Unit, Dept. Environment, Sport & Territory, Canberra.

Wilson, B.A., & Bourne, A.R. (1984), 'Reproduction in the male dasyurid *Antechinus minimus maritimus* (Marsupialia: Dasyuridae)', *Australian Journal of Zoology,* **32**:311–18.

Wilson, B.A., & Friend, G.R. (1999), 'Responses of Australian mammals to disturbance: a review', *Australian Mammalogy,* **21**:87–105.

Wilson, B.A., Bourne, A.R. & Jessop, R.E. (1986), 'The ecology of small mammals in coastal heathland near Anglesea, Victoria', *Australian Wildlife Research,* **13**:397–406.

Wilson, B.A., Robertson, D., Moloney, D.J., Newell, G.R., & Laidlaw, W.S. (1990), 'Factors affecting small mammal distribution and abundance in the eastern Otway Ranges, Victoria', *Proceedings of the Ecological Society of Australia,* **16**:379–96.

Wilson, B.A., Newell, G., Laidlaw, W.S., & Friend, G. (1994), 'Impact of plant diseases on faunal communities', *Journal of the Royal Society of Western Australia,* **77**:139–43.

Wilson, B.A., Aberton, J.G., & Reichl, T. (2001), 'Effects of fragmented habitat and fire on the distribution and ecology of the Swamp Antechinus *(Antechinus minimus maritimus)* in the Eastern Otways, Victoria', *Wildife Research,* **28**:527–36.

Woolley, P. (1966), 'Reproduction in *Antechinus* spp., and other dasyurid marsupials', *Symposia of the Zoological Society of London,* **15**:281–94.

Woolley, P. (1971a), 'Maintenance and breeding colonies of *Dasyuroides byrnei* and *Dasyuroides cristicauda*', *International Zoo Yearbook,* **11**:351–4.

Woolley, P. (1971b), 'Observations on the reproductive biology of the Dibbler, *Antechinus apicalis* (Marsupialia: Dasyuridae)', *Journal of the Royal Society of Western Australia,* **54**:99–102.

Woolley, P. (1984), 'Reproduction in *Antechinomys laniger* (Marsupialia: Dasyuridae): field and laboratory investigations', *Australian Wildlife Research,* **11**:481–9.

Woolley, P.A. (1991), 'Reproductive patterns and captive breeding of the Boullanger Island Dibbler *Parantechinus apicalis* (Marsupialia: Dasyuridae)', *Wildlife Research,* **18**:157–63.

Woolley, P.A. (1995), 'Southern Dibbler *Parantechinus apicalis* (Marsupialia:Dasyuridae)', in *Mammals of Australia* (ed. R. Strahan), pp. 72–3, Reed Books, Sydney.

PART V

CHAPTER 29

CARNIVORE CONCERNS: PROBLEMS, ISSUES AND SOLUTIONS FOR CONSERVING AUSTRALASIA'S MARSUPIAL CARNIVORES

Menna E. Jones[A], *Meri Oakwood*[B], *Chris A. Belcher*[C], *Keith Morris*[D], *Andrew J. Murray*[E], *Patricia A. Woolley*[F], *Karen B. Firestone*[G], *Brent Johnson*[D] *and Scott Burnett*[H]

[A] School of Botany and Zoology, Australian National University, Canberra, ACT 0200, Australia
[B] School of Ecosystem Management, University of New England, Armidale, NSW 2351, Australia
[C] ECOsystems Environmental Consultants, R.M.B. 7285, Timboon, VIC 3268, Australia
[D] Department of Conservation and Land Management, CALMScience Division, Woodvale Research Centre, PO Box 51, Wanneroo, WA 6946, Australia
[E] Victorian Department of Natural Resources and Environment, 171-173 Nicholson St, PO Box 260, Orbost, VIC 3888, Australia
[F] Department of Zoology, La Trobe University, Bundoora, VIC 3086 Australia
[G] Evolutionary Biology Unit, The Australian Museum, 6 College Street, Sydney, NSW 2010, Australia
[H] PO Box 324, Herberton, QLD 4872, Australia

Anthropogenic declines have been reported in all of Australasia's eight larger marsupial carnivores (genera *Dasyurus*, *Sarcophilus*, and *Thylacinus*). One species is now extinct (*T. cynocephalis*), one subspecies is Endangered (*D. maculatus gracilis*), one species and one subspecies are classified as Vulnerable to extinction (*D. geoffroii* and *D. m. maculatus*, respectively), two species are Lower Risk – Near Threatened (*D. hallucatus*, *D. viverrinus*), two are of unknown conservation status (the New Guinea quolls, *D. albopunctatus* and *D. spartacus*), and only one species is classified as Lower Risk – Least Concern (*S. laniarius*). A successful recovery program has been executed for one species (*D. geoffroii* – formerly Endangered), which will shortly be removed from threatened fauna lists. While the causes of decline are multiple and complex, the overriding factor is probably habitat loss, degradation and fragmentation, and the associated factors: loss of protective cover from predators, reduced food availability, and increased contact with humans. Introduced predators, cats (*Felis catus*), foxes (*Vulpes vulpes*) and dingoes (*Canis lupus dingo*), play a not well understood role. Other important mortality factors include persecution, non-target poisoning and road mortality. In this chapter, we conduct a critical review and summary of 1) the causes and correlates of decline, 2) the issues related to trophic level and life history that predispose marsupial carnivores to anthropogenic decline, 3) a range of solutions, some of which are being implemented, and 4) future directions for the large amount of work still to be done.

SCOPE AND CONTEXT

The conservation of Australasia's marsupial carnivores must be considered in the context of an extinction scenario unequalled in the mammalian world (50% of recent extinctions of mammals in the world are from the Australian continent, Short and Smith 1994) and matched in vertebrates only by the extinction of island bird and frog faunas in New Zealand (Towns et al. 1997). This is at a time when world extinction rates are 100–1000 times faster than background rates (Balmford 1996). The causes of mammalian extinction in Australia are several and confounded. They include habitat destruction, fragmentation, and degradation resulting from land clearance, altered fire regimes and grazing by introduced livestock and rabbits (*Oryctolagus cuniculus*). It is speculated that these processes may increase vulnerability to predation from introduced foxes (*Vulpes vulpes*) and cats (*Felis catus*) (Morton 1990; Kerle et al. 1992;

Short and Smith 1994). Extinctions and declines have been greatest in arid areas and in medium-sized terrestrial species. The extent of human impact on the Australian fauna may result from a combination of ecosystem fragility, due to soil nutrient deficiency and unpredictable climate (see Westoby 1988), and perhaps from Australia's unique faunal composition. In contrast to Australia, placental predators introduced by European colonists in North America (from Europe or other parts of North America) interacted with native wildlife that had coevolved with a suite of similar canids and felids. Impacts occurred mainly on islands, where faunas had evolved with few predators. The introduction of predators such as foxes to Australia, that are ecologically different from native predators, may have implications for their predatory impact.

By virtue of their high trophic level and obligate endothermy, which results in low natural densities and large space requirements, many conservation issues are common to both placental and marsupial carnivores. Mammalian carnivores, marsupial or placental, are charismatic animals. These features mean that carnivores are potentially useful flagship and umbrella species for the conservation of ecosystems (Simberloff 1998). The ensuing discussion will attempt to account for patterns and causes in the decline and extinction of Australasian marsupial carnivores both in terms of their vulnerability as high trophic level predators and the continent-wide extinction crisis in Australia. We discuss only the larger marsupial carnivore species, for which vertebrate prey comprise a major part of the diet.

PROBLEMS

Extent and significance of decline

All species of Australasian marsupial carnivores for which there are adequate data have experienced human-induced declines in range and abundance (Table 1). Perhaps the most significant loss was the thylacine, which was the last member of the Family Thylacinidae and the only really large marsupial carnivore to persist into historic times (see Wroe, this volume). All Australian species have declined over their continental range. Thylacines and devils disappeared before European settlement (Jones 1995b; Rounsevell and Mooney 1995) and the last known mainland population of eastern quolls became extinct in the 1960s (Nathan 1966; Maxwell et al. 1996). The chuditch contracted from a former range of 70% of the continent, including most of the arid zone, to currently 2% of its former range in south-west Western Australia (Morris 1992; Orell and Morris 1994). Also of great concern is the spotted-tailed quoll, especially the phenotypic subspecies *D. maculatus gracilis*, which is now restricted to six small and fragmented populations in tropical rainforest in northern Queensland (Burnett 1993; Maxwell et al. 1996). Except for the thylacine, declines in Tasmanian populations have been comparatively minor. The conservation status of the New Guinean quolls is poorly known. The bronze quoll was not reported in the scientific literature until 1979 and was not recognised as a separate species from *D. geoffroii* until 1988 (van Dyck 1988). Large-scale logging may be of concern for this species (Woolley, unpublished).

Patterns of decline

The patterns of decline and extinction of marsupial carnivores accord with general patterns for Australian mammals (Short and Smith 1994). Consistent and broad-scale declines have occurred across the drier, more openly vegetated parts of distributional ranges (western and northern quolls, Braithwaite and Griffiths 1994; Orell and Morris 1994) and in species with preferences for open habitats (eastern quoll, Jones and Barmuta 2000). Declines have been slower in forest-dependent species and in mesic and relatively fox-free regions (northern and spotted-tailed quolls and devil, northern Australia and Tasmania, respectively). Patterns of decline correlate temporally and spatially with geographical range expansion of the introduced carnivores, foxes (for quolls) and dingoes (*Canis lupus dingo*; for devils and the thylacine), and the introduced herbivorous rabbit (*Oryctolagus cuniculus*, the smaller quolls, Short and Smith 1994; Smith and Quin 1996), with the expansion in range of the cane toad (*Bufo marinus*, northern quoll, Burnett 1997), and with the widespread use of 1080 poison for rabbit control in the Otway Ranges in Victoria (spotted-tailed quoll, Belcher 1999).

Causes and correlates of decline

The major causes of human-induced species' declines worldwide are, most significantly, habitat loss through excessive use of resources, the impact of exotic organisms, and pollution (Diamond 1989; Coblentz 1990). The first two of these causal factors also appear to be of major importance in marsupial carnivore decline, as is direct human-induced mortality, which is characteristic of mammalian carnivores.

Habitat loss

Because carnivores live at low density, including for example spotted-tailed quolls in northern Queensland tropical forests (Laurance 1990; Laurance 1991), they are among the first species to disappear in a fragmented landscape (Diamond 1984). Habitat loss and the resultant fragmentation of suitable habitat may be the most important conservation issues for all marsupial carnivore species, as has been observed for the smaller, tropical, forest-dependent species of placental carnivores (Schreiber et al. 1989; Nowell et al. 1996). Habitat loss occurs through vegetation clearance for urban expansion, horticulture or extractive industries, vegetation alteration by farming or forestry practices, changes in fire regimes, and degradation from grazing and trampling by introduced herbivores, domestic sheep and cattle, and rabbits (Hobbs and Hopkins 1990; Short and Turner 1994).

Table 1 Body weights, diet, and conservation status of Australasian marsupial carnivores. Conservation status from (Maxwell et al. 1996). Taxonomy: *Sarcophilus laniarius* = *harrisii* (Werdelin 1987).

Species	Mean weight m/f (kg)	Former distribution, timing and extent of decline, and reason for listing	Conservation status	References
Dasyuridae				
Northern quoll, *Dasyurus hallucatus*	0.76 / 0.46	formerly broad band across northern Australia, 10–50% decline in extent of occurrence with six fragmented populations remaining	Lower Risk – near threatened	(Braithwaite and Griffiths 1994; Oakwood 1997)
New Guinea quoll, *D. albopunctatus*	0.63 / 0.50	widespread above 1000 m, decline reported	unknown	(Flannery 1995; Woolley 2002)
Bronze quoll, *D. spartacus*	1.0 / 0.68	very restricted distribution; known to scientists from only a few specimens, anecdotal decline reported	unknown	(van Dyck 1988; Flannery 1995, P. Woolley 2002)
Western quoll / chuditch, *D. geoffroii*	1.3 / 0.9	98% decline from 70% of continent, western two-thirds; now restricted to south-west	Vulnerable	(Morris 1992; Orell and Morris 1994)
Eastern quoll, *D. viverrinus*	1.3 / 0.9	formerly broad band in south-east and Tasmania; 50–90% decline in range, extinct on continent, Tasmanian populations stable	Lower Risk – near threatened	(Vertebrate Advisory Committee 1994; Jones and Rose 1996)
Spotted-tailed quoll, *D. maculatus maculatus*	4.0 / 1.9	formerly broad band in south-east and Tasmania; estimated total population size of less than 10 000 (est. 5500) which is fragmented and declining; Tasmania population is a distinct ESU and deserves subspecific status; extinct on Bass Strait Islands	Vulnerable	(Hope 1972; Belcher 1994; Belcher 1995; Jones and Rose 1996; Firestone et al. 1999; Belcher 2000)
D. m. gracilis	1.6 / 1.2	formerly all of Wet Tropics North Queensland; six small and fragmented populations remaining	Endangered	(Burnett 2001)
Tasmanian devils, *Sarcophilus laniarius*	10.0 / 8.0	formerly much of the continent; last records on mainland 5000 years south-east and 430±160 years south-west; >90% decline pre-European; secure in Tasmania	Lower risk – least concern	(Gill 1971; Archer and Baynes 1972, Jones, unpublished)
Thylacinidae				
thylacine, *Thylacinus cynocephalus*	est. 15.0/30.0	formerly much of the continent, New Guinea and Tasmania; last records on mainland 3090±90 years and 0±90 years (enigmatic); 100% decline; last definitive wild record 1933; last captive specimen 1936.	Extinct	(Brandl 1972; Wright 1972; Archer 1974; Dixon 1989; Guiler and Godard 1998)

The probable key factors in habitat loss for marsupial (and placental) carnivores are as follows:

Removal of cover and predator refuges

Grazing and fires remove vegetative cover and destroy hollow logs and trees that are used for cover or refuge from predators, increasing vulnerability to predation, although fire also generates new den sites (Newsome 1975; Braithwaite and Griffiths 1994; Friend 1994; Orell and Morris 1994; Oakwood 2000). Evidence for the importance of cover and refuges from predators for the smaller quolls includes; the most common proximate cause of adult northern quoll mortality was predation (60%, n = 15 radio-tracked to sites of death plus an additional two quolls were killed by dingoes) and all quolls that were killed by predators were in open areas of savanna where there was little or no ground cover (usually after fire). No northern quolls were killed on the rocky hills where there are abundant refuges under rocks despite the hills being preferred quoll habitat. Female northern quolls whose home ranges included the most rocky habitat had the longest lifespans (Oakwood 2000). Circumstantial evidence includes the broadscale decline of northern quolls in savanna habitats where cattle grazing and frequent fires remove understorey vegetation (Braithwaite and Griffiths 1994; Oakwood 2000), and the disappearance of eastern quolls from lowland areas in Tasmania where dense ground cover and forests (where dens are usually sited), have been extensively removed (Godsell 1983; Jones and Rose 1996).

Reduced availability of food resources

Significant habitat alteration may result in reduced food resources. Of major concern with all forms of intensive forestry is the long-term availability of hollow-bearing trees, which are essential for the persistence of arboreal prey populations (e.g. Lindenmayer et al. 1993), important in the diet of spotted-tailed quolls (Belcher 1995; Jones and Barmuta 2000). Spotted-tailed quoll populations have declined where structurally complex vegetation has been extensively removed (Jones and Rose 1996), and they use selectively logged forest only where >40% canopy cover was retained and after ground, shrub and understorey had recovered (Belcher 2000). Spotted-tailed quolls do live in reasonable densities in very steep terrain with naturally little vegetative cover, although these areas are limited in geographic extent (M. Oakwood and M. Jones, unpublished). Time since fire influences population abundances of vertebrate and invertebrate prey and of predators (e.g. eastern quolls in Tasmania, Driessen et al. 1991). Changes in fire regimes can result in homogenisation of habitat, as happened in central Australia after the cessation of Aboriginal burning practices. This alteration in burning practices may have played a role, through collapse of the prey base, in the arid-zone extinction of the chuditch (Johnson and Roff 1982), although Short and Turner (1994) failed to find evidence for influence of scale of mosaic on numbers, condition or reproduction in three species of medium-sized marsupial in spinifex grasslands on offshore islands.

Increased contact with exotic predators and human populations

Fragmentation and disturbance of forest habitat may result in increased use by exotic predators that may not prefer continuous forest (Catling and Burt 1995) and increased contact with and persecution from human populations (discussed later).

Introduced species

Placental carnivores

Three species of placental carnivores introduced to Australia have established extensive, high-density feral or free-ranging populations and have had a major impact on the native mammal fauna (Short and Smith 1994; Dickman 1996a; Smith and Quin 1996). The dingo was introduced 3500–4000 years ago by Asian seafarers and quickly spread across the continent, possibly assisted by Aboriginal peoples (Corbett 1995). Foxes were introduced successfully for hunting about 1871, reaching most of their current range in the southern two-thirds of Australia by 1930 (Rolls 1969). Cats may have arrived on the west coast as early as the seventeenth century on Dutch shipwrecks and were certainly introduced as house pets in eastern Australia in the late eighteenth century (Rolls 1969).

Predation and competitive killings are the most likely deleterious impacts of introduced carnivores on marsupial carnivores, although exploitation competition, and disease may play roles (Dickman 1996a). Direct killings, which are characteristic of carnivores and represent an extreme form of aggressive interference competition (Macdonald and Thom 2001; Van Valkenburgh 2000), are likely to have more rapid depressive effects on populations than exploitation competition for food. Primary dependence on abundant rabbit populations in Australia may enable foxes and cats to maintain artificially high population densities and even drive less abundant native prey species such as quolls to extinction, an unusual phenomenon in predator/prey systems known as 'hyperpredation' (Pech et al. 1995).

No confirmed instances of disease transmission between introduced placental and native marsupial carnivores have been recorded (see Johnson and Roff 1982; Oakwood and Pritchard 1999), although endemic infection with *Toxoplasma* has been identified in wild-caught *Dasyuroides byrnei* (Dasyuridae, Attwood et al. 1975), and cats are the definitive hosts for this parasite, which is known to affect the omnivorous eastern barred bandicoot, *Perameles gunnii* (Obendorf and Munday 1990). A sudden decline in thylacine, devil and quoll populations in Tasmania and in quoll populations on the mainland about 1910 was attributed to disease (Wood-Jones 1923; Troughton 1941; Fleay 1945; Dixon 1989; Morton et al. 1989; Guiler and Godard 1998), possibly toxoplasmosis (Shepherd and Mahood 1978), but there is no evidence in extant populations of a likely agent. At least two species of parasite introduced to Australasia with introduced predators (dogs, dingoes, foxes and cats, Coman 1972d; 1972b; 1973; Pavlov and Howell 1977; Coman et al. 1981) have been confirmed in marsupial carnivores. The cestode, *Spirometra erinacei,* is reported to have deleterious effects on northern and spotted-tailed quolls and the spiruroid nematode, *Cyathospirura seurati*, also occurs in devils, and in northern, spotted-tailed and New Guinea quolls (Coman 1972a; 1972b; 1973; Pavlov and Howell 1977; Coman et al. 1981; Beveridge and Spratt, this volume). The effects of these parasites on population levels are unknown.

Impacts of cats

Cats are not implicated as playing a major role in the decline or extinction of any quoll species, although it is thought that their effects may be significant in conjunction with other factors. All species have coexisted over their entire ranges with feral cats for perhaps hundreds of years (Serena et al. 1991, Morris, unpublished), and no substantial declines occurred before foxes and rabbits were introduced in the second half of the nineteenth century (Dickman 1996b).

The smaller species of quolls are considered to be at risk from cat predation or intraguild killings on account of their small body size and use of similar habitats (Dickman 1996b). Supporting evidence for this contention includes the identification of feral cats as the proximate cause in 13% of northern quoll deaths in Kakadu National Park (Oakwood 2000; Oakwood and Spratt 2000), feral cat kills of translocated chuditch in south-western Western Australia (Serena et al. 1991, Morris, unpublished), and behavioural anti-predator responses to cat vocalisations in the smaller juvenile but not in adult Tasmanian eastern quolls, reinforcing the importance of body size for vulnerability (Smith et al. in review). Cats also could be a problem in New Guinea, with a report of a cat killing a bronze quoll (Woolley, unpublished) and cat sightings on petroleum roads some distance from human habitation (Leary and Seri 1997). The larger, cat-sized spotted-tailed quolls and cats are known to kill each other (Troughton 1943, Museum of Victoria specimen). Resource competition for food and dens between cats and quolls, especially with the arboreal and similar-sized spotted-tailed quoll, is likely (Serena et al. 1991; Belcher 1994; Dickman 1996b; Jones and Barmuta 1998), but critical evidence is missing.

Impacts of foxes

Definitive evidence for the impact of foxes on quolls is provided by the dramatic increase in chuditch populations following fox baiting (Morris 1992, Morris et al. this volume). Circumstantial evidence includes the coincidental patterns of population declines of some quoll species with the arrival and establishment of fox and rabbit populations; in central Australia (chuditch, Johnson and Roff 1982), in South Australia (eastern quoll and chuditch, Wood-Jones 1923; Dickman 1996a), in north-eastern New South Wales (eastern quoll, 1930s – Ned Hayes 1985 pers. comm. to M. Jones), and in southern New South Wales (1890s-1930s – letter by A. Buckland 1894; Helen Bass 1978 pers. comm. to C. Belcher). Also, eastern quolls are still relatively common in Tasmania (Jones and Rose 1996), which was fox-free until 1999, and northern quoll decline in the monsoonal north of Australia, where foxes and rabbits have not become established, is more recent. The recent presence of foxes in Tasmania (one arrived on a container ship on 31 May 1999; more have been discovered in other parts of the island since early June 2001) may bode disaster for the remaining eastern quoll population. Foxes occur in moderate to high numbers in forests in southern New South Wales (Catling and Burt 1995), so forest-dependence may not offer spotted-tailed quolls protection.

Impacts of dingoes and dogs

Circumstantial evidence strongly implicates dingoes as a contributory factor in the final demise of thylacines and devils on mainland Australia. The timing of their decline roughly correlates with the establishment of dingoes (earliest fossil record 3450 ± 95 b.p., Corbett 1995), and both species persisted in island Tasmania where the formation of Bass Strait at the end of the last glaciation 12,000 years ago precluded later colonisation by dingoes. Aboriginal cave paintings in Arnhem Land in northern Australia, depicting thylacines and dingoes in Mimi and post-Mimi art styles, respectively, provide further indication that thylacines and dingoes did not coexist for a long period (reviewed in Guiler and Godard 1998). Cooperative hunting, flexible diet, and greater running speeds (see Jones, this volume) suggest that canids such as dingoes could have been dangerous killers of their solitary competitors, the devil and thylacine, as domestic dogs are to devils (see below), and multiple dogs were to thylacines (Paddle, 2000 #1466).

In conjunction with recent reductions in vegetative cover, dingoes may be significantly impacting declining populations of the northern quoll. Intraguild killings are responsible for at least 13%, but probably 26%, of northern quolls in lowland savanna (Oakwood 2000; Oakwood and Spratt 2000). That carcasses are not eaten suggests that analysis of dingo scats would underestimate impact. This raises questions about the potential for dingo predation impact on spotted-tailed quolls in eastern Australia, if they are allowed to re-establish, although the two species presently appear to coexist in undisturbed, continuous forest (Catling and Burt 1995). Quolls may derive some benefit from scavenging on dingo kills (C. Belcher, unpublished). Farm dogs can be deadly to quolls and devils, especially when individual dogs learn the skills for despatching them, and can impact significantly on a population if dogs are roaming at night or if devils or quolls are attracted to farm livestock and food stores (M. Jones, M. Oakwood, unpublished).

Impacts of cane toads

Cane toads were introduced to north-eastern Queensland in 1935 (Covacevich and Archer 1975) and are still spreading north and south on mainland Australia. Venom from the parotoid glands can be fatal to quolls and other frog-eating predators which ingest or mouth them (Covacevich and Archer 1975; Burnett 1997). Cane toads are implicated as the major factor responsible for widespread population crashes of northern quolls on the Cape York peninsula as cane toad populations advanced north between the mid-1980s to the mid-1990s (Burnett 1997). Cane toads arrived in southern Kakadu National Park in 2001 (G. Ryan pers. comm. to M. Oakwood), and dramatic impacts on the frog-eating northern quoll populations (Tyler and Davies 1986; Oakwood and Eager 1997) are expected.

Direct human-induced mortality

Persecution and hunting

Persecution is a major issue for most mammalian carnivores (e.g. Ginsberg et al. 1990; Nowell et al. 1996; Mills et al. 1998), and is a consequence of predation on humans (parts of Africa), livestock (from poultry to cattle), game species and raiding of food stores in rural dwellings. Perceptions can be dire even when predation levels are low (e.g. more sheep kills were attributed to domestic dogs than thylacines, Guiler and Godard 1998). This is especially the case for specialised scavengers, such as devils (and hyaenas, Mills et al. 1998), which occasionally eat human remains, incidents of which contribute to traditional negative attitudes towards devils in Tasmania.

All larger species of marsupial carnivores (even the little known New Guinean bronze quoll, van Dyck 1988) are deadly to improperly housed poultry of all kinds and sizes (eastern quolls kill nesting turkeys) and are sporadically trapped, shot, and poisoned for real or perceived depredations. Poultry farmers, in particular, often consider quolls to be 'Public Enemy No. 1' (Ralph Morris, Agriculture, Western Australia, pers. comm. to K. Morris; old, unpublished files, Fisheries and Wildlife, Western Australia), and large numbers of quolls or devils are sometimes killed (e.g. Franklin 1963; Dixon and Huxley 1985, Ralph Morris pers. comm. to K. Morris, Nick Mooney pers. comm. to M. Jones). Healthy sheep are thought to be safe from devils, but multiple births and weak, stressed, or penned sheep are vulnerable (Mooney 1992; Guiler and Godard 1998), as are dogs on chains. The actual impact of persecution on populations of marsupial carnivores is not known, but one of four radio-collared male spotted-tailed quolls was shot while raiding a poultry shed (Belcher, unpublished), and two devils out of a local population of about 12 were shot over a 12 month period (M. Jones, unpublished).

Poisoning programs

Marsupial carnivores are susceptible to incidental poisoning from baiting campaigns intended for wild canids (feral dogs, dingoes and foxes) and rabbits. Deleterious effects depend on the type of poison used and the method of bait delivery, but potentially result in local population sinks. Sodium fluoroacetate (compound 1080) is the most commonly used poison for wild canid control in Australia (Fleming et al. 2001). Marsupial carnivores have a higher tolerance of this poison than placental carnivores, especially the chuditch in Western Australia, where there are naturally high levels of fluoroacetate in the local vegetation (McIlroy 1981). Cyanide, which is used by farmers and pastoralists to control dingoes in Western Australia, although its use is not currently legal (Busana et al. 1998), and strychnine, which was used extensively for dingo and rabbit control in the past and is still used to poison jaws of leg-hold traps in WA, SA, QLD and NSW (Fleming et al. 2001), are nonspecific and lethal. The use of cyanide is a current problem for chuditch. Two pieces of circumstantial evidence suggest the non-target impact of strychnine poisoning programs on marsupial carnivores. Spotted-tailed quoll decline in northern New South Wales in the mid-1900s coincided with the widespread use of strychnine for dingo control (Ned Hayes 1984, pers. comm. to M. Jones). Second, some eastern quoll and devil populations in Tasmania have increased substantially since the 1960s, when the use of strychnine (for rabbit control, quolls experience secondary poisoning) was banned (Donald Fish 1996, pers. comm. to M. Jones).

Moist meat baits may pose a much greater threat than dried meat baits. Soderquist and Serena (1993) found that chuditch consumed significantly less dried meat bait than fresh meat bait in captive trials. Chuditch would need to consume two of the dried meat baits containing 1080 poison used for fox control in Western Australia to receive a lethal dose, due to their relatively high tolerance to 1080 poison (LD50 of 7.5 mg 1080 per kg body weight, King 1989). Fresh or moist meat baits, delivered via aerial baiting and inappropriate (not adequately buried) mound baiting, are still used in the eastern states and may pose an unacceptable risk to spotted-tailed quolls. Baiting at three sites coincided with dramatic declines in spotted-tailed quoll numbers (Belcher 1994; 1998; 2000; Jim Darrant 1998, pers. comm. to C. Belcher), and one male spotted-tailed quoll that died six days after aerial baiting had 1080 residues in the stomach and muscle tissues (Belcher 2000). In a recent baiting trial using non-toxic baits (containing the systemic dye Rhodamine B), 86% of adult male and female quolls consumed one or more baits (Belcher 2000; Murray et al. 2000); however, caution must be applied in extrapolating these results to toxic baits because at least one dasyurid species (*Sminthopsis crassicaudata*) (Sinclair and Bird 1984) and one population of spotted-tailed quolls (Körtner and Gresser 2002) appears to be able to detect 1080 in baits and either refuses to eat the bait or reduces intake and vomits . Although toxic baits can kill individual spotted-tailed quolls in certain circumstances, direct evidence is still required to ascertain whether baiting affects spotted-tailed quolls at the population level.

Roadkills

Evidence that animal/vehicle collisions are an important mortality factor for devils and quolls includes the local extinction of a population of 20 eastern quolls and the similar rate of decline in an initially larger Tasmanian devil population (n = 45) at Cradle Mountain National Park in Tasmania after road upgrading (Jones 2000), 20% mortality of a partially contained peninsular devil population in a single year from roadkill (M. Jones, unpublished), and 38% of male northern quoll deaths (n = 8) attributed to road mortality (Oakwood 2000). Roadkill has been identified as an important mortality factor also for the chuditch (Morris 1992; Orell and Morris 1994). Marsupial carnivores are vulnerable because they are attracted to the road surface to scavenge on roadkills and forage on insects, and they use the road for travelling (reviewed in Jones 2000). Males, particularly juveniles (except northern quolls for which juveniles are rarely hit), of all species are more likely to be killed than females. This is possibly because males are wider ranging and disperse, and is especially the case during the breeding season when males range widely to monitor females (Green and Scarborough 1990; Oakwood and Pritchard 1999; Jones 2000, Belcher, unpublished).

ISSUES

Natural rarity and predispositions to vulnerability

Trophic level and body size

By virtue of their high trophic level and obligate endothermy, many marsupial and placental carnivores, especially the larger species, are characterised by large home ranges and low population densities. These factors render them vulnerable to extrinsic forces that increase mortality or reduce population growth and survival and are major correlates of vulnerability to extinction in placental carnivores (Purvis et al. 2001).

Life history and dispersal

Some aspects of the life history of marsupial carnivores make them vulnerable to population decline. Despite early age at first breeding (in the first year after weaning), and the production of several young (potentially 4, 6 and 8 in devils, quolls, and most northern quolls, respectively), total lifetime reproductive output is low. Lifespan is short (from 12 months in male northern quolls in savanna populations, to 6 years in devils, Guiler 1983; Oakwood 2000; Oakwood et al. 2001), as is typical of dasyurids (Cockburn 1997), and females may not breed in successive years (e.g. spotted-tailed quolls, Belcher 2000). A reduced lifetime fecundity will affect the rate of population recovery and may result in more rapid decline. Semelparity predisposes savanna populations of the northern quoll to local extinction because if breeding fails in any year, males will not be present in the following mating season (Oakwood 2000; Oakwood et al. 2001), although males of this species can survive and breed in more than one year in captivity (G. Mayles pers. comm. to M. Oakwood).

Dispersal is important in influencing ability to repopulate habitat isolates and maintain genetic variation in a fragmented landscape. Males of all species are wide ranging, with daily movements of 5–7 km recorded in spotted-tailed quolls and devils (Jones, unpublished), and more spectacular movements of 180 km and 30 km recorded in translocated chuditch and in spotted-tailed quolls in the breeding season, respectively (Belcher 2000; Johnson, unpublished). Females of all species are likely to be philopatric (Belcher 2000; Soderquist and Serena 2000) which may limit their ability to recolonise patches in a fragmented environment.

Ecological specialisation

Species that are specialised in their dietary, habitat or climatic requirements are likely to be less tolerant of the matrix surrounding habitat fragments (Terborgh and Robinson 1986; Laurance 1990, 1991) and are more vulnerable to extinction. Although all of the larger marsupial carnivores are reasonably generalist in their diet, and take a variety of prey species, spotted-tailed quolls are the most specialised of this group in all three respects. They are probably obligate flesh-eaters on account of their larger body size (at least *D. m. maculatus*, Belcher 1994; Jones and Barmuta 2000). As a consequence, they may be less flexible in the environmental conditions under which they can live than other quolls or the devil. The smaller species of quolls, and the highly successful introduced fox can switch temporally and spatially between vertebrate and invertebrate prey and even utilise seasonal fruit; devils have morphological and physiological adaptations for scavenging that enable them to fully utilise carcasses (see Blackhall 1980; Godsell 1983; Triggs et al. 1984; Soderquist and Serena 1994; Jones 1995a; Oakwood and Eager 1997; Jones and Barmuta 1998). The arboreally adapted spotted-tailed quolls are more forest-dependent than the other species (Belcher 2000; Jones and Barmuta 2000), although they do use other structurally complex habitats such as heath, woodland and steep, rocky country. While forest-dependence offers them some protection from foxes, it is a major vulnerability factor in broadscale clearance, exemplified in the fragmented range of *D. m. gracilis*, which is currently restricted to tropical rainforest (Burnett 1993). Spotted-tailed quolls also are specialised in their climatic requirements. Tasmanian population distribution correlates closely with areas of highly predictable seasonality of rainfall (Jones and Rose 1996). These are very productive environments for plant growth and, therefore, for fauna but are consequently also favoured for forestry and farming. This dependence on highly productive environments leaves the spotted-tailed quoll vulnerable to environmental perturbations. In Victoria and northern Queensland, where spotted-tailed quoll populations have declined substantially, distribution is confined to high rainfall areas (Mansergh 1984; Burnett 1993).

Competition in carnivore guilds

Inter-specific competition within the guild of sympatric marsupial carnivores in Tasmania has implications for the conservation status of spotted-tailed quolls. Competition for food, both exploitative (diet overlap) and interference (dominance at carcasses and kleptoparasitism), from eastern quolls and devils appear to contribute to low population densities in spotted-tailed quolls, the middle-sized species (Jones 1995a; Jones and Barmuta 1998). This natural rarity leaves this species vulnerable to decline.

Competition, probably operating through aggressive interference, also has implications for interactions between introduced placental and native marsupial carnivores although these are very poorly understood at present. Evidence, from co-occurrence and from density changes following removal of the larger species, suggests that dingoes may displace foxes to the benefit of two native rodent species (reviewed in Smith and Quin 1996), and there are some thoughts that foxes may be able to control cat populations (Algar and Smith 1998). Removal of one of these introduced predators could result in a population increase in the smaller species and a consequent increase in predation impact on marsupials.

Climate change

Global warming may have implications for fragmented populations of marsupial carnivores, such as *D. maculatus gracilis*. If climate zones shift latitudinally or altitudinally, isolated fragments of habitat may become unsuitable. Rising sea levels may also affect isolated island populations such as those on Groote Eylandt, Northern Territory.

SOLUTIONS

Habitat management and reserve status

A high priority in the long-term conservation of marsupial carnivores is to maintain and manage sufficient area of suitable habitat and connectedness between habitat patches. Not only do the habitat requirements for each species need to be identified, but demographic responses to different types of habitat perturbation, dispersal patterns and genetic population substructure need to be quantified. Quality habitat has sufficient food resources on a year-round basis, sufficient suitable denning sites for daily movements and for breeding, and suitable structural aspects of habitat (e.g. vegetative understorey cover or rocky terrain) for refuge from predators and for foraging (e.g. structurally complex forest).

We need to understand the effects of forestry practices, fire regimes (including frequency, intensity and spatial patterns of burning), and land clearance for primary production on habitat structure, food resources and den sites. Management practices should be adjusted in light of the findings for effective conservation. Further destruction of presently high quality habitat for forestry, primary production or urban development must be addressed. In particular, intensive forestry practices such as plantation silviculture, which is currently expanding to cover vast areas, is likely to permanently alienate habitat for spotted-tailed quolls. Ability to persist in extensively cleared and fragmented habitats such as farmland also is an issue, particularly for the forest-dependent spotted-tailed quolls and devils.

Existing reserves are unlikely to be sufficiently large or well connected for the long-term conservation of these wide-ranging species. Effective conservation requires the use of a landscape approach that incorporates a combination of on- and off-reserve measures and takes into account edge effects in reserves. On-reserve measures need to examine whether the current use of aerial baiting for wild dogs is affecting quoll populations, especially in New South Wales. Off-reserve measures must include the land management issues on private and forestry lands discussed above, the provision of habitat corridors, and amelioration of localised mortality factors such as road mortality and persecution. The minimum function of a corridor is to provide sufficient continuity of habitat for interpopulation dispersal. The efficacy of corridors must be tested, to ensure that they are actually being used and that they promote animal movements between habitat patches (Simberloff et al 1992). Interpopulation dispersal can assist in local persistence within metapopulations and in long-term gene flow. The application of molecular genetic techniques to monitor both genetic depletion and interpopulation dispersal, particularly the use of microsatellites and the assignment test for studying individual dispersal, is currently the most useful tool here (Waser and Strobeck 1998; Waser et al. 2001).

Introduced predator and pest control

The other key approach in the conservation management of marsupial carnivores is to gain a deeper understanding of the dynamics between the native and introduced carnivores and to develop broad-scale and coordinated strategies for the control of introduced predators, particularly foxes, and pests such as cane toads (e.g. National Threat Abatement Plans for foxes and cats, Commonwealth of Australia, 1999). We need to know at what fox, dingo and cat densities quolls can coexist, and how exotic predator density and habitat quality interact for quoll population persistence. The most effective long-term predator control strategy may be viral-vectored immunocontraception currently being researched for rabbits and foxes (Creagh 1992; Bradley et al. 1997; Twigg and Williams 1999). In the more immediate term, high priorities are 1) the development of effective baits and toxins for controlling cat populations, which are being investigated in Western Australia, and 2) examination of the 1080 baiting system presently in use in the eastern states.

In south-western Western Australia, fox control over 3.5 million hectares of Department of Conservation and Land Management (CALM) estate, as part of the Western Shield program, has been beneficial for chuditch populations and many other medium-sized mammal species (Morris et al. 1995; Bailey 1996; Morris 1998). Translocations to the arid zone will not be undertaken until cats also can be controlled, as cats numbers increase when foxes are controlled (Christensen and Burrows 1994).

In south-eastern Australia, further research is essential to determine whether baiting directly affects quolls at the population level. Research is currently being undertaken on the effect of actual (toxic) fox baits on spotted-tailed quolls (Körtner and

Gresser 2002). Until this research is expanded comprehensively to cover a wide range of populations, a precautionary approach should be adopted, with baits buried to a depth of 15 cm to increase target specificity, and areas where quolls are known to occur avoided. Fox and wild dog control programs are usually undertaken only annually and may be ineffective in depressing population densities in the longer term (e.g. Fleming et al. 2001). This level of fox control may achieve little more than enhancement of reproductive output and recruitment to vacant territories (Macdonald 1988). Research on the effectiveness of these control programs is required.

Amelioration of human-induced mortality

Continuing public education, especially in rural areas where people are living with quolls and devils, is important because there is a long history of negative perceptions of carnivores. Conservation campaigns (school kits, media exposure, community organisations such as 'Land for Wildlife' and 'The Friends of the Chuditch') need to promote the benefits of carnivores and invest stewardship in local communities as well as effectively address ways to decrease stock losses.

Protection of livestock and food stores (where rural dwellings, caravans and shacks are insecure) from marsupial carnivores is not difficult. Secure, roofed wire mesh enclosures with concrete footings will exclude quolls, but a strong mesh is required for devils. Good stock management, including keeping lambing paddocks clean, will minimise problems between sheep and devils, which are quite easy to fence out (Mooney 1992). Control of domestic dogs and cats at night is essential where there are quoll populations.

Traffic speed seems universally to be the most important factor influencing roadkill rates, although traffic volume also has an effect, and there is a range of measures available to ameliorate road mortality of vulnerable populations (reviewed in Jones 2000).

Roles of captive breeding, translocations and genetics

Translocations and reintroductions are tools in recovery programs that are useful for expanding the range and total population size of a species and reducing the risk of extinction through the stochastic loss of small populations. Translocations, which use wild-born, wild-living animals have advantages over reintroductions of captive-bred animals. However, most quolls occur at naturally low densities. Translocations where 40–50 founder animals are required could not be sourced from wild populations without a detrimental impact on that population. Captive breeding, with the use of stud books, should be used to underpin any quoll translocation program. Given that introduced predators are a major force of decline, predator control or translocation to predator-free islands is essential for success (Burbidge 1999; Johnson 1999). Reintroduction of more than 200 chuditch, bred at Perth Zoo, since 1991, to four sites in south-western Western Australia where foxes are being controlled, together with an increase in other natural populations as a result of fox control, has resulted in downlisting of the chuditch from Endangered to Vulnerable (Morris et al. this volume). This species will be removed shortly from Western Australia's threatened fauna list but will remain "Conservation Dependant" (on continued fox control).

Molecular genetics is also a useful research tool in species' conservation. Molecular genetics can be used to determine appropriate units for conservation, examine genetic diversity in populations that may indicate past bottlenecks, current inbreeding or the risk of it, and paternity, which can be used in captive breeding programs or to infer aspects of social organisation. The application of genetics to the conservation of marsupial carnivores is discussed in Firestone (this volume).

FUTURE DIRECTIONS

Priorities for conservation of Australasia's marsupial carnivores are to complete the recovery process for the chuditch and then set in place long-term management processes, and to initiate a recovery plan for the spotted-tailed quoll (all subspecies and evolutionarily significant units, see Firestone, this volume), with the aim of downlisting those species, to Lower Risk, the Conservation Dependent sub-category if necessary. Further information is required on threatening processes for the northern quoll (in particular, the effect of cane toad colonisation) and on the conservation status of the little known New Guinean species. A nationally coordinated approach is required, with deleterious processes, habitat loss and predator control being addressed at landscape and regional scales. Broadscale, long-term monitoring of range and population densities of species at risk is required.

REFERENCES

Algar, D., & Smith, R. (1998), 'Aproaching Eden', *Landscope,* **13**:28–34.

Archer, M. (1974), 'New information about the Quaternary distribution of the thylacine (Marsupialia, Thylacinidae) in Australia', *Journal of the Royal Society of Western Australia,* **57**:43–50.

Archer, M., & Baynes, A. (1972), 'Prehistoric mammal faunas from two small caves in the extreme south-west of Western Australia', *Journal Royal Society Western Australia,* **55**:80–89.

Attwood, H.D., Woolley, P.A., & Rickard, M.D. (1975), 'Toxoplasmosis in dasyurid marsupials', *Journal of Wildlife Diseases,* **11**:543–51.

Bailey, C. (1996), 'Western Shield – bringing wildlife back from the brink of extinction', *Landscope,* **1996**:41–8.

Balmford, A. (1996), 'Extinction filters and current resilience: the significance of past selection pressures for conservation biology', *Trends in Ecology and Evolution,* **11**:193–6.

Belcher, C.A. (1994), 'Studies on the diet of the tiger quoll *Dasyurus maculatus*', MSc thesis, LaTrobe University.

Belcher, C.A. (1995), 'Diet of the Tiger Quoll *Dasyurus maculatus* in East Gippsland, Victoria', *Wildlife Research,* **22**:341–57.

Belcher, C.A. (1998), 'Susceptibility of the tiger quoll, *Dasyurus maculatus*, and the eastern quoll, *D. viverrinus*, to 1080-poisoned baits in control programmes for vertebrate pests in eastern Australia', *Wildlife Research,* **25**:33–40.

Belcher, C.A. (1999), 'The range status and distribution of the spot-tailed quoll in the Otway Ranges, West Regional Forest Agreement Area', Consultants report to DNRE and Environment Australia, Ecosystems Environmental Consultants, RMB 4269, Timboon Vic. 3268. Available on Environment Australia website.

Belcher, C.A. (2000), 'Ecology of the Tiger Quoll, *Dasyurus maculatus*', PhD thesis, Deakin University.

Blackhall, S. (1980), 'Diet of the eastern native-cat, *Dasyurus viverrinus* (Shaw), in southern Tasmania', *Australian Wildlife Research,* **7**:191–7.

Bradley, M.P., Hinds, L.A., & Bird, P.H. (1997), 'A bait-delivered immunocontraceptive vaccine for the European red fox (*Vulpes vulpes*) by the year 2002', *Reproduction, Fertility and Development,* **9**:111–16.

Braithwaite, R.W., & Griffiths, A.D. (1994), 'Demographic variation and range contraction in the Northern Quoll, *Dasyurus hallucatus* (Marsupialia: Dasyuridae)', *Wildlife Research,* **21**:203–17.

Brandl, E. J. (1972), 'Thylacine designs in Arnhem Land rock paintings', *Archaeology and Physical Anthropology,* **7**:24–30.

Burbidge, A.A. (1999), 'Conservation values and management of Australian islands for non-volant mammal conservation', *Australian Mammalogy,* **21**:67–74.

Burnett, S. (1993), 'The conservation status of the tiger quoll, *Dasyurus maculatus gracilis* in north Queensland', Queensland Department of Environment and Heritage Report.

Burnett, S. (1997), 'Colonising cane toads cause population declines in native predators: reliable anecdotal information and management implications', *Pacific Conservation Biology,* **3**:65–72.

Burnett, S. (2001), 'The ecology and conservation status of the northern spot-tailed quoll, *Dasyurus maculatus*, with reference to the future of Australia's marsupial carnivores', PhD thesis, James Cook University of North Queensland, Townsville.

Busana, F., Gigliotti, F., & Marks, C.A. (1998), 'Modified M-44 cyanide ejector for the baiting of red foxes (*Vulpes vulpes*)', *Wildlife Research,* **25**:209–15.

Catling, P.C., & Burt, R.J. (1995), 'Why are red foxes absent from some eucalypt forests in eastern New South Wales?', *Wildlife Research,* **22**:535–46.

Christensen, P., & Burrows, N. (1994), 'Project desert dreaming: experimental reintroduction of mammals to the Gibson Desert, Western Australia', in *Reintroduction Biology of Australian and New Zealand Fauna* (ed. M. Serena), pp. 199–207, Surrey Beatty and Sons, Chipping Norton, NSW 2170, Australia.

Coblentz, B.E. (1990), 'Exotic organisms: a dilemma for conservation biology', *Conservation Biology,* **4**:261–5.

Cockburn, A. (1997), 'Living slow and dying young – senescence in marsupials', in *Marsupial Biology: Recent Research, New Perspectives* (eds. N.R. Saunders, & L.A. Hinds), pp. 163–74, University of NSW Press, Sydney.

Coman, B.J. (1972a), 'Helminth parasites of the dingo and feral dog in Victoria with some notes on the diet of the host', *Australian Veterinary Journal,* **48**:456–61.

Coman, B.J. (1972b), 'A survey of the gastro-intestinal parasites of the feral cat in Victoria', *Australian Veterinary Journal*, **48**:133–6.

Coman, B.J. (1973), 'Helminth parasites of the fox (*Vulpes vulpes*) in Victoria', *Australian Veterinary Journal,* **49**:378–84.

Coman, B.J., Jones, E.H., & Driesen, M.A. (1981), 'Helminth parasites and arthropods of feral cats', *Australian Veterinary Journal,* **57**:324–7.

Corbett, L.K. (1995), *The Dingo in Australia and Asia*, University of New South Wales Press, Ithaca and London.

Covacevich, J., & Archer, M. (1975), 'The distribution Of the Cane Toad, *Bufo marinus*, in Australia and its effects on indigenous vertebrates', *Memoirs of the Queensland Museum,* **17**:305–10.

Creagh, C. (1992), 'New approaches to rabbit and fox control', *Ecos,* **71**:18–24.

Diamond, J.M. (1984), '"Normal" extinctions of isolated populations', in *Extinctions* (ed. M.H. Nitecki), pp. 191–246, University of Chicago Press, Chicago, Illinois.

Diamond, J.M. (1989), 'The present, past and future of human-caused extinctions', *Philosophical Transactions of the Royal Society of London B,* **325**:469–77.

Dickman, C.R. (1996a), 'Impact of exotic generalist predators on the native fauna of Australia', *Wildlife Biology,* **2**:185–95.

Dickman, C.R. (1996b), *Overview of the impacts of feral cats on Australian native fauna*, Report to Australian Nature Conservation Agency.

Dixon, J.M. (1989), 'Thylacinidae', in *Fauna of Australia* (eds. D.W. Walton, & B.J. Richardson), pp. 549–59, Australian Government Publishing Service, Canberra.

Dixon, J.M., & Huxley, L. (1985), *Donald Thomson's Mammals and Fishes of northern Australia*, Nelson, Melbourne.

Driessen, M.M., Taylor, R.J., & Hocking, G.J. (1991), 'Trends in abundance of three marsupial species after fire', *Australian Mammalogy,* **14**:121–4.

Firestone, K.B., Elphinstone, M.S., Sherwin, W.B., & Houlden, B.A. (1999), 'Phylogeographical population structure of tiger quolls *Dasyurus maculatus* (Dasyuridae: Marsupialia), an endangered carnivorous marsupial', *Molecular Ecology,* **8**:1613–25.

Flannery, T. (1995), *Mammals of New Guinea*, Reed Books: Sydney.

Fleay, D. (1945), 'Native cats at home; breeding carnivorous marsupials', *Wildlife,* **7**:233–7.

Fleming, P., Corbett, L., Harden, R., & Thomson, P. (2001), *Managing the impacts of dingoes and other wild dogs*, Bureau of Rural Sciences, Canberra.

Franklin, M. (1963), *Childhood in Brindabella – my first few years*, Angus and Robertson.

Friend, J.A. (1994), 'Recovery Plan for the Numbat (*Myrmecobius fasciatus*)', Department of Conservation and Land Management, Western Australia, unpublished Recovery Plan Report.

Gill, E.D. (1971), 'The Australian aborigines and the Tasmanian devil', *Mankind,* **8**:59–63.

Ginsberg, J.R., & Macdonald, D.W. (eds.) (1990), *Foxes, Wolves, Jackals, and Dogs – An Action Plan for their Conservation*, IUCN/SSC Canid and Wolf Specialist Groups, IUCN, Gland, Switzerland.

Godsell, J. (1983), 'Ecology of the eastern quoll *Dasyurus viverrinus*, (Dasyuridae: Marsupialia)', PhD thesis, Australian National University, Canberra.

Green, R.H., & Scarborough, T.J. (1990), 'The spotted–tailed quoll *Dasyurus maculatus* (Dasyuridae, Marsupialia) in Tasmania', *The Tasmanian Naturalist,* **100**:1–15.

Guiler, E., & Godard, P. (1998), *Tasmanian Tiger, A Lesson to be Learnt*, Abrolhos Publishing, Perth, Western Australia.

Guiler, E.R. (1983), 'Tasmanian Devil', in *The Australian Museum Complete Book of Australian Mammals* (ed. R. Strahan), pp. 27–8, Angus and Robertson, Sydney.

Hobbs, R.J., & Hopkins, A.J.M. (1990), 'From frontier to fragments: European impact on Australia's vegetation', *Proceedings of the Ecological Society of Australia,* **16**:93–114.

Hope, J.H. (1972), 'Mammals of the Bass Strait islands', *Proceedings of the Royal Society of Victoria,* **85**:163–96.

Johnson, K.A. (1999), 'Recovery and discovery: where we have been and where we might go with species recovery', *Australian Mammalogy,* **21**:75–86.

Johnson, K.A., & Roff, A.D. (1982), 'The western quoll, *Dasyurus geoffroii* (Dasyuridae, Marsupialia) in the Northern Territory: Historical records from venerable sources', in *Carnivorous Marsupials* (ed. M. Archer), pp. 221–6, Royal Zoological Society of NSW, Sydney, Australia.

Jones, M.E. (1995a), 'Guild structure of the large marsupial carnivores in Tasmania', PhD thesis, University of Tasmania, Hobart.

Jones, M.E. (1995b), 'Tasmanian devil *Sarcophilus harrisii* (Boitard, 1841)', in *The Mammals of Australia* (ed. R. Strahan), pp. 82–4, Australian Museum/Reed Books, Sydney.

Jones, M.E. (2000), 'Road upgrade, road mortality and remedial measures: impacts on a population of eastern quolls and Tasmanian devils', *Wildlife Research,* **27**:289–96.

Jones, M.E., & Barmuta, L.A. (1998), 'Diet overlap and abundance of sympatric dasyurid carnivores: a hypothesis of competition?', *Journal of Animal Ecology,* **67**:410–21.

Jones, M.E., & Barmuta, L.A. (2000), 'Niche differentiation among sympatric Australian dasyurid carnivores', *Journal of Mammalogy,* **81**:434–47.

Jones, M.E., & Rose, R.K. (1996), 'Preliminary assessment of distribution and habitat associations of the spotted-tailed quoll (*Dasyurus maculatus maculatus*) and eastern quoll (*D. viverrinus*) in Tasmania to determine conservation and reservation status', Report to the Tasmanian Regional Forest Agreement Environment and Heritage Technical Committee, Public Land Use Commission, Tasmania Report.

Kerle, J.A., Foulkes, J.N., Kimber, R.G., & Papenfus, D. (1992), 'The decline of the brushtail possum (*Trichosurus vulpecula*) (Kerr 1798) in arid Australia', *Rangeland Journal,* **14**:107–27.

King, D.R. (1989), 'An assessment of the hazard posed to northern quolls (*Dasyurus hallucatus*) by aerial baiting with 1080 to control dingoes', *Australian Wildlife Research,* **16**:569–74.

Körtner, G. & Gresser, S. (2002), 'Impact of fox baiting on tiger quoll populations', Report to Environment Australia and the New South Wales National Parks and Wildlife Service.

Laurance, W.F. (1990), 'Comparative responses of five arboreal marsupials to tropical forest fragmentation', *Journal of Mammalogy,* **71**:641–53.

Laurance, W.F. (1991), 'Ecological correlates of extinction proneness in Australian tropical rainforest mammals', *Conservation Biology,* **5**:79–89.

Leary, T., & Seri, L. (1997), 'An annotated checklist of mammals recorded in the Kikori Basin, Papua New Guinea', *Science in New Guinea,* **23**:79–99.

Lindenmayer, D.B., Cunningham, R.B., Donnelly, C.F., Tanton, M.T., & Nix, H.A. (1993), 'The abundance and development of cavities in montane ash-type eucalypt trees in the montane forests of the Central Highlands of Victoria, south-eastern Australia', *Forest Ecology and Management,* **60**:77–104.

Macdonald, D.W. (1988), *Running with the Fox*, Unwin Hyman Limited, London.

Macdonald, D.W., & Thom, M.D. (2001), 'Alien carnivores: unwelcome experiments in ecological theory', in *Carnivore Conservation* (eds. J.L. Gittleman, S. Funk, D.W. Macdonald, & R.W. Wayne), Cambridge University Press, Cambridge.

Mansergh, I. (1984), 'The status, distribution and abundance of *Dasyurus maculatus* (Tiger Quoll) in Australia, with particular reference to Victoria', *Australian Zoologist,* **21**:109–22.

Maxwell, S., Burbidge, A.A., & Morris, K. (1996), *The Action Plan for Australian Marsupials and Monotremes*, IUCN/SSC Australasian Marsupial and Monotreme Specialist Group, Wildlife Australia, Environment Australia, Canberra ACT Australia. Wildlife Australia, Endangered Species Program, Project Number 500 Report.

McIlroy, J.C. (1981), 'The sensitivity of Australian animals to 1080 poison. II. Marsupial and eutherian carnivores', *Australian Wildlife Research,* **8**:385–99.

Mills, G., & Hofer, H. (eds.) (1998), *Hyaenas – Status Survey and Conservation Action Plan*, IUCN/SSC Hyaena Specialist Group, IUCN, Gland, Switzerland.

Mooney, N. (1992), 'The devil you know', *Leatherwood,* **Winter 1992**:55–61.

Morris, K. (1992), 'Return of the chuditch', *Landscope,* **8**:11–15.

Morris, K., Armstrong, R., Orell, P., & Vance, M. (1998), 'Bouncing back – Western Shield update', *Landscope,* **14**:28–35.

Morris, K., Orell, P., & Brazell, R. (1995), 'The effect of fox control on native mammals in the jarrah forest, Western Australia', *10th Australian Vertebrate Pest Control Conference*, Tasmanian Department of Primary Industries and Fisheries, Hobart, May 1995.

Morton, S.R. (1990), 'The impact of European settlement on the vertebrate animals of arid Australia: a conceptual model', *Proceedings of the Ecological Society of Australia,* **16**:210–13.

Morton, S.R., Dickman, C.R., & Fletcher, T.P. (1989), 'Dasyuridae', in *Fauna of Australia* (eds. D.W. Walton, & B.J. Richardson), pp. 560–82, Australian Government Publishing Service, Canberra.

Murray, A.J., Belcher, C.A., Poore, R.N., & Darrant, J. (2000), 'The ability of spotted-tailed quolls to locate and consume meat baits deployed during a simulated aerial baiting program', Victorian Department of Natural Resources & Environment; prepared for the Australian Alps Liaison Committee and The New South Wales National Parks & Wildlife Service. East Gippsland Flora & Fauna Group Report No. 9.

Nathan, C.V. (1966), 'Quolls at Vaucluse', *Sydney Morning Herald*, Sydney.

Newsome, A.E. (1975), 'An ecological comparison of the two arid-zone kangaroos of Australia, and their anomalous prosperity since the introduction of ruminant stock to their environment', *The Quarterly Review of Biology,* **50**:389–424.

Nowell, K., & Jackson, P. (eds.) (1996), *Wild Cats – Status Survey and Conservation Action Plan,* IUCN/SSC Cat Specialist Group, IUCN, Gland, Switzerland.

Oakwood, M. (1997), 'The ecology of the Northern quoll *Dasyurus hallucatus*', PhD thesis, Australian National University.

Oakwood, M. (2000), 'Reproduction and demography of the northern quoll, *Dasyurus hallucatus*, in the lowland savanna of northern Australia', *Australian Journal of Zoology,* **48**:519–39.

Oakwood, M., Bradley, A.J., & Cockburn, A. (2001), 'Semelparity in a large marsupial', *Proceedings of the Royal Society of London – Series B,* **268**:407–11.

Oakwood, M., & Eager, R. (1997), 'Diet of the Northern Quoll, *Dasyurus hallucatus*, in lowland savanna of northern Australia', in 'The Ecology of the Northern Quoll, *Dasyurus hallucatus*', PhD thesis, Australian National University.

Oakwood, M., & Pritchard, D. (1999), 'Little evidence of toxoplasmosis in a declining species, the northern quoll (*Dasyurus hallucatus*)', *Wildlife Research,* **26**:329–33.

Oakwood, M., & Spratt, D.M. (2000), 'Parasites of the northern quoll, *Dasyurus hallucatus* (Marsupialia: Dasyuridae), in tropical savanna, Northern Territory', *Australian Journal of Zoology,* **48**:79–90.

Obendorf, D.L., & Munday, B.L. (1990), 'Toxoplasmosis in wild eastern barred bandicoots, *Perameles gunnii*', in *Bandicoots and Bilbies* (eds. J.H. Seebeck, P.R. Brown, R.L. Wallis, & C.M. Kemper), pp. 193–7, Surrey Beatty and Sons, Sydney.

Orell, P., & Morris, K. (1994), 'Chuditch recovery plan, 1992–2001', Western Australian Department of Conservation and Land Management, Western Australian Wildlife Research Centre, Wildlife Management Program Report No. 13.

Paddle, R. (2000), *The Last Tasmanian Tiger. The History and Extinction of the Thylacine*, Cambridge University Press, Cambridge, England.

Pavlov, P.M., & Howell, M.J. (1977), 'Helminth parasites of Canberra cats', *Australian Veterinary Journal,* **53**:599–600.

Pech, R.P., Sinclair, A.R.E., & Newsome, A.E. (1995), 'Predation models for primary and secondary prey species', *Wildlife Research,* **22**:55–64.

Purvis, A., Mace, G.M., & Gittleman, J.L. (2001), 'Past and future carnivore extinctions: a phylogenetic perspective', in *Carnivore Conservation* (eds. J.L. Gittleman, S. Funk, D.W. Macdonald, & R.W. Wayne), Cambridge University Press, Cambridge.

Rolls, E.C. (1969), *They all ran wild*, Angus and Robertson, Sydney.

Rounsevell, D.E., & Mooney, N. (1995), 'Thylacine *Thylacinus cynocephalus* (Harris, 1808)', in *The Mammals of Australia* (ed. R. Strahan), pp. 164–5, Australian Museum/Reed Books, Sydney.

Schreiber, A., Wirth, R., Riffel, M., Van Rompaey, H., & the IUCN/SSC Mustelid and Viverrid Specialist Group (1989), *Weasels, Civets, Mongooses, and their Relatives – An Action Plan for the Conservation of Mustelids and Viverrids*, IUCN, Gland, Switzerland.

Serena, M., Soderquist, T.R., & Morris, K. (1991), *The Chuditch* (Dasyurus geoffroii), Department of Conservation and Land Management Report.

Shepherd, N.C., & Mahood, I.T. (1978), 'The potential effect of feral dogs and cats on Australian native fauna', *Australian Advances in Veterinary Science,* **1978**:108.

Short, J., & Smith, A. (1994), 'Mammal decline and recovery in Australia', *Journal of Mammalogy,* **75**:288–97.

Short, J., & Turner, B. (1994), 'A test of the vegetation mosaic hypothesis – a hypothesis to explain the decline and extinction of Australian mammals', *Conservation Biology,* **8**:439–49.

Simberloff, D. (1998), 'Flagships, umbrellas, and keystones: is single-species management passé in the landscape era?', *Biological Conservation,* **83**:247–57.

Simberloff, D., Farr, J.A., Cox, J., & Mehlman, D.W. (1992), 'Movement corridors: conservation bargains or poor investments?', *Conservation Biology,* **6**:493–504.

Sinclair, R.G., & Bird, P.L. (1984), 'The reaction of *Sminthopsis crassicaudata* to meat baits containing 1080: implications for assessing risk to non-target species', *Australian Wildlife Research,* **11**:501–7.

Smith, A. P., & Quin, D.G. (1996), 'Patterns and causes of extinctions and decline in Australian conilurine rodents', *Biological Conservation,* **77**:243–67.

Smith, G.C., Jones, M.E., & Jones, S.M. (in review), 'Anti-predator behaviour of eastern quolls (*Dasyurus viverrinus*) in Tasmania: does it increase risk from novel, introduced predators'.

Soderquist, T.R., & Serena, M. (1993), 'Predicted susceptibility of *Dasyurus geoffroii* to canid baiting programs: variation due to sex, season and bait type', *Wildlife Research,* **20**:287–96.

Soderquist, T.R., & Serena, M. (1994), 'Dietary niche of the Western Quoll, *Dasyurus geoffroii*, in the jarrah forest of Western Australia', *Australian Mammalogy,* **17**:133–6.

Soderquist, T.R., & Serena, M. (2000), 'Juvenile behaviour and dispersal of chuditch (*Dasyurus geoffroii*) (Marsupialia: Dasyuridae)', *Australian Journal of Zoology,* **48**:551–60.

Terborgh, J., & Robinson, S. (1986), 'Guilds and their utility in ecology', in *Community Ecology: Pattern and Process* (eds. J. Kikkawa, & D.J. Anderson), pp. 65–90, Blackwell, Melbourne.

Towns, D.R., Simberloff, D., & Atkinson, I.A.E. (1997), 'Restoration of New Zealand islands: redressing the effects of introduced species', *Pacific Conservation Biology,* **3**:99–124.

Triggs, B., Brunner, H., & Cullen, J.M. (1984), 'The food of the fox, dog and cat in Croajingalong National Park, south-eastern Victoria', *Australian Wildlife Research,* **11**:491–9.

Troughton, E. (1941), *Furred Animals of Australia*, Angus and Robertson, Sydney.

Troughton, E. (1943), *Furred Animals of Australia*, Angus and Robertson, Sydney.

Twigg, L.E., & Williams, C.K. (1999), 'Fertility control of overabundant species: Can it work for feral rabbits?', *Ecology Letters,* **2**:281–5.

Tyler, M.J., & Davies, M. (1986), 'Frogs of the Northern Territory', Conservation Commission of the Northern Territory, Darwin, Australia.

van Dyck, S. (1988), 'The Bronze Quoll, *Dasyurus spartacus* (Marsupialia: Dasyuridae), a new species from the savannah of Papua New Guinea', *Australian Mammalogy,* **11**:145–56.

Van Valkenburgh, B. (2000), 'The dog-eat-dog world of carnivores: a review of past and present carnivore community dynamics', in *Meat-eating and Human Evolution* (eds. C.B. Stanford, & H.T. Bum), Oxford University Press, Oxford.

Vertebrate Advisory Committee (1994), 'Native Vertebrates which are Rare or Threatened in Tasmania', in *Species at Risk, Tasmania – Vertebrates,* Parks and Wildlife Service, Tasmania Report No. 1.

Waser, P., & Strobeck, C. (1998), 'Genetic signatures of interpopulation dispersal', *Trends in Ecology and Evolution,* **13**:43–4.

Waser, P.M., Strobeck, C., & Paetkau, D. (2001), 'Estimating interpopulation dispersal rates', in *Carnivore Conservation* (eds. J.L. Gittleman, S. Funk, D.W. Macdonald, & R.W. Wayne), Cambridge University Press, Cambridge.

Werdelin, L. (1987), 'Some observations on *Sarcophilus laniarius* and the evolution of *Sarcophilus*', *Records of the Queen Victoria Museum,* **90**:1–27.

Westoby, M. (1988), 'Comparing Australian ecosystems to those elsewhere', *BioScience,* **38**:549–56.

Wood-Jones, F. (1923), 'The Mammals of South Australia. Part I–III 1923–1925', South Australian Government Press, Adelaide, South Australia.

Woolley, P.A. (2002), 'Observations on the reproductive biology of *Myoictis wallacei*, *Neophascogale lorentzi*, *Dasyurus albopunctatus* and *Dasyurus spartacus*, dasyurid marsupials endemic to New Guinea', *Australian Mammalogy,* **23**:63–6.

Wright, B.J. (1972), 'Rock engravings of striped mammals: the Pilbara Region, Western Australia', *Archaeology and Physical Anthropology,* **7**:15–23.

CHAPTER 30

RECOVERY OF THE THREATENED CHUDITCH (*DASYURUS GEOFFROII*): A CASE STUDY

Keith Morris[A], *Brent Johnson*[A], *Peter Orell*[B], *Glen Gaikhorst*[C], *Adrian Wayne*[D] *and Dorian Moro*[A]

[A] Department of Conservation and Land Management, Science Division, Woodvale Research Centre, PO Box 51, Wanneroo, WA 6946, Australia
[B] Department of Conservation and Land Management, Wildlife Branch, Hayman Rd, Kensington, WA 6152, Australia
[C] Native Species Breeding Program, Research Section, Perth Zoo 20 Labouchere Rd, South Perth, WA 6151, Australia
[D] Department of Conservation and Land Management, Science Division, Manjimup, WA 6258, Australia

This chapter provides an account of a successful recovery program that has substantially improved the conservation status of a threatened dasyurid species. The chuditch *Dasyurus geoffroii* was once widespread throughout much of the Australian continent, and like many of Australia's medium-sized mammals, has suffered a dramatic reduction in abundance and range. Predation by the introduced European red fox has been shown to be a significant threatening process in the conservation of the chuditch and other medium-sized mammals in Western Australia (WA). Competition with the fox for food may also be involved. By the 1980s, it was estimated that fewer than 6000 chuditch remained, all confined to the south-west of WA, most being in the jarrah forest. In 1983, the chuditch was listed as a threatened species in WA and in 1991 it was listed as an Endangered species under Commonwealth legislation. Consequently a recovery plan was prepared and a recovery team appointed to co-ordinate the implementation of six recovery actions considered necessary to improve the conservation status within 10 years. This review describes progress of these recovery actions. The abundance of chuditch in the south-west of WA has increased significantly and current logging and prescribed burning practices in the jarrah forest do not appear to detrimentally impact on this. Fox control using poisoned (1080) dried meat baits over large areas of the south-west of WA has been one of the major factors in the recovery of the chuditch. The ability to breed chuditch in captivity and to translocate the progeny to areas where they once occurred has also contributed to the recovery of this species. Based on this recovery, a review in 2003 may recommend a change in its conservation status and removal from State and Commonwealth threatened species lists.

INTRODUCTION

The chuditch, or western quoll (*Dasyurus geoffroii* Gould, 1841) is a carnivorous/insectivorous marsupial belonging to the Family Dasyuridae. Chuditch is the Noongar (south-western Western Australian Aboriginal) name from the Albany and southern wheatbelt regions. It is one of four quoll species in Australia and is the largest native mammalian predator in Western Australia (WA): adults weigh between 700–1200 g (females) and 1300–1600 g (males) (Serena and Soderquist 1995).

Chuditch were the most widely distributed of all quolls and were originally found over most of the Australian mainland (Serena et al. 1991). Their physiology is especially adapted to

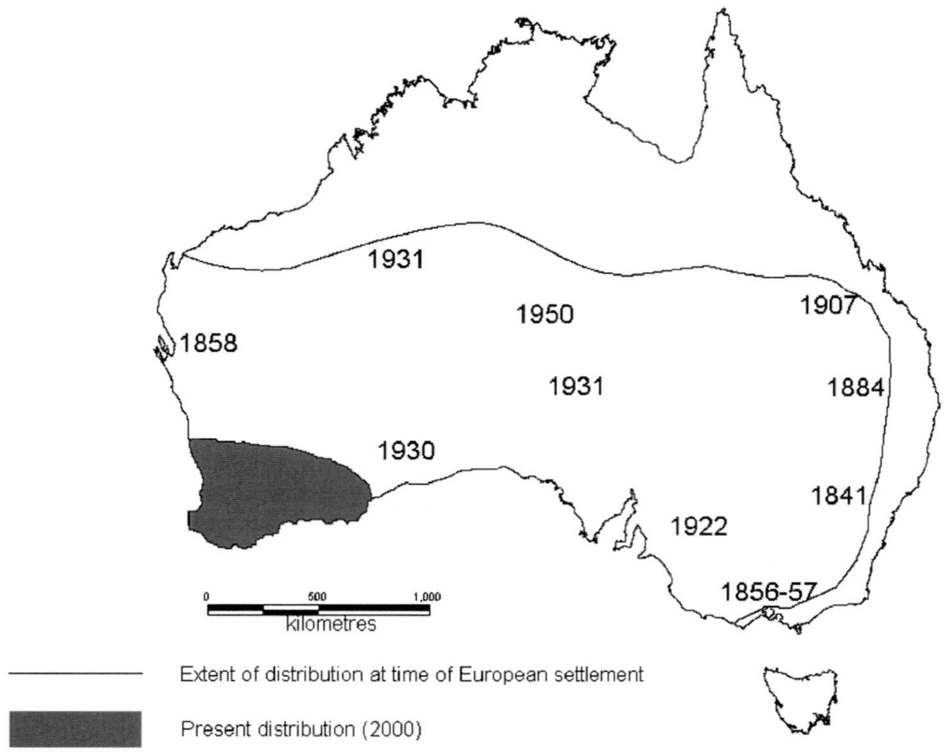

Figure 1 Former distribution of chuditch. (Dates shown are dates of last record.)

water conservation, which partly explains why this species has been so successful at colonising a range of habitats from the temperate forests of the south-west of WA to the hot, arid deserts of the interior (Arnold and Shield 1970).

Both males and females are sexually mature and breed in their first year. Females are polyoestrous and promiscuous (Soderquist and Serena 1990). Breeding is seasonal, and matings usually occur between April and June. Gestation lasts 16 to 18 days, and up to six young can be accommodated in a rudimentary pouch (Serena and Soderquist 1988). The young are deposited in a den, usually a burrow or hollow log, after 64 days (Serena and Soderquist 1989a), are weaned at about 160 days, and eventually disperse after 170 days (November to January) (Soderquist and Serena 2000). Both sexes are heavier in their second year (Soderquist 1995).

Chuditch are essentially solitary, and capable of moving several kilometres in a 24-hour period. Serena and Soderquist (1989b) estimated the home range of female chuditch to be 337 ha (core area of 90 ha), and males move up to 1.7 km from their core areas of approximately 400 ha. Within these core areas, up to 180 den sites are utilised and these are usually located in hollow logs, tree limbs, rock outcrops or burrows. In the Batalling forest, home range sizes of 278–314 ha for females and 509–791 ha were reported (Mathew 1996). In the arid-zone chuditch were known to seek refuge in termitaria and holes in the ground (Johnson and Roff 1982; Burbidge et al. 1988), and in the wheatbelt of WA, white-browed babbler *Pomatostomus superciliosus* nests are used (Redner 1999).

Chuditch are swift runners and efficient climbers, allowing them to forage over a large area in search of prey. They are opportunistic hunters and consume a wide range of food items. In both the forest and semi-arid areas, large invertebrates comprise about two-thirds of their diet, with small mammals, birds and reptiles also being taken. Some fruits, food scraps and carrion are also consumed (Soderquist and Serena 1994; Redner 1999).

Taxonomy

The chuditch was described in 1841 (Gould 1841) from a specimen collected on the Liverpool Plain in New South Wales (Iredale and Troughton 1934). Subsequently, Thomas (1906) recognised two forms based on skull size, skin and fur colour: *Dasyurus geoffroii geoffroii* from Queensland, New South Wales and South Australia, and *D. g. fortis* from WA. While it is difficult to confirm these differences because of the lack of eastern Australian specimens and extant populations, the validity of this taxonomy has been questioned both morphometrically (Serena et al. 1991) and genetically (Firestone 1999). The chuditch is currently regarded as a single taxon.

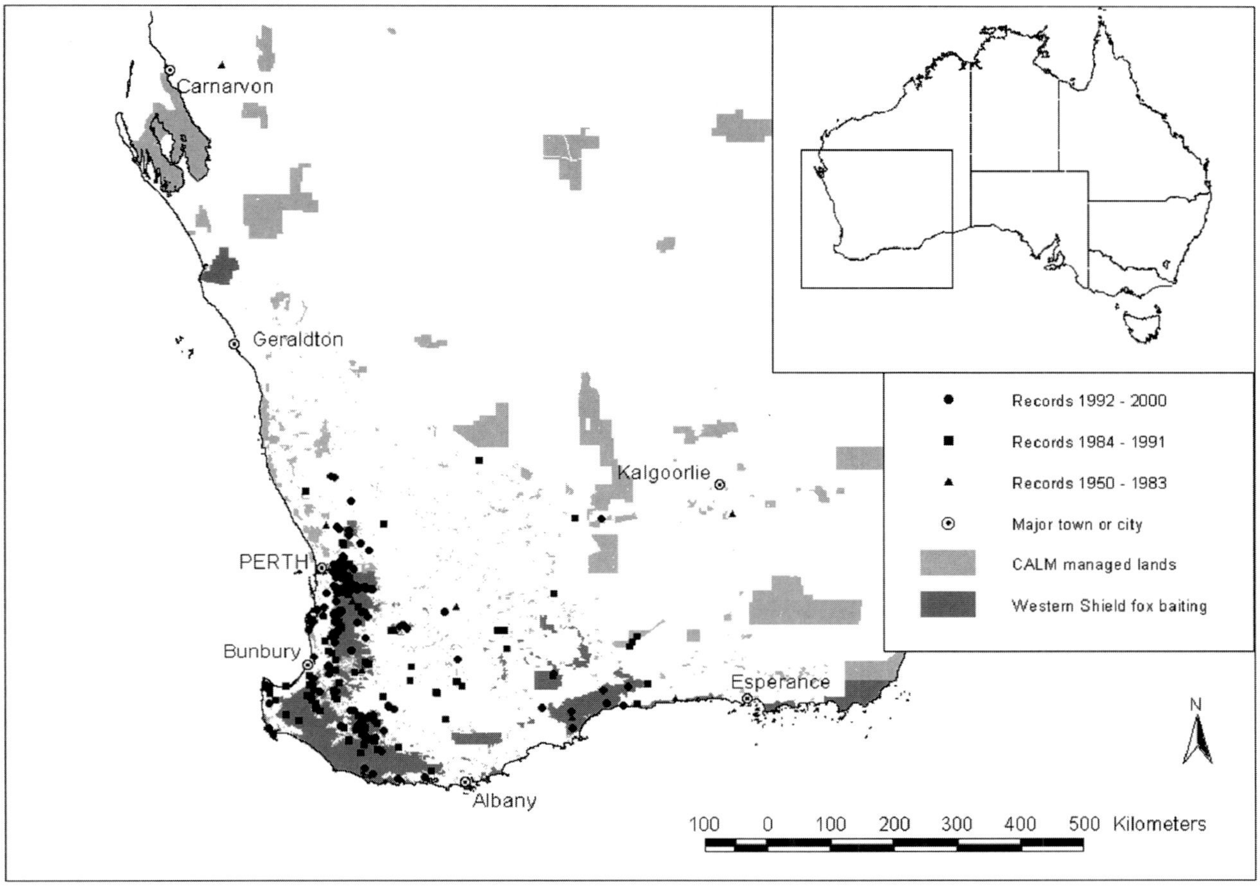

Figure 2 Distribution of chuditch in Western Australia since 1950.

DISTRIBUTION AND CONSERVATION STATUS

At the time of European settlement, chuditch were relatively abundant and occupied nearly 70% of the Australian continent, occurring in every mainland State and the Northern Territory (Fig. 1). However, a drastic decline of geographic range has occurred over the last 200 years (Collett 1887; Whittell 1954; Johnson and Roff 1982; Burbidge et al. 1988). Specimens were last collected in New South Wales (Liverpool Plains) in 1841, Victoria (near the junction of the Murray – Darling rivers) in 1857 and Queensland (Coomooboolaroo and Peak Downs) between 1887-1907 (Wakefield 1966, Krefft 1866). In South Australia, chuditch were last collected in 1931 in the north-west of that State. Chuditch were last reported in the central arid zone in the 1950s (Finlayson 1961). In arid parts of WA, the species was last collected in Shark Bay in 1858, Canning Stock Route in 1931 and on the Nullarbor Plain in the 1930s (Boscacci et al. 1987). Chuditch were still abundant in the wheatbelt in 1907 (Thomas 1906; Shortridge 1909) and persisted on the Swan Coastal Plain around Perth until the 1950s (Kitchener et al. 1978; Kitchener and Vicker 1981). They were still sufficiently common at this time to be regarded as a pest by poultry farmers in the outer metropolitan parts of Perth. Since the 1970s the chuditch has been confined to the south-west of WA, occupying a roughly triangular area bounded by Moora in the north, Cape Arid to the east and Cape Leeuwin in the south (Fig. 2). There is an unconfirmed record of chuditch along the Gascoyne River on Doorawarra Station in 1982 (McKenzie et al. 2000). By the end of the 1980s it was estimated that the wild population numbered less than 6000 and that the majority of these were confined to the jarrah forest in the south-west part of WA (Serena et al. 1991). They also persisted in low numbers in the drier woodlands and mallee shrublands of the wheatbelt.

In 1983 the chuditch was listed as 'fauna that is rare, or is likely to become extinct' pursuant to the WA *Wildlife Conservation Act* 1950, and in 1991 it was listed as an Endangered species under the Commonwealth *Endangered Species Protection Act* 1992. The 1992 Action Plan for Australasian Marsupials and Monotremes regarded chuditch as Endangered (Kennedy

1992); however, a revision of this plan in 1996 regarded it as Vulnerable using IUCN (1994) criteria (Maxwell et al. 1996). This species is currently listed as threatened (Vulnerable) under the Commonwealth *Environment Protection and Biodiversity Conservation Act* 1999.

RECOVERY PLAN

Serena et al. (1991) prepared a wildlife management program for chuditch and this formed the basis of a recovery plan for the species in 1994 (Orell and Morris 1994). This was supported by the Department of Conservation and Land Management (CALM), Alcoa, Perth Zoo, World Wide Fund for Nature, and Environment Australia. The initial objective of this recovery plan was to downlist the chuditch from Endangered to Vulnerable within 10 years (by 2001). With the revision of status by Maxwell et al. (1996), the objective became to remove the chuditch from State and Commonwealth threatened species lists (to Lower Risk – conservation dependent). The chuditch is the only quoll in Australia for which a recovery plan has been prepared and implemented.

The recovery plan objectives were to:

1 Maintain or increase average daily trap success rates above 1% at monitoring sites in the jarrah forest (trap success rates in the jarrah forest during the 1980s were 0–0.5%).

2 Maintain a chuditch population in at least one semi-arid monitoring site.

3 Establish at least one self-sustaining population outside the geographic range as known in 1992.

Six actions were considered necessary to achieve these objectives:

1 Research the impact of timber harvesting in the jarrah forest on chuditch populations.

2 Research the impact of prescribed burning in jarrah forest on chuditch populations.

3 Research the impact of the introduced fox and fox control programs.

4 Monitor existing populations in the jarrah forest.

5 Commence and maintain a captive-breeding program.

6 Translocate chuditch to areas where they once occurred.

This review examines progress against these actions from 1991–2000.

THE IMPACT OF TIMBER HARVESTING ON CHUDITCH POPULATIONS

At the time of European settlement in WA (1826), the dry sclerophyll jarrah (*Eucalyptus marginata*) forests and woodlands of the south-west covered an estimated 5,300,000 ha. This area has been reduced to about 3,300,000 ha as a result of clearing for agriculture (Dell and Havel 1989) and by the 1970s this area supported the major remaining population(s) of chuditch (Serena et al. 1991). Approximately 48% of jarrah forest is managed by the CALM.

Timber harvesting is one of the multiple uses made of State Forest and a history of logging practices in the jarrah forest since European settlement is provided by Bradshaw et al. (1991) and Heberle (1997). In 1999, approximately 976,000 hectares was available for timber harvesting. The current silvicultural practices of gap release, shelterwood creation and thinning have been in use in the jarrah forests since 1985 (Stoneman et al. 1989). Silvicultural practices also provide for the retention of hollow logs suitable as chuditch refuge sites in production forest. Since 1991, fox control through aerial baiting, has been implemented over most of the jarrah forest.

The impact of timber harvesting on chuditch, and other mammals, was studied as part of the 'Kingston Project' (Burrows et al. 1994). The design and methodology used are described in Morris et al. (2000). Given the scale of the logging treatments (e.g. approximate maximum of 10 ha for gap release) and the large home ranges of chuditch, logging impacts were studied at the landscape level (approximately 15,000 ha), rather than at the coupe level.

Population study – abundance

Chuditch abundance was assessed as trap success rates (number of chuditch trapped per 100 trapnights) before, during and after logging treatments using cage traps set along line transects covering 36 km of forest track. Trap success rates increased from less than 1% before fox control commenced in 1993 to between 2–4% in 2000, four years after timber harvesting and associated burning activities had been implemented (Fig. 3). However, the increasing number of other mammals in response to fox control has confounded determining the impacts of logging on chuditch abundance. Trap success rates for all mammals at Kingston increased from 10% in 1993 to more than 70% between 1996 and 1999 (Morris et al. 2000). The majority of these captures were of woylies *Bettongia penicillata,* which readily enter traps at dusk. This reduces the number of traps available to capture the less abundant chuditch, consequently, from August 1996 to May 1997 chuditch trap success rates declined to as low as 0.5%. In August 1997, a trial was commenced to test the attractiveness of the standard peanut-butter-based bait, which was highly favoured by woylies, with a meat-meal-based bait considered less attractive to woylies. The 'chuditch' bait reduced woylie captures by 50%, but chuditch captures increased seven-fold.

The Cormack–Jolly–Seber population estimates (Lebreton et al. 1992) of chuditch numbers within the Kingston study area

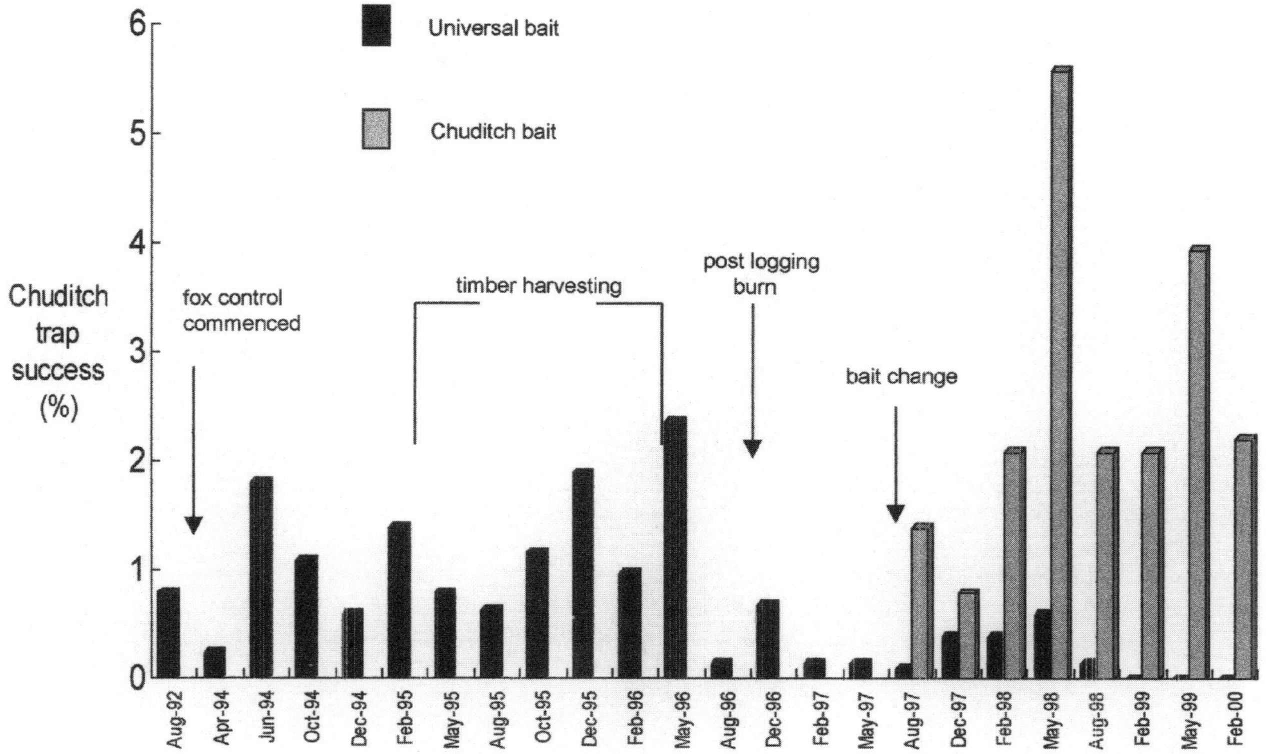

Figure 3 Chuditch trap success rates (%) in the Kingston forest using 'universal' and 'chuditch' baits. (Note: universal bait was not used after August 1998. Trap effort for each date from April 1994 was 180 cage traps set for 4 nights = 720 trapnights.)

averaged 15.6±1.3 individuals before logging commenced in 1995 and 17.9±1.8 individuals during the 18 month logging period. The post-logging population estimates using pre-August 1998 captures on 'universal' bait averaged 12.8±0.3 individuals. The population estimates derived from captures using the alternative 'chuditch' bait since August 1997 averaged 27.8±3.5 individuals. Because of the variable and confounding influence of reduced trap availability as a result of increasing woylie numbers, and because of differences between trap success rates of chuditch on the two bait types, it is not appropriate to statistically compare post logging abundances with previous measures. However, it is clear that, in the presence of fox control, chuditch are still present and persist four years after logging disturbance ceased.

Reproduction

Since 1994, at least 50% of the adult female chuditch trapped at Kingston have shown evidence of breeding (Table 1). Although the sample sizes of adult female chuditch are not large enough to statistically test the potential impacts of timber harvesting on reproduction, the capture of unmarked individuals and sub-adults each year (December to February) demonstrates that recruitment has continued post-logging. Three of the 14 sub-adults captured have subsequently been re-captured as adults. This is an underestimate of the number of sub-adults which may reach breeding age given that it is expected that most sub-adults, particularly males, would disperse away from the ranges of their mother and, therefore are likely to move beyond the range of the trapping area.

Scrotal size can be used as a measure of reproductive activity in male dasyurids (Cuttle 1982). Chuditch maximum scrotal size is achieved from February to May, coinciding with the mating season (Fig. 4). Scrotal size is at a minimum in December when there is no mating activity. This scrotal size cycle showed regular oscillations between years suggesting that the timing of male reproduction was not impaired by logging activities in 1995–6.

Injuries to females associated with mating (lesions principally to the neck and back), were occasionally observed during the regular May trapping session. During August of each year, females were observed with four to six pouch young. These young were deposited in dens before the next sampling period in December. Lactating females were observed between July and February however the majority of lactating females were captured between August and December. The number of lactating teats varied between four and six.

Table 1 The number of lactating chuditch in relation to the total number of adult females trapped during the breeding season (May to January) at Kingston forest since 1994.

	1994/95	1995/96	1996/97	1997/98	1998/99
# breeding females	3	3	1	3	3
# adult females	6	5	2	4	3
# subadult females	0	0	0	1	1
# subadult males	1	4	1	4	2

Considered together, the results of the population study between 1994 and 2000 suggest that logging has not disrupted the seasonal breeding pattern of chuditch.

Condition

Seasonal changes in body mass are often related to the condition or 'health' of individuals in a population (Bakker and Main 1980; Krebs and Singleton 1993). A condition index for chuditch at Kingston was determined by regressing body mass against other body measures of short-pes length and head length. The condition of chuditch could be predicted to decline if logging were having a detrimental impact on the population. For chuditch in the Kingston forest, the relationship between body mass and head length or short-pes length was significant: however, short – pes length was a more robust predictor of body mass ($r^2 = 0.23$) than head length ($r^2 = 0.14$) and so was used in the condition index model : CI = [1.72 log(SP) + 0.12] / log (BM), where CI is the condition index, SP is the short-pes length (mm) and BM is the body mass (g), and $r^2 = 0.23$, $F = 47.2$, df = 1, 156, $P < 0.0001$, n = 158.

Changes in condition indices over the study period for males, females and juveniles are shown in Fig. 5. There was no significant decline in the condition of chuditch after logging had occurred. A one-way analysis of variance (ANOVA) of log-transformed data found that no significant difference exists in

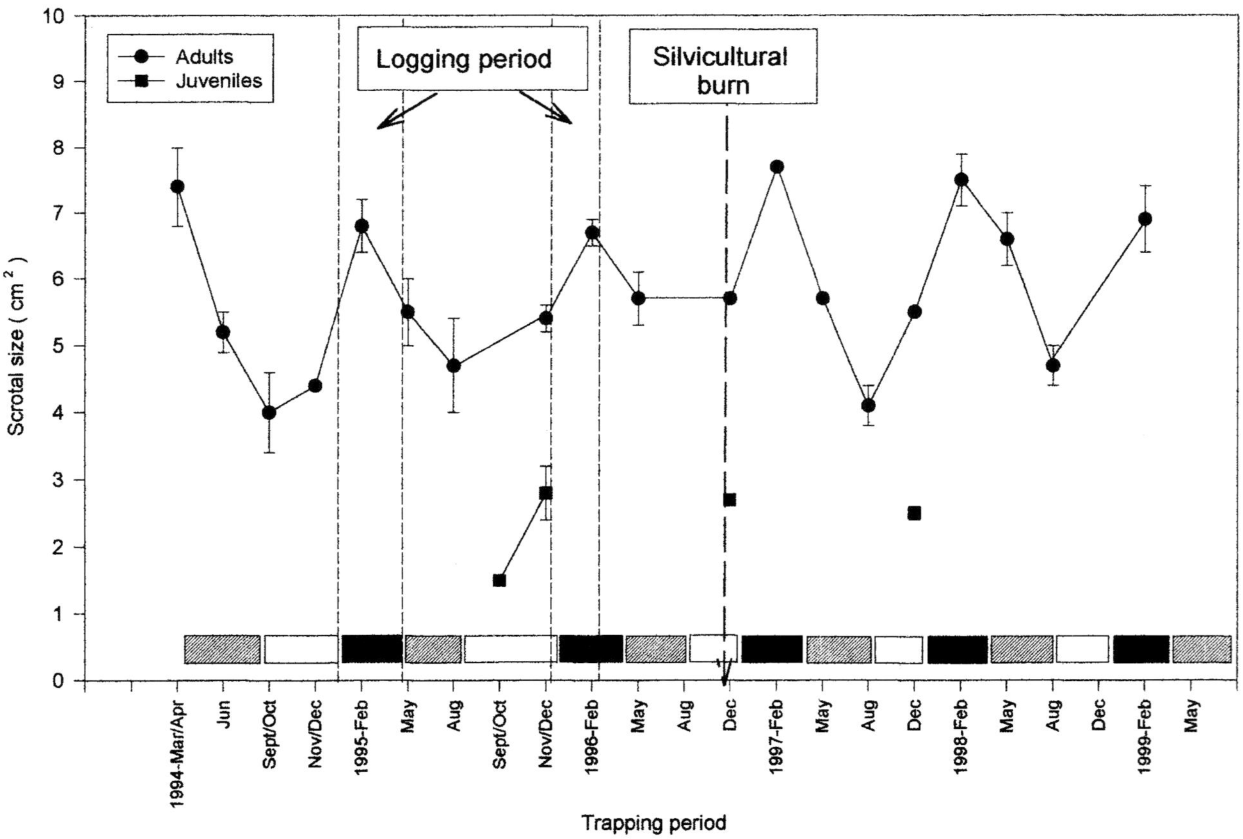

Figure 4 Seasonal variation in scrotal size of chuditch in the Kingston forest.

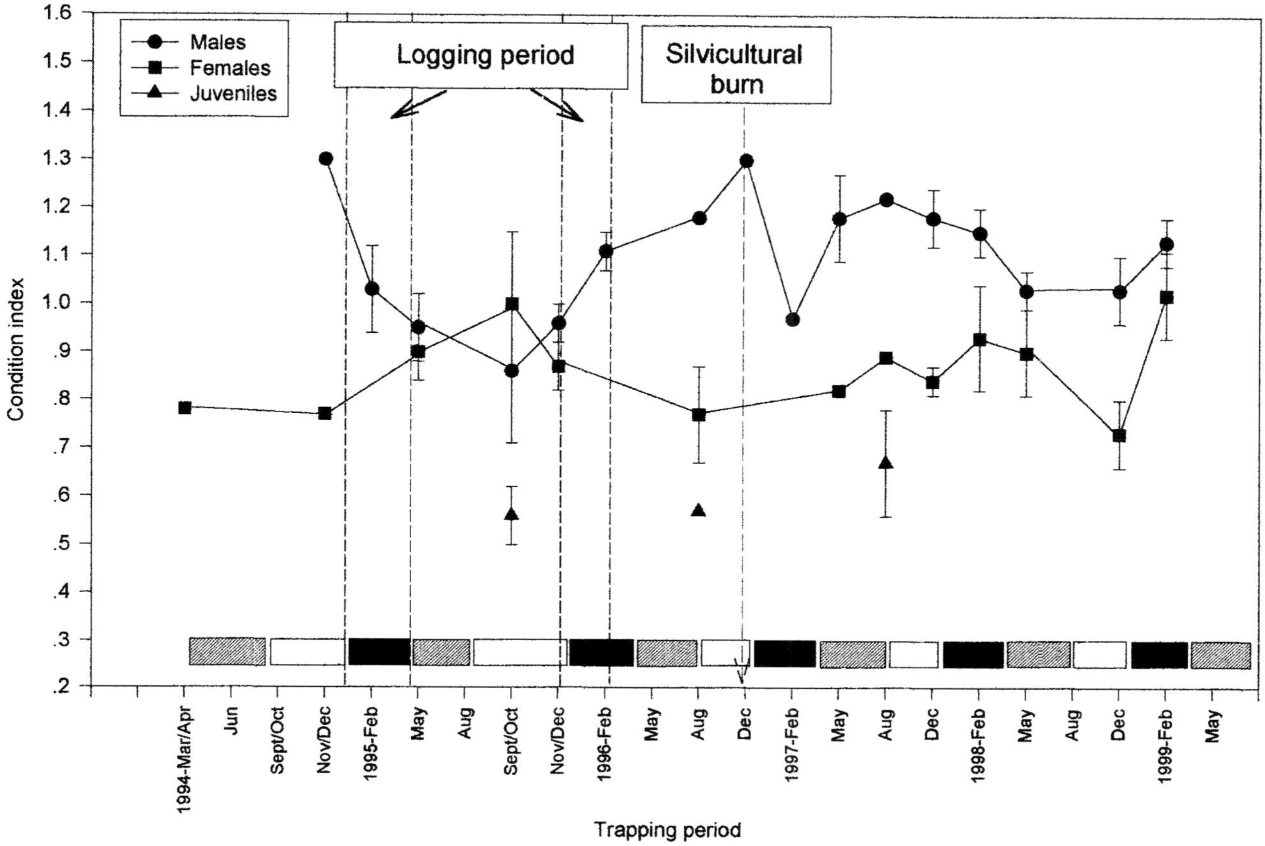

Figure 5 Seasonal variation of condition indices for chuditch in the Kingston forest.

condition indices of females (F = 1.56, df = 4, P = 0.21) between seasons. Consequently, these were grouped to give an average (±se) condition index of 0.88±0.03 for females. Condition indices for males varied with season (F = 6.78, df = 3, P < 0.001), and indices were higher in April than during the other breeding, weaning or nonbreeding months (Tukey post-hoc test, P < 0.01). Except in September 1996, male condition indicies remained higher than that for females before and after logging in the study area.

THE IMPACT OF PRESCRIBED BURNING ON CHUDITCH POPULATIONS

The jarrah forest is prescribed burnt on a rotational basis every 5–7 years to reduce the amount of forest litter build-up and the consequent risk of wildfire (Bradshaw et al. 1991). Burns are undertaken in either autumn (hot fire) or spring (cool fire) and each burn usually covers 5000–10,000 ha. Although chuditch individuals are known to survive these burns (Serena et al. 1991), the impact of burning on population dynamics was poorly understood. In 1994 a trial was commenced at Batalling forest to examine the impact of spring and autumn prescribed burns on a range of vertebrates, including chuditch. Trapping was undertaken on both grids and road transects and the results combined to provide Known To Be Alive (KTBA) estimates of population size. Radio-tracking of chuditch during the burns was also undertaken.

Population abundance

The minimum numbers of chuditch KTBA before, between, and after the spring 1994 and autumn 1996 burns are shown in Table 2. The mean number known to be alive after the burning regimes had been implemented (19.5±3.1) was significantly greater then before the burning occurred (14.7±3.2; t = 3.1, 15df, p < 0.01). Given the scale of these prescribed burns

Table 2 The mean (±SE) number of chuditch known to be alive at Batalling before (March 1993–June 1994), between and after (March 1996–December 1998) prescribed burns.

	Males	Females	Total
Before burns	6.6 ± 1.1	8.1 ± 2.2	14.7 ± 3.2
Between burns	9.7 ± 1.1	14.6 ± 2.0	24.3 ± 3.1
After burns	8.7 ± 1.6	10.8 ± 1.5	19.5 ± 3.1

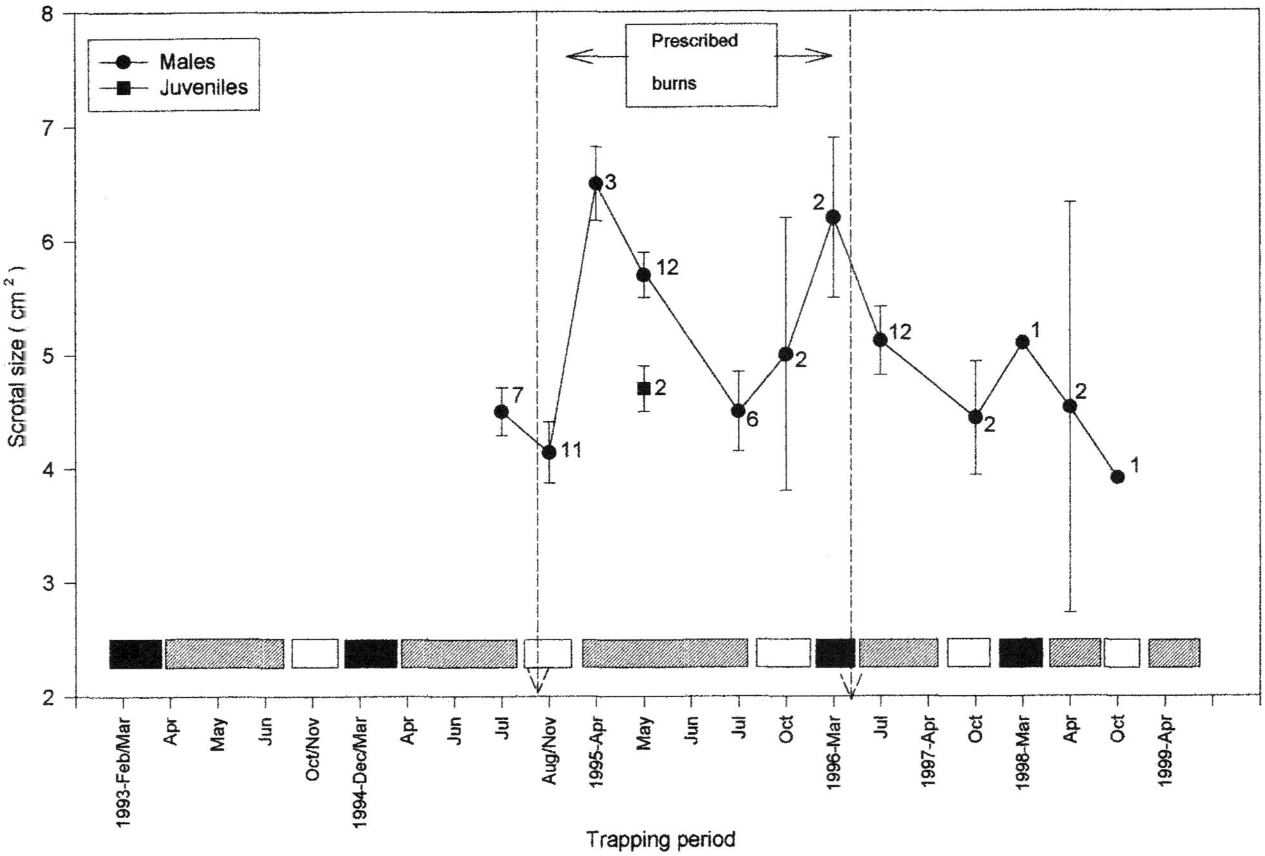

Figure 6 Variation of scrotal size for adult and juvenile male chuditch over the period 1993–1999 at Batalling forest. Breeding season (hatched), weaning/dispersal season (unshaded), nonbreeding season (shaded). Values are mean ± standard error. Sample sizes shown near data points.

(3000–4000 ha), this increase is more likely to be in response to the fox control rather than as a direct result of the burns. However, the fire may have benefited chuditch though exposure of prey items and ease of hunting in the burnt areas.

Ten chuditch were radio-collared before the spring 1994 fire and their movements followed before, during and after the fire. There were no mortalities associated with fire and none of the den logs known to be in use by lactating females were destroyed by this fire. Chuditch have continued to utilise areas burnt by the prescribed burns up to four years post-burn. Mathew (1996) radio-tracked chuditch before, during and after the autumn burn. None died as a direct result of the fire. However, five of the 13 radio-tracked chuditch died as a result of fox predation, one before the April burn and four in the six months following the fire. The clearing of the vegetation by the more intense burn may have contributed to the higher predation rate, as has been observed for woylies (Christensen 1980) and northern quolls (Oakwood 2000). Despite this mortality, trap success rates increased from 5.2% just prior to the autumn burn, to 7.6% four months after the burn. The autumn burn did destroy some refuge logs, however chuditch continued to use remaining hollow logs, and used a greater number of burrows in burnt areas than in unburnt areas (Mathew 1996).

Reproduction

All adult females trapped before, between, and after the prescribed burns during the breeding season (July to November) were either carrying pouch young or were lactating. Scrotal size for male chuditch followed a similar pattern to that seen for chuditch at Kingston with maximum size occurring just prior to mating in April May (Fig. 6). This, together with an increase in abundance, suggests that prescribed burning does not affect the timing of reproduction, or the population increasing in the presence of fox control.

Condition

A condition index was determined for chuditch at Batalling by regressing body mass against head length and short-pes length. A significant relationship was found and this was expressed as:

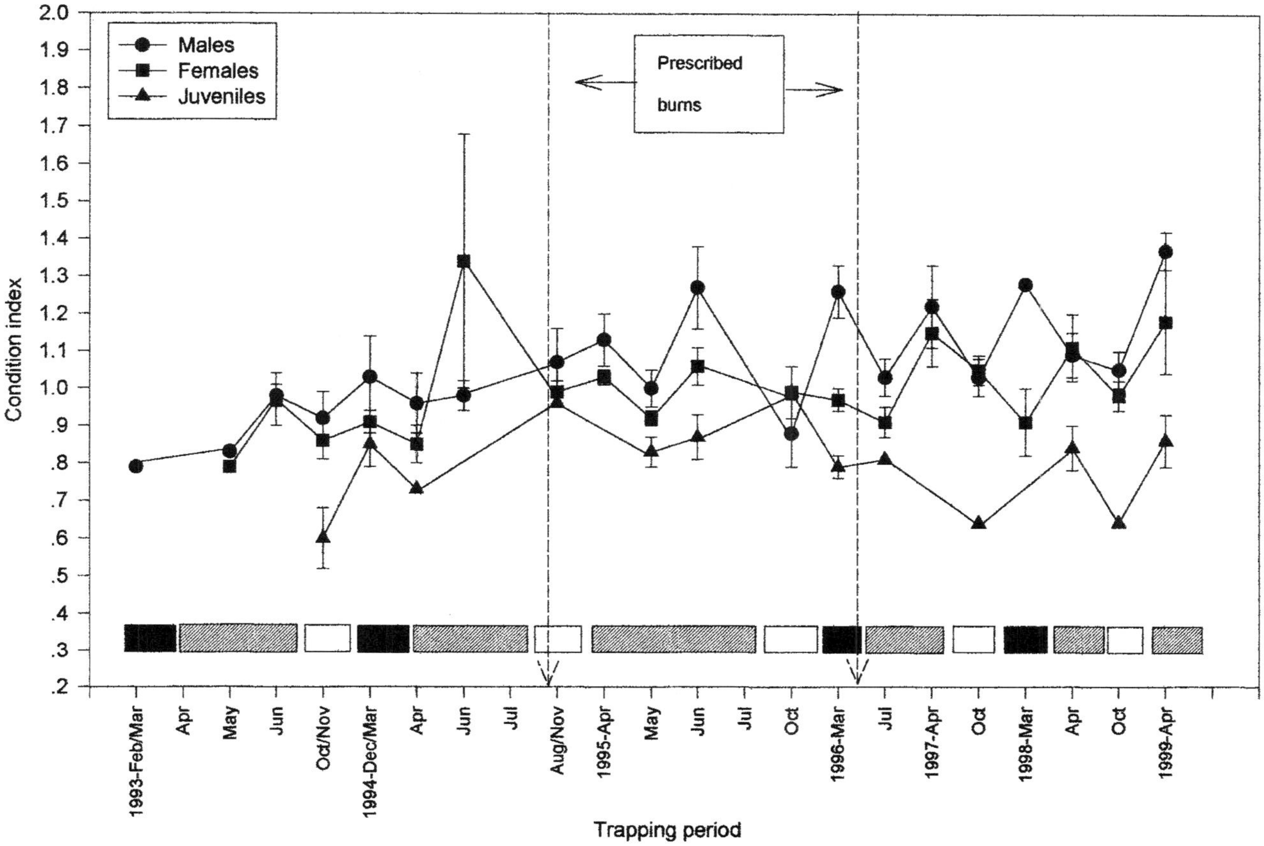

Figure 7 Seasonal variation in the condition of chuditch in the Batalling forest.

CI = [19.4 (HL) + 19.5 (SP) – 1753] / BM, where CI is condition index, HL is head length (mm), SP is short pes length (mm), and BM is body mass (g). (r^2 = 0.65, F = 263.1, df = 2, 284, P < 0.0001, n = 287).

This is a different relationship from that found for chuditch at Kingston, suggesting that the condition of chuditch in the jarrah forest may vary between populations. This may reflect the heterogenous nature of jarrah forest habitats, with some areas being 'poorer' sites than others.

Seasonal changes in condition indices for males, females and juveniles over the study period are shown in Fig. 7. There was no significant decline in condition detected after prescribed burns had been implemented. A one-way ANOVA on log-transformed data found that the condition index of male chuditch varied significantly between seasons (F = 3.60, df = 5, P < 0.01). Using condition index as a measure of the 'health' of an individual, males were in better health (Tukey post hoc test: P < 0.05) at the beginning of the breeding season (April) than at other times of the year. Similarly, the condition indices of females varied with season (F = 2.34, df = 5, P < 0.05), although pairwise comparisons (Tukey post hoc tests) failed to identify a month or season with healthier individuals. There were no significant differences in the condition indices of juvenile chuditch at Batalling throughout the year, although a lack of a difference may be subject to a low sample size of juveniles captured during each trip.

Providing fox control is maintained, it is likely that the size, frequency and timing of fuel reduction burns currently prescribed in the jarrah forest benefit the mobile and opportunistic chuditch. The resulting mosaic of vegetation ages provides a diverse food supply and ensures a supply of refuge sites. This is in contrast to the unplanned, catastrophic wildfires which would not only cause direct mortality but serve to homogenise the forest vegetation.

IMPACT OF FOX CONTROL

The European red fox (*Vulpes vulpes*) arrived in the south-west of WA in the late 1920s (King and Smith 1985) and has been implicated in the decline of several species of medium-sized marsupials (Burbidge and McKenzie 1989). Like all canids,

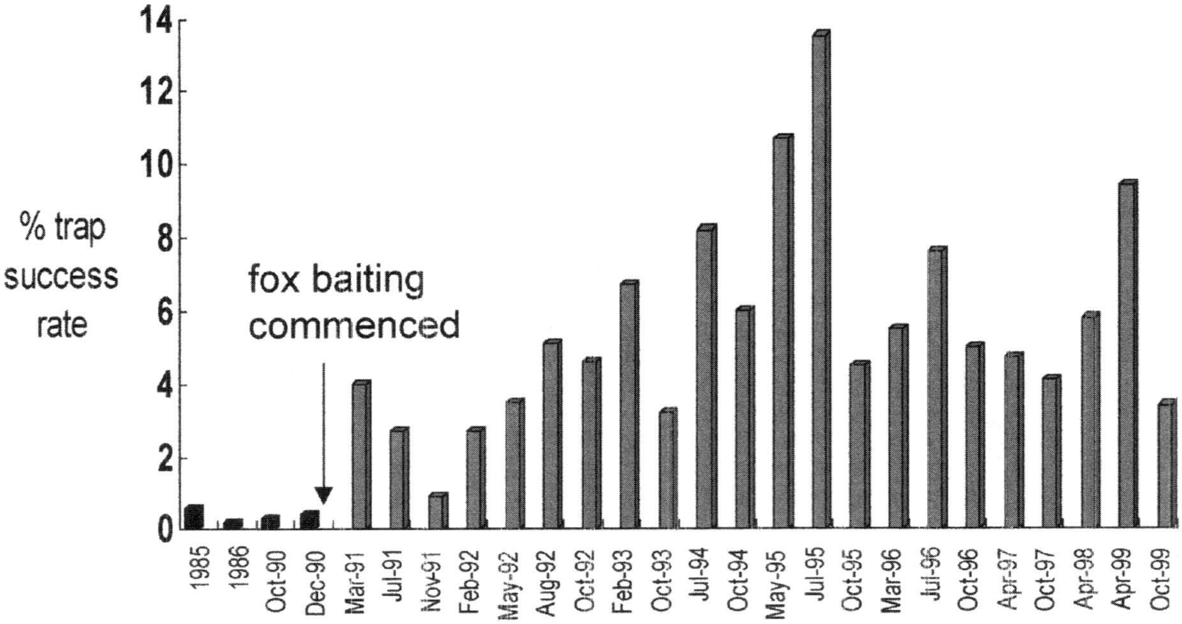

Figure 8 Trap success rates for chuditch in the Batalling forest before and after fox control. (Note: trapping after October 1997 used 'chuditch' bait rather than universal bait. Trapping effort for 1985, 1986 and 1990 was 50 cage traps set for 3 nights = 150 trapnights, from March 1991 onwards it was 120 cage traps for 4 nights = 480 trapnights).

foxes are highly susceptible to the toxin sodium monofluoroacetate, or 1080 (McIlroy 1981, McIlroy and King 1990) while most native WA mammals are tolerant to this poison (King et al. 1978, 1981; Mead et al. 1979). In areas of Western Australia where foxes are controlled with regular 1080 poisoning, populations of rock-wallabies (*Petrogale lateralis*), numbat (*Myrmecobius fasciatus*), tammar wallaby (*Macropus eugenii*), woylie (*Bettongia penicillata*) and brushtail possum (*Trichosurus vulpecula*) increase significantly (Kinnear et al. 1988, 1998, Kinnear 1990, Friend 1990, Start et al. 1998). Foxes have the potential to detrimentally impact on chuditch directly through predation, particularly of young animals. As both the fox and the chuditch have a high proportion of large invertebrates in their diet (Soderquist and Serena 1994; Lunney et al. 1990; Coman 1973) competition with the fox for food may also be involved in chuditch declines.

Broadscale fox control programs using dried meat baits impregnated with 1080 are now an effective means of managing populations of threatened mammals, including chuditch, in Western Australia. Under the Western Shield fauna recovery program (Bailey 1996), foxes are now controlled over approximately 3.5 million hectares of conservation estate. This has resulted in the recovery of several mammals and the woylie, quenda (southern brown bandicoot, *Isoodon obesulus*) and tammar wallaby, have been removed from the State and Commonwealth threatened fauna lists (Morris et al. 1998).

Before broadscale fox baiting could be commenced in the jarrah forest, the impact on the carnivorous chuditch had to be determined. While chuditch have some resistance to 1080, it is not as high as for native herbivorous species (King et al. 1989), and it was predicted that an individual would only need to consume two baits to obtain a lethal dose (Soderquist and Serena 1993). An initial trial of non-toxic baits on captive chuditch indicated that moister food was preferred to the dried meat baits. A baiting trial using toxic baits was then undertaken on wild chuditch at Batalling forest to measure the impact of 1080 baiting on abundance and reproduction (Morris et al. 1995). Radio-collars were placed on 12 chuditch and their movements were monitored before, during and for three months after toxic 1080 dried meat baits had been distributed. Longer term monitoring has continued by trapping along road transects.

None of the radio-collared chuditch died during the fox baiting trial. There was evidence that some baits were taken by chuditch but this was apparently not in sufficient quantities to kill them. Chuditch trap success has increased since fox control was commenced from approximately 0.5% in December 1990 to a peak of 13% in July 1995 (Fig. 8). Trap success rates since have been maintained at 4–6%. The issue of trap saturation by woylies at Kingston referred to earlier, is also apparent at Batalling, with approximately five times the number of chuditch being trapped on 'chuditch' bait compared to the universal peanut butter bait (Table 3). The number of empty traps

Table 3 Results of a trial at Batalling comparing trap success rates of medium-sized mammals using universal peanut butter bait and meat-meal-based 'chuditch' bait.

Species	Mean trap success rate (%) universal bait	Mean trap success rate (%) chuditch bait	Ratio chuditch / universal baits
Chuditch	1.7	9.4	5.5
Woylie	67.2	37.8	0.6
Brushtail Possum	6.7	5.6	0.8
Quenda	0.6	0.6	1.0
TOTAL	**76.2**	**53.4**	**0.7**

was significantly higher using chuditch bait (5% on universal bait *vs* 28.9% on chuditch bait).

FOREST MONITORING

In addition to the sites of more intensive forest management impact research (Kingston and Batalling), chuditch have been monitored at several sites in the jarrah forest over the last five years as part of the Western Shield program (Fig. 9). In addition to the fox control, chuditch have benefited from the considerable amount of felled timber and offcuts that have been left after logging. Hollows have been created in these through the actions of fire and these provide good quality refuge sites. At Hills Forest and Noggerup, trap success rates have increased from 0–0.5% to 4–7%. At Tone forest monitoring site, trap success rates of 0.5–2.5% are being maintained. Monitoring at the Shannon National Park commenced in September 1999 and trap success rates have been maintained at between 1–2% in that time. In the St. John forest block, chuditch were trapped for the

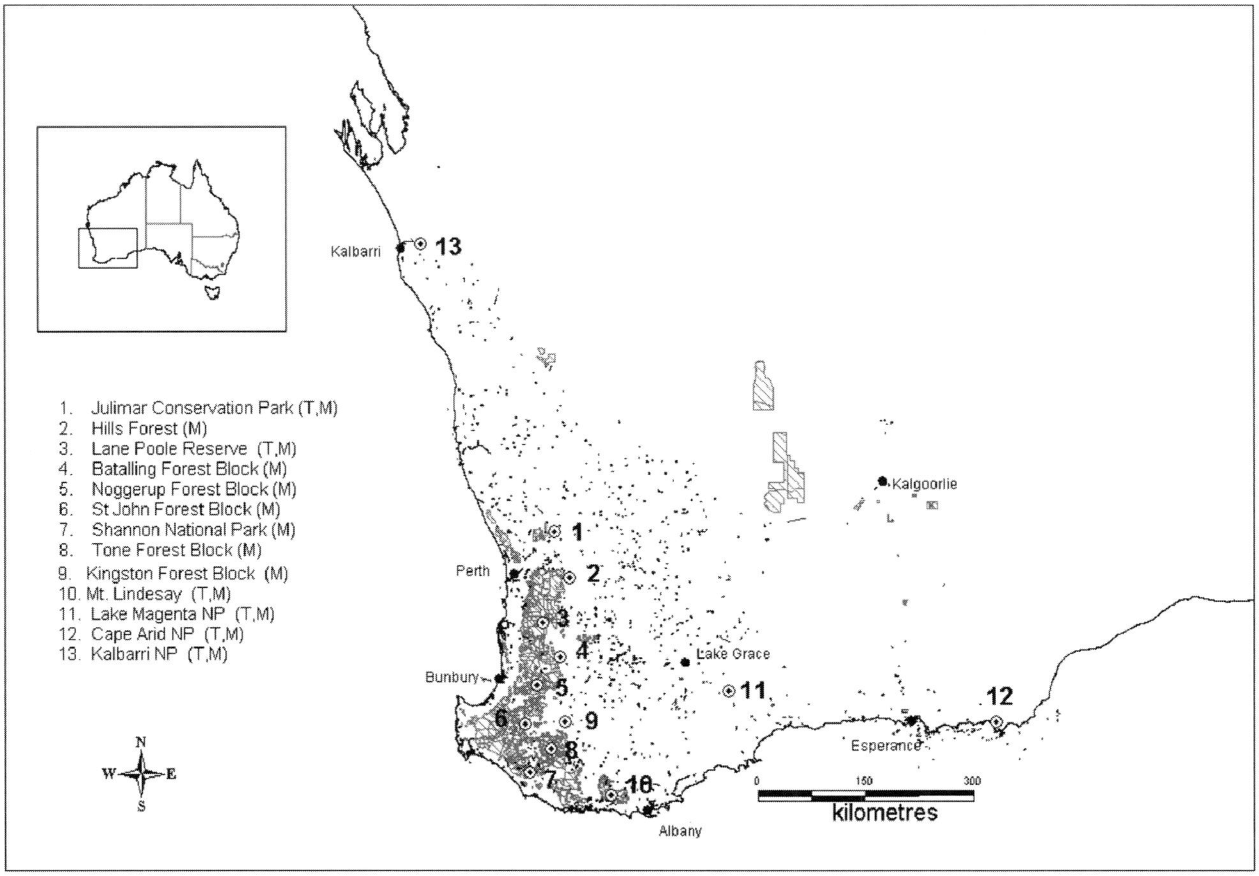

Figure 9 Chuditch translocation and monitoring sites in Western Australia.

first time in February 2000 after fox control commenced in July 1997. One of the criteria for success for chuditch recovery was the maintenance of trap success rates above 1% and this appears to have been achieved at most monitoring sites in the jarrah forest. Chuditch have also been intensively monitored at 41 sites in the northern jarrah forest as part of research into the effectiveness of fox baiting regimes and fauna recovery (Paul de Tores, Science Division pers. comm.) and increases in abundance have also been recorded at these sites.

CAPTIVE BREEDING

The first captive breeding program for chuditch commenced in 1986 at Dwellingup as part of a study into the ecology and management of chuditch (Serena et al. 1991; Serena and Soderquist 1988). In 1990, the captive-breeding colony was transferred to Perth Zoo to provide animals for CALM's proposed translocation program, and for display and education purposes. Because chuditch occur at low densities in the wild it was not possible to source translocations with wild-caught animals without some detrimental impact to these populations. Between 1990 and 2000, approximately 15 pairs of chuditch were maintained as breeding stock each year and over 300 chuditch were weaned successfully and used in translocation programs (Table 4). To ensure genetic variation was maintained in the population, three to four wild-caught animals were introduced into the captive breeding colony each year.

TRANSLOCATIONS

Chuditch have disappeared from about 90% of their former range and with the success of the captive breeding programs, translocations have been undertaken to six areas of WA where they formerly occurred (Table 7, Fig. 9).

Table 4 Births and sex ratios of all chuditch born at Perth Zoo (# males, # females, # unsexed).

Year	# births	Sex ratio	# weaned
1990	15	9.6.0	9.6
1991	14	4.8.2	4.8
1992	26	10.12.4	10.12
1993	20	7.10.3	7.10
1994	12	6.6.0	6.6
1995	37	16.19.2	11.18
1996	48	21.21.6	20.21
1997	54	27.26.1	27.26
1998	58	28.28.2	28.28
1999	53	27.20.6	27.20
2000	12	0.0.12	0.0.12
TOTAL	**349**	**155.156.38**	**149.155.12 = 316**

Lane-Poole Conservation Park

In 1987 an experimental translocation to the Lane-Poole Conservation Reserve was undertaken. No fox control was undertaken at this time. Nine captive-reared chuditch were radio-collared and released. This was unsuccessful, with at least seven of the nine being killed, or dying within a few months of release (Serena et al. 1991). The main causes of mortality were predation by cat, dog or pig and owls, illegal shooting, and road kills. The young age of the released animals was also considered a factor. A fox baiting trial at Batalling in 1990/1991 subsequently demonstrated that chuditch abundance increased following fox control. Consequently, all subsequent translocations have been undertaken to areas where fox baiting is undertaken as a normal management prescription: Julimar Conservation Park, Lake Magenta Nature Reserve, Cape Arid National Park, Mt Lindesay National Park and Kalbarri National Park.

Julimar Conservation Park

Julimar Conservation Park was selected as a translocation site because it was sufficiently large (24,000 ha), readily monitored (only 70 km from Perth) and had suitable habitat attributes for chuditch (adequate prey and refugia). The area is essentially an island of jarrah/marri/wandoo forest lying near the northern extent of the jarrah forest. Chuditch had been last recorded in 1973 (Harry Butler pers. comm), and surveys in 1983 and just prior to the translocation failed to detect them. Fox baiting commenced six months prior to the release.

In September 1992, 24 chuditch (of which 20 were fitted with radio-collars) were released. These animals were intensively tracked for six weeks and then subsequently monitored by both trapping and radiotelemetry. Four of the released animals were known to have died: one to gunshot, two to fox/cat predation and one to unknown causes. Refuge choice improved over the first two weeks and better quality sites such as burrows and hollow logs were more commonly chosen by the end of the first month. Breeding occurred within 12 months of release, during 1993, and has continued each year since. Another 39 chuditch were released between October 1993 and May 1995 to supplement the population and increase genetic vigour.

Annual monitoring of the population indicates that this translocation has been successful. In July 1998, trap success rates exceeded 14% but have since reduced to 3–8% (Fig. 10). There are now regular sightings of chuditch on roads and properties around the park.

Lake Magenta Nature Reseve

With the success of the Julimar translocation, a translocation to a semi-arid site became a priority for the recovery team, as extensive attempts to locate a substantial chuditch population within the semi-arid wheatbelt of Western Australia (where chuditch were known from earlier this century) had been unsuccessful. Lake

Table 5 Numbers of chuditch released at translocation sites in WA.

Year	# released	Source	Release sites
1987	9	Captive-bred – Dwellingup	Lane Poole Conservation Park
1992	23	Perth Zoo	Julimar Conservation Park
1993	19	Perth Zoo	Julimar Conservation Park
1994	18	Perth Zoo	Julimar Conservation Park
1995	2	Perth Zoo	Julimar Conservation Park
1996	31	Perth Zoo	Lake Magenta Nature Reserve
1997	35	Perth Zoo	Lake Magenta Nature Reserve
1998	15	Perth Zoo	Lake Magenta Nature Reserve
1998	40	Perth Zoo	Cape Arid National Park
1999	6	Perth Zoo	Cape Arid National Park
1999	48	Perth Zoo	Mount Lindesay National Park
2000	15	Perth Zoo	Mount Lindesay National Park
2000	15	Perth Zoo	Cape Arid National Park
2000	35	Perth Zoo	Kalbarri National Park
TOTAL	**311**		**6 sites**

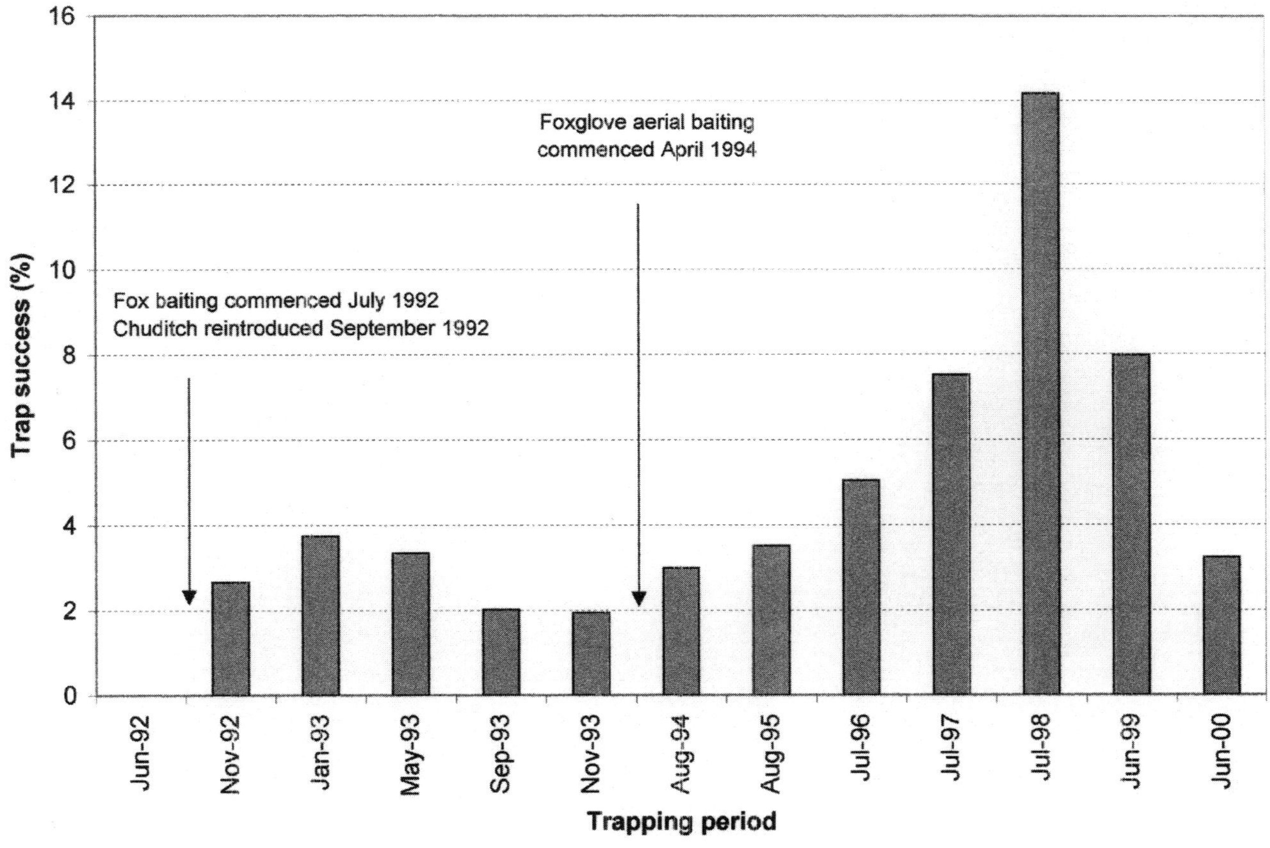

Figure 10 Trap success rates for a translocated chuditch population at Julimar Conservation Park. (Trapping effort for each date was 120 cage traps for 4 nights = 480 trapnights.)

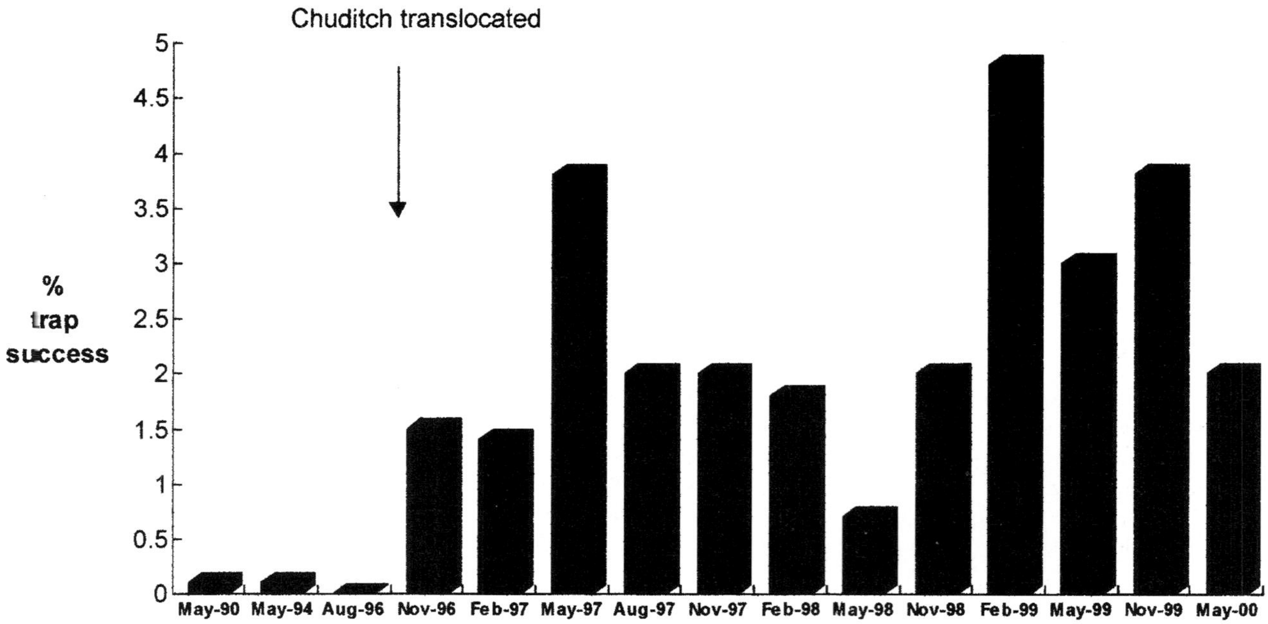

Figure 11 Trap success rates for chuditch at Lake Magenta Nature Reserve. (Trapping effort for each date was 100 cage traps for 4 nights = 400 trapnights.)

Magenta Nature Reserve, approximately 450 km south east of Perth was selected because it was sufficiently large (108,000 ha) and chuditch were known to persist there until 1994.

Fox control commenced in July 1996 and 31 chuditch were released in October 1996. Twenty of these were fitted with mortality sensing radio-collars. No feral cat control was implemented. The release site spanned salmon gum (*Eucalyptus salmonphloia*) woodlands and mallee shrublands. Intensive ground and aerial monitoring of radio-collared collared chuditch was undertaken to determine their location and status, and refuge site selection for the first six weeks, and regular monitoring continued thereafter for the six month battery life of the radio-collars.

Of the 20 radio-collared animals, six are known to have died. Four of these were females and starvation was thought to be the major contributing factor. Two other chuditch were taken by either cats or foxes, but it is not known whether they were taken alive, or dead as carrion.

An additional 50 chuditch were released in 1997 and 1998 and monitoring has been carried out by regular trapping sessions since release (Fig. 11). Breeding and recruitment were evident in the first and subsequent years. In 1999 two of the chuditch released in 1996 were trapped indicating good survivorship of these captive-bred animals. Interestingly, an unmarked chuditch was trapped in February 1997. This must have been a resident animal as breeding by the animals translocated in 1996 had not yet occurred. This indicates that a small population of chuditch must have persisted at Lake Magenta prior to the initial release and that this translocation should be regarded as a re-stocking rather than a re-introduction. Since 1996, trap success rates have increased to 3–4% suggesting that this translocation has been successful. This satisfies the second recovery plan objective to establish a chuditch population in a semi-arid area. The establishment of a population at Lake Magenta has also provided opportunities to examine the ecology of chuditch in the semi-arid parts of its range.

Cape Arid National Park

After the success of translocations to Julimar and Lake Magenta, a third translocation was undertaken to Cape Arid National Park, 150 km east of Esperance near the western end of the Great Australian Bight. There are no historic records of chuditch at Cape Arid, although they were very likely to have occurred there. The closest known records are from Hopetoun in 1990 (200 km west of Cape Arid) and at Kambalda in 1971 (300 km north of Cape Arid). The park comprises nearly 280,000 ha and contains a mix of coastal and semi-arid vegetation communities including large areas of heath with patches of mallee and eucalypt woodland. Granite outcrops occur throughout the coastal zone often with fringing *Melaleuca* thickets.

Twenty captive-bred chuditch (10 fitted with radio-collars) were released in March 1998, and a period of intense monitoring followed. It was found that, as with previous releases, the chuditch lost weight for the first four to six weeks and then stabilised body weights and gradually returned to near release condition. Refuge choice improved with time and rabbit burrows were found to be a commonly used diurnal shelter. Most chuditch stayed within the proximity of the release site for the initial monitoring period, perhaps indicating that sufficient food and shelter were available in the area. However, one individual was recaptured 10 weeks after release in a farmer's chicken pen at Salmon Gums, over 180 km from the release site. This male was found to be in excellent condition and was returned to the park.

Only one predation was recorded for this first group and that event was attributed to cannabilism of a female by an overzealous male chuditch during the breeding period. Following the release of a further 20 chuditch in late April 1998 one further death was recorded due to a roadkill. Breeding success was confirmed by the capture, in June 1998, of a female from the first release with pouch young. Subsequently both juvenile chuditch and lactating adult females were captured in December 1998. Monitoring in 1999 and 2000 confirmed that breeding and recruitment were continuing.

Only approximately one-third of Cape Arid National Park is baited for foxes and both foxes and feral cats remain in the release area. The persistence of chuditch suggests that they can withstand, to some degree, the threat posed by these introduced predators. Indeed, it was interesting to find cat fur and claws in the scat of a chuditch.

Mt Lindesay State Forest

Mt Lindesay is a proposed National Park approximately 400 km south of Perth. Although linked to other forest areas to the west, the park forms a peninsula of native vegetation surrounded by farmland. This site is near the eastern- and southern-most extremity of the jarrah forest in Western Australia. Much of the vegetation is eucalypt forest (jarrah/marri) and smaller areas of heath and woodland on lateritic and sandy soils. Several watercourses dissect the area with associated riparian zones and swamps. Numerous granite outcrops occur at the site.

Of the 48 chuditch released in 1999, five of the eight that were fitted with radio-collars are known to have died from either fox predation or from roadkill. It is too early to determine whether this translocation has been successful. The future monitoring of this translocation continues with the assistance of a local community group.

Kalbarri National Park

The final translocation undertaken as part of the recovery program was to Kalbarri National Park, approximately 650 km north of Perth, on the northern edge of the wheatbelt. The dominant vegetation type is heath with *Banksia* shrublands. Thirty-five chuditch were released in July 2000 and another 15 released in March 2001. None were radio-collared and monitoring will be by trapping transect. Success of this translocation is too early to gauge.

FUTURE DIRECTIONS

The chuditch recovery program has been extremely successful and the abundance and distribution of the chuditch in the south-west of WA has increased to the level where a review against IUCN (1994) criteria in 2003 may see the species satisfy the criteria for delisting. Two of the three recovery plan objectives have been met, i.e. the maintenance of trap success rates above 1% at jarrah forest monitoring sites, and the establishment of a population in the semi-arid wheatbelt. Because of fox control, trap success rates have increased at least five-fold at forest monitoring sites over the last eight years. Chuditch persist, and in some cases increase in abundance in the presence of logging and prescribed burning operations. The success of the captive breeding program has also been important to allow translocations to occur, particularly to the wheatbelt where chuditch populations were still declining in the mid-1990s. In addition, the implementation of the recovery plan has added significantly to the knowledge of the ecology and biology of the chuditch.

The success of the program has been due largely to a cooperative team approach of funding agencies, conservation agencies and industry, the identification of the major factors responsible for the decline of this species, and the implementation of remedial conservation actions. This program could be used as a model for recovery programs for other threatened dasyurids.

The recovery of chuditch also has implications for the recovery of other threatened species. Predation by chuditch on other translocated threatened species, such as numbats, can impact on the success of future establishment and the sequence of re-introduction needs to be determined. If fauna reconstruction is the aim, then the reintroduction of a predator such as chuditch needs to be undertaken last when other populations have become established and can withstand a greater level of predation. Feral cat control in the arid zone presently relies on trapping and the use of a small sausage bait which is palatable to chuditch (Algar 2000). Both techniques need modification before chuditch translocations can commence in the arid zone.

The increase in abundance of a predatory species that was formerly regarded as a pest by some rural communities, also raises the issue of the importance of communicating to the public, the benefits of reconstructing all components of the depauperate medium-sized mammal fauna. Solutions such as an extensive publicity program, relocating problem individuals, rescuing injured animals and providing a design for a chuditch (and fox/cat) proof poultry cage have been successful in convincing the public about the merits of such a recovery program.

ACKNOWLEDGEMENTS

Implementation of the chuditch recovery plan has been a cooperative effort between several agencies. We would specifically like to thank other members of the chuditch recovery team for their support and guidance in implementing the recovery plan: Dr Andrew Burbidge (CALM), Mr Bob Hagan (CALM), Mr Kim Williams (CALM), Mr Brett Beecham (CALM), Mr John Gardner (Alcoa), Ms Sandra McKenzie (WWF), Dr Mark Bradley (Perth Zoo), Dr Terry Fletcher (Perth Zoo) and various Environment Australia representatives. We would also like to thank the many CALM staff who assisted in the field work associated with chuditch recovery particularly Dr Gordon Friend, Mr Mal Graham, Mr Rob Brazell, Mr Colin Ward, Mr John Rooney, and Mr Ian Wheeler. Ms Amanda Mellican and Mr Matt Williams provided assistance with population abundance and statistical analysis. Ms Joanne Varley assisted in the preparation of the figures. Dr Tony Friend, Dr Andrew Burbidge and Dr Ian Abbott provided useful comments on a draft of this paper. Two anonymous referees reviewed the paper.

REFERENCES

Algar, D. (2000), 'Introduced predators in the arid zone: the WA experience – consideration of the impact on re-introductions and the need for control', in *Biodiversity and the re-introduction of native fauna at Uluru–Kata Tjuta National Park* (eds. J.S. Gillen, R. Hamilton, W.A. Low, & C. Creagh), pp. 50–2, Proceedings of a cross-cultural workshop on fauna reintroduction, Yulara, NT, Bureau of Rural Sciences, Canberra.

Arnold, J., & Shield, J. (1970), 'Oxygen consumption and body temperature of the chuditch (*Dasyurus geoffroii*)', *Journal of Zoology,* **160**:391–404.

Bailey, C. (1996), 'Western Shield – bringing wildlife back from the brink of extinction', *Landscope,* **11**:41–8.

Bakker, H.R., & Main, A.R. (1980), 'Condition, body composition and total body water estimation in the quokka (*Setonix brachyurus*)', *Australian Journal of Zoology,* **28**:395–406.

Boscacci, L.J, McKenzie N.L., & Kemper, C.H (1987), 'Mammals', in *A Biological Survey of the Nullarbor Region South and Western Australia in 1984* (eds. N.L. Mckenzie, & A.C. Robinson), pp. 103–38, South Australian Dept of Environment and Planning, Western Australian Dept Conservation and Land Management, Australian National Parks and Wildlife Service, Government Printer, Adelaide.

Bradshaw F.J., Adams R., Sneeuwjagt R., Low K., Havel J.J., Bartle J.R., & Stoneman G.L. (1991), 'The jarrah forest: a case study in multiple use', in *Forest Management in Australia* (eds. F.H. McKinnell, E.R. Hopkins, & J.E.D. Fox), pp. 1–21, Surrey Beatty and Sons, Sydney NSW.

Burbidge, A.A., Johnson, K.A., Fuller, P.J., & Southgate, R.I. (1988), 'Aboriginal knowledge of the mammals of the central deserts of Australia', *Australian Wildlife Research,* **15**:9–39.

Burbidge, A.A., & McKenzie, N.L. (1989), 'Patterns in the modern decline of Western Australia's vertebrate fauna: causes and conservation implications', *Biological Conservation,* **50**:143–98.

Christensen, P.E.S. (1980), 'The biology of *Bettongia penicillata* Gray, 1837, and *Macropus eugenii* (Desmarest, 1817) in relation to fire', *Forests Department of Western Australia Bulletin,* **91**.

Collett, R. (1887), 'On a collection of mammals from central and northern Queensland', *Zoologischen Jahrbuchen. Zeitschrift fur Systematik, Geographie und Biologie der Thiere*, Gustav Fischer, Jena.

Coman, B.J. (1973), 'The diet of red foxes, *Vulpes vulpes* L, in Victoria', *Australian Journal of Zoology*, **21**:391–401.

Cuttle, P. (1982), 'Life history strategy of the dasyurid marsupial *Phascogale tapoatafa*', in *Carnivorous Marsupials, Vol. 1* (ed. M. Archer), pp. 325–32, Royal Zoological Society of New South Wales, Mosman, NSW.

Dell, B., & Havel, J.J., (1989), 'The jarrah forest, an introduction', in *The Jarrah Forest: A complex mediterranean ecosystem* (eds. B. Dell, J. Havel, & N. Malajczuk), pp. 1–10, Kluwer Academic Publishers, Dordrecht, Boston, London.

Finlayson, H.H. (1961), 'On central Australian mammals. Part IV, the distribution and status of central Australian species', *Records of the South Australian Museum,* **14**:141–91.

Firestone, K.B. (1999), 'The application of molecular genetics to the conservation management of quolls, *Dasyurus* species (Dasyuridae: Marsupialia)', PhD thesis, University of New South Wales.

Friend, J.A. (1990), 'The Numbat *Mymecobius fasciatus* (Myrmecobiidae): history of decline and potential recovery', *Proceedings of the Ecological Society of Australia,* **16**:369–77.

Gould, J. (1841), 'Observations on *Dasyurus maugei* and *D. viverrinus* of Geoffroy, and description of a new species', *Proceedings of the Zoological Society of London* (1840), p. 151.

Heberle G. (1997), 'Timber harvesting of Crown Land in the south-west of Western Australia: an historical review with maps', *CALMScience,* **2**:203–24.

Iredale, T., & Troughton, E. le G. (1934), 'A check-list of the mammals recorded from Australia', *Australian Museum Memoirs,* **6**:1–122.

IUCN (1994), *IUCN Red List Categories: prepared by the IUCN Species Survival Commision*, IUCN, Gland, Switzerland.

Johnson, K.A., & Roff, A.D. (1982), 'The western quoll, *Dasyurus geoffroii* (Dasyuridae: Marsupialia) in the Northern Territory: historical records from venerable sources', in *Carnivorous Marsupials, Vol. 1* (ed. M. Archer), pp. 221–6, Royal Zoological Society of New South Wales, Mosman, NSW.

Kennedy, M. (1992), 'Species recovery outline. No. 4. The Chuditch', in *Australasian Marsupials and Monotremes: an Action Plan for Their Conservation* (ed. M. Kennedy), pp. 71–2, IUCN, Gland, Switzerland.

King, D.R., Oliver, A.J., & Mead, R.J (1978), 'The adaptation of some Western Australian mammals to food plants containing fluoroacetate', *Australian Journal of Zoology,* **26**:699–712.

King, D.R., Oliver, A.J., & Mead, R.J (1981), '*Bettongia* and fluouroacetate: a role for 1080 in fauna management', *Australian Wildlife Research,* **8**:529–36.

King, D.R., & Smith, L.A (1985), 'The distribution of the European Red Fox (*Vulpes vulpes*) in Western Australia', *Records of the Western Australian Museum,* **12**:197–205.

King, D.R., Twigg, L.E., & Gardner, J.L. (1989), 'Tolerance to sodium monofluoroacetate in dasyurids from Western Australia', *Australian Wildlife Research,* **16**:131– 40.

Kinnear, J.E., Onus, M.L., & Bromilow, R.N (1988), 'Fox control and rock-wallaby population dynamics', *Australian Wildlife Research,* **15**:435–50.

Kinnear, J.E. (1990), 'Trappings of success', *Landscope,* **5**:35–40.

Kinnear, J.E., Onus, M.L., & Sumner, N.R. (1998), 'Fox control and rock-wallaby population dynamics – II. An update', *Wildlife Research,* **25**:81–8

Kitchener, D.J., Chapman A., & Barron G. (1978), 'Mammals of the northern Swan Coastal Plain', in *Faunal Studies of the Northern Swan Coastal Plain: a Consideration of Past and Future Changes*, pp. 54–92, Western Australian Museum, Perth.

Kitchener, D.J., & Vicker, E. (1981), *Catalogue of Modern Mammals in the Western Australian Museum, 1895 to 1981*, Western Australian Museum, Perth, 184 p.

Krebs, C.J., & Singleton, G.R. (1993), 'Indices of condition for small mammals', *Australian Journal of Zoology,* **41**:317–23.

Krefft, G. (1866), 'On the vertebrated animals of the lower Murray and Darling: their habits, economy and geographical distribution', *Transactions of the Philosophical Society of New South Wales,* **1**:1–33.

Lebreton, J.D., Burnham, K.P., Clobert, J., & Anderson, D.R. (1992), 'Modelling survival and testing biological hypotheses using marked animals: A unified approach with case studies', *Ecological Monographs,* **62**:67–118

Lunney, D., Trigg, B., Eby, P., & Ashby, E. (1990), 'Analysis of scats of dogs, *Canis familiaris*, and foxes, *Vulpes vulpes* (Canidae: Carnivora), in coastal Bega, New South Wales', *Australian Wildlife Research,* **17**:61–8.

Mathew, H. (1996), 'An investigation into the effect of management strategies on the home range of chuditch (*Dasyurus geoffroii*)', BSc (Honours) thesis, University of Western Australia, 59 p.

Maxwell, S., Burbidge, A.A., & Morris, K. (1996) 'The Chuditch *Dasyurus geoffroii*', in *The 1996 Action Plan for Australian Marsupials and Monotremes*, pp. 83–4, Wildlife Australia, Canberra.

McIlroy, J.C. (1981), 'The sensitivity of Australian animals to 1080 poison. II Marsupial and eutherian carnivores', *Australian Wildlife Research,* **8**:385–99.

McIlroy, J.C., & King, D.R. (1990), 'Appropriate amounts of 1080 poison in baits to control foxes *Vulpes vulpes*', *Australian Wildlife Research,* **17**:11–13.

McKenzie, N.L., Hall, N., & Muir, W.P. (2000), 'Non-volant mammals of the southern Carnarvon Basin, Western Australia', in *Biodiversity of the Southern Carnarvon Basin* (eds. A.H. Burbidge, M.S. Harvey, & N.L. McKenzie), pp. 479–510, Records of the Western Australian Museum Supplement No. 61.

Mead, R.J., Oliver, A.J., & King, D.R. (1979), 'Metabolism and defluorination of fluoroacetate in the brushtail possum (*Trichosurus vulpecula*)', *Australian Journal of Biological Science,* **32**:15–26.

Morris, K., Armstrong, R., Orell, P., & Vance, M. (1998), 'Bouncing back: Western Shield update', *Landscope,* **14**:28–35.

Morris K., Johnson B., Rooney J., & Ward C. (2000), 'The short-term impacts of timber harvesting and associated activities on the abundance of medium-sized mammals in the Jarrah forest of Western Australia', in *Nature Conservation 5: Nature Conservation in Production Environments: Managing the Matrix* (eds. J.L. Craig, N. Mitchell, & D.A. Saunders), pp. 60–70, Surrey Beatty, Sydney, NSW.

Morris, K., Orell, P., & Brazell, R. (1995), 'The effect of fox control on native mammals in the jarrah forest, Western Australia', in *Proceedings of the 10th Australian Vertebrate Pest Control Conference, Hobart, May 1995* (eds. M. Stratham, & K. Buggy), pp. 177–81, Department of Primary Industry and Fisheries, Hobart, Tasmania.

Oakwood, M. (2000), 'Reproduction and demography of the northern quoll, *Dasyurus hallucatus*, in the lowland savanna of northern Australia', *Australian Journal of Zoology,* **48**:519–39.

Orell, P., & Morris, K. (1994), 'Chuditch recovery plan, 1992–2001', *Department of Conservation and Land Management, Western Australia, Wildlife Management Program,* **13**:1–25.

Redner, R. (1999), 'A study into the Chuditch *Dasyurus geoffroii* within the Lake Magenta Nature Reserve', BA – Environmental Studies (Honours) thesis, Notre Dame University, Fremantle WA, 97 p.

Serena, M., & Soderquist, T.R. (1988), 'Growth and development of pouch young of wild and captive Dasyurus geoffroii (Marsupialia: Dasyuridae)', *Australian Journal of Zoology,* **36**:533–43.

Serena, M., & Soderquist, T.R. (1989a), 'Nursery dens of *Dasyurus geoffroii* (Marsupialia: Dasyuridae) with notes on nest building behaviour', *Australian Mammalogy,* **12**:35–6.

Serena, M., & Soderquist, T.R. (1989b), 'Spatial organization of a riparian population of the carnivorous marsupial *Dasyurus geoffroii*', *Journal of Zoology,* **219**:373–83.

Serena, M., & Soderquist, T. (1995), 'Western quoll: *Dasyurus geoffroii*, Gould, 1841', in *The Mammals of Australia* (ed. R. Strahan), pp. 62–4, Reed, Sydney.

Serena, M., Soderquist, T.R., & Morris, K. (1991), 'The chuditch: *Dasyurus geoffroii*', Department of Conservation and Land Management, Western Australia, Wildlife Management Program, **7**:1–32.

Shortridge, G.C. (1909), 'An account of the geographical distribution of the marsupials and monotremes of south-west Australia having special reference to the specimens collected during the Balston expedition of 1904–1907', *Proceedings of the Zoological Society of London,* (1909) pp. 803–48.

Soderquist, T.R. (1995), 'Ontogeny of sexual dimorphism in size among polytocous mammals: tests of two carnivorous marsupials', *Journal of Mammalogy,* **76**:376–90.

Soderquist, T.R., & Serena, M. (1990), 'Occurrence and outcome of polyoestry in wild western quolls, *Dasyurus geoffroii* (Marsupialia: Dasyuridae)', *Australian Mammalogy,* **13**:205–8.

Soderquist, T.R., & Serena, M. (1993), 'Predicted susceptibility of *Dasyurus geoffroii* to canid baiting programmes: variation due to sex, season and bait type', *Wildlife Research,* **20**:287–96.

Soderquist, T.R., & Serena, M. (1994), 'Dietary niche of the western quoll, *Dasyurus geoffroii*, in the jarrah forest of Western Australia', *Australian Mammalogy,* **17**:133–6.

Soderquist, T.R., & Serena, M. (2000), 'Juvenile behaviour and dispersal of chuditch (*Dasyurus geoffroii*) (Marsupialia: Dasyuridae)', *Australian Journal of Zoology,* **48**:551–61.

Start, A.N., Burbidge, A.A., & Armstrong, D. (1998), 'A review of the conservation status of the Woylie *Bettongia penicillata ogilby* (Marsupialia: Potoroidae) using IUCN criteria', *CALMScience,* **2**:277–90.

Stoneman, G.L., Bradshaw, F.J., & Christensen, P. (1989), 'Silviculture', in *The Jarrah Forest: a complex mediterranean ecosystem* (eds: B. Dell, J.J. Havel, & N. Malajczuk), pp. 335–55, Kluwer Academic Publishers, Dordrecht, Boston, London.

Thomas, O. (1906), 'On mammals collected in south-west Australia for Mr. W.E. Balston', *Proceedings of the Zoological Society of London,* pp. 468–78.

Wakefield, N.A. (1966), 'Mammals of the Blandowski Expedition to north-western Victoria, 1856–57', *Proceedings of the Royal Society of Victoria,* **79**:371–91.

Whittell, H.M. (1954), 'John Gilbert's notebook on marsupials', *Western Australian Naturalist,* **4**:104–14.

CHAPTER 31

CONSERVATION OF THE NUMBAT (*MYRMECOBIUS FASCIATUS*)

J. Anthony Friend[A] *and Neil D. Thomas*[B]

Science Division, Department of Conservation and Land Management
[A] 120 Albany Highway, Albany, WA 6330, Australia
[B] Woodvale Research Centre, PO Box 51, Wanneroo, WA 6946, Australia

At the time of European settlement in Australia, the distribution of the numbat extended across much of southern Australia. By 1985 only two small populations survived, at sites 160 km apart in the south-west of Western Australia. The decline of the numbat coincided largely, although not wholly, with the spread of the fox (*Vulpes vulpes*) from its point of introduction in Victoria.

During the early 1980s, an experimental fox control and numbat monitoring program compared changes in numbat abundance in baited and unbaited habitat at Dryandra Woodland. This experiment showed that growth in the numbat population followed the introduction of monthly fox baiting. In 1985 reintroduction of numbats into Boyagin Nature Reserve commenced in conjunction with regular fox control and a population persists there today.

A program of fox control and numbat translocation has resulted in the establishment of six further reintroduced populations in Western Australia. Action by Earth Sanctuaries Limited has resulted in the establishment of two additional populations in fenced sanctuaries in South Australia and New South Wales. In 1994, an assessment of the status of the numbat under the existing IUCN Red List criteria recommended a change from Endangered to Vulnerable.

INTRODUCTION

Since the day in 1831 when Richard Dale and George Fletcher Moore pulled their 'beautiful animal' out of a hollow log near Brookton, Western Australia, (Dale 1833), the numbat has excited great interest due to its unusual diet, habits and appearance (Friend 1989). More recently, fascination has been increased by its rarity even in its stronghold in the south-west of the continent. Once widespread across the southern half of Australia, by 1985 the species was found only in two small populations 160 km apart (Friend 1990).

Dale's specimen was described as a new genus and species in a brief note by Waterhouse (1836) and later in more detail with a coloured illustration (Waterhouse 1841) but he was unsure of the affinities of the species. Gill (1872) placed the numbat in its own family, the Myrmecobiidae, and its distinctiveness is still recognised at that level today. Although the family is placed, together with the dasyurids and thylacinids, in the Order Dasyuromorphia, beyond that its relationships remain uncertain (Wroe and Musser 2001; Krajewski and Westerman 2003, this volume).

This article examines the conservation status of the numbat, traces the decline of the species close to extinction and relates the management actions that have moved its status towards security.

DISTRIBUTION AT EUROPEAN SETTLEMENT

Historical reconstruction of the numbat's range before European settlement in the early 1800s has been based on museum specimens collected alive, remains found in subfossil deposits, published anecdotal accounts and Aboriginal knowledge recorded through interviews (Friend 1990). Records of the species are scattered across a wide arc stretching from western New South Wales and south-eastern South Australia through semi-arid and arid lands, north to the southern border of the Northern Territory and across to the south-west of Western Australia, where it was clearly quite common.

Numbats occurred in a range of vegetation types, but the majority of sites were characterised by the presence of eucalypt species. All evidence relating to the diet of the numbat throughout its range indicates an almost complete dependence on termites (Calaby 1960) and an abundance of these insects appears to be a prerequisite for the existence of numbat populations. The other necessary factor is adequate cover near ground level, providing refuge from raptors. Cover may be provided by thickets or a combination of thickets and hollow logs. An exception to this may be the apparent existence of numbat populations in *Triodia* tussock grasslands of the arid zone, but these records may refer to sites close to woodland patches.

DECLINE IN DISTRIBUTION

The contraction of the numbat's range, traced from the latest records in any particular region, followed an east-to-west progression (Friend 1990), commencing with the recorded disappearance of the species from the Murray–Darling confluence in the 1850s (Krefft 1866). The contraction was slow until the 1920s, when the range of the introduced fox *Vulpes vulpes* expanded rapidly westward, following in the wake of colonisation by rabbits from south-eastern Australia (Friend 1990). Several observers in the 1940s predicted that the numbat's demise even in its south-western stronghold was not far off (Anon 1949), but some optimism returned in the 1950s with an increase in sighting rates at several key sites in Western Australia (Serventy 1954; Jones 1954). In the 1960s, numbats persisted only in the Gibson Desert and surrounding areas, and in the south-west (Friend 1990). By 1982 no desert populations remained (Friend et al. 1982) and numbats were found only in the south-west, at a handful of sites in jarrah (*Eucalyptus marginata*) forest and wandoo (*E. wandoo*) woodland east of Perth and in one area of jarrah forest east of Manjimup (Perup). By 1985 the northern jarrah forest numbat populations were extinct, leaving only the Dryandra and Perup populations, 150 km south-east and 280 km south-south-east of Perth respectively (Fig. 1; Friend 1990).

RESEARCH INTO CAUSES OF RECENT DECLINE

The sudden decline in numbat sighting frequencies in the late 1970s resulted in the allocation of funds to conduct research needed to formulate a recovery strategy. This work started in earnest as the species' distribution shrank to the last two populations. Preliminary research at Dryandra Woodland eliminated three possible causes of decline: shortage of food through drought, shortage of food through increased fire frequency and increased predation through increased fire frequency (Friend 1990). A study of predation of radio-collared numbats at Dryandra showed that they are taken by native predators (carpet pythons and birds of prey including brown goshawk and little eagle) as well as introduced carnivores (fox and cat).

Sodium monofluoroacetate ('1080') has been used to control introduced vertebrate pests in Australia since the 1950s. Coincidentally, high concentrations of fluoroacetate are present in some native flora species in the south-west of Western Australia and consequently many native animals from the region display a high tolerance to the toxin (Twigg and King 1991). Consequently, 1080 may be used in dried meat baits to control foxes with minimum impact on native mammal populations in Western Australia.

In the early 1980s, two Western Australian studies attempted to assess the effect of reducing fox numbers on populations of native mammals. Recovery of black-flanked rock-wallaby (*Petrogale lateralis*) populations after fox control was demonstrated by Kinnear et al. (1988), who monitored baited and unbaited populations in the Western Australian wheatbelt. A more limited study of numbat population response to fox control was carried out at Dryandra (Friend 1990). Numbats do not respond to conventional trapping techniques and population monitoring is based on changes in sighting rates. Numbat sightings were recorded on driven surveys in baited and unbaited areas within Dryandra during 1981–1985. An increase in sighting rate within the baited area was recorded, while there was no change in the unbaited area. A decade later, Caughley and Gunn (1996) suggested that the experiment should have been followed up by more rigorous trial. Their criticism of the lack of replication and the small distance of one kilometre between baited and unbaited areas was valid but overlooked the fact that there were no other areas available with a high enough numbat population to support such a study. More importantly, at this stage, however, the entire numbat population comprised less than 300 animals and the priority

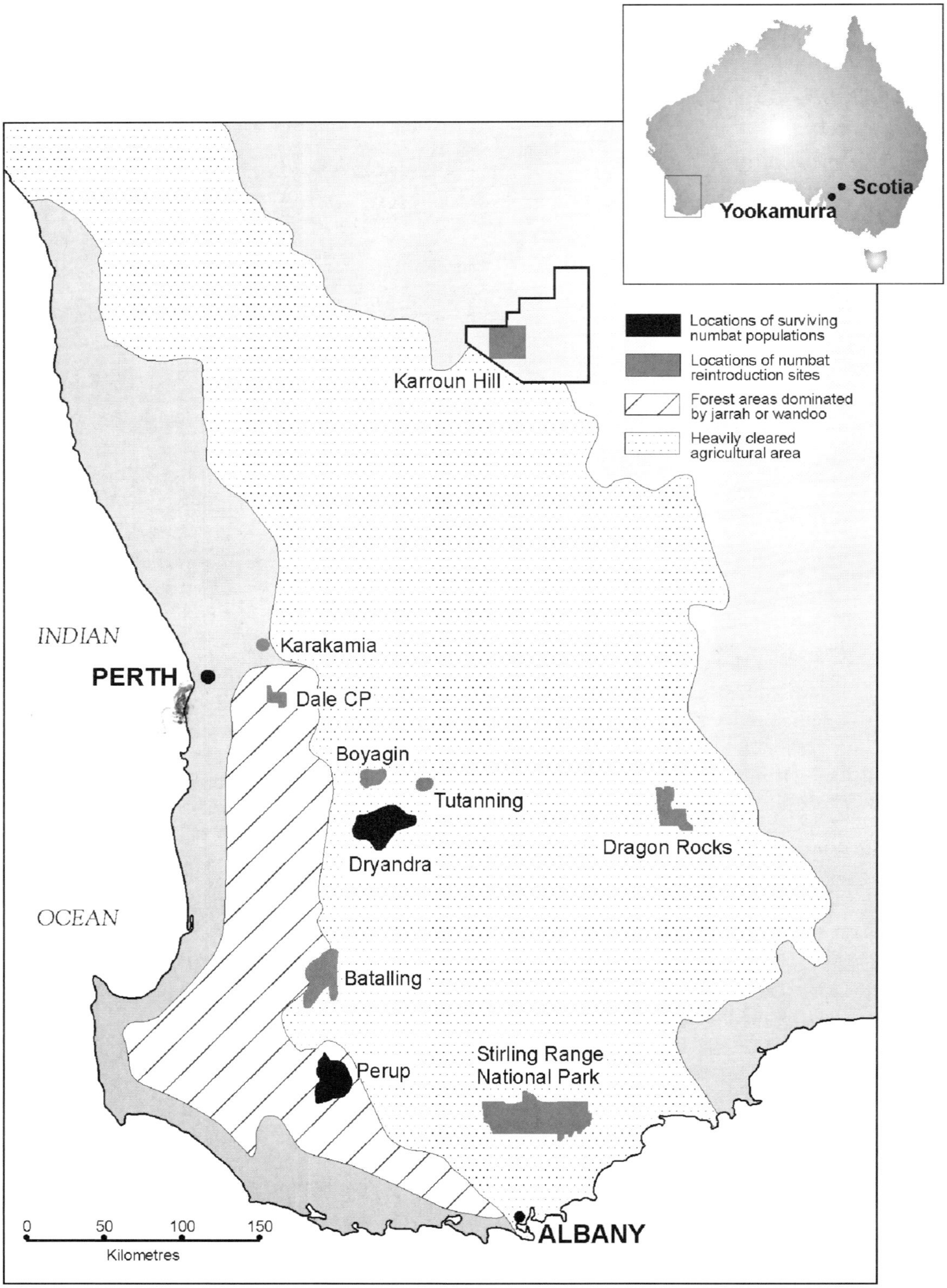

Figure 1 Locations of surviving numbat populations and translocation sites in the south-west of western Australia (main map) and elsewhere (inset).

was to increase the numbers of populations and individuals as soon as possible.

Recovery at Dryandra and experimental reintroductions

The increase in numbat numbers in the baited area at Dryandra provided the opportunity to harvest animals from Dryandra to provide stock for the establishment of additional populations. Boyagin Nature Reserve (4966 ha) was selected for the first experimental translocation commencing in 1985, because of its proximity to Dryandra and similarity in soils and vegetation (Friend 1990; Friend and Thomas 1994). This was followed in 1986 by a translocation to Karroun Hill Nature Reserve, 320 km north of Dryandra, chosen because of its large size and strategic position on the edge of the vast uncleared expanse lying east of the agricultural zone. Vegetation, soils and climate differ markedly from those at Dryandra, and this site was used to test whether a reintroduction following procedures developed at Boyagin would be successful in such a different environment. Tutanning Nature Reserve (2206 ha), 45 km north-east of Dryandra, was the site of the third experimental reintroduction. This reserve was chosen as it was considered close to the minimum area able to support a self-sustaining numbat population. It was intended that the success of this and all other translocations would be assessed after a period of at least 10–15 years using both demographic and genetic parameters (Friend 1994) but the necessary funding has not been available.

Boyagin Nature Reserve

Although the wandoo valleys of Boyagin resembled those at Dryandra, a lone radio-collared male numbat was released first to provide a guide to suitable habitat before any other releases were carried out. This animal was captured from the wild at Dryandra and released into a hollow log in wandoo woodland at Boyagin on 21 November 1985, before the commencement of fox baiting. The numbat was found freshly dead in wandoo woodland, five days after release. Its nose and some limbs had been chewed off and it was buried in a fashion typically employed by foxes. Fox control commenced on 26 November 1985 when baits of fresh kangaroo meat containing a poisoned oat (4.5 mg of 1080) were distributed at 100 m intervals along all tracks in the reserve. Although bait materials varied until the adoption of dried meat baits in 1989, this baiting regime has been in place since 1985.

All but two of the 35 numbats released at Boyagin in 1985–87 were captured at Dryandra and transferred directly to Boyagin, where they were released into hollow logs. The other two were a Dryandra male that had spent 18 months in captivity at Woodvale and one of the male progeny from the Woodvale breeding program (see below).

Boyagin comprises two blocks separated by 500 m of farmland and 35 founder animals were released into the east block during 1985–1987. In 1992 numbats diggings were first found in the west block, signifying migration from the east block. By 1995 the sighting rate in the west block exceeded the rate in the east. The successful establishment of a reintroduced population at Boyagin has been described by Friend (1990) and Friend and Thomas (1994).

Three Dryandra animals were transferred to Boyagin in 1997 in the only other release since 1987. Numbat presence is monitored annually through driven surveys and searches for fresh numbat sign in both blocks (Fig. 2). The results indicate that although population indices fluctuate substantially, a self-sustaining numbat population now exists at Boyagin.

Karroun Hill Nature Reserve

The large size of Karroun Hill NR (309 678 ha) and the lack of good access provided logistical challenges during the 8-year translocation program. Firstly, numbats dispersed more widely than at Boyagin, necessitating the use of more powerful transmitters and an improved aircraft tracking system. Secondly, as there was no track network for effective ground baiting, the translocation relied initially for fox control on a pre-existing aerial baiting program for dingoes. Between 1987 and 1990, this was supplemented by occasional distribution of dried meat baits along the main access track. Between 1990 and 1995, aerial baiting was carried out over an area of 40 000 ha twice a year and increased to four times a year from 1996 to 2001.

Most animals dispersed from the 1986 release site on the edge of the reserve and were not found. Improved radio-tracking equipment was available in 1987, and from that year on translocated animals were released 10 km from the edge of farmland near the baited granite outcrop of Karroun Hill itself. Ninety-seven numbats were released at Karroun Hill between 1986 and 1993, all sourced from Dryandra. Forty-three young born at the site were captured during monitoring trips (Friend and Thomas 1994). A population became established and numbat sign was plentiful, particularly in woodland dominated by *Callitris columellaris* with dense scrub nearby. In October 2000, 7 years after the last release, fresh numbat sign could be found in this vegetation type. Searches of the baited area in October 2001 and January 2002 revealed widespread but old numbat sign (weathered scats) and no fresh diggings, indicating a rapid population decline during the previous year or two. Aerial baiting ceased in March 2002 during a rationalisation of the *Western Shield* fox control program.

Predation of numbats by feral cats was recorded at Karroun Hill soon after the commencement of aerial baiting. In other reintroduction programs at arid and semi-arid sites, cat numbers have risen in response to fox control (Short et al. 1994, Christensen

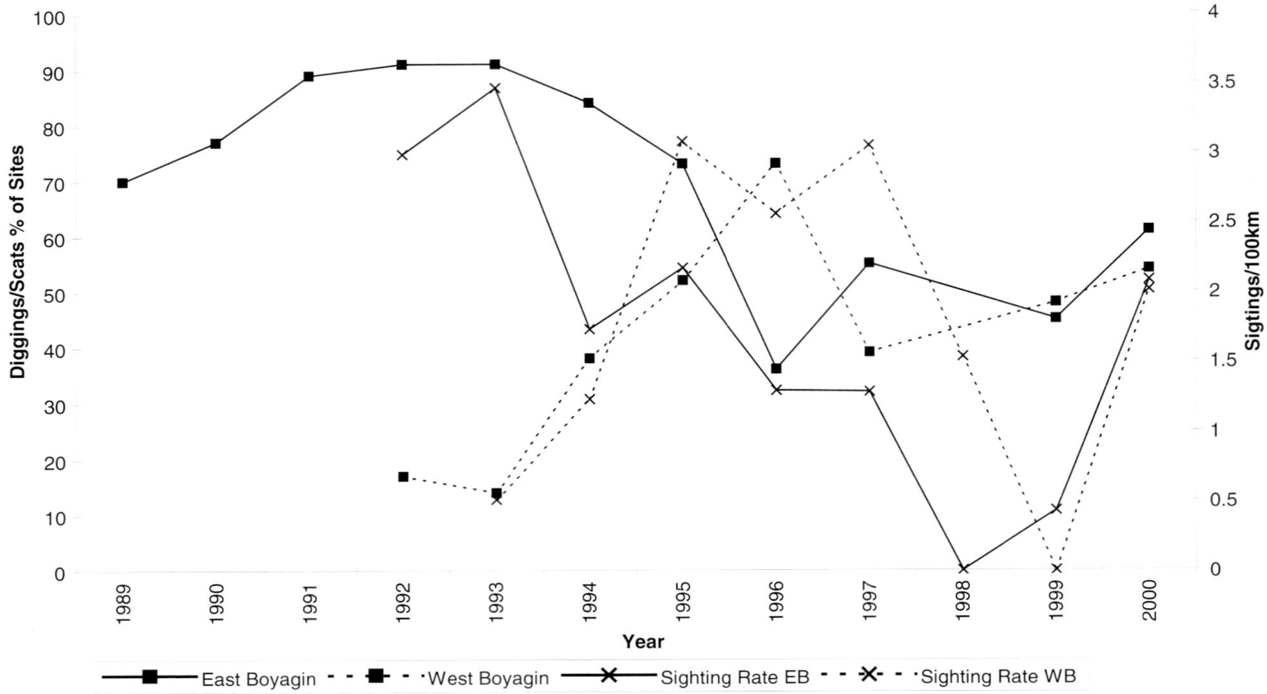

Figure 2 Results of numbat monitoring in the eastern and western blocks of Boyagin Nature Reserve by two methods: i) Searches were made for sign in October each year at 44 marked sites in the east block and 54 in the west block. Results are expressed as a percentage of the sites at which numbat diggings or scats were found. ii) Set routes were driven in October each year in each block (232 km in the east block and 195 km in the west block) and numbat sightings by two observers recorded. Results are expressed as sightings per 100 km driven.

and Burrows 1994). However, between 1986 and 1993, when numbats were monitored by radio-tracking, raptors were responsible for more predation of numbats than were foxes or cats, probably as a result of the general lack of cover (Friend and Thomas 1994). As there was no intensive monitoring at Karroun Hill during 2000–2002, it was not possible to determine the cause of the recent population crash.

Tutanning Nature Reserve

Tutanning NR is a 2206 ha vegetation remnant surrounded by cleared farmland. The Dutarning Range forms the backbone of the reserve with powderbark (*E. accedens*) woodland on the upper slopes and wandoo woodland lower in the landscape, generally close to farmland. Monthly fox baiting commenced here in 1984 so that the response of the woylie (*Bettongia penicillata*) population could be monitored (J.Kinnear pers. comm.) Numbats were reported as surviving here in the early 1980s but a search in 1985 yielded no fresh evidence. Three female numbats were captured at Dryandra, radio-collared and transferred to Tutanning in November 1987, prior to the mating season. One was found dead a week later and the signal of another was lost after release. The third female lived through four breeding seasons at Tutanning without producing young. Given the small size of the reserve and the high mobility of males during the mating season, this provided strong evidence that the original Tutanning numbat population was extinct. Between 1990 and 1996, another 32 numbats were released into the reserve. Monitoring of released animals revealed that they established home ranges in powderbark as well as wandoo woodland. Driven surveys commenced in 1996 and have been carried out in April every year or two since then. Small numbers of numbats have been sighted during every survey indicating that a population has been established at Tutanning.

Expansion of fox control in the south-west of Western Australia

In the early 1990s, the Department of Conservation and Land Management adopted fox control as a technique to reverse mammal decline in south-western Australia, based on the findings of the fox removal studies of the 1980s. Several experimental baiting programs in the wheatbelt were consolidated in 1989 and a new aerial baiting program, *Operation Foxglove*, began in 1992 in the northern jarrah forest. Results of monitoring several species of medium-sized mammal showed that fox control had a positive effect on population numbers (Morris et al. 1995). In 1996, under the name *Western Shield*, the program was extended to the southern forests, south coastal national parks and some large inland reserves. In all, 3.5 million hectares were

Table 1 Details of numbat translocations described in the text. Results of the most recent monitoring visits are shown in the last column; * only a 40 000 ha section of Karroun Hill was baited; ** only old signs were found in 2002; *** Batalling and Dale Conservation Park are included in the 2 million ha forest belt, now all under fox control.

Destination	Area (ha)	Release years	Number released	Source	Recent evidence (last check)
Boyagin, WA	4966	1985–87	35	Dryandra, Woodvale	2001 (2001)
Karroun Hill, WA	309 678*	1986–93	97	Dryandra	2000 (2002)**
Tutanning, WA	2206	1987–96	35	Dryandra, Perth Zoo, Karakamia	2002 (2002)
Batalling, WA	***	1992–95	60	Dryandra, Perth Zoo	2002 (2002)
Yookamurra, SA	1113	1993	15	Dryandra	2002 (2002)
Karakamia, WA	250	1994, 1996, 1999	6	Dryandra	2002 (2002)
Dragon Rocks, WA	32 203	1995–96	37	Dryandra	2000 (2000)
Dale CP, WA	***	1996–98	62	Dryandra, Perth Zoo, Boyagin	2001 (2001)
Stirling Range, WA	115 920	1998–2001	62	Dryandra, Perth Zoo, Yookamurra	2002 (2002)
Scotia, NSW, Stage 1	4000	1999–2001	40	Yookamurra	2002 (2002)

targeted under *Western Shield*, which aimed to remove foxes and reintroduce native mammals over much of the conservation land in Western Australia's south-west (Bailey 1996). This program included areas already baited for numbat conservation (Dryandra, Perup, Boyagin, Tutanning, Karroun Hill and Dragon Rocks – see below) and provided new areas suitable for further numbat reintroductions.

THE NUMBAT RECOVERY PLAN

The first reintroductions of numbats were carried out as research projects, and a formal reintroduction program was not approved until 1994 with the production of the *Recovery Plan for the Numbat* (Friend 1994). The strategy used in formulating this program and the basis for selection of sites are provided by Friend and Thomas (1994). All translocations were to areas in which numbats had been recorded, so all are classified as reintroductions (IUCN 1987). The details of each translocation are provided in Table 1.

A standard protocol for translocation and subsequent monitoring was developed after the experience gained at Boyagin, Karroun Hill and Tutanning and used from 1992 onwards. Under this protocol, three annual releases of 20 animals are carried out, with all animals released in late November or early December. All numbats are radio-collared before release and an aircraft is used to locate animals after dispersal. The radio-collars require replacement after 4.5 months and at this stage the numbats are captured and checked, and collars replaced before release. Females and their young are captured in October while the young are large enough to be collared but still nesting with their mother. After dispersal, the young numbats are located by aircraft in February or March, during the search for animals in the current translocation group.

Monitoring by radio-tracking continues for a year or two after the last releases. Subsequent monitoring relies initially on searches for diggings, along tracks where access is good, or at sites previously frequented by radio-collared animals where the track network is sparse. At reintroduction sites where vegetation density is low and visibility is good, driven surveys are used. Monitoring by one of these methods is carried out at each site either annually or bi-annually.

RECOVERY OF EXISTING POPULATIONS

Dryandra

John Calaby's sighting data from 1954–56 (pers. comm.) shows a relatively high sighting rate (Fig. 3). By comparison, sighting rates in the late 1970s were very low. After baiting commenced in the experimental area in 1982, then was extended in 1989, the population grew rapidly and peaked in 1992. A year later the sighting rate measured over a driven route of 400 km in the main block of Dryandra had fallen to approximately half of its 1992 value. There was no obvious new influence on the population at this stage and it appears that the population had outstripped the carrying capacity of the habitat. Although numbats (predominantly subadults) had been removed from Dryandra for the translocation program since 1985 at a rate of 17–33 animals per year, the sighting rate had continued to climb through this period. Predation of radio-collared numbats had not increased dramatically and animals captured in spring were in good condition. High numbat densities may have reduced food availability at other times of year, however. The demographic effect of fatal infestation by an acanthocephalan parasite recorded in a number of numbats at Dryandra at this time (Haigh 1994; Smales 1997) is unknown but this disease condition may have been a factor in the decline.

The numbat population at Dryandra has not recovered to pre-1993 levels despite the very low level of removals (0–5 animals per year) for translocation and captive breeding since 1998.

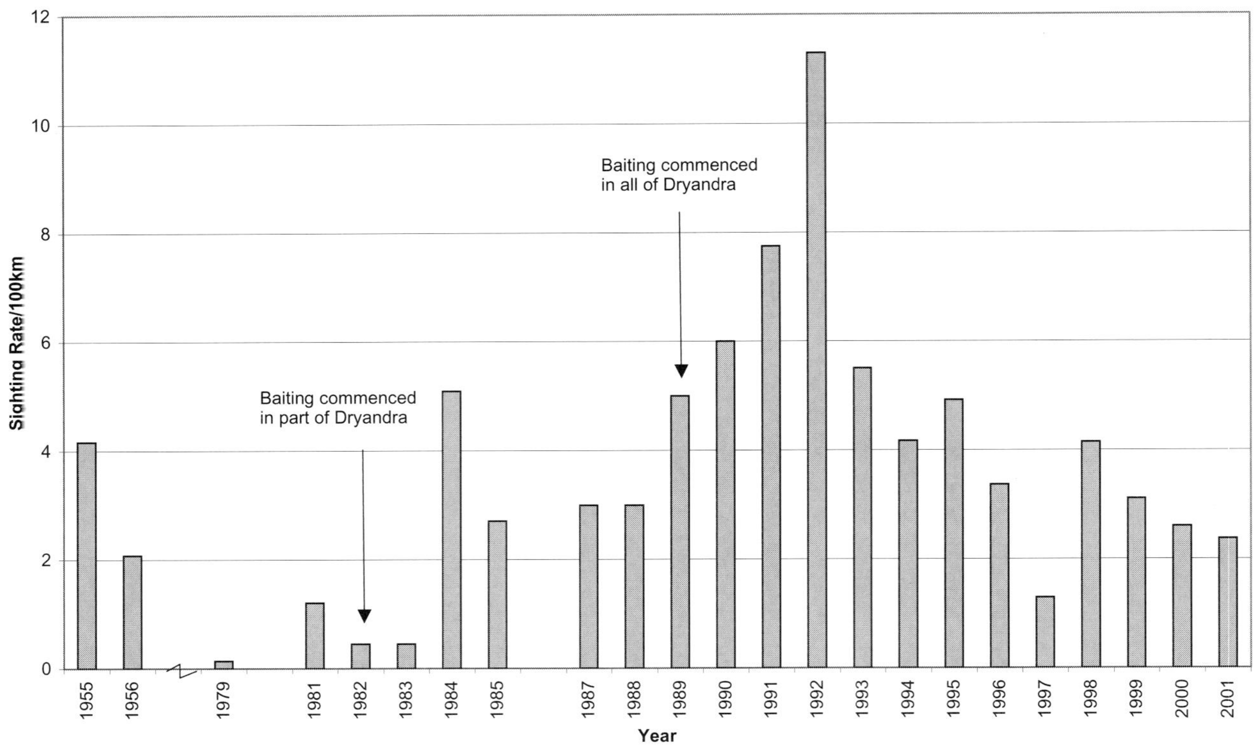

Figure 3 Numbat sighting rate, expressed as number of sightings per 100 km driven, during driven surveys at Dryandra Woodland, Western Australia. Data from 1955–56 from J. Calaby (pers. comm.); 1979 data from J. Turner (pers. comm.); 1980–88 data from Friend (1990). Sighting data for 1989–2001 are based on 384 km driven.

Perup

Quantitative monitoring of the numbat population in the Perup forest east of Manjimup requires much more effort than at Dryandra as the denser habitat restricts visibility and reduces sightability. Searches for numbat sign present difficulties here because of the high density of woylie diggings. Driven surveys were carried out in 1995 and 1996 to establish a baseline at the commencement of aerial baiting. Sightings are currently recorded by forest workers at the Department of Conservation and Land Management district office at Manjimup and this information provides distributional information on the species in the area. Christensen and others used these data to highlight the decline of the numbat in the mid-1970s (Christensen 1980; Christensen et al. 1984) several years before a similar decline at Dryandra (Friend 1987, 1990). No similar decline at Perup has been indicated by sighting report frequencies since 1980.

Sources of numbats for reintroduction

Wild populations

At Dryandra, the fox control experiment had seen 2000 ha of the 13 000 ha main block baited since 1982, and the whole block from 1989. The numbat population grew slowly at first, and then more rapidly in 1989–1992 (Fig. 3). In November 1992, a line transect survey in the main block indicated a population of approximately 800 animals (unpublished data). Over 16 years to date, a total of 325 wild Dryandra numbats have been transferred directly to translocation sites and between 1985 and 1998, stock for translocation came predominantly from Dryandra. Wild-sourced animals are likely to be more suitable than captive-bred stock for translocation on the basis of genetics, behaviour and previous exposure to disease. Translocation stock was only available from the wild until 1993 when improvement in Perth Zoo's captive breeding output presented an alternative source.

Captive breeding

When concern rose for the numbat during the late 1970s, the species had not been bred and reared to independence in captivity. The difficulty in providing termites for food prevented serious attempts at captive breeding, except at Taronga Zoo, Sydney, in the 1970s. Even there breeding failed, perhaps due to the unsuitable climate. In 1984 a small captive colony was set up at the Western Australian Wildlife Research Centre, Woodvale, with funding from the WA Department of Fisheries and Wildlife and World Wildlife Fund (Australia). A male and a female with four attached young were taken from the wild in June 1984. The adult female and her two female young produced nine young in three litters in January 1985 (Friend and Whitford 1986). The key to success was the development of an artificial diet formulated from low-lactose milk and eggs, based on a diet used by Griffiths (1977) for juvenile echidnas (Friend and Whitford 1993). All young were raised to maturity.

The numbat colony at Woodvale was transferred to Perth Zoo in 1987 but breeding did not occur there until 1993 after the construction of new enclosures. Since then, methods of numbat husbandry have been improved and four to five litters are produced there every year to provide young for release (Bradley et al. 1999). Between 1993 and 2001, 75 numbats bred at Perth Zoo were released under the translocation program. This has been very fortunate, as it has allowed the translocation program to continue despite a necessary reduction in the rate of removal of numbats from Dryandra. In order to maintain low inbreeding rates in the Perth Zoo colony, numbats are regularly brought in from the wild. A total of 19 Dryandra numbats were added to the zoo colony between 1987 and 2000.

Population genetics

The numbat recovery program has been dependent on translocation and captive breeding of stock taken from Dryandra. Perup represents a significant source of animals for the translocation program, if there are not likely to be detrimental effects from mixing stocks (e.g. outbreeding depression). A project commenced in July 1996 to compare Dryandra and Perup populations by analysis of mitochondrial DNA using small ear-tissue samples collected from numbats during monitoring exercises.

The findings indicated that the Dryandra population has been less markedly affected by bottlenecks than has the Perup population, but that both were quite recently connected (Fumagalli et al. 1999). The conclusion was that the Dryandra population is definitely the more suitable as a source for translocation, but that some mixing of stocks for translocation should not cause problems. In fact, if done correctly, it should increase the genetic variation within the translocated populations. There could also be some benefit in transfers of animals between Dryandra and Perup. This option should be approached carefully, however, due to the possibility of disease transmission and other concerns.

NUMBAT TRANSLOCATIONS 1992–PRESENT

Batalling State forest

In the late 1970s an area of jarrah forest and wandoo woodland at the eastern edge of the main forest block near Darkan was selected as a reintroduction site for the woylie because of its similarity to woylie habitat in the Perup area (P.E.S. Christensen pers. comm.). Fox control and mammal monitoring commenced at Batalling in 1990 and as the area contained suitable habitat for numbats it provided an ideal reintroduction site. The area is currently baited from the air four times a year under the *Western Shield* program.

Sixty numbats were released at Batalling over the years 1992–1995. The main difficulty encountered was the disappearance of animals after release. The forest extends 50 km west and 200 km north of the release site and although all the forest within 25 km of the release site was thoroughly searched, over half of the numbats released were never located again. It is likely that many animals dispersed well beyond the search area. Dense stands of timber in the forest may also have contributed to this loss rate, as signals are poorly transmitted from the forest floor. Fourteen numbats released at Batalling were captive-bred at Perth Zoo. There was a particularly high incidence of disappearance amongst these animals, 10 never being found after release.

Nevertheless, monitoring by diggings searches has revealed that numbats persist at a number of sites in the area. The most recent diggings survey was carried out in February 2002 and numbat sign was located at five locations within 10 km of the release site.

Yookamurra Sanctuary

Fences are being used increasingly in mammal conservation in Australia to exclude feral animals and assist their eradication within the enclosed area. The company Earth Sanctuaries Limited was set up by John Wamsley to establish and manage fenced sanctuaries to contribute towards nature conservation, particularly focussing on the reintroduction of threatened native mammals. Yookamurra Sanctuary near Sedan in the Murray Mallee area of South Australia, contains 1113 ha of remnant mallee woodland and associated vegetation surrounded by an electrified fence, completed in 1992. Much of the area is in very good condition, with trees up to 10 m in height and possessing abundant hollow logs and termites. After the fence was closed, foxes, cats and rabbits were exterminated from within the fenced area. Following a habitat assessment and in conjunction with the Department of Conservation and Land Management, 15 numbats were captured from Dryandra and transported to Yookamurra in late 1994. These animals were all radio-collared before transfer and released into hollow logs on the day after capture.

Although three animals were taken by birds of prey within the first two weeks after release, and another died as a result of the presence of an acanthocephalan parasite (Haigh 1994), the population became established. High sighting rates (10 numbats/100 km) were reported by 1996 and the population grew to the extent that 20 animals could be removed in November 1999 for translocation to Scotia Sanctuary, followed by another 20 one year later (L. Pope pers. comm.).

Karakamia Sanctuary

A fenced sanctuary was established by philanthropist and conservationist Martin Copley in 1993 in jarrah forest near Chidlow, 40 km north-east of Perth. Karakamia Sanctuary, now part of Mr Copley's Australian Wildlife Conservancy, contains 250 ha of jarrah forest and associated vegetation types. Three radio-collared numbats (a male and two females) were released in the sanctuary in December and established home ranges. The sanctuary is considered too small to support a viable numbat population and its management involves both supplementation and removal of individuals to counter stochastic loss of the population and to reduce inbreeding. The colony has been supplemented from Dryandra by a male numbat in 1996 and by a male and a female in 1999. Young have been produced on at least four occasions and one young male was transferred from Karakamia to Tutanning in October 1995.

Dragon Rocks Nature Reserve

Dragon Rocks Nature Reserve (32 203 ha), in the eastern wheatbelt 190 km east of Dryandra, is a large remnant of native vegetation in an otherwise cleared landscape. Numbats were reported there in the early 1970s (McKenzie et al. 1973) but a survey by numbat project personnel in 1983 found no diggings in woodland suitable for numbats and no sightings were reported subsequently from the area.

The vegetation of much of the reserve is classified as mallee and mallee-heath formations, amongst which high shrubland and high open shrubland have a well-developed understorey. The low open forest, which comprises stands of blue mallet (*E. gardneri*) and silver mallet (*E. ornata* or *E. argyrocaulon*) on the low ridges and covers 5–10% of the reserve, was identified as numbat habitat in the early 1970s by reserve neighbours (McKenzie et al. 1973; Coates 1992).

The cleared surroundings of Dragon Rocks NR were seen as a feature that would assist population establishment and growth. Experience had shown that in extensive areas like Karroun Hill and Batalling, released numbats and dispersing young may travel great distances before establishing a home range (Friend and Thomas 1994). On the other hand, reproductive success is enhanced in reserves surrounded by farmland because numbats rarely disperse across open fields but stay in bushland, thus remaining in contact with one other.

Fox control commenced at Dragon Rocks in October 1995 and baiting was carried out twice a year by air and six times a year from a vehicle along perimeter and internal firebreaks. In 1996 aerial baiting was increased to four times per year with ground baiting reduced to four times per year. Twenty numbats were transferred from Dryandra in December 1995 and 17 in December 1996. Survival rates were high and 10 and 27 site-bred young were captured in 1996 and 1997 respectively. This compares with six young captured at Karroun Hill in 1988 after a release of 20 numbats and 11 young captured in 1989 following a release of 17 numbats. The Dragon Rocks translocation program was curtailed after only two releases because of the high number of numbats already known to be present in the reserve.

A funding shortfall following the withdrawal of Commonwealth support from the numbat recovery program saw all radio-collars removed from numbats at Dragon Rocks during 1998. Subsequent monitoring has been by diggings searches and the latest survey, in June 2000, revealed the presence of numbats at fourteen separate locations within the reserve.

Dale Conservation Park

Numbats were recorded in the main forest belt by Calaby (1960), who mentioned sightings near Perup, Collie and Jarrahdale, stating that numbats did not live in jarrah forest but in patches of wandoo woodland within the main forest belt. In their study of the numbat at Perup in the southern jarrah forest, Christensen et al. (1984) showed that in that area, numbats were more frequently sighted in jarrah forest than in the more open wandoo valleys. Dale Conservation Park, one of several large areas of the northern jarrah forest set aside for conservation in recent years, consists of 5798 ha including jarrah and jarrah-marri forest, essentially on ridges and slopes, while wandoo and wandoo-marri woodland are the main vegetation types on the valley floors. Dale CP lies centrally in the north-south forest belt and while it was the chosen release site, it is surrounded by suitable numbat habitat under fox control, so released numbats and their progeny were expected to move into adjacent areas.

The first translocation occurred in December 1996, in the vicinity of Mount Dale, 45 km south-east of Perth. Twenty numbats from Dryandra were transported to Mount Dale and released into hollow logs at a jarrah upland site. This group of numbats dispersed across the width of the forest belt, two individuals reaching the eastern and western extremes of the forested area, 50 km apart. No numbats remained near the release site. The animals that established home ranges and survived settled in wandoo valleys north-east of Mount Dale and this area was used for the second and third releases. Coincidentally, this release site was near the location of a numbat sighting in 1985, the most recent recorded from the northern jarrah forest prior to the local extinction of the species (K. Pollock pers. comm.). Wandoo valleys were the habitat most frequently used by numbats and their progeny during this reintroduction, supporting the findings of Calaby (1960).

Stirling Range National Park

The Stirling Range National Park is a large (115 920 ha) area of remnant vegetation dominated by the Stirling Range but incorporating broad surrounding plains and foothills. The most recent museum specimen from the vicinity of the Stirling Range was collected at South Stirling in 1936, although a reliable sighting was made within the national park in the 1950s (L. Coney pers. comm.). Vegetation types suitable for numbats include wandoo woodlands in incised valleys and on heavier soils, jarrah woodlands on slopes and ridges and long-unburnt mallee-heath communities on more sandy soils. Large wildfires due to lightning strikes are a feature of the environment here, and a program of prescribed burning has been implemented in an attempt to reduce the extent of wildfires (Herford et al. 1999). The western quarter of the park contains the most extensive areas of wandoo woodland and much of this area had escaped fire for over 20 years. Fourteen numbats, including 10 from Dryandra and four adult animals from the Perth Zoo colony were released in wandoo woodland at the western section of the national park in December 1998. Subsequent releases have involved few Dryandra animals, as the bulk of the release group has originated from the reintroduced population at Yookamurra (1999) or the captive breeding colony at Perth Zoo (2000 and 2001).

The greatest source of mortality at the Stirlings has been predation by birds of prey. In 2000, Perth Zoo implemented an experimental training program whereby young captive-bred numbats are exposed to a raptor while loud noises and bird alarm calls are sounded. Early results indicate that trained animals have a higher survival rate over the first 5 months after release than untrained animals (J.A. Friend, unpublished).

Scotia Sanctuary

Scotia Sanctuary (49,000 ha) is one of Earth Sanctuaries Limited's properties in south-western New South Wales, and is situated on the South Australian border 100 km north of the Murray River. Scotia comprises two former sheep stations and the plan for development involves the progressive fencing of adjacent sections, removal of feral animals from each section and reintroduction of locally extinct native mammals. The first section, 4000 ha in extent, was cleared of feral animals (goats, rabbits, cats and foxes) by early 1999 and a habitat assessment and release site selection was carried out by the senior author in July 1999. The first section ('Stage1') included mature mallee woodland, judged the most likely community to be preferred by numbats. Release sites were chosen within the mature mallee areas with the best-developed understorey. Other major habitat types within Stage 1 were mallee over *Triodia* on sand-ridges and *Casuarina pauper* woodlands. A raptor survey carried out in August and September 1999 indicated that the density of raptors at Scotia was substantially lower than that at Yookamurra Sanctuary.

The first translocation occurred on 24–25 November 1999 when 10 numbats were captured at Yookamurra, fitted with radio-collars and taken to Scotia, four hours' drive to the north-east. The animals were released the next morning into hollow logs at the selected release sites. Individuals were located each day to check for survival, habitat use and refuge selection. Two weeks later, there had been no casualties and another 10 numbats were translocated from Yookamurra and released in the same area. Survival was very good with 80% of the released animals surviving the first year.

In November–December 2001, 20 more numbats were translocated from Scotia to Yookamurra. Monitoring during 2002 has shown that the population is expanding (Phil Harris pers. comm.).

IUCN CLASSIFICATION AND LISTING

Prior to the revision of the IUCN Red List Criteria in 1994 (IUCN 1994), the numbat was listed as follows:

- Endangered: ANZECC List of Endangered and Vulnerable Vertebrates (ANZECC 1991)
- Schedule 1, Part 1 (Endangered): Australian Commonwealth Endangered Species Protection Act 1992
- Endangered: IUCN Mammal Red Data Book (IUCN 1982)

Following extensive revision of the criteria (IUCN 1994), a workshop was held in Sydney in December 1994 to review the listing of all marsupials and monotremes. The recommendations of this workshop were essentially followed by Maxwell et al. (1996). The workshop found that the status of the numbat should be revised to Vulnerable, based on criteria A1a, D2 (observed reduction of at least 20% in the last 10 years or three generations; population characterised by an acute restriction in the number of locations, typically less than five). Downlisting must be based on the situation five years previously (IUCN 1994), in this case in 1989, when only Dryandra, Boyagin and Perup populations were considered self-sustaining. Satisfaction of the first criterion depended on the extinction of the species over the entire northern jarrah forest between 1981 and 1985 (Friend 1990).

This recommendation was adopted by the relevant listing authorities and the numbat's current listing is as follows:

- Vulnerable: 2000 IUCN Red List of Threatened Species (Hilton-Taylor 2000)
- Vulnerable: Australian Commonwealth Environmental Protection and Biodiversity Conservation Act 1999

FUTURE DIRECTIONS

Research

The limitations of the fox control experiment at Dryandra raise the question of whether the fox has been overemphasised as a limiting factor on numbat populations (Caughley and Gunn 1996). All numbat populations in Western Australia are now in areas under fox control. A replicated experiment that would produce statistically significant numbers would involve the removal of fox control from a number of areas where numbats are in sufficient density to enable the use of available techniques to assess population change. If separate colonies were used to avoid the problem of movement between control and experimental sites, this experiment would involve all eight populations in Western Australia. The effect that this experiment could have on recovery of the numbat and co-existing mammal populations could not be justified. The body of evidence indicating the beneficial effect of fox control on medium-sized mammals (e.g. Morris et al. 1995; Kinnear et al.1998) is now very large and includes the success of many reintroductions of the numbat and other Critical Weight Range mammals (Burbidge and McKenzie 1989) into areas under fox control (Morris 2000).

On a more practical note, it is useful to examine the successes and failures of the program in planning future translocations. Analysis of the factors affecting the success of establishment of new populations is confounded by the lack of similarity between the translocation sites, however. Certain factors are important at different sites. Raptor predation was particularly heavy at Karroun Hill and the Stirling Range and this is likely to be a result of a lack of understorey in woodland associations at those sites. Wide dispersal of founders from the release site at Dale CP and Batalling appears to have slowed the establishment of those colonies in comparison with reserves surrounded by farmland. The more rapid establishment of populations in fenced sanctuaries compared with baited areas could imply that predation by foxes and/or cats is still significant in unfenced baited areas. The relative effectiveness of fencing versus baiting in providing reintroduction sites needs to be examined.

Cats were identified as predators of numbats at several sites but only at low rates during the initial, radio-monitored stages of translocation. The importance of predation by cats on numbats and other fauna in areas where foxes are controlled requires urgent investigation.

Funding

Numbat recovery has necessitated intensive work over 20 years. A number of agencies and corporate entities have supported this project for various periods. Sponsorship has been provided for five years by a single donor through the World Wide Fund for Nature Australia and for three years each from McDonalds Family Restaurants and ANCA/Environment Australia. The agencies providing the necessary long-term support that recovery programs need, however, have been the institutions directly involved: WA Department of Fisheries and Wildlife/Conservation and Land Management, Perth Zoo and Earth Sanctuaries Limited.

Future funding for numbat conservation will probably follow this pattern. It is likely that another assessment of the numbat's status using IUCN criteria will see it removed from the threatened lists. Without successful biological control of foxes and perhaps cats, however, sustained recovery will depend on continued commitment to baiting programs by the Western Australian State Government, and the economic sustainability of non-government organisations such as the Australian Wildlife Conservancy, which in 2002 purchased Yookamurra and Scotia Sanctuaries from the struggling Earth Sanctuaries Limited. It is also essential that regular monitoring continues at all numbat sites as the loss of a small number of populations could once again move the species rapidly towards extinction.

Acknowledgements

We would like to thank the many enthusiastic people who have supported the numbat recovery project and those who have given financial, technical and field assistance, including: Andrew Burbidge, Gordon Wyre, Keith Morris, and Mike Scanlon, WA Department of Conservation and Land Management: Vicki Power and Cathy Lambert, Perth Zoo: Lyn Pope, Yookamurra Sanctuary: Andre Schmitz, Karakamia: members of the Numbat Recovery Team, World Wide Fund for Nature Australia, Environment Australia, and McDonalds Restaurants, and many people who volunteered their time to help. Anne Cochrane kindly reviewed the manuscript and made many suggestions that greatly improved the text, as did comments by Menna Jones and two anonymous referees.

References

Anon. (1949), 'Save our native fauna – No. 1', *Public Service Journal of Victoria* (April–May 1949), **39**:iv.

Bailey, C, (1996), 'Western Shield – bringing wildlife back from the brink of extinction', *Landscope*, **11**:41–8.

Bradley, M.P., Lambert, C., Power, V., Mills, H., Gaikhorst, G., & Lawrence, C. (1999), 'Reproduction and captive breeding as a tool for mammal conservation', *Australian Mammalogy*, **21**:47–54.

Burbidge, A.A., & McKenzie, N.L. (1989), 'Patterns in the modern decline of Western Australia's vertebrate fauna: causes and conservation implications', *Biological Conservation*, **50**:143–98.

Calaby, J.H. (1960), 'Observations on the Banded Ant-eater *Myrmecobius f. fasciatus* Waterhouse (Marsupialia), with particular reference to its food habits', *Proceedings of the Zoological Society of London*, **135**:183–207.

Caughley, G., & Gunn, A. (1996), *Conservation Biology in Theory and Practice*, Blackwell Science: Carlton.

Christensen, P.E.S. (1980), 'A sad day for native fauna', *Forest Focus*, **23**:3–12.

Christensen, P.E.S., & Burrows, N. (1994), 'Project Desert Dreaming: the reintroduction of mammals to the Gibson Desert. Reintroduction and the numbat recovery program', in *Reintroduction Biology of Australian and New Zealand Fauna* (ed M. Serena), pp. 199–208, Surrey Beatty & Sons, Chipping Norton.

Christensen, P. Maisey, K., & Perry D.H. (1984), 'Radio-tracking the numbat (*Myrmecobius fasciatus*) in the Perup Forest of Western Australia', *Australian Wildlife Research*, **11**:275–88.

Coates, A. (1992), 'Flora and vegetation survey of Dragon Rocks Nature Reserve (No. A 36128)', Unpublished report prepared for the Department of Conservation and Land Management.

Dale, R. (1833), 'Letters from Mr. Dale, giving a summary description of the country passed over in going to Mount Bakewell, and also, in an expedition to examine the country to the north and south of that place', in *Journals of Several Expeditions Made in Western Australia During the Years 1830, 1831, and 1832; under the sanction of the Governor, Sir James Stirling, containing the latest authentic information relative to that country, accompanied by a map* (ed. J. Cross), pp. 155–60, London, J.Cross.

Friend, J.A. (1987), 'Local decline, extinction and recovery – relevance to mammal populations in vegetation remnants', in *Nature Conservation: The role of remnants of native vegetation* (eds. D.A. Saunders, G.W. Arnold, A.A. Burbidge, & A.J.M. Hopkins), pp. 53–64, Surrey Beatty & Sons, Chipping Norton.

Friend, J.A. (1989), 'Myrmecobiidae', in *Fauna of Australia. Volume IB. Mammalia* (eds. D.W. Walton & B.J. Richardson), pp.583–90, Australian Government Publishing Service, Canberra.

Friend, J.A. (1990), 'The numbat *Myrmecobius fasciatus* (Myrmecobiidae): history of decline and potential for recovery', *Proceedings of the Ecological Society of Australia 1990*, **16**:369–77.

Friend, J.A. (1994), 'Recovery Plan for the numbat (*Myrmecobius fasciatus*)', Department of Conservation and Land Management, Perth.

Friend, J.A., Fuller, P.J., & Davis, J.A. (1982), 'The Numbat in Central Australia', *SWANS*, **12**:21–6.

Friend, J.A., & Thomas, N.D. (1994), 'Reintroduction and the numbat recovery program', in *Reintroduction Biology of Australian and New Zealand Fauna* (ed M. Serena), pp. 189–98, Surrey Beatty & Sons, Chipping Norton.

Friend, J.A., & Whitford, D. (1986), 'Captive breeding of the Numbat (*Myrmecobius fasciatus*)', Abstract, Australian Mammal Society, 31st Scientific Meeting, Melbourne, May 1985, *Australian Mammal Society Bulletin*, **9**:54.

Friend, J.A., & Whitford, D. (1993), 'Maintenance and breeding of the numbat (*Myrmecobius fasciatus*) in captivity', in *The Biology and Management of Australian Marsupials* (eds. M. Roberts, J. Carnio, G. Crawshaw, & M. Hutchins), pp. 103–24, Metropolitan Toronto Zoo and the Monotreme and Marsupial Advisory Group of the American Association of Zoological Parks and Aquariums, Toronto.

Fumagalli, L., Moritz, C., Taberlet, P., & Friend, J.A. (1999), 'Mitochondrial DNA sequence variation within the remnant populations of the endangered numbat (Marsupialia: Myrmecobiidae: *Myrmecobius fasciatus*)', *Molecular Ecology*, **8**:1545–9.

Gill (1872), 'Arrangement of the families of mammals with analytical tables', *Smithsonian Miscellaneous Collections*, **11**:i–vi, 1–98.

Griffiths, M. (1977), *Echidnas*, Pergamon Press, London.

Haigh, S.A. (1994), 'An investigation of the health status and occurrence of disease agents in several threatened Western Australian mammal species', unpublished report to the Department of Conservation and Land Management, Perth, Western Australia.

Herford, I., Gillen, K., Lloyd, M., Hine, C., McCaw, L. Keighery, G., & Allen, J. (1999), *Stirling Range and Porongurup National Parks. Management Plan 1999–2009*, Department of Conservation and Land Management, Western Australia.

Hilton–Taylor, C. (2000), *2000 IUCN Red List of Threatened Species*, IUCN, Gland.

IUCN (1982), *The IUCN Mammal Red Data Book, Part 1*, IUCN, Gland.

IUCN (1987), *The IUCN position statement on translocation of living organisms. Introductions, re-introductions and re-stocking*, IUCN, Gland.

IUCN (1994), *IUCN Red List categories*, IUCN Species Survival Commission, Gland.

Jones, A.D. (1954), 'The recent increase of the rarer native mammals. VII. Manjimup', *West Australian Naturalist*, **4**:139–40.

Kinnear, J.E., Onus, M.L., & Bromilow, R.N. (1988), 'Fox control and rock-wallaby population dynamics', *Australian Wildlife Research*, **15**:435–50.

Kinnear, J.E., Onus, M.L., & Sumner, N.R. (1998), 'Fox control and rock-wallaby population dynamics – II. An update', *Wildlife Research*, **25**:81–8.

Krajewski, C., & Westerman, M. (2003), 'Molecular systematics of Dasyuromorphia', in *Predators with Pouches: The Biology of Carnivorous Marsupials* (eds. M. Jones, C. Dickman, & M. Archer), pp. 3–20, CSIRO Publishing, Melbourne.

Krefft, G. (1866), 'On the vertebrated animals of the Lower Murray and Darling, their habits, economy, and geographical distribution', *Transactions of the Philosophical Society of New South Wales*, **1862–1865**:1–33.

McKenzie, N.L., Burbidge, A.A., & Marchant, N.G. (1973), 'Results of a biological survey of a proposed wildlife sanctuary at Dragon Rocks near Hyden, Western Australia', *Department of Fisheries and Fauna, Western Australia, Report* **12**.

Maxwell, S., Burbidge, A.A., & Morris, K.D. (1996), *The 1996 Action Plan for Australian Monotremes and Marsupials*, Wildlife Australia, Canberra.

Morris, K.D. (2000), 'Fauna translocations in Western Australia 1971–1999: an overview', in *Biodiversity and the Reintroduction of Native Fauna at Uluru–Kata Tjuta National Park* (eds. J.S. Gillen, R. Hamilton, W.A. Low, & C. Creagh), pp. 64–74, Bureau of Rural Sciences, Canberra.

Morris. K., Orell, P., & Brazell, R. (1995), 'The effect of fox control on native mammals in the jarrah forest, Western Australia', in *Proceedings of the 10th Vertebrate Pest Control Conference Hobart, May 1995*, pp. 177–81, Department of Primary Industries and Fisheries, Tasmania: Hobart.

Serventy, D.L. (1954), 'The recent increase of the rarer native mammals. V. Dryandra Forestry Station', *West Australian Naturalist*, **4**:128–9.

Short, J., Turner, B., Parker, S., & Twiss, J. (1994), 'Reintroduction of endangered mammals to mainland Shark Bay: a progress report', in *Reintroduction Biology of Australian and New Zealand Fauna* (ed. M. Serena), pp. 183–8, Surrey Beatty & Sons, Chipping Norton.

Smales, L.R. (1997), '*Multisentis myrmecobius*, gen. et sp. nov. (Acanthocephala: Oligacanthorhynchidae), from the numbat, *Myrmecobius fasciatus*, and a key to genera of the Oligacanthorhynchidae', *Invertebrate Taxonomy*, **11**:301–7.

Twigg. L.E., & King, D.R. (1991), 'The impact of fluoroacetate-bearing vegetation on native Australian fauna: a review', *Oikos*, **61**:412–30.

Waterhouse, G.R. (1836), 'Description of a new genus of mammiferous mammals from New Holland, which will probably be found to belong to the marsupial type', *Proceedings of the Zoological Society of London*, **2**:69–70.

Waterhouse, G.R. (1841), 'Description of a new genus of mammiferous mammals from Australia, belonging probably to the order Marsupialia', *Transactions of the Zoological Society of London*, **2**:149–54, pls 27–8.

Wroe, S., & Musser, A. (2001), 'The skull of *Nimbacinus dicksoni* (Thylacinidae: Marsupialia)', *Australian Journal of Zoology*, **49**:487–514.

PART V

CHAPTER 32

BIOLOGY AND CONSERVATION OF MARSUPIAL MOLES (*NOTORYCTES*)

Joe Benshemesh[A] *and Ken Johnson*[B]

[A] Biological Sciences, Monash University, Clayton, Victoria 3068, Australia. Email: jbenshemesh@bigpond.com
[B] Department of Infrastructure Planning and Environment, PO Box 2605, Alice Springs, NT 0871, Australia. Email: ken.johnson@nt.gov.au

Marsupial moles (*Notoryctes*) are amongst the most unusual, enigmatic and elusive vertebrates that inhabit Australia. They occupy the sandy deserts of central and north western Australia and are the most fossorial of marsupials, rarely venturing to the surface. While they have been known to Aboriginal peoples for thousands of years, and to science for over a century, very little has been documented about *Notoryctes* ecology, reproduction, or behaviour. The genus exhibits extreme morphological specialisation and shows a high degree of convergence with the unrelated eutherian golden moles (Chrysochloridae) which, like *Notoryctes*, are insectivorous and live underground in sandy deserts. While the phylogenetic affinities of marsupial moles are not yet clearly resolved, molecular and palaeontological studies confirm that they have been evolving separately from other marsupials for a very long time, perhaps from the beginning of the Tertiary, and that the genus warrants placement in its own order. Two species are currently recognised, *N. caurinus* in north-western Australia and *N. typhlops* in central Australia, although the taxonomy of the group is currently under review. Both species are regarded as rare and endangered, although so little is known about this elusive animal that this even this is uncertain. In any case, there is grave reason for concern for the conservation of marsupial moles due to high predation by introduced foxes and cats, and to changes in their habitats resulting from introduced herbivores and changed fire regimes. We conclude that studies are urgently needed on the ecology of marsupial moles, and that these would benefit from collaboration with Aboriginal people who have knowledge of these animals' habitats and habits.

INTRODUCTION

The genus *Notoryctes* is endemic to Australia and contains two named species that are the only extant representatives of this unique marsupial order. They are insectivorous predators and the most fossorial of the marsupials, only rarely venturing to the surface. Occurring in remote areas, and characterised by extreme morphological specialisation, an elusive nature and extraordinary habits, this genus continues to intrigue and baffle biologists.

Marsupial moles have a head and body length of up to 140 mm and weigh from 30 g to 60 g. They show the typical characteristics of fossorial mammals (Nevo 1979) including a tubular body form, an absence of ear pinnae, heavily keratinised skin on the snout, a reduced tail, and short dense fur. In common with most burrowing marsupials the pouch opens posteriorly as a protection against the incursion of soil. Sand plain and sand dune country of the central deserts is the preferred habitat.

HISTORY OF DISCOVERY

While marsupial moles were undoubtedly known to Aboriginal desert peoples for thousands of years, the first specimen of *Notoryctes* brought to the attention of the scientific community was collected by W. Coulthard on Idracowra Pastoral Lease in the Northern Territory. Following up on some unusual tracks Coulthard found the animal lying under a tussock. He wrapped the specimen in a kerosene-soaked rag and, in a revolver cartridge box, sent it to EC Stirling, Director of the South Australian Museum, some 1500 km to the south (Stirling 1888a).

When announced by Stirling in 1888 the discovery created great excitement in the scientific community because of its substantial likeness to the convergent eutherian golden moles (Chrysochloridae*)*. The specimen's poor state of preservation made it difficult for Stirling (1888b) to find features placing it with the marsupials and he initially suspected it was a monotreme. Its description as *Notoyctes typhlops* was published by Stirling (1891). Later the genus was tentatively placed with the polyprotodont marsupials (Ogilby 1892). Simultaneously it was considered by Cope (1892) to be unrelated to monotremes and to have arisen from the chrysochlorid stock, thereby forming a connection between the marsupials and the eutherians. Gadow (1892) secured an additional nine specimens and confirmed Ogilby's (1892) placement with the polyprotodont marsupials.

TAXONOMY AND EVOLUTION

Notoryctids are very poorly represented in the fossil record but at least one distinct genus is present in the early Miocene sediments of the Riversleigh deposit of northern Australia (Gott 1988; Archer et al. 1999). This fossil material includes postcranial and dental remains and exhibits characteristics that are much more plesiomorphic than *Notoryctes,* including pre-zalambdodont molar morphology, but Archer et al. (1999) nonetheless believe it to be ancestral to the modern species. No other fossil Notoryctids have been recorded elsewhere.

While morphologically specialised, notoryctids share some key morphological characteristics with almost every other marsupial family (Gadow 1892) and there has been considerable confusion about their phylogenetic affinities. Recent molecular studies reveal that notoryctids are not closely related to any other marsupial family (Calaby et al. 1974; Kirsch 1977; Baverstock et al. 1990; Westerman 1991). This agrees with the limited fossil evidence that marsupial moles have been evolving separately from other marsupials for a very long time and warrants the erection of a distinct order Notoryctemorphia (Aplin and Archer 1987). However, the relationship between Notoryctids and other extant marsupials is not yet clearly resolved: while some recent molecular studies suggest that marsupial moles are most closely related to the dasyurids (Retief et al. 1995; Krajewski et al. 1997; Springer et al. 1998), others (Kirsch et al. 1997; Lapointe and Kirsch 2001) suggest closer links between Notoryctemorphia, Microbiotheriidae and Diprotodontia, and with this group forming a sister-group to the Dasyuridae.

Molecular data have also suggested a date for the separation of Notoryctids from other extant marsupial lineages. Based on divergence rates of nucleotide sequences, Kirsch et al. (1997) estimate that Notoryctemorphia separated from other marsupials about 64 million years ago. At this time, South America, Antarctica and Australia were still joined (Woodburne and Case 1996), and there is no reason to place the origins of the order specifically in Australia. However, the order evolved in Australia in isolation from Antarctica for at least the past 40–50 million years or so since these continents separated (Woodburne and Case 1996). The Riversleigh fossil material suggest these notoryctids were already mole-like and well adapted for burrowing 10–23 million years ago (Gott 1988; Archer et al. 1988), and presumably lived in rainforest, which covered much if not all of central and northern Australia at that time. Decreasing rainfall from the middle to late Miocene, and the development of arid conditions at the end of the Tertiary (Martin 1998), were likely major environmental factors leading to the current, highly derived forms of marsupial mole.

Two species of *Notoryctes* are currently recognised – *N. typhlops* and *N. caurinus*. Thomas (1920) described *N. caurinus* from north-western Australia based on several features including its smaller size than *N. typhlops*. Later authors suppressed this species but it has gained widespread currency at an informal level in recent years based in part on as-yet unpublished morphological studies by Aplin at the Western Australian Museum (K. Aplin pers. comm.). We have elected to follow this more recent approach and recognise both species as this is supported by as-yet unpublished genetic analyses (S. Donellan pers. comm.). However, there is very little published information on *N. caurinus* and the following discussion is based almost entirely on *N. typhlops* unless stated otherwise.

MORPHOLOGY

The body is covered with short, dense, silky fur that is pale cream to white in colour and often coloured by oxides from the soil in which the animals reside. Vibrissae are entirely absent. The limbs are short and powerful, and digits III and IV of the manus are equipped with large spade-like claws.

The dental formula varies between individuals and much of this variation occurs anterior to the first molar. Gadow (1892) gives the dental formula as I4/4 C1/1 PM3/3 M4/4 = 48 maximum. Archer (1984) and Turnbull (1971) give slightly differing formulae having maximum of 50 teeth. Tomes (1897) noted that the molars are poorly imbedded with the root representing just one-third of the length. This was believed to impose an inability of moles to deal with hard food substances, a condition that tends to be supported by the recorded dietary intake.

There is no external evidence of eyes, and no optic nerve, but small black vestigial buds occur beneath the skin (Sweet 1906). External ear openings are covered with fur and are not accompanied by pinnae. The nostrils are small vertical slits below the horny shield of the rostrum. The brain is regarded as being very simple and extremely primitive in structure although the olfactory bulbs and tubercula olfactoria are very large (Smith 1895; Burkitt 1938). This high degree of development suggests that the olfactory sense may be central to the mole's existence. In an environment where visual senses are absent, it could be expected that other senses of marsupial moles might also be enhanced. The middle ear of *Notoryctes* shows characteristic morphological adaptations for low-frequency reception (Segall 1970). Specialised low-frequency auditory and seismic hearing has been recorded in various subterranean vertebrates including chrysochlorid moles (Mayer et al. 1995; Narins et al. 1997).

Behaviour

Apart from some limited field observations, generally made as an adjunct to a larger central program, there have been no published studies of marsupial moles in their natural habitat. Observations of captive animals have been similarly limited, firstly because the animals are not readily available from the wild, and secondly because they have proven difficult to sustain *ex situ*. Most captive animals have died within a month from what appear to be hypothermia or respiratory complications (Johnson and Walton 1989).

Locomotion

Above ground the powerfully built forelimbs with large spade-like claws haul the body over the surface while the hind limbs simultaneously propel the animal forward. Each forelimb is extended forward in unison with its opposite hindlimb so that the animal proceeds with a curving shuffle. The tail is held down so that in its sandy habitat the animal leaves a track of three parallel furrows: the outer two made by the front and rear legs of each side and the centre made by the tail.

Moles move about the surface with a frantic haste but very little speed. Captive animals move hurriedly about their cages with hectic activity, periodically stopping abruptly to commence digging. Once underground they often quickly reappear to resume patrolling of their cage.

Burrowing is initiated with rapid scratching with the forelimbs while the back is held slightly arched and the hindlimbs firmly anchored. Sand is shovelled backwards under the animal until the first half of the animal is submerged and the hindlimbs are used in unison (c/f. above ground movement) to thrust the body forwards and down. Movement into the ground can be exceedingly rapid and likened by some senior Aboriginal informants to a person jumping into water (Johnson and Walton 1989).

Feeding

There are no recorded observations of feeding by moles in the wild. In captivity they consume small eggs and larvae of invertebrates by licking from the soil surface. Presumably ingestion occurs in a similar fashion underground when burrowing moles encounter egg and larva chambers of communal invertebrates.

Larger food items such as beetle larvae are sometimes grasped in the mouth and taken underground but otherwise they are chewed progressively through the premolar part of the mouth without attempts to tear or cut the food. There is only crude manipulation of food with the forelimbs which are highly specialised for digging. The tongue is protrusible to at least two centimetres and may be used to collect small food items, or to lick out body contents of larger animals.

Observations of captive animals indicate that moles are not adapted to pursuit of prey. Their movements are slow, their forelimbs are poorly adapted for grasping, and they lack the dentition that would permit them to deliver and efficient killing bite. Olfactory senses appear to be well developed (Howe 1975) and moles are probably able to detect and follow the underground chambers and galleries of soil-dwelling animals.

Communication

Methods of communication between moles are open to speculation. The olfactory parts of the brain are highly developed and it is probable that communication is mostly olfactory. This method would deliver a very small catchment underground and it is plausible that the occasional forays by moles above ground are associated with olfactory communication. The ears have no external development and sound would be quickly attenuated underground. Nonetheless, marsupial moles produce high-pitched vocalisations when handled (Howe 1975, JB pers. obs.) and this vocal capability is probably functional and used in communication. It is unknown whether moles can produce or detect low frequency sounds which propagate more effectively underground. Much more detailed field or captive studies are required for a proper understanding of how boy meets girl in the subterranean world of the marsupial mole.

Social structure

The social organisation of marsupial moles is unknown but it is assumed that they lead a solitary life. No permanent burrows that

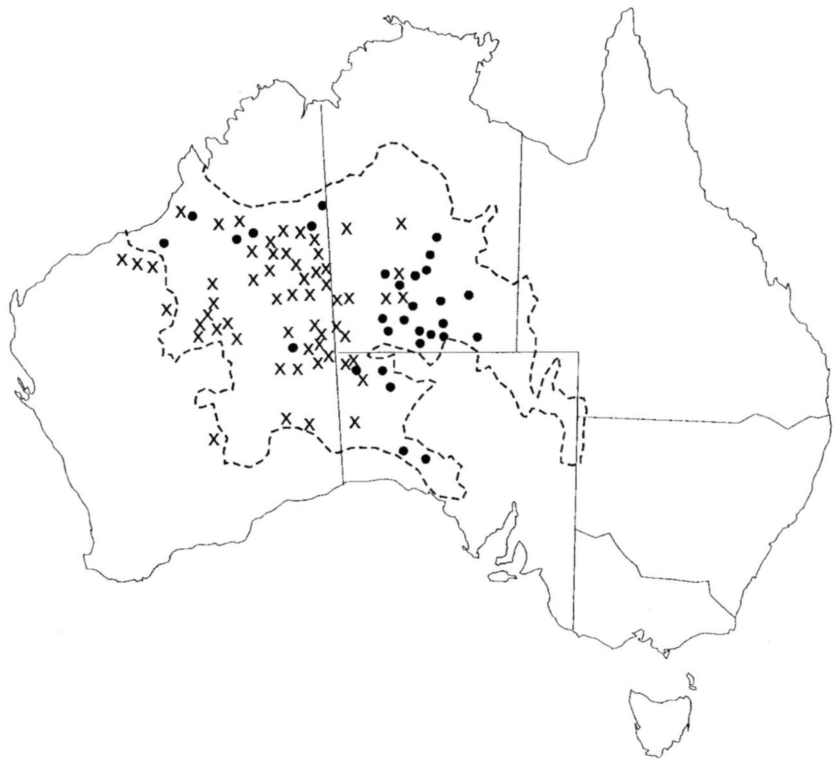

Figure 1 Distribution of Notoryctes determined from specimen records (●) and anecdotal reports from Aboriginal people (x) (after Corbett 1975, Johnson and Walton 1989, and Burbidge et al. 1988). Most points are enclosed by a line (---------) showing the extent of sandy soils in the central desert region (Hubble 1973) and probably indicates the distributional limits of Notoryctes.

might allow for communal living have been found. Visual communication is impossible and unless there is some sophisticated sense of sound or vibration, olfactory methods would seem to offer little in maintaining effective connection between two or more animals underground. On the surface, marsupial mole tracks inspected by us have appeared to be made by solitary animals.

GENERAL ECOLOGY

Distribution

The distribution of *Notoryctes* is known from scattered records throughout the sandy deserts of inland Australia. Most of these records derive from specimens (Corbett 1975) or traditional information (Burbidge et al. 1988) provided by Aboriginal people to collectors. Although there are about 240 specimens of *Notoryctes* in Australian Museums (Pearson and Turner 2001), only vague details of collecting site were recorded for most of the early specimens (many were simply labelled 'central Australia'). Johnson and Walton (1989) reviewed the range of the genus (Fig. 1) and showed that it was largely coincident with the extent of sandy soils in the central desert region (Hubble 1973). These arid regions include the Great Sandy, Little Sandy, Gibson, Tanami, Great Victoria and Simpson Deserts. Anecdotal accounts (Duncan-Kemp 1933; Johnston and Cleland 1943) suggest the species may also inhabit south western QLD, although there are no museum records and Finlayson (1961) found no knowledge of moles amongst Aborigines between the Diamantina and Barcoo rivers.

From the 1930s until recently, the two recognised species of *Notoryctes* have been regarded as synonymous and there is uncertainty about the limits of their ranges although they appear to occupy different areas. Specimens of *N. caurinus* have been collected from only about six localities in the Great Sandy and Gibson Deserts of WA (Maxwell et al. 1996). Over the past decade several specimens of *N. caurinus* have been collected in the south-western part of the Great Sandy Desert, and most recently a live specimen was collected in 1998 on the surface at Punmu in Rudall National Park (Withers et al. 2000). *N. typhlops* has been collected much more frequently than *N. caurinus* and occurs in the Great Victoria Desert in both WA and SA, in the southern NT and the Simpson Desert. Specimens attributed to *N. typhlops* have also been collected in the southern and eastern parts of the Tanami Desert in the NT. Its range extends to the north side of the Gawler ranges and at Fowlers Bay in SA where the sandy desert landform meets the coast.

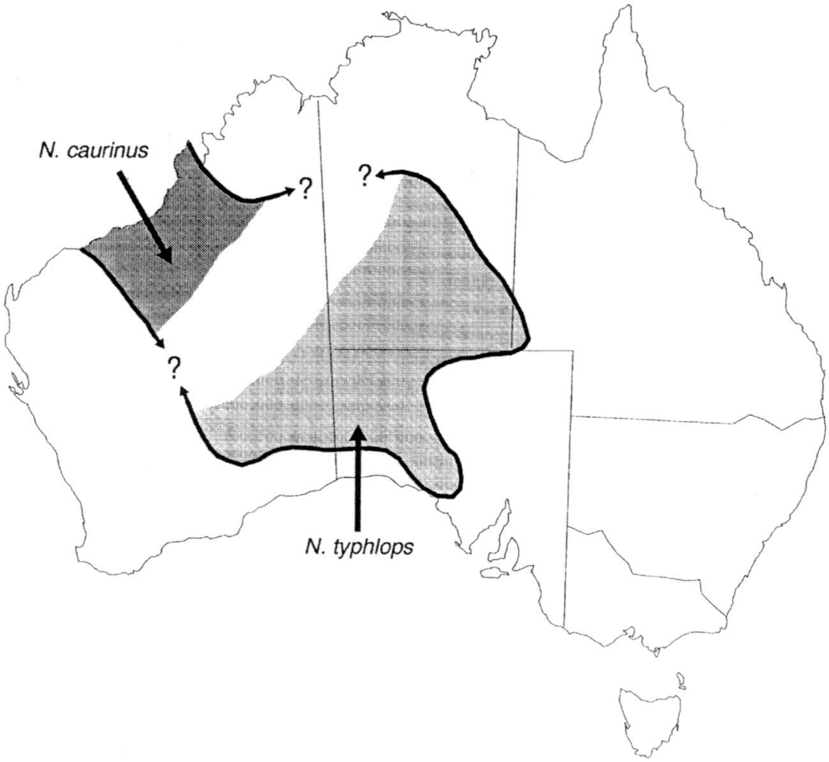

Figure 2 Distribution of *N. typhlops* and *N. caurinus* (after Maxwell et al. 1996).

These southern populations seem to be larger-bodied on average than those from the central part of the species' range, but are otherwise consistent with *N. typhlops* in cranio-dental morphology (K. Aplin pers. comm.).

Many distribution records of *Notoryctes* have relied on sightings of the animals' tracks (Burbidge et al. 1988), or their remains in predator scats (Paltridge 1998), and identification to the level of species has not been possible. In particular, it is uncertain which species inhabits the Tanami Desert as *N. caurinus* has been collected from Balgo Mission (near the WA and NT border), whereas specimens from the southern NT are *N. typhlops*. Both species appear to have been collected in the Warburton Range area of WA (K. Aplin pers. comm.) and might be sympatric there, although the eastern limits of *N. caurinus* and the western limits of *N. typhlops* are very unclear (Fig. 2).

Habitat preferences

Very little is known about the habitat preferences of *N. typhlops*. It has most often been recorded in sandy dunes or flats, usually in association with spinifex (*Triodia* spp.) and various acacias and other shrubs (Corbett 1975; Johnson and Walton 1989) (Fig. 2). Such habitat is widespread in and typical of the sandy deserts. The species may also occupy sandy river flats, such as those of the Finke River in NT where there have been several records of moles (Corbett 1975).

Aboriginal people from WA, NT and SA appear to associate marsupial moles with sand dunes and swales (Burbidge et al. 1988; Baker et al. 1993). Pitjantjatjara people from south west NT and northern SA have frequently told one of us (JB) that *N. typhlops* requires soft sand and that they are unable to tunnel through hard or loamy substrate which occurs in some areas and is often in the swales between dunes. The frequency of underground signs of marsupial moles lend support to this. Signs tend to be more common on dunes than in harder sand at their base, and have not been recorded from the hard mulga earths (JB unpublished).

Less is known about the habitat preferences of *N. caurinus* than of *N. typhlops*, and it is not known whether their preferences differ.

Diet

Information on the diet of marsupial moles has been obtained by examining the gut contents of preserved specimens and from observations of the behaviour of captive animals when presented with different foods. Stirling (1891) and Spencer (1896) noted the presence of ants and their eggs as well as other insect debris in the guts of some of the first specimens dissected. Winkel and

Figure 3 Representative spinifex habitat of marsupial moles in northern South Australia. (photo J. Benshemesh).

Humphrey-Smith (1988) examined the gut contents of 10 museum specimens and found that predatory ants (*Iridomyrmex* sp.) and seed-eating ants (Myrmeciinae) were well represented, as were termites. These authors also noted that seed material was regularly encountered in the guts. Seeds are collected and taken underground by some ant species, and it is likely that moles that intercepted or followed ant galleries underground would encounter seed caches. Whether or not seeds are intentionally ingested is unclear. The dentition and digestive tract of moles is typically insectivorous, so it would seem unlikely that they could derive much nutritional benefit from eating seeds.

Various food items have been presented to moles in captivity and large items are sometimes taken underground to be consumed. *Notoryctes* seems very partial to the eggs, larvae and pupae of insects such as ants, beetles and moths, but less so to the adult insects (Stirling 1891; Corbett 1975; Howe 1975; Johnson and Walton 1989). Moles have also been known to take centipedes, spiders and geckoes in captivity (M. Gillam pers. comm.), although such prey is only crudely manipulated. Although moles are ill-equipped to catch animals on the surface, these prey items spend much of their lives in burrows underground, and in such an environment *Notoryctes* may be a much more formidable predator.

Various Aboriginal informants have confirmed that moles eat insects, seeds and lizards, and have advised that fungi are also a dietary item (Johnson and Walton 1989; Baker et al. 1993).

Ranging behaviour

Underground

Although marsupial moles have been considered to virtually swim through the sand, their tunnels collapsing behind them, they are more accurately regarded as tunnellers that back-fill as they move along. After a marsupial mole has passed through the ground the sand-filled tunnels remain and are visible in cross-section, such as in the walls of trenches dug by hand (Johnson and Walton 1989). One of us (JB) has been involved in collaborative research with Anangu-Pitjantjatjara people in north west SA where these underground signs (of *N. typhlops*) have been examined by excavating trenches of various sizes. In areas where local Aboriginal people have seen moles (Itjaritjari) in the past, these underground signs appear common (about two tunnels per m^2 of vertical trench face), although the age of these signs varied and many were several months or possibly years old. These studies suggest that moles spend most of their active time between 20 cm and 100 cm below the surface. Mole tunnels still occurred deeper than this, but were about twice as common in the top metre as they were between one and two metres underground. Tunnels also occurred below two metres, and there is no reason to suppose that moles don't venture much deeper than this, at least occasionally. These studies have also revealed that the animals tend to tunnel horizontally or at shallow angles more often than at steep angles, although all tunnel angles were represented. No clear signs of feeding or nesting have been

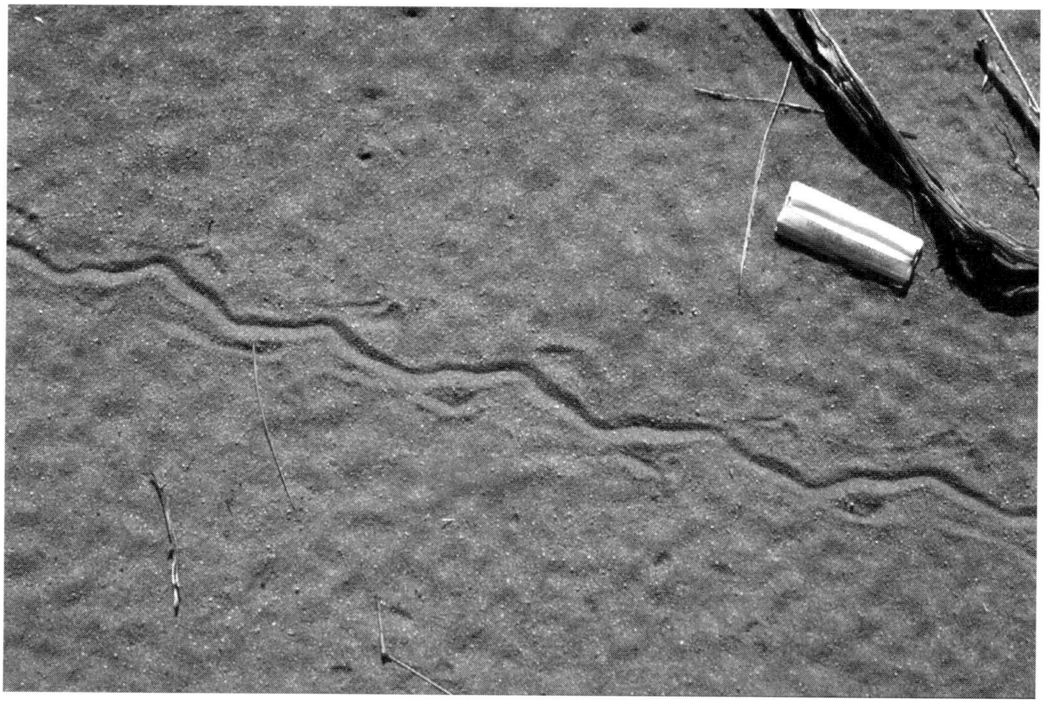

Figure 4 Tracks of a marsupial mole (*N. typhlops*). The central sinuous line is made by the animal's tail. Scale is indicated by the 7 cm long white paper. (photo J. Benshemesh).

found even though several hundred underground signs of moles have been catalogued in over 400 m^2 of exposed trench wall.

Understanding the underground ranging of marsupial moles is likely to require a better understanding of the animals' diet and the characteristics of their environment. While most foods may be expected to occur with 50 cm or so of the surface, the thermal environment at these depths is variable and is over 35°C for much of the summer and below 15°C in winter (JB unpublished). Howe (1975) observed a captive *N. typhlops* shivering when temperatures dropped below 16°C and suggested the eventual death of this animal was caused by hypothermia. He suggested that in captivity the animal needed to be kept at about 22°C. However recent studies of a captive *N. caurinus* have shown a very low metabolic rate and labile body temperature (T_b°) and that the animals are unlikely to be greatly stressed by temperatures they may encounter in their underground environment (Withers et al. 2000). In any case, marsupial moles are probably able to select their thermal environment at any time of the year by changing their depth from the surface.

Surfacing behaviour

Almost all information on marsupial moles is derived from animals observed or caught on the surface. However, it appears that marsupial moles rarely come to the surface, even in areas where they are considered common by local Aboriginal people and where their fresh signs have been found underground. Surface trails are distinctive and several that have been followed by one of us (JB) showed no obvious clues that might suggest that the animal was foraging or meeting conspecifics, or that more than one animal had been on the surface. Typically, an individual will travel only a few metres before returning underground. In most cases these visits to the surface only occur once or a few times and the trails are not seen again. However, on other occasions these visits are numerous and extensive and it is clear that one or more animals have spent many hours wandering on the surface. For example, over 200 exits from the soil were estimated in one 2 ha area, and these were probably all made within a few days or weeks. Similar such areas have also been recorded at Uluru (P. Hookey pers. comm.).

Information from Aboriginal sources suggests that while marsupial moles may surface at any time of day or year, they are more likely to do so after rain and in the cooler seasons (Spencer 1896; Bolam 1923; Russell 1934; Burbidge et al. 1988; Baker et al. 1993). The reason for this is open to speculation.

Predation

Little is known about predation on marsupial moles; however, recent studies suggest that moles may be common food items in the diets of larger mammalian predators, especially the introduced fox. Paltridge (1998) found remains of *Notoryctes* sp. in predator scats at five of her six sites in the Tanami Desert. Overall, her study revealed that 10% of fox scats, 3% of cat scats and 5% of dingo scats contained signs of moles. Signs of marsupial moles have also been found in fox and dingo scats from the A̲nangu-Pitjantjatjara Lands in SA (J. Bice pers. comm.), although not as frequently. It is uncertain whether these predators take marsupial moles on the surface or dig them up, or indeed whether they are actually killing moles or taking dead animals. Dead or severely debilitated moles have been recorded on the surface on several occasions but it seems most likely that these predators prey upon living moles that are on or just under the surface. R.T. Maurice, who travelled widely in the Great Victoria Desert between 1897 and 1903, reported that local Aboriginal people were able to capture moles after hearing them when they were under the surface (T. Gara unpublished), and larger mammalian predators may do likewise. On the surface, marsupial moles are also vulnerable to butcherbirds and corvids (Calaby 1996), birds of prey, snakes and goannas.

Breeding and life history

Virtually nothing is known about reproduction in *Notoryctes*. Their anatomy conforms to the normal marsupial pattern with the exception that there is no scrotum and the testes are prepenial lying at the anterior edge of the pubic bones between the skin and the abdominal wall (Johnson and Walton 1989). The pouch opens backwards and contains two teats (Johnson 1995).

There are no published accounts of copulation, pregnancy, birth or mother-young interactions. Single and twin pouch young have been recorded but pouch young appear very rare in museum collections, and the external characteristics of pouch young have been described from only one individual (Jones 1921). A pregnant female from Ooldea was found to contain six sub-terminal embryos (K. Aplin pers. comm.), suggesting a degree of embryonic wastage in *Notoryctes*. It is not known whether *Notoryctes* builds a nest or forms permanent burrows of any kind but none have been found and it seems unlikely that moles would be able to collect nest material.

Ecological analogues

The golden moles of Africa (Chrysochloridae) bear a striking resemblance to marsupial moles, a point remarked upon when *Notoryctes* was first described. In particular, the Namib Desert golden mole *Eremitalpa granti namibensis* resembles marsupial moles in size, morphology, habits and physiology. Like marsupial moles, this mole is blind and feeds on termites and other invertebrates in an arid sandy desert. It is highly specialised for digging in sand, travels underground without forming permanent burrows and consumes its prey underground (Fielden et al. 1990a), and has a low metabolic rate and labile body temperature (Fielden et al. 1990b; Seymour and Seely 1996; Withers et al. 2000). However, unlike marsupial moles, the Namib Desert golden mole is a capable runner on the surface where it forages for several kilometres every night, frequently stopping to bury its head in the substrate and sense its prey (Fielden et al. 1990a; Seymour et al. 1998). Although the Namib Desert golden mole occurs in more sparsely vegetated areas with lower annual rainfall (50 mm) than *Notoryctes*, it nonetheless occurs at densities of several individuals per km^2 (Fielden 1991; Seymour et al. 1998). Whether marsupial moles occur, or once occurred, at similar densities is not known.

CONSERVATION STATUS

Maxwell et al. (1996) regard both recognised species of *Notoryctes* as endangered although they note that so little is known about either species that the degree to which they are threatened could not be determined with any certainty. Nonetheless, despite this lack of data these authors believe that there is reason for urgent concern.

Some historical accounts suggest that *N. typhlops* at least was relatively common in the late 1800s and early 1900s. For example, Spencer (1896) was able to secure 48 specimens of *N. typhlops* at Charlotte Waters on the western edge of the Simpson Desert, and there are anecdotal accounts of a trade in hundreds or thousands of *Notoryctes* pelts in the early 1900s (Johnson and Walton 1989).

Aboriginal informants still regarded *Notoryctes spp.* as common in suitable areas in the 1980s and 1990s (Burbidge et al. 1988; Baker et al. 1993; Pearson and Turner 2001) although in many cases it would not have been possible to determine whether the information related to the current situation or to some more distant time in the past. Experienced mammalogists working in many parts of the species' range have located few specimens, and anecdotal reports from long-established pastoral residents suggest a decline in abundance. Recent analysis of the number of *Notoryctes* specimens held in museums and collected in each decade over the past century suggest no clear trend, with 4–10 specimens collected in most decades (Pearson and Turner 2001). Peaks in acquisition rates occurred in the 1930s and 1960s, and these may reflect eras of increased exploration by biologists and anthropologists in central Australia (Pearson and Turner 2001). Although these authors did not distinguish species in their analysis, their data suggest that endangered status applied by Maxwell et al. (1996) might not be warranted for both *N. typhlops* and *N. caurinus*. However, it is also true that the acquisition rate of specimens is difficult to interpret due to the serendipitous nature of collecting *Notoryctes* and the many changes in human activity and Aboriginal lifestyle that have occurred over the past century.

Another reason for concern about the conservation of marsupial moles is that the vast majority (about 90%) of medium-sized mammals (35–5000 g) in arid Australia have become threatened or extinct in the past century or so (Burbidge and McKenzie 1989). Marsupial moles are within this size range (30–60 g) and may be subject to the same threatening processes. In particular, introduced foxes and cats have been implicated in the decline of many other medium-sized mammals of the arid zone and recent evidence suggests they may pose a serious threat to marsupial moles (Paltridge 1998).

What other process may be threatening marsupial moles is unclear, although the past century has been ecologically turbulent in central Australia with the arrival of introduced herbivores, pastoralism and the disruption of systematic burning by Aborigines. Marsupial moles are likely to be sensitive to changes in the availability of their food, and thus indirectly to changes in vegetation caused by herbivores and fire, although too little is known about the ecology of *Notoryctes* to assess the likelihood of such a threat.

There is clearly an urgent need for studies into the conservation ecology of marsupial moles. Although these curious animals are undoubtedly elusive and difficult to study, new techniques combined with traditional knowledge and skills may bring an understanding of their ecology and conservation within reach. For example, survey techniques based on measuring the abundance of their sand-filled tunnels may provide a new quantitative tool for collecting information on *Notoryctes* distribution and abundance, and for monitoring populations and investigating habitat preferences (Benshemesh et al. 2000). Information on their distribution can also be obtained through collaboration with local Aboriginal people (Burbidge et al. 1988; Pearson and Turner 2000) and by predator scat analysis (Paltridge 1998). In addition, data on the activity and behaviour of moles underground may be obtained using seismic sensing equipment, such as geophones, to monitor tunnelling (JB unpublished) and perhaps even feeding and vocalising

Aboriginal context and knowledge

Aboriginal peoples throughout the sandy deserts have known of marsupial moles for thousands of years and know the species by a variety of names (Burbidge et al. 1988). Two of these names have been adopted as common names for the recognised species: Itjaritjari for *N. typhlops*, and Kakarratul for *N. caurinus*. Marsupial moles were eaten only during hard times and are generally regarded with sympathy (Burbidge et al. 1988), which is hardly surprising considering the animals' small size and that they are blind, clumsy, and harmless. Marsupial moles feature in Aboriginal mythology and are associated with certain sites and 'Dreaming' trails in central Australia, such as at Uluru (Mountford 1948; Baker et al. 1993) and also in the Anangu-Pitjantjatjara Lands.

In regard to scientific inquiry, Aboriginal people have collected most specimens and their involvement has been instrumental in virtually everything that has been learnt about the species. Considering the elusiveness of the species and the paucity of information on their ecology, it is likely that further efforts to understand and manage the marsupial moles will continue to rely on Aboriginal involvement. Although detailed traditional knowledge and skills are rapidly declining as intergenerational transfer diminishes (Horstman and Wightman 2001), Aboriginal people who still have good tracking skills are able to recognise and interpret subtle signs and tracks, and are often willing to teach these skills to researchers.

Conclusion

Marsupial moles are the only surviving representatives of an order of marsupials, and are phylogenetically and ecologically unique. Both species are regarded as endangered, but there is such a dearth of information that even this is uncertain. Field studies are clearly needed to understand the basic ecology of marsupial moles and there is an urgent need to clarify their conservation status. This will also require examination of the taxonomy and genetic diversity in *Notoryctes* so that appropriate population units can be determined for management.

A healthy captive population of marsupial moles would also benefit our understanding of their biology, particularly their reproduction, physiology, and behaviour. However, as every captive mole has died within a few weeks there is clearly much to learn about the animals' requirements before they can be successfully kept in captivity. Recent advances in studying in the field the distribution and abundance of marsupial moles may provide some of this detail about their requirements, as well as providing a general tool for examining the ecology of these elusive animals.

In the field, studies are required to determine and monitor the species' distribution and abundance, identify threats that may be causing populations to decline, and to better understand the species' habits and requirements. These studies would benefit from collaboration with Aboriginal people who have knowledge of the habitats and habits of marsupial moles (Pearson and Turner 2001). As these are usually older people, many of whom grew up in traditional lifestyles, much of this expertise and knowledge is fast disappearing as people age and die. There is, therefore, also an *urgent* need to learn from these people while the opportunity still exists.

Acknowledgements

We are grateful to Ken Aplin who generously provided unpublished data for inclusion, and who together with Peter Copley provided valuable comments on the manuscript. Gary Bonner, Chris Dickman and an anonymous referee also made many helpful suggestions on the manuscript. We thank Syd Milgate who prepared the distribution maps.

REFERENCES

Aplin, K.P., & Archer, M. (1987), 'Recent advances in marsupial systematics with a syncretic classification', in *Possums and Opossums: Studies in Evolution* (ed. M. Archer), pp. xv–lxxii, Surrey Beatty & Sons and the Royal Zoological Society of New South Wales, Sydney.

Archer, M. (1984), 'The Australian marsupial radiation', in *Vertebrate Zoogeography and Evolution in Australasia* (eds. M. Archer, & G. Clayton), pp. 633–805, Hesperian Press, Perth.

Archer, M., Hand, S.J., & Godthelp, H. (1988), 'A new order of Teriary zalambdodont marsupials', *Science,* **239**:1528–31.

Archer, M., Arena, R., Bassarova, M.I., Black, K., Brammall, J., Cooke, B., Creaser, P., Crosby, K., Gillespie, A., Godthelp, H., Gott, M., Hand, S.J., Kear, B., Krikmann, A., Mackness, B., Muirhead, J., Musser, A., Myers, T., Pledge, N., Wang, Y., & Wroe, S. (1999), 'The evolutionary history and diversity of Australia's mammals', *Australian Mammalogy,* **21**:1–45.

Baker, L., Woenne-Green, S., & the Mutitjulu Community (1993), 'Anangu knowledge of vertebrates and the environment', in *Kowari 4: Uluru Fauna* (eds. J.R.W. Reid, J.A. Kerle, & S.R. Morton), pp. 79–132, ANPWS, Canberra.

Baverstock, P.R., Krieg, M., & Birrell, J. (1990), 'Evolutionary relationships of Australian marsupials as assessed by albumin immunology', *Australian Journal of Zoology,* **37**:273–87.

Benshemesh, J., Bice, J., & Copley, P. (2000), 'Marsupial moles and their residual holes: A new method for investigating the distribution and abundance of *Notoryctes*', (Abstract), p. 35, Australian Mammal Society Conference 2000.

Bolam, A.G. (1923), *The Trans-Australian Wonderland,* The Modern Printing Company, Melbourne.

Burbidge, A.A., Johnson, K.A., Fuller, P.J., & Southgate, R.I. (1988), 'Aboriginal knowledge of the mammals of the central deserts of Australia', *Australian Wildlife Research,* **15**:9–39.

Burbidge, A.A., & McKenzie, N.L. (1989), 'Patterns in the modern decline of Western Australia's vertebrate fauna: causes and conservation implications', *Biological Conservation,* **50**:143–98.

Burkitt, A.N. (1938), 'The external morphology of the brain of *Notoryctes typhlops*', *Proc Koninklijke Nederlandsche Akademie van Wetenschappen,* **41**:921–33.

Calaby, J.H. (1996), 'Baldwin Spencer's post-Horn Expedition Collectors in Central Australia', in *Exploring Central Australia: society, the environment and the 1894 Horn Expedition* (eds. S.R. Morton, & D.J. Mulvaney), pp. 188–208, Surrey Beatty and Sons, Sydney.

Calaby, J.H., Corbett, L.K., Sharman, G.B., & Johnston, P.G. (1974). 'The chromosomes and systematic position of the marsupial mole, *Notoryctes typhlops*', *Australian Journal of Biological Science,* **27**:529–532.

Cope, E.D. (1892), 'On the habits and affinities of the new Australian mammal, *Notoryctes typhlops*', *American Naturalist,* **26**:121–8.

Corbett, L.K. (1975), 'Geographical distribution and habitat of the marsupial mole, *Notoryctes typhlops*', *Australian Mammalogy,* **1**:375–8.

Duncan-Kemp, A.M. (1933), *Our Sandhill Country: Nature and man in south-western Queensland*, Sydney, Angus & Robertson.

Fielden, L.J. (1991), 'Home range and movements of the Namib Desert golden mole *Eremitalpa grant namibensis* (Chrysochloridae)', *Journal of Zoology, London,* **223**:675–86.

Fielden, L.J., Perrin, M.R., & Hickman, G.C. (1990a), 'Feeding ecology and foraging behavior of the Namib Desert golden mole *Eremitalpa granti namibensis* (Chrysochloridae)', *Journal of Zoology, London,* **220**:367–90.

Fielden, L.J., Waggoner, J.P., Perrin, M.R., & Hickman, G.C. (1990b), 'Thermoregulation in the Namib Desert golden mole *Eremitalpa granti namibensis* (Chrysochloridae)', *Journal of Arid Environments,* **18**: 221–38.

Finlayson, H.H. (1961), 'On central Australian mammals. Part IV. The distribution and status of central Australian species', *Records of the South Australian Museum,* **14**:141–91.

Gadow, H. (1892), 'On the systematic position of *Notoryctes typhlops*', *Proceedings of the Zoological Society of London,* **1892**:361–70.

Gott, M. (1988), 'A Tertiary marsupial mole (Marsupialia: Notoryctidae) from Riversleigh, northeastern Australia and its bearing on notoryctemorphian phylogenetics', Honours thesis, University of New South Wales, Sydney.

Horstman, M., & Wightman, G. (2001), '*Karparti* ecology: Recognition of Aboriginal ecological knowledge and its application to management in north-western Australia', *Ecological Management and Restoration,* **2**:99–109.

Howe, D. (1975), 'Observations of a captive marsupial mole, *Notoryctes typhlops*', *Australian Mammalogy,* **1**:361–5.

Hubble, G.D. (1973), 'Soils', in *Australian Grasslands* (ed. R.M. Moore), Canberra, Australian National University Press.

Johnson, K.A. (1995), 'Marsupial Mole *Notoryctes typhlops*', in *The Mammals of Australia* (ed. R. Strahan), pp. 409–11, Reed Books, Chatswood.

Johnson, K.A., & Walton, D.W. (1989), 'Notoryctidae', in *Fauna of Australia: Volume 1B Mammalia* (eds. D.W. Walton, & B.J. Richardson), pp. 591–602, Australian Government Printing Service, Canberra.

Johnston, T.H., & Cleland, J.B. (1943), 'Native names and uses of plants in the north east corner of South Australia', *Transactions of the Royal Society of South Australia,* **67**:149–73.

Jones, F.W. (1921), 'The external characteristics of pouch embryos of marsupials. No. 2 *Notoryctes typhlops*', *Transactions of the Royal Society of South Australia,* **55**:36–9.

Kirsch, J.A.W. (1977), 'The comparative serology of Marsupialia, and a classification of marsupials', *Australian Journal of Zoology (suppl.),* **52**:1–152.

Kirsch, J.A.W., Lapointe, F.-J., & Springer, M.S. (1997), 'DNA-hybridisation studies of marsupials and their implications for metatherian classification', *Australian Journal of Zoology,* **45**: 211–80.

Krajewski, C., Buckley, L., & Westerman, M. (1997), 'DNA phylogeny of the marsupial wolf resolved', *Proceedings of the Royal Society of London B,* **264**:911–17.

Lapointe, F.-J., & Kirsch, J.A.W. (2001), 'Construction and verification of a large phylogeny of marsupials', *Australian Mammalogy,* **23**:9–23.

Martin, H.A. (1998), 'Tertiary climatic evolution and the development of aridity in Australia', *Proceedings of the Linnean Society of New South Wales,* **119**:115–36.

Maxwell, S., Burbidge, A.A., & Morris, K.D. (1996), *The 1996 Action Plan for Australian marsupials and monotremes*, IUCN/SSC Australasian Marsupial and Monotreme Specialist Group, Wildlife Australia Endangered Species Program, Canberra.

Mayer, A.V., O'Brien, G., & Sarmiento, E.E. (1995), 'Functional and systematic implications of the ear in golden moles (Chrysochloridae)', *Journal of Zoology, London,* **236**:417–30.

Mountford, C.P. (1948), *Brown Men and Red Sand*, Robertson and Mullens Limited, Melbourne.

Narins, P.M., Lewis, E.R., Jarvis, J.J.U.M., & O'Riain, J. (1997), 'The use of seismic signals by fossorial Southern African mammals: A neuroethological gold mine', *Brain Research Bulletin,* **44**:641–6.

Nevo, E. (1979), 'Adaptive convergence and divergence of subterranean mammals', *Annual Review of Ecology and Systematics,* **10**:269–308.

Ogilby, J.D. (1892), *Catalogue of Australian Mammals with introductory notes on general mammalogy,* Australian Museum, Sydney.

Paltridge, R. (1998), 'Occurrence of Marsupial Mole (*Notoryctes typhlops*) remains in the faecal pellets of cats, foxes and dingoes in the Tanami Desert, NT', *Australian Mammalogy,* **20**:427–9.

Pearson, D.J., & Turner, J. (2000), 'Marsupial moles pop up in the Great Victoria and Gibson Deserts', *Australian Mammalogy,* **22**:115–19.

Retief, J.D., Krajewski, C., Westerman, M., Winkfein, R.J., & Dixon, G.H. (1995), 'Molecular phylogeny and evolution of marsupial protamine P1 genes', *Proceedings of the Royal Society of London B,* **259**:7–14.

Russell, A. (1934), *A tramp-royal in wild Australia*, Jonathon Cape, London.

Segall, W. (1970), 'Morphological parallelisms of the bulla and auditory ossicles in some insectivores and marsupials', *Fieldiana Zoology,* **51**:169–205.

Seymour, R.S., & Seely, M.K. (1996), 'The respiratory environment of the Namib Desert golden mole', *Journal of Arid Environments,* **32**:453–61.

Seymour, R.S., Withers, P.C., & Weathers, W.W. (1998), 'Energetics of burrowing, running, and free-living in the Namib Desert golden mole (*Eremitalpa granti namibensis*)', *Journal of Zoology, London,* **244**:107–17.

Smith, G.E. (1895), 'The comparative anatomy of the cerebrum of *Notoryctes typhlops*', *Transactions of the Royal Society of South Australia,* **19**:167–93.

Spencer, B. (1896), *Report on the work of the Horn Scientific Expedition to Central Australia. Part I: Introduction, Narrative, Summary of Results and Supplement to Zoological Report*, Melville, Mullen and Slade, Melbourne.

Springer, M.S., Westerman, M., Kavanagh, J.R., Burk, A., Woodburne, M.O., Kao, D.J., & Krajewski, C. (1998), 'The origin of the Australian marsupial fauna and the phylogenetic affinities of the enigmatic monito del monte and marsupial mole', *Proceedings of the Royal Society of London B,* **265**:2381–6.

Stirling, E.C. (1888a), 'A new Australian mammal', *Nature,* **38**:588–9.

Stirling, E.C. (1888b), 'Preliminary notes on a new Australian mammal', *Transactions of the Royal Society of South Australia,* **11**:21–4.

Stirling, E.C. (1891), 'Description of a new genus and species of marsupial, *Notoryctes typhlops*', *Transactions of the Royal Society of South Australia,* **14**:154–87.

Sweet, G. (1906), 'Contributions to our knowledge of the anatomy of *Notoryctes typhlops*: Part 3 The eye', *Quarterly Journal Microscopical Science,* **50**:547–71.

Thomas, O. (1920), '*Notoryctes* in north-west Australia', *Annals and Magazine of Natural History,* **6**:111–13.

Tomes, C.S. (1897), 'Note upon the minute structure of the teeth of *Notoryctes*', *Proceedings of the Zoological Society of London,* **1897**:409–12.

Turnbull, W.D. (1971), 'The Trinity Therians: their bearing and evolution in marsupials and other therians', in *Dental Morphology and Evolution* (ed. A.A. Dahlberg), pp. 151–79, University of Chicago Press, Chicago.

Westerman, M. (1991), 'Phylogenetic relationships of the marsupial mole *Notoryctes typhlops* (Marsupialia: Notoryctidae)', *Australian Journal of Zoology,* **39**:529–37.

Winkel, K., & Humphrey-Smith, I. (1988), 'Diet of the marsupial mole, *Notoryctes typhlops* (Stirling 1889) (Marsupialia: Notoryctidae)', *Australian Mammalogy,* **11**:159–61.

Withers, P.C., Thompson, G.G., & Seymour, R.S. (2000), 'Metabolic physiology of the north-western marsupial mole, *Notoryctes caurinus* (Marsupialia: Notoryctidae)', *Australian Journal of Zoology,* **48**:241–58.

Woodburne, M.O and Case, J.A. (1996), 'Dispersal, vicariance, and the Late Cretaceous to Early Teriary land mammal biogeography from South America to Australia', *Journal of Mammalian Evolution,* **3**:121–61.

CHAPTER 33

THE APPLICATION OF GENETIC RESEARCH TO CONSERVATION MANAGEMENT IN CARNIVOROUS MARSUPIALS WITH SPECIAL EMPHASIS ON DASYURIDS

Karen B. Firestone

Zoological Parks Board of New South Wales, Conservation Research Centre, PO Box 20, Mosman, NSW 2088, Australia. Current address: Evolutionary Biology Unit, Australian Museum, 6 College Street, Sydney, NSW 2010, Australia. Email: karenf@austmus.gov.au

Conservation genetics is a relatively new field within both realms of conservation biology and molecular genetics. The application of molecular genetic principles to the conservation of any species of carnivorous marsupial is still in its infancy and there is extremely limited information available for many species. Indeed, there are probably a number of as yet unrecognised cryptic species which may only be identifiable through molecular means. In this chapter, I present an overview of some of the concepts of conservation genetics including the applicability of phylogenetics and phylogeography, population genetics, and molecular ecology to conservation. The limited number of case studies in which these principles have been applied to the conservation of carnivorous marsupials is highlighted and directions for future research are suggested.

INTRODUCTION TO CONSERVATION GENETIC CONCEPTS

Conservation biology is the science of preserving biodiversity including whole ecosystems, species assemblages, individual populations, and gene diversity (Soulé 1986). Conservation genetics is a relatively new, yet intrinsic, discipline within conservation biology that applies information obtained using molecular genetics to conservation concerns (Hedrick 2001; Ashley 1999; Haig 1998; Smith and Wayne 1996; Frankham 1995a; O'Brien 1994b; Avise 1994; Hedrick and Miller 1992). Despite the relative youth of this field, astounding progress has been made in the past few years in applying theoretical genetic concepts to real-world conservation issues for endangered and threatened taxa. This progress is partly due to the rapid technical advances in the field of molecular genetics, including the advent of the polymerase chain reaction (White et al. 1989), increasingly sophisticated computational algorithms (e.g. Swofford 1998; Schneider et al. 1997; Raymond and Rousset 1995), and non-invasive sampling techniques (Palsbøll 1999; Taberlet and Luikart 1999). Conservation genetics draws on a variety of disciplines within the broad field of molecular genetic research and addresses issues that affect small or declining populations. The various molecular techniques used in conservation genetics are best viewed as a set of tools that can provide vital information of use to agencies and managers in planning appropriate conservation strategies.

Genetic contributions to the conservation of species can be split into three areas: 1) examination of both the evolutionary relationships among species and genetic differentiation among conspecific populations as a means of determining distinct lineages

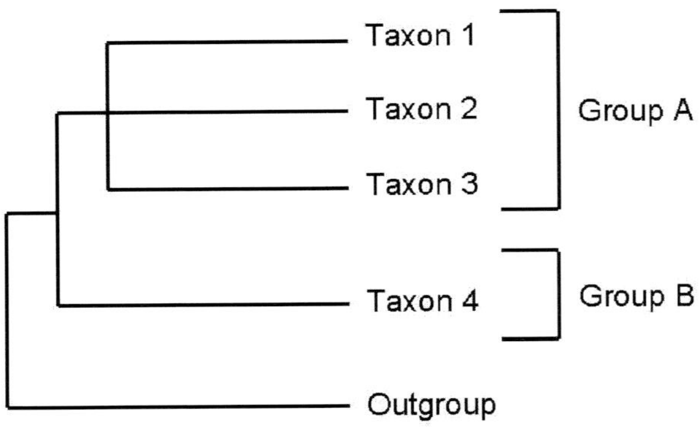

Figure 1 Hypothetical phylogenetic relationships among four taxa. Group A is comprised of three closely related taxa while group B is represented by one taxon, the sole member of its lineage. If taxon 4 and taxon 1 are both endangered, it may be considered more important to place conservation priorities towards taxon 4.

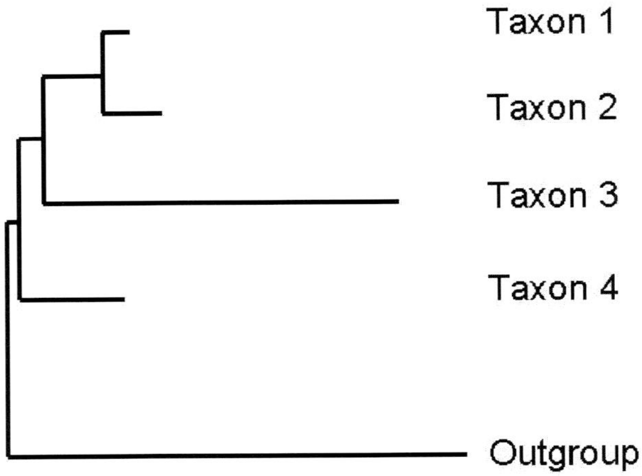

Figure 2 Hypothetical phylogenetic relationships among four related taxa. Taxon 3 is phylogenetically distinct in being the most distant of any of the four taxa and has the longest branch length. The greater the number of changes, the longer the branch length. Therefore, conservation effort may be considered more crucial for this group if priorities need to be set.

and appropriate conservation units; 2) exploration of the various aspects of population genetics such as estimation of levels of genetic diversity, inbreeding, and effective population size (N_e), the genetic consequences of population bottlenecks or founder effects in captive or reintroduced populations; and 3) the use of molecular data to infer aspects of a species' behavioural or social ecology such as localised population structure, parentage, and relatedness. Each of these points is reviewed below.

Evolutionary genetics

One of the main goals of conservation is to preserve natural genetic diversity. Yet there is much debate in the conservation literature as to what we can afford to conserve given the chronically limited resources available and the inability to conserve everything (Crozier 1997; Vogler and DeSalle 1994; Crozier 1992; Vane-Wright et al. 1991). The examination of evolutionary or phylogenetic relationships among taxa provides a means of determining distinct historical lineages which may assist conservation managers in determining what to conserve, given this 'agony of choice'. For instance, if an endangered taxon is the sole representative within a divergent lineage, this taxon may be considered more critical in terms of conservation effort than another equally endangered taxon that is closely related to more secure taxa (Fig. 1). Alternatively, a lineage may be worthy of conservation effort if it is phylogenetically very distant from its closest relatives. The greater the evolutionary time, the longer the branch lengths (Fig. 2, Polasky et al. 2001; Faith 1992).

Analysis of phylogenetic relationships may also provide useful information in instances where taxonomic status is ambiguous. For instance, the past taxonomic uncertainty of both dusky seaside sparrows (*Ammodramus maritimus nigrescens*, Zink and Kale 1995; Avise and Nelson 1989) and tuataras (the *Sphenodon* complex, Daugherty et al. 1990) had led to inappropriate management and extinctions in both taxa. Similarly, the widespread common dunnart, *Sminthopsis murina*, was shown to be comprised of a number of distinct species using molecular data (Baverstock et al. 1984). The use of genetic data to more clearly denote relationships among and within species may help to guide management decisions and prevent extinctions in other taxa. Conversely, genetics can be used to identify morphs within species that are genetically similar, thereby lowering the priority of the morphs for separate conservation (e.g. some dunnarts, Labrinidis et al. 1998).

Genetic differentiation between conspecific populations is a similar concept, in that differentiation provides a means of identifying distinct units for conservation purposes. These units generally fall within the following categories: subspecies, Evolutionarily Significant Units (ESUs) or Management Units (MUs). ESUs were originally proposed as a means of determining which of the large charismatic taxa to conserve in zoos, given the often highly subjective subspecific status of some taxa and strict limitations on space and resources within zoological institutions (Ryder 1986). Moritz (1994) further refined the genetic identity of ESUs as those units (taxa or populations) that are significantly different in allele frequencies at nuclear loci as well as reciprocally monophyletic at mitochondrial loci. He also identified management units (MUs) as those populations that are significantly different in allele frequencies (either mitochondrial or nuclear), regardless of the phylogenetic relationships of the alleles (Moritz et al. 1995; Moritz 1994; see Fig. 3). There is, however, debate as to the application of these concepts (Taylor and Dizon 1999; Paetkau 1999).

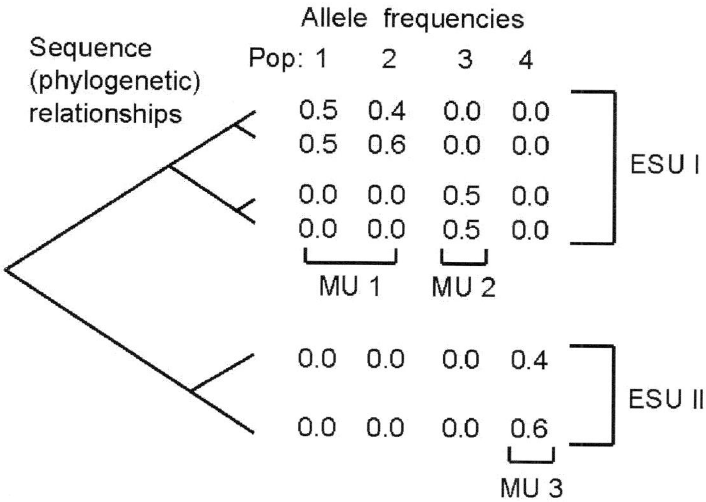

Figure 3 Diagrammatic representation of two reciprocally monophyletic clades depicting two ESUs and three MUs. Redrawn after Moritz (1994). Populations 1 and 2 have similar allele frequencies and are part of the same management unit. Population 3 has significantly different allele frequencies and forms a different MU within ESU I. Population 4 is reciprocally monophyletic to populations 1,2, and 3 and has significantly different allele frequencies; it is therefore a separate ESU and MU. Individuals from populations 1, 2, and 3 (MU1 and MU2) may be translocated or interbred since they are members of the same ESU. However, breeding or translocation of individuals between different ESUs is not recommended, therefore population 4 would be managed separately.

Although defining subspecies is often problematic, the concept of the ESU is closely related to that of subspecies and provides a framework for genetically identifying distinct units. ESUs reflect the historic phylogeographical separation of populations, and the aim of designating ESUs is to allow for their separate management so as to retain the long-term evolutionary potential of the species. MUs, on the other hand, are conservation units that are more appropriate for short-term conservation goals such as population monitoring and translocation/augmentation programs (Moritz 1999). Thus conservation managers would not want to breed individuals from two different ESUs, but rather allow for their independent evolution. It may be appropriate, however, to breed and translocate individuals between different MUs within an ESU (Moritz 1999; see Fig. 3).

Population genetics

In addition to phylogenetic considerations, various aspects of population genetics may be critical to the conservation of rare or endangered taxa. For example, knowledge of the level of genetic variability or diversity within populations is considered to be a fundamental aspect of conservation genetics (Frankham 1995c; Frankham 1995a; Frankel and Soulé 1981). The loss of genetic variation (either allelic diversity or heterozygosity), due to genetic drift, inbreeding, or other factors can reduce both individual fitness and the ability of populations to adapt to altered environmental conditions (Amos and Balmford 2001; Lacy 1997). Inbreeding depression, as a result of close consanguineous matings, may decrease individual fitness via a number of routes including decreased reproductive fitness, decreased resistance to disease, and decreased flexibility in response to environmental changes (Saccheri et al. 1998; Newman and Pilson 1997; Frankham 1995c; Miller and Hedrick 1993; Ralls et al. 1988; O'Brien and Everman 1988; Gall 1987). Conversely, relatively high levels of genetic diversity may increase qualities associated with fitness (Allendorf and Leary 1986). The long-term viability of populations also may be threatened by the loss of genetic variation due to drift. Loss of variability may affect small populations by decreasing a population's ability to adapt to changing environments and by increasing the probability of extinction due to stochastic effects (Saccheri et al. 1998; Lacy 1997). It is important to assess levels of diversity within populations as a means of determining the potential adaptability of populations as well as for future population monitoring.

A population's effective size (N_e) is relevant in conservation and population genetics as an estimate of its ability to maintain genetic diversity through time; in general, the greater an effective population size, the slower the loss of genetic diversity. N_e also provides an estimate of a population's future evolutionary potential or an estimate of the amount of genetic change within populations. The effective population size is generally defined as the size of an ideal population whose genetic composition is influenced by random processes similar to that of a real population, i.e. the same variance in allele frequencies (variance N_e) or level of inbreeding (inbreeding N_e) (Schwartz et al. 1998; Frankham 1995b; Ryman 1994; Wright 1978; Kimura and

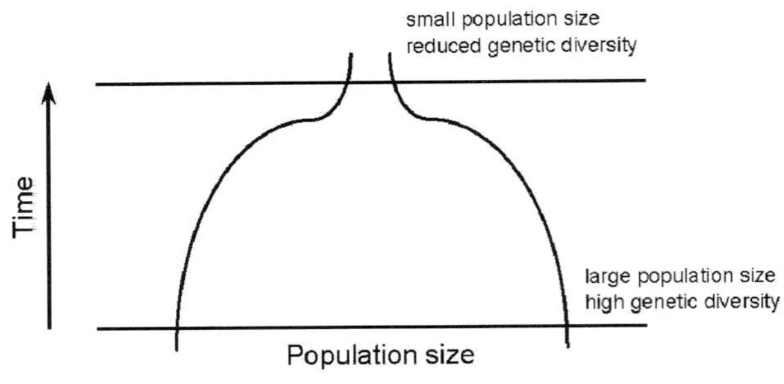

Figure 4 Diagrammatic representation of a population bottleneck. As individuals are lost from a population, genetic diversity decreases through loss of alleles and increasing homozygosity.

Crow 1963). Effective population size often may be an order of magnitude smaller than the census size (Frankham 1995b). Estimates of N_e using more traditional ecological data can be difficult to achieve for natural populations since the populations of concern are most often not 'ideal' populations, i.e. they are not in mutation-drift equilibrium. N_e is further influenced by ecological factors such as uneven sex ratios, variance in individual lifetime reproductive success, fluctuations in population census size, overlapping generations, mating systems, and population structure (Frankham 1996; Frankham 1995b; Nunney and Elam 1994; Nunney 1993; Wright 1931). While these confounding factors make estimating N_e difficult, genetic data can be used to indirectly estimate this parameter by measuring temporal changes in genetic diversity or linkage disequilibrium (Schwartz et al. 1998; Bartley et al. 1992; Waples 1989).

Another area of particular conservation concern involves those populations that have declined from very large numbers of individuals to just a few. These populations are said to have undergone a bottleneck (Fig. 4). It is important to examine population bottlenecks due to the increased likelihood of change in the genetic composition of these populations. Theory predicts that genetic variation should be higher in populations that have not gone through a bottleneck event than in those that have (Nei et al. 1975). There are many examples of natural populations known to have gone through bottlenecks that also currently exhibit low levels of genetic diversity: northern elephant seals (Hoelzel et al. 1993), Asiatic lions (O'Brien et al. 1987), black-footed ferrets (O'Brien et al. 1989), and Florida panthers (Roelke et al. 1993) all have experienced well documented population bottlenecks and exhibit depleted levels of diversity. Yet the instance and severity of population bottlenecks is sometimes assumed from current levels of diversity, not on empirical data predating a supposed bottleneck. This discrepancy can be controversial, as in the case of cheetahs where a bottleneck was inferred from the current genetic homogeneity (O'Brien 1994a; Merola 1994; Roelke et al. 1993; O'Brien et al. 1983). The analysis of museum specimens may help alleviate this problem, as specimens that were collected before an assumed bottleneck occurred can be assayed for genetic variability and compared to levels of diversity in extant populations (e.g. Goombridge et al. 2000; Matocq and Villablanca 2001).

When populations decline to severely low numbers, the only effective conservation measure may be to implement a captive breeding program for the eventual re-release of animals to the wild. Factors such as founder effects relating to captive or reintroduced populations may be assessed using current genetic techniques. Founder effects are related to bottlenecks since many populations that are captively managed are necessarily founded from a small number of individuals (Tarr et al. 1998; Brookes et al. 1997). The equalisation of founder representation in a captive colony is also important for maintenance of genetic variability (Loebel et al. 1992). Founder effects may become apparent where a low level of diversity is present in the founding individuals, inbreeding is unavoidable, and the accumulation of deleterious alleles ensues (leading to 'mutational meltdown', Zeyl et al. 2001; Gabriel et al. 1993).

Intrapopulational genetics

Another area where genetic information may assist in the conservation of species is in the field of molecular ecology, i.e. using genetic techniques to understand an individual's interactions within its environment. It is a necessary aspect of any conservation program to have an understanding of the basic biology of the organism in question. This includes the social structure, mating system, parentage and relatedness, and other behavioural characteristics of the species. However, it is often difficult to determine these traits in secretive, nocturnal, or marine species using traditional field ecology methods. Dispersal of individuals, mating systems, migratory patterns, social organisation and reproductive success of wild populations may all be examined with genetic techniques (e.g. Worthington Wilmer et al. 1999; Mossman and Waser 1999; Gompers et al. 1998; Herbers and Mouser 1998; Jones et al. 1998; Ishibashi et al. 1997; Taylor et al. 1997; Bass et al. 1996; Fleischer et al. 1994). In addition, the assessment of parentage and relatedness may be investigated in managed and wild populations. This information may be used by managers to determine genetically important individuals in captive breeding programs and to control mating strategies within such colonies with greater efficacy. Such fine-scale molecular analysis is due to the unique characteristics of particular loci such as the highly variable microsatellites or minisatellites (which allow for individual identification or DNA fingerprinting), or those loci located on the Y-chromosome (only inherited through the paternal lineage) or on the mitochondrial genome (predominantly inherited through the maternal lineage).

Review of conservation genetic research on carnivorous marsupials

Genetic management is viewed as a vital part of any modern conservation program and can play a pivotal role in guiding management objectives. Despite the wide-spread and long-term declines observed within many carnivorous marsupials, the application of genetics to the conservation of these species is only now being explored in any detail. A number of case studies that apply genetic principles to the conservation of carnivorous marsupials are reviewed.

The phylogenetic and phylogeographic relationships among carnivorous marsupials

It is important to determine the evolutionary or phylogenetic relationships among taxa before being able to identify conservation units. This is one field of study where quite extensive molecular work has been done among and within the Dasyuridae (see below), Myrmecobiidae (Krajewski et al. 2000; Krajewski et al. 2000), Notoryctidae (Springer et al. 1998), Thylacinidae (Krajewski et al. 1997; Krajewski et al. 1992; Thomas et al. 1989), and Didelphidae (Patton and Costa this volume; Palma and Spotorno 1999; Jansa and Voss 2000; Patton et al. 1996).

Determining the phylogenetic relationships within and among members of the Dasyuridae has been of great interest, yet extremely problematic, and there have been numerous molecular studies using a number of techniques over the last 25 years (Krajewski and Westerman this volume; Blacket et al. 2001; Firestone 2000; Blacket et al. 2000; Krajewski et al. 2000; Blacket et al. 1999; Armstrong et al. 1998; Krajewski et al. 1997; Krajewski et al. 1997; Krajewski et al. 1996; Painter et al. 1995; Retief et al. 1995; Painter et al. 1995; Krajewski et al. 1994; Krajewski et al. 1993; Kirsch et al. 1990b; Kirsch et al. 1990a; Baverstock et al. 1990; Baverstock et al. 1982; Kirsch 1976). The great interest in this group stems partially from the species richness of this family and the broad ecological diversity of the dasyurids.

The problems associated with determining phylogenetic relationships within the dasyurids are multifold, however, and are partly due to the variety of different characters used (molecular or morphological), the methods of analysis employed, the inclusion or exclusion of critical taxa and outgroups, and questions of homology, polarity, and independence of characters.

Flux within members of this family is apparent and the taxonomic status of many groups is not stable. Recent studies of the phylogenetic relationships among *Antechinus* species using nuclear and mitochondrial genes (Armstrong et al. 1998) revealed a deep split between the Australian and New Guinean clades and that *Antechinus* (as was then constituted) was paraphyletic. The authors recommended that these clades be recognised as distinct genera with the New Guinean species reassigned to *Murexia*.

Within most of the dasyurid genera that have been examined using molecular techniques, genetic data provide evidence for much greater diversity than was previously assumed based on morphological criteria. Within the genus *Planigale*, for example, the genetic divergence within and between the five currently recognised species suggests that there may be more species than formerly thought (Blacket et al. 2000; Painter et al. 1995). This genus will likely require further taxonomic revisions due to the enigmatic status of several morphological forms. Similarly, within *Sminthopsis macroura*, sequence variation indicates that there are a number of distinct taxa within this species (Blacket et al. 2001). It is highly likely that there are a number of as yet unrecognised species within the Dasyuridae.

The relationships among quolls have undergone a number of revisions over the last two decades with almost every study producing a different phylogeny (Wroe and Mackness 1998; Krajewski et al. 1997; Krajewski et al. 1994; Van Dyck 1987; Baverstock et al. 1982; Archer 1976). Two recent studies using molecular sequence data from the mitochondrial genome have finally brought some consensus to the question of relationships among these taxa (Firestone 2000; Krajewski et al. 1997). Both studies indicate the monophyletic nature of the members of this genus, the basal split of northern quolls from all other quolls, and the close relationship of western quolls, *Dasyurus geoffroii*, to the two New Guinean species, *D. albopunctatus* and *D. spartacus*. Furthermore, the extremely close relationship between western quolls and bronze quolls, *D. spartacus*, is within the range of intraspecific differences (Firestone 2000; Krajewski et al. 1997) and calls into question the degree of genetic differentiation between these two taxa, i.e. where are the species boundaries between members of this genus?

Despite the great effort devoted to phylogenetic and systematic relationships, there have been relatively few phylogeographical studies of carnivorous marsupials (but see Patton and Costa, this volume, for a review of phylogeography in the Didelphidae). Preliminary results from a mitochondrial DNA-based molecular phylogeny of the marsupial moles, *Notoryctes* spp., indicate three major mitochondrial lineages (S. Donnellan, unpubl. data) that accord, at least in part, with taxa delineated in preliminary morphological studies by Ken Aplin at the Western Australian Museum. Recent phylogeographical studies of brush-tailed phascogales, *Phascogale tapoatafa*, using cytochrome *b* sequence data indicate that there are three distinct lineages or ESUs within this species and that the genus is more diverse than previously thought (Spencer et al. 2001). Using microsatellite and mtDNA analyses, genetic subdivision was detected between populations of northern quolls, *Dasyurus hallucatus* from Queensland and the Northern Territory (Firestone 2000; Firestone et al. 2000), indicating that there may be at least two distinct ESUs within this species. Results from each of these studies indicate substantial genetic heterogeneity within these species

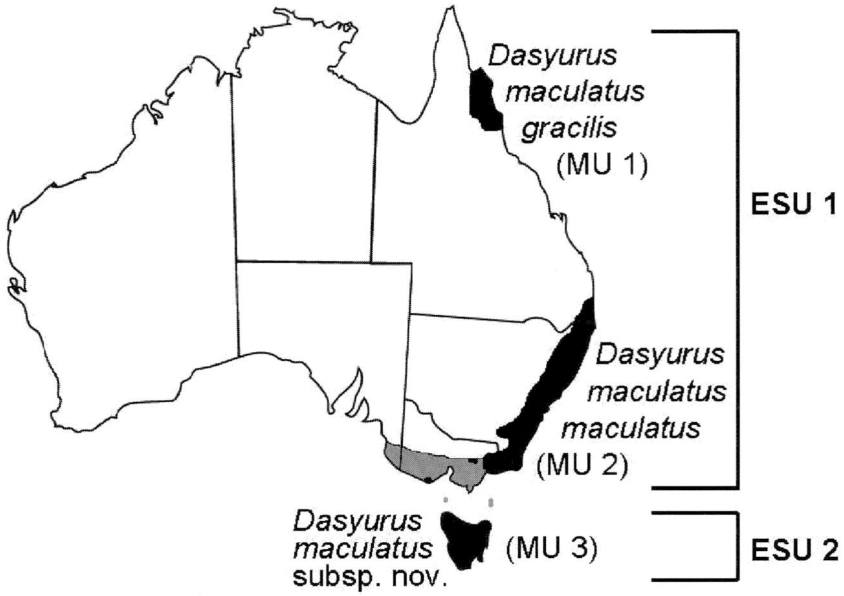

Figure 5 Current (black) and former (black + grey) distribution and subspecies designations for spotted-tailed quolls. Phenotypic differentiation exists between *Dasyurus maculatus gracilis* and *D. m. maculatus* yet no long-term historic genetic differentiation is present. Thus these subspecies are only management units within the mainland ESU. *D. maculatus* subsp. nov. forms a different ESU to that on the mainland although southern mainland and Tasmanian forms exhibit little phenotypic differentiation. From Firestone et al. (1999).

that may necessitate the designation of subspecific status for some populations.

Differences in morphological characters have often been the basis of subspecies designations. However, morphological characters often do not reflect the true evolutionary histories within species of the family Dasyuridae – for example, in grey-bellied dunnarts, *Sminthopsis griseoventer* (Labrinidis et al. 1998), fat-tailed dunnarts, *Sminthopsis crassicaudata* (Cooper et al. 2000), and spotted-tailed quolls, *Dasyurus maculatus* (Firestone et al. 1999). In some cases, morphological similarity disguises long-term historic genetic differentiation (e.g. fat-tailed dunnarts and spotted-tailed quolls). In other cases phenotypic differences are observed where no historic genetic differentiation exists (e.g. northern and southern mainland spotted-tailed quolls and island and mainland grey-bellied dunnarts).

Genetic differentiation in spotted-tailed quolls did not conform with currently recognised subspecific designations based on morphological criteria (Fig. 5; Firestone et al. 1999). In this species, the major genetic split was found to be between Tasmanian and mainland forms, not between *Dasyurus maculatus maculatus* and the smaller subspecies *D. m. gracilis* (Firestone et al. 1999). These results indicate the need for a taxonomic revision within this species and the elevation of the Tasmanian form to subspecific status, as well as a revised conservation management plan for this species. Similarly, mtDNA and allozyme data provide evidence of two distinct ESUs within the fat-tailed dunnart that do not conform to the currently recognised subspecies, indicating the need for a revision of both subspecies classifications and conservation management plans for the species (Cooper et al. 2000).

In the case of grey-bellied dunnarts, however, the lack of distinct phylogeographic subdivision was disregarded in the recent naming of a putative subspecies, *Sminthopsis griseoventer boullangerensis* (Crowther et al. 1999). In this instance, mtDNA sequence data showed no historical, long-term phylogeographical separation of mainland and island forms indicating that there was only one ESU. These genetic data were consistent with the recent geological isolation of the island from the mainland within the last 500–3000 years (Chalmers and Davies 1984).

Despite the lack of strong genetic differentiation, allozyme frequency and morphological differentiation was shown to exist between the two forms of *Sminthopsis griseoventer* (Crowther et al. 1999). However, differentiation in allozyme frequencies is consistent with recent demographic isolation. In addition, morphological characters are removed by several steps from the true level of genetic differentiation. There are many known cases where morphology follows a cline across a species' range (e.g. Houlden et al. 1996) or where different competitive environments result in morphological character displacement (e.g. Jones 1997). For example, competitive character displacement has resulted in different canine tooth and skull dimensions in Tasmanian spotted-tailed quolls, compared with quolls from the south-eastern mainland, where character release has occurred (Jones 1997). Thus, among closely related taxa, a difference in morphology alone is not a good indicator of subspecific or specific status; rather concurrence among a number of independent data sets including morphological and molecular data sets should be used in describing new taxa.

As stated above, the concept of subspecies is closely related to that of ESUs. Furthermore, the naming of subspecies has major political implications for the conservation of taxa and the allocation of scarce financial resources. While there is continued debate on the merits of ESUs for conservation purposes (e.g. Goldstein et al. 2000; Taylor and Dizon 1999; Paetkau 1999), clearly, there is a need for a consensus in applying the concepts of genetic data to the designation of subspecies and for determining units for conservation.

Population-level studies

There is no population-level information currently available for most species or populations of carnivorous marsupials. However, limited information exists on some aspects of population genetics for a few dasyurids (*Sarcophilus*, M. Jones, unpubl. data; *Dasyurus*, Firestone et al. 2000; Firestone 1999; Firestone et al. 1999; and *Antechinus*, Edwards and Wilson 1999).

Recently, genetic subdivision, or differentiation, among populations of Tasmanian devils, *Sarcophilus harrisii*, has been examined using 13 microsatellite loci developed specifically for this species. Preliminary results show that F_{ST} (a measure of genetic subdivision) is significant between eastern and north-western populations of devils, indicating that significantly different allele frequencies, and thus population differentiation, exist within this species (M. Jones, unpubl. data). The discovery of genetic differentiation among populations within this species is an important factor for managers to consider if proactive conservation measures ever become necessary.

Similarly, genetic differentiation among populations within the four Australian *Dasyurus* species was examined using six microsatellite loci as a means of determining management units (Firestone et al. 2000). Results showed that almost every pairwise population comparison was significantly subdivided indicating that each population should be considered a separate MU. Interestingly, while *D. m. gracilis* and *D. m. maculatus* were not distinct ESUs, these populations did have significantly different allele frequencies and were thus different MUs (see Fig. 5). While almost all populations examined exhibited this pattern of allele frequency differentiation, one notable exception was found within a small localised area of spotted-tailed quoll populations from the Barrington Tops region of New South Wales. These populations were from within a 50 km radius of one another and there was no apparent subdivision using these loci, indicating that these were not demographically isolated populations, but rather that there was a substantial amount of gene flow among groups (Firestone et al. 1999).

In addition to genetic differentiation, the levels of genetic variability or diversity found within species of the larger dasyurids has also been examined. As mentioned above, assessing genetic diversity is important as a means of determining baseline levels of variability for future population monitoring. In addition, genetic diversity measures are important for determining which populations may be good source populations for translocations. Preliminary analysis of Tasmanian devils, indicates that relatively low levels of allelic diversity exist within this species (M. Jones, unpubl. data). This low level of variability may be a reflection of previous bottlenecks earlier last century: several population crashes have been reported since European settlement.

Genetic diversity among and within 20 populations of the four Australian species of quolls was examined using highly variable nuclear microsatellites (Firestone et al. 2000; Firestone et al. 1999). In general, among quolls, genetic variability was lowest in small isolated populations, captively bred populations, or populations in severe decline (Firestone et al. 2000).

Despite the broad-scale decline of western quolls, *Dasyurus geoffroii*, a captive breeding colony at Perth Zoo was shown to retain significantly higher levels of diversity than any other species of quoll (Firestone et al. 2000). This high level of variability initially was surprising given the long-term history of decline for this species. However, breeding programs may greatly influence the levels of diversity and results from this study indicate that the captive management program has been successful in maintaining high levels of variability within this species. These results also indicate that the captive breeding population is a good source population for translocations or reintroductions.

Eastern quolls, *Dasyurus viverrinus*, are now presumed extinct on the mainland of Australia, yet they still occur in Tasmania. There have been a number of proposals to reintroduce eastern quolls to the mainland over the years. Until now, there has been little information available to assist managers in determining which populations would be a good source for translocation programs. Analysis of genetic variability indicates that there are significant differences in levels of diversity between populations within this species; furthermore, genetic subdivision between these populations indicates that each population should be treated as a separate MU (Firestone et al. 2000). Thus a captive breeding population comprised of individuals from each MU might be proposed as a means of increasing the level of variability for a reintroduced population.

In addition to examining levels of variability as a baseline measure, it is important to assess the effects of bottlenecks on genetic variability within populations. The effect of translocations on the genetic variability of a population of swamp antechinus

(*Antechinus minimus maritimus*) that had suffered from a bottleneck event have been examined using highly variable nuclear microsatellites (Edwards and Wilson 1999). As theory predicts (Nei et al. 1975), results from this study showed that there was a higher level of genetic variability in the source population than in the remnant bottlenecked population (Edwards and Wilson 1999).

Intrapopulational, social structure, and dispersal studies

Another area where molecular data may be of use in the conservation of carnivorous marsupials is in the field of molecular ecology. To date, however, very limited information is available for any of the carnivorous marsupials, although some preliminary data is available from a few species as discussed below.

The use of mtDNA and microsatellite markers has been of assistance in determining the social structure among spotted-tailed quolls in the Barrington Tops region of New South Wales. As mentioned earlier, data from microsatellites indicated that the populations in the Barrington Tops region have substantial gene flow between them (Firestone et al. 1999). However, other data from mtDNA have shown that there are significant differences in haplotype frequencies between populations in this region, forming different mitochondrial lineages within valley systems (Firestone et al. 1999). Thus, results from mtDNA indicate that there are different MUs within this region. The contrasting results from these two independent genetic systems provide support for sex-biased dispersal in this species. The evidence from mtDNA (which is inherited almost exclusively matrilinearly) shows that females remain within or close by to their mother's home ranges upon maturity. Males, on the other hand, appear to disperse upon maturity, traversing valley systems and taking their maternally inherited mitochondrial haplotype with them but not passing it on to their progeny. Sex-biased dispersal is further supported by preliminary trapping and radiotracking data which shows that females do remain within their mothers' home ranges whereas males disperse upon maturity (K. Firestone, unpublished data).

Currently, population structure, dispersal, and the genetic mating system of Tasmanian devils is also being examined (M. Jones, unpubl. data). One of the most interesting preliminary findings is the presence of multiple paternity among offspring within the same litter. These findings have implications for our understanding of the mating system and social structure of this species and provide information that would be otherwise unobtainable through conventional ecological methods.

Finally, Edwards and Wilson (1999) used data from microsatellite loci to examine integration of swamp antechinus from a non-bottlenecked source population in western Victoria to Airey's Inlet, a site that had experienced localised extinctions and subsequent natural recolonisation after severe bush fires. They showed that translocated swamp antechinus did not appear to contribute to the current genetic makeup of the fire-affected population, indicating that the translocated animals did not survive or breed.

Areas for future study

The application of molecular genetic research to the conservation of carnivorous marsupials is still in its infancy, yet continued declines in these species will necessitate the increased use of these techniques in future. Many of the recovery outlines that have been formulated as part of species action plans (Maxwell et al. 1996) call for genetic monitoring as a prescribed component of these species' recoveries. As demonstrated, there is a variety of ways molecular information can be used to assist managers in making appropriate conservation decisions.

The analysis of phylogenetic relationships within and among the carnivorous marsupials is necessary before we can begin to address the conservation issues facing these species. If we do not know the composition and extent of species boundaries and phylogenetic relationships among these taxa, we may be misdirecting conservation effort to taxa that are not genetically divergent. Therefore, a comprehensive examination of species boundaries within carnivorous marsupials is necessary as a first step in determining conservation efforts. Much progress has been made in this area, but there are still many genera and species that would benefit from comprehensive phylogenetic assessment. It is anticipated that there may be a number of cryptic species and subspecies within this group.

At the level of populational and intrapopulational studies, however, only limited information exists for most species. The studies reviewed in this chapter provide an example of the type of information to be gained from molecular analyses. Yet these studies are only the beginning; they open up a host of other questions for carnivorous marsupials that may assist in the conservation of these species.

For example, choose almost any species among the carnivorous marsupials and further room exists for genetic studies into: 1) examination of the phylogeographical population structure as a means of elucidating conservation units, subspecific status, and species boundaries; 2) examination of variability within and among extinct and extant populations to assess historic variation and temporal changes; 3) examination of parentage, paternity exclusion, or relatedness in populations as a means of clarifying social structure; 4) metapopulation analyses of gene flow and genetic differentiation or assessment of different dispersal models such as isolation-by-distance; 5) identification of variable source populations for the founding of new populations; 6) estimation of levels of genetic variability in translocated or captive populations to assess founder effects or levels of inbreeding and inbreeding depression; 7) examination of hybridisation and introgression where species occur in sympatry; and 8) estimation of N_e and assessing the effects of population bottlenecks. This last point is particularly important in conservation programs that seek to minimise loss of genetic diversity, yet little to no information is currently available on either the effective population size or census size for most species of carnivorous marsupials.

Furthermore, as theoretical and practical advances are made in molecular genetics, the application of these principles to the conservation of species and other units will help refine management efforts. While genetic factors may not be the crucial component when taxa are critically endangered (other factors such as establishing breeding programs, controlling predators, or restoring habitats will probably be more important, for instance), genetic data can provide insightful information that may be of great assistance in guiding managers to make accurate assessments for the conservation of these species.

EPILOGUE

As a final note, one of the most exciting, controversial, and far-reaching studies incorporating principles from molecular and conservation genetics is the recent proposal to clone the extinct thylacine (Colgan and Archer 2000). While tools currently available to the molecular biologist are not sufficient to realise this goal, genetic, cellular, and reproductive advances over the next two to three decades may enable the 'resurrection' of this iconic species. At the very least, information gained from this project will be beneficial to the conservation of other rare, vulnerable, or endangered carnivorous marsupials.

ACKNOWLEDGEMENTS

This paper benefited from discussion with a number of people. I thank Mark Blacket, Steve Cooper, Steve Donnellan, Dave Edwards, and Menna Jones for providing information prior to publication. I also thank Mark Blacket, Steve Cooper, Dave Edwards, Warwick Greville, Menna Jones, and Mike Westerman who all provided insightful comments on this paper.

REFERENCES

Allendorf, F. W., & Leary, R. F. (1986), 'Heterozygosity and fitness in natural populations of animals', in *Conservation Biology: The Science of Scarcity and Diversity* (ed. M.E. Soulé), pp. 57–76, Sinauer Associates, Sunderland, Massachusetts.

Amos, W., & Balmford, A. (2001), 'When does conservation genetics matter?', *Heredity,* **87**:257–65.

Archer, M. (1976), 'The dasyurid dentition and its relationships to that of didelphids, thylacinids, borhyaenids (Marsupicarnivora) and peramelids (Peramelina:Marsupialia)', *Australian Journal of Zoology (suppl.),* **39**:1–34.

Armstrong, L.A., Krajewski, C., & Westerman, M. (1998), 'Phylogeny of the dasyurid marsupial genus *Antechinus* based on cytochrome-*b*, 12S-rRNA, and protamine-P1 genes'. *Journal of Mammalogy,* **79**:1379–89.

Ashley, M. V. (1999), 'Molecular conservation genetics', *American Scientist,* **87**:28–35.

Avise, J.C., & Nelson, W.S. (1989), 'Molecular genetic relationships of the extinct dusky seaside sparrow', *Science,* **243**:646–8.

Avise, J.C. (1994), 'Conservation genetics', in *Molecular Markers, Natural History, and Evolution* (ed. J.C. Avise), p. 511, Chapman and Hall, New York.

Bartley, D., Bagley, M., Gall, G., & Bentley, B. (1992), 'Use of linkage disequilibrium data to estimate effective population size of hatchery and natural fish populations', *Conservation Biology,* **6**:365–75.

Bass, A.L., Good, D.A., Bjorndal, K.A., Richardson, J.I., Hillis, Z.M., Horrocks, J.A., & Bowen, B.W. (1996), 'Testing models of female reproductive migratory behaviour and population structure in the Caribbean hawksbill turtle, *Eretmochelys imbricata*, with mtDNA sequences', *Molecular Ecology,* **5**:321–8.

Baverstock, P., Adams, M., & Archer, M. (1984), 'Electrophoretic resolution of species boundaries in the *Sminthopsis murina* complex (Dasyuridae)', *Australian Journal of Zoology,* **32**:823–32.

Baverstock, P.R., Archer, M., Adams, M., & Richardson, B.J. (1982), 'Genetic relationships among 32 species of Australian dasyurid marsupials', in *Carnivorous Marsupials* (ed. M. Archer), pp. 641–50, Royal Zoological Society of New South Wales, Sydney.

Baverstock, P.R., Krieg , M., & Birrell, J. (1990), 'Evolutionary relationships of Australian marsupials as assessed by albumin immunology', *Australian Journal of Zoology,* **37**:273–87.

Blacket, M.J., Adams, M., Cooper, S.J.B., Krajewski, C., & Westerman, M. (2001), 'Systematics and evolution of the dasyurid marsupial genus *Sminthopsis*: I. The macroura species group', *Journal of Mammalian Evolution,* **8**:149–70.

Blacket, M.J., Adams, M., Krajewski, C., & Westerman, M. (2000), 'Genetic variation within the dasyurid marsupial genus *Planigale*', *Australian Journal of Zoology,* **48**:443–59

Blacket, M.J., Krajewski, C., Labrinidis, A., Cambron, B., Cooper, S., & Westerman, M. (1999), 'Systematic relationships within the dasyurid marsupial tribe Sminthopsini – a multigene approach', *Molecular Phylogenetics and Evolution,* **12**:140–55.

Brookes, M.I., Graneau, Y.A., King, P., Rose, O.C., Thomas, C.D., & Mallet, J.B. (1997), 'Genetic analysis of founder bottlenecks in the rare British butterfly *Plebejus argus*', *Conservation Biology,* **11**:648–61.

Chalmers, C.E., & Davies, S.M. (1984), 'Coastal Management Plan – Jurien Bay Area', Bulletin No. 176, Perth, Department of Conservation and Environment.

Colgan, D., & Archer, M. (2000), 'The thylacine project', *Australasian Science,* **21**:21.

Cooper, S.J.B., Adams, M., & Labrinidis, A. (2000), 'Phylogeography of the Australian dunnart *Sminthopsis crassicaudata* (Marsupialia: Dasyuridae)', *Australian Journal of Zoology,* **48**:461–73.

Crowther, M.S., Dickman, C.R., & Lynam, A.J. (1999), '*Sminthopsis griseoventer boullangerensis* (Marsupialia, Dasyuridae), a new subspecies in the *S. murina* complex from Boullanger Island, Western Australia', *Australian Journal of Zoology,* **47**:215–43.

Crozier, R.H. (1992), 'Genetic diversity and the agony of choice', *Biological Conservation,* **61**:11–15.

Crozier, R.H. (1997), 'Preserving the information content of species: genetic diversity, phylogeny, and conservation worth', *Annual Review of Ecology and Systematics,* **28**:243–68.

Daugherty, C.H., Cree, A., Hay, J.M., & Thompson, M.B. (1990), 'Neglected taxonomy and continuing extinctions of tuatara (*Sphenodon*)', *Nature,* **347**:177–9.

Edwards, D., & Wilson, B.A. (1999), 'Genetic variation in a population of swamp antechinus with a recent severe "bottleneck" event', *Newsletter of the Australian Mammal Society,* November 1999, pp. 39–40.

Faith, D. (1992), 'Conservation evaluation and phylogenetic diversity', *Biological Conservation,* **61**:1–10.

Firestone, K., Elphinstone, M., Sherwin, B., & Houlden, B. (1999), 'Phylogeographical population structure of tiger quolls *Dasyurus maculatus* (Dasyuridae: Marsupialia), an endangered carnivorous marsupial', *Molecular Ecology,* **8**:1613–25.

Firestone, Karen B. (1999), 'The Application of Molecular Genetics to the Conservation Management of Quolls, *Dasyurus* Species (Dasyuridae:Marsupialia)', PhD thesis. Sydney, University of New South Wales.

Firestone, K.B. (2000), 'Phylogenetic relationships among quolls revisited: the mtDNA control region as a useful tool', *Journal of Mammalian Evolution,* **7**:1–22.

Firestone, K.B., Sherwin, W.B., Houlden, B.H., & Geffen, E. (2000), 'Variability and differentiation of microsatellites in the genus *Dasyurus* and conservation implications for the large Australian carnivorous marsupials', *Conservation Genetics,* **1**:115–33.

Fleischer, R.C., Tarr, C.L., & Pratt, T.K. (1994), 'Genetic structure and mating system in the palila, an endangered Hawaiian honeycreeper, as assessed by DNA fingerprinting', *Molecular Ecology,* **3**:383–92.

Frankel, O.H., & Soulé, M.E. (1981), *Conservation and evolution*, Cambridge University Press, Cambridge.

Frankham, R. (1995a), 'Conservation genetics', *Annual Review of Genetics,* **29**:305–27.

Frankham, R. (1995b), 'Effective population size/adult population size ratios in wildlife: a review', *Genetical Research,* **66**:95–107.

Frankham, R. (1995c), 'Inbreeding and extinction: a threshold effect', *Conservation Biology,* **9**:792–9.

Frankham, R. (1996), 'Relationship of genetic variation to population size in wildlife', *Conservation Biology,* **10**:1500–8.

Gabriel, W., Lynch, M., & Burger, R. (1993), 'Muller's ratchet and mutational meltdowns', *Evolution,* **47**:1744–57.

Gall, G.A.E. (1987), 'Inbreeding', in *Population Genetics & Fishery Management* (eds. N. Ryman, & F. Utter), pp. 47–87, University of Washington Press, Seattle.

Goldstein, P.Z., DeSalle, R., Amato, G., & Vogler, A.P. (2000), 'Conservation genetics at the species boundary', *Conservation Biology,* **14**:120–31.

Gompers, M.E., Gittleman, J.L., & Wayne, R.K. (1998), 'Dispersal, philopatry, and genetic relatedness in a social carnivore: comparing males and females', *Molecular Ecology,* **7**:157–63.

Groombridge, J.J., Jones, C.G., Bruford, M.W., & Nichols, R.A. (2000), 'Conservation biology: "Ghost" alleles of the Mauritius kestrel', *Nature,* **403**:616.

Haig, S. M. (1998), 'Molecular contributions to conservation', *Ecology,* **79**:413–25.

Hedrick, P.W. (2001), 'Conservation genetics: where are we now?', *Trends in Ecology and Evolution,* **16**:629–36.

Hedrick, P.W., & Miller, P.S. (1992), 'Conservtion genetics: techniques and fundamentals', *Ecological Applications,* **2**:30–46.

Herbers, J.M., & Mouser, R.L. (1998), 'Microsatellite DNA markers reveal details of social structure in forest ants', *Molecular Ecology,* **7**:299–306.

Hoelzel, A.R., Halley, J., O'Brien, S.J., Campagna, C., Arnbom, T., Le Boeuf, B., Ralls, K., & Dover, G.A. (1993), 'Elephant seal genetic variation and the use of simulation models to investigate historical population bottlenecks', *Journal of Heredity,* **84**:443–9.

Houlden, B.A., England, P.R., Taylor, A.C., Greville, W.D., & Sherwin, W.B. (1996), 'Low genetic variability of the koala *Phascolarctos cinereus* in south-eastern Australia following a severe population bottleneck', *Molecular Ecology,* **5**:269–81.

Ishibashi, Y., Saitoh, T., Abe, S., & Yoshida, M.C. (1997), 'Sex-related spatial kin structure in a spring population of grey-sided voles *Clethrionomys rufocanus* as revealed by mitochondrial and microsatellite DNA analyses', *Molecular Ecology,* **6**:63–71.

Jansa, S.A., & Voss, R.S. (2000), 'Phylogenetic studies on didelphid marsupials I. Introduction and preliminary results from nuclear IRBP gene sequences', *Journal of Mammalian Evolution,* **7**:43–77.

Jones, A.G., Kvarnemo, C., Moore, G.I., Simmons, L.W., & Avise, J.C. (1998), 'Microsatellite evidence for monogamy and sex-biased recombination in the Western Australian seahorse, *Hippocampus angustus*', *Molecular Ecology,* **7**:1497–505.

Jones, M. (1997), 'Character displacement in Australian dasyurid carnivores: size relationships and prey size patterns', *Ecology,* **78**:2569–87.

Kimura, M., & Crow, J.F. (1963), 'The measurement of effective population number', *Evolution,* **17**:279–88.

Kirsch, J.A.W. (1976), 'The classification of marsupials with a special reference to karyotypes and serum proteins', in *The Biology of Marsupials* (ed. D. Hunsaker II), pp. 1–50, Academic Press, New York.

Kirsch, J.A.W., Krajewski, C., Springer, M.S., & Archer, M. (1990a), 'DNA–DNA hybridisation studies of carnivorous marsupials. II: Relationships among dasyurids (Marsupialia: Dasyuridae)', *Australian Journal of Zoology,* **38**:673–96.

Kirsch, J.A.W., Springer, M.S., Krajewski, C., Archer, M., Aplin, K., & Dickerman, A.W. (1990b), 'DNA–DNA hybridization studies of the carnivorous marsupials. I: The intergeneric relationships of bandicoots (Marsupialia: Perameloidea), *Journal of Molecular Evolution,* **30**:434–48.

Krajewski, C., Blacket, M., Buckley, L., & Westerman, M. (1997), 'A multigene assessment of phylogenetic relationships within the dasyurid marsupial subfamily Sminthopsinae', *Molecular Phylogenetics and Evolution,* **8**:236–48.

Krajewski, C., Blacket, M., & Westerman, M. (2000), 'DNA sequence analysis of familial relationships among dasyuromorphian marsupials', *Journal of Mammalian Evolution,* **7**:95–108.

Krajewski, C., Buckley, L., & Westerman, M. (1997), 'DNA phylogeny of the marsupial wolf resolved', *Proceedings of the Royal Society of London, Series B,* **264**:911–17.

Krajewski, C., Buckley, L., Woolley, P.A., & Westerman, M. (1996), 'Phylogenetic analysis of cytochrome *b* sequences in the dasyurid marsupial subfamily Phascogalinae: systematics and the evolution of reproductive strategies', *Journal of Mammalian Evolution,* **3**:81–91.

Krajewski, C., Driskell, A.C., Baverstock, P.R., & Braun, M.J. (1992), 'Phylogenetic relationships of the thylacine (Mammalia: Thylacinidae) among dasyuroid marsupials: evidence from cytochrome *b* DNA sequences', *Proceedings of the Royal Society of London, Series B,* **250**:19–27.

Krajewski, C., Painter, J., Buckley, L., & Westerman, M. (1994), 'Phylogenetic structure of the marsupial family Dasyuridae based on cytochrome *b* DNA sequences', *Journal of Mammalian Evolution,* **2**:25–35.

Krajewski, C., Painter, J., Driskell, A.C., Buckley, L., & Westerman, M. (1993), 'Molecular systematics of New Guinean dasyurids (Marsupialia: Dasyuridae)', *Science in New Guinea,* **19**:157–66.

Krajewski, C., & Westerman, M. (2003), 'Molecular systematics of Dasyuromorphia', in *Predators with Pouches: The Biology of Carnivorous Marsupials* (eds. M. Jones, C. Dickman, & M. Archer), pp. 3–20, CSIRO Publishing, Melbourne.

Krajewski, C., Wroe, S., & Westerman, M. (2000), 'Molecular evidence for the pattern and timing of cladogenesis in dasyurid marsupials', *Zoological Journal of the Linnean Society,* **130**:375–404.

Krajewski, C., Young, J., Buckley, L., Woolley, P.A., & Westerman, M. (1997), 'Reconstructing the evolutionary radiation of Dasyurine marsupials with cytochrome *b*, 12S rRNA, and protamine P1 gene trees', *Journal of Mammalian Evolution,* **4**:217–36.

Labrinidis, A., Cooper, S.J.B., Adams, M., & Baczocha, N. (1998), 'Systematic affinities of island and mainland populations of the Dunnart *Sminthopsis griseoventer* in Western Australia: data from allozymes and mitochondrial DNA', *Pacific Conservation Biology,* **4**:289–95.

Lacy, R.C. (1997), 'Importance of genetic variation to the viability of mammalian populations', *Journal of Mammalogy,* **78**:320–35.

Loebel, D.A., Nurthen, R.K., Frankham, R., Briscoe, D.A., & Craven, D. (1992), 'Modeling problems in conservation genetics using captive Drosophila populations: consequences of equalizing founder representation', *Zoo Biology,* **11**:319–32.

Matocq, M.D., & Villablanca, F.X. (2001), 'Low genetic diversity in an endangered species: recent or historic pattern?', *Biological Conservation,* **98**:61–8.

Maxwell, S., Burbidge, A.A., & Morris, K. (1996), *The 1996 Action Plan for Australian Marsupials and Monotremes*, Wildlife Australia, Canberra.

Merola, M. (1994), 'A reassessment of homozygosity and the case for inbreeding depression in the cheetah, *Acinonyx jubatus*: implications for conservation', *Conservation Biology,* **8**:961–71.

Miller, P.S., & Hedrick, P.W. (1993), 'Inbreeding and fitness in captive populations: lessons from *Drosophila*', *Zoo Biology,* **12**:333–51.

Moritz, C. (1994), 'Defining "Evolutionarily Significant Units" for conservation', *Trends in Ecology and Evolution,* **9**:373–5.

Moritz, C. (1999), 'Conservation units and translocations: strategies for conserving evolutionary processes', *Hereditas,* **130**:217–28.

Moritz, C., Lavery, S., & Slade, R. (1995), 'Using allele frequency and phylogeny to define units for conservation and management', *American Fisheries Society Symposium,* **17**:249–62.

Mosman, C.A., & Waser, P.M. (1999), 'Genetic detection of sex-biased dispersal', *Molecular Ecology,* **8**:1063–7.

Nei, M., Maruyama, T., & Chakraborty, R. (1975), 'The bottleneck effect and genetic variability in populations', *Evolution,* **29**:1–10.

Newman, D., & Pilson, D. (1997), 'Increased probability of extinction due to decreased genetic effective population size: experimental populations of *Clarkia pulchella*', *Evolution,* **51**:354–62.

Nunney, L. (1993), 'The influence of mating system and overlapping generations on effective population size', *Evolution,* **47**:1329–41.

Nunney, L., & Elam, D.R. (1994), 'Estimating the effective population size of conserved populations', *Conservation Biology,* **8**:175–84.

O'Brien, S.J., Martenson, J.S., Eichelberger, M.A., Thorne, E.T., & Wright, F. (1989), 'Genetic variation and molecular systematics of the black-footed ferret', in *Conservation Biology of the Black-footed Ferret* (eds. U.S. Seal, E.T. Thorne, M.A. Bogan, & S.H. Anderson), pp. 21–33, Yale University Press, New Haven.

O'Brien, S.J., Martenson, J.S., Packer, C., Herbst, L., de Vos, V., Joslin, P., Ott-Joslin, J., Wildt, D.E., & Bush, M. (1987), 'Biochemical genetic variation in geographic isolates of African and Asiatic lions', *National Geographic Research,* **3**:114–24.

O'Brien, S.J. (1994a), 'The cheetah's conservation controversy', *Conservation Biology,* **8**:1153–5.

O'Brien, S.J. (1994b), 'A role for molecular genetics in biological conservation', *Proceedings of the National Academy of Sciences of the United States of America,* **91**:5748–55.

O'Brien, S.J., & Everman, J.F. (1988), 'Interactive influence of infectious disease and genetic diversity in natural populations', *Trends in Ecology and Evolution,* **3**:254–9.

O'Brien, S.J., Wildt, D.E., Goldman, D., Merril, C.R., & Bush, M. (1983), 'The cheetah is depauperate in genetic variation', *Science,* **221**:459–62.

Paetkau, D. (1999), 'Using genetics to identify intraspecific conservation units: a critique of current methods', *Conservation Biology,* **13**:1507–9.

Painter, J., Krajewski, C., & Westerman, M. (1995), 'Molecular phylogeny of the marsupial genus *Planigale* (Dasyuridae)', *Journal of Mammalogy,* **76**:406–413.

Palma, R.E., & Spotorno, A.E. (1999), 'Molecular systematics of marsupials based on the rRNA 12S mitochondrial gene: the phylogeny of didelphimorphia and of the living fossil microbiotheriid *Dromiciops gliroides* Thomas', *Molecular Phylogenetics and Evolution,* **13**:525–35.

Palsbøll, P. J. (1999), 'Genetic tagging: contemporary molecular ecology', *Biological Journal of the Linnean Society,* **68**:3–22.

Patton, J.L., Dos Reis, S.F., & Da Silva, M.N.F. (1996), 'Relationships among didelphid marsupials based on sequence variation in the mitochondrial cytochrome b gene', *Journal of Mammalian Evolution,* **3**:3–29.

Patton, J.L., & Costa, L.P. (2003), 'Molecular phylogeography and species limits in rainforest didelphid marsupials of South America', in *Predators with Pouches: The Biology of Carnivorous Marsupials* (eds. M. Jones, C. Dickman, & M. Archer), pp. 63–81, CSIRO Publishing, Melbourne.

Polasky, S., Csuti, B., Vossler, C.A., & Meyers, S.M. (2001), 'A comparison of taxonomic distinctness versus richness as criteria for setting conservation priorities for North American birds', *Biological Conservation,* **97**:99–105.

Ralls, K., Ballou, J.D., & Templeton, A. (1988), 'Estimates of lethal equivalents and the cost of inbreeding in mammals', *Conservation Biology,* **2**:185–93.

Raymond, M., & Rousset, F. (1995), 'GENEPOP (Version 1.2): Population genetics software for exact tests and ecumenicism', *Journal of Heredity,* **86**:248–9.

Retief, J.D., Krajewski, C., Westerman, M., & Dixon, G.H. (1995), 'The evolution of protamine P1 genes in dasyurid marsupials', *Journal of Molecular Evolution,* **41**:549–55.

Roelke, M.E., Martenson, J.S., & O'Brien, S.J. (1993), 'The consequences of demographic reduction and genetic depletion in the endangered Florida panther', *Current Biology,* **3**:340–50.

Ryder, O.A. (1986), 'Species conservation and systematics: the dilemma of subspecies', *Trends in Ecology and Evolution,* **1**:9–10.

Ryman, N. (1994), 'Supportive breeding and effective population size: differences between inbreeding and variance effective numbers', *Conservation Biology,* **8**:888–90.

Saccheri, I., Kuussaari, M., Kankare, M., Vikman, P., Fortelius, W., & Hanski, I. (1998), 'Inbreeding and extinction in a butterfly metapopulation', *Nature,* **392**:491–4.

Schneider, S., Kueffer, J.-M., Roessli, D., & Excofier, L. (1997), *Arlequin: A software for population genetic data analysis,* Genetics and Biometry Lab, Dept. of Anthropology, University of Geneva.

Schwartz, M.K., Tallmon, D.A., & Luikart, G.H. (1998), 'Review of DNA-based census and effective population size estimators', *Animal Conservation,* **1**:293–9.

Smith, T.B., & Wayne, R.K. (1996), *Molecular Genetic Approaches in Conservation*, Oxford University Press, New York.

Soulé, M. (1986), *Conservation Biology: The Science of Scarcity and Diversity*, Sinauer Associates, Sunderland, MA.

Spencer, P.B.S., Rhind, S.G., & Eldridge, M.D.B. (2001), 'Phylogeographic structure within *Phascogale* (Marsupialia: Dasyuridae) based on partial cytochrome *b* sequence', *Australian Journal of Zoology,* **49**:369–77.

Springer, M.S., Westerman, M., Kavanagh, J.R., Burk, A., Woodburne, M.O., Kao, D.J., & Krajewski, C. (1998), 'The origin of the Australasian marsupial fauna and the phylogenetic affinities of the enigmatic monito del monte and marsupial mole', *Proceedings of the Royal Society of London, Series B,* **265**:2381–6.

Swofford, D.L. (1998), *PAUP**: phylogenetic analysis using parsimony (*and other methods)*, Sunderland, MA, Sinauer Associates, Inc.

Taberlet, P., & Luikart, G. (1999), 'Non-invasive genetic sampling and individual identification', *Biological Journal of the Linnean Society,* **68**:41–55.

Tarr, C.L., Conant, S., & Fleischer, R.C. (1998), 'Founder events and variation at microsatellite loci in an insular passerine bird, the Laysan finch (*Telespiza cantans*)', *Molecular Ecology,* **7**:719–31.

Taylor, A.C., Horsup, A., Johnson, C.N., Sunnucks, P., & Sherwin, B. (1997), 'Relatedness structure detected by microsatellite analysis and attempted pedigree reconstruction in an endangered marsupial, the northern hairy-nosed wombat *Lasiorhinus krefftii*', *Molecular Ecology,* **6**:9–19.

Taylor, B.L., & Dizon, A.E. (1999), 'First policy then science: why a management unit based solely on genetic criteria cannot work', *Molecular Ecology,* **8**:S11–16.

Thomas, R.H., Schaffner, W., Wilson, A.C., & Pääbo, S. (1989), 'DNA phylogeny of the extinct marsupial wolf', *Nature,* **340**:465–7.

Van Dyck, S. (1987), 'The bronze quoll, *Dasyurus spartacus* (Marsupialia: Dasyuridae), a new species from the savannahs of Papua New Guinea', *Australian Mammalogy,* **11**:145–56.

Vane-Wright, R.I., Humphries, C.J., & Williams, P.H. (1991), 'What to protect? Systematics and the agony of choice', *Biological Conservation,* **55**:235–54.

Vogler, A.P., & DeSalle, R. (1994), 'Diagnosing units of conservation management', *Conservation Biology,* **8**:354–63.

Waples, R.S. (1989), 'A generalized approach for estimating effective population size from temporal changes in allele frequency', *Genetics,* **121**:379–91.

White, T. J., Arnheim, N., & Erlich, H.A. (1989), 'The polymerase chain reaction', *Trends in Genetics,* **5**:185–9.

Worthington Wilmer, J., Allen, P.J., Pomeroy, P.P., Twiss, S.D., & Amos, W. (1999), 'Where have all the fathers gone? An extensive microsatellite analysis of paternity in the grey seal (*Halichoerus grypus*)', *Molecular Ecology,* **8**:1417–29.

Wright, S. (1978), *Evolution and the genetics of populations. Variability within and among natural populations*, University of Chicago Press, Chicago.

Wright, S. (1931), 'Evolution in Mendelian populations', *Genetics,* **16**:97–159.

Wroe, S., & Mackness, B.S. (1998), 'Revision of the Pliocene dasyurid, *Dasyurus dunmalli*, (Dasuyuridae: Marsupialia)', *Memoirs of the Queensland Museum,* **42**:605–12.

Zeyl, C., Mizesko, M., & de Visser, J.A.G.M. (2001), 'Mutational meltdown in laboratory yeast populations', *Evolution,* **55**:909–17.

Zink, R.M., & Kale, H.W. (1995), 'Conservation genetics of the extinct dusky seaside sparrow *Ammodramus maritimus nigrescens*', *Biological Conservation,* **74**:69–71.